HISTORY, PHILOSOPHY AND SOCIOLOGY OF SCIENCE

Classics, Staples and Precursors

HISTORY, PHILOSOPHY AND SOCIOLOGY OF SCIENCE

Classics, Staples and Precursors

Selected By

YEHUDA ELKANA
ROBERT K. MERTON
ARNOLD THACKRAY
HARRIET ZUCKERMAN

ORIGINS AND DEVELOPMENT
OF
APPLIED CHEMISTRY

J[ames] R[iddick] Partington

ARNO PRESS
A New York Times Company
New York — 1975

Reprint Edition 1975 by Arno Press Inc.

Copyright © 1935, by Longmans, Green and Co.
Reprinted by permission of Longman Group Limited

Reprinted from a copy in
  The Newark Public Library

HISTORY, PHILOSOPHY AND SOCIOLOGY OF SCIENCE:
Classics, Staples and Precursors
ISBN for complete set: 0-405-06575-2
See last pages of this volume for titles.

Manufactured in the United States of America

———◆———

Library of Congress Cataloging in Publication Data

Partington, James Riddick, 1886-
    Origins and development of applied chemistry.

    (History, philosophy, and sociology of science)
    Reprint of the 1935 ed. published by Longmans,
Green, London.
    Bibliography: p.
    Includes indexes.
    1. Chemistry, Technical--History. I. Title.
II. Series.
TP16.P3 1975   660'.09   74-26284
ISBN 0-405-06611-2

ORIGINS AND DEVELOPMENT OF
APPLIED CHEMISTRY

LONGMANS, GREEN AND CO. LTD.
39 PATERNOSTER ROW, LONDON, E.C.4
6 OLD COURT HOUSE STREET, CALCUTTA
53 NICOL ROAD, BOMBAY
36A MOUNT ROAD, MADRAS

LONGMANS, GREEN AND CO.
114 FIFTH AVENUE, NEW YORK
221 EAST 20TH STREET, CHICAGO
88 TREMONT STREET, BOSTON

LONGMANS, GREEN AND CO.
480 UNIVERSITY AVENUE, TORONTO

ORIGINS AND DEVELOPMENT
OF
APPLIED CHEMISTRY

By

J. R. PARTINGTON, M.B.E., D.Sc.

PROFESSOR OF CHEMISTRY IN THE UNIVERSITY OF LONDON, QUEEN MARY COLLEGE

LONGMANS, GREEN AND CO.

LONDON • NEW YORK • TORONTO

FIRST PUBLISHED......*April*, 1935

PREFACE

WHEN the growth of an idea, or the perfection of some technical process, is traced backwards in time, a point is often reached when the development leaves the track of formal science, and enters the wide and often ill-charted territory of the general achievement of man. For this reason, a balanced and just appreciation of the history of a science can be attained only when a complete and critical survey of its earliest origins is available. This is always a matter of considerable difficulty, and in the case of such a science as Chemistry it is particularly so, since the first narrow and ill-defined track led from that vast territory of endeavour which comprised the use by man from the most remote period of all kinds of materials. It is for this reason, no doubt, that histories of Chemistry begin with the technical arts of the Greeks and Romans, a period which has been well studied and on which information is easily accessible. For them, Egypt is represented by the accounts in Classical authors or the remains of the Eighteenth Dynasty; the Babylonian and Assyrian civilisations begin with the last days of the Assyrian Empire; the Phœnicians still occupy their undeserved position as the universal diffusers of materials and processes among the early peoples; and the Minoan and Hittite cultures, with their highly important contributions and their intimate relations with the other great centres of civilisation, are totally unknown.

The archæologist, on the other hand, feels the need for some general survey of the materials used by the ancient peoples, examples of which, in various forms, he encounters on the sites of excavation and uses for the purpose of establishing the period of the remains, and their relation to those of other regions. It is rare to find the correct technical terms used in the descriptions of remains, and the study of the origins of these materials is often, in consequence, either neglected or rendered ineffective. Exaggerated importance is sometimes attached to traces of some particular impurity found in analysed specimens, and since the analyst has usually no expert knowledge of results in other fields, he can give no useful assistance on this side of the investigation. It is only recently that the results of analyses have come to have any very definite significance.

The intention of the present work is to give a reasonably concise and systematic account of the sources, production and uses of materials in Egypt, Babylonia and Assyria, the Ægean, Asia Minor, Persia, Syria and Palestine, from the earliest times to the end of the Bronze Age. Since the further developments in applied Chemistry are mostly very recent, the ground covered is really much more extensive than the limitation in date would suggest, and in some cases the discussion has been extended to a later period. The original intention of including India and China was abandoned after much material had been collected, partly because these regions in some ways lie apart from the rest of the world in their technical processes, but more particularly because the information at present available seemed insufficient as compared with that for the other regions.

One important result of this survey, which could hardly have become clear without it, is the realisation that the knowledge of the use of materials in the Classical Period, which usually forms the starting-point for the historian of science[1], is almost wholly derived from much older cultures. It represents, in many cases, not an original and vigorous development of national genius, but a decadent form of craftmanship which had existed for a period often as long as that which now separates us from the best days of Greece and Rome. Just as the modern industrial period has ruined the traditions of craftmanship, so the irruption of the people of the Iron Age broke the continuity in a traditional use of materials which had developed almost without a break from the period of the Stone Age. The essential methods nevertheless continued with little alteration, as in some cases, such as the art of the potter, they do to the present day. In the study of the development of man no part is more significant, even if more neglected, than that concerned with the use of materials.

It was considered desirable to give brief accounts of the general history and archæology of the regions concerned, since although these can be found in standard works of reference, many of these are now incomplete and even inaccurate by reason of the rapid progress made in recent years. This great advance in knowledge made it necessary to rewrite many of the sections more than once, and as the preparation naturally extended over a period of years, some difficulty was experienced in keeping the whole work up-to-date in all its sections. It is hoped that this has, in a large measure, been achieved. In these general sections particular attention is given to the relations between the various centres of civilisation, and although an adequate account of ancient trade written by a specialist is lacking, an attempt has been made to indicate some possibilities in this field.

In the sections dealing with special materials (the arrangement of which has been varied to prevent monotony) attention has been given to the geographical distribution of ores, etc., since in some cases sites worked in antiquity have recently become of great economic importance. The section on the source of petroleum at Kerkuk, for example, was completed before the recent exploitation of this region, the occurrence of oil in which is clearly described by several older travellers. It is thought that information on older sources may not be without economic as well as historical interest.

The present work, as far as the author knows, is the first of its kind, and it must necessarily suffer from the defects of all attempts to begin something new. Since it was started, a few works of similar type have appeared for more restricted fields, references to which have been given even when the text had been completed some years before from the same original sources. A work covering a wider field, however, was still lacking. The early use of glass, for example, and a consideration of the processes and materials used in its manu- facture in early times, are now dealt with in some detail, and by combining the sections on glass in the various parts of the book (which is readily done by means of the index, the division into regions being given in the table of contents), a more complete survey will be possible than has previously been the case. In dealing with the information from very varied sources, those state- ments and hypotheses which are not possible from the point of view of the chemist have been corrected, usually without special mention, except in a few cases when it seemed desirable to draw attention to common errors which are constantly repeated. This is done without wishing to minimise in the slightest degree the value of discussions by specialists in other fields.

[1] See, e.g., Sarton's Introduction to the History of Science, 1927, i, 13, where the opinion is expressed with respect to the period before Homer that "so much research is being done . . . by a legion of specialists . . . that it is perhaps wiser to postpone the necessary synthesis until a little later"—a view which may well be true.

Since the book is intended to be of service to the specialist as well as to the reader who has merely a general interest in such subjects, the references to the literature are reasonably full, and each has actually been seen by the author in the form quoted. These references are intended to serve two main purposes. In the first place, they show the sources of each item of information, and thus enable the specialist to infer its probable accuracy. In the second place, they indicate to the reader wishing for further information a starting point for his investigations.

It seems desirable to point out that some authors convey an appearance of modernity by quoting recent works of limited accessibility. The reader who, after much trouble and expenditure of time, succeeds in getting sight of these works, often finds that they merely give in footnotes references to well-known standard works which are available in any library, or in some cases merely a reference to some Classical author. In general, it is very easy in a good library for anyone to compile in a few hours a bibliography on any subject, but in the case of a work such as the present, this cannot replace the actual acquaintance of an author with his sources of information.

There are about 7,000 numbered footnote references, each usually containing more than one citation, so that the number of single references to literature may exceed 25,000. Each of these concerns a statement of fact or opinion in the text, and each statement is usually capable of expansion into a page, chapter, book or series of volumes of text. Great pains have been taken in various ways to ensure accuracy, both in the references and in the text, but as every experienced writer knows, this is an ideal and errors are inescapable. The author hopes that these are not more numerous than he deserves, on the basis of the trouble taken to avoid them.

Complete titles are in most cases given on the page containing the first reference to a work, which is found from the index. Owing to many changes in the order of the text during revisions, this full title has sometimes gone to a later page, which is then denoted in heavy type in the index, or in rare cases it has passed through to the index. Abbreviated titles, catchwords, etc., used in the references can be traced at once in the author index. The year of publication is given first, followed by the volume number in Roman numerals, in case the work is in more than one volume, and finally the page. E.g., RV, i, 157, is a reference to vol. i, p. 157, of the Real-lexikon der Vorgeschichte. In the case of several works of the same author, these are given separately in the index, except when only a few references in all appear ; contributions of an author to periodical publications or collected works are given immediately after the name, since these publications are not indexed separately. To have given the full titles in all cases in the notes would have added very considerably to the size and cost of the book, and would have served no useful purpose. Initials of authors are not always given when the reference is to a journal or dictionary, where the article can be traced at once ; in the case of books, initials are given for authors in cases where the library catalogues are likely to contain several entries under authors of the same name. The intention is always to make the references capable of being easily traced, with a little co-operation on the part of the reader.

The task of preparing the indexes from over 5,000 cards entered from markings in the page proofs was largely carried out by the author's wife, Mrs. M. Partington, M.Sc., and since the proofs were set up in pages from the first, it was possible in this way to detect a number of inconsistencies in the spelling of names taken from various sources. The author wishes to record his sincere thanks for the great assistance thus rendered. The task of proofreading was also appreciably lightened by the unusual accuracy attained by the printers from difficult copy.

The text contains a great number of foreign words, and since most authors differ in their transliterations of these, an effort was made to ensure uniformity. The principles underlying this may be stated briefly without enlarging on their merits or defects.

The Egyptian alphabet, in the order now usually adopted[1], is transcribed as follows, the alternatives shown in brackets being used by other authors and occasionally in the present work : a, à (e, i, j), and ā (three special symbols) ; i (y), u (w), b, p, f, m, n, r, l, h, ḥ, kh (ch, χ, ḫ), kh (ḫ), s, s (s'), š (sh), k, q (ḳ, ch), g, t, th (ṯ), d (t), z (ḏ, t', tch). Vowels not included in the Egyptian alphabet are filled in as e, a or o, in order to make a word. In some cases the transliterations of authors who do not give the hieroglyphics are reproduced, although they obviously do not correspond with the usual systems. The names in Joachim's translation of the Ebers Papyrus are given unaltered. A numbered list of hieroglyphics (for which type is available), published by Messrs. Harrison[2], makes it possible to refer to any one by number. E.g., the hieroglyphic for silver (p. 42) is No. 1458. The system used for Chinese[3] could, therefore, be used for Egyptian. In many modern Egyptian and some other names of places, e is often used instead of a, e.g., Tell el-Amarna instead of Tall al-Amarna (there is no e in Arabic).

Assyrian words are usually approximately transliterated according to the system following the order[4] : a (represents several sounds, also o and u), b, g, d, z, ch (x, ḫ), t, i, k, l, m, n, s, p, ts or tz (ç, ṣ), q, r, š, t (th).

The Hebrew alphabet is usually transcribed as : ', b (bh), g, d, h, v, z, ch (ḥ), t (ṭ), j, k (kh), l, m, n, s, ', p (ph), tz (ṣ), q, r, š, t (th). Biblical names are given as in the English Bible, including shekel. Long vowels in Oriental languages are shown as â, ê, î, ô, û. In Hebrew, the evanescent vowels are represented by ă and ĕ in the body as well as at the beginning of a word, in place of the small letters often used.

Arabic words are not much used except in proper names, which are mainly given as in the British Museum Catalogue, including the use of the article al instead of the assimilation with certain consonants of the word following (ar-Râzî instead of al-Râzî), which is a peculiarity of the spoken as distinguished from the written language. Diacritical points are, however, omitted from ṣ, ḍ, ṭ and ẓ. In the case of Persian names, the forms given by the authors quoted are generally used.

The Greek alphabet has mostly been transcribed on the modern system, including $\kappa = $ k ; o, $\omega = $ o ; $v = $ y (except when sounded as u, e.g., in Amathus) ; $ov = $ u, and $a\iota = $ ai. Thus, Demokritos, Dioskurides, Diodoros, Achaian and Knossos are used instead of Democritus, Dioscorides, Diodorus, Achæan and Cnossus. Familiar names, such as Plato, Aristotle, Cyprus, Mycenæ, Crete, Crœsus, etc., are, however, retained, which makes for inconsistency. Names of Greek works are given either in English or in the Latin forms with their contractions as in international practice, e.g. Hist[oria] plant[arum], for the περὶ φυτῶν ἱστορία of Theophrastos. In quotations of Pliny, Dioskurides, Diodoros Siculus and Strabo, the following works are to be understood : Pliny, Historia Naturalis ; Dioskurides, De Materia Medica ; Diodoros Siculus, Bibliotheca Historica ; Strabo, Geographia. The citations of Pliny are to

---

[1] Budge, Egyptian Hieroglyphic Dictionary, 1920; Erman and Grapow, Aegyptisches Handwörterbuch, 1921; ib., Wörterbuch der aegyptischen Sprache, Leipzig, 1926f.

[2] Reproduced in Budge, Dictionary, 1317f.

[3] Sarton, Introduction, i, 47.

[4] Muss-Arnolt, Concise Dictionary of the Assyrian Language, Berlin, 1905.

books and chapters (corresponding with the English translation of Philemon Holland), the newer division into sections varying with the editor. Strabo is also quoted by the page of Casaubon's edition, reproduced in most modern editions. Adequate information on Classical and Oriental authors will be found in Sarton's "Introduction," and new editions or translations are often mentioned in *Isis*, a journal edited by Dr. Sarton for the History of Science Society.

With the object of avoiding expense, the use of special type, including italic, has been reduced to a minimum, the námes of journals only being in italic in order to distinguish them from books. Illustrations have also been omitted, as is the case in most other works on the history of Chemistry. It would have been possible to have given line drawings of Egyptian material, but these are easily available in standard works, and the lack of suitable material for the other countries would have produced an ill-balanced result. No adverse criticism of the absence of illustrations in recent works has come to the attention of the author, and in cases where figures are available reference has been made to them.

The above statements are mainly presented with the object of indicating the purpose of the work and the way in which it may be used. A few words of a more personal character may be said in conclusion.

The first draft of the work was almost double the length of the present text, and the revisions which followed were made with the object of presenting the material in the most concise form possible, so that the cost of the book might be the minimum consistent with the expense of production. The style is, therefore, very concise, but it is hoped that it is not obscure.

When the work was in a form more or less suitable for publication, the author had grave doubts as to whether its appearance would, in the jargon of the official scientific world, be "timely" or even "significant." Some anonymous critics reported favourably, but it was only after the leading authority on the history of Chemistry, Prof. E. O. von Lippmann, had read the greater part of the manuscript, had sent a number of helpful suggestions to the author, and had given a very definite assurance that publication at the earliest possible date was desirable, that attempts were made to achieve this end. That the appearance of the book was at last made possible by the publishers is counted by the author as an unusual favour of fortune.

In the production of an ideal book of this kind, the co-operation of a large number of specialists and an editor of unusual gifts, able to devote their whole time to the work and with abundant funds at their disposal, would be very desirable. A full knowledge of the archæology, geography and history of the various regions, and of ancient and modern Oriental and Classical languages, with facilities for consulting a great mass of periodical literature and for obtaining reprints of articles by specialists as they appear, and an adequate and trained clerical staff, would all be necessary. When, in place of this, the spare time of one person, not particularly well equipped, is all that is at command, the result must necessarily fall short of ideal.

For the purpose of consulting books and journals not otherwise available the author was permitted access to the libraries of the Royal College of Physicians and the Royal Society of Medicine, the officials of which, as well as of the Royal Asiatic Society, gave valuable and willing assistance. The London Library and the library of the Chemical Society, as well as the libraries of the British Museum and of the Patent Office, were all extensively utilised, and the invariably helpful and courteous co-operation of their staffs deserves acknowledgment.   Such a book as the present could hardly have been written in England outside of London, and this fact must be an excuse for its appearance during the period of active service of the author, who felt that if its preparation were postponed to days of leisure, then neither the opportunities nor the desire to use them might be available.

The author will be very grateful to readers who will inform him of the errors they find in the book, and to other authors who will draw his attention to their publications on subjects with which it deals.

WEMBLEY,
    MIDDLESEX.
        February, 1935.

CONTENTS

EGYPT

EGYPT I

GENERAL

CHRONOLOGY

Egyptian Chronology begins really with King Menes, although the calendar year is supposed to have been introduced at Memphis in 4241 or 4238 B.C.[1]

The following chronological scheme[2] is adopted in the present book:

1. *Predynastic Period:* to 3400 B.C.; divided by Petrie into *Sequence Dates:* earliest, S.D. 30–40; middle, S.D. 40–60; later, S.D. 60–78.

2. *Dynastic Period*, beginning with the union of Upper and Lower Egypt under King Menes, 3400 B.C.

	B.C.
(*a*) Dynasties I and II (*Thinite dynasties*)	3400–2980
(*b*) *Old Kingdom:* III–VI dyns. III 2980–2900; IV 2900–2750; V 2750– 2625; VI 2625–2475	
(*c*) *First Dark Ages:*	
VII–VIII dyns.	2475–2445
IX–X dyns.	2445–2160
(*d*) *Middle Kingdom:* XI–XII dyns. . XI 2160–2000; XII 2000–1788	
(*e*) *Second Dark Ages:*	
XIII–XVII dyns., including the rather obscure Hyksos Period, beginning *c.* 1685	1788–1580
(*f*) *The Empire* (*New Kingdom*):	
Early part rather obscure.[3]	
XVIII dyn.	1580–1350
XIX dyn.	1350–1205
XX dyn.	1200–1090
(*g*) *Period of Decadence:* beginning with end of XX dyn. .	1150– 662
α Decline of XX dyn.	1150–1090
β Tanite-Amonite Period: XXI dyn. . . .	1090– 945
γ Libyan Period: XXII–XXIV dyns. . . .	945– 745
δ Ethiopian Period: XXV dyn. 712–663 . .	745– 663
ε Assyrian Supremacy	670– 662
(*h*) *Restoration: Saite Period:* XXVI dyn.	663– 525
(*i*) *Persian Conquest:* XXVII dyn. Egypt a Persian province with short interruptions by native (XXVIII–XXX) dyns. .	525– 332

[1] Rostovtzeff, History of the Ancient World, Oxford, 1926, i, 32; Scharff, q. in JEA, 1927, xiii, 282; Glanville, *ib.*

[2] Breasted, History of Egypt, 1921, 21; *ib.*, Ancient Records of Egypt, 5 vols., Chicago, 1906–7, i, 25; Wiedemann, Das alte Ägypten, Heidelberg, 1920, 1; E. Meyer, Geschichte des Altertums, 1913, I, ii, 17 f.; *ib.*, Nachtrag to vol. i, Die ältere Chronologie Babyloniens, Assyriens und Ägyptens, 1925, 40, 68; Erman and Ranke, Ägypten und ägyptisches Leben im Altertum, Tübingen, 1923, 39 f.; Lieblein, Recherches sur l'Histoire et la Civilisation de l'ancienne Égypte, 3 fasc., Leipzig, etc., 1911–4 (pagination continuous), 176; Peet, JEA, 1920, vi, 149; 1922, viii, 5; *ib.*, CAH, i, 248; Weill, q. in JEA, 1928, xiv, 193; Hall, CAH, ii, 407; Scharff, JEA, 1928, xiv, 272; Petrie, *Ancient Egypt*, 1915, 19.

[3] Breasted, CAH, ii, 60; Budge, The Mummy[2], Cambridge, 1925, 51; Meyer, Alt., 1928, II, i, 110.

B.C.

(*j*) *Alexander the Great*, and his successors (*Diadoches*),
    Ptolemies I–XVI 332– 30
(*k*) *Roman Province* under a Prefect from 30

PREHISTORIC EGYPT

The period before Menes, perhaps from 5000 B.C., is still obscure, although much new knowledge has come to hand.[1] In the period to the end of the III dyn., sometimes called Archaic, brick buildings, hieroglyphic writing, sculpture, metal working, pottery and glazing had all developed.[2] The end of the Predynastic to the beginning of the IV dyn. is also called Protodynastic.[3] In the earliest known civilisation (Tasian) in Upper Egypt [4], the pottery mostly consists of wide bowls narrowing to the top, with small flat bases, and incised black beakers already known are Tasian. Rectangular palettes of limestone or "alabaster" (really a form of calcite) and polished stone conoid axes occur, but no corn, linen or metal. Neolithic sites at Beni Salamah, Helwân, and the Fayyûm, contain no metal, but large quantities of red and black polished and unpolished pottery, with fragments of stone vessels.[5] Next in order of date is the civilisation at Badari, and following this comes the quite different, long known, predynastic culture at Naqada.

The Badarian civilisation (S.D. 30 or earlier), a passing isolated phase known from graves at Qau and Badari [6], has resemblances in culture to those of early Predynastic Egypt (Amratian) and Nubia, and it perhaps developed and degenerated in Nubia and was in relation with Syria, the Delta, and "the North." The people wore skins, leather and linen ; wove basket work ; made excellent "ripple marked" pottery without the wheel, spreading the clay with a notched bone comb ; ate cereals ; used copper for borers. The race was perhaps mixed, although there is some evidence that the Badarians were new arrivals from Asia, perhaps related to the earlier peoples (Dravidians and Kolarians) of India.[7] They used ground stone implements, which disappeared in Egypt after them.

The hypothesis of an ancient centre of culture in Nubia, exporting bronze containing nickel, and related to Egypt, Mesopotamia and India,[8] is doubtful.[9] The ruins at Zimbabwe in Rhodesia are not earlier than the 9th century A.D.[10] A source of Egyptian culture in Mesopotamia, suggested on the basis of resemblances of predynastic and early dynastic Egyptian pottery, stone vessels, statuettes, mace-heads, cylinder seals, copper hatchets, etc., to those on old pre-Elamite sites at Susa[11] is regarded[12] as unproved and improbable, and

[1] Cf. Scharff, *Morgenland*, 1927, Heft xii ; *ib*. JEA, 1928, xiv, 272 f.

[2] BMGE, 193.

[3] Brunton, Gardiner and Petrie, Qau and Badari I, 1927, 10.

[4] Brunton, JEA, 1930, xvi, 164 ; *ib*., *Ancient Egypt*, 1930, 7, 93.

[5] Peet, JEA, 1930, xvi, 164.

[6] Petrie, *Ancient Egypt*, 1924, 33 ; *Nature*, 1927, cxix, 159 ; Brunton and Thompson, Badarian Civilisation and Predynastic Remains near Badari, 1928, 39 f.

[7] Scharff, *Morgenland*, 1927, xii, 16 ; Peake and Fleure, The Way of the Sea, Oxford, 1929, 6, 14 f., 76.

[8] Frobenius, Das unbekannte Afrika, München, 1923.

[9] Otto, Kulturgeschichte des Altertums, München, 1925, 21 f.

[10] Woolley, Digging up the Past, 1930, 22 ; *Nature*, 1931, cxxvii, 884.

[11] De Morgan, La Préhistoire Orientale, 1926, ii, 248, on " l'Origine Chaldéen de la Culture Pharaonique en Égypte " ; cf. Breasted, H., Figs. 10, 11 ; Hall, CAH, i, 263, 583 ; Woolley, Digging up the Past, 1930, 90, plate xvii ; *ib*., The Sumerians, Oxford, 1928, 6, 185 ; Frankfort, Studies in Early Pottery in the Near East, *Roy. Anthrop. Inst. Occasional Papers*, Nos. 6 and 8, 1924–7 ; Budge, History of Egypt, 1902, i, 63 ; Langdon, *Nature*, 1921, cvii, 315 ; Elliot Smith, EB[11], 149, opposite view, criticised by Capart, JEA, 1930, xvi, 101 ; S. Smith, Early History of Assyria, (EHA), 1928, 52 ; Evans, Palace of Minos at Knossos, 1928, ii, 26 f. ; Meyer, Chronol., 69 ; Scharff, *Morgenland*, 1927, xii, 16.

[12] Otto, 1925, 23.

there has been a tendency to assume too early dates for the oldest remains in Mesopotamia.

The hieroglyphic name for Egypt, supposed to represent a heap of charcoal, a crocodile's tail or a piece of fish skin, is qemi or chemi, with the meaning black—the black land ; chemia is a rare form.[1] The nature and origin of the earliest races inhabiting Egypt are very obscure and disputed, but there appear to have been incursions of Libyan tribes from the desert on the west, whose descendants, the modern Berbers, still produce hand (not wheel) made pottery of the same form as their ancestors 5,000 years ago[2]; also tribes from Nubia, Asia, perhaps from the Syrian coast, from Arabia, and from the shores of the Red Sea, including people closely related to, if not the same as, the Sumerians.[3]

The early Egyptians called their original home Punt, a place of doubtful and perhaps varying locality, which they also called "God's land." It probably meant some place on the east African coast, perhaps the Somali coast at the south end of the Red Sea, where there were pile-dwellings, although many other localities, including the coast of Arabia near Bâb al-Mandib, have been proposed.[4] Expeditions to Punt in search of incense began in 3000, but were more frequent after 2000.[5] Settlements of Semites from Mesopotamia in South Arabia and Somaliland, trading lapis lazuli from Persia, and ivory and obsidian from Somaliland with the predynastic Egyptians in exchange for gold, copper, basalt and "alabaster," [6] are problematical.

Various types of boats, at first bundles of reeds and papyrus stems, later (III dyn.) of wood (perhaps imported from the Syrian coast, since there were no suitable trees in Egypt), were used from the Predynastic Period ; they are known from representations and models, some very large. Large wooden boats, used from the IV dyn., sailed to the Syrian coast and on the Red Sea perhaps as early as 2900. Large two-masted merchant ships went to Punt, Cyprus and the Ægean in the XVIII dyn.[7] There were early relations with the Syrian coast (Byblos) from the I–II dyns. [8], and with Crete.[9]

Materials used in the Predynastic Period (4000 or earlier)[10], include ornaments of stone, bone and ivory ; bored beads of flint, rock-crystal or quartz, carnelian, agate, hæmatite, specular iron ore, other minerals and semi-gem stones, and (very rarely) of amber.[11] The flint implements were very finely worked.[12]

[1] Wilkinson, Manners and Customs of the Ancient Egyptians, ed. Birch, 3 vols., 1878, i, 10, iii, 200 ; Wiedemann, 1920, 14, 188 ; Maspero, The Dawn of Civilisation, 1894, 43 ; Plutarch, On Isis and Osiris, 33 ; Horapollo, Hieroglyphica, i, 70 ; Jablonski, Pantheon Ægyptiorum, 3 vols., Frankfurt, 1750–2, i, 97 ; Budge, The Mummy, 2 ed., Cambridge, 1925, 1 f.; ib., Egyptian Hieroglyphic Dictionary, 1920, 771—really Upper Egypt.

[2] Bates, Ancient Egypt, ii, 158, 165.

[3] Wiedemann, 1920, 13, 35, 47 ; Breasted, H.; Erman-Ranke, 34 ; Otto, 1925, 21 ; Reisner, Early Dynastic Cemeteries of Naga-ed-Dêr, Univ. of California Publ. No. 2, Leipzig, 1908, 134 ; Peake and Fleure, Priests and Kings, Oxford, 1927, 61 f.

[4] BMGE, 20, 22 ; Sethe, Der alte Orient, 1923, xxiii, 35 ; Budge, Mummy², 34, 69 ; Lieblein, Handel und Schiffahrt auf dem roten Meere in alten Zeiten, Christiania, 1886, 52 f. ; ib., Recherches, 1911–4, 38, 210 f. ; Maspero, Dawn, 395, 426, 461, 496 ; Erman-Ranke, 599 ; Meyer, Alt. II, i, 118, 140 ; Breasted, H., 127, 274 f., 319.

[5] Erman-Ranke, 599.

[6] Hornblower, JEA, 1927, xiii, 243.

[7] BMGE, 102 ; BMGE, IV–VI Rooms, 115, 248 ; Wiedemann, 1920, 212 ; Boreux, Isis, 1928, ix, 215 ; 1931, xvi, 150 ; JEA, 1927, xiii, 122 ; Egerton, Isis, 1925, vii, 264 ; Mackenzie, Footprints of Early Man, 1927, 99 ; Breasted, H., 127, 304, Figs. ; Köster, Das antike Seewesen, 1923, 10 ; ib., Seefahrten der alten Ägypter, 1926 ; ib., Beih. alt. Orient, 1923, i, 1 ; O'Leary, Arabia before Muhammad, 1927, 45 f., 51.

[8] De Morgan, Préhist. Orient, ii, 248 ; Contenau, La Civilisation Phénicienne, 1926, 26, 35, 41, 45 f., 53, 151.

[9] Pendlebury, JEA, 1930, xvi, 75.

[10] Quibell, Egyptian History and Art, 1923, 8.

[11] Wiedemann, 1920, 46 ; Breasted, H., 25 f. ; Forrer, Reallexikon der Altertümer, 1907, 87 ; Capart, Primitive Art in Egypt, tr. Griffith, 1905, 47 ; Petrie, Prehistoric Egypt, 1920, 43 f. ; ib., Diospolis Parva, 1901, 27, plate iv.

[12] Quibell, 5 f. ; Wiedemann, 1920, 47.

Ivory was known very early, although some is probably hippopotamus tooth.[1] There are bone figures with inlaid eyes of lapis lazuli ; tortoiseshell arm-bands bound with copper wire ; arm-bands of stone or sea shells ; and bone combs and hairpins.[2] Very characteristic are the carved slate palettes, perhaps for mixing cosmetics [3], or [4] sacred stones "anointed with green colour as the Israelites anointed the altar." On them, and elsewhere, is a green pigment called mafkat or mafek, used for painting under the eyes, which is malachite imported from Sinai (Mafkata, "land of the blue-green stone" [5]), Egyptian To Shûît, "land of emptiness," [6] the Horeb of the Bible.[7] These palettes went out of use in the III dyn., but painting under the eyes remained until the IV dyn., when the actual use of the green pigment, less common in Badarian than in Predynastic remains [8], ceased. The ritual texts still refer, in the Old Kingdom, to painting images with green (uat or wod) and black (mestem) pigments, used in magic ceremonies and incantations.[9] Predynastic slate palettes from Abydos were in the shape of fish, with eyes, fin, tail, etc.[10] A red pigment, called powdered "hæmatite" but probably ochre, is found in Predynastic and Badarian remains,[11] with grinding slabs of black and white porphyry and grinding pebbles of brown jasper.

Bricks of sun-dried, not baked, clay were made in the Thinite, possibly also in the Predynastic Period.[12] The pottery, not made on the wheel in this period[13], is red or black, owing to iron in different stages of oxidation[14], and unglazed but finely polished.[15] Predynastic pottery and flint are better than those of the old dynasties, but the latter excelled in polished stone vases of porphyry and diorite.[16] There is glazed ware, usually bright green coloured with copper, on beads, plaques, etc.[17] Actual glass was practically unknown, but is represented by a Predynastic bead.[18] At a somewhat later period wire and implements of copper, probably from Sinai[19], occur. Copper is found only sparingly in the oldest tombs[20], and Hall thinks Cyprus, Syria and Sinai, not Egypt, were the original foci for its use ; since there is no timber in Egypt suitable for the masts of ships, the inhabitants of the Syrian coast are supposed to have carried by sea the first known copper from the interior of Asia to Egypt and Crete from about 4000, and Sumerian copper working is supposed to be older than Egyptian. Egypt, Crete and Cyprus then developed independent copper industries, Egypt in Sinai.[21] Elliot Smith[22], on the contrary,

[1] Petrie, Researches in Sinai, 1906, 144 ; Perrot and Chipiez, History of Art in Ancient Egypt, 2 vols., 1883, i, 839 ; Rostovtzeff, i, plate 9 ; Mackenzie, Footprints, 147.
[2] BMGE, 25 ; Wiedemann, 1920, 46 ; Brunton, Qau and Badari I, 31.
[3] Capart, Art, 21 f. ; Breasted, H., 27, 40, Fig. 19 ; Erman-Ranke, 257 ; Meyer, Alt. I, ii, 52, 137.
[4] Wiedemann, 1920, 46, 146.
[5] Breasted, H., 27 ; Budge, Mummy, 1893, 12 ; Brunton and Thompson, Badarian Civ., 3, 15, 31 ; Meyer, Alt. I, ii, 52, 137 ; Petrie, Prehistoric Egypt, 1920, 43f.
[6] Maspero, Dawn, 349.
[7] Cook, CAH, iii, 370 ; Smith, DG, ii, 1003.
[8] Brunton and Thompson, Badarian Civ., 31, 46.
[9] Capart, Art, 28 ; Meyer, Alt. I, ii, 147 ; Erman-Ranke, 257 ; Brunton, Qau and Badari I, 63 ; Brugsch, Die Ägyptologie, Leipzig, 1891, 54.
[10] Frankfort, JEA, 1930, xvi, 215, plate 31.

[11] Reisner, E. Dyn. Cem., 127 ; Erman-Ranke, 257 ; Wiedemann, in Petrie, Medum, 1892, 41 f. ; Brunton, Qau and Badari I, 55 ; Brunton and Thompson, Badarian Civ., 3, 6, 7, 16, 34, 39 : some perhaps protodynastic.
[12] BMGE, 103 ; Breasted, H., 27 ; Erman-Ranke, 196.
[13] Budge, Hist., i, 92.
[14] Ib., 97.
[15] Breasted, H., 27.
[16] Reisner, Cem., 132 ; Breasted, H., 28 f.
[17] Breasted, H., 28 ; Erman-Ranke, 40 ; Wiedemann, 1920, 46.
[18] Rathgen, Sprechsaal, Berlin, 1913, xlvi, 98.
[19] Breasted, H., 28 f. ; Reisner, Cem., 132 ; Erman-Ranke, 549 ; Wiedemann, 1920, 46.
[20] De Morgan, Préhist. Or., ii, 110, 136.
[21] Hall, Ancient History of the Near East, 5th ed., 1920, ref. as NE, 90 ; ib., Civilisation of Greece in the Bronze Age, 1928, . 31 f., 37 f.
[22] EB[11], xx, 146 ; ib., The Ancient Egyptians, 1923, 77, 102, 170.

supposes that the working of gold and copper preceded the building of stone monuments by some centuries, began in Upper Egypt or Nubia, and spread from there to Palestine and Syria, Elam, Asia Minor, Cyprus and the Ægean.

Gold, silver and lead were rare in the Predynastic Age, but were all in use[1]; iron (meteoric), known but exceedingly scarce, was used for decorative beads.[2]

THE THINITE PERIOD

The technical arts of the first two (Thinite) dynasties (3400–2980) were well developed, and hieroglyphic writing was used. There were glazed plaques, some of which (II dyn.) are in the British Museum, and glazed tiles even before the I dyn.[3]; very thin diorite and also wrought copper bowls and gold work and jewellery showing mastery of work and "soldering."[4] A gold bar with the name of Menes (3400) is one of the earliest known pieces of jewellery; amethyst and turquoise set in gold were found in tombs at Abydos, e.g. the carnelian bracelets of a queen of the I dyn., now in the Cairo Museum, and jewellery of gold, lapis lazuli, carnelian, amethyst, "turquoise or blue glaze."[5]

There is mention of a vessel of asem (electrum, a natural alloy of gold and silver), with a representation of it, in a I dyn. inscription at Abydos.[6] There were vases of rock crystal and of costly stone, with wrappings of gold leaf tied with string.[7]

In the second half of the II dyn., in which there is some falling off in fine work perhaps due to the introduction of metal tools, stone temples began to be built, and the III dyn. step pyramid of King Zoser at Saqqâra is the oldest stone building in the world.[8]

Copper was very scarce in the Predynastic Age and used only for ornaments, but at the end of this period copper tools, at first chisels, make their appearance; these copper chisels, drill cutters, etc., continued in use for many centuries, flint knives, arrow heads, etc., being made long after the introduction of copper, and stone tools until 2000. Egypt seems to have passed direct from the Palæolithic to the Copper Age (including gold and knowledge of iron). In the III dyn. metal was plentiful; at its end Egypt had emerged from the Archaic Period.[9] The copper mines of Wâdî Maghâra, in the peninsula of Sinai, operated for over two thousand years, were probably worked in the I dyn.[10] They were called chetiu mafkat, "steps of malachite."[11]

THE OLD KINGDOM

The Old Kingdom (III–VI dyns.) was largely agricultural. There were granite quarries at the First Cataract, all kinds of granite being used, followed by sandstone, huge blocks of which were quarried at Silsileh. Finer and harder stones were obtained chiefly at Hammâmât, between Coptos and the Red Sea, "alabaster" from Hâtnûb, behind Amarna, where it is accompanied by calcite.[12] Masûdî (d. A.D. 957) speaks of "old" granite works at Aswân, the

[1] Breasted, H., 28; Quibell, 7.
[2] Petrie, Wainwright and Mackie, Labyrinth, Gerzeh and Mazghuneh, 1912; Wiedemann, 1920, 46.
[3] Petrie, Hist., 1903, 30; Breasted, H., 39, 43; Meyer, Alt. I, ii, 151; BMGE, 193.
[4] Petrie, Hist., 17, Fig. 8A.
[5] Breasted, H., 34, 39 f.; Petrie, Royal Tombs of the First Dynasty, Part I, 1900; Part II, 1901, passim; Quibell, 15, plate I.
[6] Breasted, H., 43, Fig. 27; Petrie, Hist. of Egypt, 1903, 15.
[7] Quibell, 15.

[8] Breasted, H., 46; Budge, Osiris, or the Egyptian Resurrection, 2 vols., 1911, i, 248; Quibell, 16, 17, 25 f.
[9] Mackenzie, Footprints, 144; Quibell, 15 f., 18.
[10] Gsell, Eisen, Kupfer und Bronze bei den alten Ägyptern, Dissertation, Karlsruhe [1910], 1; Breasted, H., 48; Quibell, 15.
[11] Meyer, Alt. I, ii, 137; Gsell, 5, 50; Brugsch, Religion und Mythologie der alten Ägypter, Leipzig, 1885–8, 155.
[12] Breasted, H., 93 f.; Maspero, Dawn, 12, 384; Erman-Ranke, 560; Petrie, Tell el-Amarna, 1894, 3.

blocks being polished by sand.[1] Limestone was found in many places, particularly at Ayan or Troia, opposite Memphis. The stones were cut with tubular copper drills and long notched copper saws with the aid of sand or emery for the hardest stones : bronze was unknown and iron tools were not used for cutting stone, although iron may have been used to a limited extent for tools.[2] A tool-box, with mallet, chisel, hatchet, knife, awl, etc., is represented in a III dyn., and a saw in a Middle Kingdom, picture : they are not often shown.[3]

The three great pyramids of Gîzah were built as tombs by Khufu (Cheops of Herodotos ; Souphis of Manetho), Khafra (Chephren), and Menkaura or Menkarā (Mencheres of Manetho), kings of the IV dyn.[4] There are two small alchemical works attributed to Sophê [5], perhaps Souphis (c. 3000 B.C.), since Manetho (c. 300 B.C.) says he wrote a book on medicine, and a copy of such a book of the IV dyn., found at Coptos, is in the British Museum.[6]

Copper, used for vessels, was obtained during the Old Kingdom in relatively large quantities from the peninsula of Sinai, where mining was said to have been "founded" by Snefru (IV dyn.), and whence also came malachite and (or) turquoise.[7] In the II dyn. the art of working copper had reached a high state of perfection, as the actual objects prove.[8] A metal-worker's shop of the Old Kingdom shows (above ; left) the weighing of precious metals and malachite ; a furnace with men blowing the fire with the mouth through tubes (centre) ; cutting and hammering metal (right) ; and putting together necklaces and costly ornaments (below). According to Breasted, bronze, an alloy of copper and tin, was not yet in use, bolts, nails, etc., being always of copper, but bronze was known in the III dyn. or earlier.[9] The early bronzes usually contain very little tin, and true bronze probably appeared about the XII dyn., after which it was freely used.[10] Tin is not found in Egypt, and its source is mysterious, Persia[11] being most probable. Silver, obtained very sparingly from Cilicia in Asia Minor, or Syria, was then rarer than gold.[12] Gold was obtained from quartz veins in the granite mountains along the Red Sea (mostly by foreigners), at Wâdî Foâkhir on the Coptos Road. It also came by trade from Nubia, where it was found in the eastern desert. Gold and copper rings, weighed against marked stone weights, were the oldest currency in Egypt.[13]

Sun-dried bricks were used for poorer buildings ; wood was scarce and expensive. Cedar wood was imported in the V–VI dyns., probably from Lebanon, Cilicia or North Syria, ebony and ivory from the south ; cedar oil was used in embalming.[14] "Alabaster," diorite and porphyry vases were

[1] Reitemeyer, Die Beschreibung Ägyptens im Mittelalter, Leipzig, 1903, 97.

[2] Breasted, H., 93 ; *Ancient Egypt*, 1915, 190.

[3] Wiedemann, 1920, 336 f.

[4] Budge, Mummy², 31 f., 405 f. ; Maspero, Dawn, 364.

[5] Berthelot, Collection des anciens Alchimistes Grecs, 3 vols., 1887–8, i, 211 f.

[6] Laqueur, PW, xiv, 1063 ; Cory, Ancient Fragments, 1832, 102 ; Müller, Fragmenta Historicorum Græcorum, Didot, Paris, 1841–70, ii, 549 ; Budge, Gods of the Egyptians (G. of E.), 2 vols., 1904, i, 524 ; Maspero, Dawn, 380.

[7] Breasted, H., 93 ; *ib.*, Ancient Records, i, 75 ; Erman-Ranke, 558 f. ; Maspero, Dawn, 349 ; de Morgan, Recherches sur les Origines de l'Égypte, 2 vols., 1896–7, i, 216 f. ; Gsell, 1 f. ; BMGE, IV–VI Rooms, 244.

[8] BMGE, IV–VI Rooms, 251.

[9] Breasted, H., 94, Fig. 41 ; Petrie, Medum, 1892, 36 ; Maspero, Dawn, 59, 393 ; *ib.*, Manual of Egyptian Archæology, tr. Edwards, 1902, 303 ; Petrie, *Ancient Egypt*, 1915, 17 ; Mosso, Dawn of Mediterranean Civilisation, 1910, 57 ; Myres, CAH, i, 103 ; Hall, NE, 136.

[10] Hall, CAH, i, 291, 319.

[11] Elliot Smith, EB¹¹, xx, 146, 151 ; Mackenzie, Footprints, 144 : N. Persia or N. Mesopotamia, about 3000.

[12] Breasted, H., 104 ; Petrie, Prehistoric Egypt, 1920, 27, 43 ; Evans, Pal. of Minos, 1928, ii, 845, suggests Crete ; Armenia was the richest source in Nearer Asia ; Schrader, Reallexikon der indogermanischen Altertumskunde, Strassburg, 1901, [RL], 764.

[13] Breasted, H., 93 f.

[14] Maspero, Dawn, 392 ; Erman-Ranke, 611.

now giving way to pottery of rich blue or green glazed faience, also coarse jars and clay pottery without glaze, but glass was not yet fully developed as an independent material.[1] Stone lamps, with bowls for the oil and floating wicks, were in use.[2] The potter's wheel was perhaps invented about 4000, within range of the ancient civilisations of Egypt and Mesopotamia, and was known to the Egyptians as early as 3500.[3]

Leather, made by curing hides with some vegetable tannins, was used from Predynastic times, and the earliest clothing was of skins.[4] Dyeing in various colours was used. Flax was plentiful, and the royal garments were of extremely fine semi-transparent linen, hardly distinguishable from silk. Papyrus had several uses—for sandals, light skiffs, and especially paper. Carbon ink was in use. The papyrus paper was an important manufacture in the Old Kingdom and it was probably exported to Phœnicia in the 12th century B.C.[5]

Medical papyri appear in the Old Kingdom, recipes being referred to old kings, such as Cheops. The medical man occupied an important position and his art was usually severely practical. The Edwin Smith Surgical Papyrus is free from magic except in a later appendix which contains charms, perhaps for use as an alternative or addition to treatment. The pharmacological knowledge in the papyri, which afterwards became incorporated into Greek medicine, is considerable. Imhotep, an official of the III dyn. under Zoser, afterwards became a god of medicine.[6]

The V dyn. began at Memphis about 2750. The symbol of the sun god, Rā, was now an obelisk exposed to the sky and capped with metal, usually copper, though electrum is mentioned in inscriptions.[7] The pyramid of Unàs, the last king of the V dyn., at Saqqâra, is the earliest containing religious inscriptions, the so-called Pyramid Texts.[8]

In a representation on a fragment of limestone, probably from Memphis (V dyn.), fringed woollen garments of Mesopotamians (Ninevites ?) visiting Egypt are clearly shown.[9] Sahurā (V dyn.) sent an expedition to Punt to fetch anti (incense), asem (electrum), ivory, and precious wood, probably ebony.[10]

In the VI dyn., which was a period of decline, gold came from the Sudan and even Central Africa, by negro traders down the Nile : the land Kash, or Kush, was Ethiopia. Iron ore occurs in North Nubia, but no workings are found. The Sinai copper mines continued to be worked to the time of Pepi II, but then declined till the XII dyn., when new deposits further north were discovered.[11]

Dynasties VII to XI represent what is still a very obscure period, the "First Dark Ages" of Egypt. At the close of the VI dyn., Nubians, negroes and Asiatics invaded the Nile Valley, the latter perhaps driven out of Mesopotamia by the Amorites, or by the southward spread of the people afterwards called Hittites in Asia Minor. This was a period of confusion : a text says that irrigation was neglected, trade was at a standstill, and cedar wood, grain, gold and charcoal

[1] Breasted, H., 93 f. ; Neumann, Z. f. angew. Chem., 1925, xxxviii, 776.
[2] Erman-Ranke, 217.
[3] H. S. Harrison, Pots and Pans : The History of Ceramics, 1928, 38 ; Mackenzie, Footprints, 144.
[4] Breasted, H., 93 f. ; Schrader and Jevons, Prehistoric Antiquities of the Aryan Peoples, 1890, 328 ; Forrer, Reallexikon der prähistorischen, etc., Altertümer, Stuttgart, 1907, 450, 498, plate 125.
[5] Breasted, H., 93 f.
[6] Breasted, H., 101, 112 ; ib., The Edwin Smith Surgical Papyrus, Chicago, 1930 ; Goodwin, TSBA, 1874, iii, 38 ;

J. B. Hurry, Imhotep, Oxford, 1926 ; Petrie, Ancient Egypt, 1924, 121.
[7] Quibell, 40 ; BMGE, 105 ; Meyer, Alt. I, ii, 203 ; Maspero, Arch., 34 ; Breasted, Anct. Rec. ii, 37, 127 f. ; ib., H., 122 f. ; Erman-Ranke, 320 ; Capart, L'Art Égyptien, Brussels, 1924, i, 166.
[8] Breasted, H., Fig. 73.
[9] W. M. Müller, Egyptological Researches, Carnegie Inst., Washington, 1906, i, 9.
[10] Breasted, H., 127 ; Maspero, Dawn, 396, 496.
[11] Quibell, 46 ; Gsell, 3 ; Maspero, Dawn, 473, 474.

were lacking. Ipuwer, in a Leyden papyrus called "Admonitions of an Egyptian Sage," complains that "gold, silver, bronze (chesmen), lapis, malachite, carnelian, and stone of Yehbet are on the necks of female slaves."[1] In 2160 Intef, a noble of Thebes, founded the XI dyn. A sea captain, Akhthoi, worked the mines of Sinai for the first time since Snefru, "punished the Asiatics in their countries" and returned with "new, shining and hard metals" (perhaps bronze ?), lapis (?), precious stones and other commodities previously unknown.[2]

THE MIDDLE KINGDOM

The Middle Kingdom, XI–XII dyns., associated with the rise to prominence of the city of Thebes (Egyptian Apt or Was), was an important period. Under Sesostris (Egyptian Senusret) I and II (1970–1887) intercourse with the Nubian gold mines improved.[3] Under Sesostris III (1887–1849) silver was still more valuable than gold. There were expeditions to Punt, now less than two months' journey, and intercourse with Minoan Crete. Sesostris III subdued Syria and invaded Palestine, where an official left a stela.[4] In the earliest period (c. 3000) Egyptian shipping plied between Crete and the Phœnician coast and Egypt ; from the beginning of the Middle Kingdom Cretan ships replaced Egyptian, and this continued until the destruction of Minoan power about 1200, when Phœnician sea traders took their place, their sea-power reaching its zenith in 1100–800, when it in turn gave place to Greek.[5] A canal dug in this period connecting the Red Sea with the Nile in the eastern Delta remained in use until it was blocked by the Arabs after the conquest of Egypt in A.D. 640.[6]

Amenemhet III (1849–1801) completed the reclamation, by draining, of 27,000 acres in the Oasis of Fayyûm, called by Greek authors Lake Moeris (Egy. mer uer, great lake).[7] He reorganised the working of the copper mines of Sinai, formerly worked only during the cool season for a few months of the year, but now in the intense heat when "the mountains brand the skin," opened new mines, and in his inscriptions at Wâdî Maghâra says they were "made for the goddess Hathor, mistress of the malachite (or turquoise) country."[8] Each mine had a foreman and there were periodical visits of treasury officials shortly after 1849. Bronze came into common use in the XII dyn. [9], when there were relations with Crete and Asia. Amenemhet III left Egypt rich and prosperous, since all the gold in the Sudan and the Eastern Desert was at his disposal, but after his death the country was soon robbed of its wealth.[10]

The close of the XII dyn. is marked by the invasion of Egypt by an Asiatic people called by Manetho the Hyksos, perhaps from the Egyptian Hq-hs-wt, the title of a king. They were a civilised people, ruling Egypt with the capital at Avaris in the Delta, bringing new mathematical knowledge, the horse (Babylonian sisû, Canaanite and Hebrew sûs, Egyptian ssm-t) and warfare on a large scale with large chariots and bronze scimitars, and introducing

[1] Gardiner, The Admonitions of an Egyptian Sage, Leipzig, 1909, 11, 31.
[2] Olmstead, History of Palestine and Syria, New York, 1931, 83 f. ; Peake and Fleure, The Steppe and the Sown, 1928, 139 f. 143, 150 ; Budge, Mummy[2], 39.
[3] Breasted, H., 180 f., 185 ; Erman-Ranke, 26 ; Budge, Mummy[2], 3 ; Quibell, 50, 95 ; Blümner, Terminologie und Technologie der Gewerbe und Künste der Griechen und Römern, 4 vols., Leipzig, 1875–87, iv, 13.
[4] Wilkinson, i, 33, 274 ; Breasted, H., 185, 396 ; Maspero, Dawn, 396, 426, 461, 496 ; Quibell, 64.

[5] Evans, The Palace of Minos at Knossos, 1921, i, 19 ; Glotz, Ægean Civilisation, 1925 ; Smith and Marinden, DA, ii, 209 ; Köster, Beiheft. alt. Orient, 1924, i, 19 f. ; Otto, 1925, 47 ; A. Rel., 1925, xxiii, 60.
[6] Breasted, Ancient Times, Boston, 1916, 79 ; Reitemeyer, 39.
[7] Meyer, Alt., 1909, I, ii, 266 ; Herodotos, ii, 149 ; Strabo, XVII, i, 37, 809 C. : Quibell, 63 ; Breasted, H., 185.
[8] Breasted, Anct. Rec. i, 314, 316 f., 318, 321 ; ib., H., 185 f. ; Gsell, 5.
[9] Erman-Ranke, 550.
[10] Budge, Mummy[2], 45.

foreign elements, including the god Tešub, afterwards the Egyptian Set.[1] In this period a form of writing was developed in Sinai, perhaps by Semitic miners using Egyptian signs, which is sometimes regarded as the precursor of the Phœnician alphabet.[2]

It is supposed [3] that the Hyksos, who entered Egypt in 1685, were essentially the autochthonous Khurri (Subaræans) of Syria, with Khatti, Hittite and Semitic elements, and perhaps led by Mitanni, but Petrie [4] thinks they came from the Caucasus, using weapons from Crete or Cyprus. The relations of the Hyksos with Crete, however, are obscure.[5]

THE NEW KINGDOM

The Hyksos were overthrown about 1580 by Theban princes, who began the XVIII dyn. with Aahmes (Ahmosis) I, with Thebes as the capital, and under this great dynasty Egypt and Crete superseded the Asiatics in the dominion of the world. Great quantities of tribute came to Egypt from Syria and Palestine. The Phœnicians were strongly influenced by Egypt, and their sea trade played a part in Egyptian commerce from about 1450.[6]

In XVIII dyn. representations the Minoans bring electrum vases, the Syrians bring gifts of "silver, as it comes from the mines, blue stone [lapis], green stone [malachite or turquoise], all precious stones, bars of copper and tin, as much as can be counted"; Nubians bring "gold in the crude state" (?), "fine gold," ebony, ivory, precious red stone, "yellow stone," etc.[7] The Egyptian king had great stores of gold, mostly from Nubia [8], and gold rings, as well as copper, were used as money. The later XVIII dyn. records speak of the Mentu of Satet, who had appeared in the XII dyn. in the Delta. Their country, which was marshy and produced copper and gold, was probably Western Asia just north of the Euphrates, between the Khabur river and the Jabal Druse.[9] Thothmes I repaired the temple of Osiris at Abydos and made a shrine boat of gold, silver, lapis lazuli, copper and precious stones.[10] Queen Hatshepsut (whose relation to Thothmes is doubtful)[11], raised monuments and built the temple of Deîr el-Baharî, west of Thebes, in which the architect Senmut has cut his name in hidden places.[12]

Thothmes I, in 1479 or 1482, captured Megiddo from the Syrians with great spoil, and conquered Syria and Palestine.[13] He took gold, electrum, silver, bronze (including armour), blocks of copper, lead, lapis lazuli, malachite, ivory, precious woods and wines from the Hittites[14]; imported gold in rings,

[1] Josephus, Contra Apion., i, 14; Maspero, Struggle of the Nations, 1896, 54 f.; Quibell, 71 f.; Breasted, H., 214, 217, 219; ib., CAH, i, 291; Evans, Palace of Minos, 1921, i, 421; Laqueur, PW, xiv, 1064 f.; Peet, Egypt and the Old Testament, Liverpool, 1922, 67 f.; ib., JEA, 1932, xviii, 116; Meyer, Alt. II, i, 41 f.

[2] Otto, 1925, 49; Cowley, JEA, 1929, xv, 200; Olmstead, H. Pal. Syr., 90.

[3] Von Oppenheim, Tell Halaf, Leipzig, 1931, 59; Meyer, Alt. II, i, 41; Hrozný, EB[14], xi, 605.

[4] Brit. Assoc. Journ., 1930, 63.

[5] Cf. Evans, Palace of Minos, i, 421; Meyer, Alt. II, i, 43; Glotz, Ægean Civilisation, 1925, 26, 44, 215.

[6] Breasted, H., 259 f., 322, 516; Meyer, Alt. II, ii, 66; Pendlebury, JEA, 1930, xvi, 76 f., 82, 89; Meyer, Alt. II, i, 138;

Olmstead, H. Pal. Syr., 143 f.; Erman-Ranke, 615 f.

[7] Müller, Egyptological Researches, 1906, i, 24; Davies, Tombs of Menkheperrasonb, etc., 1933, 6 f., 11, complete list of objects and materials; cf. Breasted, H., Fig. 118.

[8] Meyer, Alt. II, i, 66, 71.

[9] S. Smith, Early History of Assyria, 1928, 221 f.

[10] Budge, Osiris, or the Egyptian Resurrection, 2 vols., 1911, ii, 15.

[11] Budge, Mummy², 51 f.

[12] Breasted, H., 269; Meyer, Alt. II, i, 111, 116; Naville, Deîr al-Baharî, I–VI, 1908 f.; Pogo, Isis, 1930, xiv, 317.

[13] Breasted, H., 260 f., 284 f., 292, 305; ib., CAH, ii, 67 f.; Olmstead, H. Pal. Syr., 129 f.

[14] Maspero, Histoire ancienne des peuples de l'orient, 7 ed., 1905, 236; Breasted, Anct. Rec. ii, 187 f.; Meyer, Alt. II, i, 136.

bars and sheets, ebony, ivory and myrrh from Nubia and Punt[1], and silver and gold vessels from Tyre and the Ægean.[2] The city of Assur (Assyria) sent "genuine lapis lazuli," also "fine lapis of Babylon" (really cobalt blue glass), and "Assyrian vessels of many colours" (probably glazed vases).[3] Egyptian native metal work in the period from Thothmes III to Amenhotep III may have been influenced by Minoan originals.[4]

During the XVIII dyn. the relations between Crete (Keftiu) and Egypt were very close, the hypothesis that Keftiu is Cilicia being untenable. Metal ingots appear also in representations of Syrian tribute and elsewhere, but the peculiar shape of a bull's hide is so frequent in Crete that it must be regarded as a normal Minoan ingot. The ivory tusks shown in the tomb of Rekhmara are carried by men who also bear Minoan vases. Egyptian objects are found in the Ægean, all of the early or middle XVIII dyn., but nothing later than Late Minoan I in Egypt ; there was some catastrophy in Crete in the reign of Amenhotep III which put an end to the relations of the two countries. Later Mycenæan objects occur later in enormous amounts in Egypt, and Late Mycenæ also imported a large number of objects from Egypt under Amenhotep III (XVIII dyn.).[5]

Under Thothmes III (1501–1447) relations between Egypt and Asia were very extensive, Asiatic trade with the Mediterranean passing through the canal from the Red Sea to the Delta.[6] The raid of Amenhotep II (1448–1420) into Syria, which was incidental to quelling revolts, brought back 1650 deben of gold and 500,000 deben (44·75 tons) of copper. Cyprus also sent copper and silver to Egypt.[7] Paintings in tombs of the period of Thothmes III–IV show the weighing of precious materials, bars of metals, bronze, lapis lazuli, malachite, etc.[8]

Amenhotep IV, who took the name of Ikhnaton and founded the city of Akhetaton (mod. Tell el-Amarna), which was abandoned after his death, introduced a monotheistic worship of Aton, the "heat" or "power" of the actual disc of the sun, in opposition to the old sun-god of Thebes, Amen-Rā, and also a naturalistic art perhaps under Cretan influence.[9]

Tutankhaton, the son-in-law of Ikhnaton, reverted to the worship of Amen and changed his name to Tutankhamen. His tomb, discovered at Thebes in 1922 by Lord Carnarvon, contains furniture made at Amarna, inlay of polychrome faience, glass and stone, sheet gold over wood, and iron.[10]

The Tell el-Amarna tablets, written in Babylonian cuneiform script, then the language of diplomacy and commerce over a wide area,[11] refer to several nations or tribes, of which those called Danuna, Shardina and Shakalsha (or Shekelesh) have been identified "with more or less certainty" with the Greek Danai, with the men of Sardis or Sardinia, and the Sikel natives of Sicily

[1] Breasted, CAH, ii, 79 ; Meyer, Alt. II, i, 141 f.

[2] Breasted, ib. ; a "silver vase of Kefti [Cretan] work," Köster, Beiheft. alt. Or., i, 30 ; the Vizier Rekhmara received the vases.

[3] Smith, Early History of Assyria, 1928, 227, 232.

[4] Hall, NE, 292.

[5] Pendlebury, JEA, 1930, xvi, 76 f., 82 ; 87 f. 89 ; objects such as vases, animal heads, cups, ingots, etc., shown in inscriptions of Senmut, Rekhmara, etc., as coming from Crete, are collected in one plate, ib., 78, plate xx.

[6] Breasted, H., 322 ; Olmstead, H. Pal. Syr., 145 f.

[7] Budge, Mummy[2], 55 ; Olmstead, H. Pal. Syr., 143, 145 ; Meyer, Alt. II, i, 139.

[8] Wilkinson, i, plates II A–B ; Davies and Gardiner, Tombs of the Two Officials, 1923, 7, 10, 31, etc. ; ib., Tomb of Ḥuy, 1926, plates (time of Tutankhamen).

[9] Breasted, CAH, ii, 109, 211 ; Peet and Woolley, The City of Akhenaten, 1923 ; Erman-Ranke, 298 ; Budge, q. in Isis, 1926, viii, 580.

[10] Rostovtzeff, i, plates XXVI–XXVII ; Breasted, CAH, ii, 121, 127 f. ; H. Carter and A. C. Mace, The Tomb of Tut-ankh-amen, 3 vols., 1923–27–33.

[11] Meyer, Alt. II, i, 335 ; Breasted, CAH, ii, 167 ; Lieblein, Recherches, 183, 404 f.

or with the men of Sargalossos (north of Pisidia), respectively. Later come Teresh or Tursha (Tyrsenians or Etruscans) and Ekwesh or Akaiwasha (perhaps Achaians ?), all by sea ("peoples of the sea") and dangerous to Egypt. Texts of Rameses III (1198–1167) speak of tribes of the Thekel (of doubtful identity) and Peleset (Philistines, a Cretan tribe), but not Ionians (Yivana).[1]

About 1400, by intercourse between Egypt and the Hittite and related tribes living on the Black Sea, iron, known in Predynastic times as a very valuable metal in the form of meteoric iron, gradually became better known ; it was common in Egypt from about 1200, and steel from the same source, the Hittites of Kizwatna on the Black Sea, entered Egypt in the xx dyn.[2]

Seti I (*c.* 1320) went far into Africa for gold, exploring the desert route in person and having a map of the mines made.[3] Under Rameses II (1296–1230), called "Sesostris" by the Greeks, Asiatic and foreign influences became very pronounced, and in the reign of Merenptah (1230–1215) the name of Israel first occurs on a stela.[4] The account of Egypt in the Old Testament has been supposed to follow Egyptian traditions very closely[5], but the book of Exodus is usually regarded as the work of an author relying on tradition and ignorant of the geography of the Isthmus of Suez.[6]

Rameses III (1198–1167, Rhampsinitos of the Greeks) sent expeditions to the Somali and Arabian coasts, and developed the copper and turquoise mines of Sinai.[7] After Rameses XII (1090) Egypt passed under the rule of Upper Egypt, then under foreign domination by Assyria and, after an interval of native rule (xxvi dyn.), it was governed by Persia from 525 to 332, when it was conquered by Alexander the Great. Greek influence had been very strong in the xxvi dyn., and after the foundation of Alexandria it increased under the Ptolemies. Alexandria was not part of Egypt proper ; it was "a world in itself," permeated by Greek as well as Eastern culture, and the study of the Alexandrian age belongs to a later period than that covered by the present work.

---

[1] Giles, CAH, ii, 8 ; Hall, *ib.*, 276, 282, 283 ; Breasted, *ib.*, 167, 173 ; Meyer, Alt. II, i, 57 f., 334 f., 556 f. ; Olmstead, H. Pal. Syr., 158 f. ; Smith, EHA, 290.

[2] Meyer, Reich und Kultur der Chetiter, 1914, 76 ; Petrie, *Ancient Egypt*, 1915, 22 ; Erman-Ranke, 550 ; S. Smith, Early History of Assyria, 1928, 293 ; Olmstead, H. Pal. Syr., 167.

[3] Quibell, 115 ; Maspero, Struggle, 376.

[4] Breasted, H., 448 f. ; *ib.*, CAH, ii, 154, 170 ; Cook, *ib.*, 320.

[5] Yahuda, *Ancient Egypt*, 1930, 25, 28 ; *ib.*, The Language of the Pentateuch in its relation to Egyptian, Oxford, 1933.

[6] Peet, Egypt and the Old Testament, Liverpool, 1922 ; *ib.*, JEA, 1930, xvi, 157 ; Budge, Mummy[2], 67.

[7] Budge, Mummy[2], 69 ; Breasted, CAH, ii, 179 ; Meyer, Alt. II, i, 595.

EGYPT II—1

THE METALS

PART I. PRECIOUS METALS

THE METALS IN EGYPT

Egypt from the earliest times possessed considerable knowledge of the technical arts [1], and metallurgy was supposed to have been invented by gods or ancient kings.[2] In the Predynastic Age and the Old Kingdom the goldsmith and jeweller combined high technical skill with exquisite artistic taste. In the Old Kingdom he made images of the gods, e.g. a sacred hawk with a head of one piece of beaten gold, crowned with a circlet and surmounted by two tall feather plumes, the eyes being the polished ends of a single rod of obsidian ; the copper body has corroded away.[3] In the XI–XII dyns. the designations goldsmith and coppersmith appeared on the coffin ; the occupations were honourable, and these workers probably belonged to the middle class.[4] Magnificent jewellery was produced under the XII dyn., hardly surpassed by later goldsmiths of Europe.[5] Some metal workers shown in Old Kingdom pictures and statues are dwarfs[6] : the tradition of misshapen persons as metal workers is common to various races, and is best known in the lame Greek fire- and smith-god Hephaistos.[7]

The Egyptian artists, however, on the whole, were held in low esteem.[8] The temples and large houses had companies of secular workmen (stonemasons, painters, metal workers, carpenters, shoemakers, glass workers, etc.) attached to them.[9] The XVIII dyn. work was of much better quality and more artistic than the Babylonian and Assyrian of the same period, although bound by antique forms.[10] The goldsmith (nubi) was more esteemed than other

[1] Uhlemann, Thoth oder die Weisheit der alten Ægypter, Göttingen, 1855, 243 ; Lepsius, Les métaux dans les inscriptions égyptiennes, tr. Berend, 1877 ; Hoefer, Histoire de la Chimie, 2 vols., 1866, i, 30 f., 106 f., 225 f. ; Wilkinson, i, 32 ; ii, 150 ; Berthelot, Les Origines de l'Alchimie, 1885, 211 ; Maspero, History of Art in Egypt, 1912, 65 ; Garland and Bannister, Ancient Egyptian Metallurgy, 1927, 26.

[2] Diodoros Siculus, i, 15 ; iii, 13 ; Goguet, De l'origine des lois, des arts et des sciences, 3 vols., 1809, i, 152 f., 164 f. ; Lenormant, Les premières civilisations, 2 vols., 1874, i, 87 ; Gowland, *Archæologia*, 1899, lvi, 267 f.
Many coloured reproductions of early Egyptian industrial scenes are given in the second volume (Monumenti civili) of Rosellini's Monumenti dell' Egitto e della Nubia, three huge volumes, Pisa, 1832–44 ; they are more accurate than the plates in Lepsius's Denkmäler aus Ägypten und Äthiopien, twelve very large volumes, Berlin, 1849–58, or the illustrations in Wilkinson. Cf. also plates in Davies and Gardiner, Tombs of the Two Officials, plate viii, etc. (Thothmes IV). Wreszinski, Atlas zur altägyptischen Kulturgeschichte [Leipzig, 1922 f. ; not seen].

[3] Breasted, H., 40, 104 ; Fig. 58 : found at Hierakonopolis, and in the Cairo Museum ; Maspero, Guide to the Cairo Museum, 1908, 433.

[4] Breasted, H., 169.

[5] Breasted, H., 202 ; Figs. 97, 98 ; Worringer, Egyptian Art, 1928 ; Ross, The Art of Egypt through the Ages, 1931.

[6] Breasted, H., 94, Fig. 75.

[7] Schmitz, DBM, ii, 383.

[8] Erman-Ranke, 532 ; Wilkinson, i, 189 ; cf. Hübner, PW, ii, 1450 ; Maspero, Dawn, 402 ; Quibell, 38.

[9] Wilkinson, i, 158 ; Maspero, New Light on Ancient Egypt, 1908, 23 ; Wiedemann, 1920, 174 ; Meyer, Alt. II, i, 320 f. ; Otto, Priester und Tempel im hellenistischen Ägypten, 2 vols., Leipzig, 1905–8, ii, 120.

[10] Meyer, Alt. II, i, 320 f.

manual workers and the craft ran in families.[1] Goldsmiths at work with blow-pipes, balances, necklaces, vases, etc., are shown in a VI dyn. relief [2], and the great antiquity of their craft is embodied in the legend that the god Ptah, "master of gold melters and goldsmiths, caster of the golden scarab, the scarab of purest gold, master of arts and artizans," made the first statues of the gods and adorned them with gold, lapis lazuli and malachite : his temple at Memphis was the "gold smithy," his high-priest the "foreman of artists" and other priests "great wielder of the hammer" and "master of art."[3] The "double gold house" and "double silver house" were parts of the central office of the Treasury both in the Old and New Kingdoms, and not in the temple, and such titles as that of the high priest of Memphis, "he who knows the secret of the goldsmith," [4] may have represented an old title rather than a secret tradition of craftsmanship. There is a gold seal cylinder of the "foreman of the gold-casters" of Mykerinos (IV dyn.). "Foreman of the goldsmiths" occurs in the XII dyn. and New Kingdom. The "foreman of the artists," who "knew the secrets in the gold houses," perhaps made figures of the gods. "Chief gold-smiths," and "foreman of goldsmiths," occur with "goldsmiths." The "secrets of the gold houses" were handed down to the eldest son, in the case of painters for as long as seven generations : they were otherwise known only to the "foreman of the artists of Upper and Lower Egypt."[5] Plato [6] remarks on the conservatism of Egyptian artistic work ; it was the same as "ten thousand years ago, and I say ten thousand not as a word but as a fact," but the view that the Egyptians were rarely able to bring an invention to perfection [7] is too narrow. There were many goldsmiths and other metal workers in the temples under Rameses III (1198–1167): the temple at Thebes received annually 51·833 kgm. of gold, some of it, with silver and copper, apparently plundered from old graves.[8]

TEMPLE WORKSHOPS AND LIBRARIES

The older historians of chemistry were obsessed with the belief that the ancient Egyptian priests carried on metallurgical, technical and chemical operations in secret in temple laboratories, possessed books dealing with these activities, engraved their most cherished secrets on stelæ, and generally behaved in a way not unworthy of the mediæval and modern popular images of the wise men of old.[9] Actual evidence for such a belief is, of course, entirely wanting. Ipuwer, writing of the VII–XI dyns., says, however, that in this period of unrest the "secrets of the kings of Upper and Lower Egypt were divulged," and refers to the "secret art of the embalmers."[10] An official is called on his coffin "superior of the secrets and chief of the metallurgists of the house of Amen."[11]
Middle Kingdom temple industries included weaving, dyeing, and smelting,

[1] Erman-Ranke, 112 f., 504 f., 550 ; Uhlemann, Thoth, 244 ; Maspero, Arch., 316 f. ; ib., Art in Egypt, 65 ; Wilkinson, ii, 234 f. ; Blümner, iv, 268, 308 f. ; Budge, Egyptian Hieroglyphic Dictionary, 1920, 354.
[2] Wiedemann, 1920, 340 ; Erman-Ranke, 548.
[3] Brugsch, Rel. 85, 508 ; ib., Ägyptologie, 413 f. ; BMGE IV–VI Rooms, 207.
[4] Otto, ii, 121 : doubtful ; Breasted, H., 164.
[5] Erman-Ranke, 113, 117, 504 f., 550 f. ; Erman, Religion Égyptienne, tr. Vidal, 1907, 229 ; Breasted, H., 104, 245.
[6] Laws, ii, 3.
[7] Maspero, Dawn, 223.
[8] Erman-Ranke, 117, 139, 153, 341, 127 f.

[9] Borrichius, De Ortu et Progressu Chemiæ, sm. 4to., Hafniæ, 1668 ; ib., Hermetis Ægyptiorum et Chemicorum Sapientiæ, sm. 4to., Hafniæ, 1674 ; ib., Conspectus Scriptorum Chemicorum Illustriorum, sm. 4to., Havniæ, 1697 ; reprinted in Manget, Bibliotheca Chemica Curiosa, 2 vols. fol., Genevæ, 1702, i, 1 f., 38 f. ; F. J. W. Schröder, Bibliothek für die höhere Naturwissenschaft und Chemie, Marburg, 1775, pt. II, Geschichte der ältesten Chemie und filosofie, oder sogenannten filosofie der Egyptier ; lingering still in Lippmann, Alchemie, i, 275 f.
[10] Gardiner, Admonitions of an Egyptian Sage, Leipzig, 1909, 37, 55.
[11] Maspero, Guide to the Cairo Museum, 1908, 284.

and goldsmiths, silversmiths and coppersmiths, the last of lowest rank, were employed.[1] In the New Kingdom the treasurers, perhaps two, had charge of the "houses," which were magazines with workshops and workmen, partly Egyptian and partly foreign slaves. The treasurer was next in rank to the vizier and worked in close relation with him : the stores, or objects made from them in the workshops, included grain, wine, oil, cattle, clothing, instruments, tools and weapons, jewellery, etc. The masons, carpenters and builders were under his charge.[2] The temples had often very considerable treasuries of gold, silver, copper and other metals, "stamped," and presented as offerings by the king to the god, and the "treasure chests," often of very great value, were sometimes removed by foreign invaders.[3] The "red and white houses" in the texts, thought[4] to mean gold and silver treasuries, are merely the national colours of north and south Egypt[5], but there were compartments in the temples for storing stamped gold, silver and copper, fine fabrics, etc.[6] The tomb of a v dyn. "keeper of the king's treasury of gold and silver in Upper and Lower Egypt, keeper of the king's laboratory and the king's workshops," has been found at Gîzah : a single gold bracelet was found on the mummy.[7]

In a later period the temple at Denderah employed twelve artists for four months of the year in making ornaments for the gods, of gold, silver, asem and copper ; "goldsmiths of Ptah, Amen," etc.[8] There is no indication that any of these workmen were priests : an official (not a priest) of the XVIII dyn. takes pride in the fact that the king allowed him to work in the "gold house" to fashion images and figures of the gods[9], and it is unlikely that this work was carried out in secret, as Diels[10] suggests, from a late Ptolemaic text which says, "the high priests have access to the workshops of the goldsmiths," and that the making of amulets and images (*not* metal working in itself) was secret work.

The statues of the gods in the early period were of wood, from 18 in. to 6 ft. high, elaborately adorned with gold, silver and costly stones.[11] They were dressed in "sacred vestments" by a special priest called by the Greeks hierostolos, who had the privilege of entering the sanctuary. Each divinity had its own vestments : Osiris white ; Isis dyed a variety of colours.[12] A superior priest in the XII dyn. says : "I decked the body of the lord of Abydos (Osiris) with lazuli and malachite, electrum and every costly stone, among the ornaments of the limbs of a god : I dressed the god in his regalia by virtue of my office as master of secret things." He also made a shrine of gold, silver, lazuli, and fragrant woods.[13] The duties of the priest included, besides dressing the images, the application of cosmetic to their eyes.[14] The vestments or wrappings, of white, green, red and orange, were supposed to absorb some of the "divine fluid" of the statue, and were used by the priest-magicians.[15]

A curious custom in Egypt was the burying of the commemorative "foundation deposits" under temples, pyramids, or fortresses, as in Babylonia and Assyria, Nubia and the Sudan. These deposits occur in the XVIII–XIX dyns. but are commoner later. In a temple of Ptolemy II (285–246) at Naukratis there are plates of gold, silver, copper, iron, and lead ; plates of these metals, with rectangular tablets of carnelian, felspar, lapis lazuli, jasper and glazed

[1] Brugsch, Ägyptologie, 1891, 220, 417, 436.
[2] Meyer, Alt. II, i, 65.
[3] Otto, Priester und Tempel im hellenistischen Ägypten, Leipzig, 1905–8, i, 329, 259 f. ; Erman-Ranke, 341.
[4] Brugsch, Ägyptologie, 197, 214.
[5] Meyer, Alt. 1909, I, ii, 151.
[6] Brugsch, 265 f.
[7] *Daily Telegraph*, 7th April, 1932.
[8] Otto, i, 313, 326 ; ii, 20 ; Brugsch, Ägyptologie, 414 f.

[9] Breasted, H., 245 ; Meyer, Alt. II, i, 320 f.
[10] *Abh. Preuss. Akad. Wiss., Phil.-hist. Kl.*, 1913, iii, 27.
[11] Breasted, H., 62.
[12] Wilkinson, iii, 395.
[13] Breasted, Anct. Records, i, 299.
[14] Blackman, ERE, x, 300.
[15] Hopfner, PW, xiv, 334 ; Ruska, Tabula Smaragdina, Heidelberg, 1926, 22 f. ; Mme. Hammer-Jensen, Die älteste Alchymie, Copenhagen, 1921, 122 f.

ware, and small lumps of copper and lead ores were found in a deposit of Psamtik (650 B.C.), at Tell Dafnah (Daphnæ) ; tablets from tombs in the Sudan include the same materials, with bronze and glazed ware.[1]

In the Ptolemaic and Hellenistic Periods companies of workmen were attached to the temples, but this was not their sole occupation, and they did not belong to any grade of priests.[2] Officials of the king who had knowledge of, or were placed over, "the secrets of heaven," others over "the secrets of earth," others over "the depths or mines," others over "the cellars," etc., are mentioned, but it is uncertain if they were priests.[3] Silver and gold, especially "earth gold," were then still offerings to the gods, perhaps private offerings, but gilt or silvered copper, or even copper, was common, corresponding with the impoverished state of Egypt.[4] Lippmann's assumption that *imitations* now began to be made "in the temple workshops"[5] is not directly supported by any authority he quotes, and when the old texts draw distinctions between "true" chesbet and imitation, the latter is not made in Egypt, but imported from Babylon. The inventories of some Egyptian temples show that even in the Roman Period they contained great treasures of gold, silver and jewels.[6]

In the Ptolemaic Period the temples had a right to go on manufacturing a finer kind of linen cloth (byssos) according to ancient tradition, but not for sale. It was used especially for the robing of idols, and a certain portion had to be given yearly to the king. They also had industries of brewing, milling, baking, stone-cutting, brickmaking and market gardening, perhaps all for their own use. The king seems to have made large annual gifts of money to the temples.[7]

The site of the temple itself then often included various activities carried on by laymen, although the whole was enclosed within a high wall. These officials included police, bath attendants, makers of oil (which was a temple monopoly, and in which mortars were used as well as mills), bakers, brewers, makers of linen or cotton goods ('οθόνια βύσσινα), picklers (probably of fish), testers of money *by weight*, painters, stone masons, brickmakers, goldsmiths, and silversmiths.[8]

The votive offerings were probably manufactured from an early period close to the temple (as at Deîr el-Baharî, XI dyn.), which would be surrounded by a permanent fair or booths for their sale.[9] The output of the temple (or perhaps necropolis) workshops was considerable, and as it exceeded requirements many objects were sold outside.[10]

The materials from which the amulets were made were of importance, and are specified in the Book of the Dead. They were at first simple natural substances of unusual appearance, form or colour : leaves, strange roots, stones of unusual colours or having markings (banded marbles or agates) or having pyrites in them, or large shells from the Red Sea, with inscriptions (XII dyn.). In an Old Kingdom Pyramid Text, malachite is called "food for the dead," because it contains a vital essence.[11] Crystals (rock salt, alum (?), rock crystal), white, blue, red or yellow coloured things, and metals were favourite materials for early amulets, and it has been supposed that all jewellery was

[1] Budge, Mummy[2], 450 f. ; BMGE IV–VI Rooms, 208, 214 ; Petrie, Ten Years' Digging in Egypt, 1893, 45, 53.

[2] Otto, i, 287, 291 f. ; ii, 114 ff.

[3] Birch, in Wilkinson, i, 168.

[4] Otto, i, 133, 287, 333.

[5] Alchemie, i, 268.

[6] Lafaye, Cultes des divinités d'Alexandrie, 1884, 135 f.

[7] Bevan, History of Egypt under the Ptolemaic Dynasty, 1927, 150, 180, 185.

[8] Otto, i, 282 f., 292 f., 310 f. ; a newer general account, in which more papyri are used, is to be wished for.

[9] Hall, NE, 1920, 286 ; cf. Herodotos, ii, 60 f.

[10] Otto, i, 313 f., 326 f., 333 ; Maspero, Struggle, 527.

[11] Budge, Amulets, pp. xxiv, 17 f., 70 f., 87, 135 f., 202, 296, 306 f., 416 f., 423 f., 486 f. ; D. Mackenzie, The Migration of Symbols, 1929, 164.

originally magical.[1]   Other materials used for amulets are carnelian, hæmatite, glazed quartz, slate, lapis lazuli, and serpentine in the Predynastic Period : gold, glazed quartz, limestone, carnelian and ivory in the Dynastic Period : silver and electrum in the Middle Kingdom.[2]   The scarab was cut in green basalt, green granite, limestone, green marble, blue glaze or glass ; sometimes it was set in gold.[3]

In the later Hellenistic-Roman Period there were, apparently, mechanical contrivances and apparatus in the temples ; holy water was delivered from "penny in the slot" machines, and many such arrangements are described by Hero of Alexandria.[4]

LIBRARIES IN EGYPT

Before the period when the Ptolemaic library was established at Alexandria there is very little information about temple libraries.[5]   The Serapaion at Alexandria had a famous library, including general books, and the temples at Edfu, Philæ and Arsinoë also had "libraries," but Otto infers that these contained only religious works, not technical treatises. Zosimos (c. A.D. 300) [6], it is true, mentions "thousands of books" on chemical subjects in the temples, particularly in the Serapaion of Alexandria, where they were taken "when Asenau, a high priest of Jerusalem, sent Hermes to translate the Hebrew Bible into Greek and Egyptian," but these are all late traditions, and we know of no chemical books in the Ptolemaic libraries.[7]

"The House of Books" appears in the III dyn. and the temple library contained hymns, accounts of ritual, treatises on magic, medicine, geometry, etc., and mystical works, but not technological treatises or works on chemistry. The later priests probably forged "ancient documents" (some of which may have been used by Manetho) ; they made one attributed to King Zoser (III dyn.) and deceived Ptolemy with it.[8]

The temple libraries appear to have been comparatively small ; only thirty-seven works are named in the list cut on the walls of the House of Books of the temple of Edfu. They do not appear to have contained glossaries, dictionaries or grammars, as did the corresponding collections in Babylonia. The texts themselves were studied under the guidance of priests, and the Egyptians had a great respect for knowledge ("the books") : the learned class had a better life than the unlettered worker.[9]   A large number of school copy books of the XVIII dyn. are known. The scribe's work was of a purely business character : the education of the priests was different, and included a study of religious and funerary texts, cosmogonies, and legends and histories of the gods. Copies of the Book of the Dead, containing magic formulæ and with blanks for filling in the purchaser's name, were sold to the wealthy by the priest-scribes during the XVIII dyn.[10]

LABOUR AND CRAFTS

The workshops were owned by the king, the temples, and a group of wealthy merchants ; the lower artisans were either serfs or slaves. Industry was closely connected with large estates and came under the same management.[11]   Royal

[1] Budge, Amulets, 20 ; Cook, Zeus, ii, 637.

[2] Petrie, Arts and Crafts, ch. vii ; statistics are also given by Brunton, Qau and Badari II, 14 f.—mostly blue glaze and carnelian.

[3] Budge, Egyptian Magic, 90 f. ; ib., Mummy, 1893, 250.

[4] Pneumatica, i, 21, 31, 37, 49 ; cf. Otto, i, 397 ; Dümichen, Z. f. Ägypt. Sprache, 1879, 97–128 ; A. Schmidt, Drogen und Drogenhandel im Altertum,

Leipzig, 1924, 3, 41 ; H. W. Schaefer, Die Alchemie, Inaug. Dissert., Flensburg, 1887, 10.

[5] Otto, i, 338 ; ii, 21, 121.

[6] Berthelot, Coll. ii, 230 ; iii, 223.

[7] Wiegleb, Historisch-kritische Untersuchung der Alchemie, Weimar, 1777, 143 f.

[8] Maspero, Dawn, 240, 242, 398.

[9] BMGE, 68, 79 f. ; Erman-Ranke, 374 ; Breasted, H., 98, 169.

[10] Breasted, H., 99, 249.

[11] Rostovtzeff, i, 148.

monopolies in the Ptolemaic Period included at different times, certain oils (sesame, croton, saffron, pumpkin or linseed, but not olive oil, although olive trees grew in the Fayyûm), textiles (linen, wool, hemp), dyeing, fulling, goldsmith's work, stone quarrying, mining gold and emeralds, brewing beer, probably papyrus paper, salt, natron, spices and perfumes, timber, and perhaps tanning. A person desirous of exercising some craft (thick cloth maker, dyer, leather worker, goldsmith, etc.) then bought a government licence, and paid income tax. Some crafts (e.g. dyeing) were carried out both by the government and by private individuals.[1] In the Roman Period, the quarries and mines belonged to the Emperor and were usually worked directly by the State with convict labour under a military guard.[2] Every Egyptian town, even small, was then a centre of industrial activity, trades being localised in streets, and the products were sent to Alexandria.[3]

In the earliest period high officials, even princes, took charge of expeditions to mines and quarries, and metallurgy was no doubt regarded as important.[4] In the XVIII dyn. mines and quarries were the property of Pharaoh, who made personal inspections of them.[5] Copper stored in bars in the treasuries, fine linen and oil are said to belong to the king.[6] In later periods (XXI dyn.), when the priests were powerful, the control of metals perhaps passed into their hands, but the "Director of the King's Works" was an administrator only, [7] and the participation of the Egyptian priests in mundane affairs has been greatly overestimated.

Each class of workmen recognised one or more "chiefs" or "masters," i.e. shop-stewards, and indulged in strikes. The hard conditions of manual labour, even in the skilled trades, often prompted the wish to escape them by becoming a scribe. A Sallier Papyrus (B. Mus. ; XIX–XX dyn.) says : "I have seen the smith at his work at the mouth of the furnace of his forge, his fingers rough as the crocodile and stinking more than fish spawn ; as for the dyer, his fingers stink of rotten fish and his clothes are absolutely horrible" ; the unenviable lot of other workmen is also graphically described.[8]

·The Egyptian craftsman made use of all the metals and materials available, and readily adopted the use of new materials from foreign countries. Metal objects of all periods are found in museums, but the greater number belong to the XXVI dyn. and show Greek influence : doubtful objects are frequently assigned to this period. The few specimens of Coptic metal work known are usually poorly executed.[9]

THE PRODUCTION AND USE OF FIRE

An essential in the extraction of metals from their ores is fire, the use of which has been known among all nations from very remote times, at least after the Tertiary Period.[10] The use of fire, at first probably of limited application, is found in remains at Taubach and other very early stations in Europe, and the

[1] Bevan, History of Egypt under the Ptolemaic Dynasty, 1927, 148 f., from Wilcken ; Erman and Krebs, Aus den Papyrus der Königlichen Museen, Berlin, 1899, 169 ; goldsmith taxed in A.D. 169 ; the 1913 Leipzig dissertation of Reil, Beiträge zur Kenntnis des Gewerbes im hellenistischen Ägypten, was not available to me.
[2] Milne, A History of Egypt under Roman Rule, 1895, 127.
[3] M. P. Charlesworth, Trade Routes and Commerce of the Roman Empire, Cambridge, 1924, 32.
[4] Garland and Bannister, 7.

[5] Garland and Bannister, 1927, 10 ; Maspero, Struggle, 93.
[6] Erman, Die Literatur der Ægypter, Leipzig, 1923, 144, 266.
[7] Garland and Bannister, 13 ; Maspero, Dawn, 403 ; Perrot-Chipiez, Art in Egypt, i, 627.
[8] Maspero, Dawn, 310 f. ; BMGE, 69 ; BMGE IV–VI Rooms, 303 ; Erman-Ranke, 533.
[9] L. Beck, Geschichte des Eisens, Braunschweig, 1884, i, 63 f. ; Breasted, H., 573 ; Garland and Bannister, 3, 16, 19, 83.
[10] Reinach, Cultes, Mythes, Religions, 1908, iii, 82 f. ; cf. Goguet, i, 87 f.

old nations believed that fire was unknown in the earliest periods, its invention being credited to some god.[1] In Egypt, kindling fire in the dark temple was part of the daily ritual ; the flame, regarded as divine and related to the sun-god, had the power of expelling demons. The spark was probably received on a wick of linen.[2]

According to Rutot, for the first half of the Palæolithic (Old Stone) Period there are no remains of fireplaces or calcined bones, but discoloured and cracked flints (silex décolorés et craquelés) make the use of fire for the whole period probable. For the precursor of this period ("Diluvial Eolithic") there are traces of the use of fire ; blocks of flint seem to have been exposed to fire for a short time to render them more easily split. Recent evidence indicates the use of fire in the Mousterian Period (Neanderthal Man) for scaring animals and roasting bones and flesh ; smoke-blackened stones occur in the Aurignacian Period. Cro-Magnon Man probably used heated pebbles in cooking, dropping them into skin pots containing flesh and water, and in dark caves he used lamps, shallow stone basins with animal fat and wick of dried moss.[3]

Lucretius [4] suggested that fire arose from lightning and from the branches of trees blown by the wind, an old and widespread belief, but the use of fire probably arose gradually in the working, boring, rubbing and polishing of wood, horn, bone and flint, and the observation of the effects of lightning, and volcanic and natural fires.[5]

The fire drill, which is probably a Neolithic invention, was used early in Egypt and all over Africa, as well as by the Eskimos, the natives of New Zealand, the Australians, the American Indians, the Swiss Pile-Dwellers, etc.[6] It is represented by an Egyptian hieroglyphic. At Illahun there were sticks with charred blocks, wood blocks deeply drilled but not charred (? learners'), and drill-bows, with a scarab of the XVIII dyn. The drill was twirled by a cord operated by a bow, the upper end of the stick being pressed down by a stone.[7]

FUEL

For fuel, dry cow or ass dung, and straw, were mostly used ; wood and charcoal, which is definitely mentioned, were also in use but were expensive. In the Hellenistic Period, olive wood was used on some occasions (in a late gnostic papyrus).[8] Evidences of dung fuel and charcoal are found in the Predynastic Period.[9] Egyptian metallurgists are said to have used the coke of roots of sari grass, growing near the Nile, a hard fuel burning to produce an intense heat. For iron smelting, roots of papyrus, acacia charcoal in Sinai, wood of the nabak (?) tree, etc., were burnt. In ancient Babylonia, date kernels were used as fuel.[10] Grass and camel's dung are used by the modern Arabs,

[1] Forrer, RL, 221 ; M. C. Burkitt, Pre-history, Cambridge, 1925, 5 ; Hoernes, Natur und Urgeschichte des Menschen, 2 vols., Wien and Leipzig, 1909, ref. as Urg, i, 200 ; Goguet, i, 86.
[2] Wiedemann, 1920, 188.
[3] Hoernes, i, 200 ; ii, 4 ; cf. Forrer, RL, 222 ; Mackenzie, Footprints, 39 f., 61 f., 72, 81 ; W. Niemann, in Feldhaus-Klinckowstroem, Geschichtsblätter für Technik, Industrie und Gewerbe, ii, 216.
[4] De Rerum Natura, v, 1090 ff.
[5] Hoernes, Urg., ii, 4 f. ; ib., in Hoops, Reallexikon der Germanischen Altertums-kunde, 4 vols., Strassburg, 1911–9, ii, 32 ; J. Lippert, Kulturgeschichte der Menschheit, 2 vols., Leipzig and Stuttgart, 1886–7, i, 253, 319 ; Forrer, RL, 222 ; Frazer, Myths of the Origin of Fire, 1930, 219 ; Goguet, i, 88 f.

[6] Wiedemann, 1920, 187 ; Bryant and May Museum Catalogue, London, 1926, and Supplement, 1928 ; BMGE, 92 ; Maspero, Dawn, 318 ; Hoernes, Urg., ii, 8 f. ; Feldhaus, Die Technik der Vorzeit, Leipzig and Berlin, 1914, 305 ; Forrer, RL, 222 f. ; F. F. A. Kuhn, Die Herabkunft des Feuers und des Göttertränks : ein Beitrag zur vergleichenden Mythologie der Indo-germanen, Berlin, 1859.
[7] Petrie, Illahun, 1891, 11 ; Wiedemann, 1920, 188 ; Frazer, 218.
[8] Maspero, Dawn, 316 ; ib., Arch., 12 ; Wiedemann, 1920, 14, 23, 118 f. ; Wilkinson, ii, 35 f. ; F. Freise, Geschichte der Bergbau- und Hüttentechnik, 1908, 80, 85.
[9] Brunton, Badarian Civ., 1928, 11, 13, 62.
[10] Gsell, 61 f. ; cf. Pliny, xxxiii, 5.

and cow dung, collected by children and dried in the sun in flat cakes, is still burnt in open fires and baking ovens in Egypt.[1]

Charcoal appears at an early period in Sinai[2], although remains found may be charred wood, the latter being the actual fuel.[3] A charcoal-making furnace, with much charcoal, was found at Amarna.[4] Charcoal of mimosa wood, made in small meilers covered with sand, is used in modern iron smelting in Cordofan[5], and tamarisk wood makes good fires.[6]

Small portable charcoal fires blown with fans were used for roasting in the Old Kingdom[7], and in the New Kingdom a rather different type, with balls of charcoal, for heating branding irons.[8] For goldsmith's work small fireplaces with "checks" (curved pieces of metal fitted at the back), and a blowpipe and forceps were used ; bronze forceps, tongs and tweezers, "retaining their spring perfectly," have been found.[9]

BLOWPIPE AND BELLOWS

The domestic fires in Egypt were blown by fans, as in Roman times.[10] The mouth blowpipe, used in the Old Kingdom, was probably a hollow reed tipped with clay as shown in the XII dyn. at Beni Hasan (mistaken for "glass-blowing")[11] : Wilkinson[12] thought the tips were of metal ; Maspero[13] and Wiedemann[14] that the tubes were of metal, which would have been difficult to make unless of rolled sheet. They are also shown in the XVIII dyn. tomb of Tii.[15] Blowpipes continued in use till after 1500, although bellows were known at least as early as 1580.[16] In early Mexico and Peru blowpipes only, not bellows, were used.[17]

Bellows have been treated in the light of African material by von Luschan.[18] The dish bellows (Schalengebläse), indigenous to Africa, are worked in pairs, the skin or banana-leaf cover of one being slowly raised and that of the other quickly pressed down. The leaves are renewed every fortnight. There are no valves, the air entering through leaky places. The pipes and dishes are of wood and the clay blowing ends are connected with them by iron pipes. Earthenware pots with holes in the side are also used. Although modifications occur, the general form of dish bellows used in Africa is always the same. Von Luschan reproduces the figure from the XVIII dyn. tomb of Rekhmara, and an enlargement to show the mode of use, the skins being pressed out by the feet, and inflated by pulling cords attached to them.[19] The tubes between the bellows and the clay tuyères are hollow reeds or plant stems. Such bellows are still used in Bengal.[20] In the Congo two dishes are used, with leather covers raised by attached rods, and a fireclay tube.[21] Russegger[22] says a single dish was used, with a leather cover having a central hole opened and closed with the finger used to raise the leather.

[1] Layard, Nineveh and Babylon, 1867, 129 ; Wiedemann, 1920, 23 ; cf. Herodotos, ii, 36.
[2] Berthelot, Archéologie et Histoire des Sciences, 1906, 69 ; Petrie, Res. in Sinai, 52, 162.
[3] De Morgan, Or., i, 226.
[4] Petrie, Tell el-Amarna, 1894, 26 : c. 1400 B.C.
[5] J. Russegger, Reisen in Europa, Asien und Afrika, 1843-4 ; II, ii, 293.
[6] Layard, op. cit.
[7] Erman-Ranke, 526, Fig. 214.
[8] Erman-Ranke, 530, Fig. 218 ; Wilkinson, ii, 36.
[9] Birch, in Wilkinson, ii, 235 ; Fig. 415.
[10] Wilkinson, ii, 35.
[11] Newberry, Beni Hasan, 4 vols., 1893-1900 ; i, plate 11 ; ii, plates, 7, 14 ; Maspero,

Dawn, 225 ; Erman-Ranke, 548.
[12] ii, 140, 312.
[13] H. anc., 1905, 137.
[14] 1920, 341.
[15] Lexa, La magie dans l'Égypte, 1925, iii, pl. ix.
[16] Wilkinson, ii, 312, Fig. 432 ; Newberry, Rekhmara, plate 18 ; Erman-Ranke, 548 ; Maspero, Dawn, 311 ; Gsell, 65.
[17] Johannsen, Geschichte des Eisens, Düsseldorf, 1924, 5.
[18] Z. f. Ethnologie, 1909, xli, 22 f.
[19] Wilkinson, ii, 312, Fig. 432 ; Erman-Ranke, 549, Fig. 232 ; Feldhaus, Technik der Vorzeit, 1914, 369 ; Freise, 96.
[20] Gsell, 16.
[21] L. Franchet, Céramique primitive, 1911, 125, Fig. 19.
[22] Reisen, II, ii, 292 f.

The skin bellows (Schlauchgebläse) consist of a bag of sheep or gazelle skin, with a pipe at one end. The other end is open, the two sides being pressed together with two sticks by means of one hand. The skin is inflated with this part open, then closed and pressed out. These bellows are used in India and are not so common in Africa as the dish bellows.[1] The pump bellows (Pumpengebläse) are Indonesian and occur in Africa only in Madagascar.[2] They consist of two vertical wooden tubes, usually bamboo and often very large. The pistons are made tight by feathers, cotton rags, etc., which act as crude valves, and are worked up and down with long rods. At the bottom are short tubes of bamboo, or sometimes iron, leading to the clay blowing tubes. The origin is not known, but is perhaps Hindustan. Smith's bellows (Bälgegebläse) of the usual cylindrical type, of leather with wooden hoops and moved up and down, also occur in Africa, but perhaps only after Portuguese influence in the 15th century.[3] Bellows with two wooden plates joined by leather and worked by hand (modern house bellows) are first mentioned by Ausonius of Bordeaux (early 4th century A.D.).[4]

THE DISCOVERY OF METALS

An old theory [5] of the discovery of metals is that a great forest fire melted them out of the interior of mountains. According to Napier [6] this has actually occurred in the Alps and Pyrenees ; in 1762 a large mass of mixed metal (copper, iron, tin and silver) was said to have been melted out during the accidental conflagration of a wood, and the silver mines of Mount Ida are said to have been discovered in this way. No such event is known for modern times, and the old accounts must be accepted with caution ; the mixture of copper, iron, tin and silver is particularly suspicious.

Lighting a fire on ground rich in ore or exposing pieces of ore to fire by accident are possible origins of metallurgy, whilst heavy rains washing down metal or ore ; lightning splitting rocks and disclosing veins of ore (a gold mine was so discovered in Peru about 1700) ; winds blowing down trees and disclosing ore under the roots, as in the discovery of the famous Peruvian mine at Potosi ; and the unusual appearance of the soil covering metal ores, have all been suggested. The first metals used were probably native and found near the surface. Gold of great purity occurs native in various places ; simple washing of the earth only is then necessary, and sometimes even this was done by rivers and streams. Native silver and native copper were perhaps also the first forms of these metals discovered.[7]

Great masses of native copper forced themselves on the attention in America by projecting from the ground. In 1713 a crust of massive silver was found on the mountain of Ucuntaya in Peru ; even the copper is sometimes conveniently pressed by kindly Nature into sheets between masses of rock, and near Hudson's Bay the natives merely hammer lumps of copper between stones. After getting a start with native metals, primitive man perhaps noticed the ores, heavy like metals themselves, and so, by a continuous process of retrogression from the native metal, learnt how to make use of the metallic ores. Grinding, washing and mixing with fluxes were resorted to, and thus metallurgy came into being.[8]

[1] Von Luschan, 29 f. ; Figs. 9, 10, 11 ; Blümner, iv, Figs. 50, 51 ; Feldhaus, 368 ; Gsell, 17.

[2] Von Luschan, 32, 33 f. ; Figs. 12, 13.

[3] Von Luschan, 35 f. ; Figs. 14, 15 ; Gsell, 17.

[4] Freise, 99 ; Gsell, 17.

[5] Lucretius, v, 1240 f. ; Strabo, III, ii, 9, 147 C., from Poseidonios.

[6] Manufacturing Arts in Ancient Times, Paisley, 1879, 4.

[7] Alonso Barba, Metals, Mines and Minerals, London, 1739, 63 f., 74 ; Prescott, Conquest of Peru, book iv, ch. 3 ; Goguet, i, 155 f., 157, 165 ; Freise, 1 f., 9 f., 69 f.

[8] Goguet, i, 160 f. ; Napier, Arts, 4 f.

The theory of the "ages of metal" goes back to Hesiod [1] (*c.* 700 B.C.), perhaps from a Persian source.[2]   Archæologists generally recognise a sequence of stone, copper, bronze and iron ages for the materials of tools and weapons, the stone age being divided into the Old (Palæolithic) Period, when stone alone was in use, and the New (Neolithic) Period, which was a period of transition to the use of metal, generally copper in the first instance ; metal was at first used only ornamentally or for vessels.   The division into ages of gold, silver, bronze and iron [3] was revived by Borlase.[4]   That a copper age generally preceded a bronze age, when an alloy of copper and tin was used, was recognised by Wilde, of Dublin.[5]   The distinctions between the periods have been pushed too far. Conditions in different countries vary, and the actual dates of the various ages frequently differ.   In Africa an iron age was not preceded by a bronze age, and in many European cultures the transition between stone and iron was relatively short.[6]   A copper age in Britain is doubtful ; the earliest implements, formerly thought from the shape to be copper, really contain over 10 per cent. of tin, and although copper implements containing 2–4 per cent. of arsenic occur in Ireland, there was no true copper age there.[7]   Neither, again, is it necessary to assume that the discovery of a particular metal must be assigned to a definite locality ; it may have occurred in several places at once.[8]

The metals known and used in ancient Egypt were gold, silver, electrum (an alloy of gold and silver), copper, bronze, iron and lead, with tin, antimony and platinum to a limited extent.   Some metallic compounds, such as galena (native lead sulphide) and malachite (native basic copper carbonate) were also used as eye-paints from the Prehistoric Period.[9]   Although Egyptologists differed at first as to the meaning of the hieroglyphics denoting such materials[10], there is now general agreement in this field.[11]

The metals are named in a definite order of value, which was maintained by the Hebrews, Greeks and Romans, with two precious minerals interpolated : gold, electrum, silver, chesbet (lapis lazuli), mafek (malachite), copper, iron, lead, although silver is mentioned a few times before gold[12]; in a Leyden papyrus, taḥ-t means scoria, the scale or rust of a metal is called àn.[13]

GOLD

Gold was regarded in the Predynastic Period as a magic amulet, a dispeller of demons, an elixir of life, the surrogate of the goddess Hathor, "bestower of life" and "rejuvenator of men."[14]   Gold leaf was used in gilding the teeth, finger- and toe-nails, and even phalli, of mummies ; some mummies are practically plated with gold.[15]   Some of the oldest ornaments are gold models of

[1] Works and Days, 110 f. ; Pausanias, III, iii ; Hoernes, Urg., i, 383 f.

[2] Reitzenstein and Schaeder, Studien zur antiken Synkretismus aus Iran und Griechenland, Leipzig, 1926, 61 f.

[3] Lucretius, v, 1282 ; Varro, in Augustine, Civ. dei, vii, 24 ; Ovid, Fast., iv, 405.

[4] Antiquities of Cornwall, fol. Oxford, 1754, 217 f. ; F. de Rougemont, L'Âge du bronze et les Sémites en Occident, 1866, 3 ; Blümner, PW, v, 2142 f. ; de Launay, Daremberg-Saglio, Dict., ii, 1074.

[5] Hoernes, Urg., ii, 236.

[6] Hoernes, in Hoops, iii, 115 ; Rougemont, 91 ; Hoernes, Urg., ii, 236 f., 244.

[7] Gowland, *J. Anthropol. Inst.*, 1912, xlii, 240.

[8] Blümner, PW, v, 2142 ; de Launay, Daremberg-Saglio, Dict. II, ii, 1074 ; cf.

Lenormant, Prem. Civ., i, 87 f., 92 f., 99, 101.

[9] Lucas, Ancient Egyptian Materials, 1926, 214 f.

[10] Lepsius, Met., 2, 18, 36, 45.

[11] Wiedemann, 1920, 342 f.

[12] Lepsius, Met., 3, 18 ; Brugsch, Ägyptologie, 1891, 397 f. ; Kautzsch Apokryphen und Pseudepigraphen des alten Testaments, 2 vols., Tübingen, 1900, ii, 372 ; Maspero, Arch., 301 f. ; Möller, MGM, 1928, xxvii, 3 ; Schrader-Jevors, 156.

[13] Budge, Egyptian Hieroglyphic Dictionary, 1920, 58, 820.

[14] Elliot Smith, The Ancient Egyptians, 1923, 205, 206 ; Wiedemann, 1920, 343, from Moret.

[15] V. Loret, L'Égypte au temps des pharaons, 1889, 195 ; Hoernes, Urg., ii, 359 ; Wiedemann, 1920, 127 f.

shells worn as amulets.[1] Parts of statues are often covered with thin gold plates, e.g. the head-dress, eyebrows and eyelids, beard rest, neck ornaments, nipples, and nails of the hands and feet, of a XII dyn. statue at Dahshûr ; round the waist was a thin girdle of gold and the figure stood in a wooden shrine with green inscriptions on gold plates set in plaster.[2] The Amherst Papyrus (xx dyn.) contains the confession of the thief who rifled the tomb of a XIII dyn. king : "his head was overlaid above with gold, and the august body was wholly covered with gold ; his coffin was shining with gold and silver : we took the gold."[3] European alchemists almost to our own day prepared "potable gold" as an elixir of life [4], and the supposed magic properties of gold were connected with the relation of its colour to that of the sun ; certain Malay tribes consider gold, silver and tin as the seats of spirits which cause them to grow in the earth. They collect the metals with particular ceremonies, avoiding sounds unpleasant to the spirits, and use them in religious and cult ceremonies.[5]

It is often supposed that gold, with its brilliant colour, was the first metal to catch the eye of man.[6] It occurs in the sands and gravels of many rivers and in superficial alluvial deposits [7], and was probably first noticed in a few isolated places [8], e.g. after rare showers of rain in the desert [9], although much of the earliest (including I dyn.) Egyptian gold was probably imported from Nubia or S.E. Egypt, perhaps also from Asia Minor.[10]

Native gold is too soft for tools or useful applications ; it is easily beaten into sheet or even foil, and hammered into wire, but it is difficult to join small pieces together.[11] Gold sheet ornamented in repoussé was used for the handles of flint knives and vessels in the Neolithic Period in Egypt and in Europe.[12] Somewhat later the thin sheet was used in Egypt to cover wooden objects, and the metal was cold-hammered into plates, rings, and beaten vessels : soldered and brazed objects are much later.[13] The prehistoric gold is mostly worked in the forms of beads and wire[14], although foil also occurs.[15] Much of the gold in the tombs has been plundered in ancient and more modern times.[16]

Although Petrie[17] thinks copper may have been used before gold in Egypt, it is usually considered that gold was the first metal employed to any extent.[18] The hieroglyphic for gold, nub, is the oldest name of the metal.[19] Nub (Coptic noub) may have given the name to Nubia[20], where there were gold mines at least as early as the IV dyn.[21], although it is also possible[22] that gold may have

[1] Mackenzie, Footprints, 142.
[2] Budge, Mummy², 460.
[3] Petrie, Hist., 1903, 224.
[4] Darmstaedter, *Chem. Z.*, xlviii, 653, 679 ; *ib.*, *Archeion*, 1924, v, 251 ; *ib.*, MGM, 1927, xxvi, 328.
[5] Semper, Der Stil in den technischen und tektonischen Künsten, 2 vols., Frankfurt and Munich, 1860–3, ii, 480 f., 490 ; O. Stoll, Das Geschlechtsleben in der Völkerpsychologie, Leipzig, 1908, 378 f., 382 f., 413.
[6] Boyd Dawkins, Early Man in Britain, 1880, 401 ; Blümner, PW, vii, 1555 ; Gsell, 53 ; cf. Hoernes, Urg., ii, 205 f. ; Elliot Smith, Ancient Egyptians, 11, 41.
[7] Gowland, *J. Anthropol. Inst.*, 1912, xlii, 236.
[8] Meyer, Alt. II, 1893, 53.
[9] Mackenzie, Footprints, 141 f.
[10] Wiedemann, 1920, 342 ; Petrie, Arts and Crafts of Ancient Egypt, 1909, 83 f.
[11] Mackenzie, Footprints, 142 ; Semper, Der Stil, ii, 480 f., 490.

[12] Forrer, Urgeschichte des Europäers, Stuttgart, 1908, 290 ; Breasted, H., Fig. 9, p. 28 ; de Morgan, Or., i, 114 ; Erman-Ranke, 551 ; Mackenzie, Footprints, 107, 142.
[13] Semper, ii, 480 f., 490.
[14] Petrie, Prehist. Egypt, 27 ; *ib.*, Naqada and Ballas, 1896, 10, 15, 27, 28, 44, 45, 48, 66, 67.
[15] Petrie, Diospolis Parva, 1901, 25 ; Prehist. Egypt, 43 f.
[16] Petrie, Prehist. Egypt, 43 f.
[17] *Ib.*, 27.
[18] Gowland, *J. Arch. Inst.*, 1897, xxvi, 310.
[19] Lepsius, Met., 3 ; de Morgan, Or., i, 214 ; Brugsch, Ägyptol., 399 ; F. Ll. Griffith, Stories of the High Priests of Memphis, Oxford, 1900, 94, 102, 126 ; nub also meant "to smelt" (metals), Budge, Egyptian Hieroglyphic Dictionary, 1920, 353.
[20] Schrader-Jevons, 169.
[21] Gsell, 4.
[22] Petrie, Arts and Crafts, 83 f. ; *Ancient Egypt*, 1915, 14.

been named after Nubia. The metal was sacred to a special goddess, Nubt, living in the underworld and identified with Hathor.[1]

The hieroglyphic sign for gold, ꟷ, was supposed by Champollion and Lepsius to represent a bowl and a folded cloth, used for washing grains of gold from auriferous sand.[2] Birch [3] supposed it to represent a collar or necklace, probably of gold beads, and very well-formed III–IV dyn. hieroglyphics at Medum show, in fact, that this is the case, the collar being made from green, red and black beads.[4] Since gold was probably first used for collars, this hieroglyphic came to be the sign for the metal.[5] The necklaces, generally heavy gold chains or bead work, sometimes supporting gold flies, lions, etc., and hanging over the breast, were worn as symbols of office by the higher officials.[6] The suggestion of Crivelli that the sign represents a portable furnace used for the fusion of gold, the rays being flames unable to ascend because the wind inclines them horizontally [7], seems improbable. Representations of melting gold in crucibles by means of blowpipes, weighing gold in small balances and making necklaces, occur in the Old Kingdom.[8] Two of the workmen appear to be dwarfs. Washing gold with a bag is clearly shown [9], and is often mentioned.[10]

PREDYNASTIC AND EARLY GOLD

Predynastic gold beads (with garnet, a few green glaze and turquoise beads) found at Abydos were made by beating out the gold into sheet, cutting up into narrow strips, cutting these across into small fragments and curving these round till the edges met.[11]

Late Predynastic gold work was found in Nubia[12], at Abydos[13] and at Qau (Protodynastic).[14] Gold occurs in I dyn. tombs.[15] The bracelets of the queen of King Zer, at Abydos, comprise beads and a delicate flower of gold, turquoise[16], lapis and finely cut amethyst.[17] A thin gold bar with the name of Menes (I dyn., 3400, use unknown) is "the oldest known inscribed piece of jewellery."[18] There was a tax on Nubian gold in the I–II dyns.[19]

A copper rod plated thickly with gold was found in a I dyn. tomb[20]; gilt copper feathers of the VI dyn. have the copper coated with stucco (sic) for attaching the gold leaf[21]; and copper plated with gold occurs in Old Kingdom objects.[22] Copper or bronze was also gilded by applying gold leaf with "an ammoniacal solvent" which acted on the surface of the base metal.[23] Gilt bronze ladles described by Wilkinson[24] are of uncertain age and are not represented at a later period : the gilding may have been ornamental or to prevent corrosion by acid liquids. Fire gilding was unknown even in the

[1] Wiedemann, 1920, 343.
[2] Wilkinson, ii, 235 ; Blümner, iv, 140.
[3] In Wilkinson, ii, 235.
[4] Petrie, Medum, 1892, 33.
[5] Petrie, Arts and Crafts, 83 ff.
[6] Wiedemann, 1920, 65, 66 ; Figs. 5, 6.
[7] Holmyard, Makers of Chemistry, Oxford, 1931, 6.
[8] Erman-Ranke, 548, Fig. 231 ; Breasted-Jones, 56.
[9] Partington, Everyday Chemistry, 1929, 29, Fig. 30.
[10] Erman-Ranke, 553.
[11] Frankfort, JEA, 1930, xvi, 214 ; plate xxx.
[12] Archæological Survey of Nubia, Cairo, 1910, i, plate 65.
[13] Petrie, etc., Abydos I, 1902, 6, 8, 31, 34.
[14] Brunton, Qau and Badari I, 24 f.,

31 f., 34 f., 37 f.—beads, rings, jewellery, etc. ; amulets.
[15] Petrie, etc., Royal Tombs of the First Dynasty, I, 1900, 15, 33 ; ib., II, 1901, 17–19, 36—bracelets of the queen of King Zer ; 4, 21, 23, 24, 27, 36.
[16] Maspero, Arch., 357 ; ib., Guide to the Cairo Museum, 1908, 432, says really "a kind of blue glaze or paste, or this in addition to turquoise."
[17] Petrie, Arts and Crafts, 90 f.
[18] Breasted, H., Fig. 13.
[19] Lepsius, in Brugsch, Ägyptol., 399 f.
[20] Petrie, etc., Royal Tombs of the First Dynasty, II, 1901, 36.
[21] Petrie, Abydos II, 32 and plate xxi.
[22] Brunton, Qau and Badari I, 34, 66.
[23] Wilkinson, ii, 46, 229, 245 ; Maspero, Arch., 312.
[24] ii, 47.

xix dyn. [1], and gilding with gold amalgam was well known only in the Roman Period.[2] Silver was plated with gold foil laid on and soldered.[3]

Soldering and casting of gold, as well as hammering and chiselling, occur in the I dyn. Hollow spherical gold buttons are soldered to gold wire on the bracelets of the queen of King Zer, but the work was so fine that "no trace of solder can be seen" between the two : hence "the technical perfection of the soldering has never been excelled."[4] Beads of the v–viii dyns. are said [5] to have been made from very thin sheet gold "soldered" and "mounted on a core made of a mixture of a kind of resin and powdered crystal," quartz in one case and calcite in another. Borchardt [6] says soldering copper was not known in the Old Kingdom, but Petrie suggests that it was used for gold work. The details of the process do not appear to be known.

Strings of minute gold beads were worn on the ankles in prehistoric times ; larger beads were made by beating out a thin tube and drawing down the ends over a limestone core. A thin gold finger ring and a flat pendant have punched dots. Most of the prehistoric gold is found on the lips of stone vases, overlaying the handles of vases, and forming wire loops for carrying them ; gold was also used for covering handles of flint knives, but thin gold leaf on flint is not seen till the iv dyn.[7]

Gold leaf, sometimes very thin but always at least fifty times as thick as modern leaf (0·001 mm. in the v dyn.[8] ; 0·001–0·002 mm., not quite regular, in the xii dyn. [9]), was applied to various materials for decorative purposes from the Predynastic Period. Coptic gold leaf is $\frac{1}{5000}$ in. thick.[10] Good examples of gilding, of reddish colour, are known from the Old to the New Kingdoms, and it was extensively used later : specimens of gold leaf and a gilder's book are in the Louvre[11], and there are representations of gold beating from the time of the Old Kingdom, the method of beating between leaves of parchment being the same as that still employed.[12] Stone (including pyramids), wood (doors, furniture, etc.), and papier mâché were all gilded[13]: in gilding wood and papier mâché a layer of linen or plaster was first applied.[14] The Israelites probably learnt the art of gilding (e.g. covering the ark of shittim wood with pure gold)[15] from Egypt.[16] The British Museum has a scarab with a frame of gilded metal.[17]

MIDDLE AND NEW KINGDOM GOLD WORK

Egyptian gold work reached its highest stage of development in the xii dyn. On the technical side, casting was carried out by the new cire perdue process, in which a wax model is surrounded by plaster, the wax being then melted out and the metal run into the resulting mould. Moulding by pressure was used in making cowrie beads and tie beads, which were impressed in stout foil, aided by burnishing on the model so as to tool the detail. Soldering was done in a very practical manner. Wire was made by cutting strips from foil and

[1] Wiedemann, 1920, 343.
[2] Blümner, iv, 133, 308 f.
[3] Berthelot, Arch., 19 f., 33.
[4] Petrie, etc., Royal Tombs of the First Dynasty, II, 1901, 18, 19 ; Petrie, Arts and Crafts, 83 ff. ; Breasted, H., 40—Predynastic gold soldering.
[5] Beck, in Brunton, Qau and Badari II, 1928, 22 ; plate xcvii, Fig. 3.
[6] Das Grabdenkmal des Königs Sa-hu-re, Deutsche Orient Ges., 1910 [Sahurā], i, 78.
[7] Petrie, Arts and Crafts, 83 f.
[8] Feldhaus, Technik, 707.
[9] Berthelot, Arch., 22 ; Blümner, iv, 311.
[10] Petrie, Ancient Egypt, 1915, 16.

[11] Maspero, Arch., 312 ; Erman-Ranke, 551.
[12] Feldhaus, Technik, 640, 707, Figs. 433, 471 ; Wilkinson, ii, 243.
[13] Petrie, Arts and Crafts, 83 f. ; ib., Ancient Egypt, 1915, 16 ; Bucher, Geschichte der technischen Kunst, 3 vols., 1875–93, ii, 133 f. ; Lepsius, Met., 12 ; Wilkinson, ii, 243 ; Blümner, iv, 311.
[14] Pliny, xxxiii, 3 ; xxxv, 15, calls this plaster leucophoron ; Bucher, ii, 134 ; Lepsius, Met., 12.
[15] Exodus, xxv, 11, 12.
[16] Wilkinson, ii, 243.
[17] BMGE, IV–VI Rooms, 67.

soldering them together, a method afterwards used by the Jews. Ezekiel [1] refers to "fine linen with broidered work from Egypt," showing the source of the art. No drawn wire has ever been found in ancient work. Gold thread worked with linen and coloured Egyptian textiles is also mentioned by Herodotos.[2] Granulated gold work first appears in the XII dyn.[3]

The methods of the XII dyn. were continued in the XVIII. Novel designs of gold filigree jewellery are supposed to represent foreign, perhaps Cretan, influence.[4] In the XVIII dyn. plaiting of gold wire chains (the present Trichinopoly work) appears.[5] The gold and silver thread used was always solid, and this lasted until the Roman Period, perhaps to Aurelian. No plated wire or flat strip wound round silk or linen threads was found at Herculaneum and Pompeii.[6] Gold objects in the British Museum, mostly XVIII dyn., include necklace ornaments or pendants, bracelets, beads, statuettes, figure of hawk, bangles, scarabs, fibula, lion, scarab plinth (XIV dyn.), several rings (XVIII dyn.—Roman Period), ear-rings, plaques, rectangular plate, scarab settings, wire collar, and gilt bronze.[7]

COMPOSITION OF EGYPTIAN GOLD

Analyses of native gold from various localities [8] show very variable proportions of silver, e.g. in West African gold from 3 to over 20 per cent. A specimen of Russian gold contained 3 per cent. of bismuth. Attempts to deduce the source of specimens of gold from the composition seem of very doubtful utility, although Berthelot [9] considered that chemical analysis could afford a useful criterion of date. This may be true for very pure gold (although gold of 99 per cent. purity is found in the Urals), indicating purification, perhaps by heating with salt and iron sulphate[10] as used in Asia Minor from about 550 B.C., yet the criterion must be applied with caution. Gowland[11] suggests that some pure specimens may be *native* vein gold, not artificially refined. (See the section on Electrum below.) Forrer's[12] statement that Ethiopian (Nubian) gold was more or less pure, whilst Arabian gold was electrum (asem), an alloy of gold and silver, is disproved by the analyses. Napier[13] says the idea prevailed in Egypt that gold was the only true metal, other metals being different varieties of and convertible into gold, an idea which persisted as alchemy, but he gives no authority for this statement ; the idea is almost the same as that of the Arabian alchemist Abu'l-Qasim al-Irâqi in the 13th century A.D.[14]

Some analyses of Egyptian gold of various dates[15] are given in the table below.

Dyn. .	I	I	I	VI	VI	XI¹	XI¹	XII¹	XII¹	XII	XII	XII	XVIII	XVIII¹	XVIII	XVIII	Persian¹
Au . .	79·7	84·2	84·0	78·0	81·7	92·3	92·2	90·5	92·7	90·0	82·9	85·9	96·4	82·3	72·1	89·5	99·8
Ag . .	13·4	13·5	13·0	18·0	16·1	3·2	3·9	4·5	4·9	—	16·6	13·8	1·9	14·3	17·2	11·2	—
Cu . .	0	0	0	—	tr.	0	0	0	—	—	0·5	0·3	pres.	1·5	13·1	0	—
Not dtmd.	6·9	2·3	3·0	4·0	2·2	4·5	3·9	5·0	2·4	10·0	—	—	1·7	1·9	—	—	0·2

¹ Berthelot, Arch., 19 f., 22 f., 46 ; no As, Sn, Pb or Cu.

[1] xxvii, 7 ; cf. also Ex. xxvi, 36 ; xxvii, 16 ; xxxvi, 37 ; xxxviii, 18 ; xxxix, 3, 29.

[2] ii, 182 ; iii, 47.

[3] Petrie, Arts and Crafts, 83 f. ; *ib.*, History of Egypt, 1903, 177, Fig. 105 ; Hall, NE, 163.

[4] Quibell, 62.

[5] Petrie, Arts and Crafts, 83 f.

[6] Wilkinson, ii, 166 f., who refers to gold wire of Sesostris I (1970–35) and silver wire of Thothmes III (1501–1447).

[7] BMGE, IV–VI Rooms, 88 f., 100, 218 f.

[8] Leitmeier, in Doelter's Handbuch der Mineralchemie, 1926, III, ii, 188 f. ; Watts, Dict. of Chemistry, 1872, ii, 925 ; Napier, Arts, 37.

[9] Arch., 19 f., 46 ; *ib.*, Chimie des Anciens et du Moyen Âge, 3 vols., 1893 [Mâ], i, 165.

[10] Pliny, xxxiii, 4.

[11] *J. Anthrop. Inst.*, 1912, xlii, 253.

[12] RL, 200, 292.

[13] Arts, 44.

[14] Holmyard, Book of Knowledge acquired concerning the Cultivation of Gold, Paris, 1923, 4 ; A. Kent, *Proc. Roy. Phil. Soc. Glasgow*, 1932, lx, 101.

[15] Lucas, JEA, 1928, xiv, 315.

They show that the metal was essentially an alloy of gold and silver containing approximately from 72 to 96 per cent. of gold and from 3 to 18 per cent. of silver, with occasionally a little copper. The specimen of the Persian Period is exceptional and has probably been artificially purified.

VARIETIES OF GOLD

Various kinds and qualities of gold are mentioned in the texts : yellow, white, fine, gold in its stone (auriferous quartz), mountain or rock gold, gold ore (nub en set), river gold (nub en mu : alluvial gold).[1] Queen Hatshepsut (XVIII dyn.) speaks of commercial gold, and Rameses III (XX dyn.) of Asiatic, Nubian, Kushite (Ethiopian), mountain and river gold, and gold of the towns of Coptos, Apollinopolis Magna (Edfu) and Ombos.[2] Since gold does not occur in all the towns named, it was probably imported through them.[3] Thothmes III (XVIII dyn.) distinguishes between Nubian and Asiatic (Arabian?) gold [4], and his records (1501–1447) speak of gold of the best of the hills.[5]

The Harris Papyrus (13th century) and later texts (1100 B.C., etc.) refer to gold of the first, second and third qualities ; "silver-gold," good gold, or fine gold (nub ketem), perhaps purified gold [6], corresponding with the Greek washed gold ($\chi\rho\upsilon\sigma\iota\acute{o}\nu$ $\mathring{a}\pi\upsilon\rho\rho\nu$), or pure gold ($\chi\rho\upsilon\sigma\grave{o}s$ $\mathring{a}\pi\upsilon\rho\sigma s$) of Herodotos [7], though this may mean virgin gold [8] or native gold ($\nu\acute{\iota}\tau\rho\rho\nu$ $\mathring{a}\pi\upsilon\rho\rho\nu$ in the Stockholm Chemical Papyrus of A.D. 300 is either native or purified soda).[9]

There was no means of testing the purity of early gold, and fraud was not uncommon ; barter was preferred. Wholesale trade was practically a royal monopoly.[10] Tax was paid in kind even in the Ptolemaic Period, when coined money was in general use.[11] Ingots of gold and silver were, however, exchanged by weight in the earlier period.[12] Although the use of the touchstone for testing gold is supposed to be possible for early Egypt[13], there is no direct evidence that it was used.

The early Egyptian gold was always used directly without refining, and contained varying proportions of silver.[14] The Amarna letters (1375 B.C.) mention red, brown, single, double, triple, quadruple, etc., gold, and gold of Tukrish[15]: the double, triple, etc., golds may have been diluted with base metal by diplosis, triplosis, etc., as in the Leyden Chemical Papyrus.[16] Brunton[17] refers to a Protodynastic gold bead, weighing 5·8 grains, as "of pure quality, as it shows no discoloration like the others." "Red gold" leaf of the IV–V dyn. was found at Naga ed-Dêr[18], and various colours of gold in the tomb of Tutankhamen : the red and purple were stained with organic matter[19], but one kind of red gold used for sequins owed its colour to an intentional alloy with iron and copper[20], whilst other colours were due to alloys, probably natural, with silver and copper. This coloration of gold by alloy and treat-

[1] Brugsch, Ägyptol., 399, 400, 407 ; Forrer, RL, 200 ; Lepsius, Met., 6, 7, 9, 59 ; Blümner, iv, 111 ; Wilkinson, ii, 237 ; Schrader-Jevons, 169 ; Budge, Egyptian Hieroglyphic Dictionary, 1920, 353.

[2] Breasted, H., 277 ; Budge, Dict., 353 ; Brugsch, Ägyptol., 399, 475 ; Erman-Ranke, 557.

[3] Lepsius, Met., 6, 7 ; Erman-Ranke, 554.

[4] Bucher, ii, 132.

[5] Breasted, Anct. Recs., ii, 302.

[6] Erman-Ranke, 554 ; Wilkinson, ii, 285, 237 ; Lepsius, Met., 6, 7, 9, 59 ; de Morgan, Or., i, 214 ; Erman-Krebs, Aus den Papyr., 1899, 96.

[7] iii, 97.

[8] Blümner, iv, 119.

[9] Lagercrantz, Papyrus Holmiensis, Uppsala, 1913, 28.

[10] Wiedemann, 1920, 312 f., 317 f.

[11] Maspero, Dawn, 331.

[12] Wilkinson, ii, 245.

[13] Lenormant, La Monnaie, 1878–9, i, 109.

[14] Berthelot, Arch., 19 ; Wiedemann, 1920, 343.

[15] Olmstead, H. Pal. Syr., 167.

[16] Berthelot, Introduction à l'étude de la chimie des anciens et du moyen âge, 1889, 29, 32 f., 47, 56, 64, 68 f.

[17] Qau and Badari II, 1928, 21.

[18] Maspero, Guide to the Cairo Museum, 1908, 432.

[19] Lucas, in Carter, Tomb of Tutankhamen, ii, 173 f.

[20] Ridge and Scott, JEA, 1928, xiv, 190.

ment is mentioned (as βαφεῖς χρυσοῦ) by Plutarch.[1] The different colours, yellow, red, etc., of ancient gold leaf are generally due to surface films of organic matter [2], as, for example, the ear-rings with the name of Rameses XII (c. 1100 B.C.) which are covered with a rich red-brown varnish.[3]

GOLDSMITH'S WORK

Splendid gold diadems were found in the tomb of a XII dyn. princess at Dahshûr.[4] Later gold work includes the gold boat and bronze daggers inlaid with gold of Aahmes I, the first king of the XVIII dyn.[5] The gold work in Tutankhamen's tomb (1350) [6], includes a gold fan (pl. lxii), the first and second coffins of wood coated with sheet gold (pls. lxvi ; lxviii), and the third inner coffin, in the shape of a mummy, of solid gold, $2\frac{1}{2}$–$3\frac{1}{2}$ mm. thick (251, pl. lxx ff.) ; the gold mask[7] (254, pl. lxxiii); the perfume box (255, pl. lxxiv) ; the diadem (256, pl. lxxv) ; cloisonné necklets (261 f.) ; and the curious amuletic apron formed from seven gold plaques (264, pl. lxxxiii). The jewels and gold and silver weapons of Cretan style and the ornaments of Queen Aah-hotep (beginning of XVIII dyn.) [8] (which was real jewellery worn during her lifetime, not funeral jewellery as in many cases) include gilt wood ; a gilt bronze mirror with an ebony handle decorated with lotus flowers of chiselled gold ; arm- and ankle-bands of massive or hollow gold ; bracelets of gold, carneol, lapis lazuli and felspar beads on gold threads with gold fastening plates ; gold arm-bands with figured decoration on inlaid ground (not "enamel")[9] ; a diadem with sphinxes ; a necklace with scarab ; a collar with two inlaid mosaic hawks' heads and several animal heads ; a pectoral ; several chains ; miniature axes of gold and silver, and two daggers of bronze[10], either symbolical or for ordinary use, one of which was gilt.[11] The style is Minoan, but they are probably Egyptian work.[12] Two crowns of Queen Khnumuît (XVIII dyn.) are of gold decorated with carnelian, lapis lazuli, red jasper and green felspar.[13]

There is inlay of gold in wood, stone and bronze of Khufu's period (IV dyn.)[14], and inlay work ("cell mosaic") in gold goes back to the I dyn.[15], and continues till the later XIX dyn., glass and paste inlay becoming common about 100 B.C.[16] A fine collection of gold (some soldered) and gem jewellery was found in a tomb of 1400 at Lahun ("Illahun").[17] A battle axe with a bronze blade inlaid with gold and a cedarwood handle, covered with gold, in which the name of Aahmes I was inlaid with lapis lazuli, carneol, turquoise and green felspar, was found in the tomb of Aah-hotep[18], and the Leyden Museum contains many specimens of Egyptian jewellery.[19]

[1] Blümner, iv, 318.
[2] Berthelot, Arch., 22.
[3] Maspero, Guide, 440.
[4] Breasted, H., Figs. 97, 98 : Cairo Museum.
[5] Maspero, Struggle, 81, 97 ; Breasted, H., Fig. 103.
[6] Hall, JEA, 1927, xiii, 131 ; Woolley, Digging up the Past, plate xviii ; Carter, Tomb of Tutankhamen, ii, passim ; refs. in text are to this.
[7] Cf. the masks found at Mycenæ.
[8] Maspero, Struggle, 97 ; ib., Guide to the Cairo Museum, 1908, 422 f. ; Meyer, Alt. II, i, 56 ; Hoernes, Urg., ii, 220, 359 f. ; Quibell, 83.
[9] Petrie, Arts and Crafts, 83 f. ; Ancient Egypt, 1914, 87.
[10] Hoernes, Urg., ii, 220, 360 ; Lenormant, Prem. Civ., i, 242 f. ; Wilkinson, i, 340 f. ;

ii, 258 ; Erman-Ranke, 254 f. ; Maspero, Dawn, 448 ; ib., Arch., 327—"black bronze . . . formerly gilt."
[11] Hoernes, Urg., ii, 220, 360.
[12] Quibell, 83 ; Petrie, Arts and Crafts, 83 f.
[13] Maspero, Guide to the Cairo Museum, 1908, 427.
[14] Maspero, Arch., 310.
[15] Forrer, RL, 927, contradicting Petrie, in Murray, Smith and Walters, Excavations in Cyprus, 1900, 41.
[16] Petrie, Arts and Crafts, 83 f. ; Wilkinson, ii, 155.
[17] Brunton, Lahun I, 1920, 22 f., 31.
[18] Hoernes, Urg., ii, 360 ; Maspero, Struggle, 96 ; Petrie, Arts and Crafts, 83 f. ; Weigall, Ancient Egyptian Works of Art, 1924, 125 f.
[19] Wilkinson, ii, 343.

Gold chains, mostly XVIII dyn., of all sizes, single, double and triple, with large or small links, some thick and heavy and others very fine and delicate, are known ; the gold pectorals, decorated with inlaid coloured stones or paste, are also of very good workmanship (e.g. one of Sesostris II, 1903–1877, at Cairo) : they usually represent the front of a temple, with a moulded or flat border, and surmounted by a curved cornice.[1] A New Empire gilt lion of limestone was found in Nubia.[2]

Gold coin was quite unknown in Egypt until the Persian Period, when it was introduced from Persia ; in the Hellenistic-Greek Period, Egypt still used foreign coins, Athenian coins being stamped in Egypt and provided with hiero-glyphic inscriptions. Even when the silver mines of Cyprus came under Egyptian control in the Ptolemaic Period, copper was mostly in circulation as the drachma. Although weighed gold rings were used in the Old Kingdom, weighed portions of copper were the principal means of exchange in all but the latest period.[3] Gold was too valuable for coinage and in earlier Egypt silver, which was imported, was at least as valuable as gold.[4]

WEIGHTS AND MEASURES

The copper was exchanged in the forms of bars, rolled sheet or wire, the unit of weight being the deben (formerly read uten or tabnu), of 1,400–1,500 grains, or 90·959 or 91 gm., and its tenth part, the kite.[5] The spiral, or rolled sheet, of copper was the hieroglyphic for the deben.[6] The Ethiopian pek was $\frac{1}{128}$ of a deben, or 0·71 gm.[7] In the IV dyn. the so-called Phœnician shekel of 7·32 gm., or the double shekel of 14·64 gm., was in use, the deben appearing in the XVIII dyn.[5]

Weights were of hard stone (including hæmatite), tabular or loaf shaped, or in the shape of animal heads.[8]

The art of weighing, including the preparation of medicines according to complicated prescriptions, was known in Egypt from the earliest times ; a stone weight of 133·48 gm. with a cartouche of Khufu (IV dyn.) is known.[9] The standard of weight perhaps varied for different metals in the earlier period, including the XII–XVIII dyns., the deben and kite being legalised in the XVIII dyn., and the value of objects was often reckoned in a certain weight of copper, silver or gold. The unit weight of gold varied from 12·3 to 13·6 gm., the Memphite standard being 12·7 gm. and that of Upper Egypt 13·4 gm., these weights being revived in the Hyksos Period.[10] In the XII dyn. the gold standard of 12·3–13·98 gm., average 13·14 gm., was in use.[11]

Under Thothmes III (1501–1477) rings of commercial gold and silver from Asia weighed from 12 lb. to a few grains.[12] Ingots of copper, iron or lead in Asia perhaps had a fixed weight, whilst those of gold and silver were much

[1] Maspero, Struggle, 491 ; ib., Guide to the Cairo Museum, 1908, 429 f.
[2] Arch. Survey of Nubia, i, plate 72b.
[3] Lepsius, Met., 10 ; Goguet, i, 288 ; Lenormant, La Monnaie dans l'Antiquité, 1878–9, i, 97 f., 104 ; Breasted, H., 93, 97 ; Maspero, Dawn, 324 ; Wiedemann, 1920, 309 f.
[4] Lenormant, Monn., i, 98.
[5] Breasted, H., 195 ; cf. Glotz, Ægean Civ., 191 f.—88·12–101·08, average, 94·6 ; Wiedemann, 1920, 310, 415 ; Erman-Ranke, 138, 590 ; Erman, Lit., 386 ; Lepsius, Met., 11 ; Brugsch, Ägyptologie, 1891, 370 f. ; Griffith, PSBA, 1892, xiv, 323.
[6] Griffith, PSBA, 1892, xiv, 436.

[7] Lepsius, Met., 12, 60.
[8] Lepsius, Met., 10 ; Wilkinson, i, 285, Fig. 97 ; Wiedemann, 1920, 311 ; F. Hultsch, Gewichte des Altertums ; Abhandl. philol.-hist. Klasse d. K. Sächs. Ges. d. Wissensch., 1898, xviii, 1–204, 13.
[9] Griffith, op. cit., 403 f., 442 ; Walden, Mass, Zahl und Gewicht in der Chemie der Vergangenheit, Stuttgart, 1931, 16 f. ; Petrie, History of Egypt, 1903, 46, refers to a gold weight of 200 gm. with the name of Khufu.
[10] Griffith, op. cit., 435, 439, 443 ; ib., Stories of the High Priests of Memphis, Oxford, 1900, 33.
[11] Glotz, Ægean Civilisation, 1925, 191 f.
[12] Breasted, H., 307.

lighter and of variable weight. Copper rings (supposed to be Egyptian) are in the Leyden Museum.[1]

The balance (maša; mšit) for weighing metal rings is shown in the V dyn. There were "public weighers," some of them priests. The balance is of the beam type, with two pans supported by cords or chains, and a plummet, perhaps for adjustment.[2] No specimens of the large balances shown are extant, but there are some small pharmacist's (?) scales.[3]

Queen Hatshepsut (XVIII dyn.) speaks of weighing rings of commercial gold from Punt on a balance 10 ft. high. The temple balances were of gold and silver; that of Rā at Heliopolis, made for Rameses III (1198–1167), required 212 lb. of gold and 461 lb. of silver in its construction.[4]

The Babylonian sexagesimal weight system was used in Egypt, the mina (mna) occurring in texts of 1600 as a measure for wine, honey and incense.[5]

The Egyptian measure of length was the ell (450 mm., accurate to 1 or 2 mm. in extant measures); the royal ell was 525 mm., the schoinos 7·875 kilometres. The arura was the square of 100 large ells, or 2,756 sq. m. The hin was 41 to 47, usually 45·48, centilitres.[6] A smaller unit of volume was the ro, i.e. $\frac{1}{32}$ hin, or 0·0141 litre.[7] The Hebrew hin was 6·06 litres.[8]

According to Hultsch [9] the unit of $\frac{2}{3}$ kite (= 6·06 gm.) is specified in the Ebers Papyrus (1550 B.C.) for weighing drugs, and reached the Greek world as the drachma of Ægina, usually taken as 6·30 gm.[10] Hultsch[11] also considered that the Roman pound was derived from 36 kite (= 327·6 gm.); Gardner[12] gives 327·45 gm. The ordinary drachma ($\frac{1}{8}$ oz.) is supposed to be $\frac{3}{8}$ kite (= 3·41 gm.) Relations between Babylonian and Egyptian weights are also given by Hultsch[13], who regarded the old Greek κύαθος (= 0·0455 litre) as a tenth of the Egyptian hin, the Attic ἡμίνα (= 0·273 litre) as the water volume of 3 debens, the Roman amphora as the Egyptian keramion of 26·16 litres, and the English gallon (4·5435 litres) as corresponding with 10 hins (= 4·548 litres).[14]

Prehistoric, or VII–IX, or XVIII dyn., copper or bronze cylindrical cups with handles found by Petrie in a temple at Nubt were supposed to have measured gold dust on the Nubian system : 1 deben = 128 pek.[15]

In the Rhind Papyrus (Hyksos Period) the values of the deben of gold, silver and lead were in the proportion 4 : 2 : 1, whilst in the XVIII dyn. the ratio of the values of gold and silver is 5 : 3. Things represented as weighed are gold, silver, copper, lapis lazuli ; in later times incense, and in the Ptolemaic Period honey and drugs.[16] Besides the large balances, small hand-scales are shown.[17] The steelyard, an Italian invention, was not known till it was adopted from the Romans.[18]

[1] Lenormant, Monn., i, 102.

[2] Griffith, 436 ; Maspero, Dawn, 324 ; Wiedemann, 1920, 311, 415 ; Erman-Ranke, 553 ; Budge, Egy. Hierogl. Dict., 330 ; Wilkinson, i, 285 ; ii, 246 ; Davies and Gardiner, Tomb of Huy, 1926, plate xvii.

[3] Maspero, Guide to the Cairo Museum, 1908, 363.

[4] Breasted, H., 277, 491.

[5] Glotz, Ægean Civ., 191 f., says from XII dyn. ; Parthey, ed. Plutarch's Isis and Osiris, 1850, 280 ; Baumstark, PW, ii, 2714 f.

[6] Wiedemann, 1920, 414 ; Erman, Lit., 387.

[7] Walden, 1931, 17, 22.

[8] Budge, Mummy², 390 ; Smith, CDB, 1004—6·6 litres.

[9] Gewichte, 59 ; Walden, 6, 20.

[10] Glotz, 191 f. ; Gardner, DA, ii, 448.

[11] Op. cit., 5.

[12] Op. cit., 455.

[13] Table in Walden, op. cit., 20 f.

[14] See also Petrie, Ancient Weights and Measures, 1926 ; ib., Glass Stamps and Weights, 1926—none earlier than Roman ; colours on p. 11 but no analyses : one specimen perhaps remelted XVIII dyn. glass ; Peet, The Rhind Mathematical Papyrus, Liverpool, 1923, 24 f., on weights and measures ; Chace, Bull and Manning, The Rhind Mathematical Papyrus, 2 vols., Oberlin, Ohio, 1927–29, i, 31 f., on measures of capacity.

[15] Petrie, Naqada and Ballas, 1896, 67 ; Hultsch, Gewichte, 187 f., thinks they represent $\frac{1}{2}$, $\frac{1}{4}$, $\frac{1}{8}$ . . . $\frac{1}{128}$ pek.

[16] Griffith, PSBA, 1892, xiv, 436, 438, 439.

[17] Wilkinson, ii, 246 ; Wiedemann, 1920, 340, 415 : Old Kingdom.

[18] Wilkinson, ib. ; Marinden, DA, ii, 696.

Sources of Egyptian Gold

The earliest gold in Egypt was from the washings of river sand, "golden sand" (ψάμμος χρυσίτης).[1] In the correspondence of the Chalif Omar (A.D. 634–644) in a Gotha MS., "the Nile carries down gold."[2] Representations of gold washing occur in the Old Kingdom [3], and in the XII dyn. at Beni Hasan.[4] Olympiodoros (6th century A.D.) calls the seven metals "sands" (ψάμμοι), "according to the practice of the ancient [Egyptians]."[5] Psellos (1018–1078) calls the shore sand "golden" (chrysites : auri color).[6] Al Habib (8th century A.D.) says the ancient Egyptians gained immense treasures from sand.[7] The accounts of the ancient Egyptian gold mines given by modern authors are extremely confusing, since three different districts seem to have been mixed in them, and I am by no means certain that the following summary (which has cost me a considerable amount of trouble) represents the facts. The ancient mines seem to have been located in the mountainous region of the so-called Arabian Desert, which lies between the Nile and the Red Sea. In this region gold was extracted from veins in the quartz rock. There are, apparently, two sets of ancient mines in the Arabian Desert.

The first, and perhaps the oldest [8], were probably on the road now called El Foâkhir leading from the Red Sea and the old granite quarries through Coptos to the Nile, and reaching the coast in the locality of Kosseir.[9] There are remains of 1,320 stone huts used by miners in this region in the time of Ptolemy III (246–222).[10]

The old texts speak of gold from Kush (Nubia) and also gold of Edfu, Ombos and Coptos. The three last towns have no gold, so that no doubt gold brought from the Foâkhir mines and also those of the second region, in the Wâdî Allâqî, both of which lie east of this part of the Nile, is meant.[11] Although over a hundred ancient gold workings have been found in Egypt and the Sudan, there are none in Sinai.[12] Maqrîzî (d. A.D. 1441) mentions gold "like yellow orpiment" from the emerald mines of Kuft (Coptos)[13], but the Arabic geographers are usually badly informed on the localities of the Egyptian gold mines ; Qalqashandi (d. A.D. 1418) even speaks of gold found in the soil of the mountain Al-Muqaṭṭam, near Cairo.[14] The Arabs were better acquainted with the Nubian mines, which are mentioned in a letter in the Papyrus Rainer (A.D. 983–984), and Ibn al-Faqîh (A.D. 903) speaks of gold growing in Nubian sand "like yellow beet in the homeland."[15] Gold occurs more abundantly in Nubia than in Egypt proper[16], and the various journeys to West Africa in ancient and mediæval times were probably in search of gold, the outstanding attraction being the gold of the Sudan.[17]

[1] Blümner, iv, 112.

[2] Reitemeyer, 28.

[3] Breasted, H., 94.

[4] Perrot-Chipiez, Art in Egypt, i, 831 ; Blümner, iv, 13 ; Newberry, Beni Hasan, i, plate 11 ; Erman-Ranke, 552.

[5] Berthelot, Coll., ii, 106 ; iii, 115.

[6] Psellos, in Democritus Abderita de Arte Magna, Pizimentio Vibonensi interprete, Patavii, MDLXXIII, f. 68 recto ; pagination irregular ; really published in 1572.

[7] Berthelot, Mâ, iii, 101.

[8] Peake, The Bronze Age and the Celtic World, 1922, 39 f., thinks they were worked only from 2570 B.C.

[9] Maspero, Dawn, 494 f. ; Breasted, H., 93 f. ; Ritter, Erdkunde, Afrika², 1822, 673 f.

[10] Bevan, History of Egypt, 1927, 149 ;

Erman-Ranke, 551 ; Orth, PW, Suppl. iv, 110.

[11] Bucher, ii, 132 f. ; Gsell, 4, 18 ; Orth, 110.

[12] Garland and Bannister, 1927, 26.

[13] Reitemeyer, 151.

[14] Calcaschandi's Geographie und Verwaltung von Ägypten, tr. Wüstenfeld, *Abhandlungen der königlichen Gesellschaft der Wissenschaften zu Göttingen*, 1879, xxv, 31 f. ; cf. Krenkow, q. in *Isis*, 1930, xiv, 483, who says Qalqashandi describes the gold industry in Nigeria.

[15] Führer durch die Ausstellung Papyrus Erzherzog Rainer, Vienna, 1894, 204 ; MGM, ii, 439.

[16] Wiedemann, 1920, 10, 342 ; Blümner, iv, 13 f.

[17] Taylor, q. in *Isis*, 1929, xiii, 240.

The second group of mines in the Arabian Desert were on the borders of Nubia, in the Wâdî Allâqî region, seventeen or eighteen days south-east of Derow, or seventeen days east of Edfu (Apollinopolis Magna).[1] In this region there are remains of deep ancient shafts, which are well sunk, since the ancient Egyptians were good miners, and remains of tunnels 65 yards long.[2] The location of these mines was known to the Arab geographers Yaqûbî (A.D. 891), al-Idrîsî (d. A.D. 1166) and Abulfeda (A.D. 1334), who placed them in Gebel Allâqî, a mountain in the land of the Bojâ, Bujâ or Bejâ (the descendants of the ancient Blemyes), in the Bishâri mountains south-east from Bisharîjah, a village opposite to and ten days' journey east across the desert from Edfu.[3] They were still worked in 951–989, but in Abulfeda's time they were only just paying and they were afterwards abandoned, although Muhammad Ali about 1817 thought of reopening them. The small quantity of gold obtainable with immense labour, the difficulty of procuring water, and other local impediments, however, made the project impracticable. They were practically exhausted even in the ancient period ; every vein of quartz in the rocks in the deserts east of the Nile had been carefully examined by the ancient miners, samples having invariably been picked out from the fissures and broken into small fragments.[4] Maqrîzî reports that an Arabic tribe took possession of the mines in 854 and managed to enrich themselves from them.[5] The old mines were visited by Linant and Bonomi in 1831.[6]

There are remains of about 300 ancient stone huts in the Wâdî Allâqî mine region, in each of which is a granite hand mill for crushing the quartz.[7] The remains of the alluvial workings make the hillsides look as if they have been ploughed, and an area of a hundred square miles has been worked to an average depth of seven feet, so that only traces of gold in small pockets remain.[8]

The Wâdî Allâqî mines became more important after 2500 B.C.[9] There is a map of them in a Turin papyrus (XIX dyn., 1300 B.C.)[10], which is one of the two oldest maps known, the other having been made for Gudea the Sumerian.[11] An inscription of Rameses II (1296–1230) at Kuban, on the east bank of the Nile, records the construction of a tank or reservoir to supply the miners with water. They crossed the desert on asses to reach the mines and bring back the gold. Seti I, his predecessor, had bored a well 190 ft. deep at Wâdî Allâqî, but obtained no water. Rameses drove 12 ft. deeper and found water.[12] There are remains at Eshuranib of two cisterns for collecting the water of the winter rains and sloping stone tables for washing the gold dust.[13]

Gold was also obtained farther north in the same region, from the mountains of the Red Sea region of Gebal Zâbarâ[14], and there were gold mines at Meroë, the capital of Ethiopia ; the Ethiopians are reported to have had abundance of gold but little copper.[15] They did not appreciate the gold chain which Darius

[1] Wilkinson, i, 154 ; ii, 237 ; Maspero, Dawn, 480 f., 490 f. ; Breasted, H., 6, 94, 136, 181 ; Erman-Ranke, 172, 552.
[2] Maspero, Dawn, 480 ; Freise, Geschichte der Bergbau- und Hüttentechnik, 1908, 12 f. ; Orth, 110, 125.
[3] Ritter, Erdkunde, Afrika², 1822, 552 f., 666, 673 f. ; Breasted, Anct. Rec., iii, 81 ; Führer durch die Ausstellung Papyrus Rainer, 204.
[4] Wilkinson, i, 155 ; ii, 238 ; Wiedemann, 1920, 342 ; Lenz, Mineralogie der alten Griechen und Römer, Gotha, 1861, 29 ; Mining in Egypt, by an Egyptologist, n.d., 13 f. and plates ; Dareheib mines.
[5] Reitemeyer, 151 ; Wüstenfeld, Ueber die Araber in Aegypten el-Macrizi, Göttinger Studien, 1874, Abtheil. ii, 474.

[6] Wilkinson, i, 154.
[7] Dunn, MGM, xi, 485.
[8] Gowland, J. Anthropol. Inst., 1912, xlii, 255 ; Mackenzie, Footprints, 141.
[9] Wiedemann, 1920, 10 ; Roeder, RV, i, 425.
[10] Wilkinson, ii, 242 ; Maspero, Struggle, 376 and fig. ; Gowland, 255, Fig. 3 ; Feldhaus, Technik, 551 ; Erman-Ranke 556.
[11] See p. 217.
[12] Wilkinson, i, 154 ; ii, 237, 242 ; Meyer, Alt. II, i, 496 ; Breasted, Anct. Rec., iii, 81 ; ib., H., 416, 422.
[13] Gsell, 4, 18 ; Erman-Ranke, 172, 552.
[14] Breasted, H., 416 ; Ritter, 674.
[15] Strabo, XVII, ii, 2, 821 C.

sent them, and they sent Cambyses every year 200 logs of ebony, 20 elephant tusks and 2 chœnices (quarts) of virgin gold.[1]   No gold was found in Nubian graves at Karanog in Meroë, but the remains had been plundered.[2]   The rich alluvial gold deposits of Sennar, in the angle between the White and Blue Niles [3], were worked by the natives, and the pharaohs often made raids into the country from the time of the Old Kingdom.[4]

Besides the sources of Egyptian gold described another has been proposed. Since a red crust on the gold sceptre of Khasekhemui (III dyn.) contains gold antimoniate [5], and since it is alleged that "antimony will combine with gold only in the presence of tellurium," it is inferred that the Egyptians about 3000 B.C. imported gold from Central Europe, perhaps the valley of the Alt by way of the Ægean, since "there is no known source of this ore, telluride of gold and antimony, except in Transylvania." Herodotos [6] knew that the Agathyrsi in Transylvania had much gold.

The statement that "gold will combine with antimony only in the presence of tellurium," however, is incorrect, since the two metals form alloys and even a definite compound, $AuSb_2$; ordinary gold bullion may contain antimony, and gold combines with antimony on heating.[7]   Although *both* tellurium and antimony are associated with gold only in the mineral nagayagite, found in the single locality of Nagyág (Siebenbürgen) in Europe [8], there is no mention of the occurrence of tellurium in the specimen of Egyptian gold (which would be essential for the validity of Peake's hypothesis), and the antimony may have been introduced by a process of fusion with stibnite for the purpose of purification.[9]   The evidence for a European origin of *early* Egyptian gold is, therefore, very speculative and improbable.

The washing, melting, weighing and working of gold are all shown for the XII dyn. at Beni Hasan and for the XVIII dyn. at Thebes, so that the processes are known in some detail.[10]   The gold in the earliest period was washed from sand.   Native gold washers in Yezo (Japan) and elsewhere recover gold with a simple dish from sands which are too poor for treatment with modern appliances; if a sufficient amount of labour were available, vast quantities of gold could have been obtained from poor ores.[11]   Alluvial gold was washed in the XII dyn. from tracts watered by the Blue Nile and its tributaries : nuggets of considerable size were found, and the gold was stored in leather bags.[12]

The alluvial gold was won by the process described by Strabo.[13]   The gold-bearing sand was washed and allowed to settle out after the lighter parts had been carried away, and the particles of gold were caught in fleeces or cloths, as Egyptian representations show.[14]   The remains in Egypt indicate that in a later period the processes of grinding, washing and melting (slags were found) as described by Agatharchides were in use.[15]

EGYPTIAN GOLD MINING

A detailed account of the working of Egyptian gold mines, perhaps those of Wâdî Allâqî, in the Ptolemaic Period is given by Agatharchides in his work

[1] Herodotos, iii, 20, 97.
[2] Woolley, Digging up the Past, 102.
[3] Ritter, Afrika², 1822, 252 f.
[4] BMGE, 19, 73, 204 f., 228.
[5] Gladstone, q. by Peake, Bronze Age, 1922, 40, 65, 82 ; Peake and Fleure, Priests and Kings, 1927, 15.
[6] iv, 104.
[7] Giua, Chemical Combination Among Metals, 1918, 179 ; Roscoe and Schorlemmer, Treatise on Chemistry, 1923, ii, 517, 527.

[8] Slavík, in Doelter, Mineralchemie, IV, i, 848, 882.
[9] Davies, *Nature*, 1932, cxxx, 985.
[10] Wilkinson, ii, 234, 235.
[11] Gowland, *J. Anthropol. Inst.*, 1912, xlii, 255.
[12] Maspero, Dawn, 493.
[13] III, ii, 8, 146 C., from Poseidonios, with reference to Spain.
[14] Blümner, iv, 112 ; Freise, 74 f.
[15] Gowland, *op. cit.*, 257 ; Maspero, Dawn, 480.

on the Red Sea, probably written about 113 B.C. in Alexandria. The work is lost as a whole, but this part is excerpted by Diodoros Siculus.[1]

Agatharchides paints the picture of the labour in the mines in gloomy colours. The terrible conditions of slave labour which he describes may be exaggerated [2], but the old texts also speak of royal officials and slave drivers in the gold mines [3], and the mines could hardly have been worked profitably without some such system.[4] An account of an expedition to the mines of Sinai under Amemhemet III (XII dyn.) says the heat was so intense that "the mountains (the hot rocks ?) brand the skin." The hard life of the miner continued in antiquity in the Roman mines until the conditions were ameliorated under the Flavians and their successors.[5] An inscription of an Egyptian official, Sa-Menthu, says he set men, women and children at work crushing the quartz to get the gold.[6] Agatharchides says "the soil, naturally black, is traversed by veins of brilliant white rock" (ὁ μαρμάρος; quartz, not marble), and the gold actually does occur in pockets in white quartz, with oxides of iron and titanium.[7] "Out of this the overseers cause the gold to be dug by a vast multitude of people. For the kings of Egypt condemn to these mines notorious criminals ; prisoners of war, sometimes with their whole families, are bound in fetters and compelled to work day and night without intermission or hope of escape, for the guards set over them speak a foreign language. The earth [rock], which is hard, they soften by the application of fire." The splitting of rocks by fires is a method long used in mining[8] : Pliny [9] says vinegar was also used (hos igne et aceto rumpant), and Livy and Plutarch give a story of Hannibal making his way through the Alps with vinegar.[10] Vinegar was regarded as especially cold[11], and was probably thrown on rocks heated by large fires.

When the ore "has been reduced to such a state that it yields to moderate labour, several thousands of wretches break it up with iron picks. An engineer presides and directs the labourers, the strongest of whom hew the rock with iron chisels with brute force and without skill. In excavating below ground they follow the ore without keeping to a straight line. They have lamps fastened to the foreheads[12] and their bodies are soiled with the colour of the rock : they work without intermission and are lashed by the overseers. Little boys follow them and carry the fragments to the open air. Men about thirty years old pound the rock in stone mortars with iron pestles as fine as flour. At length the masters take the ground stone and carry it away for the final process. They spread it on a broad board, somewhat hollow and inclined, and pouring water upon it, rub and cleanse it till all the earthy part, separated from the rest by the water, runs off the board and the gold by reason of its weight remains. This operation is repeated frequently, the stone being rubbed lightly with the hand. Afterwards they take up the earthy part with fine sponges, gently applied, leaving clean, pure gold."

In the old period copper not iron chisels were used ; these have been found in old abandoned gold mines in Ethiopia[13], and in limestone rocks in tombs in

[1] iii, 11 f. ; De Maris Erythræi, in Müller, Geographi Græci Minores, 2 vols., Paris, 1855, i, 124 ; most modern accounts of the history of mining are imperfect, e.g. Freise's Geschichte der Bergbau- und Hüttentechnik, in 187 pages, is a mere sketch : Rickard's Man and Metals, 2 vols., New York, appeared long after my work was completed and I have not seen it.
[2] Orth, PW Suppl. iv, 141.
[3] Brugsch, Ägyptologie, 1891, 241.
[4] Erman-Ranke, 552.
[5] Charlesworth, Trade Routes, 89, 158.
[6] BMGE, 73.

[7] Maspero, Dawn, 480.
[8] Agricola, De Re Metallica, Basle, 1657, 80 ; Gowland, J. Anthropol. Inst., 1912, xlii, 238 ; ib., Archæologia, 1899, lvi, 270.
[9] xxxiii, 4.
[10] Bailey, The Elder Pliny's Chapters on Chemical Subjects, 2 vols., 1929–32, i, 199. The more rational Polybios omits this story (iii, 48 f.).
[11] Blümner, iv, 115.
[12] An Egyptian invention according to Clement of Alexandria : Gsell, 4.
[13] Blümner, iv, 41.

Thebes[1]; those in the Nubian mines were like those found in the Old Kingdom malachite mines of Sinai.[2]　A stone muller of the Predynastic or early Dynastic Period is shown by de Morgan [3], and stone mortars with metal pestles were made for public use in Egypt.　Two small Egyptian stone mortars in the British Museum (3 in. high) were probably used for drugs.[4]　The quern, an old grinding instrument, consists of two circular flat stones, the upper pivoted on the lower and turned by hand with a stick inserted into the side of the upper stone, or a circular stone basin with a circular upper stone resting in it.[5]

In the old period the gold obtained by the process just described was probably melted without further purification.　Crucibles are shown at Beni Hasan (XII dyn.); some, 5 in. high and 5 in. diameter at the mouth, are in the Berlin Museum.[6]　In the Ptolemaic Period the gold was further purified by a process described by Agatharchides.　"At last others ($ἄλλοι$) [7] take it by weight and measure, and putting it with a fixed proportion of lead, salt, a little tin ($κασσίτερος$) and barley bran into earthen crucibles well closed with clay, they leave it in a furnace for five days and nights together, after which it is allowed to cool.　The crucibles are opened and nothing is found in them but the pure gold, a little diminished in quantity.　Such is the method of extracting gold."

Alum and misy (impure copperas, or vitriol) are mentioned by Pliny [8] as additions in the purification of gold : the use of tin seems doubtful, since alloys of tin and gold are very brittle.[9]　Salt, mentioned by Agatharchides, was used in Japan in 1872.[10]　The gold from the final washing was mixed with salt and clay and piled in the form of a cone in a broad, shallow crucible, which was heated to redness for twelve hours in a charcoal furnace.　The silver was converted into chloride, which was dissolved by washing the cooled mass with hot brine and water, leaving the gold free from silver.　The silver was extracted by smelting.

Mungo Park[11] gives an account of gold washing from sands and gravels in Africa and the melting of the gold with an alkaline salt obtained from the lye of burnt corn stalks evaporated to dryness.　He describes the washings at Manding in considerable detail : no mercury was used, and both Pliny[12] and Suidas (10th century A.D.)[13] say that a process of refining with lead was used. Amalgamation is described by Pliny but not by Strabo.[14]　The extraction of gold by amalgamation in the Wâdî Allâqî, and in Sofala in Central Africa, between Abyssinia and Nubia, is described by Al-Idrîsî (A.D. 1154).[15]　The auriferous sands, taken to neighbouring pits, are washed in wooden tubs, from which the metal is taken and mixed with mercury.　"On heating the mixture of gold and mercury by means of a charcoal fire, the mercury evaporates and nothing remains except the body of the gold, pure and melted."　In Qalqa-shandi's time (d. 1418) gold (also silver) was refined in Egypt and tested by

[1] Wilkinson, ii, 252.
[2] Maspero, Dawn, 481 ; Uhlemann, Thoth, 243.
[3] Préhist. Or., ii, 499, Fig. 249.
[4] Wilkinson, ii, 203, Fig. 401 ; Yates and Marinden, DA, ii, 181.
[5] Gowland, J. Anthrop. Inst., 1912, xlii, 256 ; ib., Archæologia, 1899, lvi, 117, 273.
[6] Wilkinson, ii, 234, 236, Fig. 413.
[7] Blümner, iv, 126 f., 140, reads '$ε\psi ηται$, melters.
[8] Blümner, iv, 133.
[9] Giua, Combination among Metals, 1918, 177.
[10] Gowland, J. Anthropol. Inst., 1912, xlii, 257 ; ib., Archæologia, 1920, lxix, 121, 137, Fig. 3.

[11] Travels, Everyman ed., 218 f., 229 f.
[12] xxxiii, 6.
[13] Lexikon, s.v. $ἐξέλιπε$ $φυσιτηρ$; ed. Gaisford, Oxford, 1834, 1286.
[14] Pliny, xxxiii, 8 ; Humboldt, Examen critique de l'histoire de la géographie du Nouveau Continent, 5 vols., 1836–9, v, 93.
[15] Géographie d'Edrisi, traduit de l'Arabe en Français par P. Amédée Jaubert, 2 vols., 4to., 1836–40 ; i, 42, 67. The passage is not contained in the faulty Latin translation of Idrîsî made from an incomplete manuscript by Gabriel Sionita and published as Geographia Nubiensis, Paris, 4to., 1619, 21, 30. Al-Idrîsî was never himself in the region he describes : Ritter, Erdkunde, Afrika², 1822, 435.

keeping it molten in an earthen crucible in a furnace for twenty-four hours, the gold and crucible being weighed.[1]

STOCKS OF GOLD

Gold is represented in wall paintings in a yellow colour, in various forms : heaps (probably native, from the mine), bags (probably of gold dust), plates, bars, long bars, bricks (so named in the texts, the Greek πλίνθοι χρυσαῖ ; the modern "gold bricks" !), cast forms (from the XI dyn.), most frequently in weighed rings [2], rarely in flat discs nearly the size of the hand with a hole in the centre, perhaps from Asia.[3] Gold rings as money are still used in Sennar (Sudan), but none have been found in Egypt.[4]

Large quantities of gold were collected by the earlier kings. Sesostris I (XII dyn. ; 1950) sent an official to Nubia for gold, and the gold country east of Coptos was exploited.[5] In the XII dyn. the king had a regular income from the gold mines of both places, and received also "gold of Kush."[6] Kush, which does not commonly occur on monuments until the XI–XII dyns., denoted the whole region of the Nile below the second cataract, or the Dongola valley.[7] At the beginning of the XVIII dyn. the king received large quantities of gold and other precious materials as taxation from officials : Aahmes I obtained annually 5,600 grains from the mayor of El-Kâb, and all the officials under the Vizier of the South paid annually 220,000 grains of gold and over 1,600 grains of silver.[8] Thothmes I (1557–1501) restored the temple of Amen with electrum-tipped cedar staves, and that of Osiris at Abydos with gold and silver furniture. Thothmes III (1501–1447) also exploited the gold country of the Coptos road, appointed an official as "governor of the gold country of Coptos," and also received much gold in rings, bars and decorated foil as tribute from Nubia and from Punt, in one year 134 lb., and in his later years 600–800 lb. annually.[9]

Amenhotep III (1411–1375) received great quantities of gold from the "land of Karoy," the district around Napata in Nubia. Seti I (XIX dyn.) improved the road to the gold mines through the desert in the Gebal Zâbarâ district, digging a well on the road 37 miles from the Nile : the gold is called in his inscriptions asem, i.e. electrum.[10]

Under Tutankhamen (1375–1350) and Rameses II (1296–1230) the Viceroy of Kush received the title of "governor of the gold country of Amen" ; the mines were the personal property of this god, and great quantities of gold went into his temples. Rameses III (1198–1167) expended vast quantities of gold in temples—annually 26,000 grains—besides silver, copper, wine, etc.[11] In one year Rameses II is said to have received from his gold and silver mines precious metal to the value of 32 million minas (80 million pounds sterling).[12]

"Gold of praise" was given as presents to favourite state officials and priests from the time of the Old Kingdom[13], and "gold of valour" to military officers[14]; gold was sometimes given even to scribes.[15] An officer of Thothmes I(1557–1501)

[1] Calcaschandi's Geogr., tr. Wüstenfeld, 1879, 166 f.

[2] Erman-Ranke, 553, 554 ; Lepsius, Met., 4 f., 59, plate I ; Wilkinson, i, 150, 285 f. ; Figs. 97, 98 ; ib., ii, 3 ; Fig. 269 ; Rosellini, Monumenti, Part 2, passim ; Lenormant, Monnaie, i, 100 ; Maspero, Struggle, 490.

[3] Wiedemann, 1920, 310 ; Maspero, Dawn, 324.

[4] Wilkinson, i, 286.

[5] Breasted, H., 181 ; Wilkinson, ii, 242.

[6] Breasted, H., 163, 181, 185.

[7] Ib., 137, 255 ; Meyer, Alt. II, i, 141.

[8] Breasted, H., 238 ; Meyer, Alt. II, i, 71.

[9] Breasted, H., 265, 310, 314, 317 ; CAH, ii, 85 ; Meyer, Alt. II, i, 141 f., 146.

[10] Breasted, H., 331, 325 ; Anct. Rec., iii, 79 f. ; Meyer, Alt. II, i, 496 ; Davies and Gardiner, Tomb of Ḥuy, 1926, 11, 17.

[11] Breasted, H., 457, 490, 494—Harris Papyrus ; Davies and Gardiner, 6.

[12] BMGE, 97 f.

[13] Breasted, H., 141 ; Ikhnaton, 367 and Fig. 139 ; Thothmes III, 399, Fig. 148—necklaces ; Rameses IX, 509, Fig. 177.

[14] Breasted, H., 226 f.—Aahmes I.

[15] Ib., 241.

speaks of receiving from an Asiatic whom he captured, "gold in double measure."[1]   The gold presented by the king to high officials was usually in the form of collars or necklaces [2], such as that given by Pharaoh to Joseph ; Ikhnaton is represented as throwing these from a balcony to grateful recipients below.[3]

Great quantities of tribute, including precious metals, were obtained in the XVIII dyn. from Asia, Northern Ethiopia and Crete : the texts [4] confirm the statements of Tacitus.[5]   The gold and electrum came from Syria in bricks and rings ; from the Sudan in nuggets and dust.[6]   Some Syrian gold came as vases, and gold vases from Crete, to Thothmes III (1501–1447).[7]   In one year he received 3,311½ kilograms of gold from Asia.[8]   Thothmes III also captured great quantities of gold, silver, etc., at the battle of Megiddo in Syria (1479 B.C.), and from other towns in Syria and Palestine he collected 426 lb. of gold and silver in commercial rings or vessels.[9]   From Syria, Amenhotep II (1448–1420) took 1,660 lb. of gold in vases and vessels as tribute, and Amenhotep III (1411–1375) once received 1,657½ lb. Troy of gold.[10]   The Syrian and Babylonian kings also demanded gold from Egypt, and the metal was sent from Egypt to the Mitanni in the Euphrates region, since in the Amarna letters their king Tušratta wrote to Amenhotep III begging him to "send me so much gold that it cannot be measured, for in my brother's [Amenhotep's] land gold is as common as dust."[11]   He also mentions that Amenhotep III had sent his father "a tablet of gold as if it were alloyed with copper," or (alternative reading) "thou sentest my father a great deal of gold, a namkhar of pure (?) gold and a kiru of pure (?) gold, but thou sentest me a tablet of gold that is as if it were alloyed with copper."   Amenhotep III had sent a gift of 20 talents of gold to the king of Assyria. Burraburiaš, king of Babylon, wrote to Ikhnaton (1375–1350) that his gold was not as good as his father's : from 20 minas of gold put in the furnace, he says, only 5 minas of fine gold remained.[12]   Towards the end of the XVIII dyn., in fact, the quality of Egyptian gold deteriorated : some base gold of that period almost verges into copper.[13]   The king of Cyprus also begged silver.[14]   These letters are interesting ; they show that Egypt was then regarded by other nations as a kind of El Dorado, and also that some kind of refining was in use in Asia Minor about 1400 B.C., i.e. long before its use in Lydia in 550 B.C. Very little metal was found at Amarna, Ikhnaton's city, but there was some gold.[15]   In Tutankhamen's tomb there was abundance of gold.[16]   Two gold jugs of the XIX dyn. are at Cairo.[17]

The later Rameses lost control of Nubia and its gold mines and about 1100 B.C. this source dried up, so that with the cessation of tribute the Egyptians were obliged to go themselves to foreign countries to trade and buy goods.[18]   Even in later times much gold went into temples : Osorkon I (920–917) presented them with 73,000 lb. Troy of gold.[19]

[1] Breasted, H., 264.
[2] Maspero, Struggle, 210.
[3] Wilkinson, i, 40.
[4] Wilkinson, i, 38, plates II A and B ; ii, 239 ; Lepsius, Met., 10 ; Maspero, Struggle, 232.
[5] Ann., ii, 60 ; Quibell, 89 ; Zippe, Geschichte der Metalle, Vienna, 1857, 54.
[6] Maspero, Arch., 311 ; ib., Struggle, 267.
[7] British Museum Quarterly, i, 94, plate 52.
[8] Wiedemann, 1920, 343.
[9] Breasted, H., 292 f.
[10] Breasted, Anct. Rec., ii, 309 ; ib., H., 325.

[11] Meyer, Alt. II, i, 152 ; BMGE, 98 ; Breasted, H., 333 f. ; ib. CAH, ii, 95.
[12] Breasted, H., 334 f. ; ib. CAH, ii, 95 ; Meyer, Alt. II, i, 152 ; Meissner, Babylonien und Assyrien, 2 vols., Leipzig, 1920–25, i, 60 ; cf. Maspero, Struggle, 280, 269.
[13] Petrie, Arts and Crafts, 83 f.
[14] Meyer, Alt. II, i, 153.
[15] Petrie, Tell el-Amarna, 31.
[16] Carter, Tomb of Tutankhamen, ii, passim ; see p. 29.
[17] Maspero, Guide to the Cairo Museum, 1908, 449.
[18] Maspero, Struggle, 561 ; Wiedemann, 1920, 319.
[19] Breasted, H., 532.

Gold jewellery occurs in Nubian remains of the Meroïtic Period.[1] Gold-working and the free use of the metal were maintained in the wealthy Ptolemaic Period, when gifts of gold, especially "earth gold," to temples frequently occur. Much Greek influence appears in the patterns in the Ptolemaic and Roman Periods. Bangle bracelets were often hollow, for lightness and cheapness. Still cheaper were styles (probably of the Roman Period) of thin gold foil worked over a plaster core.[2] The Arabs believed in great stores of gold in Egypt : Yâqût (b. 1178) says Al-Mamûn (813–833) opened the Great Pyramid and found gold in a green vessel [3], and Maqrîzî [4] gives many fabulous stories of the vast stores of gold accumulated by the old kings.

ASEM; ELEKTRON; ELECTRUM

Since no method of refining gold was known in the early period, practically every specimen of Egyptian gold contains a varying proportion of silver, the alloy approximating to the metal called in Greek elektron ($\mathring{\eta}\lambda\epsilon\kappa\tau\rho\sigma\nu$) and in Latin electrum.[5] This was represented by a special hieroglyphic, now generally read asem, a pronunciation confirmed by the Greek word $\mathring{a}\sigma\eta\mu\sigma\varsigma$.[6] The modern Greek $\mathring{a}\sigma\mathring{\eta}\mu\iota$ means silver; $\mathring{a}\sigma\eta\mu\sigma\varsigma$ is "unstamped."[7] Peet [8] and Langdon [9] follow Gardiner[10] in reading ozm, not uasm, as Brugsch first read it. Gardiner concludes that "the frequently upheld identification of this metal with the Greek $\mathring{a}\sigma\eta\mu\sigma\varsigma$ is false": Birch had read tam (d'm). The identification with electrum, however, seems fairly certain.

The name asem suggests the Hebrew ḥašmal, occurring only once in the Bible[11], where its colour is compared with that of a fiery cloud : the Septuagint translates this $\mathring{\eta}\lambda\epsilon\kappa\tau\rho\sigma\nu$, and the Vulgate, electrum.[12] Many texts give both "white gold" [? silver] and asem together. The Harris Papyrus (12th century B.C.) speaks of "two-thirds gold," "fine gold" and "white gold," and another text of "copper with the colour of one-third gold" [? brass].[13] The two hieroglyphics for "white gold" and asem were not differentiated by Lepsius, but Brugsch[14] regarded them as meaning two different materials.

The hieroglyphic for asem is one of the oldest in use : an ebony tablet of King Menes, I dyn. (c. 3400), found at Abydos, represents him holding a bowl inscribed with it[15], and asem is named several times on the Palermo Stone (v dyn.), which records that King Sahurā (v dyn.) sent ships to Punt which brought back "6,000 weight" of electrum, besides anti-resin and perhaps ebony.[16] Sesostris III (1887–1849) rewarded a military officer with arms of asem, and a XII dyn. text speaks of electrum for temple work as "under the seal" of an official.[17] Electrum beads, etc., from Edfu are probably of the VI dyn.[18], and the British Museum has two electrum figures of the XII dyn. from Dahshûr.[19] The hieroglyphic was previously thought to mean pure gold,

[1] Dawson, JEA, 1928, xiv, 195.
[2] Petrie, Arts and Crafts, 83 f. ; Otto, Priester und Tempel, i, 390.
[3] Reitemeyer, 94 ; Budge, Mummy, 1893, 331.
[4] Description de l'Égypte, transl. Bouriant, Mém. Mission Archéol. Française au Caire, 1900, xvii, 50, 78, 91, 98, 116, 326, 376, 396, etc.
[5] Lepsius, Met., 64 f. ; Jacob, Daremberg-Saglio, Dict., II, i, 531.
[6] Lepsius, Met., 12 f., 60, 64 f. ; Wiedemann, 1920, 344.
[7] Schrader, RL, 765.
[8] JEA, 1923, ix, 124.
[9] Ib., 1921, vii, 150.
[10] Z. für Ägypt. Sprache, Leipzig, xli, 73, 75.

[11] Ezekiel, i, 4 ; viii, 2.
[12] Lepsius, Met., 13 ; Bucher, i, 18.
[13] Brugsch, Ägyptologie, 400, 407 ; Forrer, RL, 200 ; Erman-Krebs, Aus den Papyr., 1899, 96 ; Budge, Egy. Hierogl. Dict., 1920, 353.
[14] Ägyptol., 273, 400, 402.
[15] Breasted, Hist., 43, Fig. 27 ; cf. Petrie, Anct. Egypt, 1915, 15.
[16] Breasted, H., 127, Fig. 29 ; ib., Anct. Records, i, 70, 71, 72 ; Budge, History of Ethiopia, 1928, i, 8—"tchām or white gold."
[17] Breasted, H., 187 ; ib., Anct. Rec., i, 277.
[18] Maspero, Guide to the Cairo Museum, 1908, 433.
[19] BMGE, IV–VI Rooms, 218.

gold, bronze, or pure copper. It evidently represented some costly material, shown as weighed in rings and bags, and ranking second after gold.[1] Isis at Philæ is called "gold of the gods, asem of the goddesses, mafkat of the great cycle of gods."[2] In another text the god Rā says, "my skin is of pure asem."[3] "Real [or new] asem" is in a text of Thothmes III (1501–1447)[4], so that varieties were recognised. Asem is mentioned in the 130 ft. long Harris Papyrus of Rameses III (1198–1167)[5]; in the Westcar Papyrus (1700) in a story told of Sesostris I (1950)[6]; and in the Ebers Papyrus (1550 B.C.), which curiously does not mention either gold or silver.[7] It came from Nubia and Ethiopia.[8] Petrie[9] thinks it was imported (with obsidian and emery) from the Ægean even in prehistoric times, which is doubtful, or from Asia Minor, "since Nubian gold contains very little silver," but this statement is incorrect : the Pactolus, which he suggests as a source, does not carry gold but red mud.[10]

The metal was used for ornaments, beaten into sheets and fashioned into vessels.[11] Queen Hatshepsut (1445) speaks of goods from Punt, including ebony, pure ivory, cinnamon wood, myrrh, resin, eye-paint, etc., and "green gold of Emu,"[12] which she measured out by the peck like sacks of grain, and piled up nearly twelve bushels of it in the festival hall of the palace. "Gold of Emu" may be asemu, i.e. electrum, though the latter is mentioned separately.[13] Green gold is the modern name for electrum containing 10 per cent. of silver[14], and perhaps asem was the paler metal containing more silver : in the Harris Papyrus, "white Arabian gold" is distinguished from "yellow Ethiopian gold.[15]

The kings of the XVIII dyn. covered the pyrimidons of the great obelisks with asem.[16] Queen Hatshepsut built two obelisks 97½ ft. high, weighing 350 tons, each dedicated to the god Amen-Rā : "two obelisks of electrum, whose points mingled with heaven, their summits being of electrum of the best of every country, which are seen on both sides of the river : their rays flood the Two Lands [Upper and Lower Egypt] when the sun rises between them as he dawns in the horizon of heaven." One is still standing (without electrum) at Karnak.[17]

Asem is mentioned on the obelisk of Thothmes III, now on the Thames Embankment, which may have been plated with this metal, which was connected with Rā, the sun-god.[18] Thothmes III once received 8,943 lb. Troy of asem from Asiatics and from Nubia as tribute, which he used to cover the temple walls, and chariots mounted with electrum (perhaps from Crete).[19] The supposed use of the metal coverings of obelisks as lightning conductors (Brugsch) is impossible, since they were not properly earthed.[20]

Large quantities of electrum were used in the temples of Amen-Rā under Amenhotep III (1411–1375) ; shrines, flagstaffs, portals, obelisks and even walls and floors were made from or plated with it, and the god was presented

[1] Lepsius, Met., 12 f. ; Blümner, iv, 160.

[2] Lepsius, 13 ; Bucher, ii, 132.

[3] Erman-Ranke, 557.

[4] Breasted, Anct. Records, ii, 262.

[5] Erman-Ranke, 552 ; Breasted, CAH, ii, 188 ; ib. Anct. Rec., iv, 87 f.

[6] Wiedemann, Altägyptische Sagen und Märchen, Leipzig, 1906, 5, 8, 34, 53, 56.

[7] Joachim, Papyros Ebers, 1890, 160.

[8] Blümner, iv, 30 ; ib., PW, v, 2315 f. ; Meyer, Alt., 1909, I, ii, 258 ; Lepsius, 14.

[9] Anct. Egypt, 1915, 15 ; Dendereh 1898, 1900, 62.

[10] Schmitz, DG, ii, 508.

[11] Erman-Ranke, 552.

[12] Breasted, Hist., 127, 276, 281 ; ib. CAH, ii, 63, 65 ; ib., Anct. Records, ii, 109, 133 ;

Lieblein, Handel und Schiffahrt auf dem roten Meere, 1886, 29, 31—"asem-gold."

[13] Breasted, Anct. Rec., ii, 112, 123—"electrum of Emu."

[14] Liddell, Non-Ferrous Metallurgy, New York, 1926, ii, 998.

[15] Forrer, RL, 200.

[16] Budge, History of Ethiopia, 1928, i, 8.

[17] Breasted, H., 281 ; Anct. Rec., ii, 131.

[18] Breasted, Anct. Rec., ii, 254.

[19] Breasted, H., 307 f. ; ib., CAH, ii, 80 ; Anct. Rec., ii, § 761 ; Maspero, Struggle, 258 ; ib., Guide to the Cairo Museum, 1908, 283.

[20] Wiedemann, 1920, 413.

with "every splendid vessel of asem, without limit of number"; Ikhnaton (1375–1350) went to found his new city in "a chariot of electrum, like Aton [his god, the solar disc] when he rises in the horizon."[1]

COMPOSITION OF ELECTRUM

Asem is a natural material, taken from mines or alluvial sands in early times, although later made artificially. The name electrum was applied to the alloy when the proportion of silver varied from 20 to 50 per cent.[2] Pliny [3] says the metal was called electrum when the silver content reached a fifth. Herodotos [4] calls it pale gold (χρυσός λευκός). When the silver exceeds 40 per cent. the colour of the alloy gradually becomes silver-white ; with less silver it is bright brass-yellow.[5]

When the process of separating gold and silver became known, electrum went out of use : in the XXVI dyn. it disappears from the texts, and the mentions in the Book of the Dead are archaisms.[6] Strabo [7] says gold is obtained from electrum by calcining it with alum, no doubt by a process like that described by Agatharchides, and by this time it was known that the metal was really a mixture of gold and silver.

ANALYSES OF EGYPTIAN ELECTRUM

Dyn.	.	I[1]	I[1]	I[1]	II[2]	VI[2]	VI[2]	XII[3]	XII[3]	XI–XII[5]	XI–XII[4]	XI–XII[4]	XI–XII[4]	XVIII[5]	XVIII[5]	XVIII–XIX[5]
Au . .		79·7	84·2	84·0	78	78	82	82·94	85·92	80·1	78·7	77·3	78·2	72·9	67·0	71·0
Ag . .		13·4	13·5	12·95	17	18	16	16·56	13·78	20·3	20·9	22·3	21·1	20·5	25·0	29·0
Cu . .		0	0	0	—	—	tr.	0·5	0·30	—	—	—	—	pres.	8·0	—
Not dtmd.		—	—	—	—	—	—	—	—	—	0·4	0·4	0·7	6·6	—	—
					(light)	(dark)			(yellow)	(red)		(dark red)			

[1] Petrie, Royal Tombs of the First Dynasty, II, 1901, 24, 40 ; *Ancient Egypt*, 1915, 15—free from iron.
[2] Petrie, *Ancient Egypt*, 1915, 15 ; *ib.*, Dendereh, 1898, 1900, 61.
[3] Berthelot, Arch., 63 f. ; de Morgan, Or., i, 215—no Zn or As ; perhaps artificial [?], copper being added.
[4] Berthelot, Arch., 21 f.
[5] Lucas, JEA, 1928, xiv, 316.

The Egyptian electrum, therefore, contained approximately 70 to 80 per cent. of gold and 20 to 30 per cent. of silver, with occasionally a little copper. Analyses of 26 specimens of modern Egyptian gold from quartz gave in 15 cases more than 1 part of silver to 1 of gold, the highest being 3·3 of silver to 1 of gold. All these specimens would be white, and in appearance like silver.[8] Small additions of silver change the colour and hardness of gold to a much greater degree than small additions of gold change the properties of silver.[9]
Electrum was found at Gurob[10] and (sparingly) at Amarna.[11]

SILVER

Silver probably became known later than gold : its name in many languages means "white gold," " white" or "brilliant." It had to be got from rocks, not rivers, with greater difficulty than gold, and in the earliest period was more highly prized on account of its rarity. It does not occur plentifully on the surface but deep in the ground.[12]

[1] Breasted, Anct. Rec., ii, 358 f., 363, 367, 368, 396.
[2] Lepsius, Met., 14, 15, 60.
[3] xxxiii, 4.
[4] i, 50 ; Jacob, Daremberg-Saglio, Dict., II, i, 535.
[5] Lepsius, 15.
[6] Lepsius, 16.
[7] III, ii, 8, 146c.

[8] Lucas, JEA, 1928, xiv, 317.
[9] Lewis, Commercium Philosophico-Technicum, or the Philosophical Commerce of Arts, 1765, i, 83, 119 f.
[10] Petrie, Illahun, etc., 1891, 19—Seti I.
[11] Petrie, Tell el-Amarna, 31.
[12] Hoernes, Urg., ii, 213 ; Breasted, H., 98 ; Petrie, Arts and Crafts, 83 f. ; Brugsch, Ägyptol., 400 f. ; Wiedemann, 1920, 343.

Silver was known in the Prehistoric Period in Egypt.[1] Its hieroglyphic, nub haz [or hedj] [2], means "white gold," perhaps more correctly "brilliant or bright gold."[3] Lepsius considered the sign for gold to act merely as a determinative ; the pronunciation is preserved in the Coptic chat.[4] In the Ptolemaic Period, another sign was used, read ārq or seh, and still later a sign read ruā.[5] De Morgan [6] thinks nub haz means electrum rich in silver, not silver : the earliest specimens of silver contain 3·2 to 14·9 per cent. of gold, and the two noble metals may have been "considered originally as a single substance possessing different shades of colour."[7]

In the early period silver comes before gold in the lists, except in the old inscriptions of Sahurā (v dyn.), which refer to the "washing of silver," where it comes after gold.[8] Not only was silver placed before gold in the early period, when it was really scarcer, but it is also put first in archaising inscriptions, such as those at Denderah [9] and even in a text of Thothmes III (1501–1447).[10] In Ethiopia, which was rich in gold, silver is placed before the latter in inscriptions : on the Ethiopian stela of Dongola, gold, asem and bronze appear in this order, and silver is not even mentioned.[11] Mungo Park, in 1796, found the relative value of silver to gold in the Sudan 1 to 1½, as compared with 1 to 15 in Europe.[12] Silver is found sparingly in some Predynastic tombs.[13] Of the Second Prehistoric Period are the cap of a jar and a small spoon with a twisted handle[14], but beads and rings of silver are much less frequent in the Protodynastic Period than those of gold.[15] This early metal probably came from Asia[16] or from Cilicia in Asia Minor.[17]

The small "silver" peg in the eye of the fine III dyn. wooden statue (Shîekh el-Bel'ed) from the Pyramid of Saqqâra[18] is really of polished ebony[19], but there is silver wire inlay in a wooden palette of Men-ka-Rā (x dyn.).[20]

Silver continued to be more valuable than gold throughout the Old Kingdom to the XII dyn.[21], when it became commoner : a "superintendent of the silver house" is then mentioned[22]. A hawk about 6 in. high, some leaf, a cowrie, and a torque (very unusual in Egypt), all of silver, were found in a Middle Kingdom grave at Abydos[23], but there are only a few silver amulets of the XII dyn.[24], when the metal was also used for crowns.[25] A large silver diadem of the XIV dyn. is inlaid with stones.[26]

[1] Breasted, H., 28 ; Budge, Oriris, i, 333 ; Meyer, Alt., 1909, I, ii, 150.
[2] Lepsius, Met., 16, 60 ; de Morgan, Or., i, 214 f. ; Wilkinson, ii, 244.
[3] Erman-Ranke, 551 ; Hoernes, Urg., ii, 213 ; Griffith, Stories, 1900, 94, 102, 126 ; Budge, Egyptian Hieroglyphic Dictionary, 1920, 75, 353, 523 ; Wiedemann, 1920, 343.
[4] Schrader, RL, 767.
[5] Lepsius, 16, 17, 60 ; Budge, Egypt. Hieroglyph. Dict., 1920, 131 ; cf. ἀργός, ἄργυρος.
[6] Or., i, 214 ; Lucas, Egy. Materials, 106.
[7] Lepsius, Met., 16.
[8] Blümner, iv, 28 ; Orth, PW Suppl. iv, 111 ; Bucher, ii, 132 ; Breasted, H., 28, 94, 98 ; Schrader, RL, 767 ; Erman-Ranke, 551 ; Wiedemann, 1920, 343 ; Petrie, Arts and Crafts, 83 f. ; Lenormant, Monn., i, 98 ; Milne, JEA, 1929, xv, 150.
[9] Lepsius, Met., 17, 18, 60.
[10] Gowland, Archæologia, 1920, lxix, 135.
[11] Lepsius, Met., 13, 18.
[12] Ritter, Erdkunde, Afrika², 1822, 469.
[13] Petrie, etc., Royal Tombs, Pt. I,

1900, 28—"traces"; ib., Pt. II, 1901, 36—"rare"; Diospolis Parva, 1901, 25—rings, beads, perhaps also alloyed with copper as "white metal"; Prehistoric Egypt, 1920, 26—silver bowl to spoon of S.D. 60 ; Naqada and Ballas, 1896, 8, 10, 44 f., 48, 67 ; Gowland, Archæol., 1920, lxix, 135.
[14] Petrie, Arts and Crafts, 83 f. ; Wiedemann, 1920, 344.
[15] Brunton, Qau and Badari I, 24, 27, 38, 67 : on only 4, as compared with gold on 13, pages.
[16] Erman-Ranke, 551.
[17] Breasted, H., 94.
[18] Maspero, Dawn, 408.
[19] Maspero, Guide to the Cairo Museum, 1908, 54.
[20] Maspero, Dawn, 458.
[21] Breasted, H., 98, 185 ; Petrie, Prehist. Egypt, 1920, 27.
[22] Erman-Ranke, 559.
[23] JEA, 1930, xvi, 219, plates xxxvi and xxxvii.
[24] Petrie, Arts and Crafts, 83 f.
[25] Wiedemann, 1920, 344.
[26] Maspero, Guide to the Cairo Museum, 1908, 431.

In the early period the ratio of the values of gold and silver, according to Mariette, was 5 : 3.[1] In the Hyksos Period silver was still double the value of gold ; in the xviii dyn., 1 of gold was worth $1\frac{2}{3}$ of silver, and the value of the latter declined till the Ptolemaic Age (3rd century B.C.), when the ratio was 12 to 1 of gold. At the beginning of the xviii dyn., large payments of silver as tax were made by officials : the mayor of El-Kâb and his subordinate each paid 4,200 grains, and officials of the south over 16,000 grains, annually.[2]

The metal is represented white, in the forms of heaps (crude metal), purses, plates, vases, bricks (πλίνθοι ἀργυραῖ), chariots and rings, the latter frequently under Thothmes III, when silver wire was used.[3] It was brought as tribute by Asiatics, and was perhaps imported in spherical lumps, rings, sheets, and bricks of a standard weight.[4] The use of weighed clippings, cakes, rods, bars and tongues of silver in payment for goods lasted from a fairly early period to about the 6th century B.C.[5]

The British Museum has parts of a large chair of state found at Deîr el-Baharî with objects of Queen Hatshepsut's time : the legs are in the form of bulls' legs and the hoofs were originally covered with plates of silver and the frame was covered with plates of silver held in position by bronze nails with gilded heads. The angle supports of the back are ornamented each with a uræus inlaid with silver annules.[6] There is also a collection of silver rings, pendants, bangles, figures of gods, including silver plated with gold, etc., of various periods from the xviii dyn.[7]

Great quantities of silver were captured by Thothmes III at Kadeš, Megiddo, and other Asiatic towns, including a silver statue, 426 lb. of gold and silver in rings in Palestine, 185 lb. of silver from the Syrian coast towns, and 8 massive rings of silver weighing nearly 90 lb. from the Kheta (Hittites) as a gift (the first inscription mentioning the Hittites). The records of the king of Byblos in Syria showed that 244 lb. of "every kind of silver" had been sent to Egypt (?), probably to an xviii or xix dyn. king.[8] A rubric in the Book of the Dead describes an amulet of green jasper mounted in a frame of asem (electrum) and provided with a silver ring.[9] In the xviii dyn. silver had become commoner, as the sources in Northern Syria or Cilicia, which supplied the Hittites, became more accessible.[10] Asiatics (Rast-en-Nu ; Syrians, or perhaps Lydians ?) are represented in the tomb of Rekhmara as bringing the metal, probably as importers.[11]

Although Diodoros Siculus says Osymandias (Rameses II) obtained nearly 15,000 tons of silver annually from Egyptian mines,[12] the metal does not occur in Egypt. It was obtained from Asia[13], North Syria[14] and Crete.[15] The richest source in Nearer Asia is Armenia.[16] Silver mines at Pharnacia (Tripolis) on the Euxine[17] were rediscovered by Hamilton.[18]

Silver (and gold) vases from Keft (Crete), Cyprus and Syria are represented in the tomb of Rekhmara and mentioned in the annals of Thothmes III[19]: silver

[1] Lenormant, Monnaie, i, 98.
[2] Breasted, H., 338 ; 238.
[3] Lepsius, Met., 17, 60 ; Wilkinson, i, 150, 166f.
[4] Lepsius, 18 ; Erman-Ranke, 554 ; Maspero, Arch., 311 ; Wilkinson, ii, 237 ; Lenormant, Monnaie, i, 98.
[5] Regling, PW, vii, 972, 976, 978.
[6] BMGE, IV–VI Rooms, 55.
[7] Ib., 93 f., 215, 217 f., 219.
[8] Breasted, H., 292, 293, 302, 304, 515.
[9] BMGE, IV–VI Rooms, 66.
[10] Erman-Ranke, 551 ; Petrie, Prehist. Egypt, 1920, 27 ; ib., Arts and Crafts, 83 f. ; Breasted, H., 338.

[11] Wilkinson, i, 237, 254, 256 ; plates IIA, IIB ; Erman-Ranke, 551 ; Petrie, Anct. Egypt, 1915, 16.
[12] Wilkinson, i, 76 ; I have not found the statement in Diodoros.
[13] Birch, in Wilkinson, ii, 237.
[14] Petrie, Prehist. Egypt, 27, 43 ; Erman-Ranke, 551 ; Blümner, iv, 28.
[15] Burrows, Discoveries in Crete, 1907, 118.
[16] Schrader, RL, 764.
[17] Strabo, XII, iii, 19 : "formerly worked."
[18] Leaf, Troy, 1912, 291.
[19] Wilkinson, i, 38 ; Hall, CAH, ii, 280.

and gold rings were used for trading purposes by this king, some as heavy as 12 lb.[1] Silver vases inlaid with gold are shown in his tomb [2] and a text of Aahmes I (first king of XVIII dyn.) speaks of vessels of silver rimmed with gold[3] : such a vase of the XIX dyn. was found at Bubastis.[4] The XVIII dyn. silver dishes are rather thick and coarse.[5] A model ship in silver (with one of gold) and silver axes were found in the tomb of Queen Aah-hotep (wife of Aahmes I) [6], a silver ring (XVIII dyn.) at Medum [7], and there is another of Ikhnaton in Alnwick Castle [8], although no silver was found at Amarna.[9] Silver is not mentioned in the Ebers Papyrus (c. 1550).[10] Silver vases and a statuette of unknown date and a 63-in. silver chain are at Cairo.[11] A silver trumpet embossed with gold was found in the tomb of Tutankhamen (XVIII dyn.).[12] What is probably a decorative use of silver occurs as part of a bronze "battle axe" in the British Museum (like some represented at Thebes), with a bronze blade, 13½ in. long and 2½ in. broad, inserted into three slots in a silver tube, secured with silver nails : a wooden handle was probably inserted into the silver tube.[13]

Silver chains for women's wear became common in the XVIII dyn., some 5 ft., others only 2–3 in., in length.[14] The princes of T'nai (some place in Asia Minor ?) sent Thothmes III four silver hands and a silver vase of Keft (Cretan) work, together with iron.[15] Herodotos[16] says Rhampsinitos (Rameses III ; 1198–1167) had a great store of silver in a stone vault. A silver jug with a handle in the shape of a goat and a flat silver cup with "Phœnician" design at Cairo are of the XIX dyn.[17]

Rameses III, who received silver from Syria and Assyria[18], says he made large offerings of gold and silver to the temples, such as tablets of silver in beaten work and a great sacrificial tablet in hammered work, mounted with fine gold and inlaid with figures of ketem (fine) gold, and in making the great temple balances for weighing the offerings to Rā at Heliopolis 212 lb. of gold and 461 lb. of silver were used.[19] The "subjects" of the temple of Thebes paid 997·805 kgm., those of Heliopolis 53·351 kgm., and those of Memphis 9·359 kgm., of silver, annually, some of which they plundered from old tombs.[20] Silver was found in XIX dyn. deposits at Abydos.[21]

Petrie gives a figure of a silver bowl, with the rim turned in, which seems to have been spun, and most elaborate work is shown in silver bowls from Mendes, made entirely by hammer work, without moulds or matrices.[22] "Silver in a sack" occurs in a papyrus describing (supposed) events of 1050 B.C.[23]; Osorkon I (920–917) presented to the temples 487,000 lb. Troy of silver, with much gold, and Ašurbanipal in 661 carried off from Thebes "two enormous obelisks, wrought of bright silver, whose weight was 2,500 talents, the adornments of a temple door," together with temple furniture, images, etc.[24] A bronze box inlaid with panels of silver is of the XXVI dyn.[25]

[1] Breasted, CAH, ii, 80.
[2] Wilkinson, ii, 2.
[3] Breasted, Anct. Recs., ii, 14.
[4] Quibell, 123, pl. 14.
[5] Petrie, Arts and Crafts, 83 f.
[6] Hoernes, Urg., ii, 360 ; Maspero, Struggle, 96 ; ib., Guide to the Cairo Museum, 1908, 438.
[7] Maspero, Struggle, 27.
[8] Wilkinson, ii, 244 ; Gowland, Archæologia, 1920, lxix, 135.
[9] Petrie, Tell el-Amarna, 31.
[10] Berthelot, Arch., 240.
[11] Maspero, Arch., 314, 320 ; ib., Guide to the Cairo Museum, 1908, 405, 434, 444.

[12] Carter, Tomb of Tutankhamen, ii, 19, 30, pl. 2.
[13] Wilkinson, i, 215 ; Fig. 48.
[14] Maspero, Struggle, 491.
[15] Meyer, Alt. II, i, 130.
[16] ii, 121.
[17] Maspero, Guide, 450 f.
[18] Hoernes, Urg., ii, 296.
[19] Breasted, H., 490, 494 ; Harris Papyrus.
[20] Erman-Ranke, 127 f., 341.
[21] Petrie, etc., Abydos, Pt. I, 1902, 32.
[22] Arts and Crafts, 83 f.
[23] Wiedemann, Altäg. Sagen, 97, 105, 125.
[24] Breasted, H., 532, 559.
[25] BMGE, IV–VI Rooms, 217.

In the Persian Period silver was still uncommon : Herodotos [1] says that when Cambyses conquered Egypt in 525 B.C. the Cyreneans brought him a present of 500 minæ of silver (which they probably thought valuable), but he regarded this with so much contempt that he scattered it amongst his soldiers. The story [2] that when Cambyses burnt Thebes over sixty tons weight of silver were taken from the ashes may indicate that there were accumulated stocks of the metal in Egypt [3], but it is probably fabulous. Herodotos [4] reports that Aryandes, the Persian govern͡r of Egypt, coined money of very pure silver.

In the Ptolemaic Period silver was still imported and had a peculiarly high value relative to gold. About the xx dyn. gold was about twice as valuable as silver, which was probably imported to Naukratis by Greeks. Alexander's 10 : 1 ratio was a forced introduction of his own currency standard which was found quite unsuitable in Egypt, where the silver coins were in consequence simply reduced in weight. In 270 B.C. Ptolemy II (who is said by Athenaios to have had a silver bowl holding 5,000 gallons [5]) gave up the attempt to enforce Alexander's standard and issued gold and copper, the latter coins as tokens, and the copper drachma was recognised as a standard in Egypt from about 250 B.C. to the Roman conquest. The ratio of silver to copper in this period varied from 500 : 1 to 400 : 1, the high value of silver in Egypt making the ratio seem abnormal, gold to copper being about 1000 : 1. The second reduction of the silver tetradrachm under Ptolemy I brought it to the so-called Phœnician standard, and silver continued to be coined on this basis with a ratio to gold of about 13 : 1. The purity of the silver was maintained as long as Phœnicia was under Ptolemaic control, but in the 2nd century B.C., when this country came under Seleucid rule, the Egyptian coin was debased to contain only 25 per cent. of silver, and the mints of Cyprus, still under Ptolemaic control, followed Egypt in this reduction of standard. All late Ptolemaic silver bears their stamp, at first of Paphos, Salamis and Kition, later of Paphos only, but perhaps the stamp was merely copied in Alexandria. The monetary reform in Egypt occurred first under Diocletian. Under Roman rule the copper standard was abolished and prices quoted in silver. Nero flooded Egypt with debased silver tetradrachms of foreign metal, whilst gold was, as always, purely bullion. Bronze continued to be the popular currency and the tetradrachm was probably worth more than its nominal value in bronze.[6]

Later Egyptian silver coin contained copper, tin, zinc and lead.[7] In Egypt in Qalqashandi's time (d. A.D. 1418) the silver coinage contained $\frac{2}{3}$ silver and $\frac{1}{3}$ copper and was stamped by the Sultan ; in Syria the coinage was $\frac{1}{3}$ silver and $\frac{2}{3}$ copper.[8]

ANALYSES OF EGYPTIAN SILVER

Dyn.	Prob. early.	III	XI–XII[1]	XI–XII	XVIII	XVIII	XVIII	XIX	5–4 cent. B.C.
Ag . .	60·4	90·1	74·5	69·2	82·5	84·9	90·2	92·4	82·1
Au . .	38·1	8·9	14·9	pres.	8·7	8·4	5·1	3·2	17·9
Cu . .	1·5	1·0	—	—	8·9	4·3	4·5	3·9	—
Pb . .	—	0	—	—	—	—	—	0·2	0·5	—
Not dtmd.	—	—	10·6	30·8	—	2·4	—	—	—

[1] Berthelot, Arch., 21 f.—no As.

[1] iii, 13.
[2] Diodoros Siculus, i, 46.
[3] Gowland, *J. Anthropol. Inst.*, 1912, xlii, 269.
[4] iv, 166.
[5] Gardner, DA, i, 183; ii, 170.
[6] Milne, JEA, 1929, xv, 150 f., 169 f.

Anct. Egypt, 1914, 177 ; Revillout, PSBA, 1892, xiv, 244, 247 ; Breasted, CAH, ii, 98 ; *ib.*, H., 338.
[7] Gardner, DA, i, 184 ; Wiedemann, 1920, 343.
[8] Calcaschandi's Geogr., tr. Wüstenfeld, 1879, 145, 168.

A collection of analyses of Egyptian silver shows [1] that it was an alloy with gold containing approximately 60 to 92 per cent. of silver and 3 to 38 of gold, with occasionally a little copper, and was probably a white natural product, not obtained by smelting an ore. Argentiferous galena (lead sulphide ore containing silver) was probably first worked for silver by the Greeks about the 7th century B.C., although it occurs extensively in western Asia. In 26 specimens of modern Egyptian gold from quartz, 15 had the ratio of 1 or more of silver to 1 of gold, the highest ratio being 3·3 of silver to 1 of gold, and all these would be white metals. Old Egyptian silver (IV–XVIII dyns.) has yellowish patches due to an unequal distribution of gold. The two specimens containing lead may have been imported and obtained from argentiferous galena at a much earlier date than has been supposed.[2] Corroded silver objects from Dahshûr (XII dyn.) contained (i) 69·19 per cent. silver as chloride, with traces of gold and copper ; (ii) 55·41 and 49·90 of silver converted into chloride, 1·71 and 2·18 of copper, with no gold or arsenic.[3]

The suggestion [4] that the ancient Egyptians obtained silver by refining its alloy with gold (asem), say by roasting with salt as described by Agatharchides, is improbable.[5] Pliny [6] refers to the Egyptians as "painting the figure of Anubis on silver vases," apparently a kind of niello work, the colour being due to silver sulphide. This is a very old art and old Egyptian work of this type is known.[7]

CHESBET

Interpolated in Egyptian lists of metals are two valuable materials, chesbet (or ḫstb) and mafkat (mafek or mfke) [8], in the oldest inscriptions in Sinai mafket.[9] Chesbet, weighed like gold and silver in deben (91 gm.), was sometimes kept in blocks called "stone" and sometimes powdered : very heavy blocks (2 kgm.) were perhaps of the artificial kind, a cobalt glass.[10] The old texts describe a true chesbet (chesbet maat), identified by Champollion as lapis lazuli, since many amulets of this, and also powdered lapis, are in museums.[11] It was used for painting statues of the gods, e.g. Amen-Rā is often shown blue in wall paintings[12], but the colours were often conventional, grey hair being shown green.[13] Chesbet is represented in the XVIII dyn. as a blue material, with other precious objects ; in the Book of the Dead it is specified, mixed with gum water, in painting a scarab, and in papyri it is described as a precious material for amulets, necklaces and small objects, and for incrusting gold.[14] Specimens of lapis lazuli occur frequently in very old tombs.[15] A text of the Middle Kingdom (2200–1800) mentions [true ?] lapis lazuli ; the statue of a king is said to have a turban of lapis, the hair of the god Rā is said to be of true lapis, and an old religious poem says, "true lapis lazuli is not eaten, barley is better."[16] Chesbet is mentioned in a text of 2500 ; true chesbet in a St. Petersburg papyrus of 2000, in the Westcar Papyrus (1700), in the Ebers Papyrus (1550), in a text of 1200 ; chesbet in a papyrus of 2000–1000, and (with mafek) on a stela of 1000.[17] It is shown heaped up with gold, silver and

[1] Lucas, JEA, 1928, xiv, 316.
[2] Lucas, op. cit.
[3] Berthelot, Arch., 64.
[4] Gowland, J. Anthropol. Inst., 1912, xlii, 269 ; Orth, PW Suppl. iv, 111.
[5] Garland and Bannister, 26.
[6] xxxiii, 9.
[7] Forrer, RL, 551.
[8] Griffith, Stories of the High Priests of Memphis, 1900, 94, 102, 126.
[9] Lepsius, Met., 35, 61.
[10] Lepsius, Met., 23, 26, 34, 40 ; Darmstaedter, Studien zur Geschichte der Chemie, Festgabe E.O. von Lippmann, ed. J. Ruska, Berlin, 1927, 1 f.
[11] Lepsius, 2, 18 f., 23 f., 36, 44, 61 f.
[12] Gsell, 38, 42.
[13] Wiedemann, 1920, 28.
[14] Lepsius, 21 f. ; Wilkinson, i, pl. IIB.
[15] Petrie, Naqada and Ballas, 1896, 10, 23, 25, 28, 44 ; ib., Royal Tombs of the First Dynasty, 1901, II, 18, 37.
[16] Erman, Lit., 50, 74, 77, 195.
[17] Erman, Rel., 44 ; Wiedemann, Sagen und Märchen, 1, 19, 28, 48, 87 f., 119, 139 ; Joachim, Papyros Ebers, 91, 95.

jewels in temple treasuries.[1]   Thothmes III (1501–1447) obtained a very precious variety of chesbet with golden specks (pyrites ?) from Babylonia [2], and Amenhotep III (1411–1375) decorated his temple at Karnak with two stelæ of lapis, great quantities of gold and silver and nearly 1200 lb. of mafek (malachite) for inlay work.[3]   The Westcar Papyrus (1700) mentions an imitation chesbet.[4]   The doors of the temple of Rameses III (1198–1167) are described as inlaid with gold and silver, and incrusted with chesbet and mafek.[5]

Both true and imitation lapis lazuli came from the time of the Old Kingdom from Asia [6], by way of Babylonia and Persia [7], and the name chesbet is perhaps of Babylonian origin.[8]   The real source of the true lapis was farther east.[9]

In the New Kingdom (1600–1100) the kings, e.g. Thothmes III, are said to obtain 3, 4 or 10 "large lumps of pure real lapis lazuli" from Babylonia, also 20 seals of this and 3 pieces of (artificial ?) "lapis lazuli of Babylon," in one case 8 lb. of true and 24 lb. of artificial. The Babylonian lapis is described as packed in baskets : it was first sent to Crete and from there was shipped to Egypt.[10]   The "artificial chesbet" was, as Hoefer[11], Lepsius[12] and Berthelot[13] correctly supposed, a glass or glaze coloured with copper and cobalt, which we shall describe elsewhere (pp. 117–119).

Theophrastos[14] distinguishes between "male and female" varieties of kyanos, which Lepsius[15] thought might be glasses coloured with cobalt and copper respectively. Theophrastos, however, really distinguishes three kinds of kyanos : Egyptian, Scythian and Cyprian. The Egyptian was artificial ($\sigma\kappa\epsilon\nu\alpha\sigma\tau\delta$ς) and was probably the blue frit or glaze, containing copper or cobalt or both, which also came from Babylon. The Scythian was probably true lapis lazuli, which is found in Persia, Afghanistan, Siberia near Lake Baikal, Tibet and China. The Cyprian was probably the deep blue mineral azurite or chessylite, which is often very like lapis lazuli in appearance.[16] Theophrastos also speaks of Phœnician kyanos, "both burnt and unburnt": the burnt was probably the artificial glaze or frit ; the unburnt, azurite from Cyprus, both traded by Phœnicians.

MAFEK

Mafek, or in the oldest texts at Wâdî Maghâra, Sinai, mafket, was a green substance always put in relation with chesbet. In the Ptolemaic Period it was called ḥeb (Hebrew bâreqeth). It is represented in bricks and heaps, coloured green[17], and was a precious material used for religious purposes (as in an old book on ritual) or perhaps as a means of exchange.[18] The identification with malachite (found at Wâdî Maghâra) made by Lepsius is usually accepted, although turquoise is also possible. The meaning of mafek in hieroglyphics is usually "the region of the mines," or the material found there, a green stone,

[1] Maspero, Dawn, 302.
[2] Maspero, Struggle, 284 ; Breasted, H., 296, 304 ; Jaeger, T. Faraday Soc., 1929, xxv, 320.
[3] Breasted, H., 344.
[4] Wiedemann, Sagen, 19.
[5] Maspero, Dawn, 276.
[6] Erman-Ranke, 546 f.
[7] Budge, G. of E., i, 417 ; Lucas, Egy. Mat., 166 ; Möller, q. in JEA, 1928, xiv, 273, suggests by way of Palestine.
[8] Lepsius, Met., 20.
[9] See Babylonia and Assyria, p. 293.
[10] Meissner, Babylonien und Assyrien, i, 60, 351 ; Breasted, Anct. Rec., ii, 204, 301 ; H., 296 ; Meyer, Alt. II, i, 153, 128, 209, 240.

[11] Histoire de la Chimie, 1866, i, 64.
[12] Met., 27 f., 34.
[13] Or., 220 ; Introd., 245.
[14] De Lapidibus, 31, 55 ; Pliny, xxxiii, 13 ; xxxvii, 9.
[15] Met., 25, 32 f., 35, 61.
[16] Sir John Hill, Theophrastus on Stones, 1774, 101, 171, 220, 381 ; John, Malerei der Alten, 1836, 115 ; Blümner, iv, 500 f. ; Kisa, Das Glas im Altertume, Leipzig, 1908, 286 f. ; Lucas, Egyptian Mat., 166.
[17] Wilkinson, i, pl. IIb ; Lepsius, 2, 18f., 35 f., 40 f., 61.
[18] Brugsch, Ägyptologie, 1891, 154 ; Naville, Bubastis, 1891, 6.

"emerald, or malachite, or turquoise": the goddess Hathor, whose face was shown green (painted with mafek),[1] is the "lady of mafek" and Sinai the "land of mafek."[2] It has often been suggested that mafek or mafkat means turquoise [3], which occurs at Wâdî Maghâra in Sinai in hard nodules in the purple-brown sandstone just below a bed of iron oxide [4], and the Bedouins still trade in turquoise at Suez to about 250,000 francs a year.[5]

According to Petrie [6], mafkat is malachite, and mafkat nešau, "imperfect mafkat," is green felspar ; emerald was wholly unknown till Greek times, as likewise "root of emerald" or beryl. Malachite, a basic carbonate of copper, was probably (as now) used for ornaments, and very fine specimens were too valuable for smelting.[7] Such ornaments occur, with lapis lazuli, in very early tombs [8], and powdered malachite, used as green eye-paint (uat or wod), occurs at least as early as 3000.[9] A green eye-paint of the XII dyn. was pure malachite[10], although its general use for this purpose ceased in the IV dyn.[11] Malachite was also used for statuettes.[12] Amenhotep III (1411–1375) says he used nearly 1,200 lb. of mafek for inlay work in his temple at Karnak[13]; Rameses II (1296–1230) speaks of two "mountains" (mines ?) of it, one in Asia and one in Africa[14], and Rameses III (1198–1167) imported it plentifully from Sinai for use in temple offerings.[15]

Besides the true variety, Lepsius[16] recognised that there was an artificial mafek, which is not verdigris (as he thought possible for some varieties) but a copper-silica-lime glaze (see note 11, p. 117). The "new mafek" in the Westcar Papyrus (1700)[17] is probably the "true" variety, and a true mafek is mentioned at Denderah among eight precious substances, gold, silver, true chesbet, true chenemen (red, or ḥnmt, probably carnelian), true nešemem (light blue), true mafek, true teḥen (yellow ; topaz ?) and true hertes (white ; perhaps milky quartz, which is one of a mixture of 24 mineral substances mentioned at Denderah).[18] According to Qalqashandi (d. A.D. 1418) the "emerald" mines at Kûs (a town now vanished) were practically abandoned in his day, as they did not recoup the great cost of working.[19]

GEMS

The names of ancient Egyptian gems are very uncertain[20], and according to Mahaffy[21] the words translated "precious stones" in the Greek of the Rosetta Stone read "corn" in the hieroglyphic and demotic versions. In Egypt beads and jewellery, including arm bands of gold, silver and copper, were thought to have a magic protective influence, and were worn in necklaces as amulets, the decorative purpose being quite secondary. Burnt clay beads were worn by the poor[22], and besides gem stones, glazed ware, gold, silver, iron (very

[1] Brugsch, Ägyptologie, 398, 402.
[2] De Morgan, Or., i, 217; Petrie, Researches in Sinai, 1906, 70, 135, etc.
[3] Nature, 1928, cxxi, 375 ; Petrie, Sinai, 41 ; Brugsch, Rel., 155 ; Newberry, Studies presented to F. Ll. Griffith, Oxford, 1932, 320 ; Lepsius, Met., 36 ; Gsell, 50 ; Budge, Egy. Hierogl. Dict., 281.
[4] Seebach, in Doelter, Mineralchemie, III, i, 508 ; Petrie, Sinai, 36, 49 ; Lucas, JEA, 1927, xiii, 162.
[5] De Morgan, Or., i, 218.
[6] Quoted by Bevan, Hist., 1927, 12 ; cf. Griffith, Stories, 35 ; Erman-Grapow, 1921, 63—green precious stone or malachite.
[7] Gsell, 70.
[8] Petrie, Royal Tombs of the First Dynasty, 1901, II, 18, 37.
[9] Gsell, 42 f.
[10] De Morgan, Or., i, 215.
[11] Brunton, Qau and Badari I, 13, 39, 63 ; Erman-Ranke, 257.
[12] Wiedemann, 1920, 345.
[13] Breasted, H., 344.
[14] Albright, JEA, 1921, vii, 83.
[15] Breasted, H., 485.
[16] Met., 25, 28, 61 f.
[17] Wiedemann, Sagen, 8, 9.
[18] Lepsius, Met., 39 ; Davies and Gardiner, Tomb of Ḥuy, 1926, 24.
[19] Calcaschandi's Geogr., tr. Wüstenfeld, 13, 159 ; Benjamin of Tudela, Itinerary, ed. Adler, 1907, 69.
[20] Brugsch, Ägyptologie, 1891, 402 f.
[21] In Bevan, Hist., 266.
[22] E. W. Lane, The Modern Egyptians, 2 vols., 1846, ii, 354 f. ; Wiedemann, 1920, 128 f.

rare), striped coloured glass [1], and, very rarely, iron pyrites [2] were used as bead materials. The early Egyptian cylinder seals are of wood, copper, bone, ivory and glazed ware, not of stone [3], and were used for sealing oil and wine jars in the Old Kingdom.[4]

Bead-making developed early and bead bracelets of the queen of King Zer (I dyn.) at Abydos are very fine work.[5] Bead materials of t'1e Protodynastic Period (end of Predynastic to beginning of IV dyn.) are : carnelian (common throughout, carnelian pebbles occur in the desert) [6], porphyry (?), felspar, olivine, lapis lazuli, quartz, amethyst, garnet, green jasper, and hæmatite. The last three were not found at all in the Old Kingdom or until the XI dyn., and hæmatite not until the XII dyn. ; quartz occurs sporadically again from the IX dyn. onwards but rarely in the Old Kingdom.[7] Bead materials of the Badarian Period [8] are ivory, shell, and copper (all three scarce), carnelian, red and green jasper, various natural pebbles, slate (?), breccia, white, green and yellow calcite, varieties of limestone, "alabaster" (really a calcite), steatite (including bluish-green glazed—there was no glazed frit), soapstone and serpentine. Only two amulets were found.

In the Predynastic (not Badarian) remains at Badari were beads (some cut in facets) of blue glaze, carnelian, garnet, serpentine, soapstone, "clearest white rock crystal," olivine, calcite, coral, steatite, obsidian (?) and ostrich egg-shell ; amethyst, hæmatite and felspar were not found, but there was Badarian "branch coral," [9] and a fragment of Red Sea coral from Egypt is in the British Museum.[10]

Gem stones found in Egypt according to Pliny[11] are : emeralds, opals, amethysts, lapis lazuli, and topaz or chrysoprase, but some of these are not native products. Pliny's sardonyx was a kind of agate.[12] The "emerald" mines of Coptos[13] are mentioned by Yaqûbî (A.D. 891) and Maqrîzî (1358–1441)[14], and (as worked) in a 10th century Arabic MS., but were exhausted in the 13th century.[15] Precious and semi-precious stones actually used in ancient Egypt are : agate, amethyst, beryl, calcite, carnelian, chalcedony, felspar, garnet, hæmatite, Iceland spar, jasper, lapis lazuli, onyx, pearl, peridot, quartz, rock crystal, sardonyx and turquoise, whilst the diamond, ruby and sapphire were apparently not known[16], although small Predynastic beads believed to be of spinel ruby were found, with gold, garnet, etc., at Abydos.[17] Petrie[18] adds serpentine to this list ; Maspero[19], emerald, aquamarine and chrysoprase, with the commoner materials obsidian, granite, serpentine, porphyry and malachite for beads (round, square, oval, spindle-shaped, pear-shaped and lozenge-shaped) and amulets.

The information in the texts (as distinguished from finds) is very doubtful and the identifications are difficult. Most of the stones were obtained locally and used from the earliest period, and the same "accidents" which led to the discovery of metals could lead to the discovery of gems.[20]

[1] Wiedemann, 1920, 129 f.
[2] Erman-Ranke, 254.
[3] Budge, Amulets and Superstitions, 1930, 87 f., 293 f.
[4] Sethe, *Alte Orient*, 1923, xxiii, 10.
[5] Breasted, H., 50, Fig. 17.
[6] Erman-Ranke, 570.
[7] Brunton, etc., Qau and Badari I, 16, 22, etc.
[8] Brunton and Thompson, The Badarian Civilisation, 1928, 27, 39, 49, 56 f.
[9] Brunton and Thompson, Badarian Civ., 35, 52, 56 f. ; cf. Maspero, Arch., 244.
[10] BMGE, IV–VI Rooms, 222.
[11] xxxvii, 5, 6, 8, 9.

[12] King, q. by G. Rawlinson, The Five Great Monarchies of the Ancient Eastern World, 3 vols., 1875, iii, 162.
[13] Strabo, XVII, i, 45 ; 814 C.
[14] Reitemeyer, 149.
[15] Führer durch die Ausstellung Papyrus Rainer, Wien, 1894, 260.
[16] Lucas, Egy. Mat., 156 f.
[17] *B. Mus. Quarterly*, i, 65.
[18] Arts and Crafts, ch. vii.
[19] Arch., 244 f. ; *ib.*, Guide to the Cairo Museum, 1908, 442.
[20] Lucas, Egy. Materials, 157, 163, 171 f. ; *ib.* JEA, 1930, xvi, 94, 201 ; Goguet, ii, 102.

A large collection of bead necklaces of all periods, from Predynastic to Ptolemaic, in the British Museum includes hæmatite, garnet, rock crystal, amethyst, carnelian, agate and other hard stones.[1] Straw beads [2] are unusual : glazed beads are considered later (p. 121).

Green felspar has been wrongly called "Egyptian emerald."[3] Pearls were not used until the Ptolemaic Period, when they were obtained from the Red Sea and Persian Gulf.[4] Wilkinson [5] figures gold ear-rings set with two pearls, and Maspero [6] gives both pearls and mother-of-pearl as used in Egypt. Lieblein [7] identifies the qaš (or qa) imported by Hatshepsut as mother-of-pearl. Pink coral was found at Amarna.[8] Peridot, which is Strabo's and Pliny's "topaz," was found only once, in a scarab of the XVIII dyn. : it is a rare stone, coming from the Jazîrat Zabûgat in the Red Sea.[9] Turquoise was used from the earliest Predynastic Period (beads at Abydos), but more freely from the I dyn. (four bracelets from tomb of Zer, etc.) to the present day.[10] The Egyptians probably often confused it with malachite, which is not unlike it in colour and was obtained from the same mines at Sinai, principally at Wâdî Maghâra and at Sarâbît al-Khâdim (where old workings are found), the former of which are still worked sporadically.[11] Strabo[12] says there were mines of various kinds of precious stones at Meroë, the capital of Ethiopia. The turquoise ("Turkish stone") is still greatly prized all over Asia and in many parts of Africa ; the Arabs call it "the lucky stone," and it is said to change colour on the approach of death.[13] Another name for turquoise is callaïte (Pliny's callaïs, Greek κάλαϊς, κάλλαϊς).[14] In the tale of Snefru and the magician a mafkat (turquoise) jewel is lost.[15]

Although Terrien de Lacouperie[16] considered that real jade (nephrite) occurs only in Khorassan, Forrer[17] says it is found in Egypt. Nephrite is also found abundantly in New Zealand and is worn by the Maoris as an amulet.[18] Bauer[19] points out that several minerals have been confused with nephrite and jadeite, and quotes analyses of specimens from Burma, China, Mexico, Costa Rica, Switzerland, Piedmont, Jarkend, Bokhara, Siberia, British Columbia, Schleswig, Alaska, New Zealand, Tibet, India, Turkestan, Austria, the Hartz, etc., but not Egypt ; the main source is Upper Burma. Both jade and jadeite may have been imported.[20]

The occurrence of amber in Egypt is doubtful.[21] Maspero[22] refers to amber beads, mostly very small, found in VI, XI and XII dyn. tombs at Abydos, now in the Bulaq Museum, which were mistaken by Mariette for corroded brown or yellow glass. They are "still electric." Since glass becomes positively electrified by friction with silk, amber negatively, the matter could easily be settled. North European amber is characterised by giving a large quantity of succinic

[1] BMGE, IV–VI Rooms, 94 f.
[2] Ib., 222.
[3] Lucas, Egy. Mat., 163.
[4] Lucas, 166 ; Rommel, PW, xiv, 1685 ; Krause, Pyrgoteles, oder die edlen Steine der Alten, Halle, 1856, 18, 35.
[5] ii, 342 ; Fig. 448, No. 17.
[6] Arch., 244 f.
[7] Handel und Schiffahrt auf dem Roten Meere, 1886, 31.
[8] Peet and Woolley, City of Akhenaten, 1923, 21.
[9] Lucas, 167 ; Budge, Amulets, 321.
[10] Lucas, 168 ; ib., JEA, 1927, xiii, 162 ; Petrie, Sinai, 41 ; Mosso, Dawn of Mediterranean Civilisation, 1910, 59.
[11] Lucas, JEA, 1927, xiii, 162 ; ib., Egy. Materials, 169 f. ; Maspero, Arch., 1902, 42.

[12] XVII, ii, 2 ; 821 C.
[13] Budge, Amulets, 325.
[14] Pliny, xxxvii, 10 ; Maspero, Arch., 244 f. ; Seebach, in Doelter, Mineralchemie, III, i, 507.
[15] Petrie, Sinai, 41.
[16] Western Origin of the Early Chinese Civilisation, 1894, 312 f.
[17] RL, 549 ; cf. Feldhaus, Technik, 743.
[18] Budge, Amulets, 315.
[19] In Doelter, Mineralchemie, II, i, 649 f., 697.
[20] Mackenzie, Footprints, 139.
[21] Perrot-Chipiez, Art in Egypt, i, 840 ; Erman-Ranke, 254 ; Lucas, Egy. Mat., 12 ; ib., Analyst, 1926, 447.
[22] Dawn, 393 ; Arch., 244 f.

acid on distillation.[1]   Reinach [2] considers that Baltic amber came to Egypt.[3] True amber is the fossilised resin of some extinct species of pine and occurs only in a few localities, chiefly in deposits of the Tertiary Age on the Baltic coast in East Prussia. Varieties resembling true amber occur in Sicily, Rumania, Burma, China, Japan, the Alps, Spain, Galicia and Bukovina, but true amber is perhaps found only in Europe.[4]

[1] Schnittger, in Hoops, i, 260.

[2] *L'Anthropologie*, 1891, 107 ; 1892, 280.

[3] Cf. Herodotos, iii, 115, and Rawlinson's note.

[4] Siedler, in Ullmann's Enzyklopädie der technischen Chemie, 1915, ii, 166 ; Spencer,

Thorpe's Dict. of Applied Chemistry, 1921–7, i, 183 ; Jacob, Daremberg-Saglio, II, i, 531 ; Schrader, RL, 71 f. ; Blümner, PW, iii, 295 ; G. C. Williamson, The Book of Amber, 1932 (many misprints) ; further on bead materials see Petrie, Objects of Daily Use, 1927, 2, 14, 18 f.

EGYPT II—2

THE METALS

PART II. USEFUL METALS

COPPER

There is no metal more frequently used in antiquity than copper [1], and it is always one of the earliest metals known in any country.

Copper was represented by a sign read chomt (or ḥmt) by Lepsius; the oldest form at Wâdî Maghâra (Sinai) was a crucible: later forms (Greek and Ethiopian Periods) are different.[2] It is shown coloured red—which colour sometimes represents flint—as large discs, plates, bars and fragments, and weighed as bricks (dôbe); also fused and purified.[3] Asiatic copper is sometimes shown black.[4] A black copper, chomt chem, sometimes "inlaid with gold," [5] occurs in a text of 1200, and in a late text at Denderah is said to be used with gold, silver, etc., for making the fourteen organs of Osiris for religious ceremonies. It may have corresponded with the Greek black copper (χαλκὸς μέλας), or was perhaps simply copper as distinct from bronze (Greek χαλκὸς λευκός). That it was coarse metal (chiefly cuprous sulphide) produced in copper smelting [6] seems improbable, since Egyptian copper was made from carbonate ores. Other names which may represent copper are bâa and teḥâs-d.[7]

The sign ☥, afterwards (when inverted) the alchemist's symbol for copper, occurs on III and IX dyn. seals and scarabs [8], but represents the windpipe attached to the heart. The symbol ♀ for the ba (soul), supposed to be the same as the astronomical symbol of Venus [9], or a girdle tied together[10], may really represent an internal organ of Isis.[11]

THE EARLIEST COPPER IN EGYPT

In Egypt copper may have been the first metal known[12], since it occurs in the oldest graves, of Sequence Date 30, gold, silver and lead appearing only at the beginning of the second Prehistoric Age, of Sequence Date 42. The earliest known copper in Badarian remains came from the North, perhaps with turquoise from Sinai. The lumps of Predynastic copper found at Qau indicate that the metal was worked up locally into tools. A Predynastic copper needle

[1] Goguet, i, 166—noteworthy for the date, 1758, instead of the usual references to "bronze" or "brass."

[2] Lepsius, Met., 2, 45, 62; Petrie, Researches in Sinai, 1906, 162; Griffith, Stories, 94, 102, 126; Budge, Egy. Hierogl. Dict., 485—ḥemt; Erman-Grapow, Handwörterbuch, 110—doubtful: Coptic chomt.

[3] Gsell, 5, 38 f., 51, 68; Brugsch, Ägyptol., 400; Lepsius, 47, 62.

[4] Davies and Gardiner, Tomb of Ḥuy, 1926, 29.

[5] Gsell, 5, 51, 68; Breasted, Anct. Rec.,

ii, 65, 156—real black copper; iv, 140; Brugsch, Ägyptol., 271 f., 400; Budge, Dict., 486; black bronze in Erman-Ranke, 550.

[6] Lepsius, 48, 62.

[7] Budge, Egy. Hierogl. Dict., 210, 842.

[8] Petrie, Hist., 1903, 26, 35, 117, 118.

[9] Goblet d'Alviella, The Migration of Symbols, 1894, 187 f., 191.

[10] Wiedemann, 1920, 72.

[11] Budge, Mummy², 315.

[12] Gsell, 1 f.; Petrie, Prehistoric Egypt, 1920, 25 f.; ib., Anct. Egypt, 1915, ii, 12; Oberhummer, PW, xii, 84.

at Qau had a trace of cloth adhering to it.[1] The Badarian copper comprises two flat spiral beads of ribbon and a bent pin, the only metal implement found, which differed in shape from Predynastic specimens and was perhaps used as a borer. Predynastic copper (S.D. 44–50) at Badari included lumps and a needle, beads, hooks, etc.[2] The first copper is ornamental : needles with a bent over (hook) eye, pins, beads and bracelets ; the first tools (chisels) came at the end of the Predynastic Age.[3] Predynastic copper includes wire and strip, sometimes used for the repair of broken stone vessels and for binding tortoise-shell arm bands, and ornaments in solid and hollow castings.[4] The king carried a stone mace and a copper axe.[5]

Early copper includes :—

(1) That in Predynastic graves at Ballas and Naqada, 30 miles north of Thebes, with black pottery and flint.[6]

(2) That in Predynastic and I dyn. graves at Tarkhan (bowls, axes, armlets, tools, etc.).[7]

(3) Protodynastic and IV–XI dyn. (not analysed) at Qau [8], including tools, weapons and mirrors (very rare before the VI dyn.).

(4) Predynastic in Nubia.[9]

(5) Copper tools, vases, etc., of various dates, some prehistoric and some earlier than gold objects, at Abydos.[10] Copper sheet nailed over wood was common ; there are wire and nails, and a copper rod "plated thickly" with gold.[11]

(6) A libation vase with bent double spout (a type persistent in all periods), a small chisel, and a handleless bowl, II dyn. from Abydos.[12]

A type of hammered copper vessel with spout found at Abydos (probably VI dyn.)[13] is shown in Old Kingdom representations.[14] Finely wrought bowls were found at Abydos.[15] A "bronze" (? copper) wash basin and ewer of the type shown was found in an XVIII dyn. tomb.[16]

In the I dyn., copper wire, produced by cutting thin strips of sheet and hammering them into a round shape, was largely used.[17] It was never made by drawing through dies.[18] There was a large piece of wood with pieces of copper wire sticking out of it in I dyn. remains at Abydos, but the glazed tiles in a chamber at Saqqâra were probably fastened with gut, not copper wire.[19]

Although it has been assumed[20] that casting was earlier than hammering or forging, both methods were used from the Predynastic Period. Chisels cast

[1] Brunton and Thompson, Badarian Civilisation, 1928, 41, 46, 56 ; *Nature*, 1927, cxix, 159.

[2] Brunton and Thompson, 1928, 7, 12, 27, 33, pl. 50, 86 W₃ ; 46, 56, 60.

[3] Mackenzie, Footprints, 142.

[4] Wiedemann, 1920, 45 f., 131 ; Maspero, Arch., 351.

[5] Erman-Ranke, 624, 627.

[6] Petrie, Naqada and Ballas, 1896; Hoernes, Urg., ii, 240 f. ; BMGE, IV–VI Rooms, 238.

[7] Petrie, etc., Tarkhan I and Memphis V, 1913, 7, 8, 9, 10, 11, 13, 15, 17, 21, 22, 23 ; *ib.*, Tarkhan II, 1914, 9.

[8] Brunton, Gardiner and Petrie, Qau and Badari I, 1927, 13, 14, 17, 24, 27, 28, 30–36, 38–41, 59, 61, 66 f.

[9] Reisner, Arch. Survey of Nubia, Cairo, 1910, i, plate 65.

[10] Mosso, Mediterranean Civ., 59 f. ; Petrie, etc., Abydos I, 1902, 7, 15, 23, 30, 32, 33 ; *ib.*, Prehistoric Egypt, 1920, 25 f.,

[27] ; *ib.*, Royal Tombs of the First Dynasty, 1900, I, 13—I dyn., 27, 28 ; Royal Tombs of the First Dynasty, 1901, II, 12, 24, 27, 28, 36, 39 ; Breasted, H., 39, 93, Fig. 16 ; BMGE, IV–VI Rooms, 285, 286, 290.

[11] Royal Tombs, II, 36 f. ; cf. Petrie, etc., Diospolis Parva, 1901, 24, foil, wire, etc., of copper of Prehistoric Period.

[12] BMGE, IV–VI Rooms, 251, 285.

[13] JEA, 1930, xvi, pl. xxxvi, 217 ; Budge, Mummy², pl. vi.

[14] Erman-Ranke, 344, Fig. 155 ; Breasted, H., Fig. 41 ; Petrie, *Anct. Egypt*, 1915, 14.

[15] Breasted, H., 39, Fig. 16—I dyn. ; also vases, VI dyn., JEA, 1930, xvi, 217, pl. xxxvi ; Petrie, Abydos II, 33, pl. 21.

[16] Hall, JEA, 1928, xiv, 204 : "might well be Japanese."

[17] Petrie, *Anct. Egypt*, 1915, 12.

[18] Garland and Bannister, 70.

[19] Feldhaus, Technik, 200.

[20] Goguet, i, 164.

in pottery moulds, from I dyn. tombs, and other early copper, contained about I per cent. of bismuth, which hardens the metal[1] ; somewhat later copper contains 1–2 per cent. of arsenic and "rarely tin."[2] It is "underpoled" in the modern sense, and much cuprous oxide is left in. Strong hammering makes this metal still harder and suitable for tools for wood and limestone cutting.[3]

Very hard copper tools from a Mexican mine at Guerrero, which blunted the edge of a modern steel knife, were free from arsenic but contained nickel and cobalt, present in the ores from which they were made.[4] Some very hard copper axes from Charente (France) contained 2·8 per cent. of arsenic [5], and the presence of arsenic in some specimens of Egyptian copper and scoria from Sinai but not in others has been held to suggest that it was intentionally added, perhaps in the form of the mineral mispickel, which, however, is not found in Sinai.[6] Although Rüppell found arsenic in Sinai ore, Berthelot could not detect it.[7] Gsell [8] makes much of this arsenic, whilst Petrie [9] suggests an error in dating the Old Kingdom copper said to contain arsenic by Berthelot. Copper containing 0·15 per cent. of arsenic is rather brittle when cold and very brittle at a red heat[10], and arsenic has really little effect on hardness. Copper alloyed with bismuth is ductile at moderate temperatures, but becomes very hard and brittle when hammered ; if the bismuth exceeds 0·6 per cent. the metal cracks at the edges if hammered hot. Bismuth is found in most Egyptian copper and bronze ; although not found in copper made from malachite it occurs in crude metal from other ores. Antimony is not a common constituent ; that reported in earlier analyses may have been bismuth. Lead usually occurs in crude cast copper.[11]

Petrie's[12] theory that objects were cast in the more fusible arsenical alloy whilst plain copper was used for the beaten work is disproved by a tough copper strip from a XII dyn. wooden chest containing only a trace of tin but 4·17 per cent. of arsenic[13], and by Egyptian copper containing 2·3, 3·9 and 5·6 per cent. of arsenic.[14] A mixed smelting with arsenical ore, or a smelting of Fahl ore (a complex mineral containing sulphides of copper, arsenic, antimony, nickel, often zinc, but no tin)[15], or of copper arsenate[16], is improbable, as these minerals are not found in Sinai.[17]

OLD KINGDOM COPPER

Old Kingdom copper objects include tools, bolts, nails, hinges, mountings, wrought table vessels, files, cups, splendid weapons, and (in a later text) "copper doors with bronze bolts, made for everlasting time" are mentioned.[18] In the temple of King Sahurā (V dyn. ; 2500) at Abûsîr a section 1·02 m. long of a beaten copper water drain pipe, which had been 400 metres long and 4·7 cm. diameter, with a wall thickness of 1·4 mm., without trace of solder, was found in situ.[19]

[1] Petrie, Anct. Egypt, 1915, 14, 17, and figure.

[2] Petrie, Tools and Weapons, 1917, 7 ; Lucas, Analyst, 1926, li, 439.

[3] Gladstone, J. Anthropol. Inst., 1897, xxvi, 311 ; Petrie, Arts and Crafts, 83 f.

[4] Chem. Z., 1912, Rep. 453.

[5] Forrer, RL, 46.

[6] Berthelot, Arch., 72 ; Montelius, Die Chronologie der ältesten Bronzezeit in Nord-deutschland und Skandinavien, Braunschweig, 1900, 198 ; de Morgan, Or., i, 227 ; Gladstone, J. Anthropol. Inst., 1897, xxvi, 312.

[7] Gsell, 33.

[8] 2, 40, 68 f.

[9] Anct. Egypt, 1915, 14 f., 17.

[10] Watts, Dict. of Chem., 1872, ii, 42.

[11] Garland and Bannister, 80 f. ; Watts, op. cit., 36.

[12] Anct. Egypt, 1915, 14 f., 17.

[13] Garland and Bannister, 68.

[14] Lucas, Analyst, 1926, li, 439.

[15] Watts, Dict., 1874, v, 729.

[16] Petrie, op. cit.

[17] De Morgan, Or., i, 223 f.

[18] Breasted, H., 93 f. ; Erman, Lit., 108, 142, 155.

[19] Borchardt, Sahurā, 1910, i, 78 ; Wiedemann, 1920, 15.

A full-size copper incense-burner, with lid, VI dyn., found at Abydos with model tools of copper, is of unusual shape, resembling those of the Coptic Period.[1] Early Dynastic (II–V dyn.) copper from tombs at Naga-ed-Dêr included beads, ornaments and vessels with spouts.[2]

Copper bars with the name of a VI dyn. official [3] and a double hook of pure copper found in the Great Pyramid (2800) [4] are of the Old Kingdom, but the most famous specimens are the life-size statue of King Pepi and a smaller figure of his son, Prince Montesuphis (VI dyn. ; 2700) [5], which were found broken on wood foundations at Hierakonopolis and are now at Cairo. Copper, inlaid ("made") with gems, was also said to be used for statues of the gods.[6] The trunk and limbs of the Cairo statues are of hammered copper sheets riveted together ; the face, hands and feet are cast, probably by the cire perdue process. They indicate long practice in artistic working of copper, yet no traces of such figures are found earlier, nor for over a thousand years later. Cire perdue casting is known for gold in the XII dyn. and was perfected for bronze in the XVIII dyn.[7] ; casting of pieces of appreciable size began about 2000.[8] The suggestion [9] that cire perdue casting was used for the *whole* statue is improbable, and it is not even certain if the faces were cast by this method.[10]

There has been much confusion as to the composition of the statue of Pepi I. Gladstone found it to have the composition (corroded material) : copper 78·3, alumina (Al_2O_3) 1·5, calcium carbonate 2·5, insoluble 0·9, oxygen and loss 16·8, doubtful indication of tin.[11] Berthelot's analysis of only 24 milli-grams of material[12] showed that it is practically pure copper. Mosso[13] found in 1907 in some broken fragments 58·50 per cent. of copper, 34 per cent. of copper carbonate and 6·557 per cent. of tin, so that some mistake must have been made in furnishing his material. Recent analysis by Desch shows that the metal is almost pure copper.[14]

The so-called "sceptre" of Pepi I in the British Museum, which was found with the statues, and is a tall wooden rod overlaid with beaten copper with hieroglyphics and supporting a golden head of a hawk, is of almost pure copper, not bronze.[15]

A vase from Denderah, of the same period, showed "little more evidence of tin."[16] There is a fragment of an open-work copper brazier, coarsely tooled with a graver, of the IX–X dyn.[17]

[1] *British Mus. Quarterly*, i, 65 ; pl. xxxvi*a* ; JEA, 1930, xvi, 217, pl. xxxiii, 1, 2 ; such a burner is, however, already shown by Wilkinson, iii, 398, Fig. 599, 6.

[2] Reisner, Naga-ed-Dêr, 1908, 114 f., 127, 134 ; *ib.*, Naga-ed-Dêr III, 1932, 136, 147 f., 150, 154 f., 159 f. ; cosmetic palettes, *ib.*, 155 ; charcoal, *ib.*, 157.

[3] Wiedemann, 1920, 345.

[4] Hoernes, Urg., ii, 241.

[5] Petrie, Arts and Crafts, 90 f. ; *ib.*, Dendereh 1898, 1900, 61 ; Quibell, 46 ; Maspero, Art, 78 f. ; Rostovtzeff, i, plate v, Fig. 3 ; Mosso, Mediterranean Civ., 56 ; Hall, NE, 1920, 136 ; Erman-Ranke, 49, 501 ; Breasted, H., 104 f., Figs. 53, 54, who says the eyes are of "obsidian and white limestone," *or* "rock-crystal" ; Maspero, Guide to the Cairo Museum, 1908, 75 f. ; Weigall, Ancient Egyptian Works of Art, 1924, 66 ; the body is of sheets welded and riveted together (not soldered) and the head is cast—Maspero, *op. cit.*

[6] Erman, Lit., 118.

[7] Petrie, Tools and Weapons, 60.

[8] Maspero, Art in Egypt, 1912, 204.

[9] Garland and Bannister, 37 f.

[10] Quibell, 46.

[11] Petrie, Dendereh 1898, 1900, 61.

[12] Arch., 65 ; Mâ, i, 365 ; Sebelien, *Anct. Egypt*, 1924, 7.

[13] Med. Civ., 55, Figs. 24, 25 ; analysis also quoted from the official Catalogue of 1915 by Budge, Mummy[2], 36 ; Maspero and Quibell, Guide to the Cairo Museum, 1908, 75.

[14] Sebelien, *Anct. Egypt*, 1924, 7 ; Lucas, Egy. Mat., 77 ; Peake and Fleure, Priests and Kings, 1927, 80—Cu 94·5 ; SiO_2 3·25 ; rest Fe, Ni, S ; no Sn.

[15] Quibell, 46 ; Berthelot, Arch., 65 ; *ib.*, Mâ, i, 365 ; Maspero, Art in Egypt, 78 f. ; Perrot and Chipiez, Art in Egypt, i, 650 ; Sebelien, *Anct. Egypt*, 1924, 7.

[16] Petrie, Dendereh 1898, 1900, 61.

[17] Petrie, Hist. of Egypt, 1903, 114, Fig. 66 ; *ib.*, Arts and Crafts, 90 f. ; Maspero, Dawn, 448, says IX dyn. and of "bronze"; Petrie, *Anct. Egypt*, 1915, 48—of "beaten copper, $\frac{1}{12}$ in. thick."

A knife of uncertain age, found 13 ft. below a statue of Rameses II and analysed by Percy, contained 2·29 per cent. of arsenic, 97·12 of copper, 0·24 of "tin or gold" and 0·43 of iron.[1]

SOME ANALYSES OF EARLY COPPER

(a) Chemical and X-ray analyses of 29 specimens of metal from I dyn. tombs at Abydos are given by Sebelien.[2] The references to T.W. are to the plates in "Tools and Weapons" (Petrie); those to Abydos are to "Tombs of the Courtiers" (Petrie, 1925, 4, 6). No specimen contained any tin.

	Reference	Cu	Fe	Zn	
I dyn. objects	BRIGHT METALLIC SPECIMENS				
1. Axe, clean . . .	T.W. III 101	97·99	—	—	—
2. Axe, green . . .	,,　102	98·13	—	0·25	As tr.
3. Axe	,,　104	100·00	tr.	—	—
4. Axe, Zet . . .	Abydos 387	99·61	—	—	—
5. Adze, Zet. . . .	,,　387	99·94	tr.	tr.	Ag, Bi tr.
6. Chisel, Zet . .	,,　387	98·71	—	—	Ag, Bi tr.
7. Chisel	,,　388	98·03	tr.	tr.	As 0·25, Ag, Bi tr.
8. Tip of knife . . .	,,　508	98·50	—	0·25	As 0·6.
9. Adze, Onkh-ka . .	,,　501	97·63	—	—	—
10. Axe, broken . . .	,,　601	97·22	—	0·34	Ni tr.
11. Axe	,,　615	98·98	—	—	Ag, Bi tr.
12. Adze, bird . . .	,,　640	97·69	tr.	—	—
13. Chisel	,,　640	98·84	0·63	0·18	Ag, Bi tr.
14. Axe, with lugs . .	,,　654	98·30	—	—	—
15. Adze	,,　712	99·60	—	—	Ag, Bi tr.
16. Square bar . . .	Abydos	98·13	—	—	As 0·15.
17. Great adze . . .	T.W. XVI 66	97·01	0·50	—	Ni 0·43, S 0·3.
II and later dyn. objects					
18. Chisel, Khosekhemui, II dyn.	T.W. XXII 49	97·70	0·54	tr.	—
19. Adze, Abydos, VI dyn.	—	98·00	tr.	tr.	As, Pb tr.
20. Mason's chisel, Kahun, XII dyn. . .	,,　XXII 78	97·63	1·18	—	1·4 sand.
21. Heavy chisel, Aahmes XVIII dyn. . . .	,,　,,　68	98·53	tr.	—	0·3 sand.
	VERY CORRODED SPECIMENS				
22. Curved bar, I dyn. .	Abydos 720	88·04	0·14	—	7·96 sand.
23. Adze, very hard, I dyn.	,,　420	94·21	2·5	—	0·42 sand.
24. Adze, very hard, XX dyn. (?) . . .	T.W. XVII 92	57·97	—	—	20·6 sand.

(b) Miscellaneous early Egyptian copper objects which have been analysed include the following. The analyses are not tabulated because of the variety of constituents reported.

(1) Axe from Abydos, I dyn.[3] : copper 98·60, tin 0·38, zinc 1·55 ; (2) band from Abydos, I dyn.[4] : no tin, but 1 per cent. manganese ; (3) strip from wooden chest, XII dyn.[5] : insoluble 0·12, lead 0·29, bismuth 0·03, tin trace,

[1] Gladstone, J. Anthropol. Inst., 1897, xxvi, 311, 315 ; Montelius, 1900, 147 ; Petrie, Illahun, etc., 1891, 12—"tin about 0·5."

[2] Anct. Egypt, 1924, 7 f.

[3] Mosso, Medit. Civ., 60.

[4] Petrie, etc., Royal Tombs of I dyn., II, 1901, 12, 24, 27, 28, 36, 39.

[5] Garland and Bannister, 68.

iron 0·29, cobalt 0·06, nickel *nil*, arsenic 4·17, copper by difference 95·04 ;
(4) sheet from Denderah, XII dyn.[1] : copper 94·8, tin 1·1, arsenic trace ;
(5) pointed hammer from unopened IV dyn. tomb in Gîzah[2] : pure copper with
a little arsenic, traces of tin and antimony ; (6) dagger handle in Pyramid of
Cheops, IV dyn.[3] : copper 99·52, iron 0·48 ; (7) pipe of Sahurā, V dyn.[4] : copper
96·47, iron 0·18, arsenic trace, oxygen and chlorine, etc., from atmospheric
and soil corrosion, 3·35 ; (8) bowl of Tutankhamen, XVIII dyn.[5] : pure copper
with only a trace of lead ; (9) battle axes in Turin Museum, XIX dyn.[6] : pure
copper ; (10) a "copper" mirror from Abydos, XII dyn.[7] : not analysed ;
(11) "copper" vase from Abydos, VI dyn.[8] ; (12) XI dyn. axe, Kahun[9] : pure
copper ; (13) axe, "3,500 years old," edge hardened by cold hammering[10]:
copper 96·9, arsenic 1·5, iron 0·7, tin 0·2, traces of nickel, sulphur and oxygen ;
(14) copper tools from Medum (IV dyn.) contained, with those from Kahun
(XII dyn. ?), only traces of arsenic, antimony, etc. ; (15) adze from Medum
contained 0·38 per cent. of arsenic, traces of antimony and iron, but no tin ;
another, 0·54 per cent. of arsenic, with antimony, iron, sulphur, and possible
phosphorus, but no tin ; filings from a pick contained a possible trace of tin or
antimony.[11]

(*c*) Early Egyptian copper objects analysed by Berthelot are : (1) a prehistoric cup, 5 mm. thick, from Naqada (with flints, ivory, rock crystal, etc.)
contained no arsenic ; (2) a prehistoric axe from Abydos, weight 465 gm.,
not so old as (1), was nearly pure, with no arsenic ; (3) needles from Abydos,
cut from a sheet and folded, were pure copper ; (4) a chisel from Abydos,
from a doubly-folded sheet, was nearly pure ; (5) a canicular needle, folded
spiral, was nearly pure ; (6) another needle contained a trace of arsenic ;
(7)–(8) six objects of unknown origin, one strongly arsenical ; (9) a square
nail, V–VI dyn., was nearly pure. All these objects were free from lead, tin
and zinc. (10) Copper sheet, XII dyn. (?), copper 87·7, tin trace, oxidised
patina 12·3.[12]

Copper objects found by de Morgan at Dahshûr (near the ancient site of
Memphis) include : (11) vase (IV–V dyn.)[13], in fragments and much corroded,
71·9 per cent. copper, a "notable quantity of arsenic," and non-metallic elements
in the patina, but no tin, lead, antimony, zinc or iron : even after cleaning
with nitric acid, this metal continued to throw out an efflorescence of "malignant
patina" (atacamite), as did copper statuettes from Tello ; prolonged soaking
in frequently renewed distilled water will remove this trouble ; (12) mirror,
XI dyn., nearly pure, with trace of arsenic ; (13) foundation object of temple
of XVIII dyn., nearly pure ; (14) tablet from foundation of XXV dyn. temple
at Tanis, pure copper, no tin or lead, ? arsenic.[14]

(*d*) Copper objects in the Berlin Museum[15] are : (1) chisel, late Predynastic
or I dyn. : copper 99·90 ; (2) statuette ("Totenfigur") Rameses II (XIX dyn. ;
1300) : copper 100·0 ; (3) Isis with Horus, later than 700 : copper 97·11,
tin *nil*, lead trace, iron 0·53, nickel and cobalt 0·08, arsenic 1·72, antimony

[1] Gladstone, in Petrie, Denderah 1898,
1900, 61.
[2] Gladstone, PSBA, 1892, xiv, 224.
[3] Flight, in Gladstone, *J. Anthropol.
Inst.*, 1897, xxvi, 315.
[4] Borchardt, Sahurā, 1910, i, 78 ;
Wiedemann, 1920, 15.
[5] Hall, JEA, 1928, ix, 74, pl. 8.
[6] Mosso, Medit. Civ., 51 f.—different
shape from Cretan axes.
[7] Frankfort, JEA, 1930, xvi, 219, pl.
37.
[8] Petrie, Abydos II, 33, pl. 21.
[9] Gladstone, PSBA, 1892, xiv, 223 f.

[10] Carpenter, *Nature*, 1931, cxxvii, 589.
[11] Gladstone, *British Association Report*,
1893, 715 ; PSBA, 1891–2, xiv, 223 ;
Petrie, Medum, 1892, 36 ; Maspero, Dawn,
59, 393.
[12] Berthelot, Arch., 1906, 7 ; 8 ; 10 ; 10 ;
11 ; 12 ; 13 ; 24.
[13] De Morgan, Or., i, 211, dates it a little
later than end of III dyn.
[14] Berthelot, Arch., 51 f., 54, 56, 57.
[15] Rathgen, in Diergart, Beiträge aus dem
Geschichte der Chemie, dem Gedächtnis
von G. W. A. Kahlbaum, Leipzig and
Vienna, 1909, 212 f.

trace, sulphur 0·26. The large amount of arsenic (perhaps intentionally added to increase the fusibility) and the absence of tin in a figure of this late date are noteworthy.

Old and Middle Kingdom tools are nearly pure copper[1] ; usually only 0·5 per cent. of tin or less is present, except in a chisel, which contained 2 per cent. An axe contained 4 per cent. of arsenic, but there is little elsewhere. Copper sheet contained only 1 per cent. of tin.[2]

THE DISCOVERY OF COPPER

The bright green, blue or red ores of copper would attract attention before the dull-coloured ores of iron. Flint mining began in the Palæolithic Age before that for metallic ores, although copper mining began very early ; gold (alluvial) was on the surface.[3]

The way in which metallic copper was first reduced from the ores is quite unknown, but has provided ample scope for the exercise of the imagination. Reisner [4] and Elliot Smith [5] think copper was discovered in Egypt by the accidental reduction of a fragment of malachite, or a cosmetic prepared from it, dropped into a charcoal fire. The unfused sponge of copper so obtained is easily converted into a hard ductile mass by hammering whilst hot. [6] Smith considers that this discovery was made about 4500 B.C., not in Sinai as Breasted thinks [7], although very early working is known from crucibles, weapons and tools in Sinai [8], but in Upper Egypt or Lower Nubia [9], which Lucas[10] regards as improbable. Malachite, regarded as possessed of magic properties, an "elixir of life" compared with the Green Nile which made Egypt fertile, was known before the discovery of copper. Copper casting was not discovered until some time after the metal was known. Elliot Smith thinks the ancient Egyptians soon before 3000 spread to various lands in search of copper, gold, lapis, silver, amber, jade, etc.[11]

Since malachite partly occurs on the surface in Sinai, off the trade routes to Syria or Arabia, it may first have been observed by nomads and traded by them to the Nile Valley, where it was used as an eye paint, the source being kept secret until expeditions were sent from the time of the I dyn. An Asiatic origin of Predynastic Egyptian copper has often been suggested.[12] Naville[13] and Budge[14] think copper was brought by a Hamitic race, perhaps the ancestors of the Sumerians, from South Arabia by way of Sinai and the Delta. Peake[15] puts the discovery in Western Asia, Asia Minor, Armenia or Persia, and thinks metal was imported into Egypt about 4241 B.C. by people bringing also a knowledge of wheat and the cult of the god Osiris, entering the Delta from North Syria, somewhere between Damascus and Beyrut.

O'Leary[16] suggests that the earliest copper in Egypt and Mesopotamia came from the country around Mount Ararat, being carried by an "Armenoid" race along a specified trade route. Spielmann[17] supposes that where copper ores

[1] Breasted, H., 93 ; Petrie, *Ancient Egypt*, 1915, 14 f.
[2] Petrie, *Ancient Egypt*, 1915, 14 f. ; Berthelot, Mâ, I, 359 f. ; *ib.* Arch., 15, 24, 51 ; 54, 55.
[3] Boyd Dawkins, Early Man in Britain, 399 ; Montelius, Ält. Bronzezeit, 1900, 199.
[4] Early Dynastic Cemeteries, i, 132, 134.
[5] The Ancient Egyptians, 1923, xii, 7, 9 ; cf. Gowland, *J. Inst. Metals*, 1912, vii, 24 ; Friend, Iron in Antiquity, 1926, 16.
[6] Fournet, in Jagnaux, Histoire de la Chimie, 1891, ii, 350.
[7] Breasted and Hughes Jones, Brief History of Ancient Times, 1927, 44, 53.

[8] Erman-Ranke, 541, 549 f., 557 f.
[9] Smith, *op. cit.*, xiii, 9, 12, 194 f.
[10] JEA, 1927, xiii, 167 f.
[11] Anct. Egyptians, x f., xiv, 8, 11 f., 13, 30 f., 43, 106, 176.
[12] Montelius, Ält. Bronzez., 1900, 140 ; Hoernes, Urg., ii, 217 ; Budge, Hist., i, 41 f., 44, 63 ; Lucas, JEA, 1927, xiii, 167 ; Peet, CAH, i, 262 ; Hall, NE, 90—Cyprus, Syria, Sinai.
[13] Q. in *Isis*, 1927, ix, 545.
[14] *Op. cit.*
[15] Bronze Age, 36 f., 39.
[16] Arabia before Muhammad, 1927, 44.
[17] *Nature*, 1926, cxviii, 411.

and petroleum occur in close proximity (as in the Caucasus) ignition of the oil would produce metallic deposits liable to early discovery and use. Traces of petroleum occur in many parts of Sinai.[1] All these are merely speculations. "Copper of Shesmet land" may mean either Sinai or Asiatic copper.[2]

SOURCES OF COPPER

Since copper was freely used only at the close of the Prehistoric Age[3], Petrie[4] thinks the earliest Egyptian copper came from mines in North Syria and that Cyprus did not supply copper to Egypt until the I dyn. or later.[5]

De Morgan[6] considers the annual production of a few tons of copper from the mines of Sinai would be too small to supply the needs of Egypt, and the real object of mining at Sinai was mafek or turquoise. Knowledge of metals in Egypt, Chaldæa and Elam was imported from farther East, and Cyprus was a late source. The earliest copper perhaps came from the (rather wide) region between Smyrna and the Caspian Sea, where copper, iron, gold and silver occur, and "perhaps" tin ores; or, alternatively, perhaps from Arabia, in the chain running along the shore of the Red Sea and the Abyssinian *massif*, although the riches there are yet unknown. De Morgan's suggestions are purely speculative.[7]

Copper ore occurs both in Sinai and in the eastern desert. The amount is not large enough to pay for mining at the present day, since copper may now be obtained in much greater quantity in more easily accessible places, but there is twofold evidence of ancient copper mining in Sinai at Wâdî Maghâra and Sârabît al-Khâdim. First, the existence of ancient mines, ruins of mining settlements, mining débris, remains of furnaces, slag, broken crucibles, moulds for ingots and weapons, etc.; secondly, inscriptions in the neighbourhood of these mines, left by mining expeditions. Lucas[8] estimates that the 10,000 tons of copper corresponding with the remaining slags, covering a period of 1400 years, were sufficient for all the early needs of Egypt. The total weight of the copper objects found for this period, most of which are in museums, does not amount to more than a few tons, and the metal was probably remade from old objects and used several times. During and after the XVIII dyn. there was regular traffic between Egypt and Cyprus, and the latter supplied a large proportion of the copper used in Egypt.[9] Thothmes III in 1535 B.C. obtained 108 blocks of copper, each weighing nearly 4 lb., as well as lead and precious stones, from the king of Cyprus.[10] On one occasion the latter sent him an excuse that he could not send as much copper as was wished.[11] Thothmes III also obtained copper from Crete[12], and the mines of Sinai were also in full operation.[13] The king of Cyprus sent Ikhnaton 100 talents of copper.[14]

According to Petrie[15] prehistoric copper from Naqada contained 1·55 per cent. of zinc and only 0·38 of tin, whilst copper tools from Cyprus contain no zinc. Old copper objects from Portugal usually contain small amounts (under I per cent.) of tin, lead and zinc; one specimen contained over 3 per cent. of zinc, with no tin, lead or iron.[16] The presence of over 4 per cent. of antimony

[1] H. J. L. Beadnell, The Wilderness of Sinai, 1927, index, Petroleum.
[2] Newberry, in Studies presented to F. Ll. Griffith, Oxford, 1932, 321.
[3] Cf. Meyer, Alt., 1909, I, ii, 58, 62, 150, 748; Gsell, I.
[4] Petrie, Tools and Weapons, 5 f.; *ib.*, *Anct. Egypt*, 1915, ii, 12, 32.
[5] Cf. Brugsch, Ägyptologie, 1891, 400.
[6] Préhist. Orient., ii, 213 f., 231, 239, 247, 329 f., 336; cf. EB[11], xxx, 149.
[7] Wiedemann, 1920, 45, 48.
[8] JEA, 1927, xiii, 162.

[9] Petrie, Arts and Crafts, 90 f.; Breasted, H., 260; Hall, The Oldest Civilisation of Greece, 1901, 193 f.
[10] Breasted, H., 313.
[11] Meyer, Alt. II, i, 129, 153; Oberhummer, PW, xii, 66.
[12] Regling, PW, vii, 974.
[13] Meyer, Alt. II, i, 143.
[14] Erman-Ranke, 619; Oberhummer, PW, xii, 66.
[15] *Anct. Egypt*, 1915, 12.
[16] Bezzenberger, *Z. f. Ethnol.*, 1908, xl, 769.

and 0·32 of arsenic in a prehistoric Hungarian copper [1] is probably due to smelting of a Fahl-ore-like material. At a later period, of course, copper was mined, besides in Sinai and Cyprus, also in Syria, Palestine and Lebanon, and many places in Syria were called Chalkis.[2] The land of Arrech, which sent Thothmes III two blocks of raw copper annually as tribute, may have been in the Zagros, or the hills of Nazareth.[3] Asiatic copper (chemt Sett) and bronze, sometimes inlaid with gold, are mentioned in the XVIII dyn.[4] Amenhotep II, in 1447, obtained nearly 100,000 lb. of copper and 1660 lb. of gold from the Retenu (Asiatics ; Syrians), and Tutankhamen obtained copper ingots from them.[5] Merenptah (1225–1215) took 9,000 copper swords and weapons, and 120,000 pieces of equipment from the Libyans on the edge of the plateau on the west of the Delta.[6] Rameses III (1198–1167) obtained copper from the mines of Atika, a region somewhere in the peninsula of Sinai, sending an expedition to a Red Sea port, and displaying the great quantities of metal he received under the palace balcony.[7] He also obtained copper, silver and lead from Syria and Assyria, keeping the copper in the temple treasury, as did Rameses II (1296–1230).[8] The "subjects" of the great temple at Heliopolis under Rameses III supplied it annually with 114·66 kgm. of copper ; those of the temple at Thebes furnished 2,395 kgm. every year, and since the people could not find the 50 deben (9 lb.) of copper which each was required to pay, they resorted to robbing the tombs.[9]

EGYPTIAN MINING IN THE SINAI PENINSULA

It is generally assumed[10] that the Egyptians obtained copper in an early period, perhaps Predynastic, from the mines in the Sinai Peninsula, which has been a province of Egypt for 6,000 years, although the name Sinai dates only from the 3rd century A.D.[11]

The mine region was called collectively by the Egyptians the "land of mafkat" (malachite, or turquoise). The first region exploited was separated from the coast by a narrow plain and a single range of hills, and the minerals could be transported to the sea in a few hours. The workmen called this region the district of Baît (the mine), or of Bebît (the grottos, or tunnels), the source of the present Arabic name, Wâdî Maghâra, Valley of the Cavern.[12]

The original Semitic inhabitants were probably acquainted with copper, and the iron, copper and manganese ores, and turquoise.[13] King Semerkhat (probably I dyn. ; certainly early Dynastic) carried on mining operations in Wâdî Maghâra and military operations against the native tribes ; and Usaphais (I dyn.) records on an ivory tablet the "first smiting of the Easterners," evidently expecting the necessity for further smitings.[14] The mines were worked by Zoser (III dyn.) and by Snefru (III or IV dyn.), and actively exploited under Pepi I (VI dyn.).[15] Sahurā (V dyn.) imported copper ore from Sinai.[16] Mining in Wâdî Maghâra was actively carried on throughout the Old Kingdom, and stories of 2000 B.C. speak of treasures of copper from Sinai.[17]

[1] Gowland, *J. Inst. Met.*, 1912, vii, 28.
[2] Blümner, iv, 58 ; Job, xxviii, 2 ; Strabo, XVI, ii, 10, 18, 753 C., 755 C. ; Pliny, iv, 12.
[3] Meyer, Alt. II, i, 129.
[4] Breasted, H., 303 ; *ib.*, Anct. Records, ii, 20, 43, 72, 292 ; Budge, Egy. Hierogl. Dict., 486.
[5] Breasted, H., 325 ; Anct. Rec., ii, 309 ; Davies and Gardiner, Tomb of Huy, 29.
[6] Breasted, H., 469 ; *ib.*, CAH, ii, 168.
[7] Breasted, H., 485.
[8] Hoernes, Urg., ii, 296 ; Brugsch, Ägyptol., 268.
[9] Erman-Ranke, 127, 128, 137, 341, 590.

[10] Orth, PW Suppl. iv, 111 ; de Morgan, Or., i, 230 ; Petrie, Arts and Crafts, 90 f.
[11] BMGE, 4 ; Peet, Egypt and the Old Testament, Liverpool, 1922, 126.
[12] Maspero, Dawn, 355 ; Napier, Arts, 103.
[13] Maspero, Dawn, 354 f. ; de Morgan, Or., i, 230 ; Petrie, q. by Sebelien, *Anct. Egypt*, 1924, 10.
[14] Breasted, H., 48.
[15] De Morgan, Or., i, 230 f. ; Blümner, iv, 58 ; Budge, Mummy[2], 30 ; BMGE, 195.
[16] Budge, Hist. of Ethiopia, 1928, i, 8.
[17] Wiedemann, Sagen, 27.

After a long interval in which they were abandoned, the mines were intensively worked by the last kings of the XII dyn., especially Amenemhet III (1849–1801), and by many kings in the New Kingdom. The present Sarâbît al-Khâdim ("Servant Mountain") mines were worked first in the XII dyn., and thenceforward, e.g. under Thothmes III (1501–1447) and Hatshepsut, they became the centre of the Sinai mining industry. They, as well as the mines of Wâdî Maghâra, appear to have been exhausted in the New Kingdom : the inscriptions cease in the XX dyn.[1] Rameses III (1198–1167) does not mention copper as a product, so that working in Sinai probably ceased late in the XIX dyn., when pure copper implements are replaced by tin bronzes, perhaps imported.[2]

Although it has been said [3] that there is still much copper ore in Sinai, with considerable deposits of manganese at Um Bogma, iron ores and precious stones such as turquoise, and that only railway facilities are needed to make them worth getting [4], the copper is really exhausted except in one shaft in the Wâdî Nasb, where small quantities still remain.[5] Remains of furnaces, crucibles, slags, miners' houses and utensils were discovered on the site by de Morgan[6], and Petrie found a smeltery of the Old to Middle Kingdoms period, with much copper slag, furnaces, broken crucibles, part of an ingot mould, charcoal and an unworked charge of crushed ore, so that we can reconstruct the metallurgical processes with some precision.[7]

The copper-containing minerals collected by de Morgan [8] comprised (1) turquoise, (2) a hydrated copper silicate (*modern* chrysocolla) and (3) sandstone impregnated with basic copper carbonate (malachite) and hydrated silicate. The main ore was probably malachite, with some azurite (another basic carbonate) and a little chrysocolla.[9] The black oxide, CuO, said to occur in Sinai[10], is very rare. Although easily reduced, the impregnated sandstone would give a poor yield of metal with much labour.[11]

Besides the copper ores there was also hæmatite (iron oxide, Fe_2O_3) similar to that used in making statues, and a little crystalline pyrolusite (manganese dioxide, MnO_2).[12]

The shafts were driven horizontally into the mountain and the galleries supported by pillars of standing rock.[13] The ruined furnaces at Sinai are described[14] as 6 ft. square and $2\frac{1}{2}$ ft. high, built of stones with no trace of lining. Petrie[15] says heaps of slag were found some distance from a furnace, which was built of calcined blocks of granite, 15 ft. across and 5 ft. high, with blast holes, but in another description[16] the furnace consists of double rows of stones filled in with gravel. According to de Morgan[17] the débris of furnaces and crucibles seem to show that the former were constructed of blocks of sandstone, and the crucibles of quartz sand mixed with clay.

[1] Blümner, iv, 58 ; Berthelot, Arch., 65 f., 73 f. ; de Morgan, Or., i, 229 ; *ib.* Prehist. Or., ii, 329 ; Napier, Arts, 106 f. ; Breasted, CAH, ii, 65 ; *ib.*, H., Fig. 85 ; Erman and Ranke, 559.

[2] Sebelien, *Anct. Egypt*, 1924, 10.

[3] Petrie, Arts and Crafts, 83 f.

[4] Garland and Bannister, 28 ; Mining in Egypt, by an Egyptologist, *n.d.*, 27 f.

[5] Lucas, JEA, 1927, xiii, 162 ; Erman-Ranke, 558 ; Beadnell, The Wilderness of Sinai, 1927, 78, 133, 174.

[6] Or., i, 223 f. ; Préhist. Or., ii, 221 f., 329.

[7] Petrie, Res. in Sinai, 1906, 51, 52, 162 and plate of crucible ; cf. Berthelot, Arch., 66, 71 ; Gsell, 16 ; de Morgan, Or., i, 229, Fig. 593—mould for casting ; Napier,

Arts, 106 f.—furnace ; Lucas, JEA, 1927, xiii, 162 ; Ducros, *Ann. Serv. Ant.*, Cairo, 1906, vii, 19, 27 ; Wiedemann, 1920, 345.

[8] Berthelot, *Comptes rend.*, 1896 ; reprinted in full in de Morgan, Or., i, 223 f.; *ib.*, Préhist. Or., ii, 221 f., 329, and in Berthelot, Arch., 65 ff, 73.

[9] Lucas, JEA, 1927, xiii, 162 f.

[10] Gsell, 6.

[11] Berthelot, Arch., 73 f. ; Gsell, 1, 5 f., 68.

[12] De Morgan, *ib.*, 225 ; Berthelot, Arch., 68 f. ; Petrie, Sinai, 73.

[13] Erman-Ranke, 558.

[14] Napier, Arts, 106 f.

[15] Res. in Sinai, 18 f.

[16] *Ib.*, 242 f. ; Fig. 172.

[17] Or., i, 227.

The ḥemt ḥer-set-f, "rock copper," is probably the ore.[1] The ores from Sinai according to Rickard contain from 5 to 15 per cent. of copper, and up to 18 per cent. according to Rüppell : Desch found only 3 per cent. The ores of the Eastern Desert (where there is an old working at Hamish) are richer : two specimens from Wâdî Araba contained 36 and 49 per cent. of copper ; that of Abu Seyâl contains on an average well over 3 per cent. and in places as much as 20 per cent., but these are chalcopyrite (sulphide of iron and copper), not malachite.[2]

The Sinai ores were carried some distance, since fuel was scarce near the mines and Sinai was not wooded.[3] The reduction was effected by wood, pieces of which, carbonised to various degrees, were found [4], although the wood may have been converted into charcoal before use.[5] Napier [6] says Acacia vera was used to make the charcoal ; Petrie [7] that "desert plants" were used for fuel.

The earliest method of smelting [8] was probably simply to pile the ore with charcoal in a hemispherical cavity in the ground 10–12 in. in diameter without enclosing walls. The wind would supply the blast. The lumps of copper formed were probably cakes 8–10 in. in diameter and $1\frac{1}{2}$ in. thick, as in European hoards, having the largely columnar fracture of copper broken near its solidifying point.

One crucible with a hemispherical bottom and provided with a pouring spout, found in a temple at Sarâbît, was of friable clay and must have been tilted to pour the metal, as it was too weak to lift from the furnace[9]. A crucible of rough fireclay or ash, the surface inside vitrified in places and showing traces of copper slag, together with possible moulds, was found in Protodynastic remains at Badari.[10]

The copper smelters were called chemti, and a copper-smith repaired the miners' tools.[11] Agatharchides says the chisels in the oldest Nubian gold mines were of bronze or copper[12], and although it has often been asserted that such would be too soft, the chisels and picks found at Sinai are of practically pure copper, with only a trace of tin.[13] A fragment of a pick ("pointerolle") found by de Morgan[14] was of copper, "strongly arsenical," with practically no tin.

The scoriæ from Sinai are heavy and show unfused fragments ; perhaps sandstones and hæmatite were used as fluxes, not limestone. Some iron in the ore was driven into the slag. Some scoriæ were very fusible, as they have enveloped the bone of a small animal without causing violent calcination. The slags show crystals of fayalite ($2FeO,SiO_2$), magnetite (as iron slags), greenish pyroxenes mixed with magnetite and felspar, less basic than the first group, and in one slag crystals of cuprite (Cu_2O) mixed with metallic copper.[15] An analysis of the slag, supplied by Petrie, gives 37·9 per cent. insoluble in acid, 21·65 per cent. copper, 37·95 lead (very unusual and suspicious), 1·9 iron, traces of nickel and cobalt, 0·45 arsenic, but no antimony, silver or bismuth.[16] Both ore and slags were free from tin.[17]

[1] Budge, Egy. Hierogl. Dict., 486.

[2] Lucas, JEA, 1927, xiii, 164, 165 ; Egy. Mat., 66 ; Wilkinson, i, 155 ; cf. Gsell, 8.

[3] Berthelot, Arch., 69 ; Petrie, Sinai, viii ; de Morgan, Or., i, 226.

[4] De Morgan, Or., i, 226.

[5] Berthelot, Arch., 69.

[6] Arts, 106.

[7] Sinai, 19.

[8] Gowland, J. Anthropol. Inst., 1912, xlii, 241.

[9] Petrie, Sinai, 162, Fig. 161 ; Gsell, 61 ; on prehistoric crucibles, see Gowland, op. cit., plate 26.

[10] Brunton, Qau and Badari I, 67.

[11] Petrie, Sinai, 117, 120.

[12] Blümner, iv, 41.

[13] Petrie, Sinai, 40, 48 f., 160 f., plates, Fig. 160 ; 243 — analyses by Ramsay ; cf. Napier, Arts, 104.

[14] Or., i, 222, Fig. 592 ; Berthelot, Arch., 71.

[15] De Morgan, Or., i, 226 ; Berthelot, Arch., 70 f. ; Gsell, 6, 68.

[16] Lucas, Egy. Mat., 67 f. ; Sebelien, Anct. Egypt, 1924, 10.

[17] Gladstone, J. Anthropol. Inst., 1896-7, xxvi, 313 ; Montelius, Ält. Bronzez., 1900, 199.

A great mass of slag, estimated by Petrie at 100,000 tons, was found in the Wâdî Nasb, where the ore was probably smelted [1], and a native reported that four bars of metal, the size of an arm, had been dug from it, probably leakings from one of the several furnaces found in ruins. Heaps of slag were found at some distance from a furnace, parts of the slag (scoria ?) being rich in copper, with some smaller pieces of nearly pure metal.[2]   The estimate of 100,000 tons of slag, made by Petrie, was reduced by Rickard to 50,000 tons [3], representing 2,750 tons of copper extracted, but Lucas [4] thinks Petrie's estimate is nearer the truth. Lucas's criticism of Rickard has in turn (without any stated reason) been attacked by Glanville.[5]

AFRICAN COPPER

Although Africa seems destined in the near future to take the key position in the supply of copper, an ancient African copper industry is not established, and this metal appears to have been scarce in the countries immediately adjoining Egypt.   Herodotos[6] says the Libyan women wore on each ankle a copper or bronze ($\chi\acute{\alpha}\lambda\kappa\epsilon o\nu$) ring, and Strabo [7] that the Nubian women at Meroë in his day wore copper rings through the lips. Abulfeda [8] reports that the natives of Sofala (in East Africa) preferred copper ornaments to gold, and Herodotos [9] that the Ethiopians had so much gold that the chains in prisons were of this metal, whilst among them copper was the most valuable metal. They did not appreciate the gold chain which Cambyses (529–522) sent them. Both Diodoros Siculus[10] and Strabo[11] mention copper and gold mines at Meroë, and Strabo[12] copper mines near the North African coast. Rozière found remains of old copper and lead mines in the Baram Mountains, 40 km. east of Syene[13], and there were said to be ancient copper works in the Fayyûm, but there are none there now.[14]   The copper from the region of Bahr el-Ghazal, in the Chad territory, was worked and sent to Egypt early in the 19th century.[15]

Great quantities of copper are said[16] to have been taken from "ancient workings" in Rhodesia and the Congo, but the date of these is very doubtful. The modern natives of Rhodesia smelt copper in primitive furnaces. Budge[17] thinks copper was brought to Egypt or Nubia from Tanganyika about the XII dyn., or even in the Predynastic Period.   There is a slag heap at Kuban, where the fort was occupied not before the XII dyn.[18], and copper occurs in Predynastic, Old Kingdom and New Kingdom remains in Nubia.[19]   The copper mines in Nubia mentioned by old authors were not (as Elliot Smith assumed) in the Wâdî Allâqî but in the Wâdî Seyâl, and were probably not worked before the XIX dyn. Although there is said to be copper ore in the Wâdî Allâqî, there are no Egyptian records of its having been worked.[20]   It is quite likely that archæological investigation will establish Africa as an early source of copper, but the evidence so far available is not convincing.

THE EGYPTIAN COPPERSMITH'S WORK

In one of the earliest Predynastic graves was a small copper pin used for fastening a goatskin over the shoulders.   Later came a small chisel, then an

---

[1] Petrie, Sinai, 27 ; Napier, Arts, 104 f., 106.
[2] Petrie, Sinai, 18 f.
[3] Lucas, Egy. Mat., 60, 66.
[4] JEA, 1927, xiii, 165 f.
[5] JEA, 1928, xiv, 189.
[6] iv, 168.
[7] XVII, ii, 2, 821 C.
[8] Géogr., i, 307 ; ii, 222.
[9] iii, 20, 21, 23.
[10] i, 33.
[11] XVII, ii, 2, 821 C.

[12] XVII, iii, 11, 829 C.
[13] Gsell, 8 f.
[14] Schubart, Ägypten von Alexander der Grösse bis auf Mohammed, 1922, 255.
[15] Ritter, Erdkunde, Afrika[2], 1822, 496.
[16] Walker, Chem. Centralblatt, 1926, i, 1024 ; Nature, 1927, cxix, 43.
[17] Mummy[2], 7, 11.
[18] Lucas, JEA, 1927, xiii, 168.
[19] Archæol. Survey of Nubia, i, plate 65.
[20] Thomas, JEA, 1921, vii, 110 ; Lucas, ib., 1927, xiii, 168.

adze and a harpoon, then needles ; larger objects occur at the end of the period. All this copper was shaped with polished stone hammers.[1] A smooth stone with no handle, held in the hand, was at first used : true hammers came only just before the use of iron, although the mason's mallet was earlier.[2] The coppersmith in the Old Kingdom made knives and chisels, pieces for inlay which were chiselled and polished, large wash-bowls and copper pipes for laying underground.[3] Copper ewers and basins, examples of which were found in the royal tombs, were skilfully hammered out and cast spouts inserted. It has been supposed [4] that the hammered copper and bronze vessels are really very thin castings, since cast spouts could not have been welded on to thin beaten vessels. Annealing of metals was unknown until the Greek and Roman Periods, and spinning, brazing and soldering were unknown for copper and bronze even in the Roman Periods, the parts being simply hammered together.[5] No instance of soft solder used on copper is known till Roman times.[6]

An examination of actual objects showed that cored casting of bronze was later in use, with iron wire struts. The cores (in place in many objects), generally black or dark slate colour, are no doubt sand from deposits on the Nile bank still used for casting. Probably the best example of early hollow casting is the large XVIII dyn. statue of Horus in the Louvre.[7]

The brazier of Khety, now in Paris, is a coarse IX dyn. example of cast copper tooled with a graver. There is not much XII dyn. copper work, except tools. Moulds for casting tools, found at Kahun, were open, cut from a thick piece of pottery lined smooth with fine clay and ash.[8] Methods of casting, with illustrations of crude "ash" crucibles, are given by Petrie.[9]

A cire perdue process, used for jewellery in the XII dyn., was perfected and applied to bronze casting in the XVIII dyn., when extremely thin castings were produced by some unknown technique. An example of the XXVI dyn. is quoted[10], and a fine cire perdue casting was found at Amarna (XVIII dyn.).[11] The castings in the XXVI dyn. were very fine.[12] At first only small objects were cast, but larger castings were attempted from about 2000.[13]

In Old Kingdom pictures a metal is shown let out through a hole in the bottom of a crucible[14], and a representation of metal casting in the tomb of Rekhmara (c. 1450) shows the metal melted in fairly small flat crucibles over a fire of charcoal (emptied from baskets) blown by double foot-bellows, one on each side. The metal is poured from the crucible, which is carried with a two-handled clip, through funnels into a mould, probably of clay, to form "doors for the temple of Amen at Karnak." Three men bring metal for melting, two at the back carrying baskets with small oblong blocks, and the first one a hide-shaped bar or slab, apparently about a foot long. Different metals, probably tin and copper respectively, are no doubt intended to be represented, and an alloy, probably bronze, is being made. The text is not very clear, but[15] speaks of "bringing the bronze [?] of Asia, which his Majesty took as booty in the land of Rezen." The bars are not of lead, as these are shown of a different shape in another grave of the XVIII dyn. A "bronze furnace" is also shown in the tomb of Queen Tii (1400 B.C.).[16]

[1] Petrie, Arts and Crafts, 90 f.
[2] Petrie, Tools and Weapons, 1917, 40 ; Erman-Ranke, 549.
[3] Wiedemann, 1920, 15 ; Erman-Ranke, 221, 500, 549 f., 557 f.
[4] Garland and Bannister, 63 f., 66.
[5] Garland and Bannister, 62, 66, 68, 69, 70, 74, 121, 160—several objects examined.
[6] Petrie, Arts and Crafts, 90 f. ; Feldhaus, Technik, 636.
[7] Garland and Bannister, 37, 46, 48 : numerous figures.

[8] Petrie, Arts and Crafts, 90 ff.
[9] Tools and Weapons, 60 f. ; plate 77, Figs. 245–8.
[10] Petrie, Abydos III, 1904, 6.
[11] Petrie, Tell el-Amarna, 31.
[12] Gsell, 79 ; Maspero, Struggle, 534.
[13] Maspero, Art in Egypt, 204.
[14] Erman-Ranke, 549 ; de Morgan, Or., i, 199, 229, Figs. 527, 593.
[15] Erman-Ranke, 549, Fig. 232.
[16] Perrot-Chipiez, Art in Egypt, i, 829.

Mosso [1] sees in the hide-shaped bar a block of Cretan copper. Such blocks were found in Hagia Triada, and Cretans carrying them as tribute to Thothmes III are represented in the tomb of Rekhmara.[2] Large stocks of them are also shown as contained in Ikhnaton's temple at Amarna.[3] Large numbers of copper objects in museums include rings.[4]

HARDENING OF COPPER

It has often been supposed that the Egyptian metallurgists had some process, now lost, of hardening copper or bronze, and since the known objects of these metals are not particularly hard, the effect is assumed to have passed off with lapse of time.[5] The hardness (or strength) of Egyptian bronze is suggested in the texts by such expressions as "my heart is of bronze," and the king says he "rolled the mountain of bronze from the neck of man."[6] The commentators of Homer (Proklos, Eustathios, Tzetzes, etc.) say, from old unknown sources, that the hardening of bronze ($\beta\alpha\phi\dot{\eta}$ $\chi\alpha\lambda\kappa\sigma\hat{v}$) was effected by dipping into water, but in reality bronze becomes soft when heated and quenched, and although Plutarch says bronze could be made as hard as steel ($\sigma\tau\dot{o}\mu\omega\sigma\iota s$) he does not say how.[7] The accounts in Vergil [8], Pausanias[9] and later Greek authors of hardening bronze by quenching in the water of special springs and fountains or in oil are probably fabulous.[10] Reyer thought the hardness was due to phosphorus, produced by smelting in the presence of apatite, blood and bones, and claims to have found 0·054–0·25 per cent. of phosphorus in bronzes not of Greek or Roman origin.[11]

It has been suggested[12] that Egyptian copper is naturally hard, whilst Asiatic is soft, so that the need for alloying with tin did not arise in early Egypt but was discovered in northern Mesopotamia or northern Persia about 3000. The presence of arsenic, intentional or otherwise, has also been suggested[13] as the cause of hardness, but it is found in very varying amounts and really has very little effect.[14]

Copper may be hardened by cuprous oxide as well as by arsenic, antimony and tin, and ancient Peruvian copper is said to owe its hardness to metals of the platinum group, ruthenium, iridium and rhodium, present in the ores. Gladstone found cuprous oxide in old "hard copper" from Egypt and Asia ; some tools of the Amorite epoch at Lachiš were excessively hard and red, and an adze contained about 24 per cent. of cuprous oxide. A similar xi dyn. adze from Egypt was almost identical in character and composition.[15] Gladstone thinks the cuprous oxide was intentionally produced by a prolonged fusion in the presence of air, giving a hard and brittle "dry copper." A small amount of the oxide destroys the malleability, and the specimens described could not have been hammered into form.

Addition of more tin than the normal 10 per cent., say 22–30 per cent., is said to make bronze as hard as steel but very brittle. Egyptian dies of this composition of the 5th century B.C. are known, and the brittleness may be removed in great part by tempering.[16] Ancient surgical instruments[17]

[1] Med. Civ., 1910, 294.
[2] Ib., Fig. 166 ; Wilkinson, i, plate IIA.
[3] Erman-Ranke, 329 ; Fig. 151 : with jars, dishes, flat discs, etc.
[4] BMGE, IV–VI Rooms, 94.
[5] Blümner, iv, 333 f. ; de Morgan, Or., ii, 217 ; Lucas, Egy. Mat., 79 f.
[6] Erman-Ranke, 521 ; Erman, Lit., 341.
[7] Blümner, iv, 434 ; 333 f. ; Schrader-Jevons, 195.
[8] De Launay, Daremberg-Saglio, Dict., ii, 1077.
[9] II, iii, 3.

[10] Schrader-Jevons, 195 ; Marinden, DA, ii, 3.
[11] Blümner, iv, 333, 335, 337 ; Beck, Gesch. des Eisens, i, 46.
[12] Mackenzie, Footprints, 144.
[13] Gsell, 2, 40, 68 f.
[14] Garland and Bannister, 80 f.
[15] Gladstone, J. Anthropol. Inst., 1897, xxvi, 311 ; Montelius, Ält. Bronzez., 1900, 198 f.
[16] Zenghelis, MGM, vii, 267 ; Gsell, 5, 78.
[17] Sudhoff, in Hoops, iii, 439.

5

and files [1] of hard bronze are known, the latter containing 57–73 of copper, 18–31 of tin, 7–10 of zinc, and 7·7–8·5 of lead. Bronze razors occur early in Egypt, and in Europe in the Hallstatt Period: the Roman razors were of iron.[2] Caylus (1692–1765) discussed the hardness of Roman bronze[3] on the basis of experiments by the chemist Geoffroi (1685–1752), who found that the powder produced by attrition with stone, as well as filings, of the hard antique bronze objects was magnetic. He believed that the hardness was due to an alloy with iron, since most copper ores contain iron. An alloy of 5 parts of copper with 1 of iron was found to be hard. Copper and iron form, with difficulty, an alloy which is magnetic even when containing only 10 per cent. of iron.[4]

Iron is quite effective in hardening copper, and sufficient may have been taken up for this purpose from a ferruginous flux in Egyptian smelting.[5] Geoffroi also found that copper alone could be hardened by first heating to cherry redness and plunging into a tempering liquid composed of 1 pint of river water, a handful of common salt, 2 large handfuls of chimney soot, a pint of urine and a powdered head of garlic. Such curious mixtures are much favoured by technologists. Blümner [6], Garland [7] and Gowland [8] attribute the hardness of the cutting edges of ancient Egyptian copper and bronze implements solely to the effect of cold hammering, and this is probably correct. Græco-Roman copper razors were finished in the same way. Microscopical evidence does not indicate any secret processes for hardening or annealing.

BRONZE

Bronze, usually an alloy of about 90 parts of copper and 10 parts of tin, has three advantages over copper: (1) it has a more beautiful and permanent colour; (2) it is harder and more suitable for vessels, implements and tools; and (3) it has a lower melting-point (c. 900° C.) as compared with copper (1083° C.), and is thus more easily cast. It was used in Egypt for many purposes where iron was applied in other lands.[9] The Bronze Age of a country, in the special sense used by archæologists, begins with the use of bronze weapons and cutting tools, together with vases of the same metal[10], not merely the use of bronze for ornamental purposes.

Bronze seems to have been known and sparingly used in Egypt during the Old Kingdom[11]: that it was known in the Predynastic Period[12] is an assumption, but bronze was made by the Sumerians (p. 245).

It has been supposed[13] that bronze is shown yellowish-brown, copper being red, in pictures, but the difference in colour may be due to fading. Copper and bronze are not usually differentiated in the texts[14], but are said to have had separate names: thesed (teḥàs-d) copper, and chesmen (smen, ut, āha) bronze.[15] The usual reading for copper is ba (bâa), thesed being more doubtful.[16]

[1] Feldhaus, Technik, 515.
[2] Forrer, RL, 218, 647.
[3] Recueil d'antiquités égyptiennes, étrusques, grecques, romaines et gauloises, 7 vols., 1752–67, i, 239, 242 f., 247 f.; cf. Goguet, i, 171.
[4] Thomson, System of Chemistry, 5 ed., 1817, i, 449.
[5] Humboldt, Essai politique sur le royaume de la Nouvelle Espagne, 5 vols., 1811, iii, 307; Garland and Bannister, 80 f.
[6] iv, 335.
[7] Chemistry and Industry, 1927, 648.
[8] J. Anthropol. Inst., 1912, xlii, 244: based on experiments.

[9] Hoernes, Urg., ii, 292.
[10] Montelius, Les temps préhistoriques en Suède, tr. Reinach, 1895, 55.
[11] Perrot and Chipiez, Art in Egypt, i, 650; BMGE, IV–VI Rooms, 238.
[12] De Morgan, Or., i, 84.
[13] Gsell, 38 f.
[14] Lepsius, Met., 49, 62; Hoernes, Urg., ii, 216.
[15] Brugsch, Ägyptol., 401; Gsell, 47, 51 f.; Lepsius, Met., 52; Budge, Egy. Hierogl. Dict., 132, 188, 512, 602, 842; chesmen in Leyden Papyrus.
[16] Prof. Glanville, communication; Budge, Dict., 210.

Breasted [1] gives copper and bronze separately in XII dyn. (c. 2000) texts, e.g. copper doors with bronze bolts, and bronze was apparently plentiful about that period. Sethe [2] read by or ba on the Palermo Stone as both "metal" and "copper": it is sometimes shown blue, perhaps the patina, the bright metal being red. Copper is also called the "grown metal," perhaps because it was derived from mountains and so distinguished from the "metal of heaven," or "pure metal," i.e. iron (first mentioned in the XIX dyn.). Bronze, according to Sethe, is chomt, "an impure mixed copper."

The earliest bronzes contain very little, say 1 or 2 per cent., of tin, which may be accidental.[3] English crude copper may contain 0·2 per cent. of tin, and Spanish 0·4–0·5 per cent., and Montelius [4] thought that anything over 0·5 per cent. of tin in a bronze should be considered as intentionally added, but this is too low a proportion. The 5 per cent. limit suggested by Forrer [5], on the contrary, is much too high. About 2 per cent. is probably a safe upper limit for accidental tin content.[6] Bronze containing as much as 16–22 per cent. of tin has been found in Egypt [7], but this has no advantages over normal bronze. The mere colour of an ancient metal object gives no certain indication as to whether the metal is copper or bronze. Berthelot [8] mentions a bronze ring, containing 8 per cent. of tin and 76 per cent. of copper, which on account of strong oxidation over a long period of time looked exactly like a vase of pure copper found near it. Bronze may, in fact, assume a red colour owing to the formation of cuprous oxide.

METHOD OF MAKING BRONZE

The common ore of tin, the oxide tinstone or cassiterite, SnO_2, is usually almost pure and this makes it difficult if not impossible to say whether bronze was first formed by the smelting of mixed ores or by adding pure metallic tin to copper. Liechti [9] suggests the smelting of a mixed ore, and Montelius suggested that the small amounts of tin in early bronze might have been derived from the smelting of a copper ore.[10] Although copper ores do not usually contain tin and tinstone does not contain more than traces of copper, but iron, manganese and silica,[11] yet tin pyrites (or stannite) contains tin, iron and copper. Its formula is Cu_2FeSnS_4, although it may contain copper pyrites and thus have a variable composition. This ore, somewhat rare, is found in Cornwall (Wheal-Rock) and Bolivia.[12] It is of no importance in tin extraction, but its existence is favourable to Montelius's hypothesis, although no common copper ore contains tin, whilst copper ores rich in lead and zinc are not uncommon.[13] The smelting of stannite is still carried out on a small scale in one locality in China, yielding a metal containing almost equal proportions of copper and tin. It is a possibility for Egypt, but the amount of this ore is too small to make its use probable, and it could never have led to the discovery of the use of the only important ore, cassiterite, or to the production of metallic tin.[14]

[1] Anct. Records, i, 232, 241, 257.

[2] JEA, 1914, i, 233 f.

[3] Berthelot, Arch., 15 ; Gsell, 33 f. ; Forrer, RL, 115 ; Lucas, Egy. Mat., 70 ; ib., JEA, 1928, xiv, 106.

[4] Montelius, Ält. Bronzez., 1900, 7.

[5] RL, 115.

[6] Lucas, JEA, 1928, xiv, 106.

[7] Reinach, L'Anthropologie, 1891, ii, 107.

[8] Arch., 45.

[9] Chem. News, 1923, cxxvi, 413 ; cf.

British Museum Guide to the Antiquities of the Bronze Age, 1920, 4.

[10] Montelius, q. by Reinach, L'Anthropologie, 1892, iii, 451.

[11] Müller, in Ullmann's Enzyklopädie der technischen Chemie, 1923, xii, 266 ; Henglein, in Doelter, Mineralchemie, III, i, 178.

[12] Müller, op. cit.; Humboldt, Nouvelle Espagne, iii, 307.

[13] Gunther, in Ullmann, 1919, vii, 378 f., 423 f.

[14] Lucas, JEA, 1928, xiv, 106.

Native tin is said to occur in Bolivia but is somewhat doubtful, as lead is not a usual constituent of tin ores and the native metal, although principally tin (79 per cent.), contains lead (20 per cent.), with traces of iron, copper (0·09 per cent.) and arsenic.[1] In sulphidic copper ores [2] antimony, zinc, arsenic, iron and lead are common constituents, sometimes nickel, cobalt and bismuth occur, but in the many hundreds of analyses tin is never given, except in traces in one or two Fahl ores. The almost entire absence of tin in all copper ores is striking.

Bronze may have been discovered by accidental smelting of natural or artificial mixtures of separate copper and tin ores. Tin ore is associated with copper ore in vein ore. Whereas all ore used in the West was alluvial cassiterite, both forms may occur in the East. Lucas suggests that vein ore was first used, cassiterite only when the European deposits were discovered as a result of prospecting for gold. By smelting a very rich vein tin ore with copper ore, a white metal containing a large proportion of tin may have been produced [3], and thus tin itself could become known by heating the tin ore with charcoal, a well-known process, and the white metal tin so obtained would be recognised as a constituent of the white bronze. Alluvial tin ore, on this hypothesis, would be a later discovery than bronze : it would never be used in *mixed* smelting for bronze but always for tin, and when the latter was separately obtained a bronze of fairly definite composition, with 9–10 per cent. of tin, could be made by alloying the tin with copper.[4] This hypothesis is at least definite and chemically sound.

Bronze occurs sporadically in Egypt before the first known tin object (XII dyn.), and the tin content is variable, hence it was probably produced by mixed smelting, and possibly the tin ore was at first added as a flux [5], or as a result of experimenting with the action of fire on metals or ores, perhaps regarded as divine substances—a "primitive alchemy."[6]

Gowland [7] has disproved the statement that when a copper ore containing tin is smelted, the tin does not unite with the copper but passes into the slag. He smelted malachite with tinstone, lime, and charcoal in a hole in the ground, the charge being blown with air. The resulting alloy contained 22 per cent. of tin. The early bronzes containing little tin (1–2 per cent.) could thus very well have been produced by smelting ores containing tin.

The protagonists of the hypothesis that bronze was first made by alloying separate metals include Roberts Austen [8], who regarded the amount of tin as too definite to have been the result of haphazard smelting ; Hoernes [9], who thinks the production and casting of fairly pure copper, with the necessary metallurgical experience, must precede the making of bronze ; Kahlbaum and Hoffmann[10], who consider that tin was intentionally added to reduce the melting-point of the copper ; Napier[11], who suggested that the *small* amounts of tin (and lead) in the earliest bronzes indicate that these metals were added to produce the same effect as "poling" copper with wood, viz. removal of oxygen (which makes the metal brittle), the small excess of tin (or lead) then remaining in the copper ; and Garland[12], who points out that Egyptian copper reduced in Sinai and also received from other sources contains no tin.

[1] Doelter, Mineralchemie, III. i. 174. 178.
[2] Henglein, in Doelter IV, i, 73–220 ; 184, 187.
[3] Such as the XIX dyn. alloy of 76 tin and 16 copper; No. 15, p. 71, analysed by Berthelot.
[4] Lucas, JEA, 1928, xiv, 106 f.
[5] Petrie, Tools and Weapons, 7 ; *ib.*, *Anct. Egypt*, 1915, 17 ; Gsell, 73.
[6] Reinach, *L'Anthropologie*, 1905, xvi, 659 ; T. R. Holmes, Ancient Britain, Oxford, 1907, 121 ; Napier, Arts, 127 f.
[7] *J. Anthropol. Inst.*, 1912, xlii, 241 f.
[8] PSBA, 1892, xiv, 227.
[9] In Hoops, iii, 115.
[10] Diergart, 1909, 90.
[11] Arts, 115.
[12] Garland and Bannister, 30.

The copper ore found in northern regions, it has been stated, generally contains traces of nickel and cobalt, which remain in the metal and the bronze made from it, whilst Asiatic copper is generally free from these metals[1], and an exaggerated importance has recently been attached to the presence of traces of nickel in old bronzes. Nickel is found in the Sinai ores, but in insignificant amounts in I–III dyn. copper [2], so that a source other than Sinai has been assumed for these.

It has been supposed that the tin content of bronzes may have altered owing to atmospheric and soil corrosion, this increasing the apparent content of tin and decreasing that of copper in long buried bronzes. Experiments of Kröhnke [3] showed that a sword blade from a 3,000 year old grave in Schleswig-Holstein contained, in four samples working up towards the point, the following percentages of copper (not tin) : 63·79, 57·95, 45·91, 8·56. The metal, it is suggested, must originally have contained more than 63·79 per cent. of copper. It is, however, unlikely, except in very badly corroded specimens, that the composition as found by analysis will differ so much from the original. It has also been supposed that the low tin content of some later Egyptian bronzes, of variable composition [4], could be explained by a gradual removal of tin by oxidation caused by repeated remelting of old objects in contact with the atmosphere.[5]

Many Egyptian bronzes resist oxidation and are of good quality, e.g. daggers are very elastic. Wilkinson [6] suggests that they were rubbed with resin when hot ; Gsell [7] suggests coating with varnish. Wilkinson [8] also thought that an artificial patina was produced in some cases, then the metal covered with a substance which fills the pores, as in modern work, but these hypotheses are improbable.[9]

The hardest bronze contains about 12 per cent. of tin ; a higher percentage makes the metal brittle. Lead makes copper tougher.[10] It has been suggested[11] that other hardening constituents, such as arsenic, copper suboxide and lead, were in use before tin, small amounts of which produce hardness, hence small amounts of tin were first tried. As tin became less costly, the proportion was increased, until it was found that no improvement resulted from the addition of more than about 10 per cent., and this was standardised as the composition of good bronze.

THE OLDEST BRONZE IN EGYPT

A metal rod found by Petrie in the foundations of a III or IV dyn. mastabah (tomb ; an Arabic word meaning couch)[12] at Medum gave on analysis[13] 8·4 per cent. of tin. Maspero[14] and Hall[15] agree with Petrie that this is the

[1] Napier, Arts, pref., iii.
[2] Anct. Egypt, 1930, 90.
[3] Q. by Montelius, Ält.Bronzez., 1900, 148.
[4] See table on p. 71.
[5] Wiedemann, 1920, 346.
[6] i, 212 ; Maspero, Arch., 303.
[7] Eisen, etc., 39.
[8] ii, 256.
[9] On the treatment of corroded museum specimens, etc., see : A. Lucas, Antiques : their restoration and preservation, 1924 ; Berthelot, Arch., 47 f., 76 ; A. Scott, Cleaning and Restoring of Museum Objects, I, 1921 ; II, 1923 ; III, 1926 ; D.S.I.R., H.M. Stationery Office, London ; ib., Nature, 1932, cxxix, 339 ; J. Roy. Soc. Arts, 1932, lxxx, 487 ; H. W. Nichols, Restoration of

Ancient Bronzes and Cure of Malignant Patina, Field Museum of Natural History, Museum Technique Series No. 3, Chicago, 1930 ; Fink, Chemistry and Industry, 1934, liii, 219 ; H. J. Plenderleith, The Preservation of Antiquities, 1934 ; F. Rathgen, The Preservation of Antiquities (transl.), Cambridge, 1905.
[10] Boyd Dawkins, Early Man in Britain, 407, 409.
[11] Montelius, Ält. Bronzez., 1900, 198 ; J. Anthropol. Inst., 1897, xxvi, 312.
[12] Breasted, H., Fig. 33.
[13] Gladstone, in Petrie, Medum, 1892, 36 ; Ancient Egypt, 1915, 17—III dyn.
[14] Dawn, 393, 414.
[15] CAH, i, 29, 291, 319.

earliest specimen of bronze. A second analysis of the unaltered core of the Medum rod gave 9·1 per cent. of tin and 0·5 per cent. of arsenic.[1]

It is possible, however, that an even earlier piece of bronze than the Medum rod is known. Mosso [2] found in a piece of plate, 10 mm. wide and about 0·5 mm. thick, given to him by Maspero, 96·00 per cent. of copper and 3·75 per cent. of tin. The metal was found in a I dyn. tomb at Abydos after it had been searched by de Morgan, and is a well-worked specimen of true bronze. It is not mentioned by Petrie, but Lieblein [3] accepts the occurrence of bronze in dynasties I–III, whilst Blümner [4] accepts it only from the V–VII dyns.

A piece of thin foil of the Thinite Period (II–III dyn.) containing 2 per cent. of tin is probably derived from an accidental smelting.[5]   Copper containing a "trace of tin" was found in a III dyn. tomb by Quibell.[6]   Petrie [7] also refers to "accidental" bronze in a III dyn. tomb, and Wiedemann [8] to "accidental smelting."

The age of Petrie's Medum rod is rather doubtful, since it may have fallen into the excavations from the surface.[9]   Gladstone[10] described it as "found deep down in foundation filling" of the IV dyn. mastaba, and as consisting of an internal core and a dark outer ring. Under the microscope, the core was found to consist of miscellaneous granules, very various in colour. The outer portion was also heterogeneous, containing red cuprous oxide spotted with green and patches of blue. A piece taken right across contained the large amount of 8·4 per cent. of tin, with mere traces of antimony, arsenic and iron, and no phosphorus. A piece of the inner core contained 89·8 per cent. of copper, 9·1 per cent. of tin, about 0·5 per cent. of arsenic, and traces of antimony, iron and sulphur. There was also a residue insoluble in nitric acid, which consisted of very minute particles, some amber-coloured and semi-transparent, others almost black and opaque, and a few colourless, having the appearance of quartz. On treating these with aqua regia, the two former kinds almost wholly disappeared, leaving what appeared to be siliceous skeletons.

The "yellow bronze ring" fitted to a "pure copper loop" in a lead plug found by Borchardt[11] in V dyn. remains at Abûsîr was not analysed, and the two "bronze" vases of Unås (VI dyn.)[12] were perhaps copper, as may also be the "bronze" models of tools, implements, etc., and the "bronze" figure from Qurnah, found in VI dyn. tombs.[13] A "bronze" dagger with gold handle, XII dyn., is shown by Wiedemann.[14]

Abd al-Latîf (d. A.D. 1161) says the two obelisks at Heliopolis, one erected by Sesostris I, XII dyn., and still standing but without metal, were originally covered with bronze cones, which were eaten away in his time.[15] An Old Kingdom copper vessel, with a hole repaired "in antiquity" by a soldered bronze plate, and other similar examples from Egypt[16], are of uncertain date.

Copper was still used for tools when bronze was known ; a basket of copper tools was found at Lahun.[17]

[1] Gladstone, *B.A. Rep.*, 1893, 715; PSBA, xiv, 224 ; Petrie, *Anct. Egypt*, 1915, 17.

[2] Med. Civ., 1910, 57.

[3] MGM, xi, 178.

[4] PW, iii, 892 f.

[5] Berthelot, Arch., 15.

[6] Wiedemann, 1920, 345.

[7] Arts and Crafts, 90 f.

[8] 1920, 346.

[9] Montelius, Ält. Bronzez., 1900, 147 ; Petrie, Arts and Crafts, 90 f. ; Gladstone, PSBA, xiv, 225 ; Lucas, Egy. Mat., 75.

[10] *Ib.*, 224.

[11] Sahurā, i, 76.

[12] Maspero, Dawn, 414.

[13] BMGE, IV–VI Rooms, 208, 220.

[14] 1920, xv, 240, Plate-fig. 19.

[15] Reitemeyer, 103 ; Petrie, Hist., 1903, 157, calls the metal copper.

[16] Mötefindt, in Feldhaus-Klinckow-stroem, *Geschichtsblätter für Technik, etc.*, 1914, i, 150.

[17] Berthelot, Arch., 61 ; Petrie, Ten Years' Digging in Egypt, 1893, 115.

ANALYSES OF EGYPTIAN BRONZES

The following tables contain a collection of analyses of specimens of Egyptian bronze objects, the dynasty of each object being given in Roman numerals, and the source when stated.

TABLE I

	Cu	Sn	As	Pb	Fe	Zn	Sb
1. Thin foil, II–III . . .	56·7	2·0	o	o	o	—	—
2. Ring or bracelet (mean) v	77·0	8·2	tr.	5·5	o	o	o
3. Fragment of vase, VI .	93·8	6·2	o	o	o	o	—
4. Fragm. from Dahshûr, VI	86·23	5·68	—	—	—	—	—
5. Bowl, Thebes, XI . .	85·8	3·5	—	8·5	0·2	—	—
6. Axe, XII (?)	88·9	0·2	5·6	0·6	—	—	0·7
7. Corroded needle, Kahun, XII (second half) . .	—	10	tr.	—	—	—	tr.
8. Axe, Kahun, XII . .	93·26	0·52	3·90	—	0·21	—	0·16
9. Chisel, Kahun, XII . .	96·35	2·16	0·36	—	o	—	o
10. Part of mirror : fine *yellow* metal, bright and clear (cf. Borchardt's ring) . .	95·0	+	+	—	+	—	o
11. Bracelet, Dahshûr, XII .	68·39	16·31	o	o	tr.	tr.?	—
12. "Crochet," Dahshûr, XII	69·23	9·82	o	—	—	—	—
13. Nail, Dahshûr, XII . .	{85·02 / 84·88}	0·97	—	—	—	—	—
14. Ring, XIX (v. hard) . .	77·51	9·65	+	o	o	—	—
15. Ring (v. soft ; inverse bronze) Dahshûr, XIX	16·23	75·66	—	1·0	—	—	—
16. Vase, Dahshûr (tr. sulphur ?), XVII–XIX. .	76·79	15·18	o	o	o	o	—
17. Arrowhead, Abydos (hard), XX	81·93	12·17	—	—	—	—	—
18. Pedestal of statue, XXII.	77·86	5	—	o	—	—	—
19. Mirror : uncertain date.	78·57	11·29	o	—	—	—	—
20. Arrowhead : uncertain date	68·12	5·92	—	—	—	—	—
21. Ingot or plate, Saqqâra, Pharaonic Age . .	87·5	11·47	o	tr.	o	o	—
22. Axe, Stockholm Museum, XVIII	90·21	9·52	tr.	—	0·04	[Ni 0·08 ; S 0·03]	
23. Chisel from "ancient quarry"	94	5·9	—	—	0·1	—	—
24. Dagger	85·0	14·0	—	—	1·0	—	—

1. Berthelot, Arch., 15 f.—probably from mixed smelting with tin ore.

2. *Ib.*, 53.

3. *Ib.*, 54 ; Montelius, Ält. Bronez., 1900, 147, doubts the date, probably only on account of the tin content.

4. Berthelot, q. in de Morgan, Or., i, 212.

5. } Phillips, *Anct. Egypt*, 1924, 89—lead and arsenical bronzes.
6. }

7. Petrie, Ten Years' Digging, 115, 129, etc. ; Gladstone, PSBA, 1892, xiv, 223 ; Hoernes, Urg., ii, 241.

8–10. Petrie, Illahun, etc., 1891, 12 f. ; *Anct. Egypt*, 1915, 17 ; Gladstone, PSBA, 1892, xiv, 227 ; Reinach, *L'Anthropologie*, 1891, ii, 107.

11–21. Berthelot, Arch., 16, 54 f. ; de Morgan, Or., i, 211 f. ; Petrie, *Anct. Egypt*, 1915, 18, says the nail is the only object certainly dated.

22. Montelius, Ält. Bronez., 1900, 150.

23. Ure, in Wilkinson, ii, 255, 401 ; compares with ancient Peruvian chisel analysed by Vauquelin, copper 94, tin 6 ; Napier, Arts, 152.

24. Napier, Arts, 153 ; Watts' Dict. of Chem., 1872, ii, 46 ; Montelius, 152.

TABLE I—*continued*

	Cu	Sn	As	Pb	Fe	Zn	Sb
25. Mirror.	85·0	14·0	—	—	1·0	—	—
26. Small hatchet, Gurob, XVIII	89·59	6·67	0·95	—	0·54	—	tr.
27. Large hatchet, Gurob, XVIII	90·09	7·29	0·22	—	—	—	tr.
28. Nail, Memphis, XXVI .	74·6	0·9	—	21·3	0·3	—	—
29. "Fragment," Memphis, XXVI	92·0	6·5	—	0·8	0·3	—	—
30. Bowl, Karanog (interior tinned), early Christian	80·81	13·08	—	5·14	0·29	[Ni and Co 0·46]	

25. Wilkinson, ii, 3 ; Reinach, *L'Anthropologie*, 1891, ii, 107 ; Gladstone, *J. Anthropol. Inst.*, 1897, xxvi, 315 : analysed by Vauquelin, in Passalacqua, Catalogue raisonné et historique des Antiquités découvertes en Égypte, 1826, 238.

26–27. Petrie, *Anct. Egypt*, 1915, 18 ; Gladstone, *J. Anthropol. Inst.*, 1897, xxvi, 315 ; Montelius, Ält. Bronzez., 150 f. ; Petrie, Illahun, etc., 12, 17 f. ; *ib.*, Ten Years' Digging, 115, 120, 129.

28–29. Phillips, *Anct. Egypt*, 1924, 89.

30. Gowland, in Woolley, MacIver, etc., Karanog, 2 vols., Philadelphia, 1910, i, 67.

Berthelot concluded that pure copper continued in use till a late period, that true bronze appears about the XII dyn., and that the composition of the bronze was very variable, as in European specimens (tin 5·15–18·31, mostly 10).[1]

TABLE II

BRONZES IN THE BERLIN MUSEUM[2]

	Cu	Sn	Pb	Fe	Ni+Co	As	
1. Chisel-like tool : late Predynastic or I dyn.	99·90	—	—	—	—	—	—
2. "Totenfigur" Rameses II, XIX dyn. . . .	pure copper (100·0)	—	—	—	—	—	—
3. Isis with Horus : later than 700 B.C. . .	97·11	0	tr.	0·53	0·08	1·72	Sb tr., S 0·26
4. Figure of man, Middle Kingdom (1900 B.C.). .	92·06	6·72	0·12	0·22	0·22	0·38	—
5. Mirror, XVIII dyn. . .	92·76	5·10	0·53	0·40	0·20	0·80	—
6. Axe, 1300 B.C. . . .	85·86	12·74	0·52	0·37	0·29	0·26	Sb tr.
7. Osiris with enamel inlay, 1200 B.C. . . .	70·81	3·03	25·04	0·23	tr.	0·57	—
8. Isis with gold and niello inlay, later than 700 B.C.	83·60	14·14	1·96	0·25	—	tr.	—
9. Chons, XXVI dyn. (650 B.C.)	89·90	6·58	1·54	0·87	0·41	0·55	—
10. Kneeling king with wine vessel, after 700 B.C. .	87·61	7·39	4·72	0·20	0·07	0·07	—
11. As No. 10	88·47	6·50	4·44	0·32	0·05	tr.	S tr.
12. Neith, after 700 B.C. .	83·25	7·82	7·36	0·27	0·13	0·81	—
13. Kneeling figure of King Send, after 700 B.C. .	71·66	4·03	21·99	0·80	0·51	0·79	—
14. Neith, after 700 B.C. .	81·76	8·36	9·02	0·33	0·11	0·10	Sb tr.
15. Ibis head, after 700 B.C.	85·13	4·56	7·32	0·66	0·09	0·98	Sb 0·23, S tr.
16. Hathor, after 700 B.C. .	88·24	10·42	0·30	0·60	0·00	tr.	—
17. Chons, c. 350 B.C. . .	85·52	13·25	0·20	0·57	0·49	tr.	—

[1] Arch., 61 f. ; cf. Gsell, 34 ; Forrer, RL, 115 f., 923 ; Hoernes, Urzeit, ii, 17 f. ; Petrie, *Anct. Egypt*, 1915, 18 ; Wiedemann, 1920, 233, 236, 336, 346 ; de Morgan, Or., i, 212 ; Wilkinson, ii, 232, 401 ; Perrot and Chipiez, Art in Egypt, i, 830.

[2] Rathgen, in Diergart, 1909, 212 f.

Objects 1–3, of very different dates, are free from tin ; the third is an arsenical bronze : they should be compared with Nos. 6 and 8 in Table I. Although earlier bronzes are usually poorer in tin than later [1], this is not always so, and analysis *alone* is not reliable in deciding the date of an object. Thus No. 3, although quite late (it is true some experts dated it 1800 B.C.), might from its composition only have been put in a much earlier period, and the same applies to No. 2. The lead tends to increase in later bronzes, except Nos. 16 and 17. About 4·5 per cent. of lead was found in another statue of Osiris [2], and bronze statuettes often contain lead.[3] According to Perrot and Chipiez [4] bronze was uncommon for statuettes. Although there are bronze statuettes of the XVIII–XIX dyns. and the art of hollow-casting appeared about 1500, perhaps from the Ægean [5], the most important of such castings from the artistic standpoint are of the XXII dyn.[6] Casting large pieces and statues in bronze was well understood in the XIX dyn.[7] A bronze statue (XIX–XX dyn.) is cast over a core of sand and bitumen with an iron rod inside it.[8] Statues were sometimes cast in parts, afterwards clamped together.[9]

Bronze was both hammered and cast[10] in Egypt. The old cast bronzes often contain from 6 to 12 per cent. of lead, added to increase the fusibility[11]; the later, hammered, bronzes may contain 1–2 per cent. of iron, probably derived from the copper ores (chalcopyrite), giving the alloy considerable hardness, yet not sufficient for working hard stones such as granite, syenite, diorite and basalt.[12] Roman coins of pre-Christian times frequently had tin partly replaced by lead, especially in a later period.[13] European bronze of the Hallstatt Period contained 88 per cent. of copper, 7 of tin and 5 of lead[14]; Babylonian bronze contains up to 12 per cent. of lead[15], and old Chinese bronzes[16] over 15 per cent. When strength was not required, as in the case of statuettes, lead was no doubt often used instead of tin. A very old Hittite figure found in the bed of the Orontes contained 3·9 per cent. of lead and 3·4 of tin, and a bronze image (before 352 B.C.) found at Bubastis contained a fair amount of lead and little tin.[17]

The analyses given in Table III., p. 74 (confirmed by X-ray spectrograms, which showed traces of other metals, such as manganese), are by Sebelien.[18] The presence of antimony is supposed to lend support to one of Petrie's theories[19] that the tin (or the bronze) came from Hungary, which seems improbable, since antimony was known early in Egypt and Babylonia.

Bronze was found in small quantity at Amarna, with a "hard alloy like speculum metal"[20]; speculum metal may contain arsenic, but nothing can be said of such metals without analysis. A "white metal" found at Diospolis Parva, prehistoric, was conjectured to be an alloy of silver and copper.[21] Bronzes of the Saïte Period (663–525 B.C.) may contain considerable amounts of gold, and superb hollow-cast statues are sometimes inlaid with gold, silver and electrum wires in grooves.[22]

[1] Cf. Berthelot, Arch., 15 ; Gsell, 33 f. ; Forrer, RL, 115.

[2] Reinach, *L'Anthropologie*, 1891, ii, 107.

[3] Gladstone, B.A. Rep., 1893, 715.

[4] Art in Egypt, i, 654.

[5] Gsell, 34 ; Feldhaus, Technik, 144.

[6] Maspero, Arch., 305 f. ; Breasted, H., 573.

[7] Gsell, 34 ; Feldhaus, Technik, 144.

[8] Gsell, 79 : in B. Mus.

[9] Maspero, Guide to the Cairo Museum, 1908, 371.

[10] Petrie, Arts and Crafts, 90 f.

[11] Berthelot, Arch., 61, 62 ; Gsell, 77 ; Busch, Z. angew. Chem., 1914, 512.

[12] Gsell, 77.

[13] Gowland, in Woolley, etc., Karanog, i, 67.

[14] Forrer, RL, 115.

[15] Busch, Z. angew. Chem., 1914, 512.

[16] Dôno, Bull. Chem. Soc. Japan, 1932, vii, 347 ; 1933, viii, 133 ; 1934, ix, 120.

[17] Gladstone, J. Anthropol. Inst., 1897, xxvi, 313.

[18] Anct. Egypt, 1924, 7 f.

[19] Anct. Egypt, 1915, 16.

[20] Petrie, Tell el-Amarna, 31.

[21] Petrie, etc., Diospolis Parva, 1901, 25.

[22] Maspero, Arch., 307 ; ib., Struggle, 535, plate ; Breasted, H., 573.

TABLE III

		Cu	Fe	Zn	As	Sn	Bi / Ag	Ni	
25. Syrian axe, XII dyn.	Abyd. 51	85·92	—	—	—	12·12	—	—	Pb 0·77
26. Mason's chisel, inscribed	T.W. XVII 80	93·57	—	—	0·50	7·44	—	—	Sb tr.
27. Adze, Gurob, XVIII dyn.	T.W. XVII 85	89·82	—	0·36	0·25	3·05	—	—	Sb tr.
28. Large chisel, XVIII dyn.	T.W. XXII 89	88·02	—	0·25	0·43	11·96	—	—	{Pb 0·10 {Sb tr.
29. Winged adze, XIX dyn.	T.W. XVII 93	67·59	tr	—	—	9·59	—	0·60	—

OBJECTS OF BRONZE

Bronze daggers and razors became common in Egypt only from about 2000 ; swords and battle-axes of bronze were being adopted, but copper weapons still continued in use and the general adoption of bronze came later.[1] The bronze battle-axes with semicircular blades and tubular sockets [2] are probably Middle Kingdom.[3] Herodotos [4] says the Egyptians wore bronze helmets, although Wilkinson thought these were probably quilted and rarely of metal, whilst the shields were never of metal. The Egyptian arrows were tipped with metal or stone cemented by "a firm black paste." Spear-heads, daggers and axe-blades were of bronze, the axes being shown, however, painted blue like steel.[5] Bronze scales were used on armour from the XVIII dyn.[6] Herodotos says [7] the Egyptian naval troops in the army of Xerxes (480 B.C.) had spears, pole-axes and cutlasses, breastplates, plaited helmets and very large concave shields with rims.

A bronze knife with a wooden handle is of the New Kingdom.[8] Apep, the Hyksos king, speaks of a "gate of bronze."[9] A "bronze" ushabti of Ani (probably XVIII dyn.)[10] and a "bronze" arrow-head of Rameses II's period found at Abydos, as well as coarse "bronze" figures of the XXVI and XXX dyns.[11], are not analysed.

Miscellaneous (often rather late) bronze articles described, but mostly not analysed, are : mirrors, XVIII dyn.[12]; gilt ladles, of uncertain date[13]; large vases[14]; a vase bound with gold *represented* in a painting at Thebes[15]; a votive axe of Thothmes III[16]; a door with fittings[17]; an XVIII dyn. hinge at Gurob[18]; razors[19]; more than 1,000 statuettes of Osiris in the temple of Rameses III at Medinet-Habu[20]; bronze with silver inlay in a temple at Tanis[21]; XVIII dyn. tools at Gurob[22]; a signet ring (rare).[23] A New Kingdom (1600–1000)

[1] Mackenzie, Footprints, 144.
[2] Wilkinson, i, 215, Fig. 48.
[3] Wiedemann, 1920, 237 f., Fig. 42 ; cf. tool axe, *ib.*, Fig. 43.
[4] ii, 151.
[5] Wilkinson, i, 202, 205, 208, 211, 215, 218 ; Lepsius, Met., pl. ii, Fig. 3.
[6] Wilkinson, i, 221 ; Erman-Ranke, 652, Fig. 271.
[7] vii, 89.
[8] Wiedemann, 1920, 240, Fig. 45.
[9] Naville, Bubastis, 1891, plates xxii A ; xxv c ; transl., 23 ; Petrie, Hist. of Egypt, 1903, 241—"brass."
[10] Petrie, Royal Tombs, 1900, I, 33.
[11] Petrie, Abydos I, 1902, 25.

[12] Erman-Ranke, pl. 19, Fig. 2, opp. p. 241 ; Wilkinson, i, 350.
[13] Wilkinson, ii, 47 ; Fig. 314. 4.
[14] *Ib.*, ii, 10 ; Figs. 277, 278.
[15] *Ib.*, ii, 7 ; Fig. 274.
[16] Maspero, Dawn, 60.
[17] Wilkinson, i, 352.
[18] Petrie, Ten Years' Digging, 129, Fig. 97.
[19] Wiedemann, Sagen, 1906, 56—*c.* 1500 ; Erman-Ranke, 246—heads were always shaved in the New Kingdom.
[20] Perrot-Chipiez, Art in Egypt, i, 829.
[21] Lenormant, Prem. Civ., i, 251.
[22] Petrie, Ten Years' Digging, 129 ; ? copper.
[23] Wilkinson, ii, 341.

letter speaks of bronze weapons with heads of the "sixfold mixture."[1] Bronze arrow-heads replaced flint in the New Kingdom.[2] Thothmes I (1557–1501) says he decorated the temple of Amen at Thebes with an immense door of Asiatic bronze, with the image of the god upon it inlaid with gold [3], and a story composed in 1300 B.C. speaks of "copper doors with bronze bolts, made for eternity."[4] Records of Amenhotep II speak of 360 bronze scimitars and 140 daggers [5], of 1,475 Syrian copper chains of four rings, and 1,200 Syrian suits of armour of leather with bronze rings or scales[6]; and Libyan booty of the XIX dyn. included 3,116 vessels and 9,111 weapons.[7] Bronze locks and hinges are in the Cairo Museum.[8] Egyptian bronze objects in the British Museum, mostly of the XXII–XXVI dyns., but some earlier, include vases, jugs, censers, model boat (XXII dyn.), ladle, mortar rake, lamp, altar tongs, bells, sistra, cymbals, flute, heads of ceremonial standards, votive buckets, seals, scale pan, weights, kohl-sticks, tweezers, measures, wine strainer, incense burner, libation vase, shovel, model altars, lamp stand (Roman), combs, razors, fish-hooks, hollow castings, daggers, chisels, saws, bullets, spear-heads, knives, swords, sickles, halberds, axe-heads, and numerous statuettes.[9] Some of the statuettes are of bronze inlaid with gold and silver.[10] The bronze hypocephali, inscribed plates put under the heads of mummies, are of the XXX dyn.[11] Bronze mirrors of all periods are common.[12]

New Empire "bronze" (not analysed) was found with copper in Nubia[13], a jug and a mirror with ivory handle, late New Kingdom, at Abydos[14], and a vase of Amasis at Sidon.[15] A coin stamp of 425 B.C. (the oldest known) contained 25 per cent. of tin and 75 per cent. of copper.[16]

The analyses and the often doubtful archæological data assembled above make it probable that, although *isolated* articles of true bronze are possible for the I–VI dyns., the alloy came into use only about the XII dyn.[17], and into common use in the XVIII–XIX dyns., when the proportion of tin was increased.[18] We may accept these results and dissent from Maspero[19] and Myres[20], who say that bronze came into "actual" use not later than the V dyn.

TINNED COPPER

No tinned copper vessels have been found in Egypt proper, even in the Roman Period; a few copper vessels plated with silver, presumably "Sheffield plate," are known.[21] Dioskurides[22] and Pliny[23] mention tinned copper boilers and vessels in the Roman Period. Mme Hammer-Jensen[24] thinks the art was an invention of the Gauls. Vogel[25] points out that bronze buckles (fibulæ) tinned

[1] Erman, Literatur der Ægypter, 1923, 267; Erman-Ranke, 550.
[2] Wiedemann, 1920, 236.
[3] Breasted, H., 265.
[4] Erman, Lit., 108; Erman-Krebs, Aus den Papyrus, etc., 1899, 45.
[5] Erman-Ranke, 136.
[6] Erman, Lit., 217.
[7] Erman-Ranke, 652, 646.
[8] Guide, 1908, 370, 372.
[9] BMGE, IV–VI Rooms, 51, 53 f., 168 f., 187 f., 190 f., 195 f., 219 f., 201 f., 206 f., 208 f. (VI dyn.), 212, 214 f., 216 f., 270, 277 f.
[10] Ib., 219 f.
[11] Ib., 272.
[12] Ib., 292 f.
[13] Arch. Survey of Nubia, i, plate 65, c and d.
[14] Frankfort, JEA, 1930, xvi, 219; pl. xl.
[15] Olmstead, H. Pal. Syr., Fig. 179.
[16] Feldhaus, Technik, 727.
[17] Gsell, 34; Hall, CAH, i, 291, 319; Hoernes, Urg., ii, 241; de Morgan, Or., i, 202 f.; Petrie, Anct. Egypt, 1915, 18.
[18] Petrie, Arts and Crafts, 90 f.; Anct. Egypt, 1915, 18; Ten Years' Digging, 151; Objects of Daily Use, 1927.
[19] Arch., 303.
[20] CAH, i, 103—"not later than 2800 B.C."
[21] Wilkinson, ii, 229.
[22] i, 38.
[23] xxxiv, 17.
[24] Oversigt over det Kongelige Danske Videnskabernes Selskabs Forhandlinger, Copenhagen, 1916–7, 297 f., 300 f. (in French).
[25] Z. f. angew. Chem., 1909, 44.

only on the exposed side occur in Bronze Age remains (from 2000 B.C. ?) in Central Europe, and probably replaced decorative objects covered with tin-foil, which occur at an earlier period. They were made by dipping the copper articles into melted tin, and Vogel thinks this led to the discovery of the alloy bronze. This must have been a remote discovery, as the two metals will not form bronze at the temperature of melted tin. Some of the bronze vessels found in the Roman cemetery at Karanog, in Nubia, were plated with tin.[1]

Copper was plated with antimony as a hard and adherent film in the v–vi dyns. (basin and ewer), apparently by boiling the clean metal with stibnite and sodium carbonate solution, which produces the result.[2]

THE DISCOVERY OF BRONZE

That bronze was discovered in one locality only, viz. in Khorassan about 3000 B.C., and spread from there to the rest of the world[3] is far from certain, since it may have been discovered in several places, and its use may also have spread from some or all of these places to the rest of the early bronze-using centres, including Europe.[4] Forrer [5] limits the discovery of bronze to Egypt, Syria or "still further East," about 3000, Ægean bronze working being definitely later, but independent discoveries of the production of bronze appear to have been made in Mexico and Peru, where in the time of Cortez (A.D. 1520) it contained 6 per cent. of tin [6], and in China, where lead bronze was early in use.[7] Mosso [8] thought both copper and bronze were Egyptian discoveries; others [9] think bronze was discovered in Mesopotamia. Further suggestions as to its original home are China[10], India[11], Etruria[12], Crete or the Ægean[13], and Central Europe.[14]

In a VI or XI dyn. relief, Cretans are shown bringing copper blocks and also reddish rectangular blocks, apparently 2 ft. long and 8 in. wide, of *dhty*, tin (or possibly lead).[15] They are also shown in XVIII dyn. reliefs (see refs. 15–16, p. 64, and refs. 1–2, p. 65). Both copper and tin are also supposed to have been imported into Egypt from Cyprus and North Syria, and in a later period Phœnician and Greek colonies in Brundusium, Tarentum and Syracuse (founded 800 B.C.) may have supplied Egypt with bronze.[16]

The most attractive hypothesis of the discovery of bronze is that which locates it in Khorassan, in Persia (the Drangiana of Strabo).[17] E. K. von Baer[18], starting from a vague report of the traveller Burns that tin was found in the

[1] Woolley and MacIver, Karanog, i, 59, 64 f., 67.
[2] Fink, *Industr. and Engineering Chemistry*, 1934, xxvi, 236.
[3] Elliot Smith, EB[11], xxx, 146 ; The Ancient Egyptians, 1923 ; MGM, xi, 177.
[4] Gsell, 73 ; Hoernes, Urg., ii, 248 ; *ib.*, in Hoops, i, 329 ; Meyer, Alt. II, 53 ; Hall, Oldest Civ. Greece, 1901, 193 f.
[5] RL, 114.
[6] Humboldt, Vues des Cordillères, 1810, 117 ; Hoernes, Urzeit, ii, 123.
[7] Dôno, *Bull. Chem. Soc. Japan*, 1932, vii, 347 ; 1933, viii, 133 ; 1934, ix, 120.
[8] Dawn of Med. Civ., 59.
[9] Schrader, RL, 200 ; Schrader-Jevons, 152 ; Lenormant, Prem. Civ., i, 129 ; Montelius, Ält. Bronzez., 1900, 201.
[10] De Morgan, Or., i, p. xii ; cf. Mosso, 61 ; perhaps independent, see note 7.
[11] Perrot-Chipiez, i, 829 ; Wilkinson, ii,

229 ; G. Smith, The Cassiterides, 1863, 3 f.
[12] Wilkinson, i, 190 ; much too late, cf. Lucas, JEA, 1928, xiv, 99.
[13] W. M. Müller, Egyptological Researches, Washington, 1906, i, 5, 7 ; 1910, ii, 183 ; Petrie, *Anct. Egypt*, 1915, 18.
[14] Schulten, PW, viii, 2031 ; Petrie, Medum, 1892, 44 ; *ib.*, Arts and Crafts, 90 f. ; *ib.*, Ten Years' Digging, 153 ; *ib.*, Tools and Weapons, 7 ; Ridgeway, The Early Age of Greece, 2 vols., 1901–32 ; i, 610 ; Hall, CAH, i, 291, 319.
[15] W. M. Müller, *opp. cit.* ; Evans, Palace of Minos, ii, 176 ; Pendlebury, JEA, 1930, xvi, 76 f.
[16] Gsell, 36 ; BMGE, 98 ; Kahlbaum and Hoffmann, in Diergart, 1909, 92 ; G. Smith, The Cassiterides, 1863, 150 f.
[17] XV, ii, 10 ; 725 C.
[18] *Archiv für Anthropologie*, Braunschweig, 1876, ix, 263–7.

country beyond the Bamian Pass (perhaps the Hindu Kush), requested the officers of the Russian army in the Oxus to see if any tin mines existed there. Prof. Semenov sent him a letter from one Ogorodnikoff, which he reproduces, stating that an inhabitant of Meshed, an official of the Khorassan copper mines, reported rich deposits of tin (as well as of iron, copper, lead and sulphur) 75 miles from the town Utschan-Mion-Abot, and the tin mine Rabotje Alokaband was 22 miles from Meshed. This statement was confirmed by the "manager of the Russian Merchants' Society in Chorasan," and Ogorodnikoff himself saw tin wash-basins and dishes "of old native workmanship," which the inhabitants represented as made from local tin. The information, it will be seen, was obtained in rather an indirect way, although the result is often quoted. Tin ore is said to occur also near Tabriz and at Astrabad[1] ; it is not found in Russian Armenia, but is reported in the Kurbaba Mountains near Tillek, between Sahend and the River Axares, where it is associated with copper and hence probably in vein form ; also near Migri on the Axares, and in Hejenan.[2] It occurs in Asia Minor at Kastamuni, but although tin ore was said [3] to be found in the Caucasus (Georgia), this is a mistake.[4] If aonya in the Avesta is tin, the mines in the Paropamisus (Strabo's Drangiana) may have been worked at an early date.[5]

Although the supposed source of tin at Khorassan rests, as is clear from the above, on very insecure foundations and has been adversely criticised [6], it has been accepted by several authorities [7], and is perhaps the most probable so far suggested, although far too much superstructure has been built upon it by some archæologists.

Tin objects dated 2000–1500, reported in North Persian graves[8], are doubtful.[9] Strabo[10] says "tin is found in the country" [of the Drangæ]; the editors[11] add that "none is said to be found there at the present day." Arabic authors (Ibn Hawqal, 902–968 ; Al-Istakhrî, c. 970, etc.) also mention tin in this region[12], but Hiuen Tsiang (A.D. 629)[13] refers, not to Hilmend, but to Tarim in Chinese Turkestan, which could hardly be a source.[14] Meyer[15] definitely excludes Iran as a source of early Egyptian bronze and tin, but without giving reasons, and Montelius[16] rules out the Ural-Altai tin, which was used only locally. Lucas[17] reports the occurrence of tin ore, either alluvial or vein ore, or both, in Persia, and quotes some localities in addition to those given above, viz. between Sharud and Astrabad ; in the Kuh-i-Benan mountains and the Qara Dagh mountains : de Morgan found it about 25 km. from Tauris and at Azerbaijan, but not in Khorassan. Lucas suggests that bronze was perhaps

[1] Montelius, Ält. Bronzez., 1900, 199 ; Read, BMG Bronze Age, 8.

[2] Lucas, JEA, 1928, xiv, 100.

[3] Lenormant, Prem. Civ., i, 128 f. ; BMG Bronze Age, 8 f.

[4] Bapst, L'Étain, 1884, 7 ; Montelius, Ält. Bronzez., 1900, 201 ; Chantre, Recherches anthropologiques dans le Caucase, 4 vols., 1885–7, i, 81 f. ; Besnier, Daremberg-Saglio, IV, ii, 1459 ; Petrie, Arts and Crafts, 90 f. ; cf. Arzruni, Verhl. Berlin Ges. f. Anthropol., 1884, 58 ; Virchow, ib., 126, 503.

[5] Rougemont, L'âge du bronze, 1866, 86.

[6] Bapst, etc., q. by Berthelot, Intr., 226 ; W. M. Müller, Egyptol. Res., 1906, i, 5 f.

[7] Lenormant, Prem. Civ., i, 146; Montelius, Ält. Bronzez., 1900, 199 ; cf. Bapst, 9 f. ; Meyer, Alt. I, ii, 151, 745 ; ib., II, 157 ; Contenau, Civ. Phénicienne, 298 ; Terrien de Lacouperie, Western

Origin of the Chinese Civilisation, 1894, 322.

[8] Montelius, MGM, ii, 151 ; Gowland, J. Anthropol. Inst., 1912, xlii, 284.

[9] Myres, in H. G. Wells, Outline of History, i, 63.

[10] XV, ii, 10 ; 725 C.

[11] Falconer and Hamilton's translation, 1854–7, iii, 126.

[12] Tomaschek, MGM, ii, 152 ; Abulfeda, Géographie, tr. Reinaud, 1848–83, iii, 215.

[13] Si-yu-ki, Buddhist Records of the Western World, tr. from the Chinese by Beal, 1884, i, 19.

[14] Lippmann, Alchemie, i, 578.

[15] Alt. I, ii, 744.

[16] Ält. Bronzez., 1900, 199 ; von Sadowski, Die Handelsstrasse der Griechen und Römer . . . an die Gestade des baltischen Meeres, tr. Kohn, Jena, 1877, pref., v, xx.

[17] JEA, 1928, xiv, 98, 100 f., 107 f. ; ib., Egy. Mat., 71 f., 73.

discovered in Persia, probably by the mixed smelting of copper and tin ores, the vein ore of tin being occasionally accompanied by copper.

According to the speculations of Elliot Smith [1], bronze was discovered shortly after 3000 in the south-east corner of the Caspian, perhaps near Meshed, from which locality it spread west, south, and later east. Elizabethspol in Transcaucasia had a culture allied with the Ægean. About 2500 Asia Minor shared with the Ægean a knowledge of bronze, and long before tin was known in Europe it was brought overland through Asia Minor, and also by way of Transcaucasia and the Black Sea, from Khorassan. This is very speculative but not more so than Laufer's suggestion (quoted by Elliot Smith) that this civilisation was also taken to the Shensi province of China about 3000.

SOURCES OF EGYPTIAN TIN

Lucas considers as possible sources of Egyptian tin Bohemia, Saxony, Tuscany, Elba, Armenia, Persia, possibly Syria, and West-Central and South Africa. He dismisses Egypt, Turkestan, Mesopotamia, Arabia, Caucasia, Georgia, Asia Minor, Crete, Greece, Cyprus and Palestine, since, so far as is known, tin ore does not occur in any of them. No tin is found in Egyptian or Palestinian mines, and that reported in Syria is doubtful.[2]   The suggestion that it came from Bohemia, Saxony and Silesia, perhaps the Zinnwald in Saxony, in 2000 B.C. [3] is rejected by Lucas[4] : tin in Bohemia and Saxony occurs as vein ore, not alluvial tinstone ; the Bronze Age began rather late in this locality, and there is no evidence of early trade in tin there. These mines were, apparently, not worked before the 12th century A.D. ; Matthew Paris, in fact, says they were discovered by a Cornishman in 1241.[5] It is quite possible that they were worked locally in the Prehistoric Period and then forgotten, but there is no sound evidence that they were of any importance in the history of early Egyptian bronze. "The land of the Midianites," in north-west Arabia, has been regarded as the source of the tin which the Israelites took from the Midianites [6]: slag heaps and traces of mines were reported by Burton [7], but the suggestion [8] that this was the source of Egyptian tin is improbable.[9]

The rich mines of Malacca and Banca are hardly likely, as Bapst[10] suggested, to have been worked at an early period, and tin first reached Europe from Banca (where the deposits were discovered after a forest fire) in 1711[11], although the deposits are mentioned, as Man-li-kia, by Chinese authors of the 13th to 14th centuries.[12]

The suggestion has often been made that the ultimate source of Egyptian tin was Britain.[13] Tin mining in Cornwall, however, apparently began only about 500 B.C.[14], although Weigall[15] draws attention to Egyptian beads of Tutankhamen (1300) found at Stonehenge in Salisbury Plain, and considers that trade in tin between Egypt and Britain at this time is not impossible.

[1] EB[11], xx, 146 f.

[2] Napier, Arts, 124 ; Lucas, JEA, 1928, xiv, 97 f., 100.

[3] Petrie, Ten Years' Digging, 153 ; ib., Tools and Weapons, 7 ; Ridgeway, The Early Age of Greece, Cambridge, 1901, i, 610.

[4] Already by Wilkinson, ii, 229.

[5] Beckmann, Hist. of Inventions, 1846, ii, 226.

[6] Numbers, xxxi, 22.

[7] Montelius, Ält. Bronzez., 200 ; Lenormant, Prem. Civ., i, 129 ; Schrader, RL,

201 ; Boyd Dawkins, Early Man in Britain, 1880, 407.

[8] Tomaschek, q. by Schrader, RL, 201.

[9] Beckmann, H. Invent., ii, 208.

[10] L'Étain, 1884, 9.

[11] Blümner, iv, 84 ; Gsell, 53.

[12] Müller, in Ullmann, Enzyklopädie der techn. Chemie, 1923, xii, 265.

[13] Smith, Cassiterides, 150 f. ; Mosso, Dawn of Med. Civ., 62.

[14] Lucas, JEA, 1928, xiv, 104 f.

[15] Wanderings in Roman Britain, 1926, 23.

If British tin was available, it probably came overland through Gaul rather than by sea.[1]

Tinstone (cassiterite) is a dull and earthy but heavy mineral which would, in the Prehistoric Period, not attract attention as such, but would be separated with gold in washing, since it frequently occurs with gold in streams, e.g. in Ireland (Wicklow).[2] Borlase[3] found gold with tinstone in Cornwall in 1753. The ore is easily reduced and the metal could readily be obtained by accident.

AFRICAN TIN

Rich deposits of tinstone in Nigeria and the Congo have long been worked by the natives, the metal being used for the ornamentation of the person or of weapons[4], and Goguet[5] had already suggested that early tin may have come from African mines. Ancient tin workings exist around Rooiberg and Blaubank in the Transvaal. Nickel occurs in the same localities as the tin, whilst in Europe and Asia deposits of copper and nickel are not associated with ancient workings. The slag from the Transvaal is said to show an "ancient" smelting on a large scale for the direct production of bronze from copper ore brought to the tin and nickel-bearing areas of the Waterberg. In one area are remains of forty-three furnaces.[6]

Staudinger (op. cit.) describes the primitive modern tin-smelting in Northern Nigeria. The washed cassiterite is stamped in a stone mortar, made into cakes with water and laid on the wood-charcoal fire. The tin is collected in a channel, melted in an earthen pot and ladled into moulds of wet ashes, the forms being thin rods, easily beaten into foil, produced by laying straws in the mould material and drawing them out. The furnace is of clay, $2\frac{1}{3}$–3 ft. high, blown with skin bellows, two of which were worked by alternately pressing and expanding very rapidly. The primitive process of smelting local tinstone in the province of Zamora in Spain, used in 1853, is very similar.[7]

Vasco da Gama in the 16th century saw weapons decorated with (imported ?) tin in possession of the natives of the Congo, and the negroes of Cordofan cast small decorative objects from tin supplied from Egypt.[8] It has been suggested[9] that the discovery of large deposits of tinstone in Central Africa, and the use of tin ornaments in circulation among the negroes, made possible the entrance of the metal into Egypt along the trade routes. Tin ores occur in the Belgian Congo, Nigeria, Southern Sudan, South-West Africa, Rhodesia, Union of South Africa and Swaziland, with small amounts on the Gold Coast, Nyasaland and Portuguese East Africa. Ancient workings, remains of furnaces and ores, and tin ingots and lumps of bronze are said to occur in Rhodesia and Northern Transvaal. There is, however, no evidence for a really old working in Nigeria, and although the workings in Rhodesia and the Transvaal are admittedly old, there is no proof of intercourse with Egypt in the Bronze Age, and the Nigerian industry was quite possibly started by the Portuguese.[10] It would be a tempting speculation to suggest that both copper and tin reached ancient Egypt from the interior of Africa, but until we know very much more about the archæology of this region such hypotheses are best kept out of circulation.

[1] Mosso, 62 ; W. M. Müller, Egyptol. Res., 1906, i, 7 ; 1910, ii, 183.

[2] Boyd Dawkins, 400 f., 416 ; Henglein, in Doelter, Mineralchemie, III, i, 178.

[3] Antiquities of Cornwall, Oxford, 1754, 253.

[4] Feldhaus, Technik, 1368 ; Staudinger, Z. f. Ethnol., 1911, xliii, 147 ; Hoernes, Urg., ii, 248, 296.

[5] i, 163.

[6] Holman, Nature, 1928, cxxii, 998.

[7] Boyd Dawkins, 401 ; Gsell, 76.

[8] Rougemont, L'Âge du Bronze, 1866, 17 ; Rüppell, Reisen in Nubien, Kordofan und dem petraischen Arabiens, Frankfurt, 1829, 158.

[9] Hoover, Agricola de Re Metallica, 1912, 412.

[10] Lucas, JEA, 1928, xiv, 100 f. ; Lippmann, Alchemie, i, 579.

In the Ptolemaic and later periods tin came to Egypt from Cornwall. Stephanos of Alexandria (c. A.D. 620) calls it simply "the British metal."[1] About the 6th century A.D. ships from Alexandria regularly visited Britain with cargoes of grain, which they sold, half the value being paid for in gold and half in tin [2], and tin was an important article of Egyptian trade in the 12th to 13th centuries.[3]

TIN

Finds of tin objects have placed the knowledge of this metal in early Egypt beyond doubt.[4] The earliest is a XII dyn. (?) thin cast "pilgrim bottle" found by Petrie at Abydos [5], now in the Ashmolean Museum, Oxford. It is of pure tin.[6] A finger ring of pure tin, which crackles when bent (a characteristic property of tin), found with some glass beads of the end of the XVIII dyn. at Gurob [7], analysed by Gladstone [8], was free from copper, lead, silver, gold, arsenic and antimony, but contained a trace of black oxide insoluble in nitric acid, "indicating that the reduction of the ore had not been complete." Gladstone refers to Church's analysis of a piece of white metal having the outline of a winged scarabæus, found resting on the breast of a mummy, c. 700–600 B.C., which was pure tin.[9] This is perhaps the "rectangular thin tin plate," with the left symbolic eye engraved upon it, found by Birch[10] in a mummy : when bent it crackled. A tin finger ring[11]; two tin finger rings (analysed by Gowland), later than the XVIII dyn., found in the Roman-Nubian cemetery at Karanog[12]; a ring of an alloy of tin and silver of the XVIII dyn.[13]; and a tin vase of 1200 B.C., found in Upper Egypt[14], are evidence of the early use of tin in Egypt. No tin was found at Amarna.[15]

The Harris Papyrus (Rameses III, 1200 B.C.) twice mentions what is thought to be tin as offerings to gods from tribute, one of 95 lb. and the other of 2,130 lb.[16] The word dḥty, or teḥt-t, in the papyrus, read "tin" by Birch[17], closely resembles taḥt, the Coptic name for lead[18], but lead is separately mentioned in the papyrus.[19] "White dḥty" (white lead), which occurs in the London Papyrus 10,059[20], would correspond with the Latin plumbum candidum (tin). Erman[21] translates dḥtj (Coptic taḥt) as lead, and Budge gives no word for tin in his dictionary. Lepsius[22] thought tin was not named in the texts, and Erman and Ranke[23] still say the Egyptian name for it is unknown, but Breasted[24] gives four doubtful references to tin, three from the XX dyn. and one from the XXV dyn. Tin is not mentioned in the Ebers Papyrus (1550 B.C.)[25],

[1] Democritus de arte magna, tr. Pizimenti, Patavii, 1573, 29 ; Ideler, Physici et medici græci minores, 1842, ii, 206—Britannicum metallum, ἡ βρεττανικὴ μέταλλος.

[2] Friedländer, Darstellungen aus der Sittengeschichte Roms, 4 vols., Leipzig, 1910, ii, 83 ; Hoops, ii, 414.

[3] Qalqashandi, Geogr., tr. Wüstenfeld, 1879, 224.

[4] Lucas, JEA, 1928, xiv, 97 f.

[5] Abydos III, 50 ; Ancient Egypt, 1915, 17 ; illustrated in Evans, Palace of Minos, ii, 178, Fig. 91.

[6] Gladstone, PSBA, 1892, xiv, 223.

[7] Petrie, Illahun, etc., 1891, 12, 19 ; Arts and Crafts, 98 f., dated c. 1400 B.C.

[8] B.A. Rep. 1893, 715 ; PSBA, 1892, xiv, 226.

[9] Cf. Semper, Der Stil, ii, 482.

[10] Wilkinson, iii, 232, 474.

[11] Rathgen, Chem. Z., 1921, 1102.

[12] Woolley and MacIver, Karanog, i, 64 f. ; Lucas, Egy. Mat., 107.

[13] Lucas, JEA, 1928, xiv, 97.

[14] Weigall, Roman Britain, 23.

[15] Petrie, Tell el-Amarna, 31.

[16] Brugsch, Ägyptologie, 271, 273.

[17] Wilkinson, ii, 232,

[18] Budge, Egy. Dict., 887 ; Gladstone, PSBA, 1892, xiv, 226.

[19] W. M. Müller, Egyptol. Res., 1906, i, 6.

[20] Schrader, RL, 994 ; white = ḥetch or ḥed ; Budge, Egy. Dict., 523 ; Brugsch, 392 ; Wreszinski, Der Londoner Medizinische Papyrus, 1912, 184, not otherwise known.

[21] Erman - Grapow, Handwörterbuch, 1921, 221.

[22] Met., 59 ; Blümner, iv, 83 ; Hoernes, Urg., ii, 217.

[23] Aegypten, 1923, 550.

[24] Anct. Records, iv, 140, 160, 193 (Harris Papyrus), 471 (c. 660 B.C.) ; tyḥty.

[25] Berthelot, Arch., 240.

but it may be one of the unidentified metals in a recipe of Phœnician origin in this papyrus. In the late (A.D. 300) Papyrus of Leyden, tin is called tran ; in Coptic the name is thran.[1]

Tin oxide, perhaps obtained from the metal, was used in small quantities from the XVIII dyn. onwards for imparting a white opacity to glass.[2]

LEAD

Lead is represented by a hieroglyphic word now usually read dḥty (formerly daḥt, taḥti, deḥtu, etc.) [3], said by Lepsius to correspond with the Hebrew 'ôphereth. The metal is shown at Medinet Habu (Rameses III) as weighed in large discs and possibly bricks.[4] It is rarely mentioned in the inscriptions, but frequently occurs in the Harris Papyrus [5], although Birch [6] and Hoernes [7] considered that the metal there mentioned was tin, and a piece of lead is specified in the Ebers Papyrus (1550) for laying on a wound (for cooling ?).[8] Lead was perhaps regarded as an inferior kind of silver.[9] In Plutarch molten lead is poured as a seal over the coffin of Osiris.[10]

Lead came in bars from Asia[11], perhaps Syria in the earliest times[12], and Forrer[13] suggests also from Spain. The lead ore, a mixed carbonate and sulphide with zinc carbonate, practically free from silver, however, which occurs abundantly at Gebel Rusâs in Egypt, was almost certainly worked there.[14] Argentiferous galena, with 3 oz. of silver to the ton of lead, occurs about two miles south of Safaga Bay on the Red Sea[15], and Rozière in 1809 described old lead mines in the Thebaid in the Bahram mountains, 40 km. east of Syene.[16] Gowland[17] is, therefore, incorrect in saying that lead ore does not occur in Egypt.

Lead occurs sparingly in the Predynastic Period (before 3400), e.g. as a piece of sheet covering the wooden figure of a hawk[18], and as small figures and objects, probably from Syria.[19] A lead statuette of the I dyn. (or earlier) is in the British Museum[20], and there is a lead ear-ring of the Protodynastic Period.[21]

Galena in the form of beads frequently occurs in Predynastic and very early tombs.[22] Since galena occurs so early, it is natural that lead should also be known and used early.[23]

A lead plug with a copper loop and a bronze ring, found in V dyn. remains

[1] Brugsch, Ägyptologie, 398, 401.
[2] Lucas, Egy. Mat., 107 ; ib., JEA, 1928, xiv, 97.
[3] Brugsch, Ägyptologie, 401 ; Forrer, RL, 94 ; Lepsius, Met., 58, 63 ; de Morgan, Or., i, 215 ; Budge, Egy. Hierogl. Dict., 842, 886 f., nus = block of lead, 355 ; Blümner, ii, 97, incorrectly, says the Egyptians did not use lead.
[4] Lepsius, Met., 58, 63 ; Erman-Ranke, 549.
[5] Erman-Ranke, 550.
[6] In Wilkinson, ii, 232.
[7] Urg., ii, 213.
[8] Joachim, Papyros Ebers, 1890, 126 ; Erman-Ranke, 550.
[9] Petrie, Anct. Egypt, 1915, 16.
[10] Plutarch, Isis and Osiris, 13 ; Erman-Ranke, 306.
[11] Müller, Egyptol. Res., i, 6 ; Erman-Ranke, 550 ; Wiedemann, 1920, 347.
[12] Petrie, Arts and Crafts, 98 f. ; Prehist. Egypt, 1920, 27.
[13] RL, 94.
[14] Lucas, Egy. Mat., 102 ; Garland and Bannister, 31.

[15] Wilkinson, i, 155 ; Garland and Bannister, 31 ; Lucas, JEA, 1928, xiv, 314.
[16] Gsell, 8.
[17] J. Anthropol. Inst., 1912, xlii, 273.
[18] Breasted, H., 28 ; Petrie, Naqada and Ballas, 1896, 46 ; Gowland, J. Anthropol. Inst., 1912, xlii, 173—"3000 B.C." ; Petrie, Wainwright and Mackie, Labyrinth, etc., 1912, 17—the "only" find of Predynastic lead.
[19] Petrie, Arts and Crafts, 98 f. ; ib., Objects of Daily Use, 1927, 49.
[20] BMGE, IV–VI Rooms, 288 ; Handcock, Mesopotamian Archæology, 1912, 268.
[21] Brunton, Qau and Badari I, 66.
[22] Petrie, Royal Tombs of the First Dynasty, 1901, II, 36 ; Diospolis Parva, 1901, 25—lead and galena ; Labyrinth, etc., 20, 26, 27, 45 ; Prehist. Egypt, 43 f. ; Tarkhan I and Memphis V, 1913, 8 ; Tarkhan II, 1914, 9—in bags, perhaps for eye paint ; Brunton, etc., Qau and Badari I, 13, 33, 39, 55, 63, in pots, for paint.
[23] Petrie, Diospolis Parva, 1901, 25.

6

by Borchardt [1], which contained 99·21 per cent. of lead, with the rest chlorine, sulphuric and carbonic anhydrides, was somewhat corroded nearly pure lead. Some lead found at Amarna included a siphon with strainer end [2] and curious vases.[3]

Neither galena nor lead is common in the Old and Middle Kingdoms, but the metal was very abundant in the XVIII dyn., when it was used for sinkers for fishing nets, usually as flat strips bent over the strings, and, as in the XXVI dyn., for filling bronze weights.[4]   It is mentioned as tribute from Syria and probably came from the Taurus[5]:   Rameses III obtained it, with copper and silver, from Syria and Assyria, and in 1200 presented over 150,000 lb. of copper and 8,896 lb. of lead, with gold, silver, tin, bronze and (or) iron to the gods.[6]   Lead was not used for tanks in Egypt until about 600 B.C.[7]   Thothmes III in 1467 mentions it as coming (with 430 lb. of copper) as tribute from the king of Cyprus.[8]   The lead of a net-sinker of 1400 contained 0·0282 per cent. of silver, so that the argentiferous lead was not desilvered.[9]   The British Museum has a lead vase with cover, a massive lead jar with securely fastened cover and unknown contents, and a lead human-headed hawk with outstretched wings.[10]

Ptolemaic and later papyri mention the plumber ($\kappa o\lambda\lambda\eta\tau\dot{\eta}s$; $\mu o\lambda\nu\beta\delta o\nu\rho\gamma\acute{o}s$)[11] who makes and repairs water pipes ($\sigma\omega\lambda\hat{\eta}\nu\epsilon s$),[12] and the Arabs found much lead (perhaps Roman) in an old canal from Shanbâr to Alexandria.[13]   A finger ring in Roman-Nubian remains at Karanog was of pewter, an alloy of lead and tin much used by the Romans. ·Lead was rare in these graves, where iron and bronze were common.[14]

Lead was used for inlaying wood and for casting small statuettes, especially of Osiris and Anubis, for use as amulets (from its relation to magic).[15]   An alloy of copper and lead (pot metal) was commonly used in Egypt for statuettes in Greek and Roman times, and in the Coptic Period pewter bowls and ladles were made, the former apparently by a spinning process.[16]   Removable head decorations of lead, for placing on statuettes, some cast and others beaten, date from the Ptolemaic Period, and the wings of stone scarabs placed on the breasts of mummies were of lead, tin or silver.[17]   There is a late bronze sistrum in the British Museum in which lead, still remaining within the head, is a portion of that used in soldering the interior, coarser work or parts out of sight being soldered with lead.[18]   A bronze trap with a lead counterpoise is of the Ptolemaic Period.[19]

A grey hemispherical cup from Abu Roâsh, early IV dyn., had a patina containing 22·2 per cent. of lead, 56·96 of silica, 5·86 of chlorine, 8·94 of alumina, etc., which[20] was possibly a frit of litharge, sand and common salt, implying an early knowledge of the calcination of lead to the oxide, and the beginning of lead glaze.

[1] Sahurā, i, 76 f.
[2] Petrie, Tell el-Amarna, 31 ; Hall, JEA, 1928, xiv, 203 ; on siphons, Wilkinson, ii, 313 f., Fig. 433.
[3] Peet and Woolley, City of Akhenaten, 1923, 24.
[4] Petrie, Arts and Crafts, 98 ff. ; Anct. Egypt, 1915, 16 ; Wilkinson, i, 292, and Fig. 101, in B. Museum.
[5] Petrie, Anct. Egypt, 1915, 16.
[6] Hoernes, Urg., ii, 296 ; Brugsch, Ägyptologie, 271 f.
[7] Petrie, op. cit.
[8] Breasted, H., 313.
[9] Friend and Thorneycroft, J. Inst. Metals, 1929, xli, 105.
[10] BMGE, IV–VI Rooms, 54 f., 93— dates not stated.

[11] Liddell and Scott, Greek-English Lexicon, 1925 f., 972, 1142.
[12] Reil, 1913, 71 ; q. by Lippmann, Alchemie, i, 574.
[13] Reitemeyer, 43.
[14] Gowland, in Woolley and MacIver, Karanog, i, 67 ; 59 ; analysis of Roman pewter, Gowland, Archæologia, 1898, lvi, 17.
[15] Maspero, Arch., 302 ; Wiedemann, 1920, 347.
[16] Petrie, Arts and Crafts, 98 f. ; Anct. Egypt, 1915, 17.
[17] Garland and Bannister, 30 ; Wilkinson, iii, 486.
[18] Wilkinson, i, 499 ; ii, 258 f.
[19] BMGE, IV–VI Rooms, 207.
[20] Berthelot, Arch., 17 f.

BRASS AND ZINC

Brass is an alloy of copper and zinc. Whenever classical authors speak of χαλκός or æs, or their English translators of brass, it is usually uncertain whether copper, bronze, brass or some other alloy is meant. Birch [1] says brass was not found in Egypt, but Wilkinson [2] presented to Harrow School "a brass ring of Ancient Egypt, which perhaps had an alloy of gold like one kind of Corinthian brass," and says that "other specimens besides my ring have been found in Egypt and Greece." The ring may have been electrum. The "two vessels of fine copper, precious as gold," [3] brought from Babylon, may have been of gold-copper alloy.[4] Brass with 23·4 per cent. of zinc was known in Palestine in the period 1400–1000 B.C. [5], and prehistoric "copper" from Naqada and Ballas contained 98·60 of copper, 0·38 of tin and 1·55 of zinc, being "rather brass than bronze" : it is elastic and free from deep corrosion or change. A "base gold" [? brass] bead was also found.[6] A brass finger ring, of copper and zinc without tin, was found in late, perhaps Roman, remains at Karanog in Nubia.[7]

"Copper of the colour of one-third gold," or "of gold of the third quality," mentioned in the texts, e.g. of Rameses III (1198–1167), as coming from Etek (Sinai) [8], may have been brass [9], and the word teḥâst has been thought to mean brass.[10] A copper mixture "of the colour of good desert [? Arabian] gold," first cast and then hammered, according to the Harris Papyrus (13th century), and "a mixture of the six," used for plates, scale pans, etc., were perhaps brass.[11] Egyptian cymbals, $5\frac{1}{2}$ or 7 in. in diameter, in the British Museum, are "of mixed metal, apparently brass or a compound of brass and silver."[12]

Beads of zinc blende, the sulphide ore ZnS, occur in Predynastic graves.[13] Calamine, zinc carbonate ore, occurs in old workings at Gebel Rusâs with galena and cerussite[14], and a material translated as cadmia (zinc oxide) is mentioned several times as a constituent of eye-salves in the Ebers Papyrus (1550 B.C.)[15]; it could have been prepared by roasting calamine or blende. There is no Egyptian name for metallic zinc and the metal was probably unknown[16], all ancient brass, of variable composition and colour, being produced by a simultaneous smelting of copper and zinc ores, a method still used in Europe in the 18th century A.D.[17] Zinc oxide is more easily reduced by charcoal, and at a lower temperature, when metallic copper is present than when it is not.[18] Some copper ores contain appreciable amounts of zinc[19], and would yield brass by direct smelting. The use of such mixed ores in the direct smelting of brass is mentioned by Kircher in 1665.[20] Brass was an important article of commerce in Egypt in the 12th–13th centuries A.D.[21]

[1] In Wilkinson, ii, 350.
[2] i, 41, 452 ; ii, 341.
[3] Ezra, viii, 27.
[4] Wilkinson, i, 41 ; Napier, Arts, 96.
[5] Macalister, Excavations of Gezer, 1912, ii, 265.
[6] Petrie, Naqada and Ballas, 1896, 14, 20, 21, 22, 23, 24, 27, 29, 45, 47, 48, 54—analysis.
[7] Woolley and MacIver, Karanog, i, 62, 67 f. ; Lucas, Egy. Mat., 78.
[8] Brugsch, Ägyptologie, 402.
[9] Gsell, 8, 51.
[10] Budge, Egy. Hierogl. Dict., 842 ; a name for copper, see p. 86.
[11] Gsell, 51 ; Breasted, Anct. Rec., iv, 118, 166, 173.

[12] Wilkinson, i, 452 f. ; Fig. 222.
[13] Petrie, Prehistoric Egypt, 1920, 43 f.
[14] Garland and Bannister, 32.
[15] Joachim, 89, 90, 94, 138, 173.
[16] Lepsius, Met., 59.
[17] R. Watson, Chemical Essays, 1796, iv, 49 f. ; Gsell, 71 ; Wilkinson, i, 41.
[18] W. Rogers, *J. Amer. Chem. Soc.*, 1927, xlix, 1432.
[19] Watts, Dict. of Chem., 1872, ii, 34 ; Witter, *Chem. Z.*, 1932, 544, 763 ; Henglein, Doelter, Mineralchemie, IV, i, 75, 96, 122 f.
[20] Mundus Subterraneus, Amsterdam, 1665, ii, 218.
[21] Qalqashandi, Geogr., tr. Wüstenfeld, 224.

ANTIMONY

The common ore of antimony is the sulphide, stibnite, Sb_2S_3, used in the East as an eye paint, but the old Egyptian black eye paint called mestem (modern Arabic koḥl; Greek στίμμι; Latin stibium) was rarely stibnite, which has been found only once, in XIX dyn. remains.[1] Stibnite is easily reduced to metallic antimony, and the sporadic occurrence of this in remains could be expected, although no very early specimens are recorded. Petrie [2], who remarks that the metal was familiar to the Assyrians, thinks it was probably traded from them. It was found in the form of beads, badly reduced from the sulphide, with strange beads of iron pyrites, in XXII dyn. (c. 800) remains at Lahun.[3]

Antimony is a common impurity in many copper and lead ores, and traces occur in the nickel ore from St. John's Island in the Red Sea.[4] The early (XII dyn.) texts say that mestem came from Pitsew (in Arabia), but this was probably galena.[5] There is stibnite in Algiers, but this can hardly have been added intentionally to Egyptian bronzes.[6] Stibnite is not known to occur in Egypt, but is found in Asia Minor, Persia and possibly Arabia.[7] Ornamental plaques and buttons of badly reduced metallic antimony (still containing sulphide) were found in Transcaucasian graves at Redkin Camp and Kuban : the Kuban graves (near Tiflis) belong to the 11th–10th centuries B.C., but some are probably much older.[8] Stibnite occurs plentifully in China.

The material mentioned 36 times in recipes in the Ebers Papyrus (1550 B.C.) and translated stibium was probably mestem in the original and doubtfully stibnite : it is sometimes prescribed for internal use, and some varieties called "male" and "true" may be antimony sulphide.[9] The material in the Ebers Papyrus translated "alabaster" may have been antimony oxide[10], which is easily obtained by roasting stibnite. That Pliny's[11] variety of stimmi called larbason was white, as Berthelot[12] suggests, and hence antimony oxide, is improbable, since Pliny merely says it is brilliant (nitet) and it could have been of any colour. Ovid[13] says "diversi niteant cum mille colores," and "nitidissimus auri."

MERCURY

Mercury, the Coptic and hence perhaps Egyptian name of which was thrim[14], does not appear to have been known in early Egypt. That found in tombs in glazed earthenware bottles like pineapples is supposed to have been introduced by the Arabs[15], and the small bottle containing mercury which Schliemann brought back from Egypt, where it was said to have been found in a grave of the 16th–15th centuries B.C. at Qurnah[16], is probably of the same origin. These small bottles, also nuts, quills, etc., containing mercury, are used in the East as amulets.[17] If mercury were known in ancient Egypt (as the special name in Coptic makes just possible), it probably came from Spain, where the cinnabar mines of Sisapo (now Almaden) are very old.[18] The "stone of Tharshish" in

[1] Lucas, JEA, 1930, xvi, 42.
[2] Arts and Crafts, 98 f.
[3] Petrie, Illahun, 1891, 25 ; Arts and Crafts, 98 f. ; Gladstone, B.A. Rep., 1893, 715.
[4] Lucas, Egy. Mat., 61.
[5] Brugsch, Ägyptologie, 405 ; Gsell, 43.
[6] Sebelien, Anct. Egypt, 1924, 10.
[7] Lucas, JEA, 1930, xvi, 43.
[8] Virchow, Verhl. Berlin Ges. f. Anthropol., 1884, 126, 503 ; Forrer, RL, 32, 416, Figs. 315–324.
[9] Joachim, 19, 84, 87, 89, 95 ; Brugsch, Ägyptologie, 1891, 405.

[10] Berthelot, Arch., 243.
[11] xxxiii, 6.
[12] Intr., 238.
[13] Metam., vi, 65 ; Fasti, iii, 867.
[14] Hoffmann, in Ladenburg, Handwörterbuch der Chemie, Breslau, 1884, ii, 562 ; Peyron, Lex. Ling. Copticæ, Turin, 1835, 54.
[15] Wilkinson, ii, 213.
[16] H. Schelenz, Geschichte der Pharmazie, 1904, 41.
[17] S. Seligmann, Der böse Blick und Verwandtes, 1910, ii, 18.
[18] Dyer, DG, ii, 1014 ; Schulten, PW, viii, 2008.

the Bible is supposed by Haupt [1] to be cinnabar from Almaden, also called chrysolith (= golden stone) from its colour. Some clay dishes and pots used there for collecting mercury are probably pre-Roman, if not prehistoric.[2] Mercury was an important article of trade in Egypt in the 12th–13th centuries A.D.[3]

PLATINUM METALS

Several XII dyn. gold objects now in Cairo show small white metallic specks on the surface, consisting of platinum or one of the platinum group of metals.[4] A grey-coloured XVIII dyn. gold at Meroë contained silver and platinum[5], and gold of 1400 B.C. with grains of osmiridium on its surface was found at Lahun.[6] Berthelot[7] describes a box filled with ivory found at Thebes, perhaps Medinet Habu, with the name of Queen Shapenapit, daughter of Piankhi (XXVI dyn., 650), which is of bronze (copper 49·3, tin 4·5, lead 24·8, oxygen, etc., 21·4, no zinc, gold or silver), soldered and covered with hieroglyphics in pale and red gold and a white metal. This was partly silver, but one piece was platinum, possibly with some iridium, since it was difficult to dissolve in aqua regia. There was also gilt silver wire, gold leaf attached by glue, and lapis lazuli. The platinum probably came from alluvial deposits in Nubia or the Upper Nile, and was perhaps confused with silver. It was attached to the metal in incisions by means of a mixture of litharge (lead oxide) with some copper and tin oxides and oil, perhaps linseed oil, or soap, and then varnished. The litharge cement was probably applied hot and polished with a tooth or shell. The choice of gold and silver had a theological basis, gold corresponding with the sun and silver with the moon ; lines of stars on the box represent heaven, with the two eyes of heaven.

Petrie found at Nubt a gold-covered scarab of the XII dyn.(?), in the cover of which was contained what he calls "the first specimen of osmiridium yet known from Africa."[8] A supposed mention of platinum as aluta, or elutia, in Pliny [9] is mistaken[10], and platinum is first mentioned vaguely in the 16th and definitely in the 18th century A.D.[11]

IRON

The introduction and very early use of iron in Egypt (also metal-working in general, writing and brickmaking) have been associated with the arrival in the Predynastic Period of a special tribe, the so-called "followers" or "worshippers of Horus" (shosu Horu ; Heru shemsu), frequently called "metal workers" or "blacksmiths" (mensu or mesenu), who established him at Edfû[12] as "lord of the forge-city" (Edfû), with a forge (mesnet) in his temple behind the sanctuary.[13] Horus had also an important temple at Hierakonopolis, but his old festival, celebrated every two years in the I–II dyns., fell into disuse in the III dyn.[14]

The historical element in the legend was recognised by Maspero[15]: the "followers [worshippers] of Horus" are supposed to have entered Egypt about 4240 B.C.[16], either from Somaliland[17] or from Asia by way of the Red Sea and across the eastern desert, e.g. through the Wâdî Hammâmât, or a little

[1] MGM, i, 386.
[2] Schulten, PW, viii, 2008 ; Ganschinietz, PW, ix, 55—"aludels."
[3] Qalqashandi, Geogr., tr. Wüstenfeld, 224 f.
[4] Lucas, Egy. Mat., 93.
[5] Maspero, Dawn, 493.
[6] Brunton, Lahun I, 1920, 22 f., 31.
[7] Arch., 25 f., 35 f., 40.
[8] Naqada and Ballas, 1896, 66.
[9] xxxiv, 16 ; Hoefer, i, 140.
[10] Kopp, Geschichte der Chemie, Braunschweig, 1843–7, iv, 220.

[11] Zippe, Geschichte der Metalle, Vienna, 1857, 296.
[12] Budge, Gods of Egyptians, i, 84, 158, 476 f., 484 ; ib., Egy. Hierogl. Dict., 325, 505 ; Maspero, Dawn, 176 ; ib., L'Anthropologie, 1891, ii, 407 ; BMGE, IV–VI Rooms, 241, 283 f.
[13] Sayce, Annals of Archæology, Liverpool, 1911–12, iv, 56.
[14] Breasted, H., 46.
[15] L'Anthropologie, 1891, ii, 401 f.
[16] Meyer, Alt. I, ii, 17.
[17] Hall, NE, 91.

south of Thebes; to have conquered the indigenous peoples round Edfû, and to have made that city the centre of their civilisation.[1] The title "worshippers of Horus," however, may be only a name applied later to semimythical Predynastic kings.[2] The significance of mesenu as blacksmith or worker in metal (first read so by Brugsch) is said to be confirmed by their occupations as given in the Sallier and Anastasi Papyri [3], by the determinative sign meaning "work requiring much energy," i.e. masons, carvers of stone and metal, and by their representation as carrying a graving point in one hand and a javelin in the other. The interpretation of mesenu as "blacksmiths" is discredited by Sethe [4], who thinks it means "harpooners," carrying copper harpoons. Gardiner thinks it means "hippopotamus hunters."[5] Sethe's view is not accepted by all Egyptologists [6], but if the followers of Horus had metal weapons these were probably of copper, not iron. In a later period ironsmiths, as well as coppersmiths and other craftsmen, were employed by the temples, but their functions were purely secular.[7]

THE NAMES AND REPRESENTATIONS OF IRON

The supposed old Egyptian hieroglyphic name for iron has been given as $\bigrestroke$, men [8], but men has been translated as "metal," [9] or "copper or bronze."[10] Brugsch supposed that men represented "Nubian iron," whilst another name, teḥset, which appears in the Middle Kingdom, meant "Asiatic iron,"[11] but these translations are obsolete.[12] Lepsius[13] gives teḥset (tḥsti) as the name for iron in the Ptolemaic Period, whilst Brugsch[14] gives this as a late name for copper. The name teḥset has been supposed to be related to the tradition reported by Plutarch from Manetho (3rd century B.C.) that, whilst the magnet was "bone of Horus," iron was "bone of Set" ($\dot{o}\sigma\tau\acute{e}ov\ Tv\phi\hat{\omega}vos$).[15] These designations, said to be confirmed by Egyptian texts, e.g. "iron comes from Set" in the Pyramid Texts (v–vi dyns.)[16], are perhaps connected with Set as a god of iron and of the underworld[17], although a mention of iron so early is very doubtful.

The Coptic name for iron, benipe[18], corresponds with a hieroglyphic form bàa-en-pet, or ba-en-pet (bj-ni-pet), usually supposed to mean "iron of heaven" or meteoric iron, as opposed to ba-en-to, or bàa-nu-ta, "iron of [found in] earth."[19] This translation is uncertain[20]; ba may mean "metal,"[21] or even "hard stone."[22] Pott's suggestion[23] that $\sigma\acute{\iota}\delta\eta\rho os$ (Doric and Æolic $\sigma\acute{\iota}\delta a\rho os$)[24]

[1] Budge, Gods of Egyptians, i, xii, 485; ib., Mummy², 11.

[2] Breasted, H., 36; Maspero, L'Anthropologie, 1891, ii, 403; ib., Dawn, 182.

[3] Maspero, Art in Egypt, 230; L'Anthropologie, 1891, ii, 401 f.

[4] Q. by Hall, NE, 1920, 93; Budge, Mummy², 11.

[5] Wiedemann, 1920, 48; Thompson and Brunton, Badarian Civ., 1928, 54.

[6] Wiedemann, A. Rel., xxi, 451.

[7] Brugsch, Ägyptologie, 1891, 220, 417, 436.

[8] Lepsius, Met., 52 f., 63; Orth, PW Suppl. iv, 112; Schrader, RL, 179; Schrader-Jevons, 154.

[9] Budge, Egyptian Language, 87.

[10] Hoernes, Urg., ii, 219; Erman, q. by Gsell, 51.

[11] Orth, PW Suppl. iv, 112.

[12] Von Luschan, Z. f. Ethnol., 1909, xli, 47.

[13] Met., 52, 63.

[14] Ägyptologie, 1891, 401.

[15] Plutarch, Isis and Osiris, 62; Lepsius, 56; Brugsch, Ägyptologie, 398; Gsell, 24; Wilkinson, iii, 228; Hopfner, PW, xiii, 757.

[16] Reinach, L'Anthropol., 1891, ii, 106; Brugsch, Religion, 72; Gsell, 24; Roeder, Ro., iv, 777.

[17] Kees, PW, II Ser., iv Halbband, 1914; Schmidt, Ro., v, 1445 f.

[18] Griffith, Stories, 94, 102, 126.

[19] Lenormant, Prem. Civ., i, 88 f., 101, 155; Brugsch, Ägyptologie, 401; Budge, Egy. Hierogl. Dict., 210, 218; Beck, Gesch. d. Eisens, i, 93 f.

[20] Lepsius, 55; Schrader-Jevons, 202.

[21] Von Luschan, Z. f. Ethnol., 1909, xli, 47; Reinach, L'Anthropologie, 1891, ii, 106; Maspero, Arch., 195 f.; BMG Bronze Age, 2.

[22] Anon., in Anct. Egypt, 1915, 190.

[23] Lenormant, Prem. Civ.,i, 88 f., 101, 155.

[24] Pauli, Ro., iv, 787.

is connected with the Latin sidus, sideris, star, and thence with the Egyptian ideas of the divine origin of meteoric iron, the metal of the sky, is rejected by Schrader [1], but the connection between iron and "thunderbolts" is developed in detail by Wainwright.[2] Budge [3] says that "it is certain that iron was known to the Egyptians from the earliest times, for the oldest religious texts extant [The Pyramid Texts, v–vi dyns.], which were copied from far older archetypes, speak of the heavens being formed of a plate of iron, and the Deity is said to sit upon a throne of iron, the sides of which are ornamented with the faces of lions, and have four legs, the feet of which are in the form of the hooves of bulls. The Egyptian word for 'iron,' bàa or bàa en pet, i.e. 'bàa of heaven,' is, of course, meteoric iron, and this phrase is the exact equivalent of the old Sumerian ideographic group AN. BAR, 'iron.' The Coptic word for 'iron,' benipe, which is a direct descendant of bàa en pet, conclusively proves that this expression means 'iron' and iron only. But, in order to avoid the conclusion that iron was known to the Egyptians at this early period, it has been supposed that bàa meant 'crystal'; this, however, is disproved by the fact that the representations of weapons, knives, tools, etc., which are of a blue colour, are found upon the monuments of all periods, and, as it is clear that they cannot have been made of crystal, they must be iron."

The designation ba-en-pet, "iron of heaven" (the sky being regarded as a plate of iron falling as meteorites), may, however, date from the Hyksos Period, when iron was more valued [4], but the Egyptians probably identified meteorites of *any* kind, whether iron or not, with heaven.[5] Petrie [6] suggested that iron (meteoric or native) was discovered in a hæmatite district, the two materials being confused, and that the name ba-en-pet means hæmatite [7], being applied to iron only in the Greek Period. Dawson [8], however, says the name for Nubian hæmatite (a soft form of which he incorrectly supposes to be red ochre) is didi; statuettes of hæmatite of Rameses III are known. That ba-en [pet] = ben = men, "metal," [9] or "tree of heaven,"[10] are other suggestions.

In the Pyramid Texts (v–vi dyns.) the sky is a flat slab resting on four pillars. The material of the slab is very doubtful and has been given as iron, bronze, alabaster or crystal.[11] It is possible that it was regarded as the blue lapis lazuli, the golden specks of pyrites in this representing the stars, or else rain. The composition of the sky from iron was suggested by Deveria, and the translation of ba-en-pet as "metal of heaven" is due to Chabas.[12] According to Wainwright[13], ba-en-pet first occurs in the xix dyn., the earlier name for iron being ba. In Homer's Iliad the sky is always bronze; in the Odyssey it is usually iron.[14] It has often been supposed[15] that iron is mentioned as ba-en-pet, and shown blue, in the Pyramid Texts, e.g. of King Unàs (v dyn.), and that in Old Kingdom texts there is an "iron throne" of a god, "iron of the North" and "iron of the South."[16] Other authorities think iron is not mentioned even in the xii dyn., but only in the xviii–xix dyns.[17], about which

[1] RL, 177; Schrader-Jevons, 206.
[2] JEA, 1932, xviii, 6 f., 9 f., 159 f.
[3] BMGE, IV–VI Rooms, 238 f.
[4] Brugsch, Ägyptologie, 401; Gsell, 27, 45; Wiedemann, 1920, 344 f.
[5] De Launay, Daremberg-Saglio, Dict., ii, 1076; Wainwright, *op. cit.*
[6] Petrie, Wainwright and Mackay, The Labyrinth, Gerzeh and Mazghuneh, 1912, 19.
[7] Anon., in *Anct. Egypt*, 1915, 190.
[8] *J. Roy. Asiat. Soc.*, 1927, 497.
[9] Lepsius, Met., 55 f., 63.
[10] Hoernes, Urg., ii, 221.
[11] Budge, Gods of Egyptians, i, 156; *ib.*, Mummy², 343; C. H. S. Davis, Book of the

Dead, 1894, 75; Erman, Lit., 29; Reinach, *L'Anthropologie*, 1891, ii, 106.
[12] Maspero, Dawn, 16; BMGE, 145.
[13] JEA, 1932, xviii, 15.
[14] Cook, Zeus, i, 632; Eisler, Weltenmantel und Himmelszelt, München, 1910, 94, 545, 758.
[15] Hall, Oldest Civ. Greece, 1901, 193 f.; Reinach, L'Anthropol., 1891, ii, 106; Gsell, 11; cf. Budge, Mummy², 461; Wainwright, JEA, 1932, xviii, 3 f., 11, etc.
[16] BMG Bronze Age, 2; Budge, G. of E., i, 58, 91, 158; Gsell, 11.
[17] Hoernes, Urg., ii, 220; von Luschan, Z. f. Ethnol., 1909, xli, 48 f.; Gsell, 47.

time a blue material appears on wall pictures, e.g. in a temple at Radassiyah. These blue objects are shown as brought by Semites, who were early workers in iron.[1] The blue colour may be the colour of the metal of heaven [2], or of the sky-god Horus.[3]

What is now regarded as the first *unequivocal* mention of iron is on a stela of Rameses II (XIX dyn. ; 1250) at Abu Simbel, [4] which says that the god Ptah made the king's members of asem, his body of copper and his arm of ba-en-pet. A new translation gives : "I have wrought thy body of gold, thy bones of copper, thy vessels of iron."[5] The Ebers Papyrus (1550, from older materials) is supposed to mention iron twice : (a) iron of the town of Qesi (Apollinopolis Parva ; in Upper Egypt), and (b) art-pet, "made in heaven," meteoric iron.[6]

Other texts show that iron, including weapons, became commoner about the beginning of the XVIII dyn. (c. 1600) and probably came from the Hittites and similar tribes of the Black Sea and North-East Asia Minor, [7] and the name parthal (Hebrew barzel) is said to occur.[8] Thothmes III (1500) received from T'nai (in Asia Minor) as tribute three iron vessels, so that the metal was still valuable in Egypt.[9] Although iron from Palestine, Syria (the Retenu of Lebanon), etc., came to Egypt in his time,[10] it did not displace bronze for common purposes.[11] Tušratti, king of the Mitanni, sent steel rings and daggers to Amenhotep III (1411–1375)[12], and about 1250 the Hittite king (probably Hattusil) promised to send Rameses II iron later, since his stock was at the moment low, but forwarded a steel dagger[13]; the Hittites supplied steel to Rameses III about 1190.[14] The iron and the steel dagger found at Gerar in Palestine[15] are dated about 1350.

It is very probable that the objects, including weapons shown blue and as brought by Asiatics, in wall paintings of this period (1500–1300), represent iron and steel from Asia Minor.[16] Butchers sharpening knives on blue rods, probably "steels," are shown in later wall paintings.[17] The hair and eyebrows (presumably black) are also shown blue in some paintings.[18] No Egyptian text ever speaks of the introduction of a *new* metal, so that iron may be represented at any period[19], and common iron was perhaps shown brown or black, steel, i.e. "pure" (or "true") iron, blue.[20] The distinction in colour between steel (violet-blue) and iron (grey) is made by Homer.[21]

Gods and kings are shown with blue beards and wigs—formerly called "battle helmets."[22] That the blue colour of the beards was intended for grey[23]

[1] Schrader, RL, 179 ; Schrader-Jevons, 202.
[2] Gsell, 47.
[3] Mackenzie, Footprints, 143.
[4] Hall, q. in Lucas, Egy. Mat., 100 ; Wainwright, JEA, 1932, xviii, 14.
[5] Blackman, JEA, 1927, xiii, 191.
[6] Joachim, Papyros Ebers, 90, 168 ; Berthelot, Arch., 242 f.
[7] Erman-Ranke, 550, 615.
[8] Budge, Egy. Hierogl. Dict., 232.
[9] Olmstead, H. Pal. Syr., 145 ; Myres, CAH, i, 109 ; Meyer, Alt. II, i, 130.
[10] Gsell, 20, 47 ; Lepsius, Met., 52 f.
[11] Hoernes, Kultur der Urzeit, Leipzig, 1912, iii, 106 f.
[12] Meissner, Bab. and Assyr., i, 264.
[13] Meyer, Alt. II, i, 529 ; ib., Chetiter, 76; Hogarth, CAH, ii, 267 ; Meissner, i, 348.
[14] Gsell, 47.
[15] Petrie, Gerar, 1928, 10, 14, 29 ; Nature,

1927, cxx, 56 ; ib., 1929, cxxiii, 838 ; Brit. Assoc. Journ., 1930, 63.
[16] Jomard, in Passalacqua, Catalogue raisonné, 1826, 246 ; Wilkinson, ii, 251 ; Lepsius, Met., 56 f., 63, plate 2 ; Schrader, RL, 179 ; Schrader-Jevons, 202 ; Hoernes, Urg., ii, 296 f. ; ib., Urzeit, iii, 18 f. ; BMG Bronze Age, 2 f. ; Maspero, Struggle, 284 ; Wiedemann, 1920, 344 f.
[17] Jagnaux, Histoire de la Chimie, 1891, ii, 218.
[18] Davies and Gardiner, Tomb of Amenemhet, 32, 75 (Thothmes III) ; ib., Tombs of the Two Officials, 24 (Thothmes IV).
[19] De Morgan, Or., i, 213.
[20] Lepsius, Met., 53 f., 56 ; Birch, in Wilkinson, ii, 251 ; Gsell, 38 f., 49.
[21] Lenz, Mineralogie, 1861, 5 ; Blümner, iv, 343—somewhat doubtful.
[22] Belck, Z. f. Ethnol., 1908, xl, 64 ; Wilkinson, i, 218 ; Erman-Ranke, 66 ; Gsell, 44.
[23] Lepsius, 56.

is improbable : it represented lapis lazuli, as an amulet, just as blue beads are still used in Syria to avert the evil eye.[1]

SOURCES OF IRON

The iron used by the Egyptians and Assyrians was imported from Asia Minor ; if they worked it themselves, which is improbable for Egypt and doubtful for Assyria, it was on a very small scale only.[2] That iron in Egypt about 1200 came from Europe [3], or from the "Cretan-Philistines," who were "inventors of the technology of iron,"[4] is highly improbable.

An old argument against the early use of iron in Egypt is based on the great scarcity of finds of the metal : when iron has been used for weapons it rapidly supersedes bronze, the latter being used only for vessels or ornamental purposes.[5] Iron agricultural implements are said to be mentioned in the Sallier Papyrus (xix–xx dyn.).[6] A Bronze Age, however, does not necessarily correspond with ignorance of iron or its relative scarcity, but with a state of civilisation in which steel was either unknown or very rare, whilst the general riches have been such that arms and general utensils were made in bronze. Except in very special circumstances, iron has always been easy to procure.[7] In later tombs iron occurs with copper and bronze, which were still in extensive use.[8]

One explanation for the rarity of finds of iron in Egypt, that the metal would have been corroded away by the "strongly chloridic," [9] or "strongly nitrous,"[10] soil, or by the "strongly saline" Nile water[11], is rejected by Petrie,[12] Sir John Evans[13], de Launay[14], and Lippmann[15], but supported by Garland.[16] It is also possible that iron tools are not frequent because they were re-made when worn out.[17] The usual explanation for the scarcity of iron is that this metal was despised, hated or feared on religious grounds, as the "bone of Set,"[18] the evil opponent of the good Osiris. In the Predynastic Age, however, iron was used as an amulet, or "luck-bringing," if not sacred, material. After iron beads went out of fashion, small pieces of hammered iron were used as "lucky objects" during the early dynasties.[19] In Tutankhamen's tomb the amulet of the head-rest is of wrought iron[20], and in the later period iron was regarded as divine and possessed of magic properties.[21] Iron instruments were also used (some specimens are known) in the religious ceremony of "opening the mouth" of the mummy.[22] Although iron is still "unlucky" in Egypt, it is credited with the power of repelling demons.[23] Bronze was a "purer" metal than iron ; Hero of Alexandria (? 2nd–3rd centuries A.D.) says bronze wheels were made to revolve near the gates of Egyptian temples, "because bronze is believed to exert a purificatory influence," and one was found at Thebes.[24]

[1] E. S. Stevens, Cedars, Saints and Sinners in Syria, [1926], 195, 221.
[2] Gowland, J. Anthrop. Inst., 1912, xlii, 281.
[3] Montelius, L'Anthropol., iii, 451 ; cf. Maspero, Dawn, 59 ; Fimmen, Zeit und Dauer der kretisch-mykenische Kultur, 1909, 4.
[4] Belck, Z. f. Ethnol., 1908, xl, 45, 241 f. ; cf. Fimmen, 102 ; Gsell, 29.
[5] BMG Bronze Age, 2.
[6] Maspero, Dawn, 331.
[7] Tannery, q. in Isis, 1930, xiv, 427 ; BMGE, IV–VI Rooms, 238 ; Piehl, q. by Hall, Oldest Civ. Greece, 193 f.
[8] Maspero, L'Anthropol., 1891, ii, 105.
[9] Garland and Bannister, 99, 101.
[10] Lepsius, Met., 54 ; Wilkinson, ii, 250.

[11] Gsell, 28.
[12] Arts and Crafts, 98 f.
[13] BMG Bronze Age, 3.
[14] Daremberg-Saglio, Dict., ii, 1075.
[15] Alchemie, i, 611.
[16] Garland and Bannister, 99, 101.
[17] Ib., 85 f.
[18] Manetho, in Plutarch, Isis and Osiris, 62.
[19] Mackenzie, Footprints, 142 f.
[20] Wiedemann, A. Rel., xxvi, 340.
[21] Schrader, RL, 179.
[22] Budge, Mummy², 327 ; ib., Egyptian Magic, 192 ; Baly, JEA, 1930, xvi, 173 ; Wainwright, ib., 1932, xviii, 6, 11.
[23] Lane, Modern Egyptians, 1846, i, 306.
[24] Thorndike, History of Magic, 1923, i, 192 ; Cook, Zeus, i, 266, Fig. 191.

The red rust of iron perhaps recalled "the red (desert) land of Set," or the blood of the murdered Osiris.[1] Both Plutarch and Diodoros Siculus say the Egyptians hated red hair ; Set's weapon was an iron spear, and the executioner's axe was of iron.[2]

Plutarch [3] says from old traditions that the Egyptians called the magnet "bone of Horus" (piles of large bones, some wrapped in linen, were found at Qau), and explains that iron, the "bone of Set," is first attracted and then repelled by a "stone"; probably magnetite, a native magnetic oxide of iron, acting upon the poles of permanently magnetised steel.[4] Pliny [5] also mentions a stone which repels iron [steel]. Amulets of magnetite were used to repel evil from Horus, and two rather obscure amulets, of two different kinds of iron ore, were put in the mummy bandages to keep off evil.[6] Hæmatite beads were found in Predynastic tombs[7], and there are carved hæmatite statuettes.[8] The Book of the Dead (ch. 64) mentions the discovery, under the feet of the god at Chmunu (Hermopolis) in the time of Menkarā (IV dyn.), of an amulet written in blue letters on a tablet of hæmatite, [9] although the readings "alabaster,"[10] "metal,"[11] or "burnt clay,"[9] are also given.

Magnetic oxide of iron, Fe_3O_4 (not all specimens of which are naturally magnetic), occurs at Aswân and is perhaps Plutarch's "bone of Horus"; this author says the Egyptians knew a magnetic stone[12], which Dioskurides[13] says gives red hæmatite on burning.[14] Hæmatite, Fe_2O_3, occurs in Sinai and may have been used for iron extraction in later periods[15], although the Egyptian miners cut through rich deposits of it to look for copper ore.[16] The Sinai ore, which contains manganese and titanium, was mistaken by Lepsius for "slag heaps," an error which has often been repeated[17]; metal extraction was not done at the mines but only in the valleys of Wâdî Maghâra.[18] Meteoric iron and brilliant ores such as magnetite, iron glance, hæmatite and brown iron spar were probably first used because they would attract attention, the earthy-looking ores being unnoticed.[19]

Burton in 1822 reported traces of worked-out iron mines in the Wâdî Hammâmât, on the right bank of the Nile, where there are no hieroglyphics for iron.[20] In the Nile Valley, the Eastern Desert along the Red Sea, in Nubia, Ethiopia and the Sudan there are rich deposits of iron ores of varying quality, including magnetite and hæmatite[21], and Abulfeda[22] mentions iron mines at Dendema. The mines, if they ever existed, have been worked out and there are now no traces of old iron mining in Egypt.[23]

What is believed to be the name of iron occurs in several places in the so-called Book of the Dead. This grew out of the religious Pyramid Texts (c. 3000 B.C.) by fusion with the ritual of the worship of Osiris in the XI dyn.,

[1] Gsell, 25 ; Lenormant, Prem. Civ., i, 270.

[2] Gsell, 25.

[3] Isis and Osiris, 62.

[4] Maspero, Arch., 195 f. ; Hopfner, PW, xiii, 757 ; Wainwright, JEA, 1932, xviii, 14.

[5] xx, proem.

[6] Gsell, 25 ; Budge, Amulets, 149.

[7] Petrie, Naqada and Ballas, 1896, 10, 28, 36, 44—no iron.

[8] Wiedemann, 1920, 345.

[9] Lenormant, Chaldæan Magic, 1877, 90.

[10] Brugsch, Rel., 1885, 19.

[11] Reitzenstein, Poimandres, Leipzig, 1904, 18.

[12] Σιδηρῖτις λίθος, the usual name is μάγνης λίθος ; Strabo has σιδηρῖτις ; iron ore is γῆ σιδηρῖτις ; Latin magnes lapis' sideritis.

[13] v, 148.

[14] Gsell, 11 ; Schweinfurth, Z. f. Ethnol., 1908, xl, 61.

[15] De Morgan, Or., i, 213.

[16] Gsell, 9, 84 ; Maspero, Dawn, 354.

[17] Blümner, iv, 70.

[18] De Morgan, Or., i, 219, 227 ; Petrie, Sinai, 100 ; Maspero, Dawn, 349.

[19] Blümner, iv, 208.

[20] Wilkinson, ii, 250 ; Gsell, 10, 26.

[21] Gsell, 26 ; Breasted, H., 615—doubtful.

[22] Géogr., tr. Reinaud, ii, 225.

[23] BMG Bronze Age, 3 f. ; Wilkinson, i, 155 ; Schweinfurth, Z. f. Ethnol., 1908, xl, 61.

but assumed its present form at the end of the XII dyn.[1] The Book of the Dead was much enlarged in the XVIII–XX dyns. (the so-called Theban recension containing 180 chapters), and further in the Saïte Period (XXVI dyn.).[2] The various "layers" are distinguished by variations in particular words[3], and some parts, it will be seen, may be quite late. The Book of the Dead mentions the iron wall of heaven and an iron chain put round the neck of a serpent.[4]

PREDYNASTIC IRON IN EGYPT

Very early knowledge of iron as a precious metal in Egypt was put beyond doubt by the discovery[5] of iron beads completely turned to rust, strung with gold, agate and carnelian beads as necklaces, in two intact Predynastic graves in the cemetery of El Gerzeh. Gowland found on analysis of the rusted beads : ferric oxide (Fe_2O_3) 78·7 ; water (with traces of carbon dioxide and earth) 21·3. "They do not consist of iron ore, but of hydrated ferric oxide, which is the result of the rusting of the wrought iron of which they were originally made." Copper also occurred in the graves. The usual composition of iron rust[6], ferric oxide 85·5, water 14·5, differs somewhat from the analysis : a common ore of iron has the same composition as rust. Petrie and his co-workers, for reasons which are now known to be incorrect, at first thought the beads were of native iron, assumed to be found in Sinai as a small nugget, but in a later discussion[7], in which he dates the beads S.D. 60–63 (although 55–63 or 60–66 are also possible), Petrie admits that either native or meteoric iron is possible. Hall[8] regarded meteoric iron as more probable, since Zimmer[9] had removed Petrie's objection that meteoric iron was brittle. Meteoric iron was, in fact, cold-worked by the Eskimos as late as 1818, and many other primitive peoples use it.[10] The average weight of iron meteorites is 20 lb. and pieces for hammering can usually be detached without great difficulty. No sources of native iron are known anywhere near Egypt : the only occurrences of masses of undoubted telluric origin are those found by Nordenskiöld at Ovifak, in Greenland, which may also be buried meteorites. The Eskimos fashion knives, etc., from these, and iron was the first metal known to them. There is doubtful meteoric iron in the Toluka Valley in Mexico and small grains or nodules of telluric iron, "too small for practical use," occur in basalt or other rocks, this iron often containing large quantities (up to 65–75 per cent.) of nickel.[11] Although of the 415 iron meteorites described by Berwerth and Michel[12], there is none from Egypt (one is from Senegal), yet meteorites in Egypt are recorded, including a lump of meteoric iron as large as a cannon ball in the desert in the Wâdî Dugla, 30 km. east of Cairo.[13]

The source of the iron for the Predynastic beads appears to have been

[1] Boylan, Thoth, the Hermes of Egypt, Oxford, 1922, 49 ; Breasted, Development of Religion and Thought in Ancient Egypt, New York, 1912, 93 f. ; Erman, La religion égyptienne, tr. Vidal, 1907, 124, 143 ; Wiedemann, Die Religion der alten Ägypter, Münster, 1890, 124, 129 f. ; ib., A. Rel., vii, 481 ; Roeder, A. Rel., xv, 68.

[2] Quibell, 79 ; BMGE, 58 f. ; Peet, CAH, ii, 198 ; Davis, The Book of the Dead, 1894.

[3] Wiedemann, 1920, 393.

[4] Book of the Dead, chapters 64, 85, 130, 108 ; Davis, 103, 116, 144, 127.

[5] Petrie, Wainwright and Mackay, The Labyrinth, Gerzeh, and Mazghuneh, 1912, 15 f. ; Petrie, Ancient Egypt, 1915, 18 f. ;

Wiedemann, 1920, 344 ; cf. Lippmann, Abh., i, 259 ; Wainwright, JEA, 1932, xviii, 3.

[6] Partington, Inorganic Chemistry, 1933, 967.

[7] Prehistoric Egypt, 1920, 27.

[8] CAH, i, 572.

[9] Anct. Egypt, 1917, 45.

[10] Lenormant, Prem. Civ., i, 88, 101 ; Wainwright, JEA, 1932, xviii, 4.

[11] Gowland, J. Anthropol. Inst., 1912, xlii, 236 ; Leitmeyer, in Doelter, Mineralchemie, III, ii, 767 f. ; Vogel, ib., 562 f.

[12] Doelter, III, ii, 574 f.

[13] Wainwright, JEA, 1932, xviii, 5 ; Belck, Z. f. Ethnol., 1908, xl, 64.

finally decided by chemical analysis to be ultimately meteoric, since they contain 7·5 per cent. of nickel [1], but since we have seen that native iron may be of meteoric origin, there is still a possibility of its use.

EGYPTIAN SPECIMENS OF IRON

The difficulty of preparing an account of finds of iron in ancient Egypt is increased by the tendency of every archæologist to regard his own specimens as the only authentic ones in existence.[2] The first specimens are, as just explained, Predynastic. No iron was found in I dyn. tombs.[3] Hill in 1837 found an iron object (fragment of plate ?) embedded in the masonry of the Pyramid of Cheops (IV dyn.) when the upper layers had been blasted away, in a condition attested by English witnesses.[4] This object, now in the British Museum [5], was found by Way and Flight to be soft or wrought iron, containing combined carbon and a trace of nickel, hence Bessemer suggested that it might have been produced by carburising a meteorite.[6] A recent analysis by Plenderleith, however, showed no nickel, hence the iron is probably not meteoric.[7] Gowland [8] considers that this iron might have been obtained accidentally in the Sinai copper smelteries from a piece of the rich iron ore which crops out near the veins of copper ore.

Although Hill's specimen was, apparently, unknown to Lepsius [9], he inferred the use of iron[10] by an argument used again by Garland and Bannister[11], viz. from the working of granite even in the IV dyn., although the weapons in all Old Kingdom representations are red. There seems little doubt, however, that copper chisels were actually used, perhaps with emery[12], although Breasted[13] considers that iron tools were in use in the Old Kingdom to a limited extent before bronze, the origin of the metal being unknown.

Maspero found "five or six iron chisels, much rusted," in the pyramid of Unàs (end of V dyn.) at Saqqâra, where the inscriptions on the walls are unfinished and the chisels may have been left by workmen when the king died ; also "pieces of embedded iron" in this pyramid, as well as "several pieces of a pickaxe" in the V dyn. pyramid of Abûsîr.[14]

Iron of the VI dyn. is represented by a lump of iron rust, perhaps originally a wedge, found with copper tools by Petrie at Abydos, wrapped with copper axes of VI dyn. form and placed at the corresponding level in the foundations of the temple.[15]

Next in chronological order is the find of tools in the XII dyn. brick pyramid of Dahshûr by Maspero.[16] Then comes the XII dyn. spearhead found in

[1] Rickard, *J. Inst. Metals*, 1930, xliii, 350 ; Wainwright, JEA, 1932, xviii, 3.

[2] Garland and Bannister, 5, 85 f.

[3] Petrie, Royal Tombs of the First Dynasty, 1901, ii, 8, 36 f. ; Meyer, Alt., 1909, I, ii, 203.

[4] Gsell, 12 f. ; H. Vyse, Pyramids of Gizeh, 3 vols., 1840–42, i, 275 ; Beck, Gesch. d. Eisens, 1884, i, 85 f. ; Olshausen, *Z. f. Ethnol.*, 1909, xli, 56 ; von Luschan, *ib.*, 47 ; Meyer, Alt. I, ii, para. 258A ; Maspero, Dawn, 59 ; Petrie, etc., Labyrinth, 17 ; Foy, *Chem. Z.*, 1908, 973 ; Wilkinson, ii, 250, thought it doubtful ; Petrie, Arts and Crafts, 98, calls it a "sheet" of iron.

[5] BMGE, IV–VI Rooms, 206 : "there is no doubt that this object is contemporaneous with the building of the pyramid."

[6] Gsell, 12, 92.

[7] Balaiew, *J. Inst. Metals*, 1930, xliii, 353.

[8] *J. Roy. Anthropol. Inst.*, 1912, xlii, 284 ; Friend, Iron in Antiquity, 1926, 161.

[9] Gsell, 13.

[10] Met., 57.

[11] *Op. cit.*, 85 f. ; *ib., Chem. and Industry*, 1927, 648.

[12] Petrie, Sinai, 160 f. ; de Launay, Daremberg-Saglio, ii, 1078 ; Gsell, 95.

[13] H., 93.

[14] Gsell, 14 ; Petrie, etc., Labyrinth, 19 ; *Anct. Egypt*, 1915, 20, where the iron is said to be "cursorily mentioned" by Maspero.

[15] Petrie, Abydos II, 1903, 33 ; Labyrinth, 19 ; Arts and Crafts, 98 f. ; MGM, iii, 45.

[16] Gsell, 14 ; Feldhaus, Technik, 232 ; Petrie, Arts and Crafts, 98 f., calls them "iron ferrules said to have been found . . ." and does not mention them in Labyrinth, etc., 19.

Nubia, "the oldest iron weapon in existence."[1] From the XVII dyn. are the point of a broken chisel embedded in mortar, and a ferrule of a hoe handle found by Maspero in the pyramid at Mohammeriah, near Esna, higher than Thebes.[2] Petrie [3] states that : "In the whole town—over 2,000 rooms—of Kahun of the XII dynasty and in the town of Gurob of the XVIII–XIX dynasty I have never found the smallest trace of iron or iron rust," and no iron was found in the tomb of Aah-hotep (beginning of XVIII dyn.).[4] Several specimens of iron of the XVIII dyn., however, are known.

Gsell [5] mentions "a piece of steel-like iron" found under the obelisk of Thothmes III, now in New York, and an iron peg found in a tomb. An iron falchion or halberd, "the oldest iron weapon known," [6] was found beneath the statue of Rameses II, corresponding in shape to Wilkinson's "pole-axe,"[7] and probably a blade with a cheaper bronze stiffening. The earliest "true" iron axe is said by Petrie [8] to be one found in the store-room of the Ramesseum (800 B.C.). Petrie [9] refers to a stud from a box ; a finger ring in the Ashmolean Museum, Oxford, with no provenance but "attributed" to XVIII dyn.[10]; and a curved sickle found by Belzoni under a sphinx in an avenue leading from the temple of Mut to the temple of Karnak, now in the British Museum and regarded as "unquestionable."[11] Gsell[12] remarks that there was plenty of iron in Egypt under Thothmes III (XVIII dyn.), and Petrie[13] that it was still a foreign metal for Egyptian weapons in the XIX–XX dyns. The two statements are, of course, quite compatible, since the iron was imported in considerable quantities. A pair of iron bracelets, roughly worked with dogs' heads, are of the XVIII dyn.[14], and a lump of oxidised iron was found at Amarna in 1924.[15] Three objects of wrought iron were found in the tomb of Tutankhamen : a dagger blade with a gold sheath, a miniature head-rest amulet (usually made from hæmatite), and part of an amulet bracelet.[16]

Of the XIX dyn. are an iron knife[17], iron found at Abydos[18], and an iron sword with cartouches of Seti II (1214–1210) in the Berlin Museum, probably of Asiatic metal.[19] This type of sword is more commonly found in bronze, and at this date there is a record of the overthrow of a Libyan invasion and the capture of 9,000 bronze swords, which marks the beginning of the free use of iron weapons and coincides with the independent date of 1200 for the beginning of the use of iron in Crete.[20] An iron rod in a bronze statue of the XIX–XX dyn. is cast over a sand core[21]; three iron knives are of Ramesside date or later[22];

[1] Feldhaus, Technik, 232, correcting Gsell, 15 ; in Petrie, *Anct. Egypt*, 1915, 20, it is XIII dyn., contemporary with the II City of Troy, but in Tools and Weapons, 33, it is an "iron blade" from Nubia "unquestionably of the XII dyn."

[2] Gsell, 14 ; Petrie, Labyrinth, 19 ; in *Anct. Egypt*, 1915, 20, the date of this find is given as doubtful ; Maspero, *L'Anthropol.*, 1891, ii, 105.

[3] Q. by Montelius, Ält. Bronzez., 1900, 151.

[4] Hoernes, Urg., ii, 220.

[5] 14 f.

[6] Petrie, Abydos II, 33 ; pl. xxii; "a rarity," *ib.*, Tools and Weapons, 10, 11, 26.

[7] Wilkinson, i, 216.

[8] Tools and Weapons, 9, 25.

[9] Labyrinth, 19.

[10] Wilkinson, ii, 248, says these iron rings, of which he had a specimen with a figure of Harpocrates, were usually of the Ptolemaic or Roman Period.

[11] Petrie, Tools and Weapons, 47; *Anct.*

Egypt, 1915, 20 ; Beck, Gesch. des Eisens, i, 87 ; BMGE IV–VI Rooms, 206—"much oxidised," and "before the XIX dynasty (?)."

[12] Eisen, etc., 20.

[13] Tools and Weapons, 25.

[14] Hall, JEA, 1928, xiv, 191.

[15] Griffith, q. in JEA, 1928, xiv, 191, correcting Petrie, Tell el-Amarna, 31.

[16] Lucas, in Carter, Tutankhamen, ii, 178, 258, 268 ; plates 77, 87 ; JEA, 1928, xiv, 191 ; Wiedemann, *A. Rel.*, xxvi, 340 : iron was still valuable ; BMGE IV–VI Rooms, 24.

[17] Gsell, 15.

[18] Petrie, Abydos I, 1902, 32.

[19] Petrie, *Anct. Egypt*, 1915, 22 ; Wade-Gery, CAH, ii, 524.

[20] Petrie, *Ancient Egypt*, 1915, 22 ; Mackenzie, Footprints, 143.

[21] Gsell, 15 ; Schweinfurth, *Z. f. Ethnol.*, 1908, xl, 64, who says the bronze figures with iron stiffenings in the Berlin Museum are of doubtful period, the oldest being of the 7th century B.C., and perhaps Græco-Roman.

[22] Gsell, 14 f.

a halberd is possibly of the period of Rameses III (xx dyn.) or possibly of the
xxvi dyn.[1] ; a needle with a hole at the large end, from Nubia, is about this
period or later[2] ; and the head of a small iron hoe in the Berlin Museum, of
unusual shape, is of uncertain date.[3]   There is a xxi dyn. iron coffin nail in
the Cairo Museum [4], and iron nails of the xxii dyn. were found at Qurnah,
near Thebes.[5]   Other iron objects of uncertain date in the British Museum
are two axe-heads, a spear-head, a sickle-shaped knife, a tanged javelin-head,
a sickle blade, sickle in wooden handle and strigils (Roman Period), chisels,
spatulas, tools and weapons from Tanis, and knives (one, with a horn hilt,
of the Byzantine Period).[6]

Farther south, in Meroë in Nubia, with rich deposits of iron ore mentioned
by Strabo [7], an extensive iron industry developed in the 9th century B.C.,
iron being used instead of the copper and bronze of Egypt.[8]   Bronze in Meroë
was very scarce and its place was taken by iron for tools and weapons.[9]
Mountains of iron slag enclose the city mounds and the furnaces have been
excavated.   The whole of northern Africa might have been supplied with iron
by Meroë, "the Birmingham of ancient Africa," and the Sudan passed directly
from the Stone to the Iron Age without an intervening Copper or Bronze Age.

Sayce found iron slag mixed with clay crucibles in the Sudan at Kerma,
Kawa and the island of Argo, usually with mounds of burnt bones which
appear to have been used in making enamelled tiles.[10]   Whether iron was
worked in Nubia at an early period, knowledge of it passing thence to Egypt, is
not known.[11]

An iron spear-head found at Lahun was of c. 800 B.C.; two iron wedges of
about the same date from the Ramesseum are now in the Bologna Museum ;
two from Denderah date from c. 600, and two from Naukratis from 600–300.[12]
A group of steel tools found at Thebes are dated c. 668–666 by an Assyrian
copper helmet.[13]

The iron found by Arabs at Naukratis was from deep layers and there are
deposits of scoria and of specular iron ore on the site, showing that the objects
were made on the spot : the date is doubtful, since no jewellery earlier than
the 1st century A.D. was found.[14]

Specimens of Egyptian iron of 1200–800 B.C. were carburised and quenched,
and the introduction of iron into Europe about this period may have resulted
from its more extensive use after the discovery of these processes.[15]   The
tools found in tombs at Thebes, probably left by the Assyrian army and
comprising a sickle, flat chisel, mortice chisels, saws, punch, rasp, files, twist
scoop and two centre bits, had the edges steeled, probably by some kind of
case-hardening.   The forms are modern but the files are slight and irregular
and the bits are suitable only for hard wood, not metal.[16]   An iron adze is of
the Egyptian type (no date).[17]   At Daphnæ (O.T. Tahpanhes) there were iron
tools of about 660 B.C.[18]   The use of mild steel in Egypt in the 7th century B.C.

[1] Petrie, etc., Labyrinth, 19 ; Abydos II,
23, pl. 22.
[2] Archæological Survey of Nubia, 1910,
i, 59 f., pl. 72d.
[3] Wilkinson, ii, 252.
[4] Schweinfurth, Z. f. Ethnol., 1908, xl, 63.
[5] De Morgan, Or., i, 214.
[6] BMGE, IV–VI Rooms, 201, 203, 206 f.,
209.
[7] XVII, ii, 3 ; 821 C.
[8] Garstang and Sayce, Annals of
Archæology, Liverpool, 1911–12, iv, 45 f.,
52, 53 f. ; Wiedemann, 1920, 344.
[9] Petrie, Anct. Egypt, 1915, 22.
[10] Annals of Archæology, iv, 55.

[11] Wiedemann, 1920, 344.
[12] Petrie, Tools and Weapons, 36, 41.
[13] Petrie, Anct. Egypt, 1915, 20 ; Naukratis I, 1886, 39 ; Tanis I, 1885–6, 44 f.
[14] Petrie, Naukratis I, 39 ; Mallet, Les
rapports des grecs avec l'égypte, Mém. de
l'Inst. français d'archéol. orientale du Caire,
1922, xlviii, 71.
[15] Desch and Robertson, Nature, June
14, 1930 ; cf. Isis, 1931, xv, 437.
[16] Petrie, Six Temples at Thebes, 1896,
18 f.
[17] Petrie, Arts and Crafts, 98 f.
[18] Petrie, Ten Years' Digging in Egypt,
1893, 58.

is quite definite, since the tools found have been given an edge and can be permanently magnetised.[1]

The chief direct source of later iron in Egypt, from about 850 B.C., was probably Assyria, when the metal became common in Egypt.[2]

In the time of Herodotos (450 B.C.) [3] iron objects were used in preparing mummies. In Greek times iron ore was reduced in Egypt itself, since there are crucibles containing slag at Memphis, Daphnæ and Naukratis.[4] Scale armour found in the Palace of Apries (XXVI dyn.) at Memphis was perhaps of Persian steel.[5] There was a considerable quantity of iron, including large swords, daggers, scrapers, etc., in a graveyard of the time of Ptolemy II (283–247) at Alexandria, [6] also small objects of iron and bronze in temple foundation deposits of the early Ptolemaic Period.[7]

Iron anklets and bracelets, i.e. ornamental iron, occur in Roman-Nubian remains at Karanog, but only in the poor graves, also iron kohl-sticks, the top of a cylindrical box with an iron spring lock, several iron keys with complex wards, scissors, a chisel, an adze-head and arrow-heads of iron.[8] Iron tools were found with coins of the 2nd century A.D. in the temple of Ehnasya.[9] Iron, rare and valuable in the earliest days in Egypt, is still used for money in Cordofan[10]: thin sheet-iron is cut into the shape of rounded arrow-heads, about 3 in. long, called haschisch.[11] Iron jewellery was worn in the Coptic Period in Egypt.[12]

There seems to be no evidence of the use of cast-iron at an early period in Egypt or elsewhere, although both Gsell[13] and Feldhaus[14] think "drops of liquid iron" could be formed in the furnaces, and Feldhaus says the Harris Papyrus refers to a present by Rameses III (1165) to the temple of 6,784 statues of iron, "probably cast iron," but these were probably of hæmatite.

STEEL

It has repeatedly been asserted[15] that steel was known in Egypt in the Old Kingdom, that it was represented blue (green being merely weathered blue), iron being shown in black, and that traces of working found on very hard stones imply the use of steel chisels, iron being too soft. Maspero[16] considered that iron chisels, used one after the other and repointed by a blacksmith as they became blunt, were used for cutting the hard stone, but the *early* use of steel is not supported by the finds.[17] The *accidental* production of steely iron would be quite possible in the primitive process of iron-working still used in Africa and elsewhere.[18] Freise[19] stated that the production of steel is described by Agatharchides as carried out by heating iron with camel's

[1] Anon., in *Anct. Egypt*, 1927, 57.
[2] Ridgeway, Early Age of Greece, i, 594 f.; Cook, CAH, iii, 417; cf. Gsell, 47—1200 B.C.; Petrie, Arts and Crafts, 98 f.—not freely used till the Coptic Period; Hoernes, Urg., ii, 296 f.—rare in the Ptolemaic and even Roman Periods; Petrie, Ten Years' Digging, 152 f., used commonly only from about 800 B.C., and in actual use in Europe as soon as in Egypt; Montelius, L'Anthropol., iii, 451.
[3] ii, 86.
[4] Gsell, 15.
[5] Petrie, Palace of Apries, Memphis II, 1909, 13.
[6] Schweinfurth, Z. f. Ethnol., 1908, xl, 64.
[7] Petrie, Ten Years' Digging, 43, 46, 53, 58, 153; Gsell, 15; Montelius, Ält. Bronzez., 1900, 151.

[8] Woolley and MacIver, Karanog, i, 29, 66.
[9] Petrie, etc., Ehnasya 1904, 1905, 23, 42.
[10] Wilkinson, ii, 246—cf. Sparta.
[11] Rüppell, Reisen, 139; Russegger, Reise, II, ii, 155.
[12] BMGE, IV–VI Rooms, 330.
[13] Eisen, 47, 90 f.—c. 1200 B.C.
[14] Technik, 233.
[15] Wilkinson, i, 41; ii, 253; Lepsius, 57; Gsell, 21 f.; Garland and Bannister, 5, 85 f.; Orth, PW Suppl. iv, 111.
[16] Arch., 191, 195 f.
[17] Wiedemann, 1920, 345.
[18] Freise, Geschichte der Bergbau- und Hüttentechnik, 1908, 110 f., 121; Blümner, iv, 227.
[19] *Stahl und Eisen*, 1907, 1692, largely compiled from Beck's old Geschichte des Eisens.

dung, i.e. a kind of cementation process [1], but no such process is mentioned by Agatharchides, although it is possible by analogy with the fable of Wieland and other stories, and also the method of making Indian *wootz* steel.[2]

The production of steel from good bar iron is not a difficult operation. The bars when surrounded by charcoal powder, or even glowing charcoal, and heated for some time, take up carbon and are imperfectly converted into steel, which can be hammered at a red heat and hardened by quenching in cold water. The process is more rapid when nitrogenous organic matter is added to the charcoal. Rüppell [3], however, found that the negroes of Cordofan, who made good soft iron from inferior ore [4], did not know how to harden the metal, i.e. convert it into steel.

Garland [5] supposes that the Iron Age culture in Egypt may have preceded the Bronze Age and infers the *early* use of iron from the *large* pieces of it found in the Great Pyramid, at Abydos, etc., and from the fine work on diorite and very hard stone as early as the iv dyn., when bronze was unknown : these sculptures are very like those of the xviii dyn., when iron was in use. Iron articles in the early period may have been scarce and expensive, and few persons were skilled in making them. The sculpturing of granite, etc., could not have been done with copper chisels, and the very sharp edges of inscriptions on *vertical* granite (xii dyn.) disprove Petrie's theory of copper blades fed with emery, followed by breaking out the block between the cuts by hammering, and then hammer-dressing and grinding out the hole. Sarcophagi of blocks of granite 3 ft. high, 3 ft. wide and 6 ft. long were scooped out with well-dressed walls 6 in. thick. Garland tried to cut granite with a copper blade fed with emery and oil, but "no measurable progress could be made on the stone, whilst the edge of the copper blade was rapidly worn away and rendered useless, the bottom and sides of the groove being coated with particles of copper." The Predynastic Egyptian diorite vases, however, are scooped out of the hardest stone with excellent workmanship, and no iron was then available. It was only after Egyptian polished granite work was brought to the British Museum early in the 19th century by Belzoni that the possibility of producing similar objects from Aberdeen granite was thought feasible, and the mere objection that a process cannot be repeated by a modern workman or scientist is of no weight when the skill of the artisans of the ancient world is involved. Bottles carved from solid pieces of rock crystal, with narrow necks, and large vases of the hardest stone highly polished inside and out, are quite common museum objects which would be very difficult to make even with the latest appliances. No comparison of intelligence is, of course, instituted in these remarks. Abrasive powders, perhaps obsidian and emery grinders and cutters [6], and almost infinite patience were probably the agents employed.

The use of iron chisels in the iv dyn. is inferred by Garland from working in hard stone : copper chisels, even if hardened by hammering, would be useless, and if too much hardened they would be brittle. The making of iron presents no great difficulty as carried out in the Sudan at present, but even iron, it is admitted, would not have been hard enough for the chisels, so that steel is assumed. Garland's experiments and arguments are interesting and have been favourably commented upon by an Egyptologist [7], but they are not completely convincing and raise difficulties. An old theory that the stones were softer when taken from the quarry is disproved by the reworking of hard stone monuments by the old Egyptians after the lapse of centuries.

[1] Collier, *Mem. Manchester Lit. and Phil. Soc.*, 1798, v, 109 f.
[2] Gsell, 23, 87.
[3] Reisen, 1829, 158.
[4] Russegger, Reise, II, ii, 292, 295.
[5] Garland and Bannister, 5, 85 f., 92 f., 104 f., 111.
[6] Gsell, 23.
[7] Glanville, JEA, 1928, xiv, 190.

AFRICAN IRON

The relative ease of the production of copper and iron from the ores has been a favourite subject of discussion among archæologists, whose very definite yet completely contradictory conclusions have frequently, it is to be feared, been vitiated by their confusion of "melting" (fusion) with "smelting" (chemical reduction from the ore), and their failure to realise that the facility with which a metallurgical process is carried out depends almost entirely on the character of the ore. Thus, Gsell [1] has argued that copper is easier to reduce, even from sulphide ores, than iron, although the process is somewhat difficult. "In spite of its high melting point" (which is, of course, not involved at all), copper is more easily separated from oxide and sulphide ores than iron, and strongly sulphidic ores sometimes furnish metallic copper in appropriate circumstances. The purification of copper, it is claimed, like its mechanical working, is easier than in the case of iron. [2] The smelting of copper from a sulphidic ore, however, is really a process involving a series of stages, viz. roasting, smelting with suitable fluxes, and treatment of the resulting metal, whereas a spongy mass of iron which can be united into a compact mass by hammering, is at once obtained by merely heating a piece of a pure oxide ore in a charcoal fire. Although in most countries bronze appears in use somewhat earlier than iron, this depends on the local circumstances and in some cases iron may be in use before bronze. [3]

The problem receives some light from a consideration of the processes now in use by the natives of Africa for the production of iron. The native African tribes are often very skilled in working iron, producing decorated beads, spirals, bracelets, heavy neck-rings and thick and thin wires, which as the products of itinerant smiths who are "magicians" and members of outcast tribes, sometimes require special ceremonies of "purification" before they can be used. [4]

Iron early became known throughout Africa, but perhaps not to all tribes. The ore is found on the surface and merely requires putting on a fire to become reduced to iron. When this occurred, it was easy to look for ore in the ground. [5] The district west of the Upper Nile is rich in ore from which iron is extracted by processes far simpler than those used in the manufacture of bronze. [6] The oxide ore is reduced at 700°–800° and the malleable lump of iron only needs hammering, neither bellows nor artificial blast of any kind being necessary. The furnace was at first merely a cavity in the ground, or prolonged above it. The fuel was charcoal, piled in the furnace, sometimes above it, with alternate layers of ore. The fire could be urged by wind alone or by blowing appliances. The metal, never melted, was obtained as a solid mass, sometimes spongy, always malleable, occasionally (by chance and not under control) of a steely character. The prehistoric process is still practised in Africa and in a modified form in India, Borneo, Catalonia and Finland. In Japan the furnace is simply a V-shaped trough of clay, with holes near the bottom for the blast. [7] The African smiths who wander from place to place with the simplest tools and make very good products, form a particular class, often of different race from the tribe and usually held in great esteem. They are described as of "royal descent," "priests," "medicine men," etc., and African races which do

[1] Eisen, 55 f., 59, 67 f.
[2] Blümner, iv, 50.
[3] Lenormant, Prem. Civ., i, 88, 101; J. Percy, Iron and Steel, 1864, 873; de Launay, Daremberg-Saglio, ii, 1077.
[4] Stoll, Geschlechtsleben, 125, 395 f., 439 f., 448 f., 465; Gsell, 83; K. Faulmann,

Illustrierte Culturgeschichte für Leser aller Stände, Vienna, Pest, Leipzig, 1881, 114, Fig. 69.
[5] Hoernes, Urg., ii, 294.
[6] BMG Bronze Age, 4.
[7] Gowland, J. Anthropol. Inst., 1912, xlii, 277 f.; Wiedemann, 1920, 341.

not produce iron use bellows as fetishes. Yet in other parts of Africa to call a man a smith is punishable by death.[1]

The negro races of Africa entered into the Iron Age direct from the Stone Age. The use of iron moved from north-east to south-west and iron manufacture is most developed in the Nile region, where it is probably oldest in Africa [2], although Foy [3] thinks the Egyptian iron industry and the use of bellows came from certain Eastern races, among whom they had a unique origin. The XVIII dyn. representations [4] of double bellows, a furnace, crucible and heap of ore or fuel, supposed by Rosellini to represent iron-working, are now thought to depict the smelting of copper.[5]

The negroes may have learnt to work iron from the Egyptians, who employed them as slaves. Their present implements, weapons and bellows are like those of ancient Egypt.[6] This argument will obviously work backwards, and some [7] think iron was first known in South Africa : the Bacahapin, a Kaffir tribe, take iron (tsipi) as the fundamental name for "metal"; among them gold is "yellow iron," silver is "white iron" and copper is "red iron."[8] Lehmann-Haupt [9] suggested that iron then passed from Central Africa to the Cretan-Mycenæan culture, thence by emigration of Lycians and Philistines from Crete to Asia Minor, whence the Chalybes took it to Egypt and Assyria, but this hypothesis is very improbable.

The Periplus (1st century A.D.)[10] says iron for spear-heads and weapons was exported in the 1st century A.D. from Alexandria to Ethiopia, but this was probably steel for swords, since negro iron, which was probably well known, was too soft and copper was scarce.[11]

The Hottentots of the Cape have long smelted iron but their swords are soft and easily bent, like those of the ancient Gauls.[12] Mungo Park[13] found the African negroes most attracted by the iron of European traders and they reckoned values in terms of bars of iron. Procopius[14] says the Ethiopians possessed no iron, neither could they purchase it from the Romans, as the sale to the natives was forbidden on penalty of death. He is probably referring to steel weapons.

Large masses of iron are obtained without much difficulty, but good cutting tools and weapons are difficult to make. Even the Vikings, who forged large iron anchors successfully, could not make steel swords with good edges[15], and the negroes of Cordofan knew no method of hardening their soft pure iron and converting it into steel.[16]

The African natives use small furnaces constructed of clay, or even only pits[17], and bellows without valves, or no bellows at all. A lump of iron is formed which is worked up by the smith. This method is used in Asia by the Tartars and by the Carinthians in the Alps, and the same kind of bellows is used in Finland and Scandinavia.[18]

[1] Rougemont, 15 ; Hoernes, Urg., ii, 294 f. ; Lenormant, Prem. Civ., i, 92 ; Becker, *Islam*, iii, 261 f.

[2] Rougemont, 14f.; Hoernes, Urg.,ii, 293.

[3] *Chem. Z.*, 1908, 973.

[4] Maspero, Dawn, 311.

[5] Petrie, *Anct. Egypt*, 1915, 20.

[6] Hoernes, Urg., ii, 293.

[7] VonLuschan,*Z.f.Ethnol.*,1909,xli,22f.; Rougemont, 14; Schrader-Jevons, 154, 202.

[8] Schrader-Jevons, 154.

[9] *Z. f. Ethnol.*, 1909, xli, 55 f.

[10] Smith, Cassiterides, 1863, 8 ; W. H. Schoff, The Periplus of the Erythræan Sea, New York, 1912, 24, 79 (Berbera).

[11] Herodotos, iii, 23, vii, 69 ; Gsell, 20.

[12] Schrader-Jevons, 154, 202 ; Hoernes, Urzeit, iii, 108 f. ; Rougemont, 15.

[13] Travels, Everyman ed., 19.

[14] De Bello Persico, lib. i, cap. 19 ; ed. Bonn, 1833, i, 102 ; c. A.D. 550.

[15] *Nature*, 1927, cxix, 42.

[16] Rüppell, Reisen, 158.

[17] Gsell, 16 ; Hoernes, Urg., ii, 294 ; Beck, Geschichte des Eisens, i, 98 f. ; Friend, Iron in Antiquity, 1926, 190 f., Figs. 15 and 16.

[18] Rougemont, 15 ; Lippert, ii, 224 f. ; on primitive iron-smelting, see also Blümner, iv, 49, 69, 207 f., 211, 216; Jagnaux, Hist. de la Chimie, 1891, ii, 219 f. ; Freise, Bergbau, 94 f., 111 f. ; Feldhaus, Technik, 232, 367 f.; Gsell, 16, 19, 81 ; Schrader-Jevons, 160 ; Percy, Iron and Steel, 1864, 254, 275.

Von Luschan [1] describes with good sectional drawings and reproductions of actual photographs the iron furnaces of Africa. The iron is rarely fused and each charge is worked separately. Some furnaces are worked with bellows and some without. They are 3 to 3½ m. high, cylindrical with a slight taper at the top, and with blowing holes near the base and an opening on one side at the base. They are supported by a tree. Large numbers of furnaces (500 in one place) are operated. Charcoal and ore are put in from the top in layers from baskets, about 1·4 cu. m. of wood charcoal to 120 kg. of iron ore, some of the charcoal already ignited. The operation goes on two days, without the use of bellows. On the third day the lump of iron falls noisily to the bottom of the furnace ; it is taken out, and sand is thrown into the furnace. The product, a lump of 25–30 kg. weight of irregular shape and with adherent slag, contains only about 20 kg. of iron out of a possible 84 in the ore.

The smith breaks the lump between stones, takes out the slag and makes the pieces of iron into a cone about the size of the fist, with dry grass and wet clay. This cone is put in the fire, and when hot is hammered on the anvil. Only the requisite amount of iron for a particular job is worked up at a time. This process is used all over Togoland.

In Yoruba, near Oyo, men, women and children are engaged in mining and the metallurgy of iron. The ore, hæmatite, occurs 6–8 ft. underground. Lumps of ore are roasted over an open fire of green wood, the process lasting overnight. The roasted ore is stamped in wooden mortars, sieved, washed in running water until no more colour passes off, and put wet in the furnaces. The furnace house is 8–9 m. long, 5–6 m. wide and with walls of clay 4–6 ft. high. A roof of palm leaves is put over on posts 8 m. from the ground. In the house is the round furnace, 2–2½ m. diameter and 1·2 m. high. Burning charcoal is put in, then baskets of ore and charcoal alternately till the furnace is filled. Nine pairs of bellows, each delivering with an area of 2 sq. cm., the top of the furnace having an opening of about 20 sq. cm., are used, so that a strong blast is applied. Although it has been said that the furnace is filled 9–10 times without cooling this is incorrect, only one charge being worked at a time.

Just south of Victoria-Nyanza, in Usinja, a rough furnace is built on the ground from two lumps of clay, and a mixture of ore and charcoal is put in and blown with two to five double bellows till the ore is reduced. In other places a mere heap of ore and charcoal is used, but this can be done only with very easily reduced ores and hard work with many bellows. This primitive method is used by a few tribes near the Great Lakes. In the Congo, clay from ant-hills is used for the construction of furnaces.[2] Von Luschan[3] suggests an early origin (1500 B.C.) for the African iron industry, and a detailed discussion of the supposed primitive processes for the extraction of iron[4] contains a number of interesting technical points which, as they are not of specific historical interest, are passed over here.

In the production of iron by the negroes of Cordofan [5], smelting is carried out in a conical hole in sandy ground, 12–14 in. dia. and the same depth, filled with a mixture of broken ore and wood charcoal covered with a heap of kindled charcoal. The nozzle of the bellows is inserted at an angle of 40°–45°. They are dish bellows with a skin top having a hole in the centre in which the worker inserts a finger to close the hole on pressing out the air. The charcoal, from mimosa burnt in small heaps covered with sand, is of good quality but in small pieces. After about 10 hours' working the slags rich in iron are taken off and resmelted in the same furnace, but only a few hours' working is necessary

[1] Z. f. Ethnol., 1909, xli, 37 ; from Bellamy, J. African Soc., 1909, viii, 87, not seen.
[2] Staudinger, Z. f. Ethnol., 1909, xli, 102 f.
[3] Ib., 46 f., 104.
[4] Ib., 76–107.
[5] Russegger, Reise, II, ii, 290 f.; cf. ib., I, ii, 546.

in this case, and the iron is beaten to squeeze out slag. Only slags rich in iron and unreduced ore are obtained in the first operation. Large good pieces of metal are not often obtained, but Russegger bought one weighing 15 lb., of very good, completely soft iron. Three men work at once, two on the bellows and one superintending the furnace : 15–20 lb. of iron are obtained in 12–14 hours, and the smithying is done with the same furnaces but with the bellows nozzle at an angle of 25°–30°, and once Russegger saw double bellows used. In ordinary smelting the ores would give cold-short iron.

Mungo Park [1] also describes the iron furnaces and smelting in Africa. The production was sufficient to make iron an article of commerce from Kamalia. The furnace was a circular tower of clay, 10 ft. high and 3 ft. diameter, with seven blowing tubes of clay near the ground. The ore was dull red with grey specks, very heavy, and was broken into pieces the size of a hen's egg. Wood was first put in, then a large quantity of charcoal, then strata of ore and charcoal. The fire was blown with bellows of goats' skins.[2] The blowing was abated on the second day ; after the third day it was stopped and the furnace allowed some days to get cold. The lump of iron when broken had a granular fracture "like broken steel." It was then forged, but is "hard and brittle, and requires much labour before it can be made to answer the purpose."

The use of a limestone flux does not appear in any of these accounts and Russegger specifically says that only ore and charcoal were used.

[1] Travels, Everyman ed., 217.   Fuels, etc., 1861, 390, Fig. 107; *ib.*, Iron
[2] Also used in India; Percy, Metallurgy,   and Steel, 1864, 254, 275.

STONE; CERAMICS AND GLASS; PIGMENTS

STONE

The three predominating rocks in Egypt are granite, sandstone and limestone. The granite occurs in a low hilly region, about 50 miles broad from north to south, on the Nubian border and terminating at Aswân; granite rocks also occur on the shore of the Red Sea to near the mouth of the Gulf of Suez, reaching inland for about 30 miles. The ranges of hills on both sides of the Nile are predominantly sandstone, reaching from Aswân to Esna, where it becomes covered with limestone belonging to the upper chalk series. This continues north for 130 miles, when it is covered with tertiary nummulitic limestone.[1] Egyptian stones are classified by Petrie as (1) common: limestone, sandstone, red granite and quartzitic sandstone; and (2) less common: basalt, alabaster and diorite (used on pyramids). Limestone was probably the first stone used for building, since it is easily worked and occurs in various places, limestone cliffs fencing the Nile Valley for a distance of 400 miles. It was quarried particularly at Ayan or Troia, opposite Memphis, and was also used for constructing chambers and casing walls. A hard silicified limestone, found as nodules on the surface, was used as a tool for rough-hewing rock. The earliest stone-masonry in Egypt is the hewn limestone chamber of King Khasekhemui (end of II dyn.), surrounded by brick; the limestone and granite work of the pyramid of Khufu (IV dyn.) is very fine.[2]

Common identifications are [3]: áner-en-bâa = basalt (áner or ānu = stone in general); áner-en-benu = yellow sandstone; áner-en-bekhenu = porphyry; áner-en-ma (or mât) = granite; áner-en-rud = sandstone; áner kam = black granite; áner-hez = white limestone.

Sandstone, an impure form of silica containing oxide of iron, was used after limestone. It occurs in the Old Kingdom but is common only after the XVIII dyn., and came from quarries at Silsileh.[4] A quartzitic sandstone from Gebel Ahmar, near Cairo, was largely used in the XII dyn. and after. Flint, another variety of impure silica, was worked to perfection in the Prehistoric Period and continued in use till Roman days, all kinds of flint tools being found with metal in tombs: a XII dyn. wooden sickle still has the flint cutters in position. The stone is light brown, not dark in colour, and is really chert, since it is obtained from limestone and is harder than true flint.[5]

Much of the "red granite" described by Egyptologists is really a hard sandstone breccia [6], the quarries of which on the Kossayr road were worked in the

---

[1] Ritter, Erdkunde, Afrika[2], 1822, 697.
[2] Arts and Crafts of Ancient Egypt, chap. vii, 69–82; other sections on stone-working in Petrie, The Pyramids and Temples of Gizeh, 1883, 173 f.; Brugsch, Ägyptologie, 1891, 403 f.; Maspero, Dawn, 384, 413; Wilkinson, ii, 253 f., 301 f.; Breasted, H., 42, 93, 118; Erman-Ranke, 560 f.; Lucas, Ancient Egyptian Materials, 1926, 9 f. and 171 f.; ib., JEA, 1924, x, 131; ib., Egyptian Predynastic Stone Vessels, in JEA, 1930, xvi, 200; all these references apply also to the succeeding descriptions of stones.
[3] Budge Egy. Hierogl. Dict., 62; Erman-Grapow, Handwörterbuch, 1921, 61.
[4] Ritter, Afrika, 709 f.
[5] Birch, in Wilkinson, ii, 261; BMGE, IV–VI Rooms, 210, 279 f.; Wiedemann, 1920, 42.
[6] Maspero, Struggle, 311.

II dyn.[1] True red granite came from Aswân and was not much used after the I dyn. [2], although the quarries were worked in the Old Kingdom for pyramids, tombs, etc.[3] Masûdî says the natives called it aswanîah.[4] The so-called "grey granite" of the Egyptologists is a hornblende-biotite granite. Pliny's syenites lapis [5], quarried at Syene (Aswân), was different from modern syenite (in which quartz is usually absent) and was probably red granite, since he says it was called pyropœcilos. Granite from Elephantine, at the First Cataract, and also parallel to the Red Sea, was used for building and statuary from the earliest Dynastic Period.[6]

Smaller blocks of hard rocks came from the volcanic valleys separating the Nile from the Red Sea, especially the Wâdî Hammâmât. Such choice stone was found only in this desert region and expeditions of soldiers and workmen were organised to get it. The stone was split by means of wooden wedges driven into drilled holes, clefts or later mere pockets, then wetted, which caused the wood to swell and split the stone.[7]

Working in granite (e.g. for magnificent vases and vessels), breccia and other hard stone goes back to the Predynastic Period ; the amount of stone used decreased in the Dynastic Period, although the IV dyn. shows a full development of such work, after which softer stone (e.g. "alabaster") was more used for vessels.[8]

In stone-working in the Old Kingdom, copper chisels and wooden mallets were used.[9] The interior of a vase was cut out by a tubular copper drill, essentially a wheel with a vertical axis (the horizontal axis was, apparently, invented in Mesopotamia)[10], shown in a hieroglyphic and also represented in use, which was used with some abrasive powder such as emery. Removed cores have been found.[11] The actual operation of the drill is not very clearly described, but the borer was in use in the early Dynastic Period (c. 3000).[12]

The stone was also sawn with long copper blades carrying hard cutting points and abrasive or cutting material, sand for softer stones, emery for harder. Rozière found traces of oxidised copper from saws in cuts on stones. In general, the cutting material was used as a loose powder, but there is evidence of fixed stones, and even in prehistoric times emery blocks were used for grinding beads and later for hones.[13] An ancient Egyptian vase cut from emery itself is in the Ashmolean Museum.[14]

The emery which occurs in early remains must have been imported, since it is not found in Egypt. It probably came from the island of Naxos, from whence the Cretans obtained it for use in making stone bowls by a similar method to the Egyptians. Emery is mentioned as asmuri or smeri in a text of Thothmes III ; àsmer is perhaps emery powder, and the name is probably the origin of

[1] Wilkinson, i, 33.
[2] Petrie, Arts and Crafts, loc. cit.
[3] Erman-Ranke, 561.
[4] Reitemeyer, 97.
[5] xxxvi, 8.
[6] Breasted, H., 4, 6, 42, 49, 61, 93, 104, 135 f., 201, 266, 281, 442 ; Lucas, Egy. Mat., 10, 18, 182 f. ; cf. Ritter, Erdkunde, Afrika², 1822, 698 f.
[7] Weigall, Tutankhamen and other Essays, 1923, 153, 189, 204 ; Breasted, H., 93 ; Maspero, Dawn, 384 ; ib., Arch., 42 f.— diorite, basalt, "black granite," porphyry, breccia ; Erman-Ranke, 562 ; S. Clarke and R. Engelbach, Ancient Egyptian Masonry, 1930, 12 f. ; bricks, 207 f. ; Petrie ; Lucas.

[8] Erman-Ranke, 500 ; Petrie, opp. cit. ; Maspero, Dawn, 413 ; Budge, Mummy², 389 ; see plate xxvi, ib., 390 ; BMGE, IV–VI Rooms, 10 f.
[9] Erman-Ranke, 500.
[10] Mackenzie, Footprints, 143 f.
[11] Platt, PSBA, 1909, xxxi, 173 ; Erman-Ranke, 569, Figs. 237, 238 ; Petrie, Arts and Crafts, ch. vii ; ib., Temples and Pyramids of Gizeh, 173 f. ; ib., Ten Years' Digging, 26, Figs. 11, 12 ; Wiedemann, 1920, 349.
[12] Mackenzie, Footprints, 143, 145.
[13] Petrie, Arts and Crafts, ch. vii ; ib., Ten Years' Digging, 26 ; ib., Tools and Weapons, 44 f. ; Wilkinson, ii, 254 ; BMGE, IV–VI Rooms, 208.
[14] Platt, op. cit.

the Greek name σμύρις.[1]   In India hard stones are still drilled with a thorn fed with emery powder.[2]   Quartz crystals were drilled in Egypt, and very thin carved bowls of rock crystal (quartz) made in the I–II dyns. or earlier. Somewhat softer stones were scooped out by crescent-shaped flint drills.[3]

Diorite was used somewhat later than other stones in the Predynastic Period and remained a favourite till the Old Kingdom : there are figures of it of various types. It was used for Predynastic bowls, Old Kingdom drilled work, and statuary. According to Petrie, basalt-syenite, porphyry, "alabaster" and limestone were used in the earliest period ; somewhat later came slate, coloured limestone and serpentine ; and lastly diorite, obtained from the volcanic valleys between the Nile and the Red Sea, particularly Wâdî Hammâmât.[4]

Obsidian (black stone : àner chem) [5], used for amulets, small vases, etc., from the early Dynastic Period, is not found in Egypt itself but is plentiful in Abyssinia, Asia Minor and South Europe, e.g. in Melos in the Cyclades, its usual source in the Neolithic Age, e.g. for Crete. It may have come from Melos or from Abyssinia (Nubia)—where Pliny [6] says there were mines of it [7]— but Wainwright [8] thinks Egyptian obsidian came mainly from Armenia or Asia Minor generally.

A hieroglyphic list of stones on the Kennard tablet [9] has not yet been adequately studied, but an attempt to correct the nomenclature of Egyptian stones has been made by Lucas.[10] The Egyptologist's "alabaster," šs-t, used from Predynastic times[11], is really a variety of calcite, which in turn has erroneously been called "aragonite" ; all the specimens examined were calcite, white or yellowish-white and compact, translucent in thin sections, frequently but not always banded, fairly soft and capable of being scratched with a knife. It was used for vases, etc., from an early period.[12] "Alabaster" or "mica" (selenite ?) is prescribed in the Ebers Papyrus (1550 B.C.) in dust or meal as a constituent of cosmetics.[13] The earliest "alabaster" quarry was probably that opposite Dahshûr. There was an ancient quarry (from VI dyn.) at Hâtnûb, 10–15 miles east of Tell el-Amarna, and others from Minia to Assiût.[14]

Basalt, the "black granite" of the Egyptologists, used in the IV and XII dyns., although not plentiful in Egypt is found in small amounts in several localities, e.g. Khânqa. It was largely used in the Old Kingdom, e.g. for the pavement of the temple of the Great Pyramid at Gîzah (IV dyn.), but not much afterwards.[15]

Dolerite, in which the separate minerals of basalt are coarse-grained, is frequently called "basalt" by Egyptologists. A Predynastic diorite vase (s.D. 57–58) was found at Qau.[16] Other statuary stones[17] were breccias, diorite, granite, obsidian, porphyry (found only once before the Roman Period, in a fluted early Dynastic bowl), quartzite schist (the Egyptologist's "green

[1] Lucas, Egy. Mat., 209 ; Mackenzie, Footprints, 138 ; Maspero, Struggle, 203 ; Budge, Egy. Hierogl. Dict., 89, 602 ; Lenormant, Prem. Civ., ii, 425, thought σμίρις was from the Phœnician schamir.
[2] Petrie, in Brunton and Thompson, Badarian Civ., 1928, 27.
[3] Petrie, Tools and Weapons, 45 ; ib., Tarkhan I and Memphis V, 1913, 15, 18 ; ib., Arts and Crafts, ch. vii ; Breasted, H., 39.
[4] Petrie, Arts and Crafts, ch. vii ; Breasted, H., 28, 39, 93 ; Maspero, Arch., 251 f. ; ib., Dawn, 384 ; Lucas, Egy. Mat., 179 f.
[5] Brugsch, Ägyptologie, 404.
[6] xxxvi, 26.
[7] Lucas, Egy. Mat., 187 ; Köster,

Beiheft. Alt. Or., i, 26 ; Mackenzie, Footprints, 138.
[8] Ancient Egypt, 1927, 77 f. ; JEA, 1928, xiv, 273.
[9] Ancient Egypt, 1928, 57.
[10] Egy. Mat., 172 f. ; JEA, 1930, xvi, 200.
[11] Brugsch, Ägyptologie, 1891, 494.
[12] BMGE, IV–VI Rooms, 16.
[13] Joachim, 90, 92, 114, 155, 157, 158.
[14] Lucas, Egy. Mat., 174 ; Erman-Ranke, 560 ; Breasted, H., 93, 119, 159 ; Donne, DG, i, 81 ; Maspero, Dawn, 384.
[15] Petrie, Arts and Crafts, ch. vii ; Lucas, Egy. Mat., 175 ; ib., JEA, 1930, xvi, 203 f.
[16] Lucas, Egy. Mat., 176 ; Brunton and Thompson, Badarian Civ., 1928, 58.
[17] Maspero, Struggle, 311 ; Lucas, Egy. Mat., 176–192.

basalt"), serpentine and steatite (sometimes used for beads and glazed, and much used for scarabs). Pliny's[1] marble used in Egypt was probably often the "hard white limestone which takes a polish like marble," found on the sites of ruined temples with red granite [2], although marble was used for statuary.[3]

Stones used for Predynastic vessels, including Badarian and Neolithic, are listed below.[4] The columns give (1) the name quoted by Egyptologists; (2) the geological name, or particulars of the stone when the two names are the same; (3) the places where the stone occurs; and (4) the probable ancient source, the greater part of the stone (82·5 per cent.) being from the Fayyûm, Nile Valley and Aswân, and only 17·5 per cent. from the Eastern Desert, so that the stone industry was probably not in the Eastern Desert, as usually assumed, but in the Nile Valley.

I	II	III	IV
Alabaster.	Calcite.	Nile Valley cliffs; Cairo-Suez desert; Sinai.	Nile Valley cliffs (including adjacent plateaux and low hills).
Basalt.	Includes fine-grained dolerite.	Nr. Cairo; Cairo–Suez desert; Fayyûm; Aswân; Baharia oasis; E. desert; Sinai.	Fayyûm.
Breccia.	Chiefly red and white.	Nile Valley cliffs; E. desert.	Nile Valley cliffs.
Diorite.	Speckled, black and white.	Aswân; E. desert; Sinai.	Aswân.
Granite.	Red, black and white, and syenite.	Aswân, E. desert; Sinai; W. desert.	Aswân.
Limestone.	Amorphous, variously coloured (incl. pink and black).	Nile Valley cliffs.	Nile Valley cliffs.
Marble.	Includes all crystalline limestone except calcite.	E. desert.	E. desert.
Porphyritic rock.	Includes porphyritic diorite; usually white crystals in black matrix.	Aswân; E. desert; Sinai.	E. desert.
Schist.	Includes various metamorphic rocks, such as tuff (volcanic ash), mudstone (Petrie's "durite"?) and slate.	E. desert; Sinai.	E. desert.
Serpentine.	Incl. steatite.	E. desert.	E. desert.

There is a very fine collection of Egyptian statuary, from the III dyn. to the Roman Period, in the British Museum. The white limestone was used in all periods. In the Old Kingdom hard stones of fine texture, usually black, were used for small figures and light yellow "alabaster" was also much used. In the XII dyn. red and white quartzite and green felspar became common. In the New Kingdom black and red granite were in fashion, whilst in the XXVI dyn. black basalt, which is very suitable for delicate sculpturing of hieroglyphics, was much used.[5]

[1] xxxvi, 6.
[2] Wilkinson, i, 63.
[3] Lucas, Egy. Mat., 186 f.
[4] Lucas, JEA, 1930, xvi, 200 f., 211.
[5] BMGE, IV–VI Rooms, 118 f.

The actual mason's work is remarkable : even in the Old Kingdom, granite blocks of 50–60 tons were transported and worked, and in the pyramids (III dyn.) huge blocks were worked with "optical accuracy" to 1/10,000 in. in the joints. Enormous masses of stone were cut and polished, and hieroglyphics cut down to 2 in. or more even in the hardest stone.[1]

PLASTER AND STUCCO

In the masonry of pyramids, gypsum is used to fill joints as a bedding and to level up hollows in a face. The plaster is a mixture of ordinary lime and plaster of Paris (calcined gypsum). How it was introduced between the blocks is a mystery, since these weigh at the base 16 tons each and the stone could not be slid, yet a vertical joint five feet high and seven feet long is filled with plaster only $\frac{1}{50}$ in. thick. In rock tombs, plaster is used to fill cracks and often remains where the stone has decayed away. The gypsum of good quality used during the Old Kingdom probably came from the Fayyûm.[2] Analyses by Wallace of specimens of plaster from the exterior and interior of the pyramid of Cheops showed that they contained about 82 per cent. of calcium sulphate and about 10 per cent. of calcium carbonate, so that they were made from gypsum.[3]

Plaster was also used on brick walls, which were faced with a hard coat, $\frac{1}{16}$ to $\frac{1}{10}$ in. thick, on which paintings were executed. By the XVIII dyn. this facing became a mere white-wash over the mud-facing. A strange use of stucco was for covering a sculpture as a basis for painting, or simply left white. Although the fineness of sculpture remains, the stucco finish becomes more and more coarse, until in Ptolemaic times a smooth daub of plaster serves as a basis for the painters. Greek and Etruscan sculptures were also painted. Stucco, very extensively used, was laid for modelling on flat canvas stretched over wood, the whole relief being in the stucco.[4]

Plaster was also used for casting in moulds and making moulds : the latter are very common at Memphis and were perhaps not used for casting bronze, since plaster is reduced to powder at 260° C., but for pewter. Painted plaster statuettes were oiled and still remain waterproof. Plaster death masks were made and there is one of King Tetâ (v dyn.), but plaster casts are not common.[5]

Mortar was applied by hand, without trowels. Three kinds were used : (a) a white, friable variety, of lime (gypsum ?) only, mixed in troughs ; (b) a grey kind, rough to the touch, of lime and sand ; (c) a reddish kind, containing powdered brick.[6] "Lime" plaster was used for flooring.[7] A mortar used for setting the copper drain at Abûsîr (v dyn.) had the composition : $CaSO_4$. $2H_2O$ (gypsum) 45·54 ; $CaCO_3$ 41·36 ; silica, oxide of iron, water, etc., 13·10 per cent. Whether the calcium carbonate ($CaCO_3$) was an impurity or due to the change of lime intentionally added is uncertain.[8] Lucas[9] regards it as always an accidental impurity.

Lucas's statement[10] that the Egyptians did not burn lime till the process was introduced by the Romans from Europe, is criticised[11] on the ground that burnt lime was used for frescoes at Knossos during the period of greatest

[1] Breasted, H., 93, 117 f., 159, 199 f., 266, 281, 343, 450 f. ; Wilkinson, i, 50 ; ii, 138 ; Maspero, Arch., 247 ; Borchardt, Sahurā, 1910, i, 52.

[2] Lucas, JEA, 1924, x, 128 ; 1930, xvi, 204 ; Petrie, Arts and Crafts, ch. xiv, 142 f.

[3] Jagnaux, Histoire de la Chimie, 1891, ii, 152.

[4] Petrie, op. cit. ; Wilkinson, ii, 285 ; Wiedemann, 1920, 165, 175 f., Fig. 33.

[5] Petrie, op. cit. ; Wiedemann, 1920, 366.

[6] Petrie, Tools and Weapons, 41 ; Maspero, Arch., 1902, 50.

[7] Wilkinson, i, 356.

[8] Borchardt, Sahurā, i, 80.

[9] Egy. Mat., 25.

[10] Egy. Mat., 20 ; Wiedemann, 1920, 165.

[11] Glanville, JEA, 1928, xiv, 189.

contact between Crete and Egypt, and that the painted pavements of Amenhotep III at Medinet Habu and of Ikhnaton at Amarna show to some extent
the Cretan technique of true fresco.

Wilkinson [1] says the floors of houses were sometimes of stone, "or a composition of lime and other materials" (stucco ?), whilst roofs were of wood ;
Wiedemann [2], that the floor was usually of clay or painted stucco, and of stone
only in temples.

BRICKS

The Fayyûm and Delta abound in rich alluvial soil and in the earliest
period were sites of brickmaking. The Predynastic burial pits at Abydos
have brick linings, frequently also a second lining of wood, roofed with heavy
timbers.[3] The Egyptian name for brick, tab-t, is continued in the Coptic
tôbé.[4] Although the Nile mud, separated by the river current from the sand
and containing 48 per cent. of clay and 18 per cent. of calcium carbonate, can
be made into bricks and terra cotta [5], it is not very suitable for making bricks
of large size, hence the Egyptian bricks are relatively small and it was necessary
to mix chopped straw (teben), reeds, hair, etc., with the mud to bind it together.
Large, well-shaped bricks, such as were made by the Babylonians and
Assyrians, could not be made in Egypt, and the want of fuel prevented burning
bricks on a large scale. It is possible that the art of brickmaking was introduced in the late Predynastic Period by the western Asiatic element of the
population.[6] In making bricks [7] the clay is shown being mixed by means of
hoes with water carried in jars from a tank, kneaded, transferred to pans and
thrown down in a heap before the brickmaker, who stamps the bricks in a
wooden mould and lays them in single rows to dry in the sun. Slave or forced
labour was employed, and the stick is shown. Egyptian brickmakers and
the method of making bricks are described by Aristophanes (444–380) [8],
whose account agrees with the Egyptian representations.

The clay was often mixed with chopped straw and a little sand or, in later
periods, with broken pots. The moulds were of wood, a XII dyn. specimen
being found at Lahun with plaster spreaders, and the size varied from 22 cm. ×
11 cm. × 14 cm. to 38 cm. × 18 cm. × 14 cm.[9]

Birch[10] describes baked bricks, externally rose-red but breaking with a
deep black fracture at about $\frac{1}{8}$ in. from the surface, and Wilkinson[11] states
that Rosellini found a wall 15 ft. thick of burnt bricks at Luxor, "older than
the XVIII dyn." Although Maspero[12] says burnt bricks were first used in
Egypt in the Roman Period, burning bricks was known but seldom used, and
the only extensive use of them is in the wall built in 1000 B.C. around modern
el-Hîbe in Central Egypt.[13]

Since the seal of the king or some privileged person is stamped on bricks,
their manufacture was perhaps a state monopoly, those without stamp
being made by private individuals under licence. Those largely made by slave
labour at Thebes could be sold cheaply by the government.[14] Sometimes
bricks are stamped with the name of the temple for which they were to be
used, and brick stamps are known.[15]

[1] i, 356.
[2] 1920, 174.
[3] Breasted, H., 41.
[4] Budge, Egy. Hierogl. Dict., 819.
[5] Birch, History of Ancient Pottery,
1858, i, 12.
[6] BMGE, IV–VI Rooms, 199.
[7] Birch, i, 18 ; Wilkinson, i, 347 ;
Olmstead, H. Pal. Syr., 143.
[8] Birds, 1132 ; Moyle, DA, ii, 8.
[9] Maspero, Arch., 3, Fig. 1, tomb of

Rekhmara, XVIII dyn. ; Wilkinson, i, 36,
342 f., Fig. 112 ; Erman-Ranke, 507,
Fig. 201 ; Wiedemann, 1920, 334 ; Petrie,
Ten Years' Digging, 118, Fig. 85.
[10] Anct. Pottery, i, 11 f., Fig. 4 on p. 22.
[11] ii, 297.
[12] Arch., 8.
[13] Wiedemann, 1920, 35.
[14] Wilkinson, i, 342.
[15] Wiedemann, 1920, 335 ; Birch, i,
15 f.

Tiles were used from the oldest period to the Hellenistic Age for covering walls of rooms, grave chambers and temples, but for roofs first in the Roman Period. They were usually pressed into a coating of stucco on walls, but sometimes had "ears" at the back through which metal wires were passed to secure them. The tiles were often coloured in various shades, but these were not of clay but of glazed frit.[1]

The unburnt bricks used with stone in the mastabahs at Saqqâra are of two kinds, the earliest yellowish and the later, from the second half of the iv dyn., black. The bricks of the pyramids (xii–xviii dyns.) are rectangular, dark coloured, sun-dried and held together with chopped straw or broken pottery. Those at Saqqâra had only a little straw on the outside. All bricks in the same pyramid are of the same size, but the size varies somewhat in different pyramids.[2]

POTTERY

Egyptian pottery, of potter's clay (ast) [3] moulded when moist and then baked, differs entirely from the glazed faience on a porous base which is not clay, in that it is never glazed.[4] Predynastic and early Dynastic pottery is hand-made, even the largest vessels, but so skilfully that it is often difficult to believe that the wheel was not used.[5] The invention of the wheel has been ascribed to all the ancient nations, and it was probably used in Elam before Egypt.[6] The wheel during the Old and Middle Kingdoms was turned by hand.[7] In Egypt the art of pottery was attributed to the gods Chnum and Ptah.[8]

Making pottery is not a "spontaneous" invention, since the Australian aborigines apparently never practised it.[9] It probably had not a religious origin : "terra cotta" sarcophagi are rare but not unknown in Egypt,[10] and it never became a fine art, as in Greece. Artistic pottery was nearly always imported : Mycenæan and Syrian in the New Kingdom, later Greek.[11] Earthen pots were commoner than bronze, and the quality varied considerably, the clay being sometimes mixed with chopped straw and imperfectly fired.[12] The clay of Coptos was especially suitable, and a large variety of forms was made, indicating a high level of civilisation : at least 1,000 forms appear in prehistoric times and at least 3,000 later. Forms of the pot (åpd, or ḥent)[13] for all periods include : tall and narrow libation jars with spouts, like coffeepots ; deep dishes ; conical jars for oils and drugs ; small pots for drugs, ointments, etc. ; jugs with spouts ; open jars for honey, etc. ; crucibleshaped pots with lids for ibis-mummies ; amphoræ, which Herodotos[14] says came full of wine from Syria (Askalon jars) ; small bottles, some of the nature of crucibles, which have little spouts to pour off any liquid they contained, etc.[15]

[1] Wiedemann, 1920, 176 f., 335.
[2] Budge, Mummy[2], 398 ; Birch, i, 12 f. ; Figs. 1 and 2 ; table on p. 15.
[3] Budge, Egy. Hierogl. Dict., 10.
[4] Petrie, Arts and Crafts, ch. xi, 126–133 ; ib., 107 f., thin glaze on red ware ? ; Wilkinson, ii, 11, 150, 190 f. ; Birch, i, 9 f., 48, 54 f., 61, perhaps rubbed with resin ? ; Brongniart, Traité des Arts Céramiques, 2 vols., 1854, i, 504 ; Franchet, Céramique Primitive, 1911 ; Harrison, Pots and Pans, 1928, 51 ; Lucas, Egy. Materials, 58.
[5] Budge, Osiris, ii, 245 ; Wiedemann, 1920, 331.
[6] Birch, i, 5 ; Mackenzie, Footprints, 144.
[7] Wilkinson, ii, 192, Fig. 397—1900 B.C. ;

Erman-Ranke, 545—Wilkinson's a "poor figure" ; Harrison, Pots, 36 f., 38.
[8] Wiedemann, Rel., 71 f. ; Budge, Mummy[2], 359 ; ib., Gods of Egy., i, 500 f., 508 ; Birch, i, 10 ; Reitzenstein, Poimandres, Leipzig, 1904, 140 f.
[9] Harrison, 23.
[10] Birch, i, 24.
[11] Wiedemann, 1920, 330 f. ; Petrie, Illahun, 1891, 9, 11 ; Evans, Palace, i, 266 ; Quibell, 62.
[12] Maspero, Dawn, 320 ; Arch., 253 f.
[13] Budge, Egy. Hierogl. Dict., 42, 486.
[14] iii, 6.
[15] Birch, i, 33 f., 42 ; Wilkinson, ii, 191, 194 ; Harrison, 8 ; Petrie, Arts and Crafts, 126 f.

The Predynastic pottery is of two main types : (1) with a polished red surface, the mouth of the pot (rarely the whole pot) being black ; and (2) (made by women) with a porous, hard, reddish-buff body with white specks, with brown ornamentation of plants, animals, men, ships, etc. Type 1, with affinities to pottery in modern Nubia, in ancient Cyprus and in Iron Age graves in India, was perhaps older.[1]   The Badarian pottery is very distinctive, but has affinities with Nubian. It was made by hand from fine-grained close clay mixed with chopped straw for large cooking pots, and often has thin edges. It was burnt red and black by a process which indicates long previous experience, and was "rippled" with a comb, the incisions sometimes being filled with white. A coloured slip was applied and there are signs of burnishing. Cracked pots were carefully repaired by boring and lacing.[2]

Analyses of Egyptian pottery give[3] :

	SiO_2	Al_2O_3	$Fe_2O_3 + Mn_2O_3$	CaO	MgO	H_2O	CO_2	C
Red	56·13	18·54	9·00	5·24	1·07	5·56	4·46	0·00
Grey	52·18	15·50	2·00	23·64	0·00	1·05	5·63	tr.

The Ptolemaic pottery sometimes contained grains of sand or limestone, but was sometimes more homogeneous. The old unglazed ware was very soft and absorbent, but did not let through water. Red pottery rings as stands for water jars occur in the early dynasties.[4]

In the Pyramid Age a smooth soft brown body is usual : the bowls of this ware with a highly polished red face, of the IV dyn., are "all but equal to the best Roman." A hard drab pottery, sometimes with a half-fused surface, also appears in the V–VI dyns. In the XII dynasty the common soft brown body is general, and extends to the XVIII, by the middle of which a hard drab ware with white specks, faced with drab polish, is characteristic and continues to the XIX. Thence the brown reasserts itself, with some inferior greenish drab ware about the XXII dyn. Greek clays found during the XXVI dyn. are probably all imported, and there is soft red pottery in the Ptolemaic Period, with soft brown again in Roman times. Thin hard ware of the Constantine Period is not native, but may be due either to Nubian or Roman influence.[5]

The decoration was : (a) white slip in line patterns, copied from basket work, put on a bright red facing of "hæmatite" (the Egyptologist's name for red ochre, a clay rich in iron oxide) in the earliest Predynastic pottery, a style which persists in modern times in Algeria ; (b) black-topped ware ; (c) later Predynastic painting in dull red on a buff body ; (d) a polished red "hæmatite" facing, in the Old Kingdom only, disappearing in the XII dyn. ; (e) about the XVII dyn. a fine red polish, ceasing in the early XVIII dyn., also white on the brims, or dabbed on saucers ; (f) black or red edges next appeared and by Thothmes III there is a style of narrow black and red stripes alternating ; (g) blue paint, of copper frit, which began under Amenhotep II but was not usual till Amenhotep III, when it was common till the end of the XIX dyn. though deteriorating ; (h) after this there was no decoration till Roman times.[6]

It is doubtful if sun-dried unfired pottery was ever made. Unbaked bricks resist water to a great extent owing to their protection of one another, but pottery would not and a light firing is probable in all cases except for funerary objects, e.g. small bottles painted blue and red.[7]   A temperature of 400–500°

[1] Wiedemann, 1920, 44 f. ; Harrison, Pots, 48 f. ; Petrie, Arts and Crafts, ch. xi.

[2] Brunton and Thompson, Badarian Civ., 20 f., 25 ; cf. Wiedemann, 1920, 3, 331.

[3] Brongniart, i, 387, 502 ; Birch, i, 22, 50.

[4] Petrie, Sinai, 145.

[5] Petrie, Arts and Crafts, ch. xi ; ib., Medum, 35.

[6] Petrie, op. cit. ; Lucas, Egy. Mat., 55 ; BMGE, IV–VI Rooms, 240, 245 f., 257 f.

[7] Franchet, 12, 14, 15 ; Birch 12, 21.

would suffice, and most primitive pottery was fired at 500–700°, but for thoroughly fired ware 900° is necessary.[1] Records of unfired ware are mis-interpretations of imperfectly baked pots, softened by the action of moisture. Hardening of clay by fire may have been noticed before pottery was made, by the accidental burning of clay-plastered wicker screens or baskets, a discovery perhaps made about 5000 B.C. somewhere in a large area in the Near East.[2] The purity of the clay varied considerably; the sand in old pottery was probably mostly an impurity, although some may have been added intentionally, and powdered wood charcoal was added to many kinds of prehistoric pottery.[3]

BLACK-TOPPED POTTERY

The earliest prehistoric pots of soft body, faced with red "hæmatite," were usually baked upside down with the brim covered with fuel, the process giving a black rim and interior, partly due to smoke and partly to reduction of red ferric oxide, Fe_2O_3, in the clay to black ferrosoferric oxide, Fe_3O_4. Open dishes were also faced with "hæmatite" inside and this has been reduced to a brilliant mirror-like surface.[4] According to Petrie's "carbonyl theory"[5]: "the reason of the polish being smoother on the black than on the red is that carbonyl gas [carbon monoxide, CO], which is the result of imperfect combustion, is a solvent of magnetic oxide of iron [really metallic iron] and so dissolves and recomposes the surface facing." This theory is chemically faulty, and what Petrie calls the "old idea" that the black colour is due to carbon or a lower oxide of iron, is almost certainly correct. Except in a pot of the Ptolemaic Period, the black material has always been found by analysis to be carbon.[6]

Franchet[7] distinguished between pottery (1) blackened only on the surface by smoke; (2) blackened throughout by smoke; and (3) black pottery to which carbon has been added. The smoked pottery (poteries fumigées) is obtained by baking in an excessively reducing atmosphere, and it acquires by friction a kind of lustre. The types blackened throughout are considered to have been fired out of all contact with air in a kind of meiler, the pots being mixed with fuel. Such a process is still in use among modern gipsies, some primitive races in East Africa, viz. the Baganda, and in South America, and the Etruscan ware called buchero nero was probably baked in the manner described.[8] Pottery blackened superficially shows a core of grey or brown, and the black has penetrated only slightly; this type is represented by the Gallo-Roman Period and in modern Peru. The ware was fired at about 800°, and in the later stages was exposed to a very smoky atmosphere for 3 or 4 hours.[9] Graphite is used by the modern Bakitara of East Africa, mixed with water and the glutinous juice of a shrub, or with butter and blood.[10] In some interesting experiments made by an American potter, Mercer, pots rubbed with powdered hæmatite and burnished half-dry with bottle glass, were placed in a ring of stones and covered with chopped straw, corn-cobs, fine sticks, soft coal, coarse and fine sawdust, and chopped corn stalks, which were lighted and allowed to burn out. A variegated grey, black, reddish or buff colour was produced with an uncontrolled fire, the polish being retained. A lustrous black unglazed pot was produced with or without hæmatite by polishing the vessel by rubbing, and smothering it when red-hot with coarse sawdust, dry oats or

[1] Harrison, 13 f., 19 f.; Burton, *J. Roy. Soc. Arts*, 1929–30, lxxviii, 283.

[2] Birch, i, 2; Harrison, 18 f., 24 f.

[3] Wiedemann, 1920, 333; Franchet, 16, 29 f., 43, 76, 86.

[4] BMGE, IV–VI Rooms, 240.

[5] *Academy*, 1895, xlviii, 16; Arts and Crafts, ch. xi.

[6] Franchet, 85 f.; MacIver, *Man*, 1921, xxi, art. 51; Lucas, Egy. Mat., 56 f.

[7] Céram. Prim., 85 f.

[8] Bucher, iii, 404; Harrison, Pots, 50; Franchet, 89.

[9] Franchet, 89 f.

[10] Harrison, Pots, 51.

chopped corn-cobs. A well defined red with a sharply separated black top, never achieved except in ancient Egypt or at one time in Cyprus under Egyptian influence, was imitated by burying the mouth of the pot, rubbed with hæmatite and water, in fine sawdust in the centre of which, just under the vessel, was placed a piece of resin the size of a chestnut. Over the whole was an arch of wire netting, and over this a fire of dry rye-straw, which burned for 45 minutes. Two kinds of heat, smokeless and smoky, are required, and when the pot becomes red-hot the sawdust, igniting last and smothering its own flame, ends the baking in a smoky heat. The Egyptians may have used dry dung, reeds, pith or fibre for the upper fire, or have covered the pot with charcoal. MacIver thinks the black Etruscan ware was made in a similar way, although wax, resin or oil may have been applied to the pot taken hot from the fire.[1] In modern Nubian pottery, the body is hand, not wheel, formed from Nile mud and pounded potsherds, and the surface is coated with a paste of ochre, water and a little olive oil.[2]

In modern village pottery-making in Egypt [3], the clay is mixed with chopped straw or chaff, and pots are made either with or without the wheel. Dried maize-stalks (bûs) and dry cow-dung are used as fuel, the latter being put on the top of the pots in the kiln. The British Museum has XII dyn. Nubian pottery, Ethiopian pottery, and pottery from Meroë of the Roman and later periods.[4]

In making pottery in the XII dyn., as shown at Beni Hasan, the clay, represented grey, was kneaded with the feet, rolled out to prepare a lump, and put on the wheel, a flat circular or hexagonal table on a stand, revolving on a pin and turned with the left hand while the pottery was shaped with the right. The potter sat on the ground or on a low stool. Cups and other vessels were hollowed out with the thumb or finger, and the vase was fashioned externally with the hands. The pottery furnace shown is probably the oldest representation. It was probably about 1·25 m. high and 0·35 m. at the base, the fire was placed halfway up and air was admitted through a grating beneath.[5]

Baked clay (terra-cotta) cones found in large numbers in the XII–XXVI dyns. (mostly XVIII–XIX) tombs of Western Thebes, 6 to 10 in. in length, and 2 to 4 in. in diameter at the larger end, have the base (sometimes coloured red and sometimes white) stamped with the name of the person in whose tomb they were found. Their purpose is doubtful, but they may be models of loaves.[6] Modelling of figures in pottery occurs in all periods as subsidiary to sculpture.[7]

True porcelain was unknown everywhere before its discovery in China in the 6th–7th centuries A.D., and the Chinese vases said to have been found in tombs at Thebes [8] are probably of the early 19th century [9]; others found in the Wâdî Hammâmât, formerly thought to be of the XVIII dyn., were brought by Arab traders in the 15th–16th centuries.[10]

GLAZE AND GLAZED WARE

Egyptian glazed ware (teḥen or ḥm-t ?)[11] is neither porcelain nor glazed pottery, since the body was a white or grey, sandy, friable mass, its particles being hard but having little or no cohesion. Clay, if used at all, was only just

[1] MacIver, op. cit.
[2] Harrison, Pots, 48.
[3] Blackman, The fellāhīn of Upper Egypt, 1927, 135 f., 148.
[4] BMGE, IV–VI Rooms, 262.
[5] Birch, Pottery, i, 45 f., Fig. 16; Erman-Ranke, 345, Fig. 230; Newberry, Beni Hasan, i, plate i; Franchet, 129, Fig. 32; Bucher, iii, 412, Fig. 374; Partington, Everyday Chemistry, 19, Fig. 21.
[6] Budge, Mummy², 394 f.; BMGE,

IV–VI Rooms, 139; Davies and Gardiner, Tomb of Amenemhet, 1915, 2.
[7] Petrie, Arts and Crafts, ch. xi.
[8] J. F. Davis, The Chinese, 2 vols., 1836, ii, 261; Wilkinson, ii, 153.
[9] F. Hirth, Chinesische Studien, Munich and Leipzig, 1890, 46 f.
[10] BMGE, IV–VI Rooms, 270.
[11] Budge, Dict., 1920, 820, 842, 887; Erman-Grapow, Handwörterbuch, 1921, 109.

sufficient to hold the sand together[1] ; a small quantity of soda seems to have been added to effect the glazing. The specific gravity of the body is 2·613 and the material is not fusible even at a white heat. Egyptian clays would not combine with a siliceous glaze and lead glaze was not then known. The glaze was composed of silica, probably finely ground sand, and native sodium carbonate (natron), with certain metallic oxides to produce the colour, e.g. oxide of copper for the blue.[2] The body of the ware is sometimes called steatite, or agalmatolite, or steaschist [3], resembling soapstone, but this is a natural product. Brunton [4] describes beads of "white glazed steatite ?" found in a Badarian grave. The glazed ware is also (inaccurately) called "Egyptian porcelain," or "faience."[5]

The oldest glazed *clay* objects in Egypt are the yellowish-white, mostly six-sided, seal cylinders, perhaps derived from Mesopotamia.[6] After these come glazed scarabs, used as seals and amulets[7] ; in the IV dyn. both were in use. The clay form was dipped in a powdered glaze containing soda and fired until the latter ran into a liquid. Brunton [8] found white glaze only on I dyn. beads ; dark blue, green and black glazes occur in the Protodynastic Period and a characteristic pale blue only in the III dyn. De Morgan [9] thought Predynastic blue glazed beads were made from China clay (kaolin) which occurs in pockets in the Aswân granite, and Petrie[10] also says the prehistoric beads were blue or green glazed pottery. Attempts to glaze stone were rare, but Wilkinson[11] mentions granite sarcophagi covered with transparent green glaze, and the British Museum has a piece of crystalline limestone glazed blue.[12] Glazing began far back in prehistoric times, perhaps in the Delta.[13] The greenish-blue glaze coloured with copper is the oldest, and may have been obtained accidentally while smelting copper in the sand,[14] or by fluxing quartz pebbles with copper ore and wood-ashes in a fire, since blue or green glaze is found on quartz as early as on pottery.[15] A model boat is made from sections of quartz rock bound together with gold after glazing, and a large sphinx is of glazed quartz, 18 in. long. Fusion of the glaze corrodes the surface of quartz, and after the glaze has worn away the surface has the appearance of water-worn marble.

Glazing continued in Dynastic times, e.g. XII dyn. beads with rich blue glaze and large glazed blocks in the XVIII dyn. The oldest objects have a very thin glaze almost without lustre and perceived only in depressions where it has run together.[16] There is a blue glazed frit cylinder seal of the

[1] Lucas, in Carter, Tomb of Tutankhamen, ii, 169, found no clay at all.

[2] Boudet, Notice Historique de l'Art de la Verrerie né en Égypte ; in Description de l'Égypte, Antiquités, 1818, ii, 17 f., 34 ; J. F. John, Die Malerei der Alten, 1836, 119 ; Birch, i, 66 ; Harrison, 52. The numerous volumes of the Description de l'Égypte were written by the savants who accompanied the Napoleonic expedition in 1799 and they introduced the scientific studies of Egyptian antiquities. Berthollet, who was one of the scientists in the expedition, composed his famous Recherches sur les Lois de l'Affinité, containing the Law of Mass Action, at Cairo in that year.

[3] Birch, i, 96 f. ; cf. J. Kidd, Mineralogy, Oxford, 1809, i, 181.

[4] Brunton and Thompson, Badarian Civ., 1928, 16.

[5] Maspero, Arch., 259 ; Wilkinson, ii, 150.

[6] Kisa, Glas, 65 ; Scharff, *Morgenland*, 1927, xii, 43.

[7] Maspero, Arch., 246.

[8] Qau and Badari I, 16, 22.

[9] Préhist. Or., ii, 205.

[10] Arts and Crafts, 107 f.

[11] ii, 149.

[12] BMGE, IV–VI Rooms, 221.

[13] Petrie, Arts and Crafts, ch. x, 107–119 ; Kisa, 65 ; Blümner, PW, vii, 1382 f. ; Scharff, *Morgenland*, 1927, xii, 36.

[14] Petrie, Prehist. Egypt, 1920, 42 f. ; cf. Maspero, Arch., 263, 357.

[15] Petrie, *Anct. Egypt*, 1914, 188 ; *ib.*, Naqada and Ballas, 1896, 45 ; *ib.*, Arts and Crafts, 107 f. ; cf. Hall, CAH, i, 576 f. ; Dillon, Glass, London, 1907, 18 f. ; Wiedemann, 1920, 329, piece of quartzite vase, Amenhotep III ; Beck in Brunton, Qau and Badari II, 1928, 23 ; VI–VIII dyn. amulets of quartz rock covered with blue-green glaze.

[16] Petrie, Arts and Crafts, 107 f. ; Maspero, Arch., 259 ; Kisa, 65.

VI dyn.[1]   Green and blue glazes imitated malachite (or turquoise) and lapis lazuli : in a New Kingdom story glass [green ?] is compared with malachite.[2]

The base for glazed objects is a porous body of finely ground silica, either quartz rock or sand, slightly bound together but with all the strength of the object located in the glaze ; gum, fat or other organic matter may have been added as a binder, but all traces have disappeared in baking.[3]   The use of clay as a binding material, suggested by Brongniart [4] and Wiedemann [5], is regarded by Franchet and Lucas [6] as improbable.   The body of a statuette (XXII dyn.) [7] had the composition : SiO_2, 94·18 ; Al_2O_3, 0·59 ; Fe_2O_3, 1·64 ; CaO, 1·73 ; MgO, 1·82 ; CO_2, 0·04, i.e. nearly pure silica, not clay.   A classification of the body is given by Wiedemann as (1) white quartz sand ; (2) fine light-grey material obtained by pulverising limestone from Upper Egypt ; and (3) a reddish mass of chalk and powdered brick, together with sufficient clay to enable it to be moulded in a moist state in moulds of baked clay.

The method of making such objects was correctly recognised by Birch.[8] The object was dipped in powdered glaze with an alkaline, probably soda, flux, and fired so that the glaze formed a skin : larger objects were burnt before glazing, then dipped in fused glaze or painted with it.   Broken models, or models in sections, could be stuck together with paste before glazing.   The beautiful hard stoneware of the XXVI dyn. was apparently obtained by mixing some glaze with the body, no face glaze being used.   The colour is usually apple-green, but sometimes violet (XVIII dyn.).   The shades of green and blue are produced by copper in the cupric state, the blue demands materials especially free from iron, a trace of which gives a green.   The blue fades to white when exposed, and the green turns brown owing to decomposition of green ferrous silicate to brown ferric oxide or ferric silicate, which may go on below an unbroken polished surface.[9]

No analysis of the glaze can be found, but it is generally assumed to be similar to ancient glass.[10]   At Amarna, according to Petrie[11], quartz rock pebbles served for the floor of glazing furnaces.   After many heatings, which cracked them, they were pounded into fine chips.   These were mixed with lime and potash and some copper carbonate (malachite), and the mixture roasted in pans, the shade depending on the degree of roasting.   The half-fused mass was kneaded and toasted gradually, the colour being sampled until the desired tint was obtained.   A uniformly coloured porous (friable) mass (frit) resulted, which was ground in water and made into blue or green paint, used either with a flux for glazing, or with gum or albumin as paint for frescoes. Petrie's statement that "potash" was used is probably incorrect, since Egypt is poor in wood but rich in soda, the alkali usually added, although the earliest glazes might have been made with plant ashes containing potash.   The silica, also, was not in the form of crushed quartz, but of sand.[12]   The use of nitre and borax[13] is very unlikely, since both were probably unknown in ancient Egypt.

The ovens[14] were small, 2 or 3 ft. across.   Cylindrical pots, 7 in. dia. and 5 in. high, were set upside down and a fire lighted between them, the pans of colour, 10 in. wide and 3 in. deep, resting on the bottom edges of the pots, successive

[1] Glanville, JEA, 1928, xiv, 190.
[2] Erman-Ranke, 547 ; Erman, Lit., 312.
[3] Petrie, Arts and Crafts, 107 f. ; *ib.*, *Anct. Egypt*, 1914, 187 ; Lucas, Egy. Mat., 32 f., 34 ; Franchet, 42.
[4] Arts Céramiques, 1853, i, 505, with analyses.
[5] 1920, 329 f. ; perhaps a different variety.
[6] Egy. Mat., 34 ; *ib.*, in Carter, Tomb of Tutankhamen, ii, 169.

[7] Franchet, Céram. prim., 41.
[8] Pottery, 1858, i, 67 ; Wilkinson, ii, 150 ; Wiedemann, 1920, 330.
[9] Petrie, Arts and Crafts, 107 f.
[10] Lucas, Egy. Mat., 36 f.
[11] Arts and Crafts, 107 f. ; *ib.*, Tell el-Amarna, 25 f. ; Dillon, Glass, 1907, 24.
[12] Lucas, Egy. Mat., 48 f.
[13] Franchet, 93.
[14] Petrie, Arts and Crafts, 107 f. ; *ib.*, Tell el-Amarna, 7 f.

layers being kept apart by earthenware cones. In Roman times the glazing furnaces were 8 ft. square and deep, with an open arch to windward halfway up. By too long heating the glaze soaked through the porous body, which settled down and fell in pieces.[1] Birch [2] says the colours were sometimes laid on baked vases by water colour or tempera, the object not being baked after this process.

The black glaze may have been produced with magnetite, a native iron oxide, Fe_3O_4, either alone or, especially in the Saïte Period, with oxide of manganese. The same process was perhaps used in Greece, but not with equal parts of the oxides of iron and manganese, as proposed by Salvétat[3] : much less manganese was used and the proportion is so small that it may be accidental.[4]

The colours in the early glazes were perhaps the results of the accidental use of materials containing various metallic oxides [5], including oxides in the sand used for glass-making, and perhaps the tentative addition of "any kind of earth or metallurgical by-product which happened to be at hand." Once the colouring properties of copper became known it was used with great skill from an early period to produce definite and controllable results.

In a more specific account of Egyptian "faience" by Beck [6], it is said to consist of two parts, core and glaze, sometimes separate, sometimes mixed. The core when colourless consists of a series of varying sized particles of quartz, with a coloured vitreous cement. Some Predynastic (?) black beads are coloured throughout, probably by mixing powdered sand and manganese glaze with a binder, moulding, and firing to flux the glaze. Blue XVII dyn. (?) beads have a thick layer of glaze over a powdered quartz core, held together by a vitreous material, sometimes colourless. The sharp line of demarcation between the colourless and blue glazes, not explained by Beck, may be due to diffusion from a glaze put on outside by dipping, and then refiring. Both core and glaze are full of quartz fragments. Other Predynastic beads have a vitreous colourless core, covered with a thick layer of clear blue glaze entirely free from quartz crystals, with a melting-point considerably lower than that of the vitreous part of the core. Beads of IX–X dyn. have a "transparent" core, then a very thin layer of dense colour (probably due to iron and manganese), and outside a colourless layer, all three full of quartz crystals and the core full of bubbles : no doubt these were dipped twice. The core, of nearly pure silica, is supposed to have been made like a "sand-lime" brick : powdered quartz with 2 per cent. of lime added as milk of lime would be cemented together on heating into a "vitreous mixture," and lime would also bind the material before firing. Beck examined sections of sand-lime bricks and found that under certain conditions the quartz breaks and fuses like that in the "faience." Perhaps the fused material was ground up again before making into a bead.

Glazed beads began to replace stone beads in the I–III dyns. : glass beads came into use from the XVIII dyn. and were almost the only kind in the Roman Period. Glazed objects of the I dyn. from Abydos include models of apes, pig, flower, tiles, plaque, beads, and vases, including a large fragment of one with the name of a king inlaid in yellow glaze.[7] Besides small objects like beads and amulets, larger objects such as chests and jewel boxes were made in glazed ware, and glazed tiles for architectural purposes are very early, e.g. the small glazed tiles, some bright blue or green and others almost black, in the III dyn. step-pyramid at Saqqâra.[8] That tin oxide was used in these

[1] Cf. Wilkinson, ii, 150.

[2] Pottery, i, 50.

[3] Brongniart, i, 554.

[4] Franchet, 78, 98, 105 f.

[5] Kisa, 17, 37, 261, 276, 279 f., 282, 284 ; Rathgen in Fester, Die Entwicklung der chemischen Technik, 1923, 11 ; Partington, Inorganic Chemistry, 1933, 829.

[6] Brunton, Qau and Badari II, 1928, 23 f.

[7] BMGE, IV–VI Rooms, 288 f.

[8] Petrie, Arts and Crafts, ch. vii ; Wiedemann, 1920, 129 f., 184 f. ; Kisa, 69 f. ; Birch, i, 69.

tiles at Saqqâra [1] is improbable. They are of green glaze, set in bands with hieroglyphics in blue, red, green and yellow on a fawn-coloured ground, but may be restorations of the xxvi dyn. replacing an older decoration of the same kind.[2] There are yellow and green glazed tiles with the name of Pepi I (vi dyn. ; 2500) [3], and a greenish-blue plaque of the same king.[4] Rameses III (1198) had a temple at Tell al-Yahûdia covered with small glazed tiles set like mosaic in cement : some specimens are in the Louvre.[5]

A piece of a green glazed vase has the name of Menes (i dyn.) inlaid with a second colour, probably violet but now decomposed, and at Medum (vi dyn.) there is inlay of coloured paste in incisions in stone, mostly ground off flat but some in relief : Nefermaat in his inscription says he invented this process.[6] Such inlay otherwise occurs only later. Large Predynastic tiles covered with a hard bluish-green glaze for lining the rooms ("Dutch tiles") were stoutly made, some about 1 ft. long with dovetails on the back and holes for tying to the wall with copper wire. From the Pyramid Age there was otherwise little glaze.[7]

COLOURS OF GLAZE

The earliest glaze is greenish-blue or bluish-green, never distinctly blue or green. It is full and has not soaked into the body, and there are often pit-holes, hence it was probably not very fluid. Glazed iv dyn. statuettes are clear light-purplish blue, with dark purple stripes. In the vi dyn., dark indigo-blue appears, e.g. on a scarab and on small toilet vases. Some scarabs, probably iii–iv dyns., have a clear brilliant blue glaze, thin and well fused. Glaze of the xii dyn. is thin and hard ; on ring stands and vases it is often dry and greyish-green, but a rich clear blue was also used, best seen on scarabs and on the figures of hippopotami, made only in this period. Designs and inscriptions in the glaze are of fine black, apparently coloured with manganese. The xviii dyn. was a great age of development of glazing, which began with a close continuance of the xii dyn. style. To the time of Thothmes III (1501–1447) small pieces and blue beads are the same as of the previous age : large bowls are of bright blue and wetter glaze.[8] This king introduced a new type of scarab with a brilliant green glaze.[9] At the beginning of the xviii dyn. there was also dark green glaze on schist, mostly seen on elaborately carved koḥl-pots. Under Amenhotep II (1448–1420) was made the largest known piece of Egyptian glazing, a sceptre now in South Kensington Museum.[10] On fragments of a glazed xviii dyn. dish from Abûsîr the colours were green, red, yellow, blue and violet.[11] Birch[12] says the ground was sometimes white or yellow, with hieroglyphics, etc., in red, blue and yellow, and thinks the glaze was applied from strips of linen. There is a dark blue glazed bowl of Rameses II (1296–1230) in the British Museum.[13]

Glazing reached its perfection in the Amarna Period (1375–1350) : very large pieces were produced and a great variety of small objects. Moulds of clay, found elsewhere only from the middle of the xviii dyn. to Roman times, were used for figures, plaques and rings. Imitation jewellery, with bright

[1] Semper, Der Stil, i, 411 f. ; BMGE, IV–VI Rooms, 236.
[2] Kisa, 69 ; Maspero, Dawn, 243 f. ; ib., Arch., 276.
[3] Maspero, ib.
[4] Morin-Jean, Daremberg-Saglio, Dict., v, 936.
[5] Maspero, Arch., 267 f. ; Kisa, 70.
[6] Petrie, Arts and Crafts, 107 f. ; ib., Abydos II, 25, plate iv, 32, plate xxi ; Erman-Ranke, 546 f. ; Kisa, 70.

[7] Petrie, Arts and Crafts, 107 f. ; ib., Abydos II, plate viii ; cf. Birch, i, 67, Fig. 31, British Museum.
[8] Petrie, Arts and Crafts, 107 f. ; ib., Medum, 34.
[9] Hall, CAH, ii, 415.
[10] Petrie, op. cit.
[11] Borchardt, Sahurā, i, 131, Abb. 178, pl. xiv.
[12] Pottery, i, 30, 68.
[13] Wilkinson, ii, 11.

glazes imitating precious stones, "not bad," was made, with statues of various gods and all kinds of animal and vegetable forms. Factories for glaze and glass adjoined at Amarna, and the same process was used.[1]

Under Amenhotep III and IV (1411–1350), besides blue and green there are purple-blue, violet, brilliant apple-green, bright chrome-yellow, lemon-yellow, crimson-red, brown-red and milk-white glazes. Pendants and necklace ornaments appear in great variety. Glaze was used for architectural inlaying, the capitals of great columns being inlaid with stripes of red and blue along the palm-leaf design, separated into small squares by gilt bands, the whole capital being imitative of cloison jewellery. Coloured glaze hieroglyphics were inlaid in white limestone walls, a process carried on in a simpler way into the next dynasty, where there are many cartouches of Seti II (1214–1210), probably taken from holes of similar size found in walls at Luxor.[2] Birch describes and figures vases and cups of opaque glaze with wavy lines, and a blue glazed saucer with the figure of a fish, in the Berlin Museum.[3]

In the XIX dyn. there was much less variety, but small statuettes appear, called ushabtis, for tombs. Seti I (c. 1300) had many glazed ushabtis of blue inscribed in black, or of glazed steatite, in his tomb. Under Rameses II (1296–1230) they become usual for private persons, as many as 400 being buried in wealthy tombs. The Ramesside ushabtis are usually green with black inscriptions, rarely white with purple. The XXI dyn. gives intense blue with purple-black inscriptions roughly made and deteriorating throughout the dynasty; XXII and XXIII dyn. ushabtis are small and usually green and black; those of the XXV dyn. are mere red pottery dipped in blue wash, or little slips of mud. The XXVI dyn. started very large figures, up to 10 in. high, beautifully modelled, with incised inscriptions, back pillar and beard, always of green glaze. These deteriorated to Ptolemaic times, except some splendid blue ones and smaller ones of bright colour with ink inscriptions, of Nectanebo's time. About the XXVI dyn., glazed figures of the gods were in popular use, and by 300 B.C. they appear in vast numbers, some very well made.[4] Birch [5] mentions rare ushabti in yellow and red glaze; white is uncommon: some were of stone, e.g. red granite.[6]

Great numbers of amulets were made to be worn by the living or buried with the dead. Earlier examples are fairly modelled, of apple-green tint. In Persian times they are sharp and dry in form and olive-grey in colour, by Ptolemaic times rough and coarsely moulded. Some interesting modelled heads, covered with green or blue glaze, belong to this period.[7]

In the XXVI dyn. a light apple-green glaze is principally used for circular flasks.[8] A pale green glaze in the archaising Saïte dyn. (663–525) was an imitation of the earliest type. Persian Period glaze was thick and sugary, "like thin ice," whilst Ptolemaic glaze was thicker and glassier than the older types, and of a peculiar grey-blue colour.[9] A blue glazed Coptic plinth is a late example of the art.[10] Under Amenhotep III, grey and violet colours were especially favoured in glazes, and under Ikhnaton the yellow, green, violet, white, blue and red decorations at Amarna are small and fine; they are quite sharp and do not run into one another. A specially fine object is a mummy case in the Bulaq Museum, with rich coloured decorations of turquoise blue,

---

[1] Petrie, Tell el-Amarna, 28 f.; Kisa, 71, 74.

[2] Wilkinson, ii, plate xiv; Petrie, op. cit.; Breasted, H., 349, Fig. 138; ib., CAH, ii, 104.

[3] Wilkinson, ii, 12; Fig. 280, 2; Fig. 281, 5; ib., 42, Fig. 306, 2.

[4] Petrie, Arts and Crafts, 107 f.; Birch, Pottery, i, 29 f., 92 f.

[5] Pottery, i, 95.

[6] Maspero, Struggle, 332.

[7] Petrie, Arts and Crafts, 107 f.; Birch, i, 79 f.

[8] Birch, in Wilkinson, ii, 11.

[9] Maspero, Arch., 264; Wiedemann, 1920, 330; Hall, CAH, iii, 323 f.

[10] Gunn, Anct. Egypt, 1930, 88.

yellow and violet, on white. A very fine vase of the Ptolemaic Period from Saqqâra is 21 cm. high and 20 cm. dia., with a lapis lazuli coloured ground, the neck and foot being decorated in relief with garlands and flowers and with light green glaze.[1] A large collection of blue glazed objects in the British Museum includes bottles in the shape of fish, figures of various kinds, a wine strainer (XXII dyn.), a wig, a roll, coffin and cover, a fine glazed bowl with figures of a lake and lotus flowers in black outline, a boomerang (with the name of Ikhnaton), vases, glazed tiles from a temple of Seti I rebuilt by Rameses III, plates, dishes, jugs, ushabti figures, etc.[2] A large collection of glazed beads includes blue, green, red, yellow, white and other colours, chiefly of the XXVI dyn. Some are strung on the original linen thread.[3]

Birch suggested that oxides of chromium, tin and lead were used in the Saïte Period for green glaze [4], but this was probably always coloured with copper. A red glaze, obtained with hæmatite or cuprous oxide, was used for ear-rings, and hieroglyphics for inlaying walls, etc., were of opaque white, yellow, blue and red glaze.[5] A bright yellow glaze is said by Petrie [6] to be of unknown composition, but Lucas [7] found that a specimen of the XIX dyn. contained antimony, together with lead, as Franchet [8] had supposed. The yellow glaze appears to have been applied last when several colours were used.[9] Birch[10] had suggested the use of silver, and a yellow glaze could have been produced from titanium oxide, which occurs near the old Egyptian gold mines of Sesostris II.[11]

The violet glaze, in various depths from a faint tinge on white lotus petals to a deep, strong colour, was probably made by copper blue and one of the purples.[12] The purples, in various strengths from a rich tint on white to black-purple on blue, were probably all produced with manganese[13], gold[14] being improbable. Manganese is common on brownish-purple glazed pottery from Egypt ; a small figure of a mummy clothed in white "porcelain" very like salt-glazed stoneware, had potash in the glaze and manganese in the violet hieroglyphics applied on the glaze[15], and a black manganese glaze was found on VI dyn. beads.[16] Occasionally[17] a purplish-blue glaze contained cobalt, and Franchet[18] mentions a cobalt glaze on a XXI dyn. statuette. The dead white glaze[19] was probably produced by tin oxide.[20] A white "china" statuette (XX dyn.) with incisions filled in with blue, yellow and violet pastes afterwards vitrified in the furnace, is a remarkable object, since the pastes melt at different temperatures and several firings must have been used. An attempt to reproduce it at the Sèvres factory in 1867 was unsuccessful.[21]

A delicate thin ware of Assyrian type, in white on a slightly sunk blue ground, was made in the Persian Period and continued into the Ptolemaic Age. Large blocks for legs of furniture and stands were also made in the late periods, the characteristic colours being a dark "Prussian" blue bordering on violet and an apple-green. The Roman Age introduces an entirely new style, a purple-black body with a wreath of bright green leaves, and this continued almost to

[1] Kisa, 66 f. ; Wiedemann, 1920, 330.
[2] BMGE, IV–VI Rooms, 142 f., 234 f.
[3] *Ib.*, 210.
[4] Pottery, i, 100, 103.
[5] Birch, Pottery, i, 67, 72 ; *ib.*, in Wilkinson, ii, 149 ; Wiedemann, 1920, 133.
[6] Arts and Crafts, 107 f.
[7] Egy. Mat., 52, 60 ; *ib.*, in Carter, Tomb of Tutankhamen, ii, 171.
[8] Céram. Prim., 99.
[9] Wilkinson, ii, 150.
[10] Pottery, i, 67.
[11] Maspero, Dawn, 480.

[12] Petrie, Arts and Crafts, 107 f.
[13] Wornum and Middleton, DA, i, 486.
[14] Birch, i, 67 f.
[15] Lepsius, Met., 27.
[16] Petrie, etc., Sedment I, 1924, 6.
[17] Petrie, Arts and Crafts, 107 f.
[18] Céram. Prim., 96.
[19] Birch, i, 67 ; Petrie, *op. cit.*, "white earth."
[20] Lucas, in Carter, Tomb of Tutankhamen, ii, 170 ; Neumann, *Z. angew. Chem.*, 1929, xlii, 835.
[21] Maspero, Guide to the Cairo Museum, 1908, 397.

Coptic times. The bulk of the Roman glaze is of coarse form and bright "Prussian" blue colour.[1] Hall [2] says the Roman glaze was green, obtained by glazing blue over yellow, and gave rise to the Fostât (Cairo), Persian and Chinese styles.

Herodotos' statement that only bronze drinking cups were used in Egypt is incorrect, since there were cups of gold, silver, glass, pottery and "alabaster."[3]

EGYPTIAN BLUE

The first impression received by a visitor to a museum of Egyptian antiquities is the prevalence of a blue colour which in various shades was a favourite in Egypt and in the later Orient.[4] Although Chaptal had found alumina, lime, and oxide of copper in two out of seven specimens of colours from a painter's shop in Pompeii, the first careful examination of a specimen of Egyptian blue, obtained from the Roman so-called Baths of Titus but probably made in continuation of the Egyptian technique, was by Sir Humphry Davy [5], whose paper, combining a resumé of the information on the subject given by ancient authors, a careful statement of the origin of the materials, and accurate analyses with small specimens, constitutes a model for all investigations of this type. Davy found copper and soda in the blue frit, and obtained a similar product by strongly heating for two hours a mixture of 15 parts of sodium carbonate, 20 of powdered silica and 3 parts of copper filings. Vauquelin found copper, silica and lime in Egyptian blue.[6]

Although Theophrastos [7] probably refers to an artificial blue of this kind made in Egypt, and Dioskurides [8] mentions "Cyprian kyanos" made from copper ore and sand, the preparation of the Egyptian or "Alexandrian" blue is first correctly described by the Roman architect Vitruvius (c. 24 B.C.) [9] as follows : "Cæruleum was first made at Alexandria, then afterwards by Vestorius at Puteoli. The method of making it is curious. Sand and natron (nitri flore ; native soda) are powdered together as fine as flour, and copper (æs cyprum) is grated by coarse files over the mixture. This is made into balls by rolling in the hands. The dried balls are put in an earthen jar and this jar put into a furnace. When the copper and sand have coalesced in the intense heat and the separate things have disappeared, the colour cæruleum is made."

Russell[10] reproduced the various shades of blue and green by copper frits. The purple, produced with difficulty in a single specimen, was perhaps the so-called Alexandrian purple. A mixture of 60–80 parts of silica with 10 per cent. of sodium carbonate and the other materials gave a friable product, like Egyptian frits, when heated below fusion temperature. A delicate greenish-blue was formed with 3–5 per cent. of copper carbonate, a deeper with 10 per cent., and the purple with 20 per cent. and an equal amount of lime. Iron gave a green tint.

Laurie, McLintock and Miles[11] showed that Egyptian blues as well as those from Knossos, Palestine and Viriconium (Wroxeter), all contain a definite compound, $CaO,CuO,4SiO_2$, in tetragonal crystals, which had really been made and analysed long before by Fouqué[12] but was not obtained by Russell.

[1] Petrie, Arts and Crafts, 107 f.
[2] CAH, ii, 324.
[3] Birch, in Wilkinson, ii, 7, 42, 43.
[4] Kisa, 283 f.
[5] Phil. Trans., 1815 ; Works, ed. by J. Davy, 1839 f., vi, 131 f., " On the Colours used in Painting by the Ancients."
[6] Passalacqua, Catalogue raisonné, 1826, 239, 242 f.
[7] De lapidibus, 31, 55 ; repeated by Pliny, xxxiii, 13, xxxvii, 9, who also calls the blue colour cœlon.
[8] v, 106.
[9] vii, 11 ; note in Elzevir ed., Amsterdam, 1649, 144 ; text in Krohn's ed., Leipzig, 1912 ; Blümner, iv, 500 f. ; cf. Davy, Works, vi, 141.
[10] Petrie, Medum, 1892, 44 f. ; cf. Dillon, Glass, 1907, 18 f., 24.
[11] Proc. Roy. Soc., 1914, lxxxix, 418 ; rather inaccurate summary in Anct. Egypt, 1914, 186 f.
[12] Comptes rendus, 1889, cviii, 325.

The compound is a pure azure blue, sp. gr. 3·04, and Fouqué's samples were found to be free from alkalies, whilst some of Russell's specimens were merely blue glass. Laurie, etc., heated a mixture containing 36 gm. of fine sand, 4 gm. of fusion mixture, 8·6 of "copper carbonate," and 7·2 of calcium carbonate for several hours from 760° to over 900°. The blue compound is formed in the rather narrow temperature range of 830°–900°, but if the overheated product is afterwards maintained for a considerable time at 850°, the blue is produced by recrystallisation. Regrinding and heating is favourable, and was perhaps the Egyptian practice, but is not necessary. Above and below the critical temperature range the product is an olive-green glass, obtained by Russell : in this range blue crystals appear, but disappear again when the temperature is slightly raised. A blue can be formed, as Fouqué stated, without addition of alkali but with difficulty ; with larger amounts of alkali a green glass is formed, not crystals of blue. Addition of an artificial trona (native Egyptian alkali) made up according to Klaproth's analysis (sodium sesquicarbonate 32·6 gm., sodium sulphate 20·8 gm., sodium chloride 15·0 gm.) [1] gave a large amount of blue in 40 hours at 850°–860°. A blue was also obtained with potassium sulphate as the flux. In presence of alkali, some of the copper or calcium in the blue is replaced by alkali metals.

Le Chatelier [2] prepared a similar barium compound, $BaO,CuO,4SiO_2$, and suggested that the black coating on Etruscan pots may be the sodium ferrous silicate $Na_2O,FeO,4SiO_2$, which he did not succeed in preparing.

No ancient European people could successfully imitate Egyptian blue [3], and the secret of its manufacture was lost between A.D. 200 and 700. The technical difficulty of keeping the furnaces at the exact temperature of 900° for 48 hours, regarded as necessary for success, must have been considerable and it is not known how it was done, although the original furnaces have been found.[4] The interval 830°–900°, however, seems a normal temperature for an ordinary charcoal fire, and perhaps Laurie has overestimated the skill required.

An imitation of lapis lazuli with golden points of pyrites was made in the XVIII dyn. with blue [cobalt ?] glass sprinkled with gold.[5]

Although the principal colouring constituent is always copper, Petrie [6] says a purplish-blue glaze contained cobalt, and Franchet [7] mentions a cobalt glaze on a XXII dyn. statuette. The presence or absence of cobalt in Egyptian blue glazes and glasses has formed a subject of much discussion.[8]   Beckmann [9], whose judgments are often far from the truth, could not make up his mind as to whether cobalt was known to the ancients or not ; Klaproth found no cobalt or lead in a blue glass from Capri and Russell found no cobalt in Egyptian blue.[10]   Merimée[11], John[12], Davy[13], Boudet[14] and others[15] report the presence of cobalt in Egyptian and other ancient glass and glaze, including blue frits of the time of Rameses III (1198–1167), an amulet catalogued as "true lapis lazuli," and Ethiopian beads found by Ferlini in a pyramid at Meroë (cobalt

[1] Lunge, Alkali Industry, 1895, ii, 61.
[2] J. Soc. Glass Techn., 1926, x, 96.
[3] Mackenzie, Footprints, 166.
[4] Laurie, Chem. Trade J., 1931, lxxxviii, 232.
[5] Dillon, Glass, 1907, 32.
[6] Arts and Crafts, 107 f.
[7] Céram. Prim., 96.
[8] Cf. Watts, Dict. of Chem., 1874, i, 1039.
[9] Hist. of Inventions, i, 478 f.
[10] Kisa, 284.
[11] Passalacqua, Catalogue, 263.
[12] Malerei der Alten, 1836, 140.

[13] Works, vi, 143.
[14] Descript. de l'Égypte, Antiquités, ii, 35.
[15] Brongniart, Arts Céram., 1854, ii, 563 ; Mrs. Merrifield, Original Treatises on the Arts of Painting, 2 vols., 1849, i, p. liii ; Birch, in Wilkinson, ii, 149 ; Lepsius, Met., 27 ; Wornum and Middleton, DA, i, 486 ; Kisa, 283 f. ; Toch, Industrial and Eng. Chem., 1918, x, 118 ; Bucher, iii, 262 ; Blümner, iv, 236, 503 ; Wiedemann, PSBA, 1893, xv, 113 ; ib., Das alte Ägypten, 1920, 338, 347.

with no nickel). Parodi found it in XVIII–XX dyn. blue glasses.[1] Analyses of blue glazes and glasses in the Berlin Museum reputed to be Egyptian, made in Hofmann's laboratory [2] gave the following results :

	SiO$_2$	CaO	Al$_2$O$_3$	MgO	Na$_2$O	K$_2$O	Fe$_2$O$_3$	CuO	CoO	MnO	SnO	PbO
1. Glaze . .	70	9	—	—	4	—	1	15	—	—	—	—
2. Glaze . .	70·5	8·53	—	4·18	—	—	3·71	13·00	—	—	—	—
3. Glass beads	74·30	8·50	0·95	2·81	3·63	5·45	1·81	—	2·86	—	—	—
4. Glass beads	74·41	8·47	1·01	2·83	—	—	1·78	—	2·82	—	—	—
5. Glass beads	67·07	5·61	1·24	0·91	2·11	12·15	4·91	—	0·95	1·37	0·58	3·66

The cobalt may have been obtained from silver ores [3], or from cobalt ore from India [4] or Persia[5] ; Richmond and Off in 1892 found 1·02 per cent. of cobalt in an Egyptian mineral.[6]   The blue colouring material ("smalt") is a definite crystalline compound, $K_2O,CoO,3SiO_2$.[7]

GLASS

Common glass is made by fusing together materials containing silica (sand ; quartz ; flint), alkali in the form of sodium carbonate (native in Egypt as natron or trona) or potassium carbonate (wood ashes), and lime in the form of limestone, marble, calcined shells, etc. The process is simple and the discovery was probably made early in a place where alkali is available, as in Egypt.[8] The Egyptian name for glass, tehen (deḥen), means "Libyan," or "the Libyan thing," so that the discovery was probably made on the Libyan side of the Delta.[9]   The name ḥm-t has also been supposed to mean some kind of glass paste.[10]

Pliny's story[11], perhaps from Alexander Polyhistor (c. 100 B.C.)[12], that glass was discovered accidentally by Phœnician merchants wrecked near the mouth of the River Belus on the Syrian coast, who used lumps of their cargo of natron to support cooking pots over a fire on the sand, which melted with the alkali to glass, is fabulous, since the temperature would have been too low.[13] Although some modern authors think glass was a Syrian invention[14], there are no early remains of glass in that country and the generally accepted opinion is that glass was discovered in Egypt.[15] If invented in Egypt, glass soon reached Mesopotamia, as a lump of blue glass was found at Eridu in South

[1] Parodi, La Verrerie en Égypte, Thesis, Grenoble, 1908, q. by Duboin, Comptes rend., 1921, clxxii, 972 ; Lucas, Egy. Mat., 62, 232.
[2] Lepsius, Met., 25 f.
[3] Wornum and Middleton, DA, i, 486.
[4] Boudet, Descr. de l'Égypte, Antiq., ii, 34.
[5] Lucas, Egy. Mat., 62.
[6] Wiedemann, PSBA, 1893, xv, 113.
[7] Duboin, Comptes rend., 1921, clxxii, 972.
[8] Hoefer, Hist. de la Chimie, 1842, i, 57 ; 2 ed., 1866, i, 64.
[9] Newberry, q. by Hall, Civilisation of Greece in the Bronze Age, 1928, 71, 104 ; Budge, Egy. Hierogl. Dict., 820, 842, 887, dehen, or teḥen-t, is blue faience or crystal ; taḥen = to shine.

[10] Erman-Grapow,   Handwörterbuch, 1921, 109 ; "ein Art Glasflüss ?"
[11] xxxvi, 26.
[12] Froehner, La verrerie antique, Le Pecq, 1879, 3 ; Schwartz, PW, ii, 1449.
[13] Franchet, Céram. Prim., 91 ; Kisa, 33, 90 f. ; Kisa's work is faulty philologically, historically and technologically.
[14] Forrer, RL, 28 ; Petrie, Trans. Newcomen Soc., 1926, v, 72.
[15] De Pauw, Recherches philosophiques sur les Égyptiens et les Chinois, 2 vols., Berlin, 1773, i, 307 ; ib., English transl. by Thomson, London, 1795, i, 288 ; Hoefer, op. cit. ; Froehner, 5 ; Boudet, Descr. de l'Égypte, Antiquités, 1818, ii, 17 f. ; Franchet, 91 ; Kisa, 37, 74, 171 f. ; Lucas, Egy. Mat., 38.

Babylonia in remains earlier than 2300 B.C.[1] Herodotos'[2] λίθινα χυτά, "fused stone," may translate some Egyptian name for glass.[3]

The discovery of glass may well have been due to a gradual evolution of the art of glazing, a prehistoric technique in which the same materials were used, and opaque coloured paste was probably the first product, the colour being due to metallic oxides[4], e.g. oxides of iron, manganese and copper, either accidental impurities or, in the case of copper, intentionally added. Coloured glass frits were probably in use before glass : the Chinese and Japanese, who decorated clay and metal for centuries with glass paste, still imported window glass in the 19th century.[5] The suggestion that glass was known before glaze as slag produced in smelting ores with a flux[6] seems improbable, and the production of beads of glass by the running of thick fused glaze is more likely. This glass, like the slags, would be coloured.[7] Although, as Franchet[8] says, the earliest glaze (on seal cylinders) was not coloured, he is wrong in suggesting that coloured glazes appear only in the XVIII–XIX dyns., since blue glaze is Predynastic and coloured glass pastes go back at least to the VI dyn. and were in use in the form of bottles, jugs, vases, etc., in the Old and Middle Kingdoms.[9] Extraordinary confusion has been caused by the use of such technical terms as "glass," "enamel," "paste," "glaze," "porcelain," "faience," "cobalt blue," etc., without any appreciation of their meanings.[10]

It was formerly considered[11] that "true glass" first appeared in Egypt about 2000, but this is too late. Some authors[12] throw doubt on the use of glass, as distinguished from "paste," in the Old Kingdom, but I have not distinguished between "glass" and "paste," which are often practically identical materials.

EARLY GLASS IN EGYPT

The Berlin Museum possesses a pale green, oval, bored glass bead, found by Petrie in a prehistoric grave ("Hockergrab") (c. 3500 B.C.) at Naqada. It is 9 × 5·5 mm., 4 mm. thick at one side and 2 mm. at the other. A portion of 0·3 mgm., qualitatively analysed by Rathgen[13], consisted of lime-soda glass with a trace of potash. This bead is probably the oldest known piece of glass, although a small Hathor head of "blue glass" in the University College Museum, London, is said to be Predynastic[14], "from its colour" of Asiatic origin, and the "earliest glass known in Egypt." There is a polychrome "paste" fragment with the name of King Āḥa (I dyn.).[15] Beads in the bracelets of the queen of King Zer at Abydos (I dyn.), light blue and alternating with gold plates, were described as of "turquoise or light blue glass"[16], but Petrie, who found them, states[17] that they are of amethyst and turquoise, not glass.

Two objects in the Ashmolean Museum, Oxford, viz. a greenish-black bead

[1] Hall, Civ. Greece in the Bronze Age, 1928, 71, 104.
[2] ii, 69.
[3] Lenz, Zoologie der alten Griechen und Römer, Gotha, 1856, 419.
[4] Bucher, iii, 268 ; Hoefer, i, 57 ; Kisa, 74 ; Erman-Ranke, 547 ; Wiedemann, 1920, 328 ; M. A. Wallace-Dunlop, Glass in the Old World [1883], 3, a faulty work but rich in material.
[5] Bucher, i, 5 f. : the Japanese now make glass.
[6] Franchet, 91 f. ; Harrison, Pots, 56.
[7] Hoefer, i, 64.
[8] Op. cit., 79, 94.
[9] Budge, Mummy², 391 ; Kisa, 70.
[10] Cf. Petrie, Anct. Egypt, 1930, 6 ; Lucas, Egy. Mat., 37, 45 ; Kisa, 67, 70.

[11] Blümner, PW, vii, 1382 f.
[12] Erman-Ranke, 547 : they except the Predynastic beads ; Lucas, Egy. Mat., 40, 41.
[13] Sprechsaal, Berlin, 1913, xlvi, 98 ; Neumann, Z. f. angew. Chem., 1925, xxxviii, 776 ; cf. Forrer, RL, 290 ; Erman-Ranke, 547.
[14] Petrie, Prehist. Egypt, 1920 ; ib., Trans. Newcomen Soc., 1926, v, 72, a superficial paper ; Anon., in Anct. Egypt, 1927, 57 ; cf. Hall, NE, 119 ; Lucas, Egy. Mat., 40 f.
[15] Morin-Jean, Daremberg-Saglio, Dict., v. 936 ; Budge, Mummy², 25.
[16] Maspero, Art in Egypt, 1912, 2.
[17] Arts and Crafts, 84.

of "glassy character" found in a wooden box of the I dyn. [1], and opaque black glass inlay in the form of a figure 8 in a glazed plate, probably I dyn., may be of obsidian, a natural volcanic glass, and a cylinder of Pepi I (VI dyn.; 2600) is not glass but "clear Iceland spar or selenite [two quite different materials] lined with coloured paste."[2] What was supposed to be a semi-transparent glass tube, 2⅝ in. long and I in. outside diameter, with the name of Pepi II (VI dyn.), is not glass as formerly thought, but rock crystal.[3] Maspero [4] calls the material of the eyes of the statue of Pepi I, and those of two Old Kingdom statues (one III dyn., in the Louvre), white and black "enamel" (i.e. paste).

The eyes of mummies were of most various materials[5] : resin, wax, gypsum, obsidian, marble, talc, quartz, agate, clay, bronze, gold, gilt metal, silver, ivory, "copper enamel" and glass (perhaps also cobalt glass). The eye of Horus (uzat), an important amulet, symbolised matter, the body of the sun from which all other beings ran in the form of tears, an idea also found in the Coptic Gnostic treatise *Pistis Sophia* (A.D. 250–300 ?) [6] and in Proklos (d. A.D. 485), who attributes it to Orpheus.[7] These artificial eyes were never used for replacing lost eyes in living persons, as are modern glass eyes.[8] Herodotos [9] says the sacred crocodiles wore on their foreheads balls of glass (λίθινα χυτά), and these have been found, of millefiore glass, at Arthribis in the Delta, as well as larger and simpler ones.[10]

OLD KINGDOM GLASS

Old Kingdom glass beads are very rare in tombs, never more than one occurring in the same tomb.[11] Examples are (1) a ring bead of greenish glass (VI dyn.); (2) a ring bead of green glass (VII–VIII dyn.); (3) a spheroid of "decayed glass" (VII–VIII dyn.); (4) a spheroid of red glass (IX–X dyn., undisturbed grave, but the bead "must be ruled out as an accidental intrusion"); (5) a ring bead of bluish glass (IX–X dyn.); (6) a barrel bead of green glass (X–XI dyn.), the only one actually identified by Beck as "glass," the others are probably mostly correctly described; (7) "blue glass" spheroid beads in a necklace of the XI dyn. found by Winlock at Deîr el-Baharî. Beck[12] refers to an object not mentioned by Brunton, viz. "a barrel-shaped bead which appears to be green faience," VII–VIII dyn., a microscopic examination of which "shows that the material is a glass," impure, similar to some XVIII dyn. turquoise glass but with rather more and more finely divided colouring matter. The bead No. 6 above, of clear green glass, is dated by Beck as IX–X dyn., or not more recent than the XX dyn. [!]. The glass is full of bubbles but is only slightly corroded. Beck refers to two glass beads attributed to the XVII dyn., one blue, due "possibly to a mixture of copper and iron"; the other is corroded "but seems to have been a clear glass, and possibly to have been glazed—a process not common in Egypt, but which was used in Persia." The discussion is very indecisive.

A slab of selenite at Badari had traces of wood round it, possibly from a

[1] Kisa, 36 ; Dietz, in Ullmann, Enzyklopädie der techn. Chem., vi, 211 ; Lucas, *Analyst*, 1926, 447.

[2] Petrie, Arts and Crafts, 119 f.

[3] Budge, Mummy², 391.

[4] Arch., 215, 217, 222 ; *ib.*, Dawn, 408 ; Wilkinson, i, 409.

[5] E. Pergens, Kunstaugen aus dem alten Ägypten in technisch-chemischen Hinsicht, in Diergart, 1909, 201.

[6] Mead, Pistis Sophia, 1896 ; C. Schmidt, Koptisch-Gnostische Schriften, Leipzig, 1905, i.

[7] Legge, Forerunners and Rivals of Christianity, 1915, ii, 153, 176.

[8] Forrer, RL, 58.

[9] ii, 69.

[10] Kisa, 142.

[11] Brunton, Qau and Badari II, 1928, 21 ; correcting Erman-Ranke, 547.

[12] In Brunton, *op. cit.*, 25.

frame, and "may have been used as a mirror"[1] : no Predynastic mirrors are known but Petrie suggests that wet slate was so used.

Petrie[2] refers to "blue glass" of the VI dyn. as "unique in these tombs."

MIDDLE KINGDOM GLASS

Wilkinson[3] mentions small glass fragments (amulets ?) of the date of Antef III (XI dyn.). A blue "glass" amulet in the shape of a panther's head, in the British Museum[4], is said to be glass[5] or "soft blue paste."[6] In XII dyn. remains Berthelot[7] found three small glass rings, free from compounds of tin or other metal inside, but covered outside with a thin greenish patina containing a trace of copper, the glass being devitrified. These rings were very minute, the outer diameter being 3·05 mm. and the inside 1·55 mm. and they must have been difficult to make. They were probably originally threaded on a linen or metal thread. A blue glass frog in a silver ring, perhaps XII dyn., was found at Abydos[8], and glass objects of the XIII dyn. at Kahun.[9] A lump of solid translucent greenish glass, weathered, but found deep down with various objects of XII to XVIII dyns. at Sedment (Herakleopolis) is characterised as "a surprising object," although "clear glass beads of this age" and blue glass beads of the pre-XII dyn. period are also described.[10]

A blue glass seal cylinder of the XIII dyn., mounted in a gold finger ring, is in the British Museum[11], and a piece of square glass rod about 4 cm. long, white on three sides but with blue stripes on the fourth, with the name of Amenemhet III (1830 B.C.) in blue on a white ground, in the Berlin Museum. The design in the latter is carried through the whole length, and perhaps slices of such rods were fused on the outsides of clear glass vessels, at least in later times, for decorative purposes.[12]

NEW KINGDOM GLASS

What is described[13] as the "earliest dated piece of true glass," or "the oldest piece of vitreous paste," is the object variously called an "eye" (Petrie ; Morin-Jean) or a "bored bead" (Kisa ; Lucas), with the cartouche of Amenhotep I (1557–1501). This is a blue glass imitation of turquoise. Black and white "glass cups" of the same period are mentioned by Petrie,[14] who says all previous objects are "paste."

Sometimes referred to as the "first true glass" object is the famous bottle or vase of Thothmes III (1550 B.C.) in the British Museum, of "deep blue paste," opaque, with hieroglyphics in yellow.[15] Besides this bottle, or jug, which probably came from the tomb of Thothmes III, the British Museum has a large two-handled vase and a blue glass bowl, which certainly did.[16] Newberry,

[1] Brunton and Thompson, Badarian Civ., 1928, 3, 35.

[2] Naqada and Ballas, 1896, 45, 48.

[3] ii, 142 ; cf. Fowler, Archæologia, 1880, xlvi, 80, and Kisa, 38, who says of Antef IV, XIII dyn.

[4] Petrie, Hist., 1903, 137.

[5] Dietz, in Ullmann, 211.

[6] Petrie, Arts and Crafts, 119 f.

[7] Arch., 24.

[8] Petrie, etc., Abydos III, 50.

[9] Petrie, Kahun, Gurob and Hawara, 1890, 32 ; Kisa, 16, 74, 279, 284.

[10] Petrie, etc., Sedment I, 1924, 19 f. ; Petrie refers, for the "lump of glass," to plate xl, 5, but there is no object so numbered on the plate.

[11] Budge, Amulets, 293 f.

[12] Maspero, Arch., 260 ; Dietz, in Ullmann, 211 ; Neumann, Z. angew. Chem., 1925, xxxviii, 776 ; Erman-Ranke, 547 ; Lucas, Egy. Mat., 44, doubtful.

[13] Petrie, Arts and Crafts, 119 f. ; Hall, CAH, ii, 417 ; Morin-Jean, in Daremberg-Saglio, Dict., v, 936 ; Kisa, 36 ; Lucas, Egy. Mat., 40.

[14] Arts and Crafts, 119 f.

[15] Wilkinson, ii, 11, 142, Fig. 382, illustrated as a complete "bottle of light blue glass" ; Bucher, iii, 269 ; Erman-Ranke, 547 ; Smith, DA, ii, 972, Thothmes II ; Kisa, 38, 40 ; Newberry, JEA, 1920, vi, 155 ; Morin-Jean, 940 ; Froehner, 12 ; Petrie, Arts and Crafts, 119 f. ; ib., Illahun, 1891, 18 ; Partington, Everyday Chemistry, 21, Fig. 24, from a photograph.

[16] Budge, Mummy², 391.

in his excellent paper, with bibliography, describes a complete greenish-blue glass chalice of Thothmes III, ornamented with yellow and very dark blue, now at Munich and formerly thought to be glazed ware. About fifty perfect, or nearly perfect, examples of this type of glass are known, and Newberry had seen and noted fragments of at least two hundred and fifty broken ones. Pieces of at least a hundred vases and scores of amulets, ear-rings and broken bracelets were found among the ruins of the palace of Amenhotep III (1411–1375) at Thebes, near a glass factory similar to those found by Petrie at Amarna. A perfect bottle of this king's time, and another of the time of Tutankhamen (1350), as well as a bowl and several bottles of Rameses II (1296–1230), were found at Gurob. At Lisht there were extensive factories of this glass dating from the xx dyn. (1200–1090) and there are cups of the xxi (1090–945) dyn. at Cairo. Several xviii dyn. and later bottles, including glass (some transparent and colourless) found by Daressy in Theban tombs of Maherpre and Amenhotep II, are also described and illustrated by Kisa.[1] Ebony with "glass" inlay (including a vase containing perfume), found with glass objects in a tomb of the period of Amenhotep II, is "perhaps Sudanese work."[2]

Newberry thinks the glass itself may have been imported in the form of ingots, some known examples of which, blue and red, were probably made in the north-western Delta, where the necessary materials were found. Petrie's suggestion [3] that the work was carried out by Syrian craftsmen is dismissed [4], since no glass is found in Syria as early as Thothmes III.

There is an opaque "blue glass bead" with the name of Thothmes II (1550) [5]; a variegated blue glass bottle in the shape of a fish, now in the British Museum, found at Tell el-Amarna [6], has a decoration in wavy lines as is the case with many other specimens.

Glass objects of the xviii dyn. to the Roman Period in the British Museum include a light-blue glass koḥl-pot, a fine deep-blue glass set of Canopic jars, a deep blue glass scarab from the bead work of a mummy, a "cobalt-blue" glass head-dress of the god Bes, two two-handled white glass vases with decorated rims, and a conical cup-shaped vase of mottled black and white glass from the funerary equipment of Princess Nesi-Khensu, which fell to pieces on account of absorbed salt; much of this opaque glass is porous.[7]

All Egyptian glass except the deep red (containing cuprous oxide), the emerald green and the yellow, has some degree of transparency.[8]

A large bottle-green bead of Queen Hatshepsut had been considered as obsidian [9], the density of which is the same as that of glass, although black glass is softer and more fusible. This bead, and a turquoise-blue one of Hatshepsut found at Gurob, are probably of glass.[10] A black and white bead in the Liverpool Museum[11], supposed to be agate, may also be glass. An undated "vase at Berlin of cut glass"[12] is probably moulded. In the Tutankhamen and Rameses II remains at Gurob, Petrie found coloured glass, now in the British Museum, of surprising workmanship, including almost transparent amethyst coloured [? containing manganese]. The Seti I remains there

[1] Glas, 5, 7, 9, 11, 13, 15, 17, 19, 21, 23, 25, 42, 43, 47, 64, 67, 105, 159, 165, etc.; Maspero, Guide to the Cairo Museum, 1908, 484; Quibell, 93.
[2] Maspero, Guide, 493.
[3] Cf. Breasted, CAH, ii, 57.
[4] Newberry, op. cit., 158; Olmstead, H. Pal. Syr., 1931, 147.
[5] Petrie, Illahun, 20: Naqada and Ballas, 1896, 69.
[6] Peet, JEA, 1921, vii, 183, Pl. 30; ib., 1929, xv, 148, Pl. 28; BMGE, IV–VI

Rooms, 220; Petrie, Abydos I, 1902, 31, who dates it "by the colour," suggests that it was found at Abydos.
[7] Budge, Mummy[2], 391 f.; BMGE, IV–VI Rooms, 220 f.
[8] Kisa, 50 f., 559.
[9] Birch in Wilkinson, i, 37 f.; ii, 141, Fig. 381; Kisa, 45, 120, 288.
[10] Neumann, Z. angew. Chem., 1925, xxxviii, 776 f.; Kisa, 38, 120.
[11] Birch, in Wilkinson, ii, 141.
[12] Wilkinson, ii, 8; Fig. 275, 4.

included coarse glass or pottery beads.[1]   Other objects include a glass ushabti
(XVIII dyn.) ;   green, apparently glass, inlay ;   solid blue glass (XVIII dyn.)[2] ;
glass (XIX dyn.) and glass amulets (XXX dyn.)[3] ;   Roman glass at Denderah [4],
and finger rings at Amarna.[5]   Glass in the Cairo Museum from tombs of
Amenhotep II includes imitations of agate, onyx, marble and serpentine.[6]
By 1375 the range of colours for opaque and translucent glass had rapidly
increased and included violet, deep azure blue, bright blue, turquoise blue,
light blue, green (clear green at Amarna), yellow, orange, red (rare), golden
brown, clear white, milky-white, blue-black and black.[7]

The very thin glass vessels of Amenhotep II (XVIII dyn.) are not blown, as
might be supposed, but formed *over* (not in) a clay mould, the traces of which are
still clearly visible.[8]   Glass was never cast in Egypt, as it was too viscous to
pour ; Petrie says it was pressed as a pasty mass *into* moulds.[9]   All XVIII
and XIX dyn. glass (as well as all glass found in Greece and Italy in the 5th–4th
centuries B.C.) was formed over a mould, never blown.[10]   A lion head in relief, of
moulded glass, probably from a vase[11], is of uncertain date.   Small round plates
of glass, flat or hemispherical, colourless or coloured, transparent or opaque,
found in tombs with beads and rings, were probably made by dropping pasty
glass on a stone plate and may have been used in some game like draughts.[12]

The XVIII dyn. glass is of very fine quality ; even now it has fine colours
and is little decayed or iridescent, but Petrie's statement that Tell el-Amarna
glass (1400–1350) was better than the Ptolemaic is unjustified, since later
Alexandrian blown glass was of very fine quality and was exported even to
provinces north of the Alps in the Roman Period.[13]   The white crust on the
surface and the iridescence of ancient glass are due to the solvent action of
water.   The glass contains more alkali than modern glass, is often porous and un-
even, and is less resistant.[14]   Semper's idea[15] that the iridescence was artificially
produced, as in modern forgeries, is wrong ; it is a natural result of decay.[16]

Several cylindrical glass beakers or tumblers, of light green, yellow, blue,
black or variegated paste, were found at Deîr el-Baharî (XXI dyn. ; 1100–1000).[17]
Inlaying coloured glass paste hieroglyphics in wooden coffins goes back to the
XII dyn., and was much used in the XVIII and later dyns., the Saïte and Ptolemaic
(e.g. the inlaid throne seats in the Turin Museum) work being especially good.
Paste was also set in cement as coloured mosaic.   Canopic jars in opaque blue
glass came from Thebes : these jars were not used after the XXI dyn.[18]   Lenormant[19] refers to a vase of Amenhotep III, the grey-white glaze of which has
been incised and filled in with hieroglyphics of red and blue paste, "fused in" (?).
The cutting, grinding, polishing, and engraving of glass, probably with the use
of emery, date from the XVIII dyn., and the "cutting" (grinding) with emery
on a wheel was well known in Pliny's time.[20]

[1] Petrie, Illahun, 1891, 17, 18
[2] Petrie, etc., Royal Tombs, I, 1900, 33 ;
II, 1901, 38.
[3] Abydos I, 1902, 32, 38.
[4] Dendereh 1898, 1900, 35.
[5] Kisa, 140.
[6] Guide, 1908, 485.
[7] Hall, CAH, ii, 104 ; Kisa, 19, 47, 58,
60, 64 ; Wiedemann, 1920, 328 ; Petrie,
Arts and Crafts, 119 f. ; *ib.*, Illahun, 1891,
17 f. ; Peet and Woolley, City of Akhenaten,
22 ; Angus-Butterworth, *Glass Industry*,
New York, 1933, xiv, 21.
[8] Kisa, 48, 49, 139 ; Erman-Ranke, 547.
[9] Arts and Crafts, 119 f. ; cf. *Anct.
Egypt*, 1915, 32.
[10] Griffith, quoted by Kisa, 50 ; Morin-
Jean, *op. cit.*

[11] Wilkinson, i, 354.
[12] Kisa, 141 ; Wilkinson, i, 32 ; ii, 56.
[13] Kisa, 54, 55, 57.
[14] Wiedemann, 1920, 328 ; Neumann and
Kotyga, *Z. angew. Chem.*, 1925, xxxviii,
857 ; Kisa, 302 f.
[15] Der Stil, ii, 188 f., 195 f., 203.
[16] Forrer, RL, 388 ; Kisa, 264, 273 f.
[17] Kisa, 13, 41.
[18] Kisa, 57, 60 f., 70 ; Maspero, Arch.,
259 ; Brunton, Lahun I, 1920, 14 ; BMGE,
IV–VI Rooms, 221, 234.
[19] Prem. Civ., i, 251 ; cf. p. 116, ref. 21.
[20] Wilkinson, ii, 8, 151 ; Kisa, 60, 631,
692 ; Petrie, Tell el-Amarna, 27 ; Newberry,
JEA, 1920, vi, 158 ; Beckmann, H.
Invent., ii, 84.

LATER GLASS

The very good Saïte glass work was under Greek influence in Naukratis and other coast towns, co-operation from the 6th century B.C. reaching high development in the Ptolemaic Period. There is not much definitely Alexandrian glass known ; an example is a glass mosaic in the form of a fish at Vienna. The technique was copied throughout the whole Roman Empire from India to Britain.[1] Strabo [2] says the sand at Alexandria was the best for glass-making, and although Thebes (Diospolis) and Sidon were still important [3], Alexandria, with modern business methods including exhibitions, was the main centre [4] and Septimius Severus (193–211) taxed the Alexandrian glass in order to build free baths for the "people" of Rome. Imitation of earthenware of every kind was made in glass in Alexandria.[5] Careful methods of packing glass were used to minimise breakage in transport, including wrapping in papyrus and grass. Oriental oils and perfumes were exported in glass bottles, the larger jars being packed in wood boxes.

Large glass bottles (up to 2 gallons) shaped like carboys with longer necks, enclosed in a wickerwork of papyrus, are of uncertain date, but there are some of the XXVI dyn. of clear dark green glass.[6] Smaller bottles "of old Egyptian form" with the original wrappings, and square bottles blown in moulds, are of uncertain date, probably mostly 5th century A.D. [7], but transparent green or coloured bottles of the XXVI dyn. are in the British Museum.[8]

In the letter of Hadrian (2nd century A.D.) to Servian, which is perhaps apocryphal and of the 3rd century [9], Alexandria is said to be full of glass-blowers, papyrus and linen makers, etc. The emperor says, "I send you vases of various colours (calices tibi alassontes versicolores transmissi) given me by the priest of the temple." Alasontes or alassontes (ἀλασσοντες) perhaps means opalescent glass[10], but more probably millefiore glass, i.e. variously coloured glass vessels.[11] That the vessels contained gold purple and potassium chlorate, "like the modern iridescent glasses"[12], is impossible, both these materials being then unknown.

In the Roman Period glass was an imperial monopoly at Alexandria[13], and glass was still made there after the Arabic conquest. The head of the kiblah in the mosque of Ibn-Tûlûn still has the 10th century Egyptian glass mosaics set round with a purely classical border[14], but in a later period glass-making died out.[15]

A picture found at Hawâra (XII dyn. to Ptolemaic Period) had probably been glazed : a piece of glass which fitted it was found at Tanis, and there is in the Turin Museum a transparent, colourless glass mirror of the Ptolemaic Period backed by a metal foil (perhaps tin) in a sycamore wood frame.[16] Glass "lenses," which occur in Egypt (Kahun ; Roman Period), Asia Minor, North Africa and elsewhere, some apparently of the 3rd century B.C., were probably ornamental, not optical.[17]

[1] Kisa, 21, 76, 86 f.
[2] XVI, ii, 25 ; 758 C.
[3] Pliny, xxxvi, 26.
[4] Kisa, 62, 75 f., 86, 109 ; Büchsenschütz, Hauptstätten des Gewerbfleisses, Leipzig, 1869, 28.
[5] Athenaios, Deipnosophistæ, xi, 28.
[6] Kisa, 87 f. ; Wilkinson, ii, 152, Fig. 383.
[7] Kisa, 19, 25, 129, 131, 325 f.
[8] Birch, in Wilkinson, ii, 11, 39 ; Figs. 279, 305 g ; Fowler, Archæologia, 1880, xlvi, 84.
[9] Mommsen, Römische Geschichte, 1885,

v, 485, 576 ; Harnack, Mission und Ausbreitung des Christentums in den ersten drei Jahrhunderten, Leipzig, 1902, 448.
[10] Semper, Der Stil, ii, 199 ; Marquardt, Privatleben der Römer, Leipzig, 1879, 730.
[11] Uhlemann, Thoth, 247.
[12] Kisa, 306.
[13] Charlesworth, Trade Routes, 29.
[14] A. J. Butler, The Arab Conquest of Egypt, Oxford, 1902, 103.
[15] Lane, Modern Egyptians, ii, 3.
[16] Kisa, 360 f.
[17] Wiedemann, 1920, 412 ; Meyerhof, MGM, 1929, xxviii, 303.

THE EGYPTIAN GLASS INDUSTRY

The manufacture of glass on the large scale began under the great Theban dynasties : remains of glass factories and small heaps of rubbish have been found near the Ramesseum at Thebes (Rameses II), at El-Kâb, Tell Achmunein, and also in the Libyan Desert near the Natron Lakes.[1] The curious "natural glass" recently found in the Libyan Desert is a kind of silica, perhaps formed by a meteorite.

The most important known site of an Egyptian glass factory, however, is that at Amarna, a town founded by Ikhnaton (1375–1350) and abandoned soon after his death.[2]　Much glass was found, including blue "engraved" glass, inlays of glass rod, and a large dump of so-called "Phœnician" glass really made in the town. There was a considerable amount of gorgeous glazed ware, used for decoration of rooms, inlaying pillars, etc. (along with gold), tiles (with waves, fish, daisies, lotus, thistles, etc.), and other naturalistic art ; glass finger rings, decorations to be sewn on clothing, pendants, hieroglyphics, etc.[3] Abundant remains of melting pans and glass and glaze factories were found, but no actual glass furnaces.[4]　One furnace had been used for making charcoal. The alkali and quartz sand were first made into a frit in fritting pans, as previously described, and the frit then melted in deeper pots. The lime necessary for the glass was probably present as an impurity in the sand, on which the whole site is built.[5]　Lime is also omitted by mediæval authors. The glass was fused in the crucibles, 2 to 3 in. deep and diameter, samples being taken out with round-tipped copper tongs. It was then allowed to cool, the upper frothy scum (found on some specimens) and the lower sediment were broken off (as in modern optical glass-making), and the clear block of good glass broken up, softened by heating, laid on a smooth plate and rolled diagonally to a thick rod. Signs of this rolling are seen on the pieces of glass found. The glass was drawn into rods or thin ribbons, flattened into bands, polished and used for inlay. Tubes and beads were found.[6]

A broken pan showed broken white pebbles free from iron, taken from the Nile and coming down from the disintegration of primitive rocks farther south.

The "charcoal furnace," found with much charcoal, was an irregular square, 43–57 in. sides, 35 in. high, the north door to admit wind was 29 in. high × 15 in. wide ; the exit for gases on the south was 16 in. high × 13 in. wide. The roof was destroyed. In a later account [7] Petrie says that straw only, not charcoal, was used. Pliny [8] says dry wood was used for the glass furnaces ; other authors (Cassius Felix ; Olympiodoros) that they were fired with papyrus roots.[9]　Petrie's account, which is by no means clear, seems to imply that the frit made in pans was melted down in crucibles. Remelting glass is not an easy operation, as it tends to devitrify in the process, but in Mesomedes (c. A.D. 130) a lump of glass is said to be broken up with a hammer and melted in a furnace.[10]

Tubes were possibly made by heavy rolling of rods so as to make them hollow inside : they were used only for beads, never bent to make ornaments or siphons. Beads were usually made by wrapping glass fibres around a wire ; later Coptic beads from tube drawn and nicked. Drawn rod was bent into

---

[1] Wiedemann, 1920, 328 ; Kisa, 43 f. ; Wallace-Dunlop, 4.

[2] Breasted, Anct. Rec., ii, 402 f. ; ib., CAH, ii, 128.

[3] Petrie, Tell el-Amarna, 12 f. ; Kisa, 73.

[4] Petrie, ib., 25 f. ; Kisa, 17 f. ; Maspero, New Light on Ancient Egypt, 1908, 70.

[5] Lucas, Egy. Mat., 48; Petrie, Tell el-Amarna, 7 f. ; cf. Herodotos, ii, 8.

[6] Petrie, ib., 25 f. ; Kisa, 17.

[7] Trans. Newcomen Soc., 1926, v, 72.

[8] xxxvi, 26.

[9] Blümner, iv, 389.

[10] De Vitro, fragm. in Anthologia Græca, xvi [Anthol. Planudea], 323 ; Tauchnitz ed., 3 vols., Leipzig, 1829, iii, 306.

ear-rings. Variegated glass vases were made by dipping a sand core attached to a tapering metal rod ; on cooling, the metal rod contracted loose and the sand core was rubbed out. The glass was then hand-worked, the foot formed by pressing out, the brim turned outwards and the pattern applied by winding thin threads of coloured glass around, rolling into the mass and then "dragging" to get a wavy pattern : single drag U U U ; double drag W W W.[1] The wavy design was produced by pulling, the coloured rods being held in place by metal pins, then the final surface was rubbed down with emery and the object re-heated to obtain a smooth polished surface.[2] The twisted margin of the brim or foot was formed by winding one thread round another and bending the two round the vase. This type of glass was afterwards largely used in Magna Græcia, but was not so brilliant as the early Egyptian.

GILT GLASS

The gilding of glass was an Egyptian invention. From the XVIII dyn. gilt glass beads alternate with coloured beads in some necklaces, and some gilt ribbed cylindrical beads were found at Gurob. In the Slade collection (British Museum) there is a completely gilt small spherical Egyptian bottle. The gilding was executed by covering the hot vessel with gold leaf and dipping in molten colourless glass.[3] Froehner considered the gilt glasses (ὑάλινα διάχρυσα) of Ptolemy Philadelphos (284–246) [4] as such vessels, but they may have included vessels merely "sprinkled" with gold, perhaps made by blowing [in later times ?] a gilt vessel so as to break up the gold leaf. An Egyptian dish of colourless glass with an inscription surrounded by stars in gold leaf, is also in the Slade collection, but most gilt glasses are of the Roman Period. The tradition of their manufacture lingered in the Middle Ages in the East, in Alexandria and in Byzantium, and such objects are mentioned in the mediæval treatise of Eraclius.[5]

Glass beads with gold leaf inlay (covered with colourless glass) were made in Egypt in the 4th century B.C. and are common in the Roman Period.[6] Some were found at Badari. Gilding by fusion was replaced in the 17th century A.D. by a mere sticking on of gold leaf by adhesives.[7]

COLOURLESS GLASS

Colourless glass first appears at Thebes about 1400 B.C.[8] The earliest was perhaps made with very pure sand free from iron, although the glass found at Amarna contains manganese. Since manganese had been used before in brown glazes, the use of pyrolusite (manganese dioxide) in decolorising glass was perhaps made accidentally, and artificially decolorised glass was more used for the later thin blown vessels. The general use of pyrolusite perhaps began only in the Roman Period. Excess of manganese gives a violet glass, and manganese occurs in violet and amethyst glass from Memphis as well as in Roman specimens.[9] There is colourless glass as beads, bottles and vases of the

[1] Newberry, JEA, 1920, vi, 158 ; Petrie, Tell el-Amarna, 26 f.

[2] Cf. Wiedemann, 1920, 329 ; Kisa, 695, 772.

[3] Kisa, 834 f.

[4] Athenaios, v, 30.

[5] Kisa, 96, 838, 841 f. ; Merrifield, Original Treatises on the Arts of Painting, 1849, i, 186.

[6] Kisa, 128, 292, 834, Saïte Period ;

de Pauw, i, 308 ; Budge, Mummy², 267.

[7] Kunckel, Ars Vitraria Experimentalis, Franckfurt and Leipzig, 1679, 7 ; misread by Kisa, 834, 897 f.

[8] Wiedemann, 1920, 328 ; Kisa, 103, 105 ; Blümner, PW, vii, 1383 ; Fowler, Archæologia, 1880, xlvi, 79, put it too late, 8th century B.C. to 6th century A.D.

[9] Kisa, 262 f., 282, 294 f. ; Blümner, iv, 392.

Saïte Period (663–525), including a vase from Memphis and specimens from Thebes in the British Museum, and the torso of a statuette of crystal glass in Paris is probably Ptolemaic or early Roman[1] ; a colourless tall vase from Hawâra, of the Roman Period, has been cut with the wheel.[2]   Colourless glass was not very common, even at Pompeii, and its large scale production may have been evolved in Alexandria as a reply to the invention of glass-blowing in Sidon about 20 B.C. to A.D. 20.[3]

BLOWING GLASS

Glass-blowing was quite unknown in ancient Egypt. The XII dyn. representations at Beni Hasan, formerly thought to show glass-blowing [4], occur in a series of representations of metal work, the so-called "blowing tubes" being blowpipes for the fire, and the supposed lumps of glass, coloured bright greenish-grey, are the fireclay (not metal) ends of these pipes. In later representations, e.g. in the tomb of Rekhmara (XVIII dyn.), the ends of the blowpipes are coloured yellow to represent clay.[5]

No early Alexandrian or other Ptolemaic remains are of blown glass[6] : both Seneca [7] and Pliny [8] speak of glass-blowing as something new and it was certainly unknown in the 6th century B.C.[9]   The process of casting or moulding glass is much older.[10]   It has been supposed that glass-blowing originated in Egypt not before, or very little before, the Roman Period[11], but the modern view is that glass-blowing began in Syria during the 1st century B.C., or in the period 20 B.C. to A.D. 20., and perhaps in Sidon.[12]   Kisa thinks it developed from blowing soap-bubbles, but although soap was known there is no evidence that it was used for washing in the Roman Empire. Rathgen thinks glass-blowing was imitative of metal working. It soon spread among the glass factories of the Eastern Mediterranean, chief among which were those at Sidon, Tyre and Alexandria. There is a *blown* glass figure of a whale, possibly Greek, of the 1st century A.D.[13]

All glass vessels found in Roman-Nubian graves at Karanog[14] are blown. The quality varies, some being very thin and transparent, others thick and nearly opaque ; the colours are green, blue-greenish, "approaching white" [colourless], purple-brown decorated with white opaque spiral ; some specimens were probably once gilded, and one was blown in a mould. Two large vases were very thin, and the transport of vessels so large and delicate must have been difficult. All the glass was probably imported from Lower Egypt or some other place in the Roman Empire, since very fragile vessels found along the Roman wall in Britain were not made there. Many glass beads, probably of local make, were found at Karanog, some transparent colourless, others opaque in every shade of red, blue and yellow, and also *millefiori* beads. They are mostly made from glass rods cut and rolled, the variegated beads from a rod composed of a bundle of coloured threads lightly fused together.

[1] Kisa, 292 f. ; Budge, Mummy[2], 391 ; Fowler, *Archæologia*, 1880, xlvi, 84 ; Marquardt, Privatleben, 724.
[2] Petrie, Ten Years' Digging in Egypt, 1893, 101, Fig. 74.
[3] Kisa, 298, 300.
[4] Wilkinson, i, 32, 38 ; ii, 140 ; Smith, DA, ii, 972 ; Maspero, Histoire ancienne, 1904, 137 ; *ib.*, Arch., 257 ; Rodwell, The Birth of Chemistry, 1874, 48 ; Blümner, iv, 380.
[5] Kisa, 3, 34 f., from Griffith ; Petrie, Arts and Crafts, 119 f. ; *Anct. Egypt*, 1914, 33.

[6] Kisa, 174, 292, 295 ; Wiedemann, 1920, 327.
[7] Epist., 90.
[8] xxxvi, 26.
[9] Kisa, 36.
[10] Semper, ii, 194.
[11] Woolley and MacIver, Karanog, i, 82 : cf. Erman-Ranke, 547.
[12] Rathgen, *Sprechsaal*, 1913, xlvi, 98 ; Kisa, Glas, 292 f. ; *ib.*, MGM, viii, 34.
[13] *B. Mus. Quarterly*, i, 105.
[14] Woolley and MacIver, Karanog, i, 72 f. ; ii, plates 37–40.

ANALYSES OF EGYPTIAN GLASSES

Several analyses of Egyptian glasses of various periods are available.

EGYPTIAN GLASSES FROM THEBES, ABOUT 1500 B.C. DARK BLUE, OPAQUE.[1]

SiO_2	CaO	MgO	Al_2O_3	FeO	Mn_2O_3	K_2O	Na_2O	CuO	SO_3	Total
67·82	4·03	2·30	4·38	1·08	1·12	2·34	13·71	1·96	0·98	99·72
62·48	5·57	4·16	1·56	1·73	0·76	2·24	17·80	2·72	1·39	100·41

The first specimen in the table resembled good Alexandrian glass in high silica and low alkali content, and these "dolomitic" glasses (containing magnesia) with high alumina content are probably typical of the oldest Egyptian products. With the firing facilities available, they must have been extremely difficult to melt. The dark blue colouring in these glasses was due to copper and manganese oxides only.

Analyses of glasses from Amarna, c. 1400 B.C., gave [2] the following results, the analysis of a modern bottle [3] being added for comparison :

	Opaque			Transparent	
	Dark blue	Leaf green	Turquoise blue	Colourless	Bottle (modern)
SiO_2 .	61·7	62·44	62·58	63·86	62·54
Al_2O_3 .	2·45	1·00	0·82	0·65	4·42
Fe_2O_3 .	0·72	0·84	0·58	0·67	1·34
MnO .	0·47	—	—	trace	—
CaO .	10·05	9·23	9·33	7·86	20·47
MgO .	5·14	3·05	4·37	4·18	5·41
Na_2O .	17·63	18·08	18·19	22·66	4·73
K_2O .	1·58	2·76	2·75	0·80	0·94
CuO .	0·32	2·00	0·52	—	—
Cu_2O .	—	—	—	—	—
PbO .	—	0·47	—	—	—
SnO_2 .	—	—	0·47	—	—
SO_3 .	—	0·72	0·45	—	0·10

Another dark blue opaque glass contained 0·44 Fe_2O_3 and 0·45 CuO, with a trace of MnO ; a colourless transparent glass 0·54 Fe_2O_3 ; a violet opaque glass 0·57 Fe_2O_3 and 0·89 MnO. A black glass from Amarna (1400 B.C.) contained 0·50 Fe_2O_3, 0·32 MnO, 0·20 CuO ; a dark yellow 1·08 Fe_2O_3 ; a light yellow 0·96 Fe_2O_3 ; a "hæmatine" (sealing-wax red) glass contained 0·75 Fe_2O_3 and 12·02 Cu_2O, the colour being due to cuprous oxide. A honey-yellow transparent glass contained 0·80 Fe_2O_3. The proportion of alkali is large as compared with modern glass, and the proportion of lime smaller, so that the Egyptian glass is more fusible and less resistant. Softening points of Egyptian glasses [4] are : (1) 1500 B.C. (Thebes), 880° C. ; (2) 1400 B.C. (Amarna), 785° C. ; (3) 2nd century B.C. (Elephantine), 744° C. ; (4) 1st century B.C. (Alexandria), dark blue transparent, 770° C.

The small amount of copper required to produce the colour is striking and the lead and tin in some specimens should be noted. Jackson [5] implies that lead is frequent in Egyptian glass, but gives no authority. The potash is

[1] Neumann, Z. angew. Chem., 1927, xl, 963.
[2] Neumann and Kotyga, Z. angew. Chem., 1925, xxxviii, 776 ; W. Turner, J. Chem. Soc., 1926, 2091 ; Anct. Egypt, 1930, 90.
[3] Verneuil, in Moissan, Traité de Chimie, 1904, iii, 751.
[4] Neumann, Z. angew. Chem., 1927, xl, 966.
[5] Nature, 1927, cxix, 400; cxx, 264, 301.

small and, except in the case of the colourless glass, in practically constant ratio to the soda, indicating a possible impurity in the latter, although potash has not been found in native soda from Egypt.[1] The addition of nitre to ~lear the glass is improbable, since the proportion of potash is lower in the colourless glass. Petrie[2] states that the glass was free from lead and borates, usually not quite colourless, but sufficiently so to take up various colours.

Analyses of glasses from Gurob Medinet, of 1500 B.C., gave[3] the following results :

	Dark blue trans-lucent ; grey and yellow inlays	Dark blue	Violet, nearly transparent	Rich blue trans-lucent
SiO_2 . .	67·80	62·70	62·90	67·03
CaO . .	3·80	8·80	8·87	7·83
MgO . .	2·89	3·29	5·49	4·93
Fe_2O_3 . .	0·92	1·07	1·29	1·88
Al_2O_3 . .	3·22	3·82	2·58	2·48
Mn_2O_3 . .	0·54	0·83	1·71	2·26
CuO . .	1·51	1·00	0·46	0·79
Na_2O . .	16·08	15·21	12·83	10·12
K_2O . .	2·08	2·12	1·86	1·82
SO_3 . .	1·01	0·94	1·51	0·75
SnO_2 . .	0·51	0·41	0·42	0·39

The tin oxide was present in the white opaque stripes on the transparent or translucent blue ground. No cobalt was found in these glasses. Neumann refers to analyses by Lucas of glasses from Tutankhamen's tomb (c. 1350 B.C.)[4], one of which probably contained tin oxide (it was not analysed), whilst cuprous oxide was found by analysis in the red glass, and cobalt in one dark blue glass, although usually the blue was coloured by copper, and a XIX dyn. yellow glass contained lead and antimony. The dark blue glass showed signs of deterioration ; the light blue and red glasses were well preserved. The analytical results leave no doubt that cobalt occurs occasionally in *old* Egyptian blue glass and glaze, as Wiedemann[5] had correctly stated.

Rathgen found only cobalt as colouring material in a very thick dark blue glaze, with a violet tone, of 1375 B.C.[6] Some at least of the blue cobalt glass was imported from Babylonia.

Gmelin, who analysed some specimens of ancient blue glass, found the colour was due to iron[7] ; Kisa[8] suggests along with arsenic, Sprengel[9] "the blue scum which floats in the smelting of hæmatite." The usual colours given by iron are bottle-green, brown or almost black, but some blue Chinese glazes are said to owe their colour to ferrous oxide[10]. Bancroft and Cunningham say the blue colour is apparently due to an unstable blue modification of ferric oxide, stabilised by ferrous oxide.[11]

The use of the word "cobalt" as an adjective ("cobalt blue") to describe a full blue colour, without reference to chemical composition, should be discontinued.[12]

[1] Lucas, Egy. Mat., 49.
[2] Arts and Crafts, 119 f.
[3] Neumann, *Z. angew. Chem.*, 1929, xlii, 835.
[4] Carter and Mace, Tomb of Tutankh-amen, ii, 170 f.
[5] PSBA, 1893, xv, 113 ; *ib.*, Das alte Ägypten, 1920, 338, 347.
[6] Fester, Entwicklung der chemischen Technik, 1923, 11.

[7] T. Thomson, System of Chemistry, 1831, i, 536.
[8] Glas, 284.
[9] Histoire de la Médicine, 1815, i, 64.
[10] H. Jackson, *Nature*, 1927, cxix, 400.
[11] *J. Phys. Chem.*, 1930, xxxiv, 1.
[12] Examples of incorrect uses in Trow-bridge, Philological Studies in Ancient Glass, Urbana, Illinois, 1928, 14 : "cobalt paste formed of pounded kyanos," etc.

Early analyses of ancient red glass by Klaproth, Quicheret and Hilburg indicated two main varieties : (1) Pliny's hæmatinum, described as dark purple-red and opaque, and (2) a sealing-wax red. According to Tischler, type 1 shows under the microscope dendritic crystallisations of cuprous oxide in a colourless glass ; type 2 in thin layers shows fine opaque particles of cuprous oxide or copper which appear metallic red in reflected light. The hæmatinum was reproduced in the 19th century by Pettenkofer, but was perhaps known before in the Vatican mosaic works, and there are earlier examples of the Ptolemaic Period.[1]

Jackson [2] says the brilliant scarlet early Egyptian glass owes its colour to large crystals of cuprous oxide ; an orange-yellow glass found in Saxony, which contains cuprous oxide, has only recently been reproduced. It is said [3] that not all red glass, at all events during the xviii dyn., is of the cuprous oxide type which shows green breaks when corroded, but it is not clear whether the "cuprous oxide" or the "green breaks" constitute the peculiarity.

OPAQUE COLOURED AND TURBID GLASSES FROM ELEPHANTINE, SECOND–FIRST CENTURIES B.C.[4]

	Turquoise blue	Reed green	Black	Black	Milk white	Hæmatine red			Dark blue transparent	Colourless transparent
						(a)	(b)	(c)		
Fe_2O_3	0·67	0·57	0·81	9·98	0·51	0·86	1·57	1·30	0·78	0·28
MnO	—	—	0·51	0·25	—	0·53	0·74	0·31	0·61	0·97
CuO	2·73	3·01	0·16	—	—	—	—	—	0·95	—
Cu_2O	—	—	—	—	—	2·09	2·52	4·40	—	—
PbO	—	0·93	1·27	—	—	1·28	3·02	6·28	—	—
SnO_2	—	—	—	—	0·54	—	—	—	—	—

Late Egyptian glass from Alexandria, nearly transparent and deep blue, contained : (a) Fe_2O_3 0·89, MnO 0·75, CuO — ; (b) Fe_2O_3 0·14, MnO 0·51, CuO 0·31. All the glasses examined contained 60 to 70 per cent. of silica, 14 to 23 per cent. of alkali (as oxides) and 4 to 11 per cent. of lime, i.e. less silica and more alkali than good modern glass (72 silica, 15 alkali, and 13 lime). This was probably because ancient glass was worked at a lower temperature. Sulphate was present, and small air bubbles which give a milky appearance. A dark blue colour was generally produced by iron, copper and manganese ; a pale blue, pale violet, yellow and green by iron and manganese, though certain Egyptian glasses were coloured green by copper and lead. Black was produced with iron and manganese, and sometimes copper. Brilliant red opaque glass was coloured with varying amounts of cuprous and lead oxides.[5]

The presence of cuprous oxide in Egyptian red glass is indicated by the "pieces of crimson glass, resembling sealing-wax but turned green by decomposition," in the Ashmolean Museum, Oxford.[6] The cuprous oxide in the glass, by the action of air and moisture, would be converted into a green surface deposit. Red glass was used instead of jasper for amulets.[7]

[1] Kisa, 277 f., 282.
[2] *Nature*, 1927, cxix, 400 ; cxx, 264, 301.
[3] Glanville, JEA, 1928, xiv, 190.
[4] Neumann and Kotyga, *op. cit.*
[5] Neumann and Kotyga, *op. cit.*

[6] G. J. Chester, Catalogue of Egyptian Antiquities in the Ashmolean Museum, Oxford, 1881, No. 307 ; cf. also No. 145, an object of uncertain use, of red glass.
[7] BMGE, IV–VI Rooms, 97.

Some earlier analyses of Egyptian glasses[1] gave :

	SiO_2	Na_2O	CaO	Fe_2O_3	Al_2O_3	MnO	MgO
1. Colourless rod .	72·30	20·83	5·17	0·51	1·19	—	—
2. Disc, used in games	70·58	20·70	6·54	0·99	1·19	—	—
3. Bottle-green disc .	71·15	18·76	8·56	0·25	0·84	0·44	tr.
4. Brown rod . .	65·90	22·33	8·42	0·97	1·44	0·94	—

IMITATION GEMS

The imitation of gems by coloured glass is an old tradition in Egypt, going back at least to the XVIII dyn.[2] Pliny [3] says emeralds were easily imitated and the art of counterfeiting gems, of which he refuses to give details, was "the most profitable devised by the ingenuity of man, the counterfeit being very difficult to distinguish from the real." Imitation gems, marble, onyx, etc., with veinings, were made at Alexandria. The older fragments show as principal colours many blues and yellows, pure opaque white and whites with casts of yellow and blue, orange, brown, black, purplish azure-blue, turquoise blue (the last two very fine), green, red and violet.[4] Pliny [5], who says emeralds were found at Thebes, mentions an emerald in the form of a statue of Serapis in the Labyrinth, nine cubits (12 ft. 9 in.) in height ; another, sent to Egypt by a king of Babylon, was four cubits high and three cubits broad, and there were other pillars. Pliny says these must have been of imitation emerald (pseudosmaragdos), and Wilkinson [6] thinks they may have been green glass. Similar statues, pillars, coffins, etc., are also reported for Babylonia, Tyre, Ethiopia, etc. It would have been very difficult to make such large pieces of glass, and old authors had a very imperfect knowledge of the glass industry, which was probably kept secret. In all probability the objects were of malachite, some green stone, or glazed tiles.[7]

The beginnings of glass painting go back to later Egypt : fragments were found especially at Oxyrhynchus, not earlier than the Ptolemaic Period and probably early Roman, with a black border and filled (without modelling) with red and yellow.[8]

VASA MURRINA

The fragments of coloured glass picked up during the Napoleonic expedition to Egypt were thought by Rozière to be pieces of ancient vasa murrina (or murrhina).[9] These are mentioned only three times by rather late Greek authors[10], but by many Latin authors, first by Propertius[11], but most fully by Pliny.[12] Their nature is very obscure, but the most reasonable hypothesis appears to be that there were two kinds, a natural one, probably fluorspar (although stone, shell, amber, meerschaum, onyx, sardonyx, carneol, alabaster, agate, jade, chalcedony, gum such as myrrh, obsidian, and Chinese soapstone have been suggested), and an artificial, which was probably coloured glass

[1] Benrath and Schüler, in Muspratt, Chemie, Brunswick, 4 ed., iii, 1366.

[2] Hoefer, i, 64 ; Froehner, 45–49 ; Kisa, 62.

[3] xxxvii, 12.

[4] Kisa, 59 f. ; Maspero, Arch., 259.

[5] xxxvii, 5.

[6] i, 63 ; ii, 146.

[7] Wilkinson, ii, 146 ; Bucher, i, 6 ; Kisa, 44, 56, 72, 92, 101, 273, 301.

[8] Kisa, 811 f.

[9] Hoefer, Hist. de la Chim., 1866, i, 64.

[10] Periplus Maris Erythræi, written c. A.D. 60, ed. Müller, Geographi Græci Minores, Paris, 1855, i, 261, 293, μυρρίνη, made in Thebes ; Pausanias, VIII, xviii, 5, c. A.D. 170, μορρία, "glass, crystal, murrina and other things made by men from stone"; Pollux, x, 69, c. A.D. 190.

[11] Eleg., IV, v, 26, murrea in Parthis pocula cocta focis.

[12] xxxvii, 2.

(although glazed pottery and porcelain—not known in antiquity—have been proposed).[1] Kisa [2] thinks they were *all* of coloured glass, and no other kind is known in actual specimens.

Bowls of coloured glass are shown in Egyptian paintings.[3] The method of making them [4] was probably to form a glass rod by fusing a bundle of coloured rods together. This was cut into thin slices (perhaps after drawing out to any required thinness) to give concentric rings, or if cut sloping, ovals. Such plates were found by Petrie at Tell el-Nebesheh [5] of the Saïte, Ptolemaic and Roman Periods, and this type of glass is described by Caylus, who gives attempts to reproduce it [6], as like agates. Birch considered that this technique was late Alexandrian or Roman, and there is a late specimen in the British Museum.[7] The Romans used plates and vessels of millefiore glass from the 1st century A.D.[8]

To make the dishes, the slices were perhaps laid in a terra-cotta bowl, were softened by heat and joined by blowing a transparent, usually coloured, glass inside them. The bowl could then be finished by pressing and grinding. Bands were formed from longitudinal sections of the rods, and filigrane glass by radial arrangement. An outer colourless layer, however, is never found on antique bowls, but only on Venetian specimens.[9] The colour penetrates throughout ; the ground is usually transparent deep blue, violet-red, golden brown, or green ; the rings, bands, and spots are opaque white, yellow, dark brown, red, blue, or emerald green. Some pieces contain gold leaf or gold fragments. Most of the mosaic bowls were found in Egypt, the East, Greece, South Italy and Etruria ; the South Italian are perhaps native, the Etruscan imported from Egypt. Most museum specimens are Italian. The best period is A.D. 0–50. Imitation of agate and onyx, unknown in ancient Egypt, began in the Ptolemaic Period, perhaps because coloured walls were replaced by glass plates ; it was developed in the first period of the Roman Empire at Alexandria and Rome.[10]

ENAMEL

Franchet[11] defines enamel as identical with glaze in composition, but containing an added opaque body : it is always opaque, owing to oxide of tin in the white ; antimony (in presence of lead) in the yellow ; oxide of chromium in the green, rose, purple-red and coral red ; and oxide of iron in the bright red and brown-red. This definition of enamel includes "paste."[12]

Although Semper[13] saw specimens of the three different kinds of enamel from Egypt, including a purple enamel on gold from Canopus, and Forrer[14] thinks true fused enamel ("émail champlevé") was used in ancient Egypt, it is usually stated[15] that all so-called Egyptian "enamels", including the supposed XII dyn. specimens found by Maspero at Dahshûr, are not true enamels. The vitreous paste was never united with the metal support by firing, as in Chinese cloisonné, but separate bits of gems or stones (carnelian, amethyst, lapis

[1] Thiersch, *Abhl. kgl. Bayerischen Akad. d. Wiss., Philos.-Philol. Kl.*, 1835, i, 439–510 ; Trowbridge, Studies in Ancient Glass, 1928, 83 f. ; Marquardt, Privatleben, 740 ; Blümner, iv, 396.

[2] Glas, 180, 273, 420, 501, 532, 551, 555, 565.

[3] Erman-Ranke, 195, Fig. 49.

[4] Wilkinson, ii, 144 ; Kisa, *locc. cit.*

[5] Kisa, 501, 504 ; BMGE, IV–VI Rooms, 221.

[6] Recueil d'antiquités, i, 293 f., plate cvii.

[7] Wilkinson, ii, 145, 288, plate 14.

[8] Woolley and MacIver, Karanog, i, 75.

[9] Semper, Der Stil, ii, 201 ; Kisa, 509, 513 ; Figs. 203, 204, in South Kensington Museum ; 205, plate iv ; Figs. 211–15 ; Forrer, RL, plate 70, Figs. 1, 2 ; Dillon, Glass, 1907, 49 f. and plate.

[10] Kisa, 510, 512, 518 f., 522, 834.

[11] Céram. Prim., 79.

[12] Eyer, in Ullmann, Enzyklopädie der techn. Chem., 1916, iv, 543.

[13] Der Stil, i, 425, 475 ; cf. Kisa, Glas, 148.

[14] RL, 306.

[15] Kisa, 146 f. ; Petrie, Arts and Crafts, 83 f., 94 ; *ib., Anct. Egypt*, 1914, 87.

lazuli, turquoise, jasper, etc.), as in those in the tomb of Queen Aah-hotep (XVIII dyn.) [1], or else coloured glass pastes, were mounted in gold, silver, etc., by a cold process. This work, really a kind of mosaic, therefore resembles the process of hammering gold and silver threads into grooves in bronze, which is also an Egyptian method.[2] The inlaying of floors and walls in mosaic with glass cubes began only about the commencement of the Roman Empire.[3] Labarte [4] considered that ἤλεκτρον (elektron) in Homer meant enamel, as in some Mediæval authors [5], but this is doubtful, and the material was probably either amber or the alloy of gold and silver.[6] F. de Lasteyrie [7] considered that true enamel was not known to the Greeks and Romans before the 3rd century A.D., and Blümner [8] that true cell enamel (émail cloisonné), in wire cells soldered on the metal, was also not known till a late period.

Opaque white glasses containing oxides of lead and tin [9] which occur in Egypt are, however, "enamels" in one sense, and a set of gold cloisonné enamel necklets was found in the tomb of Tutankhamen.[10] The variety called émail champlevé (Grubenemail), in which the paste is melted into engraved depressions, is said by Forrer[11] to be "old Egyptian," but especially used by the Kelts of the later La Tène Period.

The "Ethiopian jewellery" discovered in a pyramid at Meroë, which has been supposed to show Ptolemaic-Greek influences, includes very fine cell enamel in light and dark blue, emerald green and red, also blue glass beads containing cobalt and gilt glass beads.[12] Birch doubted whether this jewellery in Berlin, which was purchased in London, was genuine, but the Berlin authorities[13] maintain that it is. True enamel was probably known in Greece from about the 4th century B.C. and also in Alexandria and Rome, but there it was rare. There is a crown from a 4th century B.C. grave in Italy, with inlays of blue enamel of Greek workmanship, in Munich and azure blue enamel on gold of the Alexandrian Period in Berlin. The first literary mention of true enamel is said to occur in the 3rd century A.D. in the *Imagines* of Flavius Philostratos[14], who says of certain coloured metal decoration : "the barbarians of the ocean pour out these colours (τὰ χρώματα) over the red hot bronze (τῷ χαλκῷ διαπύρῳ) so that these congeal and adhere, become hard as stone (λιθοῦσθαι) and preserve the image." The "barbarians" are probably the Kelts of the French and British coasts, the latter being very expert in enamel work[15] which is quite different from the Græco-Egyptian type.[16]

The constituents of antique enamels according to Cohausen are copper (blue, green, red) ; cobalt (blue) ; antimony and uranium (yellow and orange) ; iron and chromium (green). Egyptian glazes analysed by K. A. Hofmann contain soda and ferric oxide in the brown-red ; cobalt with alumina, silica

[1] Said by Hoernes, Urg., ii, 395, and Feldhaus, Technik, 265, to be true enamel.

[2] Mariette, quoted by Perrot-Chipiez, Art in Egypt, i, 837, 839 ; Bucher, i, 5 f. ; ii, 566 ; Hall, CAH, ii, 416 ; Forrer, RL, 201 ; Maspero, Arch., 322 f. ; Blümner, iv, 407 f. ; Kisa, 147.

[3] Kisa, 129, 372 f. ; Pliny, xxxvi, 25 : the frieze inlaid with blue glass found at Tiryns, q.v., however, is a beginning of this style.

[4] Recherches sur la peinture en émail dans l'antiquité et en moyen âge, 1856; *ib.*, Histoire des arts industriels au moyen âge et à l'époque de la renaissance, 1872-5, iii 377.

[5] Theophilus, iii, 12 ; electra in auro : Hendrie, Essay upon Various Arts by Theophilus, 1847, 130.

[6] Bucher, i, 3, 7 ; Kisa, 146, 163.

[7] L'Electrum des anciens était-il de l'émail ? 1857, 81, 91 ; similar view in Petrie, Arts and Crafts, 94 ; *Anct. Egypt*, 1914, 87.

[8] iv, 407, 413.

[9] Wiedemann, 1920, 328.

[10] Carter, Tomb of Tutankhamen, ii, 261 f.

[11] RL, 306.

[12] Kisa, 129, 148 f., 372 f., 502 f.; Blümner, iv, 411 ; Lepsius, Met., 27.

[13] Schäfer, Möller and Schubart, Ägyptische Goldschmiedearbeit, 1910, 95.

[14] i, 28 ; ed. Jacobs, Leipzig, 1825, 44.

[15] Kisa, 150 ; T. R. Holmes, Ancient Britain, 237 f.

[16] Blümner, iv, 412 ; Kisa, 151 f.: Semper, Der Stil, i, 475 ; ii, 567.

and soda in the blue ; manganese and iron in the black ; and manganese in the violet.[1]

The taste for coloured glass died out about A.D. 50, after which colourless transparent glass was preferred.[2] A transparent yellow glass is apparently referred to in Revelations.[3]

PIGMENTS

The Egyptians used pigments for statues and walls at a very early period (IV dyn.), in a variety of colours.[4] Ceilings, wood and linen were painted.[5] Pliny [6] refers to the painting of stone (lapidem pingere), perhaps to look like porphyry, as something new in Rome, and also says that walls were painted with plants (herbis tingi lapides). The Egyptians, like the Greeks, painted stone statues : one of limestone at Memphis of the V dyn. shows colours which are still fresh, and Yâqût (b. A.D. 1178) says the head of the sphinx had been painted red, which lasted a long time.[7] The painter is shown holding the brush in the right hand and the pot of paint in the left.[8] The Egyptian loved colour and used it whenever possible.[9] The small number of colours used is thought[10] to have had a religious significance : they were prescribed and then used without change, but the range of available materials was small. Old V dyn. palettes show seven colours, red, yellow, brown, black, blue, green and white.[11] The colours were kept in small bags[12] or made up into cakes, found in tombs along with the small rubbing stone (šed-t) and stone or pottery mortar (meger ?) and pestle, probably also used for mixing drugs as well as colours.[13] The seven pigments : black (qemi), dark blue (chesbet), red, light blue, green (wod, wd, or uaz), yellow and white, probably correspond with the precious minerals and stones, lapis lazuli, ruby, turquoise (bright blue), emerald (green), topaz (yellow) and rock-crystal (white) ; a relation with the planets is improbable.[14] In the later periods, flesh-colour, cinnabar-red, purple, and gold are used.[15]

Prehistoric (Cro-Magnon ; Aurignacian ; Magdalenian) artists in Europe used red, yellow, blue-black, black, and white pigments, kept as dry powders in hollow bowls and mixed with grease for use. Red and yellow were ground nodules of iron and manganese ores (also blue black) ; black was calcined bones, and white was kaolin.[16]

The effect of "light and shade" (never used otherwise in Egypt) was achieved at Amarna, in the earliest known example, by dark shading and dusting on "orpiment" [?] for the light part.[17]

---

[1] Blümner, iv, 413 ; cf. Merimée, in Passalacqua, Catalogue raissonné, 1826, 243, 263.

[2] Kisa, 196.

[3] xxi, 21 : "pure gold like transparent glass," ὡς ὕαλος διαυγής ; Bucher, i, 62.

[4] Perrot-Chipiez, Art in Egypt, i, 784 ; Budge, Mummy, 1893, 350 f. ; Wilkinson, ii, 285 f. ; Erman-Ranke, 505 ; Maspero, Arch., 1889, 197 ; ib., 1902, 202 f. ; BMGE, 109.

[5] Wilkinson, i, 363 ; Berger, Die Maltechnik des Altertums, Munich, 1904, frontispiece, 11 f., 18 f.

[6] xxii, 2 ; xxv, 1 ; Semper, Der Stil, i, 460, 478, 497.

[7] Maspero, Arch., 1889, 197 ; ib., Dawn, 247 ; Gsell, 42 f. ; Lenormant, Prem. Civ., i, 256 ; Reitemeyer, 99.

[8] Wilkinson, ii, 294, Fig. 424.

[9] BMGE, 110.

[10] Berger, Maltechnik, 22 f.

[11] Berger, 5, 22 f. ; Maspero, Arch., 1889, 197 ; Perrot-Chipiez, Art in Egypt, 1883, i, 784, and plate opp. 807 ; Wilkinson, i, 363, coloured plate viii, of XVIII dyn. ; BMGE, 175, plates xv, xvi and Fig. on p. 175.

[12] Maspero, Arch., 202 f.

[13] Budge, Egy. Hierogl. Dict., 331, 757 ; BMGE, IV–VI Rooms, 9 ; Perrot-Chipiez, i, 784 ; Maspero, Arch., 170, Fig. 160 ; ib., PSBA, 1892, xiv, 312 ; Davies and Gardiner, Tomb of Amenemhet, 1915, 12.

[14] Brugsch, Ägyptologie, 1891, 83 ; Budge, Egy. Hierogl. Dict., 150.

[15] Berger, 5, 22 f.

[16] Mackenzie, Footprints, 72 ; Hoernes, Urg., ii, 340 ; M. C. Burkitt, Prehistory, Cambridge, 1925, 205.

[17] Petrie, Tell el-Amarna, 15 ; Breasted, H., 378, 608, Fig. 144.

In the XVIII dyn. the colours are three varieties of yellow, three of brown, two of red, two of blue and two of green, as well as white and black. The colouring was highly conventional and conservative and never quite true to nature.[1] The medium was white of egg (geese were kept in ancient Egypt) [2], gum arabic or mucilage[3], perhaps milk, glue[4], gum tragacanth[5], and honey, which is mentioned in a Bulaq papyrus : "the picture of the god Chem is painted with green colour (chenti) mixed with honey." Plutarch and Vitruvius both say the purple dye was mixed with honey [6], which is still used for certain water colours, and it was probably mixed with egg or gum.[7]

When walls were painted, the stone if of poor surface was first covered with stucco put on over a coating of clay [8], or sometimes limewash from lightly burnt shell lime, or a coat of chalk or gypsum mixed with glue solution.[9] In some cases the clay was merely distempered white with chalk, or with chalk tinted red or yellow.[10] The walls at Beni Hasan were stained and sprinkled with colour to give the appearance of red granite.[11] The pigments were applied by means of a reed, or a more or less fine hair brush, and when well prepared are stable : the reds have darkened, the greens faded, and the blues turned somewhat green or grey, all superficially only, the colour underneath being brilliant and unchanged.[12] Wooden coffins were "whitewashed" inside and out.[13] The British Museum has a series of "frescoes" from XVIII dyn. tombs at Thebes, the colours being painted on a thin layer of plaster laid upon a backing of coarse mortar.[14]

Egyptian wall painting is never true fresco, executed on a damp surface of caustic lime and water only, but always tempera painting on gypsum plaster with some medium, probably gum (which has been detected). Painting on wood is nearly always on a preliminary coat of whiting and glue called gesso by Egyptologists (although this is properly plaster of Paris and glue), or stucco, and this in turn is sometimes put on coarse canvas glued to the wood.[15] Painting on linen canvas is represented by XVIII dyn. specimens.[16]

ANALYSES OF PIGMENTS

Egyptian pigments from the tomb of Perneb (2650 B.C.) in the Metropolitan Museum, New York, were[17] : red hæmatite, yellow ochre or ferruginous clay, blue frit, pale blue azurite, green malachite, black charcoal or bone-black (easily distinguished chemically) ; a pigment in a pot was a mixture of hæmatite, limestone and clay. Other analyses are given in the table opposite.[18]

SOURCES AND COMPOSITION OF PIGMENTS

The pigments used were very probably all found in Egypt. Deposits of ochres, red oxide, green earth, barytes, chalk, and yellow clay, perhaps those used in ancient Egypt, occur near Aswân.[19]

[1] Maspero, Arch., 1889, 197 ; ib., Dawn, 412 ; Breasted, H., Figs. 155, 156 ; Worringer, Egyptian Art, 1928.

[2] Maspero, Arch., 1889, 197 ; Newberry, Beni Hasan, ii, 2.

[3] Maspero, Arch., 1902, 202 ; Perrot and Chipiez, Art in Egypt, i, 785.

[4] Found by John, Malerei der Alten, 1836, 156, 213, in Egyptian specimens ; Berger, Maltechnik, 6, 8, 11 f., 13.

[5] Maspero, Arch., 1889, 197 ; Berger, 11 f., doubtful.

[6] Maspero, ib. ; Perrot and Chipiez, i, 785.

[7] Berger, 12.

[8] Erman-Ranke, 502.

[9] Berger, 7.

[10] Maspero, Arch., 192 ; Dawn, 319, 524 ; Berger, 6, 9 ; Hall, CAH, ii, 414.

[11] Wilkinson, ii, 292.

[12] Maspero, Arch., 1889, 197.

[13] Brunton, Qau and Badari I, 30, 47.

[14] BMGE, IV–VI Rooms, 79 f.

[15] Maspero, Arch., 192 ; Lucas, Egy. Mat., 148 f.

[16] Rustafjaell, The Light of Egypt, 1909, 46.

[17] Toch, J. Ind. Eng. Chem., 1918, x, 118.

[18] Cf. also Prisse d'Avennes, Histoire de l'art égyptien, 1879, 276, 293 f. ; Merimée, in Passalacqua, Catalogue raisonné, 1826, 258 f.

[19] Anon., in Chem. Trade Journal, 1927, lxxx, 241.

	(1) III dyn. "frescoes": tomb of Nefermaat.	(2) Thebes: ? XVIII dyn.	(3) Amarna; 1350 B.C.	(4) Hawâra: Greek-Roman Period.
White.	Powdered "efflorescence" of gypsum (calcium sulphate).	Very pure chalk.	—	Powdered gypsum.
Black.	Lampblack (not charcoal or graphite); a circular cake, 1¾ in. dia. and ⅜ in. thick, was found in a well.	Ground bone black (grey ash soluble in HCl with little effervescence) and gum. No bitumen: nothing soluble in petrol.	Soot.	—
Yellow.	Ochre and clay; the best nearly pure ferric oxide. Different tints used, also different thicknesses of film. Lighter by mixing with gypsum.	Yellow iron ochre.	Raw ochre.	Ochre: hardly any alumina.
Orange.	Yellow ochre with faint wash of red haematite (really red ochre).	—	—	—
Red.	Haematite ground with water in a curved vessel, hollow rock or potsherd. Coarse and dark reds from ochreous clays, some containing manganese? One sample 79·11, another 81·34 per cent. Fe_2O_3.	Earthy bole, mostly soluble in HCl; contained iron and aluminium.	Burnt ochre.	(a) Dark red: burnt ochre: ferric oxide with a little sand. (b) Minium (red lead), with some alumina? (c) Rose-red: an organic pigment (? madder) in gypsum medium.
Brown.	Thin wash of haematite over black, or unburnt ochre over haematite.	—	—	—
Green.	Pure malachite, often overlaid on yellow. No frits used.	Blue "glass," with colourless glass [? sand or quartz] and yellow ochre; probably a frit.	Copper frit.	—
Grey.	Pale yellowish earth and a little lampblack.	—	—	—
Blue.	Finely ground chessylite (impure copper carbonate mineral); no malachite or frits used.	Powdered blue copper frit; some iron present.	Copper frit.	Copper frit: very permanent and not attacked by hydrochloric acid. Note red (c): a cochineal (not madder) lake was suspected in a pink pigment from the Baths of Titus by Davy, Works, vi, 147.

(1) Spurrel and Russell, in Petrie, Medum, 1892, 28 f., 44.
(2) Ure, in Wilkinson, ii, 287 f.; plate xiv.
(3) Petrie, Tell el-Amarna, 14.
(4) Russell, in Petrie, Hawara, Biahmu and Arsinoe, 1889, 11; Kisa, 290 f.; C. Smith, DA, ii, 393; Berger, 26.

The black pigments are soot (lampblack) or burnt bones[1] ; the white gypsum, plaster or chalk[2] ; powdered white enamel, given by Perrot and Chipiez on the authority of Prisse d'Avennes and Merimée [3], does not appear. The blue is chessylite (azurite) in the earliest work [4], later copper frits or powdered blue glass.[5] Cobalt is rare, but occurs in a v dyn. pigment.[6] Green was malachite in the earliest work, later copper frits, sometimes with yellow ochre. Some greens have faded to olive.[7] It was not green earth (terre verte).[8] The violet is supposed to be the result of the falling away of gold from a place once gilded, and formed from the mordant or mixture applied to receive the gold[9], but some ochres have a violet colour when burnt.[10]

The yellow, red and brown are almost always iron ochres of various shades, both unburnt (yellow) and burnt (red and brown)[11], the Greek miltos ($\mu\iota\lambda\tau\sigma s$). Red ochre (qenà-t ?), *not* hæmatite, is the only red pigment used in early Egypt for painting and writing,[12] is found in the northern oasis[13], and occurs in pre-historic tombs.[14] Ochre was one of the oldest colours used by man[15]; many native tribes, including African negroes, still paint themselves red with burnt ochre or clay, sometimes mixed with black or white spots and stripes.[16] Tacitus[17] reports this of the Germans, and Herodotos[18] says the Ethiopians going into battle painted their bodies half white with gypsum or chalk ($\gamma\upsilon\psi\sigma s$) and half red with ochre ($\mu\iota\lambda\tau\sigma s$) ; they covered with gypsum and painted the bodies of their dead, placed inside a hollow "glass" pillar, the glass being dug in abundance in the country, and very easy to work. This "glass" may have been rock salt, since Ibn Batûta[19] (14th century A.D.) reports that the North African tribes still used rock salt for this purpose. Herodotos[20] says the Maxyañs and Gyzantians, two Libyan tribes, painted themselves red ; also[21] that the early Greeks painted their ships red. The red colour represented blood, and hence had magic properties among primitive peoples.[22] The XXI dyn. mummies in Egypt are painted red (men) or yellow (women), but these were used to represent the natural complexions.[23]

The red colour afterwards called sandyx or serikon by the Greeks was probably rouge (iron oxide), although in a later period red lead (minium) and cinnabar (HgS), or a red vegetable dye, were used.[24] Minium and cinnabar were important articles of trade in Egypt in the 12th-13th centuries A.D.[25]

The Egyptians did not rouge the lips except of mummies[26], although a Theban papyrus shows a woman with a mirror and a stick applied to the lips.[27] A kind of rouge was used for the cheeks, also on mummies[28], and the

[1] Maspero, Arch., 202 ; Dawn, 362.
[2] Petrie, Medum, 28 f. ; Maspero, *opp. cit.* ; Ure, in Wilkinson, ii, 287 f. ; Perrot and Chipiez, Art in Egypt, i, 785.
[3] In Passalacqua, Catalogue, 258 f., 263.
[4] Petrie, Medum, 28 f. ; John, Malerei der Alten, 1836, 115.
[5] Wilkinson, ii, 287 f. ; Perrot and Chipiez, i, 784.
[6] Toch, *J. Ind. Eng. Chem.*, 1918, x, 118 ; Lucas, Egy. Mat., 63 ; *ib.*, in Carter, Tomb of Tutankhamen, ii, 179 ; Berger, 24.
[7] Petrie, Medum, 28 f. ; Wilkinson, ii, 287 f. ; Perrot and Chipiez, i, 785.
[8] Berger, 23.
[9] Perrot and Chipiez, i, 785, from Champollion.
[10] Berger, 26.
[11] Perrot and Chipiez, i, 785 ; Berger, 23 f. ; Schrader, RL, 233 ; Wagenaar, *Pharmaceutisch Weekblad*, 1933, lxx, 894.
[12] Lucas, JEA, 1930, xvi, 44.

[13] Brugsch, Ägyptologie, 1891, 406 ; Budge, Egy. Hierogl. Dict., 773.
[14] Forrer, RL, 661.
[15] Schrader, RL, 233.
[16] Lippert, Kulturgeschichte, i, 375 f. ; Gsell, 44.
[17] Germania, 33.
[18] vii, 69 ; iii, 24.
[19] Travels, tr. by S. Lee, 1829, 231.
[20] iv, 191, 194.
[21] iii, 58.
[22] E. Wunderlich, Die Bedeutung der roten Farbe im Kultus der Griechen und Römer, Giessen, 1925, 4, 12, 24, 109 ; Dawson, Magician and Leech, 1929, 7 f.
[23] Dawson, 26.
[24] Blümner, i, 245 ; Berthelot, Mâ, ii, 8, 12, 161, 331 ; *ib.*, Intr., 261.
[25] Qalqashandi, Geogr., tr. Wüstenfeld, 1879, 224 f.
[26] Dawson, 26.
[27] Erman-Ranke, 285, Fig. 101 ; Wiedemann, 1920, 48.
[28] BMGE, 81 ; Dawson, 26.

Egyptians, from the oldest period to the present day, used eye-paints. The red ochre found in graves with palettes, and as stains on the palettes and stones used for grinding, was perhaps used for painting round the eyes and possibly sometimes for the cheeks, as described in the Old Testament [1], the pigment (Hebrew puch ; Greek φῦκος, phykos ; fucus) being a red vegetable dye, not stibium, as in St. Jerome's translation.[2]

In the New Kingdom a pink pigment was made simply by mixing red and white.[3] White and pink pigments in Leyden Museum are powdered sea shells.[4] The use of face-powder is uncertain, but Greek women painted with white lead and red anchusa (from the root of the Anchusa tinctoria).[5]

It will be seen from the table that orpiment (As_2S_3), cinnabar (HgS), and Naples yellow, given by Merimée [6] and mentioned by Maspero [7], do not occur: the yellows and reds are exclusively ochres, or ferric oxide minerals. In the Græco-Roman Period pigments found at Hawâra, minium (Pb_3O_4) has come into use, but not cinnabar or orpiment. The latter, however, is said to occur at Amarna (XVIII dyn.).[8]

John [9] found the "brilliant sulphur yellow" from boxes from Thebes and Abydos to be a vegetable product, indistinguishable from the extract of yellow berries called in Germany Schüttgelb, and not orpiment : a bright yellow pigment in the Leyden Museum consists of ochre and some organic pigment.[10] There is, apparently, no record of the use of true cinnabar (mercuric sulphide, HgS) in ancient Egypt, but some archæologists use the name loosely, meaning an iron oxide red, or minium, etc., or as an adjective : "cinnabar or vermilion red," in the same way as they speak of "cobalt blue" without meaning cobalt. Merimée[11], although he did not find it in Egyptian colours, assumed that cinnabar was a possible pigment. Even in the Greek and Roman Period, e.g. in Dioskurides and Pliny, the nomenclature of red pigments (ochre, red lead, cinnabar) was very confused.

USE OF PIGMENTS ; MEDIA

An experiment peculiar to the III dyn. tomb of Nefermaat at Medum is the inlaying of inscriptions of coloured pastes in deeply incised undercut hollows. The paste has perished by exposure and salt efflorescence.[12] Inlays of coloured glaze (which have largely fallen out) in gilt gypsum occur in a late period at Abûsîr.[13] This inlaid work is different from "fresco." The pastes at Medum contained a white pigment and what was later recognised as gelatin, also, except the white, "grains of gum" or "resin," or perhaps "segregated turpentine," which "comes nearest to mastick by tests, and especially imitative preparations," and was "probably used instead of gum arabic, or ignorantly mixed with it." The inlays were "polished by oil or liquid turpentine," and the turpentine "may have been liquefied by strong wine"[14]; as the gum was probably found useless, glue was perhaps put on over the surface.

Varnish of acacia gum was applied in the XX dyn. only : it was soluble in water and has cracked and darkened from the effects of age. This fault was

[1] Jeremiah, iv, 30 ; Lucas, JEA, 1930, xvi, 44.
[2] Gesenius, q. in Smith, DA, i, 880 ; I. Löw, Flora der Juden, Vienna, 1926, i, 20.
[3] Glanville, JEA, 1928, xiv, 190.
[4] Wagenaar, Pharm. Weekblad, 1933, lxx, 688, 894.
[5] Lucas, JEA, 1930, xvi, 44 ; Gsell, 44.
[6] Passalacqua, Catalogue, 258 f.
[7] Arch., 202 f.
[8] Lucas, Egy. Mat., 79, 144 ; Peet and

Woolley, City of Akhenaten, 32, 34, 86 ; ? by chemical analysis.
[9] Quoted by Berger, 25.
[10] Berger, 26 ; Rübencamp, in Ullmann, Enzykl. techn. Chem., v, 269 ; Wagenaar, Pharm. Weekblad, 1933, lxx, 688, 894.
[11] In Passalacqua, Catalogue, 258 f. ; Berger, 23.
[12] Petrie, Medum, 25, 50 ; ib., Hist. of Egypt, 1903, 35 ; Maspero, Dawn, 362.
[13] Borchardt, Sahurā, i, 127, Fig. 172.
[14] Impossible : the whole of Petrie's description is confused.

probably soon discovered and the method abandoned. Before the Theban Period no precautions were taken to protect work from air and light.[1]

Varnish applied as a liquid natural oleo-resin (not "dissolved in strong wine") was used from the XVIII dyn.[2] Turpentine, drying oil and naphtha were not known or used till Roman times, although John suggested the use of turpentine, Berger a solution of resin in fatty oil applied hot on a linen ground [3], and Merimée a solution of resin in Persian naphtha.[4] The use of varnish over painting is described by Prisse d'Avennes : its use on mummy cases belongs to the XIX–XX dyns.[5] The wall paintings in the tomb of Ḥuy (c. 1350 B.C.) are heavily coated with varnish, which has turned deep red or black and opaque, bringing away the colour when removed.[6] Some kind of black lacquer varnish was also used : the kinds known at present are Chinese, Japanese or Indo-Chinese.[7]

Analyses by John and by P. L. Geiger [8] gave indications of the use of a medium other than glue, and both suggested wax, perhaps with resin.

ENCAUSTIC PAINTING

The process of encaustic painting (with a wax medium) was not in use in ancient Egypt [9] but appeared in the Ptolemaic Period for painting on wood[10], although Herodotos[11] says Amasis (559–525) sent a portrait of himself to Cyrene. The encaustic paintings on wood on mummy cases are Greek and Roman.[12] The encaustic technique may have originated in Egypt ; preparations of wax for preserving paintings are said to go back to the XVIII dyn., and the names of most of the encaustic painters of antiquity appear to be Alexandrian or Egyptian.[13] The first literary mention is after Alexander : a reference in a supposed ode of Anacreon (c. 550 B.C.) is of doubtful date.[14] Eusebius (264–340) calls the process κηρόχυτος γραφή ("drawing in liquid wax") : it continued in use till the Middle Ages, but had declined after the 9th century A.D. The pigments (now in the British Museum) found at Hawâra by Petrie are really water colours, but it is probable that they would be similar to the pigments used by the encaustic painter. The process, according to Petrie[15], was as follows. The colours were ground in the wax, previously bleached by heating to its boiling-point, and fused in the sun in hot weather or in a hot-water bath, which is mentioned by Theophrastos.[16] The portrait was made on a wood panel, previously primed with distemper, the wax colour being put on from a pot with a lancet-shaped spatula or (more probably) with a brush, pressed out at the end of the stroke.[17] Petrie says wax is softened and actually melted on the surface by the sun in Egypt, so that no other solvent such as oil or turpentine would be required, although he does not explain how the picture hardened.

A wooden colour box with a pivoted lid in the shape of a figure 8, containing sand coloured with ferric oxide and found in V dyn. remains at Abûsîr with

[1] Maspero, Arch., 203.
[2] Lucas, Egy. Mat., 150 f. ; Perrot and Chipiez, Art in Egypt, i, 786.
[3] Pliny, xiv, 20, already says all oils dissolve resin.
[4] Passalacqua, Catalogue, 262 ; Berger, 15, 26 ; Lucas, Egy. Mat., 130, 150 f.
[5] Berger, 9 f., 14, 15, 26 f.
[6] Davies and Gardiner, Tomb of Ḥuy, 1926, 2, 22.
[7] Lucas, Egy. Mat., 154.
[8] Chemische Untersuchung alt-ägyptischer und alt-römischer Farben, Karlsruhe, 1826, q. by Berger, 7 f., 308.

[9] De Pauw, i, 212.
[10] Perrot-Chipiez, Art in Egypt, i, 786.
[11] ii, 182.
[12] Forrer, RL, 501 ; Wilkinson, ii, 294.
[13] C. Smith, DA, ii, 393 f.
[14] Schmitz, DBM, i, 157.
[15] Ten Years' Digging ; frontispiece, 97, 100 ; ib., Hawara, 1889, 18, 37.
[16] De odoribus, v, 22.
[17] John, Malerei der Alten, 1836, 157 f., 197 f. ; Rich, Dict. of Roman and Greek Antiquities, 1874, 501.

a coil of string used to mark out the lines on granite for cutting, is of the Roman Period, when the columns of the temple were cut into millstones.[1]

A terra-cotta bottle carried on the thumb was perhaps used to hold the water for painting.[2]

The early Pyramid Age is represented by a picture of geese. There is no very fine example of painting of the Middle Kingdom[3] ; many are excellent but not so good as Old Kingdom work. The great age of painting in tombs and palaces is in the XVIII–XIX dyns., when it replaced sculpture.[4]

EYE PAINT

Eye paint has always had a special magical and religious value in the Near East, sustained in later days by reported sayings of Muhammad.[5] Besides its supposed effect in strengthening the sight and in averting the evil eye [6], eye paint was supposed to ward off disease [7], and was no doubt of use in treating the inflammation of the eyes which is endemic in the hot and dusty climate of Egypt.[8] The pots and boxes containing eye paints have statements such as : "to lay on the lids and lashes," "good for the sight," "to stop bleeding," "best eye paint," "to cause tears." Various kinds of pots were used, including glass or ivory tubes.[9]

Different eye paints for persons of different ages, and different parts of the year, were kept in divided boxes. Sometimes the kohl pots are in sets of four, apparently containing different materials : in one set the tubes are labelled : powder "for daily use," powder which "opened the eyes," powder which "cleansed the eyes," and powder which "removed rheum from the eyes." Double and quadruple tubes are common, but triple tubes are rare. The kohl stick ("needle," as the Arabs call it) for applying the pigment was of wood, ivory, bronze, hæmatite, glass, etc. ; the pear-shaped end dipping into the powder was moistened with water or scented unguent and drawn along the eyelid, under the eye, or over the eye.[10]

Two kinds of eye paint for ritual use are mentioned in the IV dyn. or earlier texts : a green, called uaz or wod, rarely šesmet ; and a black, mestem.[11] The green was probably always malachite (mafkat or mafek), and was in use before the black.[12] It occurs in beads and powder in prehistoric tombs[13] and Badarian remains, and continued in use till at least the XIX dyn. It was applied under the eye.[14] The black mestem (Coptic stem, Greek στίμμι, Latin stibium)[15] was used by all classes.[16] There were several kinds : "real mestem" and "good mestem"[17]; "true" and "male" mestem, and another kind (perhaps "female" ?) in the Ebers Papyrus (c. 1550).[18] One kind, not used till the late Predynastic Period but continuing till Coptic times, was galena (native lead sulphide, PbS,

[1] Borchardt, Sahurā, 107 f., Abb. 133.

[2] Wilkinson, ii, 19, Fig. 295 ; Petrie, Arts and Crafts, 55–61 ; Russell, in Petrie, Medum, 1892, 44 f.

[3] Petrie, op. cit. ; Breasted, H., 202.

[4] Breasted, H., 308, 348, 378, 417, Fig. 155.

[5] Hornblower, JEA, 1927, xiii, 243.

[6] Birch, in Wilkinson, ii, 349 ; Lane, Modern Egyptians, 1846, i, 51 f. ; Maspero, PSBA, 1892, xiv, 315.

[7] Wiedemann, 1920, 146.

[8] Dawson, Magician and Leech, 1929, 97.

[9] Wilkinson, ii, 347 f., Fig. 451 ; Budge, Mummy², 259, 260 ; ib., Hist., i, 55 ; ib., Osiris, i, 247 f. ; Predynastic.

[10] Budge, Mummy², 260 f., 265.

[11] Wiedemann, in Petrie, Medum, 1892, 41 f. ; Wiedemann, 1920, 146 f. ; Brugsch, Ägyptologie, 154 ; Lucas, JEA, 1930, xvi,

41 ; Newberry, in Studies presented to F. Ll. Griffith, Oxford, 1932, 320.

[12] Maspero, Arch., 252 ; Lucas, 41 ; Newberry, op. cit.

[13] Petrie, Naqada and Ballas, 1896, 6, 10, 15, 19 f., 26 f., 43, 45.

[14] Maspero, Guide to the Cairo Museum, 1908, 40.

[15] K. B. Hofmann, Über Mesdem, Mittheil. d. Vereins der Ärzte in Stiermark, 1894, Nos. 1, 2 ; q. in E. von Meyer, History of Chemistry, 1906, 19 ; Budge, Mummy², 259 ; ib., Egy. Hierogl. Dict., 329.

[16] Maspero, Arch., 252 ; Budge, op. cit.

[17] Erman-Ranke, 258 ; Brugsch, 411 f.

[18] Joachim, Papyros Ebers, 1890, 84, 87, 89, 95 ; Pliny, xxxiii, 6, has male and female kinds of stibium, and there are male and female stones and plants in Theophrastos and Pliny.

which gives a black powder). Charcoal was used, including soot, and the poorer classes perhaps used charred sticks and Nile mud.[1] A piece of graphite (black-lead) found at Gurob [*not* "in a tomb"] [2] may have been used as a pigment or as eye paint.

Malachite and galena occur in various forms : fragments of raw material ; stains on palettes and grinding stones ; prepared, either as a compact mass of finely ground material made into a paste (now dry) or more frequently as a powder, and often placed in graves in small linen or leather bags, shells (also used for other pigments), segments of hollow reeds, wrapped in leaves, and in small vases, sometimes reed shaped. The paste was probably (although not certainly) made up with water and gum, or water alone. Fat, said by Barthoux to be used, is not found.[3]

After the XI dyn. the pigments were obtained from dealers ready powdered for use [4], and were not triturated by the user as in the earliest period, when mestem is shown in bags (I dyn. and before).[5] On an Old Kingdom stela, "eye paint" and "stibium" (?) are mentioned as separate things.[6] The true imported mestem, perhaps from the South Arabian coast, was supposed to be antimony sulphide, galena was an Arab substitute, and manganese dioxide, copper oxide and magnetite were Egyptian substitutes.[7] In view of the great rarity of finds of antimony sulphide, this seems doubtful. The texts make eye paint come in King Sānkh-ka-Rā's time (2000) in bags or purses from, or through, Pitsew (Arabia ?).[8]

Malachite (in Sinai and the eastern desert) and galena (near Aswân and on the Red Sea coast) both occur in Egypt, but antimony compounds do not, but are found in Asia Minor, Persia and possibly Arabia. Eye paint was obtained in the XII dyn. from Asiatics, and in the XIX dyn. from Koptos (perhaps galena from the Red Sea coast), while eye cosmetics (? the same) were received in the XVIII dyn. from Naharin in western Asia and from Punt.[9] The eye cosmetic from Punt may have been malachite or galena traded from Arabia, where both occur ; antimony sulphide[10] is not very probable.

An official of Seostris III in 1880 obtained a present of eye paint from a desert chief, perhaps from Syria.[11] Rameses III (1198–1167) gave a present of 50 lb. of mestem to the god, so that it was valuable.[12]

Black eye paints from Medum (III dyn. on) analysed by X. Fischer (32 specimens) consisted[13] of : frequently galena, gently roasted and used as such or mixed with gum ; pyrolusite (manganese dioxide) ; copper oxide, from roasted malachite ; magnetite (Fe_3O_4), perhaps partly reduced hæmatite ; carbon was perhaps also added and much metallic lead, from roasted galena, occurs ; one specimen (perhaps Coptic) appears to have been made by gently roasting impure mineral molybdenite, molybdenum sulphide, MoS_2 (very like graphite in appearance), since in 70·39 parts it contained 6·01 of molybdenum, 58·10 of lead (probably from admixed galena), 5·06 of sulphuric acid and 1·22 of sulphur as sulphides. Crude stibnite, antimony sulphide, Sb_2S_3, quite free from lead, occurred in *one* XIX dyn. specimen only.

[1] Lucas, JEA, 1930, xvi, 41 ; Wiedemann, in Petrie, Medum, 1892, 41 f. ; Forrer, RL, 418 ; Erman-Ranke, 257 ; Wiedemann, 1920, 146.
[2] Lucas, *Analyst*, 1926, li, 447 ; cf. *Ancient Egypt*, 1927, 58.
[3] Lucas, JEA, 1930, xvi, 43.
[4] Brunton, Qau and Badari I, 63.
[5] Petrie, etc., Tarkhan I and Memphis V, 1913, 8 ; Tarkhan II, 1914, 9 ; Capart, Primitive Art in Egypt, 1905, 29 ; Wilkinson, i, 480.
[6] Budge, Mummy², 440.
[7] Wiedemann, in Petrie, Medum, 1892, 44.

[8] Brugsch, Ägyptologie, 405 ; Gsell, 43 : probably galena.
[9] Lucas, JEA, 1930, xvi, 43 ; Wiedemann, Sagen, 31 ; Brugsch, 389, 399 ; Lieblein, Handel und Schiffahrt auf dem roten Meere, Christiania, 1886, 29, 31, 35, 70.
[10] Erman-Ranke, 610 ; Wiedemann, 1920, 145 ; Brugsch, 405, thinks from XVIII dyn. ; Gsell, 43.
[11] Meyer, Alt., 1909, I, ii, 260 ; Olmstead, H. Pal. Syr., 89.
[12] Brugsch, Ägyptologie, 273.
[13] Petrie, Medum, 41.

Black galena (*or* manganese) with a palette, was found in Badarian remains.[1] Green eye paint from a XII dyn. tomb at Dahshûr was pure malachite ; the black was a mixture containing 23 per cent. of galena, 5 of ferruginous clay, 5 of manganese dioxide, 1 of ferrous oxide, 1 of antimony sulphide and 1 of copper oxide.[2]

Florence and Fischer [3] give some earlier analyses and two of their own. Barthoux found most of the specimens were galena ; the rest (which appear in some cases to have been doubtfully cosmetics) included lead carbonate (white lead); a compound of lead and antimony (the only one in which any antimony occurred), vegetable black (soot), compounds of arsenic with and without iron pyrites (some orange coloured), chrysocolla, and "bitumen impregnated with aromatic essences" (?). Some specimens contained resin but were probably medicaments.

Of 58 black and green specimens analysed [4] (excepting those of Barthoux), 37 were galena (2 of these contained traces of antimony sulphide, and 5 carbon) ; the rest included 2 of lead carbonate (which is white when pure : 1 specimen contained a trace of antimony sulphide), 1 of black oxide of copper [5], 5 of brown ochre, 1 of magnetic oxide of iron, 6 of oxide of manganese, only *one* of sulphide of antimony (XIX dyn.) : of the green pigments 4 were malachite, and 1 chrysocolla.

Black eye paints now used in the East include : charcoal and lampblack, charred almonds and frankincense[6] ; the soot of the oily qurtum plant[7] ; galena or lead sulphide, or lead preparations, supposed to have medicinal properties[8] ; pyrolusite or black oxide of manganese (from the Sudan) ; black copper oxide, from roasted malachite ; true stibnite, antimony sulphide, and mixtures of various materials.[9] Stimmi was imported from East Africa to India in the 1st century A.D.[10] It is still used in Abyssinia for painting the eyes, and by women for tattooing the gums.[11] "Paint" (cosmetic), eye salve and "eye-black," were important articles of Egyptian trade in the 12th–13th centuries A.D.[12]

The later use of two kinds of black pigments, galena or lead sulphide, and stibnite or antimony sulphide, and the known fact that two metals, lead and antimony, could be extracted from them, no doubt gave rise in Egypt to the curious statements about the "two leads," and "our lead," i.e. antimony, in later Egyptian alchemy.[13]

In a Leyden papyrus magnesia, called manesia in Coptic, a mineral like stem, i.e. stimmi, and grey when powdered[14], was probably pyrolusite.

The British Museum has a number of eye-paint tubes of various materials and toilet outfits ; an interesting wooden toilet box (XVIII dyn.) with four compartments contains a terra-cotta and two alabaster vases with unguents ; a piece of pumice stone ; a double "stibium" tube with a wood and an ivory stick, one tube containing powdered eye paint and the other an unguent ; an ivory comb ; a bronze shell for mixing the unguents ; a pair of pink gazelle skin sandals, and three red cushions.[15]

[1] Brunton, Qau and Badari I, 43, 62.
[2] De Morgan, Or., i, 215.
[3] Q. by Lucas, JEA, 1930, xvi, 41 f.
[4] Lucas, *op. cit.*
[5] Cf. Budge, Mummy[2], 259.
[6] Wilkinson, ii, 347 ; Lane, Modern Egyptians, 1846, i, 51 f.—interesting details of use in Cairo ; soot, αἰθάλη or λιγνύς, was used for ophthalmic remedies by Dioskurides, i, 84, etc.
[7] Brunton, Qau and Badari I, 63.
[8] Lane, *op. cit.* ; Maspero, PSBA, 1892, xiv, 315 ; Wilkinson, ii, 347.

[9] Wilkinson, ii, 347 ; Budge, Mummy[2], 259 ; Gsell, 44.
[10] Periplus Maris Erythræi, ed. B. Fabricius [i.e. H. T. Dittrich], Leipzig, 1883, 90, 96.
[11] Stoll, Geschlechtsleben, 376, 288 f.
[12] Qalqashandi, Geogr., tr. Wüstenfeld, 1879, 224 f.
[13] Lippmann, Alchemie, i, 47.
[14] Brugsch, Ägyptologie, 397.
[15] BMGE, IV–VI Rooms, 263, 266, 268.

EGYPT III—2

NON-METALLIC MATERIALS

SALT

The early Egyptians used common salt (ḥemai-t or chemai-t; Coptic ḥmou or chmou) for preserving fish and birds [1], but not meat, which was dried. Packets of salt were found with XX dyn. remains (c. 1150) at Abydos [2], but salt, used in the very early period for preserving dead bodies before mummifying was introduced, was obtained long before this from lakes on the sea-coast and from the Natron Lakes, and rock salt from several places in the Western Desert [3]. Herodotos [4] mentions the salt pans near a temple at the Canopic mouth of the Nile, and says solid salt was found in the Libyan Desert, where, in the midst of it, were springs of fresh water.

Various kinds of salt are mentioned in the Ebers Papyrus (1550 B.C.), e.g. merely "salt"; "mine salt" (probably rock salt); "sea-salt" (the "Typhon's foam," $\tau\upsilon\phi\hat{\omega}\nu\sigma\varsigma$ $\alpha\phi\rho\acute{o}\varsigma$, of Plutarch); and "salt of the North."[5]

The salt of the Oasis of Ammon in the Libyan Desert near Sîwah (where houses of rock salt have been found) covers the soil with an incrustation through which camels' feet break as through a thin coat of ice. It contains, besides sodium chloride, often large amounts of the carbonate and sulphate, and is used, as in early times, for glass-making.[6] This salt efflorescence was pure white and had a special sanctity: it was dug up ($\H{a}\lambda\epsilon\varsigma$ $\dot{o}\rho\upsilon\kappa\tau oi$) in large crystals, wrapped in palm leaves, and taken by certain priests of Ammon to Egypt as a gift to the Persian king, being preferred as an article of sacrifice to the impure sea-salt ($\theta\alpha\lambda\acute{a}\sigma\sigma\eta\varsigma$ $\dot{a}\lambda\hat{\omega}\nu$). It is still exported.[7]

The Egyptians seem to have regarded sea-salt as impure and forbidden, and the Hindus also reckon salt workers of low caste. The Egyptians may have considered sea-salt as impure because of the numerous insects and small animals, drowned in the sea, which occur in salt efflorescences (e.g. in Elton Lake, in Astrakhan).[8]

In Abyssinia salt is scarce, and at one time slabs of rock salt, 10–12 in. long, $1\frac{1}{2}$–$2\frac{1}{2}$ in. wide and thick, were used as coins : such a slab, weighing about 12 oz., was called 'amôlê.[9]

NATRON

Natron, a naturally occurring compound of sodium carbonate and bicarbonate, with varying amounts of common salt and sodium sulphate as impurities, is mentioned in old Egyptian records as ḥosmen or hesmen (Coptic qosen); alkaline salt is bsm or bsn and salt from the Natron desert

---

[1] Wilkinson, i, 292; ii, 30, 111, 117, 118; Budge, Egy. Hierogl. Dict., 484; Erman-Grapow, Handwörterbuch, 1921, 110; Wiedemann, 1920, 259, 265, 294.
[2] Petrie, Abydos I, 1902, 39.
[3] BMGE, 5, 83, 161.
[4] ii, 113; iv, 181, 185.
[5] Joachim, 29; 146; 31, 172; 116.
[6] Donne, DG, ii, 178; Rawlinson, note on Herodotos, iv, 185.

[7] Arrian, Anabasis of Alexander, iii, 4; Cook, Zeus, i, 380; Parthey, Plutarch über Isis und Osiris, 1850, 158 f.
[8] Chwolson, Die Ssabier, St. Petersburg, 1856, ii, 734; Parthey, Isis und Osiris, 159; Schrader, RL, 699; Zycha, in Hoops, iv, 75.
[9] C. Ritter, Erdkunde, Afrika², 1822, 199; Budge, Amulets and Superstitions, 1930, 323.

in west Upper Egypt is šrp.[1] It occurs in Lower Egypt in the Wâdî Natrûn, in the Behêra province about 30 miles north of this (Barnugi and Harrâra lakes), and in Upper Egypt at El-Kâb. The deposits at Wâdî Natrûn and El-Kâb are the only ones named in old texts ; Thothmes III also obtained natron from Syria.[2] Eisler[3] suggests that the Hebrew neter, natron, is connected with the Egyptian neter, god, perhaps because of a resemblance of the crystalline soda with hailstones falling from heaven, but the resemblance in name is purely accidental : the derivation from "the pure" (as cleansing material) in relation to "divine" is improbable, and the word neter is perhaps Libyan.[4] Natron was, however, used in daily ceremonies connected with the rebirth of the sun-god and may have been regarded as a purifying and divine substance. Salt and natron were perhaps both considered as having magic properties.[5] The names nether and bsn in Egyptian also meant natron ; neter also meant wine or strong beer ; another name for natron is ḥesmen, ḥesmen tešer is red natron.[6] The natron of the Libyan lakes was probably used at a very early period for making glass, remains of very old glassworks being found there[7], and natron (nitrike) is often mentioned in the Book of the Dead.[8] A I dyn. text speaks of "mouth-wash of the royal house,"[9] which may have contained natron. Natron is also mentioned in Egyptian stories of the Middle or New Kingdom[10], and the natron of Nekhabît (El-Kâb) was probably largely used for embalming during the XVII–XVIII dyns.[11] Rameses III (1200 B.C.) presented the temples with 1,843 artabes (a measure of capacity represented in hieroglyphics as a flat tub with liquid flowing from it) of natron and salt, including bricks of these materials.[12]

Various kinds of natron are mentioned in the Ebers Papyrus (1550 B.C.), merely natron, or "red natron"; natron of the North ; of the South ; sa-seeds of Upper Egypt, perhaps finely crystalline natron (?) or saltpetre (?) ; bedet, perhaps very pure natron (?) ; and natron water, used to drive away vermin.[13]

Strabo[14], Pliny[15], and Vitruvius[16]—who mentions the solid crust formed on the surface (ut habeant insuper se salem congelatum)—refer to the Natron Lakes in Egypt, perhaps those at Barnugi, not in the Wâdî Natrûn, Barnugi being the modern representative of the famous Nitria. Old authors say that natron was obtained in the N.W. Delta near Naukratis, but the accounts are confused and often unintelligible.[17] Qalqashandi (d. A.D. 1418) describes two other natron deposits, now unknown, one of about 100 acres at Tarâbiya, near Behnesah in Upper Egypt, worked from the time of Ibn Tûlûn (835–884) and yielding an annual revenue of £50,000 ; and the other in the Fâqûs district in the E. Delta.[18] Egyptian natron or trona contains sodium sesquicarbonate ($Na_2CO_3,NaHCO_3,2H_2O$), sulphate, and chloride, but no potassium salts.[19] The use of plant ashes, consisting largely of potassium

[1] Brugsch, Ägyptologie, 406, 426 ; Woenig, Die Pflanzen im alten Ägypten, Leipzig, 1897, 373,
[2] Brugsch, op. cit. ; Lucas, JEA, 1932, xviii, 62 f.
[3] Z. Assyriologie, 1926, xxxvii, 117.
[4] Wiedemann, A. Rel., xxi, 442; xxvi, 332.
[5] Dawson, Magician and Leech, 29 ; Lucas, JEA, 1932, xviii, 136.
[6] Budge, Egy. Hierogl. Dict., 408, 512 ; Erman-Grapow, Handwörterbuch, 1921, 50.
[7] Fowler, Archæologia, 1880, xlvi, 81 ; Kisa, 44.
[8] Chapters 17, 20, 79, 133, 161 ; Davis.
[9] Petrie, Hist. of Egypt, 1903, 21.
[10] Erman, Lit., 159, 262 ; Wiedemann, Sagen, 38.

[11] Maspero, Struggle, 84.
[12] Brugsch, Ägyptologie, 274, 379.
[13] Joachim, 15 ; 140 ; 144; 81, 85, 144 ; 120, 157 ; 179 ; on red natron, τὸ πυρρὸν νίτρον, nitrum rubrum, see Salmasius, Plinianæ Exercitat., Utrecht, 1689, 778.
[14] XVII, i, 23 ; 803 C.
[15] xxxi, 10.
[16] viii, 3.
[17] Lucas, JEA, 1932, xviii, 63.
[18] Qalqashandi, Geogr., tr. Wüstenfeld, 161 ; Lucas, 63 ; Wüstenfeld gives the second as "near Chitara, a day's journey from the first."
[19] BMGE, 162 ; Lucas, Egy. Mat., 47, 237 ; Partington, The Alkali Industry, 1925, 80.

carbonate, in washing clothes is probably very old [1], but it is not known if they were used in Egypt. What is called borith in the Old Testament [2] was perhaps natron or the ash of the soda plant (Salicornia herbacea), which is very common in Egypt as well as in Syria, Judæa and Arabia.[3]

The springs feeding the soda lakes do not contain any appreciable amount of sodium carbonate, but only sulphate and chloride, and are full of algæ. In the lake, sodium sulphide is produced by reduction near the place of origin of the springs. The water then shows an alkaline reaction, which increases in strength, whilst the mud, coloured black by ferrous sulphide or red by algæ, yields a large number of bubbles of carbon dioxide, developed by a micrococcus. This changes the sodium sulphide into carbonate. The sodium sulphate originally present is supposed to be due to the filtration of water containing sodium chloride through beds of gypsum [4], and the sand around the Natron Lakes is strongly impregnated with sodium chloride, carbonate and sulphate.[5] Although this biochemical theory has been regarded as the most likely one, there are at least six other alternative suggestions for the origin of the soda, of which one postulates the interaction between sodium zeolites and calcium carbonate, and another assumes that the Nile water, rich in calcium bicarbonate, percolates through soil rich in common salt and gypsum (calcium sulphate), forming sodium sulphate and calcium chloride, the sodium sulphate then reacting with more calcium bicarbonate to form sodium carbonate.[6]

The soda lakes of Lower Egypt, the oldest known source, which vary in number according to the season, occur in the Libyan Desert between Memphis and Hermopolis, in a valley called the Wâdî Natrûn, 38 km. to the west of the Nile and N.W. of Cairo. There are now twelve principal lakes. They are fed by a large number of small salt springs, and the liquid in the lakes is concentrated by the heat of the sun. They are partly coloured red (hence, probably, the name "red natron") and contain sodium chloride and sometimes magnesium sulphate, but very few of them sodium carbonate. The crusts forming on the sides, 15–18 in. thick, are pushed off with poles and spades, spread out to dry on the banks, and carried in baskets to the Nile, where the native soda is shipped. It is called trona, natron or atrun, or latroni in the Levant and Greece; it is the impure sesquicarbonate, and is used by the Egyptians in softening their very hard drinking water. It was principally used in Crete for soap-making and 2,500 tons were annually exported in the period 1850–1892 from Alexandria. In the Oasis of Fezzan in the Sahara (Libya) there are also natron lakes, the largest of which contains islands of solid trona. The natron in the El-Kâb region occurs in five different places and is easily accessible.[7]

Analyses of Egyptian trona show no appreciable amounts of potassium salts, although the solid residue from Nile water contains 3·3 per cent. K_2O and 14·852 per cent. Na_2O. The Nile water also contains fairly large amounts of calcium and magnesium bicarbonates.[8]

[1] Hoefer, Hist. de la Chimie, i, 58.
[2] Malachi, iii, 2; Jeremiah, ii, 22.
[3] Goguet, Origines des lois, etc., 1809, i, 144; Löw, Flora der Juden, Vienna, 1924–8, i, 637 f., 644, who suggests that O.T. neter = natron, qali (Arabic) = ash of soda plant, O.T. borith = vegetable ash.
[4] Hooker, q. by Lunge, Sulphuric Acid and Alkali, 1895, ii, 56 f.
[5] Donne, DG, ii, 441.
[6] Wegscheider, in Doelter, Mineral-chemie, 1912, i, 144 f., 157 f.; Chatard, Bull. U.S. Geol. Survey Rep., 1890, 27 f.,

89 f.; Clarke, Data of Geochemistry, 5 ed., U.S. Geol. Survey Bull. 770, 1924, 237 f.; Gmelins Handbuch der anorganischen Chemie, 1928, Syst. Nr. 21, 16, 33 f.
[7] Lunge, op. cit.; Lucas, JEA, 1932, xviii, 63; ib., Natural Soda Deposits in Egypt, Min. of Finance, Egypt, Survey Dept. Paper No. 22, 1912; BMGE, 5.
[8] Popp, Liebig's Ann., 1870, clv, 344; contradicting von Œfele, in Puschmann, Geschichte der Medizin, Jena, 1902, i, 84, who says Nile water is very pure and can be used instead of distilled water.

The quality of the trona varies considerably, as is seen from Popp's analyses :

	I. Good crystals.	II. Crystalline crust with obvious crystals.	III. Compact mass, no crystals.
$Na_4C_3O_8$.	64·3	32·2	26·15
Na_2SO_4 .	1·5	24·0	66·66
NaCl . .	8·4	33·3	2·63
$CaSO_4$.	1·3	—	—
Water .	22·5	8·87	4·05
Insoluble .	1·65	1·35	0·40

The formula $Na_4C_3O_8=Na_2CO_3+2NaHCO_3-H_2O$, assumed by Klaproth to be that of trona, corresponds with no known compound.[1] Natron from El-Kâb contained 16–23 of natron ; 12–57, or 25–54 of common salt ; 11–70, or 12–54 of sodium sulphate ; that from Wâdî Natrûn contained 2–27 of common salt and a trace to 39 of sodium sulphate, all in percentages.[2] The Egyptian soda lakes are worked by the Egyptian Salt and Soda Co., which makes caustic soda and also soap. In 1911, 10,000 tons of caustic soda and 1,500 tons of natron were exported.[3]

The soda used for mediæval Egyptian glass was made by burning the plant Mesembryanthemum Copticum, which the Venetians bought from Alexandria. It was called qali by the Arabs and grew abundantly in the desert between Cairo and Alexandria, as well as in Syria, but the best soda was made in Narbonne until about 1825.[4] An Arabic name for natron was baurach.[5]

Petrie's statement [6] that Laurie, McLintock and Miles [7] found potassium in Egyptian natron is incorrect : these authors make no such statement, and neither Popp nor Lucas [8] found potassium in any of various specimens. In view of the large quantities of natron available in the lakes, plant ashes were probably not used in making glass in Egypt, although they may have been used for later Alexandrian glass ; the earlier glass contains very little potassium, whilst vegetable soda, especially from marine plants such as Salsola clavifolia and Salsola soda, may contain large quantities of potassium salts.[9]

Natron is said to have been used as a constituent of incense, with which it has been found.[10] During the Roman Period in Egypt it was subject to a tax, and that shipped to Rome was much adulterated.[11] In the Arabic Period, the 10,000 qantars (about 450 tons) of red or green natron produced in western Egypt was a government monopoly.[12]

SAL AMMONIAC

There is considerable obscurity surrounding the early history of sal ammoniac, i.e. ammonium chloride, which was made in later Egypt by subliming the soot from burning dung.[13] Ghâfiqî (d. 1165) distinguishes an artificial sal ammoniac (milh ammoniya) from the natural (nušadir) found in Khorassan,

[1] Lunge, 53.
[2] Lucas, JEA, 1932, xviii, 66, 125 ; ib., Nat. Soda Deposits, 5 f., 15 f.
[3] Lucas, Nat. Soda Deposits, 22 f.
[4] Salmasius, Plinianæ Exercit., 1689, 771 ; de Pauw, i, 304 ; Neri, in Kunckel, Von der Glasmacher-Kunst, Frankfurt and Leipzig, 1679, 9 ; Fowler, Archæologia, 1880, xlvi, 95, from Sandys, 1610 ; Lunge, op. cit.
[5] F. Adams, The Seven Books of Paulus Ægineta, 1848, iii, 232.
[6] Anct. Egypt, 1914, 186 f.

[7] Proc. Roy. Soc., lxxxix, 418.
[8] Egy. Mat., 47, 237.
[9] Watts, Dict. of Chem., 1874, v, 324, 792 ; Muspratt, Chemistry, 1853–61, 918.
[10] BMGE, 5 ; Lucas, JEA, 1932, xviii, 65.
[11] Wilkinson, i, 337 ; Lucas, op. cit. ; Charlesworth, Trade Routes, 25, 247.
[12] Maqrîzî, d. 1441 ; Description de l'Égypte, transl. by Bouriant, Cairo, 1900, 315.
[13] De Pauw, i, 331 ; Muspratt, Chemistry, 183.

and says it is made from the soot of baths.[1]   Although the name sal ammonia-cum is in Pliny[2], Dioskurides[3] and other early medical authors, it is generally supposed to mean a variety of ordinary rock salt[4], since some of its uses correspond with those of culinary salt.   Some years ago, however, the author saw blocks of compressed sal ammoniac made in England for export to the East, where they were used instead of common salt for flavouring food : the taste is very sharply saline.

Sal ammoniac was an article of trade in Egypt about A.D. 1410, but its use and the method of preparation at that time are not known.   An account of its manufacture in the Delta was given in 1720 by a Jesuit, Sicard[5], and in Gîzah in 1750 by Hasselquist.[6]   Dried animal dung, including but not exclu-sively camel's, and never camel's urine, when used as fuel evolves a thick smoke, and the soot, rich in sal ammoniac, was carefully collected in March and April and taken to factories, where it was heated in fifty large glass globes set in a gallery furnace, when cakes of crude sal ammoniac sublimed on the upper parts of the globes as a dull, spongy, greyish mass.

It seems probable that the Egyptian nitrum mentioned by Pliny[7] as giving off a strong smell when mixed with quicklime, and was calcined (? sublimed) in earthen pots, and also his sal hammoniacum, which had an unpleasant taste and formed long opaque or transparent pieces (characteristics of sal ammoniac), may at least have contained sal ammoniac.   Arrian[8], who seems to be describing the same material, calls it a mineral salt, which corresponds with a kind of rock salt ; he says it came from the Libyan Desert.

Ruska[9] denies that sal ammoniac was known to the Greek alchemists and physicians of the 7th century, and supposes that it first became known in Persia, where it is used under the name nûšâdir by the physician Sahl ibn Rabbân al-Tabarî (early 9th century).   It is also mentioned in the Book of Poisons attributed to Jâbir ibn Hayyân (? 8th century).   Ruska thinks nûšâdir is not a Greek or Persian word, but probably Bactrian.   The alleged "volcanoes" of Central Asia, emitting sal ammoniac, were perhaps really burning coal measures.   Ruska's investigations must be regarded as of a preliminary character only, and not conclusive.

ALUM

Alum ('ibn ; ybn ; abenu ; Coptic oben) was produced in large quantities in Egypt.[10]   Herodotos[11] says Amasis, king of Egypt, sent a present of 1,000 talents of alum for the rebuilding of the temple at Delphi, which had been destroyed by fire in 548 B.C.   Lenz[12] thought it was used for fire-proofing the timber, but[13] it was perhaps to be sold to dyers to provide revenue for rebuilding the temple.   The united Greeks of the Nile Valley, it is specially emphasised, sent only 20 minas of alum as their quota.   Archelaos, a general of Mithridates, fire-proofed a wooden tower with alum in the war with Rome in 87 B.C.[14] and the Roman siege-engines were similarly treated in the war of Constantine and the Persians in A.D. 296.[15]   Egyptian alum was the best kind in Rome.[16]

[1] Ibn al-Baitâr, Traité des Simples, cc. 2167, 2241 ; tr. by Leclerc, Notices et Extraits des MSS., xxvi, 337, 380.

[2] xxxi, 7.

[3] v, 126.

[4] Beckmann, H. Inventions, ii, 396 ; Lippmann, Abh., ii, 230.

[5] Muspratt, 183 f.

[6] Voyages and Travels in the Levant, London, 1766, 55, 67, 304 f.

[7] xxxi, 10.

[8] Anabasis of Alexander, iii, 4 ; A. Schmidt, Drogen, Leipzig, 1924, 59.

[9] Sal ammoniacus, Nušadir und Salmiak, Heidelberg, 1923 ; Z. angew. Chem., 1922, xxxv, 719 ; 1928, xli, 1321.

[10] Maspero, Dawn, 219 and refs.

[11] ii, 180.

[12] Mineralogie, 11.

[13] Maspero, Passing of the Empires, 1900, 646.

[14] Aulus Gellius, Noctes Atticæ, xv, 1.

[15] Ammianus Marcellinus, XX, xi, 13 ; aliæ unctæ alumine ; Aldrovandi, Musæum Metallicum, Bologna, fol., 1648, 329.

[16] Charlesworth, Trade Routes, 25 ; Blümner, ii, 331.

Alum was used in magic in the Old Kingdom [1] and is still used in Cairo to avert the evil eye, and in magic ceremonies; it is also worn as a charm with common salt dyed red, yellow, and blue.[2] Pieces of alum used as amulets against the evil eye are common in Palestine, Syria, Persia, Egypt and westwards along the whole of the north coast of Africa, and alum is used in Morocco, mixed with salt, by Jewish and Arab magicians.[3]

Alum (ybn) often occurs in the Ebers Papyrus [4], e.g. in a recipe of a "Phœnician of Byblos." The substance called imru (ymrw) which frequently occurs in the Edwin Smith Surgical Papyrus [5], is written with an Old Kingdom determinative sign which shows that it is a mineral product. Imru is not translated by Breasted, but has been identified with alum.[6] The word 'ibn (ybn) in the Ebers and the Berlin medical papyri (but not in the Edwin Smith Papyrus), and known in the Coptic oben, had been identified as alum through an Arabic form in Ibn al-Baitâr by Athanasius Kircher (1602–80), and this is one of the very rare cases in which this laborious German Jesuit's extensive excursions into Egyptian antiquities have any permanent value. Loret [7] showed that the uses of 'ibn mentioned in the Ebers and Berlin papyri agree with those which could be made of alum. In the Berlin papyrus it is a remedy in skin diseases for which alum is specified by Dioskurides. The Edwin Smith Papyrus abounds in archaic terms, and Prof. Glanville informs me that, although it is unlikely, it is by no means impossible that ymrw also means alum ; ymrw (known only in the Edwin Smith Papyrus) and ybn are possibly two words for the same thing from different stages of the language. Alum is frequently mentioned in Egyptian papyri written in Greek in the 1st–3rd centuries A.D., also as "Egyptian alum" and "rock alum."[8] Qalqashandi (d. A.D. 1418) says alum was produced in pits in Upper Egypt and the oases (it was extracted in the Fezzan in the 19th century) [9], and brought to Alexandria : it was much used to dye red [with madder, q.v.] by the Greeks.[10]

SULPHUR

Wilkinson[11] says, without further details, that the sulphur which abounds in the deserts of the Red Sea was "not neglected by the ancient Egyptians" ; sulphur is said to be mentioned in the Ebers Papyrus[12], and "yellow sulphur" was an important article of Egyptian trade in the 12–13th centuries A.D.[13] It is obtained in the Oasis of Fezzan.[14]

NITRE ; SALTPETRE

The crystalline salt called sa in the Ebers Papyrus is supposed to be salt-petre (potassium nitrate), and bed has been translated saltpetre or natron.[15] Napier[16] thought the ancient Egyptians were acquainted with both nitric and sulphuric acids, and knew that these dissolved silver. The writings and markings upon ancient mummy cloth have also been said to have been made with a solution of silver, probably the nitrate[17], and Wiedemann[18] refers to

[1] Lexa, La Magie dans l'Égypte, i, 49.
[2] Lane, Modern Egyptians, i, 342 f.
[3] Budge, Amulets, 307.
[4] Joachim, 99 ; Budge, Egy. Hierogl. Dict., 5, 39.
[5] Ed. by Breasted, 2 vols., Chicago, 1930, i, 57, 98, 101, 264, 336, 352, 357, 583 ; see especially 264 ; Isis, 1930, xv, 357.
[6] Lippmann, Chem. Z., 1931, lv, 933.
[7] Recueil des travaux relatifs à la philologie et à l'archéologie égyptiennes et assyriennes, 1893, xv, 199.

[8] Dawson, Magician and Leech, 1929, 121 f.
[9] Ritter, Erdkunde, Afrika², 1822, 998.
[10] Geogr., tr. Wüstenfeld, 1879, 160.
[11] i, 155.
[12] Joachim, 90.
[13] Qalqashandi, 224 f.
[14] Ritter, Erdkunde, Afrika², 1822, 998.
[15] Joachim, Papyros Ebers, 81, 85, 144 ; Budge, Egy. Hierogl. Dict., 227, 349.
[16] Arts, 55.
[17] Herapath, Phil. Mag., 1852, [iv], iii, 528.
[18] 1920, 83.

the corrosion of papyrus and linen by some kinds of ink, especially the red, and supposes "an acid" was sometimes used.

There is no *direct* evidence that nitre was known as such in Egypt, although it is formed on the soil in the Sinai Desert, where Petrie's expedition made gunpowder for blasting from the natural "saltpetre taken from the rocks."[1] Saltpetre is also obtained from the soil in the Oasis of Fezzan.[2]   Mungo Park[3] found the negroes near Gomba (Nigeria) making gunpowder from sulphur supplied by the Moors and nitre recrystallised from the white efflorescence on the mud in dried-up cattle ponds. A native showed him a bag of very white but small crystals. Natron is said to have been used in mixtures for incense, though saltpetre would have been a more suitable constituent and was perhaps actually used.[4]   Wilkinson[5] says the fine sand of Egypt is impregnated with "nitre," and the alluvial soil used as a top-dressing for land was "nitrous earth," but the analysis he quotes from Regnault does not give nitre as a constituent. Archæologists frequently speak of "nitrous soil" without meaning saltpetre and any efflorescence is, to them, usually "nitrous." There is no reason why saltpetre should not have been known quite early in Egypt, but we cannot say that it was. The soil on the sites of old villages is rich in nitrates and is used as a fertiliser ; the saline efflorescence on such soil may well have attracted attention at an early date, and its extraction with water is not a difficult operation. The shale (Blättermergel) adjoining the Nile in Upper Egypt contains sodium nitrate (6–60 per cent.) with common salt and a small quantity of nitrite.[6]

[1] Petrie, Res. in Sinai, 257.

[2] Ritter, Erdkunde, Afrika[2], 1822, 998 ; other products are soda (trona), rock salt, alum (shub), and gypsum.

[3] Travels, Everyman ed., 88.

[4] BMGE, 5.

[5] ii, 377, 395 f. ;   cf. Maspero, Dawn, 26 f., and references.

[6] Gmelins Handbuch der Chemie, 1928, Syst. Nr. 21, Natrium, 17.

EGYPT III—3

ORGANIC MATERIALS

TEXTILES

Linen (àau) [1] was known in Predynastic Egypt.[2] Pliny and Athenaios say that weaving was invented in Egypt [3], and spinning was said to be an invention of the goddess Neith. Clothing was regarded as decorative and served mainly magic purposes.[4]

Linen cloth (men), probably invented in the Delta where flax grew [5], is found in prehistoric tombs, some of it very fine.[6] It is usually fine, silk-like and almost transparent (as Pliny says), even in the I dyn., although linen mummy cloths were generally coarse. The linen of Pepi I (VI dyn.) and that of Thothmes III (XVIII dyn.) were especially fine, and mosquito nets were used in the III dyn.[7] Woven linen of a high order of quality was found next to the skin on all bodies in Badarian graves, skins being worn over it.[8]

In the Old Kingdom a noble of rank had the control of the Pharaoh's flax harvest [9], and the cultivation, soaking and beating of flax, the production of rope and cloth, and representations of women weaving and using the spindle are found in the XII dyn. pictures at Beni Hasan; large ropes were also made from the fibres of the date palm.[10] The Ebers Papyrus (1550 B.C.) mentions the flax plant, its seeds and the infusion of them in water, rich in mucilage, also linen lint.[11] Large stocks of linen cloths were found in a V dyn. cemetery: the very fine varieties had 93 × 44 and 72 × 36 threads to the inch, and the open kinds 13 × 10; others with 152 × 71 threads to the inch, and even much finer, are mentioned, and some linen seems to have been "starched" by some material.[12]

Mummies of the II and III dyns. are wrapped in sheets of linen about 4 ft. square, the pads between the limbs being of pieces from 18 in. to 2 ft. square. In the XVIII dyn. sheets 9–12 ft. square were used. The length and breadth of mummy bandages vary from 3 ft. × 2½ in. to 13 ft. × 4½ in. The saffron-coloured pieces of linen dyed with Carthamus tinctorius, with which mummies were finally covered, measure about 8 ft. × 4 ft. Usually two or three different kinds of linen were used in bandaging. The wrappings are usually plain

---

[1] Budge, Egy. Hierogl. Dict., 110.

[2] Wiedemann, 1920, 323 f.; Kroll, PW, xii, 602.

[3] Wilkinson, ii, 160.

[4] Wiedemann, 1920, 118, 325.

[5] Budge, Dict., 300; Hall, CAH, ii, 422.

[6] Petrie, Tarkhan II, 1914, 6, 10; Erman-Ranke, 233; Elliot Smith, The Ancient Egyptians, 55; Maspero, Arch., 299.

[7] Pliny, xix, 1; Wilkinson, i, 185; ii, 161 f.; Breasted, H., 96; Petrie, Anct. Egypt, 1930, 5; BMGE, 164; J. Yates, Textrinum Antiquorum, London, 1843, i:

Raw Materials—all published—255 f., 261 f.; A. Scott, T. Faraday Soc., 1924, xx, 226; Wagenaar, Pharm. Weekblad, 1933, lxx, 688, 894; Wiedemann, 1920, 181, 326.

[8] Brunton and Thompson, Badarian Civ., 1928, 19; 54, plate 48, 6; 64 f. and plates with supposed representations of Predynastic linen workers.

[9] Breasted, H., 96.

[10] Wilkinson, i, 317, Fig. 110; ii, 172, 177, Fig. 389.

[11] Joachim, 28, 114, etc.

[12] Maspero, Arch., 299; Erman-Ranke, 233; Wilkinson, i, 162, 165, 185.

until Greek times, when the outer covering of fine linen is decorated with figures of gods, etc., in gaudy colours. The hope of resurrection given by Christianity practically killed the art of embalming, though the Copts still dressed the body in remarkably beautiful garments, many specimens of which occur in 2nd–9th century A.D. graves at Akhmîm. This town (Egyptian Àpu ; Greek Panopolis) was a famous centre of linen manufacture, as it is at present, but there must have been other centres of manufacture, as Egypt exported large quantities of linen, for example to Western Asia. The cartonnage cases of mummies were sometimes of linen with plaster run in between the folds.[1]

Linen thread was made and wound on wooden reels, as at present [2], and a bronze needle found at Amarna had two eyes at right angles, so that two coloured threads could be used.[3] Some linen was made by temple weavers [4], perhaps for the garments of the priests only, who, as Herodotos [5] and Pliny [6] say, wore garments entirely of linen. The priests and votaries of Isis in Rome still wore only linen, [7] which was also an Orphic custom.[8] Pliny gives four kinds of Egyptian linen, all made in the Delta, viz. at Tanis, Pelusium, Buto and Tentyra, and says linen was the only textile used in Egypt. Wool, which was known in Egypt somewhat later than linen, was used for garments in the Hellenistic Period only.[9]

Linen stained bright green [with copper ?] was found in a Badarian grave.[10] Copper salts are still used in impregnating canvas against rotting and for waterproofing ("Willesden canvas"), and an ancient Egyptian canvas bucket so treated was shown by Petrie in Manchester about 1910. Perhaps a solution of malachite in (ammoniacal) putrefied urine had been used.

Herodotos[11] calls mummy cloths σινδόνος βυσσίνης ; Wilkinson[12] thought σινδών was the Hellenised Egyptian shenti, garment, and byssos from the Egyptian hbos, clothes or to clothe, although byssos is usually derived from the Hebrew bûts.[13] Herodotos is not clear as to this material, for he says in another place[14] that Amasis sent presents of linen embroidered with gold and "tree wool" (εἰρίοσι, or εἴριον, ἀπὸ ξύλου), and it has been thought that this "tree wool," which is undoubtedly cotton[15], was an Egyptian product. Herodotos, however, says[16] that it came from India, and Philostratos[17] that byssos (here cotton) was imported to Egypt from India. Pollux[18] explains that "tree wool" is distinct from byssos, which he calls a kind of Indian flax, and that the old Egyptian name for coarse linen or sailcloth was phōsson (φώσσων). The Egyptian name for fine linen, however, was pek.[19] The name khet-en-šen has been translated as cotton.[20] Wilkinson[21] suggested that βύσσος and λίνον meant, respectively, the flax plant and its product ; Yates[22] that λίνον was common flax and βύσσος a finer variety. The very confused statements in Greek and Roman authors puzzled historians, and the confusion was made worse by Rouelle in 1750, who alleged that mummy cloths which he examined were of cotton.[23] Ure, and Thomson[24], however, showed that they were of

[1] Budge, Mummy[2], 216 f., 221 ; BMGE, 97, 100, 164 ; BMGE, IV–VI Rooms, 223 f., 227 f., 274.
[2] Wilkinson, ii, 176 ; Fig. 392.
[3] Petrie, Tell el-Amarna, 30.
[4] BMGE, 97, 99.
[5] ii, 37.
[6] xix, 1.
[7] Apuleius, Metamorphoses, xi ; Wilkinson, ii, 159.
[8] Marquardt, Privatleben, 465.
[9] Birch, in Wilkinson, ii, 157 ; Erman-Ranke, 537 ; Budge, Mummy, 1893, 189 f. ; Wiedemann, 1920, 125.
[10] Brunton and Thompson, Badar. Civ., 15.
[11] ii, 86.

[12] ii, 158.
[13] Smith, DA, i, 319.
[14] iii, 47.
[15] Pliny, xix, 1 ; Marquardt, Privatleben, 470 f. ; Smith, op. cit.
[16] iii, 106.
[17] Apollonios of Tyana, ii, 20.
[18] Onomastikon, vii, 71, 75.
[19] Marquardt, 465, from Brugsch.
[20] Budge, Dict., 566.
[21] ii, 159.
[22] Textr. Antiquor., 267.
[23] J. Thomson, Phil. Mag., 1834, v, 355; Budge, Mummy[2], 215.
[24] Op. cit. ; Dingler's Journal, 1835, lvi, 154.

linen and never cotton, and this is undoubtedly correct.[1] Lucas [2] says cotton was probably not known or used in Egypt before 200 B.C., when it is first found, but cotton lint (? cotton wool) is said to be mentioned in the Ebers Papyrus (1550 B.C.) [3], whilst Forrer [4] says cotton fabrics first occur in the Byzantine graves of Akhmîm.

Egyptian linen was made from Linum usitatissimum, Linn., true flax, not from wild flax, Linum angustifolium, from which linen found in Italian Stone Age remains and Swiss pile-dwellings was made.[5] Seed pods of Linum humile occur in Egyptian tombs.[6] The true Egyptian flax may be derived from the wild flax, the seeds of which were used as food in Europe : the plant did not occur in Scandinavia in the Stone Age.[7]

Carpet-weaving (from flax) is said to be represented [8], but generally rush mats were used for hanging on walls and carpets were not used : the tapestry-like material woven from coloured linen threads about 1500 B.C. was very expensive and was probably used only for valuable clothing. Some ceilings were apparently of coloured leather, also used for walls.[9] Carpet-weaving from wool is probably an old Persian invention.[10]

In sharp contrast to its extensive use in Mesopotamia, wool was not used for clothing by the Egyptians, but woollen clothing is shown worn by foreigners : the material was regarded as unclean by the Egyptians.[11] Wigs, specimens of which occur in XII and XVIII dyn. tombs, were of human hair (not horse-hair), curled and fixed in position by beeswax.[12] Lucas (opp. cit.) states that the wigs are "probably never of wool," as Wilkinson supposed, but there is a wig of human (?) hair mixed with black sheep's wool at Cairo.[13] Although Pliny[14] says felt was used by the "ancient peoples," it does not seem to have been used in Egypt. Silk first became known in Egypt, as in Greece and Rome, about the 2nd century A.D.[15] Lucas[16] says silk was not known in Egypt until the 4th century A.D., but Feldhaus[17] mentions silk objects of the 1st–2nd centuries. No silk was found in the Roman-Nubian remains at Karanog.[18] Silk cloth was used as wrappings for mummies of wealthy people in the 1st century A.D. : some are in the British Museum. It was not used in tapestries till the 8th century A.D.[19]

DYES AND DYEING

Although savage peoples prefer music in the minor key, their colours are always bright. The first dyes used were probably organic colours from plants etc., in the earliest times no doubt without mordants and thus very fugitive.[20] Most tissue materials are dull in colour; the crushing of fruits and plants and the effects of rain on some minerals perhaps suggested a means of colouring them. Dyeing may be carried out in hot or cold solutions : it is not known which was first in use.[21]

[1] Wilkinson, ii, 158 ; Yates, 254 f. ; Marquardt, 465 f. ; Budge, Mummy[2], 215 ; Wiedemann, 1920, 327 ; Keimer, Die Gartenpflanzen im alten Ägypten, Hamburg, 1924, i, 59.

[2] Egy. Mat., 196.

[3] Joachim, 115, 117, 124.

[4] RL, 83.

[5] Forrer, RL, 246.

[6] Wiedemann, 1920, 323.

[7] Schrader, RL, 246 f., 249 ; F. Woenig, Die Pflanzen im alten Ägypten, Leipzig, 1897, 181.

[8] Maspero, Arch., 295.

[9] Wiedemann, 1920, 177 f.

[10] Orth, PW, xii, 605.

[11] Hall, CAH, ii, 423 ; BMGE, 80.

[12] Wilkinson, ii, 329 f. ; BMGE, IV–VI Rooms, 264 f. ; Lucas, Ann. Service des Antiquités de l'Égypte, Cairo, 1930, xxx, 190 ; ib., Egy. Mat., 195.

[13] Maspero, Guide to the Cairo Museum, 1908, 491—XVIII dyn.

[14] viii, 48.

[15] Budge, Mummy[2], 213 and refs. ; BMGE, IV–VI Rooms, 227 ; Yates, op. cit.

[16] Egy. Mat., 197.

[17] Technik, 1016.

[18] Woolley and MacIver, Karanog, i, 27.

[19] Budge, Mummy[2], 213 f., 218.

[20] Hoefer, Hist. de la Chimie, i, 59.

[21] Goguet, i, 142 f. ; Berthollet, Art of Dyeing, tr. by Ure, 2 vols., 1824, i, 4 f.

Dyeing was used very early in Egypt, where red, yellow and green clothing is represented in the oldest pictures.[1] The name àun (colour) probably means pigment rather than dye.[2] Coloured sails, red and green, are shown in the tomb of Rameses III [3], and Nubian women from the Sudan in the tomb of Ḥuy (xviii dyn.) wear skirts striped in blue, red and yellow.[4] Specimens of brown and salmon-coloured mummy linen and coloured wearing apparel, of the period of Amenhotep I (xviii dyn.), are in the British Museum.[5] Old Kingdom mummy linen was quite plain, but in the Middle Kingdom it began to be dyed with a blue stripe at the edge[6] ; red (rose colour) and blue stripes or fringes were most frequent in the xxi dyn. Bright blue clothes were worn in mourning.[7]

Ancient Egyptian dyes established by analysis are indigo and safflower (Carthamus).[8] Woenig and Schelenz thought that indigo was imported from India and used as a dye in Egypt as early as the vi dyn. (2500 B.C.) ; this seems improbable, but indigo is apparently mentioned in the Ebers Papyrus (1550 B.C.), [9] and its use is proved by actual examination of mummy cloths.[10] One of these from Thebes (now in Harrow Museum) was dyed blue at the edges with indigo[11], and another mummy wrapping of linen was found by chemical tests and by microscopic examination to be dyed in the skein with indigo.[12] A mummy cloth of the time of Alexander the Great was dyed with indigo and gave off the characteristic purple vapour on heating.[13] Woenig[14] says indigo was used under Amenhotep I.[15] A dyer's shop of the time of Ptolemy Physkon (146–117) contained pots stained with indigo and red dye (safflower ?)[16], and linen shrouds with "an angle woven in blue, with occasional greens and yellows and red thread," were found in the Roman remains at Karanog.[17]

Indigo is not mentioned by Herodotos[18]; it is included by Dioskurides[19] and by Pliny[20] among minerals. Although the ancient Egyptians and probably the Hindus used it as a dye, the Romans were not acquainted with the method of getting it into solution by a reduction process, although this is not very difficult, and therefore used it only as a pigment.[21] Pliny[22], indeed, says : "at in diluendo misturam purpuræ cæruleique mirabilem reddit," which *could* mean dissolved, and is so given in Philemon Holland's translation.[23] Pliny also says, quite correctly, that indigo on heating gives a beautiful purple vapour ("flame" by mistake).[24]

The woad plant (Isatis tinctoria) is different from the indigo plant (Indigofera tinctoria) although both yield the same indigo blue[25], and the variety of indigo growing wild in India, North Africa and Arabia was probably not the true indigo (Indigofera tinctoria) but the Indigofera cœrulea, which would yield a

[1] Erman-Ranke, 241, 244 ; Wiedemann, 1920, 326.

[2] Budge, Egy. Hierogl. Dict., 34.

[3] Wilkinson, ii, frontisp., pl. 13.

[4] Davies and Gardiner, Tomb of Ḥuy, 1926, plates xxiii, xxx ; Quibell, 90, 91.

[5] BMGE, 101 ; Maspero, Struggle, 104.

[6] BMGE, 164.

[7] Wiedemann, 1920, 113, 117.

[8] Loret, L'Égypte, 177 ; Woenig, Pflanzen, 353.

[9] Schmidt, Drogen, 3 ; Joachim, 21, 157.

[10] Merimée, in Passalacqua, Catalogue, 1826, 258 f.

[11] Thomson, *Phil. Mag.*, 1834, v, 361 ; Wilkinson, ii, 152, 163, Fig. 383 ; Woenig, Pflanzen, 353.

[12] Herapath, *Phil. Mag.*, 1852, iii, 528 ; "not so deep nor so equal as the work of the modern dyers."

[13] Rathgen, *Chem. Z.*, 1921, 1102 ; cf. Fester, Entwicklung der chem. Technik, 1923, 10.

[14] Pflanzen, 353.

[15] Lippmann's doubts, Abh. i, 93 ; Alch., i, 271 ; cf. Abh. ii, 20, are based on an obvious misprint in Woenig.

[16] Petrie, Arthribis, 1908, 11.

[17] Woolley and MacIver, Karanog, i, 27.

[18] Wilkinson, ii, 164.

[19] v, 107—ἰνδικόν.

[20] xxxiii, 13—indicum.

[21] John, Malerei der Alten, 1836, 120 ; K. C. Bailey, The Elder Pliny's Chapters on Chemical Subjects, 1929–32, i, 236 ; Berthollet-Ure, ii, 37.

[22] xxxv, 6.

[23] The Historie of the World of C. Plinius Secundus, fol., 1634, 531.

[24] Cf. John, *op. cit.* ; Wilkinson, ii, 164.

[25] Gmelin, Handbook of Chemistry, tr. Watts, 1859, xiii, 36.

true indigo on treatment. Considerable confusion exists in the nomenclature of the species of Isatis and Indigofera, about 50 species of Isatis being recognised, whilst 250–300 species of Indigofera have been reported. Woad was cultivated in Egypt early in the Christian Era, there being eight references to it in the Oxyrhynchus Papyri.[1]

The indigo used by the ancient Egyptians may very well have been derived from the Indigofera argentea, which grows wild in the desert west of the Thebaïd, and in Abyssinia, Nubia, Kordofan and Sennar.[2] True indigo was cultivated in Egypt in the 19th century and was probably imported from India.[3] Mungo Park found the African natives on the banks of the Gambia dyeing cloth an excellent blue with indigo [4], and Arabic authors speak of indigo cultivation in Egypt.[5]

According to Loret [6] the old Egyptian name for indigo was terenken, dinkon or tinkon, from which ἰνδικόν and indicum were derived. In all probability, however, these names are derived from the name of India (as Pliny says) ; the Sanskrit name of indigo was nîl. Loret's identification is based on the occurrence of tinkon in texts dealing with dyeing and in medical papyri, where it is called "the plant which expels the gripes," whilst Dioskurides [7] says indikon "breaks inflammation and œdema." The "Indian black" (μέλαν ἰνδικόν) of Greek authors was not indigo but Indian (really Chinese) carbon ink.[8] Indigo was an important article of commerce in Egypt in the 12th–13th centuries A.D.[9] ; Maqrîzî[10] says it grew near Alexandria.

The undyed part of the mummy cloth found at Thebes in which Thomson detected indigo[11], was stained a yellow colour soluble in water. The solution contained no tannin but gave a cloud with lead subacetate, indicating the presence of extractive matter, which was probably not used for impregnating the threads before weaving[12] but was the yellow or brown colouring matter of unbleached linen, perhaps produced by the process of steeping the flax, and rendered soluble by oxidation on long keeping.[13]

Another piece of linen from Thebes examined by Thomson was apparently dyed by safflower (Carthamus tinctorius, κνῆκος; Sanskrit kâñcana, "golden")[14], which is perhaps the nesti or nas of the Ebers Papyrus (c. 1550), in which senau is perhaps Carthamus lanatus, and it was used for dyeing mummy bandages from about 2500.[15] The usual names for safflower (Carthamus tinct.), however, are ma-t and katha.[16] Qalqashandi (d. 1418) says it was cultivated in Egypt.[17] Safflower seeds, from which an oil was also extracted, occurred in a Theban tomb.[18] The powder of Carthamus, used as a cosmetic which gives a

[1] G. Watt, Dictionary of Economic Products of India, 1890, iv, 383, 389 f. ; ib., Commercial Products of India, 1908, 660 f., with full bibliographies ; Hurry, The Woad Plant and its Dye, Oxford, 1930, 8 f., 10.

[2] Loret, La Flore Pharaonique, d'après les documents hiéroglyphiques et les spécimens découverts dans les tombes, 2 ed., 1892, 90 ; ib., L'Égypte au temps des Pharaons, 1889, 177 ; Beckmann, History of Inventions, 1846, ii, 258 ; Woenig, Pflanzen, 353.

[3] Woenig, 353 ; Lucas, Egy. Mat., 199.

[4] Travels, Everyman ed., 7.

[5] Reitemeyer, 44.

[6] La Flore, 90 ; L'Égypte, 177.

[7] v, 107.

[8] Salmasius, Plinianæ exercitat., 1689, 181.

[9] Qalqashandi, Geogr., tr. Wüstenfeld, 1879, 224 f.

[10] Tr. Bouriant, 201.

[11] Thomson Phil. Mag., 1834, v, 361; Wilkinson, ii, 152, 163, Fig. 383 ; Woenig, Pflanzen, 353.

[12] Wilkinson, ii, 164.

[13] Donovan, Treatise on Chemistry, 1832, 181 ; Berthollet-Ure, i, 120 f.

[14] Wilkinson, ii, 165 ; Blümner, i, 111 ; Woenig, Pflanzen, 351 ; Keimer, Gartenpflanzen, i, 7, 80, 127 ; Schrader, RL, 698 ; Berthollet-Ure, ii, 189.

[15] Joachim, 23, 124 ; Loret, L'Égypte, 177 ; Fester, 9.

[16] Budge, Hierogl. Dict., 268, 791.

[17] Reitemeyer, 25.

[18] Schrader, RL, 698 ; Wilkinson, ii, 399 ; Woenig, Pflanzen, 351.

red colour with water, also occurred in tombs [1], and the use of the dye has been confirmed by analysis.[2]

An iron-buff was dyed, perhaps with iron vitriol.[3]   In later Egypt at least, Anchusa tinctoria (alkanet) [4] and fucus ($\phi\hat{v}\kappa o\varsigma$) (archil) [5] were used. A red dye sandyx (a name also applied to mineral red pigments) is mentioned for the later period.[6]   Scarlet was dyed at Assiût in the Arabic Period.[7]

Dyed and embroidered woollen fabrics were imported from Babylonia fairly early [8], and two specimens of wool, one (XII dyn.) as yarn, dyed blue, red and green, and one of the XVIII dyn., were found in graves.[9]

Pliny says the Egyptians had a strange way of dyeing. They first impregnated the fabric with chemicals (medicamenta) which did not cause the appearance of any kind of colour, then plunged the cloth into a cauldron of boiling dye, and drew it out showing various colours (picta) although there was only one in the vat. This probably refers to the use of mordants, i.e. metallic salts fixing the dye, often with various colours. Parti-coloured fabrics, some printed, of the early Christian Period were found in graves at Akhmîm, confirming Pliny's account, and the process may go back in Egypt to an earlier period.[10]

Two roots, nstj (Coptic Opion) and jp' (Coptic Apia), used in dyeing, are identified by Loret[11] with alkanet (anchusa) and madder (Rubia tinctorum). Madder, mentioned (as $\dot{\rho}\dot{\iota}\zeta a$, rhiza, "the root") by Strabo[12], was an important article of commerce in Egypt in the 12th–13th centuries.[13]

Wigs of sheep's wool dyed blue or black, or of horse-hair or human hair, were used with the head shaved from a very early period by all classes, including priests.[14] The kings wore blue and green wigs (formerly thought to be "steel helmets")[15] and blue beards, and the people coloured their hair blue or green, or powdered it white, and depilatories were probably in use. The modern kind of depilatory in Egypt is a mixture of quicklime with one-eighth the weight of orpiment.[16]

An old custom in the East and in Egypt, which is still used, is the staining of the hands, feet and nails with henna, a dye from the leaves of the Lawsonia inermis (L. alba). The plant, twigs of which were found in Hawâra (Ptolemaic Period), was called $\kappa\acute{v}\pi\rho o\varsigma$ by the Greeks.[17] The best plants, and an oil extracted from the flowers are said by Pliny to come from Egypt, the next best from Cyprus, Sidon, Askalon and Canopus in Egypt.[18] The plant probably came from farther East and was introduced into Egypt about 1400, or at latest about 1300–1000.[19] The hair and nails of a few later mummies are said by Olck to be dyed with henna (doubted by Lucas), and its use is indicated in an Old Kingdom statue.[20]

[1] Loret, L'Égypte, 195.
[2] Hübner, q. in Lucas, Egy. Mat., 199 ; BMGE, IV–VI Rooms, 225—purple linen.
[3] Lucas, 200 ; Wiedemann, 1920, 326.
[4] Blümner, i, 246.
[5] Stadler, PW, vii, 196 ; Blümner, i, 246.
[6] Blümner, i, 245.
[7] Yaqûbî, A.D. 891 ; Reitemeyer, 146.
[8] Maspero, Dawn, 370.
[9] Lucas, Egy. Mat., 195.
[10] Pliny, xxxv, 11 ; Wilkinson, ii, 168 f. ; Berthollet-Ure, i, 5 ; Forrer, RL, 213 ; Fester, 17 ; Führer durch die Ausstellung des Papyrus Erzherzog Rainer, Wien, 1894, 228 ; gives manufacture in Persia about A.D. 550.
[11] JEA, 1932, xviii, 186 ; cf. C. Neumann, Chemical Works, tr. Lewis, 1759, 338.
[12] XIII, iv, 14, 629 C.

[13] Qalqashandi, Geogr., tr. Wüstenfeld, 1879, 224 f.
[14] Wilkinson, i, 163, 185, 424 f. ; ii, 329 f.; Maspero, Dawn, 54, 278 ; Erman-Ranke, 245, 247, 251 ; specimens in British Museum ; Wiedemann, 1920, 138.
[15] Wilkinson, i, 218.
[16] Lane, Modern Egyptians, 1846, ii, 50.
[17] Olck, PW, vii, 805 ; Engel, Kypros, 1841, i, 65 ; Fester, Technik, 9 ; Perrot and Chipiez, Hist. of Art in Phœnicia and Cyprus, 1885, ii, 89 ; Keimer, Gartenpflanzen, i, 51, 107 ; BMGE, 81 ; Wiedemann, 1920, 148 ; Woenig, Pflanzen, 349 ; Lane, Mod. Eg., i, 54 f., with illustrations ; Gsell, 44 ; Lucas, JEA, 1930, xvi, 45 f.
[18] Pliny, xii, 22 ; xiii, 1 ; Olck, op. cit. ; Engel, Kypros, 1841, i, 13 f., 64 f.
[19] Maspero, Struggle, 200 ; Olck, op. cit.
[20] Wiedemann, 1920, 148.

Budge [1] speaks of linen dyed "pink or purple" in a XXI dyn. coffin on which is a "mauve or purple" varnish (perhaps red and blue mixed).

POMADE AND OINTMENT

The use of ointment (baq ; àbr ; àmzart) for the body was common to all classes in ancient Egypt as a protection against heat, and it is still largely in use.[2] A pomatum of scented fat (probably animal) was used, and Egyptian white and black ointments, mentioned by Pollux and Theophrastos, have been found at Thebes.[3] Petrie found what he calls "palm oil (?)" in jars in prehistoric tombs, as well as a "scented fat" common in the Predynastic Period and the I dyn. [4], but the composition of the solid ointments used has probably altered with time.[5]

Although Theophrastos [6] says some Egyptian ointment was not strongly scented, the texts refer to the "smell of unguents," "sweet oil of gums," "ointment of gums" (probably gum resins) and to seven (later eight to ten) sacred oils, the names of which were (?) finest oil of Libya, finest oil of cedar, tuatu, nemnen, sefth (bitumen), heknu, and festival oil [7], and scented pomade was put on the head.[8] Chapman [9] says an *odour* of coconut oil does not mean anything in an ancient fat, and although he was unable to form a definite opinion of the nature of the fat from Tutankhamen's tomb he thought it probably of animal origin, and the chemical evidence generally seems to exclude coconut or palm-kernel oils. The pomade probably contained about 10 per cent. of resin, as previously suggested by Ure.[10] Sometimes the fatty matter consists of mixed palmitic and stearic acids, probably from animal fats : one specimen was stearic acid, probably from some fat.[11] Gowland[12], from an examination of five specimens of material from a toilet box of unknown date, concluded that it was beeswax mixed with an aromatic resin and a little animal oil. Traces of perfume are said to have been found in a blue glass bottle in the tomb of Amenhotep II.[13]

Oils and fats used in Egypt probably included animal fats, butter, castor oil (used for anointing the body, as by negroes at present), olive oil, linseed oil, sesame oil and safflower oil.[14] The material qemi, used medicinally and coming from the south coast of the Red Sea, is not, as Erman once thought, a salve but is gum arabic ($\kappa \acute{o} \mu \mu \iota$).[15]

Unguents in museum jars are brown and viscous, some are semi-liquid and in one case the ointment has percolated through the alabaster pot, sticking it tightly to the mount.[16] There is no evidence of the use of animal perfumes (musk, civet, ambergris) in ancient Egypt, but plant products, including certain resins and gum resins, were used, and perfumes were perhaps extracted from the odoriferous plants, flowers, etc., with oils and fats, a process (enfleurage) still used, but now the perfume is extracted from the fat by alcohol. In one

[1] Mummy², 220, 429 ; BMGE, IV–VI Rooms, 7.
[2] Wiedemann, 1920, 148 ; Budge, Egy. Hierogl. Dict., 39, 55, 205.
[3] Wilkinson, i, 426 ; ii, 345 f. ; analysis by A. Chaston Chapman, *J. Chem. Soc.*, 1926, 2614.
[4] Naqada and Ballas, 1896, 39 ; Tarkhan I and Memphis V, 1913, 9, 11, 12, 18.
[5] Wiedemann, 1920, 151.
[6] De Odoribus, 55.
[7] Wilkinson, ii, 345 ; Lucas, JEA, 1930, xvi, 46; Wiedemann, 1920, 151 ; BMGE, IV–VI Rooms, 213.
[8] Wilkinson, i, 426, Fig. 203 ; Davies

and Gardiner, Tomb of Amenemhet, 1915, 37, 64.
[9] *J. Chem. Soc.*, 1926, 2614 ; cf. Lucas, Egy. Mat., 126 f.
[10] In Wilkinson, ii, 401.
[11] Lucas, JEA, 1930, xvi, 46 f.
[12] PSBA, 1898, xx, 268.
[13] Kisa, Glas, 48, 318, 565.
[14] Lucas, Egy. Mat., 126 f. ; Wilkinson, ii, 399 f. ; Wiedemann, 1920, 149 ; Budge, Mummy², 265 f.
[15] Erman, Lit., 9, 268, 311, 387 ; Erman-Ranke, 599 ; Wiedemann, 1920, 151.
[16] Budge, Mummy², 266 ; BMGE, IV–VI Rooms, 19.

case the odour of the fat is merely due to valeric acid formed by chemical change.[1] Theophrastos (d. 287 B.C.) [2] describes the extraction of perfume by means of Egyptian or Syrian balanos, or ben oil (oil of Balanites ægyptiaca), though olive and almond oils were also used, and mentions one Egyptian ointment containing cinnamon and myrrh with several other (unspecified) ingredients, and another one made from quince. Pliny says the oil was heated. That a corresponding process was used in Egypt is indicated by Pliny's [3] enumeration of various oils among the constituents of Egyptian ointments. Pliny says the Mendesian ointment consisted originally of ben oil, resin and myrrh, but later contained oil of bitter almonds (metopium), olive oil (omphacium), cardamoms, sweet rush, honey, wine, myrrh, seed of balsamum, galbanum and turpentine resin. Constituents of ointments which he mentions are oils of ben nut (myrobalanum), growing in the country of the Troglodytes, in the Thebaïd and in Ethiopia ; the Egyptian elate or spathe, and the fruit of a palm called adipsos. His cyprinum, an Egyptian tree, was probably henna.

Eight specimens of unknown date, thought to be perfumes by Reutter, were said to contain all or most of the following : storax, incense, myrrh, turpentine resin, bitumen of Judea perfumed with henna, aromatic vegetable material mixed with palm wine or the extract of certain fruits such as cassia or tamarind, and grape wine. The analyses, made in 1914 with very small quantities of material (0·498–2·695 gm.), are said [4] to require confirmation, the presence of bitumen, in particular, being doubtful. Rather elaborate recipes for the preparation of unguents contained in inscriptions on the four walls of the "laboratory" (àsit) in the temple at Edfu specify many resins and plants.[5] The presence of vegetable fibres in fat or pomade does not, according to Chapman (op. cit.), necessarily mean that the fat is of vegetable origin, since animal fats were wrapped in coverings of vegetable material.

PLANT PRODUCTS

The plants, trees and vegetable products of Egypt, both ancient and modern, are very numerous.[6] Vegetable products in Predynastic graves at Badari (S.D. 37–44) include tamarix wood, pine, cedar, cypress or juniper, "a Rubiaceous wood"—some of these did not grow in Egypt but probably came by trade from Syria—pods and seeds of gilban (Lathyrus sativus), bark of cinnamon or sandal (?) tree—"foreign to Egypt"—tubers of Hab-el-aziz (Cyperus esculentus), and a complete fruit of dûm palm (Hyphæne thebaica), similar to that of the wild tree.[7] The date palm and dûm palm were indigenous to Egypt.[8] Weinleitner [9] says the Egyptians knew 190 varieties of plants, of which 110 were cultivated in gardens and most of the rest on the large scale in fields. Some representations are stylised, but many are very true to nature and indicate good observation : in a temple at Karnak an arum is shown accurately in various stages of development. The importation of foreign plants and trees and the establishment of botanical gardens by Thothmes III and Queen Hatshepsut (on the walls of whose temple at Deîr el-Baharî the expedition to the Red Sea is shown, with good drawings of incense trees)[10] indicate an interest in natural history. Zoological gardens and natural

[1] Lucas, JEA, 1930, xvi, 45 f.
[2] De Odoribus, 29 f.
[3] xiii, 1.
[4] Lucas, JEA, 1930, xvi, 47.
[5] Dümichen, Z. f. ägypt. Sprache, 1879, 97–128 ; Woenig, Pflanzen, 372.
[6] Wilkinson, ii, 398–417 ; F. Hartmann, L'Agriculture dans l'ancienne Égypte, 1923, 17–70 ; Brugsch, Ägyptologie, 395 f. ; Wiedemann, 1920, 273 f. ; Loret, L'Égypte,

1889 ; ib., Flore Pharaonique, 1892 ; Newberry, in Petrie, Hawara, 1889, 46 ; Woenig, Pflanzen, 1897 ; Keimer, Gartenpflanzen, 1924 ; E. Rohde, Garden Craft in the Bible, 1927.
[7] Brunton and Thompson, Badarian Civ., 1928, 62 f.
[8] Wiedemann, 1920, 275.
[9] Isis, 1927, ix, 16 f.
[10] Maspero, Struggle, 251.

representations of animals indicate an interest in this branch of natural history.[1] A fragment of a herbal, dealing with the virtues of various parts of the Ricinus plant, occurs in the Ebers Papyrus.[2] In one part of the kingdom of Osiris, the blessed ones are said to cultivate the mythical divine plant, Maāt, on which both they and Osiris lived : in this way they became one with him, divine, incorruptible and immortal.[3]

Plant products were indigo[4], mustard[5], colocynth, caraway, cummin (seeds sprinkled on bread), coriander, onions and garlic[6], myrobalans[7], almond, and the fig (after the XII dyn.)[8], which classical authors say was the first cultivated tree.[9]

Pliny[10] says the peach (persica), which he also confuses with the persea, grew in Egypt ; he rejects the statement that it is poisonous, although in fact the crushed kernels produce prussic acid ; a [gnostic ?] papyrus in the Louvre says : "Do not pronounce the name of Iao under the penalty of the peach,"[11] and Abd al-Latîf[12] says the Egyptians relate that the kernel of the persea (lebakh) produces unconsciousness : although described by Arabic authors it became extinct in the Middle Ages.[13] The persea (περσεα) in Plutarch[14], sacred to Isis[15], has been identified with the Cordia myxa or Balanites Ægyptiaca[16], and also[17] with the mimosa, the ἄκανθος of Herodotos[18] and Strabo[19], producing a gum and seed-pods used in tanning.[20] The bamboo is supposed to be represented in XVIII dyn. tombs.[21]

There are also the palms[22]; olive[23]; papyrus[24]; the white or violet-blue lotus, "typical of the Sun" according to Proklos[25]; the pomegranate, the ancient rhodon or rose flower, used for dye, which gave its name to the island of Rhodes and is found in tombs[26]; the castor oil tree[27]; urtica (used for medicinal oil, perhaps a nettle)[28]; calamus (a reed, used for arrows); flax ; gossypion or xylon, mentioned by Pliny[29], i.e. cotton, which Woenig thinks was known in Egypt, although not cultivated in ancient times[30]; poppy, the opium of which is perhaps the nepenthes of Homer and Pliny[31], though this may have been hashish[32]; the poisonous strychnos[33]; and cnicus (carthamus).[34]

The Papyrus Salt (B. Mus.) of the Saïte Period describes various vegetable and other products (myrrh, cedar and cedar oil, incense, wax, honey, linen ?, papyrus, laurel, asphalt) as tears or other excretions of gods.[35]

[1] D. Paton, The Animals of Ancient Egypt, Materials for a sign list of Egyptian hieroglyphs, E, Princeton, 1925 (not seen).
[2] Dawson, Magician and Leech, 1929, 116.
[3] BMGE, 143, 151.
[4] Wilkinson, ii, 403 ; Brugsch, Ägyptologie, 1891, 390 f., 407 f. ; Wiedemann, 1920, 275 f. ; Woenig, 353.
[5] Woenig, 224 f. ; doubtful according to Wiedemann, 1920, 278.
[6] Woenig, 192 f.
[7] Wilkinson, ii, 404.
[8] Wiedemann, 1920, 276 ; Woenig, 292 f.
[9] Goguet, i, 128.
[10] xiii, 9 ; xv, 13.
[11] Hoefer, Hist. de la Chimie, 1866, i, 232.
[12] Relation de l'Égypte, transl. by S. de Sacy, Paris, 1810, 17, 47 f.
[13] Reitemeyer, 64.
[14] Isis and Osiris, 68.
[15] Not Harpokrates, as Hoefer, i, 232, seems to infer, misled, no doubt, by Squire in the earlier edition of Wilkinson, quoted by Parthey, Isis und Osiris, 258 f.
[16] Parthey, op. cit. ; Gubernatis, Mythologie des plantes, 1878-82, ii, 284.
[17] Wiedemann, 1920, 276.
[18] ii, 96.
[19] XVII, i, 35 ; 808 C.
[20] Cf. Reitemeyer, 64.
[21] Davies and Gardiner, Tomb of Amenemhet, 1915, 64.
[22] Wiedemann, 1920, 276 ; Woenig, 304, 315 ; Wilkinson, ii, 405.
[23] Woenig, 327 ; Petrie, Hawara, 1889, 46.
[24] Wilkinson, ii, 406.
[25] Wilkinson, op. cit. ; Woenig, 17 f. ; Steier, PW, xiii, 1516.
[26] Wilkinson, ii, 407 ; Woenig, 323 ; Wiedemann, 1920, 277 ; Dawson, Magician and Leech, 1929, 112.
[27] Woenig, 337 f.
[28] Wilkinson, 408.
[29] xix, 1.
[30] Wilkinson, ii, 402, 409 ; Woenig, 346 f.
[31] Woenig, 224 f. ; Schrader, RL, 49 ; Pliny, xxi, 21 ; xxv, 2.
[32] Wilkinson, ii, 410, 412.
[33] Pliny, xxi, 15, 31 ; Wilkinson, ii, 411.
[34] Pliny, xxi, 15, 32 ; Wilkinson, ii, 411.
[35] Lexa, Magie, ii, 65.

The mandrake (dedmet ; tha-en-unš), used in ancient medicine as an aphrodisiac [1], is known in specimens and representations in XVIII dyn. tombs, and was perhaps imported from Asia.[2]   Dawson says it was not used medicinally in Egypt and that the reading of didi is hæmatite. The mandrake has a strong narcotic action and, although very poisonous, was used from the 1st to the 17th century A.D. in surgical operations, but more and more sparingly, and its supposed magic properties date from the Classical Period.[3]

A cake of a kind of conserve, like the modern agweh, made from pressed dates, was found in a tomb at Thebes with dried dates [4], and Herodotos [5] says the Gyzantians, a Libyan tribe, made more honey (date honey ?) than bees.

Many plants were used medicinally, but the identifications of the names, e.g. in the Ebers Papyrus, are often doubtful.[6]   They probably include : castor oil plant (qiqi) ; caraway and fennel (besbes) ; dill (Anethum graveolens) ; fenugreek ; melilot (chebu) ; peppermint (frequently) and coriander ; juniper in a recipe "from Byblos" ; cassia (?) (genti) ; wormwood (saam) ; aloes and aloe-wood [two quite different things].[7]   Juniper was probably not indigenous to Egypt, but came from Syria and Asia Minor.[8]

Water melons, melons, beans, artichoke (?), cucumber, lettuce, lentils, endive (qatsut), and radishes (nn, or nnt ; Coptic noune) [9] were used as food[10], mostly from before the XII dyn. Onions are clearly shown in the representation of temple offerings and (with leeks and garlic) seem to have been freely eaten and used : their supposed prohibition as food perhaps applied only to priests. The eyes of the mummy of Rameses IV had been replaced by small onions.[11] Benjamin of Tudela (A.D. 1173) praises the cherries, pears, cucumbers, gourds, peas, beans, chickpeas, purslane, asparagus, pulse, lettuce, coriander, endive, cabbage, leek and cardoon of Egyptian gardens.[12]

Apricots are said to have been introduced into Egypt only between A.D. 400 and 640[13]; although Theophrastos[14] is said by Olck to mean apricots when he says the Egyptians of the Thebaïd made jam from the fruit of the κοκκυμηλέα, this probably means the common plum.[15] Pliny[16] mentions the oil of bitter almonds (oleum amigdalis amaris) of Egypt, and in the preparation of this, hydrocyanic acid would be formed.[17]   Brunton[18], from finds of horns and spoons, thinks the Predynastic Egyptians used snuff, not tobacco but some pungent plant.

Gum arabic (qemi or qami ; Greek κόμμι), which came from Nubia or the south coast of the Red Sea, was used in making papier mâché for mummies (really made from linen gummed together) and for varnishing wood.[19]   Hero-

---

[1] Steier, PW, xiv, 1031 ; Budge, Egy. Hierogl. Dict., 850.

[2] Davies and Gardiner, Tomb of the Two Officials, 11 (Thothmes IV) ; Dawson, Magician and Leech, 113, 147.

[3] Steier, 1032, 1035 ; Lippmann, Abh., i, 190, where many of the old references are collected.

[4] Wilkinson, i, 398 ; ii, 43.

[5] iv, 194.

[6] Wilkinson, ii, 417 ; Woenig, 224 f.

[7] Joachim, 5, 11, 70 ; 18 ; 142 ; 111 ; 164 ; 62 ; 88 ; 36, 51, etc. ; 21, 129 ; 30, 35, 122, 142, 180, 196 ; useful summary of Joachim's rather imperfect work in Lippmann, Abh., ii, 1 f., who does not, however, quote all the places in Joachim where the material occurs.

[8] Woenig, Pflanzen, 362.

[9] Loret, Studies presented to F. Ll. Griffith, Oxford, 1932, 304.

[10] Wilkinson, ii, 25 f., 398 f. ; Joachim, 9, 13, 129 ; Wiedemann, 1920, 278.

[11] Wilkinson, i, 181 ; ii, 25 f. ; iii, 350, 419 ; Brugsch, 394 ; Wiedemann, 1920, 277 ; Woenig, Pflanzen, 192 f., 344, 375, 380, who translates St. John's bread berries in the Ebers Papyrus instead of "onions."

[12] Itinerary, ed. Adler, 1907, 73.

[13] Olck, PW, ii, 270.

[14] Hist. plantarum, IV, ii, 10.

[15] Liddell and Scott, Greek-English Lexicon, Oxford, 1930, 971.

[16] xiii, 1.

[17] Cf. Hoefer, Hist. de la Chimie, i, 232 ; Baslez, Les poisons dans l'antiquité égyptienne, 1933, which I have not seen.

[18] Badarian Civ., 1928, 60.

[19] Erman, Lit., 9, 268, 311, 387 ; Erman-Ranke, 599 ; Wiedemann, 1920, 151 ; Budge, Egy. Hierogl. Dict., 771 ; Loret, L'Égypte, 118.

dotos [1] mentions a gum exuded by a thorny tree, ἄκανθος, growing in Egypt, and this acanthus gum (sont) is frequently mentioned in the Ebers Papyrus.[2] The gum was probably true gum arabic from the Acacia nilotica, which was brought as tribute to Rameses III and perhaps used also in making papyrus.[3] Gum was an important article of Egyptian trade in the 12th–13th centuries A.D., as were also saffron and gall nuts.[4]

The lotus tree, on the fruit of which the Lotophagi, a North African tribe, lived, and from which they made a sort of wine [5], is a Rhamnus, the Zizyphus Lotus, the modern Arabic sidr ; the fruit (which looks and tastes like a bad crab-apple) is called nebk.[6]   The tree still grows in the oasis of Fezzan.[7] The water lily (sšn), Nymphæa Lotus, which grows in Egypt about 2 ft. above the surface of the water of the Nile and adjacent rivulets (magnificent plants in Kew Gardens), is quite a different plant.   There was another species of lily growing in the river, the fruit springing up alongside the blossoms on a separate stalk, in appearance like a comb made by wasps, and containing a number of seeds the size of olive stones, which are good to eat, both green and dried.[8] Herodotos says the Egyptians collected the lotus flowers (from the Nymphæa Lotus, L.), dried them in the sun, and extracted from the centre of each a substance like the head of a poppy, which they crushed and made into bread. The root was also edible.[9]   It is still roasted and eaten in Senegal.[10]

OILS

Although Diodoros Siculus[11] makes the god Hermes-Thoth the discoverer of olive oil, Mahaffy[12] says it was not used in Egypt until 260 B.C. Strabo[13] reports that in his time (1st century B.C.) only the Arsinoïte nome and gardens near Alexandria were planted with olive trees, which gave no oil. The tree is shown in early representations and the leaves are found in tombs.[14] The olive was probably known early in Egypt, as it grows with very little attention.[15] If the oil were carefully extracted it might be obtained good, but a large amount of it obtained in Egypt has a disagreeable smell.[16] Qalqashandi (d. 1418 A.D.) says only a few olives were grown in Egypt and they were not used for oil but for eating with salt.[17] The principal edible oil was, as in Assyria[18], sesame oil, and other oils were[19] castor (qek, qiqi) oil and oils of safflower seed, pumpkin, linseed, lettuce seed, rape seed (Brassica olifera, probably Pliny's raphanus) and radish seed.[20] Linseed oil was used to make a cement with litharge[21], and is perhaps the "thick oil" of the Ebers Papyrus, the "tree oil" being perhaps castor oil.[22] Dawson says the Egyptian name for castor oil was degam, and that k'ak'a (qiqi) was not the Ricinus (κίκι), the seeds of which were used medicinally[23], but a different plant. There was an important oil

[1] ii, 96 ; Wilkinson, ii, 207.
[2] Joachim, 48, 110, etc.
[3] Woenig, Pflanzen, 114, 298 f. ; Lieblein, Handel, 1886, 49, 64.
[4] Qalqashandi, Geogr., tr. Wüstenfeld, 1879, 224 f. ; Reitemeyer, 50.
[5] Herodotos, iv, 177.
[6] Rawlinson's note to Herodotos ; cf. Homer's myth of the lotus eaters : Gubernatis, Mythologie des plantes, ii, 202 f. ; Woenig, Pflanzen, 1 f., long account.
[7] Ritter, Erdkunde, Afrika², 1822, 998.
[8] Herodotos, ii, 92.
[9] Hartmann, L'Agriculture, 44.
[10] Adams, Paulus Ægineta, 1847, iii, 236.
[11] i, 16 ; cf. Goguet, i, 126.

[12] Empire of the Ptolemies, 1895, 146.
[13] XVII, i, 35, 808 C.
[14] Wilkinson, ii, 413 ; Wiedemann, 1920, 278; Hartmann, 66.
[15] Schrader, RL, 589 ; Newberry, Anct. Egypt, 1915, 97.
[16] Wilkinson, ii, 407 ; C. Dubois, L'olivier et l'huile d'olive dans l'ancienne Égypte, Rev. de Philol., 1925, xlix, 60.
[17] Reitemeyer, 16, 26.
[18] Herodotos, i, 193.
[19] Wilkinson, ii, 399 ; Mahaffy, Emp. of Ptol., 146.
[20] Pliny, xv, 7 ; xix, 5.
[21] See p. 85.
[22] Joachim, 9, 23, etc. ; 34.
[23] Magician and Leech, 114.

industry in Diospolis (Thebes) in 140 B.C.[1] Gardiner [2] thought qiqi meant "plants or shrubs in general."

Castor oil was probably obtained (from 3000 B.C. ?) both by pressing [3] and, as described by Herodotos [4], by heating the seeds with water over a fire and skimming, which is the process still used in Egypt.[5] Castor oil was used from the Predynastic and Badarian Periods by the poorer people for anointing the body (as is still the case in Nubia) [6] and for lamps.[7] A Predynastic lamp from Qau is the oldest known [8], and Clement of Alexandria says lamps were an Egyptian invention.[9] In the time of Herodotos[10] these consisted of saucers containing oil and salt, with a wick above (perhaps stuck in a heap of salt), and he mentions a festival of lamps held at Saïs. The non-smoking lamp used in tombs appears to have been a kind of night-light, the thin wick being supported in a heap of salt, so that it just touched the bottom of the vessel and projected a little at the top. The salt was then soaked in oil and the wick lighted. Wax candles appear also to have been used[11], but torches are rarely represented. Other types of lamp gave a large flame, and sometimes lamps were put on stands like candelabra. Lanterns appear first in the Hellenistic Period. At first only one wick was used in lamps (an exception has four burners in Amenhotep IV's time), but in the Hellenistic Period several burners were used, and the Copts used lamps with seven burners in the ceremony of extreme unction. Only in the later period (Demotic ; Coptic-Greek ; Arabic) were lamps used for magical purposes, although the use of 365 lamps in a festival of Osiris is mentioned.[12] No actual specimens of candles have been found.[13] Naphtha was used in the Arabic Period[14], and Benjamin of Tudela (A.D. 1173) says fish oil was used for lamps.[15] The Egyptian lamp passed to Crete and Mycenæ : with the fall of these civilisations it disappeared in Europe, but it had also passed to the Syrian coast and Palestine, from where the Greeks obtained it with so much which disappeared at Mycenæ. The Greeks used olive oil in lamps, and Pliny says castor oil was too fatty and did not give a clear flame ; in Babylon and Agrigentum naphtha or petroleum was used.[16]

MANNA

Manna descended on the Israelites in the desert of Sinai[17], where Josephus[18] says it fell in his day : Whiston in his note says it fell usually in Arabia, and that Artapanos (100 B.C.) describes it as like oatmeal. The nature of this heavenly gift has given rise to much discussion.[19] Modern manna is produced

[1] Erman-Krebs, Aus den Papyrus, etc., 1899, 174.
[2] Admonitions of an Egyptian Sage, Leipzig, 1909, 86.
[3] Bellwood, J.Soc.Chem.Ind.,1922, 213 R.
[4] ii, 94.
[5] Wilkinson, ii, 400 ; Wiedemann, 1920, 279.
[6] Lucas, JEA, 1930, xvi, 44.
[7] Diodoros Siculus, i, 18 ; Strabo, XVII, ii, 5, 823 C.
[8] Brunton and Thompson, Badarian Civ., 1928, 41, 61 ; cf. Wiedemann, 1920, 151.
[9] Goguet, i, 127.
[10] ii, 62.
[11] Davies and Gardiner, Tomb of Amenemhet, 1915, 97 ; Thothmes III.
[12] Wiedemann, 1920, 189 f. ; on legends of ever-burning lamps, see Hargrave Jennings, The Rosicrucians, 6, 13, 63.
[13] Anct. Egypt, 1915, 141.

[14] Reitemeyer, 49.
[15] Itinerary, ed. Adler, 1907, 72.
[16] Hug, PW, xiii, 1572 f. ; 1606 f.
[17] Exod., xvi ; Numb., xi.
[18] Antiq., III, i, 6.
[19] Pilter, PSBA, 1917, xxxix, 155, 187 ; D. Hanbury, Science Papers, 1876, 355 f. ; F. Vigouroux, Dictionnaire de la Bible, 1891 f., iv., 656 ; Neumann's Chemical Works, tr. Lewis, 1759, 324 ; Salmasius, De Manna, in his Homonymis Hyles Iatricae, Trajecti ad Rhenum, 1689, 245–254, usually bound with the Exercitationes Plinianæ ; R. James, Medicinal Dictionary, London, 1745—containing much of interest—unpaged, art. Manna ; Lippmann, Geschichte des Zuckers, 2 ed., 1929, 145 f., quoting Kaiser, Der heutige Stand der Mannafrage, 1924, not seen ; ib., Z. d. Vereins d. Deütsch. Zucker-Industrie, Tech. Teil., 1934, lxxxiv, 806, 833 ; Berthelot, Mâ, i, 385 f.

by two species of ash, the Fraxinus Ornus, Linn., and the F. rotundifolia, Lamarck, which are cultivated in Sicily and Calabria, although many other sweet exudations have been called manna, as many as eight being given by Landerer[1]: Manna larcinia, from leaves of Larix Europæa (Briançon manna) ; Manna cedrina, from branches of Pinus cedrus, gathered on Mount Lebanon ; Manna calastrina ; Manna quercina ; Manna australis, produced by Eucalyptus ; Manna cistina or labdanifera, a rare variety met with in Greece, an exudation from several species of Cistus and really ladanum ; Alhagi Manna, an exudation from the Hedysarum Alhagi, a plant covering the plains in Arabia and Palestine, used as food by travellers in the desert, and probably the mel ex aëre of Aristotle[2] and Pliny[3], and the μελιτώδη χυλός (humor melleus) of Theophrastos[4]; Manna tamariscina, from the puncture by a coccus insect of the trees of Tamarix mannifera growing about Mount Sinai, and probably the Biblical manna.[5] The last variety was found by Berthelot to contain 55 per cent. of cane sugar, 25 per cent. of invert sugar and 20 per cent. of dextrin.[6] The name manna (μάννα) was also applied to incense.[7] The manna of Persia is exuded by a prickly shrub [8], the leguminous Camel's thorn (Alhagi Maurorum and A. camelorum) growing especially there but also in Egypt, the Arabian Desert, Syria, Palestine, etc. ; it is not known if this kind, which forms reddish gum-like grains, is a physiological or pathological product : it has also been regarded as the Biblical manna.[9] The Sinai tree was seen by Ebers, and Tischendorf had actually seen it dispersing manna on May 23rd, 1838.[10]

The manna issues from the tender twigs of the tamarisk (Tamarix gallica mannifera, Ehrenberg) as a sweet, sticky, honey-like substance, and falls to the ground in hard drops, which are gathered by the Bedawa and put in leather bottles, partly for use as food and partly for sale in Cairo. The exudation is caused by the puncture of the tender peel by a coccus. The Arabs call this manna, man : on Egyptian monuments it is called mannu or mennu, and "white mannu" is named in the Edfu laboratory with mulberry-fruit, figs, milk, pure olive oil (tuat' hét), wines of Upper and Lower Egypt and beer. About 700 lb. of manna are produced annually in Sinai.[11]

Another theory of the Biblical manna, advocated by Kerner and adopted by Lippmann[12], is that it was the manna-lichen, called by old botanists Lichen esculentus, but now called Lecanora or Parmelia esculenta, L. desertorum, and L. Jussufi, which forms wrinkled, warty crusts on stones, greyish-yellow outside but white inside, and is found in South-west Asia, extending to south-west Europe and north Africa, including the Sahara Desert. The loose parts are rolled and swept about by wind, and lie piled up behind bushes and under-growth. It is collected for food, being ground and baked as bread. There is no direct evidence that it has been found in Sinai, but it may have been in ancient times. Pilter, who shows that the properties of manna in the Biblical accounts agree with neither form separately, but with both together, concludes that both tamarisk manna and lichen manna formed the manna of the Israelites.[13]

The Portuguese physician Garcia da Orta who was born in 1490 at Elvas, was at Goa from 1534 until his death in 1570. His "Coloquios dos

[1] Watts, Dict. of Chem., 1873, iii, 821 ; Hanbury, 355 ; Neumann, 325 ; EB[11], xvii, 588.

[2] Hist. Animal., V, xxii, 4.

[3] xi, 12.

[4] Hist. Plant., I, xii, 1 ; Salmasius, 247.

[5] Ebstein, Die Medizin im alten Testament, Stuttgart, 1901, 35.

[6] Watts, op. cit.

[7] Salmasius, 245 f.

[8] Neumann, 325.

[9] Lippmann, Gesch. d. Zuckers, 145.

[10] Pilter, 156.

[11] Pilter, 156, 157, 162 ; Budge, Egy. Hierogl. Dict., 300.

[12] Gesch. d. Zuckers, 150.

[13] Pilter, 165 f., 188 f., 206.

simples e drogas he cousas mediçinais da India"[1] mentions four kinds of manna :

(1) Shirqest (Persian schîr-khist, hardened milk), probably on one side identical with Dioskurides' elaiomeli[2], a pathological gummy exudation from the olive trees of Syria[3], but mainly the "sweet honey-dew" (drosomeli ; æromeli) mentioned by classical[4] and Arabic authors. Leeuwenhœk and Réaumur first explained it as an exudation from an insect, which often occurs in great numbers and produces very large amounts of exudation.[5]

(2) Tiriamjabim (Persian turanjabin, freshly fallen honey), like gum and in reddish grains : this is the exudation from the camel's thorn.

(3) Manna from Basra, in large pieces mixed with leaves ; probably the same as European manna, obtained from insects on larches in Briançon, and ash trees, especially the manna ash. This was collected as early as the 9th century in Calabria and Sicily for Venetian trade, and called by Arabic authors alaschar, etc. : the real al'uschar is the concreted juice of the Eurphorbia asclepias gigantea, etc., not manna.

(4) A kind of thick honey from Ormuz ; rarely mentioned but referred to by Salmasius.[6]

An expedition to Sinai from the Hebrew University of Jerusalem in 1927 visited the manna district and found that the appearance of manna is the same phenomenon as is known in other countries as "honey-dew," a sweet excretion of plant lice (Aphidæ) and scale insects (Coccidæ). Two scale insects mainly responsible for the production of manna were : (i) Trabutina mannipara, Ehrenberg, occurring in the lowlands, and (ii) Najacoccus serpentinus, var. minor Green, which replaces the former in the mountains. Two other Hemipterous insects, Euscelis decoratus, Haupt, and Opsius jacundus, Leth., also produce manna, but to a lesser extent. All these insects live on the Tamarix nilotica var. mannifera Ehrenb. ; no manna was observed on any other species of Tamarix.[7] Manna is, therefore, similar to the sweet excretion of the aphids which is the food of ants. It consists of sugars (glucose, fructose, saccharose) with no trace of protein.[8] The alchemists used the name manna for many metallic salts[9], and it also meant incense in some Greek authors, such as Galen.[10]

INCENSE AND PERFUMES

The word "incense" (from the Latin incendere, to burn or kindle) has the same literal meaning as "perfume," the aroma given off with the smoke (per

---

[1] First ed., Goa, 8vo., 1563, 131 v. ; The Simples and Drugs of India, tr. by Sir Clements Markham, London, 1913, 280 f. ; epitomised in Latin by Clusius (de l'Escluze) in his Exoticis, Antwerp, 1567 ; cf. Sprengel, Geschichte der Botanik, 2 vols., Altenburg and Leipzig, 1817–8, i, 347.

[2] i, 37.

[3] Salmasius, Hyles iatr., 245 f.

[4] Cf. Athenaios, Deipnosophistæ, xi, 102.

[5] Lippmann, 145 f.

[6] Cf. also Gorræus (Jean des Gorris, 1505–77), Definitionum medicarum libri xxiiii, literis Græcis distincti, fol., Frankfurt, 1578, 279 ; Prosperi Alpini, De Medicina Ægyptiorum libri quatuor, 4to., Venice,

1591, 127 ; Prosper Alpinus, an Italian botanist and physician (1553–1617), was in Egypt in 1580–86 ; his book, which appeared in several editions, is usually bound with the De Medicina Indorum of the Dutch physician Jacob Bontius, who lived for several years from 1627 in Java, as well as the Liber de Balsamo of Alpinus and other of his works according to the edition ; Macalister, Hastings DB, iii, 236 ; E. Bibl., iii, 2929 ; Reiske and Fabri, Opuscula medica ex monimentis Arabum et Ebræorum . . . recensuit C. G. Gruner, Halle, 1776, 83 f.

[7] Bodenheimer, q. in Isis, 1930, xiv, 529.

[8] Anct. Egypt, 1930, 64.

[9] James, Medicinal Dict., art Manna.

[10] Gorræus, 279.

fumum) of any odoriferous substance when burnt.[1] The "perfume" of flowers is, therefore, rather a derived expression. The word incensum for incense occurs only in later authors[2] ; the Egyptian name was senter neter (divine smell)[3] ; the classical name is tus or thus, θύος, frankincense. Incense is sometimes called in Egyptian "the sweet of the god" [Osiris].[4] The Egyptians were particularly fond of flowers : real and artificial flowers were used as offerings to the dead, and fresh flowers were used in living rooms.[5] Many kinds of perfume and incense were used.

The aromatic called in the texts anti was imported from Punt as early as the v dyn., and the trade lasted into the Hellenistic Period. The expedition of Queen Hatshepsut (c. 1550 B.C.) brought back from Punt not only ivory, precious woods, "eyepaint," "green" gold (electrum) and gold rings (still used as currency in East Africa), but also large quantities of anti and anti-trees, which are shown at Deîr el-Baharî.[6] These trees are said not to be myrrh trees, and anti probably not myrrh : Wiedemann says they are the Boswellia thurifera : Schoff [7] and Lieblein say neither myrrh nor frankincense meets the case, and it is supposed that the trees with foliage represented are of the Boswellia Carteri, the frankincense tree of Dhofar in Southern Arabia : the trees without foliage, also shown, could just as well have been true frankincense.[8] The anti resin was used in the cult of the dead ; with common resin in the worship of Amen-Rā in the XXII dyn., and also for embalming.[9] Thothmes III imported over 223 bushels of anti from Punt.[10] The tree could not be acclimatised in Egypt.[11] According to Joachim, "myrrh," including "dried myrrh," is frequently mentioned in the Ebers Papyrus.[12] Other varieties of resins said to be mentioned in this Papyrus, sometimes described as "fats" of the trees, are turpentine, pistacia gum, acanthus and sycamore resins, Nubian storax, lotus resin (?), and mastic (?).[13] In the Book of the Dead the "pure man" is anointed with septen and anti ; a text is said to be written on a strip of new papyrus with a spar pen dipped in water of anti, and there is mention of anointing with "anti from the sacred limbs."[14]

The use of incense and incense burners, and the offering of incense, are frequently mentioned in ancient records and pictured in temples and graves, as well as in illustrations in the Book of the Dead. There are references to incense in the v and vi dyns. and an incense burner of the viii dyn. is known. The earliest certain specimen of incense mentioned by Lucas is of the XVIII dyn., in small balls as shown on monuments. These balls are said by Wiedemann to be of turpentine resin. This would cause much black smoke, and although Lucas says some specimens giving an aromatic odour have been found,[15] its use was probably restricted to ordinary purposes and poorer people. Whether the soda (natron) said to be used with incense would diminish the smoke is doubtful : saltpetre would, of course, have had this effect.[16]

A fragment of "perfume" from a XIII dyn. tomb was a fragile yellowish-

[1] Lucas, JEA, 1930, xvi, 47.

[2] Isidore of Seville, Etymologiæ, IV, xii, 3 ; W. Smith, Latin Dict., 545.

[3] Wiedemann, 1920, 152.

[4] Shorter, JEA, 1929, xv, 137 f.

[5] Wiedemann, 1920, 97 f. ; Woenig, Pflanzen, 234 f.

[6] Breasted, H., 127, 274 ; BMGE, 21 ; Budge, History of Ethiopia, 1928, i, 8 ; ib., Egy. Hierogl. Dict., 29, 128 ; Wiedemann, 1920, 153, 321 ; Maspero, Struggle, 253 f. ; Lieblein, Recherches sur l'Histoire et la Civilisation de l'ancienne Égypte, 3 fasc., Leipzig, etc., 1910–14 (pagination continuous), 221, 227 ; Woenig, Pflanzen, 354 f., 375 ; Schmidt, Drogen, 3.

[7] Periplus, 1912, 218.

[8] Lieblein, 227 ; Lucas, JEA, 1930, xv, 49 ; cf. Schmidt, Drogen, 3.

[9] Wiedemann, 1920, 153 ; BMGE, 21, 211 ; Breasted, Anct. Rec., iv, 136.

[10] Breasted, H., 305.

[11] Wiedemann, 1920, 154.

[12] Joachim, 21, 30, 83, etc. : ? anti.

[13] Joachim, 21, 27 ; 55 ; 51 ; 114, 180 ; 34, 51 ; 180.

[14] Book of the Dead, caps. 124, 129, 145 ; Davis, 139, 143, 161.

[15] Lucas, JEA, 1930, xvi, 48.

[16] Lucas, JEA, 1930, xvi, 48 ; Wiedemann, 1920, 152 f.

brown compact resin, with a conchoidal fracture. On heating it behaved like colophonium (pine resin), with a similar odour. Round balls from the same tomb, supposed to have been originally perfumed, consisted of quartz or crushed glass cemented together with calcium carbonate, and also contained a small percentage of magnesia and a little organic matter.[1] Analyses by Ure of resins which may have been used for incense showed that one had a specific gravity of 1·204, was red in colour, intumesced on heating and then burned like amber, and was insoluble in oil of turpentine but soluble in alcohol ; another specimen of sp. gr. 1·067, soluble in oil of turpentine and alcohol, was probably mastic.[2] Incense of the Ptolemaic Period from Philæ was partly in balls and partly in discs. Some finds supposed to be incense, however, such as that in the foundation of the tomb of Aahmes I (1557) at Abydos, may be common resin, dark brown lumps of which occur frequently in graves, especially of the earlier periods.[3] Five pots and a glass jar, all covered with linen, in an XVIII dyn. tomb, contained perfumes, also large sealed jars "containing what remains of the perfumes and salts employed in embalming." Some balls made from powdered glass mixed with perfume are still faintly odorous.[4]

The incense burners were either tall standing vessels or portable burners with handles carried in the hand, the handle being shaped like an arm. They were provided with lids which could be put on from time to time. Wiedemann thinks the flame-like projections shown were wicks, the vessel containing oil or fat : Wilkinson calls them flames.[5] The object of burning incense in the religious ceremonies was probably to purify the room where the image of the god was situated, and the ritual meal was served up soon after to be eaten as physical food : it is also used in magic ceremonies, when it no doubt served to attract the god or spirit.[6] The use of incense came into religion from profane life ; it is a very old custom in the East, where fumigation is still in full use.[7]

FRANKINCENSE AND MYRRH

The two principal constituents of modern incense are the gum resins myrrh and frankincense, which are found in tombs and are said to be mentioned in records, although the same word has been translated sometimes "frankincense" and sometimes "myrrh." A mere enumeration of the records under separate headings of these materials would thus be misleading.[8] Breasted's translations give myrrh in V, XI, XVIII, XX and XXI dyn. texts [9], in which it comes usually from Punt, but in one XVIII dyn. text from Retenu in western Asia, perhaps ultimately from Arabia. Reutter identified myrrh in "ancient" (undated) Egyptian perfumes ; Lucas in specimens of gum-resin from royal and priestly mummies of the XVIII–XXI dyns. inclusive[10]; and Verneuil in the contents of a small ancient Egyptian medicine chest in the Berlin collection.[11]

Myrrh (khari ; kher) comes from Somaliland and Southern Arabia, which are possible sources for ancient Egypt, is derived from various species of Balsamodendron and Commiphora, and occurs in yellowish-red masses of agglutinated "tears," often covered with its own fine dust ; it is never white or green, so that it cannot be the incense of these colours mentioned in ancient

[1] Berthelot, Arch., 23.
[2] Wilkinson, iii, 398.
[3] Lucas, JEA, 1930, xvi, 48.
[4] Maspero, Guide to the Cairo Museum, 1908, 493, 495, 524.
[5] JEA, 1930, xvi, pl. xxxiii, p. 216 : from Abydos ; Wilkinson, iii, 398, Fig. 599 ; B. Mus. Quarterly, i, 65, pl. xxxvi, a ; VI dyn. ; Wiedemann, 1920, 152.
[6] Budge, Osiris, i, 250 ; Moret, Le Rituel du Culte Divine, 1902 ; cf. Davies and

Gardiner, Tomb of Amenemhet, 1915, 76 ; Shorter, JEA, 1929, xv, 137 f. ; Dawson, Magician and Leech, 28 f.
[7] Ganszyniec, PW, xii, 53, 56.
[8] Lucas, JEA, 1930, xvi, 48 f.
[9] Lucas, 50 ; Schmidt, Drogen, 3 ; Steier, PW, xvi, 1134, 1142.
[10] Lucas, JEA, 1930, xvi, 50 : "probably" myrrh.
[11] Parthey, Isis und Osiris, 1850, 277.

records.[1] Its use in Egyptian ointments is mentioned by Theophrastos [2] and Pliny ; Plutarch [3] says it was used in Egypt as incense (i.e. for fumigation) of the Sun along with resin ($\dot{\rho}\eta\tau\dot{\iota}\nu\eta$) and kyphi ($\kappa\hat{\upsilon}\phi\iota$) (see below), and that the Egyptian name was bal ($\beta\dot{\alpha}\lambda$), which Parthey thinks is a manuscript mistake for sal ($\sigma\dot{\alpha}\lambda$), or schal ($\sigma\chi\dot{\alpha}\lambda$), the Coptic name. In Coptic, bal means eye. Solar, lunar and other varieties of kyphi are specified by Paulos Ægineta (c. A.D. 640).[4]

The incense par excellence is frankincense, often regarded as the only genuine kind, which occurs in large "tears" almost colourless when pure, one of its characteristics according to Pliny [5], who says there were male and female kinds [large and small pieces], and hence it is probably the "white incense" in the Harris Papyrus (xx dyn.). It is usually light yellowish-brown, translucent when fresh but after attrition in transport it becomes covered with its own dust and is then opaque. Most other incense materials are dark yellow, dark yellowish-brown, or in a few cases grey or black. Frankincense exudes from certain trees of the genus Boswellia, growing, like myrrh trees, principally in Somaliland and South Arabia, though one variety is obtained from the Commiphora pedunculata in the eastern Sudan near Gallabat and in the adjoining parts of Abyssinia. The statements in records that it came from certain Negro tribes in the vi dyn., from Punt in the xviii and xx dyns., and from Western Asia in the xviii dyn. (the original source being then probably South Arabia), would agree with these facts, although the Asiatic kind may have been a different variety of incense. Both African and Arabian frankincense were imported to Egypt in the Roman Period, being prepared for sale, probably by cleaning and sorting, at Alexandria.[6] The cultivation of the balsam tree at Matarîya, just south of 'Ain Schems (Heliopolis) goes back at least to Cleopatra. Abd al-Latîf (1203) and Yâqût (1224) describe it, but it disappeared about 1615. In the 13th century about 500 kgm. of incense was obtained annually and sold for twice its weight of silver.[7]

OTHER KINDS OF INCENSE

Various kinds of incense other than myrrh and frankincense are mentioned, according to Dümichen fourteen kinds in an inscription in the "temple laboratory" at Edfu.[8] Several are specified in Joachim's translation of the Ebers Papyrus [9], such as Nubian storax; mastic; lotus resin (?) ; pistacia ; and sycamore resin. The fumes were supposed to drive away demons and have a curative influence[10]. Joachim's identifications are doubtful and perhaps misleading, the only materials of which there is evidence of use being galbanum, ladanum and storax, although they are not certainly identified in old Egyptian remains.[11] Benzoin and camphor from the Far East, and (in the earlier periods) the products of India, are unlikely. Schmidt thinks cassia bark (cinnamon) is mentioned in the xvii dyn. ; as it grew in China, where it is mentioned in 2700 B.C. as kwei, he assumes an early sea traffic between Egypt and China, which is improbable.

True storax, from Palestine and other places, is a balsam obtained from a tree, Liquidambar orientale, belonging to the natural order of Hamamelideæ indigenous to Asia Minor, although at one time the name was applied to the

[1] Budge, Egy. Hierogl. Dict., 561 ; Lucas, JEA, 1930, xvi, 50 ; Loret, L'Égypte, 199 ; Woenig, Pflanzen, 354.
[2] De Odoribus, vi, 28.
[3] Isis and Osiris, cc. 52, 79 ; Parthey's ed., 94, 277.
[4] Hypomnema, vii, 22 ; tr. Adams, 1847, 599 ; on myrrh in wine, murrina, in the Classical Period, see Schneider, PW, xvi, 669.

[5] xii, 14 ; Breasted, Anct. Rec., iv, 131.
[6] Lucas, JEA, 1930, xvi, 49.
[7] Reitemeyer, 67 f., 71.
[8] Schmidt, Drogen, 3.
[9] Joachim, 114, 180 ; 180 ; 34, 51 ; 55 ; 51.
[10] Wiedemann, 1920, 154.
[11] Lucas, JEA, 1930, xvi, 50 f. ; cf. Schmidt, Drogen, 4.

resin of the Styrax officinalis which somewhat resembles benzoin. Reutter claimed to have identified storax in undated mummy material and perfumes.[1]

Galbanum, a fragrant gum-resin generally occurring in light yellowish-brown to dark brown, sometimes greenish, masses of agglomerated "tears" of a greasy appearance, usually hard but occasionally semi-solid, is the product of several species of a Persian umbelliferous plant, Peucedanum, the most important being P. galbaniflorum. Although there is no record of its occurrence in Egyptian graves, Pliny [2] says it was a constituent of Mendesian ointment, and it is mentioned as a constituent of Jewish incense in the Bible [3], so that it may have been the "green" incense in an XVIII dyn. record.[4]

Ladanum, a true resin and not a gum resin, which occurs in dark brown or black masses, often viscid or easily softened on handling, is obtained from the short glandular hairs on the leaves of various species of Cistus, e.g. C. ladaniferus, growing in Asia Minor, Crete, Cyprus, Greece, Palestine, Spain, and other parts of the Mediterranean region, but not at present in Egypt.[5] The earliest certain references to its use in Egypt appear to be those in the Bible [6], although Newberry thinks it was called the "plant of Osiris" and mentioned as incense in the Old Kingdom. It is used in modern Egypt in chewing to sweeten the breath [7], and in Europe in perfumery. The Turks wear it as an amulet. It was considered in older medicine to have stimulant and expectorant properties and was used internally and externally, at one time being regarded as a specific against plague.[8] Its only recorded use in older Egypt is its supposed presence in a late (7th century A.D.) specimen of Coptic incense from Faras, containing 31 per cent. of mineral matter.[9] Of the sixteen species of Cistus spread throughout the Mediterranean region, five are ladanum bearing, and the best ladanum is said to be that from Cyprus.[10]

The so-called "whip" or "flail" held by Osiris, shown on a mace-head as held by King Menes (I dyn.), and also in a representation of a god at Ur, is the ladanisterium, a stick with leather thongs used in the Lebanon district to collect ladanum. Middle Kingdom examples resemble modern ones from Crete.[11] Herodotos[12] says the Arabians gathered ladanum from the goats' beards, it having come from the bushes on which they browse ; this method was mentioned by 17th century travellers to Cyprus[13], and a specimen of such a beard is in the Kew Museum. Herodotos gives the Arabian name as ladanon ; Dioskurides[14] calls the plant kistos and the gum ladanon, sometimes incorrectly translated "gum mastic" (e.g. in the Loeb Classical Library edition of Herodotos). It is the Hebrew lôt. The origin of the name may be the Assyrian ladanu (Newberry), or laduna.[15] References to the plant in the classical, medical and other authors, including its confusion with ivy, are given by Newberry, whose article is a model of such discussions and is of great interest.

Herodotos[16] says Arabia is the only country which produces frankincense ($\lambda\iota\beta\alpha\nu o\varsigma$), myrrh ($\sigma\mu\upsilon\rho\nu\alpha$), cassia ($\kappa\alpha\sigma\iota\alpha$), cinnamon ($\kappa\iota\nu\nu\alpha\mu\omega\mu o\nu$), and ladanum ($\lambda\alpha\delta\alpha\nu o\nu$ or $\lambda\eta\delta\alpha\nu o\nu$), a resin or gum ; which they do not get, except the myrrh, without trouble. The frankincense is procured by means of the Phœnician gum storax ($\sigma\tau\upsilon\rho\alpha\xi$), the smoke of which dislodges from the frankincense trees vast numbers of small winged serpents of various colours. These gum resins

[1] Schmidt, Drogen, 4 ; Lucas, JEA, 1930, xvi, 52 ; Hanbury, Science Papers, 1876, 8, 127 f., 175.
[2] xiii, 1.
[3] Exod., xxx, 34.
[4] Lucas, JEA, 1930, xvi, 51.
[5] Newberry, JEA, 1929, xv, 86 f. ; Lucas, op. cit.
[6] Gen., xxxvii, 25 ; xliii, 11 (Revised Vers. only).
[7] Lane, Modern Egyptians, i, 259.

[8] Newberry, 87 f. ; Stadler, PW, xii, 380.
[9] Lucas, JEA, 1930, xvi, 51.
[10] Newberry, op. cit. ; Stadler, PW, xii, 380.
[11] Newberry, 85 f. ; S. Smith, Early History of Assyria, 1928, 135, 379.
[12] iii, 112.
[13] Newberry, 92.
[14] i, 128.
[15] Stadler, PW, xii, 379.
[16] iii, 107, 120 ; ii, 86.

were sent as gifts on embassies, and used in embalming. The Phœnix bird (which he never saw, except in pictures) came every 500 years from Arabia to Egypt, bringing the parent bird plastered over with myrrh to the temple of the Sun, where he buried him. In Tacitus [1] the phœnix, on the approach of death, builds a nest and "sheds a generative power" on it, so that a young bird appears, whose first duty is to burn the body of his father on the altar of the sun. Herodotos says the body was buried inside a lump of myrrh like an egg. The resurrection of the phœnix first appears in Pomponius Mela (1st century A.D.).[2] Herodotos [3] says fine bread, honey, raisins, figs, frankincense, myrrh, etc., were also used in Egypt for stuffing the body of the sacrificed bullock, which was afterwards burnt, pouring on it a great quantity of oil.

Cassia fistulæ Willd. ("manna") is indigenous to Egypt, but cassia bark was imported by Hatshepsut from Punt about 1500.[4] The Ebers Papyrus mentions calamus as imported from T'abi (in Asia ?).

The Egyptian incense, at least in the later periods, was a mixture of several materials. Recipes for Egyptian perfumes to be used in religious ceremonies were found, as copies, in temples of the Ptolemaic Period, and Loret, with the assistance of the French perfumers Rimmel and Domère, prepared specimens of these perfumes, which were deposited in the office of the Académie des Inscriptions in 1886.[5]

KYPHI

The incense kyphi ($\kappa\hat{v}\phi\iota$, $\kappa o\hat{\iota}\phi\iota$, $\kappa v\phi\iota\omega\nu$), on which books were composed by Manetho (3rd century B.C.) [6] and Julian (? 1st century A.D.) [7], is said by Plutarch [8], perhaps from Manetho, to consist of sixteen ingredients, the translation of which has been somewhat variously given[9] : honey, wine, raisins, galingale ($\kappa v\pi\epsilon\rho os$; Cyperus rotundus ?), resin, myrrh, asphaltum, cardamoms and sweet flag, with the less easily identified $\dot{a}\sigma\pi\dot{a}\lambda a\theta os$ (? camel's thorn ; Alhagi maurorum), $\sigma\epsilon\sigma\epsilon\lambda\iota$, $\sigma\chi\hat{\iota}\nu os$ (? mastic), $\theta\rho\dot{v}o\nu$ (? rushes, Parthey ; saffron, Wilkinson), $\lambda\dot{a}\pi a\theta os$ (? sorrel)[10] and $\dot{a}\rho\kappa\epsilon v\theta os$ (a juniper).

Ganszyniec thinks probably only one kind of kyphi was used in ancient Egypt, honey and wine being later additions. The recipes (like those for the Greek theriac) became numerous and more complicated as time went on, the ten ingredients of kyphi in the Ebers Papyrus[11] (myrrh, juniper berries, incense, cypress, aloes, šebet-resin, sweet flag, mastic, storax, sesel) being probably retained and others added, so as to make total numbers connected with astrological or superstitious beliefs : solar kyphi with 36, the number of dekans and parts of the body, and lunar kyphi with 28, the days in the lunar month. The medicinal use of kyphi is very old in Egypt, as is seen from the mention in the Ebers Papyrus and also from temple inscriptions. Kyphi occurs in a temple inventory of the Greek Period.[12]

Dioskurides[13], Aëtios[14] and Galen[15], like the Ebers Papyrus, give only ten ingredients ; Nicholas Myrepsos ("the ointment maker") (c. A.D. 1280) gives

[1] Ann., vi, 28 ; cf. Achilles Tatius, tr. R. Smith, 1889, 418.
[2] De Situ Orbis, iii, 8, ex se rursus renascitur ; Schmitz, DBM, iii, 344.
[3] ii, 40.
[4] Woenig, Pflanzen, 343, 359.
[5] Maspero, Études de Mythologie, i, 317.
[6] Schmitz, DBM, ii, 915 ; Müller, Fragmenta historicorum græcorum, Paris, 1841–70, ii, 511 ; Ganszyniec, PW, xii, 52, 54.
[7] Beausobre, Histoire critique de Manichée et du Manichéeism, Amsterdam, 1734–39, i, 322 ; Kircher, Œdipus Ægyptiacus, 4 vols., fol., Rome, 1652–53, II, ii, 539 ;

Fabricius, Bibliotheca Græca, Hamburg, 1718–28, ii, 498.
[8] Isis and Osiris, 52, 81 ; Parthey, 94, 143, 277.
[9] Wilkinson, iii, 398 ; Wiedemann, 1920, 155; Parthey, op. cit., 143, 277 f.—long note —E. Rimmel, Le livre des parfums, 1870, 27.
[10] Cf. Dioskurides, i, 19 ; iii, 60 ; ii, 140, respectively.
[11] Joachim, 180.
[12] Erman-Krebs, Aus den Papyrus, etc., 1899, 178.
[13] i, 24.
[14] Tetrabiblos, I, i, 99.
[15] Parthey, Isis und Osiris, 278.

twenty-eight or fifty ingredients.[1]   Paulos Ægineta (7th century A.D.) [2] and
Aëtios give twenty-eight and thirty-six.

Voigt, of Berlin, made up for Parthey three prescriptions of kyphi, according
to Dioskurides, Plutarch and Galen [3], the first being the best ; the preparations
when added to wine gave it an astringent taste, something like the modern
Greek vino resinato. When smeared on a hot plate they evolved a sharp
aromatic smell, the first again giving the best result.   Plutarch [4] says a "sacred
book" was read during the preparation of kyphi, and Nicholas Myrepsos
in describing the preparation says : "arida contere una cum liquidis, et hæc
conficiendo dic ἂ ἒ υῖ οὐ ὦ," these being the "seven vowels" of the magic
papyri.[5]   Loret also gives from a "hieroglyphic text" a recipe for kyphi
with sixteen ingredients, made up by Rimmel : the perfume was "plutôt étrange
qu'agréable."[6]

Other resins and gum resins (some not identified) were used, or are suitable
for use, as incense.[7]   Powdered incense mixed with honey and made into balls
was chewed, especially by women, to sweeten the breath[8] : if myrrh, it would
have a beneficial action on the gums.

Common resin was used in mummification, as a varnish—a "kind of gummy
varnish, light brown in colour, which gave it a rich lustrous appearance," was
found on a coloured wood box [9]—as a cementing material[10] and for beads and
personal ornaments[11]: the "scented beads, perhaps resinous," which still smell
when warmed[12], may be amber. Syrian turpentine resin, mentioned in the
Ebers Papyrus[13], was used for incense[14] : resin is mentioned in the Book of the
Dead[15] and was melted and poured to secure metal in stone in the pyramid of
Cheops (2800).[16] Lucas, who promises further communications, thinks this
resin, a common material in ancient graves of all periods and perhaps the
earliest kind of incense, càme from coniferous trees in Asia.

THE MUMMY

We now interrupt our account of organic materials with two long digressions,
one on mummies and the other on medicine.

The belief in the resurrection held in relation to the religion of Osiris made it
essential that the body should be preserved, since the body was to live again on
earth as a spirit-body, while the soul dwelt in heaven.[17] From the earliest
times the body, perhaps only of the king, who could rise again "as Osiris,"
was preserved, and in somewhat later periods the practice of mummifying
became a complicated ritual.[18] The essential thing was recognised as dryness :
"those who are upon their sand" are safe, the temple is "founded on sand,"
and a little sand was always symbolically sprinkled on the foundations. If
the body decomposed, the soul died a "second death," from which there was no
survival.[19] The good preservation of mummies is largely due to the very dry

[1] Greenhill, DBM, ii, 1193 ; Hopfner, Fontes Historiæ Religionis Ægyptiacæ, 1922 f., 757 ; Berthelot, Arch., 239 f. ; Parthey, Isis und Osiris, 278.

[2] Hypomnema, vii, 22.

[3] Parthey, Isis und Osiris, 279.

[4] Isis and Osiris, c. 81.

[5] Parthey, 280 ; Ganszyniec, PW, xii, 54.

[6] L'Égypte, 201 f.

[7] Lucas, Egy. Materials, 120 f. ; ib., JEA, 1930, xvi, 52 f.

[8] Wiedemann, 1920, 155 ; Lane, Modern Egyptians, i, 259.

[9] Woolley and MacIver, Karanog, i, 71—Roman-Nubian Period.

[10] Lucas, JEA, 1930, xvi, 52 f.

[11] Brunton and Thompson, Badarian Civ., 1928, ii, 35, 46, 56, 63—Badarian and Predynastic : analysed.

[12] Brunton, Qau and Badari I, 39.

[13] Joachim, 21, 27.

[14] Wiedemann, 1920, 153.

[15] Cap. 17, Davis.

[16] Petrie, Hist., 1903, 48.

[17] BMGE, IV–VI Rooms, 241 ; BMGE, 154.

[18] Sethe, Alte Orient, 1923, xxiii, 11.

[19] Maspero, Dawn, 113, 116, 118.

climate of Egypt.[1] In the Predynastic Period removal of the intestines, followed by drying in the sun, rubbing with salt [2] or smoking [3], were alone used. Burial in the garments, packed round with salt (free from natron), was in use in Nubia and in the Coptic Period, when the bodies are well preserved.[4] Wiedemann [5] summarises the types of mummifying at various periods as follows :

(1) *Earliest period* : a superficial treatment, so that the remains crumble rapidly and effloresce on exposure to air. Bodies of the IV–XI dyns. are not often mummified ; there is perhaps some salting and drying, and the use of "pitch," but in Badarian graves there is no sign that the bodies had been preserved in any way, and no coffins either of wood or of clay.[6] A limb wrapped in linen, perhaps soaked in oil or resin, was found in a I dyn. tomb. Mummification is mentioned in the II dyn. and traces of it are found in bodies of this period, involving wrapping in linen and perhaps crude natron (which has caused corrosion). It was perhaps suggested by the observation that bodies buried in the sand, which was impregnated with natron, became preserved, whilst those in brick vaults decayed. The vaults were still used, but attempts were made to preserve the body.[7] The first mummy is represented by the well-preserved head of Menkarā (Mykerinos), IV dyn., in which resin was used.[8] Another early mummy is that of the son of Pepi I and elder brother of Pepi II (VI dyn.) at Gîzah [9], and the graves at Aswân, Beni Hasan, and other places prove that the Egyptians mummified their dead under the IV–VI dyns. The earlier mummies, and all preserved with natron only, have a hard skin hanging loosely from the bones, which are white and somewhat friable.[10] Wiedemann says the bodies at Beni Hasan were not embalmed and attempts to preserve the body with natron began in the III dyn. Such mummies deliquesce on exposure or become covered with crystalline sodium sulphate.[11]

(2) XI dyn. mummies are yellowish and very brittle ; less resin was used and the cartonnage head-piece first appears.[12]

(3) XII dyn. mummies are black.

(4) XVIII to XXI dyn. mummies at Memphis are black ; at Thebes mostly yellowish. The art reached its zenith at Thebes, especially with royal mummies, but later it declined. The brain was now removed.[12]

(5) *Saïte and Hellenistic* mummies are black, extremely hard, heavy and with few exceptions ill-shaped, so that they are more repulsive than those of earlier periods.

Some mummies have been re-wrapped more than once in ancient times (XIX–XXI dyns.).[13]

The actual process of mummification is still unknown : Ipuwer, writing of the VII–XI dyns., calls it a "secret art," although he also mentions "cedar oil for mummies."[14]

The process of treating the body with medicaments and bandaging the mummy was called uta, the embalmer utu ; "to embalm" was also setekh, "to wrap in bandages" qes (which may be a short form of qeres, the mummy and

[1] Hoefer, Hist. de la Chimie, i, 69.
[2] BMGE, 161.
[3] Hall, NE, 84.
[4] Dawson, JEA, 1928, xiv, 195 ; Lucas, ib., 1932, xviii, 136 f.
[5] Ägypten, 1920, 111 f. ; cf. Wilkinson, iii, 482 f. ; Budge, Mummy², 207 f.
[6] Brunton and Thompson, Badarian Civ., 1928, 19, 20 ; Brunton, Qau and Badari I, 47 f.
[7] Budge, Mummy², 176 ; Elliot Smith, JEA, 1914, i, 189 f. ; Peake and Fleure, Priests and Kings, 1927, 91 ; Dawson,

Magician and Leech, 1929, 17 f., 19 f., 31 f.
[8] Budge, Mummy², 210 ; Breasted, H., 77 ; Elliot Smith, op. cit. ; Dawson, 22.
[9] Budge, Mummy, 1893, 184.
[10] Budge, Mummy², 208, 210 ; Dawson, 22.
[11] Wilkinson, iii, 482 f.
[12] Dawson, 23, 27.
[13] Quibell, 129 ; Dawson, 24 f.
[14] Gardiner, Admonitions of an Egyptian Sage, Leipzig, 1909, 37 ; 10 ; Davies and Gardiner, Tomb of Amenemhet, 1915, 113.

coffin). The Coptic forms kes, kos, koos, etc., translate the Greek ἐνταφιασμός, ταφή, θάττειν, etc., and the Coptic mioloe is given by Kircher. The accounts of the process given by Herodotos and Diodoros Siculus are correct in a broad way, but many actual details are still unknown.[1]

The operations were carried out under the direction of mortuary priests and the whole constituted a religious ceremony, said to have been invented by the god Anubis, a mask representing the jackal-head of this god being worn by a priest. Thoth, Isis and Horus endued the process with new powers by magic amulets and paintings on the walls of the tomb and on the coffin, the various faculties being supposed to be restored by incantations : this ritual enabled Osiris to live again, and when carried out by the priests it was hoped and believed that it would enable the dead man also to become "an Osiris."[2]

The process lasted a considerable time, 40–70 days, and at least six operators were probably engaged at once. Remains of embalmers' workshops have been found, and a wooden embalming platform of the xi dyn., encrusted with natron and resin, was found at Thebes [3], where in the xviii dyn. and in the Greek Period there was a separate quarter for embalmers, undertakers and makers of coffins and tomb furniture.[4] Dawson [5] thinks the place used for embalming (w'bb) was a temporary structure, tent or booth (syh), erected in the necropolis.

Herodotos [6] says that in the first, expensive, way the brain, liquefied by certain drugs (φάρμακα) to make it soft, and the viscera, were removed, after which the body was washed with palm-wine (οἴνῳ φοινικηΐῳ : perhaps strong Syrian wine), and the cavities filled with powdered myrrh [7] (σμύρνης), cassia and other aromatics, but not frankincense. The body "they lay in natron" (λίτρῳ ταριχεύουσι) not longer than 70 days, when it was washed and smeared with gum (κόμμι), which the Egyptians use instead of glue, and was then ready for the wooden coffin. In the second method the viscera are removed by an injection with cedar oil (? ; κέδρου ἀλείφαντος), which after the prescribed number of days (during which the body is kept in natron) dissolves the intestines and vitals, whilst the natron removes the flesh, leaving only the skin and the bones. In the third method, used only for the poor, the intestines are washed out and the body then kept in natron for seventy days.

Diodoros Siculus [8] says the scribe or chief priest marks the body on the left side, then the paraschistes opens it with an Ethiopian stone ; he then runs away whilst being pelted with stones, "since they look upon one who wounds the body as a hateful person." The embalmers are the taricheutæ, or priests' companions, whose office was hereditary.[9] They were admitted to the temples. One of them drew out with his hand all the intestines but not the heart or reins (νεφρός ; kidneys), another cleansed the bowels and washed them in Phœnician wine mixed with various aromatic spices. The body was then anointed all over with cedar oil and other precious ointments for 40 days, after which it was rubbed with myrrh, cinnamon and such things, both preservative and imparting a pleasant odour, and was finally delivered so that every part, even the eyebrows, seemed unaltered. Diodoros does not refer

[1] Sprengel, H. Med., i, 60 f. ; Kircher, Œdipus Ægyptiacus, 1654, iii, 395 f. ; Budge, Mummy, 1893, 161 f., 182 ; ib., 2 ed., 1925, 203 ; BMGE, 158 ; Lucas, Egy. Mat., 110 f. ; Dawson, JEA, 1927, xiii, 40 f. ; T. J. Pettigrew, History of Mummies, 4to., 1834, illustrations by Geo. Cruikshank ; Elliot Smith and Dawson, Egyptian Mummies, 2 vols., 4to., 1924.
[2] Dawson, Magician and Leech, 33 ; ib., JEA, 1927, xiii, 40 f.; Blackman, ERE, x, 301 f.; Maspero, Dawn, 179 f. ; Budge, Egyptian Magic, 45 ; Wilkinson, iii, 475,

plate 72 ; Lexa, Magie, iii, plates 33, 54, 67.
[3] Lucas, JEA, 1932, xviii, 126.
[4] Maspero, Struggle, 508 ; Breasted, H., 251.
[5] JEA, 1927, xiii, 40 f.
[6] ii, 86 f.
[7] Lucas, Egy., Mat., 114 f., says the use of myrrh is doubtful.
[8] i, 83, 91.
[9] As is confirmed by the texts, Blackman, ERE, x, 301 ; they were perhaps physicians, Otto, Priester und Tempel, i, 100 f.

to the treatment with natron or bandaging in his description of human embalming, but in his account of the embalming of animals he says the body was wrapped in fine linen and carried away to be salted (ϵἰς τὰς ταριχείας), after which it was anointed "with oil of cedar and other things, which both give the body a fragrant smell and preserve it a long time from putrefaction."

It is generally assumed that the body was immersed in a solution of natron, perhaps containing some common salt, and was then washed to remove the fatty acids, etc., produced by the action of the solution on the flesh.[1] Dawson thinks the body was immersed, except for the head, in a vertical jar of the liquid, which jar he sees represented, with the body, in some pottery models otherwise difficult to explain : the Pyramid Texts say the king Unás "has come forth from his jar." After soaking and washing and whilst the body was pliable, the embalmers of the XXI dyn. packed padding material under the skin and moulded on this a life-like form. In the New Kingdom this had not been done, and such packing as the body received was done externally by padding the cavities before the bandages were applied. The cavity of the body, when the viscera were removed, was filled up with sand mixed with amulets, or with linen bandages (XVIII–XIX dyns.), or with sawdust, lichen, mud, etc. (XX and later dyns.).[2]

According to Pettigrew [3] and Dawson [4], an important stage in the process, omitted by Herodotos but apparently mentioned by Lukian [5] and suggested by Rouelle in 1754 from an examination of mummies, was the washing and drying of the body, "perhaps in an artificially heated chamber or fire heat with some unknown apparatus," as is suggested by remains found at Thebes in 1924. Dawson assumes that the body was then rendered supple by a liberal application by a brush of "a paste consisting of resin [kind not specified], mixed with natron or salt and animal fat," or in Ptolemaic times molten unmixed resin was poured into and over the previously desiccated body. Possibly also fire heat was used to render the stream of resin (which Dawson thinks was the "gum" of Herodotos) more mobile ; it penetrated into every cavity and crevice, even into the structure of the bones. Abd al-Latîf (d. 1231) noticed that the flesh and bones of mummies were coloured red, which he thought had penetrated from the tar and plant juices in which the bandages were soaked.[6] In some cases the resinous or bituminous material has penetrated so completely that it is difficult to distinguish it from bone : "the arms, legs, hands and feet of such mummies break with a sound like the cracking of chemical glass tubing ; they burn very freely, and give out great heat. Speaking generally, they will last for ever."[7]

The Rhind Papyrus says the "paste" applied to the bandages was heated, and this hot resin treatment is thought by Dawson to be shown in XIX dyn. tombs at Thebes, but these seem to me to represent the painting of the cartonnage.[8] Lucas [9] considers that a mixture of fat and resin was applied *before* treatment with natron.

An entirely different hypothesis of the mummifying process is offered, on the basis of experiments, by Lucas[10], who maintains that *dry* powdered natron, which may contain over 50 per cent. of common salt, was put round the body and acted as a desiccating agent. Neither Herodotos nor Diodoros Siculus

[1] Sprengel, H. Med., i, 60 ; Pettigrew, 43 f. ; BMGE, 159 f. ; Wiedemann, Rel., 124 ; Dawson, JEA, 1927, xiii, 43 f. ; 1932, xviii, 125 ; *ib.*, Magician and Leech, 25, 31 f., 41 f.
[2] Wiedemann, 1920, 32 ; Dawson, JEA, 1927, xiii, 49.
[3] History of Mummies, 1834, 62.
[4] JEA, 1927, xiii, 45.
[5] Budge, Mummy², 204.

[6] Relation de l'Égypte, tr. de Sacy, 1810, 200, 268 ; Reitemeyer, 139, 141.
[7] Budge, Mummy², 208.
[8] Dawson, JEA, 1927, xiii, 47 ; *ib.*, Magician and Leech, 48 ; Forrer, RL, 496 ; Maspero, Struggle, 510.
[9] JEA, 1914, i, 123 ; *ib.*, in Carter, Tomb of Tutankhamen, ii, 183 f.
[10] JEA, 1932, xviii, 125 f., 131 f., 139.

mention "steeping in solution." The meaning of the verb ταριχεύω which they use for embalming was originally to preserve, dry or salt fish, etc., which was carried out in ancient Egypt by opening, salting and drying in the sun.[1] Athenaios also uses the same word for salted fish and mummies, and (except in the case of mackerel) implies that dry salting, not pickling, was used. The fact that the epidermis is frequently missing is not a proof of maceration, since it is sometimes absent (from putrefaction) without embalming and sometimes present with embalming. Lucas also does not think that drying with artificial heat was used after the washing, which he assumes followed the treatment with dry natron, and the body was sometimes wrapped whilst still damp. In the New Kingdom the body, after drying, was sprinkled with solid natron, which is said to have "corroded the bandages in such mummies,"[2] but the colour of the bandages is really due to a fungus.[3]

The description in the Ritual of Embalmment Papyrus throws little light on the actual process. As each bandage was put on, magic words were uttered and each bandage had a magic name. The body was anointed with a mixture of ten perfumes, especial care being taken with the head ; the intestines were placed in "liquid of the children of Horus" and a chapter read over them ; the backbone was immersed in "holy oil," a "secret liquid," an emanation of the gods Shu and Seb ; the finger and toe nails were gilt—"thy fingers are of gold and thy thumb of asem" ; and there was a further elaborate ritual of bandages.[4]   Herodotos says the treatment with natron lasted 70 days, which is confirmed for the total time by the texts [5], whilst Diodoros gives 40 days, which is the time taken by the Egyptian "physicians" (really "priests," Hebrew rephem) in the Bible to embalm the body.[6]

MATERIALS USED IN EMBALMING

The use of palm wine, or Phœnician wine, mentioned by Herodotos and Diodoros, is supposed to be confirmed by "the presence of alcohol in some of the tissues" of mummies [7], but the suggestion that palm wine was used as an alcoholic solvent for embalming resins, such as cedar and Pistacio terebinthus resins [8], is wrong, since these resins do not dissolve in strong wine.[9] The "cedar oil" of Herodotos is said to correspond with the Egyptian hatet aš.[10] Liquid resin[11] and "distilled oil of turpentine" (which would not be known then)[12] have been suggested. Modern cedar oil, which would not dissolve the viscera, could hardly have been meant[13], but Paulos Ægineta[14] says oil of κέδρος corrodes soft flesh. Pliny[15] says the first liquid running from the carbonising ovens in making charcoal is as clear as water, is called cedrium by the Syrians and is used in Egypt for embalming : this could be either turpentine or wood vinegar. The gum (κόμμι) mentioned by Herodotos may be acacia gum[16]; Dawson[17] thinks it meant resin or a paste of resin and fat.

Bags of natron, of the same composition as that of the lakes, have been found in places where mummies were made or in collections of embalming materials ; natron has also been found on the body, wrapped with viscera,

[1] Wilkinson, ii, 117 f.
[2] Maspero, Struggle, 510 ; Dawson, JEA, 1927, xiii, 48.
[3] Lucas, JEA, 1914, i, 119 ; 1932, xviii, 139.
[4] Thorndike, History of Magic, i, 8 ; Budge, Egyptian Magic, 185 f. ; Dawson, JEA, 1927, xiii, 49.
[5] Wiedemann, 1920, 114 ; Dawson, JEA, 1927, xiii, 41, 43.
[6] Gen. l, 2 ; Hoefer, i, 65 ; Lucas, JEA, 1932, xviii, 133.
[7] Dawson, op. cit., 49.

[8] Offord, Anct. Egypt, 1914, 180.
[9] Lucas, Egy. Mat., 150 f. ; wine was used by surgeons in washing wounds from antiquity until the 19th century.
[10] Anct. Egypt, 1915, 97.
[11] Sprengel, H. Med., i, 60 f.
[12] Hoefer, H. Chim., i, 68.
[13] Lucas, Egy. Mat., 112.
[14] vii, 3 ; tr. Adams, 1847, iii, 164.
[15] xvi, 11 ; cf. Adams, 452.
[16] Lucas, 125.
[17] JEA, 1927, xiii, 40 f. ; 1929, xv, 187.

and on bandages, etc. [1], and a 3 per cent. solution, coloured brown by organic matter, in IV and XII dyn. Canopic jars.[2] The natron of Nekheb (El-Kâb) is said in a text to have been used for embalming, and the constituents of natron, sodium sesquicarbonate, sulphate and chloride in variable proportions, with traces of calcium, have been found both in mummies and in crystals on their exterior [3], yet Dawson [4] maintains that "common salt mixed with various impurities," not natron, was used. The Coptic mummies, which are without incisions, are packed in solid common salt, free from natron, and are well preserved, but this purified salt was not used till the early Christian Period. The white crystalline patches on such mummies are not salt but fatty acids derived from the body.[5]

If we could read the λίτρον (νίτρον) in Herodotos as nitre or saltpetre, which is used together with common salt in preserving meat, the process might be more intelligible. Granville [6] found saltpetre both in the crystals on the exterior and also in the interior of a mummy, although he says its effects would be different from those produced by mummification, and Ure did not find it in a Greek-Egyptian mummy. There is no proof that saltpetre was ever used as such, and the small quantities found may have been produced from the body.

Egyptologists usually refer to "bitumen" in speaking of mummies, but the identity of this material is very imperfectly known. The word mummy (momie, Mumie, mummia) is not derived from any Egyptian [7] or Coptic word, but from the Arabic mumia, bitumen, through the Byzantine Greek μουμία or μώμιον. It occurs in the Latin form mumia only about A.D. 1000. The Arabic name came in turn from the Persian mum, meaning originally wax, then substances similar to wax in appearance, such as mineral wax (ozokerite) and asphalt (bitumen), then any material used in embalming, and finally the complete mummy. The Syriac name for embalming material was mûmyâ, the Greek πιττάσφαλτος.[8] Authorities differ considerably both as to the identity of the "bitumen" used and the period when it was first applied. Dawson [9] says "bitumen" was not used till the Græco-Roman Period, and then not universally; Budge[10] that it was not used extensively till the XXII dyn. (900 B.C.). Lucas[11] thinks it was never ozokerite but sometimes wood pitch in the Ptolemaic Period, and sometimes resin, which has become black with age. Wiedemann[12] speaks of the body as being immersed in "hot asphalt." Lucas says that, although Reutter claimed to have identified "bitumen" in mummies, he himself never found it, but Spielmann[13] thinks the tests he made point to the use of bitumen, with resin and wood tar. Petroleum is said to be mentioned in the Ebers Papyrus (1550 B.C.).[14]

Ibn al-Baitâr (d. 1248), quoting Dioskurides[15], says mumia flows from mountains and becomes hard and thick, forming the bitumen of Judæa. It is also found in Egyptian tombs in large quantities and was a mixture formerly used by the Byzantine Greeks for embalming.[16] Abd al-Latîf[17], quoting Galen,

[1] Wiedemann, 1920, 112 ; Lucas, Egy. Mat., 114 ; ib., JEA, 1932, xviii, 125 f.

[2] Brunton, Lahun I, 1920, 20 ; Meyerhof, A. Nat., 1927, x, 336 ; Lucas, JEA, 1932, xviii, 127.

[3] Granville, in Pettigrew, 62 ; BMGE, 162 ; Lucas, JEA, 1914, i, 122 ; 1932, xviii, 126.

[4] JEA, 1927, xiii, 49.

[5] Granville, in Pettigrew, 53 ; Lucas, JEA, 1932, xviii, 128 f.

[6] In Pettigrew, 62, 83.

[7] Erman-Grapow, Handwörterbuch, 1921, 65, give the hieroglyphic mnnn = asphalt ; mnḥ = wax.

[8] Kircher, Œdipus Ægyptiacus, 1654,

iii, 396 f. ; Pettigrew, 1 f., 7 f. ; Wiedemann, 1920, 111 ; Budge, Mummy[2], 201 ; Netolitzky, MGM, 1927, xxvi, 276.

[9] JEA, 1927, xiii, 45 ; 1929, xv, 187.

[10] BMGE, 158.

[11] Egy. Mat., 124 ; JEA, 1914, i, 241 f.

[12] Ägypten, 1920, 33.

[13] JEA, 1932, xviii, 177.

[14] Joachim, 15, 168, 170.

[15] i, 100.

[16] Ibn al-Baitâr, c. 2190 ; tr. Leclerc, Notices et Extraits des MSS., xxvi, 346 ; Budge, Mummy, 1893, 174.

[17] Relation de l'Égypte, tr. de Sacy, 1810, 200 f., 268 f. ; Reitemeyer, 139 ; Budge, Mummy[2], 202.

says mumia is a mixture of pitch and myrrh, but is really like asphalt or mineral wax, which can be used instead if one takes the trouble to procure it. Masûdî (d. 956) describes a mixture of pitch and medicaments found in Canopic jars, which had no odour until thrown on a fire.[1]

Reddish resin found in the tomb of Tutankhamen (XVIII dyn.) was thought by Plenderleith [2] to be Angora copal. This resin may, however, have been used as incense.[3] Diodoros Siculus does not mention the use of bitumen in his section on mummifying, but in another place [4] he says it was sold in Egypt and used for embalming mixed with other spices, since otherwise the body would not keep long. Offord [5] suggests that this was Judæan balm, otherwise identified with resin, etc.[6] Reutter, in materials in a XXVI dyn. (663–525) coffin regarded by him as constituents of incense, identified some aromatic wood (probably juniper), Chios turpentine, cedar resin, storax (from Liquidambar orientale), Aleppo resin and mastic (from Pistacia lentiscus).[7] All these identifications are doubtful. Two vases containing "bitumen" (Thothmes III) are in the Cairo Museum.[8]

In Upper Egypt the priests probably carried out a lucrative business in mummifying, the sale of funeral papyri and amulets and in conducting funerals.[9] The number of bodies mummified must, however, have been comparatively small.[10] Each mummy was provided with a copy of the Book of the Dead, more or less complete[11], and all kinds of amulets.[12]

The Egyptians embalmed various sacred animals, birds, fish, and even flies and shells.[13] Herodotos[14] was struck by the embalming of cats, and there are numerous mummified cats in museums. The human intestines, etc., were cleaned, wrapped in linen with powdered spices, salt, etc., and placed in four jars, later called Canopic jars, each dedicated to one of the four sons of Horus, the gods of the four cardinal points.[15] Plain stone jars appear at least as early as the V dyn., but during the VI–XII dyns. they had covers representing four different heads, and were sometimes put into wooden Canopic chests. The heart was always left in the body, the kidneys sometimes.[16] The use of Canopic jars was discontinued in the XXI dyn., the viscera being wrapped in linen and put back in the body.[17] The contents of the jars have been differently given.[18] They probably contained natron solution.[19] The opening in the body was sometimes (not always) sewn up and covered with a wax or silver plate stamped with the sacred eye (uzat) to scare off demons, and these eyes were painted on the coffin.[20] The face of the mummy was covered through the Dynastic Period with a beaten gold, gilt wood or gilt papier mâché portrait mask. In Ptolemaic times the gilding gave way to coloured portraits. Portrait masks of gold occur also in Nineveh, Mycenæ, and Etruria ; in

[1] Reitemeyer, 143.
[2] In Carter, Tomb of Tutankhamen, ii, 214.
[3] Lucas, JEA, 1930, xvi, 48 ; 1932, xviii, 126.
[4] xix, 99 ; Lucas, JEA, 1914, i, 214 f.
[5] Anct. Egypt, 1914, 180.
[6] W. Smith, Concise Dict. of the Bible, 1865, 92 ; cf. Pettigrew, 75 f.
[7] Wiedemann, 1920, 113 ; Lucas, JEA, 1930, xvi, 41 f.
[8] Maspero, Guide to the Cairo Museum, 1908, 307.
[9] Budge, Gods of the Egyptians, i, 101.
[10] Budge, Osiris, i, p. xx.
[11] Maspero, Hist. anc., 1905, 46.
[12] Wiedemann, Rel., 124 ; BMGE, 150 ; Erman-Ranke, 356.
[13] Wilkinson, iii, 248 ; BMGE, IV–VI Rooms, 31 f. ; Abd al-Latîf, in Reitemeyer,

140 ; Mackenzie, The Migration of Symbols, 1926, 140.
[14] ii, 66, 67.
[15] BMGE, 161 ; BMGE, IV–VI Rooms, 36 ; Wilkinson, iii, 219, pl. 48 ; Budge, G. of E., i, 490 ; Mummy², 240.
[16] Dawson, JEA, 1927, xiii, 42 f. ; 1930, xvi, 163 ; Shorter, ib., 1929, xv, 138 ; cf. Diodoros Siculus, i, 91, whose statement that the incision was made on the left side is correct.
[17] Dawson, Mag. and Leech, 49, 89.
[18] BMGE, 161 ; Quibell, 56 ; Elliot Smith, in Mackenzie, Symbols, 37.
[19] Lucas, JEA, 1932, xviii, 127.
[20] Dawson, JEA, 1927, xiii, 46 ; Wiedemann, 1920, 33 ; Thorndike, History of Magic, 1923, i, 8 ; Lexa, Magie, iii, plate 63 ; Erman-Ranke, plate 24.

Sardinia and Tunis they are of polychrome clay.[1] Flat gold plates were laid on the tongue of the mummy only in the Roman Period.[2]

Mummies were used until the 17th century A.D., perhaps later, as medicine and sold in pharmacists' shops. They were provided mainly by Alexandrian Jews, and specially prepared for the purpose.[3] Boerhaave does not mention mummy; Neumann[4] censures its use. Mummies are still imported from Egypt for sale as cattle medicine in Bavaria.[5]

EGYPTIAN MEDICINE

Medicine in Egypt is of great antiquity, and medical literature is mentioned in 2800 B.C.[6] What are supposed to be "drug pots" from an "apothecary's shop" of the VI dyn. (2500 B.C.) are in the British Museum[7] ; a wicker case of bottles on a stand, supposed to have belonged to a medical man, but probably a lady's toilet set, found at Thebes, is in the Berlin Museum.[8] A very small ladle-shaped spoon found with it *may* have been used to measure out doses, or, if the bottles were for toilet purposes, to take out cosmetics. Small bronze "strainers"[9] *may* have been used with medicines.

A modern detailed account of early Egyptian medicine is lacking and Dawson's sketch[10] hardly touches the chemical aspects. Egyptian medicine, which may have owed something to Asia[11], had an important influence[12], descending through Greece, Arabia, Syria and other parts of Western Asia down to the Middle Ages.[13] Some remedies in the works attributed to Hippokrates, a standard authority until recent times, are contained in the Egyptian medical papyri[14], and even psycho-therapy, the most modern branch of the healing art, is represented by the ancient Egyptian physician who remarks : "I say to myself every day," repeated.[15] Unpleasant drugs played an important part, since the patient feels that a medicine with a violent taste must have an energetic action. The partiality for nasty medicines of animal origin was continued in the popular Roman medicine preserved for us in Pliny[16], which continued through the Middle Ages and still flourishes in a modern scientific form in gland and vaccine therapy. In its older form this branch of the healing art made use of roast dog, boiled earthworms and preparations from toads, spiders, flies and woodlice, as well as animal and human excrements[17], and was based on an immense mass of superstition about occult properties.[18] Perhaps as useful drugs were accidentally discovered in their participation in magic ceremonies the latter declined in importance ; the unpleasant medicines were probably originally intended to be unpalatable to the spirit causing the

[1] Wilkinson, ii, 244; Hoernes, Urg., ii, 426 f. ; good collections of portraits in British Museum and in Vienna.

[2] Wiedemann, 1920, 31.

[3] Budge, Mummy, 174 ; Schmidt, Drogen, 159 ; Lefévre, Traité de la chimie, Paris, 1674, i, 159 f. : "La mumie qu'on prepare avec la chair du microcosme, est un des plus excellens remedes qui se tire des parties de l'homme," but "est en horreur à quelques-uns" ; Pettigrew, 9 f.

[4] Chemistry, tr. Lewis, 1759, 551.

[5] Netolitzky, MGM, 1927, xxvi, 276.

[6] Sarton, Isis, 1931, xv, 357.

[7] BMGE, IV–VI Rooms, 257.

[8] Wilkinson, i, 428 ; ii, 47, Fig. 206.

[9] Wilkinson, ii, 48.

[10] Science Progress, 1927, xxii, 275 : "Medicine and Surgery in Ancient Egypt" ; ib., Magician and Leech, 1929.

[11] Budge, The Divine Origin of the Craft of the Herbalist, 1928, 11.

[12] Maspero, Dawn, 220 ; Meyerhof, Isis, 1926, viii, 200 ; Sarton, ib., 1931, xv, 357.

[13] Budge, The Syriac Book of Medicines, 1913, i, cxxx f. ; Wiedemann, 1920, 418.

[14] M. Neuburger, History of Medicine, tr. Playfair, 1910, i, 154.

[15] Maspero, Dawn, 399 ; Ganszyniec, PW, xii, 54.

[16] Books 28–30.

[17] Budge, Syriac Med., i, p. cxxxiv ; Hausrath, History of New Testament Times, the Time of Jesus, tr. Poynting and Quenzer, 1878, i, 131 ; Lenz, Zoologie der alten Griechen und Römer, Gotha, 1856, 27 f.

[18] Eitrem, Papyri Osloensis, Oslo, 1925, i, 90 f., 103 f., etc. ; Stemplinger, A. Med., 1920, xii, 33 ; ib., Antiker Aberglaube, Leipzig, 1922.

disease, although honey is said to be "sweet to the living and bitter to the dead."[1]

A main principle of Egyptian medicine was to expel the materia peccans by emetics, purgatives, enemata, blood-letting, diaphoretics, diuretics and sternutatories ; unwholesome air was eliminated by exciting ructus and flatus with onions, leeks and beans. Most diseases were thought to arise from over-feeding—an early recognition of the importance of dietetics—and the principal therapeutic agents were pure air, fumigations and draughts of liquids.[2]

Egyptian Physicians

Diodoros Siculus [3] says the physicians (perhaps army physicians) practised without fee during wars ; the ordinary physicians had a public stipend. In Zosimos (c. A.D. 300) [4] the "superior priest" is contrasted with the ordinary physician who works "by books," a distinction also made in the Ebers Papyrus, which adds the "common magician" to these two classes. According to Gardiner [5], Egyptian medicine is the direct offspring of magic. The god Thoth, the first physician, was also the first magician and the distinction between physician and magician was, if anything, that the physician worked according to the book, whilst the magician was a priest under divine guidance. Lexa [6], who otherwise follows Gardiner closely, considers that the Egyptians distinguished clearly between medicine and magic, and physicians from magicians, although a "royal physician, interpreter of a difficult science," is also a magician.[7] The magic formulæ in the Edwin Smith Surgical Papyrus are separate and perhaps intended for use with patients who required them.

Physicians and Priests

There were different grades, perhaps three, of Egyptian physicians as well as veterinary surgeons.[8] Von Oefele [9], Sprengel[10] and Uhlemann[11] considered that medicine was confined to the temples, and that the physician was the hierogrammateus ($\iota\epsilon\rho o\gamma\rho a\mu\mu a\tau\epsilon\upsilon\varsigma$) or "sacred scribe," mentioned by Clement of Alexandria[12], who carried the books of Hermes-Thoth. This official was the Egyptian kheri heb, or lector priest, the "man with the book," who recited the magic formula, the "words of power" (hekau), written by the god Thoth, the first magician and physician. These words when spoken in the correct tone and with proper emphasis were of irresistible power, could compel even the gods to obey the commands of the magician, and could even stop the course of the sun-god Rā sailing in his boat across the sky by day and in the underworld at night.[13] Magic in Egypt goes back to the oldest period[14], but its importance increased considerably from the time of the xix dyn.[15]

[1] Dawson, Mag. and Leech, 58, 73 ; Thorndike, Magic, i, 10.

[2] Dawson, 62 ; Neuburger, i, 26, 29 ; Wilkinson, ii, 354; von Oefele, in Puschmann, Geschichte der Medizin, Jena, 1902, i, 83.

[3] i, 82.

[4] Berthelot, Collection des anciens Alchimistes Grecs, ii, 223 ; iii, 226.

[5] ERE, viii, 267 ; cf. Wiedemann, Rel., 146 f. ; Maspero, Dawn, 214 f. ; Dawson, Mag. and Leech, 61.

[6] Magie, i, 17, 133—a mediocre book.

[7] Dawson, Mag. and Leech, 59, 60.

[8] Maspero, PSBA, xiii, 501 ; Wilkinson, ii, 452, Fig. 487.

[9] In Puschmann, Gesch. Med., i, 78 f. ; cf. Budge, Syriac Med., i, p. cl.

[10] H. Med., i, 45 f., 50.

[11] Thoth, 150 f.

[12] Stromateis, vi, 4.

[13] Erman-Ranke, 330, 406 ; Maspero, Dawn, 281 ; Budge, Osiris, i, 199, 282 ; ii, 169 f. ; ib., Egyptian Magic, 3 f., 27, 144 ; ib., Gods of Egy., i, 401 f., 409 ; Wiedemann, 1920, 411, 420; Gardiner, ERE, viii, 263 ; Aram, JEA, 1930, xvi, 108 ; Dawson, Mag. and Leech, 69 ; Lexa, Magie, i, 43, 46, 61 f. ; Hopfner, PW, xiv, 307 f. ; Peet, CAH, i, 354 ; Origen, Contra Celsum, i, 24 ; v, 45 ; Iamblichos, de Mysteriis, vii, 4, 5 ; Green, JEA, 1930, xvi, 34 ; Pietschmann, Ro., v, 835 ; ib., Hermes Trismegistos, Leipzig, 1875, 12 f., 43 ; Boylan, Thoth the Hermes of Egypt, Oxford, 1922, 92 f., 124 f., 131 f.

[14] Budge, Gods of Egy., i, pp. xiii, 2, 10, 27, 152 ; Gardiner, ERE, viii, 262.

[15] Meyer, Alt. II, i, 417; Hall, CAH, vi, 164.

Otto [1] thinks the priests may have practised medicine—although the medical knowledge required for embalming may have misled Clement of Alexandria— but it was not their monopoly. The divine origin of medical knowledge was, however, often claimed by the Egyptians : the Ebers Papyrus (*c.* 1550 B.C.) gives a recipe "made by Rā," [2] and a book on medicine was "revealed" to a priest watching the altar at night in an Isis temple at Coptos. [3]

Great stress was laid on personal cleanliness, especially by the priests, who frequently washed the hands and fingers. Apparatus for washing, consisting of a jug and bowl (represented with "a square object, certainly not a piece of soap") and towel is shown in the Old Kingdom. [4]   The sign of an oval basin containing the water line, used for the name of the god Amen from the time of the Libyan dynasties, represents the Berber word aman, "water," which coincides with the pronunciation of the name of the god at that period. [5]

The Bible [6], Homer—who calls Egypt "a country whose fertile soil produces an infinity of drugs, some salutary and some pernicious" [7]—and Pliny represent the Egyptians as using many drugs. [8]   Herodotos [9] says "every place is full of doctors."

Pliny[10], who calls the embalmers physician-priests, says the Egyptians claimed to have invented medicine, and the medical schools of Alexandria had a great reputation : Ammianus Marcellinus (d. A.D. 400) reports that a doctor recommended his skill by saying that he had studied there.[11] Much of the superstition of early Egypt persisted in Alexandria.[12] Herodotos[13] implies that there were specialists, and he is supported by texts from the time of the Old Kingdom.[14] Collections of old recipes were made and annotated, especially in the beginning of the New Kingdom.[15]

IMHOTEP

Osiris, Rā and Isis all had medical functions[16]. The Ebers Papyrus opens with an invocation to these divinities, in which Isis is called "powerful in magic,"[17] and Diodoros Siculus[18] says Isis when on earth discovered many medicines and raised her son Horus from the dead by a medicine of immortality, still curing men in sleep. The most important gods of medicine, however, were Thoth, who included medical knowledge in his all-embracing wisdom[19], and Imhotep, originally human, a vizier of King Zoser (early III dyn.). Imhotep left a great reputation for wisdom in architecture, medicine and magic, "knew the wisdom of Thoth," and was "acquainted with the books of magic."[20] He was the architect of the terrace or step pyramid of King Zoser : the first true pyramid was built by Zoser's successor, Snefru, as his tomb, "the horizon," in which he was "swallowed up," as his father the sun every evening.[21]

Imhotep's career is divided[22] into three stages : (1) a contemporary of

[1] Priester und Tempel, i, 96 ; ii, 194.
[2] Joachim, 59.
[3] Maspero, Dawn, 224.
[4] Wilkinson, ii, 48 ; Wiedemann, 1920, 95 f., Figs. 10, 11.
[5] Griffith, JEA, 1922, viii, 275.
[6] Jerem., xlvi, 11.
[7] Od., iv, 229.
[8] Wilkinson, ii, 355 ; medicinal plants in Egypt, *ib.*, ii, 417 f.
[9] ii, 84.
[10] vii, 56 ; xix, 5.
[11] Wilkinson, ii, 358 ; Freind, Opera omnia, 1733, 471.
[12] Wehrli, in Singer and Sigerist, Essays on the Early History of Medicine, presented to Karl Sudhoff, Zürich, 1924, 384 f. ; Neuburger, i, 18 ; Wiedemann, 1920, 246.

[13] ii, 84.
[14] Maspero, Dawn, 216 ; Erman-Ranke, 409 ; Sarton, *Isis*, 1931, xv, 359.
[15] Erman-Ranke, 413 ; Joachim, Papyros Ebers, xii.
[16] Budge, Divine Origin of the Craft of the Herbalist, 1928, 12.
[17] Joachim, 2.
[18] i, 25 ; Lafaye, in Daremberg-Saglio, Dict., III, i, 581 ; Sprengel, H. Med., i, 34.
[19] Roeder, Ro., iv, 854.
[20] Reitzenstein, Poimandres, 1904, 120.
[21] Breasted, H., 112 ; Quibell, 34 f. ; Maspero, Dawn, 370.
[22] Hurry, Imhotep, Oxford 1926, 4 f., 20 f., 31, 39, 42, 65 ; largely based on Sethe, Imhotep : der Asklepios der Aegypter, Leipzig, 1902 ; the name is really Imhutep, Ij-m-ḥtp.

King Zoser (2980 B.C.) ; (2) a medical demigod in the reign of Mykerinos (c. 2850 B.C.) ; (3) a full deity of medicine in the Persian Period (c. 525 B.C.). Under Zoser he was grand vizier, architect, chief lector priest (kheri-ḥeb), sage and scribe, astronomer and magician-physician, associated with Hermes-Thoth in making astronomical observations, a famous priest and magician of Memphis. During the period when he was regarded as a demigod, he "appeared in sleep," carrying a book, to a sick person in his temple and effected a cure.[1] In the Persian Period he is called a "son of Ptah," perhaps under Hellenistic influence.[2]

Although it has been suggested that Imhotep wrote the Edwin Smith Surgical Papyrus [3], there is no direct evidence that he wrote any medical works [4], and instead of being a man afterwards deified [5], he may have been an old god afterwards specially honoured.[6] Imhotep's name may mean "he who comes in peace," an old epithet of the god Ptah of Memphis, where there was a temple of Imhotep with an important medical school. He also had temples at Philæ (Ptolemaic Period) and Thebes, and shared several temples with other gods. As a god Imhotep did not take shape till the New Kingdom ; his popularity was greatest in the Saïte and early Greek Periods and declined on the rise of Greek medicine.[7] In Greek works (e.g. in Zosimos) he is called Imuthes (᾽Ιμούθης) or Imuth (᾽Ιμούθ) [8], identified with Asklepios [9] or the Phœnician god Ešmun (῎Εσμουνος), the chief of the eight Kabiri.[10] The Asklepios serpent staff, however, is shown in a representation of the Sumerian god Ningišzida, son of Ninazu[11], and Asklepios may, therefore, have an Asiatic origin.

MAGIC IN MEDICINE

Egyptian medical texts all contain elements of magic in varying amount. The Ebers Papyrus opens with a collection of spells (ro) to give efficiency to the prescriptions (pakhret) which follow, some for external and some for internal medicines (spells for "placing" or "drinking" the prescription), ending "say the words."[12] In the Pyramid Texts, the Book of the Dead and in magic papyri of all periods, on stelæ, etc., the parts of the body are represented as under the care of different gods : Nu the hair, Rā the face, Hathor the eyes, Anubis the lips, Thoth all the limbs : "there is not a limb of him without a god"[13]. In a later theory reported by Origen[14] these are under the control of the 36 dekans or signs of the Egyptian zodiac[15], whose hieroglyphics were engraved on plants and stones and could be invoked when the part was diseased. Mention of death is usually avoided : Pharaoh when he dies "becomes a god."

[1] From a late papyrus of Oxyrhynchus, 2nd century A.D., where Imhotep is identified with Asklepios ; Schubart, Einführung in die Papyruskunde, 1918, 157.
[2] Breasted, H., 575 ; Budge, Dict., 580.
[3] Isis, 1931, xv, 359.
[4] Dawson, Mag. and Leech, 87.
[5] Budge, Gods of Egy., i, 522 ; Hurry, Imhotep, 31 f.
[6] Wiedemann, 1920, 87 ; ib., A. Rel., 1922, xxi, 459.
[7] Maspero, Dawn, 106 ; Meyer, Alt. I, ii, 170 ; Wilkinson, iii, 205 ; cf. Sethe, Imhotep, 1902, 4, 6 ; Reitzenstein, Poimandres, 120 ; Haeser, Grundriss der Geschichte der Medicin, Jena, 1884, 4 ; Hurry, Imhotep, 43 f., 48, 57, 71 f.
[8] Berthelot, Or., 9, 184 ; Sethe, Imhotep, 1902, 25 ; Hurry, 68 ; Reitzenstein, Poimandres, 120.

[9] Brugsch, Rel., 526 f. ; Wiedemann, Rel., 77.
[10] R. Asmus, Das Leben des Philosophen Isidoros von Damaskios aus Damaskos, Leipzig, 1911, 124 ; J. H. Baas, Outlines of the History of Medicine, tr. Handerson, New York, 1889, 14 ; Lepsius, Über die Götter der vier Elemente bei den Ægyptern, Abh. Berlin Akad., 1856, 181 f.
[11] Budge, Herbalist, 17 ; but cf. Prinz, PW, vii, 1908 f.
[12] Dawson, Mag. and Leech, 71.
[13] Wiedemann, Rel., 147 ; Lenormant, Chaldean Magic, 1877, 36 ; Dawson, 60 f.
[14] Contra Celsum, viii, 58.
[15] Wiedemann, Rel., 147 ; Murray, Five Stages in Greek Religion, Oxford, 1925, 45 ; Wilkinson, ii, 356 ; Schulzius, Historia Medicinæ a Rerum initio ad annum urbis Romæ DXXXV deducta, Lipsiæ, 4to., 1728, 11.

The "breath of life" and the "breath of death" (? contagion in the air) entered through the right and left ear respectively.[1]

The Ebers Papyrus implies that the "spirits" of drugs entered the patient and protected him as an amulet. The evil spirit causing illness fed on the nourishment taken by the person and could be expelled by taking things repugnant to him. On the other hand, by rubbing the body with oil of cypress or myrrh, the virtue of these sacred materials entered the body, and the evil spirit dare not attack it.[2] Although it is generally stated [3] that the Egyptians, like the Babylonians, believed that disease was always caused by evil spirits, which entered the body and against whom only magic could prevail, Budge [4] says there is good reason for thinking that although some diseases were attributed to evil spirits entering bodies, "the texts do not afford much information on the matter," a cautious statement which is likely to be true.[5] The Pyramid Texts (v–vi dyn.) imply that death was a state specially created at the same time as the gods and man[6]; they (and also the Harris Papyrus) contain spells against poisonous snakes and scorpions, etc.[7] A text from Byblos says that if the poison has entered the body it is better to use a magic formula than to sacrifice to the gods.[8]

The medical papyri contain magic formulæ to be spoken during the preparation or use of drugs, and some mixtures are "of divine origin."[9] Dawson[10] thinks the prescriptions are merely elaborations of the manual rites of magic, the names of drugs punning with the words of incantations; the Egyptian was very fond of puns, which probably had more of a magic than a humorous character. In some cases a formula could be written on papyrus, the ink washed off with beer or water and drunk as medicine[11], or the patient touched the statue of the god, which transmitted its vital fluid (sa en ankh) to him (or to the priest or inanimate objects), the gods renewing their sa from greater gods or from a magic pond in the northern sky.[12]

Divine bodies continued by means of sa for very long periods; as they grew old their bones changed to silver, their flesh to gold; their hair, "piled up and painted blue, after the manner of great chiefs," into lapis lazuli, and the god was changed to "an animated statue."[13] The Harris Papyrus says of Amen: "his bones are of silver, his limbs of gold, his hair real lapis lazuli."[14] In other papyri the blood of Rā changes into salt[15], or the blood of Horus is said to be of gold.[16] In a Pyramid Text the god Sokar "smelts the bones" of the dead Pharaoh from copper; Ptah told Rameses II: "I have wrought thy body of gold, thy bones of copper, thy vessels of iron," and in the Westcar Papyrus (17th century B.C., but including older material c. 2000 B.C.) a prince is "born with limbs of gold and a head-dress of real chesbet" (lapis lazuli).[17] The substances of plants were the same as the substances of the bodies of gods: anti (myrrh) was "the tears of Horus;" bees (giving honey) sprang from the "tears of Rā," etc.[18] A text of 2500 B.C. says "gold is the body of the

[1] Dawson, 6, 93.
[2] Budge, Osiris, i, 284; Lexa, Magie, i, 74; ii, 106; Berlin Papyrus.
[3] Maspero, Dawn, 215; Breasted, H., 102; Dawson, Mag. and Leech, 2 f., 55, 62 f.
[4] Egy. Mag., 206.
[5] Thorndike, Hist. of Magic, i, 11.
[6] Dawson, 5.
[7] Wiedemann, 1920, 248 f.; ib., Rel., 151.
[8] Wiedemann, A. Rel., xxvi, 360.
[9] Neuburger, H. Med., i, 26 f.; Wiedemann, Rel., 147 f.
[10] Mag. and Leech, 58, 63.
[11] Wiedemann, Rel., 32, 149; Hopfner, PW, xiv, 336; cf. Ezekiel, iii, 3; Haeser

Gesch. d. Med., Jena, 1884, 110; later uses, Lippmann, Alch., ii, 218.
[12] BMGE, 117; Budge, Amulets, 164; ib., Gods of Egy., ii, 39; Maspero, Les contes populaires de l'Égypte ancienne, 1905, 221; ib., Dawn, 119 f., 257, 267; Hopfner, PW, xiv, 333 f.
[13] Maspero, Dawn, 110, 164.
[14] Chabas, Mélanges, III, ii, 254; Erman, Rel., 44.
[15] Amélineau, Ann. Musée Guimet, xiv, 304.
[16] BMGE, 116.
[17] Blackman, JEA, 1927, xiii, 191; Erman-Krebs, 30 f., 40.
[18] Budge, Syriac Med., i, p. cxxxii; ib., Herbalist, i, 24 f.

gods" and the skin of Rā is said to be of "pure asem" (electrum) ; in one of
1600 B.C. it is said : "for love of Rā the desert brought forth gold, silver and
lapis lazuli."[1]

MEDICAL TEXTS

Old medical books are often mentioned. In a story found in the tomb of
Ptahuash, chief architect of King Neferārikarā (v dyn.), the latter was inspecting
some works when Ptahuash was injured. The king's own physicians (senu) were
summoned and the king ordered his case of medical papyri to be brought ;
the chief physician examined the man and told the king that his case was
hopeless. The king, after giving orders that everything possible should be done,
retired to his chamber and prayed to Rā, but Ptahuash died, as the physician
said he would.[2] Manetho [3] says the king Athothis (Teti I) wrote a book on
anatomy : his name was confused with that of Thoth.[4] Horapollo [5] mentions
a sacred book used by the priests, called Ambres (ἀμβρὴς), which decided
from the recumbent position of the sick person whether he would live or not.
The name is reminiscent of the magician Mambres, but Baas [6] says it is the
Egyptian ha em re em per em hru, "here begins the book of the preparation
of drugs for all parts of the human body." Some recipes from books in the
temple of Imuthes at Memphis are given by Galen[7] : they include litharge,
squamæ æris, æruginis rasæ, burnt copper, colophonium, etc. The extensive external use of copper preparations seems to have been derived from
ancient Egypt.

Although it is often assumed that the Egyptian physicians were not specially
skilled, practised empirically, could make no proper diagnoses and had no
knowledge of anatomy derived from dissections until the Greek Period [8],
they must have derived much knowledge of the organs of animal and human
bodies from the preparation of animals for sacrifice and from mummifying,
respectively. The hieroglyphics representing these (heart, head and wind-pipe,
ear, uterus, etc.) are of animal origin, those of external parts are human. The
old Egyptians had more than a hundred anatomical terms, which show that
they differentiated and named a great number of organs ; they failed to understand the nerves, muscles, arteries and veins, although the pulse felt in various
parts was connected with the beating of the heart.[9]

In the Ebers Papyrus the physician makes a diagnosis ["say to this," i.e. the
symptoms] and then says "I will act" [according to this diagnosis][10], but he
was expected to diagnose a complaint by merely looking at the patient.
The oldest medical papyrus, the Kahun Papyrus (c. 2000) is practically free
from magic ; the Ebers Papyrus (c. 1550) contains it to some extent, and the
Berlin Papyrus (15th–14th centuries) is full of it.[11] The main part of the
Edwin Smith Papyrus is free from magic, but it contains an appendix of magic
formulæ added by a later user.[12]

THE EGYPTIAN MEDICAL AND SURGICAL PAPYRI

The Egyptian medical and surgical papyri are of different dates and have
been differently described by various authors : in the following account I have

[1] Erman-Ranke, 557 ; Erman, Religion
Égyptienne, 1907, 44, 86.

[2] Budge, Syriac Med., i, p. cxxxiii.

[3] Cory-Hodges, Ancient Fragments, 1876,
112.

[4] Budge, Herbalist, 10 ; M. B. Lessing,
Gesch. der Medizin, 1838, 3 ; Maspero,
Dawn, 786 ; Sprengel, H. Med., i, 38.

[5] Hieroglyphica, i, 38 ; ed. and transl.
by Cory, London, 1840.

[6] H. Med., 1889, 14.

[7] De compos. med. sec. gen., lib. v,
cap. 2 ; Opera omnia, fol., Venice, 1562, v,
120 verso ; Maspero, H. anc., 1905, 88.

[8] Wiedemann, Rel., 146 f. ; cf. BMGE,
73 ; Sprengel, H. Med., i, 62.

[9] Wilkinson, iii, 474 f. ; Dawson,
Magician and Leech, 90 f.

[10] Joachim, 192, etc. ; Maspero, Dawn,
218 ; Erman-Ranke, 412.

[11] Meyer-Steineg and Sudhoff, Geschichte
der Medizin im Überblick, Jena, 1922, 21.

[12] Dawson, Mag. and Leech, 74, 100.

Papyrus	Date of Papyrus B.C.	Columns	Lines	Recipes	Provenance	Present Location
I. (a) Kahun[1] (gynæcological).	2000–1850.	3	154	34	Illahun by Petrie, 1889; in fragments.	—
I. (b) Kahun[1] (veterinary).	do.	—	—	—	do.	—
II. Gardiner[1] fragments.	c. 2000.	—	29	—	—	—
III. E d w i n Smith[2] (surgical).	17th cent.	[21½ 17	469] 377	—	Thebes? With Ebers Pap.; purchased by Edwin Smith at Luxor, 1862; presented by his daughter in 1906.	New York Historical Society.
IV. Ebers.[3]	1550–1500, early 16th century.	110	2289	877	See III; purchased by Ebers, 1872–3.	Leipzig University.
V. Hearst.[4]	Later than IV; c. 1500?; partly damaged.	15–16	273	260+8 incantations.	1899 at Deîr el-Ballâs; purchased in 1901 by Reisner.	California University.
VI. B e r l i n[5] 3027.	c. 1450.	15 pages	—	20 incantations+ 3 prescriptions.	Acquired in 1886 from Miss Westcar.	Berlin Museum.
VII. B e r l i n[6] 3038.	Time of Rameses II (1296–1230)	24	279	204	Passalacqua, early 19th century, from tomb in Saqqâra.	Berlin Museum.
VIII. London[7] 10,059.	XVIII–XIX? dyn.	19	253	63	Unknown; acquired by British Museum from Royal Institution in 1860.	British Museum.
IX. Oxyrhynchus (Greek).[8]	Early 1st century A.D.	fragment.	—	—	Oxyrhynchus.	British Museum.
X. C o p t i c (Chassinat).[9]	9–10th cent. A.D.?	—	—	237	Meshaïkh.	French Archæol. Institute, Cairo.

[1] Griffith, PSBA, 1891, xiii, 526; Griffith, The Petrie Papyri: Hieratic Papyri from Kahun and Gurob, London, 1898, 5, 12; Weinleitner, Isis, 1927, ix, 18; Wreszinski, Der grosse medizinische Papyrus des Berliner Museums, 1909, glossary, etc.

[2] Breasted, The Edwin Smith Surgical Papyrus, 2 vols., 4to., Chicago, 1930, with Engl. tr. and notes; Peet, CAH, ii, 220; Elliot Smith, introd. to Bryan, The Papyrus Ebers, 1930; Rey, Archeion, 1928, ix, 10; Lippmann, Chem. Z., 1931, lv, 933; Sarton, Isis, 1931, xv, 361.

[3] G. Ebers: Papyros Ebers, das hermetische Buch über die Arzeneimittel der alten Äegypter, in hieratischer Schrift, 2 vols., fol., Leipzig, 1875 (includes in vol. ii. hieroglyphic-Latin glossary by L. Stern); Wreszinski, Der Papyrus Ebers, Umschrift, Übersetzung und Kommentar, I Theil, Umschrift (all publ.), Leipzig, 1913; German translation by H. Joachim, Papyros Ebers, Berlin, 1890; English transl. of parts of Joachim's German, with childish commentary, by A. P. Bryan, The Papyrus Ebers, 1930; R. H. Major, Annals of Medical History, 1930, ii, 547 (not seen). The papyrus was discovered by Edwin Smith and was at first called the Smith Papyrus.

[4] Reisner, The Hearst Medical Papyrus, Leipzig, 1905 (facsimile, introduction and glossary); see also note[7]; summary in Maspero, New Light on Ancient Egypt, 1908, 266; Peet, CAH, ii, 219.

[5] Erman, Abh. Königl. Preuss. Akad. Wiss., 1901, with transl. ("Zaubersprüche für Mutter und Kind"), 28 f.; Hieratische Papyri aus den Königl. Museum, Berlin, 1911, iii, text only; Erman-Krebs, Aus den Papyrus, etc., 1899, 76f.

[6] Lithogr. facsim. (unsatisfactory) in Brugsch, Recueil des Monuments Égyptiens, 1863, ii, plates 85–107, text pp. 101–120; Wreszinski, Der grosse medizinische Papyrus des Berliner Museums, Leipzig, 1909 (photogr. facsim., hieroglyphic transcript, and translation, commentary and glossary); Erman and Krebs, Aus den Papyrus, 1899, 63 f., copious extracts; date c. 1300.

[7] Birch, Z. f. Ägypt. Sprache, 1871, ix, 61–64; a palimpsest in damaged condition; Wreszinski, Der Londoner medizinische Papyrus (B.M. 10,059) und der Papyrus Hearst, Leipzig, 1912, dates c. 1200 B.C., p. xiv; Budge, Syriac Med., i, p. cxxxi; Herbalist, 26, dates XVIII dyn.; Dawson, Magician and Leech, 78, XIX dyn.; Sarton, Isis, 1931, xv, 357, 11th century B.C.; Joachim, Papyros Ebers, p. ix, XVIII dyn.

[8] A. S. Hunt, The Oxyrhynchus Papyrus, 1911, viii, 110.

[9] Chassinat, Mém. Inst. Franç. Cairo, 1921, xxxii; summary in Dawson, P. Roy. Soc. Med., 1924, 51.

generally followed the latest authorities available.[1]   The table on p. 183 summarises the information. The division into sections and recipes is modern : the dates are those of the MS. ; the contents are much older in every case.[2] The references give the publications of the papyri, translations when available and summaries or critical observations when these have appeared. The Kahun Veterinary Papyrus is unique, although there is a tomb painting showing balls of drugs being administered to cattle.[3]   Coptic medicine was to a large extent the continuation of the old Egyptian.[4]

THE EBERS PAPYRUS

The Ebers Papyrus, said to have been found in a terra-cotta jar between the legs of a mummy at Thebes buried 10 ft. deep, is 20 m. long and 30 cm. wide, comprising a collection of 39 small treatises compiled without order from older sources and sometimes accompanied by notes of the copyist, such as "found destroyed," "good," etc.[5]   A slight notice of it, pointing out its chemical interest, was published in 1875 by H. Carrington Bolton [6], and its importance in the history of medicine was emphasised in 1888 by Lüring.[7]   Although Joachim, a medical man, had the assistance of the Egyptologist Lieblein, of Christiania, his translation is not satisfactory, the meanings of the technical terms being, as in other medical papyri, very obscure.[8]   The originals used in compiling the Ebers Papyrus seem to be of the XII–XIII dyns., although their subject-matter may be many centuries older. It is perhaps not a "Hermetic" book (as Ebers called it), or a book at all, but a collection of prescriptions interspersed with spells and incantations.[9]   Ebers's identification of it with the "books on medicine" mentioned by Clement of Alexandria[10] as the books of Hermes-Thoth carried by one of the priests in processions, of which six dealt with medicine and anatomy, viz. the structure of the body, diseases, organs, drugs, eyes, and diseases of women, is now regarded as improbable.[11] Lieblein[12] thinks the Ebers Papyrus was compiled for Bnon, the second king of the xv (Hyksos) dyn.

The Ebers Papyrus is not exclusively medical but contains prescriptions for cosmetics, hair-dyes, hair-restorers, mixtures for driving out rats, etc., still in use in Egypt.[13]   There is a recipe for restoring hair in a Greek papyrus of the 3rd–4th centuries A.D.[14]   Most of the drugs in the Ebers Papyrus are of organic, especially vegetable, origin, since the Egyptians were acquainted with a large number of plants, but some are mineral, such as copper sulphate (?), salt,

[1] Wreszinski's three publications—see refs. to table on p. 183 ; Dawson, 73 f. ; somewhat different account in Sarton, *Isis*, 1931, xv, 357, partly from Breasted, The Edwin Smith Surgical Papyrus, Chicago, 1930, 2 vols., 4to.

[2] BMGE, 32, 72, 91 ; Breasted, H., 101 ; Peet, CAH, ii, 219 ; Dawson, 75 f. ; Erman-Ranke, 409, 411 f. ; Budge, Herbalist, 26 f. ; *ib.*, Syr. Med., i, p. cxxxi.

[3] Maspero, Guide to the Cairo Museum, 1908, 524.

[4] Erman-Krebs, Aus den Papyren, etc., 1899, 252 f.

[5] Joachim, xii ; Berthelot, Arch., 235 ; Dawson, Mag. and Leech, 81.

[6] *American Chemist*, 1875, vi, 165.

[7] Die über die medicinischen Kentniss der alten Ägypter berichtenden Papyri, Leipzig, 1888 ; title from Brugsch, Ägyptologie, 1891, 408, and Budge, Syriac Med., i, p. cxxxi ; Joachim, 5, gives Strassburg as the place of publication : the original is not available to me.

[8] Berthelot, Arch., 232, 234, 238 ; Wiedemann, 1920, 417 ; Budge, Herbalist, 11 ; Dawson, pref. vii f., 75, 96, 109 ; *ib.*, JEA, 1932, xviii, 150 ; Wreszinski, Der Papyrus Ebers, 1913, preface, criticising Joachim severely ; Wreszinski's own method, in his translations of the Hearst, London and Berlin Papyri, of merely transcribing most of the Egyptian names does not, howevc.r, take us very far, and Joachim did his best.

[9] Dawson, 76 ; Wiedemann, *A. Rel.*, xxvi, 337; Joachim, x ; Berthelot, Arch., 234 f.

[10] Stromateis, vi, 4.

[11] Mme. Hammer Jensen, *Oversigt over det Kongelige Danske Videnskabernes Selskabs Forhandlinger*, Copenhagen, 1916, 279 f., in French.

[12] Recherches, 1911–4, 112 f.

[13] Erman-Ranke, 417, 418.

[14] Möller, *Isis*, 1930, xvi, 122.

natron, and aner sopdu (Memphite stone ?).[1] Prescriptions resemble those in
European medical books of two or three hundred years ago, and there are
alternative prescriptions. Powders and decoctions from plants and seeds, the
piths of certain trees, dates, sycamore-figs and other fruits, salt, magnesia (?),
oil, honey and sweet beer formed the main ingredients, mixed with materials
such as bone dust, rancid fat, droppings of animals, pig's gall and milk of the
mother of a male child, used with incantations repeated four times.[2] There
are over fifty vegetable products, including herbs and trees (e.g. cedar shavings).
The mandrake of Elephantine (not yet endowed with the human form) gave
an intoxicating narcotic drink; Isis gave it to Rā, when he felt himself
growing old.[3] Magical processes are frequently used with remedies. Some
remedies were used by gods when they reigned on earth, such as a salve of
sixteen ingredients used by Rā to cure himself of boils. A prescription good
for inflamed eyes contains equal parts of myrrh, "Great Protector's" seed,
copper oxide, citron pips, northern cypress flowers, antimony, gazelle droppings,
oryx offal and white oil, placed in water, allowed to stand one night, strained
through a cloth, and smeared over the eye for four days or painted on with a
goose-feather.[4] To prevent hair falling out, fat was rubbed in.[5] A cosmetic
for removing wrinkles consists of "ball of incense," wax, fresh oil, and cypress
berries, equal parts. "Crush and rub down and put in new milk and apply it to
the face for six days. Take good heed [of this]"[6] : this would form a good
skin cream, similar to modern preparations. In the Berlin Papyrus, iron
(ben-i-pet) with Nile water [rusty water ?] is prescribed for fever ; iron rust
was also used by the Romans for various diseases.[7] Materials are beaten in
stone mortars, powdered, filtered through linen and sieved.[8] As a solvent
pure water is generally used—Nile water, well water, "rain of heaven," "water
of the bird pond," etc.—but various kinds of beer, milk, oils and even urine
are specified.[9] Drugs are boiled in a cloth[10] (76, 123) and fermented with
yeast (13), and draughts are sucked through a tube (14). A surgical instrument
is of copper, and a rod called "man of metal" (perhaps of copper or bronze) is
used for cauterising (190–194). Liquids are heated by hot stones thrown into
them and the vapours are breathed through a tube passing through the lid
of the vessel (78, 79). There is no reference in Joachim's translation to the
water bath or to distillation. Some recipes specify 17, 18 and even 37 ingredients (51, 132, 145), a taste which, descending from Egypt (and, as will be
seen, Babylonia and Assyria), passed through later Greek, Byzantine and
Arabic medicine almost to the modern times.[11] There are some "universal"
remedies.[12]

No directions are given in the papyrus for the preparation of medicines, and
the book seems to be the manual of a physician rather than of a pharmacist,
the two being probably separate persons.

Dümichen[13] translates the text on the four walls of what he calls the
"laboratory" of the temple at Edfu, dealing with the preparation of unguents.

[1] Berthelot, Arch., 236, 238 f., 242 ;
Maspero, Histoire ancienne, 1905, 91 ;
Loret, L'Égypte, 225 f. ; Woenig, Pflanzen,
364 f., 394 f.
[2] Maspero, H. anc., 91 ; Dawson, Mag.
and Leech, 58, 119 ; Lexa, Magie, i, 73 ;
ii, 108—Isis and Horus ; BMGE, 32 f.
[3] Maspero, H. anc., 91 ; Dawn, 166 ;
Dawson, 113, says it was never used in
medicine.
[4] Maspero, H. anc., 92 ; Wilkinson, ii,
356 ; BMGE, 32 f.
[5] Wiedemann, 1920, 137.
[6] BMGE, 34.
[7] Gsell, 48 ; Pliny, xxxiv, 15.

[8] Joachim, 13, 14, 15.
[9] Joachim, 67, 82, 89, 106, 124, 133, 157,
160, 170 ; Maspero, H. anc., 91 ; ib.,
Dawn, 220 ; Erman-Ranke, 414 f. ; Hurry,
Imhotep, 80.
[10] The numbers in brackets in the text
are the page references to Joachim,
Papyros Ebers, 1890.
[11] Thorndike, H. Magic, i, 172 ; Ganszyniec, PW, xii, 54.
[12] Erman-Ranke, 414.
[13] Ein Salbölrecept aus dem Laboratorium
des Edfutempels ; Z. f. ägypt. Sprache,
1879, Heft 3 and 4, 97–128.

The materials are weighed, boiled over a fire of acacia wood, strained, ground, etc., and the quantities are specified, including fractions. Various resins and plants are named. The process usually takes six days ; a hieroglyphic in these texts may mean "liquid extract" as well as water. In the Ebers Papyrus also (in contradistinction to Assyrian texts, where weights are not usually given), the weights of the drugs are minutely specified, with a notation of small fractions, so that considerable care was used in dispensing.[1] The weight unit was $\frac{2}{3}$ kite = 6·06 gm., and fractions are $\frac{1}{2}$, $\frac{1}{4}$, $\frac{1}{8}$, $\frac{1}{32}$, $\frac{1}{64}$; the volume unit, the water volume of $\frac{2}{3}$ kite, was 0·01 dnat.[2] In the older Kahun Papyrus the quantities are given by volume, not by weight.[3] In the Coptic Period the medical scales were kept in wooden boxes with two divisions.[4]

Among various vegetable seeds and products identified by Joachim [5] are aloes (44, 126), linseed (?) (114), cummin, anise, fennel (besbes) (18), dill and fenugreek (111). Fenugreek is identified by Dawson with a plant called "hairs of the earth", and it is still an officinal drug, although it is used chiefly as flavouring in curry powder and in veterinary medicine : throughout the Middle Ages and in later herbals and dispensatories it is prescribed for the same purposes as in the Egyptian medical papyri.[6] Pomegranate skin was already used as a vermifuge and other bitter materials as endive (qatsut) and wormwood (?) (saam) [7] (122, 142, 180 ; 129 ; 21, 70, 129). Chebu is melilot (164) ; other vegetable materials are currants or raisins (sa-sa), grapes, figs and sycamore-figs (fresh, dried or roasted), and dates (fresh or green, paste or meal—probably date honey)[8] ; also peppermint and coriander (62), "juniper berries of Byblos" (88), saffron of various kinds (from the south, from the north, from the mountains, from the Delta)—perhaps used as a collective name [9], although Ebers identified matet with Chelidonium majus—indigo (21, 157), safflower (23), Carthamus lanatus (senau) (124), mandrake as berries and "flour" from Elephantine (44, 126), opium (4, 14, 64, 112, 114, 120, 131, 143) as chesit-gum, also the plant (25, 74, 133, 153)[10] and as a "soothing drink" (seter-seref) (33, 106, 139), and hyoscyamus (sepet) (108). A curious find in an XVIII dyn. tomb was an oil or ointment containing iron and morphine (un grasso di natura vegetale, continente ferro ed oppio).[11] Tamarisk, calamus from T'abi (Asia), and aloe-wood (122, 142, 180), are aromatic materials, perhaps also cassia (gentet plant ; genti-seeds) (15, 59, etc., 213 ; 36, 51, etc., 213) : Woenig[12] says Cassia fistulæ Willd. ("manna") is indigenous to Egypt ; cassia bark was imported from or through Punt by Hatshepsut.

Oil appears 87 times (204), including "pure" and "white," especially castor oil "pressed from the fruit" (26, 63) ; the plant (qiqi) also yielded fruit (qesebt) and berries (deqm, degam) as well as leaves (55), pith and the skin of the root for medicinal purposes (5, 11, 70, etc.). A "dried oil" (34) is used in "pieces" for salves (perhaps soap ?). Sesame and ben (qebu, baq) oils are also specified (26, 98, 111, 158, 162, etc.).[13]

The fats from definite animals, no doubt supposed to have specific virtues, are prescribed : ox, cow, ram, goat, ass, gazelle, antelope, ibex (Ibex Nubiana)

[1] Dawson, Mag. and Leech, 110, 129.

[2] Hultsch, Gewichte des Altertums, Abh. Kgl. Sächs. Ges. Wiss., phil. hist. Classe, 1898, xviii, No. 2, 59.

[3] Griffith, The Petrie Papyri, 1898, 5 f.

[4] BMGE, IV–VI Rooms, 329.

[5] Page references in brackets in text ; transliterations as in Joachim ; cf. Bryan, 25 f., 32 f.

[6] Mag. and Leech, 112 ; Isis, 1931, xv, 265 ; Erman-Krebs, 69.

[7] Woenig, Pflanzen, 375, identifies saam with Vitex agnus, "Keuschbaum."

[8] Cf. Woenig, 312, 315—dûm palm ; Brugsch, Ägyptologie, 220—"maker of date-sweetmeat" at the court.

[9] Lippmann, Abh., ii, 20.

[10] Budge, Egy. Hierogl. Dict., 563 : khesi, some fruit or plant ; khesait = cassia ?

[11] Schiaparelli, q. in JEA, 1928, xiv, 205.

[12] Pflanzen, 343, 359.

[13] Budge, Dict., 767, qeb-tree oil.

(74, 116) [1], hippopotamus, crocodile, serpent, fish, ostrich, geese ; also tallow (71, 134) and even lard (118, 142), although swine were regarded as unclean.[2]

Various curious oils and fats are oils of worms (144), thersu animal, lion, mouse and cat (116) ; also excrements of the wasp (116) and crocodile (uric acid?) (81). Clement of Alexandria [3] complains of the use of crocodile excrement as a cosmetic in Egypt. An ass's claw (?) in oil is a hair restorer (105, 106) ; blood—including that of a vulture and of a black calf (100, 101, 104, 160)— urine (101, 119, 123, 155, 180, etc.), and "milk of a woman who has had a son" (23, 89, 92, etc.) appear. Numerous parts of animals, including blood and brains, and their excrements are frequently specified[4] ; hartshorn is very frequent.[5] Spanish flies (Cantharides) were apparently known in Egypt.[6] The fats from animals which were rare or difficult to obtain (lion, oryx, lizard, etc.) were probably supposed to have special virtues, but the pharmacist very likely simply made up these prescriptions with goose grease, which is given in the Ebers Papyrus.[7] It is noteworthy that raw liver is prescribed in the old Kahun Papyrus.[8]

Among the metals and minerals in the Ebers Papyrus [9] copper, lead, iron and electrum (asem) are specified in Joachim's translation (to which the numbers in brackets in the text refer), but gold, silver, zinc, tin and mercury are not. Asem is mixed with myrrh and yeast as a salve (160), which seems a peculiar use of a malleable metal. Five kinds of "metal" [probably "mineral"] are specified in a recipe from "an Asiatic of Byblos" (99), viz. abennu-mineral [really ybn = alum], anch-mineral[10], netri-tit-mineral, hunnut-mineral[11] and asem-metal ; other constituents of the recipe are "granite" (?) and (possibly) powdered sulphur as a constituent of an eye-salve (90). The anch-mineral (160), the netri-tit-mineral (64) and asem (160) are again mentioned. Netri-tit occurs both as "metal" (?) and "grains" : Brugsch translated it "divine excrement" (65, 135) : Lippmann[12] suggests "residues from temple services," but perhaps it means incense. No attempt is made to identify these "metals" by Joachim, Berthelot or Lippmann, but the latter suggests that, since "sulphur is also given as a metal," these may not be metals at all but mineral products. The translation "sulphur," however, seems doubtful, and the five "metals" exclude sulphur and granite. Sulphur is said to be called hunnut-mineral once as a constituent of an eye-salve (90). The netri-tit-mineral is perhaps[13] natron, or native carbonate of soda, but natron is a mere guess from a comparison with netri. Natron often appears specifically, including red natron (15), natron of the North (140), natron of the South (144) and bedet (120, 157, 179), used in solution to drive away vermin. A material translated "green earth of lead" ("grüne Bleierde") by Joachim[14] is mentioned with lapis lazuli.[15] The name in the papyrus is hntj, or chenti, some medicine for the eyes or nose and perhaps malachite[16], although chenti is also called "seeds" and described as a material for "mortar" (68, 139, 189), perhaps a paste with water, and a "deposited green earth of lead" (82) may also be washed malachite.

[1] Loret, L'Égypte, 91.

[2] Loret, 97.

[3] Pædagogus, ii, 11 ; iii, 2 ; Dioskurides, ii, 98.

[4] Berthelot, Arch., 238, 239.

[5] Dawson, 110 f.

[6] Parthey, Isis und Osiris, 1850, 268.

[7] Dawson, 65, 66.

[8] Griffith, The Petrie Papyri, 1898, 5 f.

[9] Berthelot, Arch., 240 ; Bryan, 19 f. ; Lippmann, Abh., ii, 4–14.

[10] Anch was a personal name in the

v dyn. ; Lieblein, Handel und Schiffahrt auf dem roten Meere, 1886, 13.

[11] Hunnut, "a kind of metal," Budge, Egy. Hierogl. Dict., 487.

[12] Abh., ii, 12, 17.

[13] Lippmann, 13 ; cf. Wiedemann, A. Rel., xxi, 442 ; xxvi, 332.

[14] See his Index.

[15] Joachim, 95.

[16] Von Oefele, in Lippmann, Abh., ii, 7 ; in the Hearst Papyrus hnty is an "earth," Wreszinski, 1912, 79.

Copper compounds are said to have been used internally, as emetics, in Egypt.[1]

Copper is used as filings for plasters ; lead in fragments with cat's and dog's excrements for the same purpose (126). Scoria of copper (155) [2] ; verdigris (?) (50, 89, 90, 110, 135, 160, etc. ; probably malachite; once "verdigris of the ship" (?), 50) ; perhaps blue vitriol (gesfen) (92)[3] ; chrysocolla (64, 82, 87, 92, 103, 112, perhaps malachite) ; "real" chesbet (lapis lazuli) (91, 95) ; salt (frequently), sometimes mountain (? rock) salt (146), sea salt (31, 172) and salt of the North (116) ; the sa-grains of Upper Egypt are perhaps fine-grained crystalline salt, or saltpetre (81, 85, 144).[4] "Minium" (134, 168) is sometimes used internally and thus could not have been red lead : it is perhaps $\mu i \lambda \tau o s$, i.e. ochre ; Berthelot [5] thinks, improbably, that it is sometimes litharge. A black "knife-stone" ("schwarzer Messerstein") (90, etc.) suggests obsidian, which was used for knives. Berthelot [6] identifies it, improbably, with magnetite. Stibium (87 ; probably mestem, not "stibium") is a collyrium (198), including "male" (87, 95) and "real" (87, 89) ; "alabaster" frequently occurs as flour for cosmetics (114, 155, 157, 158), but in one place it is doubtful if iron or alabaster is meant (90). Berthelot [7] thinks alabaster might sometimes mean oxide of antimony. Other minerals are gypsum *or* mica (92) (perhaps selenite) ; crocus, perhaps orpiment [8], but more likely yellow ochre ; hæmatite (92, 115, 120, 158)—"granite" proposed by Joachim and Lieblein (92) is probably incorrect[9] ; red jasper for stopping bleeding (170, 177), a "sympathetic" amulet[10]; uat-stone for eye pigment (65 : malachite) ; powdered flint (?) for tooth powder (161 : chalk would be better for the teeth) ; charcoal (123, 179), "charcoal" of walls (144 ; ? lime plaster) ; soot (20, ? containing sal ammoniac) ; "cadmia" (as an eye-salve, perhaps zinc oxide) (89, 90, 94, 138, 173) ; "stone of Memphis"[11] (18, 19, 22, 23, 102, 103), and petroleum[12] (15, 168, 170). Although petroleum and asphalt were known, there do not appear to be any representations of the use, as in Assyria, of incendiary weapons.[13]

Joachim's "lead alum" (17, 20) is perhaps the material called sory ($\sigma \omega \rho \upsilon$), which Dioskurides[14] says is found in Egypt, whilst a similar material, misy ($\mu i \sigma \upsilon$), was found in Cyprus ; it was probably a basic iron sulphate.[15]

In all probability a large number of the identifications just given are incorrect, and the Ebers Papyrus awaits a critical study.

Sprengel[16] and von Oefele[17] suspected a system of cover-names (urine, fæces, etc.) in the *old* Egyptian medical papyri, whereas Wiedemann[17] thinks cover-names first appear in the Greek Period. Budge[18] suggests that the names of plants, etc., in the Ebers Papyrus ("of which no satisfactory translation has been made") are ancient native names known throughout the Nile Valley, but not to the later Dynastic Egyptians. Recipes from foreign sources include plants from Crete (Keftiu)[19]; there is a formula in the Cretan language

[1] Von Oefele, in Puschmann, Gesch. d. Med., 1902, i, 87.

[2] Berthelot, Arch., 240 f.

[3] Budge, Egy. Hierogl. Dict., 813 ; gesfen, only in Ebers Papyrus, some kind of seed ?

[4] Bryan, 22.

[5] Arch., 240.

[6] Arch., 240 f.

[7] Introd., 238 ; Arch., 240 f.

[8] Berthelot, Arch., 240 f.

[9] Lippmann, Abh., ii, 8.

[10] Lepsius, Met., 38 ; Dawson, Mag. and Leech, 8.

[11] Dioskurides, v, 158, says it is "fatty"

and of various colours ; Lippmann, Abh., ii, 14, identifies it with asphalt.

[12] Berthelot, Arch., 243.

[13] Cf. Wilkinson, i, 244.

[14] v, 116, 119 ; cf. Galen, de Simpl., ix.

[15] Berthelot, Intr., 242 ; Lippmann, Abh., ii, 6 ; Adams, Paulus Ægineta, 1847, iii, 253, 367, 399, Arabic, zaj, vitriol ; still used in the East and known to Boerhaave, Elementa Chemiæ, Leyden, 1732, i, 54.

[16] H. Med., i, 46.

[17] In Puschmann, i, 87.

[18] Herbalist, 11.

[19] Evans, Palace of Minos, ii, 748.

in the London Papyrus for exorcising a disease [1], and one recipe in the Ebers Papyrus is due to "an Asiatic [Phœnician] of Byblos."[2] Probably drugs imported by Phœnicians were used.[3] Maspero thought a transcription in hieroglyphics of a cuneiform conjuration occurs: Semitic words occur repeatedly in other texts of this period (XVIII–XIX dyns.), and the Phœnician Astarte is mentioned in the Amherst Papyrus IX (XIX–XX dyn.).[4]

Horapollo [5] says anemone flowers stood in Egyptian for "disease." Late classical authors (4th century A.D.) imagined that the hieroglyphics, when knowledge of them was lost in Egypt, were still known in Ethiopia, and that there were books on plants and animals in which they had secret names. Plutarch and Neoplatonic authors give such cover-names: ivy was "plant of Osiris" ($\Sigma\chi\eta\nu\delta\sigma\iota\rho\iota\varsigma$); verbena "tears of Isis"; a kind of lily was "blood of death"; a kind of artemisia was "heart of Bubastis"; saffron was "blood of Herakles"; and the squill was "the eye of Typhon."[6]

The blood of a black cow or black calf, containing the "blackness" of the animals, was used in ancient Egypt to dye grey hair black.[7] Recipes in the Ebers Papyrus deal with forcing the growth of the hair, dyeing grey hair— "formula of Lady Ses, mother of King Tetá"[8], and for the teeth, nose, eyes and ears. Vermin are killed ("he creeps not") or expelled from houses by natron water and mice by cat's grease.[9] Red stones (carnelian, jasper) formed amulets supplying the place of the blood, in which the life resided, for the mummy.[10]

Some of the formulæ in the later part of the Ebers Papyrus are of the chimerical character appearing later in Pliny, Dioskurides, the Geoponica and Marcellus Empiricus of Bordeaux[11]: brains, excrements, parts of animals and reptiles ("blood of the lizard," "scarabs," etc.); a double sense is possible, but is without positive proof.[12]

Near the end of the Ebers Papyrus is a book on anatomy[13], describing "vessels," often imaginary, various parts of the body and a kind of pathological theory. The Book of Usaphais (King Semti, I dyn.), "found in the temple at the feet of the statue of Anubis, in the town of Letopolis"[14], is a theory of vessels, with remedies. The vital spirits, conveyed like air or breath along the vessels, enter by the right ear or nose, mingle with the blood, are distributed to all parts of the body, and are collected and redistributed by the heart, "the perpetual mover."[15] The Ebers and Brugsch Papyri describe the air vessels, and distinguish between good and bad air, air of life and air of death, supposed to circulate differently, perhaps inspired and expired air[16], but there is no notion of a circulation of the blood.[17]

The Ebers Papyrus is practical and on the whole has a scientific form[18]; in particular it shows no trace of the so-called Iatromathematics, an alliance of

[1] Glotz, Ægean Civ., 385; Gordon, JEA, 1932, xviii, 67.
[2] Maspero, Struggle, 495; Erman-Ranke, 414; Berthelot, Arch., 237.
[3] Neuburger, H. Med., i, 27.
[4] Maspero, op. cit.; Wiedemann,1920, 88; Newberry, Amherst Papyri, 1899, 14, 47.
[5] Hieroglyphica, ii, 8.
[6] Sprengel, H. Med., i, 51; Plutarch, Isis and Osiris, c. 62; Lobeck, Aglaophamus, Königsberg, 1829, 887; Schmidt, PW, iii, 2106; Berthelot, Intr., 11, quotes several such curious names of plants from Dioskurides.
[7] Wiedemann, 1920, 137; use of blood, Dawson, Mag. and Leech, 9 f.; such recipes were used in the Arabic Period.
[8] Joachim, 106.
[9] Joachim, 179; Berthelot, Arch., 237.
[10] Dawson, 8.
[11] De medicamentis liber, ed. Helmreich, Leipzig, 1889, 299, 310, 339, 352 f., 378, etc.
[12] Berthelot, Arch., 236 f.
[13] Joachim, 180 f.
[14] Joachim, 185; Berthelot, Arch., 233, 238; Budge, Mummy², 25.
[15] Maspero, Dawn, 216 f.; Dawson, 92 f.
[16] Neuburger, H. Med., i, 25; Dawson, 6 f., 93.
[17] Peet, CAH, ii, 220.
[18] Erman-Ranke, 261; Berthelot, Arch., 233 f.

medicine and astrology which appears in the Hellenistic Period in works attributed to Hermes Trismegistos.[1]

THE HEARST PAPYRUS

In the Hearst Papyrus, which is only half complete and with the chapters in a disconnected order, the treatment of diseases resembles that in the Ebers Papyrus, but the arrangement is different. Massage, purgatives and methods of treatment are interspersed with prayers and magical incantations. Diseases were caused by the anger of the gods, or the presence in the body of one or more evil spirits, which incantations could expel. There is an attempt at grouping of prescriptions, the composition of some being attributed to gods, and of others to kings of the earliest dynasties. Many remedies are like those of the Ebers Papyrus, containing milk, saliva, urine, excrements, worms, insects, horn, gall, etc., but some are useful and there is an indication of the beginning of scientific order.[2]

Materials mentioned in the Hearst Papyrus [3] are : honey (61, and often), salt (often, including "salt of the North," 74, 96 f., 105, 128), natron (74, 99, 102) and red natron (105), bedet (63), sycamore gum (64), sweet beer (67, 70, 89), pig's blood in wine (68), myrrh (71), incense (71), degam (Ricinus) (73), dill (75, 85), yeast of date wine (82), yeast of sweet beer (97), mandragora (95, 127), wax (95), "stibium" (sic ; 95, 127), "mennig" and "fresh mennig" (this would be red lead, which is improbable—ochre is meant, 112, 130), residue ("Abfall ?") of coppersmith (109), sory (112). Medicines are made for Rā by Tefnut or Isis (82, 83).

EGYPTIAN SURGERY

The Egyptians were probably good operative surgeons.[4] Cupping was practised by means of cow's horns sawn off near the point, as in later Egypt and India.[5] Although an undisturbed tomb of a physician of the Middle Kingdom was found at Beni Hasan, there were no surgical instruments or special apparatus in it [6], but Egyptian medical and surgical instruments (knives, scarifiers) are known, and are represented (balance, cupping glasses, etc.) in reliefs, especially those in the temple at Kom Ombos, which are actually of about 100 B.C. but are copies of much older drawings.[7] Early Egyptian surgery is known from the Edwin Smith Papyrus.

THE EDWIN SMITH SURGICAL PAPYRUS

This papyrus is an incomplete copy, made in the 17th century B.C., of an original by an unknown author of about 3000–2800 B.C. The vocabulary and style are very archaic and there is a very detailed and valuable commentary of 69 glosses, probably added in 2500 B.C., in explanation of words or expressions which had then fallen into disuse.[8] The papyrus, discovered at Thebes in the same tomb as the Ebers Papyrus [9], is beautifully written in black and red.

[1] Ideler, Physici et Medici Græci Minores, 2 vols., Berlin, 1841–42, i, 387, 430 ; Riess, PW, ii, 1808 f. ; Bezold, Boll and Gundel, Sternglaube und Sterndeutung, Leipzig, 1931, 54, 129, 134, 207 ; Berthelot, Arch., 233.

[2] Peet, CAH, ii, 219 ; Dawson, 76 f. ; Budge, Syriac Med., i, p. cxxxi ; ib., Herbalist, 26 f. ; Maspero, New Light on Ancient Egypt, 1908, 266–271.

[3] Wreszinski, Der Londoner Med. Papyr. und der Hearst Papyr., 1912, page references to this in text.

[4] Baas, Outlines of the Hist. of Med.,

New York, 1899, 15 ; Budge, Syriac Med., i, p. cxxxiii ; Neuburger, i, 26 f.

[5] Stapleton, Mem. Asiat. Soc. Bengal, 1927, viii, 362 ; plate in Alpinus, De Medicina Ægyptiorum, Venice, 1591, 63 v.

[6] Erman-Ranke, 410.

[7] Meyer-Steineg and Sudhoff, 28 f., Figs. 16–18 ; Singer, Short History of Medicine, Oxford, 1928, 40 ; Feldhaus, Technik, 704, bronze knife of 1500 from Thebes.

[8] Breasted, The Edwin Smith Surgical Papyrus, Chicago, 1930, pref. xiii ; 9, 73, 595 ; 10, 61 ; references in text are to this.

[9] Ib., 20 f.

The commencement is lost and the copyist broke off for some reason before it was finished. The present form comprises 17 columns of 377 lines ; the title is not known, since although Breasted thinks it is probably the "Secret Book of the Physician" named in the Ebers Papyrus, this is doubtful.[1] The subject-matter is surgery, arranged in the form *de capite ad calcem* (from head to heel) so popular in the compendia of the Middle Ages [2], and the fragment contains 48 cases of injuries and wounds from the bones of the skull as far as the bones of the spine, where the text breaks off.[3] The crisis of a disease is recognised.[4] The knowledge shown is surprising ; the papyrus contains the oldest known description of the skull and its sutures, of the brain ('ysh), called "the marrow of the skull," and of its convolutions ; and of the disturbance of the functions of specific parts of the body caused by injuries to certain parts of the brain.[5] The heart is the organ which drives the blood through the arteries to the various parts of the body and the rôle of the pulse is well observed and described in the 26th century glosses ; the first Chinese references to the pulse go back only to the 2nd–3rd centuries A.D. [6], and the earliest Greek references are in the Hippokratic works and Demokritos.[7] The timing of the pulse could not have been well carried out, since the shadow-clocks of the 13th century B.C.[8] would not serve, and water-clocks are a later, Greek, invention.[9] Each case has a description of the observation, then the making of the diagnosis, good, doubtful or bad (when the physician would not take the case, "not to be treated"), then the treatment, mostly by bandaging and treatment of wounds.[10] The materials described are much fewer than those in the Ebers Papyrus, which is more medical. Wounds are closed by adhesive plaster, held by clamps and loops, and stitched (the first description) (54, index 589), with previous treatment with cooling, disinfecting and astringent substances. The bandages are of linen, as used on mummies ; wooden splints wound with linen were found by Elliot Smith on two Old Kingdom mummies (8, 57 f., 98). They were used dry or as lint for drainage (9, 53, 57, 316) or soaked in solutions, or in glue and plaster so that they would harden (8). Instruments for cauterisation and opening abscesses include the "hot point," also in the Ebers Papyrus, the fire drill made hot by friction (9, 365). Although metal instruments must have been used, perhaps with wood and stone, they are not specially named (9, 53), and copper is the only metal mentioned, the wrinkled slag (or oxide ?) on the surface of molten copper before pouring into the stone mould being compared with the markings on the skull (10, 65, 167, 173, 174, 208). The "green pigment" (w'dw ; wod) for the purification of infected and necrotic wounds is malachite (9, 59, 382, 411). Black mestem (msdmt ; Coptic stem) is used for treating eye diseases and for blackening the lashes, and is perhaps stibnite (489), but more probably galena. Northern salt (common salt from the N.W. Delta) is used for wounds (9, 59, 379, 383, 411, 491). Other minerals are common and red neter (sodium carbonate) (9, 59, 412, 491), a "glassy mineral" (382 ; hardly "vitriol," perhaps selenite), alabaster kernels (491), mason's mortar for "cooling" (60, 407, 410, 412 ; also in the Ebers Papyrus, but rare), and imru (ymru), which, from the archaic determinative used with it, was a mineral substance, used sometimes alone and sometimes with honey as a disinfectant and styptic (57, 98, 101, 264, 336, 352, 357, etc.). Lippmann[11] has plausibly identified imru with alum.

[1] Breasted, The Edwin Smith Surgical Papyrus, Chicago, 1930, 10, 33, 65.
[2] Dawson, Magician and Leech, 99.
[3] Breasted, pref. xvii ; 6, 14, 33.
[4] *Ib.*, 9.
[5] *Ib.*, pref. xiv, xv ; 10, 12, 65, 166.
[6] Sarton, *Isis*, 1931, xv, 364.

[7] Sarton, *ib.* ; Breasted, pref., xvi ; 13, 64, 104.
[8] Breasted, Ancient Times, 1916, 91 ; Thothmes III.
[9] Breasted, Edwin Smith Pap., 104, 106 ; *ib.*, Anct. Times, 359, 467.
[10] Breasted, E. Smith Pap., 6 f.
[11] *Chem. Z.*, 1931, lv, 933.

Fresh meat is used to heal wounds (9, 57, 98, etc. ; index) and fresh ox-dung for drawing out inflamed places and for cooling (60, 411) ; milk and cream (60, 287) are (doubtfully) named. Ointments often used are made with animal (e.g. ibex) grease or vegetable oils and honey (perhaps wild) (57, 96, 98, 146, 379, 411, 488, etc. ; index). Infusions of plants include that of the bark and leaves of willow, which is rich in strongly antiseptic salicin (9, 59, 60, 379, 380), also of acacia and sycamore leaves, containing much tannin (60, 379, 411). The many plants named include (380 f., 411 f., 422, 489) the carob (383), coriander (422), and incense (482, 488, 490). A "sweet beer" (488) may, according to Lippmann, be wine from grapes, but is probably fresh, as opposed to sour, beer. The unit of measure is the hekat (300 cu. in.) (489). The papyrus contains the first mention of pest-laden winds (470). There is no dental surgery or tooth-stopping, which were probably unknown in ancient Egypt.[1]

At the beginning of many prescriptions in the Edwin Smith Papyrus the word *ky*, "otherwise," occurs (476), resembling the ἄλλως or ἄλλο of the Greek papyri.[2]

At the end of the incomplete text of the Edwin Smith Papyrus, the 17th century B.C. copyist has added extracts from two works on magic, entitled "Incantations for Driving Out the Wind in the Year of Pest," and "The Book for Transforming an Old Man into a Youth of Twenty," the second containing the earliest known prescriptions for this favourite speciality of later quacks.[3] Magic formulæ are thus supplied in a separate appendix, probably because they were regarded as useless in surgical cases [4], although Lippmann suggests that they were for use only with such patients as required them. Evil demons, the goddess Sekhmet and the god Set, flies (sometimes swallowed) (482), pest of the year (an epidemic), etc., can be scared off by the formulæ or the help of Isis, Osiris and other favourable gods and spirits enlisted (14, 19, 59, 73, 411 f., 469, 501 f. ; 220).[5] Some superstitious practices occur in the preparation of drugs and ointments, as when the addition of the powder of the (very hard) shell of an ostrich egg helps the cicatrisation of wounds (219 f.), or when the activity of a medicine conferring youth is enhanced by keeping it in vessels of costly stones such as lapis lazuli, jasper or possibly "alabaster or aragonite" (495 : aragonite is unknown in Egypt). The transformation of the old man into a youth, a recipe added in the 17th century, depends on the use of face cream.[6]

THE LONDON PAPYRUS (10,059)

The contents of this papyrus (which is a badly damaged palimpsest) are mostly spells and incantations, but some materials are mentioned [7]: "mennige" (177, 209 ; literally red lead, probably ochre), salt of the North (177), a quartz bead for a spell (196), safflower (207), "tree oil" (207), "stibium" (*sic*, 209), copper scales and "green stone" (210 ; malachite ?). A "white lead" (perhaps tin ?) (184) is not otherwise known.

THE BERLIN PAPYRUS (3038)

This papyrus, first described by Brugsch in 1853 and published in lithographed facsimile by him in 1863 [8], very frequently mentions emetics, clysters, salves,

[1] Dawson, Magician and Leech, 101.
[2] Lagercrantz, Papyrus Græcus Holmiensis, Uppsala, 1913, 96 ; Dieterich, Abraxas, Leipzig, 1891, 189, 193.
[3] Dawson, 99.
[4] Dawson, 74, 100 f.
[5] Gods in Ebers Papyrus, see Joachim, 59, 60, 174, 185.
[6] Lexa, Magie, i, 30.
[7] Wreszinski, Der Londoner Med. Papyrus, Leipzig, 1912, refs. in text are to this.
[8] Recueil des Monuments Égyptiens, 1863, ii, pl. LXXXV–CVII ; explanatory text, 101–120 ; photographic reprod., transcription, etc., in Wreszinski, Der grosse Med. Papyr. des Berliner Museums, 1909, references in text are to this ; Erman-Krebs, Aus den Papyr. der Kgl. Mus., 1899, 63 f.

and fumigations ; the prescriptions are often made up with honey and beer, including sweet beer (49 f., 97, etc.), date wine (78), sea salt (54, 57, 64, 72, etc.), ball of salt (70), natron and "pure" natron (62, etc.), also natron with tallow (soap ?, 79), "stibium" (74), and many vegetable products not identified, one being "hairs of the earth."[1]

THE COPTIC AND GREEK MEDICAL PAPYRI

The long anonymous Coptic Papyrus, found in 1892 at Meshaîkh and now at Cairo [2], contains old Egyptian material but also information from Greek (Dioskurides, Alexander of Tralles) as well as Arabic (Avicenna) sources. The date is supposed to be mostly 9th–10th centuries A.D., but Avicenna is 11th century. It is a collection of 237 prescriptions, some claiming to be new, and is free from spells and incantations. Coptic medicine, known also in other fragmentary works, usually followed the old Egyptian.[3] The Coptic Papyrus, which is sometimes difficult to read on account of the use of peculiar contractions and unusual signs, deals mainly with diseases of the eye, and mentions a large number of materials, including : Armenian borax (paurak armenei) (ll. 24, 26 ; p. 89 f.)[4] ; sal ammoniac (alos ammoniakon ; nousater) (ll. 24, 47, 106, 168, 246, 267, 357, etc. ; pp. 89, 91, 159, 193, 238, 257, 295, etc. ; see full Coptic and Greek indexes) ; quicklime (asbestos) (ll. 293, 343 ; pp. 269, 291) ; gum ammoniac (l. 209 ; p. 223) ; alum (obne, oben) (ll. 18, 281 ; pp. 64, 265), including "liquid" alum (l. 338 ; p. 290), and "round" alum (l. 67 ; p. 134), the latter called obne esishp ; yellow and red arsenic sulphides (arseni-kon, arsenigon, santaraches) (ll. 22, 23, 42, 118, 293, etc. ; pp. 81, 88, 105, 170, 269, etc.), perhaps also as "red" or "native" *sulphur* (l. 70 ; p. 142 f. ; ? alchemical) ; azurite (armenion) (l. 220 ; p. 226) ; antimony sulphide (stimeos, kalliblepharon) (ll. 27, 84, 199, 205, 209, 212, 290, etc. ; pp. 91, 153, 219, 222, 223, 225, 268, etc.) ; copper scales (lepitos chalkou) (ll. 51, 158, 251, 283, etc. ; pp. 128, 190, 240, 266, etc.) ; burnt copper (ll. 155, 178, 180, 205, 257 ; pp. 189, 204, 222, 243, etc.) ; verdigris (usu [*los*]) (ll. 165, 306, etc. ; pp. 192, 273, etc.) ; lead (molebos) (l. 339, p. 290) ; burnt lead (tathe efroch) (ll. 349–351 ; pp. 293 f.) ; litharge (ll. 68, 70, 250, etc. ; pp. 137, 139, 240, etc.) ; minium (silikon [σίρικον]) (ll. 68, 69, 180 ; pp. 137, 138, 204) ; white lead (psimuthion) (l. 120 ; p. 170, etc.) ; hæmatite (l. 81 ; p. 150) ; mercury (Coptic, "water of silver," moou nchat) (l. 224 ; p. 230), "killed" by trituration with vegetable acids ; cinnabar (kennabereos) (l. 53 ; p. 130) ; marcasite (marka shithe) (l. 102 ; p. 158) ; soda (nitron) (ll. 148, 170 ; pp. 187, 195, etc.), including Arabic nitron (sosm narabikon) (l. 135 ; p. 176) ; common salt, including mountain salt (l. 306 ; p. 273) and andrani salt (salt of Al-Anderâ in Syria) (l. 111 ; pp. 166, 168) ; tutia (thoutie, thouthia ; oxide of zinc) (ll. 27, 84, 88, 111, 116, 162, 173, etc. ; pp. 91, 95, 153, 154, 166, 169, 191, 196, etc.), and cadmia (kadmias, katmias, ketmie, aklemia, i.e. zinc oxide) (ll. 60, 73, 102, 106, 120, 157, 178, 199, 215, 220, 254, 257, 267, etc. ; pp. 131, 146, 158, 159, 170, 190, 204, 219, 225, 226, 240, 243, 257, etc.), including calcined and washed (l. 212 ; p. 225), and "golden" (with an Arabic symbol for gold) (ll. 82, 99 ; pp. 150, 157, cf. p. 164) ; the Alexandrian-Coptic names for tutia are supposed by Chassinat (98) to have been akrenken and oikobin. Various kinds of vitriols are named, including copper vitriol (kalkanthos, kalkanth) (ll. 18, 51, 60, 97, 120, 240, etc. ; pp. 64, 126, 131, 156, 170, 235, etc.) ; "stepterias," which, however, is probably alum (ll. 53, 186 ; pp. 130, 206) ; "white vitriol" (?)

---

[1] Erman-Krebs, 69 ; see p. 186.
[2] Chassinat, *Mém. Inst. Franç. Orient. Cairo*, 1921, xxxii, pp. 392 + 20 plates ; short summary in Dawson, *P. Roy. Soc. Med.*, H. Med. Sect., 1924, xvii, 51 ; *ib.*,

Magician and Leech, 1929, 84 f.
[3] Chassinat, pref. ; Lexa, Magie, i, 146 ; ii, 181 f. ; Dawson, *Proc.*, 52.
[4] All references are to lines and pages in Chassinat, 1921.

13

(chalkiteos ; probably not white vitriol, zinc sulphate, but crude green vitriol) (ll. 51, 199, 205, etc. ; pp. 126, 219, 222) ; "yellow vitriol" (miseos ; impure iron sulphate) (ll. 53, 122, 270 ; pp. 130, 172, 260) ; "green vitriol" (k[alkanthos] amgam, i.e. anikam) (ll. 60, 150, 236 ; pp. 131, 187, 233, 330), with blue, white and yellow vitriols in the same recipe (l. 279 ; p. 265). Opium (opion) is very often specified (ll. 20, 73, etc. ; pp. 74, 146, etc.).

Distillation is perhaps described (l. 70 ; p. 140), and heating substances in an oil-flask suspended in a stone (earthenware ?) pot, perhaps in an apparatus similar to those figured in the Greek chemical MSS. (pp. 196, 220 f., and Figs.). Some cover names appear to be used, as when orpiment (?) is given as "yolk of egg," with the symbol ⊙ (gold or the sun in Greek texts) for egg (l. 109 ; p. 162). Some prescriptions contain animal materials, as in the old papyri (p. 214 f.). The publication of a Coptic treatise on alchemy is promised by Chassinat, whose notes on the treatise just dealt with are very copious and interesting : the first copy of his manuscript went down with a torpedoed ship.

Some Egyptian medical papyri found at Oxyrhynchus were written in Greek. In one of these (early 1st century A.D.), calamine (καδμήας), white lead (ψιμιθίον), antimony (στίμεως), copper scale, "purified schist" (λίθος σχι[στος] πεπλ[υμένος]), saffron (κρόκος), opium (ὀπίον), cantharides, yellow orpiment (ἀρσενικὸν), aphronitron, and rock alum (?) (χαλκίτιδει; as a styptic) are specified.[1]

WOODS

Egypt in the Dynastic Period was poor in timber, but remains of fossil wood show that there was at a very early period a great forest in what is now the desert to the east of Wâdî Ḥalfah.[2] In the Old Kingdom the chief trees were the date palm, sycamore, acacia and tamarisk, none of which gave good timber. The common wood (khet) [3] used for building, furniture, etc., was acacia (at least as early as 2700) or sycamore, the better woods such as ebony (heben ; iban ; hbry) (from Nubia) and cedar (from Syria, probably Lebanon, in the Old Kingdom) being imported.[4] Cheaper woods (also stone) were often stained (XVIII dyn.) (usually on stucco) or veneered to imitate expensive foreign kinds. The identifications of the woods, except date palm, acacia and sycamore, are doubtful. Remains of several species of acacia, tamarisk, dûm palm, olive, and pomegranate wood, all from trees which grow in Egypt, occur in tombs [5], and there is a fine cypress or cedar coffin, bound with copper, of the XI–XII dyn.[6] Aahmes I (1580–1557) built a temple barge of cedar ; Thothmes I (1520) says he decorated a temple with cedar staves tipped with electrum ; Thothmes IV and Seti I received it as tribute from Lebanon but Rameses IX (1142–1123) had to send an expedition to Syria to beg for it.[7] Sawdust in a mummy packing of 1000 B.C. was from the Atlas cedar (in Morocco) and not from the cedar of Lebanon : the Gaunches of the Canary Islands mummified their dead by methods very like those of the XXI dyn. Egyptians (1090–945).[8]

Ebony, an offering to the gods, was also used in medicine and (inlaid with ivory) for Predynastic furniture ; it came from Punt and the Sudan.[9] Wilkinson[10] refers to a walking-stick of cherry wood from Egypt in the British

[1] A. S. Hunt, The Oxyrhynchus Papyri, London, 1911, Part viii, 110 f. ; Dawson, Magician and Leech, 87, 123.
[2] BMGE, IV–VI Rooms, 135.
[3] Budge, Egy. Hierogl. Dict., 566.
[4] Breasted, H., 95 ; Honigmann, PW, xiii, 1 ; Kanngiesser, A. Nat., iii, 82 ; Budge, Dict., 142 ; Wilkinson, i, 352, 356, 376, 378, 402, 411 ; ii, 194 f. ; Lieblein, Handel, 1886, 35, 71 ; Erman-Grapow,

Handwörterbuch, 1921, 100; Wiedemann, 1920, 215, 338.
[5] Wilkinson, loc. cit.; Lucas, Egy.Mat., 27 f.
[6] Maspero, Guide to the Cairo Museum, 1908, 25.
[7] Breasted, H., 168, 252, 265, 328, 410, 513f.
[8] Mackenzie, Footprints, 177.
[9] Wiedemann, 1920, 339 ; Breasted, H., 39, 127, 136, 276, 277.
[10] ii, 352.

Museum : Pliny [1] specially says the cherry would not grow in Egypt, and it is not now found there. "Cinnamon wood" is among the materials brought from Punt by Hatshepsut.[2]

Small tubs were used as measures for grain. Polished wood is frequently found in tombs. Carpenters' tools of bronze with wooden handles included the axe, adze, handsaw, chisels, wooden mallet, drill, two kinds of planes, ruler, plumb-line, square, nails in a leather bag, hone, and horn of oil. Dovetailing was used in the early period. Boards were joined by flat pins inserted horizontally in the edges, with round pins then driven through these from the flat of the board, and glue was used.[3] Papier mâché was a substitute for wood : some New Kingdom coffins are of sheets of linen glued together, whilst in the Greek Period papyrus was used.[4]

DRINKS

The usual drink in Egypt was Nile water, kept in skin (leather ?) bags and cooled in large porous earthenware jars.[5] Opinions differ as to its purity. Von Oefele [6] says it is so pure that it can be used at present in chemical works instead of distilled water. Plutarch [7] reports the Egyptian opinion that it was sacred yet fattening, and water was, in fact, a divine element.[8] Parthey [9] says Nile water is, if anything, purgative and is very pure : 4·89 hektograms, according to Regnault, contain only 5·4 centigrams of foreign matter. Benjamin of Tudela (A.D. 1173) says it was medicinal.[10] It is generally foul, of a muddy colour, and full of bacteria[11], but its quality varies according to the season, and it is sometimes limpid.[12] Heliodoros[13] and Achilles Tatius[14] say it was clear and sweet when drunk direct from the river with the hand, in Egyptian fashion. Whether it was clarified with alum, etc., is not stated. Maqrîzî[15] says this was added and the water filtered. The only water found in an Egyptian tomb is a saline solution in the tomb of Hetepheres (c. 3000), the mother of Cheops.[16] There are mineral springs near Cairo; the only spring of sweet water in the Nile Valley is near Heliopolis (Matarîya).[17]

Milk (àar-t), grape wine and beer[18] were also drunk in Egypt from early times, beer being regarded as an invention of gods. In Egyptian texts, goddesses make beer for the king.[19] Herodotos says the Egyptians drank "a wine from barley"; the Emperor Julian wrote an ode on this drink.[20] Diodoros Siculus[21] says that Osiris taught men to make a drink called xythus, of barley, which was little inferior in taste and strength to wine itself, and Strabo[22] calls the drink zythus. This was beer.

BEER

Beer, really kwas[23], was made from the earliest times both in Egypt and Babylonia ; the accidental similarity between its names, ḥqt or ḥenq in Egypt, hîqu in Babylonia[24], led Meyer[25] to suppose early relations between the two

[1] xv, 25.
[2] Breasted, H., 276.
[3] Wilkinson, ii, 204 ; 40 ; 195 f.
[4] Erman-Ranke, 544.
[5] Erman-Ranke, 225 ; Wiedemann, 1920, 295.
[6] Puschmann, Gesch. der Med., i, 84.
[7] Isis and Osiris, 5 ; Wilkinson, iii, 93.
[8] Budge, Herbalist, 20 f.
[9] Isis und Osiris, 161.
[10] Itinerary, ed. Adler, 1907, 72.
[11] Borlase, Antiquities of Cornwall, 1754, 234 ; Wiedemann, 1920, 297.
[12] Maspero, H. anc., 1905, 2 ; ib., Dawn, 22 f. ; Wilkinson, ii, 428.
[13] Æthiopica, ii ; tr. R. Smith, 1889, 54.
[14] Tr. R. Smith, 1889, 434.

[15] Tr. Bouriant, 174 f., 182.
[16] Meyerhof, q. in Isis, 1928, xi, 215.
[17] Wiedemann, 1920, 16.
[18] Herodotos, ii, 60, 77.
[19] Wiedemann, 1920, 297, 299 f. ; Budge, Egy. Hierogl. Dict., 21.
[20] J. Harrison, Prolegomena to the Study of Greek Religion, Cambridge, 1903, 416.
[21] i, 18.
[22] XVII, i, 14 ; 797 C.
[23] Brugsch, Ägyptologie, 1891, 67 ; Feldhaus, Geschichtsblätter für Technik, Industrie und Gewerbe, ii, 96 ; ib., Technik der Vorzeit, 85.
[24] Wiedemann, 1920, 299 ; Budge, Egy. Dict., 491.
[25] Alt. I, ii, 167.

countries, probably through Syria [1], but the meaning of hîqu is somewhat uncertain.[2] The modern Egyptian name for beer is busa.[3]

Beer from red barley is mentioned in the IV dyn., and several other kinds in the Book of the Dead. These were perhaps beer flavoured with various plants in the manner of modern hops, such as chickpeas, lupins, skirret (Sium Sisarum), rue, safflower and the root of an Assyrian plant.[4]

Although hops are said to be mentioned, as lupus salictarius, by Pliny[5], the meaning is much disputed and they were, apparently, not used in brewing until the 8th century. They are mentioned by Hildegard of Bingen in 1079. Their introduction into Europe by Finns, Letts or Slavs has been assumed but is not certain.[6] An infusion of plants in beer was used for sprinkling the floors of rooms to drive away snakes and scorpions.[7]

The Golenischeff Mathematical Papyrus (c. 1850) has problems on how much beer can be made from a given amount of grain (a hekat or gallon of grain makes 2 des of beer), and in one problem a beer of great "alcoholic" (? intoxicating) strength is derived from two others.[8]

Beer is mentioned a great number of times in the Ebers Papyrus and in various forms: bitter, fermenting, foaming, flat, stale, cool, strong, sweet, etc.[9] The Pyramid Texts (V–VI dyn.) refer to "dark beer" (? stout), iron (?) beer, and hes beer : other texts to sweet beer, perhaps mixed with honey, or mead,[10] imported beer, beer from red grain (deser-t) and beer made in Egypt by foreigners. Queen Hatshepsut sent a foreign prince a present of Egyptian beer.[11] "Beer" and "sweet beer" appear in the old Kahun Medical Papyrus.[12] Beer was drunk by all classes and ages in ancient Egypt, and the king had a brewery ; it was used only by lower classes in Alexandria.[13]

Goguet[14] considered the invention of beer difficult, but the process actually used in Egypt was quite simple. There is a wooden model of a brewer of the Old Kingdom.[15] The grain was moistened and allowed to sprout, then ground in a mortar, moistened with water, formed into lumps with leaven and baked superficially into a sort of bread, the inside being raw.[16] Bread is still used in Egypt for making busa ; malt is a much later invention. This bread (really a kind of malt) was then broken up and allowed to ferment in tubs for about a day, sieved, and the liquor pressed through a strainer. The beer was kept in pitched jars stopped with Nile mud. The loaves are represented in the Old Kingdom model, together with a small vat, and the process of making them in a XII dyn. model.[17] The description of making beer in the Alexandrian chemical texts[18] would give a similar product to the modern Egyptian whitish, frothing drink which does not keep.[19]

[1] Schrader, RL, 90 ; Richter, A. Nat., iv, 428 ; Wiedemann, Sagen, 21—close relations from about 1700.

[2] De Morgan, Préhist. Orientale, ii, 316 f.

[3] Erman-Ranke, 225 ; Lane, Mod. Egyptians, i, 133.

[4] BMGE, 83 ; Lippert, Kulturgeschichte, i, 620 ; Sprengel, H. Med., i, 55 ; doubtful according to Woenig, Pflanzen, 170 f. ; Goguet, i, 122 ; Wiedemann, 1920, 300 ; Wilkinson, i, 395 f.

[5] xxi, 15.

[6] Schrader, RL, 90, 376 f. ; Feldhaus, Geschichtsblätter, i, 84 ; ib., Technik, 85 f., 533 ; Mesue, Opera, Venice, 1562, 16v. ; Salmasius, Homonymis Hyles Iatrici, 1689, 80 ; Lippmann, Abh., ii, 246 ; Huber, Nature, 1927, cxix, 404.

[7] Wiedemann, 1920, 175.

[8] Archibald, Isis, 1931, xvi, 150.

[9] Joachim, Index.

[10] Cf. Feldhaus, Technik, 85.

[11] Lieblein, Handel und Schiffahrt auf dem roten Meere, 1886, 28.

[12] Griffith, The Petrie Papyri, 1898, 5 f.

[13] Wiedemann, 1920, 300 f. ; Budge, Dict., 889.

[14] i, 121.

[15] Erman-Ranke, 209, pl. xvi, Fig. 1.

[16] Wiedemann, 1920, 299 ; Capart, Art, 174.

[17] Erman-Ranke, 225 ; Wiedemann, 1920, 174 ; Maspero, Guide to the Cairo Museum, 1908, 525.

[18] Berthelot, Coll., ii, 372 ; C. G. Gruner, Zosimi Panopolitani de zythorum confectione fragmentum, Sulzbach, 1814, 10 f., 29 f., 40.

[19] Grässe, Bierstudien, Dresden, 1872, q. by Wiedemann, 1920, 299, not seen.

Malt is generally said not to have been used as such ; its origin is doubtful, but it is supposed to occur as βύυι in the Stockholm Chemical Papyrus (c. A.D. 300) [1], and as βύνη in Aëtios (c. A.D. 540).[2] The late Latin name is said to have been brace or braisium.[3] Wilkinson [4] quotes "red barley or malt," bet teser, for making beer, from a IV dyn. text. The Egyptian word dqw is supposed to mean malt.[5]

Various kinds of yeast mentioned in the Ebers Papyrus are : wine yeast (Joachim, 34), beer yeast and mesta-yeast [6] (67), growing yeast (109), bottom yeast (160), yeast juice (36), and yeast water (77, 78). The suggestion that Iceland moss (stocks of which were found at Deîr el-Baharî) was used in place of yeast for "raising" dough [7] seems unnecessarily improbable.

Millet beer (from the Andropogon sorghum vulgaris) is still the chief drink of native Africans : in Egypt it is called durra beer, in Central Africa pombe.[8] The African negroes are particularly fond of intoxicants, the more so the blacker their skins, and the modern Nubians distil brandy from grape wine.[9] Mungo Park[10] found the negroes near Gomba drinking good beer made from corn. Potent manawa is made from fermented oatmeal, and banana wine distilled with a bamboo and gourd still in Victoria Nyanza gives a very potent alcoholic liquor.[11]

Egyptian beer was drunk from flat dishes, rarely provided with feet, and, as the descriptions and representations of its effects show, it was strongly intoxicating. The intoxicating quality was thought to be due to a demon, and a papyrus of 1200 B.C. contains a magic formula to get rid of its effects.[12] The famous strong beer of Pelusium in the Delta produced exhilarating effects.[13]

Beer was offered to the gods : Rameses III provided 466,303 jugs of it for temples, and it was used in ceremonies of magic. In the Roman Period there was a tax on beer.[14]

Jars in the Roman-Nubian cemetery at Karanog contained a solid sediment of vegetable remains, apparently from some form of beer[15] ; a similar jar, with remains of yeast (?), was found in an XVIII dyn. tomb at Thebes.[16]

VINEGAR

Although Berthelot[17] is wrong in saying that vinegar (ḥum'za)[18] as such is frequently mentioned in the Ebers Papyrus, since Joachim does not once give it, there is a sour beer ("abgestandenes Bier," perhaps only flat or stale beer)[19], as well as sweet and bitter. Diluted vinegar was drunk in the Ptolemaic

[1] Lagercrantz, Papyrus Græcus Holmiensis, Uppsala, 1913, 25, 203.
[2] Aetii Tetrabiblos, tr. Cornarius, Basle, 1549, 588 ; cf. Hesychios, Lexikon, s.v. βύνη ; Palladius, de febribus, ed. Bernard, Leyden, 1745, 124.
[3] Maigne d'Arnis, Lexicon mediæ et infimæ latinitatis, Paris, 1890, 349, 351.
[4] i, 396.
[5] Erman - Grapow, Handwörterbuch, 1921, 216.
[6] Joachim thought mesta was a fermented drink ; Budge, Egy. Hierogl. Dict., 328, says it means "some medicinal solution."
[7] Wreszinski, q. in Anct. Egypt, 1927, 29.
[8] A. E. Crawley, Hastings ERE, v, 72 ; Speight, Pharm. J., 1931, lxxii, 478 ; 1933, lxxvi, 26, on African folk-chemistry

and drugs—vegetable poisons, alcoholic drinks, hair dyes, detergents, tanning, etc.
[9] Wilkinson, i, 397.
[10] Travels, Everyman ed., 89.
[11] Speight, opp. cit.
[12] Wilkinson, i, 396 ; Wiedemann, 1920, 298, 301 f.
[13] Columella, De Re Rustica, x, 116 : Pelusiaci proritet pocula zythi, "provoke fresh bumpers of Pelusian beer" ; Wilkinson, i, 396.
[14] Wiedemann, 1920, 301 f. ; Lenormant, Chaldean Magic, 1877, 92 f.
[15] Woolley and MacIver, Karanog, i, 30.
[16] Winlock, Illustr. London News, 1932, clxxxi, 322.
[17] Arch., 239.
[18] Budge, Egy. Hierogl. Dict., 471.
[19] Joachim, 40.

Period [1], vinegar from sycamore-figs was made in the Arabic Period, and the modern Egyptians prepare it from dates.[2]

WINE

Fruit juices, including unfermented grape juice, were drunk mixed with water. Grape wine, e.g. under Rameses II, was probably expensive and was also used medicinally.[3] It is mentioned in the IV dyn. as àrp (Coptic erp), of various kinds.[4] It was kept in earthenware jars with stoppers of Nile mud stamped with seals [5] as in the Roman Period [6], or in leather bags coated with a resinous substance ; even when jars were used it is possible that some resin or bitumen was put in the wine, or it was kept in pitched jars, as was the Roman wine or modern wine in southern countries. Wine was fermented in large jars (perhaps with some addition) from grape juice trodden and also squeezed from bags of grapes in hand-presses. The use of grape wine by the higher classes is certain from the earliest dynasties, and it was also an offering to the gods, wine and grapes being regularly placed in tombs.[7]

Palm wine (àama ?) and date wine (benà) were drunk by the lower classes.[8] Palm wine is said [9] to have been used in mummification, and traces of alcohol have been found in the tissues of mummies.[10] Pomegranate wine was little used[11]; perhaps it was called shedeh, a "sweet intoxicating drink."[12] The very intoxicating palm wine (Indian toddy ; mod. Egy. lowbgeh)—different from date wine—is much used in West Africa and Madagascar and is made from the sap of the Raphia vinifera, etc.[13] Pigafetta[14] found palm wine used in most of the East Indian islands ; other plant juices are fermented in various parts of the world.[13] The date wine in Egypt was made when the Nile water was clearest, in January.[15]

At least six kinds of wine are mentioned, including white, red, black, spiced (perfumed) and wine of the Delta (famous in Roman times as the Mareotic, Sebennytic and Teniotic wines).[16] Much wine was also imported from Syria and Palestine, and in the later period (6th century B.C.) from Greece.[17] Herodotos[18] says wine was imported twice a year from Greece and Phœnicia in earthen jars, "all of which are collected, sent to Memphis, where they are filled with water, and then go to the desert tract of Syria." The wine was cooled by fanning the jars.[19] Siphons were used for drawing off and mixing wines, and one was found at Amarna.[20]

The Egyptian wine was strong and its effects are shown.[21] In feasts at Denderah so much wine was drunk that the town was called a "place of

[1] Schubart, Ägypten von Alexander bis auf Mohammed, 1922, 156.
[2] Reitemeyer, 63, 66 ; Wilkinson, i, 399.
[3] Loret, L'Égypte, 122 ; Wiedemann, 1920, 303 f., 308, 309.
[4] Wilkinson, i, 166, 380 f., Figs. 152, 166 ; Richter, A. Nat., iv, 428 f. ; Parthey, Isis und Osiris, 1850, 163 ; Maspero, Guide to the Cairo Museum, 1908, 28 ; Budge, Egy. Hierogl. Dict., 72, àrp might also mean mead.
[5] Examples in British Museum ; BMGE, IV–VI Rooms, 249 f.—I dyn.
[6] Pliny, xiv, 20.
[7] Erman-Ranke, 65, 216, 225, 227—Old Kingdom ; Breasted, H., 41 ; Wiedemann, 1920, 303 f., 308 ; Wilkinson, i, 380 f., 384, 386, Figs. ; ii, 186 ; Woenig, Pflanzen, 254 f., 268.
[8] Woenig, Pflanzen, 304, 315 ; Wilkinson, i, 397 f. ; BMGE, 83 ; Budge, Egy. Hierogl. Dict., 20, 217.

[9] Herodotos, ii, 85.
[10] Dawson, JEA, 1927, xiii, 49.
[11] Wiedemann, 1920, 298 ; Woenig, 323.
[12] Erman, Lit., 388.
[13] Wilkinson, i, 397 ; Crawley, ERE, v, 72 ; Balfour, Cyclopædia of India, 1885, iii, 906 ; see p. 304.
[14] Voyage autour du Monde, Paris, An ix, passim.
[15] Masûdî, d. 956 ; in Reitemeyer, 34.
[16] Wilkinson, i, 390, 394 ; Erman-Ranke, 226 ; Lucas, Anct. Egypt, 1928, 1.
[17] Wiedemann, 1920, 307.
[18] iii, 6.
[19] Davies and Gardiner, Tomb of Amenemhet, 1915, plate xxvi (Thothmes III).
[20] Erman-Ranke, 228 ; Wilkinson, ii, 33, 313 ; Griffith, Isis, 1931, xv, 266.
[21] Wilkinson, i, 392 f., Figs. 167 f. ; Wiedemann, 1920, 302 ; Richter, A. Nat., iv, 428 f. ; Hartmann, L'Agriculture, 156.

drunkenness" and its goddess Hathor "mistress of drunkenness." Suidas
reports that the Egyptians thought they could avoid the effects of wine by
eating boiled cabbage.[1]

Although Chabas thought alcohol was known and called set in ancient Egypt,
this is improbable. It is commonly assumed that distillation was unknown
before the 1st century A.D. [2], and that the date of the discovery of alcohol is
much later.[3] Brandy from dates was made in Egypt in the 17th century.[4]

BREAD

Hekataios of Miletos (500 B.C.) calls the Egyptians the "bread eaters," [5]
and no other people has ever cultivated the art of bread-making to such a
degree.[6]   Egypt was, especially in the Roman Period, a great source of corn.
In the Ptolemaic Period its rich stores had an important political significance :
in the Roman Empire Egyptian corn, mostly from temple land, fed both Rome
and the provinces and enabled the Emperor to keep them under control.[7]
Wheat (su-t) was extensively grown in ancient Egypt.[8]   Large stocks of barley
and emmer wheat were found in Predynastic remains in Fayyûm [9], and
bread and porridge of cereals, perhaps including emmer wheat, were used in
the Badarian Period. Predynastic bread, not very homogeneous, found at
Qau was porous, so that probably yeast was used. Emmer wheat (Triticum
dicoccum) grains were found of the Predynastic Period.[10] Emmer, or split,
wheat has a kernel split in two halves which are still held together by the hull
when threshed ; it requires hard rubbing to break the hull and obtain the
kernels. It is raised in parts of Europe and used to make starch ("starch
wheat").[11] As Archibald[12] points out, the German dictionaries do not contain
the word "Emmer," which is said to be the spelt called bodet by the Egyptians
and "much cultivated in Europe." Maspero[13] calls the grain used for Egyptian
bread spelt (Egy. kurishtit or kulishtêt ; Herodotos' kyllestis ; modern spelt
—Triticum spelta—is a variety of ordinary wheat).[14]

Charred remains of what is supposed to be ordinary wheat which occur in
Predynastic pots of S.D. 40 perhaps came from Mesopotamia, and drawings
of emmer wheat occur in the tomb of King Zer (1 dyn.).[15] Barley, grown in
the Predynastic Age[16], and millet, grains of which, of a species now cultivated
only in the East Indies, have been found in mummies, are native to Egypt :
they were cultivated with grasses which had been growing for thousands of
years in the Nile Valley, the seeds of which increased in bulk on cultivation.
They had, however, been cultivated first by the Nile. Each year this rises
and floods the valley, then gradually retreats, leaving lakes and pools, just at
the beginning of the cool season. Barley and millet wild grasses shoot up in
the damp soil and grow ripe at the beginning of the hot season, when the seeds
fall off and are buried in the dry soil, where they remain until the river floods
again and causes them to sprout. The seeds had, by this process, become plump
enough to attract attention.[17] "Mummy wheat" is a fiction : the really old

[1] Wiedemann, 1920, 308, 316 ; 302.
[2] Wilkinson, i, 394 ; Lucas, JEA, 1930, xvi, 45.
[3] Lippmann, Beitr., 56 ff.—11th cent.
[4] Vansleb, q. in Reitemeyer, 63.
[5] Wiedemann, 1920, 292.
[6] Duncan, *Anct. Egypt*, 1930, 26.
[7] Rostovtzeff, PW, i, 134, 157.
[8] Wilkinson, ii, 42 ; Wiedemann, 1920, 267 f., 271 f. ; Erman-Ranke, 513 f. ; Woenig, Pflanzen, 164 f.
[9] Peake and Fleure, The Way of the Sea, Oxford, 1929, 14.

[10] Brunton and Thompson, Badarian Civ., 1928, 14, 25, 41, 63 ; Wiedemann, 1920, 271.
[11] Breasted, Anct. Times, 1916, 38.
[12] *Isis*, 1931, xvi, 151.
[13] Empires, 797.
[14] Cf. Kanngiesser, *A. Nat.*, iii, 98 ; Herodotos, ii, 36, ζειά or ὄλυρα, "spelt."
[15] Peake, *Nature*, 1927, cxix, 158, 894 ; cf. Elliot Smith, *ib.*, 81.
[16] Breasted, Anct. Times, 1916, 38 ; H., 92.
[17] Mackenzie, Footprints, 91.

specimens never germinate, and those which are repeatedly said to do so are modern.[1]

The grain was stored in bins with sliding doors, enclosed in a building, or in tall, tower-shaped granaries filled through doors at the top, but with doors below for emptying, the top being reached by a ladder.[2]

In the earliest period roasted grain, not bread, was used, but baked bread (tiû ; àu, àb-t) was well known in the Old Kingdom, when there are representations and also wooden models of bakers (including women) at work. The grain was crushed between stones (specimens of I–III dyns. known), wheat (sûo ; the Semitic name kamhû is in the Pyramids) for the best bread and barley (iati, ioti) and sorghum for the inferior.[3] Various kinds of grain, including wheat, barley and millet, are understood by the names àm, ānkh, but, pert, hi, hez-t, khai, khend, sud, ta, etc.[4] Leaven or yeast was used [5], the dough was kneaded with the feet, as Herodotos [6] was surprised to find, and the flat or other cakes were laid on the flat stones of the oven.[7] Representations of baking in the time of Sesostris I(1970–1935) show the white dough lifted from a large jar and filled into conical shapes, which are then laid horizontally in the fire : the baked loaf is shown yellow.[8]

As many as thirty kinds of bread and cakes mentioned in the Harris Papyrus (1180 B.C.) are perhaps only different shapes, shown in the Middle Kingdom[9] : cakes and biscuits of various shapes (resembling those from Turfan, A.D. 650–750) were found in a tomb of the XVIII dyn.[10], and bread made from coarse wheat and barley meals has also been found.[11] The bread was baked in the house, not in public bakeries.[12] The sizes of the loaves varied, e.g. some of 300 gm. and of 1,500 gm. are mentioned in the XIX dyn.[13] Although Pliny[14] says the Egyptians used sieves of stalks of papyrus and reeds, and there is an Egyptian sieve in the Louvre[15], the bread was usually of poor quality and full of grit[16], so that it wore down the teeth, although "white" bread is mentioned, with "flat loaves," in XII dyn. texts.[17] In the Book of the Dead the deceased says : "let my bread be made of white grain and my beer from red grain."[18] Some loaves and cakes were sprinkled with seeds[19], some bread was imported from Asia[20], and some kinds were perfumed.[21]

Milk and cream (for a medical plaster) are mentioned in the Ebers Papyrus, but not butter or cheese, which were perhaps not known[22], although "much butter" is said to be mentioned in a papyrus of 2000 B.C.[23], and Diodoros Siculus says cheese (tser ?) was eaten.[24] Tallow, "thick fats" of various

[1] Petrie, *Anct. Egypt*, 1914, 78 ; Feldhaus, Technik, 418 ; Wiedemann, 1920, 271 ; BMGE, IV–VI Rooms, 222 ; *Nature*, 1933, cxxxii, 271, 469.

[2] Erman-Ranke, 520, Figs. 209–10 ; BMGE, IV–VI Rooms, 22.

[3] Budge, Mummy[2], 327 ; *ib.*, Egy. Hierogl. Dict., 31, 37 ; Wiedemann, 1920, xv, 290 f., Plate-fig. 21 ; Goguet, i, 108 f. ; Wilkinson, i, 359 ; ii, 33 f., 41 ; Maspero, Dawn, 66 ; Erman-Ranke, 224, Fig. 71 ; Woenig, Pflanzen, 174 f. ; BMGE, IV–VI Rooms, 20 f.

[4] Budge, Egy. Hierogl. Dict., 50, 126, 215, 227, 242, 468, 523, 529, 559, 592, 611, 648, 821.

[5] Maspero, Dawn, 320, 338 ; Joachim, 34, 36, 67, 77, 78, 109, 160 ; Berthelot, Arch., 239.

[6] ii, 36.

[7] Wiedemann, 1920, 290.

[8] Davies and Gardiner, Tomb of Ante-

foker and Senet, 1920, 14 f., plates XI *A*–*B* ; XII, XII *A*.

[9] Wiedemann, 1920, 289 ; Erman-Ranke, 223, Fig. 70.

[10] Hall, JEA, 1928, xiv, 205.

[11] Wiedemann, 1920, 289—XI to XII dyn.

[12] Wilkinson, i, 358.

[13] Wiedemann, 1920, 292.

[14] xviii, 11.

[15] Wilkinson, ii, 179 ; Hoefer, i, 39.

[16] Maspero, Dawn, 320, 338 ; Wiedemann, 1920, 291.

[17] Breasted, Anct. Rec., i, 268, 270.

[18] Budge, Mummy[2], 460.

[19] Wilkinson, ii, 33, 41.

[20] Wiedemann, 1920, 289 : ? cake, as bread would become too stale.

[21] Schmidt, Drogen, 31.

[22] Joachim, 33, 76, 77 ; Wiedemann, 1920, 297, 298.

[23] Maspero, Dawn, 471.

[24] Wilkinson, ii, 24.

animals, lard and fat are mentioned in the Ebers Papyrus for medical uses.[1]

The main food was bread, although roast goose, boiled (not roasted, and not salted) beef, and baked or salted fish—an important food [2]—were eaten, the meat being boiled in large (copper ?) pots over an open fire.[3] Fish were split, salted and dried [4], smoking and preserving in oil being unknown.[5] Specimens of foods and fruits from tombs, some in reed baskets, in the British Museum, include dates, nuts of dûm palm, pomegranates, castor-oil berries, dried fish, crushed wheat, bread, cakes and pastry, and specimens of wheat and barley which may be ancient.[6]

LEATHER

Capart [7] and Budge [8] indicate the use of leather (àmeska) (? tanned) in the Predynastic or Early Dynastic Period. "Well tanned" leather, both with the hair removed and also with the fur, was found plentifully in Badarian graves. Goat and antelope skins are commonest, but there is also a fine black fur like a cat's. Leather fringes were used. The skin was not decorated with bead work, and sandals do not seem to have been in use.[9] Prehistoric *white* leather shoes have been found.[10] Petrie[11] found leather in early tombs (some prehistoric), some coloured black, white, yellow and with zigzag patterns, etc. Coloured (green and red) leather was used for sandals, and leather was used for chair seats, bottles, wine skins[12], also as nets, bags, tyres for chariot wheels and sides of chariots.[13] It is mentioned in the Westcar Papyrus.[14] Fine soft-cured hides were used in the Old Kingdom for furniture, etc., dyed in all colours.[15] Rose-coloured, green and red leathers occur on XXI dyn. mummies[16], and black, white, red and green leather sandals in Roman-Nubian remains at Karanog.[17] Shoes of goat or gazelle skin, well tanned and stained a pink colour, were made[18]; leather sandals (papyrus was also used) go back to the Old Kingdom, and silver, leather and reed models of sandals were put in tombs.[19] Leather was also used for shields and tunics.[20] A shoemaker is represented in the XVIII dyn., and a shoemaker's shop, with leather, was found at Amarna.[21] There is a painted leather (vellum) roll of the XIX dyn. in the British Museum.[22] Valuable documents were written on leather rolls: the museum specimens are from the Middle and New Kingdoms.[23] The laws are described as being written on forty leather rolls.[24]

Tanning was common in Egypt, but the process used is not known.[25] The

[1] Joachim, 71, 117, 134, 144.
[2] Dawson, Mag. and Leech, 18.
[3] Wiedemann, 1920, 263, 293 f., 295.
[4] Wilkinson, ii, 117 f.
[5] Lucas, JEA, 1932, xviii, 131 f., 133.
[6] BMGE, IV–VI Rooms, 222.
[7] Art in Egypt, 1905, 48, 57, 68, 101, 173.
[8] Osiris, i, 28 ; ib., Egy. Hierogl. Dict., 54.
[9] Brunton and Thompson, Badar. Civ., 1928, 19, 40, 41.
[10] Mosso, Dawn of Mediterran. Civ., 1910, 45.
[11] Naqada and Ballas, 1896, 16, 29, 48, 49.
[12] Cf. Herodotos, ii, 121—perhaps for transport only.
[13] Wilkinson, i, 232, 283, 411, 414, 422, 466 ; ii, 19, 20, 185, 189 ; BMGE, IV–VI Rooms, 80, 88, 205, 213, 222—red from Syria ; Maspero, Guide to the Cairo Museum, 1908, 492 f. ; Hall, CAH, ii, 423.

[14] Wiedemann, Sagen, 21.
[15] Breasted, H., 96.
[16] Wiedemann, 1920, 127.
[17] Woolley and MacIver, Karanog, i, 28.
[18] BMGE, 81.
[19] Wiedemann, 1920, 126, 127.
[20] BMGE, 120.
[21] Erman-Ranke, 539 ; Wilkinson, ii, 187 f. ; Feldhaus, Technik, 613 ; Peet and Woolley, City of Akhenaten, 1923, 33.
[22] Budge, Osiris, i, 43 ; BMGE, IV–VI Rooms, 294.
[23] Erman-Ranke, 378, 538 ; Birch, in Wilkinson, ii, 182 ; Breasted, H., 197, 313—xii dyn. and Thothmes III ; ib., Ancient Records, ii, 164 ; Maspero, Dawn, 504 ; Glanville, JEA, 1927, xiii, 232 ; Isis, 1930, xiv, 508.
[24] Meyer, Alt. II, i, 64.
[25] W. E. Austin, Fur Dressing and Fur Dyeing, London, 1922, 30 ; Wiedemann, 1920, 126 ; Goguet, i, 133 f. ; Kobert, A. Nat., vii, 185 f.

preliminary treatment of skins is shown ; a man dips a hide into a vase, probably containing water for soaking, after which lime was perhaps applied (as in modern Egypt) to remove hair.[1] Alum (plentiful in Egypt) may have been used for preparing leather, as in early Europe and the Classical Period.[2] The pods of the acacia (Acacia nilotica ; Egy. sont ; Strabo's acanthus), cultivated also for its wood and gum, were probably used in tanning.[3] The acacia wood was used for boat-building [4], asphalt and pitch being applied to stop leaks.[5]

The modern Arabs use the bark of the Acacia Seyal, which grows wild in the desert but not in the Nile Valley [6], although both A. nilotica and A. Seyal are cultivated in Lower Egypt.[7] The Egyptians perhaps used the bark and wood of the Rhus oxycanthoides, a native of the desert. The Arabs use the acrid juice of the desert plant Periploca secamone, which winds round shrubs, etc., and is shown in Egyptian pictures : it is not known if the Egyptians used it for curing skins. The Arabs use the Periploca (Ghula) as follows. The skins are put into flour and salt for three days, and the fat, etc., removed from the inside. The stalks of the plant, pounded between large stones (in pictures at Thebes the curriers have vases or mortars), are put into water and applied to the inside of the skin for one day. The hair having fallen off, the skin is left to dry for two or three days and the process is complete.[8] Leather was made at Ikhmîm (Panopolis) in the Arabic Period.[9]

The leather roll No. 10,250 in the British Museum, a mathematical document, although in a very brittle condition was successfully unrolled by Dr. A. Scott by first impregnating it with a 2 per cent. solution of celluloid in equal volumes of amyl acetate and acetone. According to Scott the roll was of preserved skin, not tanned, but treated by some unknown process.[10]

Hippopotamus hide was used to make shields, whips, javelins and helmets, but whether tanned or not is not stated. Catgut (not wire) was used for harp strings.[11]

DETERGENTS

Certain plants have marked detergent properties. Pounded lupins (doqáq of modern Egypt) were probably used for washing the hands after eating : the Egyptian or Coptic word for lupin is tharmos, the Arabic termes. The later Greeks used natron and hyssop. The Egyptians do not appear to have been acquainted with soap, unless a mixture of natron and clay or steatite was used.[12] In demotic papyri a word corresponding with the hieroglyphic anzir is found, and in Coptic anchir is generally translated by soap.[13] A small quantity of soap was found at Pompeii, and Aretaios of Cappadocia (c. A.D. 50) says the Greeks learnt its medicinal properties from the Romans, who according to Pliny[14] got to know it from the Gauls, who used it only as a pomade.[15]

The Egyptians had some unknown means of removing grease from clothing[16], perhaps natron, or a kind of fuller's earth. A supposed soapy mixture of fat and natron found with mummies was really formed by the action of natron on fat derived from the body.[17] Since in the Ebers Papyrus (1550 B.C.) oils and fats are frequently directed to be boiled with large quantities of alkaline

[1] Wilkinson, ii, 186 f., 417 ; Austin, 31.
[2] Schrader, RL, 498 ; Austin, 32.
[3] Wilkinson, ii, 195, 224, 416, 417 ; Austin, 33 ; Lucas, Egy. Mat., 198 ; M. Nierenstein, Natural Organic Tannins, 1934, 220 ; Woenig, Pflanzen, 302 ; Kanngiesser, A. Nat., iii, 82.
[4] Wilkinson, ii, 190, 195, 224, 417.
[5] Wiedemann, 1920, 217.
[6] Wilkinson, op. cit.
[7] Hasselquist, Travels, 1766, 250.

[8] Wilkinson, ii, 186 f., 417.
[9] Yaqûbî, A.D. 891 ; Reitemeyer, 148.
[10] JEA, 1927, xiii, 239.
[11] Wilkinson, i, 469, 471, 483 ; ii, 126.
[12] Wilkinson, ii, 48 f.
[13] Budge, Mummy², 265.
[14] xxviii, 12.
[15] Wilkinson, ii, 49 ; Adams, The Seven Books of Paulus Ægineta, 1847, iii, 326.
[16] Erman-Ranke, 245.
[17] Lucas, JEA, 1932, xviii, 127, 128.

substances, it is very probable that soaps and plasters were obtained [1], but these are never specially mentioned.

The body was anointed, but was washed with water afterwards if a visit was made to a house : the Egyptians had "refreshing habits of cleanliness, even when society was in its earliest stage," and the modern natives, who are not obliged by religion to wash at meals, are as particular as the Moslems themselves.[2]

IVORY

Ivory (ab, àab, abu), already used in Predynastic times [3], was imported from Punt, Kush (one tusk 10 ft. long), Asia or Libya, although some was probably from the native elephants shown on Predynastic pottery. Sesostris I (1970–1935) refers to the capture of an elephant in Kush, Thothmes II (xviii dyn.) received elephants from Syria, and Thothmes III (1501–1447) hunted them there.[4] The early ivory may have been hippopotamus tusk.[5] Ivory was used for beads, and in the construction of furniture, from the earliest times [6], sometimes coloured, e.g. red or blue, and inlaid in wood, e.g. for chests.[7] Chair legs of carved ivory were used in the early dynasties, and crude statuettes of ivory made in the Predynastic Period.[8] The upper portion of an ivory figure of a 1 dyn. king, found at Abydos, is of very fine workmanship and is "the most important object of archaic Egyptian art hitherto discovered."[9] Maspero[10] suggests that the Latin name ebur for ivory was derived, through aburu, from the Egyptian name.

VARIOUS ANIMAL PRODUCTS

Ostrich feathers and eggs were used ornamentally and the capture of the bird is shown. Ostriches (later imported from the Sudan) are shown on Predynastic pottery, and ostrich egg shells, also used for jars, occur in various forms in Predynastic and Badarian graves.[11]

Although fowls are supposed by Sethe to have been known under Thothmes III[12], they are usually supposed to have been introduced from India in the Persian Period.[13] The albumen used as a medium in painting, etc., was no doubt obtained in the earlier periods from the innumerable birds' and reptiles', including crocodiles', eggs.[14] The egg *may* have been mixed with fig juice, which makes it liquid and enables it to work easily with pigments.[15] The eggs of geese are clearly shown[16], and clay models of eggs were put in tombs.[17] Diodoros Siculus[18] describes the artificial incubation of chickens in Egypt, both fowls and geese, and the Copts are still very skilful in this process[19], which is described by Maqrîzî and Abd al-Latîf.[20]

The use of glue was well known. There is a picture of a glue-pot and brush

[1] Lippmann, Abh., ii, 15.
[2] Wilkinson, i, 425 ; ii, 48.
[3] Petrie, Tarkhan I and Memphis V, 1913, 8 f., 22, 25 ; ib., Tarkhan II, 1914, 9 f. ; ib., Illahun, 1891, 19—Seti I ; Brunton and Thompson, Badarian Civ., 58—amulets, with black paste eye inlay ; ib., 4, 13, 15, 28, 31, 58, etc.—Badarian spoons, statuettes, etc., Predynastic, including jars ; Budge, Egy. Hierogl. Dict., 4, 19 ; BMGE, IV–VI Rooms, 244, 250, 286.
[4] Breasted, H., 30, 136, 141, 181, 271, 276, 280.
[5] Mackenzie, Footprints, 147 ; Hornblower, JEA, 1927, xiii, 240.
[6] Erman-Ranke, 254, 544, 592 f. ; Lucas, Egy. Mat., 12 ; Maspero, Arch., 270.
[7] Wilkinson, ii, 200 ; BMGE, IV–VI

Rooms, 84, 88, 266, 285; Birch, Anct. Pottery, 1858, i, 72.
[8] Breasted, H., 15, 39 ; 28.
[9] BMGE, IV–VI Rooms, 289.
[10] Dawn, 494.
[11] Wilkinson, i, 282 ; Breasted, H., 136, Fig. 11 ; Brunton and Thompson, Badarian Civ., 3, 28, 40.
[12] Wiedemann, 1920, 285.
[13] Breasted, Anct. Times, 1916, 188.
[14] Shrewsbury, *Analyst*, 1926, li, 624 ; *Anct. Egypt*, 1927, 57 ; Reitemeyer, 74 f.
[15] *Anct. Egypt*, 1924, 96.
[16] Wilkinson, ii, 448, Fig. 485.
[17] Wiedemann, 1920, 261.
[18] i, 74.
[19] Wilkinson, ii, 449 f.
[20] Reitemeyer, 73.

(Thothmes III), together with a piece of glue shown with its characteristic curved fracture.[1] A piece of glue which had originally been rectangular but had dried up was found in the tomb of Queen Hatshepsut (XVIII dyn.).[2]

Beeswax (menḥ ; Coptic moulch) was used extensively for closing the apertures of the embalmed body, for luting vases, for covering painted surfaces (XVIII dyn.), and in the Ptolemaic and Roman Periods for encaustic painting.[3] It was also used for making amulets and magic statuettes.[4] Wax is frequently mentioned in the Ebers Papyrus [5], and was used with honey in embalming.[6]

Honey (ȧbȧ-t ; baȧa ? ; bȧt ? ; Coptic ebiô), whether of the wild or cultivated kind is not certainly known—although bee-keeping is prehistoric in most countries [7]—is often mentioned as used as a medium for drugs in the Ebers Papyrus (1550 B.C.), in raw, boiled and solidified forms [8], and its general use was extensive in the Middle Kingdom.[9] It appears in the composition of an amulet, "sweet to the living but bitter to the dead."[10] Honey is said to have been used to preserve the body of Alexander the Great[11], and Abd al-Latîf tells a story of some Arabs who plundered an Egyptian tomb finding a body preserved in a jar of honey.[12]

WRITING MATERIALS

Early writing materials[13] were stone, bricks and tiles[14], and clay in Babylonia ; wood, leather and paper in Egypt ; the bark of trees and linen in Italy ; and birch bark in early India.[15] Etruscan documents on linen are known[16]; leaves of trees are still in use among the Eastern people (waraka in Arabic means either a leaf or paper), and even the shoulder blades of sheep were used among the Arabs. Linen was used for note-books in Rome in Aurelian's time. Other writing materials were plates of bronze, lead and other metals, and wooden tablets, bare or covered with wax or a glazed composition.[17] Wood tablets covered with glazed composition (stucco ?) from which the ink could be washed off after use, were used early in Egypt, for example at Lahun in the XII–XIII dyn. ; they continued in use long after papyrus was common, and are still used in Cairo schools.[18] The references quoted in the section on leather[19] show that vellum or parchment was used in Egypt before the date 1400 B.C. given by Feldhaus.[20] Pliny[21], from Varro, says it was not used by the Greeks till the time of Eumenes II, king of Pergamos (197–159), from which place the name parchment (charta pergamena) is derived. A fragment of parchment of excellent quality, dated 195 B.C., the oldest Greek text on parchment known, which is written in a small and fine cursive script similar to that on contemporary Egyptian papyri, was found at Euporos (founded about a century before).[22]

[1] Wilkinson, i, 205 ; ii, 19, 196, 198, 199, Fig. 398.
[2] Lucas, in Carter, etc., Tomb of Tutankhamen, ii, 166.
[3] Lucas, Egy. Mat., 133 f. ; Erman-Grapow, Handwörterbuch, 1921, 65.
[4] Wiedemann, 1920, 287.
[5] Joachim, 206 ; in 38 recipes according to the index.
[6] Wiedemann, 1920, 287.
[7] Hahn, in Hoops, i, 277.
[8] Joachim, Index, 201—130 times.
[9] Erman-Ranke, 229 ; Wiedemann, 1920, 286 ; Erman-Grapow, Handwörterbuch, 1921, 46 ; Budge, Egy. Hierogl. Dict., 39, 201, 202, 209.
[10] Lexa, Magie, ii, 28—XVIII dyn.
[11] BMGE, 162.
[12] Reitemeyer, 138.

[13] Wilkinson, ii, 183 f. ; Schrader, RL, 734 f. ; Marquardt, Privatleben, 778 f. ; Dziatzko, PW, ii, 554 f. ; Smith, DA, ii, 57 f. ; Erman-Krebs, Aus den Papyrus d. Königl. Mus., 1899, 36 f.
[14] Pliny, vii, 56.
[15] The inner bark of trees was liber, which became the name for book ; Pliny, xiii, 11.
[16] Schubart, Papyruskunde, 1918, 47 ; Livy, iv, 20, 23, says linen pieces were used about 440 B.C.
[17] Smith, DA, ii, 57 ; Smith and Marinden, ib., 753 ; Marquardt, 778 f.
[18] Petrie, Illahun, 1891, 13.
[19] P. 201.
[20] Technik, 782.
[21] xiii, 11 ; Smith, DA, ii, 58.
[22] Cumont, Isis, 1925, vii, 564.

The principal writing material in Egypt was papyrus (thuf, thufi, or thef; qem), also used for many other purposes, e.g. the large cabbage-shaped head was boiled and eaten, and the roots were burnt as fuel.[1] Herodotos [2] says the priests wore papyrus sandals and it was used for boats, made watertight with pitch, as in the story of Moses.[3] Papyrus paper was already in use in the III dyn. (2980 B.C.), but the oldest specimen known is of the XI dyn.[4] Earlier specimens have no doubt perished.[5] Several authorities date the earliest papyrus in the V dyn.[6] The manufacture of papyrus, a flourishing industry in the Old Kingdom, is represented in a wall painting at Saqqâra, the stalks of the plant being plucked, tied in bundles and carried away.[7] At first perhaps a temple monopoly, papyrus was largely exported in later times [8], and was introduced into Syria under Rameses III (1198–1167).[9]

The plant is the Cyperus papyrus, although several varieties of the genus Cyperus were called papyrus.[10] The ancient Egyptian name of the plant was pa apu[11], perhaps (according to Bondi) derived from pa-p-ior, "that which is of the river."[12] Strabo[13] says the plant was not cultivated or abundant around Alexandria in his time, but abounded in the lower part of the Delta. There were two kinds, the hieratica being better, and in places its cultivation was restricted so as to keep up the price. The plant grew in the Delta and marsh lands near the Nile to a height of 20–25 ft. and the largest diameter of its almost triangular stalk was $3\frac{1}{2}$–6 in.[14] Lotus and papyrus plants no longer grow in Egypt : Abd al-Latîf (d. 1231) does not mention them, although papyrus probably grew in Egypt in his time. In the earlier centuries of Arab dominion it probably grew plentifully, and was made into paper (fâfyr) in Bûrah and Wasîmah, etc.[15] Ibn Hawqal (c. A.D. 950) says it was cultivated at Palermo in Sicily, being transplanted there by the Arabs.[16] It still grows in Sicily in the River Anapos, near Syracuse[17], also in the Sudan, the plants being 25–30 ft. high, with very thick stems.[18]

The colour, quality and width of papyrus paper varied in different periods. The oldest is darkest, but some early specimens are much lighter and of a silky texture, and some XVIII dyn. papyri are remarkably white.[19] In the Middle Kingdom it is dark brown, in the Theban Period light to dark brown, in the Saïte Period light brown, and later light yellow to almost white.[20] Papyrus was superseded by parchment in the 7th century A.D., although it was occasionally used at a later date, e.g. for a Papal Bull of A.D. 1100.[21]

[1] Budge, Mummy[2], 171 ; ib., Egy. Hierogl. Dict., 776, 854.
[2] ii, 37.
[3] Wilkinson, ii, 120, 205.
[4] Spanton, Anct. Egypt, 1917, 6 f. ; Hartmann, L'Agric., 155–v dyn. spec. at Cairo.
[5] Breasted, H., 85 ; Lucas, Egy. Mat., 202.
[6] Deissmann, Light from the Ancient East, tr. Strachan, 1910, 22 ; Blümner, i, 297 f. ; Maspero, H. anc., 1904, 10 f. ; Marquardt, 784 f. ; Erman-Ranke, 534 ; Lenormant, Prem. Civ., i, 193, 241 ; Schrader, RL, 613 ; Feldhaus, Technik, 778 ; Kanngiesser, A. Nat., iii, 85.
[7] Birch, in Wilkinson, ii, 182 ; Breasted, H., 97, Fig. 44 ; Woenig, Pflanzen, 57, 85.
[8] Schubart, Papyruskunde, 1918, 36 f., 428.
[9] Breasted, H., 484, 517.
[10] Wilkinson, ii, 120 f., 179 ; Budge, Mummy, 1893, 349 ; Woenig, Pflanzen, 57, 74 f., 85, 88 f., 93 f., 123 f. ; Pliny, xiii, 11 f. ; illustration of the plant in Deissmann,

Light from the Ancient East, 1910, 22, Fig. 2.
[11] Wilkinson, ii, 179.
[12] BMGE, 53 ; E. Bibl., iii, 3556 ; peḥ = swamp : Budge, Egy. Language, 76.
[13] XVII, i, 15 ; 799 C.
[14] Budge, Mummy[2], 171.
[15] Woenig, Pflanzen, 123 f. ; Reitemeyer, 63 ; A. von Kremer, Culturgeschichte des Orients unter den Chalifen, Vienna, 1875–7, ii, 305.
[16] A. F. von Schack, Geschichte der Normannen in Sicilien, Stuttgart, 1889, i, 292.
[17] Wilkinson, ii, 406 ; Bunbury, DG, i, 722.
[18] BMGE, 53.
[19] Birch, in Wilkinson, ii, 182 ; Woenig, Pflanzen, 93.
[20] Wiedemann, 1920, 82.
[21] Birch, op. cit. ; Schubart, Einführung in die Papyruskunde, 1918, 45 ; Spanton, Anct. Egypt, 1917, 16.

The method of making papyrus is described by Pliny [1] ; Strabo merely mentions the plant. The pith was sliced, a layer of slices was laid on a board with a second layer in a transverse direction, and the whole was pressed, beaten and smoothed with a tooth or shell, forming a single page. Pliny says the layers were stuck together with Nile water, and almost every modern writer emphasises that he is wrong. Schubart [2], however, found that no binder is really necessary, the juice of the plant being sufficient, although binders were probably used, perhaps a paste from the sifted crumb of sour bread, or flour with hot water and vinegar.[3] Perhaps gum arabic was used. The name protocol ($\pi\rho\hat{\omega}\tau\sigma$; $\kappa\acuteo\lambda\lambda a$) originally meant "first gummed," or the top leaf.[4] Pliny mentions nine kinds of papyrus, the chief manufactures being at Alexandria and Rome.

Pieces of papyrus were stuck together to form long rolls : the Harris Papyrus (Rameses III) in the British Museum is over 135 ft. long and 16½ in. wide, and others in the British Museum are 123 ft. long and 18½ in. wide (1100–1000 B.C.), 78 ft. by 15 in., and 77 ft. by 13½ in.[5] The rolls ("books," etc.) were kept in large jars, sycamore-wood boxes or stone chests in the temple libraries[6].

The papyri for ordinary purposes were 10–20 ft. long and 7–9 in. wide, those for liturgical and magical compositions a little wider. Papyri with bilingual contracts in Demotic and Greek are 2–10 ft. long, and 10–14 in. wide. The Copts wrote on small sheets, bound up into books : two in the British Museum (6th century A.D.) have 156 folios, 11¾ × 8¼ in., forming 20 quires, each signed by a letter ; and 175 leaves, 11 to 12¾ × 8½ to 9⅜ in., 22 quires, each signed by a letter, respectively.[7]

Saline efflorescences on papyri are supposed to originate from the soil.[8] Papyrus when old and dry is brittle : to unroll it, the roll must be moistened by wrapping slightly damp paper around it, or holding it over steam from warm water.[9]

INK

Ink (nau ; sešu ?) was probably first used in Egypt,[10] but may have been invented independently in other countries. Egyptian ink was a fine powder of carbon (like lampblack) mixed with gum water to keep it in suspension.[11] Various kinds of carbon are specified for medicinal use in the Ebers Papyrus (1550 B.C.).[12] Earths such as ochre, and green copper preparations were probably also used for inks, the solids being rubbed down on stone slabs with little mullers.[13] Budge[14] thinks ink was also made from vegetable matter made up with gum water into cakes. The ink pot, called pes, was of glazed ware, the palette, mestha, was usually of wood, with a sliding lid, oval or circular hollows for the ink and a groove for the pens.[15] The reed or rush

[1] xiii, 11 ; cf. Perrot-Chipiez, Art in Egypt, i, 578 f. ; Blümner, i, 313 ; Marquardt, 786 f. ; Wilkinson, ii, 180 ; Budge, Mummy², 171 f. ; Smith, DA, ii, 58.
[2] Papyruskunde, 37, 45.
[3] Wiedemann, 1920, 81.
[4] Woenig, Pflanzen, 114.
[5] Wiedemann, 1920, 82 ; Feldhaus, Technik, 778 ; BMGE, 53 ; Budge, Mummy², 171.
[6] Wiedemann, 1920, 83 f. ; older literature in H. A. Zimmermann, De Papyro, Dissertat., Vratislaviæ, 1866 ; Melch. Guilandinus, Papyrus, hoc est commentarius in tria C. Pliniis majoris de papyro capito, Venice, 4to, 1572, Lausanne, 1576 ; Lenz, Botanik der alten Griechen und Römer, Gotha, 1859, 271 f. ; Woenig, 88 f. ; Schubart, 44 f. ; Smith, DA, ii, 57.

[7] Budge, Mummy², 172.
[8] Wiedemann, 1920, 81.
[9] Wilkinson, ii, 180 ; Davy, Works, 1840, vi, 160 ; Wiedemann, 1920, 81.
[10] Laufer, q. in Isis, 1927, ix, 137 ; Budge, Egy. Hierogl. Dict., 346, 620.
[11] Hendrie, Theophilus, 1847, 74 ; Woenig, Pflanzen, 105 ; physical chemistry of carbon ink, see Terada, Yamamoto and Watanabe, Scientif. Papers of Inst. of Phys. and Chem. Research, Tokyo, 1934, xxiii, 173.
[12] Joachim, Index.
[13] BMGE, 55.
[14] Mummy, 1893, 352.
[15] BMGE, 53, 55 ; BMGE, IV–VI Rooms, 48 f. ; Glanville, JEA, 1932, xviii, 53 ; Budge, Dict., 247 : specimens in British Museum.

pen was not pointed but frayed out at one end into a sort of brush, but after the XXVI dyn. a reèd cut like a quill pen was used.[1] Some black ink was found by Petrie at Amarna [2], and there are several specimens on palettes. In the Book of the Dead written in hierogiyphics, the rubrics (instructions for the performance of the rites and ceremonies which follow the chapters) are in red ink made from ochre ; in the Book of Overthrowing Apep it is ordered that the name of this fiend shall be written in green ink (probably from a copper preparation).[3] Inks for writing, however, were practically only black and red ; the yellow, blue, green (reà, ri-t) and white inks being for the artist.[4] Red ink was used on stone blocks in the pyramid of Cheops (IV dyn.), green ink in the pyramid of Tetà (VI dyn.).[5] The black ink is always described as carbon ink ; the red and yellow as ochres, the blue and green as copper compounds. Analyses by Wiener of black ink of the 9th–13th centuries A.D. and those of Lucas [6] show that early black inks were of carbon.

Carbon ink was probably used in the I dyn., certainly on the XI dyn. Prisse Papyrus, and it was still used in the 5th century A.D. on papyrus, not parchment.[7] Hoefer [8] says the preparation of ink from green vitriol and oak bark (tannic acid) "does not go back beyond three or four hundred years before the Christian era," and Lippmann [9] that iron-gall ink was known to the "ancient Egyptians." Vitriol was added to *carbon* inks in Dioskurides' time (1st century A.D.).[10] Iron-gall ink was known in later Egypt, since it is mentioned in a recipe in one line in the Leyden Papyrus V : 4 drams of misy, 2 drams of green vitriol, 2 drams of gall nuts, 3 drams of gum and 4 drams of an unknown substance represented by a symbol.[11] According to Lucas[12] it occurs on Egyptian parchments of the 7th–12th centuries A.D. A Coptic MS. written in Arabic and dated A.D. 1312 is partly in carbon and partly in iron ink. The modern Syrians and Arabs boil shavings of the root of the arta tree in wine, strain the liquor and add a little copperas (green vitriol) and gum to make an ink hard to erase from leather or parchment ; another kind was made from wine in which galls had been steeped and gum water, with a little copperas (recipes in a British Museum Syriac MS.).[13] Isidore of Seville (A.D. 624)[14] mentions goose-quill pens and iron-gall ink. From the 4th century A.D. a reddish-brown, perhaps metallic, ink occurs, which has corroded the parchment on which it was used.[15] Exceptionally, the black ink has corroded papyrus and linen bandages, and the red ink has frequently corroded papyrus, hence "sometimes an acid was used."[16]

The use of gold on papyri is very rare : the only known example mentioned by Budge is in a copy of the Book of the Dead in which the solar disc encircled by a serpent, carried on the head of a hawk, is gilded ; the papyrus was painted yellow before the gold leaf was laid on it.[17]

Iron-gall ink may have been invented for use on parchment, on which it is "almost impossible to write with carbon ink."[18] The use of galls is mentioned not only by Martianus Capella (5th century A.D.)[19], but also by Hippolytos[20]

[1] Birch, in Wilkinson, ii, 296 ; Budge, Dict., 777 ; BMGE, 55.
[2] Tell el-Amarna, 30.
[3] Budge, Amulets, 28 ; *ib.*, Mummy², 177.
[4] Lucas, Egy. Mat., 203 f. ; Budge, Dict., 417, 419.
[5] Budge, Mummy², 407, 413.
[6] Egy. Mat., 205 f.
[7] Lucas, *op. cit.* ; C. A. Mitchell and Hepworth, Inks : their Composition and Manufacture, 1924, 1 ff.
[8] Hist. de la Chimie, 1866, i, 62.
[9] Abh., i, 91.
[10] v, 183.

[11] Berthelot, Intr., 12 ; Fester, Entwicklung der chem. Technik, 1923, 31.
[12] Egy. Mat., 205 f.
[13] Budge, Mummy², 176 f.
[14] Etymol., VI, xiv, 3 ; XVII, vii, 38.
[15] Schubart, Papyruskunde, 44 ; Mitchell and Hepworth, Inks, 4.
[16] Wiedemann, 1920, 83.
[17] Budge, Mummy², 177—date not given.
[18] Flather, DA, i, 244 ; Mitchell and Hepworth, 1 f. ; Budge, Mummy², 176 f. ; *ib.*, Amulets, 33 ; Kobert, *A. Nat.*, i, 103 f.
[19] Flather, 244.
[20] Refutationis omnium hæresium, lib. iv, cap. 28 ; Göttingen, 1859, p. 92.

(A.D. 222), and much earlier by Philo of Byzantium (250 B.C. ?).[1] The use of iron-gall ink had been established by Davy's experiments [2] for some early (2nd–3rd centuries) *parchment* manuscripts. Philo [3] describes a type of sympathetic ink in which the writing was done with an infusion of galls, and after drying was then sponged over with a solution of vitriol (chalcanthon), when the letters appeared; the Latin translation in Thevenot reads: "scribuntur autem occultæ intra causium seu pileum novum, vel in cute galla confracta et aquis madefacta : ubi enim siccatæ fuerint litteræ, amplius non comparent : chalcantho autem, seu æris flore contrito, quemadmodum atramentum in aqua teritur, et spongia in illo madefacta, postquam ea detersæ fuerint litteræ, apparebunt." Philo gives other types of sympathetic ink but not, of course, the cobalt ink, which is said to have been introduced about 1705.[4]

A very remarkable find reported by W. Herapath [5] seems to indicate that marking ink containing dissolved silver was used in ancient Egypt. On unrolling a mummy which he dates 4500 B.C. (on the old dating this would be about v dyn., or what is now taken as 2750 B.C. ; if we suppose the mummy 4500 years old, the date would be 2648 B.C.) it was found that : "on three of the bandages were hieroglyphical characters of a dark colour, as well-defined as if written with a modern pen ; where the marking fluid had flowed more copiously than the characters required the texture of the cloth had become decomposed and small holes had resulted. I have no doubt that the bandages were genuine, and had not been disturbed or unfolded : the colours of the marks were so similar to those of the present 'marking ink' that I was induced to try if they were produced by silver." He obtained a button of this metal with the blowpipe from the material, and considered that this silver had been dissolved by nitric acid, obtained in turn by sulphuric acid. We need not follow him in his further speculation that Moses dissolved the golden calf in aqua regia (a mixture of nitric and hydrochloric acids). J. D. Smith [6] suggested that the silver had been dissolved, as chloride, in ammonia. T. J. Herapath [7] thereupon made a fresh investigation of the material by the microscope and showed that the appearance of the fibres was such as could be caused experimentally by silver nitrate, but not by a solution of silver chloride in ammonia. No suggestion was made by Smith that the stains were not caused by some form of silver in solution and Herapath's results are difficult to understand.

Mitchell [8] finds that the supposed "marking ink" on the edge of mummy linen in the British Museum, the winding sheet of Tehutesat (*c.* 1550 B.C.) [9], is not made with "indelible ink" : it is not a carbon ink, contains no iron or copper, and is probably an organic pigment like bistre.

[1] Diels, in Wiedemann, *Sitzungsber. Phys. Med. Soc. zu Erlangen*, 1920–21, lii–liii, 219 ; usual date is 2nd century B.C.
[2] Works, vi, 175.
[3] Thevenot, Veterum Mathematicorum Opera, fol., Paris, 1693, 102 ; cf. Nierenstein, *Isis*, 1931, xvi, 439 ; *ib.*, The Natural Organic Tannins, 1934, 5, 291 ; Kobert, *A. Nat.*, i, 106.
[4] Thomson, System of Chemistry, 5 ed., 1817, ii, 245 ; Beckmann, Hist. of Inventions, 1846, i, 109.
[5] *Phil. Mag.*, 1852, iii, 528 ; Napier, Arts, 55.
[6] *Phil. Mag.*, 1852, iv, 142.
[7] *Phil. Mag.*, 1853, v, 339.
[8] *Anct. Egypt*, 1927, 18.
[9] BMGE, IV–VI Rooms, 223 ; Budge, Mummy[2], 217.

BABYLONIA AND ASSYRIA

BABYLONIA AND ASSYRIA I

GENERAL AND HISTORICAL

THE ANCIENT CIVILISATION

Mesopotamia, the "land between the rivers" Tigris and Euphrates, rising in the mountains of Asia Minor and flowing down to the Persian Gulf, was the site of an ancient civilisation rivalling that of Egypt. Of the four rivers of Paradise [1], which Philo Judæus (30 B.C.–A.D. 50) thought were allegorical, it is generally agreed that Hiddekel is the Tigris and P'rath the Euphrates; various identifications of Pison and Gihon have been proposed [2], of which the Nile and the Wâdî Hamd in Arabia are probable.[3] The Tigris and Euphrates now unite to a single stream, the Shatt al-Arab, but once flowed separately into the Persian Gulf, and many sites now inland, such as the old towns of Eridu, Lagaš and Nippur, once stood near the sea.[4]

Many of the dates for the early period are probable estimates only; the foundations on which they are based are uncertain and their interpretation is problematical.[5] Some of the sites, e.g. the very large mound at Warkâ (ancient Erech or Uruk), have only recently been systematically excavated.[6]

The Babylonian records [7] begin with the Creation and include dynasties before the Flood [8], which seems to have been an historical event at Ur and Kiš.[9] The last king in the lists before the Flood is Zi-û-sut-ra (Xisuthros, or "the Babylonian Noah," of Berossos), and these kings are said to have ruled over Šubaru, Larsa, Sippar, Šuruppak), Eridu, etc.[10] The earliest chronological lists are artificially based on multiples of sars, cycles of 360 years, which may have some astronomical significance.[11] Although the oldest remains known probably go back to 4000–3500, the first ascertained dates are about

---

[1] Gen. ii, 11.

[2] W. Smith, Concise Dict. of the Bible, 1865, 222.

[3] Miller, *Isis*, 1927, ix, 459; Barton, JE, ix, 518; Yahuda, *Daily Telegraph*, 30 Aug., 1933.

[4] Contenau, La Civilisation Assyro-babylonienne, 1922, 10; Meissner, Babylonien und Assyrien, 2 vols., Leipzig, 1920–25, i, 2.

[5] Rogers, History of Babylonia and Assyria, 2 vols., New York and London, i[6], 1915, ii[2], 1901; i, 494 f.; Peake and Fleure, Priests and Kings, 16 f.; Unger, RV, viii, 162; CAH, i, 156; Meissner, i, 22; Contenau, Civ. Assyr., 144; *ib.*, Manuel d'Archéologie Orientale, 3 vols., 1931, iii, 1616 f.; Delaporte, La Mésopotamie, les civilisations babylonienne et assyrienne, 1923, 65; Weidner, MGM, 1918, xvii, 44; Meyer, Chronol., 25, 30, 32 f., 37 f., 39 f.; L. W. King, History of Sumer and Akkad [S. and A.], 1916, 334; Hall, Civilisation of Greece in the Bronze Age, 1928, 37; Woolley, The Sumerians, Oxford, 1928

[quoted by date], 21 f., 27 f., 62, 185 f.; *ib.*, Ur Excavations, 1934, ii, 208 f., 392 f.; interesting account of earlier excavations on various sites by H. V. Hilprecht, Excavations in Assyria and Babylonia, Philadelphia, 1904; *ib.*, Exploration in Bible Lands during the Nineteenth Century, Edinburgh, 1903.

[6] Cf. *Abh. Preuss. Akad. Wiss.*, 1930 f.

[7] D. D. Luckenbill, Ancient Records of Assyria and Babylonia, Chicago, 2 vols., 1926–7.

[8] Meissner, i, 23.

[9] Woolley, 1928, 31 f., 45, 62; Contenau, Manuel, iii, 1506; C. Marston, New Knowledge about the Old Testament, 1933, 43 f.; Hilprecht, The Earliest Version of the Babylonian Deluge Story, Philadelphia, 1910.

[10] Berossos, in Müller-Didot, Fragm. Histor. Græc., ii, 503; Meissner, ii, 113, 115 f., 439 f.

[11] Woolley, 1928, 29, 62; Delaporte, 252, gives $60^2 = 3,600$ as the sar, and $60^4 = 12,960,000$ as a great sar.

3100–3000, before which no monument is known with certainty. The date 4600 given for the I dyn. of Ur in the Sumerian lists of kings, which were compiled about 2000 by scribes from earlier documents but are known only in excerpts in much later Babylonian chronicles, must be reduced to about 3100.[1] If the I dyn. of Ur is dated 3100–3000, the earlier remains at al-'Ubaid (near Ur) and at Ur probably go back to 3500.[2] The dates adopted by Contenau [3] are (all B.C.) :

3500 : Al-'Ubaid ; oldest layers at Warkâ and Kiš.
2950 : I dyn. at Ur ; Tello before Ur Nina.
2880 : Ur Nina and his successors.
2725 : Sargon of Akkad.
2500 : The Guti.
2400 : Gudea.
2350 : III dyn. of Ur (2278 also given) ; dyns. of Isin and Larsa ; Cappadocian tablets.
2100 : I dyn. of Babylon.
2000 : Hammurabi.
1750 : The Cassites.
1500 : The Mitanni.

The first known historical king is Mesannipadda (c. 3100 B.C. ?), founder of the I dyn. of Ur, a seal cylinder of whose queen was found. Between this and the next well-established dynasty, that of Akšak (Opis), are several others which probably did not last long, although the native lists give 4000 years. A gold scaraboid and a marble tablet from the foundation of the stage tower at Ur, with the name of A-annipadda, son of Mesannipadda, establish the I dyn. of Ur, and the preceding dynasty of Kiš is probably also historical.[4]

Although finds at Tello (Tell-Lô), a provincial town, led to the earlier opinion [5] that the Sumerian culture is later but more primitive than the Egyptian, the finds at Ur show a still earlier highly artistic culture which may be as old as, if not older than, the Egyptian.[6] Woolley, in fact, suggested that the Dynastic Period in Egypt beginning with Menes was due to Sumerian influence, and although the two civilisations may have developed quite independently, an early Sumerian settlement in Egypt is often assumed.[7] The early Sumerians were a cultivated race, and their traditions of a "period before the flood" show a speculative and pseudo-scientific cast of thought.[8]

A little later than the I dyn. of Ur are the patesi or priest-rulers (not kings) of Lagaš, beginning with Ur Nina (2700), Eannatum, Entemena (2600), Urukagina and Gudea (2450).[9] Several readings of the names of kings have been altered in recent years (Ur Nammu instead of Ur Engur ; Sulgi instead of Dungi—both possible—etc.), and their dates have also fluctuated.

In the earliest period the dead were buried lying on the side in brick pits or in (or under) clay pots or jars.[10] Cremation proper was not used, but at

[1] Meyer, Chronol., 37, 39, 40 ; Woolley, 1928, 27, 29, 62.
[2] Contenau, Manuel, 1616 ; Hall and Woolley, Ur Excavations, al-'Ubaid, Oxford, 1927, i, 10, 140.
[3] Manuel, iii, 1616 f.
[4] Woolley, Times, 19 Jan., 11 Feb., 1924 ; ib., Sumerians, 1928, 33, 122 ; Hall and Woolley, Ur Excav., 1927, i, 128 f. ; S. Smith, Early History of Assyria [EHA], 1928, 13 ; Peake and Fleure, Priests and Kings, 24 f.
[5] Meyer, Chronol., 40 ; ib., Sumerer und Semiten in Babylonien, 1906, 75 ; Elliot Smith, EB11, xx, 149.

[6] Langdon, Nature, 1921, cvii, 315 ; MGM, 1923, xxii, 269 ; Woolley, 1928, 185, 188 f. ; ib., Ur Excavations, 1934, ii, 392 f. ; Hall, Civ. of Greece in Bronze Age, 1928, 37 ; ib., NE, 1920, 30 ; Contenau, Manuel, iii, 1509 f.
[7] King, S. and A., 334 ; Rostovtzeff, i, 17 ; cf. Woolley, Ur Excav., 1934, ii, 395.
[8] Smith, EHA, 15, 25.
[9] Woolley, 1928, 21 f. ; ib., Ur Excav., 1934, ii, 208 f. ; Meyer, Chronol., 32 ; ib., Sumerer, 75 f.
[10] Meyer, Alt. I, ii, 447 f. ; Jeremias, 319 (see p. 2161) ; Delaporte, 189, Fig. 15.

Ur a fire was lighted near the head and allowed to burn out before the earth was filled in, producing the so-called "burnt burials." In a later period (III dyn.) at Ur the bones of the skeletons were burnt, producing a result ("calcined bones") quite different from cremation of bodies.[1]   The prehistoric graves at al-'Ubaid near Ur, and the very early remains at Lagaš, both of about 3500 B.C., yielded clay vases, stone bowls, gold, silver, copper as pots, beaten plates covering walls, pillars and beams, nails, beaten copper reliefs of cattle with copper heads, and also at al-'Ubaid a frieze of figures of men and animals in fine white stone silhouetted against a background of black paste or bitumen, the whole framed in copper.   There is inlay in mother-of-pearl, lapis lazuli and red paste.[2]

Perhaps the most important result of recent archæological investigation is the disclosure that the rather crude artistic work found on what were regarded as very early sites is really the product of a later, decadent, period, when the splendour and majesty of the early dynasty at Ur had passed away. The culture of Ur Nina's time represents, not the beginning, but the end of the *old* Sumerian Period, which had reached its zenith about 3000 B.C.[3] The early Sumerian Period may be provisionally divided into : (1) the earliest, represented by objects from Kiš ; (2) the period of the tablets, etc., from Šuruppak (mod. Fâra) and the Predynastic cemetery at Ur (*c.* 3500 B.C.) ; (3) the advanced culture of the I dyn. at Ur (*c.* 3000 B.C.), followed after an interval by (4) the period formerly known as " Early Sumerian." The earliest known king of Kiš is Mesilim (*c.* 3100 B.C. ?).[4]

Agriculture was very early, implying a skilful irrigation system which is represented in very old pictures [5], but the statement that wheat grows wild in Mesopotamia is erroneous.[6]   The grain was cut in the early period (3000) by sickles of flints set with bitumen in wooden handles.[7]

SUMERIANS AND AKKADIANS

Early Mesopotamian racial and cultural origins are still very obscure.[8] From very early times two races lived side by side, a Semitic race in the north, speaking a language now called Akkadian (after the town of Akkad), and a non-Semitic race in the south called the Sumerians, speaking a language now wholly extinct.[9]   This was superseded about 2000–1500 in official documents by the Semitic Akkadian, then still written in Sumerian script or in bilingual texts. Akkadian was superseded about the 8th century B.C. by the Aramaic language, which had been in common use elsewhere much earlier.[10] This circumstance, and the fact that the earliest monuments show men with non-Semitic features, led to the view that the Sumerian was the earliest civilisation.[11] Although the Akkadians may not have appeared till about 2800[12], it is

[1] Meissner, i, 425 ; Woolley, Ur Excav., 1934, ii, 142, 407.

[2] Woolley, 1928, 13 f., 185 f. ; Hall and Woolley, Ur Excav., 1927, i ; BMGB, 57 f., 81 f. ; Meyer, Alt. I, ii, 483 ; Langdon, Excavations at Kish, Paris, 1924 ; *ib.*, *Alte Orient*, 1928, xxvi, 1 f.

[3] Woolley, Ur Excav., 1934, ii, 217.

[4] Smith, EHA, 15 ; Peake and Fleure, Peasants and Potters, Oxford, 1927, 81 f. ; Contenau, Manuel, ii, 658 ; Woolley, Ur Excav., 1934, ii, 226 ; *ib.*, 1928, 65.

[5] M. Jastrow, The Civilisation of Babylonia and Assyria, Philadelphia, 1915, 7.

[6] De Morgan, Préhistoire Orientale, 1926, ii, 298.

[7] Frankfort, *Illustr. London News*, 1932,

clxxxi, 529 ; Hall and Woolley, Ur Excav., 1927, i, 151.

[8] Langdon, *Isis*, 1933, xix, 560.

[9] Jastrow, Civ., 120 f. ; Meyer, Alt. I, ii, 433 ; Maspero, Dawn, 550 ; Delaporte, 17 f.

[10] Lenormant, Magic, 369 ; Sayce, Babylonian and Assyrian Life and Customs, 1900, 56 ; Kugler, Im Bannkreis Babels, Münster, 1910, 117 ; von Oppenheim, Tell Halaf, eine neue Kultur im ältesten Mesopotamien, Leipzig, 1931, 44 ; A. Jeremias, *Der Alte Orient*, 1929, xxvii, Heft 4.

[11] Lenormant, Magic, 352 ; Sayce, B. and A. Life, 4 ; Meissner, i, 15 ; Baumstark, PW, ii, 2706 ; Smith, EHA, 10 f., 43 f. ; Otto, 1925, 24 f.

[12] Unger, RV, viii, 166.

now generally accepted that the first population was Semitic, their country being invaded by the Sumerians with a more advanced civilisation.[1] The southern half of the country was once a sea, and the absence of Palæolithic finds (stone without metal) indicates that it was comparatively late, in the Chalcolithic Age (copper with stone), that the Sumerians arrived, probably by sea from Oman in the Persian Gulf, and subdued the primitive Semitic population. The south country was called after them Sumer ; in the north the Semitic majority was dominated by them.[2] Alternative hypotheses that the Sumerians came from east of the Tigris [3] or from Central Asia by way of the Caspian [4] are less probable.[5]

RELATIONS WITH THE INDUS VALLEY

In the Indus Valley at Mohenjo-daro and Harappā (between Lahore and Karachi) there are sites of cities with a relatively mature Chalcolithic civilisation similar to the Sumerian and quite different from the Vedic [6], with a high standard of art and craftsmanship and a developed pictographic system of (undeciphered) writing on seals. The remains in these two widely separated Indian sites include houses and temples, massively built of burnt brick and provided with well-constructed water conduits covered with marble slabs. The smaller antiquities include a quantity of painted and plain pottery, terra-cotta, toys, bangles of blue glass-paste and shell, new types of coins or tokens, curious stone rings, dice, various gem stones, a cylindrical vase and cover of silver, large quantities of gold, electrum and jadeite jewellery, and a bronze figure. The tin bronzes usually contain 6 to 13 per cent. of tin, but exceptionally 27 per cent. Tinstone occurs in the Western Presidency, Bihār and Orissa. The bronzes may also contain up to 15 per cent. of lead, but this is unusual, and they also contain nickel in proportions usually resembling those in the Sumerian bronzes but sometimes considerably higher, one specimen of copper containing 9·38 per cent. of nickel and bronzes 0·5 and 1·9 per cent. Another alloy was of copper with 3 to 4·5 per cent. of arsenic. Lead (sparingly used), white lead (basic lead carbonate) and cinnabar (mercuric sulphide) were also found. Weights are supposed to be on the Sumerian standard.[7] The finds represent a widespread culture which must have flourished for many centuries in the plains of the Indus [8], and there was a close trade connection between the Tigris and the North-Western Provinces of India at least as early as 3000 B.C. A fragment of a steatite vase at Mohenjo-daro is of the same unusual form as at Susa II, indicating Elamite import at least as early as 2800 ; the upper layers are dated 2500.[9] A cylinder seal (2600) from the Akkadian site of Tell Asmar in Irâq shows elephants and a rhinoceros, and resembles stamp seals at Mohenjo-daro.[10] A possible trade route in 2900–2700 is via the Persian Gulf ; another is a land route from Bokhara westwards through Khorassan and Qumis to Hamadan and the Tigris Valley, which is supported by finds of painted pottery at Seistan. The two civilisations may have spread from

[1] King, S. and A., 51 f. ; Meyer, Alt. I, ii, 437 ; ib., Sumerer und Semiten in Babylonien, 1906, 107 f. ; Jastrow, Civ., 106, 121 ; ib., Hebrew and Babylonian Traditions, New York, 1914, 8 ; Hall, NE, 1920, 171 f. ; Rogers, i, 447 f.
[2] Woolley, 1928, 1 f., 4 f., 12, 13 ; Meissner, i, 15 f. ; Barton, Isis, 1931, xv, 439.
[3] Delaporte, 20.
[4] BMGB, 3.
[5] Smith, EHA, 11.
[6] J. Marshall, Mohenjo-Daro and the Indus Civilisation, 3 vols., 1931 ; JRAS,

1932, 453 f. ; Chanda, q. in Isis, 1931, xv, 277.
[7] Ancient Egypt, 1930, 90 ; Desch, B.A. Report, 1931, 69 ; Marshall, Mohenjo-Daro, i, 29 ; ii, 674 f., 691.
[8] The Times, 10th Dec., 1924 ; Woolley, 1928, 8, 46 ; ib., Ur Excav., 1934, ii, 335 ; Smith, EHA, 49 ; Gadd, Discovery, 1924, v, 325 ; E. Mackay, q. in Ancient Egypt, 1930, 30 ; ib., Sumerian connexions with ancient India, JRAS, 1925, 697.
[9] Nature, 1932, cxxxi, 339.
[10] Frankfort, Illustr. London News, 1932, clxxxi, 504, 510.

the hills of Baluchistan.[1] Indian tribes as such influenced Mesopotamia only about 150 B.C.[2] Remains very like the Sumerian also occur in Anau, near Merv in the oases of Turkestan, and also at Elam in north Persia, and influences from these sites seem to have been felt in Asia Minor and even in Central Russia. Early connection between Egypt and Mesopotamia, proved by finds of Sumerian mace-heads, cylinder seals, vases, etc., in Egyptian remains of 3100 B.C., probably had its focus in Syrian coast-towns. Sumerian ships appear to have engaged mainly in coastal trade and did not make long sea voyages, although Ur was a large manufacturing town in the 4th millennium B.C. Sumerian work is found at Astrabad and there were trading settlements at Ganeš before 2500 B.C. The country was in constant touch with the north, the Halys basin and the country east of the Caucasus, and drew on the mineral wealth of these regions from the earliest period. The relations with Egypt are now generally accepted, so that former scepticism as to a Sumerian element there (see p. 4) is perhaps unjustified.[3]

THE SEMITES

The ultimate home of the Semites was probably Arabia; in the north they entered Babylonia and Assyria, in the north-west Palestine.[4] King[5] thinks they entered by way of the Syrian coast. The Syrian Desert is an integral part of Arabia and the Semites are supposed to have descended from it into Syria and Mesopotamia.[6] Jastrow[7] and Smith[8] suggest that the Semites came from Amurru, the land of the wandering tribe of Amorites who rose to power during the dynasty of Opis (a town near the present Seleucia).

Woolley[9] arranges the early races of Mesopotamia as follows:

(1) Semites, called Martu, later Amurru, living in the north, and as the alluvium formed occupying Sippar and Opis.

(2) A Caucasian race living in the Zagros hills and across the plain to the Tigris, who occupied Assyria.

(3) The Bedouin Semites, nomads, different from (1).

(4) The Sumerians, who came last by way of the Persian Gulf from an unknown place.

In Assyria, the Semitic colonies were modified by mixture with the Mitanni and Hittites. Other races in relation at various times with the Babylonians and Assyrians, often as conquerors, were: (1) the Elamites, perhaps mixed Negroid-Mongolian, whose country was afterwards inhabited by the Indogermanic Medes and Persians; (2) Kossæans, from the Persian border; (3) Lulubæans and Gutæans, from the mountains of the North; (4) Mitanni, who came as far south as Nippur; (5) Hittites of Asia Minor; and (6) in 720 B.C. the Gimirri (Cimmerians) and Aschkuza (Scythians). Farther north lay the countries of the Na'irî and Urartu (Biblical Ararat; modern Armenia). The inhabitants of Babylonia and Assyria also knew the Syrians, Cappadocians, Lydians, the people of Alašia (Cyprus, producing copper in the 15th century B.C.), Jawan (Ionians) and Egypt.[10]

[1] Smith, EHA, 51; Woolley, 1928, 9; Jastrow, *Alte Orient*, 1925, xxv, Heft 1.

[2] Hillebrandt, q. in *Isis*, 1927, ix, 555.

[3] Woolley, 1928, 47, 115 f.; *ib.*, Ur Excav., 1934, ii, 379 f.; Delaporte, 21, 396 f.; Contenau, Assyr., 26 f.; *ib.*, Manuel, iii, 1582 f. (Egypt), 1564 f. (Caucasus); Scharff, *Morgenland*, 1927, Heft xii; Maspero, Dawn, 616.

[4] Meyer, Alt. I, ii, 388, 435.

[5] S. and A., pref. viii.

[6] O'Leary, Arabia before Muhammad, 1927, 42 f.; Stapleton, etc., *Mem. Asiatic Soc. Bengal*, Calcutta, 1927, viii, No. 6.

[7] Trad., 9.

[8] EHA, 43.

[9] Sumerians, 4 f., 12 f.; cf. Unger, RV, viii, 166.

[10] Meissner, i, 15 f.

The Sumerians

The Sumerians were acquainted with the use and sale of gold, silver, copper and certain precious stones from the earliest period, when they also represented them by signs and regarded their brilliance as beneficent and divine. They also worshipped gods occupied with metallurgy, and (like the Egyptians) they regarded the setting sun passing through the lower hemisphere as the god of the treasures of the underworld. In later Babylon the gods had different metals as attributes.[1] Inlil (or Ellil) was "lord of gold," a title afterwards taken by the god Marduk (Bêl) of Babylon ; Ea of Eridu was also (with his human-fish form Oannes) lord of gold, protector of smiths, including goldsmiths, and in a later period this function was assumed by Gibil, the fire god, the "divine smith."[2]

Decorative cast copper occurs in the oldest Sumerian remains along with gold.[3] The free use of copper, earlier by "many centuries" than the I dyn. of Ur, perhaps came with the "new people" into the alluvial plain[4] ; although the metal was used by the "Pre-Sumerian" marsh-dwellers (before 3500 B.C.) it was very rare.[5] The noble metals, gold and silver, were used from the earliest Sumerian times as standards of value in the extensive commerce with other nations [6], silver being relatively scarce in the oldest remains (except at Ur).[7] Gold, silver and copper were imported from some mountainous country, perhaps the Zagros, or copper from Cappadocia.[8]

The old idea, based on excavations at Tello, that the Sumerians had little artistic or literary ability [9] is disproved by recent excavations in the royal city of Ur.[10] The Sumerians in the earlier period were a military people, fighting heavily armed in phalanxes, wearing helmets formerly thought to be of leather[11] but really of copper[12], and carrying shields and maces with stone heads, sharp copper "sickles" or scimitars, and lances (as shown on the Stela of the Vultures, of Gudea's time).[13] The Sumerians shaved, probably with copper razors, and had the head cropped. They wore wool skirts covering the legs, but no shoes. The priests wore beards but shaved the cheeks and upper lip ; they wore no clothes, but the priestesses wore a mantle over the left shoulder.[14] Wool was largely worked in the country.[15]

A very old, practically independent Sumerian town is Lagaš, composed of several parts, the oldest being Girsu. The earliest remains are a massive

[1] Lenormant, Prem. Civ., i, 113 f., 121 ; ib., Magic, 184 f., 363 f. ; A. Jeremias, Handbuch der altorientalischen Geisteskultur, Leipzig, 1913, 295.

[2] Jeremias, 15, 28 ; 69, 282 ; Maspero, Dawn, 653.

[3] Hall, JEA, 1922, viii, 247 ; Meyer, Alt. I, ii, 448 ; Heuzey and de Sarzec, Découvertes en Chaldée, ed. Amiaud and Thureau Dangin, 2 vols., 1884–1912, i, plates ; BMGB, 84, 86, 89 f. ; Contenau, Assyr., 93 ; King, S. and A., 74 ; Rostovtzeff, i, plate viii—3000 B.C. ; Hall and Woolley, Ur Excav., 1927, i, 33 ; Woolley, Ur Excav., 1934, ii, 411 f.

[4] Smith, EHA, 25.

[5] Woolley, 1928, 14.

[6] Meyer, Alt. I, ii, 448 f.

[7] Gowland, J. Anthropol. Inst., 1912, xlii, 268 ; Banks, Bismya or the Lost City of Adab, New York, 1912, 314 ; cf. Budge, Rise and Progress of Assyriology, 1925, 251, 297.

[8] Banks, 275, 308, 314 ; Meyer, Alt. I, ii,

447 ; BMGB, 82 f. 86, 90 ; Perrot and Chipiez, History of Art in Chaldæa and Assyria, 2 vols., 1884, i, 124.

[9] Meyer, Alt. I, ii, 448 f., 472 f.

[10] Woolley, 1928, 45, 122, 191 ; ib., Ur Excavations, 1934, ii, The Royal Cemetery.

[11] Meyer, Alt. I, ii, 450 ; King, S. and A., 138.

[12] Woolley, 1928, 52, 56 ; Meissner, i, 81 f.

[13] Meyer, 450 f. ; Delaporte, 75, Fig. 8, 77, Fig. 9 ; Woolley, 1928, 54 ; Miessner, i, 81 ; King, S. and A., 138—scimitars were "curved maces."

[14] Meyer, 451 ; King, S. and A., 111 ; Meissner, i, 407 f. ; Langdon, Excav. at Kish, 1924, i, 6 ; ib., Times, Jan., 1924 —mother-of-pearl inlay representations ; Frankfort, Illustr. London News, 1932, clxxxii, 526—copper statue of c. 3000 B.C. at Khafaje.

[15] Sayce, B. and A. Life, 100.

brick building, with stone and clay inscriptions.[1] The oldest buildings in
Mesopotamia were of reed matting, also used for baskets, waterproofed with
bitumen or mud.[2] The earliest historical ruler of Lagaš is Ur Nina or Ur
Nanše (2700 B.C.), who speaks of his buildings, and his name appears on a
temple of burnt brick cemented with bitumen. The earliest sculptures at
Lagaš are crude, like those of the Mexican Indians or the Hittites of Zinjirli.[3]
A little later (c. 2450) comes the patesi or priest-ruler (not king) of Lagaš,
Gudea, whose town was excavated at Tello above the ruins of Lagaš. He built
a large temple with a brick stage-tower, probably with seven stages, and a
plan of a town is cut on his statue. Akkadian influences were evident in the
art of his period.[4] The oldest representations of Sumerians show them with
large beak-shaped noses and round eyes, whilst the type under Gudea was quite
different. They are shown with slanting eyes and fine straight noses.[5] In
the earliest period the patesi (also read as isag or issakku) or the priest-ruler
received his authority direct from the town-god, for whom he acted as a sort
of tenant-farmer.

At this period there were continual wars with the Elamites, a people with a
high state of civilisation similar to the Sumerian ; their capital at Susa had a
stone-copper civilisation about 3000.[6]

THE AKKADIANS

The capital of Semitic Akkad was originally Sippar (mod. Abu Habba),
but King Sargon (2630 or 2528 ?) built (or rebuilt) the town Akkad (or Agade),
the site of which is now unknown.[7] His empire is known from a geographical
text based on an original of about 2000 B.C.[8] Before his time the old pictographic or hieroglyphic writing (c. 3500) had given way to cuneiform (Latin
cuneus, a wedge), made up of wedge-shaped marks impressed on soft clay
tablets, afterwards baked, by means of a writing stylus, a unique specimen of
which was found at Kiš.[9] The Sumerians made lists of words at least as early
as 2600. The linear script on the oldest Sumerian tablets is the earliest writing
known.[10] The earliest tablets are oval and small, covered on both sides with
fine writing. These were soon replaced by rectangular tablets, sometimes
enclosed in burnt clay envelopes, even in the early period.[11] The Akkadians
used rectangular tablets with clearer inscriptions than the Sumerian.

The Akkadians wore longer hair and beards, were clothed in coats with girdles
and wore sandals. The troops wore copper helmets with long horns, fought
with bows and lances and appear to have predominated over the Sumerians,
whose pre-eminence began to wane about 2500, if not earlier.[12]

Akkad adjoined the plains of Mesopotamia, then inhabited by several
peoples called the Subari or Subarians.[13] Semitic peoples had settled there at

[1] King, S. and A., p. x ; Maspero, Dawn, 603 f. ; Woolley, 1928, 66 ; Meyer, Sumerer, 1906, 34.
[2] Woolley, 1928, 185 f. ; Frankfort, Illustr. London News, 1932, clxxxi, 529.
[3] Woolley, Ur Excav., 1934, ii, 214, 319 f., 322 ; ib., 1928, 66 ; Meyer, Alt. I, ii, 485 ; BMGB, 242.
[4] Meyer, Sumerer, 35 ; ib., Alt. I, ii, 542 f. ; Jeremias, 12 ; Heuzey and de Sarzec, Découvertes en Chaldée, i, 129 f. ; de Morgan, Délégation en Perse, i, 38 f. ; King, S. and A., 265, 361 ; Jastrow, Civ., 138.
[5] Jeremias, Der Alte Orient, 1929, xxvii, Heft 4.
[6] Smith, EHA, 45 f. ; Woolley, 1928, 64, 128 ; Gadd and Legrain, Ur Excav., Texts,

1928, i, 2 ; BMGB, 3 ; Baumstark, PW, ii, 2707 ; Meyer, Alt. I, ii, 488, 499.
[7] Smith, EHA, 79 f. ; Meyer, Alt. I, ii, 503 f. ; ib., Chronol., 33 ; King, S. and A., 220 f. ; Delaporte, 30 ; Woolley, Ur Excav., 1934, ii, 210, 334.
[8] Albright, q. in Isis, 1927, ix, 547.
[9] De Morgan, Délég. en Perse, ii, 129 ; Meissner, ii, 341 ; Smith, EHA, 11 f. ; Roeder, RV, i, 56 ; Langdon, Excav. at Kish, 1924, i, 95 f., correcting earlier theories of writing on clay.
[10] Woolley, Ur Excav., 1934, ii, 393.
[11] Meyer, Alt. I, ii, 467 ; Meissner, ii, 342, 344 ; BMGB, 98, 108, 186.
[12] Meyer, Alt. I, ii, 506 f. ; Meissner, i, 83 ; Jastrow, Trad., 12.
[13] Meyer, 507.

an early date, and occupied the town of Aššur (Assur), where the earliest settlers, non-Semitic and non-Aryan, were closely related to the Sumerians.[1] Many authorities think the Amorites appear as a new people about 3000–2600 in the west ("the land of sunset in its totality") and that they are mentioned by Gimil-Sin (c. 2200).[2] Their country, Amurru (Sumerian Martu), has been identified with Palestine [3], Syria [4], the country between Palmyra and Ana-Hit, or the west in general as far as the Mediterranean and east as far as Palmyra, but indefinite in the north and south, including the great land south-west of the upper reaches of the Euphrates.[5] In texts of the 3rd millennium it may include places much farther east, but not Syria, although the Amorites are generally called East Canaanites and came thence to Mesopotamia. It is doubtful if they were known to Gudea, that the 1 dyn. of Babylon was Amorite, or that the Amorites were then in Babylonia at all, although this was generally assumed.[6] From 1400 B.C. they appear more definitely in the Amarna letters; for the Assyrians the name was again geographical, the country extending to Palestine. The Amorites were similar to north European races. They were settled in Lebanon about 1500. They cut the hair and shaved, their speech was allied to Canaanitish (Phœnician and Hebrew) and they worshipped the storm-god Hadad, the Subarian Tešub, carrying a hammer and lightning fork, who was afterwards adopted by the Hittites, Mitanni and Akkadians.[7]

Under Sargon the Sumerians and Akkadians tended to fuse into one people. His influence extended over the whole of Mesopotamia, now becoming Semitised, also Palestine, Sinai, part of Egypt and possibly Cyprus. A commercial settlement at Ganeš in Cappadocia, however, was probably Sumerian.[8]

MAGAN AND MELUCHCHA

Sargon speaks of materials brought from two places, Magan (or Makan) and Meluchcha (Meluḫḫa), the identity of which has been much discussed. "Ships from Magan" are mentioned in a very old inscription at Ur [9], and the place was reached by sea. It was a country from which copper came and from which Narâm-Sin (2550 B.C.) imported diorite stone.[10] Magan was formerly identified with Sinai[11] or with East Arabia.[12] The first identification however, relies on a description of Magan as "mountain of copper," which occurs only in a late (7th century B.C.) text[13], and the modern view is that Magan was some place on the west coast of the Persian Gulf, probably Oman, where copper, diorite and dolerite still exist.[14] De Morgan[15], however, located

[1] King, History of Babylonia, 1915, 139, 227.

[2] Jastrow, Civ., 133; ib., Trad., 7; BMGB, 61, 85; Meyer, 509; King, History of Babylonia, 119 f.; Sayce, Ancient Egypt, 1924, 72; Barton, Isis, 1926, viii, 576.

[3] Jastrow, Trad., 7; W. H. Worrell, A Study of Races in the Ancient Near East, Cambridge, 1927, 102.

[4] Meyer, 509.

[5] Von Oppenheim, Tell Halaf, 47.

[6] Meissner, i, 15 f.

[7] Alt, RV, i, 155; Reche, ib., 157; Meyer, Alt. I, ii, 509 f.; II, i, 28, 100; ib., Reich und Kultur der Chetiter, 1914, 94; King, H. Bab., 139; von Oppenheim, Tell Halaf, 47, 85, 93, 222.

[8] Meyer, Alt. I, ii, 516; Woolley, 1928, 49, 79.

[9] Gadd and Legrain, Ur Excavations, Texts, 1928, i, 11.

[10] Sayce, Religion of the Ancient Babylonians and Assyrians, 1898, 31 f.; Meyer, Alt. I, ii, 527; Heuzey-de Sarzec, 122.

[11] Amiaud, Rec. of the Past, N.S., i, 42; ii, 73 f.; Sayce, Religion, 1898, 31, who suggests a connection with mafkat; Maspero, Dawn, 564, 600; King, S. and A., 241 f.; Heuzey-de Sarzec, i, 122, 172.

[12] Meyer, 527; Hommel, in Hilprecht, Explorations in Bible Lands, Edinburgh, 1903, 737; E. Schrader, Keilinschriften und das Alte Testament, 3rd ed., by Zimmern and Winckler, 1903, 15.

[13] O'Leary, Arabia before Muhammad, 1927, 48; Langdon, JEA, 1921, v, 143; but cf. Hall, ib., 40; Albright, ib., 80 f.

[14] A. T. Wilson, The Persian Gulf, Oxford, 1928, 27; Woolley, 1928, 46, 82; Peake, q. by Dussaud, La Lydie et ses Voisins, Babyloniaca, 1930, xi, 81.

[15] Préhist. orient., ii, 323; Délég. en Perse, i, 38 f.

Magan on the Khabur, a tributary of the Euphrates, and it has been suggested [1] that the name meant vaguely "the land to which ships go."

From his expeditions to Asia Minor, Sargon brought home specimens of foreign trees, vines, figs and roses, for acclimatisation in his own country.[2] He speaks of Mari (Amor or Amurru), Iarmuti and Ibla (probably in North Syria), of the cedar forest (Lebanon) and the silver mountain (in Asia Minor, just north of the Cilician gates), and obtained building material and stones from Tidanum (Syria) [3], Magan, Martu, etc., gold from the mountains of Ghaghum; from Kimaš (Elam), he obtained copper, bitumen, silver, perhaps bronze, etc.[4]

Meluchcha has been identified with Sinai [5], Central and Western Arabia as far as the Sinai Peninsula [6], the desert south of Wâdî el-Arish [7] and Ethiopia.[8] In the Amarna letters (c. 1400 B.C.) Meluchcha is equivalent to Kâši, the Egyptian Kosh, Hebrew Kush, i.e. Nubia; in Ašurbanipal's time it was Ethiopia, reached by sea round the south of Arabia. We can only conclude that the names Magan and Meluchcha were later extended to countries reached by ships which on their way touched and traded with the places to which those names were originally applied. Dilmun, Magan and Meluchcha probably lay, in this order, along the Arabian coast of the Persian Gulf, the first corresponding with the largest of the Bahrain islands; at a later period Magan was extended to Sinai and the Red Sea coast of Egypt, and Meluchcha to Ethiopia. The whole east side of Arabia was perhaps familiar to the Sumerians and Akkadians, who reached it both by land and sea from 3000 B.C.[9]

THE THIRD DYNASTY OF UR

The dynasty of Akkad ended after 2571 B.C. by the invasion of the people of Su (Subarians) and Gu (Guti or Gutium) from the north; the Guti from the Zagros conquered the country, although the southern cities of Sumer held out for 30 years under the leadership of Erech. The Guti then ruled the whole country for 125 years from Arrapkha (? Kerkuk). Gudea (2450) was patesi of Lagaš under the Guti dynasty, although he enjoyed autonomy in a period when there was a reversion to "city states." During this period a trading settlement, probably Sumerian, was established at Kara-Euyuk in the heart of Asia Minor.[10]

The Third Dynasty of Ur, "when chronology begins to be certain," was founded in 2278 by Ur Nammu. Its kings first called themselves "kings of Sumer and Akkad." The second king, Sulgi (2276–2231), was the first to claim divinity in Babylonia.[11]

Sumerian influence was then predominant. Ur Nammu and Sulgi rebuilt and renovated the land, which had fallen into ruin under the Guti, and vast quantities of material were imported from Elam. They rebuilt the temple of the kings of Akkad at Nippur; the chief god was the moon-god Nannar (Sumerian) or Sin, whose great temple at Ur was rebuilt by Ur Nammu.[12]

[1] Hall, NE, 187.
[2] Woolley, 1928, 79.
[3] Olmstead, History of Palestine and Syria, 47, 79, 533; Ebeling and Meissner, Real-Lexicon der Assyriologie [RA], 1929, i, 99.
[4] Amiaud, Rec. of the Past, N.S., i, 42; ii, 73 f.; Meissner, i, 53; Meyer, Alt. I, ii, 539; Price, J. Amer. Oriental Soc., 1923, xliii, 45; BMGB, 32, 56.
[5] Meyer, Alt. I, ii, 543.
[6] Hommel in Hilprecht, Explor., 737; Schrader, Keil., 15.
[7] Sayce, Rel., 32.
[8] Hall, NE, 187; King, S. and A., 261;

Meissner, i, 33; Schroeder, RV, vii, 348; Price, op. cit., 41.
[9] O'Leary, 49 f.; cf. Maspero, Dawn, 562, 564, 600, 614, 616.
[10] Meyer, Alt. I, ii, 546; Peake and Fleure, Steppe and Sown, 128 f., 130, 138.
[11] Delaporte, 18; King, S. and A., 274, 283, 291 f., 298; Woolley, 1928, 134 f.; Ält. Chronol., 30; Meissner, ii, 445; Woolley, 1928, 24.
[12] Meyer, Alt. I, ii, 552 f.; King, S. and A., 283, 291 f., 298; Woolley, 1928, 134 f.; L. Legrain, Le temps des rois d'Ur, Bibl. de l'École des hautes Études, Philolog. et Hist. Cl., vol. 199, with plates, Paris, 1912.

The rich cities of the south country then controlled the head of the Persian Gulf with its shipping. One trade route led up the eastern bank of the Tigris to the north, and goods in transit for Syria passed up the Euphrates to the fords at Carchemish and so made their way south or north-west. By the same route, floated downstream in wooden boats broken up for their timber at the journey's end, came silver and copper ore, cedar from Lebanon, lye [?] and cystus gum. The Gulf traffic was more important than that from the north. A bill of lading shows that a ship in 2048 unloaded at Ur from the Gulf a cargo of gold, copper ore, ivory, precious woods and fine stone. The raw materials were worked up by Sumerian craftsmen and re-exported by the land routes to the west and north. There were trading colonies in Cappadocia and Syria.[1]

The old town rulers, the patesi, who had often assumed the title lugal (king), were now degraded to officials. The kingdom was much extended. Copper and gold were imported from Meluchcha and bitumen or naphtha (or both) from the spring of Madqa (Ibisin).[2] The Third Dynasty of Ur ended with the rise of the Elamites, who sacked and looted the city in 2180 or 2170, when the independent history of Sumer ceases. Conditions in the following period are very obscure, but an Amorite dynasty was founded at Babylon in 2049 B.C.[3]

After the downfall of Ur there were again city-states ; those of Isin, Larsa, Erech (Uruk), Sippar, Babylon and Kiš had independent kings. Under the Larsa dynasty (2170–1910) the scribes composed historical works, collected books of hymns, omens, signs of divination from the liver, old legends, etc., and reduced the Sumerian religion to a system.[4] In this period there was active commerce in the larger towns (not "cities") with metal money, the standard of value being silver. Gold was more valuable than silver, but copper, used for tools, etc., was very cheap as compared with its value in Egypt.[5]

Weights and Measures

Coinage proper did not appear until the Persian Period and silver was measured by weight in scales according to a sexagesimal system : 1 talent (Sum. gun ; Akkad. biltu) = 60 minas ; 1 mina (Sum. mana ; Akkad. manu or mâneh ; $\mu\nu\hat{a}$ first in Herodotos)[6] = 60 shekels (Sum. gin ; Akkad. šiqlu).[7] A crude representation of weighing metal in a balance, apparently of the Egyptian type, was found in the palace of Sargon II at Khorsabad.[8] The origin of the sexagesimal system may be connected[9] with the numbers 7 (a sacred number) and 5 (the number of fingers and thumb) : $5 + 7 = 12$; $5 \times 12 = 60$; or the obvious mass divisions $\frac{1}{3}, \frac{1}{2}, \frac{2}{3}, \frac{1}{10}$, being extended at each end, gave 60, divisible by 2, 3 and 10, the older decimal system being perhaps retained in ordinary life.[10] Neugebauer considers that a minus sign existed and postulates borrowing by India from the Babylonians, perhaps the Sumerians.

The standard mina was fixed by Sulgi at 505 gm., about 2 oz. more than 1 lb. avoirdupois (453·6 gm.), the shekel being 8·4–8·5 gm., and the talent 30·3–30·505 kgm.[11] The actual weights of three specimens of Gudea's mina, in

[1] Woolley, 1928, 115 f.
[2] Meyer, Alt. I, ii, 554, 556 ; Legrain, 8 ; Schrader, Keil., 15.
[3] Woolley, 1928, 24, 168, 170 ; Meyer, Chronol., 25, 28, 30 ; King, S. and A., 304 ; Jastrow, Civ., 141.
[4] Woolley, 1928, 171, 178 f.
[5] Meyer, Alt. I, ii, 579 ; Meissner, i, 355 ; BMGB, 105 f. ; Woolley, Ur Excav., 1934, ii, 210 f. ; ib., 1928, 117.
[6] ii, 168, etc. ; Baumstark, PW, ii, 2714 f.

[7] Delaporte, 256 ; Meissner, i, 357.
[8] Contenau, Manuel, iii, 1267.
[9] Jeremias, 164.
[10] Neugebauer, Isis, 1928, xi, 525 ; Cruickshank, ERE, ix, 414 f.
[11] Sayce, B. and A. Life, 159 ; Meyer, Alt. I, ii, 579 f. ; Meissner, i, 357 ; Lenormant, Monnaie, i, 104, 111 ; Luckenbill, ii, 499 ; Delaporte, 248 f. ; Gardner, DA, ii, 444 f.

diorite, were 489·6 gm., 492·9 gm. and 491·2 gm., the mean being 491·2 gm.[1] Babylonian weights from Nineveh show that two principal standards were used in Mesopotamia, Syria and Asia Minor in the Assyrian Period, one based on the light mina of 505 gm. and the other on the heavy mina of 1,010 gm.[2] The corresponding shekels of 8·4 and 16·8 gm. were the weights of the earliest gold coins of Asia Minor, and Brandis suggested that the heavy shekel was adopted in Phœnicia and the light in Lydia and thence the Greek colonies of Asia Minor. The relation between the weight of the light shekel and the Egyptian kite (9·09 gm.) is probably accidental.[3]

Weights were of stone, including diorite, agate and onyx, or of baked clay, later of bronze or other metals. They had various shapes : geometrical (rings, cones, pyramids, etc.), or animal forms (lion, duck, etc.), or human heads.[4] Weights varied from time to time and from place to place, and there were fraudulent weights, a "heavy" set for buying and a "light" set for selling. The old Babylonian unit, "the barleycorn" (grain) (še'u = 46·75 mgm. ; cf. Greek $\kappa\rho\iota\theta\dot{\eta}$, "crith"), $\frac{1}{180}$ shekel, was later replaced for small weights by the subdivisions of the shekel into $\frac{2}{3}$, $\frac{1}{2}$, $\frac{1}{3}$, $\frac{1}{4}$, $\frac{1}{5}$, $\frac{1}{6}$, $\frac{1}{8}$, $\frac{1}{10}$ and $\frac{1}{24}$ parts, each with a special name.[5]

Gold, silver, lead and copper, in bars, as well as corn, fruits, animals, etc., were used in payments. Maspero says the silver was weighed in ingots, not rings ; Meyer, Woolley and Sayce that both bars and (as in Egypt) rings were used, those of one-third of a shekel being common, and smaller amounts of silver were cut off standard bars or rings and weighed.[6]

In Canaanitic countries copper was the coinage metal.[7] A fixed ratio of value of equal weights of gold and silver was adopted in 2300 B.C., and later between silver and copper : the final value of the first ratio, 11 : 1, remained until a late period the norm for Lydian, Persian, Phœnician and Carthaginian coinage.[8]

Measures of length were very variable : Gudea's ell (ammatu) was 0·495 m., the Nippur ell 0·518 m., etc., and in ancient Assyria buildings were measured in "brick courses" (tibku). Assyrian units of length were the "finger" (ubânu) of 16·5 mm., cubit (ammatu) of 24 ubânu = 396 mm., "reed" (qânu) of 6 cubits = 2·376 m., and the gar of 2 "reeds."[9] The division of a foot into twelve inches is of Babylonian origin : the Babylonian foot was about 330 mm., the ell 450 mm., and the old Babylonian "double ell" 990–996 mm., which Lehmann thought was the length of the seconds pendulum (992·5—993·1 mm. for latitude 30°–40° N.), or else a multiple of the "finger" (16·5 mm.) of Gudea's time (60 × 16·5 = 990 mm.).[10]

The superficial measure in old Babylonia was the "bed" [of grain] (Sum. sar : Akkad. musarû) of 35·284 sq. m.[11] The "field" (Sum. iku ; Adkka. ikû) was 100 sars ; 18 fields was a "hole" ["Loch"] (Sum. bur ; Akkad. bûru). The unit of volume, the sila, formerly thought to be 0·4 litre, but really 0·842 litre, was constant in Mesopotamia for a long period. The Assyrian sûtu was

[1] Walden, Mass, Zahl und Gewicht in der Chemie der Vergangenheit, Stuttgart, 1931, 13.

[2] 982·4 in Feldhaus, Technik, 691 ; 982·4–985·8, Lehmann, q. by Walden, 13.

[3] Gardner, DA, ii, 445 ; BMGB, 171 ; Sayce, B. and A. Life, 159 ; Meyer, Alt. I, ii, 580.

[4] Walden, 15 ; J. Brandis, Das Münz, Maass und Gewichts-System in Vorderasien, 1866, 53, 78 ; Sayce, B. and A. Life, 265 ; Lenormant, Monnaie, i, 104, 116 ; Layard, Discoveries in the Ruins of Nineveh and Babylon, 1853, 600.

[5] Meissner, i, 357 ; BMGB, 25, 214 ; Luckenbill, ii, 499 ; Baumstark, PW, ii, 2714 f.

[6] Meyer, Alt. I, ii, 580 ; Maspero, Dawn, 721, 749 ; Woolley, 1928, 117 ; Sayce, B. and A. Life, 160, 226.

[7] Jeremias, 87.

[8] Baumstark, PW, ii, 2715.

[9] Meissner, i, 358 ; Luckenbill, ii, 500.

[10] Walden, 12 f. ; BMGB, 24 ; Gardner, DA, ii, 444. ; cf. Petrie, Isis, 1934, xxii, 377.

[11] Luckenbill, ii, 500, says 22·58 sq. m.

10 sila, the pi was 50·52 litres. The old Babylonian kurru (or gur) was 300 sila (252 litres), the late Babylonian kurru 180 sila, and the Assyrian imêru (homer) was 100 sila = 84·2 litres.[1]

Lehmann thought the tenth of the Babylonian double ell (99 to 99·6 mm.) was the basis of the volume unit, the water weight of which, $(9·9)^3$ to $(9·96)^3 =$ 971 to 988 gm., was the heavy mina [2], whilst Balaiew [3] concludes that the (light) mina introduced by Sulgi, possibly in relation with Egypt, was represented by three values : 491·14 gm., 502·10 gm. and 522·8 gm., surviving as the Russian pound, the Persian mina and the Russian bezmen respectively. According to Hoppe [4] the oldest Sumerian time unit was one-sixth of a day, called sussu ; half this was a kasbu. There was also a division according to a decimal system, giving a day of 24 hours, and a sexagesimal system. In the Assyrian Period the sussu was divided into 60 parts, each $\frac{1}{360}$ of a day. The circumference of the circle was divided in the Sumerian Period into 360 degrees, each of 60 minutes.[5]

ASSYRIA

Assyria arose about the town of Assur (more correctly Aššur), now Qal'at Šarqât, near Môsul, which was perhaps called after the god Aššur.[6] From Assur dominion spread along the Tigris Valley about 2000, although incursions from the north-west and settlements from the south date to a much earlier period.[7] The early Assyrians (2750–2650) were a distinct people, acquainted with metals, and in relation with the Sumerians and Akkadians.[8] They are sometimes regarded as very closely related to the Sumerians, if not the same people [9]; Meyer[10] thought they came from Asia Minor. A great conflagration marked the end of early Sumerian domination in Assur ; a new temple was then built and the Semitic (Akkadian) influence became powerful.[11]

It is not proved that Assur was originally Sumerian, and Sumerian influence was perhaps imposed on it only later (c. 2700 B.C.).[12] The chief element in the early population of Assyria was Subarian (Subaraean), the district being then called Subartu.[13] The Subarians, whose culture has been supposed to go back to 4000 B.C., were quite different from the Sumerians, with whom they were constantly in conflict.[14] The so-called Cappadocian tablets, found at Kültepe in 1920 and now in the British Museum, were written about 2250–2150 in cuneiform in a dialect of the town of Assur, and give information of trade in metals, etc. They show that Assyrians lived among the Semitic people round Cæsarea (Mazaca), and their caravans transported large quantities of lead, copper, silver and cloth to Assur. The settlement in Cappadocia probably existed as early as 2400 B.C.[15]

Subartu adjoined Elam ; for the Akkadians it was the great tract of country

[1] Meissner, i, 357, 358, 466 ; ii, 492 ; Luckenbill, ii, 500 ; BMGB, 25 ; Maspero, Dawn, 773 f. ; older ideas of metrology have been considerably revised, according to Otto, 1925, 29, who refers to O. Viedebantt, Antike Gewichtsnormen und Münzfüsse, 1923, which I have not seen.

[2] Walden, 13.

[3] Q. in Isis, 1929, xiii, 203 ; 1931, xv, 439.

[4] Q. in Isis, 1928, xi, 491 ; cf. Jeremias, Der Alte Orient, 1929, xxvii, Heft 4, 12.

[5] Delaporte, 251 ; cf. Gandz, Isis, 1929, xii, 457.

[6] Jastrow, Civ., 17 ; Sayce, Rel., 123 ; Smith, EHA, 102 ; Delaporte, 345 ; E. Unger, Der Alte Orient, 1929, xxvii, Heft 3, Das Stadtbild von Assur.

[7] Jastrow, Civ., 228 ; Smith, EHA, 1 ; Unger, RA, i, 170 ; Weissbach, ib., 228 f.

[8] Delaporte, 264 f., 364 ; Smith, CAH, iii, 89 ; ib., EHA, 8.

[9] King, H. Bab. 137 ; Woolley, 1928, 47.

[10] Alt. I, ii, 607.

[11] Smith, EHA, 77, 79.

[12] Smith, EHA, 70, 139 ; Jastrow, Civ., 115.

[13] Smith, EHA, 72 f. ; Knudtzon, Die El Amarna Tafeln, Leipzig, 1915, 1071.

[14] Von Oppenheim, Tell Halaf, 42 f., 53, 214 ; Smith, EHA, 72, 74, 102 f.

[15] Landsberger, Alte Orient, 1924, xxiv, Heft 4 ; Contenau, Manuel, ii, 812 ; BMGB, 10, 11, 20, 160, 163 f. ; Jastrow, Civ., 160 ; Meyer, Chetiter, 76.

in the west of south-west Persia, Middle and Upper Mesopotamia and Syria ; the mountains near Kermanshah, Kurdistan and Armenia to the Mediterranean; and Palestine in the south. The Subarian culture was prehistoric and took its place next to the Egyptian and Babylonian. In the 3rd millennium Kültepe (old Kaneš) and Ursu were important industrial centres trading with Assur, gold and silver in varying qualities being given in payment for goods from Cappadocia. These goods were transported not only on pack animals and wagons but also on boats on the Tigris.[1]

The Assyrians were probably a mixed race, an Armenian element related to the Cassites of Asia Minor (perhaps the Zagros) being combined with a Semitic strain. Asarhaddon (680–669) was perhaps the first truly Semitic ruler of Assyria, of which he was represented as the "founder." After the earliest period there is a blank until an Assyrian embassy is represented as bringing lapis lazuli to Thothmes III (1500 B.C.).[2]

The Assyrian god Aššur is represented as an archer inside a winged circular disc, usually regarded as the Egyptian solar symbol of the god Horus of Edfu [3], but perhaps really Subarian.[4] The symbol was afterwards adopted by the Hittites.[5] Aššur has been regarded as related to the Egyptian Osiris ; he was a sun-god as early as 2400 B.C., and was afterwards the counterpart of the Babylonian god Marduk (Bêl).[6] As "the great lord" he is associated with various forms of a sacred tree, perhaps representing the "tree of life," the later elaborate and stylised forms of which are shown along with genii carrying sacred baskets. Several sacred trees, e.g. date palm, cedar, cypress and tamarisk (which appears in the Egyptian myth of Osiris) are depicted.[7]

In the old (c. 2700 B.C.) temple of the goddess Ištar at Assur are water pots, stone statues, terra-cotta cult models of houses, incense and libation pots, two broken mirrors (one of copper and one of silver), many beads of frit dipped in glaze, a copper sickle, stone moulds for casting metal ornaments in the shape of animals and rosettes, statuettes of gypsum and ivory and gypsum plaques, some painted in black, white and red. The pottery has curious painted marks near the lip, generally circles with spots.[8]

HAMMURABI AND BABYLON

The complete fusion of Sumer and Akkad was accomplished by Ḥammurabi (1924–1913), an Amorite or East Canaanite(Achlamâ) who ruled from Babylon.[9] From his time we can speak of a Babylonian culture, which incorporates both Sumerian and Semitic-Akkadian elements.[10]

Ḥammurabi's influence probably extended to Syria, from whence he obtained stone, metals and resinous woods, although this country is not actually named. A series of letters from him to a high official in the south indicate a high organisation of political business, and enforce a just treatment of the peoples concerned. His famous code of laws was probably based on Sumerian originals. It is written in Akkadian on a diorite stela found at Susa.[11]

[1] Von Oppenheim, Tell Halaf, 47, 52 ; Landsberger, Alte Orient, 1925, xxiv, Heft 4, Assyrische Handelskolonien in Kleinasien.

[2] Meyer, Alt. I, ii, 607 ; Smith, EHA, 102 f., 151 ; Delaporte, 266 f.

[3] Handcock, Mesopotamian Archæology, 1912, 393 ; Rawlinson, Mon., ii, 2 ; Smith, EHA, 12.

[4] Von Oppenheim, 85, 93, 222.

[5] Delaporte, 346 f., 369.

[6] Delaporte, 345 ; Smith, CAH, iii, 91 ; ib., EHA, 123 f.

[7] Rawlinson, Mon., ii, 3 f., 7 f., 29 f. ;

BMGB, pl. xiv ; Smith, CAH, iii, 92 ; Schrader, Keil., 526 ; Sayce, Rel., 240.

[8] Smith, EHA, 62, 65, 67 f.

[9] Meyer, Alt. I, ii, 628 f. ; ib., Chron., 25 ; Weidner, MGM, 1918, xvii, 44 ; Jastrow, Civ., 148 ; ib., Trad., 13 ; Unger, RV, viii, 167 ; King, H. Bab., 119 f., 162 f.

[10] A. Jeremias, Der Alte Orient, 1929, xxvii, Heft 4, Die Weltanschauung der Sumerer.

[11] Delaporte, 44 ; Meyer, Alt. I, ii, 634 f., 638 ; de Morgan, Délégation en Perse, iv, 11 ; Woolley, 1928, 91.

The Invasion of Mesopotamia

The first Babylonian Empire ended in 1750, when the Ḫatti or Hittites appeared from Cappadocia, Asia Minor beyond the Taurus, and sacked and burnt Babylon.[1]   Another race, the Cassites (Kaššû), had come perhaps once as early as 1896 B.C., but certainly again about 1750. They were a non-Semitic people, probably the ancestors of the Kurds, either Aryan or (more probably) under Aryan influences, and came from the mountains of Iran, the wildest parts of the Zagros.[2]   There was also an Aryan invasion with the horse, which was used for drawing wagons and chariots and is first mentioned definitely by Ḫammurabi, but is represented at Khafaje about 3000 B.C.[3]

During the Cassite Period (1746–1171 B.C.), horses and lapis lazuli were the chief exports from Babylonia, and there was contact with Assur, perhaps even with the Mediterranean.[4]   The Assyrians during the Cassite rule were kept in check by the Hittites and the non-Semitic Mitanni, a branch of the Khurri, whose kingdom flourished for about 75 years about 1450. The Mitanni, who have been called Subarians, were established between the Hittites on the west and Assyria on the east, their kingdom extending from the Euphrates in the south to the mountains in the north ; their fall opened the way to the Assyrians, who made repeated campaigns into the hill countries of the Euphrates. The Mitanni, who were settled in modern Erbil, Kerkuk, Samarra and on the Khabur, came from the Taurus about 2000, and became powerful in the 15th century. They adopted the Subarian culture of the land in which they settled and brought with them extensive knowledge of the care and breeding of the horse. Their language was sub-Aryan, and von Oppenheim regards the Mitanni as East Indogermanic, though different from the Hittites.[5]

In this period (1450–1400) Babylonia had close relations with Egypt, as is shown by the Tell el-Amarna letters.[6]   About 1300 come the tablets from the Hittite capital of Boghazköi ; Assyrian was then the language of diplomacy in Egypt, Syria, Cyprus and Asia Minor.[7]   About 1400 the Achlamê, or Aramæans, who were Semitic nomads or half-nomads from South Arabia, appeared in Syria and Palestine, penetrating to Mesopotamia and adopting the Subarian culture ; their language (Aramaic) was soon adopted over a wide area.[8]

The Aramaic Kaldu, or Chaldæans, appeared in south Babylonia, perhaps from east Arabia, in the 14th century B.C. and in large numbers about 1100, at first confining their settlements to the banks of the lower Euphrates, a marshy district which had gradually become derelict from the Cassite Period.[9]

Assyrian Supremacy

The Assyrian Tiglathpileser I (1115–1089) extended his kingdom to the sources of the Euphrates, subdued the Hittites, and controlled the regions of

[1] Jastrow, Civ., 150 f. ; BMGB, 118.

[2] Meyer, Alt. I, ii, 648 f., 653 ; II, i, 40 ; ib., Chronol., 25 ; Jastrow, Civ., 150, 153, 157 ; King, H. Bab., 84, 197, 210 f., 214 f. ; BMGB, 7 ; Unger, RV, viii, 167 ; Meissner, i, 17 ; Contenau, Manuel, ii, 876.

[3] Hrozný, EB[14], xi, 604 ; von Oppenheim, Tell Halaf, 61, 133 f., 137 f. ; Langdon, Alte Orient, 1928, xxvi, 35, corrected in Peake and Fleure, Merchant Venturers in Bronze, Oxford, 1931, 114, 117 ; Ur Excav., 1934, ii, 271 f. ; Frankfort, Illustr. London News, 1932, clxxxi, 529.

[4] King, H. Bab., 224 ; Meyer, Alt. I, ii, 668 ; Jastrow, Civ., 159 ; Garstang, The Land of the Hittites, 1910, 369.

[5] BMGB, 8 ; Meyer, Alt. II, i, 29 ; Unger, RV, viii, 167 ; Schroeder, ib., 217 ; Gustavs, ib., 226 ; von Oppenheim, Tell Halaf, 59, 61, 82, 133 f., 138.

[6] King, H. Bab., 219 f. ; Schrader, Keil., 192 f. ; BMGB, 122 f.

[7] BMGB, 160.

[8] Meissner, i, 15 ; Breasted, Anct. Times, 1916, 144 f. ; von Oppenheim, Tell Halaf, 62, 65 f.—from the 13th century ; Meyer, Alt. II, ii, 366, 386 f.

[9] Schrader, Keil., 22 ; Meyer, Alt. II, i, 342 ; ib., II, ii, 388, 390 ; Smith, EHA, 303 f. ; Maspero, Dawn, 563 ; Unger, RV, viii, 162.

Asia Minor beyond the Taurus. Egypt acknowledged his independence. He reached the Mediterranean, captured Babylon and controlled the caravan routes of Western Asia ; Syria and the Phœnician cities were then freely accessible and minerals from Asia Minor came to Assur across the Upper Euphrates.[1]   The queen Semiramis, whose exploits are described by Herodotos and Diodoros Siculus, has been (doubtfully) identified with Sammurâmat, the wife of Šamsi-Adad V (1055–1050).[2]   On the death of Tiglathpileser I the extensive Assyrian empire came to an end, but under Tukulti-Urta (or Ninib) II (889–884) Assyrian power revived and reached a culmination about two centuries later. The Assyrians were then a great military nation, using all kinds of equipment, including great metal-capped battering rams mounted on four or six wheels, heavily armoured like a "tank " and carrying a high tower from which archers shot arrows over the city wall whilst the latter was mean-time demolished by the battering-rams. An officer in a metal-domed observation tower with spy-holes directed the attack on a fortress, various kinds of engines, including catapults, being used by the defence, and burning pitch or petroleum was poured over the heads of the attackers storming the walls with scaling ladders. With all their military, business and technical skill the Assyrians were and remained always a half-civilised nation.[3]

Ašurnasirpal II (885–860), often called Ašurnasirpal III, and Šalmaneser III (860–824), obtained gold, silver, copper, bronze, lead, iron (not from Phœnicia), ivory, coloured garments and valuable stone from the Hittites, Palestine and Phœnicia.[4]

Šalmaneser III brought Syria and Palestine under Assyrian power, and Jehu, king of Israel, sent him tribute. Tiglathpileser III (745–727) (not IV, who never existed), the "Pul" of the Bible, was a usurper who headed a new line of rulers under whom Assyria entered upon her last and most glorious phase.[5]   In his time, caravans transported to Nineveh, along five or six trade routes between Mesopotamia and Media, the products of Central Asia : gold, copper, iron, fabrics, precious stones (including carnelian, agate and lapis lazuli), and curious animals.[6]

URARTU

The civilisation of Urartu centred about Lake Van, the Ararat of the Bible, the land of the 'Αλαρόδιοι of the Greeks, and modern Armenia. It is first known under Šalmaneser I (1276–1257). The city of Van (ancient Tuspâ ; Θωσπία) was probably founded about 840 by King Šarduris I.[7]   The language of the inscriptions is related to Georgian and not connected with Hittite, and went out of use in the 7th century B.C. The people are supposed to have brought with them from their unknown original home, perhaps in part Thrace, a know-ledge of iron, and they were influenced by the Assyrian culture.[8]   They hewed out rock dwellings with iron tools and had as gods the Khaldi (perhaps Khaldis as supreme god), associated by some with the Chaldians (Χαλδοί, or Χαλδία) of the Pontos, perhaps borrowed from the Hittites and including Tešub. The name Chaldian seems also to apply to the Urartians ; Strabo's "Chaldæans"

[1] Meyer, Alt. II, ii, 377 ; Jastrow, Civ., 166 ; Meissner, ii, 451.
[2] Contenau, Manuel, iii, 1211 ; Hil-precht, Excavations in Assyria and Baby-lonia, Philadelphia, 1904, 130.
[3] Breasted-Jones, 105 ; Meissner, i, 39, 108 f., 110 f.; Feldhaus, Technik, 309 ; Delaporte, 305 ; Maspero, The Passing of the Empires, 1900, 8 f., 11 ; Meyer, Alt. II, ii, 383, 399 ; Luckenbill, i, 163 ; Worrell, Study of Races in the Ancient Near East, 1927, 104.

[4] Sayce, Rec. of the Past, N.S., 1889, ii, 128 f. ; Rogers, ii, 46, 64 f.
[5] Jastrow, Civ., 173 f.; Luckenbill, i, 200 ; Delaporte, 277, Fig. 41 ; BMGB, 13, 46, 183 ; Smith, CAH, iii, 32 f.
[6] Rawlinson, Mon., i, 553 f. ; Maspero, H. anc., 1905, 467.
[7] Maspero, Empires, 52 ; Meyer, Alt. II, ii, 418 f. ; Herodotos, vii, 79 ; Sayce, CAH, ii, 240 f. ; iii, 19, 169, 173.
[8] Sayce, CAH, iii, 19, 172 ; Rostovtzeff, i, 110, 117 ; Meissner, i, 18.

15

were Chalybes from near the Black Sea, and have nothing to do with the Chaldæans of Mesopotamia.[1] There was, however, no connection in the earlier period between the Black Sea and the people of Lake Van, the Alarodians, who were neither Aryans nor Semites.[2]

Pliny [3] and Strabo [4] refer to the minerals of Armenia. Strabo says the water of Lake Van ('Αροηνή) contains nitron (soda), which makes it suitable for cleaning cloth (the scum is still so used) [5] but unfit for drinking. The gold mines at Hyspiratis (probably north Armenia) were worked at the time of Alexander's expedition, and Pompey obtained a contribution of 6,000 talents of silver from Armenia. There was a mine of sandyx, "the Armenian colour." The Urartians executed fine gold and silver work similar to that found at Carchemish, and bronze, copper and iron were worked, the bronze being excellent.[6] It has been suggested that much of the metal work attributed to Assyria was actually made in Van, the artistic traditions of which were transmitted by the Medes to Achæmenid Persia.[7] There are remains of very old copper mines in Armenia [8], and the earliest copper of Egypt and Mesopotamia may have come from this region.[9] Iron was plentiful and was early introduced from mines of north-east Asia Minor. The extensive replacement of bronze by iron under Sargon II (722–706) was probably due to his campaigns in the north, when he demolished the temple of the Khaldis and took immense spoil of gold, silver, bronze, precious woods, stone, ivory, iron and scarlet linen, the metals being in the form of cups, daggers, shields, etc. During the period of the Vannic kingdom there was a general resemblance between the inhabitants of the Armenian plateau and those of Asia Minor.[10]

ASSYRIANS AND CHALDÆANS

Sargon II built a great palace at Dûr-Šarrukin (mod. Khorsabad), from which many antiquities have been excavated.[11] His records are unusually ferocious and brutal even for an Assyrian, and he usually calls himself "king of the universe"—a title already assumed by Tukulti-Urta I (1260–30). The brutality, however, reaches its zenith with his successor Sennacherib (705–681), who usually finishes his records with the words : "I besieged, I conquered, I despoiled, I destroyed, I devastated, I burned with fire ; with the smoke of their conflagration I covered the wide heavens like a hurricane," and whose detailed descriptions of the mutilation of prisoners have probably never been excelled in horror and barbarity.[12] Sennacherib defeated the Chaldæan pirates in 694 by means of a fleet built on the Tigris, and drove them to Gerrha, where it was so hot that people lived in houses built of blocks of salt.[13]

Ašurbanipal (668–626) restored the temples and royal houses ; it is possible that his library was public.[14] His records are still brutal, but he promoted art and learning, which reached a high level, and had the old inscriptions and the works of reference, grammar, lexicography, poetry, history, science and

[1] Sayce, CAH, iii, 19, 170, 185 ; Maspero, Empires, 56 ; Meyer, Alt. II, ii, 420.
[2] Maspero, Empires, 55 ; Lehmann-Haupt, PW, xi, 400 ; Karst, Grundsteine zu einer mittelländisch-asianischen Urgeschichte, Leipzig, 1928, vi, 132, etc.
[3] xxxvii, 6.
[4] XI, xiv, 8, 529 C.
[5] Tozer, History of Ancient Geography, Cambridge, 1896, 269.
[6] Sayce, CAH, iii, 19, 185 ; Maspero, Empires, 57 ; Meissner, i, plates, Fig. 74—bronze statue.
[7] Herzfeld, q. by Dalton, The Treasure of the Oxus, 1926, 8.

[8] Montelius, Ält. Bronzezeit, 1900, 197.
[9] O'Leary, Arabia before Muhamm., 44.
[10] Sayce, CAH, iii, 180, 186 ; Herzfeld, Islam, 1921, xi, 128 f., 136 ; cf. Maspero, Empires, 59 f. ; O. M. Dalton, The Treasure of the Oxus, 2 ed., 1926, p. xlii.
[11] V. Place, Ninive et l'Assyrie, 3 vols., 1867–70—Place thought Khorsabad was Nineveh ; Smith, CAH, iii, 45 ; BMGB, 14, 32.
[12] Luckenbill, ii, 50, 124, 127.
[13] Strabo, XVI, iii, 3, 765 C. ; H. G. Rawlinson, Intercourse between India and the Western World, Cambridge, 1926, 6 ; ? El-Khatir, Williams, DG, i, 998.
[14] Sayce, Rel., 11.

religion from old libraries in Babylonia copied and annotated.[1] The tablets in his library at Nineveh (mod. Quyûnjiq), discovered in 1842 by Layard and deciphered by G. Smith and Sir H. Rawlinson, deal largely with magic and are in the British Museum.[2] In his reign successive waves of Aryan immigration washed across the Caucasus, made up of Cimmerians, Mannai, Ašguzeans, etc. These wild hordes of the north-east, to which the general name Manda was applied, were only kept in check by vast armies.[3] On his death Nabopolassar, a man of humble birth from Chaldæa, the "sea land" in Southern Babylonia, became king of the Kaldu or Kalda. Kaldu, or Chaldæans, was the original name of the people of South Babylonia, or the "land of the sea-kings," which was less affected by invasion than the more northerly regions. Berossos (c. 280 B.C.) calls the oldest inhabitants of Babylonia or Chaldæa (he uses both names) always "Chaldæans," whilst the "Babylonians" are the inhabitants of the city of Babylon.[4] Nineveh was captured in 612 B.C. by a confederation of Cyaxares, king of Media, Nabopolassar and (possibly) the Scythians (perhaps Bactrians). The Medes then turned on the Scythians, and the latter disappeared for ever as a nation. Nineveh fell, for the first time in history, after three unsuccessful assaults : the Babylonian chronicles say : "a great havoc took place . . . they carried off the booty of the city, a quantity beyond reckoning, they turned the city into a ruined mound." This ended the Assyrian power for ever : the remnants of the defeated garrison took refuge in the fortress of Harran, but this in turn was devastated in 610, and Chaldæan Babylon became the leading power in the world.[5]

Nebuchadnezzar II (604–562), who founded the Second Babylonian (or Chaldæan) Empire with the capital at Babylon, was a vigorous ruler. Syria and Palestine were subject to him. Jehoiakim, king of Judah, revolted, but he was defeated and Jerusalem captured in 597, a thousand "workers in iron" being carried off.[6] Babylon then rose to great importance.[7]

Nabu-na'id (Nabonidus), who took the throne of the Chaldæan Empire in 556, was a scholar and antiquary and restored the great temple of the Moon (Sin) at Harran, as well as the temples of the Sun at Sippar and Larsa. He was succeeded in 539 by his son Belshazzar, but in the same year Babylon was taken without difficulty by Cyrus the Persian, who had begun a war in 547.[8]

The fall of Babylon produced an impression even more powerful than that resulting from the sack of Nineveh.[9] Owing to the tolerance of the Persians, who formed only a very small fraction of the total population of the Persian Empire (500,000 in 50 millions in the period of the Achæmenidæ and no doubt fewer in the earlier period), the Babylonians even after the conquest retained most of their old beliefs and customs, although an interchange of Babylonian and Persian elements of culture occurred.[10]

Babylon was captured by Alexander the Great in 331 ; on his death it became a province of Seleukos I (d. 280 B.C.), who ruled it from Antioch and built Seleucia near the site of the modern Bagdad. The Parthian (Arsacid) dynasty lasted from about 250 to 226 B.C. ; after it the Persian Sassanid kings ruled Babylonia until the Arab conquest in A.D. 635, when the country once more was under a Semitic power.

[1] Rogers, ii, 279 f. ; Luckenbill, ii, 379.
[2] Jastrow, Civ., 21 ; Handcock, 47 ; BMGB, 25, 28, 162, 212, etc.
[3] Jastrow, Civ., 180.
[4] Sayce, B. and A. Life, 7 ; Meyer, Alt. I, ii, 434 ; III, 132 ; King, H. Bab., 200, 213 ; Rogers, i, 457 ; ii, 300 f. ; Schnabel, Berossos, 1923, 6, 10 ; Laqueur, PW, xiv, 1063.
[5] Minns, CAH, iii, 189 f. ; Smith, CAH, iii, 129 ; Luckenbill, ii, 417 ; cf. Contenau, Manuel, iii, 1353.

[6] Minns, CAH, iii, 212 f. ; King, H. Bab., 277 ; Rogers, ii, 319.
[7] Daniel, iv, 30 ; Josephus, Contra Apion., i, 19 ; Herodotos, i, 178 f. ; Diodoros Siculus, ii, 7 f. ; Grote, History of Greece, 1869, iii, 298.
[8] Rogers, ii, 364 f. ; Minns, CAH, iii, 218, 221 f.
[9] Isaiah, xiii, 7 ; xxi, 9 ; Herodotos, i, 190 f. ; Rawlinson, India, 16.
[10] Meyer, Alt. III, 91, 94 ; Lenormant, Magic, 195, 217.

BABYLONIA AND ASSYRIA II

THE METALS

ARTS AND CRAFTS IN MESOPOTAMIA

Even in the earliest Sumerian Period the arts and crafts were developed : the workmen included caulkers, carpenters, perfumers, curriers, metal casters, masons, statuaries, gem engravers, gardeners and navvies.[1] Manual work, which was taught to men by Ea (Oannes in Berossos) and other gods, was held in esteem and well executed, and the craftsman, who was respected even by kings and in some cases excused taxation, had masses of illiterate labourers under his direction, although (as in Egypt) even artists were merely workmen. Foreign craftsmen were either invited or after wars compelled to enter the country. The work was very conservative : "as in olden time," or even "as before the Flood," was an ideal. In Hammurabi's time (1924–1913) artisans had smaller wages than farm workers, the average being six shekels (less than 2 oz.) of silver a year.[2]

Craftsmanship was hereditary and workers were organised in guilds. Apprenticeship lasted for varying times : bakers $1\frac{1}{2}$ years, stonemasons 4 years, weavers 5 years. There was no premium, but the master could receive a "gift" : he was bound to teach his trade thoroughly.[3] The skilful goldsmith worked precious metals and gems. In Assyria a "gate of the metal workers" is mentioned ; perhaps they lived in a special part of the town.[4] The later Assyrians were highly skilled technologists.[5]

METAL WORKING

The metal worker, qurqurru, was perhaps a coppersmith, also called nappâch siparri, and the smith is nappachu, perhaps a blacksmith, also called nappâch parzilli. (All names given are Semitic-Babylonian unless it is stated that they are Sumerian.) The smith took his name from the reed blowpipe, nappachu. The goldsmith was nappâch churâsi. Metals were hammered and also cast in moulds of limestone and soapstone, examples of which have been found.[6]

The melting-furnace (Sum. gir ; Akkad. kiru ; Babylonian utunu, from the Sum. udun, utunu or tinuru ; Arabic tannur ; Mediæval Latin athanor, i.e. at-tannur)[7] was heated with the expensive wood charcoal (pêntu)[8], "abbas tree wood"[9] or charred date kernels.[10] The smith obtained the metal in bars (libittu). Casting metal was šapâku in the time of Ur Nina and Gudea ; alloying was bullulu. The bellows was nappachtu or nappachu. To work in precious metals was sarâpu.[11]

Other terms related to mining and metallurgy were : bursu, mountain ;

---

[1] Delaporte, 70.
[2] Meissner, i, 163, 229 f., 274 f.
[3] Meissner, i, 231 ; Smith, CAH, iii, 97.
[4] Meissner, i, 269, 359 ; plates, Figs. 148–151.
[5] Rawlinson, Mon., i, 405.
[6] Meissner, i, 264, 266, 269, 270, Fig. 78.
[7] Meissner, ii, 347 ; Schrader, Keil.,

648 f. ; Gadd and Legrain, Ur Excav., Texts, 1928, i, 32.
[8] Meissner, i, 414.
[9] Handcock, 107.
[10] Meissner, i, 206 ; Rawlinson, Mon., i, 35 ; Maspero, Dawn, 555.
[11] Meissner, i, 246, 266 ; ii, 492 ; Schrader, Keil., 648 f.

irsitu, earth ; išâtu, fire ; mû, water ; illu or namru, bright or pure.[1] The name tibira in Badtibira, the unidentified town of an antediluvian king, may mean metal workers.[2] Scenes on very old (proto-Elamite) seal cylinders at Susa are supposed to represent metallurgists at work with blowpipes making vases.[3]

Babylonia had no metals or even stone ; the kings imported both for their own use, and it is doubtful if the Assyrians mined the deposits of copper, lead, silver, antimony and salt in their country although an inscription of Sargon's may imply this. Gold, silver, copper, bronze, lead and antimony were worked in the Sumerian Period ; iron was known, perhaps meteoric, and was used very sparingly for decorative purposes.[4] The Sumerians brought their knowledge of metal working with them, perhaps from the upper country of the Caucasian highlands. Metal work at al-'Ubaid in 3100–3000, with beginnings probably in 3500, is very skilful ; the socketed cast copper axes are much in advance of the Egyptian, in which the hafting did not progress beyond the primitive tang and rivet till the xii dyn. (2000 B.C.). The gold and silver work at Ur under King Mes-kalam-dug (3500 B.C.) is very elaborate and finely executed ; there was a decline in art from 3500 to 3100, when design had become conventional. Besides garish designs there are plain gold bowls and goblets of exquisite shape in the earliest period.[5] A chain of links of *drawn* gold wire occurred in early remains at Kiš.[6]

The Sumerian name for precious metal was kutimmu ; the oldest texts have signs for copper and precious metal (gold or silver) and in old legends the god Ea created, as emanations, gods connected with brickmaking, carpentering and metal working, including gods of goldsmiths and coppersmiths, although in some legends gold is the metal of the demons of the underworld and thus an evil thing. This legend of Ea (Sum. Enki) is probably Sumerian, although the source is late. A god of copper, Nindara, who shone like it, came out of the earth (or the mountains) where the metal was found and "covered solid copper like a skin."[7]

GOLD

Gold, Sum. guhsgin or azaggi [8], Akkad. hurâsu or churâsu, "the brilliant" [9], the latter name dating from the time when the Semites formed one people[10], was highly prized.[11] It is not found to any extent in the remains as a whole, apart from Ur, partly owing to depredations and also because it was used only for exceptional purposes.[12] Prehistoric and Sumerian gold objects from Ur (3500–3000) include a thick, hollow soft gold bull's horn from al-'Ubaid, found filled with bitumen.[13] Gold was used from the earliest Sumerian Period for ornaments, statues and ritual weapons[14], and was buried in tombs, but it was not used generally with the profusion common in Egypt.[15] Although engraving

[1] King, Babylonian Magic and Sorcery, 1896, 152 ; 140 ; 160 ; 135, 163.
[2] Peake and Fleure, Priests and Kings, 1927, 39 ; Pantibibla in Berossos ; Maspero, Dawn, 565.
[3] Contenau, Manuel, ii, 598 ; cf. Egyptian representations, p. 8.
[4] Meissner, i, 53, 233, 264 ; Rawlinson, Mon., i, 38 ; Perrot-Chipiez, Chaldæa, ii, 310 ; King, S. and A., 73 ; Place, Ninive, ii, 263 ; Woolley, Ur Excav., 1934, ii, 284 f., 299 f.
[5] Woolley, 1928, 22, 38, 42 f., 44, 45 ; ib., Ur Excav., 1934, ii, 30, 150, 288, 398 ; Hall, Civ. of Greece in the Bronze Age, 1928, 31 f., 37 f.
[6] Langdon, *Alte Orient*, 1928, xxvi, 64.

[7] Lenormant, Magic, 176, 178, 364 ; Maspero, Dawn, 653 ; Meissner, i, 229, 265, 269, 345 ; ii, 13.
[8] Orth, PW, Suppl. iv, 112 ; Unger, RV, i, 428 f.
[9] King, Magic, 152.
[10] Schrader-Jevons, 170.
[11] Meissner, i, 60, 268, 345.
[12] Handcock, 103, 263.
[13] BMGB, 57 f., 82 ; Hall, JEA, 1922, viii, 247, pl. 32 ; Hall and Woolley, Ur Excav., 1927, i, 16, 30, pl. 5 ; Partington, Everyday Chemistry, 13, Fig. 15.
[14] Bucher, ii, 131 ; Woolley, 1928, 37, 55.
[15] Meissner, i, 345, 427 ; Hall, CAH, i, 586.

on gold was rare [1], a single piece of gold sheet with cuneiform inscriptions and some gold beads were found at Tello[2] ; a narrow strip inscribed with the name of Narâm-Sin of Akkad at Bismaya ; gold fragments, gold-headed (or gilt-headed) nails and thin gold sheet on the site of the ziqqurat of Eridu ; and a gold lightning fork belonging to a statue of Adad at Assur.[3] A considerable amount of decorative gold work, "the oldest gold jewellery," including delicate granulated work for the later period, was found in the Sumerian remains at Ur (3500–3000), including a fine complete helmet of King Meskalam-dug, a lamp, vases, tools and weapons, a copper pin with a gold and lapis head, gold lions' masks on the sledge chariot of Queen Shub-ad, a thin gold tube decorating the wooden shaft (which had disappeared) of a spear with a copper head, and headbands made up of lengths of gold or silver chain and heavy beads of lapis lazuli and gold secured behind by a string.[4] The gold objects found at Ur are far too numerous to specify in detail.[5] They include a strainer, a goblet (p. 159, pl. 160), fluted bowls (p. 159, pls. 160, 162), vessels (one with upturned spout and one with horizontal spout) (pls. 41, 161, 164), fluted tumbler (pls. 157, 162), cup, cockle shell, adze and pin (pl. 165), ostrich shell (p. 283, pl. 170), beads (pl. 145)—some composed of three double conoids soldered together (p. 158)—daggers (pls. 13, 60, 146, 151, 155, 157), spear-heads and butt (pl. 149), dice (pl. 158), saw and chisels (pl. 158), chain (pl. 159), lyre (pl. 114), harp (pls. 109–110), bull's head (pls. 115, 117), finger ring, various ornaments and jewellery (pls. 138 f.), gold-plated copper (pl. 309) and gold imitation of bamboo (pl. 154). The regalia of Queen Shub-ad was very magnificent (p. 84 f.). Gold objects were found in tombs of ordinary people and the city was wealthy. Gold ornaments, viz. solid ear-rings with wire hanger and a solid finger ring, were found in old remains at Kiš [6], a gold bracelet encrusted with lapis lazuli and carnelian at Tello, and a gold chain with lapis lazuli beads at Ašara (3100–3000 B.C.).[7] Gold was, apparently, used profusely at Babylon. Sumu-la-ilu (2035–2000) speaks of building a throne adorned with gold and silver for the god Marduk ; Nabû-apal-iddin (885–852) built for building an image of Šamaš of pure gold and lapis lazuli ; and the statue of Marduk at Babylon seems to have been of pure gold. The images of the gods had wheeled cars of gold, silver, bronze, etc.[8]

Gold appears to have been used to some extent in commercial transactions [9], and as a money standard more frequently in the Cassite Period (1746–1171), although actual payments were in silver. Its value changed with time ; Hammurabi (c. 1950 B.C.) regulated the exchange of gold by laws, and payment in gold (sometimes large sums) is specified.[10]

GOLD JEWELLERY

Apart from that at Ur, much of the gold jewellery found in Babylonia is of uncertain date, as that found by Loftus at Warkâ, which is perhaps of the

[1] Handcock, 103.

[2] Heuzey-de Sarzec, 345 ; Cros, Heuzey and Thureau-Dangin, Nouvelles Fouilles de Tello, 1910–14, 113.

[3] Handcock, 103, 263 ; Rawlinson, Mon., i, 81.

[4] Woolley, 1928, 37, 46, 53, 55, 162 ; ib., Nature, 1927, cxix, 610 ; cxx, 124, 927 ; 1928, cxxi, 110, 331 ; ib., Times, Feb. 11, 1924 ; ib., Digging up the Past, 1930, 110, 115, 136 ; pl. xxv.

[5] Described in Ur Excavations, 1934, ii, inadequately indexed ; the objects are best found in the "Analysis of Objects," pp. 411–509 ; the page references in the volume of plates often do not correspond with the volume of text. The references in our text are to the pages and plates of this work ; cf. Contenau, Manuel, iii, 1512 f., who describes and illustrates most of the objects.

[6] Langdon, Excavations at Kish, 1924, 89 f.

[7] Contenau, Manuel, ii, 600 f.

[8] Meyer, Alt. I, ii, 618 ; ib., Chronol., 25 ; Handcock, 264 ; Jastrow, Civ., 415 ; Meissner, ii, 73, 133.

[9] Meyer, Alt. I, ii, 448 ; Handcock, 103, 263.

[10] Meissner, i, 154, 167, 174, 356, 363.

Sassanian Period [1], and a gold ring with a peculiar inlay of carnelian and lapis lazuli found at Tello is perhaps Persian or even Greek.[2]   The king of Assyria wore a gold chain hung with gold religious emblems, and bracelets of beaten gold as much as an inch broad were worn by the Babylonians.   In addition to the Babylonian and Assyrian finds, there is also a small gold Elamite statuette from Susa which appears to have been cast.   Besides beaten gold work there was also cast gold in Babylonia, as steatite moulds for small articles, one perhaps representing a Mycenæan altar, are in the British Museum [3], and there is a fine cast pendant representing Gilgameš and two bulls, of about 2500 B.C., in the Louvre.[4]   Full and hollow casting were probably continued in Hammurabi's time.[5]

Gold was beaten into very thin sheets at Ur, probably on flat hard stones and not between goldbeater's skin. A small hæmatite hammer or burnisher found *may* have been used in the process. The thin sheet ("leaf") was used to cover beads of bituminous shale, wood, etc., and the heads of copper pins. The sheet was moulded into shape by rubbing over a model of wood or bitumen, and large objects such as helmets were made from the sheet metal, sometimes of pieces riveted together. Small objects only were cast. Wire was not drawn [6] but cut in square section from thin plates ; it was coiled into a spiral with the strands touching and the inside soldered, apparently with a white alloy of gold and silver. Metals were both sweated (by heating to incipient fusion) and soldered, and great skill was shown in fusing together (not soldering) granulated work : there is a gold ring only 2 mm. diameter composed of six balls so sweated together. The colours were deliberately varied by the use of gold and electrum in the same object. True cloisonné work was produced by inlaying stones in cells formed of soldered or sweated wires. Gold objects of very uniform size (ear-rings, wads), some found with hæmatite and pebble weights, may have been used as money.[7]   The gold and silver weapons found in the old graves were state objects carried by the king or the chief officers ; the gold tools (saws, chisels, adze) may be either the attributes of the king as "master craftsman" of an industrial city or symbols of royalty used in the inauguration of temples.[8]

Gold jewellery in Babylonia and Assyria included studs, buttons, ear-rings, beads, necklaces, bracelets and finger rings. An emerald set in a gold ring by the goldsmiths Ellil-ach-iddin, etc., in a commercial house at Nippur in the time of Artaxerxes (465–424), was guaranteed (under a penalty of 10 minas of silver) not to fall out in twenty years. Granulated gold work, which appears in the oldest remains at Ur, also occurs in a chain of Hammurabi's time and in soldered granules on a foundation in Elamite work from Susa. Assyrian and Babylonian work were similar, but the later Assyrians used heavy and showy forms. Gold was sometimes used instead of bronze on furniture. The lightning fork of the god Adad at Assur is of pure gold sheet 0·5 mm. thick hammered over wood.[9]   Two fragments of copper plated with gold and a piece of alabaster inlaid with carnelian, turquoise and "gilt copper," of the period of Gudea (2450), were found at Tello.[10]

[1] BMGB, 237 ; Koldewey, Excavations at Babylon, 1914, 264, 266 ; Handcock, 265.
[2] Heuzey-de Sarzec, 391, 392.
[3] Meissner, i, 70, 270, 272 ; plates, Fig. 149 ; BMGB, 189 ; Rawlinson, Mon., i, 107 ; ii, 567.
[4] Contenau, Manuel, ii, 695.
[5] Blümner, PW, vii, 1571.
[6] Cf. p. 229, wire at Kiš.
[7] Plenderleith, in Ur Excav., 1934, ii, 295 f. ; Woolley, *ib.*, 299 f. ; E. A. Smith, *Discovery*, 1930, xi, 20, on gold solders— no analyses ; Smith refers to C. R.

Williams, Cat. of Egy. Antiq. of the New York Historical Society, New York, 1924, on solders, but this was not available to me.
[8] Ur Excav., 1934, ii, 303.
[9] Rawlinson, Mon., i, 106 ; Meissner, i, 270 f., 247, 273 ; plates, Fig. 150 ; Hilprecht, Babylonian Exped. of the University of Pennsylvania, Series A, Cuneiform Texts, 1898, ix, 30.
[10] Heuzey-de Sarzec, 345, 346 : "ce qui ouvre toute une vue nouvelle sur les procédés de l'art Chaldéen"; Delaporte, 200 ; gilt copper also at Ur, see p. 241.

SOURCES OF GOLD

Gold is not found in Babylonia and Assyria and most, if not all, of the Babylonian gold was probably imported as dust in bags, and as rings, bars, tongues, "hands" and clippings from Egypt and Nubia.[1] Gold in bars came from XVIII dyn. Egypt, where the Asiatics thought it was "as common as dust."[2] Other sources suggested besides Africa are India, Arabia [3], the borders of China and the Altai and Northern Asia.[4] In the Sumerian Period gold (mostly alluvial) probably came from Elam, the Khabur district, the Antioch region of Syria and Cappadocia [5], also (with silver) from mines in Cilicia, afterwards visited by Šalmaneser III (859–824).[6] In the Ur dynasty gold came from (or through) Meluchcha [7], and in Sargon's and Gudea's times this country sent much gold dust.[8] The "gold of Meluchcha" mentioned by Gudea was perhaps African gold, since there is no gold in the place he suggests.[9] The Babylonians and Assyrians believed that gold came from the north-east, a fabulous "dark" land called Aralli or Arallu, inhabited by gods and spirits, the "lower world of the dead."[10] Gold may have come from the Kurdish mountains : it is regularly mentioned in the early period as a product of mountains, such as Mount Chachum (not identified), the mountains of Meluchcha in Gudea's time (as "mountain dust"), and also in later texts of Tiglathpileser I.[11] Although Layard thought gold mines were worked in Assyrian dominions, no gold is known there and the metal was probably imported, either from the Babylonians or (later) by the Phœnicians from the West African coast by way of Carthage, or from Media and countries farther east. A part of Media was called "Šikraki of gold."[12]

QUALITIES OF GOLD

The gold found at Ur was of varying composition, some being relatively pure and some containing silver and fairly large quantities of copper. The purest specimen analysed contained : gold 77·84, silver 21·74, copper 0·42 ; another specimen contained : gold 83·95, silver 13·95, copper 2·10. The least pure specimen contained : gold 30·30, silver 59·37, copper 10·35. All the specimens were really electrum, the percentage of gold varying from 30·30 to 83·95, of silver from 7·69 to 59·37 and of copper from 0·32 to 10·35. The metal is less pure than some specimens of Egyptian gold.[13] The quality of the gold from Ur could not be judged from the colour alone. Some specimens of different colour had practically the same composition, and the base specimen with 10 per cent. of copper had a rich yellow colour suggesting a purer metal. It is considered that some surface treatment involving the removal of baser metals and followed by burnishing was used, since the outer surface is sometimes deeper in colour than the mass of the metal.[14]

"Pure gold" is mentioned frequently in the texts, as also in the Bible and in Egypt[15], and some method of purifying and assaying gold was known in Babylonia

[1] Hall, CAH, i, 586 ; Meissner, i, 345.

[2] Maspero, Struggle, 287 ; Meissner, ii, 346.

[3] Rawlinson, Mon., i, 102, 108, 556.

[4] Bucher, ii, 131 f. ; in Job, xxxvii, 22, written before 400 B.C., "gold cometh from the north."

[5] Woolley, 1928, 46 ; ib., Ur Excav., 1934, ii, 394.

[6] Smith, EHA, 92, 99.

[7] Schrader, Keil., 15.

[8] Meyer, Alt. (1909) I, ii, 476 ; ib., 1913, 539 f.

[9] Langdon, JEA, 1921, vi, 143 ; O'Leary, Arabia before Muhammad, 49.

[10] Lenormant, Magic, 152 ; Luckenbill, ii, 261 ; Siecke, A. Rel., i, 125 ; King, Magic, 140.

[11] Rawlinson, Mon., i, 219 f. ; Meissner, i, 53, 345, 346 ; Amiaud, Rec. of the Past, N.S., ii, 81.

[12] Rawlinson, Mon., iii, 16 ; Meissner, i, 53, 345.

[13] See p. 27.

[14] Plenderleith, in Ur Excav., 1934, ii, 292, 294, 298 ; JEA, 1928, xiv, 317.

[15] Talbot, TSBA, 1874, iii, 437.

as early as 1500 B.C. In the Amarna letter of Burnaburiaš, king of Babylon, to Amenhotep III of Egypt (1411–1375), the metal is said to have been "put in a furnace," when it was found "not of full weight," in fact so poor that "of the 20 minas put in the furnace only 5 minas came out." In Nabonidus' time (555–538) gold and silver were regularly assayed and the officials report that the metals lost varying amounts in the fire, in some cases up to 25 per cent. : e.g. 1 mina $2\frac{1}{2}$ shekels of gold lost 2 shekels ; $4\frac{5}{6}$ minas of gold lost 1 mina 2 shekels ; and $34\frac{1}{2}$ minas of silver lost 1 mina. Pure refined gold was churâsu sagiru, and the quality was obviously variable.[1] Gold leaf (yellow) from the palace of Sargon II (722–705) at Dûr-Šarrukîn, modern Khorsabad, contained a notable amount of silver, but no copper, lead or iron ; it was a variety of electrum.[2] The lightning fork of Adad from Assur is practically pure gold sheet.[3] Varieties of gold for money named in the texts are : full value (ša ginnu ; babbanu), purified or refined (qalû), white (pisû), poor quality (nuch-chutu) and Akkadian.[4] Nebuchadnezzar II (604–562) speaks of "pure and mixed gold," thought [5] to mean gold and electrum, but probably refined and impure gold. From gold, silver and copper various artificial alloys were produced : red (sarîru or sarîtu), a product of the underworld ; "mountain dust" ; "shining" (probably an alloy of gold and copper) ; bright white (zachalû) and white (ešmarû), probably two varieties of electrum and orich-alcum.[6] The Cappadocian tablets specify payments in rings (riksu) of $\frac{1}{2}$–$\frac{2}{3}$ mina of "fine gold" (hurasum pašallum), as well as two other commoner varieties of gold.[7]

TRIBUTE OF GOLD

Many gold articles have been found on Assyrian sites.[8] Gold as tribute is frequently mentioned in the texts of Ašurnasirpal II (885–860).[9] He received from the Hittite king Sangara at Carchemish beads and a chain of gold ; from Lubarna, ruler of the state of Khattin (of Hittite origin), in his city of Kunalua (Kinalia, east of the Orontes), one talent of gold ; gold, etc. (but not iron), from the Phœnicians ; and he made a stone image of his god, Ninurta, plated with gold.[10] In his campaign of 854 Šalmaneser III (859–824) obtained gold as tribute from these kings, from Jehu and a king of the Sukhu (Khabur and Middle Euphrates).[11] In the late Assyrian Period 12 minas of gold are said to be used to make ear-rings and a cap for the statue of the goddess Šarpanitu.[12] Sargon II (720 B.C.) obtained 11 talents 30 minas of gold as tribute from the Hittites and 34 talents 18 minas from the Armenians, gold from temples of his enemies, and large fines in gold.[13] Sennacherib (705–681) presented the god Ea with a ship, a fish and a crab of gold[14]; Ašurbanipal (668–626) paid for the corpse of a dangerous rebel its own weight of gold.[15] Nabonidus offered 100 talents 21 minas of silver and 5 talents 17 minas of gold to Marduk[16], whose gold statue in the temple at Babylon in the time of Herodotos weighed 80 talents.[17] Some of the temple images were probably covered with gold plates hammered into shape (Diodoros Siculus's σφυρήλατα) and were not solid gold.[18]

[1] Meissner, i, 60, 269 ; BMGB, 128, 182.
[2] Berthelot, Arch., 81.
[3] Meissner, i, 273 ; plates, Fig. 151.
[4] Meissner, i, 356.
[5] Lippmann, Alchemie, i, 531.
[6] Meissner, i, 346, 269.
[7] Landsberger, Alte Orient, 1924, xxiv, 21.
[8] Rawlinson, Mon., i, 554.
[9] Rogers, Hist., ii, 46, 64 f. ; Sayce, Rec. of Past, N.S., 1889, ii, 131, 140, 144, 148, 151, 154, 155, 162 f., 169, 170, 172 ; Luckenbill, i, 144 f.

[10] Meyer, Alt. II, ii, 372, 403.
[11] Rogers, ii, 76 ; Luckenbill, i, 211, 217, etc. ; BMGB, 46 ; Maspero, Empires, 28.
[12] Meissner, ii, 85.
[13] Meissner, i, 111, 139, 181 f. ; Luckenbill, ii, 9 f.
[14] Meissner, i, 101 ; Luckenbill, ii, 146.
[15] Meissner, i, 113.
[16] Meissner, ii, 86.
[17] Meissner, ii, 61.
[18] Rawlinson, Mon., ii, 567.

ELECTRUM

Electrum (išmerû ; zahalû ; sarîru), a natural product, was known in the earliest period.[1] There were two electrum axes with much gold and silver in a tomb of 3500, and a solid cast electrum ass surmounting a silver rein-ring on the chariot of Queen Shub-ad (3100 B.C.), at Ur. There were also two bowls, ribbon sweated to a silver bowl, and handles of a silver bowl, all of electrum, and all the "gold" found was really electrum, as it contained notable amounts of silver.[2]

Decorative work in electrum is mentioned by Tiglathpileser III, Sargon II, Asarhaddon and Ašurbanipal [3], who sometimes describe it as a product of the underworld and as "ruddy," so that gold-copper alloys may be meant.[4]

SILVER

Silver (Sum. ku babbar, gu babbas, or gu, "the white or shining" ; Akk. kaspu or sarpu) [5] was known and used in the early Sumerian Period. Except at Ur, it is rather rarely found on the sites. None was found in the oldest graves at Uruk[6] ; one piece was found at Adab (Bismaya) [7], and it occurs in the oldest remains at Kiš [8], at Nippur [9], at Tello[10], as a piece of three-ply wire at al-'Ubaid (3500–3000)[11], and as finger rings with copper objects at Šuruppak (Fâra) in tombs of 3500–3000.[12] A silver arm-band was found at Nippur, and the eyelashes and eyebrows of a Sumerian marble head from this site are inlaid with silver.[13] Silver occurs plentifully in early graves (3500–3000) at Ur, not only as ornaments but also as vessels (including nested beakers), weapons and even tools[14]: a belt (pls. 12, 13), comb (pl. 20), bowls (pls. 33, 42, 167, 171, 172), ewer (pl. 172), fluted bowl (pl. 172), tumblers (pl. 172), lyres (pls. 75–76, 111–112), cow's head (p. 310, pl. 120), head of lioness (pls. 127–130), hair ornaments (pls. 136–137), cockle shell for cosmetic paint (p. 245, pl. 137), binding for spear-heads (pl. 149), pin (pls. 159, 189), boat (p. 71, pl. 169), lamp (?) (pp. 71, 293, pl. 170), libation vessel (pl. 171), vases with electrum handles and electrum ribbon sweated on (p. 297, pl. 173), tubing with holes at regular intervals and probably a musical instrument (double pipe) (p. 258), nose-rings of oxen (p. 64, pl. 35), rein-ring (pp. 78, 301, pls. 166–167). The silver rein-ring (3100 B.C.) contained : silver 93·5, copper 6·10, gold 0·08, zinc 0·15.[15] White metal as an ingot and turnings at Tello contained 96·3 silver and 3·5 copper, and no lead.[16]

A fine example of Sumerian metal work is the silver vase of Entemena, patesi of Lagaš (2600 B.C.) which (with its original copper stand) is in the Louvre. It was found at Tello and cleaned by Heuzey. It is dedicated to Ellil (Bêl) of

[1] Unger, RV, i, 428 f.
[2] Woolley, 1928, 38, 43 ; ib., Ur Excav., 1934, ii, 78, 89, 156, 271, 297, 302 ; plates 140, 156, 159, 166, 171, 173 ; analyses, Plenderleith, ib., 294.
[3] Luckenbill, i, 289 ; ii, 38, 53, 261, 275, 276, 296, 309, 343 f., 353.
[4] Meissner, i, 269 ; ii, 73 ; Luckenbill, ii, 38.
[5] Schrader-Jevons, 180 ; Orth, PW, Suppl. iv, 112 ; Schrader, RL, 767 ; Weissbach, MGM, v, 502 ; King, Magic, 156; Fossey, La Magie Assyrienne, Bibl. de l'École des hautes Études, Sciences relig., 1902, xv, 184, 226, 426 ; Muss-Arnolt, Concise Dictionary of the Assyrian Language, Berlin, 1905, 417.
[6] Reinach, L'Anthropologie, 1892, iii, 453.

[7] Banks, Bismya, 314.
[8] Langdon, Excav. at Kish, 1924, 77, 90.
[9] Hilprecht, Explor., 1903, 544.
[10] Meyer, Alt. (1909) I, ii, 416 ; Heuzey-de Sarzec, 361, 419 ; Nouv. Fouilles, 128, 133.
[11] Hall and Woolley, Ur Excav., 1927, i, 103.
[12] Gowland, Archæologia, 1920, lxix, 133.
[13] Hilprecht, Ausgrabungen im Bêl-Tempel zu Nippur, Leipzig, 1903, 20, 67.
[14] Woolley, 1928, 37 f., 46, 48, 52 f., pl. 13 ; ib., Ur Excavations, 1934, ii, 411–509, summary of objects, references to this in text.
[15] Plenderleith, Ur Excav., 1934, ii, 293.
[16] Berthelot, Arch., 78.

Nippur.[1] The inscription states that it is "of pure silver."[2] The vase is hammered from a single piece of silver except for the interior of the neck near the orifice, which is composed of a second piece of thicker silver soldered with copper.[3]

Various silver objects in the British Museum (bracelet, bell with bronze clapper, ring set with garnet, thin embossed plate found wrapped with linen in a jar at Warkâ) are of uncertain date.[4]

An inscription on a statue of Gudea (2450 B.C.) at Tello says : "of this statue, neither in silver, nor in copper, nor in tin [?], nor in bronze, let anyone undertake the execution," [5] and tablets of Sulgi (2276 B.C.) and later kings refer to silver in minas and shekels.[6] An old Sumerian code of laws (known only in fragments) speaks of payments in silver.[7]

SOURCES OF SILVER

Gudea speaks of "silver from the mountains," [8] probably the Taurus range, or the mines of Bulgar Maden in North Cilicia, Asia Minor [9] or Zagros.[10] Sargon I and Erimuš (soon after Gudea) speak of the silver mountains and the silver mines as somewhere in the extreme western limit of their territory[11], probably Cilicia[12], the Taurus[13], or Elam.[14] Maništusu, king of Akkad (2572–2558), went to Elam for silver and captured the mines.[15] Suggested sources are also Bactria and Carmania (Luristan), and for a later period the Armenian and Caucasian mountains near the Black Sea, and Armenia—which Marco Polo[16] says was rich in silver.[17] The Kurdish mountains, a suggested source, still supply silver.[18] In the earliest Sumerian Period the metal probably came from Bulgar Maden and from the hills south of Elam, from "South Persia and the mountains of the North."[19] The Taurus (Tunni) mountains on the Cilician coast were worked in the Sumerian Period for silver and gold[20]: Šalmaneser III (858–824) still obtained silver from them, and they are also mentioned in late Egyptian and Hittite texts.[21] The mountain Saršu, producing silver, is not identified[22], and in general the source of all early silver in Mesopotamia is uncertain.[23] After 1500 the metal was perhaps imported from Spain, by Phœnicians, and became cheaper.[24]

That silver was more valuable than gold in the earliest period[25] is doubtful. The ratio of value of gold to silver fluctuated at different periods :

Sargon I (2725 B.C.) 8 : 1.   Ḥammurabi (1924–1913) 6 : 1.
Gimil Sin (2150 B.C.) 7 : 1.   Sargon II (722–705) 13 : 1.
Bur Sin (2060 B.C.) 10 : 1.   Nabonidus (555–538) 12 : 1.

[1] Meyer, Sumerer und Semiten, 104 ; Maspero, Dawn, 757, before cleaning ; Handcock, 265, Fig. 45 ; King, S. and A., 167 f., Fig. 52 ; Meissner, i, 24 ; Partington, Everyday Chemistry, Fig. 12, after cleaning.
[2] Rogers, i, 380.
[3] Contenau, Manuel, ii, 601.
[4] Loftus, Travels and Researches in Chaldæa and Susiana, 1857, 211, 219 ; Handcock, 267 ; BMGB, 278.
[5] Boscawen's translation ; Gowland, Archæologia, 1920, lxix, 134.
[6] Gowland, op.cit., 134; BMGB, 138, 145 f.
[7] Meissner, i, 150.
[8] Meissner, i, 53, 346 ; Amiaud, Rec. of the Past, N.S., ii, 84 ; Meyer, Alt. I, ii, 447.
[9] Langdon, CAH, i, 428, 545 ; Hall, CAH, i, 586.
[10] Perrot and Chipiez, Chaldæa, i, 124 ; Woolley, 1928, 46.
[11] Meissner, i, 25, 346.
[12] Gadd and Legrain, Ur Excavations, Texts, 1928, i, 79.
[13] BMGB, 5 ; Meissner, i, 25, 37, 53.
[14] King, S. and A., 262.
[15] Handcock, 267 ; Delaporte, 132 ; Woolley, 1928, 80 ; Meissner, i, 365.
[16] i, 4.
[17] Blümner, iv, 31 ; Schrader, RL, 765 ; Schrader-Jevons, 181 ; Dussaud, La Lydie et ses Voisins, Babyloniaca, 1930, xi, 81 ; Gowland, Archæologia, 1920, lxix, 133.
[18] Rawlinson, Mon., i, 219 f.
[19] Woolley, 1928, 46 ; ib., Ur Excav., 1934, ii, 394.
[20] Delaporte, 16.
[21] Smith, EHA, 92, 99 ; Luckenbill, i, 246 ; Meissner, i, 37, 346.
[22] Meissner, i, 34.
[23] Handcock, 267 ; E. Meyer, Histoire de l'antiquité, 1926, iii, 262—"totally unknown."
[24] Movers, Die Phönizier, Bonn, 1841–56, II, iii, 27 f. ; Blümner, iv, 29.
[25] Schrader, RL, 541.

In the late Babylonian Period the ratio appears to have been 10 : 1 ; the ratio 13 : 1 was common in antiquity.[1] A supposed relation of the ratio $1 : 13\frac{1}{3} = 27 : 360$ to the periods of the moon and sun (planets associated with silver and gold respectively) [2] seems fanciful.

Taxes, fines and commercial debts were paid in Ḥammurabi's time in silver, which was stamped or sealed.[3] Ḥammurabi speaks of decorating a temple throne with gold and silver.[4] Singašid of Uruk (c. 2000 B.C.) had laid down that 1 shekel (8·4 gm.) of silver was equal in value to 10 minas (5 kgm.) of copper.[5]

On some old Babylonian sites silver is very scarce.[6] The metal is rarely found native and its extraction from the ores is rather difficult. It is noteworthy that the silver found at Ur, which contains traces of gold, is generally of good quality and is rarely coloured green, as is the case when the metal contains much copper. The conversion of the metal to chloride, owing to salt in the soil, causes warping and deformity owing to changes of volume.[7]

Silver was known at an early period in Assyria. The Cappadocian tablets (3 millenn.) mention it [8], and a king of the Mitanni (c. 1450 B.C.) took silver and gold from Assur to decorate his palace.[9] Small silver tablets were put in the foundations of the Ištar temple at Assur (13th century B.C.)[10], and clippings in 9th–7th centuries B.C. deposits at Assur.[11]

Silver is very frequently mentioned in texts, e.g. of an early period at Kiš[12], and from Gudea's time downward[13]. Bur-Sin, king of Ur (c. 2220), speaks of adornment in gold, silver and lapis lazuli.[14] In Babylon the consort of the god Marduk was called in Akkadian Šarpanitu (i.e. "silvery," white or shining), and the "skull" of Marduk is said to be of silver.[15]

A very early text mentioning silver is on a granite obelisk found at Susa, giving title-deeds of property purchased by Maništusu, king of Kiš (3000 B.C.), in which a silver monetary standard of talent, mina and shekel, arranged according to a sexagesimal scale, appears. The values of bronze wedges, cleavers, etc., are given in shekels of silver.[16] A text of 2575 B.C. speaks of the looting of silver from the destruction of the town of Lagaš.[17]

The later Assyrian kings took great quantities of silver as tribute from their vassal princes and as spoil in wars. Ašurnasirpal II (885–860) obtained it from the Hittite king of Carchemish and from Lubarna, ruler of the state of Khattin in his city of Kunalua, he obtained 20 talents, also silver from the Phœnicians.[18] Šalmaneser III, in his campaign of 854 B.C., took tribute of silver from the same kings and from Jehu, Suâ and Marduk-apal-usur of the land of Sukhu.[19] Tiglathpileser III (745–727) took 1,000 talents of silver from Menahem, the king of Israel.[20] A document specifies payments of 15–100

[1] Thureau Dangin, in Delaporte, 132 ; Meyer, Alt. (1909) I, ii, 606 ; ib., Hist. ant., iii, 264 ; Woolley, 1928, 117 ; Weissbach, PW, ia, 2506, 2510 ; Meissner, i, 363 f.

[2] Winckler, in Walden, Mass, Zahl und Gewicht, 1931, 15.

[3] Meissner, i, 58, 86, 124 f., 154, 157, 161, 162 f.

[4] Meyer, Alt. (1909) I, ii, 548.

[5] Meissner, i, 361 ; BMGB, 246.

[6] Gowland, J. Anthropol. Inst., 1912, xlii, 268 ; Rawlinson, Mon., i, 98.

[7] Plenderleith, Ur Excav., 1934, ii, 293.

[8] Landsberger, Alte Orient, 1924, xiv, 13 f., 21.

[9] Meyer, Alt. II, i, 133.

[10] Andrae, Mitteil. Deutsch. Orient. Ges., 1914, liv, 36 ; Feldhaus-Klinckowstroem, Geschichtsblätter, i, 66.

[11] Regling, PW, vii, 976.

[12] Langdon, Excav. at Kish, 1924, i, 9.

[13] Meissner, i, 25, 37, 53, 177, 181, 259, 306, 346, 355, 360, 363, 402, 427 ; BMGB, passim ; Handcock, 267 ; Amiaud, Rec. of the Past, N.S., ii, 84—Tello.

[14] Gadd and Legrain, Ur Excavations, Texts, 1928, i, 16.

[15] Meissner, ii, 16, 133.

[16] Gowland, Archæologia, 1920, lxix, 132 ; cf. Meissner, i, 365.

[17] Meyer, Alt. (1909) I, ii, 457 ; Woolley, 1928, 71.

[18] Handcock, 267 ; Sayce, Rec. of the Past, 1889, N.S., ii, 128 f., 131, 140, 144, 148, 151, 162 f., 169 f., 172 ; Rogers, ii, 46, 64 f. ; Meyer, Alt. II, ii, 371 f. ; Jastrow, Civ., 172 ; Luckenbill, i, 144 f.

[19] Rogers, ii, 76 ; BMGB, 46 ; Luckenbill, i, 211, 217, etc.

[20] Rawlinson, Mon., ii, 122.

talents of silver from various towns.[1] Sargon II (722–706) took 2,100 talents 24 minas from the Hittite king at Carchemish ; also 33 silver cars of the gods, 167 talents 2½ minas of silver and 18 minas of gold from a temple in Mussasir (Lake Van, in Armenia).[2]

When Babylonia became a Persian province it paid 1,000 talents of silver ; Egypt only 700. Nebuchadnezzar II of Babylon (604–561) speaks of over-laying the walls of a chamber and the doors of a temple with silver, and over-laying with gold a silver statue of Bêl. Marduk-apal-iddina II (Merodach-baladan) (721–702) had a silver umbrella.[3] Josephus even speaks of a silver robe, a very rare mention of this use.[4]

QUALITIES AND USES OF SILVER

The Sumerian ku-luh-ha, "washed silver," was purified silver.[5] In later times silver was refined in a furnace, no doubt by cupellation, and a text says that 34½ minas lost 1 mina in the process, so that the metal was no doubt of varying degrees of purity and was "tested" (chatu).[6]

The Cappadocian tablets (3 millenn.), of Assyrian origin, mention "purified silver" (kaspum sarrapum) in weights (nêpešu) of 4–15 minas and also impure "silver of Kaneš."[7] A Neo-Babylonian text speaks of money [silver] of inferior (nuchchutu) and full (ša ginnu ; babbanu) value ; white (pisû), purified (qalû) and Akkadian silver, in bars and weighed portions. Lead, copper and tin [?] were separated from silver by "washing" (misû), "boiling" (bašlu) and refining (barru ; sarpu) : thus were obtained fine (damqu) and superfine (watru) silver. The meaning of these terms can hardly be as given by Meissner, except "refining," unless "boiling" means the "spitting" of silver. Silver came into commerce as rings (unqu), bars (libittu)—often stamped—wire and clippings. As early as the archaic king Enkegal of Lagaš and the contracts of Šuruppak, money was in use, but of copper, which remained the standard till the Assyrian Period. In the old Assyrian code of laws, lead is the ordinary currency. Copper and lead were soon displaced by silver in clippings, bars, spirals, rings and sheets, which were weighed in the balance (zibânîtu), very like the Egyptian balance, with conical, duck and animal-head (lion) shaped weights. Genuine and refined silver, of correct weight, was stamped (kanku), various towns having their own stamps.[8]

Copper or bronze was plated with silver as well as gold : Neriglissar (559–556) set up eight serpents of silvered bronze in the gateways of Babylon.[9] Silver was also inlaid in bronze, as in Egypt[10]: there is no reason to suppose[11] that similar Egyptian work is probably of Assyrian origin. Philostratos[12] calls inlaid bronze "Indian work."

Sennacherib of Assyria (705–681) speaks of "coins" cast by him, weighing a shekel, or half a shekel, but none have been found and their use must have been very limited.[13] A silver ear-ring seems to be mentioned (the passage is not clear), and moulds for casting ear-rings (probably in gold or silver) were found both at Nimrûd (Calah) and Quyûnjiq (Nineveh).[14]

In the Babylonian Period silver was freely used for statues, furniture, and utensils, and in plates, especially when applied to walls or internally to the

[1] Meissner, i, 142.
[2] Meissner, i, 139 ; ii, 73, 111, 129 ; Weissbach, PW, ia, 2506, 2510 ; Lucken-bill, ii, 9 f., 96 f.
[3] Lenormant, Prem. Civ., ii, 250 ; Rawlinson, Mon., iii, 29 ; Meissner, i, 130, 306.
[4] Antiq., XIX, viii, 2 ; Blümner, i, 157.
[5] Nouv. Fouilles, 55.
[6] Meissner, i, 269, 356.
[7] Landsberger, Alte Orient, 1924, xiv, 13 f., 21.

[8] Meissner, i, 346, 355 f., 360, Fig. 360, plate fig. 209 ; Rawlinson, Mon., i, 476.
[9] King, Babylonian Religion and Mytho-logy, 1899, 104.
[10] Handcock, 251, 266, Fig. 40 D.
[11] Blümner, iv, 270.
[12] Apollonios of Tyana, ii, 20.
[13] Rostovtzeff, i, 131.
[14] Rawlinson, Mon., i, 372.

woodwork of palaces, but silver images, ornamental figures and utensils were probably solid.[1] The connection of silver with Sin, the moon god, and the description of both as "green" are probably old.[2]

Although platinum was found in Egyptian remains [3], it has not been found in Babylonia or Assyria.[4]

COPPER

Copper was known in the earliest period and was the first metal worked.[5] It may have been known before the invention of cuneiform writing [6], and was, apparently, worked by the Sumerians at least as early as in Egypt.[7] The Sumerian name for copper, urud or urudu, is the only name of a metal not written with a compound ideogram.[8] Lenormant [9] thought urud was connected with the Finnish rauta and Slavonic ruda, which are names for iron ; the Basque name for copper is urraida.[10]

The Akkadian (Semitic) name for copper is erû or urû, but in magic texts it is called kêmassi.[11] Although the Akkadian names erû and sipparu are said to be properly used for copper and bronze respectively[12], the Sumerian urudu has been translated as "copper," as "bronze," and as both in the same text.[13] The Euphrates at Siparra was called "river of copper (urud)," perhaps because the buildings there were decorated with copper[14], or perhaps from some connection between Siparra and siparru. According to Eisler[15] the Chaldæan name for copper, hal-hi (chal-chi), is the source of the Greek χαλκός and means "belonging to heaven (hal)" ; he thence develops a whole series of mythical relations to alchemy, which are quite improbable.[16] The Assyrian name for copper is said to be kipar[17], and Ludwig[18] suggests that the name copper is from the Hebrew k'pôr, lid or plate, which is doubtful. The usual derivation of the Latin cuprum is from Cyprus, although Wilser[19] thinks it was adopted from the northern people, the metal being obtained in the Alps and in Scandinavia in an early period. A derivation of χαλκός from Khaldi (Chaldik), the name of the people of Lake Van, a region rich in copper, is proposed.[20]

Deep in the underworld, in the bowels of the earth, it was supposed, lived many gods who were independent of the gods living in heaven and were ministered to by a special priest called kalu. Their leader was Enmešarra (or Nindara), a warrior god whose special mission was to combat demons, monsters and plagues, perhaps a form of Ellil. Enmešarra, "lord of the ghost-world," was the night-sun hidden during half his course like Rā in the form of Atum in Egypt, and emerging triumphant each morning. He was the god of hidden metallic and mineral treasures, waiting like the sun to come out of the earth and shine with splendour. He came from the mountains where copper was found, "covered solid copper like a skin," and had gems with talismanic powers.[21] Pliny[22] refers to a book on these gems by Zachalios of

[1] Rawlinson, Mon., ii, 567.
[2] Jeremias, 240.
[3] See p. 85.
[4] Rawlinson, Mon., i, 98.
[5] Meissner, i, 264.
[6] Handcock, 242.
[7] Gowland, J. Anthropol. Inst., 1912, xlii, 247 ; Langdon, Excavations at Kish, 6—not before 4000 B.C.
[8] Meissner, i, 265 ; Hoernes, Urg., ii, 216 ; Schrader-Jevons, 186, 192.
[9] Magic, 254.
[10] Hoernes, 216.
[11] Fossey, Magie, 178, 184, 226, 264, 359 ; Sayce, Rel., 450 ; Meissner, i, 265.
[12] Meissner, i, 53, 265, 346.
[13] Gadd and Legrain, Ur Excav., Texts, 1928, i, 64, 68 ; 52, 59 ; 92.

[14] Sayce, Rel., 418.
[15] Die chemische Terminologie der Babylonier, Z. f. Assyriologie, 1926, xxxvii, 109—131, 113.
[16] Ruska, ib., 1927, xxxviii, 273 ; MGM, 1926, xxv, 82 ; ib., Chem. Z., 1925, xlix, 965 ; Darmstaedter, ib., 967, quoting Zimmern.
[17] Beck, Gesch. d. Eisens, i, 126 : ? sipparu.
[18] MGM, vi, 163.
[19] Q. in F. Dannemann, Die Naturwissenschaften, 4 vols., Leipzig, 1920—3 ; i, 43.
[20] Dussaud, La Lydie et ses Voisins, 1930, 82.
[21] Lenormant, Magic, 174 f., 176 f. ; Sayce, Rel., 145, 147 ; King, Rel., 29 ; BMGB, 29 ; Maspero, Dawn, 638.
[22] xxxvii, 10.

Babylon, dedicated to King Mithridates. The fire-god Gibil (the Assyrian Nusku) [1], perhaps connected with Tubal Cain [2], was the "fire of the reeds with pure and brilliant flame," who resides in all kinds of fires, of the hearth and of the sacrifice, and in the cosmic fire spread throughout the universe and shining in the stars. He is said in a hymn to "purify gold and silver" and to "mix copper and tin" [or lead ?]. In the night on the earth the substitute of the sun, Gibil the fire-god, a friend of man, scared away wild beasts, spells and pestilences.[3]

EARLY SUMERIAN COPPER

Copper implements, weapons, tools and ornaments (including a copper hairpin with a knob of lapis lazuli) were found in the prehistoric remains (3500–3000) at Kiš [4], at Nippur [5], at Ašara (3100–3000) [6] and at Eridu with flint, obsidian and pottery of Susian type.[7] Copper dishes or drinking bowls occur in the oldest tombs at Ur.[8]

Copper at al-'Ubaid (c. 3500 B.C.) is rather scarce and is hardly used for tools (those of the "pre-Sumerian" marsh dwellers are still of stone), although there are socketed axes and daggers.[9] There is abundance of copper from the earliest period at Ur.[10] Analysis shows that many of the early objects from the Royal Cemetery at Ur are not of copper but of tin-bronze.[11]

"Thin copper hammered over wood" is Woolley's description of the bull's head filled with bitumen from al-'Ubaid (near Ur), of 3500 B.C.[12], although Hall supposed it to be cast. The inlaid eyes are of lapis lazuli, mother-of-pearl, "paste," etc.[13] Inlaying the eyes of ivory (not metal) amulets with "black paste" and shell also occurs in Predynastic Egypt.[14] Copper statues of bulls at al-'Ubaid (c. 3500 B.C.) were apparently made by covering a carved wooden body with bitumen and hammering over this thin sheets of copper, secured with copper nails, and wooden pillars were covered with nailed copper sheets.[15] A copper goat's head with spiral horns, of very good workmanship, was found in Sumerian remains of Ur Nina's time at Šuruppak (Fâra).[16]

A copper-rimmed frieze of 3500–3000, representing cows' milk strained into jars, etc., was found at al-'Ubaid[17], also a relief with stags, etc. (3500–3000), of very advanced technique. The relief is of copper on wood and was probably placed over the door of a temple. The rectangular frame is composed of lengths of copper, 4 in. by 6 in., nailed to wood, and the group inside is also of copper, fastened by nails and three holdfasts of twisted lengths of copper bar ½ in.

[1] BMGB, 28 ; Meissner, ii, 165, says the two are different.
[2] Friend, Iron in Antiquity, 170.
[3] Lenormant, Magic, 184 f., 186 f., 191 ; Fossey, Magie, 260 ; Sayce, Rel., 478, 488 ; Ragozin, Chaldea, 1889, 173 f.; Maspero, Dawn, 635 ; Meissner, ii, 165.
[4] Langdon, Excav. at Kish, i, 77 f., 89 ; Peake and Fleure, Priests and Kings, 1927, 42.
[5] Peters, Nippur, or Explorations and Adventures on the Euphrates, New York, 1897, ii, 384.
[6] Contenau, Manuel, ii, 601.
[7] Hall, NE, 4.
[8] Rawlinson, Mon., i, 88 ; Woolley, 1928, 36 ; Maspero, Dawn, 87.
[9] Woolley, 1928, 13 f. ; Hall and Woolley, Ur Excav., 1927, i, 15 f., etc., 210, pl. 46 ; Smith, EHA, 25.
[10] Woolley, 1928, 14, 37, 39, 53, pl. 10 ; ib., Ur Excav., 1934, ii, 285, 303, 412, and passim ; Smith, EHA, 58 ; Meissner, i, 81.

[11] See p. 246.
[12] Woolley, Digging up the Past, 1930,118.
[13] Hall, JEA, 1922, viii, 248, pl. 32 ; Hall and Woolley, Ur Excav., 1927, i, 29 f., 33 f., 84 f., 110, pl. 5 f., 30 ; BMGB, 60, pl. 23 ; Partington, Everyday Chemistry, 1929, 15, Fig. 13 ; the idol of Bêl in Daniel, ii, 34, is of "brass without and clay within," a late reference ; King, S. and A., 74 ; Contenau, Civ. Assyr., 93.
[14] Brunton and Thompson, Badarian Civ., 1928, 59.
[15] Hall and Woolley, Ur Excav., 1927, i, 84, 102 ; pls. 7, 37, 38, 3 ; Contenau, Manuel, ii, 591 f.
[16] Contenau, Manuel, ii, 596 ; Hilprecht, Ausgrabungen im Bêl-Tempel zu Nippur, Leipzig, 1903, 67 ; ib., Excavations in Assyria and Babylonia, Philadelphia, 1904, 539 f.
[17] Woolley, The Times, Feb. 11, 1924 ; Mackenzie, Footprints, 114 ; Hall and Woolley, Ur Excav., 1927, i, 91 f., 110 f., pl. 31.

thick. Parts of the objects were of sheet copper roughly hammered over wood. The stags' heads may be hammered or cast ; the antlers are of hammered copper bar of square section, $\frac{1}{2}$ in. diameter, and "each branch is brazed on to that from which it rises." They were fixed into the head by lead poured into the root-holes.[1]

Although there has been some discussion as to whether the bulls' heads are cast or beaten work [2], it appears that the early arrow-heads at Ur were cast and then sometimes finished by hammering, whilst the socketed bronze and copper axes were cast.[3] In the early Sumerian remains are cast copper idols with nail-shaped extensions, which after magic consecration were buried in the foundations of buildings.[4]

Although hammering sheet copper may be older than casting [5], casting in the more fusible alloy bronze seems to have been earlier at Ur than working copper with the hammer.[6] Less than 1 per cent. of tin makes the alloy more fusible and easily cast.[7] A statuette at Khafaje, of 3000 B.C., with a stand of forged copper plates is, however, of nearly pure copper (99 per cent.), with only 0·63 per cent. of tin, traces of iron and lead and no nickel. It was perhaps made from Persian ore.[8] The earliest metal object found at Ur is a spear-head of nearly pure copper : copper 99·69, iron 0·01, arsenic 0·16, zinc 0·12, with no tin, nickel, lead, antimony, silver or manganese. A specimen of native copper from Angora contained 99·83 of copper, 0·17 of iron and a trace of tin ; and one from Arghana 97·08 copper, 0·27 tin, 2·13 iron, 0·03 nickel, 0·49 sulphur, all in percentages. Neither specimen contained lead, arsenic, antimony or bismuth.[9]

A nail and lion fragment from al-'Ubaid (3000 B.C. ?) were practically pure copper, without tin, but containing 0·109 per cent. and a trace of nickel respectively. A plate from the same source, probably later, contained 7·95 per cent. of tin and a trace of lead, but no nickel. It is supposed that the copper of the spear-head was smelted from malachite or oxidised outcrops above a layer of sulphide ore. The presence of sulphur in the Arghana native copper and its absence in the Angora specimen are noteworthy, in view of the often repeated statement that the presence of sulphur denotes the use of a sulphide ore, and so enables the conclusion to be drawn that a specimen free from sulphur must have been derived from a malachite ore.[10]

The fine cast statuettes of Ur Nina's period are solid ; in Gudea's time there is also hollow casting.[11] The remains of a smelting (?) furnace, and a clay crucible filled with scrap copper, were found at Ur ; probably the ores were imported and worked up *in situ* into metal.[12] A large flat, dish-shaped furnace or fireplace of baked clay, provided in front with two tubes joined together, for the insertion of bellows, was found at Tello and may have served for melting copper[13], and the remains of a copper furnace (*c.* 2300 B.C.) were found at Tell Asmar (Ešnunna).[14]

[1] B. Mus. Quarterly, i, 85 f., pl. xlvi ; Rostovtzeff, i, pl. 8 ; King, S. and A., 74 ; Hall and Woolley, Ur Excav., 1927, i, 22 f., pls. 5, 6 ; Hall, JEA, 1922, viii, 248.

[2] Hall and Woolley, Ur Excav., 1927, i, 33 f.

[3] Ur Excav., 1934, ii, 304 f.

[4] Meyer, Alt. I, ii, 448—"bronze" ; Heuzey-de Sarzec, i, plates ; BMGB, 84, 86, 89—one found *in situ* at Ur : 1990 B.C. ; Woolley, Digging, 65, pl. 8 ; Contenau, Civ. Assyr., 93 ; Maspero, Dawn, 643 ; also at Tell Halaf, "Subarian" ; von Oppenheim, Tell Halaf, Leipzig, 1931, 188 f., pl. 57.

[5] Perrot-Chipiez, Chaldæa, ii, 116, 410.

[6] Ur Excav., 1934, ii, 286, 304 f., 307.

[7] Desch, B.A. Rep., 1931, 271.

[8] Frankfort, *Illustr. London News*, 1932, clxxxi, 528 ; Frankfort, Jacobsen and Preusser, Tell Asmar and Khafaje, *Univ. Chicago Orient. Inst. Publ.*, No. 13, 78.

[9] Desch, B.A. Rep., 1928, 437 ; 1930, 267 ; Ur Excav., 1934, ii, 286.

[10] Plenderleith, Ur Excav., 1934, ii, 286.

[11] Meissner, i, 246, 267.

[12] Woolley, 1928, 149, 116.

[13] Nouv. Fouilles, 151, Fig. D.

[14] Frankfort, *Illustr. London News*, 1932, clxxxi, 505.

The copper (and bronze) objects from Ur, where the metal came into general use before 3300 B.C., are far too numerous to mention in detail.[1] They include numerous pins, needles, razors and toilet instruments (pls. 230–31), fish-hooks (p. 310, pl. 230), tools (including a three-pronged fork, chisels and adzes) (pls. 225–31), daggers and knives (pls. 78, 225 f.), spear-blades and pike-heads of square section and sometimes 30 in. long (pp. 63 f., 68, 304 ; pls. 149, 189, etc.), very cleverly made cast bronze axes and flat hatchets with sockets for fixing to the handles (p. 305, pls. 147, 153, 223), hammered copper axes (pls. 225 f.), helmets with reliefs and decorative figures (p. 63, pls. 148 f.), reliefs (p. 69, pl. 169), heads of stags, bulls and other animals, and a head of a horned deity (pls. 74, 113, 116–21, 143), various vessels such as pots, patens, vases, pans, beakers, bowls and fluted bowls, buckets and trays (pls. 12, 20, 41, 48, 50, 56, 62, 78, 172, 184–5, 232–40), a collar of an ass (p. 78, pl. 39), friezes, inlay in wood and furniture, wood pillars overlaid with sheet copper, curved strips in pairs for use as cymbals (p. 127), hollow cones made from sheet (perhaps from drums) (p. 259), shields, a cylindrical mace-head with knobs (p. 47, pl. 224) and gold-plated copper (p. 57).

The copper objects found at Lagaš and Tello include human heads filled inside with bitumen, with the eyes of mother-of-pearl and pupils of lapis lazuli (3500–3000 ?), and hollow "cast" bulls' heads with eyes of coloured stone, found near objects of Ur Nina's time.[2] A bull had a nail-like extension.[3] A large blade, spear or lance, $31\frac{1}{2}$ in. long, with the name of King Lugal and the side engraved with the figure of a lion, with remains of Ur Nina, is of nearly pure copper with a trace of antimony only, no tin, lead, zinc or arsenic, but a trace of lead in the patina.[4] A hollow beaten pipe, 10 ft. long and 4 in. diameter, is of similar date.[5] There are also a beaker with a very long spout [6] (410) ; a sword with the handle engraved with two lions, probably a sword of sacrifice and very old (387) ; cast serpents interlaced (390) ; a "magic eye" with inlay of shell, black stone, or lapis, and "enamel or bitumen" [sic], to avert the evil eye (390 : perhaps from a head of copper ?) ; dagger blades (410) ; a large "machine" of copper plates (235) ; plates with rivets (237) ; many nails [7], and fish hooks [8], and a curious vase with a long, trough-shaped tapering spout.[9] Numerous votive cast statuettes with long nail-shaped extensions are very characteristic of the remains and were used for driving away demons, although copper as a product of the underworld was sometimes regarded as unlucky.[10] Many were found with statuettes of fine baked clay at Tello, of various periods from Ur Nina (2700), built into foundations with alabaster tablets giving the names of the rulers.[11] These figures were analysed. A statuette of Rim Sin (2000 B.C. ?) was of nearly pure copper : copper 92·9, sand 0·2, oxidised patina 6·9 ; second analysis of interior part gave copper 95·7, iron 3·5, sulphur and oxygen 1·2, no tin, lead, zinc, antimony or arsenic. An adze with a circular socket, of Ur Nina's period, was pure copper with a little cuprous oxide [which would confer

[1] Ur Excav., 1934, ii, 299, 412 ; references below are to pages and plates of this ; Woolley, 1928, 14, 37, 39, 53, plate 10 ; dagger of 3000–2500 with riveted haft, B. Museum Quarterly, i, 57 ; Smith, EHA, 58 ; Meissner, i, 81.

[2] Heuzey-de Sarzec, i, 61, 69, 70, 238; Contenau, Manuel, ii, 595 ; Hilprecht, Explor., 1903, 540 ; Meyer, Alt. I, ii, 544 ; Handcock, 249.

[3] Maspero, Dawn, 757.

[4] Heuzey-de Sarzec, i, 410 ; Handcock, 243 ; Meyer, Alt. (1909) I, ii, 445 ; ib., Sumerer, 82, 87 ; Maspero, Dawn, 756 ;

Petrie, Tools and Weapons, 33 ; analysis in Berthelot, Arch., 76 f.

[5] Handcock, 243 ; cf. the Egyptian one of King Sahurā.

[6] The numbers in the text refer to Heuzey and de Sarzec.

[7] Handcock, 243.

[8] Meissner, i, 226.

[9] Contenau, Manuel, ii, 596.

[10] Fossey, Magie, 112 f. ; Meissner, i, 265, 267 ; Contenau, Manuel, ii, 586, 705, 757, 789, 790.

[11] Heuzey-de Sarzec, i, 238, 240, 246, 248, 259, 414, 420 ; Handcock, 244 ; Meissner, i, 267 ; ii, pl.-figs. 3 and 4.

hardness], but no lead, tin, zinc, arsenic or antimony ; a hatchet from the oldest remains was pure "hard copper" with traces of arsenic (whether present in the ore or added intentionally is not known) and phosphorus, but no tin, lead or zinc.[1]

Sharp copper sickles and perhaps curved maces were carried as weapons by the Sumerian soldiers.[2]

Many other objects from Tello, "of bronze or at least of copper," [3] are most safely included as copper : they comprise a lance blade (67), statuettes (58, 67, 72), a saucer (59), a bull's horn of metal plates on wood (61 : "bronze") and a large vase, possibly of the time of Gudea (26). Of especial interest is a fragment of a very old Sumerian helmet, rising to a conical pointed piece at the top, formed by hammering out[4] : similar helmets and representations of them were found at Ur [5], and are shown in representations of Sumerian troops.[6] There were also razors, a vase, bracelets [7], a dagger blade with two rivets (110), a dagger blade with three rivets and central rib (112 : ? lance head) [8], an adze cast in a two-piece mould (115) [9], an article with two prongs (117), cups (127, 132), an arrow-head (117), a goblet (139), a blade (141), "articles" (134), an implement like a shepherd's crook (149), a dagger (255), a hatchet (261), a bill-hook (301), a rod or bar (300), chisels (80), a saw with curved handle (99 : very old), and a tubular cast copper mace handle, half decorated, with wood inside (77 and Fig.). An example in copper of the votive staff, with an oval handle attached at the side, frequently shown in religious scenes, was found at Tello.[10] "Bronze" weapons, scraps of "bronze" in a treasure chest and a "bronze" stag's head cast over a clay core, of the first Babylonian dynasty (c. 2000), found at Kiš[11], do not appear to have been analysed. A "bronze" sabre of Adadnirari I (1310–1281)[12] is the same in make as one found in Canaan with a long sword of Cretan-Mycenæan shape.[13] Copper fish-hooks and a butcher's axe, of 3500–3300, were found at Jemdet Nasr, near Kiš.[14]

SOURCES OF COPPER

Some metals were probably discovered independently in various places[15], although it has been supposed[16] that copper working was a specifically Sumerian invention. This suggestion is just as speculative as the alternative hypotheses that knowledge of copper was brought to Mesopotamia by South Arabian Hamites[17]; that it was brought both to Egypt and Mesopotamia by an Armenoid race from the region of Mount Ararat[18]; or that Asiatic copper was taken by sea to Egypt and Crete about 4000 B.C. by the inhabitants (pre-Phœnicians) of the Syrian coast, these countries and Cyprus then developing their own supplies of copper.[19] Hall[20] suggests that the Greek name πελέκυς for axe may be derived through the Assyrian pilakku from the Sumerian. Cretan axes of 2000 B.C. are also socketed like the Sumerian.

The assumption of a migration of the Sumerians from Central Asia would

[1] Berthelot, Arch., 75, 78, 80 ; ib., Intr., 225 ; ib., Mâ, i, 391 f. ; Hoernes, Urg., ii, 241 ; Jeremias, 59 ; BMGB, 104 ; cf. Mallowan, Discovery, 1930, xi, 259.

[2] Meyer, Alt. I, ii, 451 ; Delaporte, 77, Fig. 9 ; Nouvelles Fouilles, 129, planche viii ; King, S. and A., 138 ; curved knife, Partington, Everyday Chem., 1928, 14, Fig. 11, c. 3000 B.C.

[3] Heuzey-de Sarzec, i, 237 ; numbers in text are to pages of this work.

[4] Nouv. Fouilles, 43, 44, 114 ; Meissner, i, 81, 82, Fig. 17.

[5] Woolley, 1928, 52 ; cf. ib., 56.

[6] Maspero, Dawn, 606, 722.

[7] Nouv. Fouilles, 133, 152 ; references in text to this.

[8] Cf. Maspero, Dawn, 722 ; Rawlinson, Mon., i, 96, Fig.

[9] Cf. Rawlinson, Mon., i, 97.

[10] Contenau, Manuel, ii, 588.

[11] Langdon, Excav. at Kish, i, 78 f., 89, 92.

[12] Maspero, Struggle, 607.

[13] Hoernes, Urzeit, ii, 24.

[14] Langdon, Alte Orient, 1928, xxvi, 74.

[15] Orth, PW, Suppl. iv, 109.

[16] Montelius, Die älteste Bronzezeit, 1900, 195.

[17] Naville, q. in Isis, 1927, ix, 545.

[18] O'Leary, Arabia before Muhammad, 1927, 14.

[19] Hall, Civ. of Greece in the Bronze Age, 1928, 31, 37.

[20] Op. cit., 90.

involve their bringing a knowledge of copper with them [1], but it has been suggested that foreign craftsmen may have settled on the Lower Euphrates and the king of Ur, who had the copper statues and reliefs made, imported the copper-workers from a district, probably Cappadocia, where the technique was indigenous. The similarity of the early Copper Ages in Babylonia, Syria and Egypt suggests a common point of origin for copper working. Weapons and implements in Babylonia include all the types known in Egypt and Syria, whilst typically Sumerian shapes never occur in Egypt. The copper-workers in Babylonia and Egypt, it is suggested, shared a common knowledge, perhaps from Cappadocia, but this developed differently.[2]

The copper was certainly imported, either as metal or ore, since none occurs in Babylonia, which is an alluvial region. The old copper mine at Arghana Ma'adan in Kurdistan (which was worked by the Germans during the Great War) and that of Sabzawar in Afghanistan, would be possible sources.[3] The presence of nickel and the absence of sulphur in practically all specimens of copper and bronze found at Ur have been supposed [4] to indicate that the ore came from the state of Oman (Jabal al-Ma'adan in Wâdî Ahin, inland from Sohar) in the Persian Gulf, where copper ore is still found at Jabal Akhdar, but ore containing nickel and manganese also occurs in Sinai and the conclusion seems to me too hasty. A large proportion of sulphide ores of copper contain nickel, and if a desulphurising process was known the copper containing nickel might have come from anywhere. The ore from Oman contains 0·19 per cent. of nickel, but two specimens of old slag, containing 1·50 and 4·30 per cent. of copper, contained no nickel. Five specimens of metal from tumuli in the Bahrain Islands (in the Persian Gulf), about 1200 B.C., contained 77·53 to 94·69 per cent. of copper and 3·18 to 19·27 per cent. of tin. They mostly contained traces of lead and sulphur, but only two contained nickel.[5] Import of copper ore (?) by sea is mentioned in a tablet at Ur.[6] Copper from an old mine in Kurdistan contained cobalt, and could not, it is said, have been used, although a copper nail from a Sumerian statuette of 3000 B.C. in Copenhagen contains traces of cobalt and arsenic, and some iron, lead and nickel.[7] The copper imported to Ur from Meluchcha [8] probably came from the Persian Gulf. The Caucasus and the Kushite Mountains have been suggested as sources of supply [9], also the Black Sea region (Pontos), where copper ore occurs in the mountains.[10] Copper ore also occurs in Persia, in Assyria and on the upper reaches of the Tigris, but it is doubtful if it was mined[11]; it is also found abundantly in the Tiyari Mountains, near Nineveh, where it forms surface outcrops, and also near Diarbekir.[12] Deposits of native copper still exist at Arghana, 75 miles north-west of Diarbekir, and several others, some known to have been worked at an early date, near Diarbekir and Erzerum, whilst other deposits are still worked between Sinope and Trebizond. Copper may have been first discovered in this region (the ancient Amida).[13] The copper found at Šuruppak may have come from Diarbekir, Sinai, the Tiyari Mountains or (probably) the Caucasus.[14]

[1] BMGB, 3.
[2] Smith, EHA, 25 f., 56 f., 372.
[3] Contenau, Manuel, iii, 1510.
[4] Peake, The Copper Mountain of Magan, Antiquity, Gloucester, 1928, ii, 452.
[5] Desch, B.A. Rep., 1928, 437; ib., Nature, 1928, cxxii, 886; Petrie, Anct. Egypt, 1929, 49; Marples, ib., 100; Woolley, 1928, 45; ib., Digging up the Past, 1930, 137; ib., Ur Excav., 1934, ii, 287, 394.
[6] B. Museum Quarterly, i, 58.
[7] Sebelien, Anct. Egypt, 1924, 11, 14.

[8] Schrader, Keil., 15.
[9] Orth, PW Suppl. iv, 112.
[10] Sebelien, op. cit.; Maspero, Dawn, 614; Rollin, Ancient History, ed. Bell, Glasgow, 1845, ii, 696.
[11] Meissner, i, 53; Sebelien, Anct. Egypt, 1924, 14.
[12] Rawlinson, Mon., i, 219 f.
[13] Peake and Fleure, Priests and Kings, 1927, 12; von Oppenheim, Tell Halaf, Leipzig, 1931, 189.
[14] Gowland, Archæologia, 1920, lxix, 133; Dussaud, La Lydie et ses Voisins, 1930, 81 f.

Much copper was imported in Gudea's time from a mountainous country called Magda [1], perhaps the foothills of the Taurus range [2], or Zagros [3], Elam [4], or Cappadocia.[5] Copper also came from a place called "Abullat in the mountain of Kimaš," [6] from "Albadh in the depth of the mountain," [7] and from Dilmun, perhaps on the Persian Gulf.[8] Although Hilprecht [9] and Amiaud[10] suggest that it was Nejd, the Mash of the Bible[11], and Sebelien that it was Sinai, Kimaš in Harsi (Hursitum) is generally understood to mean the mountains of Elam and Iran sloping towards Assyria.[12] The name Kimaš may be the origin of the Assyrian name kêmassi for copper.[13]

Ašurnasirpal II (885–860) obtained 100 talents of copper from the Hittite king Sangara, of Carchemish, and much copper from other Syrians.[14] In the later periods copper probably came to Babylonia and Assyria by Phœnician trade, perhaps by way of Palmyra and Thapsacus.[15] Šalmaneser II (860–825) and Asarhaddon (681–669) also obtained copper and bronze as tribute.[16] Nabonidus (550 B.C.) imported copper from Mount Amanus[17], perhaps in North Syria[18] or Lebanon.[19] Cyprus (Alašia), even in the 15th century B.C., exported much copper, which was "prepared" there : it mostly went to Egypt, and in one letter the king of Alašia excuses himself to Pharaoh because he has sent only the small amount of 500 talents (about 15 tons) of the metal.[20] Sayce[21] thought some copper came to Mesopotamia from Cyprus, which is mentioned by Sargon, but most of it from Sinai ; O'Leary[22], that both Egypt and Mesopotamia obtained copper from an Armenoid race near Mount Ararat. There appears to have been an extensive metallurgical industry centred around Lake Van ; there are remains of very old copper mines in Armenia, and in the Assyrian Period Urartu (Van) was a principal source of copper.[23] When Sargon II captured the town in 714 B.C., he took 3,600 talents (about 108 tons) of "unworked" copper (probably in bars).[24]

USES OF COPPER

It is impossible to give a detailed account of the numerous later copper objects from Babylonia and Assyria, but a brief survey of the technology may be attempted.[25] Early Sumerian copper figures of lions and bulls with "cast" (?) heads filled with bitumen, and the bodies formed of plates hammered over wood, are said to combine both casting and hammering.[26] Castings filled with bitumen found by Layard at Quyûnjiq (Nineveh) are of much later date, so that the technique was handed down.[27] The copper articles in the British Museum from Tell Sifr ("mound of copper") near Senkereh, the ancient Larsa, are probably of 2500 B.C.[28] Copper articles of all kinds—daggers,

[1] Meyer, Alt. I, ii, 447.
[2] Langdon, CAH, i, 427.
[3] Perrot and Chipiez, Chaldæa, i, 124.
[4] Meyer, Alt. I, ii, 545 ; ib., 1909 ed., 492, 499 ; Meissner, i, 347.
[5] BMGB, 90 ; Meissner, i, 53, 346.
[6] Meissner, i, 347 ; Sebelien, Anct. Egypt, 1924, 14.
[7] Meissner, i, 265.
[8] Maspero, Dawn, 616 ; ib., Hist. anc., 1905, 167.
[9] Excav. in Assyria and Babylonia, 1904, 238.
[10] Rec. of the Past, N.S., ii, 81.
[11] Gen. x, 23.
[12] Meyer, Alt. I, ii, 545 ; King, S. and A., 261 ; Herzfeld, Islam, 1921, xi, 125.
[13] Amiaud, op. cit. ; Meissner, i, 53.
[14] Jastrow, Civ., 172 ; Sayce, Rec. of the Past, N.S., 1889, ii, 128 f. ; Rogers, ii, 64 f. ; BMGB, 51 ; Luckenbill, i, 144 f.

[15] Rawlinson, Mon., iii, 16.
[16] Luckenbill, i, 211 f., 288.
[17] Contenau, Civilisation Phénicienne, 1926, 298.
[18] Amiaud, op. cit., 79.
[19] De Morgan, Délégation en Perse, i, 40.
[20] Meissner, i, 18, 346 f.
[21] B. and A. Life, 6, 131.
[22] Arabia before Muhammad, 1927, 44.
[23] Sayce, CAH, iii, 185 ; Montelius, Alt. Bronz., 1900, 197.
[24] Meissner, i, 111 ; Weissbach, PW, ia, 2506, 2510 ; Luckenbill, ii, 9 f. ; Dussaud, La Lydie et ses Voisins, Babyloniaca, 1930, xi, 78.
[25] Cf. Rawlinson, Mon., i, 99.
[26] Hall, CAH, i, 291 ; Partington, Everyday Chemistry, 15, Fig. 13.
[27] Rawlinson, Mon., i, 367.
[28] Loftus, 1857, 268, 269 ; Maspero, Dawn, 747 ; Hilprecht, Explor., 1903, 156.

knives, fish-hooks, vases and kitchen utensils—were largely used in Babylonia and Assyria, along with bronze, to the latest times.[1] Copper was also used in Assyria for the manufacture of colour.[2] A copper bell and a large ring, probably Assyrian, were found at Arban, on the Khabur.[3] There are two copper mace-heads with the name of Ḥammurabi (1924–1913) and one (perhaps older) with seven entwining snakes, in the British Museum.[4] Under Ḥammurabi copper was cheap and was used for tools, surgical instruments and door-keys.[5]

Objects mentioned in texts are two "palms" and a large statue of copper for the temple of the sun god of King Gungunnu of Larsa (2094–2068) ; "strong copper" [perhaps a gong ?] for scaring demons ; and a "lightning flash" symbol of the god Ramman made for Tiglathpileser I (1115–1093).[6]

King Singašid of Uruk (2000 B.C.) is said to have fixed the price of copper at 600 to 1 of silver or 700 of wool, which is supposed to be too cheap to be real [7], although King [8] gives the ratio 600 : 1 for Babylon about 1900 B.C. In the Sumerian Period at Tello the value of copper to silver was 1 : 240.[9] In Hammurabi's time 1 shekel of silver bought 2 to $2\frac{1}{3}$ minas of copper, so that the value of copper had increased to 120–140 to 1 of silver[10], or 3,600 to 1 of gold[11], whilst in the Assyrian Period a slave was worth 20, 30 or 32 shekels of silver, or 50 to 100 minas of "bronze", the values of silver and bronze being about 1 : 185.[12] Sargon II's statement (722–705) that copper and silver were "equal in value" is exaggerated.[13] Copper was used as money from the oldest Sumerian to the Assyrian Period, and copper and bronze were of sufficient value in 1450–1400 B.C. to be exchanged with Egypt along with gold and silver.[14] In the so-called Cappadocian tablets (3 millenn.) an "inferior" copper (êrum chaburatai) is distinguished from a "good" or "better" (šikum) copper [perhaps bronze ?], the ratios of value to silver being 1 : 55 and 1 : 25, whilst the usual ratio for ordinary copper was 1 : 45 to 1 : 60. These metals were traded with Assur.[15]

BRONZE

The suggestion, made in an early period of archæological investigation in Mesopotamia, that copper and bronze occur together in the earliest Sumerian Period[16] was based on a mere inspection of objects thought to be bronze but afterwards found on analysis to be copper.[17] The next representation was a Copper Age preceding a Bronze Age, the transition occurring gradually and the first bronzes being poor in tin, or containing lead or antimony instead of tin.[18] Analysis of objects from Ur and Eridu shows, however, that true tin-bronze occurs in the earliest periods (3500–3000 B.C.) along with copper, and that bronzes poor in tin become more frequent in the later period (from 2700

[1] Perrot and Chipiez, Chaldæa, ii, 410 f. ; Maspero, Dawn, 747 ; Sayce, B. and A. Life, 133.
[2] Handcock, 253.
[3] Layard, Nineveh and Babylon, 1867, 122, 125.
[4] BMGB, 87, 92 ; Smith, JEA, 1927, xiii, 277.
[5] Meyer, Alt. (1909) I, ii, 517 ; Meissner, i, 247 ; ii, 318.
[6] Meissner, i, 26, 141 ; ii, 209, 238, 363 f., 445 ; Maspero, Dawn, 662.
[7] Meissner, i, 54, 149, 361 ; BMGB, 246 ; Meyer, Alt. (1909) I, ii, 412 ; ib., Hist. antiq., 1926, iii, 256 ; Delaporte, 257.
[8] H. Bab., 211.
[9] Meyer, Histoire, iii, 264.
[10] Meissner, i, 361.

[11] Thureau-Dangin, Amer. J. Archæology, 1910, xiv, 211.
[12] Delaporte, 335.
[13] Meissner, i, 363.
[14] Meissner, i, 355 ; BMGB, 128.
[15] Landsberger, Alte Orient, 1924, xxiv, Heft 4, 23.
[16] Meyer, Alt. (1909) I, ii, 416, 744 ; ib., 1913, 418 ; ib., Sumerer, 56 f., 75, 77, 79, 85, 92, 115 ; Delitzsch, MGM, v, 11 ; Feldhaus, Technik, 1032 ; Hoernes, Urg., ii, 242.
[17] Berthelot, Intr., 225 ; Maspero, Dawn, 756 ; Meissner, i, 265 ; Orth, PW, Suppl. iv, 112.
[18] Contenau, Assyr., 92 ; Meissner, i, 265, 348.

B.C.), when the supplies of imported tin ore were becoming exhausted or inaccessible.[1] At al-'Ubaid, near Ur, on the contrary, the metal of *c.* 3000 B.C. is copper, not bronze.

The analytical results for the Ur specimens, which are very numerous, may be summarised as follows :

(1) *Earliest Period* : 3500–3200 B.C.—Royal Cemetery, Ur. The alloys contain 70·24 to 97·2 per cent. of copper and 0·56 to 20·2, but mostly about 11 to 12 per cent., of tin (one 5·8 per cent. and one 8·3 per cent.) ; also traces to 2·2 per cent. of nickel (mostly 0·2 to 1·3 per cent.), some iron (up to 1·7 per cent. exceptional, mostly below 1 per cent.), and also lead up to 1·62 per cent., but mostly below 1 per cent. Sulphur and antimony are absent, or present in traces only, in the bronzes of *all* periods at Ur, but sulphur occurs in alloys from al-'Ubaid, Kiš and the Bahrain Islands. Arsenic occurs (0·16 per cent.) in metal from al-'Ubaid (pre-3500 B.C.), but not at Ur.

(2) *Pre*-3000 B.C. Four specimens contain a trace, 1·45, 8·8 and 15·1 per cent. of tin, and all contain nickel, up to 1·1 per cent. with 15·1 of tin, but traces only or no lead. There is no relation between the tin and nickel contents.

(3) *Sumerian* : 3200–2700 B.C. Four specimens were practically pure copper (mostly containing some nickel) and two contained 14·3 per cent. of tin. Only traces of lead were present.

(4) *Sargonid* : 2700–2500 B.C. The alloys are poor in tin (0·1–2·4 per cent.), containing only traces of lead and iron, but appreciable amounts of nickel (up to 1·77 per cent.).

(5) *Sumerian Revival* : 2500 B.C. The alloy consists of copper 95·2, tin 0·22, lead 0·36.

Early bronzes (before 3000 B.C.) at Kiš contained 3·0 to 13·2 per cent. of tin (3·0, 3·27, 6·1, 7·4, 11·0, 12·2, 13·2) and some nickel (0·006 to 0·54) ; metal of 3000 B.C. was 94·01 copper, 0·43 tin, 0·58 lead, 1·31 iron, 3·34 nickel and 0·17 sulphur ; metal of Nebuchadnezzar's time was 88·16 copper, 4·65 tin, 6·16 iron and 0·15 lead.

Metal from al-'Ubaid before 3500 B.C. was sometimes nearly pure copper (99·69 copper, 0·01 iron, 0·16 arsenic, 0·12 zinc), metals of the 1 dyn. period contained 0·16–7·95 per cent. of tin, traces of lead, traces to 0·23 per cent. of nickel, traces or no arsenic, and no zinc. Some specimens contained small amounts, up to 2·8 per cent., of tin, in one case 4·4 per cent. of lead, usually nickel (up to 0·8 per cent.), often sulphur, and arsenic (up to 1·6 per cent.). The percentage of nickel is smaller than in the Egyptian copper of Pepi's statue.[2] A bronze from Irâq (2000 B.C.) contained 9·77 per cent. of tin, 0·68 of lead, 0·17 of sulphur and 0·28 iron, but no nickel.

The older socketed axes and adzes at Ur are very skilfully cast in bronze, whilst the later axes are of copper, hammered, not cast. The change in technique results from the change of material, since copper is difficult to cast, but is malleable and gains in hardness on hammering. At first, rather feeble attempts were made to imitate the old socketed shapes, but these were soon abandoned for simpler forms better suited to the purpose. One of these has remains of a wooden handle and the butt has been passed through a slot in the handle and bent over, a copper nail being driven through the wooden shaft and butt from the opposite side. Closed moulds for casting were never used in Predynastic Egypt, and even when Egypt and Mesopotamia were afterwards in relation through Syria, the technique was never adopted in

[1] Desch, *B.A. Rep.*, 1928, 437 ; 1929, 264 ; 1930, 267 ; 1931, 271 ; Contenau, Manuel, ii, 598 ; Ur Excav., 1927, i, 36 f.; 1934, ii, 284 f. ; Peake and Fleure, Merchant Venturers in Bronze, Oxford, 1931, 13, 44.

[2] See p. 55 ; Hall and Woolley, Ur Excav., 1927, i, 36, who say the "sceptre" of Pepi is a seal cylinder.

Egypt.[1] The Syrian axes of a somewhat later period are also socketed, but the shape of the blade is oval. The socketed axe is represented on a stela of Hammurabi.[2]

The complete analyses of four specimens of metal from the earliest period (3500–3200 B.C.) at Ur are (in percentages) :

Copper	.	.	.	84·18	85·13	85·01	—
Tin	.	.	.	12·00	11·78	14·52	20·2
Lead	.	.	.	1·62	1·13	0·47	nil
Nickel	.	.	.	2·20	0·25	tr.	0·7
Iron	.	.	.	—	1·71	—	tr.

Free from arsenic, antimony, silver, zinc and manganese.[3] These are true bronzes, the last being unnecessarily rich in tin, since too much tin makes the metal brittle. *White* alloys very rich in tin (speculum metal) do not seem to occur.

The compositions of early copper alloys from Ur are strikingly different from those found for metals from most other Sumerian sites. Of the objects from Tello analysed by Berthelot [4], a votive statuette of Bur Sin (2060 B.C.) contained 76·0 of copper, 18·1 of lead, a "notable amount" of sulphur, and some iron and oxygen, in one analysis ; a second analysis gave 77·4 copper, 17·0 lead, 2·3 sulphur, no tin, zinc, arsenic or antimony. A Babylonian statuette of uncertain date contained 79·5 copper, 1·25 tin, 0·8 iron, 9·75 oxygen and 8·3 calcium carbonate (the last two undoubtedly in the patina), but no lead, zinc, antimony or arsenic. The centre of Bur Sin's statuette was composed of copper with 3 per cent. of iron. A statuette of Rim Sin (2000 B.C.) was of practically pure copper. The pedestal of a Babylonian bronze bull incrusted with silver contained 82·4 copper, 11·9 tin, 4·1 iron, no lead, antimony, arsenic or zinc.

Very early Sumerian bronzes found at Nippur by Hilprecht contained copper and antimony as principal constituents ; one piece contained a considerable amount of tin [5] :

	Cu	Sb	Fe	Ni	O, etc.	Pb	Sn	S
Fragment of sword . .	96·38	1·73	0·24	0·22	1·43	tr.	—	—
Stylus	80·52	3·05	0·35	0·55	9·90	—	5·45	0·18
Copper nail . . .	98·27	0·30	0·17	0·39	0·87	—	—	—
Dish	80·35	2·24	0·40	0·87	14·93	1·15	—	0·06

From these results it was generally assumed that, before Sargon's time (2600 B.C.) when tin-bronze was actually in use, lead (anaku) or antimony (guchlu) were added to copper to increase the fusibility and harden the alloy. Later, perhaps about 2000 B.C., tin (also called anaku) replaced lead or antimony. Bronze (Sum. zabar ; Akk. sipparu) was still valuable in Hammurabi's time, but became common in the Assyrian and later Babylonian Periods, the alloy then containing about 8·5 of copper to 1 of tin.[6] Sayce, however, had suggested from references in Hittite texts at Boghazköi, which show that Asia Minor was invaded by "Babylonians" at an early date for the sake of the copper, lead and silver mines, that it would be necessary to refer the origin

[1] Ur Excav., 1934, ii, 30, 150, 396 ; plates 223–6, 229.
[2] Contenau, Manuel, ii, 587, 837.
[3] Desch, *B.A. Rep.*, 1928, 438 ; 1931, 271.
[4] Arch., 78 f., 81 ; BMGB, 104 ; Handcock, 253.
[5] Helm, *Verh. Berlin Anthrop. Ges.*, 1901, 157 f. ; Hilprecht, Explor., 1903,

252 ; *ib.*, Excav., 1904, 454 ; *Chem. Z.*, 1901, Rep. 250 ; Peters, Nippur, 1897, ii, 384.
[6] Sayce, B. and A. Life, 131 ; Schwenzner, MGM, xv, 51 ; Handcock, 253 ; Hall, Oldest Civilisation of Greece, 1901, 193 f. ; Unger, RV, i, 428 ; Petrie, *Ancient Egypt*, 1929, 49.

of bronze to an earlier date than was hitherto supposed [1], viz., about 3000 B.C. [2], and this is confirmed by the recent discoveries at Ur.

The North Mesopotamian and Persian sites have not as yet been sufficiently studied to enable us to draw any very certain conclusions as to the origin of bronze, and in particular as to whether the new discoveries at Ur require any modification of the theory given in a preceding section, written before the information was available.[3] They might be supposed to lend support to the older theory that bronze was a Sumerian invention, afterwards taken to Egypt [4], in support of which there is also some literary evidence.

A Sumerian hymn to the god of fire, Gibil (Bilgi, Gira, Nusku), one of the emanations of Ea, which is extant in a bilingual version (Sumerian-Akkadian), says :

" Fire (Gibil) with thy [pure and] brilliant flame (fire)

. . . . . .

Thou mixest copper and anaku
Thou purifiest gold and silver."[5]

That anaku is necessarily tin, and hence the alloy mentioned in the hymn is tin-bronze, is uncertain, since anaku and annamu originally mean lead, although anaku is a *later* name for tin.[6] There is no satisfactory name for tin in Assyrian [7], but anaku is generally accepted.[8] Since anakum in the Cappadocian tablets (3 millenn.) stood in the ratio of value of $1 : 3\cdot5$ to $1 : 6$ to silver [9], it was perhaps there tin, not lead.

In a pre-Sargonic Sumerian text bronze (zabar) is made from 1 mina 4 shekels of pure copper (urudu luchcha), $10\frac{2}{3}$ shekels of lead (anna), $4\frac{2}{4}$ shekels of ne-ku and $\frac{1}{2}$ shekel 21 grains [*sic*] of su(g)-gan. Thureau-Dangin thinks ne-ku means antimony, but does not specify su-gan. The proportions are : copper 80·05, lead 13·34, antimony 5·8 and su-gan 0·77. In another text from Tello, ku-babbar, abridged as ku(g), i.e. silver, is also called ku(g) zabar, so that zabar does not always mean bronze.[10] It is usually supposed, however, that the Sumerian names urudu and zabar meant copper and bronze respectively, zabar being "fiery red" (which is not the usual colour of bronze), and that the Akkadian names eru and sipparu, which are obviously derived from the two Sumerian words, also meant copper and bronze respectively. Thus in Sumerian texts at least as old as Gudea's time there are separate names for copper and bronze. This is doubtful for Egypt[11], and in the Hebrew (něchôšeth) and Indo-Germanic languages (æs, áyas, aiz, etc.), as well as in Greek ($\chi\alpha\lambda\kappa\delta s$), copper and bronze have only one name.[12] One Elamite name for bronze, zubar, corresponds with the Sumerian zabar ; another, sahi, is probably derived from the name of a place.[13]

In a copy of a text attributed to Sargon (2600 B.C.) in the Royal Library

[1] *Man*, 1921, xxi, Art. 97.
[2] Mackenzie, Footprints of Early Man, 144.
[3] See p. 76.
[4] Tomaschek, 1888, q. by Much, Kupferzeit in Europa, Jena, 1893, 363 ; Hall, Oldest Civ. Greece, 193 f. ; Meyer, Alt. I, ii, 424 ; Schrader-Jevons, 192 ; Schrader, RL, 199 ; Hoernes, Urg., ii, 216 f.
[5] Meissner, i, 229 ; ii, 15, 165, 207 ; Lenormant, Magic, 185 ; Maspero, Dawn, 635 ; Hoernes, Urg., ii, 217.
[6] Unger, RV, i, 428 ; Fossey, Magie, 426 ; Meissner, i, 266 ; ii, 165.
[7] Thompson, Chemistry of the Ancient Assyrians, 1925, 40.

[8] Zimmern, in E. Schrader, Keilinschriften, 1903, 624.
[9] Landsberger, *Alte Orient*, 1924, xxiv, Heft 4, 22.
[10] Thureau-Dangin, in Nouv. Fouilles de Tello, 55 ; Amiaud, *Rec. of the Past*, N.S., ii, 84 ; Sayce, *Man*, 1921, xxi, Art. 97.
[11] See p. 66.
[12] Schrader, RL, 199 ; Baumstark, PW, ii, 2714 ; Sayce, *Man*, 1921, xxi, Art. 97 ; Hall, Oldest Civ. Greece, 1901, 193 f. ; Unger, RV, i, 428 ; see also Gadd and Legrain, Ur Excav., 1928, Texts, 52, 59, 64, 68, 92.
[13] De Morgan, Délégation en Perse, 1904, v. 47.

at Assur, this king is made to say [1] that his conquests extend "from the lands of the setting sun to the lands of the rising sun, viz., to the tin land (Kû-Ki) and Kaptara (Crete), countries beyond the Upper Sea (the Mediterranean ?)." Kû-Ki has been identified with Spain or Brittany, and Sayce considered that the western extension of the empire ended with the Syrian coast, beyond which were Kaptara or Crete and "the tin land." Peake suggests that Kû-Ki is Cyprus or some other island in the Mediterranean, or some region easily accessible from it, and that the discovery of tin and bronze was made before Sargon's time somewhere in Western Asia. Although ancient tin workings have been found in Khorassan [2], this local source was insufficient and merchants from the Persian Gulf are supposed to have traded in tin with a place "in the Mediterranean region." Peake develops this hypothesis into regions where we need not follow him. Meyer [3] quotes the passage as follows : "The tin land (a-na-azag-ki, in case this is the correct pronunciation of anaku) and Kaptara, the lands beyond the upper sea, Dilmun and Magan, the lands beyond the lower sea, the lands of the rising sun to the setting sun, have been thrice conquered by the hand of Sargon, king of the world." Meyer identifies Kaptara with Kaphtor, the first mention in cuneiform ; the "upper sea" is the Mediterranean, the "lower sea" is the Persian Gulf ; if Ku means copper, as Langdon suggested [4], the "tin land" is perhaps Cyprus and not Tartessos in South Spain, but Meyer doubts if Sargon I is mentioned, and the text is probably based on old legends of this king. King's new translation gives the Persian Gulf, not the Mediterranean, and Cyprus is doubtful.[5]

LATER ASSYRIAN AND BABYLONIAN BRONZE

Later Assyrian tin-bronze was still of rather variable composition and often contained some lead, as did the early European [6] and Chinese [7] bronzes. It was cast in clay, stone or bronze moulds, examples of which are known. The cast objects are usually small and are sometimes strengthened with iron. Repoussé and engraving work in bronze were well executed.[8] Some analyses of later Assyrian bronzes are given in the table on p. 250.

Sennacherib (705–681) says he made forms of clay, on the command of "the master of wisdom, Ea," and cast bronze in them [9], and Nabonidus (555–538) mentions ½ talent 5 minas ⅓ shekel of copper and 4 minas 5 shekels of tin, probably for the composition of bronze.[10]

Tukulti-Urta I (1260–1230) and Tiglathpileser I (1125–1100) speak of hewing rock with bronze picks.[11] Tiglathpileser I obtained great quantities of copper and bronze from Asia Minor, which was then a very important centre of metal industry.[12] Copper plates, rings, cups, etc., and bronze picks are mentioned in texts of Ašurnasirpal II (885–860)[13]: from the Hittite king Sangara at Carchemish he obtained 100 talents of copper, also copper, etc. (but not iron) from the Phœnicians. Šalmaneser III in 854 obtained copper from the

[1] Sayce, Man, 1921, xxi, Art. 97 ; Peake and Fleure, Priests and Kings, 1927, 56 ; Peake, The Bronze Age and the Celtic World, 1922, 41 f., 43, 57.
[2] Gowland, J. Anthropol. Inst., 1912, xlii, 252.
[3] Chronol., 23.
[4] Peake and Fleure, Priests and Kings, 57.
[5] Dussaud, Civilisations préhelleniques dans le bassin de la mer égée, 1914, 225.
[6] Blümner, iv, 188 f.; ib., PW, iii, 561 f.; Hoops, Reallexikon, i, 314—1·72 and 5·25 per cent. of lead.
[7] Dôno, Bull. Chem. Soc. Japan, 1932,

vii, 350 ; 1933, viii, 133 ; 1934, ix, 120; Desch, B.A. Rep., 1930, 268—mostly free from nickel.
[8] Perrot and Chipiez, Chaldæa, ii, 413 ; Rawlinson, Mon., i, 99, 366 f., 370 f. ; Maspero, Dawn, 756 ; Handcock, 258 f.
[9] Belck, Z. f. Ethnol., 1908, xl, 47 f.
[10] Meissner, i, 265, who gives 1 tin to 8½ copper.
[11] Luckenbill, i, 56, 81.
[12] Smith, EHA, 299 ; Luckenbill, i, 76.
[13] Sayce, Rec. of the Past, N.S., 1889, ii, 140, 142 f., 148 f., 153 f., 157, 160, 162 f. ; Luckenbill, i, 144 f.

Hittites and the kings Jehu, Suâ and Marduk-apal-usur of the land of Sukhu [1], and Sargon II (722–705) obtained great quantities of bronze, including "white bronze" (? tin) as tribute from Armenia and elsewhere.[2]

	Cu	Sn	Pb	Fe	Sb	As	Ni
1. Grey thick rod	88·03	0·11	3·26	4·06	3·92	0·60	—
2. Bent rod . .	86·84	12·70	0·28	tr.	—	—	0·18
3. Decoration of furniture . .	86·99	12·33	0·38	tr.	—	—	0·30
4. Rim of dish (or cup)	80·84	18·37	0·43	0·16	—	—	0·20
5. Tripod . .	{83·96 {83·76	8·18 8·20	7·64 7·70	0·26 0·28	—	—	—
6. Figure . .	{85·0 {85·14	5·30 5·37	8·50 8·45	0·7 0·8	—	—	—
7. Vessels . .	—	{10·63 { 9·78	—	—	—	—	—
8. Bell . . .	—	14·1	—	—	—	—	—
9. Gate fasteners .	90·0	{8·88 {8·63	0·41	tr.	nil.	0·12	nil.

1–4. Bronzes from Nineveh (Quyûnjiq) in the British Museum, found by Layard and carefully analysed by L. R. von Fellenberg, q. by E. von Bibra, Die Bronzen und Kupferlegirungen der alten und ältesten Völker, Erlangen, 1869, 94 f., 154—the book contains a large number of analyses of old bronzes from various sources ; the place of publication is not Stuttgart, as Desch, B.A. Rep., 1928,: 437, says, by an oversight. The numerous analyses of bronzes in Fellenberg's papers in the Mitteilungen der Naturforschenden Gesellschaft in Bern, 1860–65, do not include those of the Assyrian bronzes, yet von Bibra gives no other reference ; Gowland, J. Inst. Metals, 1912, vii, 37 ; Busch, Z. angew. Chem., 1914, xxvii, 512 ; Feldhaus, Geschichtsblätter, i, 153.
    5. Assyrian tripod at Erlangen, analysed by Busch and Leuze ; refs. 1–4.
    6. Figure in Assyrian tripod No. 5 ; refs. 1–4.
    7. Vessels found by Layard at Nimrûd, Ašurnasirpal II, 885–860 B.C., analysed by Percy ; Layard, Discoveries in the Ruins of Nineveh and Babylon, 1853, ch. viii, 191, 670 f. ; ib., Nineveh and Babylon, 1867, 62 ; Handcock, 255 ; Perrot and Chipiez, Chaldæa, ii, 312 ; Rawlinson, Mon., i, 371, 585 ; Montelius, Ält. Bronzezeit, 1900, 138.
    8. From Nimrûd ; ref. 7.
    9. From gates of Šalmaneser, 860–824, in British Museum, analysed by Sebelien, Anct. Egypt, 1924, 13 ; free from antimony, cobalt, zinc, silver and bismuth.

ASSYRIAN BRONZE OBJECTS

There is a large (damaged) bronze statuette with the name of Ašurdân (933–912) in the Louvre.[3] The bronze objects, covered with a green crystalline crust, found in Ašurnasirpal II's palace at Nimrûd (885–860) include [4] small bells with iron clappers, tapering rods bent into hooks and ending in a kind of lip, small rosettes (probably ornaments of harness and chariots), cups, bowls, plates and dishes, crown-shaped ornaments, long ornamented bands of copper, a grotesque mace-head, an elegant wine-strainer, various "metal vessels of peculiar form" and a bronze ornament. Many of the objects are embossed or incised by a punch, probably by laying the vessel on a bed of mixed clay and bitumen and punching from the outside, and the graver was sparingly used. Bronze shields have iron handles : from the style of ornamentation it has been suggested that they are imported Phœnician [5], but they probably came from

[1] Sayce, op. cit., 169, 172 ; Rogers, ii, 64, 76 ; Luckenbill, i, 211 f. ; BMGB, 46.
[2] Luckenbill, ii, 9 f., 95 f., 109.
[3] Contenau, Manuel, iii, 1173.

[4] Layard, 1867, 58 f., 62 f.
[5] Layard, op. cit., 62 ; Meissner, i, 97, 269 ; Rawlinson, Mon., i, 370 ; Contenau, Manuel, iii, 1339, 1352.

Lake Van, in Armenia [1], which had a very advanced bronze industry.[2] Bronze objects from Van are a shield and plate in the British Museum [3], a throne and footstool in Assyrian style and a statue of bronze and ivory.[4] The royal throne at Nimrûd was of wood cased in bronze, elaborately engraved and embossed with winged deities, griffins, etc. ; the legs were, apparently, partly of ivory. The footstool was also of wood cased in bronze. Near the throne was a circular band of bronze, 2 ft. 4 in. in diameter and studded with nails, perhaps from a wheel.[5]

The Babylonian bronze work must have been remarkable. Town and palace gates of great weight were said to have been cast, and "the metallurgists must have been able to run into a single mould vast masses of metal." The Assyrian town gates are always shown as solid, but there is a representation of a gate composed of a number of perpendicular bars united above and below by horizontal bars.[6] These doors were, however, probably not cast from bronze or copper, but were of wood, probably cedar, decorated with strips of beaten and ornamented copper or bronze sheet, as in the case of Šalmaneser's gates and the decoration of Ašurnasirpal's throne.[7] Door-sockets were of bronze : two massive ones, more than 6 lb. each, were found at Nimrûd.[8] The "bronze gates of Nebuchadnezzar" in the British Museum [9] are probably the bronze decorations of Šalmaneser's gates, since no gates of Nebuchadnezzar are mentioned in the British Museum Guide. The gates of Šalmaneser were found at Balâwât near Môsul, the ancient Imgur-Ellil.[10]

Copper and bronze were used in Babylonia and Assyria at all periods for a great variety of objects : axes, hammers, knives, cups, cauldrons, dishes, jugs, funnels, dish handles, jars, ladles, chains, nails, door and coffer keys and mountings, statuettes, ornaments, ear, toe, finger and ankle rings, covering for pillars, armlets, bracelets, fish-hooks, weapons such as arrow and lance heads, swords, daggers, helmets, shields and temple objects such as horns and the large laver or "sea" (apsu) set up before the sanctuary. All these are found in actual specimens or in representations.[11]

The bronze bowls from Nimrûd (885–860) are excellent examples of advanced Assyrian metal work. Some designs are wholly Egyptian in character, others show no Egyptian influence but are based on native styles similar to that on Entemena's silver vase. Those of Egyptian style have hieroglyphics which are meaningless, and the style is Phœnician, even if the bowls were not made in Phœnicia : they were perhaps brought as tribute, though many bowls (e.g. one with a silver boss) are probably native work.[12] There are in the British Museum an Assyrian mirror[13], a doorstep of the Chaldæan Nebuchadnezzar II (604–561) from Borsippa[14] and other miscellaneous bronze objects.[15]

Wooden pillars were overlaid with bronze, and the king's tomb or sarcophagus was "sealed with bronze"[16]; bronze plates were used for decorating doors[17], and door locks or bolts were generally of copper or bronze, but could

[1] Dussaud, La Lydie et ses Voisins, Babyloniaca, 1930, xi, 140 f.
[2] Sayce, CAH, iii, 185.
[3] BMGB, 169, 213.
[4] Meissner, i, 268, plate-fig. 118 ; Contenau, Manuel, iii, 1332.
[5] Layard, 1867, 65.
[6] Rawlinson, Mon., ii, 567.
[7] Hilprecht, Exploration, 1903, 208 ; Meissner, i, 267 f.
[8] Rawlinson, Mon., i, 582.
[9] Perrot and Chipiez, Chaldæa, i, 241 f.
[10] Hilprecht, Explor., 208, who, with Sebelien, op. cit., calls the king Šalmaneser II ; Meissner, i, 267, Luckenbill, i, 200 f., 224, and the BMGB, 12, say III.

[11] Meissner, i, 94 f., 247, 282, 312 ; Maspero, Dawn, 756, with details of discoveries ; Rawlinson, Mon., i, 99, 106, 584 f.
[12] Perrot and Chipiez, Chaldæa, ii, 327, 333 ; Layard, 1867, 62 ; Meissner, i, 354 ; Handcock, 255 f., plate 29 ; 256, Fig. 42 ; 258 ; Rawlinson, Mon., i, 370 ; BMGB, 169.
[13] Rawlinson, Mon., i, 575.
[14] BMGB, 76, 77, plate 29 ; Handcock, 262.
[15] BMGB, 80, 170 f., 212.
[16] Meissner, i, 283 ; 79.
[17] Handcock, 261.

be of gold.[1] Lion weights in "copper or bronze" of the period of Tiglathpileser III in the British Museum have the numbers of minas marked in strokes : the large ones have handles cast to the bodies, the smaller have attached rings ; some weights were conical.[2] Bronze objects from the palace of Sargon II include a lion with ring (probably attached to a wall by a chain and not a weight), a bracelet, a statuette, and a plaque with mythological scenes. A cast bronze bell in the Berlin Museum has also mythological figures. A similar bronze lion, of the Persian Period, was found at Susa.[3]

Assyrian bronze casting of the periods of Šalmaneser IV (781–772), Tiglath-pileser III (745–727) and Sennacherib (705–681), found at Quyûnjiq, Nimrûd, etc., includes weights in the shape of animals, apparently Phœnician work, or made by Phœnician immigrants, since they bear Phœnician stamps.[4] Sennacherib says he made, on the command of the god, forms of clay (moulds) and cast works of bronze such as great pillars weighing 6,000 talents, colossal lions of 11,400 talents, bulls, etc., "as in making half-shekel pieces," and he emphasises that no king before him had attempted such work, which he pondered deeply and invented himself.[5]

Bronze cubes inlaid with gold scarabs of Egyptian design found at Nimrûd are probably Egyptian [6], although some Assyrian swords are inlaid with gold.[7] A bronze head of a dragon in the Louvre, probably of the 9th–8th centuries B.C., is like that shown on boundary stones.[8]

Even after the introduction of iron, bronze was still largely used for military equipment.[9] The pioneers in the army of Sargon II in 714 used large picks of this alloy, not iron.[10] Bronze cast over an iron core is in the British Museum.[11] In the Assyrian and Neo-Babylonian Periods parts of houses and furniture, as well as statues, were of bronze.[12] Philostratos[13] mentions bronze (or copper) roofs at Babylon. Rawlinson[14] calls the Assyrian maces used in battle "iron or bronze" ; bronze arrow-heads (with iron and glass) found at Babel[15] are of doubtful date, since the objects are of all periods.

BRASS

There seem to be no definite discoveries of Babylonian or Assyrian brass. Layard[16] says much of the metal called gold in antiquity was really copper alloyed with other metals, aurichalcum or orichalcum, "used in the bowls and plates discovered at Nimrûd," but these are said to be of bronze. The Vulgate Bible also refers to the casting of brass (aurichalcum) in clay moulds[17]. Loftus[18] mentions "bangles or rings of brass" on skeletons at Warkâ (Erech) and "brass rings" of the Parthian Period. A mixture of gold with silver or copper may have been the source of several new alloys similar to aurichalcum : the bright red sarîru, the brilliant white zachalû and ešmarû, "diluted" (sagiru) and "white" gold, not pure. The boat of Marduk in the temple was inlaid with "orichalcum" and gems "like the stars in heaven,"[19] and Nebuchadnezzar II (600 B.C.) says he offered to a goddess dogs of gold (churâsu), silver (kaspu) and "shining copper" (eru namru), the last perhaps bronze, brass or polished

[1] Meissner, i, 247.
[2] BMGB, 170 f. ; Meissner, i, 360.
[3] Contenau, Manuel, iii, 1304 f., 1311, 1445.
[4] BMGB, 170 ; Handcock, 254 ; Perrot and Chipiez, Chaldæa, ii, 413.
[5] Luckenbill, ii, 162, 169, 176.
[6] Rawlinson, Mon., i, 372 ; Layard, 1867, 64 ; BMGB, 237.
[7] Perrot and Chipiez, Chaldæa, ii, 345.
[8] Contenau, Manuel, iii, 1169.
[9] Sayce, B. and A. Life, 180.
[10] Delaporte, 301.
[11] Semper, Der Stil, i, 235.
[12] Meissner, i, 247, 282, 283, 293, 306—Nebuchadnezzar.
[13] Apollonios of Tyana, i, 25.
[14] Mon., i, 429.
[15] Layard, 1867, 285.
[16] 1867, 388.
[17] III Regum, vii, 46, 49 : fudit ea rex in argillosa terra ; Gsell, 78.
[18] Travels, 1857, 202.
[19] Meissner, i, 269 ; ii, 73 ; these are sometimes translated electrum—see p. 234.

copper.[1]   The traces of zinc found in many old bronzes are probably derived from the copper ores.[2]

LEAD

Lead was known in the early Sumerian Period, when it is mentioned as abar or anna (Akkadian and Assyrian abaru or anaku), the "metal of Ea," the god of Eridu.[3] Erimuš (2581–2573) takes pride in first making a lead statue ; Gudea (2450) had "jewels, precious metals and lead" in his sealed treasure-house, and presented a "beautiful basin of lead and stone" to a temple.[4] In later texts lead comes as a manufactured product, perhaps from the refining of silver, from the otherwise unknown mountains of [Ch]archa or Chachua and Mašgungunnu.[5]   An old Assyrian code of laws shows that about 2000 B.C. lead was used as currency, gold, silver (in bars, rings and discs) and bronze being also used in the time of Sargon I, but to a less extent than lead. In 1400–1050 lead in the form of animal heads was the commonest means of exchange in Assyria.[6]   The lead was generally used for this purpose in lumps ; some very large ones (500 kgm.), with some kind of stamp, were found in a temple of Ištar and Denitu at Assur (1250 B.C.) with lead ornaments and curious roundels, perhaps used as tokens for payment of the temple women.[7] The Cappadocian tablets (3 millenn.) speak of various qualities of lead (anakum, perhaps sometimes tin) under seal : "pure," "good," etc., and "loose lead" (anak qâtim) in limited use as a means of exchange.[8]

Lead occurs sparingly in the early remains ; pieces of pipe and a vase found at Ur and a piece at Tell Sifr (Larsa) [9] are of uncertain date. Early Sumerian metal from Kiš contained 98·29 of lead, 1·30 of tin and 0·41 of iron, and had probably been smelted from a simple ore.[10] Lead tumblers were found in very early graves at Ur, deep below the Royal Cemetery, but in the latter there was only one piece of corroded sheet, containing practically no silver.[11] The copper antlers of the copper stags' heads found at al-'Ubaid (3000 B.C.) were soldered in place with lead, which has expanded by corrosion and burst the heads.[12] This lead contained 0·0131 per cent. of silver[13], that of a tumbler from al-'Ubaid, 0·07 per cent.[14] Three lead pot-menders for a stone bowl, before 2450 B.C., were found at Khafaje.[15] There is a bronze pedestal with a lead core from Lake Van.[16] Layard[17] found at Nimrûd one piece of lead which had been melted, with an iron axe-head embedded in it ; an inscription on lead was found at Quyûnjiq (Nineveh)[18]; fragments of thin sheets on which amuletic texts were written, also thin leaves with Greek inscriptions, at Babylon.[19] An Assyrian bronze gate socket set in lead is in the British Museum[20], a piece of

[1] Weissbach, MGM, v, 502 f.
[2] Von Bibra, Bronzen and Kupferlegirungen, 17, 36 f.
[3] Schrader, RL, 97 ; Schrader, Keilinschriften, 648 f. ; Meissner, i, 266, 348 ; Unger, RV, i, 428 ; Handcock, 253 ; Berthelot, Arch., 79 ; Muss-Arnolt, Dict. Assyr., 1905, 70, 'KAT ; R. C. Thompson, The Chemistry of the Ancient Assyrians, 1925, 69.
[4] Meyer, Sumerer, 47 ; Unger, 428.
[5] Meissner, i, 54, 176, 265, 346, 347, 355.
[6] Delaporte, 332, 393 f. ; Orth, PW, Suppl. iv, 112 ; Meissner, i, 355 ; Landsberger, Alte Orient, 1924, xxiv, Heft 4, 15 f., 21 f.
[7] Smith, EHA, 324, plate 14 ; Feldhaus-Klinckowstroem, Geschichtsblätter, 1914, i, 66 ; Andrae, Mitteil. Deutsch. Orient-Ges., 1914, liv, 36 ; CAH, iii, 76.
[8] Landsberger, Alte Orient, 1924, xxiv, Heft 4, 22.
[9] Perrot-Chipiez, Chaldæa, ii, 309 ; BMGB, 213 ; Rawlinson, Mon., i, 98 ; Loftus, Travels, 1857, 269.
[10] Desch, B.A. Rep., 1928, 440.
[11] Woolley, Digging up the Past, 91 f. ; ib., Ur Excav., 1934, ii, 148, 293, 295, pl. 10.
[12] Hall, JEA, 1922, viii, 248 ; Hall and Woolley, Ur Excav., 1927, i, 23, 28, pls. 5, 6.
[13] Friend and Thorneycroft, J. Inst. Metals, 1929, xli, 105.
[14] Ur Excav., 1934, ii, 295.
[15] Frankfort, Jacobsen and Preussen, Tell Asmar and Khafaje, 1930–31, Univ. Chicago Orient. Inst. Publ., No. 13, 100.
[16] Belck, Z. f. Ethnol., 1908, xl, 49.
[17] Discoveries, 1853, 357.
[18] Smith, TSBA, 1874, iii, 463.
[19] Budge, Amulets, 34.
[20] Handcock, 131, 267.

almost pure lead wire was found at Tell el-Hesy in Palestine [1], and there is some lead of the Persian Period in the British Museum.[2] Lead, as we know from texts, was used in the form of sheet as a damp-course in the walls of Babylon [3], and also in fixing blocks of stone: Herodotos[4] and Diodoros Siculus [5] say the stones in the piers of the bridge at Babylon were clamped by iron set in lead. The metal was also used for adjusting bronze weights, sometimes fraudulently.[6] In 679 B.C. a talent of lead was dedicated as a gift to the goddess when a slave was sold, and payment of a talent of lead replaced the punishment of mutilation in the Assyrian Period.[7]

Tukulti-Urta II (889–859) obtained large quantities of lead in "bricks" from the countries on the Khabur, with gold, silver, copper and iron.[8] Lead as tribute is frequently mentioned in the texts of Ašurnasirpal II (885–860) [9], who obtained 100 talents (with as much iron, etc.) from Lubarna, ruler of the state of Khattin in his city of Kunalua[10], and also from the Phœnicians. Šalmaneser III, in his campaign of 854, took lead from the same kings and from Jehu, Suâ and Marduk-apal-usur of the land of Sukhu.[11] Sargon II (722–705) obtained lead and "white bronze" (tin ?) as tribute from Armenia, Sennacherib cast lead capitals for bronze pillars, and Asarhaddon obtained lead as tribute from the Sutû at Memphis.[12]

Lead ores occur in great abundance as an outcrop in the Tiyari Mountains near Quyûnjiq (Nineveh) and are also obtainable near Diarbekir.[13] It has been suggested[14] that the metal was imported into Ur from regions bordering on the Black Sea. There are remains of lead mines in Mount Nich in Khorassan and at Segend near Lake Urmia.[15] Assyrian lead of the 7th (?) century B.C. contains 0·11 per cent. of silver.[16]

TIN

Tin was used for the preparation of the earliest and the later bronzes, but was rare and expensive.[17] The bilingual texts put tin between silver and bronze and before iron ; in Hebrew lists tin (or lead ?) comes at the end.[18] The metal tin as such seems to have been known fairly early[19], but the early Sumerian bronzes were probably derived from the mixed smelting of oxide ores of copper and tin. Although the unhomogeneous nature and the impurities in some early Babylonian bronzes have been held to suggest the smelting of a single mixed ore[20], these ores would not have given such pure alloys as the early Sumerian bronzes.[21]

It is commonly assumed that the Babylonian and Assyrian anaku, Akkadian anag, naga, or anna, at first meant lead (also called annamu), and later tin[22], the two being perhaps regarded as two varieties of one metal, as plumbum (lead) and plumbum candidum (tin) were in the Roman Period. In Syriac

[1] Gladstone, B.A. Rep., 1893, 715.
[2] BMGB, 236.
[3] Rawlinson, Mon., ii, 517.
[4] i, 186.
[5] ii, 8.
[6] Handcock, 255, 268 ; Meissner, i, 156, 360.
[7] Delaporte, 329, 337.
[8] Luckenbill, i, 130 f.
[9] Sayce, Rec. of the Past, N.S., 1889, ii, 140, 144, 149, 155, 160, 162 f., 170, etc. ; Luckenbill, i, 144 f.
[10] Rogers, ii, 65 ; Meissner, i, 138 ; Meyer, Alt. II, ii, 371.
[11] Rogers, 76 ; BMGB, 46 ; Luckenbill, i, 211 f., 218 ; in Lippmann, Alchemie, i, 578, this is "tin," from Hommel—no doubt the translations vary.

[12] Luckenbill, ii, 9 f., 95 f., 109 ; 183 ; 229.
[13] Rawlinson, Mon., i, 219 f. ; cf. i, 98.
[14] Maspero, Dawn, 614.
[15] Unger, RV, i, 428.
[16] Friend and Thorneycroft, J. Inst. Metals, 1929, xli, 105.
[17] Sayce, B. and A. Life, 132.
[18] Hoernes, Urg., ii, 217.
[19] Rawlinson, Mon., i, 98 ; Baumstark, PW, ii, 2714 f.—3000 B.C. ; Elam, J. Inst. Metals, 1932, xlviii, 97.
[20] Liechti, Chem. News, 1923, cxxvi, 413.
[21] Desch, B.A. Rep. 1930, 268.
[22] Schrader, Keilinschriften, 648 f. ; Fossey, Magie, 184, 260, 426 ; Schrader, RL, 995 ; Unger, RV, i, 428 ; Muss-Arnolt, Dict. Assyr., 1905, 70 ; Meissner, i, 348.

and Arabic, anak or ânuk means tin, and this word (ănâk) is used once in the Hebrew Bible.[1] The "white bronze" which Sargon II (722–709) took as tribute from Armenia [2] may have been tin. According to Delitzsch [3] the Babylonians had a special name, anâk siparri, for tin. A late Babylonian name for tin, kastira (Arabic qasdir, Central African kesdir), has been derived from a supposed Assyrian (or Georgian) kâsazatira and an Akkadian id-kasduru [4], but these words probably never existed and the name kastira is derived from the Greek kassiteros (used by Homer).[5] Lenormant's suggestion that the Arabic qal'iyun, Georgian gala, Armenian klajek, Ossetic kala, Turkish kalai and modern Greek καλάϊ, all names for tin, point to "a vestige of the diffusion of tin in Caucasian Iberia among the Semites" is also to be rejected, since these names are derived from that of Malacca, viz. Qalah, and the Malay Peninsula is a very recent source of tin.[6] Equally improbable is the supposed connection of the Greek kassiteros with the Kaššû, or Cassites, of the Zagros, who first appear about 2100 B.C. and are kept back by a wall, kar-Kaššû [7] ; or with the name of the Caspian, or with the town Kaspatryos [8] (perhaps on the borders of India and Persia or Bactria) [9] ; or Haverfield's[10] suggestion that the name goes back to Kassitira, "the land of the Kassi" (Elamites), and denotes the metal which came from there or was sold by that people.

Tin itself occurs in Egypt, but I have not found any notice of metallic tin being found in Mesopotamia, although the presence of tin oxide in later Babylonian glazed tiles[11] shows that it may have been known at least in the later period.

The sources of tin are, as in the Egyptian case, matters of conjecture.[12] Sayce[13] suggests India (which imported tin), the Malay Peninsula (which is a late source) and Cornwall (which is just possible).[14] Layard[15] and Rawlinson[16] suggest Cornish tin traded by Phœnicians by way of Palmyra and Thapsacus in exchange for textiles, but this could apply only to later periods, since the Phœnician traffic began only about 1250 B.C.[17] The metal perhaps came from the north or north-east[18], possibly from the now exhausted mines of Caucasian Iberia (Georgia), whichιare thought to have supplied tin for the bronzes of the Chalybes, or of Iran or Susiana[19], although tin is no longer found in these districts.[20] The favourite hypothesis is that it came from the supposed mines of Khorassan and Transoxania, Paropamisus, Drangiana, etc., south-east of the Caspian.[21] Drangiana, the modern Seistan[22], was said by Strabo[23] to have had tin mines, and deposits of tin were said to occur in north Khorassan at Meshed and Kutschau.[24] Rawlinson[25] suggested the Kurdish mountains, and Orth[26] the Medic-Elamite mountains, as the source. Whatever the source of the tin used for the earliest Sumerian bronzes, it seems to have been exhausted in a few centuries, or the supply of metal was otherwise cut off, since in a

[1] Amos, iv, 78.
[2] Luckenbill, ii, 9, 95 f., 109.
[3] Q. in Meissner, i, 348 ; ii, 492.
[4] Lenormant, Prem. Civ., i, 147 ; Bapst, L'Étain, 1884, 2.
[5] Besnier, Daremberg-Saglio, Dict., IV, ii, 1458 ; Schrader, RL, 993, correcting Schrader-Jevons, 216 ; Meyer, Chronol., 23.
[6] Schrader, RL, 995.
[7] Jastrow, Civ., 157 ; Lehmann-Haupt, PW, xi, 407 ; Weissbach, ib., 521, 1501.
[8] Herodotos, iii, 102.
[9] Smith, DG, i, 558 ; Hüsing, Der Zagros und seine Völker, Der alte Orient, ix, Heft 3–4, 24.
[10] PW, x, 2330.
[11] Semper, Der Stil, i, 356 ; ii, 156.

[12] Meissner, i, 348 ; Hall, Oldest Civ. Greece, 193 f.
[13] B. and A. Life, 132.
[14] Hoernes, Urg., ii, 217.
[15] Discoveries, 1853, 191.
[16] Mon., i, 556 ; iii, 16.
[17] Köster, Beihefte zum alten Orient, 1924, i, 18, 19, 21, 37.
[18] Baumstark, PW, ii, 2714, 2715.
[19] Lenormant, Prem. Civ., i, 129 f.
[20] Bapst, 7.
[21] Bapst, 8 ; Gsell, 36 ; Lenormant, Prem. Civ., i, 128 f. ; Hoernes, Urg., ii, 217.
[22] Vaux, DG, i, 787 ; Meissner, i, 348.
[23] XV, ii, 10, 725 C.
[24] Meissner, i, 348 ; see p. 77.
[25] Mon., i, 219 f.
[26] PW, Suppl. iv, 112.

somewhat later period bronzes poor in tin, or containing lead or antimony, were in use. The later supplies of tin probably came from the same source as furnished tin to Egypt.

ANTIMONY

Antimony occurs as the native sulphide, stibnite, supposed to be mentioned as guchlu, "eye-paint," [1] which came from "the mountains of antimony (?)." This locality was probably the modern Afschâr near Takht-i-Suleimân, where it is found in layers 4–5 m. thick in chalk.[2] It also came from the town of Kinaki in Media.[3] The Kurdish mountains still supply stibnite.[4] Antimony occurs in some pre-Sargonic alloys with copper, as in some old European bronzes.[5] The metal is fairly easily reduced from stibnite. A vase of Gudea's time (c. 2450 B.C.) found at Tello consists of antimony free from copper, lead, bismuth and zinc, but with a trace of iron. Antimony otherwise occurs only in graves of about 1000 B.C. at Kuban near Tiflis in Transcaucasia, as a material for vessels and ornaments.[6] The metal was used in the Assyrian Period.[7]

Stone or alabaster pots containing a black eye-paint, also "eye brushes" for its application, occurred in early remains (3000 B.C.) at Ur, Adab (Bismaya ; with henna ?) and other sites [8], but the assumption that the pigment was stibium (antimony sulphide) is always unjustified without analysis.[9] Burnt almond shells and soot are still used in the East[10], and the actual black cosmetic found at Ur contained manganese dioxide and turquoise, with some lead (perhaps from galena), and not antimony sulphide.[11] According to Muss-Arnolt[12] the meaning of guchlu is "precious tribute," rubies or a *red* (not black) colour for the eyes, the Hebrew koḥl[13], and the translation "stibium" is very arbitrary and misleading. Most ancient eye-paints have been assumed to be antimony sulphide on very little evidence. Sennacherib obtained "antimony" from Jerusalem, and Asarhaddon got it (also "magnesium," first discovered in A.D. 1808 !) from Memphis.[14]

Since antimony oxide occurs in Assyrian glazes as lead antimoniate[15], the method of roasting stibnite to form the oxide was probably known, but the method by which the Assyrians made lead antimoniate is not known.

IRON

Hæmatite, an oxide of iron (Fe_2O_3) which occurs in a compact crystalline state, was used in making cylinder seals. An "oval object," and smaller olive-like objects [weights] of the Sumerian Period found at Tello, consisted of hard, crystalline, black and brilliant oxide of iron[16], perhaps hæmatite or magnetite (Fe_3O_4). Either oxide is easily reduced by heating in a charcoal fire to spongy iron. There are polished hæmatite weights from Uruk (Warkâ) in the British Museum, and Balaiew[17] thinks early Sumerian iron was reduced from hæmatite, whilst Rickard[18] says it is meteoric, since it contains nickel.

In the earliest period when it was known in Babylonia and Assyria, perhaps

[1] Meissner, i, 244, 266 ; Schrader, Keil., 648 f.

[2] Meissner, i, 347 f.

[3] Unger, RV, i, 428.

[4] Rawlinson, Mon., i, 219 f.

[5] Meissner, i, 265 ; Nies, PW, i, 2438 ; Forrer, Urgeschichte des Europäers, Stuttgart, 1908, 410 ; Hoernes, Urg., ii, 235.

[6] Meissner, i, 347 ; Berthelot, Intr., 223 ; Forrer, RL, 32 ; *ib.*, Urgeschichte, 410.

[7] Perrot-Chipiez, Chaldæa, ii, 310 ; King, S. and A., 73 ; V. Place, Ninive et l'Assyrie, ii, 263—no details.

[8] Handcock, 327 ; Meyer, Alt. (1909) I, ii, 416 ; M. H. Farbridge, Studies in Biblical and Semitic Symbolism, 1923, 215 ; Frankfort, Studies in Early Pottery, 1924, i, 65.

[9] See p. 141.

[10] Smith, CDB, 664 ; Thompson, The Assyrian Herbal, 1924, 256.

[11] Ur Excav., 1934, ii, 248.

[12] Concise Dict. of Assyrian Language, Berlin, 1905, 215.

[13] See Index, *puch.*

[14] Luckenbill, ii, 121, 136, 229.

[15] Semper, Der Stil, i, 356 ; ii, 123.

[16] Berthelot, Arch., 78.

[17] *J. Inst. Metals*, 1930, xliii, 352.

[18] *Ib.*, 350.

later than its appearance in Egypt [1], iron was rare and expensive, and was used for rings, bracelets, bangles and personal ornaments.[2] Iron rings and ornaments occur even in rather late Babylonian graves [3], although in the still later period the metal was used both in Babylonia and Assyria on a much more lavish scale than in Egypt.[4] An iron rod decorated with polished ornaments, found at Babylon, perhaps formed part of a throne.[5] In Ḥammurabi's time (1924–1913) the metal was still costly, iron and silver being valued in the ratio 1 to 8.[6] In the Neo-Babylonian Period, when it was imported from the Hittites and Chalybes, iron was as cheap as, if not cheaper than, copper, and raw iron and silver were valued in the ratio of 1 to 225.[7]

The first mention and use of iron in Mesopotamia are doubtful. Gudea (c. 2450 B.C.) says : "from the mountains of Meluchcha I get gold and girzanum, and from Kimaš copper," and girzanum has been translated iron [8], but the translation may be faulty, stone being meant [9], although the Assyrians thought that the Sumerians were acquainted with iron.[10] Iron, including knives, was found in all layers at Nippur.[11] An iron instrument ("tang or blade"), oxidised and crumbled, found in an early grave in the lowest deposits (3500–3000) at Ur[12], contains 89·1 per cent. of iron and 10·9 per cent. of nickel, and is therefore probably of meteoric origin.[13] Too much stress has recently been laid on the supposed meteoric origin of *all* early iron[14], and a dagger blade found at Khafaje is of terrestrial iron.[15] Iron knives with bone handles found at Tello may be Sumerian, but their date is uncertain.[16] Several *small* finds of Assyrian iron of 3000–2000 are known.[17] Hall suggested that the Sumerian iron was meteoric in the first instance, since the Sumerian an-bar has practically the same meaning as the Egyptian ba-en-pet, "heavenly metal," but according to Johns[18] the identification of an-bar with ba-en-pet and iron is doubtful, since the ideogram an-bar, which he takes to denote "divine weight," is of uncertain meaning. The reading an-bar = iron, however, is adopted in the translation of Elamite-Anzanite texts from Susa by de Morgan[19], and Sayce[20] believes that iron is represented in pre-cuneiform picture-writing[21] by an-bar, two characters meaning "heaven" and "metal," also that the use of bar, "shining," with the determinative of divinity, indicates that the first iron worked in Babylonia was meteoric. The meteoric origin of early Sumerian iron, in fact, seems probable from the analyses.

The pronunciation of an-bar is unknown : another equivalent possibly reads bar-gal, taken over as the Semitic-Babylonian parzillu, which name

[1] Handcock, 269.
[2] Rawlinson, Mon., i, 95, 99 ; Maspero, H. anc., 1905, 154 ; Koldewey, Excavations, 265.
[3] Hoernes, Urg., ii, 243 ; Reinach, L'Anthropologie, 1892, iii, 453.
[4] Perrot-Chipiez, Chaldæa, ii, 310 f. ; Beck, Geschichte des Eisens, 1884, i, 128 f.
[5] Handcock, 269.
[6] Meissner, i, 265, 363 ; Schwenzner, MGM, xv, 51 ; Meyer, Histoire, iii, 262.
[7] Meissner, i, 364.
[8] Hilprecht, *Verhl. Berlin Ges. für Anthropol.*, 1901, 164 ; retracted in *ib.*, Excav. in Assyr. and Bab., 1904, 238 ; Scheil, in de Morgan, Délégation en Perse, 1905, viii, 338 ; Sebelien, *Ancient Egypt*, 1924, 14.
[9] Hilprecht, Exploration, 1903, 238 ; Handcock, 268 ; Meissner, i, 265.
[10] Hall, Oldest Civ. Greece, 1901, 193 f. ; Boson, q. by Contenau, Manuel, iii, 1247, says iron is mentioned by Gudea.

[11] Peters, Nippur, 1897, ii, 241, 381.
[12] Woolley, *Nature*, 1927, cxx, 124 ; *ib.* JEA, 1928, xiv, 191 ; *ib.* Ur Excav., 1934, ii, 49.
[13] Desch, *B.A. Rep.*, 1928, 440; Wainwright, JEA, 1932, xviii, 3.
[14] For example, the authenticity of the object found in the Great Pyramid has been doubted on this ground; see p. 92.
[15] V. G. Childe, New Light on the Most Ancient East, 1934, 189.
[16] Heuzey-de Sarzec, i, 35 ; Hall, Oldest Civ. Greece, 1901, 193 f. ; Hilprecht, Excav., 1904, 238.
[17] Orth, PW, Suppl. iv, 112.
[18] Quoted by Cook, Zeus, i, 632.
[19] Délégation en Perse, 1907, ix, 63, 102, 133, etc.
[20] B. and A. Life, 135 ; Rel., 153, 186 ; cf. Muss-Arnolt, Dict. Assyr., 1905, 828.
[21] BMGB, 19.

may have come to the Semites with the metal itself from the Chalybes, Tubal and other Armenian tribes.[1]   Dussaud [2], from Zimmern, also thinks parzillu is not Semitic but a word from Asia Minor, as indicated by the termination -ill. Schrader [3] thought iron, known to the Sumerians not later than 3000 B.C., was called barza. But parzillu, barzel or barza are the oldest Semitic names for iron, and barzil, Aramaic parzilia, which is used in the Bible [4], is almost certainly the Babylonian-Assyrian parzillu.

In the Gilgameš epic "a meteor (?) falling from heaven" and "death from iron" occur.[5]   The date of this epic is not accurately known, but Gilgameš is shown on seal cylinders at Ur of the time of Sargon I, and the story may be from the old Subarian people.[6]

USEFUL IRON

Iron was certainly well known in Babylonia and Assyria about 2000 B.C., but first became plentiful about 1400, when steel was used.[7]   Scheil [8] thinks iron was used as early as 2500 : it is mentioned as parzillu in a text of 1750–1170 (sic). Babylonia had no useful iron in the early period, and although Assyria used iron occasionally from the 13th century B.C., it was then without an iron industry, the metal being obtained from the highland district of Commagene, between North Syria and Asia Minor. In North Syria an iron-using culture intruded from the north-west in the 12th century, and about the same time, after the collapse of the Minoan sea-power, iron weapons suddenly became rather common in Cyprus. Egypt also received Syrian iron in the XIX dyn., but made no general use of the metal until Greek times.[9]   Iron is named in the texts about 1700, and about 1500 we hear of Canaanite iron chariots and of ironsmiths in Palestine. This period probably represents the transition from bronze to iron in Babylonia.[10] Rings and daggers of iron, or probably steel (chabalkinu ; cf. χάλυψ), were sent by Tušratti, king of the Mitanni, to Amenhotep III (1411–1375) of Egypt, and the Hittites of Kizwatna, on the Black Sea, supplied Rameses II (1296–1225) with iron and steel.[11]

Iron ore occurs in Armenia near Urmia and north of Tabriz[12]; also (with surface outcrops) in the Tiyari Mountains to the north-east of Quyûnjiq (the site of Nineveh) ; near Khorsabad ; a few days' journey from Môsul ; and in the neighbouring part of Kurdistan. It is worked at present and may have been worked on a small scale by the ancient Assyrians, but in 881 B.C. the metal was brought as tribute from the iron district of the Euxine (the country of the Chalybes, Tibareni and Moschi of classical writers), and in Ašurnasirpal's time from the Hittites of Carchemish.[13]

Iron ores occur extensively in two important districts in Western Asia, where remains of early iron manufacture are found. (1) The region south-east of the Euxine, the ancient Paphlagonia and the Pontos, extending from the modern Yešil Irmak to Batum and comprising a series of mountain ranges not far from the coast, along the lower slopes and foothills of which the iron deposits are scattered.[14] (2) In the Taurus and Anti-Taurus region in the

[1] Hall, Oldest Civ. Greece, 1901, 193 f.
[2] La Lydie et ses Voisins, 1930, 82.
[3] RL, 173 ; Schrader-Jevons, 202.
[4] Smith, Concise Dict. Bible, 347.
[5] Jeremias, 64, 197 ; ib., Alte Orient, 1925, xxv, Heft 1, 22 f. ; Meissner, ii, 147, 192.
[6] Meissner, ii, 191 ; Otto, 1925, 41 ; Woolley, Ur Excav., 1934, ii, 358 f.
[7] Unger, RV, i, 428.
[8] In de Morgan, Délégation en Perse, 1905, viii, 338.

[9] Myres, CAH, i, 109 ;   Contenau, Manuel, iii, 1247.
[10] Sayce, B. and A. Life, 131, 135 f.
[11] Meissner, i, 264, 348 ; Unger, RV, i, 428.
[12] Unger, RV, i, 428.
[13] Rawlinson, Mon., i, 219 f. ; Gowland, J. Anthropol. Inst., 1912, xlii, 281 ; Perrot-Chipiez, Chaldæa, ii, 312 ; BMGB, 32.
[14] Strabo, XII, iii, 9, 544 C. ; XII, iii, 19, 548 C., says "on the mainland there are at present mines of iron."

south-east of Asia Minor, extending on the west from Cape Anamur to the borders of Syria, and in Syria to Aleppo, the Euphrates and Lebanon. The former was probably the region where iron was first regularly produced.[1] Nabonidus (550 B.C.) speaks of iron coming (with copper) from Mount Amanus and from Mount Lebanon.[2]

The Amarna tablets (c. 1400 B.C.) mention "rings of iron covered [plated ?] with gold";[3] ; the tablets of a banking firm of 1395–1242 mention gold, silver, copper, bronze and lead, but not iron.[4]

There appears to have been an early iron industry in Armenia, centred about Lake Van, the people being acquainted with iron "long before they migrated" from "an unknown place," perhaps Thrace. The use of iron may, however, have been introduced from Asia Minor. The extensive replacement of bronze by iron under Sargon II (722–705) may have been on account of his campaigns in Armenia.[5] Iron came into *use* in Mesopotamia about 1300–1200, when the Hittites of Asia Minor were working iron mines. The Assyrians in northern Mesopotamia adopted iron weapons and armour, probably getting the metal in ingots from north-east Asia, and their use of the new metal enabled them to make extensive military conquests.[6]

Šalmaneser I (1276–1257) recorded that he laid an iron tablet in the foundations of the temple at Assur which he restored, and iron tablets were discovered in the foundations of the temple of Ištar at Assur.[7] A text in which a woman is compared with a "sharp iron dagger" is known in Neo-Assyrian and late Babylonian copies, but the original was older.[8] Tiglathpileser I (1115–1103) speaks of slaying four wild bulls with mighty blows sukuth parzilli, the last two words of which may, or may not, mean "with arrows of iron" (or "tipped with iron"), since the ideogram of parzilli is the same as that of Ninib, the god of war and hunting.[9]

Tukulti-Urta II (889–884) says he hewed the mountains of the Upper Euphrates with iron picks and he obtained iron daggers from Dûr-Katlimmu (on the Khabur).[10]

Ašurnasirpal II (885–860) obtained 250 talents of iron from the Hittite king Sangara at Carchemish, and 100 talents from Lubarna, ruler of the Hittite state of Khattin in his town of Kunalua ; also iron from the state of Bit-Hallupi on the Khabur and from Bit-Zamani between the Tigris and Euphrates above Diarbekir, etc., but none from the Phœnicians. He also speaks of tribute of gold, silver, copper and iron bars from the city of Šadikanni (now Arban on the Khabur) and of hewing his way through forests by bronze picks and iron axes.[11]

A considerable amount of iron was found in Ašurnasirpal's palace at Nimrûd (Calah) : bronze bells with iron clappers in two large copper cauldrons, with other bronze objects, tripod stands consisting of iron rings bound in some places with copper and feet of iron cased in bronze, with the iron rusted away, which are also frequently represented in bas-reliefs ; remains of iron armour, instruments, spear-heads and arrow-heads, falling to pieces soon after exposure to the air ; iron handles on bronze shields, the head of a pick, a double-handled

[1] Gowland, *op. cit.* ; de Morgan, Délég. en Perse, 1905, viii, 338.

[2] Contenau, La Civilisation Phénicienne, 1926, 298 ; Ridgeway, Early Age of Greece, Cambridge, 1901, i, 594 f.

[3] Ridgeway, 594 f.

[4] Thompson, CAH, i, 566.

[5] CAH, iii, 19, 185 ; Rostovtzeff, i, 110, 117.

[6] Mackenzie, Footprints, 149.

[7] Meissner, i, 265 ; Andrae, *Mitt. Deutsch. Orient Ges.*, 1914, liv, 36.

[8] Meissner, ii, 432, 434.

[9] Ridgeway, i, 616 ; Scheil, in de Morgan, Délég. en Perse, 1905, viii, 338 ; Luckenbill, i, 86—"iron spear"; Meissner, i, 73 ; Friend, Iron in Antiquity, 1926, 178.

[10] Luckenbill, i, 130 f.

[11] Sayce, *Rec. of the Past*, N.S., 1889, ii, 139, 151, 157, 160, 169, 170, 172 ; Luckenbill, i, 144 f., 153, 157, 179, etc. ; Rogers, i, 64 f. ; Meissner, i, 138 ; Rawlinson, Mon., ii, 100 ; Ridgeway, i, 594 f., 615 ; Friend Iron in Antiquity, 179.

saw (3 ft. 6 in. long), several objects resembling the heads of sledge-hammers, and a large blunt spear-head. Iron rods found were probably not arrow shafts.[1] The armour scales are 2–3 in. long, rounded at one end and square at the other, with a raised or embossed central line.[2] Hilprecht [3] suggests that this iron is really of the time of Sargon II, who restored the palace at Nimrûd. Iron inlaid with bronze was found at Nimrûd. A bronze casting over an iron core was probably formed by pouring molten bronze over the iron, a process not in modern use.[4] Although the shields were actually of bronze with iron handles, shields and arrow- and spear-heads could probably be of either bronze or iron, and perhaps even silver or gold shields (as in the Bible) may have been used.[5]

Assyrian tribute lists of 881 B.C. mention iron and cows from Moscher (Meshek, or Meshed, on the upper Tigris or Euphrates, or more probably the Pontos), and gold, silver and iron from Niri (Kurdistan) ; 5,000 talents of iron, 3,000 of copper, 2,300 of silver and 20 of gold were paid as tribute to Adadnirari III (810–782) by Mari, king of the Philistines, in Damascus.[6] Tiglathpileser III (745–727) took 3,000 talents of copper and 5,000 of iron from Damascus.[7] Assyrian steel is said to be mentioned as paldâh in the book of Nahum [8], which was written in the 8th century B.C. in Assyria.[9]

With Sargon II we reach the use of iron in Assyria on a most lavish scale. This king wrote to the town of Ulchu (east of Lake Urmia in Armenia) : "I have broken down your walls with iron picks and crowbars and have smashed them like pots." He took tribute of iron from Armenia. There were excellent steel picks in his palace at Khorsabad, but "how they have vanished is unrecorded" (some were used by the excavators).[10] This palace was built by Sargon and abandoned soon after his death in 705 B.C., so that its date is accurately known.[11] There was also a large quantity of iron scale armour, sometimes inlaid with copper lines[12], also iron chains, rings, nails, pikes, tools (?), etc.[13] The buried store of 150 tons of excellent soft iron found was tested in Paris and found not to be steel as Place at first believed, and it is free from nickel and manganese.[14] The iron is in the form of rough bars, 12–19 in. long, $2\frac{3}{4}$–$5\frac{1}{2}$ in. thick, roughly tapered at each end and pierced by a single jagged hole, probably for convenience in transport, each weighing from 8 to 44 lb. Similar pieces and bars survived to Roman times, and in Sweden and Finland to about 1870.[15] Iron bits and bridles for horses and iron fastening rings in stables were in use in Sargon's time. He probably obtained iron from the Hittites, although there is no mention of the metal in a list of tribute from Pisiri, king of Carchemish, which includes 11 talents 30 minas of gold and 2,100 talents 24 minas of silver.[16]

[1] Layard, 1867, 58, 63 ; Perrot-Chipiez, Chaldæa, ii, 312 ; BMGB, 13, 212 ; Handcock, 255 ; Rawlinson, Mon., i, 454, 456.
[2] Friend, 186 f. ; Beck, Geschichte des Eisens, i, 138.
[3] Excav. in Assyr. and Bab., 1904, 239.
[4] Percy, in Layard, Discoveries, 1853, 670 ; Perrot-Chipiez, Chaldæa, ii, 312 ; Rawlinson, Mon., i, 371 ; Contenau, Manuel, iii, 1247.
[5] Rawlinson, Mon., i, 446, 456 ; Meissner, i, 95 f., Fig. 25.
[6] Hoernes, Urzeit, 1912, ii, 114 ; ib., Urg., ii, 222 f. ; Beck, i, 109 f., 133 f. ; Maspero, Empires, 102 ; Luckenbill, i, 263 ; Dussaud, La Lydie et ses Voisins, 1930, 80 f.
[7] Hoernes, Urg., ii, 222 ; Bucher, iii, 5 ; BMGB, 275.
[8] ii, 3.
[9] Rougemont, L'Âge du Bronze, 1866, 195 ; Smith, CDB, 900.

[10] Meissner, i, 266 ; ii, 369 ; Luckenbill, ii, 9 f., 87, 96 f., 110 ; Petrie, Tools and Weapons, 1.
[11] Smith, CAH, iii, 59.
[12] Petrie, Tools and Weapons, 38 ; Rawlinson, Mon., i, 443, 444.
[13] Place, Ninive, i, 84–89 ; iii, 70 ; Meissner, i, 266 ; Beck, Geschichte des Eisens, i, 136, says no tools ; Gowland, J. Anthropol. Inst., 1912, xlii, 281 ; Contenau, Manuel, iii, 1246.
[14] Beck, 136 ; Desch, B.A. Rep., 1929, 265.
[15] Place, Ninive, i, 84 ; Maspero, Dawn, 722 ; Hilprecht, Explor., 1903, 238 ; Beck, i, 134 ; Meissner, i, 266, 294 ; Bucher, iii, 5 ; BMGB, 32 ; Perrot-Chipiez, Chaldæa, ii, 311 ; Blümner, iv, 68 ; Forrer, RL, 402 ; Gowland, J. Anthropol. Inst., 1912, xlii, 281, Fig. 7 ; Friend, Iron in Antiquity, 183, Fig. 14.
[16] Meissner, i, 139, 218 ; Forrer, RL, 197.

Hezekiah sent as tribute to Sennacherib (705–681) at Nineveh iron daggers, bows, arrows and spears, with 30 talents of gold, 800 talents of silver, also "antimony," jewels, etc. [1], and Sennacherib speaks of quarrying hard stones with iron picks. Asarhaddon (681–669) mentions iron as tribute from Van.[2] An Assyrian bronze helmet and iron tools found at Thebes in Egypt probably date from 672–670, and the iron came from the Chalybes.[3] Ašur-banipal (668–626) mentions an iron dagger, and speaks of supplying vessels of iron, gold, silver, copper, wood and stone to the temple at Babylon.[4] Assyrian riveted conical iron helmets, perhaps of his time (the shape varied with the period) [5], inlaid with copper or with copper fittings and found at Quyûnjiq (Nineveh), are in the British Museum.[6] An iron dish was also found there.[7] Nebuchadnezzar II (604–561) constructed an iron sluice at Babylon to supply water to a moat. King Kandalanu (647–626) and Nabonidus (555–538) gave receipts for iron.[8] Iron arrow-heads of uncertain date, with bronze and glass, were found at Babylon [9], and an Assyrian iron comb in the British Museum is $3\frac{1}{2}$ in. long and 2 in. broad in the middle and has teeth on both sides.[10]

Herodotos[11] says the Assyrian army with Xerxes (480 B.C.) had bronze helmets and iron maces with wooden handles. Loftus found at Susa iron spear-heads with copper and clay objects inscribed in Babylonian characters, and "alabaster" vases with the name of Xerxes, hence this iron is perhaps Persian.[12] In the late Babylonian Period 8 shekels of silver bought 10 minas of iron, so that silver was 225 times as valuable as iron, which was now a common metal.[13]

[1] Olmstead, H. Pal. Syr., 481.
[2] Luckenbill, ii, 178, 289.
[3] Petrie, Six Temples at Thebes, 1897, 18, pl. xxi ; Meissner, i, plate-figs. 113, 114.
[4] BMGB, 49, 50, 74.
[5] Meissner, i, 96.
[6] Rawlinson, Mon., i, 441 ; Meissner, i, 95, Fig. 24 ; plate-fig. 58.
[7] Smith, TSBA, 1874, iii, 463.
[8] BMGB, 139, 145, 153.
[9] Layard, 1867, 285.
[10] Rawlinson, Mon., i, 574.
[11] vii, 63.
[12] Travels, 1857, 409.
[13] Meissner, i, 364.

BABYLONIA AND ASSYRIA III—1

NON-METALS, INORGANIC

STONE

In Babylonia proper there was no stone, and every pebble was valuable, whilst Assyria was much better supplied with stone (sigarru) and minerals.[1] Stone of good quality (limestone, sandstone and conglomerate) was at hand and basaltic rock could be obtained from Mons Masius on the northern border. Various marbles and ores of useful metals occur in the mountains of Kurdistan.[2] Flint and obsidian were used for arrow-heads, axes, grindstones, etc., and for knives in the earliest period, and even after metal was known.[3] The stone for statues was, as in Egypt, diorite or dolerite (ušu) in blocks [4], which came in Gudea's time from Magan [5], and since it is different from that of the Egyptian statues [6], the theory that Magan was Sinai is doubtful.[7] This stone was polished with some abrasive powder.[8] Building stones were diorite, "alabaster," magnesite [dolomite ?], limestone (pîlu pisû), breccia and basalt (kašurrû).[9] Gudea also imported stone from the land of the Amorites.[10]

Stone vessels, dishes and vases (perhaps made by imported Egyptian craftsmen)[11] occur in the early Sumerian Period, when they were rare and costly gifts (with blocks of stone and lumps of metal) to the gods.[12] Clay vessels were more common in later times, yet stone vases were made in all periods. Those at al-'Ubaid and at Ur (3500–3000) were of calcite (the Assyriologist's "alabaster"), more rarely of soapstone, diorite or limestone, and as rarities at Ur there were cups or bowls of obsidian and lapis lazuli. Stones used for vases at all periods included white calcite, stalagmites, "alabaster," basalt, diorite, marble (doubtful, according to Meissner), limestone, onyx, porphyry, green porphyry, blue freestone, sandstone and green steatite.[13] Many archaic stone vases were found at Bismaya and Ur.[14] Tiglathpileser I (1115–1103) imported rare stones from the Na'irî for the temple; Sennacherib (705–681) boasts that he used such stones from Syria for large works never before attempted.[15] As early as Nebuchadnezzar (604–562) the alabastron of pottery or white "alabaster" was common.[16]

Stone employed sparingly in the earliest work at Lagaš was limestone, granite and marble (doubtful): at Eridu, exceptionally, limestone, granite and marble (?) were extensively used in the Sumerian Period, and as there

---

[1] Sayce, B. and A. Life, 102 ; Delaporte, 15—no metals or stone; King, Magic, 169 ; Meissner, i, 259, 349.
[2] Rawlinson, Mon., i, 219 f.
[3] Meyer, Alt. I, ii, 447 f.; Meissner, i, 260.
[4] Meissner, i, 349 ; Thompson, The Chemistry of the Ancient Assyrians, 1925, 79, 121.
[5] King, S. and A., 262 ; Handcock, 127.
[6] Thompson, 122.
[7] Sayce, Rel., 33 ; Handcock, 127.
[8] Thompson, 128.

[9] Ib., 83 f.—full account.
[10] Meyer, Alt. I, ii, 529 ; Meissner, i, 281 f., 285, 349.
[11] Smith, EHA, 55.
[12] Jastrow, Civ., 447 f.
[13] Woolley, 1928, 13 f., 37 ; ib., Ur Excav., 1934, ii, 379, pl. 165 ; Handcock, 325, 327, 330 ; Meissner, i, 261, 466.
[14] Banks, Bismya, passim ; Meissner, i, 261 ; Ur Excav., 1934, ii, 378 f.
[15] Luckenbill, i, 89 ; ii, 168, 179.
[16] Koldewey, Excav., 252.

is no stone on the site these were imported. At Nippur there were stone gate-sockets. In the Neo-Babylonian Period stone was extensively used for important purposes where expense was no objection, but was applied over and over again after reshaping.[1]   Stones found at al-'Ubaid include diorite, dolerite, jasper, jadeite, serpentine, breccia, black marble, chert, limestone and "alabaster."[2]

The materials used in the construction of the tombs at Ur are a coarse white limestone in rough unshaped blocks of various sizes laid in plain mud mortar ; burnt and crude bricks, and *terre pisée* ; as mortar there are "cement," bitumen, clay and mud. Wall surfaces were finished with plaster, either mud or a "fine white lime cement made from gypsum" (which would give neither "lime" nor "cement"). The arched vault was used in the early period : the arch is known at Tell Asmar in 2500 B.C., in private houses at Ur in 1900 B.C. and in the later period at Babylon.[3] "White calcite" which on burning was "reduced to plaster of Paris" [4] is, of course, chemically impossible and gypsum must be meant. This "calcite" came from the west coast of the Persian Gulf, diorite from "the East."[5]   Shell inlay and engraving were cleverly executed from the earliest period.[6]

For ordinary purposes in Babylonia bricks were used and stone was generally of secondary importance ; even Nebuchadnezzar's palace in Babylon was of brick.[7]   In Assyria the buildings were generally of limestone, with some "alabaster" (abundant in Assyria but precious in Babylonia) and basalt.[8] "Alabaster" and marble (?) were imported from the mountains of Armenia [9], although "soft grey alabaster" abounds on the banks of the Tigris.[10]

A competent study of stone objects from Mesopotamia, such as has been attempted for Egypt[11], is still lacking. The stone gišširgallu, formerly incorrectly read *parûtu*, is never either marble or the modern alabaster, which is a variety of gypsum[12], but is probably[13] that variety of limestone which Theophrastos[14] calls chernites ($\chi\epsilon\rho\nu\acute{\iota}\tau\eta s$), "like ivory," and found near Thebes in Egypt. Later classical authors call it chermites, and Hill[15] calls it a "white marble." Pliny[16] repeats from Theophrastos that it was like ivory and that Darius was buried in a sarcophagus of it. The stone can be burnt into lime[17] and is thus a variety of calcite. Vases of it were made in the royal factory at Babylon at an early period, and in the Persian Period were sent as presents by the king.[18] A fossiliferous limestone full of shells, which is mentioned by Xenophon as used for building[19], occurs near Môsul.[20]

Stone, including millstone grit, could be procured from the Euphrates Valley at almost any point above Hît, from the mountain region of Susiana and from the western provinces. Near Babylon the commonest was limestone, but round Haddisah a gritty siliceous rock alternates with ironstone, and sandstone and granite occur in the Arabian Desert. Abundance of grey alabaster is found near Nineveh and in much of the Mesopotamian region, and a better kind was quarried near Damascus.[21]

Limestone in the early Sumerian Period could be got from Jabal Simran,

[1] Handcock, 127 f., who adds sandstone, "alabaster" and agate at Eridu ; Meissner, i, 281.

[2] Hall and Woolley, Ur Excav., 1927, i, 51 f.

[3] Ur Excav., 1934, ii, 229 f.

[4] *Ib.*, 142.

[5] *Ib.*, 379 f.

[6] *Ib.*, 263 f.

[7] Maspero, Dawn, 623 ; Meissner, i, 281.

[8] Handcock, 129 ; Jastrow, Civ., 368 ; Perrot-Chipiez, Chaldæa, i, 121 f.

[9] Jeremias, 295.

[10] Rawlinson, Mon., i, 219 f.

[11] See p. 103.

[12] King, Magic, 167 ; *ib.*, H. Bab., 41 ; Meissner, i, 349, 466.

[13] Birch, in Loftus, Travels, 1857, 411.

[14] De Lapidibus, c. 6.

[15] Theophrastus on Stones, 1774, 38.

[16] xxxvi, 17.

[17] BMGB, 32.

[18] King, H. Bab., 41 ; *ib.*, Magic, 60.

[19] Anabasis, III, iv, 10.

[20] Rawlinson, Mon., i, 257.

[21] *Ib.*, ii, 489 ; iii, 160 f.

or finer varieties from the Upper Euphrates Valley ; Gudea imported it from Meluchcha. Diorite came by sea from Magan, and the "alabaster" was partly Persian and partly stalagmitic calcite from cave deposits on the eastern shore of the Persian Gulf.[1] Ašurnasirpal II (885–860) obtained a stone called umu [2] as tribute from Syria.

Small pieces of polished agate, alabaster and marble (?), $\frac{1}{2}$ in. to 2 in. long, and $\frac{1}{4}$ in. or less broad, with copper pins for fixing, were used to decorate the interior walls of the ziqqurat at Eridu, and agate beads occur in tombs.[3] Stones and half-gem stones were used for mace-heads, parts of sceptres, feet of thrones, etc.[4]

Unidentified stones in the texts are cha-u-na, na-lu-a (basalt ?) (Gudea) ; saggilmut stone (Asarhaddon) ; musa stone (Šamsi-Adad III, c. 1630) ; ašnan stone (compared with cucumber seeds or grains of wheat, and perhaps a Fusuline limestone) ; the precious stones dušû, chulâbu, mušgarru, etc. In a hymn the stone schammu is cursed, since it "tried to lay hold of Ninurta in the mountains," whilst dolerite is assigned a favourable position as a "stone from Magan, which cuts the mighty bronze like leather."[5]

SEAL CYLINDERS

The seal was in common use from the earliest period : "every Babylonian had a seal."[6] An early physician's seal dedicated to a god Edina-mugi, a "messenger of Gir" (a designation of Nergal, god of pestilence), shows a scalpel, knives and cups.[7] In Ḥammurabi's time (1924–1913) ready-made seals which required only the addition of the owner's name were available, and many scenes engraved were standardised.[8]

Seals engraved on small stone cylinders, with figures of demons, gods, animals and mythological scenes, together with texts in cuneiform, were perhaps originally used as amulets, the scenes often representing struggles between good and evil genii, and also as votive offerings.[9] In their use as seals they were rolled on soft clay.[10] Small cylinders of "meteoric stone" were found, with copper and pieces of "bamboo" (reed ?), in an old tomb at Ur.[11] The best seal work was of the dynasty of Akkad (2637–2457).[12] An Egyptian origin of the art has been suggested[13], but the early occurrence of such seals in Egypt has also been used as an argument that the art was imported into that country by the Sumerians.[14] The seal, also used in Assyria from the earliest times, was later adopted by the Persians.[15] Most Assyrian seals in museums are of Sargon II's time (722–705) ; their characteristic feature is the extensive use of ball-shaped incisions.[16]

The seal cylinders in the British Museum range in date from 3000 to 300 B.C.[17] The very varied materials used include shell, the engraving of which was very early[18]; hæmatite, which is most common, especially in the

[1] Woolley, 1928, 46 ; Meyer, Alt. I, ii, 529.
[2] Sayce, Rec. of the Past, N.S., 1889, ii, 144.
[3] Rawlinson, Mon., i, 81, 106.
[4] Meissner, i, 262.
[5] Meissner, i, 350, 351 ; ii, 173 ; Hilprecht, Babyl. Exped. of University of Pennsylvania, Series D, vol. 5, fasc. 2, 1910, 30 f., 58.
[6] Herodotos, i, 195.
[7] Meyer, Alt. I, ii, 425 ; Contenau, Manuel, 1931, ii, 691.
[8] Meissner, i, 263.
[9] Budge, Amulets, 87 f., 123 f. ; Fossey, Magie, 108 f.
[10] Meyer, Alt. I, ii, 470.

[11] Rawlinson, Mon., i, 88, 93.
[12] Handcock, 309 ; Jeremias, 295 ; King, S. and A., 78 ; Sayce, B. and A. Life, 102 ; Heuzey-de Sarzec, i, 275 f. ; Meissner, i, 262 f.
[13] Kisa, Glas, 572 f. ; Erman-Ranke, 174.
[14] Furtwängler, Die antiken Gemmen, iii, 1 f. ; BMGB, 231—a hieroglyphic sign is supposed to represent a cylinder seal with string attached ; Frankfort, Studies in Early Pottery in the Near East, 1924, i, 130 f.
[15] Handcock, 304 ; Perrot-Chipiez, Art in Persia, 1892, 453 f.
[16] Meissner, i, 264.
[17] BMGB, 222 f., 231 f. ; Perrot-Chipiez, Chaldæa, ii, 251 f.
[18] Handcock, 309 ; Heuzey-de Sarzec, i, 259, 265 f. ; mother-of-pearl was later.

Ur dynasty[1] ; chalcedony, obsidian, agate, jasper, lapis lazuli (imported from Persia in Gudea's time) (2450) [2], marble (?), black and green serpentine, quartz (especially in the Assyrian Period) [3], carnelian, steatite, occasionally jade, emerald, root of emerald, amethyst, topaz, onyx, green glass (later period ?), flint (rare), aragonite, dolomitic marble, limestone, syenite, amazon stoņe (a very hard green felspar), steatite, sandstone and green calcite. The amuletic properties of the seal depended on the material from which it was made (kagina, hæmatite ? ; lapis lazuli ; dušia, rock crystal ; tuudaš ? ; zatumušgir, green serpentine ; gug, red jasper, or carnelian).[4]

Gold caps [5] were used, and a wire of copper, sometimes silver or gold, and in a later period iron, was used to secure the drilled cylinder [6], but drilling was not used in the earliest period.[7] Metal seals were probably never in general use, although there is one rough silver cylinder in the British Museum, and the stamp seal is much later (from the 9th century B.C.) than the cylinder.[8] In the Persian Period and somewhat before (612–350 B.C.) cylinders were replaced by cone seals in sard, carnelian, agate and especially chalcedony. Sassanian amulets, and seals used in Western Iran or Persia in the period A.D. 226–632 ("Pahlawî gems"), are in agate, lapis lazuli, sard, carnelian, chalcedony, etc., and were used as amulets.[9] Gems and stones for beads are considered later.[10]

Softer stones were probably engraved with flint, harder with corundum or emery (which is found on the islands of the Greek Archipelago). Hard stones such as porphyry, basalt and hæmatite were engraved in Gudea's time (2450), and the very hard stones such as agate, jasper, garnet and rock crystal at the end of his period. The lapidary's wheel was probably not in use till the 8th century B.C., and before this a bow-drill with crushed emery was used.[11]

BRICKS

Two types of bricks were used from the early Sumerian Period (3300 B.C.). The first are unbaked or sun-dried bricks ("crude bricks"), in which the clay is often mixed with chopped straw, and the second are kiln-fired ("baked bricks"). Crude bricks are still made on the banks of the Tigris in large flat wooden frames, and the process continued practically unchanged from the Sumerian to the late Assyrian Period.[12]

The invention of bricks was ascribed to the twin gods Sin and Ninib, in the months of May and June, the "month of brick" (Sum. murga ; Akkad. simanu), when the river had subsided and exposed the clay, and the sun was hot enough to dry the bricks and yet not crack them.[13]

Babylonia is essentially a flat region of alluvial clay deposited by the two rivers, and much of the present country towards the Red Sea has been formed in historical times.[14] This clay was the earliest building material, and was mixed with reeds and straw to form walls from 4000 B.C. The poorer huts had reed walls plastered with clay, but for palaces and temples clay bricks were formed by hand on a board or in a mould (nalbantu). They were thicker in the middle ("plano-convex bricks"), were set in clay as mortar and were not

---

[1] Meyer, Alt. I, ii, 418 ; Sayce, B. and A. Life, 104 ; Rawlinson, Mon., i, 88, 93, 382 ; cf. Krause, Pyrgoteles, Halle, 1856, 124.

[2] Sayce, 94 ; Rawlinson, 382 ; Woolley, 1928, 46.

[3] Rawlinson, 405 ; Sayce, 104.

[4] Meyer, 418 ; Thompson, 84 f., 97, 109 ; Handcock, 287 ; Rawlinson, 382, 405 ; BMGB, 81 ; Budge, Amulets, 87, 293.

[5] Meissner, i, plate-fig. 148.

[6] Handcock, 286 f.

[7] Meyer, 419.

[8] Perrot-Chipiez, Chaldæa, ii, 277, 280.

[9] Budge, 124 f.

[10] P. 288.

[11] Meyer, 418 ; Perrot-Chipiez, Chaldæa, ii, 259 f.

[12] Rogers, i, 409 ; Heuzey-de Sarzec, i, 13 f. ; Langdon, Alte Orient, 1928, xxvi, 72 ; Maspero, Dawn, 624.

[13] Maspero, Dawn, 624, 753.

[14] Sayce, B. and A. Life, 1 f.

commonly burnt until a later period.[1]   The bricks were baked in kilns fired with reeds cultivated for use as fuel.[2]

Stone and wood, both scarce, were obtained from outside by war or trade. Much damage was done to buildings by heavy rains and by fires ; the crude clay bricks settled and artificial mounds were formed on the sites.[3]   For the sites of temples a thick foundation of bricks was laid down[4] ; vertical drains made up from sections of burnt clay drain-pipes, with perforated clay heads like colanders, were set in the foundations, and there were also horizontal clay drain-pipes with T-shaped connections, those at Nippur (before 3000 B.C. ?) being laid in cement in the floor of a vaulted culvert.[5]

The earliest Assyrian buildings were copies of the Babylonian, i.e. of brick raised on wholly unnecessary mounds (since there was no danger of floods), although stone was readily obtainable.[6]   Assyrian bricks are large, square and flat.[7]

There were also at Tello burnt clay cones, similar to those found at Uruk (Warkâ) and Eridu, curved and triangular bricks, clay drain-pipes and brick cavities coated inside with bitumen on a wicker-work of reeds, probably used for keeping beer or date-wine[8] : Strabo [9] says such vessels were used in Babylonia for holding water. Clay cones and pots were set in mortar in walls, for decorative purposes.[10]

Plano-convex sun-dried bricks, with two finger-marks to give lodgment to the mud mortar, were used in the earliest period at Kiš during the reigns of Ur Nina and Eannatum, but were replaced by flat bricks under Entemena. Plano-convex bricks also occur in early graves at Nippur and Ur, although the very earliest bricks at al-'Ubaid are flat, and at Kiš some of the earlier burnt bricks are nearly flat. In walls of early large buildings (stage towers) at Ur and Uruk, the plano-convex bricks are laid in alternate courses slanting in opposite directions to produce a herring-bone effect. The brickwork at Ur is probably unsurpassed and all the basic architectural forms were worked out.[11]

The usual Babylonian bricks are nearly 1 ft. square and $3\frac{1}{2}$ in. thick, although the size varied. Some are oblong. The burnt bricks from Birs Nimrûd are generally dark red, those from Babylon light yellow and very hard. The best baked brick is yellowish-white, resembling Stourbridge fire-brick ; a second kind is blackish-blue, very hard but brittle ; a third, the coarsest and earliest type, is slack-dried and pale red. Crude and baked bricks were sometimes used in alternate layers, and sometimes the inside of the structure was of crude brick with a face of burnt brick.[12]

Crude bricks, often bonded with reeds, when used for interiors were sometimes painted white, black, purple, blue and orange, as well as silvered and gilded.[13]   Early Sumerian houses (e.g. at Eridu) were decorated in bands of red, white and black, or plain red and white paint on stuccoed crude brick walls, and this primitive colour scheme seems characteristic of early Sumerian

[1] Handcock, 120 f., 124 f. ; Hilprecht, Explor., 1903, 401 ; Meissner, i, 275 ; cf. Genesis, xi, 3 ; Meyer, Alt. I, i, 444 f.

[2] Sayce, B. and A. Life, 138.

[3] Jastrow, Civ., 10, 367, 378 ; Perrot-Chipiez, Chaldæa, i, 114 f.

[4] Meyer, 444 f.

[5] Delaporte, 193 ; Feldhaus, Technik, 208 ; Hilprecht, Ausgrabungen im Bêl-Tempel zu Nippur, Leipzig, 1903, 40, 65 ; ib., Excavations in Assyria and Babylonia, Philadelphia, 1904, 396 f.

[6] Meyer, Alt. I, ii, 609 ; Sayce, B. and A. Life, 93.

[7] Jastrow, Civ., 368.

[8] Loftus, Travels, 1857, 186 ; Rawlinson, Mon., i, 82, 90 ; Heuzey-de Sarzec, 53, 60, 62, 413, 415.

[9] XVI, i, 9, 739 C. ; Pliny, xiv, 3.

[10] Birch, Anct. Pottery, i, 142.

[11] Hilprecht, Ausgrabungen, 64 ; ib., Excav., 1904, 372 f., 389 ; Ur Excav., 1934, ii, 223 f., 228 f., 393 ; Woolley, 1928, 34 ; the dimensions are very variable.

[12] Rawlinson, Mon., i, 72 f. ; ii, 524 f., 555 ; Birch, Ancient Pottery, i, 108 f., 139 ; Maspero, Dawn, 624 ; Layard, 1867, 296 ; Heuzey-de Sarzec, i, 13 f. ; Handcock, 126.

[13] Birch, 106 f., 141.

art.[1] A gypsum plaque from Assur (early period) is painted in black, white and red.[2] The walls of Gudea's palace were not decorated or coloured ; the bricks may have had a coat of white plaster on which scenes were painted in one or two colours. Wood panelling, glazed bricks and decorations were reserved for temples. Magic copper nails with heads representing gods were put in the ground or some part of the masonry to preserve by magic the bricks from destruction.[3]

The Assyrians employed baked bricks less frequently than the Babylonians. The clay was mixed with loam and sand, stubble and vegetable fibre, before burning, apparently to hold the bricks together before they were sent to the kiln. In thinking of "sun-dried" bricks we must remember that the temperature in Babylonia can reach 50° C. in the shade. The terra-cotta (burnt clay) prisms used for texts were of very fine clay, sometimes glazed with a vitreous white coating. Curious round pillars built closely together in groups of four with shaped bricks, resting on a brick foundation, occur only in Gudea's buildings.[4] Callisthenes (4th century B.C.) and Epigenes (early 3rd century B.C.) knew that the Babylonians wrote astronomical observations on baked bricks or tiles (coctilibus laterculis).[5]

Various kinds of mortar were in use : frequently clay or mud only, as at Babylon [6], or mud with chopped straw or reeds, as at Kiš and other places.[7] Layers of reeds were sometimes put between those of unbaked brick.[8] Besides clay mortar, asphalt or bitumen and lime mortar (the modern mortar) were used. Bricks were set in bitumen in the Sumerian Period, e.g. at Tello in Ur Nina's time, in pre-Sargonic remains at Ur, etc. In later buildings (Nebuchadnezzar) at Babylon, asphalt was replaced by white lime mortar as good as the best Roman.[9] In some cases, e.g. in a wall of Nebuchadnezzar, both kinds of mortar were used at once. In the new palace of this king at Babylon an extremely durable fine, pure white, lime mortar was employed, and lime mortar with clay and bitumen was also used at Nippur, Kasr and Birs Nimrûd. At Ur a mixture of lime and ashes (the modern kharour) was used, or bitumen with a thin coating of gypsum or "enamel plaster" on arches. In Nabonidus's (555–538) building, the very hard, bright red burnt bricks were laid in bitumen only.[10] In Assyria, although bitumen was used for under-work or basements to prevent leakage, including perforated bricks and drain-pipes set in bitumen[11], mortar was generally used sparingly, the stones being carefully dressed to fit, and burnt bricks were often cemented with clay.[12] Of two specimens of mortar from the palace at Ctesiphon (A.D. 600), one was pure gypsum and the other a fat dolomitic lime mortar containing powdered brick as aggregate in the proportion of 1 : 1.[13] Gypsum (gassu) plaster was used in the late Babylonian Period.[14]

Beside mud mortar for the construction of walls and roof, a fine lime plaster was applied to the whole inner face of the chambers of royal graves at Ur (from 3500 B.C.), and in some cases to the floor. The arch, vault and (probably) the dome were then in use.[15] Remains of an old lime-kiln (before 2450 B.C.) with limestone (analysed) were found at Khafaje.[16]

[1] Hall and Woolley, Ur Excav., 1927, i, 17 ; pls., 34, 38.
[2] Contenau, Manuel, ii, 645.
[3] Maspero, Dawn, 716 f.
[4] Birch, i, 107, 113 ; Meissner, i, 186, 275, 280.
[5] Pliny, vii, 56.
[6] Handcock, 124 f.
[7] Langdon, Excav. in Kish.
[8] Herodotos, i, 179 ; Maspero, Dawn, 624 ; Meissner, i, 277.
[9] Birch, i, 138 ; Handcock, 125 ; King, S. and A., 91 ; Rawlinson, Mon., ii, 556.

[10] Handcock, 125 f. ; Loftus, 1857, 133 ; Delaporte, 198 ; Rawlinson, Mon., ii, 524, 529.
[11] Birch, i, 138, 142 ; Rawlinson, Mon., ii, 556 ; Perrot-Chipiez, Chaldæa, i, 157 f.
[12] Handcock, 126 ; Meissner, i, 277.
[13] Biehl, Building Science Abstracts, London, 1930, iii, 290.
[14] Meissner, i, 277.
[15] Woolley, 1928, 37.
[16] Frankfort, Jacobsen and Preussen, Tell Asmar and Khafaje, 1930–31, Univ. Chicago Orient. Inst. Publ., No. 13, 90.

Abundant remains of bricks set in bitumen were found at Warkâ : Perrot and Chipiez [1] say burnt bricks, Loftus [2] unburnt. A damp-course of sheet lead was sometimes used in walls [3], and elaborate sewers were found at Nimrûd.[4] The lowest stage of the ziqqurat at Nippur was coated with bitumen.[5]

Traces of colouring occur on sculptures, in red and yellow and sometimes blue and black.[6]   They were visible on the large stone bas-reliefs from Nimrûd (red, blue and green) when they arrived at the British Museum in 1850–56, and originally these were probably brilliantly painted.[7] In the Assyrian Period a thin (4 mm.) lime plaster with remains of fresco painting occurs, e.g. at Khorsabad.[8] Interior decorations were then frequently in painted stucco [9], which was used early in Babylonia, before glazed tiles.[10] Painting was not used in Hammurabi's time.[11]

Layard says green and yellow occur continually on bas-reliefs at Khorsabad, whilst Botta found only red, blue and black, the green and yellow being used only on glazed bricks. The red was bright and far more brilliant than that of Egypt ; at Khorsabad it approaches vermilion, at Nimrûd it inclines to a crimson or lake tint ; perhaps it was cuprous oxide, which the Assyrians used to colour glass red.[12] Blue and green were rare in the Assyrian Period for bas-reliefs.[13] The blue is an oxide of copper glaze with a trace of lead[14], or powdered lapis lazuli.[15] White rarely occurs on reliefs but frequently on later pottery, where it is an oxide of tin enamel ; the black is of uncertain composition, but is perhaps bone-black with gum.[16]

A curious feature of some ziqqurats is that certain layers of bricks (the fifteenth, twentieth, etc.) are covered with crossings of white filaments, falling to powder on a touch. The powder (analysed by Webster) is largely alumina, "no doubt calcined reed or matting which has absorbed alumina, probably from the brickwork."[17]

BITUMEN OR ASPHALT

Asphalt or bitumen (kupru or iddû ; "slime" in Genesis, ix, 3 ; ἄσφαλτος in the Septuagint ; bitumen in the Vulgate ; Arabic mumia) is found in several places in Mesopotamia. Gudea and the kings of Ur obtained "earth pitch" from the mountain Madqa (Ibis) by ship.[18] Model boats of bitumen occur in tombs at Ur.[19] The source of bitumen at Qijâra, north of Assur, was probably used, but Hît (Babylonian Id, Greek Is), called by Tukulti-Urta II the "place of the ušmeta stone, in which the gods speak," was the most famous source and is still productive.[20] Hît is in the country of the Subartu and supplied the Sumerians with bitumen in 3500 B.C.[21]; it is on a river of the same name which flows into the Euphrates about a hundred miles north from Babylon. Herodotos[22] describes the building of the walls of Babylon in Nebuchadnezzar's time (604–561) from bricks baked in kilns and cemented with hot bitumen

[1] Chaldæa, i, 155, 156.
[2] Travels, 1857, 167 f.
[3] Rawlinson, Mon., ii, 517.
[4] Perrot-Chipiez, Chaldæa, i, 228.
[5] Handcock, 125.
[6] Rawlinson, Mon., i, 360, 363.
[7] BMGB, 31 ; Meissner, i, 329.
[8] Delaporte, 362 ; Meissner, i, 329, plate-fig. 200.
[9] Jastrow, Civ., 371 ; Handcock, 279.
[10] Sayce, B. and A. Life, 91 ; Meissner, i, 278.
[11] Breasted, Anct. Times, 137.
[12] Rawlinson, Mon., i, 363 f.
[13] Delaporte, 370.
[14] Rawlinson, Mon., i, 363.
[15] Delaporte, 370.

[16] Rawlinson, 382.
[17] Langdon, Excav. at Kish, 46, 47, 51 ; Hilprecht, Excavations, 1904, 16—fifth to seventh layers.
[18] Birch, Ancient Pottery, i, 138 ; Meissner, i, 53, 260, 348 ; King, S. and A., 261 ; Muss-Arnolt, Dict. Assyr., 18, 423— kupru = pitch or asphalt, iddû = naphtha ; Schrader, Keil., 648 ; Meyer, Alt. I, 1i, 545, 556.
[19] Ur Excav., 1934, ii, 145.
[20] Meissner, i, 348 ; Peters, Nippur, i, 159 ; Luckenbill, i, 129.
[21] Woolley, 1928, 45.
[22] i, 179 ; Ragozin, Chaldea, 1889, 44 ; Handcock, 125 ; other old accounts in Bochart, Geographia sacra, 1712, 40.

from Is, with layers of wattled reeds between the courses. Between Môsul and Bagdad bitumen oozes from cracks of the soil. Kugler [1] thought the material from Hît was petroleum, which is actually found there, and was imported by Thothmes III.[2] The pits were described by Ainsworth [3] and by Chesney.[4] Kugler suggested that the mortar was made from the lime-stone of Hît, "burnt in an asphalt fire," but there is no doubt that bitumen was used.[5]

A picturesque description of the bitumen springs at Nimrûd, which emit tongues of flame and jets of gas, is given by Layard.[6] The modern name Muqayyar for Ur is derived from the Arabic qir, bitumen.[7] Bitumen is also found in the Dead Sea in Palestine, and at Kerkuk, which was nearly as celebrated as Hît.[8]

Other sources of bitumen and petroleum mentioned by Pliny, Strabo, Vitruvius, Dioskurides, Herodotos, etc., include many places on the Tigris, Susiana, Arderikka, Cilicia, India, Ethiopia, Carthage, Zakynthos (Zante), Akragas (Girgenti, in Sicily) and Apollonia near Epidamnos (Dyrrachium).[9] The information on bitumen in Strabo and Dioskurides is probably from Poseidonios (135–51 B.C.), that in Diodoros Siculus from Poseidonios or Ktesias (c. 415 B.C.).[10] Diodoros Siculus[11] says the bitumen, which flows from the earth as a liquid, afterwards hardens and is used as fuel : a well emitted a sulphurous suffocating vapour.

Besides its use in building, asphalt was used at various times for plastering reed baskets and reed boats[12], for varnishing wood in building, as a mortar, in fumigating and as incense, mixed with sulphur for fumigating birdcages and nests, for salves, for putting round trees to prevent ants climbing them, medicinally (externally and internally), in warfare as an incendiary (maltha, "poix minerale," according to Haüy) and probably also in magic.[13] Black hair on Sumerian stone heads is represented by bitumen.[14]

In the Deluge Tablets, which form part of the Gilgameš Epic, the Ark, as in Genesis, is said to be smeared outside with bitumen.[15] Nicholas of Damascus (c. 40 B.C.) says[16] the remains of the Ark were to be seen in his day on the top of Mount Baris in Armenia[17], where they were seen by Benjamin of Tudela (A.D. 1170)[18] and many later travellers. Discoveries of charcoal and bitumen on Gebel Judî, one of the mountains identified as that on which the Ark rested, probably gave rise to the tradition.[19] Another tradition put the resting-place of the Ark at Apamea in Phrygia, and the Ark is shown on 3rd century A.D. medals from the mint there.[20] In Berossos's account (c. 250 B.C.) of the Flood it is said that there are remains of the Ark on the Gordyæan mountains, and the bitumen coating of the ark was scraped off for use as an amulet.[21] Some late Chaldæan amulets are pieces of bitumen with inscriptions

[1] Sternkunde und Sterndienst in Babel, Münster, 1907, II, 117 f.

[2] Rogers, i, 427 ; Rawlinson, Mon., i, 39 ; ii, 487.

[3] Researches in Assyria, Babylonia and Chaldæa, 1838, 85 f.

[4] Narrative of the Euphrates Expedition, 1868, 76, 280.

[5] King, H. Bab., 25.

[6] Nineveh and Babylon, 1867, 68.

[7] Maspero, Dawn, 561.

[8] Rawlinson, i, 219 f.

[9] Nies, PW, ii, 1728.

[10] Reinhardt, Poseidonios, Munich, 1921, 114 ; Schmitz, DBM, i, 899 ; Ctesiae Cnidii Operum Reliquiae, ed. Baehr, Frankfurt, 1824, 397.

[11] ii, 9, 12.

[12] Meissner, i, 245 f., 250.

[13] Geoponika, XVIII, ii, 4 ; XIV, ii, 4 ; XII, viii, 1 ; Nies, 1729.

[14] BMGB, 190.

[15] Ib., 220.

[16] Müller-Didot, Fragmenta Historicorum Græcorum, iii, 415.

[17] James, DG, i, 379.

[18] Itinerary, ed. Adler, 1907, 33.

[19] Maspero, Dawn, 572.

[20] Lenormant, Hist. ancienne de l'Orient, 1869, i, 25.

[21] Cory, Ancient Fragments, 1832, 27 f. ; Müller-Didot, Fragm. Hist. Græc., ii, 502 ; Josephus, Contra Apion., i, 19 ; ib., Antiq., I, iii, 6 ; Bochart, Geographia Sacra, Leiden, 1712, 13 f., 18, 38 f.

in Greek letters [1], and the use of amulets of black coral [2] is perhaps related to this tradition. Ear-studs of bitumen, as well as bone and baked clay, occur at al-'Ubaid (3500 B.C.).[3]

A very hard natural or artificial mixture of clay and bitumen was used for seal impressions [4] and as an artificial stone.[5] Two enormous blocks of a mixture of asphalt, earth and lime, very hard on the outside, which were found at Tello [6], were probably the same material as that analysed by Berthelot [7], which could become very hard. It was used by Entemena for bathroom floors and for plastering baths [8], and a room floored with asphalt was found at Siparra.[9]

From the bitumen springs was also obtained crude petroleum or naphtha (naptu ; šaman iddi ; mazût), used for lamps.[10] In some Babylonian temples the walls were varnished with a solution of asphalt [in naphtha ?].[11] Plutarch[12] and Strabo[13] report that Alexander found at Ekbatana a spring of eternal fire and a lake of mineral oil ($\acute{\rho}\epsilon\upsilon\mu\alpha$ $\tau o\hat{\upsilon}$ $\nu\acute{\alpha}\phi\theta\alpha$), which inflamed so easily that it began to burn as soon as fire was brought near it. The inhabitants sprinkled a whole street with it one evening and set fire to one end, when the flame flashed in an instant to the other end. A bath attendant then soaked a boy in naphtha and lit him, but Alexander threw water over the boy, putting him out only with great difficulty.[14] Hall[15] suggests that the bricks in the ziqqurat at Ur may have been burnt by a fire of brushwood soaked in pitch or petroleum, and reeds soaked in bitumen were used as torches.[16] A kind of Greek fire, probably petroleum or pitch, is shown poured on the heads of soldiers attacking Assyrian fortresses[17], and melted asphalt was poured over the head as a form of torture in Assyria. The supposition that corpses were burnt in reed mats soaked in bitumen is improbable.[18]

Strabo says (from Eratosthenes of Cyrene, or from Poseidonios) that the white naphtha of Babylonia, "which attracts flame, is a liquid sulphur ; the second kind, the black naphtha, is liquid asphaltos and is burnt in lamps instead of oil. The liquid asphaltos, which is called naphtha ($\nu\acute{\alpha}\phi\theta\alpha$) and is found in Susiana, is of a singular nature. When it is brought near the fire, the fire catches it, and if a body smeared with it is brought near the fire it burns with a flame, which it is impossible to extinguish except with a large quantity of water ; with a small quantity it burns more violently, but it may be smothered and extinguished by mud, vinegar, alum and glue." Pliny[19] mentions (perhaps from Ktesias) "naphtha flowing like liquid bitumen. Burning springs exist in Bactria, Media, on the borders of Persia and at Susa. The plain of Babylon throws up flame from a place like a fish-pond, an acre in extent. There is a lake at Samosata, a city of Commagene, which exudes an inflammable mud called maltha [naphtha ?]. It sticks to any solid body which it touches and its power of adhesion enables it to follow those who try to escape it. With this they defended their walls when besieged by Lucullus, and set his soldiery on fire, arms and all. Water makes it burn more fiercely (aquis accenditur) and experience shows that earth (terra) is the only thing which will put it

[1] Maspero, Dawn, 572.
[2] Gardiner, *Nature*, 1921, cviii, 505 ; Hickson, *ib.*, 1922, cx, 217 ; Nierenstein, *ib.*, 313.
[3] Woolley, 1928, 13 f.
[4] King, S. and A., 166.
[5] Perrot-Chipiez, Chaldæa, i, 383.
[6] Nouv. Fouilles, 153.
[7] Heuzey-de Sarzec, 204.
[8] Meissner, i, 260, 278, 412.
[9] Perrot-Chipiez, Chaldæa, ii, 401.
[10] Rawlinson, Mon., i, 39, 219 f. ; Hug,

PW, xiii, 1573 ; Hall, JEA, 1923, ix, 180 ; Muss-Arnolt, Dict. Assyr., 18.
[11] King, H. Bab., 70.
[12] Alexander, c. 25.
[13] XVI, i, 15, 742 C.
[14] Lenz, Mineralogie, Gotha, 1861, 85, 176 ; Nies, PW, ii, 1729 ; Wilson, The Persian Gulf, Oxford, 1928, 45.
[15] JEA, 1923, ix, 180.
[16] Meissner, i, 246.
[17] Delaporte, 305.
[18] Meissner, i, 177, 425.
[19] ii, 104, 105, 106.

out.[1] Naphtha is an exudation like liquid bitumen occurring near Babylon and among the Astaceni in Parthia. It is closely related to fire, which leaps upon it from any quarter as soon as it beholds it." Pliny, who deals with bitumen under sulphur [2], also calls the white naphtha bitumen liquidum candidum ; Dioskurides [3] calls it "filtered asphalt," ἀσφαλτος περιήθημα, and liquid pitch, πισσασφαλτος, Pliny's pissasphaltum. It is difficult to see how a white naphtha could have been obtained from crude petroleum except by distillation, although filtration is possible.

Sulphur, found in large quantities in mines at Môsul, was quarried in great blocks conveyed to considerable distances.[4] Great quantities of hydrogen sulphide gas are evolved from the oil wells in Mesopotamia.[5] Sulphur was used in medicine in ancient times and was perhaps also obtained from the Dead Sea, on the shores of which balls of nearly pure sulphur occur.[6]

THE TEMPLES

The temples, even in Sumerian times, were orientated [7], the *corners* facing the four cardinal points, except at Babylon and the Assyrian ziqqurat at Birs Nimrûd, of which the *sides* face them. In the early period, north-east was taken as north, and so on, in maps. The orientation, in theory, agrees with that of the Egyptian pyramids, but, as it is usually only approximate, it is sometimes denied for both countries.[8]

The temples were largely adorned with silver and gold. Asarhaddon (680–669) says they were made "as splendid as the sun," or "as the day," by the profuse use of precious metals [9], and Bur Sin (2060 B.C.) adorned his temple with gold, silver and lapis lazuli. The temple at Ur had extensive workshops and store chambers, in which vast quantities of goods brought as tribute to the god were stored. There were factories within the precincts, where women spun and wove wool brought from the country. Remains of a smelting furnace, with a clay pot containing scraps of copper, were found : the temple servants apparently melted down metal, brought in by the city merchants, into ingots for storage, such ingots being found in another part of the building.[10] The temple, in fact, resembled less a church than a mediæval monastery, in which industries of all kinds were practised, the workshops and schools being scarcely less important than the chapel.[11] Their store-houses contained accumulations of gold, silver, copper, lapis lazuli, gems and precious woods, and the priests, who maintained great numbers of workmen, made great profits out of corn and metals.[12] The Babylonian temples, even in the earliest period, were centres of learning, trade, industry and agriculture. The priests ruled the land, the oil, date and cotton plantations, the crops, herds and hand-workers. Peasants and workmen were connected with them (as in Egypt) and many classes of workmen were "priests."[13] Technicians who belonged to the college of priests (kiništu) were scribes, directors of archives and of stores, doorkeepers, builders and craftsmen. There were also temple slaves: Nabonidus gave 2850 prisoners of war for the service of Bêl, Nebo and Nergal.[14]

[1] This is a property of the later "Greek fire."

[2] xxxv, 15.

[3] i, 101.

[4] Rawlinson, Mon., iii, 160.

[5] Information from Prof. J. F. Thorpe.

[6] Meissner, ii, 309 ; Rawlinson, Mon., i, 219 f. ; ii, 488.

[7] Loftus, Travels, 1857, 128 ; Ragozin, Chaldea, 284 f. ; King, S. and A., 91 ;

ib., H. Bab., 69 ; Jeremias, 53 ; Ur Excav., 1934, ii, 140 f.

[8] Koldewey, Excav., 225, 243.

[9] Rawlinson, Mon., ii, 37.

[10] Woolley, 1928, 145 f., 147, 148 f., 199.

[11] Eastlake, Materials for a History of Oil Painting, 1847, 1 f.

[12] Maspero, Dawn, 677, 679, 750.

[13] Jeremias, 288.

[14] Meissner, ii, 67; Diodoros Siculus, ii, 29 f.

From the early Sumerian Period to the last days of Assyria, the temples were the law courts and the schools in which writing, astrology, medicine and botany were taught.[1] In later times the temple at Borsippa became a university and perhaps a medical school, and in Babylonia even some slaves could read.[2] The Babylonian scribe, though not so well off as the Egyptian, was important and well educated.[3] The schools taught prophecy, astronomy and medicine, with regular text-book tablets and probably examinations. In Hammurabi's time (1924–1913) the old text-books of Šuruppak, in use even in Susa, were replaced by new ones, the tablets of which are still useful to Assyriologists.[4] The temples probably had offices for adjusting weights and measures, and libraries, perhaps the only ones in existence, but probably not public. In one text the reader is told to "refer to the library."[5] An important collection of books (tablets) was made by Ašurbanipal, especially from Babylonia, Kûtâ and Nippur ; 30,000 fragments (including some catalogues) were taken from his palace at Nineveh. The tablets in the royal library (which had a "chief librarian") are mostly copies from much earlier texts, well classified and numbered.[6]

The images in the temples were usually of baked clay (terra-cotta), wood covered with metal (usually bronze, but also silver, and in a few cases gold), stone, or sometimes, but less frequently, of metal only, including iron. In some cases images of one metal were overlaid with plates of another ; one of the great images of Bêl, originally of silver, was overlaid with gold by Nebuchadnezzar. Bronze hammered work over a model of clay mixed with bitumen was used in Assyria and in Babylonia, e.g. at Ur and Tello ; copper was used over the same material. The idols were provided with costly clothing, often of gold and jewels.[7]

The statues of gods buried or hidden in temples or erected on stage towers were supposed to become imbued with the spirit or "actual body" of the god on consecration, when his name was written on them, as was also the case in Egypt : a text says Marduk (Bêl) "inhabits the image." In some cases the statues or images of foreign gods were broken up on conquest and their temples despoiled, to destroy their power. The spirit could also animate stones, trees, votive lances and talismans.[8] The ceremony of "opening the mouth," Sumerian ka-tuchuda, Akkadian pit-pî, was performed with the idol, perhaps by putting magic materials in the mouth of the statue.[9] It was performed in Egypt with the mummy. The idols were probably contrived to emit sounds by pulling strings operated by a hidden priest.[10] The offerings made to the gods were in the nature of meals and consisted of food of all kinds[11], which was secretly removed by the priests.[12]

The altars and stands were of brick and stone ; the incense burners, lustration vessels and lavers (apsu) were of copper or bronze ; some vessels (e.g. salt cellars) were of silver or gold. Nets, torches, trowels and other implements, tables, sacred cars and sacred boats adorned with electrum and jewels were used in the temples.[13]

[1] Jastrow, Civ., 273, 275 ; cf. Langdon, Alte Orient, 1928, xxvi, 43.
[2] Sayce, B. and A. Life, 54 f.
[3] Maspero, Dawn, 723.
[4] Meissner, ii, 326, 328.
[5] Jeremias, 292 ; Sayce, B. and A. Life, 54 ; ib., PSBA, 1874, iii, 160 ; Meissner, ii, 330 ; Maspero, Dawn, 771.
[6] Jeremias, 294 ; Meissner, ii, 330 f. ; Fox Talbot, TSBA, 1874, iii, 432.
[7] Birch, Ancient Pottery, i, 143 ; cf. the "idol with feet of clay" in Daniel, ii,

33 ; v, 4 ; Rawlinson, Mon., ii, 33 ; iii, 25, 28 ; Maspero, Dawn, 679 ; Meissner, ii, 72, 85, 95 ; Heuzey-de Sarzec, i, 238 ; Hall, CAH, i, 291 ; Delaporte, 184.
[8] Maspero, Dawn, 641, 679 ; Meissner, ii, 129, 131.
[9] Ball, PSBA, xiv, 160 ; Meissner, ii, 237 f.
[10] Maspero, Dawn, 642.
[11] Woolley, 1928, 125.
[12] Kautzsch, Apokryphen, i, 190.
[13] Meissner, ii, 73 f., 77 f.

THE TOWER OF BABEL

Associated with the Babylonian and Assyrian temples, but perhaps not with all the later ones (that of the goddess Nin-makh at Babylon had no tower)[1], was a very characteristic structure placed behind or on one side of the temple proper and separate from it, called the stage tower or ziqqurat (Sum. ekur ; Akk. ekurru), the "mountain house," or "link between heaven and earth" (Sum. dur-an-ki ; Akk. rikis-šamê u irsiti), which perhaps represented a mountain in miniature and was intended, probably at first by the Sumerians, to keep the god of the atmosphere, Ellil, and later other gods, on the level plain inhabited by their human subjects.[2]

The idea seems to have been to reach into heaven by the tall structure of the tower[3] : Tiglathpileser I (1115–1103) speaks of rearing two towers "up to heaven," like the tower of Babel.[4] The old Jewish idea was of three, or four, heavens, and the division into seven superimposed heavens associated with the seven planets occurs in the Bible only under Babylonian influences, e.g. a house is built on seven pillars [5], and in the Qabbalah (e.g. Sephir Jetzira, 9th century A.D.) the usual astrological significance of the seven heavens also appears.[6] The Elamites also built stage towers ; one at Susa, "of blue bricks, with summits of bronze," was destroyed by Ašurbanipal.[7]

The ascent of the stage tower, perhaps in some cases by a spiral ramp representing the revolution of the planets, was part of the ritual[8] : the sun god Šamaš is shown stepping up a series of seven stages, and a spiral ramp is both shown on old boundary stones and described by Herodotos [9] for the tower at Babylon. On most of the old towers, however, the stair from stage to stage was on inclined planes going round the tower.[10] The ladder of Mithra ($\kappa\lambda\hat{\iota}\mu\alpha\xi$ $\dot{\epsilon}\pi\tau\acute{\alpha}\pi\upsilon\lambda\sigma$) with steps of seven metals, each associated with a planet (lead with Saturn, tin with Venus, copper with Jupiter, iron with Mercury, mixed metal with Mars, silver with the Moon, gold with the Sun), which is described by Origen from a lost work of Celsus (c. A.D. 178)[11], and probably taken from a Persian original, would be more appropriate for this type of tower.[12]

The later towers, e.g. one at Samarra (9th century A.D.), had a spiral ramp like a screw[13], and possibly there were these two main types[14], the Sumerian step-tower and the Assyrian spiral ramp, which were both combined in the tower and upper part of the tower at Babel, which was in seven stages.[15]

[1] Handcock, 136.

[2] Jeremias, 41 ; ib., Ro., iii, 52 ; Sayce, TSBA, 1874, iii, part 1, 151 ; Rawlinson, Mon., ii, 543 f. ; Jastrow, Civ., 4, 32, 49, 123, 187, 270, 375 ; Meissner, i, 302, 310, 314 ; Maspero, Dawn, 624, 674 ; M. H. Farbridge, Studies in Biblical and Semitic Symbolism, 1923, 180 ; Lagrange, Religions Sémitiques, 2nd ed., 1905, 192 ; Meyer, Alt. I, ii, paras. 370, 408, 413 ; Woolley, 1928, 141 ; Dombart, in Reitzenstein, Das iranische Erlösungsmysterium, Bonn, 1921, 207.

[3] Jeremias, Der Alte Orient, 1929, xxvii, Heft 4, 9.

[4] Genesis, xi ; Berossos in Cory, Ancient Fragments, 1832, 34.

[5] In Proverbs, ix, 1 f., so understood by Reitzenstein, Iranische Erlösungsmysterium, 208.

[6] S. Karppe, Études sur les Origines du Zohar, 1901, 154 ; Schrader, Keilinschriften,

268 ; Jeremias, Handbuch, 42 ; Bousset, A. Rel., iv, 268 ; Woolley, 1928, 144.

[7] De Morgan, Délégation en Perse, iii, 33.

[8] Jeremias, 41, 43, 47, 327 ; Meyer, Alt., I, ii, paras. 370, 380 ; Handcock, 133 ; Meissner, i, 312.

[9] i, 181.

[10] Maspero, H. anc., 159 ; Hilprecht, Explor., 1903, 551 ; Meissner, i, 310.

[11] Origen, Contra Celsum, vi, 22 ; Keim, Celsus wahres Wort, Zürich, 1873, 84 ; Lobeck, Aglaophamus, Königsberg, 1829, 936 ; Eisler, Weltenmantel und Himmelszelt, Munich, 1910, 61, 209.

[12] Jeremias, 43, 47, 87, 89 ; A. Dieterich, Eine Mithrasliturgie, Leipzig, 1903, 186 ; Cook, Zeus, ii, 128.

[13] Layard, Nineveh and Babylon, 1867, 263 ; Jeremias, 44, Fig. 26.

[14] Maspero, Dawn, 628.

[15] Unger, RV, xiii, 254 ; ib., Z. f. Alttest. Wiss., 1927, xlv, 162, 171.

18

Koldewey's [1] idea that the tower was not stepped but consisted of a single stage decorated by coloured bands and provided with a triple stairway, the whole surmounted by a temple, is incorrect.[2]

The tower at Ur, originally built in 2350 B.C., was perhaps in two stages, as it was reconstructed by Nabonidus ; in its present state it has three stages, which are (1) painted black, (2) composed of bright red brick and (3) built of brilliant sky-blue bricks, respectively, from the base. It is oriented (like all other ziqqurats) with its corners to the cardinal points, the whole being surmounted by the shrine.[3]   The stage tower of Tell al-Uhaimir is supposed, from a seal cylinder found in it, to date to 3200 B.C.[4]   The tower at El-Hibba has two circular (unusual) stages ; that at Nippur (remains of Ur Nammu, 2278 B.C.) probably had three stages, or perhaps five, which were apparently square.[5]   A blue stone vase found in the very old (c. 3000 B.C.) tower at Bismaya (Adab) has a design of a tower with four stages.[6]   There is a representation of a four-stage tower (at Borsippa ?) on a 13th century B.C. boundary stone [7]; the Assur temple, built about 1140 B.C., had two four-stage square towers [8]; that at Nineveh, represented in a relief, had four or five stages [9]; and the best preserved ziqqurat, that at Khorsabad, has distinct traces of exterior colouring (red, black, white) of four stages.[10]   The temple of Nebo at Nimrûd was rectangular and probably had five or seven stages.[11]

Unger[12] quotes remains or representations of twelve towers in all, with two, three, four and five stages, and supposes that the one at Babel had seven stages, the eighth, mentioned by Herodotos[13], being the imaginary underground tower (kigallu) or reflection of that above ground. The stages were usually of different heights, diminishing upwards, although the four at Khorsabad were equal in height.[14]

Baumstark's idea[15] that the earliest towers had only two stages, afterwards increased to four, then seven, the number depending on the number of countries ruled over by the king, is incorrect, although later kings extended the number of stages of old towers. Nebuchadnezzar II (600 B.C.) raised the three stages of the tower at Borsippa or Babel to seven, "as it was in ancient times," and replaced the silver wainscoting and altar with gold.[16]

In an inscription of Gudea the temple of Ningirsu at Lagaš is called the "house of the seven tubuqati,"[17] and Jastrow[18] thinks that it had seven stages probably representing the planets.[19]   The word tubuqati has been translated

---

[1] Excavations at Babylon, 1914, 188, 195 ; ib., Tempel von Babylon und Borsippa, Deutsche Orient-Ges. in Bab., Leipzig, 1911, 58.
[2] King, H. Bab., 77 ; Unger, opp. cit.
[3] Woolley, The Times, 14 May, 1924 ; ib., Sumerians, 1928, 142, plates 23, 24 ; British Mus. Quarterly, i, 82 ; cf. Maspero, Dawn, 630 ; Langdon, Der Alte Orient, 1928, xxvi, 32.
[4] Ancient Egypt, 1930, 83.
[5] Handcock, 64, 133, 146 ; King, S. and A., 89 ; Hilprecht, Explor., 1903, 551 ; Meyer, Alt. I, ii, 477.
[6] Langdon, CAH, i, 390 ; Jastrow, Civ., 52 ; King, H. Bab., 81 ; L. H. Vincent, Canaan, Paris, 1907, 264.
[7] Jastrow, Civ., 376 ; King, H. Bab., 79, Fig. 30 ; Unger, Ebeling's RA, i, 422.
[8] Handcock, 142 ; Meissner, i, 309 ; W. Andrae, Anau-Adad-Tempel in Assur,

Deutsche Orient Ges., 1909 f., No. 10 ; Jastrow, Civ., 56 ; Delaporte, 346.
[9] Rawlinson, Mon., i, 314, Figs. VIIa and b ; Meissner, i, 311.
[10] Jastrow, Civ., 370, 376 ; Perrot-Chipiez, Chaldæa, i, 387.
[11] Rawlinson, Mon., i, 315.
[12] RV, xiii, 254 ; Z. f. Alttest. Wiss., 1927, xlv, 162.
[13] i, 181 f.
[14] Jastrow, Civ., 270, 376 ; Handcock, 138.
[15] PW, ii, 2676.
[16] Baumstark, op. cit. ; Unger, Z. f. Alttest. Wiss., 1927, xlv, 447 ; Jastrow, Civ., 30 ; Meissner, i, 306.
[17] Schrader, Keilinschriften, 616 ; Nouv. Fouilles, 28.
[18] Aspects of Religious Belief and Practice in Babylonia and Assyria, New York, 1911, 287.
[19] Cook, Zeus, ii, 128 ; cf. Jastrow, Civ., 371, 375—doubtful.

"divisions" [1] which is probably more correct than "spheres," [2] "lights," [3] "rulers," [4] or seven superimposed spaces or regions [5], but the usual significance is "planets."[6] Although separate planets were known at an early date, the idea of a group of *seven* planets is relatively late. It apparently arose in Babylonia, where the study of the stars and their supposed influence on human affairs goes back to a very early period, but only in the 10th–8th centuries B.C. in the period of Chaldæan domination.[7] The belief in the significance of the number seven [8] and the identification of planets with gods (at least as early as 2000 B.C.) are much older. Nebuchadnezzar II (600 B.C.) in the "India Office" text from Borsippa [9] says (text rearranged) : "I repaired and rebuilt the wonder of Borsippa, the temple of the seven spheres (?) of heaven and earth. In the midst of Borsippa I built the tower, the eternal house, the brilliance of which is increased by gold, silver and other metals and by glazed bricks. I rebuilt the tower, the stage tower of the temple, the base of which is covered with beaten gold, and the summit with copper, lead (not given by Ball) and stones." It lay in ruins in Alexander the Great's time, but was repaired again in the following century.[10]

What Sir H. Rawlinson actually found at Birs Nimrûd[11] was a first stage set in black bitumen, a third (?) stage of red half-burnt bricks set in crude red clay mixed with reeds, and "what seemed to be the sixth" stage of blue bricks, vitrified by great heat to a dark blue mass of slag. Four fragments, one fluxed, were sent to the British Museum[12], and when I examined them, looked like yellowish-red bricks and ordinary "blue" bricks respectively. Jensen[13] also regarded the black and red bricks as ordinary building materials, and the blue were thought to have been fluxed by an accidental, not intentional, fire.[14] Doubts as to the existence of coloured stages[15] are unreasonable in view of the remains of red, black and white colouring on the stages of the tower at Khorsabad[16] and the black basement, red bricks and blue tiles on the tower at Ur.[17]

H. Rawlinson was convinced that the tower at Borsippa had originally seven distinct rectangular stages one above the other, each having a definite colour, and also that these coloured stages represented the seven planets according to the system preserved in the descriptions in Arabic authors of the temples of the Sabians of Harran, who were descendants of the old Chaldæan starworshippers.[18] Rawlinson's reconstructions, however, varied, as can be seen from lists I–IV in the following table :

[1] Jeremias, 45 ; Jastrow, Civ., 30 ; Baumstark, PW, ii, 2676 ; King, H. Bab., 73.

[2] Loftus, Travels, 1857, 28.

[3] Lenormant, Prem. Civ., ii, 176.

[4] BMGB, 140.

[5] Schrader, Keil., 615.

[6] Hilprecht, Explor., 1903, 184, 232.

[7] Meissner, ii, 130, 404 ; Meyer, Alt. I, ii, para. 427 ; ib., Ursprung und Anfänge des Christentums, Stuttgart, 1921, ii, 55, 373 ; Riess, PW, ii, 1805 ; Boll, PW, vii, 2556 ; Boll, Bezold and Gundel, Sternglaube und Sterndeutung, Leipzig, 1931.

[8] Wundt, Völker-Psychologie, Leipzig, 1909, II, iii, 540 ; Farbridge, Symbolism, 114.

[9] Tr. by Oppert, in Wollheim, Die National-Literatur sämmtlicher Völker des

Orients, 1873, ii, 447 ; Ball, PSBA, 1887, x, 87 f., 362 f.

[10] Meissner, i, 315.

[11] BMGB, 34, 36, plate x.

[12] BMGB, 91 f.

[13] Die Kosmologie der Babylonier, Strasburg, 1890, 143.

[14] Loftus, Travels, 1857, 31 ; G. Rawlinson, Mon., ii, 548.

[15] Jastrow, Aspects, 287 ; ib., Civ., 370 f.

[16] Place, Ninive, i, 142 ; Jastrow, Civ., 370, 376 ; Perrot-Chipiez, Chaldæa, i, 387.

[17] Woolley, The Times, 14 May, 1924 ; cf. Langdon, Alte Orient, 1928, xxvi, 33.

[18] Hilprecht, Explor., 1903, 186 ; Rawlinson, Mon., ii, 548 ; Schrader, Keilinschriften, 616.

Stage	1	2	3	4	5	6	7
I. .	Black	Red-brown or orange	Red	Golden; gold-plated ?	Pale yellow or yellowish-white	Dark blue	Silver; silver-plated ?
II. .	Black	Bright red	Red	Yellow	Green	Blue	White
III. .	Black bitumen	Red brick	Pink	Yellow brick (perhaps gilded)	—	Blue vitri-fied clay	Green-grey
IV. .	Black bitumen	Orange	Blood-red	Gold-plated	Pale yellow	Blue vitri-fied brick	Silver-plated
		burnt bricks			burnt brick		
Planet	Saturn	Jupiter	Mars	Sun	Venus	Mercury	Moon
V. .	White	Black	Purple	Blue	Yellow or red	Silver-plated	Gold-plated
VI. .	Black	Green	Red	Golden yellow	Cornflower blue	Brown	Silver

I. H. Rawlinson, in Schrader, Keil., 616 ; G. Rawlinson, Mon., ii, 548.

II. Loftus, 28 ; for Birs Nimrûd.

III. Birch, Anct. Pottery, i, 135, from a lecture by H. Rawlinson at the Royal Institution.

IV. Ragozin, Chaldea, 282, cf. 280 ; traced back to H. Rawlinson.

V. Ekbatana (circular walls), from Herodotos, i, 98 ; cf. Polybios, x, 27 ; Rawlinson, Mon., iii, 362.

VI. Harran, from a Sabian account in Dimešqî ; Manuel de la cosmographie du moyen âge [of Dimešqî], tr. Mehren, Copenhagen, 1874, 41 f., 71 ; Chwolson, Die Ssabier und der Ssabismus, St. Petersburg, 1856, ii, 381, 658, 671, 828 ; Carra de Vaux, Les Penseurs de l'Islam, 1923, iv, 97.

Oppert [1] gave still another order and the "reconstructed" schemes of colours seem very speculative : the *ascertained facts* are as follows[2] :

Stage	1	2	3	4	5	6	7
Babylon	Clay colour	Yellow-green	—	—	—	—	Various colours
Borsippa	Black	—	Red	—	—	Blue	—
Khorsabad	White	Black	Red	White or blue	—	—	—

It is not certain that the red and blue stages at Borsippa were the third and sixth, and the interpretation of the white stage at Khorsabad as faded blue [3] is uncertain.

The two inner walls of Ekbatana (built in 607 B.C.), which were gold and silver plated [4], no doubt represented the sun and moon respectively, and the same holds for the walls of Nineveh, which were destroyed at this time but

[1] Schrader, Keilinschriften, 617.

[2] (1) Babylon : Geo. Smith, q. in Jeremias, 45, 86 ; ib., Ro., iii, 54 ; (2) Borsippa : H. Rawlinson, *opp. cit.* ; also Bouché Leclercq, L'Astrologie grecque, 1899, 41 ; Chwolson, ii, 840 ; Perrot-Chipiez, Chaldæa, i, 273 ; (3) Khorsabad : Place, Ninive, i, 137 f. ; ii, 79 f. ; iii, plates 36, 37.

[3] Berger, Maltechnik, 29.

[4] Herodotos, i, 98.

later rebuilt. The planetary associations of the remaining walls at Ekbatana are doubtful [1], but the following are suggested :

I[2]	Sun	Moon	Jupiter	Mercury	Mars	Saturn	Venus
	Gold	Silver	Yellow or orange	Blue	Purple or scarlet	Black	White

II[3]	Sun	Moon	Mars	Mercury	Jupiter	Venus	Saturn
	Golden	Silver	Red	Blue	Purple-red	Black	White

Bouché Leclercq says the series has been confused in the case of Borsippa, in one case being in the reverse order of the planets, but these must be doubtful, since Nebuchadnezzar altered the silver plating to gold.[4] He also provided Sin, the god of the *silver* moon, with a *blue* beard, and the sceptre and beard of tne king are often shown as blue in the pictures.[5]

The Persian poet Nizâmî (d. A.D. 1198) describes a seven times built temple of the Sassanid king Bahrâm Gûr (Vahram V, A.D. 420), in seven parts in seven colours, dedicated to the seven planets[6] :

Saturn	Jupiter	Mars	Sun	Venus	Mercury	Moon
Black	Sandal-wood colour (white)	Scarlet	Golden (green)	White (blue or varie-gated)	Azure-blue (greenish-yellow)	Green

Colours given in other authors are quoted in brackets. There is said to be no *direct* confirmation of the relation between colours and planets in the old Babylonian texts.[7]

THE ASSOCIATION OF METALS WITH PLANETS

The association of certain gods with specific metals goes back to the Sumerians.[8] Ellil of Nippur was "master of gold," a title afterwards given to Marduk [9], and Ea, the ancient god of Eridu, was also the god of gold and the protector of goldsmiths and metal-workers generally.[10] The fire-god Gibil afterwards assumed the duty of god of smiths, but the idea of a "divine smith" is not old Babylonian. In the old texts silver = Anu, gold = Ellil, copper (erû) = Ea.[11]

The "sympathetic" association of planets and stars with animals, plants, metals and gems[12], although ascribed by Hellenistic writers to the Egyptian

[1] Bouché Leclercq, L'Astrologie grecque, 1899, 41, 73.
[2] Jeremias, Handbuch, 1913, 86.
[3] Bousset, A. Rel., iv, 240 : last two interchanged.
[4] Boll, PW, vii, 2547 ; Bousset, A. Rel., iv, 238 f. ; Roscher, Ro., iii, 2529 ; Baumstark, PW, ii, 2676.
[5] Jeremias, 172, 177, 244.
[6] Lenormant, Magic, 227 ; Ritter, *Islam*, 1925, xv, 111 ; cf. Horovitz, *ib.*, 1919, ix,

159 f. ; Jeremias, 82 f., 84, 86 ; Thompson, Chem. Assyr., 98.
[7] Meissner, i, 312.
[8] Lenormant, Prem. Civ., i, 113, 121.
[9] Jeremias, 237.
[10] Maspero, Dawn, 653 ; Jeremias, 15, 28.
[11] Jeremias, 282, 69, 85.
[12] Meissner, ii, 130 f., 132—brief ; Berthelot, Intr., 73.

Hermes-Thoth, is perhaps really Babylonian in origin.[1] Ninib ruled over plants and stones and Ninurta, son of Ellil, "determined the fate of stones," whilst the plants "praised his name." The skull of Marduk was of silver and his sperm of gold.[2] In a text translated by Sayce [3] there is mention of stars of (1) the stone absia, (2) the stone of the bronze fish, (3) alabaster, (4) silver, (5) gold, (6) bronze ; in another text Saturn is called "husband of the queen of copper" (Ištar ?). A magic tablet from Nippur, of the Cassite Period (1600–1400) but probably based on older sources, "not to be shown to the uninitiated," contains sixty symbols with their divine implications.[4] "Silver is the great god" (the Moon ; Sin, or Anu of Heaven), "gold is Enmešarra" (the Sun : in other texts gold is associated with Ellil and the latter frequently is Šamaš, the Sun), "copper is Ea," "lead is Ninmah" (the Mother Goddess), [?], probably a metal, is Ninurta (Saturn), and a series of materials, perhaps other metals, are referred to other planets. "The copper kettledrum is Ninsar" (Nergal), "gypsum is the storm god"(Ninurta), "the torch is Gibil."

FOUNDATION DEPOSITS

The *old* texts are confirmed by the so-called "foundation deposits" of temples, in which plates of various metals and stones are buried as amulets in stone chests.[5] Mr. Sidney Smith, of the British Museum [6], states that "the custom was at least as early as 2000 B.C.," and connects it with the burying of copper in the deposits at Ur. These deposits are not confined to Babylonia or Assyria for the later period, but also occur in Egypt, Nubia and the Sudan.[7]

A stone chest found in 1854 by Place under a building of Sargon II (710 B.C.) at Khorsabad, contained five (or six) bars of material, the inscriptions showing that there were originally seven. One bar, the heaviest, of lead, was lost in transport :. the others had long inscriptions.[8] That the seven tablets were connected with the seven planets is regarded by Zimmern as doubtful, since other texts describe the burying of eight or nine tablets. The names of the materials of the bars given in the inscriptions were read as ḥurasi, kaspi, urudi or er, anaki, kasazatiri or abar, sipri or zakur, and gis-sin-gal, and were translated by Oppert as gold, silver, copper, lead, tin, marble and alabaster. The materials of the four bars analysed by Berthelot were (1) gold (8 cm. ; weight 167 gm. ; no notable alloy) ; (2) silver (11 cm. ; 435 gm. ; nearly pure, somewhat blackened) ; (3) bronze (19 cm. ; 952 gm., *not* "copper" : tin 10·04 per cent., copper 85·25 per cent.), and (4) pure white crystalline magnesite (10 cm. ; 185 gm.). The last had been called "antimony" and "oxidised tin" by the Assyriologists, and the reading "antimony" for aburu, originally proposed by Oppert, is maintained by R. C. Thompson.[9] Gis-sin-gal, now read giš-šir-gal, is now taken as "alabaster" ; Jeremias translated it "lapis," but the name of this is uknû.[10] Luckenbill gives lapis for Oppert's marble.

An early king of Assyria, Irišum (*c.* 2000 B.C.), buried tablets of silver, gold, bronze and lead ; Šalmaneser I added an iron tablet to these four materials.

[1] Hopfner, PW, xiii, 750 ; xiv, 316 ; Eisler, *Chem. Z.*, 1925, xlix, 578.

[2] Meissner, ii, 9, 133.

[3] TSBA, 1874, iii, part 1, 167–172, 177, 200, 207 ; cf. Meissner, ii, 132 ; Berthelot, Intr., 219 f.

[4] Langdon, Sumerian Liturgies and Psalms, Univ. of Pennsylvania Museum, *Publ. of Babylonian Section*, x, No. 4, Philadelphia, 1919, 334–339 ; Cook, Zeus, i, 626, 632 ; Bousset, *A. Rel.*, iv, 241 ; Maury, La Magie et l'Astrologie dans l'Antiquité et au Moyen Âge, 1860, 29 ; Hopfner, PW, xiii, 749.

[5] Perrot-Chipiez, Chaldæa, i, 319.

[6] Communication, 16 February, 1927.

[7] Budge, Mummy², 450 ; see p. 16.

[8] Place, Ninive, i, 63 ; Berthelot, Intr., 80, 219. Oppert, Expédition scientifique en Mésopotamie, Paris, 1859, ii, 343 f. ; Perrot-Chipiez, Chaldæa, i, 319 f. ; Ragozin, Chaldea, 116 ; Jeremias, 87 ; Zimmern, in Schrader, Keilinschriften, 624 ; Contenau, Manuel, iii, 1303 ; Luckenbill, ii, 57, 59.

[9] Chemistry of the Ancient Assyrians, 1925, 37.

[10] Zimmern, *op. cit.* ; L. W. King, Babylonian Magic and Sorcery, 1896, 60.

Tukulti-Urta I and Šalmaneser III also buried tablets. After Sargon II, Asarhaddon buried tablets of silver, gold, copper, lapis lazuli, marble, salamdu (a black stone), "wheat stone" (Fusuline limestone), elalu stone and white limestone.[1] Foundation deposits containing gold, silver and bronze (c. 2000 B.C.) were also found at Susa.[2] A deposit similar to that at Khorsabad was found under a large stone block weighing 4,500 kgm. in the old Ištar temple at Assur.[3] The three lead tablets with inscriptions had been strewn with glass beads, shells, pieces of lead, copper and iron. Over this layer was clay mortar, on which plates of silver and gold, with inscriptions, were laid, then a limestone block and a lead tablet. Inlaying walls with seven tablets according to the order of the planets is mentioned by Nonnos of Panopolis (4th century A.D.).[4]

The Babylonians, especially the Chaldæans and under their influence the later Persians, had idols of wood, stone and metals such as gold, silver, copper, bronze and iron, which have been considered (perhaps incorrectly) as representing planetary gods. The idea that images of the gods must be made of materials "sympathetic" to them (e.g. Hermes of ebony) occurs repeatedly in the Neo-Platonic authors, and probably came from a Chaldæan source.[5] In the account of Nebuchadnezzar's dream in Daniel [6], written under strong Iranian influences, the four kingdoms are represented by idols of gold, silver, copper and iron.[7]

POTTERY

The oldest Sumerian pottery is not older than 4000 B.C. [8], about which date the potter's wheel may have been invented in Elam.[9] It is possible that before pottery was made, wooden vessels or baskets of straw or wicker-work plastered with clay were in use.[10] The earliest pottery may have been sun-dried[11], but the pottery from the very old site of Adab (Bismaya) is both sun-dried and baked black[12], and the oldest (unglazed) vases at Warkâ (Uruk ; Erech) and Muqayyar (Ur) are baked and contain chopped straw.[13] Straw, sand, crushed pottery, etc., are added as dégraissants to make the clay less sticky and to diminish the tendency to crack when the vessel is dried or fired.[14] Somewhat later vases from the same sites are wheel-made and have a finer body, sometimes covered by the potter (pacharu) with a slight glaze (perhaps really a slip). It had been supposed[15] that the earliest pottery was hand, not wheel, made, but wheel-formed pots occur in the lowest strata at Fâra (Šuruppak) and Surghul (although those at Warkâ and Muqayyar are said to be hand-made), and so-called hand-made pottery is perhaps sometimes a copy of contemporary ware made on the wheel. Koldewey thought that in Babylonia "pottery and the potter's wheel were invented at the same time."[16] Painted hand-made pottery, however, occurs in the lowest layers (below the "Flood stratum") at Ur, dated about 4000 B.C.[17], and also at al-'Ubaid[18] with black

[1] Luckenbill, i, 14, 41, 63, 251 ; ii, 248.

[2] Contenau, Manuel, ii, 923.

[3] Andrae, Mitteil. der Deutschen Orient Ges., 1914, liv, 36, Fig. 8.

[4] Reitzenstein, Das iranische Erlösungsmysterium, 184.

[5] Jeremias, 86 f. ; Hopfner, PW, xiii, 748 f., 767 f.

[6] v, 4, 23.

[7] Meyer, Ursprung und Anfänge des Christentums, 1921, ii, 179, 184, 189, 333 ; cf. Berthelot, Coll., ii, 167 ; Josephus, Antiq., X, x, 4.

[8] Langdon, Excav. at Kish, 1924, 6.

[9] Harrison, Pots and Pans, 38 ; Smith, EHA, 55 ; Frankfort, Studies in Early Pottery, 1924, i, 7.

[10] Harrison, 24, 28, 38 ; Meyer, Alt. I, ii, 447 ; Lippert, Kulturgeschichte der Menschheit, Leipzig and Stuttgart, 1886–7, i, 332 f.

[11] Handcock, 333 f. ; Meissner, i, 233.

[12] Banks, Bismya, 175, 261, 349.

[13] Rawlinson, Mon., i, 91 ; Perrot-Chipiez, Chaldæa, ii, 298.

[14] Frankfort, Pottery, 1924, i, 5.

[15] Meissner, i, 233 ; Perrot-Chipiez, ii, 298 f.

[16] Jastrow, Civ., 380, pl. 41 ; Koldewey, Excav. at Babylon, 253 ; Maspero, Dawn, 747 ; Meissner, i, 233 ; Perrot-Chipiez, ii, 299.

[17] Mallowan, Discovery, 1930, xi, 256.

[18] Hall and Woolley, Ur Excav., 1927, i, 151 f., 161 f., 165.

or brownish-red iron oxide colouring. Among old pottery at Tello is a grey "tournassés à la main" and a red "fabriqués au tour."[1]

Decoration of pottery began very early. At Tello in the oldest layers there was black (smoked ?) pottery with incised lines filled with red paste and some small amount of painted ware, and in the oldest layers at Nippur, vases and pottery painted with green and yellow stripes occur [2], although the very oldest Sumerian pottery has only black, white and red colours, not blue or green, and even the later Babylonians did not paint the ware before firing.[3] The later pottery at al-'Ubaid (3000 B.C.) has a blue colour.[4] Painted clay dogs (blue and green) were found at Kiš and painted pots, with coarser clay ware, sometimes with incised decorations, also baked clay ear-studs, at al-'Ubaid (near Ur) : these objects are of c. 3500 B.C.[5] Other painted dogs found at Nineveh have blue, red and black colours (now very faint) laid on in plaster.[6] "Artificial flowers" in painted cream-coloured clay (red, white and black) on conical clay "stems" were found at al-'Ubaid (3500 B.C.).[7] These were probably for wall decorations.

Vases painted red and black, like those of Susa II (q.v.) were found in old layers (before Narâm-Sin, 2550 B.C.) at Tello, with a curious circle of small pots like some found in Cyprus.[8] The old pottery at al-'Ubaid is made from a creamy-white clay, fired to a dull green, dark grey or greenish-grey, and contains small red lumps, apparently of crushed pottery. No slip is used. The greenish colour is probably due to the admixture of chalk with a clay containing iron. The black or smoky-grey ware has been fired in a smoky and reducing atmosphere.[9]

All the pottery in the royal tombs at Ur was wheel-made and unpainted, with one exception, which was painted in red and black on a buff slip in a way typical of Susa II pottery. Painted ware, characteristic of the earlier period at Jemdet Nasr and al-'Ubaid[10] was not found, and even handles, which were used in the early period, are absent later until the Neo-Babylonian and Persian Periods. The colour of the ware at Ur is various (grey, red, pink, brown, only one black), and the pots are sometimes covered with a red "hæmatitic" layer, sometimes matt, but usually rag-burnished. No terra-cotta figures were found. Some pots have incised decorations. The shapes are very various, including flat dishes, conical and cylindrical jars, bottle and alabastron shapes, spouted jugs, egg-cups with feet, libation vases, tall jars with bases or pointed below, round-bellied pots, etc.[11] Babylonian and Assyrian pottery is difficult to date, since many forms were used in all periods : the unglazed pale yellow saucers, plates and vases, and the terra-cotta figures found on the sites are of various periods.[12]

Pottery was more popular than glass when the latter was known, and was used also for objects which we should construct in wood or stone, such as tubs, chests, pipes, ovens, coffins, spindles, etc. Sometimes models in pottery were afterwards cut in stone. Votive figures for the common people and for burying in foundations, and drain-pipes were also made in pottery.[13] The Assyrian

[1] Nouv. Fouilles, 35 ; cf. Meissner, i, plate-fig. 95.
[2] Handcock, 282 ; Jastrow, Civ., 369 ; Nouv. Fouilles, 35.
[3] Hall, CAH, i, 587.
[4] Hall and Woolley, Ur Excav., 1927, i, pl. 49.
[5] Langdon, Excav. at Kish, i, 91 ; Woolley, 1928, 13 f.
[6] Birch, Pottery, i, 125 ; Rawlinson, Mon., i, 342 ; BMGB, 221, 239.
[7] Hall and Woolley, Ur Excav., 1927, i, 17, 81, 117 ; pls. 12, 30, 34.

[8] Nouv. Fouilles, 145, 310.
[9] Hall and Woolley, Ur Excav., 1927, i, 162 ; Frankfort, Pottery, 1927, ii, 64.
[10] BMGB, 57, 81 f.
[11] Ur Excav., 1934, ii, 387 f. ; pls. 251–267.
[12] Birch, Pottery, i, 119 f., 124 f., Fig. 77 ; Meyer, Alt. I, ii, 447, with bone implements ; Handcock, 325 ; Meissner, i, 234 ; Hall, CAH, i, 587.
[13] Rawlinson, Mon., i, 89 f. ; Meissner, i, 235.

unglazed pottery was rarely painted, although there are some red and black geometrical designs in the very early period. The Assyrian pottery is finer in paste, brighter in colour and used in thinner masses than the Egyptian. The Babylonian is similar to the Assyrian ; the clay was light red, sometimes greenish-yellow, and unpainted.[1]

The remains of an old potter's kiln (*c.* 2300 B.C.), similar to the modern kilns used in the East, were found at Nippur. It was a chamber 4 m. long, 2·13 m. wide and 1·2 m. high, with nine arches and a horizontal flue closed except at one end, the other end of the kiln being open. The pots were supported in it on three-cornered clay stands.[2]

The potter also made and burnt the clay writing tablets, prisms and cylinders, with the clay envelopes for the tablets. These were usually unglazed, but some are covered with a vitreous silicate glaze or white coating. The colour of the clay varies from pale yellow or pink, through a bright brown to a very dark brown, almost black.[3]

Pottery hands clasping battens were used as architectural ornaments by Ašurnasirpal II (885–860), and terra-cotta figures of gods, priests, demons, etc., were made at all periods.[4] Besides ordinary clay, a "porcelain-like paste" was used for vessels, beads and similar things.[5]

A curious form of decoration used from the Sumerian Period consisted of terra-cotta cones, sometimes with inscriptions, set in patterns in walls with the circular bases facing outwards ; some are dipped in red and black colour, some are white and composed of limestone and marble (?), and sometimes with a rim round the edge filled with copper. There were also cups fitted with the mouths facing outwards.[6]

GLAZED OBJECTS

Glazed beads occur very sparingly in the Predynastic Period at Ur with decorative blue glaze and more commonly about 2400 B.C. Vases of glazed frit appear as a new technique about 2000 B.C. The usual household vessels were of unglazed pottery, stone and copper.[7]

The decoration of the earliest pottery consists of incised patterns and glazing of pottery begins later.[8] Coloured and glazed pottery appears in Assyria in the 15th–12th centuries. The ware is very thin ("egg-shell" pottery) and is brightly coloured. Although it has been suggested [9] that this pottery was Minoan, perhaps based on Egyptian originals, this seems improbable in view of the great development of glazing in Assyria. The art passed to Persia on the decline of Assyria and influenced the art of the Caucasus and Scythia, perhaps ultimately even Siberia and China : it died out in later Assyria and the pottery of Sargon II is coarse and plain buff ware.[10] An Assyrian pottery furnace of the 8th–7th centuries B.C. is figured by Andrae.[11] The composition

[1] Birch, i, 105, 130, 144 ; Handcock, 282.

[2] Hilprecht, Ausgrabungen im Bêl-Tempel zu Nippur, Leipzig, 1903, 21 f. ; ib., Excavations in Assyria and Babylonia, Philadelphia, 1904, 489 f. ; Meissner, i, 233 ; Figs. 55, 56, plate-fig. 96.

[3] Sayce, B. and A. Life, 51, 138 ; Birch, i, 113, 115 ; Breasted-Jones, 1927, 96.

[4] BMGB, 187 f.

[5] Meissner, i, 234 ; BMGB, 237.

[6] Loftus, Travels, 1857, 187 f., 2 Figs. 191 ; Handcock, 156 ; Hall and Woolley, Ur Excav., 1927, i, 49 ; Sayce, B. and A. Life, 91 ; Ragozin, Chaldea, 1889, 88 ; Langdon, *Alte Orient*, 1928, xxvi, 35 ;

Heuzey-de Sarzec, 53 ; Meissner, ii, 343, plate-figs. 49, 51, 52 ; Jordan, etc., *Abh. Preuss. Akad.*, *Phil.-Hist. Kl.*, 1931, 1932.

[7] Woolley, 1928, 160, 162 ; ib., Ur Excav., 1934, ii, 121, 366 f., pls. 87–90 ; Thompson, Chemistry of the Anct. Assyr., 12.

[8] Meissner, i, 234.

[9] Hall, CAH, ii, 430 ; Smith, ib., iii, 110.

[10] Hall, CAH, iii, 332 ; Smith, ib., 110 ; BMGB, 82 ; Thompson, Chem. Anct. Assyr., 12 ; Contenau, Manuel, iii, 1322 f.

[11] Farbige Keramik aus Assur, 1923, 4 f. ; faulty English translation, Coloured Ceramics from Assur, 1925, 4, Fig. 37.

of the Assyrian glazes still awaits chemical investigation, but according to Rathgen the white probably contains bone-ash, the black manganese, the blue and green cupric oxide, and the yellow lead antimoniate. About 1300 B.C. black and white were most commonly used for pots, and there are two sherds with the name of Adadnirari I (1310–1281) the yellowish glaze of which is the oldest tin glaze known : it is "known without chemical investigation that it is a tin glaze" (!) since it was applied to potter's clay.[1] The first really satisfactory surface glaze for pottery was the lead glaze, used in Mesopotamia at least as early as 600 B.C.[2] An early Assyrian vase [3] has the neck light blue and the body painted in equal stripes of white, yellow and dark blue, the latter having "now become a sulphur yellow" (by decomposition ?).

Small vases with a white, yellow or blue glaze, sometimes with a blue edge, occur in Babylon in the old Cassite Period (c. 1750 B.C.), when they are also made of coarse frit.[4] Vessels lined with a coarse blue glaze, on brick clay except in one or two instances when fine white clay was used, were found on Assyrian sites. The prevalent colour is a fine bright blue, verging to green when the surface is slightly decomposed. Other fragments are pale lilac or have a yellow pattern on a blue ground. In some cases the ground was white with brown and purple stripes. Although not analysed, the glaze was probably a vitreous silicate coloured with metallic oxides, principally of copper, the brown being iron silicate. Glazed pottery of the Assyrian Period is painted in blue, yellow and white, occasionally in brown, purple and lilac. The colours were probably metallic oxides, covered with a vitreous silicate glaze.[5]

Great numbers of blue-glazed coffins of friable frit, found at Nippur by Layard, with pottery, broken glass bottles, jars, etc., and several highly glazed dishes, probably belong to the Parthian Period [6] ; they are in the British Museum [7] and were used in the period 200 B.C.–A.D. 200. Earlier coffins of similar shape at Kiš were of sun-baked clay with an imperfect blue glaze.[8] The blue-glazed ware is much inferior to the corresponding Egyptian, being coarse and dull, with lack of cohesion between the body and the glaze, the latter being unsuitable for earthenware.[9] A fragment from Quyûnjiq with the name of Asarhaddon (681–669) has a resemblance to some Mycenæan fragments.[10] There is "gilded terra cotta" of uncertain date in the British Museum[11], and vases with thick blue and yellow "faience" glaze, of uncertain date, probably native, were found at Birs Nimrûd (near Môsul).[12] Painted and glazed earthenware knobs and plaques were used in decorating the buildings of Ašurnasirpal II (885–860) at Nimrûd.[13]

GLAZED BRICKS

Glazed coloured bricks were largely used in Assyria in the period 900–600 B.C. The colours were usually pale, but occasionally possess some brilliance. Pale blue, olive green, dull yellow and white are common, brown and black less so, and red comparatively rare. The ground was generally blue, in imitation of lapis lazuli.[14] Many bricks of Nebuchadnezzar's time from Babylon are covered with thick glaze. The colours, which preserve their original brightness,

[1] Andrae, Ceramics, Engl. tr., 9.
[2] Harrison, Pots and Pans, 53.
[3] Andrae, Engl. tr., 41 c.
[4] Koldewey, Excavations, 248.
[5] Birch, Pottery, i, 130, 144 ; Rawlinson, Mon., i, 365, 389 ; ii, 588 f.
[6] Layard, Nineveh and Babylon, 1867, 320 ; Birch, i, 149 ; Loftus, 1857, 203— "late period."
[7] BMGB, 78.
[8] Langdon, Excav. at Kish, 89.

[9] Birch, i, 105, 126 ; Perrot-Chipiez, Chaldæa, ii, 305 ; Handcock, 283 ; Harrison, Pots and Pans, 53.
[10] Perrot-Chipiez, Chaldæa, ii, 301, 302, 304.
[11] BMGB, 236.
[12] Layard, Disc., 1853, 282 ; Perrot-Chipiez, Chaldæa, ii, 113, 306.
[13] BMGB, 214.
[14] Rawlinson, Mon., i, 331, 376 f. ; Meissner, i, 236.

are principally brilliant blue, red, deep yellow, white and black.[1] Blue-glazed
tiles were also found at Ur and Uruk.[2]

The art of making glazed bricks was apparently first developed in Assyria,
but was perhaps originally discovered in Babylonia or Elam [3] or Egypt.[4]
The late Babylonian glazed tiles and bricks were inspired by Assyrian and
Hittite models.[5] The Arabs transmitted the art to Europe through Persia.[6]
At the beginning of the 18th century A.D. the art in Persia was as prosperous
as when the great mosque at Ispahan (16th–17th century) was built [7], but it
died out in the 19th century. Similar glazed tiles are also made in India.[8]

The Babylonian glaze ("enamel") was firmer and harder than the Assyrian,
and the older bricks in Assyria are more thickly glazed.[9]   A fine example of
the art of glazed bricks is the doorway arch of Sargon II (722–705). The
ground is deep azure, the bodies of the lions are yellow, the manes, muscles
and feathers are blue, horns of bulls white, leaves green. There were also glazed
plaques.[10] The art reached its highest form under Nebuchadnezzar II (604–
561) ; the sacred way or processional street in Babylon past his palace (Ištar
Gate), along which the statues of the gods were carried, was lined with magnificent figured glazed tiles[11], and the three circular walls of Nebuchadnezzar's
palace are described (from Ktesias) by Diodoros Siculus[12] as decorated with
glazed ware representing animals. The front of the throne room was entirely
of glazed bricks.[13] Black, white, blue and red glazed bricks were found at
Borsippa and white, yellow, blue and green, with seven-ray stars, at Warkâ.[14]

The bricks were probably first shaped according to a model and lightly
burned. An outline of black glass thread was fixed on and filled with the
coloured paste, the whole being fired ; the designs were drawn on separate
bricks, placed together and covered with a layer of glaze, afterwards burnt.[15]

According to Thomas, who accompanied Oppert, a large mass of clay was
first taken and the design modelled in low relief. The mass was then cut up
into bricks, each being then painted separately, placed in the furnace and
baked. The glaze has sometimes run down the adjoining faces. Then the
baked bricks were restored to their original places in the design, a thin layer of
the finest mortar keeping them in place.[16] The bricks were probably lightly
baked before the colour was applied, and returned to the kiln afterwards,
and the melting-point of the glazes shows that a high temperature, such
as that attained in a muffle furnace, must have been used.[17]

Glazed bricks found by Layard at Borsippa were analysed by de la Beche
and Percy.[18] The principal colours were yellow (lead antimoniate—"Naples
yellow," with tin oxide and sodium silicate), blue (copper silicate with some

[1] Layard, 1867, 288 ; cf. Rawlinson,
Mon., ii, 525.
[2] Maspero, Dawn, 754.
[3] Meissner, i, 236 ; Perrot-Chipiez,
Chaldæa, i, 272, 281 ; Delaporte, 198.
[4] Hall, CAH, iii, 331 ; Semper, Der Stil,
i, 411 ; ii, 152.
[5] Delaporte, 195, 198, 214, Fig. 37.
[6] Jeremias, 295.
[7] Dussaud, Deschamps and Seyrig, La
Syrie antique et médiévale illustrée, 1931,
93.
[8] Perrot-Chipiez, Chaldæa, i, 287 ; ib.,
Art in Persia, 439 ; Jastrow, Civ., 370 ;
Birdwood, Industrial Arts in India, 1880,
ii, 140.
[9] Place, Ninive, ii, 253 ; Handcock, 279 ;
Perrot-Chipiez, Chaldæa, i, 281 ; ii, 247.
[10] Meissner, i, 236 ; Contenau, Manuel,
iii, 1240, 1242, 1323.
[11] Jastrow, Civ., 369 ; Handcock, 270,

280 ; Rogers, i, 318 ; Koldewey, Excavations, passim ; Contenau, Manuel, iii,
1358.
[12] ii, 8 ; Birch, i, 149 ; cf. Ezekiel,
xxiii, 14, 15.
[13] Meissner, i, 236.
[14] Jastrow, Civ., 32 ; Rawlinson, Mon., i,
363, 376 ; Birch, i, 128, 141.
[15] Meissner, i, 237 ; Jastrow, Civ., 369 ;
Handcock, 277 ; Maspero, Dawn, 754.
[16] Rawlinson, Mon., ii, 565.
[17] Birch, i, 128 ; Berger, Maltechnik,
32.
[18] Layard, Discoveries in the Ruins of
Nineveh and Babylon, 1853, 166 ; Jastrow,
Civ., 369 ; Perrot-Chipiez, Chaldæa, i, 280 ;
Birch, i, 128 ; Thompson, Chem. of
Assyrians, 10 ; Semper, Der Stil, i, 356,
123, 156 ; ii, 129 ; Rawlinson, Mon., i,
331, 363, 376, 382 ; ii, 129 ; Berger, Maltechnik, 32 ; Contenau, Manuel, iii, 1241.

lead as a flux), red (cuprous oxide) and white (stannic oxide). The glaze of the tiles of the processional street in Babylon contains copper, lead anti-moniate and antimony oxide, but no tin oxide or bone ash.[1]

The use of tin oxide in the glazes at Borsippa is probably the first true enamel technique, an enamel being produced by adding tin oxide to a lead glaze, in which the particles of oxide are insoluble. Coloured enamels are made by the use of compounds of metals such as chromium, cobalt, iron, copper, and others, which dissolve in the vitreous part of the mass. Knowledge of enamels survived in the Near East until brought to Spain by the Moors, and several kinds of tin oxide glazed ("enamelled") wares were developed in Europe from the 14th century A.D. (majolica, faience, delft).[2]

Rawlinson's account [3] of the Assyrian glazed bricks is very complete. The bricks are of two classes : (1) those merely patterned, and (2) those which contain designs representing men and animals. The colours chiefly used in the patterns are pale green, pale yellow, dark brown and white. Now and then an intense blue and a bright red occur, generally together, but on patterned walls the colours are usually pale and dull. On the second type of bricks the colours are almost exclusively pale yellow, greenish-blue, olive-green, white and brownish-black. There is no evidence that the colours have faded, although they may have become dull.[4] Red was used very sparingly, except in painting on sculptures, where it may have been intended to receive gilding. Olive-green was used for grounds, and occasionally other half-tints. A pale orange and a delicate lilac or pale purple were found at Khorsabad, while brown is far more common on the bricks than black. The glazed bricks from Nimrûd are $13\frac{1}{4}$ in. square and about $4\frac{1}{2}$ in. thick, glazed on one narrow edge only, the colours being blue, black, yellow, red and white. The glaze has scaled off, and the colours have become dull[5] : Layard says he found purple and violet, but he does not represent these colours.

The tints used in a single composition vary from three to five : combinations of five contain brown, green, blue, dark yellow and pale yellow ; or orange, lilac, white, yellow and olive-green. Combinations of four are more common, e.g. red, white, yellow and black ; or deep yellow, brown-black, white and pale yellow ; or lilac, yellow, white and green ; or yellow, blue, white and brown ; or yellow, blue, white, and olive-green. Sometimes three colours are used, with the ground generally of the colour used in the figures : yellow, blue and white on a blue ground ; the same colours on a yellow ground ; or white and yellow on a blue or olive-green ground. Intense red occurs on patterned bricks, balanced by intense blue and accompanied by a full brown and a clear white, in one case further accompanied by a pale green, which has a very good effect. A similar red appears on a design, accompanied by white, black and full yellow. Where lilac occurs, it is balanced by its complementary colour yellow, or by yellow and orange, and further accompanied by white. Bright hues are not usually placed one against the other, but are separated by narrow bands of white, or brown and white. Sometimes the intention is to be true to nature, but sometimes local colour was purposely neglected, the artist limiting himself to certain hues. The red was cuprous oxide, the yellow ferric oxide or lead antimoniate, the blue either cupric oxide or possibly cobalt. The scales of deep purplish-blue glaze from bricks from Babylon, examined by Salvétat and Lenormant, may have contained cobalt, but the quantity of material available was too small to make this certain.[6]   The light blue and

[1] Rathgen, quoted by Fester, Entwick-lung der chemischen Technik, 1923, 12.

[2] Harrison, Pots and Pans, 55.

[3] Mon., i, 376, with thirty-two references ;

W. D. Bancroft and R. L. Nugent, *J. Phys. Chem.*, 1929, xxxiii, 729.

[4] Birch, Pottery, i, 27.

[5] Birch, i, 127.

[6] Brongniart, Traité des arts céramiques, 1854, ii, 89 ; cf. Birch, i, 148.

yellow glazes contain no tin or lead, as did the Assyrian specimens analysed by Percy, but consist of aluminium and alkali silicates, coloured with copper in the blue and oxide of iron in the yellow glaze. The composition of the white glaze was not given. Babylonian work in cloisonné enamel, although very rare, is not unknown.[1]

GLASS

Glass paste beads were found sparingly with a vessel supposed to be of glass in the oldest remains (3500–3000) at Ur [2] and bluish paste beads at Tello [3], and the use of glazed frit and glass for beads was very early in Babylon.[4] A small glass bottle of the early Sumerian Period (before 3000 B.C.) was found at Nippur.[5] Glass beads occur fairly plentifully in III dyn. remains (2450 B.C.) at Ur.[6] No true glass was found at Mohenjo-daro, although there were vitreous pastes coloured with copper and iron, perhaps cobalt.[7] Colourless glass may have been early in Mesopotamia, since there are quartz pendants covered with colourless glaze supposed to date "from very early times to 2000 B.C."[8] A lump of blue glass of c. 2400 B.C., found by Hall at Abu Shahrain (Eridu), is in the British Museum.[9] Glass beads of 2000 or earlier occur in temple foundations at Assur.[10] A number of red and dark glass beads practically free from corrosion, found at Ur, are probably older than 1600[11], but blue glass found at al-'Ubaid is perhaps Arabic.[12] The Sumerian sudam and the Semitic elmesu have been supposed to mean glass[13]; the Assyrian-Babylonian name for common glass is sirsu or dušû (Syriac sištha).[14]

Babylonian and Assyrian glass has been considered as imported Syrian or Egyptian[15], or at least of doubtful origin.[16] There is now no doubt that glass was made from an early period both in Babylonia and Assyria.

Glass objects of uncertain date abound in remains of Assyrian palaces, but there is little or no glass in most of the *oldest* Babylonian tombs.[17] Small glass bottles, which occur in "tolerable abundance," were probably intended for the toilet, and the Babylonian mounds are covered with fragments of [recent] glass.[18]

The famous oval glass vase with the name of Sargon II (722–705) now in the British Museum was found by Layard at Nimrûd, but was lost in Bombay and then found later in Devonshire.[19] It has been incorrectly described as transparent[20], but is really of opaque greenish glass; it is $3\frac{1}{4}$ in. high, with the king's name and the figure of a lion on the neck, and is of Assyrian, not Egyptian, manufacture.[21] It has either been scooped out of a solid piece of

[1] Meissner, i, 270.
[2] Langdon, *Der Alte Orient*, 1928, xxvi, 41; Woolley, Ur Excav., 1934, ii, 366.
[3] Heuzey-de Sarzec, 67.
[4] Koldewey, Excavations, 246.
[5] Peters, Excavations at Nippur, 1897, ii, 160, 374; E. Bibl., ii, 1737.
[6] Morey, *Discovery*, 1930, xi, 61; Woolley, Ur Excav., 1934, ii, 366—not very common.
[7] Marshall, Mohenjo-Daro, 1931, ii, 469, 574, 582.
[8] Beck, in Brunton, Qau and Badari II, 1928, 25.
[9] Hall, Civ. of Greece in the Bronze Age, 1928, 71; Glanville, JEA, 1928, xiv, 189.
[10] Feldhaus - Klinckostroem, *Geschichtsblätter*, i, 66.
[11] Beck, in Brunton, Qau and Badari II, 1928, 25.
[12] Hall and Woolley, Ur Excav., 1927, i, 53.

[13] Sayce, Rel., 246, 490.
[14] Thompson, Chem. Assyr., 26, 67.
[15] Deville, Histoire de l'art de la verrerie dans l'antiquité, 1873, 15, 17; Kisa, Glas im Altertum, 102, 104; Hall, Civ. of Greece in the Bronze Age, 1928, 71.
[16] Bucher, iii, 268; Koldewey, Excavations, 255—glass of 1500 at Babylon is "not necessarily Egyptian."
[17] Layard, Discoveries, 1853, 196, 672, 674; Perrot-Chipiez, Chaldæa, ii, 306; Handcock, 330; Sayce, B. and A. Life, 98.
[18] Rawlinson, Mon. i, 574; ii, 566, 569.
[19] Layard, Discoveries, 1853, 196; BMGB, 196; Perrot-Chipiez, Chaldæa, ii, 306; Kisa, 102; the object shown by Meissner, i, 235, is of stone.
[20] Layard, 196; Perrot-Chipiez, ii, 306; Rawlinson, Mon., i, 390.
[21] BMGB, 196; Smith, EHA, 396, correcting Kisa, 102.

glass [1] or more probably moulded over a clay form, and is not blown.[2] Opaque glass objects, believed to be much older than the bottle, were found with it [3], and round and flat beads, of uncertain age, have frequently been found. A "cylinder or tube" of glass found at Nineveh (Quyûnjiq) is probably of 705–626 B.C.[4] Two entire glass bowls, the larger 5 inches in diameter and $2\frac{3}{4}$ inches deep, the other 4 inches in diameter and $2\frac{1}{2}$ inches deep, found at Nimrûd and probably of Sennacherib's period (705–681), are covered with iridescent pearly scales, due to decomposition.[5] With the bowls was the so-called "rock crystal lens" (not glass), plano-convex and supposed to be "the earliest specimen of a magnifying and burning glass," used for reading very small tablets, but probably (like the fragments of blue opaque glass found with it) either part of some decorative work, such as inlay in wood or ivory, or a wall decoration, or a button.[6]

A fragment of a glass vase from Nimrûd (700 B.C. ?), which is sealing-wax red inside and coated outside with green copper patina, is coloured with cuprous oxide.[7] Glass bottles of various dates were found by Layard at Babel and Nimrûd with numerous bricks of Nebuchadnezzar (604–561), and by Loftus at Nippur and Warkâ. Some are as late as the 6th–7th or even 11th centuries A.D.[8] Some of the bottles are coloured, others are ribbed and otherwise ornamented ; those found at Babel, some of the clear ones probably Seleucid-Parthian, were accompanied by vases of various sizes and forms, sometimes with a rich blue glaze.[9]

Loftus[10] found at Warkâ (Uruk) "pottery, vitrified and inscribed bricks, scoria, and glass in abundance on the surface of the ruins," along with "several large mounds, covered with black slag and scoria, like the refuse of a glass factory."

Glass lachrymatories, found in or near "slipper" clay coffins at Warkâ are probably late (Parthian and Sassanian). Loftus also found seven different forms of fragile coloured glass bottles, probably also late, since they occur with Parthian circular copper coins having two projections.[11] The Babylonian and Assyrian bottles are small, usually coloured, ribbed and often mis-shapen.[12] A pottery vase of the 7th–6th centuries B.C. found at Ur is roughly painted with a design in yellow, green and black under the glaze, and is made in imitation of glass. Similar vases, of the same date, occur in Greece, Egypt and Sicily and the place of origin is not known.[13]

At Arban, on the Khabur, Layard found pottery and glass ; some highly glazed earthenware had become iridescent by corrosion ; there were funeral urns of highly glazed blue pottery. With these objects were Egyptian scarabs (some XVIII dyn.) and a small Chinese bottle of doubtful date, etc. There was a considerable Jewish settlement in the cities of the Khabur until the end of the Arab Period.[14] Assyrian glass bracelets and finger rings and an Assyrian necklace of light blue glass beads, square and flat with horizontal flutings, are known.[15]

The "great emerald," 6 ft. long and $4\frac{1}{2}$ ft. broad, said by Pliny[16] to have been

[1] Perrot-Chipiez, ii, 306.
[2] Kisa, 102.
[3] Perrot-Chipiez, ii, 306 ; Meissner, i, 235.
[4] Layard, Discoveries, 1853, 596 ; Perrot-Chipiez, ii, 307 ; Rawlinson, Mon., i, 391, 573 ; in B. Mus., also glass finger rings.
[5] Layard, Nineveh and Babylon, 1867, 65 ; Smith, TSBA, 1874, iii, 463.
[6] Layard, Discoveries, 1853, 197, 674 ; ib., Nineveh and Babylon, 1867, 65 ; Smith, TSBA, 1874, iii, 463 ; Perrot-Chipiez, ii, 308 ; Sayce, B. and A. Life, 51 ; BMGB, 196 ; Meissner, i, 269 ; ii, 309.

[7] Percy, in Layard, Discoveries, 1853, 672 ; Rawlinson, Mon., i, 390.
[8] Layard, Discoveries, 1853, 503 ; BMGB, 196 ; Meissner, i, 235.
[9] Layard, 1867, 285 ; Koldewey, Excav., 255.
[10] Travels, 166 f., 185.
[11] Loftus, 203, 211 f.
[12] Rawlinson, Mon., i, 390 ; ii, 569, Fig.
[13] British Mus. Quarterly, i, 58.
[14] Nineveh and Babylon, 1867, 122, 125.
[15] Rawlinson, Mon., i, 573.
[16] xxxvii, 5.

sent by a king of Babylon to an Egyptian pharaoh, may have been a plate of green glass [1], or (more probably) glazed ware or native malachite.[2]

ANALYSES OF GLASSES

About two-thirds of the "lapis lazuli" found by Peters or Hilprecht in a temple of the Cassite Period (1746–1171) at Nippur, probably 14th century B.C., including votive offerings in the form of hammers, rings and tablets, and also stocks in bars weighing about 1 mina (2 lb.), consisted of an artificial blue glass coloured with copper and cobalt.[3] Darmstaedter's analysis of the material of three dark blue pieces of oval section, 3 cm. long and wide and 2 cm. high (perhaps from the handle of a vase), covered with a yellowish crust containing magnesium and aluminium, differs from analyses by Neumann [4] of blue artificial lapis lazuli and turquoise, also found at Nippur, of 1400 B.C. Peters [5] thought the cobalt ore came from China.

Later glass (c. 250 B.C.) from Nippur analysed by Neumann [6] was all transparent, well melted and almost without seed, and the fragments were those of large vessels, which Neumann thinks had been made with the blow-pipe, in contradistinction to older vessels formed over a core of clay. The blowpipe is not otherwise known before 20 B.C. [7], and Neumann's assumption is not accepted by other experts.

GLASSES FROM NIPPUR[8]

Cassite, 14th century B.C.			Babylonian, 250 B.C.					
Artificial :		D.	No. 89	No. 90	No. 91	No. 92	No. 93	No. 94
lapis	tur-quoise		Deep rose	Faint rose	Dark green	Pale green	Dark blue	Pale blue
SiO$_2$ 65·03	64·41	57·78	71·14	69·82	63·10	61·18	64·41	65·38
CaO 5·65	6·19	3·04	5·26	5·79	7·66	7·10	7·53	6·10
MgO 2·52	5·59	—	5·40	4·09	3·42	4·92	4·59	5·50
PbO 0·19	0·00	15·83	—	—	—	—	—	—
Al$_2$O$_3$ 2·13	1·52⎫	2·92	2·48	1·40	2·75	2·06	2·58	1·50
Fe$_2$O$_3$ 0·97	1·36⎭		0·72	1·80	1·76	2·15	1·44	1·54
Mn$_2$O$_3$ 0·65	0·00	—	1·04	0·41	5·92	4·37	2·75	0·99
CuO 1·94	2·60	1·19	0·55	0·36	1·02	0·52	1·20	1·48
CoO 0·93	0·00	0·42	—	—	—	—	—	—
K$_2$O 1·68	2·37⎫	16·98	1·30	2·18	3·18	2·88	2·08	2·48
Na$_2$O 17·37	13·98⎭		10·81	13·51	9·35	12·90	12·75	13·55
SnO$_2$ 0·00	0·32	—	—	—	—	—	—	—
SO$_3$ 1·70	1·28	—	0·87	0·96	1·52	1·67	1·00	1·19
100·76	99·62	98·16	99·57	100·32	99·68	99·75	100·33	99·71

The sulphate content partly represents sulphide and, with the fairly high iron and aluminium oxides, is peculiar to ancient glass. The low alkali content of most of the glasses indicates a noteworthy technical achievement for the time, since such glasses are not easily fusible. All six later glasses, even the faint rose (No. 90), contained both copper and manganese, the rose tints

[1] Rawlinson, Mon., ii, 570.
[2] Kisa, 101.
[3] Peters, Nippur, 1897, ii, 134, 240, 374; Darmstaedter, in Ruska, Studien zur Gesch. d. Chemie, 1927, 5 f.
[4] Z. angew. Chem., 1929, xlii, 835.
[5] Nippur, ii, 134.
[6] Z. angew. Chem., 1928, xli, 203; J. Soc. Glass Technol., 1930, xiv, 104 Abstr.
[7] See Egypt, p. 128.
[8] Analyses by Neumann; Darmstaedter's analysis is given in the fourth column.

being traced to the latter. The manganese content was unusually high in the green glasses, the colour of which was due to copper or manganese oxide (blue) in conjunction with alkali sulphide (yellow).

Babylonian glass of about 100 B.C.[1] had the normal composition. Late (9th century A.D.) Mesopotamian glass from Samarra[2] is well made and in a very good state of preservation. Samarra (which has the remains of a ziqqurat or stage tower) lies north of Bagdad on the left bank of the Tigris ; the ruins excavated were of an Abbâsid town inhabited only from A.D. 838–883.[3] The silica corresponds with that in Roman or Rhenish glass, but the alkali is lower and much smaller than that in Egyptian glass. The lime, magnesia and (with one exception) the alumina contents are high. The composition of the emerald-green glass shows that the colour is due to copper (blue) and alkali sulphide (yellow), the latter calculated as SO_3.

MESOPOTAMIAN GLASSES FROM SAMARRA, NINTH CENTURY A.D.

	No. 82 Window glass	No. 83 Dark blue plate	No. 84 Emerald-green window glass	No. 85 Colourless small dish	No. 86 White hollow glass
SiO_2 .	68·48	66·93	65·86	67·22	67·44
CaO .	5·71	3·62	5·95	5·84	4·80
MgO .	5·28	5·42	4·55	5·64	5·64
Al_2O_3	0·70	2·08	2·16	1·80	2·98
FeO .	0·91	1·44	1·19	1·03	0·51
Mn_2O_3	—	0·34	1·09	1·07	0·73
K_2O .	2·83	2·62	2·70	2·68	1·93
Na_2O .	14·95	15·26	12·84	13·28	13·94
CuO .	—	1·76	2·66	—	—
PbO .	0·95	—	—	0·82	1·01
SO_3 .	0·54	0·52	1·72	0·39	0·84
	100·35	99·99	100·72	99·77	99·82

GEMS AND MINERALS

The gem-stone (Sumerian šubû or zadimmu ; Akkadian sasînu), highly prized by the Babylonians and Assyrians as in the case of all Eastern nations, was worn in jewellery and known in several varieties.[4] The name enâte (lit. "eyes") is often used for jewels in Assyrian, as well as šubû, uknû (or ugnû) and elmesu.[5]

Precious stones were used and appreciated by the Sumerians, who had many varieties now hardly distinguishable, and the cutting of gems was a very early art in Mesopotamia.[6] Babylonian trade collected gems from various lands : Susiana, Arabia, the Pamirs, Further Asia, Egypt and India, which (as Ktesias says) was a great source of precious stones.[7] Agates, beryls and sards came from the bed of the Choaspes, amethysts from near Petra, and alabaster from near Damascus ; jasper is found near Zenovia on the Euphrates.[8]

[1] Rathgen, Chem. Zeit., 1921, 1101.
[2] Neumann, Z. angew. Chem., 1927, xxxiv, 963.
[3] Herzfeld, Die Ausgrabungen von Samarra, 1923, i, p. vii ; vols. ii and iii, 1927, deal only with wall decoration and painting, without technical information ; vol. iv, 1928, by C. J. Lamm, deals with the glass, with coloured plates.

[4] Schrader, Keilinschriften, 648 ; Meissner, i, 269, 272, 350.
[5] Muss-Arnolt, Concise Dict. of Assyr., Berlin, 1905, 47, 73, 1000.
[6] Schrader, RL, 149, quoting Hommel.
[7] Garbe, Die indischen Mineralien, Leipzig, 1882 ; Scheil, Revue d'Assyriologie, 1918, xv, 115.
[8] Rawlinson, Mon., ii, 488 ; iii, 162.

In the sand left as a deposit by "the Flood" at Ur were two beads of amazonite, a green stone for which the nearest known source is said to be in the Nilgiri Hills of Southern India, or the mountains beyond Lake Baikal, indicating caravan trade across a thousand miles of mountain and desert in the antediluvian period.[1] Jewellery in the earliest Sumerian Period included necklaces of amethyst, coral, lapis lazuli, mother-of-pearl and agate ; there is also an axe-head of agate slightly later than Gudea's time (2450 B.C.). Jewellery was also much used in the Assyrian Period and the later forms had astrological symbols[2] : it probably had a magic and religious rather than a decorative significance.

Bead necklaces at al-'Ubaid (3500 B.C.) contain roughly chipped rock-crystal, carnelian, lapis lazuli, jasper, garnet, chalcedony, obsidian and shell[3] ; those at Kiš, agate, carnelian and lapis lazuli.[4] Bead shapes and materials at Ur are various.[5] In the Predynastic Period only four materials are regularly used for beads : gold, silver, lapis lazuli and carnelian ; wood occurs occasionally, glazed frit is not common and glass paste is very rare. In the Sargonid Period the range of materials is much greater : agate is common, sard sometimes occurs ; steatite, hæmatite, marble, pebble and crystal occasionally ; glazed frit is not infrequent and there are two instances of glass paste. At the same time the four old materials were commonest. The beads, even the oldest when they have not suffered by corrosion, have a very fine and brilliant polish. The technique of piercing, grinding and polishing is not clearly made out, but probably fine drills, flat grindstones, bow drills, rotating polishers and emery (found at Ur, but not on the cemetery site) were used. White inscriptions on red carnelian beads found at Ur and somewhat later at Kiš and (set in silver) at Tell Asmar in Irâq (2600 B.C.) were produced by applying sodium carbonate and roasting. They may be Indian work, since the bleached carnelian beads also occur at Mohenjo-daro and on Greek, Scythian, Parthian and Kushan sites in north-west India, and the technique is still used in India.[6]

Although the early use of pearls in Babylonia and Assyria has been questioned [7], a text of 2000 B.C. from Ur speaks of "fish-eyes," which are probably pearls.[8] Other texts mention "precious stones from the sea" and pearls were probably known and used for jewellery, e.g. ear-rings, at an early date, the source being (as at present) the Bahrain Islands in the Persian Gulf (first definitely mentioned by Nearchos, 326–325 B.C.). Shell, mother-of-pearl, real amber and ivory (šinni pîri) were also used.[9]

Inlays of shell and mother-of-pearl in slate occur in the very old temple at Kiš (before 3500 B.C.), and in the so-called "standard" from a grave of 3500 B.C. at Ur there is a panel of mosaic in shell and lapis lazuli, showing the Sumerian army with chariots, copper helmets, short-handled spears, axes, scimitars, daggers, etc. There is at Ur inlay of lapis lazuli and red stone (carnelian ?) in shell, shell inlay and shell carving, shell lamps, a mother-of-pearl knife handle (?), and other decorative use of shell. Lapis lazuli is extensively used,

[1] Woolley, Digging up the Past, 1930, 138.

[2] Handcock, 340, 346.

[3] Woolley, 1928, 13 f. ; Hall and Woolley, Ur Excav., 1927, i, 52.

[4] Langdon, Excav. at Kish, 1924, i, 89.

[5] Woolley, 1928, 37, 40, 43, 44 ; ib., Ur Excav., 1934, ii, 366 f., 373.

[6] Langdon, Der Alte Orient, 1928, xxvi, 64 ; Marshall, Mohenjo-Daro, 1931, 509, 515, 526, 583 ; Frankfort, Illustr. London News, 1932, clxxxi, 504, 510 ; ib., Tell Asmar, Khafaje and Khorsabad, Univ.

Chicago Orient. Publ., 1933, No. 16, 47 ; Woolley, Ur Excav., 1934, ii, 373, pls. 133–4.

[7] Ball, PSBA, 1887, x, 99.

[8] Wilson, Persian Gulf, Oxford, 1928, 5, 28 ; Muss-Arnolt, Dict. Assyr., 73.

[9] Meissner, i, 270, 351 f. ; Rawlinson, Mon., i, 372, 559, 573 ; ii, 15 ; Rommel, PW, xiv, 1682, 1686 ; Dakin, Pearls, Cambridge, 1913, 2 ; Lenz, Zoologie der alten Griechen und Römer, Gotha, 1856, 631 ; Heyd, Commerce du Levant, Leipzig, 1885–6, ii, 648 ; Benjamin of Tudela, Itinerary, tr. Adler, 1907, 63.

19

e.g. for inlay, beads, amulets, dice, whetstone, cups, mosaic and also for beards of bulls' heads, etc.[1]

Some stones were regarded as male (the larger and more brilliant) and others as female. Stones used as amulets were tied to different parts of the body by knotted cords of different lengths as protection against the dreaded demoness Labartu or Lamaštu, who is shown in a boat on one side of a bronze plaque.[2]

One of the earliest precious stones known was lapis lazuli, the identification of which with the uknû of the texts, proposed by Steindorff and Hilprecht, is now generally accepted.[3] The names of agate, onyx, chalcedony, rock crystal, topaz, serpentine, garnet, etc., all of which stones occur in the remains, are not known with certainty, and conversely a number of stones named in the texts cannot be identified unequivocally.[4] There is mention of twelve "holy jewels," as in the Bible.[5] The superstition that each kind of stone possessed a living personality, could experience sickness (as pearls are supposed to do even to-day) and disease, become old and powerless and die seems to have been common in Babylonia in 3000–2000.[6]

Stones (mostly not identified) named in the magic texts are : chulalini, sirgarru, chulâlu (a porous stone), sandu, uknû, dušû, elemêšu, šimeluchcha, gabî, eye-stone of Meluchcha, gubsu and sapingu.[7] Attempts at identification have been made by Boson [8] and R. C. Thompson [9], but are still tentative. A "meteoric" stone is thought to be nodular pyrites.[10] A remarkable find in old layers (2nd millenn. B.C.) under the temple tower at Assur contained amber, stone and glass beads.[11]

The jasper (iašpu, ašpu), probably fairly certainly identified, came from Mount Zimur, east of Lake Urmia ; it occurs in Cyprus, Asia Minor and Persia.[12] The sipru is identified as the very hard sapphire, the name coming from the Assyrian sapâru, to scratch, although it may have been corundum, which is the Hebrew šâmîr stone and probably the hardest stone known to the ancients.[13] The sapphire and corundum are chemically the same, and come next only to diamond in the hardness scale. The Semitic šâmîr is probably connected with the Greek σμύρις, emery ; without emery the Assyrian gems could probably not have been engraved.[14] An obsidian bowl (c. 3000 B.C.) was found at Ur.[15]

The burallu is probably beryl, which may have come from India, although it is also found in Armenia. The Sanskrit name is vāiḍūrya, the Prakriti verulia and Pali veluriya.[16] The barraqtu is probably the emerald, which occurs in Egypt, Media and Cyprus.[17] The Bactrian emeralds (smaragdi Bactriani) of Pliny[18] may be green rubies.[19]

[1] Woolley, 1928, 35, 50, Fig. 14 ; ib., Ur Excav., 1934, ii, pls. 91–97, 99–105, 107, 142 f., 158, 174, etc.
[2] Budge, Amulets, 108 f. ; Maspero, Dawn, 690 f.
[3] Darmstaedter, in Ruska, Studien zur Gesch. der Chemie, 1927, 5 ; Meissner, i, 269, 350; Ball, PSBA, 1887, x, 99, translated uknû as "onyx," probably relying on the sound of the word; the Syriac aqna, he says, is the lapis Lydius or touchstone.
[4] Rawlinson, Mon., iii, 162 ; Meissner, i, 269 f.
[5] Exod., xxviii, 17 f. ; Josephus, Antiq., III, vii, 1 ; Zimmern, in Schrader, Keilinschriften, 628, 629.
[6] Budge, Amulets, 87, 423.
[7] Fossey, Magie, 300, 362 ; Sayce, Rel., 490, 491.
[8] Les métaux et les pierres dans les inscriptions Assyro-babyloniennes, Dissert.,

Munich, 1914; ib., Revista degli Studi orientale, 1918, vi, 969, not seen ; q. in MGM, xix, 135.
[9] Chemistry of the Anct. Assyr., 1925, 79–128 ; ib., JRAS, 1933, 885.
[10] Thompson, JRAS, 1933, 885.
[11] Meissner, i, 352 ; Feldhaus-Klinckowstroem, Geschichtsblätter, i, 66.
[12] Meissner, i, 270, 351 ; Rawlinson, Mon., iii, 162.
[13] Thompson, Chem. Assyr., 100 f.
[14] Rawlinson, Mon., i, 558.
[15] Ur Excav., 1934, ii, 379, pl. 165.
[16] Meissner, i, 270, 351 ; Rawlinson, Mon., iii, 163 ; Finot, Les lapidaires indiens, Bibl. de l'École des hautes études, Sciences philol. et hist., 1896, cxi, pp. xlv, 260, says vāiḍūrya is cat's eye, not beryl.
[17] Schrader, Keilinschriften, 648 ; Meissner, i, 270, 351.
[18] xxxvii, 5.
[19] Rawlinson, Mon., iii, 161.

Although the name elmesu has been translated diamond [1], this stone is not found in ancient remains and the word may mean glass [2], or perhaps rock crystal. The stone sâmtu was identified by Jensen as malachite, the Egyptian mafkat, but it was a red stone and has been identified by Thureau Dangin with carnelian [3], which was used in the earliest Sumerian Period for beads [4], although porphyry [5] and red coral [6] have also been suggested.

The "eye-stone" (ênâti) from Meluchcha was perhaps agate with circular markings, found in Carmania, Susiana and Armenia.[7]

The stone KA-, the brilliant šâdanu "as valuable as gold," coming from Armenia, is supposed to be rock crystal [8] or hæmatite [9], which, however, is fairly common and not as valuable as gold. A dark brown paint used in Assyria (for pottery ?) was iron oxide.[10] Hæmatite, a hard material, was used as an amulet for virility and fertility, also (for similar purposes) in the so-called "Gnostic" amulets. A "hæmatite which grasps" (šâdanu sabitu) is supposed to be the magnet (naturally magnetic ferroso-ferric oxide), and its attraction for iron was known. It was black, softer than hæmatite and used as a drug. "Living hæmatite" (šâdanu baltu) is Pliny's ferrum vivum, a species of lodestone ; a "living copper" is in Ur texts. Other varieties of iron ore are described as bright, fortunate, strong, black, white and yellow (i.e. ochres), also prescribed in medicine.[11]

Stones coming as tribute from Arabia (?) and places not identified are ud-aš, dušû, chulâlu, mušgarru "and others."[12] The mušgarru is perhaps malachite (Egyptian mafkat), possibly named after Meluchcha, but may also be serpentine, smaragdus or "practically any green stone." The identification with serpentine is not improbable, since mušgarru is "some kind of reptile" ; serpentine is probably the Greek ὄφιτις stone.[13] Another kind of "reptile stone" is the Greek βατράχιον (batrachion), mentioned by the Greek alchemists Synesios and Dioskuros[14], which in an old Greek chemical glossary is said to be chrysokolla : βατράχιον ἐστι χρυσοκόλλα.[15] Hoefer's[16] identification of batrachion with mountain green (vert de montagne) would, according to Thompson, be "exactly what is wanted," although malachite is "rarely (or never ?) found in Assyrian ruins." The molochitis of Pliny[17], a green stone used for making seals, has been identified with malachite.[18]

Abaru is magnesite[19], which was buried with metals in foundations (p. 278). Sum. im-bar = gassu, is supposed to be gypsum : the Sumerian god Enurta was identified with this stone.[20]

Oval green stone talismans were found at Tello[21], and Layard[22] describes a seal cylinder of "translucent green felspar, called amazon." The king of

---

[1] Muss-Arnolt, Dict. of Assyrian, 1905, 47.

[2] Sayce, Rel., 246.

[3] Albright, JEA, 1921, vii, 83 ; Meissner, i, 270, 351.

[4] Woolley, 1928, 37, 40, 43 f.

[5] Langdon, JEA, 1921, vii, 150.

[6] Meissner, ii, 493.

[7] Ib., i, 351 ; Rawlinson, Mon., iii, 162.

[8] Meissner, i, 351.

[9] Thompson, Chem. Assyr., 122 ; ib., JRAS, 1933, 885.

[10] Olmstead, History of Assyria, New York, 1923, 569 ; Thompson, Chem. Assyr., 124.

[11] Thompson, Chem. Assyr., 125, 127 ; Budge, Amulets, 210, 318.

[12] Meissner, i, 350 ; Luckenbill, ii, 109, 150.

[13] Albright, JEA, 1921, vii, 83 ; Thompson, 94, 95 ; Blümner, iii, 25.

[14] Fabricius, Bibliotheca græca, Hamburg, 1718–28, viii, 241.

[15] Marcianus MS. 199, 10th century, q. by d'Orville, in Bernard, Palladii de Febribus, Lugduni Batavorum, 1745, 123 ; B.N. MS. Grec 2,325, 13th century, fol. 2 v. ; Berthelot, Coll., ii, 6.

[16] Histoire de la chimie, 1866, i, 257.

[17] xxxvii, 8.

[18] W. Smith, Latin-Engl. Dict., 1904, 694 ; Krause, Pyrgoteles, Halle, 1856, 243.

[19] Meissner, i, 266 ; Schrader, Keilinschriften, 648.

[20] Gadd and Legrain, Ur Excav., Texts, 1928, i, 41.

[21] Heuzey-de Sarzec, i, 41.

[22] Nineveh and Babylon, 1867, 174 ; cf. Rawlinson, Mon., i, 382, 405.

Assyria's regalia comprised an apotropaic breastplate composed of seven different precious stones.[1] Mosaic work and inlay of coloured stones, including lapis lazuli, goes back to a very early Sumerian Period[2] ; mosaic pavements probably go back to the Assyrian or Persian Period, but do not seem to appear in Greece until about Alexander's time.[3]

A Sum. prefix za-tu in the names of stone is supposed to mean that these types effervesced with acids.[4] Marchasi or marchušu is translated pyrites or marcasite (pyrites and marcasite are different minerals), occurring in Marchašu in the east of Assyria, and a "spangled form" (?) of it is said to have been imitated in a kind of Aventurine glass [5], i.e. a glass containing golden crystalline spangles, invented in its present form by Briani at Venice in 1280.[6]

PIGMENTS

The pigments used in mural decoration, which are found in a poor state of preservation, very friable and often faded, were probably frequently renewed. The decorations copied (with difficulty) by Place at Khorsabad (Sargon II) comprise human figures with bands, palms and rosettes in black, green, red and yellow on a white ground (used for representing the skin). Layard published some fragments of wall painting in yellow on blue grounds, with red and blue borders separated from the middle part by white lines, and Smith found coloured bands painted on stone plates. It is not known how the colours were applied, whether as tempera or fresco, or with a wax or oil medium : the use of vegetable colours has been inferred from the fact that most of the decorations on limestone or stucco walls have almost completely faded.[7]

Thompson suggests that sandu, which came from Meluchcha (Arabia or Sinai), means cinnabar, vermilion, carnelian or red jasper. Cinnabar is suggested because sandu occurs in medical texts as a drug and is found less than 75 miles from Quyûnjiq (Nineveh). The name sandu is suggested as the original of σάνδυξ ; Sum. im-kal-gug is thought to mean "vapour of cinnabar," i.e. mercury.[8] Since, as far as I know, cinnabar has never been found in Babylonian or Assyrian remains, this identification is dubious. Cinnabar was found at Mohenjo-daro.[9] Cakes of pigment in Sargon II's palace at Khorsabad were : blue, lapis lazuli ; red, oxide of iron ; yellow, lead antimoniate ; white, tin oxide.[10]

A red cosmetic paint called šaršerru and a yellow paste (šipu, lêru, damatu) were also used for colouring walls. Boxes contain rouge, henna and stibium. The remains of colours (šimtû) found have rarely been analysed. They include white, black, yellow, red, green and blue.[11] A "morsel of purple colour" at Tello[12], a purple-red wall distemper in a temple, a black plinth, and horizontal stripes of red, green and yellow have been found.[13] Cockle shells at Kiš (c. 3000 B.C.) contained remains of black, white, red, light green and blue pigments.[14] Similar shells (from the Persian Gulf) of the same date at Ur contained white, yellow, red, blue, green and black pigments, now in the form of hard pastes, which are supposed to have been used as cosmetics. The blue was powdered turquoise, the black a mixture of manganese dioxide (pyrolusite) and turquoise, with some lead carbonate which may have been added as

[1] Meissner, i, 70.
[2] Woolley, Ur Excav., 1934, ii, 126, 252, 262 f. ; pls. 73, 95 f., 113, 116, etc.
[3] Blümner, iii, 324 f. ; cf. Esther, i, 6 : pavimentum smaragdino et pario stratum lapide.
[4] Thompson, Chem. Assyr., 105.
[5] Thompson, 117, 119, 119A ; Gadd and Legrain, Ur Excavations, Texts, 1928, i, 4.
[6] Feldhaus, Technik, 450.
[7] Meissner, i, 329 ; Berger, Maltechnik, 29 f.
[8] Chem. Assyr., 59, 80, 86.
[9] Marshall, Mohenjo-daro, 1931, 691.
[10] Contenau, Manuel, iii, 1241.
[11] Meissner, i, 244, 329.
[12] Nouv. Fouilles, 152.
[13] Meissner, i, 330.
[14] Woolley, Digging up the Past, 137 ; Peake and Fleure, Priests and Kings, 1927, 42.

white lead or originally as lead oxide (perhaps red lead). Cuttlefish bones found were probably used as depilatories.[1] Although guchlu is usually translated "eye-paint or stibium" [2] it is doubtful if antimony sulphide was used as eye-paint.[3]

LAPIS LAZULI

The Sumerian za-gi-in, the Akkadian uknû or ugnû [4], is now usually identified with lapis lazuli, the Greek kyanos, formerly thought to be sapphire [5], although Scheil's [6] identification with lazurite (azurite, a basic copper carbonate) is also quite possible as an alternative.

The sapphirus of Pliny[7] , the best kind of which "with golden points" (of pyrites) came from Media, was probably the Babylonian lapis lazuli.[8] The source, "a mountain Bilki, or Bikni in Media" [9] (probably Demawend), is given in a text of Asarhaddon (680–669), which was perhaps the original source of the information in classical authors. Demawend, however, was not the ultimate source of the stone but a trading station for lapis lazuli coming from farther east, from Badakshan, north of the Hindu Kush[10], or Central Asia.[11] There is no sign of old working in Persia or of lapis lazuli in Demawend or in Media.[12] Rawlinson[13] suggested that lapis lazuli came from Bactria, or mines near Fyzabad east of Balkh, or the upper Jihun river, or from near Lake Baikal, or from Tibet and China. It probably came from the Pamirs, north-west of India, the "mountain" mentioned in Assyrian texts being a commercial station where caravans from the east and west met. This trade would be consistent with the existence in the Indus Valley of the civilisation related to the Sumerian.[14]

Lapis lazuli was used in the earliest Sumerian Period[15], for example as tablets for inscriptions[16], and a panel at Ur of mosaic in shell and lapis lazuli, of 3500 B.C., represents chariots drawn by asses[17] and foot soldiers. The wheels of the chariot are represented as made in two semicircles fastened together by strips, and with leather tyres, which were found in graves.[18] Lapis lazuli was also used for seals or in powder as a pigment as early as the Third Dynasty of Kiš (c. 3500 B.C. ?)[19], and it is found freely in graves of 3500–3000 at Ur.[20] Lapis lazuli cylinders built into a kitchen fireplace (?) were found at Tello.[21] A marble (?) head with eyes inlaid with shell and lapis lazuli, of c. 2000 B.C., was also found at Ur[22], and a kilogram of powdered lapis lazuli in the palace of Sargon II at Khorsabad.[23]

The Egyptian import of lapis lazuli from Babylonia became important in the Cassite Period. Egypt also obtained it from Tefrert (Asia Minor ?). A ruler of Assur sent Thothmes III (c. 1500 B.C.) three large lumps (8 lb.) of "true lapis lazuli" and three pieces (24 lb.) of "blue stone of Babel" (the

[1] Woolley, Ur Excav., 1934, ii, 245, 248, 394 ; pl. 137.
[2] Schrader, Keilinschriften, 648 f. ; Meissner, i, 244 ; cf. Muss-Arnolt, Dict. of Assyrian, 1905, 215.
[3] See p. 256.
[4] Maspero, Struggle, 284 ; King, Magic, 133 ; Meissner, ii, 351 ; cf. Syriac quna, Thompson, Chem. Assyr., 22 ; Muss-Arnolt, 37, uknû = precious stone or crystal, or lapis lazuli ; Uknu was also the name of a river in Elam, Maspero, Dawn, 563, 751.
[5] Rawlinson, Mon., ii, 488.
[6] Revue d'Assyriol., 1918, xv, 121.
[7] xxxvii, 9.
[8] Rawlinson, Mon., iii, 162.
[9] Mackay, JRAS, 1925, 700.
[10] Meissner, i, 351.

[11] Perrot-Chipiez, Chaldæa, ii, 294.
[12] Ruska, in Studien zur Gesch. d. Chem., 1927, 3 ; Meissner, i, 350.
[13] Mon., i, 558 ; ii, 488.
[14] Woolley, 1928, 46 ; ib., Digging up the Past, 138 ; ib., Ur Excav., 1934, ii, 394 ; Meissner, i, 350 ; Mackay, JRAS, 1925, 700.
[15] Meissner, i, 351.
[16] Langdon, Excav. at Kish, 1924, 4.
[17] Horses ? ; von Oppenheim, Tell Halaf, 1931, 137.
[18] Woolley, 1928, 50, plate 14, 52 ; ib., Digging, 115 f.
[19] Thompson, Chem. Assyr., 85, 99.
[20] Woolley, 1928, 38.
[21] Heuzey-de Sarzec, i, 42.
[22] Woolley, 1928, 164, plate 28.
[23] Meissner, ii, 385.

artificial kind). In the Amarna letters (14th century B.C.) lapis lazuli is as valuable as gold ; Burraburiaš sent his "brother" Amenhotep III (1375) four minas of "beautiful azure" and ten pieces and ten signet rings of "beautiful lapis lazuli."[1]

The statues of the gods Sin, Šamaš, etc., and of the androgyne Ištar, were actually provided with beards of lapis lazuli, and the statue of Marduk stood in a chapel roofed with this material to symbolise the sky or vault of heaven. The Sacred Way of the gods in Babylon was paved with lapis lazuli, other gems and ebony. Lapis had also medicinal uses.[2]   In many periods, including the late Babylonian, it is frequently mentioned in documents and found in remains.[3]   Sargon II obtained it as tribute from Armenia.[4]   In the Sumerian text of the Descent of Ištar into Hades (3rd millenn. B.C.) there is mention of a flute of lapis (or decorated with lapis ?).[5]   Two large bars of lapis (imitation ?), with images of Marduk and Adad, found at Babylon, probably served as "seals" for the gods.[6]   Wall paintings at Assur (1260–1238) show a very pure blue colour like uknû.[7]

COLOURED GLASS AND IMITATION GEMS

Beside beads of amber and precious stones found at Assur in the foundations of the temple, there were beads of coloured glaze and glass, perhaps imitating precious stones.[8]   Much interesting information on Assyrian glass has come to light from the translation of the so-called "chemical tablets" from Ašurbanipal's library (668–626), now in the British Museum, which contain lists of stones, liquids, etc.[9]   These resemble the Egyptian chemical papyri (c. A.D. 300) in some respects, and although they have long been known, their translation has been the most interesting event in Assyriology in recent years, making several older accounts [10] of the history of glass quite out of date.

Meissner[11] gives a list of the texts, which except for one fragment in Virolleaud's *Babyloniaca* were then unpublished.[12]   The chemical treatises called "The Door of the Furnace" deal with the production of false gems and enamels. The interpretation is often impossible and it is always difficult to identify the materials named : I have omitted a large number of marks of interrogation in Meissner's text.

" When you will lay the foundation of the furnace for the stone [or preparation], you must seek out a propitious month and a propitious day and then lay the foundation of the furnace. Whilst you look upon the furnace and make it you shall number the (divine) fœtus. A stranger must not enter, neither an impure person stand opposite you. Constantly heap up sacrifice before it. When you place the stone in the furnace you must bring sacrifice before the fœtus, set up a vessel of fumigation with cypress, pour out libation of intoxicating liquor, make a fire under the furnace and carry the stone into the furnace. The men whom you bring to the furnace must purify themselves, and then only shall you bring them to the furnace. The wood which you burn under the furnace is a thick peeled mulberry tree, and palm charcoal without knots, laid in a bag and gathered in the month of Ab. This wood shall be put under your furnace."

[1] Meissner, i, 351 ; Meyer, Alt. II, i, 128, 153 ; King, H. Bab., 224.
[2] Woolley, 1928, 44, plate 11 ; *ib.*, Ur Excav., 1934, ii, pl. 107 ; Meissner, ii, 19, 21, 120, 131, 165 ; ii, 73, 108, 309.
[3] Meissner, i, 60 f., 137, 269, 272, 305, 350 f.; ii, 19, 21, 73, 120, 131, 165, 184, 309, 351, 383 f.
[4] Luckenbill, ii, 95 f., 109.
[5] Meissner, ii, 184 ; Jeremias, *Der Alte Orient*, 1925, xxv, Heft 1, 16.

[6] Meissner, i, 272.
[7] Andrae, Ceramics from Ashur, 1925, 41.
[8] Meissner, i, 352.
[9] BMGB, 180.
[10] E.g. in Kisa, Glas im Altertum, 101 f.
[11] ii, 383 ; cf. *ib.*, 212, 223.
[12] Zimmern, Assyrische chemisch-technische Rezepte für Herstellung farbiger glasierter Ziegel im Umschrift und Übersetzung, Z. f. Assyriol., 1925, xxxvi, 177–208 ; *Isis*, 1927, ix, 180.

Thompson [1], who refers to Meissner's "attempt" at translation, practically repeats it, including the placing of the embryo under the foundations of the furnace, and the sacrifice made to it then and also when the materials for the glass are put in. He translates the last part as "styrax wood billets, cut in a hot month, which have not lain but have been kept in leather coverings." In a tablet of Ḫammurabi (1924–1913) abba trees are felled for the use of metal smelters.[2] An incantation for driving evil spirits from a furnace is preserved.[3]

The introduction of the fœtus (which Zimmern [4] says is really a miscarriage) has given rise to a considerable amount of rather inconclusive discussion, notably on the part of Eisler [5], based on the very brief section in Meissner. Probably the embryo (kûbu) (which is found in Canaanitish graves) [6] took the place of a human sacrifice.[7] A "stone with an embryo in it" (i.e. a hollow stone with a loose kernel) was also used in medicine [8], and such stones were used as birth-charms in Europe at least as late as the 18th century A.D.[9] The burying of "mascots" under the floors of houses, with bones of a sacrifice offered, was an Assyrian custom and did not appear in Babylonia till the Assyrian Period.[10]

Meissner's translation of the part of the tablets dealing with the preparation of an imitation lapis lazuli (uknû) is :

" To make bright lapis lazuli take 10 minas of immanakku stone, 15 minas of ashes of salt plant, $1\frac{2}{3}$ minas of white plant, pound them together, and mix them and bring them into a furnace with its 4 eyes cold and take great care. Then make a good fire which gives no smoke, until the mass is white-hot. Then take it out, break it up, grind it and collect it in a good (form ?). Then bring it into a furnace cold inside, make a good fire till it is [white] hot [an annealing furnace ?], then bring it out on the brick as your enamel."

" To make merku (otherwise an unknown word) lapis : To 1 mina of brilliant enamel (?) add $\frac{1}{3}$ mina of ground sand, $\frac{1}{3}$ mina of amnaku-stone, 5 half-shekels [kisal : some subdivision of the shekel] of mother-of-pearl, dry, stir, grind, fill in the form, shut it up and by repetition observe exactly ; then you will find merku lapis."[11] Another recipe for lapis contains more ingredients : enamel, tarabanu, sand, "mother-of-pearl of the sea from the middle of the pearl" (oysters ?), salt, gold paste (?), alum, gold perfume (?) and white plant. The agate (dušû) is made from 20 minas amnaku stone, 1 talent horned salt plant, 2 minas salt, 10 shekels mother-of-pearl, 1 mina tuskû, 6 shekels colour (?) ; then the agate stone comes out [after fusion ?].

Although no recipes for the preparation of gold have come to us, they "must have existed," and different kinds of copper and silver produced by synthesis were attempted.[12]

The Assyrian-Babylonian common glass, sirsu or dušû (Syriac, sištha)[13] was composed of 60 parts of sand, 180 of alkali, 2 of chalk (?) and 5 of the salt mil'u, improbably identified by Thompson with borax or saltpetre.

Dušû was, apparently, sirsu with 1 part more mil'u, $1\frac{1}{2}$ parts less chalk (?)

[1] Chemistry of the Ancient Assyrians, 1925, 57, 70, 72 ; cf. JRAS, 1925, 726.

[2] Handcock, 107 ; BMGB, 99.

[3] Sayce, Rel., 446 f.; Maspero, Dawn, 636.

[4] Z. f. Assyriol., 1925, xxxvi, 177.

[5] Der Babylonische Ursprung der Alchemie, Chemiker Zeit., 1925, xlix, 577, 602 ; ib., Z. f. Assyriol., 1926, xxxvii, 109.

[6] Vincent, Canaan d'après l'exploration récente, 1907, 51, etc.

[7] Schrader, Keil., 597.

[8] Meissner, ii, 309.

[9] S. F. Geoffroy, Treatise of the Fossil, Vegetable and Animal Substances that are made use of in Physick, tr. from the author's manuscript by G. Douglas, 1736, 68 : lapis ætites, Dioskurides, v, 161.

[10] Brit. Mus. Quarterly, i, 58.

[11] Cf. recipe for imitation purple in the Greek Stockholm Papyrus, 3rd century A.D. ; Lagercrantz, Papyrus Græcus Holmiensis, Uppsala, 1912, 204—"you will find purple."

[12] Meissner, ii, 384, 385 ; with reference to British Museum texts.

[13] Thompson, Chemistry of the Anct. Assyrians, 26, 67 ; Sayce, Rel., 490, gives the Sumerian name for glass as sudam, the Semitic elmesu, ib., 246.

and the addition of 3 parts of tuskû and three-tenths of pearl-oyster shell (?). The dušû was a material for seal cylinders, perhaps "crystal glass." The tuskû is supposed to be "oxide of tin," but this would give an opaque milk-glass, not a "crystal." "Oxide of tin," according to Thompson, occurs at Qara Dagh, "less than 300 miles from Nineveh," and was used in the 13th century B.C. for a yellow glaze [at Assur ?]. If tuskû was added in excess, a "pink enamel" was formed.[1] (Perhaps pink is the colour of light transmitted by an opaque milk-glass.)

The name tuskû was thought [2] to be the origin of the name tutia, which, however, was oxide of zinc, not of tin [3], so that Thompson [4] later thought it meant cadmia or calamine, zinc oxide or carbonate. These identifications are very hypothetical.

The alkali (uchûlu) was perhaps crude soda, since one form is made by burning the "horned plant" (salicornia).[5] The sand was perhaps a special kind, that of the River Belus (modern Na'man) being mentioned by Pliny.[6] The lime is supposed to be represented by namruta (Arabic nûra, lime) "of the sea" (not oyster shells, which are lulû), the "bright thing," a soft, whitish mineral and probably chalk. Lead was probably not added, since its name anaku does not occur in the texts.[7]

A fusible glass is composed of the same materials as simple glass [8] : the large proportions of alkali in the glass would make it impure and liable to attack by water.

In the Assyrian method of making glass [9], a simple glass ("glaze," Thompson), a frit (achussa) is composed of 30 of sand, 45 of alkali and 5 of "storax gum," the latter perhaps to serve the purpose of charcoal in modern glass making, in preventing the formation of scum.

A ruby glass, made from 7,200 parts of zukû-glass, 32 parts of tuskû, 20 parts of abaru (antimony ?), some mil'u (saltpetre ?) and 1 part of gold [which could also mean silver][10], imitated coral, since the ruby was not known to the Assyrians or Egyptians, whilst Pliny[11] says coral came from the Persian Gulf, where it was known as iace.[12] I do not see how "gold" as such could be incorporated in glass and the production of gold ruby glass is very difficult : perhaps cuprous oxide is meant, as in the "reddish-purple" glass.[13] Green glass contains aš-ge-ge and barummu, the first perhaps arsenic, since a lump of orpiment was found at Zinjirli[14], and in medical texts as-char is a medicine for the eyes, "a mineral smelling of assafœtida."[15] The barummu is supposed to be "iron rust," oxide of iron[16], which would normally give a yellow or brown colour. The use of manganese, "the earthy product which attracts" [? really magnetite], is suggested, and a black glass (?) was obtained with hæmatite.[17] The chemistry of these materials is much confused by Thompson, whose difficulties must have been considerable, as is clear from such translations as the composition of a "bronze inlay (?)" from simple frit, "some mineral and perhaps something else."[18] By heating copper both cuprous oxide, giving a red glass, and cupric oxide, giving a blue glass, could be made.[19]

[1] Thompson, 29 f., 31, 38, 65.
[2] Ib., 30, 38.
[3] Kopp, Geschichte der Chemie, Braunschweig, 1843–47, iv, 114 f.
[4] JRAS, 1929, 813.
[5] Thompson, Chemistry of the Assyr., 12.
[6] xxxvi, 26 ; Thompson, 14.
[7] Thompson, 16 f.
[8] Ib., 22, 26, 63, thinks it was perhaps "poured into water": perhaps the meaning is "soluble in water," i.e. a kind of waterglass, or "poured like water."
[9] Ib., 19 f., 21, 26.

[10] Ib., 32 f., 37.
[11] xxxii, 2.
[12] Thompson, 34.
[13] Ib., 56, 63.
[14] Meissner, i, 244.
[15] Thompson, 42, 44, 45, who thinks, ib., 25, that arsenic was added to ordinary glass, as in modern practice.
[16] Ib., 46 f. ; Fossey, Magie, 393, translated abnu anbar as "ironstone": perhaps magnetite.
[17] Thompson, 52, 54, 55.
[18] Ib., 56, 62.
[19] Ib., 50, 58.

ALKALIES 297

Tersitu is a blue glaze containing copper to be "poured on" (applied to ?) burnt bricks. The imitation lapis is uknû ibbu : ibbu is "clear or translucent," and uknû "blue." The material is used either as "the stone" (bars or pieces for transport or inlay ?) or powdered as a pigment, and was a "blue frit" or "blue glaze," as in the slabs 3 in. by 2 in. in the Louvre [1] and perhaps the blue beards of statues.

Many of Thompson's identifications are necessarily provisional.[2] His conclusion [3] that the Assyrian technicians of the 7th century B.C. had a good empirical knowledge of what we may call pre-chemistry and also of mineralogy is fairly justified. The level of attainment was probably similar to that in contemporary Egypt. When we add to this the extensive knowledge of plants possessed by the physicians, it must be admitted that the foundations for the beginning of Greek mineralogy and botany in Theophrastos may well have been laid in Mesopotamia and Egypt.

SALT

The names tâbtu and mil'u are generally supposed to mean common salt. This was always used with food and also in sacrifices, since it was regarded as a sacred material, a gift of the god Ellil. It was also used in preserving fish. Salt was supposed to have great magic power in purification from demons and pieces of salt were attached to the body in rites of purification. A text says : "Thou art salt, produced in a pure place and destined by Ellil for eating by the great gods. Without thee no meal is taken in the temple." It was invoked to break the spell of witchcraft and to cure fever.[4] The body of a Chaldæan emperor preserved in salt was sent to Ašurbanipal by the Elamites.[5]

Salt brine occurs in the bitumen pits of Hît, there is rock salt in Assyria and at Kerkuk, and salt was obtained, as at present, from the "saline and gypsiferous soil of Assyria" and from the "saltpetre containing soil of Babylonia."[6]

A peculiar kind of salt, tâbat amâni, used in medicine and formerly regarded as sal ammoniac[7], is perhaps the rock salt of the Oasis of Ammon in the Libyan Desert [8], although this is merely a guess as to its identity.

ALKALIES

Uchûlu, šikku, or nitiru (the Syriac achla, lye) is perhaps crude potassium carbonate, or alkali from the soda plant or from reeds.[9] Even in the Sumerian Period a kind of soap was made from alkali and oil.[10] If this were a true soap, its preparation would imply a caustification of the alkali with quicklime, which may have been used. In a Tello text the god Uri-zi was in charge of the "mixing of oil," perhaps making soap[11]. Thompson[12] thinks soap

[1] E. Dillon, Glass, 1907, 40 ; Thompson, 22 f., 47, 59.
[2] Darmstaedter, Archiv für Gesch. d. Math., Naturw. u. Technik, 1927, x, 72–86 ; Isis, 1928, xi, 216 ; Z. f. Assyriol., 1926, xxxvi, 302 ; Chem. Zeit., 1925, xlix, 967 ; Isis, 1927, ix, 180.
[3] Chem. Assyr., 5, 12, 53, 79.
[4] Meissner, i, 349, 415, 226 ; ii, 228, 240.
[5] Ib., i, 113.
[6] Rawlinson, Mon., i, 213, 219 f. ; ii, 488 ; Meissner, i, 256, 349.
[7] Meissner, i, 349 ; ii, 309 ; von Oefele, in Puschmann, Geschichte der Medizin, Jena, 1902, i, 100.

[8] Ruska, Isis, 1926, viii, 197.
[9] Löw, Flora der Juden, Vienna, 1928, i, 637 f., 644 ; Nouvelles Fouilles, 175 f. ; Meissner, i, 244, 255, 413 ; ii, 306, 384 ; Thompson, Chem. of Assyr., 12 ; Muss-Arnolt, Dict. Assyrian, 29, says uchûlu is a plant for rubbing a sick person ; Whittaker and Lundstrom, Review of Patents and Literature on the Manufacture of Potassium Nitrate, U.S. Dept. Agricult. Misc. Publ., No. 192, Washington, 1934, 3.
[10] Meissner, i, 244.
[11] Nouv. Fouilles, 175 ; Meissner, i, 413.
[12] The Assyrian Herbal, 1924, 191, 270.

was made from castor oil and alkali, and says that rue is still burnt near Carchemish to make lye.

ALUM

Alum (sikkatu ; the word is perhaps foreign) probably came mostly from Egypt [1], although it occurs plentifully in the hills round Kifri.[2] Synkellos [3] states that it was found in the water of the Dead Sea, together with salt "a little different from the common kind."

NITRE

According to Rawlinson [4] the soil near and on the sites of old towns in Babylonia is "deeply impregnated with nitre," and Loftus [5] thought this was the cause of the cold of winter ! Meissner [6] also speaks of the soil of Babylonia as encrusted with or "containing saltpetre." There seems no reason to doubt the accounts of travellers, such as Loftus, Mignan, [7] Ainsworth [8] and Banks [9], which describe the soil of Mesopotamia as "impregnated with nitre or marine salt," both being clearly distinguished[10], but whether this nitre has been formed from the decomposing organic matter of old sites or was present in ancient times is not made clear. That mil'u in the chemical texts is "provisionally and not certainly saltpetre,"[11] or borax, which occurs near Urmia, is doubtful ; mil'u is generally read as meaning common salt. The existence of saltpetre in the incrustation on the soil in the neighbourhood of Babylon is proved by its use in making gunpowder by the Arabs.[12] The salt halmyrrax (ἁλμύρραξ), which Pliny says was found in caves in Media, might have been saltpetre, his nitrum being soda (natron). In Assyrian medicine black and white mil'u are prescribed, once in a venereal case with arsenic.[13]

OTHER SALTS

Gypsum, which may possibly have been regarded as a kind of salt, was used in medicine and (perhaps in the\form of selenite) for magical purposes.[14] According to Thompson, chulâlu is white lead, something which on roasting becomes red (?), "the acetated thing (?)," although there is no Assyrian word for vinegar corresponding with that translated "acetated" ; it is even suggested that sugar of lead (lead acetate) was known and used in medicine, which seems improbable.[15]

Šubû is compared with Arabic and Syriac words meaning brass and vitriol, and supposed, with its synonym sichru, to mean green vitriol (ferrous sulphate ; cf. "copperas," the name for iron, not copper, vitriol).[16] Thompson connects the word with sory[17], an impure green vitriol[18], but according to Muss-Arnolt[19], šubû is some precious stone. In an Assyrian text kammu is "vitriol (?) of the

[1] Meissner, i, 255, 413, 460 ; ii, 309.
[2] Rawlinson, Mon., i, 219 f.
[3] In Hippolytos, Ante-Nicene Library tr., vol. VI, pt. ii, 175.
[4] Mon., ii, 521, 529.
[5] Travels, 1857, 146 ; Maspero, Dawn, 553.
[6] i, 187, 349.
[7] Quoted in Langdon, Excav. at Kish, 1924, 51.
[8] Researches in Assyria, Babylonia and Chaldæa, 1838, 118.
[9] Q. by Rogers, Hist., i, 311.
[10] Thompson, Chem. Assyr., 27 f. ; nitre is accompanied by natron ; cf.

Whittaker and Lundstrom, op. cit., 4, 11 f., and bibliography.
[11] Thompson, 27, 28.
[12] Koldewey, Excavations, 108.
[13] Pliny, xxxi, 10 ; Smith, Latin Dictionary, 1904, 491 ; Thompson, Chem. Assyr.
[14] Meissner, ii, 208 f., 309.
[15] Thompson, Chem. Assyr., 89, 90, 91, 93.
[16] Ib., 111 ; cf. ib., JRAS, 1933, 885.
[17] Pliny, xxxiv, 12.
[18] Berthelot, Intr., 242.
[19] Dict. Assyr., 1000.

shoemakers," perhaps "green vitriol," or "copper scales."[1] Šamaitu, the "sky-blue mineral," is thought [2] to be blue vitriol, copper sulphate : perhaps it is the same as sâmtu, which has been translated as malachite [3], carnelian or red coral.[4] The entirely provisional character of nearly all Thompson's identifications must be emphasised and is admitted by him ; many of them are, in fact, very improbable, and conclusions drawn from them as to the early history of chemical materials must be accepted with the greatest reserve.

[1] Thompson, Assyrian Herbal, 1924, 274 f., who confuses iron vitriol with copper oxide.

[2] Thompson, Chem. Assyr., 116 f. ; *ib.*, JRAS, 1933, 885.
[3] Albright, JEA, 1921, vii, 83.
[4] Meissner, ii, 493.

BABYLONIA AND ASSYRIA III—2

NON-METALS

ORGANIC MATERIALS: FOODS, PLANTS, WOOD, ETC.

WHEAT

Agriculture, held in high esteem in Babylonia, was said to have been invented by the gods Ninurta and Ningirsu and the god Tammuz also represented it.[1] The Sumerians had a goddess of wheat (Nidaba) and a god of barley (Ašnan).[2] There was an extensive irrigation system of canals and the shaduf, for raising water, was used as in Egypt.[3] Just as Egypt was "the gift of the river [Nile]," [4] so Mesopotamia was a fertile country because of its two great rivers.[5] Although the soil naturally varied in fertility in different parts of the country [6], the account in Genesis (i, 9–11) is probably very true.[7] This fertility, owing to neglect of the irrigation system, has long since disappeared and the south country is now a parched and sandy plain in autumn and winter, and during the spring and summer almost a continuous marsh, on which a rank vegetation of reeds and coarse grass flourishes.[8] These sterile conditions are already described in the Old Testament. The plough (epinnu ; epittu), which was used from the earliest times and is shown on seal cylinders of the early period at Ur (before 3000 B.C.), was furnished with a tubular seed-drill (still used in Syria) and drawn by a yoke of oxen. The soil was thinly sown. Grain was ground to flour between flat stones, or was parched and bruised for porridge.[9] At al-'Ubaid (c. 3500 B.C.) barley was cut with burnt clay sickles, and ground in querns or pounded in mortars to make a kind of porridge, which, with fish, was the chief food.[10] Remains of lentils and barley were found at Khafaje (before 2450 B.C.).[11]

A painted Sumerian jar of about 3500 B.C., found at Jemdet Nasr (near Kiš), contained blackened wheat.[12] The "wild wheat" of Mesopotamia and Palestine, once thought to have been cultivated by the Sumerians, sheds each grain from its ear as it ripens, could not be harvested, and when crossed with other wheats becomes sterile. Modern wheat is not a cultivated variety of wild Mesopotamian or Palestinian wheat, and does not appear to have been naturally cultivated, as was barley : Cherry suggests that early seafarers discovered it in Delos in the Cyclades. The Egyptians and Sumerians had the same name for wheat.[13]

Wheat does not grow wild in Mesopotamia, although Berossos said that wheat, barley, sesame and a kind of leguminous plant which he calls ὦχρος

---

[1] Meissner, i, 185 ; Woolley, 1928, 71, 114.
[2] Ur Excav., 1934, ii, 333.
[3] Meissner, i, 191 f. ; Rawlinson, Mon., iii, 16.
[4] Herodotos, ii, 5, 10.
[5] Contenau, Assyr., 9.
[6] Meissner, i, 187.
[7] Woolley, 1928, 4.
[8] Hilprecht, Excavations in Assyria and Babylonia, Philadelphia, 1904, 4 f.
[9] Woolley, 1928, 114 ; ib., Ur Excav.,

1934, ii, 336, 350 ; Breasted-Jones, Brief History of Ancient Times, 1927, 80 ; Meissner, i, 193, 195, plate-figs. 780–81.
[10] Woolley, 1928, 13.
[11] Frankfort, Jacobsen and Preussen, Tell Asmar and Khafaje, 1930–31, Univ. Chicago Orient. Inst. Publ., No. 13, 98.
[12] Langdon, Alte Orient, 1928, xxvi, 73.
[13] Mackenzie, Footprints, 116 ; Peake, Nature, 1927, cxix, 158, 894 ; cf. Elliot Smith, ib., 81.

grew wild in Babylonia.[1] J. Percival [2] considers that emmer wheat was the first kind grown and gives an illustration of grains of it from a Sumerian house site of 3500 B.C. Peake [3] thinks Predynastic Egyptian wheat may have come from Mesopotamia, but according to Percival the grains of Mesopotamian wheat belong to more highly developed races of this cereal. The identification by the grains only is always difficult and sometimes impossible, but in this case the grains seem to belong to a variety of Rivet wheat (Triticum turqidum), unknown to the ancient Egyptians, and to be the first authentic samples known. The Sumerians were thus in possession of an advanced type of wheat at a very early date.

The miller (kaziddaku) and baker (nuchatimmu) were separate, but the baker was also a cook-shop keeper : most of the grain, however, was ground at home. The oven was a large earthen pot, heated by flames, and the thin cakes were plastered on the inside.[4]

CULTIVATED PLANTS AND TREES

Cultivated plants included several kinds of grain (še'u), viz. barley (ašnan), which was the principal grain ; wheat (Sum. gig ; Akk. kibtu) ; emmer wheat (Sum. zizna ; Akk. kunâšu) for bread and especially beer ; millet (duchnu), the modern durra (as in Egypt), which was found in a late Assyrian sarcophagus ; sesame (šamaššammu) for oil ; lentils (halluru) ; perhaps beans and mustard (šachlû) ; linseed, from the flax (kitû ; pištu) ; and perhaps truffles (kamtu : Berossos's gongæ roots, still favourites). Rice was not known till the Persians brought it from India just before Alexander's time. Beets (mušarû), lucerne (aspastu) from the Persians (μηδική, medica, of the Greeks and Romans), garlic, onions, cress, dill, cardamoms, coriander, thyme, hyssop, gherkins, kummel, chicory, poppy, sweetwood, rose, lotus, etc., and various trees were cultivated.[5] Tiglathpileser I (1115–1103) imported foreign trees and plants for his garden.[6]

Reeds (Sum. gi or gin ; Akkad. qanu, the origin of "cane"), both wild and cultivated, were much used.[7] The papyrus plant is mentioned as "reed of Magan" in the Sumerian Period, and is also called "grass of guiding" or "vegetable of knowledge" ; Egyptian papyrus was used in Babylonia and called "vegetable skin." The Assyrians, besides clay tablets, had books on parchment or papyrus.[8]

The date palm (Sum. gišimmar ; Akkad. gišimmaru) was characteristic of the whole of Babylonia from the oldest period. The tree furnished fruit, honey (i.e. date-sugar syrup), wine and vinegar from the sap. It was sacred and used in magic, and grew freely in Babylonia, where it is represented more naturally than in the stylised pictures in Assyria and Subartu : it is not indigenous to North Mesopotamia, Syria, Asia Minor or Subartu.[9] Gathering the fruit of the palm tree is shown in Sumerian times.[10] The ripe dates (suluppu ; uchûnu) were great favourites. Xenophon said they were

[1] Maspero, Dawn, 555 ; Meissner, i, 185, 198.
[2] The Wheat Plant, 1921, 178 f., 186 f., 335 f. ; ib., Nature, 1927, cxix, 280 ; Isis, 1928, ix, 219.
[3] Nature, 1927, cxix, 894 ; Peake and Fleure, Peasants and Potters, 1927, 16.
[4] Meissner, i, 238.
[5] Ib., i, 198 f., 209 ; Rawlinson, Mon., i, 566 ; Maspero, Dawn, 554 ; E. Bonavia, The Flora of the Assyrian Monuments, 1894, is useless.

[6] Luckenbill, i, 87.
[7] Meissner, i, 212, 246.
[8] Sayce, Rel., 9, 32 ; ib., TSBA, i, 343 ; Fox Talbot, ib., 1874, iii, 430, 432, 434 ; Rawlinson, Mon., i, 268.
[9] Meissner, i, 202, 206 ; Rawlinson, Mon., i, 35 ; von Oppenheim, Tell Halaf, 147 ; Contenau, Manuel, iii, 1153 ; Popenoe, Scientific Monthly, New York, 1924, xix, 313, "The Date-Palm in Antiquity"—indigenous to India ?
[10] Ur Excav., 1934, ii, 333.

like amber [1], and a kind of treacle (the modern Arabic dibs) was made from them[2] : dibis in Syria is the syrup from boiled grape juice.[3]

The earliest honey was that from dates, but bee-honey and wax are mentioned about 750 B.C. by a large landowner who is proud of having first introduced bees. The use of honey for preserving bodies, mentioned by Herodotos, is not confirmed by the texts.[4]

Herodotos [5] says the plains of Babylonia were planted with palm trees, the fruit (dates) of which furnished bread, wine and honey. "They are cultivated like the fig tree in all respects, among others in this. The natives tie the fruit of the male-palms, as they are called by the Greeks, to the branches of the date-bearing palm, to let the gall-fly (ψῆν) enter the dates and ripen them, and to prevent the fruit from falling off." In Egypt and Palestine the fig is pierced with a sharp instrument a few days before gathering.[6] The procedure, correctly described by Herodotos, is called caprification, and depends on a transfer of pollen by the insect.

The artificial fertilisation of the date palm by shaking the male flowers over the female (described in texts of Ašurbanipal's time, 626 B.C.) was an invention of the gods, who are frequently represented with a bunch of flowers in the right hand for use in the process.[7] There is mention in religious texts of a sacred or cosmic tree, with a root of white crystal, and there was probably such a tree in the temple at Ur which pronounced oracles.[8] The "tree of life" in Genesis was perhaps a date palm.[9] The sexes of the palm were known to the old Babylonians[10] and the sexes of palms and their fertilisation and the caprification of figs are fully described by Michael Glykas (A.D. 1150)[11], yet Jungius (A.D. 1587–1657) still thought flowers were of one sex only : the two sexes of plants were finally differentiated by Camerarius of Tübingen in 1694.[12]

Other fruit trees in Babylonia were the fig (tittu), vine[13], pomegranate and perhaps mulberry, pistachia, pear, almond, etc., and edible plants such as St. John's bread and strawberry (girgiššu).[14] A list of plants which grew in the garden of Merodachbaladan (c. 700 B.C.)[15] contains the oldest mention of the Beta vulgaris as silqu, "the Sicilian (?)," a foreign plant perhaps introduced into Syria about 800 B.C.[16] The citron tree was cultivated in Pliny's[17] time and the fruit was supposed to be an antidote against poison.[18] Pips, perhaps of lemons, were found with barley grains at Nippur.[19]

The "hanging gardens" of Nebuchadnezzar at Babylon, described by Diodoros Siculus[20] and Josephus[21], have been regarded as fabulous[22], although such

[1] Anabasis, II, iii, 15 ; Philostratos, Apollon. Tyana, i, 21.
[2] Meissner, i, 206 ; Woolley, 1928, 114.
[3] E. S. Stevens, Cedars, Saints and Sinners in Syria, 231.
[4] Herodotos, i, 193, 198 ; Meissner, i, 223 f., 425 ; King, Magic, 148 ; Lippmann, Geschichte des Zuckers, 1929, 8 ; CAH, iii, 31.
[5] i, 193.
[6] Smith, Concise Dict. Bible, 906.
[7] Maspero, Dawn, 555 f. ; Meissner, i, 205 ; ii, 381 ; Thompson, Assyrian Herbal, 1924 ; ib., Isis, 1926, viii, 506 ; Sarton, ib., 1934, xxi, 8.
[8] Sayce, Rel., 238, 410, 471 ; Maspero, Dawn, 642.
[9] Enoch, 24, Kautzsch, Apokryphen und Pseudepigraphen des Alten Testament, Tübingen, 1900, ii, 254 ; Popenoe, 318.
[10] Meissner, i, 205 ; ii, 381.
[11] E. H. Meyer, Geschichte der Botanik, Königsberg, 1856, iii, 380 ; sexes of palms described again by Ibn Alwardi, d. 1232.

[12] Hoefer, Histoire de la botanique, de la minéralogie et de la géologie, 1872, 40, 161, 190.
[13] Herodotos, i, 193, is wrong in saying the fig and vine were not grown.
[14] Meissner, i, 193, 208 ; ii, 304 ; Maspero, Dawn, 556 ; Rawlinson, Mon., i, 36, 216, 352.
[15] Meissner, i, 210.
[16] Lippmann, Geschichte der Rübe (Beta) als Kulturpflanze, 1925, 33, 168 ; appendix in Z. d. Vereins d. Deutschen Zucker-Industrie, 1934, lxxxiv, Techn. Teil, 15.
[17] xii, 3.
[18] Rawlinson, Mon., i, 216 ; Matthiolus, Commentarii in Dioscoridis, Venice, 1570, 182.
[19] Meissner, i, 209.
[20] ii, 10.
[21] Antiq., X, xi, 1 ; ib., Contra Apion., i, 19.
[22] Budge, Rise and Progress of Assyriology, 1925, 63.

terraces raised on arches are mentioned by Sennacherib [1], and Koldewey, whose habitual scepticism towards classical accounts of Babylon has not always been justified by archæology, considered that he had found traces of the "hanging gardens" in vaulted structures in the ruins.[2]

WOOD

Wood was very scarce in Babylonia ; palm wood was generally used and was better when seasoned : remains of palm-wood beams were found at Ur.[3] Ebony combs occur [4], and Indian teak logs of the 6th century B.C. were found at Ur and in Nebuchadnezzar's palace.[5] Teak and ebony (ušu ?) were imported and the later trade in these, also sandalwood and blackwood between Barygaza (Baroche ; Gulf of Cambay) [6] and the Euphrates is mentioned in the Periplus of the Erythræan Sea (1st century A.D.) ; boxwood (urkarînu) came from Armenia.[7]

A much prized tree was the cedar (erinu) : the mast and ceiling of the ark of the Babylonian Noah were of cedar wood, supposed to counteract sorceries. Arrian says that in Alexander's time cypress (burâšu) wood was used even for ship-building ; its real home was Armenia, near Lake Van. The extreme south of Babylonia produced poplar or willow (chuluppu).[8] Gudea (2450 B.C.) imported a wood (probably cedar) from Amalum or Amanum (Mount Amanus in North Syria, or Lebanon), and coniferous trees appear frequently on seals of Sargon of Akkad, probably because the mountainous country of Zagros was subject to him.[9] The "pieces of bamboo" found in a tomb at Ur[10] were perhaps reeds, since the latter were very large[11], but a gold representation of bamboo was also found at Ur.[12]

Other trees growing in Babylonia and Assyria were the poplar, tamarisk and (in the later period) cedar. The carpenter (naggaru) used the saw (sassaru), axe (pasu), hammer (chasinu), chisel, rasp, file, borer, ferrule, etc. The house doors were portable and belonged to the furniture, which was largely of wood. Wooden ships were made at an early period.[13]

The oak growing in Assyria yields galls.[14] Manna, perhaps supalu, "earth of the moon crescent,"[15] is gathered from the dwarf oak and is deposited on various trees (oak, tamarisk, etc.), shrubs, rocks, sands, etc., "falling" plentifully in wet or especially foggy weather but not in dry. The natives catch it on cloths under the trees, which are shaken, and it soon putrefies. The modern manna is mostly obtained from a species of ash cultivated in Sicily and Calabria.[16]

FERMENTED DRINKS

Wine and beer were consumed in large quantities, but water, the commonest drink, was held in great esteem. The usual expression for intoxicating liquor was sîkaru and there were many kinds. The vine was grown, but wine, including "poor wine," was also made from dates and sesame. Much wine of various kinds was also imported from Armenia and Syria, that from Lebanon being

[1] Luckenbill, ii, 161.

[2] Contenau, Manuel, iii, 1356.

[3] Meissner, i, 282 f., 352 ; Rawlinson, Mon., i, 84.

[4] Perrot-Chipiez, Chaldæa, ii, 350 ; Handcock, 349 ; Sayce, B. and A. Life, 108 ; ib., Rel., 137.

[5] Rawlinson, Mon., i, 109.

[6] Tozer, History of Ancient Geography, Cambridge, 1896, 177.

[7] Meissner, i, 353 ; Rawlinson, India, 1926, 3.

[8] Lenormant, Magic, 160 ; Meissner, i, 211, 352 f.

[9] Jastrow, Civ., 41 ; Meyer, Alt., I, ii, 529 ; ib., Sumerer, 60 ; King, S. and A., 261 ; Honigmann, PW, xiii, 1 ; Meissner, i, 352.

[10] Rawlinson, Mon., i, 88.

[11] Maspero, Dawn, 552, 561.

[12] Ur Excav., 1934, ii, pl. 154.

[13] Meissner, i, 247 f., 250, 283, 352.

[14] Rawlinson, Mon., i, 217.

[15] Thompson, The Assyrian Herbal, 1924, 268 f.

[16] Rawlinson, Mon., i, 219 ; Hanbury, Science Papers, 1876, 355.

especially good.[1] Thompson's "alcohols" (sîkaru) [2] are really various kinds of intoxicating drinks, including date wine, grape wine and beer.

Wine (Sum. gestin ; Akk. karanu ; gapnu) was used in the earliest period, and viniculture, certainly known for the later period and perhaps introduced from Armenia (wine from Lake Van being celebrated), was perhaps known to and mentioned by Gudea, or known before Sargon I. The vine was especially cultivated in Assyria and the neighbouring country. Unfermented grape juice mixed with water was drunk and spiced wine was made by the perfumer.[3]

Date wine (dašpu), not "brandy," as Meissner calls it, was a favourite drink which would keep for a year. According to a recipe preserved in the Talmûd, it was made by extracting the dates thirteen times with water. The addition of cassia leaves or sesame oil (?) improved it, when it was called kurunnu.[4] The date wine which keeps for a year before turning sour is quite different from the rapidly fermented sap ("toddy") of the palm tree which turns sour in a day or two. The date wine is no longer used in Mesopotamia, but is made in Nubia by boiling the fresh, mature fruit for three or four days in an earthenware boiler, straining and allowing to ferment for 10 to 12 days in well-closed jars buried in the ground. It is now distilled to make brandy.[5] In later Babylonian times 1 kur (121 litres) of dates gave 1 cask of wine.[6] Vinegar (ensu), including "fresh" and "strong," perhaps from date wine, was drunk and used in the kitchen.[7] A syrup ("honey," not "sugar") was also made from dates, and the fibres of the palm were twisted into ropes.[8]

Beer (Sum. kaš ; Akkad. hîqu) was known in early Sumerian times and was made from barley, emmer and other grains ; there are recipes of Lugalanda's period (2630 B.C.).[9] The ingredients were grain, "beer bread" (as in Egypt) and buqlu (perhaps malt)[10], which is perhaps mentioned in Hittite texts.[11] Hops were unknown, but rue and safflower were perhaps used as flavours.[12] Beer was also flavoured with cinnamon.[13]

One recipe gives for 100 sila (40 litres) of beer, 72 sila of emmer without husk, 12 sila of beer bread and 96 sila of buqlu (malt). There were many kinds of beer : black (porter ?), good black (stout ?), red, fermented, strong and diluted. The beer was strong and intoxicating, but when one had got used to the taste was pleasant. It was kept in leather bags or clay barrels or jars, and before use was strained to free it from sediment and floating grains. Otherwise it was drunk through straws, with the grains floating on the top of the liquor.[14] This method of drinking through straws from a pot is shown on early Sumerian cylinders at Ur, on a Hittite cylinder of 2000 B.C. and on an Egyptian stela of 1400 B.C. representing a Syrian in Egyptian service.[15] Xenophon[16] says the Armenians drank sweet barley wine (beer) from jars through canes to prevent the floating grains getting into the mouth. Archilochos,

[1] Rawlinson, Mon., i, 578 ; Sayce, B. and A. Life, 139 f. ; Meissner, i, 71, 239, 242, 417 ; ii, 317 ; Muss-Arnolt, Dict. Assyr., 1905, 1033.
[2] Assyrian Herbal, 1924, v, 195.
[3] Meissner, i, 207 f., 242 ; Rawlinson, Mon., i, 567 ; King, Magic, 156 ; Sayce, B. and A. Life, 139 ; Jeremias, 338.
[4] King, Magic, 148 ; Meissner, i, 190, 239, 241 ; Ruska, Isis, 1926, viii, 196 ; Rawlinson, Mon., i, 35, already says clearly that it was not distilled.
[5] Rollin, Ancient History, ed. Bell, Glasgow, 1845, i, 461.
[6] Meissner, i, 239.
[7] Ib., i, 241 ; Thompson, Assyrian Herbal, 1924, 195 f.

[8] Sayce, B. and A. Life, 127.
[9] Jeremias, 338 ; Meissner, i, 241 ; Meyer, Alt., I, ii, 167 ; Sayce, Rel., 534.
[10] Budge, Amulets, 115.
[11] Friedrich, Alte Orient, 1925, xxv, Heft 2, 17.
[12] E. Hahn, Bier und Bierbereitung bei den Völkern der Urzeit, 1927, ii, 102, q. in Nature, 1927, cxix, 404.
[13] Jeremias, 338 f.
[14] Meissner, i, 241 f. ; Jeremias, 338 f.
[15] Ur Excav., 1934, ii, 328, 337 f. ; Meyer, Chetiter, 58 ; ib., Alt., II, i, 17 ; Jastrow, Civ., 422 ; Wiedemann, Das alte Ägypten, 1920, 302 ; Schrader, RL, 88 ; Forrer, RL, 92 ; Feldhaus, Technik, 85.
[16] Anabasis, IV, v, 26 : 401 B.C.

as quoted by Athenaios, reports that the Thracians and Phrygians drink the brew (βρῦτος) through a reed : βρῦτος, mentioned in a fragment of Aischylos, is perhaps a Thraco-Phrygian word.[1]

The Babylonian beer could be brewed at home or in breweries, when the client provided the materials, the vats (namcharu), fermenting tubs (namzitu), barrels (dannu) and jars (kannu) for the product ; the works, and especially the "brewer's furnace"(boiler, or malt kiln ?), belonged to the brewer (barippu), who must supply the right amount of product from the materials.[2]

There are "feasts of malt" in a religious text, a god of intoxicating drink, Siris, was created by Ea and intoxicating drink was an offering to the gods Ea, Šamaš and Marduk. There were regular public houses, with barmaids (one of whom became a queen, Ku-Bau), in the old period (3000 B.C.). The brewers were not held in great esteem : for selling adulterated beer they or the barmaids could be ducked in water. The Babylonian and Assyrian intoxicating drink (sîkaru) was strong : its effects, the texts say, produced an unsteadiness of the legs and the faculty of seeing things several times over. A cure consisted of the seeds of five kinds of plants ground up in wine and drunk off "without tasting."[3]

FOODS

A large temple kitchen at Ur had an open court with a well, a fireplace for heating water and a brick table for cutting up meat. The grain was ground on flat saddle-shaped querns of hard stone ; opening off the yard were two roofed chambers, one containing the beehive-shaped clay bread oven and the other the cooking-range of fireclay with a flat top and circular flues.[4] Fuel was largely desert thorn (ašâgu) [5], dried dung of various kinds, date kernels and the expensive wood charcoal (pêntu). The fire, in a fixed (kinûnu) or portable (tinûru muttaliku) brazier was kept always burning. The large cooking-pots were set over the fire on stones.[6]

Fire was probably kindled (as in Egypt) by the friction of a pointed wooden stick—the fire-drill—represented by two ideograms meaning "the wood of light."[7] Pottery lamps are represented on kudurru (boundary stones) of the Cassite Period as the symbol of Nusku, the fire god, and the same kind of lamp has been found in Kiš.[8] The earliest lamps were unglazed, the later glazed.[9]

The staple Babylonian and Assyrian foods were vegetable, although fish, including dried, smoked and salted, was largely eaten ; meat was only sparingly used. Salt (tâbtu) was added to food : the god Ellil had ordained its use, and no meal could be taken without it.[10]

Fruits and garden vegetables (onion, date, radish, cucumber, pineapple, pomegranate, grape, and in Assyria the fig, apple and pear) were eaten. Corn, crushed and baked, often by women, flour (qêmu), "fine flour" (siltu), groats (chašlâti) and barley bread were the chief foods. A favourite breakfast dish was porridge (zisurrû ; pappasu) of thick milk and honey ; other dishes were a mush (mirsu) of thick milk, honey, oil, etc., and a marzipan (muttaqu) of flour, sesame oil and date honey.[11]

[1] Schrader, RL, 88 ; Liddell and Scott, Lexicon, 1926, 332.
[2] Meissner, i, 240.
[3] *Ib.*, i, 239, 417 f. ; ii, 94, 171.
[4] Woolley, 1928, 154 ; *ib.*, Ur Excav., 1934, ii, 4 ; on cooking, etc., in the later periods, see Meissner, i, 414, and illustrations of baking and roasting ovens, boiler, etc.
[5] Cf. the "crackling of thorns under the pot" in Eccles., vii, 6.

[6] Meissner, i, 206, 414 f. ; Handcock, 107.
[7] Handcock, 100 ; Sayce, Rel., 180.
[8] Langdon, Excav. at Kish, i, 90, plate xxv, 2—Neo-Babylonian.
[9] Rawlinson, Mon., i, 92.
[10] Maspero, Dawn, 768 ; Meissner, i, 413, 415 f.
[11] Meissner, i, 413 f. ; Rawlinson, Mon., i, 107, 576 ; iii, 18 ; Maspero, Dawn, 739.

OILS AND FATS

Lard was used as food and was offered to the gods.[1] In the oldest and latest periods alike, as Herodotos [2] and the Talmûd say, the only oil in use was sesame oil. Gudea speaks of great stores of oil in summer time and large oil jars were found in a temple at Ur (2080 B.C.). Sennacherib's (705–681) attempt to cultivate the olive in Assyria was a failure. The yield of one-third of oil from sesame (šamaššammu) was good : it was used for food, anointing and illumination, and for religious and medical purposes.[3]

Butter (definitely mentioned by Ašurnasirpal II) was probably used by the Sumerians ; the "juice of Enurta," used in metal buckets for anointing, was a mixture of oil, butter and honey.[4] Round cheeses are shown on a Sumerian seal at Ur.[5] Oil for anointing (pašašu) was called piššatu : it appears to have been perfumed with flowers. The hair, hands and feet were coloured with henna.[6]

AROMATIC MATERIALS

Perfumes and aromatic oils and resins were largely used[7] : a text mentions four different kinds of resin.[8] The aromatics were favourite offerings to the gods and Herodotos [9] reports that a thousand talents of incense were burnt annually in the temple of Bêl at Babylon. The aromatic materials included oils of cedar, cypress, myrtle, etc., myrrh (murru), bdellium (budulchu), Indian myrrh, nard (lardu), ladanum (ladanu), sweet flag, saffron, cyperus grass (su'âdu), hemp and myrtle (asu), and a composition of cosmetics and drugs called "life oil," which has not been identified, was used as an offering. The myrrh, bdellium, nard and ladanum resin for use as incense and cosmetics were imported, the latter from Damascus under Tiglathpileser III (745–727), although Sennacherib succeeded in growing myrrh at Nineveh.[10] In a text of 698 B.C. a criminal drinks as punishment "blood of the cedar," perhaps turpentine.[11]

Although classical authors mention "Assyrian" nard, amomum and perfumes generally, and "Syrian" myrrh and frankincense, yet in reality no spices are produced in Assyrian territory and they must have been imported. Assyria may have traded in these spices between Arabian and Indian producers and the Syrian coast.[12] Arrian says the town of Eridu was an entrepôt for incense and spices from Arabia.[13]

Cinnamon, usually supposed to have been known to the Jews from the time of the Exodus, was probably imported into Assyria from the Arabians, who may have obtained it from Ceylon or Malabar, the most accessible countries producing it, as there is no true cinnamon in Arabia. It was imported into Greece by Phœnicians.[14]

TEXTILES

Spinning and weaving were known very early, flax (kitû ; pištu) being grown in the Sumerian Period for making linen and probably muslin (called after

[1] Meissner, i, 221.
[2] i, 193.
[3] Meissner, i, 54, 198 f. ; Luckenbill, ii, 173—"oil of the fruit tree"; Woolley, 1928, 152.
[4] Fossey, Magie, 345, 347, 349 ; Hilprecht, Excav. in Assyr. and Bab., 1904, 249 ; BMGB, 43.
[5] Ur Excav., 1934, ii, 333.
[6] Meissner, i, 238, 411 f.
[7] Ib., i, 243, 353, 411 ; E. Rimmel, Le livre des parfums, [1870], 91 f. ; Egyptian perfumes, ib., 27 f.

[8] Fossey, Magie, 393.
[9] i, 183.
[10] Meissner, i, 201, 209, 243, 309, 353 ; ii, 84 ; Luckenbill, ii, 171.
[11] Delaporte, 336.
[12] Bochart, Geographia Sacra, lib. ii, cap. 18, ed. Leyden, 1712, 103 ; Rawlinson, Mon., i, 562.
[13] Meissner, ii, 85.
[14] Herodotos, iii, 107 and Rawlinson's note ; Pliny, xii, 19 ; Rawlinson, Mon., i, 560.

TEXTILES 307

Môsul).[1] Seeds of flax were found at Khafaje (before 2450 B.C.).[2] Pieces of linen were found attached to skeletons in tombs, but few traces of actual fabrics remain. Specimens of "fine linen" (bûsu ; Greek byssos) were found, at Susa, where the texts say that it was made.[3] Balls of string of hemp or flax are represented.[4] Although linseed, the seed of the flax plant, was used medicinally, there is no mention of its oil.[5] There was a large linen factory at Borsippa [6], and native accounts say the linen of Eridu was especially good.[7] In a list of Sargon II's time (722–706) linen and byssos (cotton) garments appear together, the latter being more numerous. Linen was the dress of the priests.[8]

In a bilingual description of Nineveh under Sennacherib (who imported foreign plants), written in 694 B.C. on a clay prism, it is said that "the trees that bore wool (sindhu) they clipped and they shredded it for garments," and the reference is to cotton. The mention of vegetable cloth in old Akkadian ideograms is doubtful, and could also mean linen.[9] Indian cotton is mentioned as "tree wool" by Herodotos[10] and was perhaps imported into Babylonia and Assyria by way of the Persian Gulf[11], although cotton may have come from Egypt.[12] The small scraps of cotton with a purple dye (probably madder) found at Mohenjo-daro were a coarse kind, produced by a plant closely related to Gossypium arboreum, whilst the fine Indian cotton (sindhu) is supposed to have been a true cotton.[13] Early intercourse between Babylonia and the west coast of India had been assumed by Sayce.[14]

In Sargon II's time a cotton dress and a linen shirt each cost 1 mina (1 lb.) of silver.[15] The finely woven muslins of Babylonia, dyed in brilliant colours, were famous in foreign countries. A room in Nero's palace was hung with Babylonian tapestries which cost 4,000,000 sesterces (over £32,000), and Cato sold for 800,000 sesterces a Babylonian tapestry because it was too splendid for a Roman to wear.[16] According to Maspero[17] the work was needle embroidery or painting on stuffs, and neither in Babylonia nor Egypt was *true* tapestry ever known. Working in gold thread is also mentioned by Pliny as an invention of an Asiatic king Attalus, and this type of work is a very old Eastern art, mentioned in the Bible[18] and practised also in Persia, Lydia, etc.[19] Fringed woollen shawls are represented in the sculptures, as well as rich woven stuffs and turbans (μίτρα) worn with shaven heads.[20] The characteristic flounced skirt (kaunakes of Greek authors), used in Ur Nina's time and in the Akkadian Period, is clearly shown on statuettes from Ur of 2700 B.C.[21]

The wool came from flocks on the banks of the Euphrates and was woven into tapestries, curtains, robes and carpets in brightly coloured patterns

[1] Woolley, 1928, 115 ; Meissner, i, 199, 254 f.
[2] Frankfort, Jacobsen and Preussen, Tell Asmar and Khafaje, 1930–31, *Univ. Chicago Orient. Inst. Publ.*, No. 13, 98 f.
[3] Rawlinson, Mon., i, 100 ; Meissner, i, 135, 256 f.
[4] Rawlinson, Mon., i, 524.
[5] Meissner, i, 199.
[6] Strabo, XVI, i, 7 ; 739 C.
[7] Rawlinson, Mon., ii, 570 ; Meissner, i, 255.
[8] Meissner, i, 135 ; ii, 55.
[9] BMGB, 225 ; CAH, iii, 773 ; Budge, Herbalist, 50 ; Luckenbill, ii, 170, 172 ; Rawlinson, India, 1926, 2, who says Ašurbanipal instead of Sennacherib.
[10] iii, 106 ; vii, 65 ; Schrader, RL, 61.

[11] Meissner, i, 209 ; Sayce, B. and A. Life, 108 ; Rel., 137 f.
[12] Feldhaus, Technik, 73 f.
[13] Marshall, Mohenjo-daro, 1931, i, 33.
[14] Rel., 137 ; B. and A. Life, 108.
[15] Meissner, i, 135.
[16] *Ib.*, i, 256 ; Sayce, B. and A. Life, 108 ; Rawlinson, Mon., ii, 570 f. ; Perrot-Chipiez, Chaldæa, ii, 365 ; Pliny, viii, 48 : colores diversos picturæ intexere Babylon maxime celebravit ; picta was an embroidered robe.
[17] Dawn, 758.
[18] Ezek., xxvii, 24.
[19] Blümner, i, 155.
[20] Herodotos, i, 195.
[21] Maspero, Dawn, 718 ; Woolley, 1928, 56 ; B. Mus. Quarterly, i, 38, plate xix.

(Joseph's "coat of many colours").[1]   The later Babylonian carpets, dyed in various colours and representing griffins and monsters, were very famous and were largely exported.[2]   Wool was treated by the fuller (qâsiru), who trod it with the feet in a pit, beat it with a mallet (mazûru) and carded it with thistles. The washerman (ašlaku) beat the woollen or linen clothes in water with a mallet, with soap or potash (soda ?) (nitiru), or with potash (soda ?) lye (uchûlu), or with alum (?) (šikkatu) : the strong sun then bleached the fabric. Black, white and brown wools were of natural colours : grey was obtained by spinning black and white together.[3]   The Sumerians, who shaved the head, sometimes wore ceremonial wigs.[4]

The Babylonian purple and embroidered robes are mentioned by Roman poets[5] : the wool garments were dyed blue and red, like those of Asiatics shown in Egyptian pictures, who wear clothes with red and blue designs in rings, dots, zigzags, stripes, etc.[6]   A text speaks of "linen and coloured clothes," and Assyrian records from the 12th century B.C. repeatedly speak of wool dyed the colour of blood.[7]

DYES

The dyes, which were mainly vegetable, included safflower and henna (found in stone pots) in the early period [8], but according to Ktesias (401 B.C.) kermes was imported from near the Indus and Indian lac may also have been used as a dye.[9]   A "robe of many colours" is mentioned in a hymn to the gods, as well as a "dark blue dress,"[10] perhaps dyed with indigo. The dyeing of cloth, although not actually mentioned in the texts, was, of course, practised[11], and purple, black, violet (or blue), bright yellow and white wools are described or represented. "Brilliant purple" is argamânu and violet or blue is takiltu. The king's tiara was probably of red felt or cloth ; felt was used by Scythian tribes as a tent covering.[12]

Ašurnasirpal II (885–860) and Šalmaneser II (860–825) received red and purple garments as tribute, and coloured and linen garments from Lubarna, a Hittite king.[13]   Imitations of purple appear in old Babylonian and Assyrian recipes.[14]   Red (šaršerru) and yellow or golden pastes (šipu ; lêru ; damatu) were used as cosmetics and as colours for walls, and the priests wore some red garments.[15]

TANNING

Leather was tanned, apparently in the Sumerian Period and certainly later, both with alum and galls and was of good quality at least as early as 700 B.C.[16] The oaks growing in Assyria yield galls.[17] Tušratta, king of the Mitanni, sent Amenhotep III of Egypt shoes of sheep leather made up with coloured (blue-purple) linen and decorated with beads of gold and silver. The leather was sometimes dyed red and blue. The tanner was called susikku, the shoemaker aškapu ; the sârip tachšê probably dyed the leather. The shoemaker tanned

[1] Sayce, B. and A. Life, 108 ; Meissner, i, 67 f.
[2] Rawlinson, Mon., ii, 570 ; Meissner, i, 257 ; mosaic representing a carpet, ib., Fig. 69.
[3] Meissner, i, 254 f.
[4] Woolley, Digging up the Past, 1930, 136.
[5] Rawlinson, Mon., i, 396 ; Contenau, Manuel, iii, 1188.
[6] Maspero, Dawn, 719, 759, calls these persons "Babylonians"; Meyer, Sumerer, 13, "Syro-Palestinians."
[7] Meissner, i, 136 ; Luckenbill, i, 80, etc.
[8] Sayce, B. and A. Life, 108 ; Handcock,

327 ; Maspero, Dawn, 554 ; Meissner, i, 244.
[9] Rawlinson, Mon., i, 560.
[10] Sayce, Rel., 491, 492.
[11] Smith, CAH, iii, 102.
[12] Meissner, i, 255 ; Rawlinson, Mon., i, 465, 486, 496.
[13] Sayce, Records of the Past, N.S., 1889, ii, 143, 144, 155, 170, etc. ; Luckenbill, i, 144 f., 217 f. ; Rogers, ii, 65.
[14] Darmstaedter, MGM, 1928, xxvii, 28.
[15] Meissner, i, 244 ; ii, 55.
[16] Ib., i, 52, 257, 466 ; Handcock, 349 ; Sayce, B. and A. Life, 131.
[17] Rawlinson, Mon., i, 217.

skins sent to him and besides shoes made saddlery, hoses, etc. Parchment (ni'âru) of good quality was made and used by the later Babylonians and Assyrians for Aramaic documents, instead of clay tablets.[1]

IVORY

Ivory (šinni pîri) was not found at Tello [2], but Loftus [3] found one piece of uncertain date at Warkâ, and it occurs sparingly in Predynastic and III dyn. graves at Ur.[4] Ivory was hardly known in the earliest period, when shell and mother-of-pearl were used for inlays[5] ; that used extensively in the Assyrian Period, e.g. for combs, may have been native (see below), Indian, or African, obtained by way of Egypt and Syria or Phœnicia, where it was also carved before export.[6] Ašurnasirpal II and Šalmaneser certainly obtained ivory as tribute from the Phœnicians [7], and Hezekiah of Judah sent it to Sennacherib (686 B.C.).[8] The carved ivory of Ašurbanipal (668–626) was either imported from Egypt or made after Egyptian models. In the early periods the elephant was hunted on the northern frontiers : Tiglathpileser I (1115–1093) killed ten and caught four alive in Harran, and Thothmes III (16th century B.C.) had hunted it near Carchemish.[9]

The large collection of carved ivory found in Ašurnasirpal's palace at Nimrûd shows unmistakable Egyptian influence, although some specimens are purely Assyrian and some show mixed influences, probably Phœnician. This king hunted elephants and once killed thirty himself. Among this ivory were panels inlaid with lapis lazuli or with blue glass and blue and green pastes, and gilded. These were probably for use in furniture, perhaps by mortice or glue attachment to wood.[10] With this ivory were several hundred studs and buttons in mother-of-pearl (used for amulets in modern Palestine and on the Red Sea coast)[11] and ivory, also various objects in ivory, the workmanship, subjects and treatment of which are more Assyrian than Egyptian. They were perhaps copied from Egyptian models or made by a native of a country (perhaps Phœnicia) under the influence of Egypt. The most interesting ivory relics were a carved royal sceptre and several entire elephants' tusks, the largest 2 ft. 5 in. long, also several figures and rosettes, and four oval bosses with copper nails still remaining ; the legs of a chair were partly ivory and partly wood overlaid with bronze.[12]

[1] Rawlinson, Mon., i, 488 ; Meissner, i, 133, 257, 259.
[2] King, S. and A., 78 ; Meissner, i, 270.
[3] Travels, 223.
[4] Ur Excav., 1934, ii, 439, 471, 489.
[5] King, S. and A., 333 ; Meissner, i, 271.
[6] Meissner, i, 271 ; Perrot-Chipiez, Chaldæa, ii, 320 ; Layard, Discoveries, 1853, 195 ; Rawlinson, Mon., i, 557 ; BMGB, 46, 128, 227.
[7] Sayce, Records of the Past, N.S., 1889, ii,

172 ; Luckenbill, i, 166, 175, 189, 232 ; Rogers, Hist. Bab. and Assyr., ii, 65.
[8] BMGB, 227 ; Luckenbill, ii, 121.
[9] Meissner, i, 73, 270, 273, 323, 354 ; Sayce, B. and A. Life, 136 ; Luckenbill, i, 86.
[10] Handcock, 313 f., Fig. 84 ; Contenau, Manuel, iii, 1318, 1333 ; BMGB, 165, plates 41 and 42 ; Meissner, i, 73 ; Rawlinson, Mon., i, 375.
[11] Budge, Amulets, 70.
[12] Layard, Nineveh and Babylon, 1867, 58, 62 f.

BABYLONIA AND ASSYRIA III—3

MEDICINE

MEDICINE AND MAGIC

The physician in Mesopotamia was generally a magician, since medicine was regarded as a branch of magic, and he is named as one of the three types of priests : conjurers, physicians and theosophists.[1] Sickness and disease were caused by evil spirits, which could be expelled, or more easily passed on to another person, only by incantations, or when these failed, by medical treatment. Specific diseases were due to, or actually were, specific (named) demons, and the use of incantations against these goes back to the old Sumerian Period, when magic was developed.[2] The old Sumerians apparently regarded man as a microcosm in which the macrocosm of the universe was reflected, and all evil, including disease, was originally regarded as the result of sin, the cure for which was ultimately to be sought in religious rites of purification rather than in the use of drugs.[3]

In some cases the cause of illness was regarded as an evil force, and all the results of these evil forces were called the bann (Sum. sagba ; Akkad. mâmîtu) : it was treated by exorcism like a demon, several demons being named so that none should be missed. The great number of ways in which the demons might enter are stated.[4] The transference of diseases to animals, etc., also features in Pliny's "popular" medicine, based on magic. These include the cure of a boil by rubbing nine grains of barley round it, each grain three times with the left hand, and then throwing them all on a fire. Similar cures for warts still persist in English villages. Great numbers of examples of the use of amulets occur in Pliny : roots suspended from the neck by a thread, a fox's tongue worn in a bracelet, coloured strings, cloth, skin, a box, a nail, etc., are specified.[5]

The technique of Babylonian and Assyrian magic comprised a number of steps.[6] (1) The divination of the cause of the trouble or disease, e.g. lack of attention to the gods, impurity or sorcery. (2) The purification of the affected person by ablution, or fumigation, or both, often accompanied by an incantation, the waters of the Tigris and Euphrates preserved in bronze "seas" (apsu) being especially efficacious. Herbs could be added to the water, or else plants, seeds, woods, milk, cream, butter, copper, silver, gold, stones and gems, gypsum, asphalt, salt, etc., were used as active agents of purification. The metals, the "pure things," e.g. "strong copper," were perhaps touched or handled. Fumigations with cypress and resinous woods (charru) were also used. (3) Destruction of the evil when drawn into and collected in water, fruits, onions, etc., by burning (Akk. maqlû). Statuettes of clay, dough,

[1] Fossey, Magie, 135 ; Lenormant, Chaldean Magic, 1877, 14 ; Maury, La Magie et l'Astrologie dans l'Antiquité et au Moyen Âge, 1860, 24 ; Sayce, Rel., 329 ; Meissner, ii, 231, 343 f. ; Pinches, ERE, x, 284.
[2] Jastrow, Civ., 240 f. ; Meissner, ii, 198 f., 201, 214, 217, 235, 241, 293 ; Lenormant, Magic, 12, 70, 126, 141 ; Fossey, Magie, 18, 144 ; King, ERE, viii,

253 ; ib., Magic, 134 ; Sayce, Rel., 315, 345, 441 ; Budge, Herbalist, 36 ; Maspero, Dawn, 690 ; Dhorme, La Religion Assyro-Babylonienne, 1910, 246; Woolley, 1928, 126.
[3] Jeremias, Der Alte Orient, 1929, xxvii, Heft 4, 16.
[4] Meissner, ii, 217, 225.
[5] Thorndike, Hist. of Magic, 1923, 88 f.
[6] Fossey, 65 f. ; Meissner, ii, 210 f.

etc., could be burnt, when Gibil the fire god destroyed the demons, or they could be buried. The purifying action of fire destroys the "germs" of evil. (4) The transmission of evil to animals or images, magic knots and rings, minerals, plants, saliva, ointments, etc. (5) The use of drugs and minerals, which probably acted by "sympathy," e.g. of colour. Some of these materials are known from Assyrian lexica and are mostly quite common and without "mystery." (6) The use of incantations and imprecations. (7) The use of amulets and talismans. (8) The invocation of gods, who were first placated by offerings and sacrifices, and when this was done the incantation was uttered.[1]

The incantation (šiptu), which must be recited in the correct form with strict adherence to detail as well in the text as in the accompanying ceremonies [2], was generally combined with the use of talismans, pieces of textiles, statuettes, amulets, etc., and with prayers and mysterious hymns intended to invoke the gods and spirits. They were very effective : in one case "the earth gaped" and crowds of spirits rushed out during the ceremony.[3] Several texts of incantations in British Museum tablets (1900 B.C. and later) [4] include prayers against evil demons with various names, e.g. utukkê, lemnutî, etc., whose home is the desert and who are especially grouped in sevens and related to the spirits of the dead. A "good demon" (lamassu damku) was represented outside temple doors.[5] Efficacious spells depended on the "supreme name" of the divinity, which the god Ea alone knew, but which he communicated to his son Marduk.[6]

MEDICINE AND PHARMACY

Whereas Egyptian medicine gradually left off its magic components as it progressed, that of Babylonia and Assyria, which was permeated with magic and sorcery, never freed itself from them and never became a scientific system.[7] The Babylonian physician was unimportant, if we can credit the testimony of Herodotos [8], who says (incorrectly, but perhaps correctly if he meant "apart from priests") : "they have no physicians, but when a man is ill they lay him in a public square and the passers-by give him advice."

The physician (âsû) was at the same time a sukkallu priest, since anointing formed part of his duties : the name âsû (Sum. azu) means "water knowing," perhaps of medicinal waters or the bodily juices [9], but perhaps also he who knows the use of lustration and the ritual of Ea. The physician is either âsû, "one who understands water," or iazu, "one who understands oil," i.e. lecano-mancy. Sprinkling with holy water, connected with Ea, continued in medicinal use till the latest periods. Medicine, like all other arts and sciences, was regarded as a gift of the gods, especially of Ea, the "master of the physicians." His element water was especially useful in healing, and he stood in close relation to the physician and determined the fate of the curative plants.[10]

According to Sayce, Babylonian medicine improved by contact with Egypt during the XVIII dyn., and drugs were also in use as alternatives or additions to incantations.[11] Jastrow[12], however, considers that medicines were chosen

---

[1] Fossey, Magie, 70, 72, 74, 79, 82, 88, 90, 101, 104, 122 ; Meissner, ii, 77, 208, 223, 225, 227, 230, 238 ; King, Magic, p. xxx.

[2] King, Magic, 149 ; Sayce, Rel., 319 ; Fossey, 13 ; S. A. Pallis, Mandæan Studies, 1926, 45, calls šiptu "exorcism."

[3] Maspero, H. anc., 1904, 172 f. ; Dawn, 696.

[4] BMGB, 122.

[5] Lenormant, Magic, 17, 25, 204, etc. ; Schrader, Keil., 455, 458 f. ; Fossey, 21 ; Maspero, Dawn, 634 ; Jastrow, Civ., 240.

[6] Maspero, H. anc., 172 f. ; as in Egypt and in the Jewish Qabbalah.

[7] Jastrow, Civ., 211, 246 ; Meyer, Alt., I, ii, 587 ; Breasted, Isis, 1931, xv, 356 ; Maspero, Dawn, 780 ; Meissner, ii, 292 ; Sayce, Rel., 304, 317.

[8] Herodotos, i, 197 ; Lenormant, Magic, 35 f. ; Meissner, ii, 283.

[9] Pinches, ERE, x, 287.

[10] Budge, Herbalist, 26 ; Jastrow, Civ., 211, 246 ; Meissner, ii, 284.

[11] Rel., 317 ; B. and A. Life, 163 ; cf. Meissner, ii, 283.

[12] Civ., 251 ; cf. Fossey, 63, 104 ; Meissner, ii, 233.

merely because, by their bitter, pungent or malodorous properties, they were unpleasant to the evil demon and drove him out ; or else, when things like milk, honey and butter were used, their pleasant properties coaxed out the demon causing the disease.

The goddess Ninchursag created the eight lower healing gods, each of which cured a special disease. The most potent divinities in medicine were Ninurta, the dissolver of spells, and his wife Gula, "who made the dead (the very ill) live" by the "touch of her hand" ; her emblem was the dog (cf. the clay dogs buried in foundations). Gula, a "great physician," had temples in Babel, Borsippa and Assur : in Isin she was called Ninkarrak or Nin-isinna.[1] The god Ninazu was "master of physicians" and his son Ningišzida had as emblem a staff with two serpents twined about it (shown on a steatite vase of Gudea), which is still the sign of the physician as the so-called "staff of Asklepios." The serpent is typical of healing because he casts his skin and "becomes young again," and the snake, as the epicene god Sachan, was worshipped at Der as "master or mistress of life."[2]

A physician is mentioned in the period of Gudea and the names of several Assyrian royal physicians are known. In Hammurabi's time there were veterinary surgeons.[3] In Sippara in the same period the druggists lived in a separate street.[4] The Babylonian physicians seem to have been renowned, since Mutallu, the brother and predecessor of the Hittite king Hattusil, sent for a physician and an exorcist, but as compared with magicians they were not very numerous. Assyrian physicians also went to Egypt, but the Egyptian doctors were generally preferred in foreign countries, although Assyrian goddesses were "lent" (in the form of their statues) to Egypt for medical purposes. In the Persian Period the royal physicians were Greek or Egyptian.[5] Ancient medicine and pharmacy were decidedly international, and drugs such as cypress oil, myrrh, cedar oil and incense gums were imported from foreign countries through agents.[6]

The payment of physicians depended on the monetary circumstances of the patient when the treatment was successful : if it failed, the physician could have his hand cut off, hence he did not operate where the result was in doubt. The medical fees charged to a nobleman were double those of a burgher (muškinu).[7]

Scientific medicine developed from popular medicine and long remained under the influence of superstition. Some descriptions of symptoms remain. Mental disease was due to "possession," and toothache, as is often explained in the texts, was due to a "worm." Poisoning (šimmatu) was recognised, including that by bites of snakes and scorpions, the Babylonian name for poison being intu.[8]

MEDICAL TEXTS

The medical texts, perhaps parts of an encyclopædia, are mostly late, but some tablets from Boghazköi which are Hittite copies of Babylonian (perhaps Sumerian) originals go back to the 13th century B.C.[9] From the "temple library" at Nippur, 23,000 literary and 28,000 business tablets, the oldest from about 2500–2200 B.C., were recovered, and these included mathematical,

[1] Meissner, ii, 31, 33, 207, 284.

[2] Jeremias, 277, Fig. 178 ; Meissner, i, 261 ; plate-fig. 126 ; ii, 284.

[3] Sayce, B. and A. Life, 162 ; Meissner, ii, 285, 321.

[4] Schmidt, Drogen, 2.

[5] Delaporte, 51 ; Meissner, ii, 126, 285 ; Maspero, Dawn, 780.

[6] Herodotos, iii, 1, 129 f. ; Schmidt, Drogen, 2 f., 6.

[7] Meissner, ii, 286 ; Meyer, Alt., I, ii, 640—Hammurabi's code of laws ; Woolley, 1928, 96.

[8] Meissner, ii, 234, 287, 291, 296, 314 ; King, Magic, 148.

[9] Budge, Herbalist, 37 ; Meissner, ii, 291, 296.

astronomical and medical texts, very few of which have been published.[1] The numerous fragments from Assur are late Assyrian and carelessly written : most of the texts (including those in the British Museum) are from Ašurbanipal's library, or are late Babylonian. Hammurabi's code shows, however, that medicine was practised at a much earlier period. Anatomical knowledge was weak, since dissection was forbidden and no surgical operations are described. In one incomplete text the parts of the body are associated with gods, perhaps an astrological speculation on the relation of the macrocosm and microcosm.[2]

The medical texts are very numerous ; over 450 are catalogued in the British Museum. A useful preliminary account of them was published, with some translations, by Küchler [3] and Ebeling [4], but their first satisfactory study [5] is by R. C. Thompson [6], who has published practically all of them. Some of these texts were copied from originals of about 2200 B.C., and these in turn refer to traditions "before the Flood"—which was itself an historical event revealed by excavations at Ur and Kiš.[7] Vegetable products and drugs are mostly specified : onions [8], "dog's tongue" (a plant) rubbed in oil (13, 31, 57), "plant of Dilbat in oil" [9], good wine or diluted wine (13, 17, etc.), intoxicating drink (5, 23, 25, 31, 33), date juice (37), honey (45), cypress and cedar preparations (very often), oil of ma-pin-ma (3), myrrh (59), aloes (?) (95), salt (3, 9 21, 23, 31, 47, 72) of different kinds and gypsum (51) for "an accident" (possibly plaster of Paris for setting a limb). In some cases the treatment is to be applied when the stars have a certain appearance (7, 27, 67 f.), probably with an astrological significance, and the unlucky patient would have to wait till the proper constellation turned up. Magic formulæ are given for repetition (11). There are already anticipations of the Greek theory of four humours, e.g. sualu = phlegma (65), which Küchler wrongly thought indicated a late date for the texts. Filtration through a cloth is described (28, 80, 81, 142), as in the Egyptian Ebers Papyrus.

Various methods of preparing drugs are given, but not distillation : drying, powdering, uniformly grinding, stamping fine in a mortar (madakku) and pestle (elîtursi)[10], mixing together, pressing out juices, peeling, filtering, decanting, shaking, mixing, dropping (liquids), moistening, kneading, heating over a fire of thorn branches in a small kettle or large pot or a lustration vessel on the hearth, heating in a furnace, warming, roasting, boiling and seething. The operation may be repeated, often several times. The quantities are not usually given, but sometimes the number of barleycorns, shekels, or minas is stated for weights, or silas for liquids.[11]

The smell of copper dust is enough to cure some diseases ; making a fire before the sick person, fumigation, application of poultices, powders and

[1] Hilprecht, Die Ausgrabungen der Universität von Pennsylvania im Bêl-Tempel zu Nippur, Leipzig, 1903, 17, 53 ; ib., Babylonian Expedition of the University of Pennsylvania (numerous vols.), Philadephia, 1906, Series A, vol. xx, pt. 1 (several parts of this) ; ib., The Earliest Version of the Babylonian Deluge Story and the Temple Library of Nippur, Philadelphia, 1910, 15 ; ib., The Excavations in Assyria and Babylonia, Philadelphia, 1904, 511, 524.

[2] BMGB, 160 ; Meissner, ii, 291, 317 ; Sarton, Isis, 1931, xv, 356.

[3] Beiträge zur Kentniss der assyrisch-babylonischen Medizin, Leipzig, 1904.

[4] A. Med., 1921, xiii, 1, 129 ; 1923, xiv, 26, 65.

[5] Meissner, ii, 283 f., 381 ; Budge, Rise and Progress of Assyriology, 1925, 180.

[6] Assyrian Medical Texts from the Originals in the British Museum, Oxford, 1923.

[7] Budge, Herbalist, 37 f. ; ib., Assyriology, 180 ; Thompson, Isis, 1925, vii, 256 ; 1926, viii, 506.

[8] Küchler, 7 ; the references in the text are to this work.

[9] Fossey, 374.

[10] Meissner, i, 261.

[11] Ib., ii, 310 ; Sayce, An Ancient Babylonian Work on Medicine (B. Mus. text), Z. f. Keilschriftforschung, 1885, ii, 1, 13.

lotions, massage, anointing, snuffing, etc., are used externally; draughts or "eating" internally, and emetics, siphoning and blowing in drugs through tubes all appear. There are complicated recipes like those in the Egyptian Ebers Papyrus, e.g. one with 75 ingredients.[1]

Many of the medical tablets are pure recipes without magic and in this form they passed to Syrian and Hebrew medicine.[2] Surgical instruments were used even in Hammurabi's time : the physician carried them in a case (takâltu) with bandages (sindu), and a seal of an old physician, Ur-Lugal-edinna, found at Tello represents two probes and two ointment pots.[3] A copper lancet is mentioned in a legal text [4], and bronze surgical instruments (scalpel, saw, trepan, two knives) and a small obsidian knife were found at Quyûnjiq.[5]

A tube (uppu) of bronze was used for blowing a mixture of myrrh, storax and "salt of Akkad" into the eyes, or storax in oil into the urinary organs, and a bronze spatula for applying drugs to the eyes.[6] Cloth and parchment were used for poultices, and copper, arsenic and antimony compounds in treating diseases of the scalp, skin and eyes.[7]

DRUGS

Babylonian and Assyrian authors made lists of drugs, plants, animals and minerals, with synonyms and equivalents [8] : about 250 drugs were vegetable, 120 mineral (including antimony) [9], and 180 "others."[10] Vegetable products (not all free from doubt) are most numerous and the name for herb, šammu, was used generally for medicine.[11] They include rosewater (as a medium), gums, oils or other products of cedar, cypress, laurel, myrtle, tamarisk and juniper ; gums and gum-resins such as storax, galbanum, myrrh (murru), liquidambar, asafœtida and opoponax ; narcotics such as hemp, opium[12], lolium temulentum, mandragora (namtar-ira plant) and some of the solanaceæ ; and the stomachic value of many umbelliferæ and peculiar uses of anemones and chamomiles, liquorice and mustard are recognised. The fruits, juices or skins of pomegranate, lemon, apple and medlar are used, and "various alcohols" (wines), oils and fats are specified as vehicles or solvents. According to von Oefele[13], several hundred plants used for medicinal purposes by the Babylonians passed into Greek pharmacology.

Seaweeds were burnt to a kelp containing iodides[14]; gallnuts and oak galls, pomegranate, acacia and sumach were used as astringents and also probably for tanning.[15] Sweetwood root (perhaps liquorice)[16], sunflower root, purgatives and emetics are specified.[17]

There are some interesting old Babylonian plant fables similar to those in the Bible.[18] The "plant of life" is eaten by gods to prolong their existence and used as snuff.[19] Curative plants mentioned in an incantation are saffron (Akk. asupiru), mustard (sichlu), cassia (kašu) and thyme (chašû). Other

[1] Meissner, ii, 123, 311, 313, 317.
[2] Thompson, CAH, iii, 240.
[3] Meissner, ii, 285, plate-fig. 43 ; Contenau, Manuel, ii, 691.
[4] Woolley, 1928, 112.
[5] Meyer-Steineg and Sudhoff, Geschichte der Medizin, Jena, 1922, 19, Fig. 10.
[6] Meissner, ii, 213.
[7] Thompson, CAH, iii, 241 ; ib., Proc. Roy. Soc. Med., Hist. Med. Sect., 1924, xvii, 10.
[8] Meissner, ii, 304 f., 352, 381.
[9] Budge, Herbalist, 44.
[10] Thompson, Assyrian Herbal, 1924, p. v. ; ib., CAH, iii, 240 f., 250 f. ; Isis, 1926, viii, 506.

[11] Budge, Herbalist, 40 ; Meissner, ii, 304.
[12] Assyrian Herbal, viii, xviii, 41 f., 46.
[13] Q. by Schmidt, Drogen, 6.
[14] Thompson, Assyr. Herbal, 99 f. ; CAH, iii, 240, 249 f.
[15] Ib., Assyr. Herbal, xi, 122, 124 ; xxiv, 109 ; ib., Proc. Roy. Soc. Med., Hist. Med. Sect., 1926, xix, 37, 43 ; Nierenstein, Incunabula of Tannin Chemistry, 1932, 159 ; ib., The Natural Organic Tannins, 1934, 174, 203, 220, 245 f.
[16] Theophrastos, Hist. Plantarum, IX, xiii, 2 : γλυκεῖα ῥίζα.
[17] Meissner, ii, 295, 306, 313 f.
[18] Judg., ix, 5 ; Jeremias, Alte Orient, 1925, xxv, Heft 1, 6.
[19] Meissner, ii, 123, 317.

plants are mixed with oil or wine, or boiled with water, etc., and salt is also held in the mouth, the saliva being swallowed.[1]

Various herbs were used, including sapru, thorn-weed, sammu, cassia, khaltappan, kitmu, ararû, mukhurtu, each with an appropriate incantation. The white cedar was a defence against evil, cone-bearing plants were prophylactic, and many plants were no doubt regarded as possessed of magic properties.[2] Some plants have names which can be translated but not identified: fish-plant, sun-plant, fox wine (nightshade ?), bitter corn, field clumps, "plantain," field stalk, tongue (ground ivy ?), serpent's ear, horned alkali (a kind of salicornia ?), sea tooth, bitter plant and "plant of life." Plants which "have stood on graves or were gathered out of the sun" are mentioned (superstition ?).[3]

Special parts or products of plants for medicinal use are roots, stem, skin, bark, resin, green, blooms, creeper, branch and blooms, "blood" (sap ?), fruit, cones of conifers, seeds, grains, flour, malt (?), dough, porridge, mixed food, oil, mucilage, sherbet, beer, intoxicating drink, wine and tartar of wine.[4]

The designation of plants and minerals as male or female is old Babylonian.[5] The "male" plants were often the stronger or darker varieties, although sex proper was recognised only in the date palm.[6] The term male as applied to ašlu (cyperus) and namtar (mandrake), and male and female to nikibtu (liquidambar orientale), referred merely to outward appearances, the latter to the shape of the gum as in the description of frankincense in Pliny.[7] The "human form" of the mandrake was early recognised.[8]

PLANTS NAMED IN ASSYRIAN TEXTS

An early study of the plant lists in the British Museum tablets was made by Boissier.[9] The following list is compiled from Meissner[10] and Thompson[11], some of whose alternative identifications are given in square brackets in the list. Thompson also consulted Syriac works and compared the texts with the modern flora of Mesopotamia, with the drugs of Oriental and Classical writers and of modern pharmacopœias, and with descriptions in other Semitic languages. He remarks that the Assyrians did not arrange plants according to modern ideas of botany, but began with grasses, followed by rushes and Euphorbiaceæ, and then grouped the Papaveraceæ and Cucurbitaceæ together merely because the names for the principal plants began with the sign hul. Compositæ were scattered throughout the series. The arrangement is that of a rather superficial but laborious cataloguer, but "the doctors and chemists [pharmacists ?] of Nineveh" possessed great knowledge.

almond	šiqdu	bdellium	budulchu
"ammi"	nînû	bean	pulilu
apple	chašchûru	beet	silqu
asafœtida (?)	surbu	belladonna	šag-šar
asphodel (?)	urânu	black cummin	zibû
barley	še'u or ašnan	box (?)	urkarînu or šimeššalu

---

[1] Meissner, ii, 229, 314 f., with specimen recipes.

[2] Jastrow, Civ., 252; Farbridge, Symbolism, 31, 33; Ragozin, Chaldea, 162; BMGB, 201, 221.

[3] Meissner, ii, 306.

[4] Ib., ii, 307.

[5] Zimmern, Z. f. Assyriol., 1925, xxxvi, 187.

[6] See p. 302.

[7] Thompson, Assyrian Herbal, p. xix f., 142, 189; Isis, 1926, viii, 507, 508.

[8] Lippmann, Abh., i, 195; old Egyptian representations in Keimer, Gartenpflanzen im alten Ägypten, 1924, i, 173.

[9] Revue Sémitique d'Épigraphie et d'Histoire Ancienne, ii, 135–145, q. by Maspero, Dawn, 782.

[10] i, 210 f.; ii, 304.

[11] The Assyrian Herbal, 1924, unknown to Meissner.

caper baltu
caraway kamûnu
cardamoms qaqûlu
cassia kasû [Persian rose]
cedar erinu
chaste tree (Vitex agnus) šunû
chicory kukru [fir turpentine]
coriander kusiburru
cress šachullatu [lolium]
crocus kurkânû [turmeric]
cummin kamunu
cyperus grass su'âdu
cypress burâšu or dupranu
date palm gišimmaru
dill šibêtu
dog's tongue lišân kalbi
emmer bututtu
fig tittu
flag qanu
flax kitu
garden rocket girgirû or engengirû
garlic šûma
gherkin qiššû
hemlock (?) gi-šul-šar
henbane (?) šakiru
hyssop zûpu
incense (?) lubbânu
ladanum ladunu
laurel (?) urnu
leek karâšu [cherry]
lettuce chassu
linseed kitû
lotus tree (?) musukkânu [1]
lucerne aspastu
mountain ash sindu or balachchu [ferula communis]
mulberry (?) sarbatu [1]
mullein bûsînu
mustard šachlû [lolium]
myrobalans ammaluga

myrrh murru
myrtle asu
nard lardu
oleander (?) pallukku or chaldappanu [mustard]
olive (?) sirdu
onion bisru
opoponax (?) ararû
pea challûru
pear kameššaru or chachchu (?)
pine šurmînu
pistacia butnu
plantain dulbu
pomegranate armannu or nurmû (?) [apricot]
poplar (?) e'ru
poppy irru [or hul gil]
pumpkin (?) abtachu
purslane zuriratu or purupuchu (?)
radish puglu
rape laptu
reeds urbatu or šišnu
rose (?) amurdînu
saffron azupiru
St. John's bread tree charûbu
sesame šamaššammu
spurge anunutum or kamti eqli
squill uššurati
squirting cucumber piqqûti or tigilû
strawberry tree girgiššu
sweetwood šûšu or sillibânu
tamarisk bînu or ašlu
terebinth šurmênu
thorns ašâu or baltu or puquttu
thyme zambûru or chašû
truffle kamtu
wheat kibtu
willow (?) chaluppu
xanthium strumarium [2]

MINERAL DRUGS IN ASSYRIAN TEXTS

Mineral substances supposed to be mentioned in the medical texts are[3] :

alum (mostly from Egypt) (xix, 65 f., 69 ; M)
antimony, including "needles," per-

haps crystals of stibnite [4] (xvii, 22 ; xix, 31, 35, 36, 42, 48)
arsenic (?) (xvii, 22 ; xix, 35, 43, 49, 63)

---

[1] Luckenbill, ii, 36, 117, 132, etc., 494 (Sargon II), and Thompson say the mulberry is musukkânu, from which by transposing the first two consonants the Greek name συκάμινος, previously of uncertain origin, is derived ; Dawson, Magician and Leech, 1929, 148.

[2] A black dye growing near Basra.

[3] Meissner, ii, 309, 312 = M ; Thompson, CAH, iii, 240, 249 f. = C. ; ib., Proc. Roy. Soc. Med., Hist. Med. Sect., 1924, xvii, 1-34 ; 1926, xix, 29-78, reference in text by vol. and page only ; ib., J. Roy. Asiat. Soc., 1929, 801 f. = J.

[4] Or, perhaps more probably, merely the sticks for applying kohl.

arsenic sulphide [sandarach or realgar, As₂S₂?], šindu arku (xvii, 22 ; xix, 35, 65 ; C)

arsenic trisulphide (orpiment, As₂S₃) (xvii, 22 ; xix, 35, 65)

borax (?) (xix, 77 ; J)

bronze (xvii, 22)

chalk (M)

cinnabar (sâmtu or sandu)[1] (xix, 77 ; C)

clay, of yellow and other colours (M)

copper (including a saucepan, "hard," "male" and "dust"—perhaps verdigris ?) (xvii, 22 ; xix, 43)

dust from "the gate" or "graves" or "a deserted town," or a "deserted house or temple," or "a river" (M)

furnace shards or bits of old pot (with adhering slag ?) (M)

gold (xvii, 22)

gypsum (gassu) (C, M)

jasper (iašpu) (C)

lapis lazuli (M)

lead (xvii, 22)

limestone (M)

magnetic oxide of iron (xix, 34 ; J)

mercury (?) (xix, 31, 64, 78)

misy (basic sulphate of iron) (xix, 39, 42, 43, 47)

naphtha (naptu) (C)

ochre (xix, 53)

quicklime and almond oil (xvii, 10)

rock salt (M)

rock or mountain stone (valuable) (M)

sal ammoniac (?) [2] (tâbat amânim) (C, M)

salt, salt of Akkad, asallu salt, and various salts and alkalies (xix, 36, 61 ; C, M)

sand from the river (M)

scum from river water (M)

"stone of conception with the embryo in it" (M) [3]

sulphate of iron (green vitriol)(J)

sulphur (xvii, 22 ; xix, 42, 76 ; J, M)

verdigris [4] (xvii, 22 ; M)

vitriol (šubû) (xix, 43)

water from a river or spring (M)

zinc oxide (J)

Drugs of vegetable origin not mentioned above are cantharides (?) (xix, 31, 49), opium (xix, 39, 50, 56), vinegar (xvii, 22) and yeast (šuršummu) (xix, 71).

To restore the dead to life the water of life, closely guarded and given out only by order of Ea, as well as the plant of life hidden in the infernal regions, was necessary.[5] The water of life (apsu) was an old Babylonian idea. The relation of the "water of life" to the "water of release" is uncertain. The plant of life (šammu balâtu), the water of life (me balâti) and the saliva of life (imat balâti) were reserved for the gods, the water of life being kept in the underworld in a leather bag. They made the old young again, but were denied to mankind.[6] The great copper basin containing the water used in ritual ablutions in the temple was also called apsu ; one was found at Tello, and a text describes the setting up of such a basin supported by bronze oxen.[7] The water of life and the plant of life at first belonged to all the great gods. Their special attribution to Ea of Eridu, who had risen from the position of a South Babylonian river god to that of the particular god of the deep and of the waters of the ocean, appeared about 2000 B.C. Both the plant and water of life could, however, still be acquired by other gods and also by heroes such as Gilgameš.[8]

---

[1] Compared by Thompson with the Greek sandyx, which was not usually cinnabar (mercuric sulphide) but red iron oxide or red lead ; cf. Rhousopoulos, in Diergart, 1909, 172—cinnabar known in Greece about 600 B.C.

[2] It is doubtful whether sal ammoniac, i.e. ammonium chloride, is meant ; see p. 297.

[3] Probably a hollow stone with a kernel, which rattled on shaking ; ætites lapis,

Dioskurides, v, 161 ; Adams, Paulus Ægineta, 1847, iii, 227.

[4] Verdigris is properly a basic acetate of copper, but the name is sometimes loosely applied to any green tarnish on copper.

[5] Maspero, Dawn, 693, 697.

[6] Meissner, ii, 113, 123, 149, 184, 420.

[7] Maspero, Dawn, 675 ; Sayce, Rel., 63.

[8] Prinz, Altorientalische Symbolik, 1915, 102, 105, 114, 137, 139, 141 ; Meyer, Chetiter, 148.

318

The gods had sacred animals, both real and fabulous, and there were many animal medicines [1], including birds, reptiles, fish and insects, and their parts or products, such as honey, and castoreum (garidu), presumably brought by caravans from the region of the Black Sea, in the 7th century B.C.[2] Most of these remedies, like similar ones in the Ebers Papyrus, were of no value.[3]

[1] Meissner, ii, 131 f., 307 f.
[2] Thompson, *Isis*, 1928, xi, 405.

[3] Breasted, *Isis*, 1931, xv, 356.

ÆGEAN CIVILISATION; TROY;
CYPRUS

THE ÆGEAN. TROY. CYPRUS.

THE ÆGEAN CIVILISATION

With the completion of the descriptions of the Egyptian and the Mesopotamian cultures, a choice of material for further consideration presents itself. It would be possible to select a treatment based on ethnology and to follow the Semitic culture into Syria and Palestine, then to take up a consideration of Asia Minor and Persia with populations having important "Aryan" elements, ending with the Ægean civilisation based on Crete, which preceded the European Iron Age and the Classical Period—falling outside the scope of the present work. Alternatively, it would be possible to follow a sequence in time, when the early civilisation at Knossos in Crete and the whole Ægean culture dependent upon it would claim the first place. The second choice has been taken, and although other considerations might well weigh against it, the selection is deliberate and has been made after due thought.

The first settlement of man in the Ægean region was probably at Knossos in the later (Neolithic) Stone Age. Other ancient sites in Crete which have been excavated are Gournia, Phaistos, Hagia Triada, Palaikastro and Zakro. Phylákopi in Melos is also an old site, but not so old as Knossos. No remains of the old (Palæolithic) Stone Age have been found.[1] Pottery of the Neolithic period occurs in quantity ; the earliest is unornamented, that of the next period has incised lines often filled with a white substance. At Phaistos, where remains of a city were also discovered, the second kind only was found, hence Phaistos was perhaps settled from Knossos. The Bronze Age (2800–1200), characterised by advanced metal work and polychrome pottery, has been called by Evans Minoan, after Minos, a legendary king of Crete in Homer.[2] It has also been called the Ægean or the Mediterranean Civilisation (although the latter name would, strictly speaking, include Egypt), the early and late periods being called Minoan and Mycenæan respectively. Cycladic refers to a corresponding culture on island sites other than Crete, and Helladic to the mainland of Greece.[4] Although sometimes regarded as a prehistoric Greek civilisation [3], the Ægean culture is non-Aryan and a branch of the Mediterranean culture (perhaps autochthonous) more related to Egypt than to the European Greeks, who were of a different race.[5]

The dating and arrangement of the periods into Early (EM), Middle (MM) and Late (LM) Minoan, adopted in this section, are essentially those of Evans, slightly modified by Hawes.[6] Evans's division of EM is not very obvious in

[1] Homer, Odyssey, xix, 179 ; Homer's Odyssey, Books XIII–XXIV, ed. Monro, Oxford, 1901, 156—the Dorian and Pelasgian passage is perhaps interpolated ; Sir Arthur Evans, The Palace of Minos at Knossos, i, 1921 ; ii, 1928 ; iii, 1930 ; iv, announced in 1934 (all references to Evans are to this work unless otherwise indicated), i, 12 ; Peake and Fleure, Priests and Kings, 1927, 97 ; Meyer, Alt., II, i, 212 ; F. H. Marshall, Discovery in Greek Lands, Cambridge, 1920, 1 f., 112 f.—bibliography.

[2] Evans, i, 13.
[3] Otto, 1925, 55 f., 61 ; Marshall, 26 f.
[4] Glotz, Ægean Civilisation, 1925, p. vi, 13.
[5] Hall, NE, 5 ; ib., Oldest Civ. Greece, 1901, 20 ; Karo, RV, i, 29 ; Ridgeway, The Early Age of Greece, Cambridge, 1901, i, 66, vol. ii appeared in 1932.
[6] C. H. and C. B. Hawes, Crete, the Forerunner of Greece, 1911 ; alternative arrangements in Meyer, Alt. I, ii, 761 ; ib., II, i, 165 ; Glotz, 22 ; Fimmen, Zeit und Dauer der kretisch-mykenische Kultur, 1909, 38 ; ib., Die kretisch-mykenische

the remains [1] and EM III and MM I*a* (some writers subdivide into *a*, *b*, etc.) are difficult to separate [2], but MM II is a well-defined period and Meyer and Fimmen separate it from the rest [3], whilst Glotz does not. Meyer [4] rejects Evans's date of 10,000 B.C. for the earliest Neolithic settlement.

Finds of African ivory in the earliest remains at Phaistos (similar to those at Knossos) indicate an early relation of Crete with Africa, and at the close of the Neolithic Period at Knossos a flat copper axe indicates a connection with Egypt.[5]

Some catastrophe occurred at the end of MM II, perhaps due to an earthquake, and the earlier palace at Knossos was destroyed, either by fire [6] or by an attack by people of the islands anxious to capture Egyptian trade.[7]

THE MINOAN CIVILISATION

	B.C.		B.C.		B.C.
EM I	2800–2600	EM II	2600–2400	EM III	2400–2200
MM I	2200–2100	MM II	2100–1900	MM III	1900–1700
LM I	1700–1500	LM II	1500–1450	LM III	1450–1200

The Mycenæan Period, which begins on the mainland with LM II, has been divided into Myc. I, etc., or E. Myc. (= MM II—LM II) and L. Myc. (= LM III)—which lasted in Greece to the period of the xx dyn. in Egypt and in Cyprus perhaps longer : a scarab of Rameses III (1200) occurs in Cyprus and Mycenæan vases are depicted in his tomb.[8] More recent dating of MM III by comparison with Grave VI at Mycenæ gives 1600.[9]

The *Cycladic civilisation* has the same classification (EC I, etc.) as the Minoan : there are no Neolithic remains. There is no sign of a Neolithic culture in the Peloponnesos, but it existed for a long time in Thessaly, where it had two periods : Th. I (till the end of EM II) and Th. II (middle of MM I) ; the Chalcolithic Age (Th. III) follows with the use of bronze. Th. IV came with relations with the Ægean (LM). The *Helladic civilisation* on the mainland of Greece began with that of the Cyclades, but developed more slowly : EH I continued till the last quarter of EM II, and EH III commenced at the end of EM III. With MH II the great days of Mycenæ began.[10]

EGYPT AND THE ÆGEAN

Relations between early, perhaps Predynastic, Egypt and the Ægean were intimate.[11] The Hanebu in early Egyptian texts may have been Cretans, and there are also indications of early relations with Anatolia. The "Hanebu people of the papyrus lands" have also been identified with the inhabitants of the northern Delta and unlocalised foreigners in the Delta, the place of entry into Egypt by sea. No definite name for the Cretans occurs in Old Kingdom texts.[12]

Kultur, ed. G. Karo, 1921, 210 ; Mackenzie, Footprints of Early Man, 1927, 126 ; Karo, RV, i, 29 ; *ib.*, PW, xi, 1718 ; Peake and Fleure, Priests and Kings, 1927, 36, 101 ; *ib.*, The Steppe and the Sown, 1928, 99, 102 ; *ib.*, The Way of the Sea, 1929, 56.
[1] Fimmen, 1921, 210.
[2] Hall, Bronze Age, 1928, 67.
[3] Cf. Evans, i, 261.
[4] Alt., I, ii, 773.
[5] Peake and Fleure, Steppe and Sown, 99, 104.
[6] Wace, CAH, i, 597 ; cf. Hall, Bronze Age, 109, 199 ; CAH, ii, 431.
[7] Pendlebury, JEA, 1930, xvi, 89.
[8] Glotz, 23, 27 ; Hall, NE, 37.

[9] Hall, Bronze Age, 146 ; Evans, ii, 172.
[10] Glotz, 23 f.
[11] Pendlebury, JEA, 1930, xvi, 75 ; Roeder, RV, i, 39 ; Glotz, 202, 402 ; Evans, i, 17 ; ii, 22, 30, 53, 60 ; Hawes, 42 ; Fimmen, 1909, 58 ; Hall, CAH, ii, 277 ; *ib.*, NE, 34 ; *ib.*, Bronze Age, 69 ; Köster, *Beih. alt. Orient*, i, 24 ; *ib.*, Seefahrten der alten Ägypter, 1926, 17 ; Mosso, The Dawn of Mediterranean Civilisation, 1910, 21.
[12] Glotz, 202 ; Meyer, Alt., I, ii, 778 ; Evans, i, 20 ; Hawes, 42 ; Dussaud, Les Civilisations préhelleniques dans le Bassin de la Mer Égée, 1914, 282.

From the end of the Old Kingdom to the beginning of the XI dyn. the relations were relaxed but not completely broken off ; in the XII dyn. they were again very active. Sesostris II (1903–1887) and Amenemhet III (1849–1801) collected gangs of workmen at Lahun (abandoned about 1765) and in this village there is Cretan pottery, including fragments of good MM II Kamares ware, and Middle Kingdom scarabs were found in Crete.[1] It is possible that Cretan materials, including raw metals, ivory, gold and silver vases and jewellery, were imported into Egypt in the XII dyn., and went up the Nile as far as Abydos: some Cretan vases have been found even in Nubia.[2] Relations slackened during the Hyksos Period, but an alabastron of a Hyksos king was found at Knossos. During the XVIII dyn., after a period of piracy, the Cretans resumed regular relations with Egypt, where they are often represented and named Keftians.[3] The Cretan script, apparently containing also numerical signs, is undeciphered.[4]

In the EM Period there seems to have been influence, perhaps indirect, from Syria and Mesopotamia. This is shown by a Babylonian cylinder seal (perhaps really MM I) and a supposed correspondence between Cretan weights and the light Babylonian talent.[5] Intercourse between Crete and Sicily, South Italy, Sardinia and Libya has also been regarded as probable, and finds similar to those in Crete occur on other Ægean islands and as far afield as Etruria.[6]

Many objects found at Knossos have been extensively "restored," a practice which must be regretted from the archæological point of view.[7]

CERAMICS AND WORK IN STONE

In the Neolithic strata at Knossos were stone axes, serpentine maces, obsidian knives (probably from Melos) and stone and clay spindle whorls, but the only trace of metal was one copper axe, perhaps imported from Egypt or Cyprus.[8] Pottery decoration indicates no break between the Stone and Bronze Ages. At the end of the Neolithic Period the pottery was made of cleaner clay and is black, perhaps produced by smoke or by carbonaceous material such as graphite, mixed with the clay to prevent cracking in firing in the closed ovens which were used.[9] Hard stone for tools replaced limestone. The pottery vessels were smoothed by stones or bones and a linear ornamentation filled in with white, probably chalk or gypsum, began.[10]

The Early Minoan Period begins in Knossos with pottery covered with a brilliant black paint with white lines or points painted on, or vice versa ; or with a red glaze of varying tone and brown stripes. The decorations remained

[1] Glotz, 203 f. ; Dussaud, 283 ; Evans, ii, 192 ; Hall, Oldest Civ. Greece, 67, 154 ; Ridgeway, EAG, i, 64 ; Petrie, Illahun, Kahun and Gurob, 1891 ; Furtwängler, Die antiken Gemmen, Leipzig and Berlin, 1900, iii, 19.
[2] Glotz, 204, Fig. 34 ; Dussaud, 284 ; Köster, Beih. alt. Orient., i, 32 ; ib., Seefahrten, 18.
[3] Glotz, 205 f. ; Dussaud, 285 ; Fimmen, 1909, 82 ; ib., 1921, 183 ; Meyer, Alt., II, i, 106, plates ii, iii ; Hall, Bronze Age, 199 ; Pendlebury, JEA, 1930, xvi, 75 ; amulets thought to represent mummies were found at Hagia Triada ; Peake and Fleure, Priests and Kings, 99.
[4] Glotz, 371 ; Evans, i, 278, 612, 638, 646 ; Hall, Bronze Age, 90, 134 ; Dussaud, 425 ; Contenau, Civ. phénicienne, 1926, 310.

[5] Evans, i, 15 ; ii, 28 ; ib., Palace of Knossos, Report in Annual of the British School of Archæology at Athens, 1900, 43 ; Wace, CAH, i, 592, 596 ; Glotz, 25 ; Hall, Bronze Age, 107 ; Meyer, Alt., I, ii, 778 ; II, i, 163 ; Thomsen, RV, i, 44 ; Otto, 1925, 34, 59—doubtful ; Peake and Fleure, Way of the Sea, 61.
[6] Evans, i, 66, 312 ; Peake and Fleure, Steppe and the Sown, 112, 116, 118, 127 ; ib., Way of the Sea, 34 ; Meyer, Alt., I, ii, 779.
[7] Berthelot, Archéologie, 25.
[8] Hawes, 14 ; Evans, ii, 14.
[9] Meyer, Alt., I, ii, 776 ; Glotz, 30, 348 ; Perrot-Chipiez, Histoire de l'Art, ix, 159 ; Mosso, Dawn of Med. Civ., 80, 93—carbon by analysis.
[10] Evans, i, 37 ; Meyer, op. cit. ; Mosso, 81, 95.

linear till the end of the period, when very characteristic spiral decoration began. The forms of the pottery resemble those of Troy II and Cyprus, but there were no urns with human faces, although rude stone and clay idols appear.[1] Numerous vessels of coloured stone, extremely well worked, which were copied from but not imported from Egypt, occur with Egyptian seals, ivory and fragments of diorite vases of the III dyn.[2]

A large amount of obsidian, a black volcanic glass, was exported from Melos and played the part of flint : flint weapons are rare in Crete and obsidian knives were used throughout Greece, Crete, Troy and even Egypt. The town now called Phylákopi was an early settlement in Melos, where the painted and ornamented pottery was better than that of Crete.[3]

In EM II a characteristic mottled pottery appears in Crete, probably made on small discs of wood or stone turned by hand [4], and the brilliant orange, red and black colours produced by firing resemble those produced in Egypt by the same method.[5] A direct influence of Egypt is suggested, lasting to the end of MM.[6] Although the ivory seals also suggest trade with Egypt [7], there is no evidence of direct relations of Crete with Cyprus or the Syrian coast till the MM Age.[8] Glazed paste in native faience of a pale bluish-green colour and a bowl of the same material resemble Egyptian products.[9] There are large bowls of porphyry, diorite and of the volcanic obsidian glass liparite (identified by Sir H. Miers), which is found only at Lipari in the Æolian islands.[10] This liparite, indicating early intercourse with Sicily, was apparently exported to Crete, the mainland of Greece and Troy, and Egypt probably obtained it from Crete.[11] Engraved seals, which did not begin till EM III, are of soft material, although beads of carnelian, amethyst and crystal were made at the same time, and in the EM Period there is a "collar stud of chrysocolla." Some at least of these seals, and some from Mycenæ, were probably really amulets.[12]

In EM III the mottled pottery gave way to a painted ware showing a light decoration on a dark ground, and there were beginnings of polychrome decoration, a matt "Indian red" bordered with dull white on a black-brown glaze ground. The spiral decoration appears through Cycladic, not Egyptian, influence, and the red and black varnished glazed ware (Urfirnis) was also developed in the Cyclades, not on the mainland. Vases and idols of Parian marble, alabaster and steatite were imported, and ivory seals with representations of ships show early maritime enterprise.[13]

In MM I, when the slow wheel was perhaps first used, there was polychrome pottery. An exceptional effect was produced by a thick hand-burnished wash of dark reddish-brown colour, on which are hatched medallions in a new clear white. There were also jugs with a dark brown decoration on a pale buff slip.[14] The "Indian red" of the earlier period is now superseded by a much more brilliant "vermilion red" and very thin "egg-shell" polychrome pottery imitating metal.[15] There is lapidary work in rock crystal, agate, carnelian, etc., bowls of diorite and liparite, and small beads of deep "cobalt" blue (no

[1] Meyer, 776 ; Evans, i, 63.
[2] Meyer, 777 f. ; Hall, NE, 35 ; Glotz, 24 ; Evans, i, 16, 26.
[3] Meyer, Alt., I, ii, 779 f. ; ib., II, i, 162 ; Mosso, 89.
[4] Hall, Bronze Age, 47, says not, although "they look as if they were."
[5] Evans, i, 78 f.
[6] Ib., i, 83, 291.
[7] Hall, Bronze Age, 69.
[8] Evans, i, 83 : cf. Wace, CAH, i, 592, 596.
[9] Evans, i, 85.
[10] Ib., 87 f. ; ib., Knossos Excav., Brit. Sch. Arch. Athens, 1902, 122 ; Mosso, 363,

365 f. ; Frankfort, Studies in Early Pottery, 1927, ii, 190.
[11] Hall, Bronze Age, 49 ; Meyer, Alt., II, i, 162 ; Köster, Beih. alt. Orient, i, 24 ; Dussaud, 98, on the method of working up into knives.
[12] Hopfner, PW, xiv, 303 ; Evans, Palace, i, 55, 93 : the ancient chrysocolla was malachite.
[13] Evans, i, 108, 111, 114, 120 ; Hall, Bronze Age, 58 ; Fimmen, 1909, 15.
[14] Hall, Bronze Age, 47 f. ; Evans, Palace, i, 165 f.
[15] Evans, Palace, i, 168 ; Hall, Bronze Age, 75 ; Meyer, Alt., II, i, 163.

analysis : colour only, not material) with larger beads of a green tone resembling those of the VI dyn., and probably derived from the Egyptian technique. There are traces of inlay work, votive clay balls, prickly decorations, and representations of birds and animals in pottery, imitations of silver and copper ware in clay and imitations of Egyptian scarabs. The quick potter's wheel was probably not yet in use.[1] Ancient Peruvian pottery (early Christian era) represents the imitation of men, animals, fruit, etc., at its highest level.[2] Attempts by the Egyptians to imitate Cretan pottery were unsuccessful, as the clay was unsuitable.[3]

The acme of polychrome pottery, red, white and blue on a black ground, was reached in MM II, when the quick potter's wheel was in use. This is the so-called Kamares, or egg-shell, ware, belonging to this period only, and exported to Egypt. The white is creamy, the red (beginning in MM) has a touch of orange and the crimson a cherry tint. The black has sometimes a purple tone and often a brilliant metallic lustre. Large stores of this pottery, probably for the king's use, were found. There is sometimes a brilliant white glaze resembling enamel (tin glaze ?).[4] The pottery imitations of gold and silver cups are similar to the actual metal cups found in shaft graves at Mycenæ.[5] The "riot of colour" in the pottery, which reached its zenith in MM II, afterwards suddenly died down.[6]

MM seals and gems are in rock crystal, white and red carnelian, green jasper and banded agate, with suggestions of Babylonian influence. Remains of a MM III lapidary's workshop contain unfinished marble, blocks of "Spartan basalt," steatite, jasper and beryl (?).[7]

The MM III pottery is often coarse and badly made and badly glazed as compared with the preceding, with a great falling-off in polychromy. Broad white spirals on a dark ground are the most frequent ornamentation, as at Mycenæ. There are very large decorated store jars, double-walled pottery charcoal braziers, clay candlesticks similar to those of IV dyn. Egypt, large jars for burials, inscriptions in ink on jars, imitations of breccia on pottery and vases with shell inlay in imitation of Babylonian products.[8]

LM I pottery is naturalistic, with designs mostly of marine life and plants : the severely simple designs of MM III give way to the "fruition of naturalism," which in LM II becomes stylised.[9]

BUILDINGS

The ordinary MM II houses at Knossos were of two or three stories with roof attics and windows with four or six panes containing some substitute for glass (?), possibly oiled parchment ; the material is shown bright red on the coloured faience plaques representing the houses.[10] Hall[11] maintains that the windows were open spaces. Remains of baked brick Minoan houses were found at Gournia and Roussolakkos.[12] Large quantities of wood were used in buildings.[13] The Cretan bricks were 40 × 30 × 10 cm., and fired.[14] Although Glotz suggests that the Greek name for brick (πλίνθος) is pre-Hellenic, the sun-dried bricks in the Little Palace at Knossos are more like the Babylonian

[1] Evans, i, 170, 175, 179, 182, 193, 199, 242 ; Hall, Bronze Age, 49, 70, 90 ; Glotz, 350.
[2] Harrison, Pots and Pans, 8.
[3] Hall, Bronze Age, 73.
[4] Evans, i, 231, 237, 240, 259 ; Fimmen, 1909, 17 ; Glotz, 38 f. ; Hall, NE, 41 ; ib., Bronze Age, 74.
[5] Evans, i, 244 ; Meyer, Alt., II, i, 163.
[6] Hall, Bronze Age, 76.
[7] Evans, Knossos Excav., Brit. Sch. Arch. Athens, 1900–1901, 20 ; ib., Palace, i, 274 ; iii, 269.
[8] Evans, i, 552, 557, 562 f., 568, 578, 585 f., 601 ; Hall, Bronze Age, 123 ; on pottery, cf. Pottier, in de Morgan, Délégation en Perse, xiii, 67 ; Fimmen, 1921, 69.
[9] Hall, Bronze Age, 167, 180.
[10] Evans, i, 303, 308 ; ib., iii, 342 ; Glotz, 107, 114.
[11] Bronze Age, 103.
[12] R. M. Burrows, Discoveries in Crete, 1907, 26, 27.
[13] F. Baumgarten, F. Poland and R. Wagner, Hellenische Kultur, Leipzig and Berlin, 1913, 36.
[14] Glotz, 107.

(square form) than the Egyptian, were little used in Crete and bricks were not a native invention in Greece.[1]

The ruins of the (unfortified) palace at Knossos reveal three structures [2] : (1) MM I, burnt down perhaps owing to earthquake and revolution and (2) rebuilt, c. 1800, (3) enlarged c. 1500. It is built of limestone blocks and white gypsum covered with lime wash ; the stone blocks were cut by copper saws, one of which, 2 m. long, was found.[3] Gypsum occurs at Sitia.[4] Remains of earlier buildings erected before the use of metal tools are of mud and wickerwork with clay floors. The mortar was of lime with clay and pebbles and is now extremely hard. The gypsum of the Palace site is very homogeneous, with no shell or organic impurities, while that of the houses is crystalline, micaceous, large-grained and grey or greyish-brown. Besides gypsum and limestone of a dull sandy colour, bluish-black or light blue slate was used for pavements, etc.[5] Decorations are in fresco and plaster relief, with brilliant colours ; the earliest (all EM) was red ochre (burnt clay), then yellow (natural clay) and black. Blues, from pale greenish-blue (MM) to dark blue (LM), followed. Cretan paintings were on wet plaster and were less detailed than Egyptian of the same period, which were executed on fine limestone.[6] There was an elaborate water supply with a very advanced drainage system of cemented stoneware pipes (MM I, c. 2000 B.C.) recalling the v dyn. copper pipework of King Sahurā at Abûsîr.[7]

Huge stone vases were for the storage of food ; one is 28 in. high with three handles, with spiral bosses and had metallic (probably gold) inlay. The great pottery jars (pithoi) [8] with many handles, one 7 ft. high and 15 ft. girth, were decorated with trickle ornament and some rope decoration.[9] There are also stone chests lined with lead, but supposed baths are religious objects.[10] The palace doors were provided with locks and keys.[11] There is a representation of an artificial fountain, unknown in Ancient Egypt, Mesopotamia and Classical Greece, where they appeared only in the Hellenistic Period.[12]

The MM culture spread all over the Ægean and extended to the mainland ; vases are found as far north as Bœotia and relics in' Mycenæan tombs were perhaps carried there by workmen.[13] MM III opens with some catastrophe which coincides with the break-up of the Middle Kingdom in Egypt about 1800 B.C. The palaces of Knossos and Phaistos were reduced to ruins, probably by fire.[14] The pillars of the *restored* MM III palace show for the first time the slope dwindling towards the base : those of the later Lion Gate at Mycenæ, always seen as of this type, are now said to be straight and the sloping form is characteristic of MM III. The material was cypress wood : the walls were of blocks of stone with layers of timber.[15]

FRESCOES

Painted red stucco decoration began in MM III.[16] The elaborate wall paintings in true fresco[17] are probably a Minoan invention[18] related to *buon*

[1] Hall, Bronze Age, 293.
[2] Forrer, RL, 413, plates 106–7 ; Glotz, 40 ; Fimmen, 1921, 39 ; Phylákopi, in Melos, was fortified.
[3] Dussaud, 17.
[4] Bürchner, PW, xi, 1738.
[5] Evans, Knossos Excav., Brit. Sch. Arch. Athens, 1899–1900, 53 ; report by Fyfe.
[6] N. Heaton, J. Roy. Inst. of British Architects, 1911, xviii, 697 ; Hall, JEA, 1914, i, 197 ; ib., Bronze Age, 118 ; Evans, i, 265.
[7] Glotz, 115 ; Evans, iii, 253 ; Hall, Bronze Age, 101 ; Borchardt, Sahurā, i, 29.
[8] Smith, DA, i, 650.
[9] Evans, iii, 264.

[10] Hawes, 55 ; Evans, Knossos Excav., Brit. Sch. Arch. Athens, 1899–1900, 21 ; Hall, Bronze Age, 102.
[11] Glotz, 184.
[12] Evans, ii, 461 ; iii, 254.
[13] Ib., i, 23 ; Discovery, 1921, ii, 35.
[14] Evans, i, 315 f., 419 f. ; Hall, Bronze Age, 123.
[15] Evans, i, 342, 344, 349 f.
[16] Ib., 356 ; Glotz, 309, and Hall, Bronze Age, 14, say MM II.
[17] Evans, i, 524 f. ; Wace, CAH, ii, 434.
[18] Heaton, J. Roy. Soc. Arts, 1932, lxxx, 427 ; Meyer, Alt., II, i, 178, thinks under Egyptian influence.

fresco with only a slight assistance of tempera and carried out in pure lime plaster.[1] They began crudely in the Neolithic Period with the object of protecting loose rubble and sun-dried brick. In MM the surface is practically pure [what ? : CaCO$_3$?] and reached perfection in LM. When applied to a gypsum backing it was very thin. The material was nearly pure lime from chalk dug on the site. In EM only red ochre, afterwards burnished by hand, was applied, but by the close of MM I there were three colours, red, white, and black, and a stencil was used in MM II.[2] Of the MM III colours, Heaton [3] believes the white was calcium carbonate applied as slaked lime, the black a carbonaceous shale or slate, the red and yellow ferruginous earths [ochres], calcined and uncalcined, respectively ; deep red was probably hæmatite ; a deep natural blue was used in earlier MM, but in MM III a deep blue was copper silicate (*kyanos*), probably imported from Egypt.[4] The green was a mixture of blue and yellow, and ground malachite was only later used at Tiryns, whilst in Egypt it was used in prehistoric times for face-paint.[5] The colours were applied to the moist lime and have proved extremely durable. No marble dust was mixed with the lime, as at Pompeii or in mediæval frescoes. The designs are very natural and living (cats, birds, fish in water, plants). Shells were also painted.[6]

FAIENCE. GLAZE. GLASS

Faience objects made in moulds and probably based on Egyptian originals reached great perfection in MM II, although the technique perhaps goes back to EM.[7] Faience plaques for inlay are sometimes covered with gold foil or minute grains of melted gold.[8] The work represents a true faience technique, as is shown by chemical analysis. The body was almost pure quartz sand and clay moulded into shape, dried, coated with glaze and fired at a temperature just sufficient to soften the glaze. The colour was mainly due to copper, nearly related to Egyptian blue ; the dark browns and black are probably due to iron. Great quantities of faience beads in imitation of Egyptian are of home manufacture.[9] Segmented paste or perhaps glass beads of MM III to LM III are of a type widely diffused over England and Spain, but the latter are probably indigenous. Similar Egyptian beads of the XI-XII dyns. are much earlier.[10] The most remarkable faience work is represented by the figures of the "snake-goddess," with very modern clothing.[11]

Church's report, as given by Evans,[12] states that the glaze contained silica, lime, a little magnesia, some soda and a larger amount of potash [plant-ash alkali ?]. The friable and rather porous body contained : moisture and "other" [organic ?] matters 1·22 ; matter soluble in concentrated hydrochloric acid, 2·22, consisting chiefly of lime and oxides of iron, alumina and copper [from the glaze ?] ; the part insoluble in acid, "96·56 per cent. (= 100)" [sic] was quartzite sand with traces of mica, felspar and clay. The matter insoluble in acid consisted chiefly of silica (97·01 per cent.) with 1·33 per cent. of alumina, 0·17 of lime, "with traces of lime [sic : iron ?], magnesia, copper [from the

[1] Evans, i, 528 ; Hall, Bronze Age, 103 ; cf. Mrs. Merrifield, The Art of Fresco Painting, 1846, with a valuable account of ancient pigments.
[2] Evans, i, 530 f.
[3] J. Roy. Inst. British Architects, 1911, xviii, 697 ; ib., J. Roy. Soc. Arts, 1932, lxxx, 411, 415, 419, 427.
[4] Egyptian blue was really a copper calcium silicate : see p. 117.
[5] Evans, i, 534 ; iii, 30, plate 15.
[6] Evans, Palace, i, 543 ; ib., Knossos

Excav., Brit. School Arch. Athens, 1903, 43.
[7] Evans, i, 487 f. ; Glotz, 352 ; Karo, PW, xi, 1761.
[8] Evans, Knossos Excav., Brit. Sch. Arch. Athens, 1903, 33, 34.
[9] Evans, Palace, i, 489 f. ; ib., Knossos Excav., 1903, 64.
[10] Evans, i, 492 ; Peake and Fleure, Merchant Venturers in Bronze, 1931, 90.
[11] Hall, Bronze Age, 127, 128, 171.
[12] Knossos Excav., 1903, 64.

glaze ?] and alkalies." This account is chemically almost unintelligible, and could hardly have been furnished in this form by Church.

"Blue glass" (imitation lapis lazuli) eyes inserted in gold tubes were inset in a white faience bull's head of MM I᾿I.[1] Lapis lazuli was used for statuettes.[2] The kyanos (blue paste) technique, of interest in connection with its later application at Tiryns, etc. [3], is the blue glaze described in Homer[4] and Hesiod.[5] It was used for inlaying along with crystal, precious metals, ivory and Baltic amber [6], and a kind of draught-board of MM III (much restored) is inlaid in a very elaborate manner with gold, silver, ivory, crystal and kyanos.[7] The "enamel" [faience] plaques used for inlaying a large mosaic are LM I, but the most remarkable specimens of faience work, which came from treasure chests below the floor of a small chamber (perhaps a shrine of the Great Goddess), belong to the later Palace before its re-modelling.[8] "Steatite" moulds for the paste were found with shells and rosettes.[9] Egyptian glass vases similar to those made under Amenhotep II (1448–1420) were found at Enkomi[10], and a spherical glass bead (? 1400 B.C.) at Knossos.[11] It seems probable that all true glass objects which occur in Crete are of Egyptian origin, including the beads already mentioned.

METALS IN CRETE

Although Mount Ida is covered with timber there are no signs of metal working on it and there are no notable deposits of metallic ores in Crete, which in antiquity, however, had the reputation of being peopled by skilled metallurgists.[12] A cycle of legends, now preserved only in obscure fragments in Hellenistic authors, made Crete the home of the Dactyls, but the accounts confuse Crete with Asia Minor and the Dactyls with other mythological beings in a way which it is now impossible to disentangle. The Dactyls were dwarf magicians, skilled in the metallurgy of bronze, but especially of iron, in the service of the Great Mother of the Mountain ($M\acute{\eta}\tau\eta\rho\ \acute{o}\rho\epsilon\iota\acute{\eta}$), Mount Ida, either in Crete or Phrygia.[13]

GOLD

The precious metals appeared early in Crete, but probably (as in Egypt) not so early as copper.[14] Al-Idrîsî (A.D. 1154) says gold was found at Chania in Crete.[15] Gold, silver, copper and lead were used in the EM Period, but not bronze.[16] A miniature gold axe-adze, EM II, was found at Knossos[17]; at Mochlos there

[1] Evans, Palace, iii, 434.
[2] Meyer, Alt., II, i, 177.
[3] Ib., II, i, 187, 209 ; see p. 355.
[4] Odyssey, vii, 87 ; θριγκὸς κυάνιοι, a frieze of blue ornament.
[5] Shield of Herakles, 143, 167 ; cf. Burrows, 20.
[6] Burrows, 33.
[7] Hawes, 130 ; Evans, i, 472.
[8] Hawes, 121 f.
[9] Evans, Knossos Excav., Brit. Sch. Arch. Athens, 1903, 65 ; the name steatite is rarely used correctly by archæologists.
[10] Fimmen, 1921, 179.
[11] Evans, Knossos Excav., 1902, 38 ; Meyer, Alt., II, i, 209.
[12] Hawes, 116 ; James, DG, i, 703 ; Meyer, Alt., 1909, I, ii, 658.
[13] Diodoros Siculus, v, 64 ; xvii, 7 ; Hesiod, fragm. in Loeb ed., 1914, 76 ; Dussaud, La Lydie et ses Voisins, 1930, 86, 94 ; Tümpel, PW, i, 406 ; Kern, ib., iv,

2018 ; Rzach, ib., viii, 1223 ; Jenssen, ib., 969 ; Gudemann, ib., 115 ; Bürchner, ib., xi, 1728 f. ; Pohlenz, ib., 1996 ; Schwenn, ib., 2209 ; Eitrem, ib., 2342 ; Strabo, X, iii, 20 f., 472 C. ; Lobeck, Aglaophamus, Königsberg, 1829, 1156, 1250, 1282 ; Meyer, Alt., 1919, I, ii, 658 ; ib., II, ii, 120 ; Schrader, RL, 728 ; Schrader-Jevons, 164, 205 ; Maury, Religions de la Grèce antique, 1857–59, i, 202 ; iii, 246 ; Immisch, Ro., ii, 1587 ; Bapp, ib., iii, 3040 ; v, 990 ; Wäser, ib., iii, 3314 ; Friedländer, ib., iv, 236 ; Orth, PW, Suppl. iv, 115 ; Cook, Zeus, i, 109 ; ii, 313, 664, 947 ; Lagarde, q. by Klinkenberg, MGM, xv, 339 ; Farnell, ERE, vii, 628.
[14] Evans, i, 68 ; Hall, Bronze Age, 53 ; Thomsen, RV, i, 45.
[15] Bürchner, PW, xi, 1719, 1732.
[16] Karo, PW, xi, 1747 f.
[17] Evans, ii, 629.

is EM II gold work, mostly punched sheet, chains, etc.[1] ; and a gold cylinder, perhaps earlier than MM I, with a spiral decoration, was found at Kalanthiana.[2]

EM III goldsmiths' work is very fine and similar to the silver work of Troy II and the Cycladic tombs, e.g. the EM II dagger, repoussé work, etc., in graves of Mochlos.[3] The Cretan gold work includes[4] admirable reproductions of flowers and foliage in thin gold, which recur in the Mycenæan Period, gold death masks like those of ancient Bolivian mummies and at Mycenæ, rock crystal, amethyst, etc., set in gold and silver, necklaces of gold and jewels, a gilt fish and gold animal heads, also found at Delphi and Mycenæ.[5] There was a traditional connection between Crete and Delphi, and Sparta had also many customs like, and perhaps derived from, the Cretan.[6] In this period there is more gold than silver.[7]

In MM I there is gold plate, thicker than the foil of later periods, and copper on which gold had probably been overlaid.[8] In MM III gold foil and crystal are used as inlays, also silver foil covered for protection with gold foil when applied to wood or ivory ; there is crumpled gold foil (repeatedly found on the Palace site), a gold signet ring, fish, pendant, lion, duck with granulated work and a solid bead.[9] A small vessel of blue glaze with a foot, collar and thimble-like inner receptacle of gold plate is perhaps LM.[10] Carved steatite vases were coated with gold leaf[11] and MM III bronze was plated with gold.[12] "Very elegant fern-like sprays of thin gold-plated wire" of an early period were found at Knossos.[13]

In the "cup-bearer" fresco[14] two heavy bands on the arm and a signet on the hand are shown yellow (gold) : the long conical vase is shown as yellow (gold) and blue (silver or kyanos ?). The goldsmiths' work about 1450 was very good[15] and was exported to Egypt.[16] The gold bulls' heads represented, as well as the small gold bars found at Mycenæ and Enkomi, are thought to have been used as money, but as similar bulls' heads of bronze filled with lead were found at Psychro, they may have been used as weights.[17] Electrum occurs in the Ægean at first (c. 3000) as pellets (φθοῖδες) and later in rods, cakes and bars.[18]

The gold came partly from Ægean mines and rivers, but mostly from Nubia by way of Egypt, perhaps through Libya, where there were one or two Minoan settlements and an established intercommunication. Mount Tmolos in Lydia and perhaps Transylvania, Macedonia and Thrace may also have furnished some gold, the sources being probably the same in the Mycenæan Age. Remains of old workings occur in many places in Asia Minor, notably in the back country to the south, south-east and south-west of Trebizond, in north Armenia, Anatolia and near Balia on the north-east of Mount Ida. Several of these have been re-opened for working.[19]

According to Hall, Cretan and Trojan work in gold and silver came with the

[1] Dussaud, 1914, 39, 41, Fig. 21.
[2] Evans, ii, 194.
[3] Ib., i, 95, 97 ; Hall, NE, 39 ; ib., Bronze Age, 52 ; Dussaud, 40, 41, Fig. 22.
[4] Evans, i, 97 ; ib., Knossos Excav., 1902, 25 ; Baumgarten, Hellenische Kultur, 1913, 44 f. ; Peake and Fleure, Steppe and Sown, 103 ; Meyer, Alt., I, ii, 777.
[5] Glotz, 272.
[6] Wade-Gery, CAH, iii, 563.
[7] Evans, Palace, i, 99.
[8] Ib., i, 170 ; ib., ii, 641 ; ib., Knossos Excav., Brit. Sch. Arch. Athens, 1902, 69 f.
[9] Evans, i, 469 ; ii, 557 ; iii, 401, 410 ; ib., Knossos Excav., Brit. Sch. Arch. Athens, 1900–01, 46 ; ib., 1902, 38 : the last two objects perhaps LM.

[10] Evans, Knossos Excav., 1902, 24, Fig. 11.
[11] Burrows, 33.
[12] Evans, Palace, iii, 401, 414.
[13] Hawes, 116.
[14] Hall, Bronze Age, frontisp. ; Hawes, 119.
[15] Glotz, 333.
[16] Meyer, Alt., II, i, 209.
[17] Fimmen, 1921, 121, Fig. 113.
[18] Regling, PW, vii, 978.
[19] Meyer, Alt., II, i, 210 ; Köster, Beih. alt. Orient, 35 f. ; Fimmen, 1921, 121 ; Gowland, J. Anthropol. Inst., 1912, xlii, 261, 266 ; Davies, Nature, 1933, cxxx, 985.

metals from Anatolia, and hence is perhaps of Babylonian origin, although some gold may have come from Egypt. The Spanish silver was not worked in the earliest period. The spiral decoration occurs first at Troy on gold, spread to the Cyclades and then Crete, where it is used first on stone, and went to Egypt (XII dyn. scarabs) from the Ægean. It was perhaps ultimately of Sumerian or Elamite origin, since it occurs in appliqué wire work at Ur in 3000, which is "the oldest true spiral known."[1] Spirals appear early also on Black Earth and Danubian pottery, perhaps an independent origin.[2] Two golden eggs (7th century B.C. ?) found at Siphnos have been lost. Ægean gold contains 8–25 per cent. of silver, but some pieces from Mochlos and Leukos are of a good colour and some Bronze Age lumps are yellower than the foil ; a suggested treatment with a solvent [3] is improbable, but some process of fire refining by heating with minerals containing iron sulphates (sory and misy) may have been used.

SILVER

Silver occurs and is imitated in pottery at an early period.[4] Silver daggers (EM) at Kumasa [5], votive axes (EM II) [6], a cylinder at Mochlos (EM II) [7], a small hoard of bowls, dish, jug, etc. (MM or LM I) at Knossos [8], a cup at Gournia (MM II) [9] suggesting an indirect trade connection with Mesopotamia or perhaps with Troy II[10], and cast weights at Knossos[11] are early examples of Ægean silver. This metal was perhaps more valuable than gold in the earliest period[12]. In the dumps of stamped metal at Knossos the earliest "coins" (?) are silver, not gold.[13]

A silver rhyton of bull's head shape in the fourth shaft grave at Mycenæ (MM III–LM I) is perhaps Cretan work.[14] It is suggested that the Syrian silver work of Amenemhet III's time (1849–1801) at Byblos, when there was a very vigorous intercourse between Egypt and Crete, is Cretan and points to trade between Crete and Syria-Palestine[15], but this work was probably either native or else imported from Crete through Egypt.[16] Gold and silver occur sporadically on the Cyclades about 2500 and there is a beautiful diadem of sheet silver from Syros, presumably early.[17] The alleged "comparative absence" of silver objects at Knossos[18] is only relative, since a fair amount of silver was found, although the metal was more used in the Cyclades.[19]

Suggested sources of Cretan silver are Egypt (doubtful), Spain, Sardinia, Anatolia, Laurion in Greece, or the wider region of Macedonia, Thrace, Laurion, Siphnos, Sardinia, south-east of the Euxine and western Asia Minor. Mycenæan silver was probably from the same sources.[20] Traces of silver occur in Crete itself.[21]

[1] Hall, Bronze Age, 55, 58, 67, 206.
[2] Childe, Dawn of European Civilisation, 1925, 27 ; Hall, Bronze Age, 60.
[3] Davies, Nature, 1932, cxxxi, 985.
[4] Evans, i, 191, 193, 242 ; Glotz, 333, 350 ; Fimmen, 1909, 17 ; Hawes, 116 : MM.
[5] Meyer, Alt., II, i, 210 ; Fimmen, 1921, 121 ; Hall, Bronze Age, 52, 54 ; Mosso, Med. Civ., 371.
[6] Glotz, 231.
[7] Wace, CAH, i, 592.
[8] Hawes, 116 ; Evans, ii, 387.
[9] Karo, PW, xi, 1780 ; Wace, CAH, i, 592, 596.
[10] Peake and Fleure, Way of the Sea, 1929, 60.
[11] Glotz, 196.
[12] Ib., 333.
[13] Burrows, 118.

[14] Evans, ii, 640.
[15] Glotz, 215 ; Breasted, CAH, ii, 97 ; Hall, ib., 280 ; Evans, Times, 19 Sept., 1923 ; ib., Palace, ii, 654.
[16] Meyer, Alt., II, i, 209.
[17] Meyer, Alt., 1909, I, ii, 695, 697, 713 ; Peake and Fleure, Priests and Kings, 116, Fig. 70.
[18] Gowland, J. Anthropol. Inst., 1912, xlii, 267.
[19] Peake and Fleure, op. cit.
[20] Köster, Beih. alt. Orient, i, 32, 35 ; Fimmen, 1921, 121 ; Meyer, Alt., II, i, 210 ; Mosso, Med. Civ., 372 ; Hall, Bronze Age, 52, 54 ; Peake and Fleure, Way of the Sea, 1929, 60 ; Gowland, J. Anthropol. Inst., 1912, xlii, 266 ; ib., Archæologia, 1920, lxix, 140.
[21] Bürchner, PW, xi, 1732 ; Karo, ib., 1747.

The silver buttons found at Knossos and Enkomi, many of them marked, may have served as money.[1] Inlaid metal work similar to the daggers found at Mycenæ and in the tomb of Aah-hotep in Egypt is supposed to have been "a specially Minoan art"[2] : a bronze MM II dagger inlaid with silver was found at Psychro.[3] The inlaid silver is filled in cavities with black niello [4] (which Evans calls an alloy of iron and silver), electrum and copper, but not tin, which is mentioned by Homer[5], although there is stone inlay in metal.[6] Plates of crystal, "painted at the back, or backed with gold or silver foil," were decorative.[7]

COPPER

Although it was first thought that no metal occurs in the earliest Neolithic remains at Knossos[8], a copper axe, perhaps imported from Egypt or Cyprus, was found.[9] Meyer's date of 3000 B.C. for decorative wire spirals, etc., and the suggestion that this early metal was bronze[10], are incorrect. Copper (probably pure) occurs in EM I (c. 2800), in which evidence for gold and silver is defective.[11] The earliest (EM) finds are toilet articles and weapons, vessels appearing somewhat later.[12]

Weapons such as daggers were of copper before EM III.[13] EM II copper includes models of oxen and a small waggon drawn by oxen, daggers, spearheads and axes, including votive axes, the small specimens of which may have been used as weights[14], since Hesychios says the Greeks made them 10 minas in weight, and the Minoans *may* have had a currency of copper ingots.[15] A currency of gold, silver and bronze[16] is doubtful. Copper bars in multiples of the supposed Cretan mina of 618 gm. are very widespread, indicating Crete as a depôt for Cyprian sea trade about 1500 B.C. with Greece, Italy, Sardinia, etc.[17] Whether the Phœnicians[18] or Minoans then took copper from Cyprus to Egypt and whether it went at all in the earlier period are unsolved problems.[19]

Large MM copper cooking vessels were found at Gournia[20] and a hoard of four enormous riveted cauldrons of MM III or LM I with a copper bar ("talent") at Tylissos, the largest vessel being 1·4 m. in diameter and weighing 52 kgm. ; similar cauldrons occur in Cyprus, Mycenæ, Chalkis and Sardinia.[21] Copper (?) hoards (axes, cauldrons, tripods, etc.) of MM III–LM I occur at Knossos and Mochlos.[22] An obvious source for Cretan copper is Cyprus, which had an early copper industry[23], but an importation from Italy or from Siphnos and Seriphos is suggested. Daggers and flat axes found in Sicily resemble those of

[1] Fimmen, 1921, 121.
[2] Evans, iii, 111, 131.
[3] Karo, PW, xi, 1780.
[4] Cf. Hesiod, Shield of Herakles, 300.
[5] Schuchhardt, Schliemann's Excavations, an Archæological and Historical Study, tr. Sellars, with an appendix on recent discoveries at Hissarlik and an introduction by W. Leaf, 1891, 232.
[6] Karo, 1780.
[7] Evans, Knossos Excav., Brit. Sch. Arch. Athens, 1899–1900, 41 ; ib., 1903, 46.
[8] Meyer, Alt., 1909, I, ii, 689, 693 ; Hoernes, Kultur der Urzeit, 1912, ii, 49.
[9] Evans, Palace, ii, 14.
[10] Alt., 1909, I, ii, 691, 744, 748 ; ib., 1913, 777 ; Myres, CAH, i, 103 ; Hall, NE, 33 ; cf. Hoernes, Urzeit, ii, 49.
[11] Evans, i, 68 ; Karo, RV, i, 35.
[12] Karo, PW, xi, 1748, 1758 ; Ganszyniec, ib., xii, 289.
[13] Glotz, 93.

[14] Evans, i, 99, 101 ; Meyer, Alt., II, i, 184.
[15] Lexikon, s.v. πέλεκις ; Petrie, Tools and Weapons, 14 ; Glotz, 192.
[16] Burrows, 15.
[17] Forrer, Urg., 361 ; Thomsen, RV, i, 45.
[18] Blümner, iv, 60.
[19] Meyer, Alt., II, i, 209 ; Petrie, Tools and Weapons, 6.
[20] Hall, Bronze Age, 164.
[21] Evans, ii, 569, 570, 624, Figs. 355–356, calls them MM III "bronze" ; Hall, 164, LM I "copper."
[22] Evans, ii, 392 f., 624, 629, "bronze" ; Köster, Beih. alt. Orient, i, 36, "copper."
[23] Rougemont, Âge du Bronze, 1866, 92 ; Hall, NE, 34 ; Fimmen, 1909, 14 ; ib., 1921, 120 ; Evans, The Times, 16 Sept., 1924, via Asia Minor ; Blümner, iv, 60 ; Meyer, Alt., II, i, 210 ; Glotz, 33, 211, 214 ; Davies, Nature, 1932, cxxx, 986 ; Petrie, Tools and Weapons, 6, "from the mainland, the most recent view."

Crete.[1]   Hall [2] thinks a knowledge of copper and the first specimens of metal reached Crete (and Egypt) from 4000 B.C. by ship from the Syrian coast.

Evidence has also been put forward of copper mining and smelting in Crete itself, in Sphakia, Gavdos, Prase and Phourne (Kydonia), Selinos and Hierapetra.[3]   There are said to be remains of ancient mining at Phourne and scoria at Hierapetra, but the statements of Hall that Bambakas found "silicic acid" in all specimens of Cretan copper and that ore from many parts of Crete— Gavdos, Kydonia, Selinos, etc.—"all contained this acid in large quantities," do not convey anything tangible to me and are chemically meaningless.

The supposed prehistoric or EM [4] copper mines of Chrysokamino near Gournia (χρυσοκαμινο=gold furnace) are described by Mosso.[5]   Pieces of scoria and one piece of "metal" containing cuprous oxide (Cu_2O) 45·05, ferric oxide 2·40, gangue 23·80, water, carbon dioxide and oxygen 28·75 per cent., were found. The ore was probably mined in a cave in which were found an almost intact EM II cup and a small MM III pot, showing a long period of working. Outside the cave and close to the sea, Hazzidaka found pieces of "crucibles" with irregularly disposed holes 20 mm. dia., showing the effects of fire. Pieces of scoria in some of the holes contained iron and copper. Since no copper ore was found, it had probably been worked out and was then brought from the island of Gavdos, where there was supposed to have been a copper mine and there is still a "vast deposit of rich ore" consisting of "sulphate [? sulphide] of copper," partly modified into malachite and basic carbonate. The analysis gave : oxide of copper with traces of metallic copper 50·00, sulphur 0·137, arsenic and antimony 0·080, gangue, silica, alumina and lime 27·00, ferric oxide 5·10, water and carbon dioxide 17·683. Gavdos is opposite Mount Ida, the home of the traditional Curetes, who are constantly represented on Etruscan monuments, and Gournia is rich in bronze finds.[6]

An object found at Zakro (not far from Chrysokamino) by Dawkins [7] and called by him a "portable brazier" is regarded by Mosso as a conical "crucible" with perforations at the top, which has been subjected to a high temperature, since slag was found adhering to it. The ore was put in through the large aperture, the crucible was "heated in a vertical position and then lifted by means of green branches and the fused metal poured out by the small holes. It could then be filled afresh." Bellows were probably used and "if this was the case, the form of the crucible appears to be well adapted for keeping up a current of air."[8]   Chrysokamino, however, is said [9] to have been too small to have supplied all the copper to Crete during the Bronze Age, and Cyprus remains the probable source. The process described by Mosso does not seem practicable.

Copper daggers from the great tholos of Hagia Triada, "the most ancient tomb known in Crete" (EM II according to Hawes) were analysed. One was pure copper, the other contained copper 98·617 and tin 0·158 ("a negligible quantity, which may be regarded as an impurity"). A contemporary dagger from Palaikastro contained : copper 99·54, zinc 0·16, lead 0·13. The weapons are very small (6 cm.–15·6 cm.), indicating the high value of the copper. Daggers at Porti near Kumasa, of the same period, 10 cm. long, contained copper 96·500, tin 0·197, lead 0·170, iron, zinc, and traces of nickel, 2·400, loss, etc., 0·733.[10] Mosso reports no arsenic, which Davies[11] suggests was "sometimes added to colour the surface"—an obscure statement.

---

[1] Mosso, 310.

[2] Bronze Age, 1928, 31, 37.

[3] Wace, CAH, ii, 442 ; Hall, NE, 34 ; ib., Bronze Age, 31 ; Mosso, 289; Bürchner, PW, xi, 1732 ; Meyer, Alt., II, i, 210 ; Davies, Nature, 1932, cxxx, 986.

[4] Karo, PW, xi, 1747.

[5] Med. Civ., 289 ; cf. Peake and Fleure, Way of the Sea, 1929, 59.

[6] Mosso, 292, 296.

[7] J. Hellenic Studies, 1903, xxiii, 258.

[8] Mosso, 292, Fig. 165.

[9] Davies, Nature, 1930, cxxx, 987.

[10] Mosso, 105 f., 110.

[11] Op. cit., 986.

Nineteen large blocks of copper in the shapes of double axes [1] or hides without head and tail were cast in open forms. They have a wrinkled surface, weigh *about* 29 to 30 kgm., and are supposed to represent the weight of the Babylonian talent, 28·8 kgm.[2] A large stone weight with engraved octopus feelers, found in Crete, was 28·6 kgm.[3] These copper ingots, found at Hagia Triada, were probably cult-objects, and had characteristic signs stamped on them which are also found on stones in the palaces at Phaistos and Knossos. Analysis by Mosso gave : copper 98·606, zinc and iron 0·630, lead 0·034, sulphur 0·445. They are of the shape carried by the Cretans in the tomb of Rekhmara, who bring gifts to Thothmes III (1500 B.C.). Copper pieces broken off from ingots showing a fibrous structure (perhaps for melting with tin to form bronze) were found in Cannatello and Candia, and an ingot ("Phœnician talent") of copper weighing 26·5 kgm. at Tylissos (west of Candia), where there was a MM I palace. Copper ingots marked with Cretan signs also occur in Sardinia and a piece of one in Sicily, so that commerce in copper was carried on by Crete with the countries of the Mediterranean.[4]

The weight of these and other Cretan bars has been thought by Forrer to indicate the use of weights of $\frac{1}{4}$, $\frac{1}{2}$ and 1 Babylonian talent of 30 minas (29 kgm.).[5] The large bars, however, vary in weight from 27 to 47 kgm., and seventeen bars with weights from 6·93 to 17·64 kgm. were found in the sea off Eubœa. The supposed relations are doubtful and the Minoan weight system is still obscure.[6] Forrer regards a copper double axe which weighed 3,040 gm., and lead and tin bars with bronze hooks, found in a Swiss pile-dwelling, as weights for Cretan and Phœnician minas. The axe has a very narrow orifice and could not have been used as a tool ; some others weigh only 540 grams, so that the theory of Forrer is doubtful.[7]

Scales for weighing the ingots of copper shown with them appear on a tablet at Knossos. Two small toy balances were found at Mycenæ with stamped gold discs as scales. Cast pieces of silver at Knossos have the marks H and Ⱶ, one being just double the weight of the other.[8]

Numerous small double axes, the early of copper, the later perhaps of bronze, occur in various places in Crete, and these, with the larger ones, were probably used in religious ceremonies.[9] Some are decorated, e.g. with a "conventional butterfly,"[10] said to represent really the death's-head moth symbolising the soul[11], and some (in Crete and Mycenæ) are intentionally broken. An ornamented axe at Phaistos is thought to have been cast by the cire perdue process. Large double axes raised above pyramidal bases on long staves are frequently represented in use as cult objects and have been found, e.g. four at Nirou Khani (LM I ?), one in the Palace at Knossos and one each in the caves at Mount Lasithi and Arkalochori.[12]

[1] Mosso, 293, Fig. 167.
[2] Evans, ii, 624, Fig. 391 ; Köster, *Beih. alt. Orient*, i, 35 ; Fimmen, 1921, 123 ; Mosso, 293, Fig. 167; Karo, PW, xi, 1774 ; Glotz, 191 ; Meyer, Alt., II, i, 188, 210.
[3] Evans, Excav., 1901, *Brit. Sch. Arch. Athens*, 42, Fig. 12 ; Glotz, 191 ; Dussaud, 1914, 81, 83, Fig. 60.
[4] Mosso, 293, 296 ; Glotz, 191 ; Köster, *Beih. alt. Orient*, i, 36 ; Meyer, Alt., II, i, 189.
[5] Dussaud, 81 ; Nilsson, *A. Rel.*, xiv, 425 ; Forrer, RL, 78 ; *ib.*, Urg., 361.
[6] Fimmen, 1921, 123 ; Meyer, Alt., II, i, 211 ; Köster, *Beih. alt. Orient*, i, 36 ; Dussaud, 250.

[7] Dussaud, 251.
[8] Glotz, 195 f., Fig. 33 ; Fimmen, 1921, 123 ; Köster, *Beih. alt. Orient*, i, 36 ; the weight standards in Glotz, 192, 196, are in disagreement.
[9] Mosso, 136, 309; Peake and Fleure, Steppe and Sown, 104, Fig. 59 ; Meyer, Alt., II, i, 184 ; Hall, Bronze Age, 107 ; Cook, Zeus, ii, 617, 628 ; Evans, Knossos Excav., 1901–2, *Brit. Sch. Arch. Athens*, 101.
[10] Mosso, 318, 347.
[11] Cook, Zeus, ii, 644.
[12] Mosso, 136, 139, 309, 318, 347 ; Cook, Zeus, ii, 637 ; Meyer, Alt., II, i, 188.

BRONZE

Bronze appears in Crete and the Ægean in the EM III Period (2400–2200) and was common in the MM Period (2200–1700). The composition of the earliest alloy was very variable.[1]

ANALYSES OF ÆGEAN BRONZE AND COPPER OBJECTS

Object	Place of origin	Composition, per cent.	
		Copper	Tin
1. Large double axe	Sitia	100	—
2. Do., same form	Hagia Triada	—	18
3. Double axe	Palaikastro	—	3·71
4. ,,	Psychro	"apparently copper"	
5. ,,	Selakano (period doubtful)	84·60	4·169
6. Dagger, EM III	Platanos (Mesara)	—	nearly 2
7. ,, ,,	,,	100	—
8. Vase, ? MM I	Phaistos	89·5	3·146
9. Double axe, MM I	,,	89·50	3·146
10. Fused metal	—	89·40	1·57
11. Fused and spongy piece of metal	—	63·80	2·35
12. Nail	—	84·00	3·16
13. Piece of lebes	—	100	—
14. Daggers and knives, Mycenæan	Hagia Triada	88·70	9·480
15. Knife, Mycenæan	Tourlotti di Sitia	90·88	8·65
16. Tools	Gournia	—	9·6–10·45

1–5 : Mosso, 313, 316 ; No. 2 was of poorer quality metal than 10 per cent. bronze ; No. 3 was much damaged by use, the edge being bent over and worn ; No. 4 was damaged.
6–7 : Peake and Fleure, Steppe and Sown, 105 ; both found together.
8 : Evans, i, 195, apparently confused with No. 9.
9–15 : Mosso, 305, 308 ; No. 9 was 105 mm. long and 30 mm. thick at centre ; No. 11 was attached to a piece of gypsum (mould ?) ; No. 14, blades riveted with copper ; Nos. 14–15 are "several centuries later" than the early specimens Nos. 10–13, containing very little tin.
16 : Hawes, 38, 100 ; long and short saws, light and heavy chisels, awls, nails, files and axes, all of "modern type" and much battered by use.

Objects described (without reference to analyses) as bronze but some possibly copper, are the following : (1) A very large sword and a dagger with the hilts decorated with thin gold plate (EM III–MM I) from the palace of Mallia, east of Knossos, the sword being much longer than the earliest European swords : the long sword was a characteristic Cretan weapon,' the greaves and circular shields being also of bronze [2] ; (2) hoards of cauldrons at Tylissos (MM III), Cyprus, Mycenæ, Chalkis and Sardinia, and hoards of axes, cauldrons, tripods, etc. (MM III–LM I), at Knossos and Mochlos ; (3) large decorated

[1] Glotz, 21, 34, 36 ; Hall, Bronze Age, 51, 87 ; Frankfort, Studies in Early Pottery, 1927, ii, 150 and refs.　　　[2] Evans, ii, 272 ; Meyer, Alt., II, i, 205.

vases or ewers and basins, with the necks or rims soldered on (LM I), found at Knossos ; (4) a cup (MM III) and razors (LM III) from tombs.[1] The MM bronze axes of Crete and the Ægean are socketed for the haft like the Sumerian axes of 3000 B.C., whilst the Egyptian axes were always stuck into the haft, which is supposed to indicate that Cretan bronze work was a continuation of the Sumerian rather than the Egyptian technique.[2]

The common opinion, which seems on the whole insufficiently supported by analyses, is that the Bronze Age began in Crete in the MM Period (2200–2000) with vessels which were also imitated in pottery, and a few axes. In the LM Period (1600–1250) bronze vases, cups and utensils are common and weapons finally appear in greater numbers. The bronze at Mycenæ (LM III) is supposed by Karo to have come from Crete as plunder.[3] The statements that bronze was known in Troy II "about 2000 B.C., before Crete," [4] and that knowledge of bronze reached Crete from Troy II [5] are incompatible with the dates for EM III. Thomsen's date [6] of 1200 for bronze in Crete is much too late, and Hawes's of 2800, with a transition period of copper [7], is probably too early. Although Crete was possibly the first centre of bronze manufacture in the Ægean [8], the assumption that the alloy was discovered there [9] is improbable. The Cretans no doubt first obtained it by commerce, perhaps from Asia by way of Asia Minor, and then made it themselves.[10] Bronze weapons in Sicily and South Italy are of Late Mycenæan type, although there was early influence from Crete.[11]

Cretan bronze working was very skilled and Gournia (1700–1500) was a centre of the industry : slag, moulds, pure copper and bronze with 10 per cent. of tin, but no tin itself, were found there.[12] Tin buttons, of a conical shape widely disseminated through the Keltic and Iberian West, were found in EM III (c. 1800) Crete with metal toilet instruments of Egyptian fashion[13]; two such buttons with an "early Minoan" dagger occur in an Etruscan grave[14], so that Evans thinks the bronze was made by alloying the separate metals. Hammering, stamping, punching, graving and casting in soft stone moulds and casting boxes and perhaps by the cire perdue process were all practised.[15] Small moulds for casting double axes (perhaps of lead, q.v.) have been found.[16]

The usual sources for the supply of tin have been proposed: "Keltic" or British (Cornish) as early as the EM Period[17], Spain[18], Asia Minor,[19] Etruria or Gaul[20], Bohemia or the Erzgebirge[21], and Khorassan [22] ; traces of tin have

[1] (1) Evans, ii, 569, 624, Figs. 355 f. ; (2) ib., 624, 629, Figs. 392 f. ; (3) ib., Knossos Excav., 1903, 121, 125 ; Hall, Bronze Age, 164, 182, Fig. 234 ; (4) Evans, Palace, ii, 480 ; Wace, CAH, ii, 446 ; C. T. Tsountas and J. I. Manatt, The Mycenæan Age, 1897, 146, 166, 386, Fig. 60 ; Köster, Beih. alt. Orient, i, 36.

[2] Childe, Dawn of European Civilisation, 1925, 34 ; ib., The Aryans, 1926, 190 ; Hall, Bronze Age, 90.

[3] Karo, PW, xi, 1758, 1767, 1771, 1780, 1790 ; ib., A. Rel., 1909, xii, 361 ; Forrer, RL, 413.

[4] Hall, Bronze Age, 52.

[5] Peake and Fleure, Way of the Sea, 60, 64.

[6] RV, i, 45.

[7] Hawes, 17.

[8] Besnier, Daremberg-Saglio, Dict., IV, ii, 1461.

[9] Glotz, 225 ; Peake and Fleure, Priests and Kings, 12 ; Childe, The Aryans, 134 ; ib., Nature, 1927, cxx, 138 ; Hall, Bronze Age, 206.

[10] Hawes, 39 ; Childe, op. cit. ; Mackenzie, Footprints, 126.

[11] Meyer, Alt., II, i, 219.

[12] Glotz, 181 ; Hawes, 38.

[13] Evans, i, 101.

[14] Mackenzie, Footprints, 126.

[15] Glotz, 182 ; Mosso, 318 ; Meyer, Alt., I, ii, 782 ; Cook, Zeus, ii, 643.

[16] Dussaud, La Lydie et ses Voisins, 1930, 96.

[17] Evans, ii, 176 ; Mosso, 62, 270, 396 ; Glotz, 36, 225.

[18] Fimmen, 1921, 121—England also possible ; Köster, Beih. alt. Orient, i, 35; Spain doubtful—Frankfort, Studies in Early Pottery, 1927, ii, 135.

[19] Mackenzie, Footprints, 126.

[20] Glotz, 36, 225.

[21] Glotz, ib. ; Childe, Nature, 1927, cxx, 138 ; Hall, Bronze Age, 87, 206 ; W. M. Müller, Egyptological Researches, 1906, i, 5 ; 1910, ii, 183—Britain possible.

[22] Hawes, 42 ; cf. Thomsen, RV, i, 45 ; Kyrle, ib., 409.

been reported in Crete itself.[1]   The Cretans seem to have known tin at an early date, since they are shown in a VI or XI dyn. relief bringing it (dḥty) to Egypt.[2] Ashes of human bones from Palaikastro in Crete, as well as some (c. 2000 B.C.) from Gortan Kalembo in Pergamos, are rich in tin, which may have been derived from cremation in bronze vessels.[3]

LEAD

Lead occurs in the form of small votive axes with some of copper and silver in EM II tombs. It has been suggested that these were used as weights, but in all probability they, as well as the numerous lead double axes of Cretan shape found at Pantikapaion and Sarmatia, were votive objects or amulets.[4]   Lead is also found as a lining of stone chests in the MM palace at Knossos [5], as weights of MM III to LM I at Kairatos [6], as a statuette of Cretan form in a tholos at Kampos [7] and as bullets for slings in a palace, very late Minoan and perhaps even Achaian, crudely cast and pointed by paring with a knife.[8]   Meillet has suggested that the Greek names of some metals, such as lead (μόλυβδος), were borrowed from the [unknown] Ægean language.[9]

IRON

A cube of iron was found in 1927 in a MM II (c. 2000) tomb at Mavro Spelio at Knossos.[10]   A magnetic Neolithic piece of "iron" weighing about 1 lb., found by Mosso at Phaistos and probably an idol or cult object, not a hammer[11], is not meteoric iron or even iron at all but unsmelted native magnetite (the oxide ore Fe_3O_4), which is quite strongly magnetic and easily mistaken for iron.[12] Iron was used for rings, which occur in LM I (c. 1500 B.C.) tombs at Kakovatos and Vaphio.[13] A finely shaped iron nail with a flat ornamental top decorated with a typical Mycenæan rosette belongs to the period of transition between bronze and iron (c. 1375 B.C.).[14] An Egyptian wall painting of c. 1450 B.C. (Thothmes III) represents Minoans bringing a rhyton in the form of a jackal's head painted "a somewhat indistinct black, probably dark blue," which may represent iron or "black copper."[15]

Iron weapons at the end of LM (1200–1100), probably brought by intruders from the Greek mainland, appear first in East Crete in the early transitional period, soon after this in Thessaly in the Dotian plain and at the end of the transitional period abundantly at Tiryns and Athens.[16]   Although Hesiod[17] makes the Idæan Dactyls in Crete the first workers of iron, which was said to have been discovered by the veins of ore in Mount Ida being melted by a forest fire[18], there is no doubt that iron was a foreign metal in the early Ægean. The Hittite use of iron suggests that Greece obtained it from north-east

[1] Bürchner, PW, xi, 1732.
[2] Müller, opp. cit. ; Hall, Bronze Age, 200 ; Dussaud, 1914, 249, Fig. 179.
[3] Kobert, in Diergart, 1909, 118.
[4] Evans, i, 99 ; Glotz, 231 ; Karo, PW, xi, 1748 ; Meyer, Alt., II, i, 184 ; Cook, Zeus, ii, 540 ; Peake and Fleure, Steppe and Sown, 104.
[5] Evans, Knossos Excav., 1899–1900, Brit. Sch. Arch. Athens, 21.
[6] Evans, Palace, ii, 536.
[7] Tsountas-Manatt, 229, plate 17; Evans, iii, 461.
[8] Evans, ii, 345.
[9] Dussaud, 439.
[10] Evans, ii, 557, Fig. 353 ; Hall, Bronze Age, 86, 253: "the oldest worked

iron in Greece" ; correcting Karo, PW, xi, 1790 ; Glotz, 42.
[11] Mosso, 71, 243 ; Glotz, 228.
[12] Hall, Bronze Age, 253.
[13] Glotz, 42 ; Karo, 1790 ; Hall, 253.
[14] Evans, Knossos Excav., 1899–1900, Brit. Sch. Arch. Athens, 66.
[15] Müller, Egyptological Researches, 1910, ii, 19, 25 ; Evans, ii, 746 ; Hall, Bronze Age, 201, Fig. 262 B—no mention of iron ; on black copper, see p. 52, and Breasted, Anct. Rec., iv, 87 f.—translation of the Harris Papyrus.
[16] Mackenzie, Footprints, 148 ; Wade-Gery, CAH, ii, 524.
[17] Fragm., in Loeb ed., 1914, 76.
[18] Goguet, i, 156 ; Much, Kupferzeit, 1893, 286.

Anatolia. The use of iron weapons is traditionally supposed to have come, not with the Achaians, but with the so-called Dorian invasion, which appears to have spread from the region of Illyria about 1150 B.C. and to have broken up the remains of the old Minoan civilisation which lingered in the Mycenæan sites on the mainland of Greece. By this time the Minoan rule at Knossos had long disappeared.[1]

The Egyptian records of the 13th–12th centuries B.C., from Rameses III, speak of invasions of the "peoples of the sea," who seem to have included Achaians, Etruscans, Lycians, Sardinians and the (unidentified) Šakaruša. Part of this movement seems to have affected Crete, and the Philistines who appeared in Palestine in the same period are usually supposed to be dispersed Cretans. The Mycenæan culture on the mainland was apparently largely Achaian, although it continued Minoan traditions. The Dorian invasion put an end, in its turn, to the Achaian rule.[2]

The oldest Greek iron swords are of two types[3] and an iron sword from Gournia (Iron Age) is like those of Central Europe, which disproves the hypothesis of Belck[4] that the Cretan-Philistines were the discoverers of iron : there is no reason to suppose that the Philistines had iron weapons when they entered Syria.[5]

Traces of old iron mines are said to exist in Crete.[6]

ORGANIC MATERIALS

Although the discovery of the famous purple dye was formerly attributed to the Phœnicians, a legend preserved by Suidas says it was first made in Crete.[7] Clement of Alexandria[8] and Horace[9] say the purple fish (murex) was found in the Gulf of Lakonia and the purple of the Spartans, who had colonies in Crete[10], was the best European variety in the Roman Period.[11] The Cretan origin of murex purple seems established by the existence of the murex on the Cretan coast at present[12], by its representation on vases and by the vast heaps of its shells with MM II pottery on the island of Kouphonisi and near Palaikastro ; Mosso suggested that the lead-lined vats at Knossos were used for the dye.[13] Herodotos[14] also mentions a Cretan dealer in purple. The use of purple in Crete in 2000 B.C. would thus precede the Phœnician industry by some centuries.

According to Evans[15] the flowers of saffron (Crocus Sativus Græcus) are represented and "its dye, so much prized by the later Phrygians and Persians, may have coloured the yellow robes" seen on MM II frescoes. Mosso[16] also speaks of the "saffron flower, with its bulb," for colouring the skin. Perhaps safflower is meant : I can find no record of the use of saffron as a dye. The marine dye plant Rocella tinctoria, still growing in north-west Crete and exported to Egypt, may have been used in antiquity.[17]

[1] Wace, CAH, ii, 466 ; Wade-Gery, ib., 518, 531 ; Childe, The Aryans, 118.
[2] F. Bilabel, Geschichte Vorderasiens, Heidelberg, 1927, 117, 127, 231, 380 ; Evans, i, 24 ; Glotz, 49 ; Hall, CAH, ii, 284 ; Wace, ib., 447 ; cf. Bury, ib., 473.
[3] Hall, Bronze Age, 254, 257; Daremberg-Saglio, ii, 1602, figs. 3604, 3605.
[4] Z.f. Ethnol., 1908, xl, 45, 241.
[5] Dussaud, 1914, 55, 300.
[6] Meyer, Alt., II, i, 210 ; Glotz, 30—"rich in iron ore"; James, DG, i, 703—"no traces of mines."
[7] Lex., s.v. Ἡρακλῆς ; Goguet, ii, 88.
[8] Pædagogus, ii, 11 ; Ante-Nicene Library tr., iv, 263.
[9] Odes, ii, 18, 7.
[10] W. Smith and G. E. Marinden, History of Greece, 1902, 41.
[11] Charlesworth, Trade Routes, 125, 265.
[12] Bürchner, PW, xi, 1743.
[13] Bosanquet, J. Hellenic Studies, 1904, xxiv, 321 ; Glotz, 171, 177 ; Bailey, Pliny's Chapters on Chemical Subjects, 1929–32, i, 155.
[14] iv, 151 ; Dussaud, 419.
[15] Knossos Excav., 1899–1900, Brit. Sch. Arch. Athens, 30 ; Palace, i, 281.
[16] Med. Civ., 260.
[17] Glotz, 163 ; Bürchner, PW, xi, 1739, 1741 ; on the flora of Crete, ib., 1738.

22

Pliny[1] says the chief product of Chios was gum mastic (mastiche)—still one of the most important exports of the island and obtained by making incisions in the Lentiscus [2]—and also that storax was found in Crete.[3] The Cretans cultivated the vine, fig (ὄλυνθος), olive, date-palm, plum, quince, cypress, flax (for ropes : μήρινθος), poppy and sesame (σήσαμον) seeds, crocus or saffron, mint (μίνθα), catmint (καλαμινθος), wormwood (ἀψίνθος), lichens and various flowers, and kept bees for honey and wax.[4] Evans [5], from the forms of candlesticks, thinks wax candles were used in LM I Crete : earlier candles were perhaps of tallow. The olive was grown and its oil extracted in large quantities in the MM II Period and there was probably extensive export to Egypt. The silphium plant was important : in Classical times it was exclusively cultivated in Cyrene and may have been brought to Crete from there.[6]

There are indications that beer was brewed (MM) before wine was made (LM).[7] There do not seem to have been grain mills, the barley being cooked whole, and as there were no sickles the grain may have been plucked instead of cut.[8] Baltic amber occurs in EM III–MM I tombs near Gortyna, from at least 1600 on the mainland of Greece, and as beads in LM II tombs at Knossos.[9] Leather[10] and ivory[11] were known.

MALTA

Remains at Hagiar Kim and Mnaidra (near Valletta) in Malta resemble those of the Ægean.[12] Malta was the bridge between Europe and Africa and there were also early relations between Crete and Sicily. There are in Malta Neolithic (Chalcolithic) temples of huge slabs with standing stones and arches of limestone and floors of white clay earth over stone beds. The rooms, with small windows, were probably used for (fictitious) oracles. Stone decoration consists of spiral scrolls and pitting.

Stone objects include hammers, mortars, balls (possibly for grinding), axes, flint and obsidian knives and stone statuettes. Jade and jadeite were imported for making axe-head amulets and all varieties of pottery occur. Copper is found in remains of a later civilisation (c. 2000), the older remains being dated about 3000. Tarxien is the only place in Malta where Bronze Age metal implements and characteristic pottery occur. Later Malta is LM III or Phœnician.[13]

TROY (HISSARLIK)

A ship sailing into the narrow Hellespont from the Ægean passes on the starboard side a broad heart-shaped peninsula which, with the ancient Thracian Chersonese to port, forms the mouth of the Dardanelles, of sad memory. On this peninsula, in the large and fertile valley of the Skamander, and between this river and its ancient tributary the Simois, there rises from the level plain a low spur, 52 ft. in height, covered with ruins. This is Hissarlik, the ancient citadel of Troy.

---

[1] xii, 17, 25.
[2] Long, DG, i, 610.
[3] Bürchner, PW, xi, 1739 ; Salmasius, Homonymis Hyles Iatricæ, Utrecht, 1689, 151.
[4] Glotz, 163, 169.
[5] Palace, ii, 127.
[6] Evans, i, 283, 462 ; Forrer, RL, 738 ; Birch, Ancient Pottery, i, frontisp. ; weighing of silphium represented on a vase.
[7] Evans, i, 414.
[8] Mackenzie, Footprints, 118.

[9] Mosso, 368, 371 ; Meyer, Alt., II, i, 209, 226, 229 ; Evans, ii, 174 ; Hall, Bronze Age, 198 ; Karo, PW, xi, 1750.
[10] Glotz, 178.
[11] Meyer, Alt., II, i, 209.
[12] Evans, Palace, ii, 180, 190 ; T. Zammit, Discovery, 1922, 202 ; Peet, Ashby and Leeds, CAH, ii, 575—Neanderthal man at Ghar Dalam ; Mackenzie, Footprints, 138 ; Hall, Bronze Age, 213 ; Dussaud, 1914, 206 ; Meyer, Alt. II, i, 219, denies Cretan influence on Malta.
[13] Perrot-Chipiez, Phœnicia, i, 301.

The site was excavated in 1870–90 by Schliemann.[1] In his time the technique of excavation was in its infancy and he worked under very unfavourable conditions. For these reasons his results are often defective, objects from different layers having been mixed, and his accounts are sometimes contradictory. When he reported his work at Athens he had no notes and relied largely on memory [2], and the remains which he thought were the ruins of Homer's Troy (which is really Troy VI) were those of an earlier city (Troy II).[3]

According to Dörpfeld [4] the following nine strata represent the successive sites at Hissarlik :

I : The earliest settlement, sometimes regarded as two and numbered I and II. On this site are pieces of primitive hand-made, open-fired, usually black, pottery and some copper.

II : The prehistoric fortress or Burnt City (Troy II), on the site of which was kiln-fired monochrome pottery made on the wheel, much gold, silver and bronze, with moulds for casting ; lapis lazuli, amber, ivory, one axe-head of white jade and a small lead idol. The pottery is sometimes made in the form of human faces and animals ; the silver includes the forms of jars and knives ; the bronze is in the forms of spear-heads, daggers and cups. Troy II was a manufacturing town much larger than any other in the Eastern Mediterranean district at that time, and its importance depended on its central position between Europe and the regions to the east and south of the Black Sea.[5]

III, IV, V : Prehistoric villages, ephemeral and unimportant.

VI : Homer's Troy, on the site of which was a developed monochrome pottery and imported Mycenæan vases. This town was destroyed by fire.

VII : Two pre-Greek settlements with rude pottery and metal objects of a Northern tribe (? Treres), dated c. 1000–700 : a decadent period.

VIII : The Greek Ilion, 700–0 B.C.

IX : The Acropolis of the Roman Ilium, A.D. 0–500.

Troy is not an outpost of the Ægean civilisation proper but one of the principal seats of an alien culture which came into close contact with it.[6] The so-called Dardanians who had crossed the Hellespont and settled at Troy had become "Hellenised" about 1200 B.C. by contact with the original inhabitants, the "Pelasgians."[7] Troy may be Indo-European, since its walls are similar to those of the Hittite town Boghazköi.[8] The prehistoric culture, especially of Troy II, was part of that of a large region having common features extending through the north of the Balkans to the Danube Basin and Hungary.[9] Troy is particularly important because of its position on the boundaries of several nations and its close relations with peoples far in the interior of Asia Minor, whose culture it served to diffuse.[10] Graves and cemeteries of the so-called Trojan type of about 2000 B.C. in the interior of Asia Minor are rich in copper tools, knives and needles.[11] The Trojans were probably a Thraco-Phrygian people with natural relations to the interior of the Asiatic and the

[1] Meyer, Alt., 1909, I, ii, 656 ; Schuchhardt, Schliemann's Excavations, tr. Sellars, 1891, 17, 26 ; Leaf, Troy, 1912 ; Dussaud, 1914, 118 ; Dörpfeld, Troja und Ilion, 2 vols., Athens, 1902 ; Montelius, Ält. Bronzezeit, 1900, 155 ; Heinrich Schliemann's Selbstbiographie, bis zum seinem Tode vervöllstandigt, ed. Sophie Schliemann, Leipzig, 1892.

[2] Reinach, L'Anthropologie, 1892, 455 ; Montelius, 155.

[3] Dussaud, 119.

[4] i, 31 ; summary in Forrer, RL, 849, and plate 254 ; Marshall, Discoveries in Greek Lands, 19.

[5] Perrot-Chipiez, Histoire de l'Art, vi, 154 ; Ridgeway, Early Age of Greece, i, 45 ; Hall, Bronze Age, 53 ; Meyer, Alt., I, ii, 744 ; Hoernes, Urg., i, 69 ; Köster, Beih. alt. Orient, i, 26 ; Frankfort, Pottery, 1927, ii, 147.

[6] Wace, CAH, ii, 472.

[7] Bury, CAH, ii, 488.

[8] Otto, 1925, 55.

[9] Drerup, Ro., v, 1248 ; Peake and Fleure, Steppe and Sown, 1928, 91, 97.

[10] Baumgarten, Hellenische Kultur, 1913, 46 ; Köster, Beih. alt. Orient, i, 18, 26.

[11] Meyer, Alt., 1909, I, ii, 668.

littoral of the European peninsulas, and Troy II was in relation with Cyprus and Crete [1], perhaps with Babylonia.[2] The curious pottery with face decorations at Troy II resembles the face-decorated handles of pots found in Sumerian remains at Kiš, which are much earlier.[3]

Troy II is contemporary with EM II, i.e. 2800 B.C., and lasted to 1900 B.C.[4] The so-called Minyan pottery, of delicate grey colour and made on the wheel, was found in the VI city contemporary with MM (or Middle Helladic) ; it appears suddenly in Greece but not in Crete and is perhaps of north-west Anatolian origin or originated in Phokis and Bœotia. It appears to have been produced by first firing at a good temperature and then closing the air-holes of the furnace, so as to produce reducing gases, which converted the red ferric oxide into black ferrous oxide, giving a grey colour which is supposed to imitate silver.[5] The principal contact of the Ægean civilisation with the Sumero-Akkadian seems to have occurred both by sea and land by way of Troy II [6], and an extension of the Ægean culture from the same origin to El Argar in Spain, where there are flat copper axes and jewellery in gold (rare), silver and copper or bronze, has been suggested.[7]

In the early layers (Troy I) there is a mixture of Stone and "Bronze" cultures. The lowest layers (c. 3300), almost completely of Stone Age character, contain a few copper objects such as thin knives. The later layers (Troy II, etc.) frequently contain copper objects, including cast, some containing a small amount of tin, also bronze objects such as razors.[8]

The dates of the various layers at Hissarlik have been variously given[9] :

Troy	Forrer	Dörpfeld	Dussaud
I	3500–3000	3000–2500	3000–2400
II . . .	3000–2000	2500–2000	2400–1900
III–V . . .	2000–1500	2000–1500	1900–1500
VI . . .	1500–1000	1500–1000	1500–1180
VII . . .	—	1000–700	—

BUILDING MATERIALS

The building materials[10] of the various strata (35 f.) consist of clay-earth, porous limestone, wood and bricks, the earliest unburnt, and in some cases marble, syenite and stone. Stone was sawn and drilled with solid and hollow drills, perhaps of soft material aided by sand, or sometimes of flint (378). Burnt bricks occur only in the Roman Period (35), previous ones being two kinds of sun-dried (i) containing chopped straw, (ii) (rare) without it (37). Layers of wood often alternate with bricks, as in Cheshire to-day (40, 88, 91). The size of brick most used was that common in Egypt (40). Earth roofs with a slight fall from the middle to the edges were used (41) ; tiled roofs occur only, but then in abundance, in the last stratum (42). Lime mortar is found in the

[1] Glotz, 217.
[2] Meyer, Alt., II, i, 163 ; Hall, Bronze Age, 54.
[3] Peake and Fleure, Steppe and Sown, 97, Fig. 53.
[4] Hall, Bronze Age, 53.
[5] Hall, Bronze Age, 80 ; Frankfort, Pottery, 1927, ii, 141, 144.
[6] Dussaud, La Lydie et ses Voisins, 1930, 99, 110.

[7] Peake and Fleure, Merchant Venturers in Bronze, 15.
[8] Meyer, Alt., 1909, I, ii, 661, 665 ; ib., II (1893), 122 ; Forrer, RL, 850.
[9] Dörpfeld, i, 31 ; Forrer, RL, 849 ; Dussaud, 1914, 120 ; C. H. and C. B. Hawes give Troy II 2500, Troy VI 1350.
[10] References in the text are to pages of Dörpfeld, Troja und Ilion, i—no index.

earliest strata (36). The buildings of Troy VI were of limestone, wood, unburnt bricks, clay earth and chalk (109). Burnt chalk used for flooring cement was (by analysis) nearly pure calcium carbonate, probably imported from the mainland of Greece or elsewhere (112).

CERAMICS

The pottery in the various strata is divided into coarse and fine (243 f.). The coarse in Troy I is dark grey, contains bits of granite, mica, quartz, etc., and is irregularly burnt in open flames. The surface layer (slip ; engobe ; Überzug) is of fine washed clay smoothed by stones and of a grey, yellowish and (rarely) brownish colour. It is perhaps blackened by smoke and coloured by more or less complete oxidation of the iron in the clay (245). That the black was produced by soaking in liquid pine resin after firing and then heating [1] is improbable ; it is said to be due to carbonaceous matter, since when the ware has been exposed to a bright red heat in an oxidising atmosphere, the yellow or red colour will not change to black again in a reducing atmosphere, as is the case with ware in which the black colour is due to ferrous oxide.[2] The fine pottery of Troy I, of washed clay regularly fired and uniform in colour (grey, yellowish, brownish, deep black and fine red) (246), is sometimes ornamented by wavy lines filled with white "chalk," perhaps from the limestone soil.[3]

The pottery of Troy II–V is divided into three groups according to technique, form and decoration (252). The earliest types are still of badly washed clay containing large pieces of mica, are not formed on the wheel and are burnt irregularly and incompletely in an open flame. Attempts were made to get a definite colour on an outer fine clay by firing. No Troy II pottery is painted.[4] The quality is poorer than in the first stratum, but the kiln and potter's wheel appear (253). The ware is more uniform and of a brighter colour [5], yellow, red and traces of a clear grey, and the clay is better washed. There is curious and characteristic pottery in grotesque human and animal forms and Schnabel-kannen.[6] In Troy III the fine red ware of Troy II gives way to a fine grey or grey-black, or a beautiful yellow or brown. Up to now the surface had been polished mechanically, but attempts are now made to produce this effect by firing, and traces of painting occur. The ornamentation consists of horizontal grooves, sometimes filled with white material, and the shape is various and characteristic (255 f.).

In Troy VI the pottery is grey (mostly), yellow or red, the red being scarce as compared with the preceding period (281). The fineness of the clay varies according to the kind of vessel ; the polish is mostly obtained by burning, but in some cases still mechanically, and a surface layer is still used. Mycenæan ware was imported and imitated (283). The chief ornamentation consisted of wavy and spiral lines, with painting in black.

In Troy VII (296 f.), immediately below the Roman stratum, is a gradual introduction of old Greek glazed pottery. Trojan monochrome pottery probably came first, then early geometrical painted and Bügelkannen (300 f.).

The pottery in Troy VIII–IX (304 f.), the old Greek Period, is of very fine well-washed clay burnt red or light brown, the surface being light brown and smooth with a brown-black, bright red or chocolate glaze of very good quality. On one piece is a swastika (308). This was imported ware and was

[1] Hostmann, q. by Schuchhardt, 39.
[2] Doulton, q. by Frankfort, Studies in Early Pottery, 1924, i, 10, whose criticism of Franchet shows a lack of technical knowledge ; see the discussion of black Egyptian pottery, p. 109.
[3] Schuchhardt, 39, 41 ; Tubbs, DA, ii, 919 f.
[4] Montelius, Ält. Bronzez., 1900, 160.
[5] Cf. Schuchhardt, 41 ; Frankfort, Pottery, 1927, ii, 66.
[6] Schuchhardt, 68 ; Dussaud, 1914, 132 f., 134, Figs. 96, 97, 98.

imitated in Ilion (310). In the Roman Period very little pottery is found. In the post-Roman Period lead-glazed vessels first occur. Large pithoi (jars) were for storage (314 f.).

JADE AND IVORY

The white jade in Troy II is said (31 f.) [1] to be Chinese, but it is found also in Asia Minor (where Troy is situated) as well as in Syria, Egypt and Europe.[2] An ivory knife-handle in Troy II–V [3] is doubtful [4], but there was an ivory comb in Troy VI (399). Bone implements occur in Troy I, stone articles (in jade, greenstone, serpentine and a little hæmatite and porphyry) plentifully in I and II (324) [5] and wool and linen in Troy II (340), but there is no amber on the site.[6]

GLASS

Glass and faience beads were found in Troy II–V [7], including two stick knobs in green paste with white stripes, and polished rock crystal, including a "lens" (?) or button (338, 340). In Troy VI was one small button of blue glass (399).

GOLD

Gold as jewellery, as pellets and flat cakes ($\phi\theta o\ddot{\imath}\delta\epsilon\varsigma$) [8], small rods, tongue-shaped bars [9], nicked bars 10 cm. long with notches for breaking off equal pieces, wire ornaments, beads, sheet and objects of pressed and worked sheet occurred in abundance in Troy II in the remains called the "Treasury of Priam" by the enthusiastic but incautious Schliemann.[10] It is suggested by Forrer that the bars were weighed in Egyptian units, each segment weighing 0·188 gm. or approximately $\frac{1}{36}$ of the shekel (6·817 gm.) of the Egyptian mina (409 gm.). There are doubtful clay weights.[11] The gold is generally of 96 per cent. (= 23 carat) purity.[12] The wire is sometimes fine and the leaf thin, and the objects include ornaments partly of "barbaric" character, some spirals and individualistic work, such as the gold hair-pin with an oblong head comprising spiral work in panels surmounted by six miniature gold jugs, and an eagle of gold sheet partly soldered with gold.[13] The work is often elaborate and connected and there are spun vessels of gold as well as of silver and copper.[14] It has been thought that gold was worked before copper[15], but copper occurs in the oldest layers where gold is not reported. The gold may be Asiatic; the silver and gold of Troy II, where metal work was very advanced, perhaps Anatolian.[16] The silver of the Hellespont and the gold of Pactolus are traditional, and although the actual source of the silver is doubtful[17], this metal occurs in the mountains behind Hissarlik.[18] A gold "sauce boat" of very modern form with two spouts[19] came from Troy II. Some at least of the metal work seems to have been made with stone hammers in Troy II–V.[20]

[1] Also by Schuchhardt, 38, "or Central Asian," and Rawlinson, India, 8.

[2] Forrer, RL, 549; cf. Dussaud, La Lydie et ses Voisins, 1930, 98.

[3] Schuchhardt, 67.

[4] Perrot-Chipiez, vi, 175.

[5] Schuchhardt, 38.

[6] Perrot-Chipiez, vi, 947; Montelius, 160.

[7] Dörpfeld, 340, 385, page references in text; Perrot and Chipiez, Hist. de l'Art, vi, 944, say "no glass was found at Troy," but on the same page, "six glass objects were found."

[8] Hesychios $\phi\theta\acute{o}\ddot{\imath}\varsigma$, Lexikon, s.v.

[9] The "tongues of gold" in Jos. vii, 21, 24.

[10] Meyer, Alt., 1909, I, ii, 665; Regling, PW, vii, 978—thinks bars and tongues used as money; Blümner, PW, vii, 1555; Schuchhardt, 55; Dörpfeld, i, 328, 365; Wroth, DA, i, 324; Forrer, RL, 78.

[11] Dörpfeld, i, 389.

[12] Ib., i, 366.

[13] Schuchhardt, 56, 61, 64, 66; Figs. 57, 59, 61; Feldhaus, Chem. Z., 1910, 1133; Dörpfeld, 370.

[14] Hoops, i, 315; Blümner, iv, 236; Feldhaus, Technik, 639; Forrer, RL, 78, 850; ib., Urg., 290.

[15] Meyer, Alt., 1909, I, ii, 665.

[16] Blümner, PW, vii, 1555; Hall, Bronze Age, 54.

[17] Meyer, Alt., 1909, I, ii, 665.

[18] Peake and Fleure, Steppe and Sown, 91.

[19] Schuchhardt, 63, Fig. 47; Dussaud, 1914, 142, Fig. 105.

[20] Dörpfeld, i, 371.

SILVER

Silver occurs very plentifully in Troy II, mostly as vessels, but also as tongue-shaped bars weighing 40 Egyptian shekels of the heavy mina of 818 gm.[1] which Perrot and Chipiez [2] thought were Homer's "talents," as ornaments, beads [3], needles [4], goblets and jugs [5], a dagger and vases.[6] Some rust was found on silver articles, and "something which might be slag." The silver of some vessels contained 5 per cent., that of the bars 2·5 per cent., of copper. Silver was welded and soldered.[7] The silver pin, ear-ring and piece of wire said to have been found in Troy I, which would then be some of the oldest wrought silver in the world [8], are very doubtfully of this layer.[9]

In Troy II were crucibles in which, probably, gold and silver were melted and the metal workmanship is good.[10] Gowland[11] considers, from the results of analyses by himself (G) and Roberts-Austen (RA), that (in opposition to the view of "a Continental author") the silver was purified by cupellation with lead. The Roman patera was known to have been cupelled and Roman objects in the British Museum contain from 92·5 to 95·6 per cent. of silver, "which was of the nature of a definite standard."

	Silver	Gold	Copper	Lead	Iron
Troy II Bar (RA) . .	95·61	0·17	3·41	0·22	0·38
Vessel from Mycenæ (G) .	95·59	0·30	3·23	0·44	0·12
Roman patera (G) . .	95·15	0·47	3·44	0·33	0·07

Electrum occurs in Troy II as vessels, ornaments and bars, the metal containing gold and silver in the ratios 2 to 1 and 4 to 1, either natural or artificial.[12]

COPPER AND BRONZE

Although the presence of metal in Troy I seems certain,[13] whether it is copper or bronze is undecided. In 1893 what is described as a bronze pin was found in Troy I.[14] Objects which Schliemann (whose statements must be accepted with caution) reports from this layer are knives and pins—probably hair-pins and not brooch pins[15]—but the silver and gilt plate which he reports probably do not belong to this layer.[16] Earlier analyses of copper or bronze objects from Troy I by W. Chandler Roberts gave[17] :

Gilt knife : copper 97·4, with copper carbonate and cuprous oxide.

Needle 1 : copper 97·83 ; tin 0·21 ; nickel, tr. ; iron 0·90 ; cobalt, tr. ; sulphur, nil.

Needle 2 : copper 98·20 ; tin, tr. ; nickel, nil ; iron 0·75 ; cobalt, nil ; sulphur, nil.

[1] Dörpfeld, 327, 366 ; Schuchhardt, 62, Figs. 42–44 ; Forrer, RL, 78, 737, 850 ; ib., Urg., 288 ; Gowland, Archæologia, 1920, lxix, 139, Fig. 5.
[2] Histoire de l'Art, vi, 204.
[3] Meyer, Alt., 1909, I, ii, 664.
[4] Dörpfeld, 342 f.
[5] Gowland, J. Anthropol. Inst., 1912, xlii, 267.
[6] Schuchhardt, 62, 67, Figs. 45, 46.
[7] Dörpfeld, 366 f., 370.
[8] Gowland, 267.
[9] Dörpfeld, 324.
[10] Gowland, J. Anthropol. Inst., 1912, xlii, 265 f.
[11] Archæologia, 1920, lxix, 139 ; cf. also Friend and Thorneycroft, J. Inst. Metals, 1929, xli, 105 ; Davies, Nature, 1932, cxxx, 985.
[12] Dörpfeld, 332, 339, 366; Schuchhardt, 76.
[13] Fimmen, 1909, 21.
[14] BMG Bronze Age, 1920, 163.
[15] Schuchhardt, 37.
[16] Dörpfeld, 324.
[17] Montelius, Ält. Bronzez., 1900, 157 ; Schrader, RL, 731.

The metal is much harder than modern copper, probably because of impurities (the iron would produce this effect), and the small amount of tin in Needle 1 is probably "accidental." The results seem normal and reasonable. Newer analyses, however, are said to show that the metal is really bronze containing 10 per cent. of tin, and it is suggested that Troy passed from a Stone to a Bronze Age without an intermediate Copper Age.[1] In all probability both sets of analyses are correct, and the samples submitted seem to have been mixed, as Dörpfeld hints. The suggestion [2] that the earlier analyses are faulty seems quite impossible, since no analyst could miss 10 per cent. of tin or return it as copper, and the further assumption that the tin was removed by corrosion is equally improbable.[3]

If the copper knives were really found in Troy I (which is now considered probable), the metal was worked there at least as early as in Cyprus [4], but import of copper from Cyprus to Troy II, probably overland, is proved by the occurrence of Cypriote daggers in an unfinished form in the town.[5]

There is less doubt of the occurrence of both copper and true bronze in Troy II, which perhaps traded in copper and bronze with Asia Minor and the Balkans, and was probably also under Babylonian and Minoan influences [6], but the assumption by Hall that bronze was earlier in Troy than in Crete must be accepted with caution, since the chemical analyses indicate confusion among the objects.

The earlier analyses of Troy II bronzes by the chemists named gave [7] the following results : no lead is reported except a trace in No. 9 :

	Copper	Tin	Iron
1. Flat axe (Damour)	95·80	3·84	—
2. ,, ,, ...	90·67	8·64	—
3. ,, (Roberts)	95·41	4·39	—
4. ,, ,, ...	93·80	5·70	—
5. Large axe (Th. Schuchhardt) . .	93·50	5·80	0·70
6. Lance-head ,, . . .	90·96	9·04	—
7. Nail ,, . . .	98·65	0·45	0·85
8. ,, (largest) ,, . . .	99·55	tr.	tr.
9. Axe (Rammelsberg) . . .	97·10	2·90	,,
10. ,, ,, ...	97·11	2·89	,,
11. ,, ,, ...	95·38	4·11	,,
12. Chisel ,, ...	99·16	0·84	,,
13. Lance-head ,, ...	94·57	5·43	,,
14. Needle ,, ...	93·73	6·27	,,

These results would be normal for a civilisation passing from a Copper to a Bronze Age culture. The bronzes usually contain from 3 to 6 per cent. of

[1] Dörpfeld, 324—doubtful ; Hall, NE, 33 ; Meyer, Alt., 1909, I, ii, 665, 744—confidently.
[2] Montelius, Ält. Bronzez., 161.
[3] Cf. Perrot-Chipiez, Histoire de l'Art, vi, 178 ; Maspero, Struggle, 362, who refers to "a kind of bronze in which the proportion of tin was too slight to give the requisite hardness to the alloy."
[4] Gowland, J. Anthropol. Inst., 1912, xlii, 247 ; Frankfort, Pottery, 1927, ii, 7.

[5] Dussaud, La Lydie et ses Voisins, 1930, 85.
[6] Dörpfeld, 326 ; Myres, CAH, i, 103, 105 ; Hoernes, Urg., ii, 243 ; Meyer, Alt., II (1893), 120 ; II, i, 163 ; Schrader, RL, 731 ; Hall, NE, 39 ; ib., Bronze Age, 52, 54.
[7] Dörpfeld, 367 ; Montelius, Ält. Bronzez., 161 ; Blümner, iv, 188 ; Hoernes, Urg., ii, 243.

tin, less than in normal bronze (10 per cent.) and variable, although the lance-head [1], No. 6, contains 9 per cent. of tin. Newer analyses, again [2], give much larger amounts, from 8 to 11 per cent. of tin, and again the earlier analyses are doubted and the existence of a Copper Age called into question.[3] Dörpfeld's suggestion that there was no copper in Troy I and yet a fully developed Bronze Age in Troy II seems dubious, since no tin was found as such and the source of the tin for the bronze is doubtful.[4] Schrader [5] also comments on the remarkable fact that not a single sword was found in any level at Troy.

The mica-schist and granite moulds for casting are of the open ("hearth-casting") type, by means of which only one side of the object could be given the required form, the exposed side being then hammered.[6] That ordinary casting with full moulds was used in the II–V periods is probable but doubtful. Crucibles, funnels of clay and stone for molten metal, hammering, spinning of vessels, wire-drawing, chisel work and punching all appear.[7] There are bronze ornaments and some soldered and spun vessels in Troy II–V [8], but bronze utensils, implements, weapons and vessels become more numerous in Troy VI.[9] In Troy II there were bronze needles, razors and beads, some cast.[10]

In Troy V–IX bronze also occurs, in VIII–IX a bronze surgeon's knife.[11] The analyses are given below.[12] It is very remarkable that Troy VI, which should have been in the full Mycenæan culture, should have bronzes so poor in tin or even (in the case of the knife) practically pure copper, whilst those of Troy V, regarded as a decadent village, should contain so much tin. Dörpfeld, who was also puzzled by this anomaly, does not suggest that the objects have been mixed, but it almost seems as if this might be the case.[13] In Troy VII the bronze is rich (11–13 per cent.) in tin, although there is a pure copper needle. There are also hollow clay moulds, probably built round a wax model afterwards melted out (cire perdue), in this layer.

	Copper	Tin	Lead	Iron	Cobalt and Nickel	Arsenic
Troy V						
1. Dagger	86·93	10·62	0·68	0·25	0·49	0·77
2. Celt	88·67	9·70	0·74	0·62	—	tr.
3. Borer	96·17	2·69	0·35	0·49	—	0·30
Troy VI						
4. Axe (double)	94·11	4·15	0·85	0·26	—	0·63
5. Knife	98·88	—	—	0·40	0·38	tr.
6. Arm ring	89·48	9·34	0·77	0·41	—	,,
7. Knife	90·88	7·04	0·91	0·70	0·24	,,
Troy VII						
8. Axe (double)	85·17	12·92	1·08	0·33	tr.	,,

[1] Which has become a "celt" in Dörpfeld, 367.
[2] Dörpfeld, 367 ; Schuchhardt, 63.
[3] Meyer, II (1893), 120 ; ib., I, ii, 666 ; Schrader, RL, 731 ; cf. Frankfort, Pottery, 1927, ii, 7.
[4] Dörpfeld, 367 ; Meyer, Alt., 1909, I, ii, 665 ; Dussaud, La Lydie et ses Voisins, 1930, 85, suggests Bohemia.
[5] RL, 748.
[6] Perrot-Chipiez, Histoire de l'Art, vi, 204 ; Schuchhardt, 70, Fig. 65.

[7] Dörpfeld, i, 368 f. and plate ; Blümner, iv, 281.
[8] Schmidt, in Hoops, i, 315.
[9] Forrer, RL, 850.
[10] Dörpfeld, i, 343, 347.
[11] Ib., 394, 407, 408, 413.
[12] Ib., 421 ; in all cases "Pl," unknown as a chemical symbol, has been read "Pb," i.e. lead.
[13] BMG Bronze Age, 163.

Object No. 1 contained 0·32 per cent. of antimony, and No. 5 contained 0·05 per cent. of sulphur. Both these elements (and silver throughout) are reported as absent from all other specimens.

A glance at the table, keeping in mind the facts that good normal bronze contains 10 per cent. of tin and that lower and higher proportions of tin affect the quality prejudicially, leads to the conclusion that the Trojan metallurgist was neither particularly advanced nor very skilled, and to the suspicion that the bronze of Troy V was imported from some centre of higher culture, probably Egypt or Mesopotamia.[1] The suggestion [2] that tin-bronze was first invented in Troy II with tin imported from the Erzgebirge must also seem unlikely, since the varying proportions of tin in these much later alloys do not suggest to the chemist any long period of previous manipulation. If the early use of true bronze in Troy II is accepted [3], we are reminded of the very early tin-bronzes of Ur (q.v.). I am, however, by no means satisfied with the archæological material, in view of the very discordant analyses, and in any case the use of bronze in Ur precedes that at Troy according to present dating, so that an invention of bronze in Asia Minor seems to be ruled out.

LEAD

Gowland [4] refers to "shapeless lumps of lead" found in "the Lowest City (3000–2500)" as "perhaps the most ancient specimens of lead in the world," but I can find no mention of these in Dörpfeld or Schuchhardt. A small lead idol of a naked goddess is a remarkable object. It was formerly said to have a swastika cut in it [5], then to be without a swastika.[6] It was first said to have been found in the "burnt city" (Troy VI ?), then Troy II [7] or Troy II–V [8], or of 3000–1500.[9] It has been said to be indigenous[10], or as showing Babylonian influences[11], or to have come from Crete or Cyprus.[12] A lead wheel of Troy IV[13] is not mentioned by Dörpfeld or Schuchhardt.

Lead occurs with silver and "bronze" in the Cyclades from about 3000 B.C.[14], but although Perrot and Chipiez[15] say lead "seems to have been very common in Troy," the one doubtful idol is the only object described by Dörpfeld and Schuchhardt and Gowland says the metal was found in very small amounts. It probably came from the mines in the mountain ranges containing Mount Ida and Mount Olympos, where considerable deposits of argentiferous galena in ancient workings have recently been exploited. The mine at Balia, probably now the most important in Asia Minor, produced 7,600 tons of lead containing 63 oz. of silver per ton in 1903.[16]

IRON

Information as to the occurrence of iron at Troy is very confused. Maspero[17] speaks of a few specimens in early layers ; Dörpfeld[18] and Regling[19] of a tongue-shaped bar said to have been found in Troy II but of uncertain age ; and Schuchhardt of the two lumps of iron found in Troy II in 1890 by

[1] Cf. also Dörpfeld, 392.
[2] Peake and Fleure, Way of the Sea, 1929, 60, 64.
[3] As it is, e.g. by Frankfort, Pottery, 1927, ii, 7.
[4] J. Anthropol. Inst., 1912, xlii, 271.
[5] Dörpfeld, 362 ; Schuchhardt, 66, Fig. 60, showing the swastika ; Perrot-Chipiez, Hist. de l'Art, vi, 204.
[6] Dussaud, 1914, 364, Fig. 269, without the swastika.
[7] Ib., 364.
[8] Dörpfeld, 362 ; Forrer, RL, 94 ; Gowland, op. cit., 271.

[9] Feldhaus, Technik, 104.
[10] Dussaud, 364.
[11] Meyer, Alt., II, i, 163 ; cf. ib., 1909, I, ii, 670 ; Hall, Bronze Age, 54.
[12] Meyer, Alt., II, i, 163 ; further discussion by Dussaud, La Lydie et ses Voisins, Babyloniaca, 1930, xi, 71 f.
[13] Gowland, 271.
[14] Hoernes, Urzeit, ii, 43.
[15] Hist. de l'Art, vi, 953.
[16] Gowland, Archæologia, 1920, lxix, 140.
[17] Struggle, 363.
[18] i, 362.
[19] PW, vii, 978.

Schliemann and "probably rarer and more precious than gold," [1] which may really have come from Troy III.[2] Montelius [3] mentions three iron objects, one doubtful. No iron was found in Troy VII although it was probably used, but it is said to occur in Troy VIII–IX.[4] Sudhoff [5] mentions an iron "razor blade" found at Troy (no date). Useful iron does not appear until after the destruction of Troy VI.[6]

A piece of iron said by Götze to have been found in 1897 in Troy II was considered after analysis to be merely a mineral containing iron.[7] Von Luschan reports that Körte in 1899 found a piece of slag in an old Phrygian tumulus, "which appeared to be derived from a true smelting process," but this is isolated and it is unsafe to postulate Phrygian metallurgy in the Mycenæan Period. According to de Launay [8] the only iron found at Hissarlik consisted of two balls.

The reader may now be able to form or obtain a clearer idea of the occurrence of iron at Troy than I have found possible.

THE CYCLADES

The Cyclades provide the stepping-stones between Crete and the mainland of Greece as well as between the two sides of the Ægean. Melos, Paros, Naxos, Syros, etc., show a civilisation approximately synchronous with EM, with figurines, pottery with spirals, glaze, bronze, gold, some silver and model boats in lead. Designs stamped on clay with movable types of clay or wood probably represent the first use of such types. In the Middle Cycladic Period, ending about 1600, glaze on pottery was abandoned for a matt paint.[9]

Until the fall of Knossos, intercourse was only on the direct line between Crete and Egypt and even Cyprus was hardly touched till the end of the Palace Period. In the Mycenæan Period (LM III ; after 1400 B.C.) Cyprus was colonised, Curium, Citium and Salamis founded, and intercourse with Palestine established. This period ended abruptly with the Early Iron Age.[10]

In Late Cycladic II (close of 15th century B.C.) the influence of mainland Mycenæ became more important than that of Crete. Ægean culture had passed from the Cyclades to the mainland of Greece about 2800 B.C., but this is distinct from the foundation of the so-called Mycenæan or mainland-Cretan culture, which came later, perhaps in the MM III and LM I Periods. The newcomers brought varnish-painted pottery.[11]

On the volcanic island of Santorin in the Sporades, with its chief town Thera and the ruins of Eleusis, there were painted vases and a copper saw.[12] The Cyclades were rich in bronze implements, weapons and vases as early as 250C B.C., but the arrow-heads were of obsidian from Melos.[13] A saw from Naxos is made from arsenical copper, not tin-bronze.[14] Although the Cyclades probably knew metal early, there is little evidence of its use before EC III,

[1] Schuchhardt, 332 ; Schliemann, Bericht über die Ausgrabungen in Troja im Jahre 1890, Leipzig, 1891, 20 ; Tsountas-Manatt, 321.
[2] Perrot-Chipiez, Hist. de l'Art, vi, 204, 953.
[3] Ält. Bronzez., 156.
[4] Dörpfeld, 408, 413.
[5] In Hoops, iii, 439—no reference ; cf. Schrader, RL, 1016.
[6] Myres, CAH, i, 109 ; Wade-Gery, ib., ii, 524.
[7] Olshausen, Verh. Berlin Anthropol. Ges., 1897, 500 ; accepted by von Luschan, Z. f. Ethnol., 1909, xli, 49, and Montelius,

Ält. Bronzez., 156 ; reply by Götze, Verh. Berlin Anthropol. Ges., 1897, 504.
[8] Daremberg-Saglio, Dict., ii, 1076.
[9] Wace, CAH, i, 600 ; Tsountas-Manatt, 265.
[10] Myres, CAH, iii, 635.
[11] Wace, CAH, ii, 448 ; Hall, Bronze Age, 61, 140.
[12] Perrot-Chipiez, Hist. de l'Art, vi, 149, 153.
[13] Meyer, Alt., 1909, I, ii, 695 ; II (1893), 123.
[14] Zenghelis, q. by Frankfort, Studies in Early Pottery, 1927, ii, 150.

the use of obsidian being encouraged. Gold, silver, copper and lead were probably obtained in exchange for obsidian, marble and emery, the islands only later working their own copper (at Paros, Seriphos and Siphnos) and lead ores. When obsidian knives were replaced by copper, Melos declined until it was settled by Minoans in MM I–MM II.[1]

MYCENÆ

Mycenæ (Μυκῆναι), an old walled town in the Argolis, may have been built as an outpost to secure the trade routes to the Corinthian Gulf. In Homer it is the city of King Agamemnon and "abounding in gold," a description amply confirmed by the rich finds of artistic works in gold in graves on the site by Schliemann in 1876.[2] In its present form the city dates back to the 14th century B.C.[3] Egyptian objects of about 1400 B.C. were perhaps brought by way of Syria and Asia Minor.[4] Mycenæ is supposed to be an offshoot of Cretan culture by colonisation of the mainland and the description in Homer is post-Mycenæan.[5] The metal work especially, including the moulds of dark soapstone (Speckstein) in Crete and Mycenæ, makes it very probable that both come "from the same forms, the same workshop."[6] The ruins of the town, which was sacked in 463 B.C., were described by Pausanias (A.D. 160–180) [7], who mentions the Lion Gate, the beehive-shaped tombs (tholoi) and that strange circle of shaft graves which still offers problems to the archæologist. The tholoi resemble others at Menidi in Attica, at Orchomenos in Bœotia, at Pharis in Lakonia, at the Heraïon near Argos and at Volo in Thessaly : the objects found in them are all of Mycenæan type. The rich treasures of gold were found in the six shaft graves containing stelæ and bodies, not burnt and one partly mummified, with ashes, probably burnt offerings.[8] The famous Lion Gate has a relief of two lions carved in anhydrite facing each other, with their paws resting on the support of the remarkable column, always seen to be thicker towards the top (as is the case at Knossos) [9], but now said not to be.[10] It shows influences of the art of Asia Minor. The walls of the citadel are of limestone blocks, in the oldest part unhewn or only roughly dressed, in the somewhat later part of rectangular masonry, and both are bonded with clay ; much later are polygonal close-fitting blocks.[11] The stone reliefs (stelæ) (perhaps later than the graves), of a much cruder character than the Lion Gate and the metal work, are probably native work, as opposed to that of Cretan artists.[12]

The dates of the finds have been the subject of much discussion, with a tendency to lower them to 1600–1500 B.C., instead of 2000 B.C., as formerly

[1] Peake and Fleure, Priests and Kings, 116, 117 ; ib., Way of the Sea, 63 ; Marshall, Disc. in Greek Lands, 10.
[2] Schliemann, Mycenæ, Engl. tr., 1878 ; Schuchhardt, 134 ; Dussaud, 1914, 146 ; Perrot-Chipiez, H. de l'Art, vi, 131, 303 ; Forrer, RL, 528 ; Tsountas-Manatt, 67 ; Hall, Bronze Age, 8, 20, 140, 153, 207, 215, 233, 248 ; Marshall, Discovery in Greek Lands, 13.
[3] Wace, CAH, ii, 457.
[4] Breasted, H., 337 ; Meyer, Alt., II (1893), 133, 200 ; II, i, 166 ; Baumgarten, Hellenische Kultur, 61.
[5] Hall, Bronze Age, 140, 154 ; Karo, PW, xi, 1774, 1795 ; Meyer, Alt., II, i, 221 ; Peake and Fleure, Merchant Venturers in Bronze, 1931, 104.

[6] Baumgarten, 60.
[7] Description of Greece, II, xvi ; the German theory that Pausanius was a compiler who did not travel is completely antiquated and incorrect.
[8] Schuchhardt, 15, 158, 162 ; Meyer, Alt., II, i, 226 ; Forrer, RL, 526 ; Herodotos, i, 67, 68, and other authors preserve a pre-Homeric tradition of burial, not cremation.
[9] Schuchhardt, 138, 142, Fig. 137 ; cast in South Kensington Museum.
[10] Evans, Palace, i, 342.
[11] Schuchhardt, 136, Figs. 134–136 ; Forrer, RL, 526 ; Tsountas-Manatt, 25.
[12] Schuchhardt, 170, Figs. 145 f. ; Hall, Bronze Age, 141, Figs. 175 f. ; Meyer, Alt., II, i, 229.

assumed.[1]   Wace considers the tholos and Acropolis tombs at Mycenæ not to be LM I but LM III (Late Mycenæan), when Knossos and Crete had ceased to be important. The quite different Neolithic culture on other parts of the mainland, which remained much longer in Thessaly, used hand-made but fine pottery, often with painted cross-hatched designs, but no metal. The *early* Ægean culture on the mainland is called Helladic.[2]   From this period there is a gold "sauce boat" in the Louvre.[3]

A modern dating of the graves is as follows : vi, the oldest, contents imported Cretan, 1600 B.C., MM III ; graves ii, iv and v, MM III–LM I ; the rest LM I. The tholoi of Mycenæ and Orchomenos ("Treasury of Minyas") are later ; some, including the "treasuries" of Atreus and Clytæmnestra, are perhaps Late Mycenæan, *c*. 1400 B.C. That at Vaphio is LM I (*c*. 1500 B.C.) and Evans puts all the tholoi as LM I and the "treasury of Clytæmnestra" as early as MM III. The tholos probably originated in Crete, where it is known for LM I.[4]

The numerous finds at Mycenæ are here described briefly from the point of view of material only : the artistic forms may best be appreciated from a glance at the illustrations in Schliemann [5] and in Schuchhardt.[6]   Cretan artists and craftsmen probably introduced art into Mycenæ, but by the end of the 16th century the native workers had developed the Minoan technique for themselves and Mycenæ gave its name to this stage of culture in the Ægean.[7]

CERAMICS

The vases [8] are of two main types : (1) those painted with lustrous dark brown varnish glaze resisting the action of acid, with designs of sea-plants and animals ; (2) those with matt painting, in dull brown, purple or red (some with green), with linear and spiral ornaments ; the types of decoration are characteristic of the character of the varnish or colour (dull), with one exception. The body is of fine yellowish-brown clay, the vases thin and made on the wheel. The two types occur with the same decorations and material in widely separated localities, including Rhodes, Sicily and Sardinia, and thus probably formed articles of export.[9]

Analyses of red and black glaze from Mycenæan ware by Foster[10] showed that the colours were due to ferric and ferrous oxide respectively from the clay, with alkali in the glaze, and that manganese was not essential. The black glaze was made with a white clay, the body containing less ferrous iron than the glaze. In the red ware the body contained more ferrous iron than the glaze, which contained considerably more ferric iron than the body. Attempts to reproduce the red glaze by fusing pipe-clay with an alkaline flux (sodium nitrate), with free admission of air, gave a black, since ferric oxide on heating turns black. A mixture of a red-burning yellow clay with pipe-clay and sodium nitrate (3 yellow clay + 1 pipe-clay ; 1 part of mixture + 8 parts sodium nitrate), thinly applied and fired at 980–990° C., gave a red glaze, but

[1] Hall, Bronze Age, 146 ; Fimmen, 1909, 38 ; Meyer, Alt. II (1893), 130 ; *ib*., II, i, 221 f.

[2] Hall, Bronze Age, 11, 61, 64 f., 117, 247.

[3] Childe, *J. Hellenic Studies*, 1924, xliv, 163, whose conclusions are criticised by Hall, Bronze Age, 66.

[4] Hall, Bronze Age, 146 f. ; Dussaud, 150, 196 ; Evans, Palace, ii, 172 ; Schliemann's numbering of the graves is different : Schuchhardt, 156 ; cf. Casson, *Discovery*,

1929, x, 262 ; shafts and tholoi are contemporary ?

[5] Mycenæ, Engl. tr., 1878.

[6] Schliemann's Excavations, Engl. tr., 1891.

[7] Wace, CAH, ii, 451, 454 ; Perrot-Chipiez, Hist. de l'Art, vi, 133.

[8] Tubbs, DA, ii, 921 ; Dussaud, 154.

[9] Schuchhardt, 186 f., 190, 209, 213, 270, 273.

[10] Tonks, *Amer. J. Archæology*, 1910, series 2, xiv, 417 ; black glaze, Tonks, *op. cit*., 1908, xii, 417, 421, with bibliography.

not easily. An analysis of Mycenæan clay gave : SiO_2 40·60, Al_2O_3 17·07, Fe_2O_3 6·93, FeO 0·56, CaO 19·80, MgO 4·42, K_2O 2·96, Na_2O 0·21, CO_2 5·40, H_2O 2·95 : total 100·90. Another specimen gave : Fe_2O_3 (total Fe) 8·89, CaO 13·82, Al_2O_3 20·40, SiO_2 47·51, MgO 4·41, loss on ignition 2·78.

GOLD

The precious metals, which occur in the tombs in large quantities, were worked skilfully but artistically rather crudely by local craftsmen under Cretan influence. Egyptian elements probably also came from Crete. The forms of gold objects include death-masks, diadems, breastplates, vases, jewellery, etc. ; the work is very profuse, since the craftsmen had enormous quantities of metal to work upon.[1] Examples are (the figures in the text are pages in Schliemann, 1878) : diadems piped with copper wires (79, 115, 153, 155) ; thin gold leaves with repoussé, probably done on a lead base, butterfly, tree-crickets and griffin (Figs. 156, 239–52, 165, 176) ; vases (185, 206) ; signet rings (215, 218, 220, 223) ; wood covered with gold (227, 229, 231, 232 f., 236, 246 f., 258 f., 259) ; buttons (260 f.) ; gold leaves in large numbers (266) ; wire and rings (267 f., 278, 287 f., 292, 301 f., 309–16, 318, 332, 350, 353). The wire, of round and square section (142, 354 f.), may be "cut from sheet," although Blümner[2] says it is drawn, as is the case with Greek and Roman wire (wire is often mentioned by Homer). The round discs of foil with spiral and octopus decorations have been supposed to be coins.[3] The foil was 1/600 to 1/500 inch thick, the sheet 1/100 inch thick. The foil had apparently been lacquered reddish-yellow, since on heating it gave off organic matter and turned greenish-yellow. There were two colours of gold, a red with copper alloy and a light with silver.[4] Brownish-black spots on gold rosettes contained manganese oxide.[5] The "red" gold may have contained some iron, as did the Egyptian red gold in the tomb of Tutankhamen[6], but the specimens of metal analysed contained only a little iron.

The gold objects are also described by Schuchhardt.[7]

ANALYSES OF GOLD FROM MYCENÆ[8]

	Gold	Silver	Copper	Lead	Iron
Foil . . .	73·11	23·37	2·22	0·35	0·24
Sheet . .	89·36	8·55	0·57	—	0·20

The first specimen is really electrum, whether natural or artificial is uncertain. Percy suggested that the alloy was artificially made from purified gold, cupelled silver containing a little lead, and traces of copper, the last in some cases perhaps added. The use of the "lead bath" in purifying gold is, however, mentioned by Suidas.[9]

[1] Wroth, DA, i, 324 ; Glotz, 334 ; Perrot-Chipiez, Hist. de l'Art, vi, 865, 877 f. ; Tsountas-Manatt, 219 ; Forrer, RL, 528.
[2] iv, 250.
[3] Baumgarten, Hellenische Kultur, 53.
[4] Schliemann, 367 ; Schuchhardt, 231, 265.
[5] Rhousopoulos, in Diergart, 1909, 190.
[6] Hall, Bronze Age, 143.

[7] Schliemann's Excav., 1891, 176 f., 191 f., 196 f., 201 f.—typical art forms ; 207 f., 211, 215 f., 221—engraved rings ; 223 f.—masks ; 231—red and pale gold ; cf. 236 f., 252 f., 265, 269, 271, 275 f., 277—engraved ring.
[8] R. Smith, in Schliemann, Mycenæ, 367.
[9] Lexikon, ed. Gaisford, Oxford, 1834, 1286.

The Mycenæan Age was rich in gold and had a highly developed gold industry. The use of gold of different colours in combination with alabaster, ivory, malachite, lapis lazuli, blue glass paste (kyanos), etc., the metal being sometimes in the form of spun or pressed sheet and sometimes soldered, indicates a high degree of skill.[1]

Two interesting figured gold cups found in 1889 by Tsountas at Vaphio, a hamlet in Lakonia, with representations of bulls, etc., are probably of the 16th century B.C., and were imported from MM III–LM I Crete (perhaps to Mycenæ, from whence they were stolen). An outer repoussé gold sheet is fastened over another plain sheet for the interior of the cup.[2] Two decorated gold cups which belonged to a king and queen, the latter cup ornamented with niello, have been found at Mideia near Mycenæ.[3]

Electrum was much used at Mycenæ for ornaments.[4] A band (1690 B.C.) contained 7 gold to 1 silver ; a diadem (1600 B.C.), 3 gold to 1 silver ("true electrum"). Mycenæan gems are sometimes found covered all over with gold leaf, including the engraved face, and were probably not seals but amulets.[5]

SILVER

Although Mycenæ used silver freely and had developed an artistic silver industry, this metal does not occur so frequently as gold.[6] The objects include a cup, vases, a cow's head, a flagon, an object of gold and silver, an ox head inlaid with gold and "enamel," some niello work, a silver vase with a battle scene, perhaps on the coast of Asia Minor, and a sceptre plated with gold.[7] A silver spoon and *plain* silver cups found at Vaphio with the figured gold cups (see above) were probably also MM III–LM I Cretan work obtained by way of Mycenæ.[8] The silver vase with the battle scene was found in rather a decomposed state[9] : the metal contained 71·60 of silver, 0·22 of gold, 2·42 of copper, 0·33 of lead, 0·09 of iron and a trace of chlorine ; the crust contained 21·97 of silver chloride and 3·16 of gold, silver, copper oxide and calcium carbonate. From these results the original composition of the vase metal was deduced as : 95·59 of silver, 0·30 of gold, 3·23 of copper, 0·44 of lead and 0·12 of iron.[10]

A silver arm-band was practically pure, with a trace of gold[11], although the silver at Mycenæ was usually impure and contained gold, antimony and copper.[12] A clumsy cast model stag with antlers from Schliemann's Grave iv contained 2 parts of silver to 1 of lead, perhaps an imperfectly cupelled argentiferous lead.[13] The silver cow's head had gold-plated horns, made by first plating the silver with copper and then plating the copper with gold ; the neck of a silver vase was similarly plated.[14] The cow's horns were soldered. Schliemann says that borax was used for soldering gold : Landerer "was lucky enough to

[1] Forrer, RL, 528 ; Meyer, Alt., II (1893), 156, 175 ; *ib.*, II, i, 234 ; Blümner, iv, 239, 271.
[2] Perrot-Chipiez, Hist. de l'Art, vi, 409, 770, 784, plate 15 ; Tsountas-Manatt, 227 ; Baumgarten, Hellenische Kultur, 54, Figs. 72, 77 ; Hall, NE, 55, 61 ; *ib.*, Bronze Age, 154, Fig. 102 ; Meyer, Alt., II, i, 170, 231.
[3] Casson, *Discovery*, 1929, x, 262.
[4] Blümner, iv, 161 ; *ib.*, PW, v, 2315 ; Rhousopoulos, in Diergart, 1909, 182, 184.
[5] Ridgeway, Early Age of Greece, i, 331.
[6] Meyer, Alt., II (1893), 156 ; Forrer, RL, 528 ; Glotz, 336.
[7] Schliemann, 58, 210, 215, 243, 308, 316 ; Schuchhardt, 194, 207, 244, 259 f.,

265, 278, 297 ; Perrot-Chipiez, Hist. de l'Art, vi, 811, 821, Fig. 381 ; Tsountas-Manatt, 212 ; Meyer, Alt., II, i, 232, plate vii ; Anderson, DA, ii, 611.
[8] Tsountas-Manatt, 145, 227 ; Forrer, RL, 857, and Meyer, Alt., II, i, 234, speak of "figured" cups of silver.
[9] "Restored" in Evans, Palace, iii, 89, Fig. 50.
[10] Schliemann, 367 f.
[11] Rhousopoulos, in Diergart, 1909, 193.
[12] Blümner, iv, 151.
[13] *Ib.*, iv, 285 ; Gowland, *Archæologia*, 1920, lxix, 143 ; Schliemann, 257 ; Schuchhardt, 245.
[14] Schliemann, 158, 218, 228 ; Schuchhardt, 249.

discover this salt on the border of an ancient false [silver-plated copper] medal from Ægina."[1] Borax was used in glazing Roman terra sigillata of the 1st century B.C.–1st century A.D., but not older Etruscan ware.[2] Perrot and Chipiez say that Mycenæan metal, unlike that at Troy, was not soldered, but soldering was found in beehive tombs in later Greece.[3]

Copper plated with gold was also found at Mycenæ.[4] One of the cows' heads had a double axe between the horns.[5]

COPPER

Copper was found in various forms. A large vase (1600 B.C.) was pure copper.[6] Copper boxes (Sn., 207, St., 160) and cauldrons (Sn., 215, St., 244, 256) were not soldered, but the plates joined with copper pins (Sn., 207, 215, St. 160: on copper objects see Sn., 255, 274, 277—tripod ; St., 185, 207, 222, 244, 246, 251, 269). (See p. 354.[8])

Analyses of objects found by Schliemann at Mycenæ are[7] :

	Copper	Tin	Lead	Silver	Iron	Nickel	Arsenic
1. Kettle	98·47	0·09	0·16	0·013	0·03	0·19	0·83
2. Bowl	99·4	—	0·2	—	—	—	0·2
3. Sword handle	99·4	0·1	—	—	—	—	—
4. Thin sheet	95·6	0·1	—	—	0·9	—	—

No. 1 contained a trace of bismuth ; the large amount of arsenic is noteworthy.

Mycenæ had a well-developed early copper industry. The work of the smith (χαλκεύς) included cold hammering, casting, turning, soldering, beating into foil and drawing into fine wire. Some objects, such as the large cauldron and the daggers covered with black and inlaid with gold and electrum (see below), are objects of advanced art.[8]

BRONZE

Bronze was found as daggers, tools, implements, razors and knives, but not as arrow-heads (mentioned in Homer), which are all of stone or obsidian.[9] The bronze (or copper) worker (χαλκεύς) and the goldsmith (χρυσοχόος) are one person in Homer, who also gives bronze with gold, silver, electrum and ivory as materials for wall decorations, so that bronze was perhaps expensive in the early period.[10] The early bronzes at Mycenæ are poor in tin and it has been suggested that those rich in tin (10 per cent. or more)[11] are of foreign origin[12], or were obtained as piratical booty from Crete[13], but the variation in

[1] Schliemann, 231 ; Schuchhardt, 249 ; cf. Feldhaus, Technik, 637.
[2] Nasini, Chem. Z., 1930, liv, 985.
[3] Hist. de l'Art, vi, 974.
[4] Schliemann, 278.
[5] Ib., 218 ; Schuchhardt, 249.
[6] Rhousopoulos, in Diergart, 1909, 182 f.
[7] (1) by Percy, Schliemann, 375 ; Blümner, iv, 186 ; (2)–(4) by G. B. Phillips, quoted in Anct. Egypt, 1924, 89.

[8] Meyer, Alt., II (1893), 156, 167, 173 ; Blümner, iv, 50 ; Davies, Nature, 1932, cxxx, 986 ; Rhousopoulos, in Diergart, 1909, 184 f., Fig. 4 f.
[9] Hoernes, Urg., ii, 243 ; Tsountas-Manatt, 166 ; Schuchhardt, 237.
[10] Perrot-Chipiez, Hist. de l'Art, vi, 558, 880.
[11] Tsountas-Manatt, 73.
[12] Meyer, Alt., II (1893), 156.
[13] Karo, PW, xi, 1767.

composition may be accidental. Some analyses (in percentages) are given below[1] :

	Copper	Tin
1. Kettle	—	1
2. Nail	—	2
3. Sword (corroded) . .	46·4	5·5
4. Sceptre . . .	90·56	9·44
5. Sword . . .	86·36	13·06
6. Vase handle . . .	89·69	10·08
7. Nails	88	12

No. 5 also contained 0·11 lead, 0·17 iron, 0·15 nickel and traces of cobalt, bismuth and silver. No. 6 was a very pure bronze. Miscellaneous bronze objects include knives, rings, safety pin and a dagger with the blade soldered in the middle, etc.[2]

The bronze swords or daggers inlaid with gold, silver, electrum and a black substance, all on a bronze plate let into the bronze blade, are similar to late Cretan work and are mentioned in Homer.[3] The inlay of gold on iron or bronze, or gold on silver or bronze, by hammering the noble metal in the form of thin sheets into engraved designs, has been considered [4] to be a Mycenæan art, but [5] the inlaid daggers are perhaps Egyptian, not Cretan work. On one is a spot of red gold, perhaps involving the Egyptian technique of alloying with iron.[6] Bronze very rich in tin, or "an alloy of lead and silver," found at Vaphio (Lakonia) was not analysed[7]. Theban bronze contained 18 per cent. of tin.[8] The castings in Mycenæ are solid, not hollow.[9]

LEAD

Lead occurs frequently in the form of jars for storing grain, as rings, wire, discs, etc. This metal and lead at Tiryns probably came from Laurion, where there are three domed tombs containing Mycenæan remains. Lead was also found at Vaphio, at Thoricas and at Gha in Lake Copais (Bœotia).[10]

IRON

Schliemann found no iron at Mycenæ[11], but Tsountas found some chains along with gold chains in late tombs.[12] Iron was very scarce ; it appeared at Mycenæ and Tiryns only at the end of the Mycenæan Period and was used for ornamental purposes.[13] In weapons of 1400 B.C. iron is used with strips of gold, copper and bronze inlay.[14] A fragment of an iron ring of the Mycenæan

[1] (1)–(2) Blümner, iv, 188 ; Lippert, Kulturgeschichte, 1887, ii, 231 ; (3)–(4) Rhousopoulos, in Diergart, 1909, 185–192 ; (5) Schliemann, 372 ; Blümner, iv, 188, analysis by Smith ; (6) Schliemann, 375 ; Blümner, iv, 188, analysed by W. F. Ward ; (7) used for fastening bronze plates to walls, Schliemann, 44.
[2] Schliemann, 75, 111 ; 142, 144 ; 158, 219, 279 f., 306 ; Schuchhardt, 185, 206, 212, 222, 229, 231 f., 269, 296.
[3] Schuchhardt, 229, 231, 232 ; Perrot-Chipiez, Hist. de l'Art, vi, plates 17–19 in colour—silver and gilt, but no black shown ; Glotz, 338.
[4] Forrer, RL, 799.
[5] Meyer, Alt., II, i, 177.
[6] Hall, Bronze Age, 143.

[7] Perrot-Chipiez, Hist. de l'Art, vi, 975.
[8] Glotz, 52.
[9] Blümner, PW, vi, 607.
[10] Schrader, RL, 95 ; Forrer, RL, 94 ; Blümner, iv, 286 ; Feldhaus, Technik, 104 ; Tsountas-Manatt, 73, 381 ; Gowland, J. Anthropol. Inst., 1912, xlii, 271.
[11] Schuchhardt, 229, 332.
[12] Perrot-Chipiez, Hist. de l'Art, vi, 954 ; Tsountas-Manatt, 72, 145, 146, 165, 183, 321—five or six times only, as rings.
[13] Perrot-Chipiez, Hist. de l'Art, vi, 990 ; Blümner, iv, 286 ; ib., PW, v, 2143 ; Hoernes, Urg., ii, 220 ; ib., Urzeit, ii, 33 ; Forrer, RL, 525, 528 ; Meyer, Alt., II (1893), 379 ; Schrader, RL, 176 ; de Launay, Daremberg-Saglio, Dict., ii, 1076.
[14] Gsell, 30 ; Hoernes, Urzeit., iii, 11.

Period was found at Kakovatos in Triphylia in Greece, with lapis lazuli and Baltic amber.[1] The early iron at Mycenæ may have come from Egypt.[2] The first iron sword found in the Ægean is from Mouliana, at the end of LM III.[3] It was a foreign people armed with such swords which ended the Mycenæan culture, and the plentiful use of iron in Greece for tools and weapons began only after the so-called Dorian invasion from the North (Illyria).[4] Another—modern—view is that the transition between the Bronze and Iron Ages was not a sharp one but occurred gradually from about 1200 B.C.[5] Gowland [6] traces the form of iron furnace used in Southern Europe east of the Apennines to Central Europe, and thence to Central Asia ; it has no affinity with the peculiar form used in Egypt, Etruria and the western Mediterranean. The iron tools, arrow-heads, bits of scale armour, etc., in late remains in Egypt (Naukratis and Daphnæ) are also of Greek or European, not Egyptian, form.[7]

LAPIDARY WORK

There were engraved gems and cylinders at Mycenæ [8] (Sn. 112) ; lapis ollaris (Pliny's Siphnian stone) (Sn. 110) ; rock crystal as perforated hollow beads for pin-heads (Sn. 200, 210, St. 195), a jar (Sn. 300) and inlay (St. 234, 250) ; alabaster as a votive spoon (St. 206), a vase (St. 242), a knot (St. 251) and a cup (St. 263) ; obsidian knives and arrow-heads (Sn. 272 ; St. 209, 237) ; fluorspar (or agate) beads (Sn. 121 ; St. 196) ; granite and basalt moulds for casting metals (Sn. 107 f., St. 279) ; and a block of porphyry with decoration (St. 285).

GLASS

A fine white paste head ("Egyptian porcelain") is perhaps "lead glazed."[9] Translucent blue or white glass paste, e.g. a white bead[10], weathered true glass beads from Pylos (1600–1200 B.C.) and a fragment of a vase of potash glass coloured with copper[11], show that glass was known in the Mycenæan Period.

Schliemann refers to a "lead glassy substance" found with "blue glass," and beads of glaze, such as a blue paste bead "coloured with cobalt," cast in stone moulds found on the site.[12] The centres of some spiral decorations probably had a coloured bead, since traces of the fixing cement remain. There are imitation gems in glass paste and a small tube of cobalt glass, analysed.[13] The source was probably Egypt, since a small figure of an ape in glass paste with the name of Amenhotep II was found.[14] No progress was made towards true glass working[15]; that glass was blown in the late period[16] seems improbable. Wall painting was carried out in white, grey, yellow, red, blue and black.[17]

ORGANIC MATERIALS

Ivory at Mycenæ[18] is probably African imported through Egypt[19], and there was an ostrich egg-shell, which effervesced with acid.[20] Pausanias[21] says that

[1] Dussaud, 1914, 172.
[2] Hall, Oldest Civ. Greece, 1901, 193 f. ; Tsountas-Manatt, 321.
[3] Glotz, 389.
[4] Hall, Oldest Civ. Greece, 193 f.
[5] Fimmen, 1909, 21.
[6] Archæologia, 1899, lvi, 315.
[7] Ridgeway, Early Age of Greece, i, 594 f.
[8] References in text to Schliemann = Sn., and Schuchhardt = St.
[9] Schliemann, 330 ; Schuchhardt, 207, 213, 265.
[10] Schliemann, 114 ; Tsountas-Manatt, 77.
[11] Rhousopoulos, Archiv für die Geschichte der Naturwissenschaften und der Technik, Leipzig, 1909, i, 288.

[12] Schliemann, 114, 121, 157 ; Schuchhardt, 185 ; Perrot-Chipiez, Hist. de l'Art, vi, 943, Fig. 502 f.
[13] Schliemann, 153, 157 ; Schuchhardt, 145 ; Tsountas-Manatt, 182.
[14] Meyer, Alt., II, i, 177.
[15] Schliemann, 158.
[16] Perrot-Chipiez, Hist. de l'Art, vi, 944.
[17] Schuchhardt, 288 ; Perrot-Chipiez, Hist. de l'Art, vi, 544 f., Figs. 240, 241, 245 ; ib., 698, plate 13 ; ib., 883 f., Figs. 437 f.
[18] Schliemann, 152—"? bone," 329 ; Schuchhardt, 295, 296—"ivory."
[19] Wace, CAH, ii, 459.
[20] Schuchhardt, 268.
[21] I, xii.

although ivory was known to all nations from the earliest times, the elephant was not known outside India, Libya and adjacent lands until Alexander crossed into Asia. Homer, he says, mentions ivory beads, etc., used by the wealthier kings, but never the elephant. Amber, [1] which was analysed, was probably Baltic amber, known in EM III or MM I Crete, and on the mainland in 1500–1400 B.C.[2] There was also linen [3], but the supposed "spindle-whorls" of clay, found also in gold, are probably beads.[4]

TIRYNS

Tiryns, excavated by Schliemann [5], whose work was completed by Dörpfeld, is a second royal seat nearer the sea than Mycenæ. It was built later than Mycenæ, in the 14th century B.C., and has been ruined since 468 B.C.[6] The stone for the strongly fortified massive walls, of native workmanship [7], was quarried by boring, inserting wood and wetting ; the blocks are roughly faced and arranged and set in clay mortar.[8] A palace was excavated, in the construction of which a kind of clay concrete was used : in no cases was lime, freely used for plaster and fresco work, employed as mortar. The stone was probably cut with a thin copper blade and emery.[9] The wall frescoes, which are cruder than those at Mycenæ, are really later and not before the late 14th century B.C., perhaps LM III (1350–1250 B.C.). The style and details are Minoan, with some differences.[10] They were painted in white, black, blue, red, yellow and (later) green, on lime laid over clay.[11]

The friezes were of alabaster inset with blue glass paste, also used at Mycenæ, the kyanos (κύανος) of Homer[12] and the chesbet of Egypt. It is a lime-copper-silica composition without cobalt ; a "cobalt blue" bead is probably described only by its colour. In early editions of Homer, kyanos was often regarded as steel.[13] The suggestion of Landerer[14] that kyanos was copper sulphide, beads of which were found at Troy, is improbable. The soda glass at Tiryns contains considerable amounts of lead.[15] Mycenæan Age gilded glass paste cast in moulds for decorative purposes at Spata (Attica) was whitish, rarely blue.[16]

Of fragments of vases and very crude terra-cotta figures, the earliest resemble pottery of Troy I–II, and the pottery has the characteristic "Mycenæan" decoration. There are terra-cotta female idols and crude attempts at figure decorations, sometimes with the curious narrow waists found in Crete. The clay is coarser than at Mycenæ and the paint is always a lustrous brown on a light yellow ground.[17]

[1] Schliemann, 245 ; Schuchhardt, 195, 219.
[2] Mosso, Med. Civ., 244, 368 ; Stoll, Geschlechtsleben, 419, 444 ; E. Speck, Handelsgeschichte des Altertums, 4 vols., Leipzig, 1900 f., i, 94, 103, 463.
[3] Schuchhardt, 233.
[4] Mosso, 200, 203.
[5] Tiryns, Engl. tr., 1885 ; Schuchhardt, 93–133 ; Tsountas-Manatt, 44 ; Perrot-Chipiez, H. de l'Art, vi, 258 ; Marshall, Discovery in Greek Lands, 16 ; Baumgarten, Hellenische Kultur, 48.
[6] Hall, Bronze Age, 11 ; Meyer, Alt., II, i, 245 ; Schuchhardt, 94.
[7] Meyer, Alt., II, i, 240, 242.
[8] Schuchhardt, 98.
[9] Ib., 104 f., 107, 113 f.
[10] Hall, Bronze Age, 11, 103, 146, 152, 230 ; Dussaud, 1914, 197 ; Meyer, Alt., II, i, 239, 242.

[11] Tiryns, 338 ; Blümner, iv, 415 ; Perrot-Chipiez, vi, 296—"on rough-cast" ; decorations, ib., 533, Figs. 209–19, 222, 239 ; Schmitz, DA, i, 656 ; Dussaud, 160 f., plate.
[12] Odyssey, vii, 87.
[13] Tiryns, 117 f., Figs. 106–7 ; ib., 82, 285, 291 ; Tsountas-Manatt, 47 ; Forrer, RL, 826 ; Perrot-Chipiez, vi, 559, 1004 ; Baumgarten, Hellen. Kultur, 49 ; Rossbach, PW, vii, 1065 ; Blümner, PW, vii, 1385, c. 1500 B.C. ; Meyer, Alt., II, i, 243 ; Merry and Riddell, Homer's Odyssey, Oxford, 1886, i, 289.
[14] A. Rössing, Geschichte der Metalle, Ergänzungsheft, Verhandl. Verein zur Beförderung des Gewerbfleisses, 1901, 146.
[15] Tiryns, 82 ; Blümner, iv, 390.
[16] Dussaud, 176.
[17] Tiryns, 122, 126, 128, 130, 132 ; Forrer, RL, 825 ; Schuchhardt, 128.

METALS

Gold (especially an important find in the lower town) and silver occur at Tiryns, and bronze with 10–13 per cent. of tin was abundant.[1] Lead was common, e.g. numerous fragments of large vases and jars bound with clamps of lead, large lumps, half a pig and fragments of sheet lead. A lead statuette was found in a beehive tomb at Lakonia.[2] Iron occurs in 10th century B.C. graves, but was rare before in the Ægean, and was probably Asiatic. Iron of this period occurs also at Athens and in the Dotian plain in the modern village of Marmariani.[3]

Mycenæan type gold objects were also found in graves at Pylos, in the remains of a Mycenæan palace on the Acropolis at Athens and in several graves in the plain of Attica. Salamis in Cyprus, Ægina, Bœotia, especially Orchomenos ("treasury of Minyas"), the island of Gha (Homer's Arne ?) on Lake Copais, Thebes, Delphi, the island Kephallenia, Dimini (old Iolkos), etc., all have Mycenæan remains. The islands of the Eastern Mediterranean and Sicily were all under Mycenæan influences.[4]

The mainland of Greece was settled comparatively late by a few intruders from the northern region, the people being ignorant of metal but using pottery like that made in Thessaly. This culture was distributed in Central Greece, at Chæronea, Orchomenos and Hagia Marina ; later traces of it occur near Corinth and in Arcadia and Argolis. The rest of the Peloponnesos seems to have been uninhabited in these times.[5] The archæology of Classical Greece belongs to a later period than that with which we are concerned.

CYPRUS

THE CYPRIAN CULTURE

Nothing certain is known of the earliest population of Cyprus, since there are no Palæolithic and very few, if any, Neolithic remains, but it appears to have been similar to those of Troy and Asia Minor.[6] A stone-copper (Chalcolithic) culture appeared very early, but the earliest tombs contain neither copper nor stone implements. The datings given in the table below have been

	Dussaud	Contenau	Walters	
Copper Age . .	3000–2200	} 3000–1580	Copper Age (pre-Mycenæan) .	2500–1500
1 Bronze Age .	2200–1550		Bronze Age (Mycenæan) .	1500–900
2 Bronze Age (Mycenæan)	1550–1100	1580–1100	Greek-Phœnician .	900–550
1 Iron Age (Greek-Phœnician)	1100–600	—	Hellenic . .	550–200

---

[1] Perrot-Chipiez, Hist. de l'Art, vi, 954 ; Fimmen, 1921, 13 ; Hoernes, Urg., ii, 293 ; Lippert, Kulturgeschichte, 1887, ii, 231.

[2] Perrot-Chipiez, vi, 954 ; Gowland, J. Anthropol. Inst., 1912, xlii, 272.

[3] Wade-Gery, CAH, ii, 524, 529 ; Forrer, RL, 197.

[4] Baumgarten, Hellenische Kultur, 55 ;

Fiechter, PW, ia, 965 ; Hoernes, Urzeit, ii, 76.

[5] Peake and Fleure, Priests and Kings, 1927, 118.

[6] Oberhummer, PW, xii, 82 ; Dussaud, Civ. préhelleniques, 1914, 216, 222 f. ; Meyer, Alt., II, ii, 88 ; Myres and Richter, Catalogue of the Cyprus Museum, Oxford, 1899, 13 ; BMG Bronze Age, 166.

proposed, but there is no doubt that the Copper Age goes further back than 2500 B.C. Modestov [1] put it back to 4000 B.C., but on the basis of the date 3800 for Sargon I, which is now thought to be a thousand years too early. Meyer [2] took 3000 B.C. for the earliest use of copper, and this seems now generally accepted. Myres does not date the older periods, but puts the Greek-Phœnician Period from the beginning of the Iron Age to 295 B.C. The Mycenæan Age in Cyprus, especially on the site of Enkomi, probably corresponds with a Minoan migration after the destruction of the Palace at Knossos by Achaians.[3]

There was some relation of Cyprus with Egypt in the XII dyn., and Cyprus was called Asi or Isi, the Alašia of the XVIII dyn. Amarna tablets. Alašia exported copper, oil, ivory and woods to Egypt, and is probably the country called Jatnana in Assyrian documents.[4] The Old Testament Elisha, however, may have been Carthage, and the Hebrew Kittim, formerly identified with Cyprus, is now read Kasdim, the Chaldæans.[5] Trade with Egypt and Palestine is probably at least as early as 1700 B.C. In the 6th century B.C. Cyprus was occupied by Egyptians.[6]

Although an early colonisation from Crete is uncertain, there was probably strong Ægean-Mycenæan (LM III) influence from the mainland of Greece about 1450–1400 B.C., and this was more important than the early Oriental-Phœnician.[7] Others [8] think the Ægean influence was less important, that Cyprus was out of touch with the mainland of Greece and that the Oriental influence came from the Cilician frontage of Asia Minor or through the intermediary of Rhodes rather than from the Syrian ("Phœnician") coast. An early influence of Babylonia is doubtful and old Babylonian goods, especially pottery, may have come by way of the Syrian coast.[9] The argument that the earliest bronzes of Cyprus contain tin whilst those of Mesopotamia contain lead or antimony is no longer valid after the discovery of very early Sumerian tin-bronzes. In Cyprus, however, the forms of copper and bronze weapons are very primitive and characteristic until the Mycenæan Period, especially the daggers with stout midrib and tang bent round at the end to serve as a handle.[10] There was certainly trade with Assyria in the 8th–7th centuries B.C.[11] Relations with Troy are indicated by the similarity of early pottery with linear ornaments of Cyprus and Troy II, and there are similar finds in Phrygia.[12] A black ware with dotted lines appearing about the middle of the Bronze Age in Cyprus was probably imported from the coast of Palestine : it was also sent to XII–XVIII dyn. Egypt.[13]

The Iron Age culture of Cyprus shows that mixture of Egyptian, Syro-Phœnician and Greek elements which is usually regarded as typical of Phœnician influence, but the early and strong Phœnician influence on Cyprus formerly assumed has been over-estimated. The mixed styles may be really Cyprian and the objects could have been exported from Cyprus (where

[1] Introduction à l'Histoire Romaine, 1907, 85.

[2] Alt. (1909), I, ii, 670.

[3] BMG Bronze Age, 160.

[4] Hall, Bronze Age, 205, 212, 218 ; ib., CAH, ii, 281, improbably thinks Cilicia ; Bilabel, Geschichte Vorderasiens, Heidelberg, 1927, 10, 90, 429—Asi = Alašia is doubtful ; Meyer, Alt., II, i, 139 ; Oberhummer, PW, xii, 59.

[5] Meyer, Alt., I, ii, 752 ; cf. Oberhummer, op. cit., 59, 60, 85.

[6] Meyer, Alt. I, ii, 752 ; Oberhummer, 85 ; Myres, CAH, iii, 641 ; Hall, ib., 306, 327.

[7] Hall, CAH, ii, 279 ; ib., Bronze Age,

205, 212 ; Pottier, in de Morgan, Délégation en Perse, xiii, 86.

[8] Myres, CAH, iii, 645 ; Giles, CAH, ii, 18 ; Olmstead, H. Pal. Syr., 367 ; Pottier, op. cit. ; Fimmen, 1921, 105 ; Hall, Bronze Age, 218.

[9] Meyer, Alt., I, ii, 752 ; Köster, Beih. alt. Orient, i, 18 ; Dussaud, 1914, 226, 228.

[10] BMG Bronze Age, 166.

[11] Myres, CAH, iii, 145 ; H. Berger, Geschichte der wissenschaftlichen Erdkunde der Griechen, Leipzig, 1903, 176.

[12] Forrer, RL, 163, 400 ; Meyer, Alt., I, ii, 748 ; Glotz, 270 ; Dussaud, 364 ; Maspero, Struggle, 201 ; Köster, Beih. alt. Orient, i, 18.

[13] BMG Bronze Age, 167.

they were actually found) rather than imported from Syria. Early metal work in Cyprus is certainly pre-Phœnician and the Phœnicians may have been attracted to Cyprus in the 11th century B.C. by its indigenous metallurgy.[1] Later Phœnician influence on Greece by way of Cyprus, however, must not be under-estimated.[2]   The Phœnicians certainly reached Cyprus later than the Ægeans, probably about 1000 B.C., although most Phœnician inscriptions in Cyprus are as late as the 4th century B.C.   Idalion and Tamassos were annexed only in the Persian Period.[3]   Phœnician influence was mainly concentrated in Paphos, Amathus and Kition, a Tyrian colony perhaps being established first at Kition.   From there its influence spread to Idalion, Tamassos, Lapethos and Narcana, the island being girdled with establishments which were not solely mercantile depôts, since 4th–3rd cent. B.C. inscriptions mention Phœnician kings of Kition and Idalion.[4]   Ionian Greeks were also in Cyprus, and the names of 7th century Cyprian princes tributary to Assyria are not Phœnician.[5]

Cyprus would early attract attention because of its fertility and its mineral and other natural wealth. It produced the ash, pine, cypress, oak, olive and vine ; sweet-smelling woods for incense ; wheat and barley ; medicinal plants such as the poppy and ladanum ; henna ; sweet-smelling flowers such as the violet, anemone, lily, hyacinth, crocus, narcissus and wild rose ; and mineral wealth such as iron, precious stones and, in the hills of Tamassos, abundance of copper at an early period (c. 3000). Strabo says the forests were cleared as fuel for smelting copper and silver.[6]

POTTERY

Although Copper and Bronze Ages are well established for metal in Cyprus, there is no break in the development of associated objects.[7]   The pottery may be classified as follows[8] : (1) Sub-Neolithic, hand-made from coarse ill-washed clay, like that of Troy I. (2) Copper Age, hand-polished slip with incised or relief decorations and unpainted. (3) First Bronze Age, with rectilinear decoration matt painted in black or red-brown on white slip : this pottery occurs in Athens, Troy, Egypt and Syria. (4) Second Bronze Age, Cyprio-Mycenæan, with brown simple decoration on fine pale yellow clay with brilliant slip.

The pottery which occurs before metal is coarse and clumsy, hand- not wheel-made, except at the very end of the period, and hence could not have been introduced from Egypt. Though the clay is coarse, the characteristic slip is fine, bright red (with an ebony-black variety) and polished with stone or horse-tooth burnishers. It is often only slightly baked, so that the fine surface layer is inclined to separate and flake off, and was probably introduced just before the Copper Age : the forms are probably indigenous and, if anything, only the technique was imported. It resembles the Predynastic Egyptian except that it is based on gourd forms, not on stone. The use of black paint for pottery, a native umber which is still worked, is earlier in Cyprus than in Mycenæ, Troy and Phœnicia, but the red paint may have been introduced from Egypt. The black glaze is lustrous, and on it are painted in lustreless red paint groups of short parallel lines. The black becomes red when over-fired.[9] The wheel was introduced only in 1550–1100, perhaps from Mycenæ.[10]

[1] Engel, Kypros, 2 vols., Berlin, 1841 ; Hall, Bronze Age, 227 ; Meyer, Alt., II (1893), 220 ; II, ii, 86 ; Dussaud, 304, 310.
[2] Olmstead, H. Pal. Syr., 368.
[3] Meyer, Alt., I, ii, 751 ; II, ii, 86 ; Myres, CAH, iii, 641 ; Dussaud, 226 ; Oberhummer, PW, xii, 93.
[4] Engel, i, 165 ; James, DG, i, 730.
[5] Contenau, Civ. phénic., 1926, 90.
[6] Meyer, Alt., I, ii, 673 ; II (1893),

219 ; Maspero, Struggle, 200 ; ib., Hist. anc., 1905, 280 ; James, DG, i, 729 ; Engel, i, 42 ; Strabo, XIV, vi, 5, 683 C. ; Pliny, xxxiv, 1.
[7] Myres-Richter, 14 ; Dussaud, 240.
[8] Dussaud, 231, 239.
[9] Myres-Richter, 15, 16 f., 36, 39 ; Dussaud, 234, puts the early pottery in the Bronze Age.
[10] Dussaud, 228 ; Myres-Richter, 21.

The Græco-Phœnician pottery is wheel-made, most vessels having a foot or base-ring, absent in the Bronze Age pottery, and is mostly decorated with lustreless black paint, much of which is applied on the wheel; relief and incised ornaments are almost wholly absent.[1] At certain periods Hellenic vases appear, particularly black figured Attic of 600–450 B.C., and red figured (red clay, black glaze ground) Attic in 500–200. The best period is 450–400.[2]

LAPIDARY WORK

The inscriptions on seal cylinders at Salamis are probably an intermediate stage of art from the Babylonian, probably not Phœnician but Hittite (some in plate XII in Cesnola seem to show Minoan affinities, e.g. the thin waist with band); the date is doubtful: one is c. 2000 B.C., some are Cassite and one is 9th century B.C.[3] The cylinders are of steatite, jasper and dark green or black magnetite or hæmatite. The scarabs, seals, etc., were of steatite with blue, green or red glaze, or of agate, carnelian and other hard materials. Engraved gems and stones of various dates include deep purple amethyst, transparent paste (Roman, 1st century B.C.), onyx or niccolo, carnelian, bloodstone, iridescent glass or paste, hyacinth, sard, yellow and red jasper, sardonyx, agate, garnet, spinel ruby (or coloured paste), pale green paste, greenstone, basalt, green and white mosaic glass and "calcined steatite." A few objects in chalcedony include a symbolic eye amulet and there are a few objects in crystal, amber beads and lapis lazuli beads at Enkomi.[4]

The "diamonds" (adamas vergens in aërium colorem) said by Pliny [5] to be found in Cyprus were not true diamonds, since they could be bored with the latter [6], but probably sapphires [7] or rock crystal.[8] Other gems said (some probably incorrectly) [9] to be found were amethyst, emerald (malachite ?), aquamarine, opal (pæderos), agate, heliotrope (?), jasper, chalkosmaragdos (turbida æris vena)[10], lapis lazuli and coral.[11] Red coral was set in a ring.[12] What has been called chrysoprase is agalmatolite[13], but jasper, agate, heliotrope and possibly opal and amethyst actually occur in Cyprus.[14]

"Alabaster," commonly found[15], is plentiful in Cyprus and of fair quality, although the white at Enkomi and the vases of the finest banded variety were probably imported from Egypt; the earliest alabastra are of the Mycenæan Period at Salamis.[16] Other stones said by Pliny to occur in Cyprus are talc, grey chalk, gypsum, marble, grindstones and emery (naxium).[17] What Engel calls antit, with a reference to Pliny, is probably the unidentified anachites[18], also mentioned (ἀναχίτης; perhaps diamond) in the *Lithica* of the pseudo-Orpheus.[19] Dioskurides[20] mentions asbestos (ἀμίαντος), which is mined in Cyprus at present.[21]

GLASS

Glazed objects and rarely transparent glass, e.g. a spiral ear-ring, occur in Bronze Age remains; that glass was suggested by metallurgical slag, since fragments of glass and furnace slag occur together at Tamassos, is possible

[1] Myres-Richter, 21; Tubbs, DA, ii, 920.
[2] Myres-Richter, 22, 59.
[3] A. P. di Cesnola, Salaminia, 1882, 118; Dussaud, 1914, 273.
[4] Cesnola, 115, 118, 136, 163, 167; Dussaud, 272; Myres-Richter, 121, 183, 186.
[5] xxxvii, 4—adamas refers to the great hardness.
[6] Engel, i, 54; James, DG, i, 729.
[7] Rawlinson, Mon., iii, 162.
[8] Oberhummer, PW, xii, 67.
[9] *Ib.*, 66; *ib.*, Cypern, Munich, 1903, i, 185.
[10] Pliny, xxxvii, 5; perhaps malachite, Smith, Lat. Dict., 1904, 170.
[11] Engel, i, 55; James, 729; Rawlinson, 162.
[12] Myres-Richter, 139.
[13] *Ib.*, 55.
[14] Oberhummer, *opp. cit.*
[15] Cesnola, Salaminia, 111; Murray, Smith and Walters, Excavations in Cyprus, 1900, 25.
[16] Myres-Richter, 99.
[17] Engel, i, 55; James, DG, i, 729; Rawlinson, Mon., iii, 162.
[18] xxxvii, 5.
[19] Krause, Pyrgoteles, Halle, 1856, 7.
[20] v, 156.
[21] Oberhummer, PW, xii, 67.

but not very probable, and knowledge of glass probably came from Egypt or Syria. The large, clumsy, spherical or cylindrical Greek-Phœnician beads of variegated glass, probably native, have been supposed to be a by-product of the copper industry : "the glass (or rather vitreous slag) is highly vesicular, and almost pumice-like in texture." Coarse glass beads found in 9th–8th century B.C. Greek and Italian tombs are very like those found in Cyprus and were perhaps carried by Phœnician traders.[1] Some pottery is in imitation of glass ware.[2]

The opaque vessels of various colours (pale citron, green, orange, yellow, white, dark maroon, blue and dark violet) resemble XIX dyn. glass from Gurob and probably date to about 1400 B.C.[3] They were cast over a sand core in the Egyptian technique and have "dragged" designs made with threads of coloured glass. Myres and Richter [4] believe that they are as late as the 6th century B.C. and were made in Rhodes and Naukratis ; they suggest that such vessels occur with blown glass, rather infrequently ; that glass does not become common in Cyprus until the later Ptolemaic Period and that the remains of a glass factory found at Tamassos are of the Hellenistic-Roman Period. Late Mycenæan glass and glass paste ornaments are always cast, whilst those in Cyprus are nearly always rolled or modelled. The Hellenistic-Roman amphoræ, found with blown glass, are of coarser fabric than the earlier specimens, are less opaque and have much less brilliant and tasteful colouring, a black or brown ground being frequent.[5] The welded millefiore bowls and saucers, made from particoloured rods of amber, yellow, white, green and blue glass ; the coloured blown glass bowls, cups and bottles, of dark blue, green, amber, amethyst or mixed colours ; and the colourless blown glass bottles, bowls, jugs, tumblers and plates (replacing pottery), of thin glass, which are now often (but not always) iridescent, with a flaky or granular corrosion, are probably all of the Roman Period.[6] An analysis of the colourless glass [7] gave : silica 68·18, alumina 2·70, iron oxide 0·82, "oxide of manganese" 0·92, lime 7·73, soda 18·46, magnesia slight trace [total 98·81]. A thin bowl of colourless blown glass with distemper painting on the under side, which flakes away very easily, was cut from the bottom of a blown bottle.[8] Toilet articles (stirring rods, needles, hair-pins, finger rings) of variegated glass, glasses like watch-glasses, and a glass egg cup, with remains of egg-shell, were also found.[9]

Coloured pastes were used in jewellery and a free use of vitreous enamel of bright blue, bright green, white and dark blue colours, placed in the scales of gold work as in cloisonné work, but now mostly dropped out, was found at Amathus in 5th century B.C. work.[10] The art of making coloured glass continued rather late in Cyprus : Qalqashandi (d. A.D. 1418) says the windows of the mosque at Cairo contained coloured Cyprian glass.[11]

GOLD

Gold and electrum occur very rarely in Bronze Age remains, mostly as cylinder seal mountings closely resembling the Babylonian.[12] Although

---

[1] Myres-Richter, 15, 23, 100.
[2] Tubbs, DA, ii, 921.
[3] Murray, Smith and Walters, 23, 69 ; Cesnola, Salaminia, 44, 173, 181.
[4] 12, 15, 23, 26, 100.
[5] Myres-Richter, 100; Murray, Smith and Walters, 69, 115 ; L. P. di Cesnola, Cyprus, its Ancient Cities, Tombs and Temples, 1877, 73, plate iii ; Forrer, RL, 164, 288.
[6] Myres-Richter, 100 f. ; Blümner, iv, 382 ; Cesnola, Salaminia, pp. x, 172, 182, and plates.

[7] Sandwith, Archæologia, 1877, xlv, 140.
[8] Myres-Richter, 102, 105.
[9] Cesnola, Salaminia, 173, 178, 181.
[10] Myres-Richter, 121 f. ; Murray, Smith and Walters, 101.
[11] Reitemeyer, Beschreibung Ägyptens, 203.
[12] M. Ohnefalsch-Richter, Kypros, the Bible, and Homer, 1893, 35, 337 ; cf. Meyer, Alt. (1909), I, ii, 671 ; II (1893), 829 ; Myres-Richter, 15, 33—native work ; Dussaud, 1914, 276—imported.

Myres and Richter say that gold appears only after copper and silver in the Mycenæan Age, and (except for Mycenæan sites such as Salamis and Kurion) very rarely until the Græco-Phœnician Age, Regling [1] states that it appears somewhat rarely about 2500 B.C., at first in pellets and flat cakes (φθοΐδες), later in thin rods and bars : the earliest form, however, appears to have been thin stamped sheet.[2] Homer [3] makes Kinyras, king of Cyprus, give Agamemnon on his departure for Troy a breastplate containing 10 bars of dark kyanos (μέλανος κυάνοιο), 12 of gold and 20 of tin (κασσιτέροιο). It is doubtful, according to Regling, if the pieces of gold bars and nodules were used as money.

Native gold occurs in Cyprus, especially at Akamas, but the Venetians ceased to work it.[4] The gold in the early objects is purer and softer than that after 600 B.C.[5] A considerable variety of gold was found, especially on later sites (5th–4th centuries B.C.), including cast work, armlets weighing over 2 lb. (7th century ?), small nuggets, of average weight 72 grains, cut from a rough ingot with a chisel, wire, sheet and leaf, beads, small gold bottles for necklaces, a gold stopper to a crystal bottle and a gold bowl with repoussé work.[6] Granulated soldered gold work with diagonal patterns, probably 1400 B.C., occurs at Enkomi.[7] The fine filigree and chased work disappear after about 295 B.C. Spiral decorations were produced by laying sheet over a spiral of wire soldered to a surface and hammering. Gold ornaments of the 5th–4th centuries B.C. with wire cell-enamel and inserts of glass or paste, and a pectoral with inlays of blue, pink and white pastes, were perhaps Egyptian work of about 1400 B.C.[8]

Bronze is covered with gold leaf or plated with gold and silver, gypsum or clay is covered with gold. There is heavily gold-plated bronze in the Greek Period, hollow gold plate beads of the 5th century B.C. filled with sulphur, hollow gold rings, those of the Hellenistic Period filled with sulphur, and gold tube. Some gold objects have a red or deep red colour.[9]

Electrum, which occurs as jewellery from 1400 B.C. in a few thin plates and spirals, two engraved rings (one perhaps Babylonian) and two ear-rings, is freely used with gold and silver from the Greek Period.[10] "Touchstones" of the Bronze Age and later may be stones for sharpening copper or bronze knives.[11]

SILVER

Richter[12] says that silver is rare in the Copper and Bronze Ages but is commoner than gold, which appears later ; Myres and Richter[13] that silver appears after copper and bronze in the early or middle Bronze Age ; and Forrer[14] that it was an important decorative metal in the Copper Age. The king of Cyprus begged silver from Amenhotep III of Egypt (1411–1375).[15] According to Richter the earliest silver is pure, but that of the Iron Age is

[1] PW, vii, 978.
[2] Dussaud, 1914, 276.
[3] Iliad, xi, 35.
[4] Engel, i, 54 ; James, DG, i, 729.
[5] Richter, Kypros, 452.
[6] Wroth, DA, i, 324 ; Richter, Kypros, 338, 359, 367, 374, 447, 477, 492 ; Cesnola, Salaminia, pp. xii, 12, 18, 24, 27, 32, 44 ; Dussaud, 277 ; Cesnola, Cyprus, 282, 298, 307, 309, 315, 325 ; Murray, Smith and Walters, 65, 102.
[7] Murray, Smith and Walters, 18 ; Meyer, Alt., II, i, 552 ; Richter, Kypros, 338 ; Hall, Bronze Age, 210 ; Cesnola, Salaminia, 19.
[8] Myres-Richter, 27 ; Richter, Kypros,

339, 368 ; Meyer, Alt., II, i, 552 ; Murray, Smith and Walters, 19, 21, 41, pl. v— 800 B.C.?
[9] Dussaud, 277 ; Richter, Kypros, 368 f., 408 f. ; Myres-Richter, 34 f., 125 f., 130 ; Cesnola, Cyprus, 309 ; ib., Salaminia, 32, 44 ; Murray, Smith and Walters, 101.
[10] Cesnola, Salaminia, 24 ; Myres-Richter, 33, 127 ; Richter, Kypros, 362, 493 ; Murray, Smith and Walters, 19 f.
[11] Myres-Richter, 52 ; Dussaud, 256.
[12] Kypros, 451, 492.
[13] Catalogue, 33 ; cf. Schrader-Jevons, 181.
[14] Urgeschichte, 1908, 288.
[15] Meyer, Alt., II, i, 153.

alloyed with lead ; Myres and Richter [1] and Dussaud [2] state that it is always largely alloyed with lead and has a light-coloured and powdery rust, whilst later silver is refined, dark and compact. The presence of lead indicates the use of some kind of imperfect cupellation process, as at Mycenæ.

Silver objects at Enkomi (1400 B.C.) include "Mycenæan" vases and two rings with engravings, one of Assyrian type and the other of Egyptian.[3] No silver (and very little gold) was found at Kurion in the Mycenæan Period, but later sites (5th–4th centuries B.C.) were rich in silver, e.g. studs, spirals, rings, etc., at Amathus, with small lumps, of average weight 90 grains, broken from a thin ingot, also pieces of gold, lumps of bronze and a bronze mirror plated on both sides with silver. Silver ornaments all copy the very earliest copper spirals and rings.[4]

Several authors mention silver mines in Cyprus, but although Strabo [5] reports that silver was smelted on the island, there was probably only very little produced.[6] In the Greek Period, especially in the 6th–5th centuries, silver was far commoner than gold, and was probably imported from Spain and from Laurion in Greece, the latter source being extensively worked about 600 B.C. This pure metal is easily distinguished from the early silver-lead alloy. In the Ptolemaic and Roman Periods silver rapidly went out of use and was replaced by mean and tasteless gold work, probably because in the 3rd century B.C. the Spanish silver was diverted to Rome.[7] Gilded silver objects occur [8] and Cyprian metal pateræ are largely silver.[9]

Other examples of silver work in Cyprus are girdles (partly gilt), embossed leaf, plates, spirals, a mask and other objects.[10] The origin of the silver objects described by Cesnola[11], including spoons, bowls, hairpins, wire, rings, pendants, bottles, ear-rings, bracelets and a dagger, is doubtful, since it is not certain whether he actually excavated at Kurion.[12]

COPPER

Old legends associate copper with Cyprus. Servius, in the commentary on Vergil's Æneid which he wrote in the 4th century A.D. with the use of much older sources[13], says the Corybantes or priests of Cybele were called after copper (ab ære appellatos) because there was a mountain in Cyprus rich in copper, called by the natives Corium (apud Cyprum mons sit æris ferax, quem Cyprii Corium vocant).[14] Legend attributes the discovery of copper to a mythical king of Cyprus, Kinyras.[15] Homer[16] makes Mentes take iron to Temese, either Tamassos in Cyprus or Temese in south Italy[17], to exchange for copper, and Nikokreon, king of Cyprus (4th century B.C.), sent a present of copper for the winner of the Argolid games.[18] Copper was also said to have been mined in Cyprus by Agapenor and the Arcadians wrecked there on the return from Troy.[19] The name of the metal (χαλκὸς κύπριος; æs cyprium; cuprum) is

[1] Catalogue, 15, 33 f.
[2] Civ. préhellenique, 278.
[3] Murray, Smith and Walters, 17 ; Hall, Bronze Age, 210.
[4] Murray, Smith and Walters, 65, 101 f. ; Myres-Richter, 33, 54, 118.
[5] XIV, vi, 5, 683 C.
[6] Engel, i, 54 ; Orth, PW, Suppl. iv, 115 ; E. Oberhummer, Die Insel Cypern, Munich,1903, i (with bibliography),183, says no silver or lead is found in Cyprus.
[7] Myres-Richter, 21, 24, 27, 34 f.
[8] Ib., 133 ; Cesnola, Cyprus, 329.
[9] Richter, Kypros, 45.
[10] Ib., 52, 360, 408 f., 451, 494 f.

[11] Cyprus, 297, 315, 325, 329 ; ib., Salaminia, pp. xiii, xvii, 46 f.
[12] Dussaud, 220, 233.
[13] Ramsay, DBM, iii, 794 ; Seyffert and Sandys, Dict. of Classical Antiquities, 1891, 578.
[14] Engel, i, 189.
[15] Pliny, vii, 56 ; Homer, Iliad, xi, 35 ; Engel, i, 43.
[16] Odyssey, i, 184.
[17] Dyer, DG, ii, 1084 ; Bunbury, ib., 1123.
[18] Perrot-Chipiez, Art in Phœnicia and Cyprus, ii, 90.
[19] Cesnola, Cyprus, 219.

probably derived from that of Cyprus.[1] The great antiquity of copper in Cyprus is proved by its use as in Egypt and Mesopotamia for ornamental purposes in the earliest period, for jewellery at least as early as 3000 B.C., although the Neolithic Age may have lasted much later than this.[2]

Spirals, open rings and pins, no doubt of copper, are represented on the necks of the very old anthropomorphic vases [3], but the metal is rare in the earliest tombs [4], one of which contained a seal cylinder of 2750 B.C. [5], and in that period it may have come from the mainland of Europe.[6] Gowland [7] suggests 2500 B.C. for the first copper working in Cyprus. Native copper was perhaps the first form mined in Cyprus, since the early specimens of metal are of great purity.[8]

Import of copper to Egypt from Cyprus (Asi), which may go back to the 3rd millennium B.C., is shown in the 16th century B.C. and is mentioned (probably as coming by commerce, not as tribute) in the Amarna letters (1400 B.C.).[9] Copper may have been first worked in the Mediterranean in Cyprus.[10] Whether Cyprus was the original home of copper working, whether it was taught by Egypt or whether (as Hall[11] thinks probable) the knowledge of copper was brought both to Cyprus and Egypt from Asia by maritime trade from the Syrian coast, are questions which remain undecided. The Boghazköi texts (1300 B.C.) show that the Hittites imported copper from Cyprus.[12] When the metal was known there, the native ore was exploited.[13] In one year the records say the king of Cyprus could send no copper to Amenhotep III since there was a pestilence and the miners died.[14]

The copper mines were royal monopolies under the native kings, Ptolemies and Cæsars.[15] The king had a monopoly of bronze during the XVIII dyn. in Egypt : he retained all payment made by Pharaoh for the metal, whether made on the island or imported.[16]

Ancient copper mines occur in Cyprus near Lithrodónta (Larnaka district), Marion (Arsinoë, in the Paphos district, the headquarters of the ancient copper trade) and in the hills of Tylliria (perhaps the ancient Mount Tyrrhias).[17] The richest mines were said to have been at Tamassos, Amathus, Soli (where Richard Pococke [18] in 1738 saw a hill of *iron* ore or slag), Kurion and Krommyon (where there were said to have been numerous smelteries). The old tombs at Amathus and Kurion contained more copper objects, of crude type, than those on other sites.[19] The Tyrian copper ($T\nu\rho\rho i\alpha\nu \chi\alpha\lambda\kappa\delta s$) in ψ-Aristotle[20] is probably metal from Kurion.[21]

Tamassos, Tainaron in Lakonia, and Seriphos (Sarepta) have been connected with supposed Phœnician words : temes or themæs, smelting ; tannur, a smelting-furnace, or tinar, rock ; and zar(e)phat, to smelt, respectively. There

[1] Engel, i, 42 ; Blümner, PW, xi, 2194 ; Oberhummer, *ib.*, xii, 66 ; Dussaud, 247 ; Strabo, XIV, vi, 5, 683 C. ; Pliny, xxxiv, 1 ; the name probably often means brass.
[2] Hall, Bronze Age, 33, 37 ; Meyer, Alt. (1909), I, ii, 673 ; II (1893), 219 ; Myres, CAH, i, 143.
[3] Myres-Richter, 33.
[4] Dussaud, 236.
[5] Montelius, Ält. Bronzez., 1900, 152— doubtful ; Dussaud, 225 ; Meyer, Alt., I, ii, 749 ; Reinach, *L'Anthropologie*, 1892, iii, 452 ; Forrer, RL, 163.
[6] Hall, Oldest Civ. Greece, 1901, 193 ; Much, Kupferzeit, 1893, 303—the two regions were independent.
[7] *J. Anthropol. Inst.*, 1912, xlii, 246.
[8] Rickard, q. in *Isis*, 1932, xvii, 593.
[9] Dussaud, 247 ; Oberhummer, PW, xii,

66 ; Hall, Oldest Civ. Greece, 1901, 193 ; Köster, *Beih. alt. Orient*, i, 23 ; *ib.*, See-fahrten der alten Ägypter, 1926, 17.
[10] Myres, q. by Hall, NE, 1920, 33.
[11] Bronze Age, 1928, 33, 37.
[12] Köster, *Beih. alt. Orient*, i, 19.
[13] Dussaud, 249.
[14] Meyer, Alt., II, i, 153 : "the hand of Nergal" was over the land.
[15] Oberhummer, PW, xii, 66.
[16] Maspero, Struggle, 287.
[17] Dussaud, 249 ; Myres-Richter, 9 ; Perrot-Chipiez, Phœnicia and Cyprus, ii, 90.
[18] Travels, in J. Pinkerton, Collection of Voyages and Travels, 1811, x, 589.
[19] Engel, i, 44 ; Cesnola, Cyprus, 281, 297.
[20] De mirabilibus auscultationibus, c. 43.
[21] Aristotelis Opera, ed. Müller, Didot, Paris, 1889, iv, 81 ; Engel, i, 44.

are remains of old iron and copper mines, not mentioned by classical authors, at Seriphos.[1]

The copper mines of Cyprus were intensively worked in the Roman Period, but are not mentioned after the 4th century A.D. They probably continued in the Byzantine Period, but under Frankish rule (1192–1517) they were neglected. They were reopened by an English company in 1887 without success, though in 1912 a new company was founded at Limni, near the site of the ancient mines.[2]

Mosso argues that the Mediterranean copper industry did not originate in Cyprus, which he thinks is named after κύπρος, the henna plant (Lawsonia inermis).[3] Cesnola [4], Myres and Richter [5] and Dunstan [6] do not mention copper mines in Cyprus, and "there is no evidence that copper was worked there before its use in Egypt or Crete."[7] The deposits may have been secondary, on or near the surface, and produced by the decomposition of pyrites deeper in the earth, and these surface deposits were completely worked out.[8] There are, however, still remains of old mines.[9] That the copper ingots of Cyprus and Crete are "identical in shape," as Mosso says, is also incorrect, since the shapes are different.[10] The evidence for copper mining in Crete is not very conclusive and much less impressive than that for early copper working in Cyprus.

The copper ore of Cyprus, called by Aristotle[11] χαλκῖτις λίθος, was probably the sulphide ore chalcitis[12] or copper pyrites[13] : it is free from tin.[14]

Analyses of two kinds of scoria are said to show two different processes of copper extraction, an ancient method in which soluble salts were added and a later, conjecturally attributed to Kinyras, in which ferric oxide was used as a flux.[15] The analyses are as follows, traces only of soluble chlorides being found in each case :

	SiO_2	Al_2O_3	FeO	Fe_2O_3	Mn_2O_3	CuO	Rest
1. Summit of Mt. Olympos	5·00	10·84	–	80·18	tr.	tr.	4·63
2. Other localities (mean)	28·85	1·16	28·75	tr.	33·72	0·56	6·96

The red scoriæ (1) are coloured by ferric oxide (not copper), and since the silica is low it was probably not added as a flux but (with the alumina) was probably derived from the gangue. The scoriæ (2) show a more advanced technique, ferric oxide being either used as a flux or more probably derived from the copper pyrites. The manganese carbonate or peroxide associated with the ores would have been separated in washing the ore, and was perhaps

[1] Oberhummer, PW, xii, 66 ; Schrader, RL, 69 ; Schrader-Jevons, 159, 204 ; Dyer, DG, ii, 1086 ; Smith, ib., 968, 1084 ; Orth, PW, Suppl. iv, 115 ; Bochart, Geographia Sacra, Leyden, 1712, 459.
[2] Oberhummer, 66 ; Richter, Kypros, 504 ; Perrot-Chipiez, Phœnicia and Cyprus, ii, 90 ; Orth, PW, Suppl. iv, 115.
[3] Olck, PW, vii, 805.
[4] Cyprus, 284.
[5] Catalogue, 12.
[6] Q. in Mosso, 301.
[7] Mosso, 299 ; Petrie, Tools and Weapons, 6.
[8] Dunstan, Bull. Imperial Inst., 1906, iv, 213 ; Mosso, 302.

[9] Fimmen, 1921, 120 ; Oberhummer, Die Insel Cypern, 1903, i, 176.
[10] Forrer, Urg., 361 ; Dussaud, 249, Fig. 179 ; Regling, PW, vii, 794.
[11] Hist. animal., v, 19.
[12] Engel, i, 46.
[13] Orth, PW, Suppl. iv, 115.
[14] Gladstone, PSBA, xii, 232 ; Montelius, Ält. Bronzez., 199.
[15] Fournet, De l'influence du mineur sur les progrès de la civilisation, Instruction Publique, Académie de Lyon, Lyons, 1861, 67 ; Blümner, iv, 91 ; Rougemont, L'Âge du Bronze, 1866, 93 ; Oberhummer, PW, xii, 66 ; Jagnaux, Histoire de la Chimie, 1891, ii, 352.

intentionally added as a flux, since manganese ores occur in Cyprus. This addition would give a more fusible and more basic slag, with less tendency to take up copper oxide. Common salt may also have been added. The materials contain very little copper, and if they are scoriæ, the copper extraction must have been efficient.[1] Dunstan [2] found appreciable amounts of cobalt in the copper pyrites of Cyprus. Ancient scoria or slag contained only 1·4 per cent. of copper oxide (CuO) but large amounts of manganese (36·09 per cent. MnO and 1·33 per cent. MnO$_2$). Dunstan considers it possible that a manganese ore was added during the smelting, or that the copper ore contained manganese, e.g. as lampadite or pelokonite, an abnormal form of psilomelane (the latter when pure being a compound of manganese dioxide and a basic oxide, usually barium oxide) in which copper oxide is present.[3]

BRONZE

Since in the Mycenæan Period Cyprus had a primitive bronze industry metallic tin was perhaps imported ; the quantity of tin in the bronze was very variable and bronzes rich in tin were perhaps also imported, those poor in tin being made locally.[4] It has been supposed that bronze poor in tin began soon after 3000 B.C., and true bronze about 2200 B.C., the objects for a long time having the same forms as those of copper. The forms are like those of Troy and the early Bronze Age in Central Europe, especially in Hungary, and there are no arrow-heads, fibulæ or safety-pins. Hittite bronze of Zinjirli (North Syria) has been regarded as Cyprian.[5] If the Phœnicians worked copper and bronze in Cyprus [6], which is doubtful, they were rapidly outstripped by the Greeks.[7]

Remains of a copper or bronze factory at Salamis (Enkomi) include several tools, such as rakes, pincers and hammers, and fire shovels, with a pig of copper of the Minoan Period shaped like a double axe and stamped with a sign like an anchor [8], thought to be the Cyprian syllable si.[9] Numerous objects from Cyprus, the earliest of hammered (not cast) copper and the later of bronze, are in museums and include weapons, armour (6th–5th centuries B.C., with iron weapons), cauldrons, tripods and small objects such as pins.[10] About a thousand specimens at Salamis consisted of vases, mirrors (Greek and Roman), bowls, strigils, weapons (lance-heads, daggers), pins, dishes, horse-trappings, fragment of a tripod and miscellaneous objects. According to Birch some weapons are of copper, "which may have preceded the use of bronze." A Phœnician-Egyptian bowl appears to represent the ceremonies described by Herodotos.[11] A pin and needles of an "alloy of copper and tin, or silver," in a bronze box, three bronze mirrors, one with an engraving of the temple of Venus at Paphos, forty oxidised bronze coins "of great antiquity" and bronze buckets with traces of enamel inlay, not very old, were found.[12] Objects at Enkomi[13] included a pair of greaves and a stamped slab of 81 lb. 10 oz., of copper. The analysis of the copper by Church gave : Cu 98·05, Pb 0·31, Zn 0·05, Ag, Bi,

[1] Orth, PW, Suppl. iv, 115.
[2] *Bull. Imperial Inst.*, 1906, iv, 213.
[3] Koechlin, in Doelter, Mineralchemie, III, ii, 874.
[4] Blümner, iv, 187 ; Meyer, Alt. (1893), II, 156, 219.
[5] Myres, CAH, i, 143 ; Hoernes, Urg., ii, 248 ; *ib.*, Urzeit, ii, 30 ; Contenau, Civ. phénic., 385 ; Forrer, RL, 163 ; Myres-Richter, 17.
[6] Blümner, iv, 60.
[7] Büchsenschütz, Hauptstätten des

Gewerbfleisses, Leipzig, 1869, 44 ; Meyer, Alt. (1909), I, ii, 671.
[8] Dussaud, 1914, 249, Figs. 179, 180.
[9] Forrer, Urg., 361 ; Mosso, 303.
[10] Cesnola, Cyprus, 87, 335, 338, pl. 30 ; *ib.*, Salaminia, pp. xiv, 51, 56 ; Murray, Smith and Walters, 66 ; Myres-Richter, 115 ; Richter, Kypros, 47, 377, 410, 455 ; Dussaud, 257.
[11] i, 196.
[12] Cesnola, Salaminia, 55, 59 f.
[13] Murray, Smith and Walters, 15 f., 17 ; Dussaud, 253.

Fe, S, traces, Sn, Ni, nil. There were also two scale pans with four loops for cords.[1]

Analysed objects found by Cesnola, of doubtful origin and date [2], are mostly copper (Table, Nos. 1–4), although two iron axes were found with them. A dagger blade of uncertain date, a small pair of forceps and a spiral ring analysed by Naue in 1888 [3] are all bronze (Table, Nos. 5–8). Reinach [4] considers these objects to be early.

	Cu	Sn	Ni	Fe	Au	Pb	As	S
1. Dagger (knife) .	97·23	tr.	—	1·32	0·28	0·08	1·35	—
2. ,, (lance-head) . .	98·40	—	0·15	0·73	0·305	—	tr.	0·31
3. Dagger (lance-head) . .	99·47	—	0·08	0·38	—	—	tr.	—
4. Axe . .	97·95	0·08	—	1·02	—	—	0·49	0·09
5. Dagger with 2 rivets . .	88·77	8·51	tr.	0·48	—	1·50	—	—
6. Forceps . .	91	9	—	—	—	—	—	—
7. Spiral ring .	93·8	6·2	—	—	—	—	—	—
8. Lance-head .	—	6·0	—	—	—	—	—	—

No. 5 contained 0·304 per cent. of cobalt ; traces of phosphorus were found in Nos. 1–3 and 5 ; No. 4 contained 0·25 per cent. of oxygen and 0·12 per cent. of zinc.

The analysis of the dagger No. 5 is said by Gladstone to be by Flight ; that of the lance-head No. 8 (in Dussaud) is by Nilson.

Curious copper needles with an eye halfway down were found at Gurob in XVIII–XIX dyn. remains, in Cyprus and in Europe in Stone and Bronze Age remains. Cyprian graves as late as the XXVI dyn. (7th century B.C.) in Egypt at Tell-Nebesheh, near Tanis, include objects of bronze, and a bronze axe found in Cyprus is as late as the beginning of the 1st century B.C.[5]

Analyses of objects made by Schuchhardt (Nos. 9–11) and by Boudouard (Nos. 12–15) for Ohnefalsch-Richter are given for the first time by Dussaud :

	Cu	Sn	CaO	Fe	P_2O_5	CO_2	SiO_2
9. Ring-spiral, Phœnikiais	65·73	23·11	3·89	0·62	0·93	4·24	1·48
10. Dagger, Alambra .	58·98	13·18	2·67	2·90	1·69	14·26	3·52
11. Axe, Alambra . .	60	35	tr.	tr.	—	—	tr.

	Cu	Sn	Fe	As	Pb
12. Axe . .	96·33	—	1·88	1·27	0·47
13. Forceps .	97·4	—	0·51	1·45	—
14. Forceps .	93·42	—	1·53	4·70	—
15. Lance-head .	86·73	11·59	1·00	tr.	tr.

[1] The Egyptian had three ; Wilkinson, ii, 246.

[2] Flight, *Ber. Deutsch. Chem. Ges.*, 1874, 1460 ; Gladstone, *J. Anthropol. Inst.*, 1897, xxvi, 316—whose descriptions are in brackets in the table ; Sebelien, *Anct. Egypt*,

1924, 7 ; Montelius, Ält. Bronzez., 1900, 153 ; Mosso, 303 f. ; Dussaud, 1914, 253 f.
[3] Montelius, 154 ; Dussaud, 253.
[4] *L'Anthropologie*, 1892, iii, 452.
[5] Montelius, 1900, 17, 145, 154 ; Cesnola, Salaminia, plates III, IV, 8A.

Mosso quotes analyses of objects found by Richter containing 10–11 per cent. of tin, including a sword "evidently not very early." No antimony was found. The small proportion of tin in early bronzes, according to Dussaud, may be due to its rarity or to the necessity (for metallurgical reasons) of using only small amounts of tin ores in smelting mixed ores ; it may be due to remelting of the objects.

LEAD

Lead objects include oval urns, small cylindrical boxes with flat covers, a vase handle, a fragment of water pipe (from Salamis), weights (?), a miniature cart-wheel [1], a lump of lead in a bronze casing, small cylindrical rods "perhaps for solder," a mass apparently run between stones, and net sinkers.[2] Meyer [3] makes Cyprus export a considerable amount of lead to Egypt in the period 2000–1500, whilst actually no lead ore is found in Cyprus.[4]

The remains of lead at Salamis, which are usually late, are mainly toys and votive offerings, but also plates with inscriptions, some probably 4th century B.C., seals attached to merchandise, Roman and Greek sling bullets and small vases (for eye-ointment). In tombs of Cyrenaica sham jewels as well as boxes of lead were often found. The lead rolls were brittle, but could be unrolled after exposure to gentle heat. Lead seals were Byzantine or early Venetian.[5] Early Bronze Age pins, bracelets, etc., are sometimes of "ill-refined silver lead" (cupelled ?). A bronze weight (?) with lead adjustment was found at Salamis.[6]

IRON

As a precious metal for jewellery, Cyprus, like Rhodes, Crete and the Minoan area generally, had known iron since about 1400 B.C. It was perhaps through Minoan intercourse that iron finger rings became customary in parts of peninsular Italy. Iron does not appear in actual use at Troy till after the destruction of Troy VI (12th century B.C.) and there is no reason to believe that Asia Minor obtained its iron from Europe or that it was brought to Europe direct, *via* the Hellespont.[7] Montelius [8] conjectures that iron was known in Cyprus, Egypt and Western Asia soon after 1500 B.C. An iron arrow-head was found in an old grave at Tamassos with gold, iron plated with silver (a cabinet ; a ring) at Salamis [9] and two iron knives with ivory handles in a tomb of 1200 B.C. at Enkomi.[10]

The iron industry of Cyprus was not so important as that of copper, although fable made the Telchines the discoverers of iron and Pococke saw remains of iron mines at Paphos, Soli and Bole (where there is a mineral spring).[11] In the Greek-Phœnician Period bronze was still the commonest metal, but iron frequently replaced it for knives and swords. Lance-heads of both iron and bronze have tubular sockets, borrowed from Mycenæ. The very early appearance of iron and its great frequency in this period probably indicate intercourse with the Syrian coast or the Pontos region. Cyprus has also much iron ore of fair quality, discovered and worked as soon as knowledge of the metal extended. The iron objects include knives, double-edged sword-blades, spear-heads, arrow-heads, long cylindrical spits, a fire-rake, shield bosses, tweezers, nails and strigils.[12] Iron swords of the 6th century B.C., others

[1] Cf. tin wheels from pile dwellings, Forrer, Urg., 316.
[2] Myres-Richter, 120.
[3] Alt., II, i, 129 ; Breasted, H., 313.
[4] Oberhummer, Cypern, 1903, 3, 183 ; Orth, PW, Suppl. iv, 115.
[5] Cesnola, Salaminia, pp. xiii, 61, 63, 66.
[6] Myres-Richter, 15, 117.

[7] Myres, CAH, i, 109 ; Wade-Gery, *ib.*, ii, 524 ; Wilkinson, ii, 340.
[8] Ält. Bronzezeit, 1900, 154.
[9] L. P. di Cesnola, Cyprus, 282 ; A. P. di Cesnola, Salaminia, 47.
[10] Murray, Smith and Walters, 25 ; Meyer, Alt., II, i, 565; Dussaud, 1914, 279 f.
[11] Pococke, Travels, 585 f. ; Engel, i, 53.
[12] Myres-Richter, 21 f., 119; Dussaud, 265.

not later than the 5th century B.C. and an oxidised iron chain were found.[1] A saw at Phœnikias, in the period of transition from the Bronze to the Iron Age, was of nearly pure iron. A double axe of iron is Greek.[2] Iron implements and objects, e.g. a box of weapons and rings, although traced to the 9th–8th centuries B.C. at Salamis, are not numerous, owing to loss by oxidation.[3] An implement in a Bronze Age grave at Nicosia which "looked like iron" was found to be iron oxide.[4] A supposed mention of a "caster of iron" in a Phœnician inscription [5] and the suggestion that cast iron was known in Cyprus [6] are very doubtful. The crucibles found [7] were used for bronze.

BRASS

One of the most interesting industries of Cyprus was the manufacture of brass, an alloy of copper and zinc. The zinc ore was probably called cadmia (καδμία), a general name also given to varieties of zinc oxide obtained as by-products from the brass furnaces, viz. botrytis, pompholyx, diphryges and spodos. The information is all of the Græco-Roman Period and the date of origination of the industry is not known.[8]

Galen [9] says the best cadmia, botrytis (βοτρῦτις; zinc oxide), came from Cyprus, where pompholyx was obtained by washing spodos, prepared in brass furnaces.[10]

The brass (not "copper," as in Engel) smelting-furnace described by Dioskurides was built into a house of two storeys, the roof being open to the air. In the wall was a hole for bellows and a door for charging. The bellows and the men operating them were contained in an adjoining building. In smelting, the fuel was put into the furnace and kindled, then the cadmia broken into small pieces was fed in through a funnel from above along with more fuel. The lighter sublimate of zinc oxide (produced by the oxidation of zinc vapour) collected in the upper storey on the walls, at first like bubbles and then like fleeces of wool ; the heavier part sank down below. When Galen visited Cyprus there was no brass furnace in operation, but he was shown the manufacture of pompholyx direct from cadmia in a furnace. The same type of furnace was used for making brass in Rammelsberg in the 17th century, and as small quantities of metallic zinc were then obtained without distillation,[11] this may well have been the case also in antiquity.

SALTS

The Cyprian misy, said to be the best[12], is specified by Galen[13]; Cyprian sory was second only to the Egyptian.[14] Sory and misy were probably impure sulphates of iron and copper obtained from the oxidation of pyrites.[15] Basic iron sulphate occurs in Cyprus and could have been used as a mordant. The alum which occurs is not ordinary alum (potassium aluminium sulphate) but magnesium aluminium sulphate, $MgSO_4,Al_2(SO_4)_3,22H_2O$, known as

[1] Richter, Kypros, 48, 445 f., 465.
[2] Schuchhardt, in Dussaud, 254, 258.
[3] Cesnola, Salaminia, pp. xiii, 58, 60 ; objects in plate V.
[4] Myres-Richter, 10.
[5] Movers, Phönizier, II, iii, 68.
[6] Feldhaus, Technik, 234.
[7] Richter, Kypros, 453.
[8] Berthelot, Intr., 239 ; Oberhummer, Cypern, i, 182—no zinc ore known in Cyprus.
[9] De simplic. med., c. ix ; Opera, Venice, 1562, v, fols. 68 v., 70 r.
[10] Engel, i, 47, 51 ; cf. Strabo, III, iv, 15, 163 C. ; καδμεία λίθος σποδίον in

Dioskurides, v, 85 : Hippokrates, de morb. mulier. i, 63 ; in Opera omnia, ed. van der Linden, Leyden, 1665, ii, 455 ; μελάνθιον καὶ κύπειρον is translated spodium cyprium, but in other places, e.g. i, 100, melanthion is charcoal.
[11] Löhneyss, Bericht von Bergwercken, Leipzig, 1690, 83 ; Geoffroy, Of the Fossil, etc., Substances used in Physick, tr. by Douglas, 1736, 211.
[12] Pliny, xxxiv, 12.
[13] De compos. med. per gen., v, 13.
[14] Dioskurides, v, 117 ; Pliny, xxxiv, 12.
[15] Berthelot, Intr. 14, f., 47 ; Engel, i, 48.

feather alum ("plumous alum") or pickeringite [1], which could no doubt also serve as a mordant.[2] Black and white alum are mentioned as products of Cyprus, also sea salt at Salamis and rock salt at Kition, kyanos [3] (azurite, not lapis lazuli), sil and ochre [4] (oxides of iron). Cyprian salt is praised by Hippokrates and Dioskurides : salt working reached its maximum under Frankish occupation, and in the 16th–17th centuries A.D. was the main industry.[5] Verdigris [6] was made and chalkosmaragdos was probably malachite.[7] Chalkanthon ($\chi \acute{a} \lambda \kappa a \nu \theta o \nu$), copper sulphate or blue vitriol, was made in Cyprus by crystallising the drainage from copper mines[8] ; one variety is called by Dioskurides $\sigma \tau a \lambda a \kappa \tau \iota \kappa \acute{o} \nu$ (stalactite-like crystals). Copper scales ($\chi a \lambda \kappa o \hat{v} \, \check{a} \nu \theta o s$; flos æris), cuprous oxide, not to be confused with chalkanthon, were made by quenching heated copper.[9] Strabo[10] says the rich copper mines at Tamassos produced chalkanthon and verdigris ($\grave{\iota} \acute{o} s \, \tau o \hat{v} \, \chi a \lambda \kappa o \hat{v}$), the latter perhaps made from copper.

IVORY AND AMBER

Ivory occurs in Bronze Age remains, but the numerous objects at Salamis are principally Greek and Roman, although an ivory box, inside a lead box and two pateræ, is older. Bone and shell antiquities are probably later than ivory.[11] Ivory was found at Enkomi, etc.[12], and in a tomb of the Mycenæan Period near Athens in Greece, with bronze, glass and glazed faience.[13] Although Cesnola[14] says amber does not occur in tombs in Cyprus, it is occasionally found.[15]

[1] Watts, Dict. of Chemistry, ed. Morley and Muir, 1894, iv, 574.
[2] Dunstan, *Bull. Imperial Inst.*, 1906, iv, 213.
[3] Dioskurides, v, 106.
[4] Engel, i, 53, 55.
[5] Oberhummer, PW, xii, 67.
[6] Pliny, xxxiv, 11 ; Dioskurides, v, 104.
[7] Engel, i, 53.
[8] Dioskurides, v, 114.

[9] Dioskurides, v, 88.
[10] XIV, vi, 5, 683 C.
[11] Myres-Richter, 15 ; Cesnola, Salaminia, pp. xiv, 69, 71, 76, 80, Fig. 70.
[12] Murray, Smith and Walters, 9 ; Dussaud, 1914, 278.
[13] Perrot-Chipiez, Hist. de l'Art, vi, 417.
[14] Salaminia, 28, 35.
[15] Myres-Richter, 139, 184 ; Murray, Smith and Walters, 43, 123 : ring, beads, e.g. at Amathus, small cylinder at Enkomi.

ASIA MINOR

ASIA MINOR

THE HITTITES

The Hittites appear in the records of Egypt and Mesopotamia as one of the races inhabiting Asia Minor. Their culture is not definitely known, and the name "Hittite" has been too vaguely used,[1] but the suggestion of Jensen in 1898 that they were originally an Aryan race, although doubted, if not actually rejected, is now generally accepted.[2]

The Hittites appear to have migrated from their Indo-European home by way of the Caucasus (less probably the Bosporos), to have reached Asia Minor (already the seat of a very old and rich civilisation) some time in the period 2500–2000 B.C., probably nearer 2000 B.C., and to have established the Hittite kingdom about 1900 B.C. in the towns of Kaneš and Kušar.[3] Their earlier homes were in north-west Mesopotamia and the Taurus ; their ultimate place of origin perhaps in the heart of Asia.[4] They are perhaps the White Syrians (Λευκόσυροι) of the Greeks, living in the Pontos, and the later Assyrians have called Hittites who adopted the civilisation of Babylon.[5]

The representations of the Hittites [6] show that they were a mixed race, the Indo-European element on entering Asia Minor having come into contact with Assyro-Babylonian colonies, from which were borrowed cuneiform writing, many words and many gods.[7] The records at Boghazköi contain texts in the language of the Khatti, which is not Indo-European and is of uncertain origin, perhaps a Caucasian language spoken by the original inhabitants of Asia Minor who were of the same "Armenoid" type as those of Syro-Palestine, Mesopotamia, Armenia and Persia. Other languages involved are Lûish, an Indo-European language of west Cilicia, and Khurrish, non-Aryan and non-Indo-Germanic, related to that of the non-Aryan and non-Indo-Germanic but sub-Aryan Mitanni, the two being probably only two dialects. The Khurri apparently inhabited north Mesopotamia and the Armenian mountains much earlier than the true Hittites (before 3000 B.C.), and part of their country was that of the Mitanni, the later Mygdonia.[8] The Mitanni kingdom was wealthy : it sent to XVIII dyn. Egypt precious metals and stones, gilt bronze axes, decorated weapons for offerings to gods (including

[1] Meyer, Alt. (1909), I, ii, para. 472 ; ib., II, i, 3, 333, 512 ; ib., Reich und Kultur der Chetiter, 1914, 12 ; Maspero, Struggle, 351, 650 ; Hall, NE, 326 ; Otto, 1925, 35 f. ; Delaporte, La Mésopotamie, 1923, 267 ; Garstang, The Hittite Empire, 1929 ; ib., The Land of the Hittites, 1910, 2, 319, 321 ; Sayce, JRAS, 1920, 49 ; King, H. Bab., 227.

[2] Hrozný, EB¹⁴, xi, 598 ; Friedrich, RV, i, 126 ; Reche, ib., v, 321 ; Frankfort, Studies in Early Pottery, 1924, i, 74.

[3] Meyer, Alt., II, i, 20 f., 38 ; Olmstead, H. Pal. Syr., 116 f. ; Hogarth, CAH, ii, 259 ; Dussaud, La Lydie et ses Voisins,

1930, 24 ; Hrozný, EB¹⁴, xi, 602, 604 ; Contenau, Manuel, ii, 874 f. ; Bilabel, Geschichte Vorderasiens, Heidelberg, 1927, 134, 244—both somewhat out of date ; A. Götze, Das Hethiter-Reich, Der Alte Orient, 1928, xxvii, Heft 2.

[4] Hogarth, CAH, ii, 259 ; iii, 152 Roeder, RV, v, 317 ; Meyer, Alt., II, i, 24.

[5] Meyer, Alt., II, ii, 362 ; Giles, CAH, ii, 6.

[6] King, H. Bab., 227 ; Maspero, Struggle, 353.

[7] Hrozný, EB¹⁴, xi, 602.

[8] Ib. ; Reche, RV, v, 321 ; Otto, 1925, 36 ; Meyer, Alt., II, i, 6.

iron plated with gold), steel daggers and iron swords, iron rings covered with gold, bronze armour, purple garments and purple shoes decorated with gold beads, perfumes and spices and fine oil; in return it asked for gold and sometimes received gold of poor quality or even wood plated with gold.[1]

The Mitanni and Khurri have been called Subarians, the inhabitants of Subartu, the land north and north-west of Babylonia, the Armenian mountains. The Subarians were a non-Aryan race older than the Semitic and the Indo-European (Armenian) populations of these regions and were quite different from the Hittites.[2] Tablets from Boghazköi (ancient Hatti) in the Hittite language written in Babylonian cuneiform script and collected about 1300 B.C. to form an official library, contain a speech identical with that of Arzawa (Cilicia Tracheia, etc.) which was already known, and this is believed to be the earliest Indo-European language.[3] The Hittites had a picture-writing on stone monuments (the oldest of which may be Subarian) which is still undeciphered, but for correspondence they used Babylonian cuneiform.[4]

Archæological evidence has been held to show that the Khurri were known to the Sumerians of 3000–2500 B.C.[5] Cuneiform inscriptions and the so-called Cappadocian tablets, written in the old Assyrian or the Akkadian language at Kültepe-Kaneš, are of about 2250–2000 B.C.[6] The earliest specific reference to the Khurri is by Sargon I (2600 B.C.).[7] They were known to the Egyptians of the XII dyn. (c. 1950 B.C.), when figured red "Hittite" (really Cappadocian) pottery went to the Ægean and to the Semites of south Palestine, and commerce along the route from Egypt through Palestine over Mount Carmel and northwards to the trade routes leading down the Euphrates to Babylon had begun. The natural barriers of Anatolia preserved the Hittite characteristics and prevented direct contact between Mesopotamia and the Ægean in the period 2000–1000; such relation between these two civilisations as existed was probably by way of Cilicia, the Hittites, not the Phœnicians, being the intermediaries. The Khatti who are supposed to have been in early relations with Susa, Crete and Troy, could not have been the Indo-Germanic Hittites, who appear only about 2500–2000 and come more distinctly into history about 1800, when they had been mainly responsible for the overthrow of the First (Amorite) Dynasty at Babylon.[8] It has been said that there was no "Hittite race" as such, but only a confederation of Asiatic tribes with common affinities of race, language and religion, under the leadership of one, the Indo-European Khatti.[9]

The Hittites emerge into full light about 1500–1400 B.C. as Kheta in Egyptian records of Thothmes III (1500–1450), who received from them silver rings, costly wood and some precious stone not identified.[10] About 1380 the town

[1] Bilabel, Gesch. Vorderasiens, 1927, 91.
[2] Hrozný, EB[14], xi, 604; Unger, RV, á, 135; viii, 167; Peake and Fleure, Way of the Sea, 1929, 143; Dussaud, La Lydie et ses Voisins, 1930, 93; von Oppenheim, Tell Halaf, Leipzig, 1931; Contenau, Manuel, ii, 877; Forrer, Die Nachbarländer des Hatti-Reiches, 1929, was not available.
[3] Winckler, Die Völker Vorderasiens, Mitteil. Deutsch. Orient-Ges., 1907, No. 37; Sayce, J. Roy. Asiat. Soc., 1920, 49; Hogarth, CAH, ii, 6, 253, 256, 259; ib., iii, 148, 156; Meyer, Alt., II, i, 5, 336; Unger, RV, i, 135; Hrozný, EB[14], xi, 600, 602; Garstang, Nature, 1927, cxix, 819, 860; Wesendonk, A. Rel., 1929, xxvii, 62.
[4] King, H. Bab., 227; Hogarth, CAH,

iii, 151; Breasted, H., 262, 380; von Oppenheim, Tell Halaf, 55.
[5] Woolley, Sumerians, 1928, 47; Giles, CAH, ii, 6; Hogarth, ib., 257.
[6] Hrozný, EB[14], xi, 604; Dussaud, La Lydie, 21.
[7] Meyer, Alt., II, i, 12; Garstang, Nature, 1927, cxix, 819, 860.
[8] Hall, NE, 326; ib., CAH, i, 588; Wace, ib., 594; Giles, ib., ii, 6; Breasted, H., 188, 379; Maspero, Struggle, 365; Köster, Beih. alt. Orient, i, 18; Garstang, Land of the Hittites, 1910, viii, 316, 323; King, H. Bab., 210; Götze, Der Alte Orient, 1928, xxvii, Heft 2, 44.
[9] Contenau, Manuel, iii, 1060.
[10] Garstang, Land of the Hittites, 1910, 322; Breasted, H., 304, 315; Meyer, Alt., I, ii, 755; Sayce, B. and A. Life, 189.

of Qatna (near Homs) in Syria was sacked by the Hittite king Šubbi-luliuma. Cuneiform tablets at Qatna (containing one Hebrew, some Hittite, but no Egyptian words) mention gold, silver, copper, bronze and iron (then rare).[1] The earlier remains at Qatna (end of 3rd millennium) show strong Sumerian and some Ægean influences.[2]

The Hittites exerted a profound influence : they destroyed the Egyptian Empire in Asia, overthrew the 1 dyn. of Babylon, checked the desolating advance of the Assyrians and were important in Palestine. They are said to have founded the Heraklid dynasty of Lydia, and to have acted as intermediaries in the transmission of Asiatic culture, religion, art and astronomy to Europe, including the Ionian Greeks.[3] They were acquainted with Babylonian astronomy at least as early as 1400 B.C., including the planets Mars, Jupiter, Saturn and Mercury, and the Babylonian divination by inspection of the liver passed through (if not with) them to Etruria.[4] The double axe (clay models of which occur in a very old site (? 4000 B.C.) near Nineveh)[5] is probably of Subarian origin, like the storm-god Tešub of the Hittites, who carries a lightning-fork and double axe.[6] Many elements of so-called Hittite culture were probably Subarian. Von Oppenheim[7] considers that the principal Subarian divinities were Tešub (= heaven), the highest god ; the Great Goddess, Hepet (= earth) ; and their son, the Sun god, who is represented as a winged disc.[8] The veiled sphinx of Tell Halaf represents the Great Goddess, is older than the Egyptian sphinx and embodies the oldest known representation of a veil.[9] A "sun god in heaven" and a "sun god in water"[10] remind us of the alchemists.[11]

There is a representation of a sacred marriage typifying the growth of vegetation in spring, which is a myth well known in later Asia Minor.[12] The composite gigantic monsters of the Assyrian sculptures and reliefs probably came from the Subarians, and frequently occur on Hittite monuments[13], although small human-headed bulls occur in Sumerian remains at Tello.[14]

Parts of the Babylonian Gilgameš Epic written in Akkadian, Babylonian, Hittite and Khurri were found at Boghazköi, and the work may be of Subarian origin : it may have influenced the Greek epics. The medical texts at Boghazköi are in Hittite and Babylonian.[15]

The power and extent of the Hittite Empire is one of the surprises of the history of 2000–1000 B.C. which archæology has revealed during the last fifty years.[16] The principal dates are now supposed to be[17] : Mursil I (Hittite ; 1780–1740), overthrew 1 dyn. of Babylon and took Aleppo, c. 1750 ; Saušatar (Mitanni ; c. 1500) ; Šubbi-luliuma (Hittite, c. 1380) ; Boghazköi, Euyuk,

[1] Contenau, Manuel, ii, 878, 881.
[2] Dussaud, Revue de l'Art, 1931, lix, 10.
[3] Garstang, Land of the Hittites, 1910, viii, 151, 170, 354 ; Strong and Garstang, The Syrian Goddess, 1913—Syria Dea, Cybele ; Dussaud, La Lydie et ses Voisins, 1930, 14, 108—Hittite influence only towards the East.
[4] Hrozný, EB14, xi, 607 ; Otto, 1925, 40 ; Jeremias, 134, 142 ; Dussaud, La Lydie et ses Voisins, 1930, 10.
[5] Illustr. London News, 16 Sept., 1933.
[6] Meyer, Chetiter, 90 ; ib., Alt., II, i, 32 ; Giles, CAH, ii, 7 ; von Oppenheim, Tell Halaf, 85, 93, 229 ; cf. Dussaud, La Lydie, 52, 92 f., 96.
[7] Tell Halaf, 1931, 85, 93, 222, pl. 9.
[8] Farbridge, Symbols, 1923, 35.
[9] Von Oppenheim, 98, 102, 154 ; Jeremias, 260.
[10] Friedrich, Alte Orient, 1925, xxv, Heft 2, 21.

[11] J. Ferguson, Bibliotheca Chemica, Glasgow, 1906, ii, 264.
[12] Meyer, Chetiter, 90—Attis.
[13] Garstang, Land of the Hittites, 1910, 158, 254 ; Jeremias, 296 ; von Oppenheim, Tell Halaf, 73, 151, 165 ; Unger, RV, viii, 195.
[14] King, S. and A., 76.
[15] Otto, 1925, 41 ; Hrozný, EB14, xi, 607 ; Jeremias, Der Alte Orient, 1925, xxv, Heft 1, 27 ; Friedrich, ib., Heft 2, 30.
[16] CAH, ii, pref., v ; King, H. Bab., 210 ; Meyer, Alt., II, i, 333, 448 ; Olmstead, H. Pal. Syr., 116 ; H. von der Osten, Explorations in Hittite Asia Minor, Oriental Inst. Communications, Nos. 2, 6, 8, Chicago, 1927–9–30—much space taken up by descriptions of his meals ; W. H. Worrell, A Study of Races in the Ancient Near East, Cambridge, 1927 ; A. Götze, Das Hethiter-Reich, Der Alte Orient, 1928, xxvii, Heft 2.
[17] Contenau, Manuel, iii, 1618 ; Götze, 15.

ancient sculptures of Zinjirli and Carchemish, c. 1300 ; Hattusil III (1283–1260) ; "peoples of the sea," c. 1250 ; Carchemish, Zinjirli, Sákjé-Geuzi, 1000–800 ; end of Hittite Empire, 709. Under Šubbi-luliuma the Khatti kingdom was the first military and political power in the East ; the Khurri kingdom was then decaying and was dissolved into the enemy states of the Khurri and of the Mitanni, probably in Râs al-Ain, in north Mesopotamia.[1] This king annexed north Syria and invaded Mesopotamia, defined the limits of the Amorite territory, and exacted an annual tribute of 300 shekels of fine gold. Under Mursil II (1340), the kingdom was very powerful and in touch with an Achaian state in Asia Minor. About 1200 Boghazköi was overthrown by "sea peoples," with Indo-European Thracians, Phrygians and Armenians following, and a second Hittite Empire was founded in Cilicia, with Tyana as capital.[2] Shortly before this the Mitanni kingdom was conquered by the Assyrians, who captured subsequent small Hittite states in Syria (Carchemish ; Kunalua), the Hittites disappearing from history about 718.[3]

The geographical area of Hittite civilisation embraced the eastern half of Asia Minor with southern Phrygia and possibly Cilicia, and all north and north-central Syria, with extensions across both the middle and upper Euphrates on the one hand, and into lands west of the central plain of Asia Minor on the other. It occupied at a certain epoch the inter-continental bridge between Asia and Europe.[4]

HITTITE METALLURGY

The Hittites and related tribes (Chalybes, Tubal, etc.) of Asia Minor were important in the early development of the metallurgy of iron, and the Šardana are shown in Egyptian reliefs as carrying blue (steel ?) weapons.[5] Cilicia and Cappadocia, which were important centres of metallurgy in the III dyn. of Ur (2275) continued to be so under Hittite dominion until the 9th century B.C. The Hittites controlled the mines of Asia Minor, providing nearly every kind of mineral, and supplied metals to the ancient world.[6] In the Roman Period some workings had been exhausted, but King Archelaos (34 B.C.) had mining engineers (μεταλλευτί) in Galatia.[7] The regions of the Pontos and Euxine and the valleys of the rivers Thermodon and Iris were rich in minerals, and the tribes on the shores of the Black Sea, called Chalybes by the Greeks and famous as metallurgists, especially of iron, were closely related to the Hittites.[8]

The legend of the Argonauts, which does not seem anterior to 1000 B.C. [9], appears to imply that the Pontic coast had been visited by Ægeans from the 13th century B.C.[10], and the roads from Syria, Mesopotamia and Egypt met in this region.[11]

GOLD

Gold, some with blue "enamel" paste inlay, occurs as plaques, thick foil, beads, curious rings and other forms in graves of c. 3000, and especially of

[1] Hogarth, CAH, ii, 265 ; Cook, ib., 299 ; Meyer, Alt., II, i, 447 ; Hrozný, EB[14], xi, 606.

[2] Hrozný, 606 ; Giles, CAH, ii, 15 ; Hogarth, ib., 268.

[3] Garstang, Land of the Hittites, 1910, 390 ; Hrozný, op. cit. ; cf. Sayce, J. Roy. Asiat. Soc., 1929, 895, who has the divisions Proto-Hittite to 2000, Hittite 2000–1200 and Moscho-Hittite 1200–600.

[4] Hogarth, CAH, ii, 252 f.

[5] E. Reyer, Z. d. Deutschen Morgenländischen Ges., 1884, xxxviii, 149, 151.

[6] Smith, EHA, 163, 165 ; Garstang, Land of the Hittites, 1910, ix.

[7] Strabo, XII, ii, 10, 540 C. ; ib., XIII, i, 23, 591 C. ; Charlesworth, Trade Routes, 87, 258.

[8] Maspero, Struggle, 361, 363 ; Gowland, Archæologia, 1920, lxix, 159 ; Perrot and Chipiez, Hist. of Art in Sardinia and Judæa, 1890, ii, 268.

[9] Dussaud, La Lydie et ses Voisins, 1930, 77.

[10] J. R. Bacon, The Voyage of the Argonauts, 1925, 43 f., 64 ; Myres, CAH, iii, 661.

[11] Maspero, Struggle, 361, 365 ; on Egyptian objects in Asia Minor, von der Osten, Explor., No. 8, 1930, 162.

c. 2000, B.C., at Tell Halaf.[1] A Boghazköi text (14th century) says the Hittite kings "bring gold to the city of Bi . . ."[2] There was practically no gold in the regions of Asia Minor under Roman control, but there were rumours of gold mines near Caballa in Armenia, and gold and silver among the Suani in Colchis.[3] The legend of the golden fleece probably represents the collection of gold dust in greasy sheepskins in the coast torrents of Colchis, and the process survived in this region within living memory.[4] The graves of Colchis have long been violated but show traces of offerings of gold.[5] *ψ*-Aristotle [6] says the Chalybes collected gold in a small island lying off their coast : in the island of Gyaros (or ? Cyprus) mice eat iron[7] : Ælian [8] reports the same peculiar taste of the mice of Teredon in Babylonia, as does Theophrastos [9], adding that the mice devoured the iron and steel of the forges of the Chalybes and even stole gold from the mines, so that when caught they were disembowelled to recover the metal !

SILVER

The Hittites were particularly important in the development of the metallurgy of silver, the ores of which occur in large amounts in Asia Minor. The Hittite name for silver was khattus, and the old name of Boghazköi meant "town of silver."[10] Lubarna, king of the Khattin, sent Ašurnasirpal 20 talents of silver (with gold, etc.) from Kunalua (Kinalia, east of the Orontes)[11], and Thothmes III had received presents of silver, including eight rings weighing 41 lb., from the Hittite king, and much silver, including four silver hands and a vase of Cretan work, from north Syria.[12] In a Boghazköi text silver comes from kuzza (the mines ?).[13] Silver found at Tell Halaf[14] includes a ring and a deep dish, also a stylus in a rouge box. Early silver was also found at Carchemish.[15] In a Hittite book of laws (*c.* 1300 B.C.) 4 minas of copper cost 1 shekel of silver, i.e. the ratio is 240 : 1, and higher than in Babylonia, and payments of $1\frac{1}{2}$ or 1 mina, 3, 5, 6, 10, 12 or 20 shekels of silver for specified offences are prescribed.[16]

A copy of a treaty made in 1272 B.C. between Hattusil and Rameses II was engraved on a plate of silver. A silver boss with cuneiform and Hittite inscriptions of King Tarkondemos (or Tarkuntimmi) "of the land of Irmi," about 1100 B.C., is now lost, but there is an electrotype of it in the British Museum.[17] What is described as the oldest stamped silver bar, for use as money, is one weighing about a mina from Zinjirli in private possession, shaped like the lump from the bottom of a crucible, with an incised inscription of a Hittite king of 700 B.C.[18]

A Boghazköi text (14th century B.C.) refers to wooden beams plated with gold and silver[19], and bronze and copper were also plated with these metals.[20]

[1] Von Oppenheim, Tell Halaf, 65, 75, 192, plate III.
[2] Sayce, *Man*, 1921, xxi, Art. 97.
[3] Charlesworth, Trade Routes, 88, 258.
[4] Myres, CAH, iii, 661 ; Lenormant, Prem. Civ., i, 117 ; Strabo, XI, ii, 19, 499 C.
[5] Blümner, iv, 16 ; Schrader-Jevons, 176.
[6] De Mirabil. Auscultat., 26.
[7] *Ib.*, 25.
[8] De Nat. Animal., v, 14 ; Cook, Zeus, i, 632.
[9] Q. by Pliny, viii, 57.
[10] Sayce, *J. Roy. Asiat. Soc.*, 1920, 68, 78.
[11] Rawlinson, History of Phœnicia, 1889, 437 ; Meyer, Alt., II, ii, 371 ; Sayce, *Rec. of the Past*, N.S., 1889, ii, 128, 170.

[12] Meyer, Alt., II, i, 128, 130 ; Wainwright, Balabish, 1920, 39.
[13] Sayce, *Man*, 1921, xxi, Art. 97.
[14] Von Oppenheim, 192, 194.
[15] Garstang, Land of the Hittites, 1910, 81 ; *ib.*, The Hittite Empire, 1929, 290 ; Olmstead, H. Pal. Syr., 80, 117.
[16] Meyer, Alt., II, i, 517 ; Friedrich, *Der Alte Orient*, 1923, xxiii, 5, 28.
[17] Perrot-Chipiez, Art in Sardinia and Judæa, ii, 30, 268, Fig. 261 ; Breasted, CAH, ii, 149 ; Cesnola, Salaminia, 128 ; *B. Mus. Quarterly*, 1926, i, 42 ; Contenau, Manuel, ii, 944.
[18] Regling, PW, vii, 979 ; cf. Herodotos, iii, 96.
[19] Sayce, *Man*, 1921, xxi, Art. 97.
[20] Maspero, Struggle, 357 ; Meyer, Chetiter, 44—late silvered copper seal.

A Hittite seal of unknown origin in the Louvre is nearly pure silver, with traces of copper and iron [1], and a text from Kültepe speaks of "purified silver."[2]

There were celebrated silver mines in Armenia and on the Pontic coast; Tigranes, king of Pontos, paid Pompey a reparation of 6,000 talents of silver [3], and the mines near Trebizond are mentioned by Marco Polo.[4] Those near Tripolis, exhausted in Strabo's time [5], were seen by Hamilton.[6] There is a Hittite inscription near the argentiferous lead mines of Bulgar Maden, which are again in operation, as well as several sites of ancient argentiferous lead mining in the Pontos and Euxine region, all probably worked by the Hittites.[7]

The Hittite Khalywa, "land of the Halys," was perhaps the land of the Alybes, the silver in the Taurus Mountains being the chief ancient source of the metal [8], although Armenia [9] or somewhere east of Paphlagonia[10] is also possible. Alybe ('Aλύβη) is the "birthplace of silver" in Homer[11]: it is otherwise unknown.[12]

COPPER AND BRONZE

The copper and bronze objects from Asia Minor have not been well differentiated, but in all probability the use of copper preceded that of bronze.[13] The supposition by Virchow[14] that bronze was invented by the metal-working tribes of the Black Sea coast is without proof. A large quantity of copper was found on the Subarian site at Tell Halaf, which passed through a stone-copper period (3000 B.C.), including idols, a beaker, a half-moon, statuettes (like old Sumerian); a rectangular matrix with figures of Tešub and Hepet, of the oldest period, $8\frac{1}{2}$ cm. × 6 cm., which was used for stamps and has holes for casting beads; dishes, punches, spatulas for colours or salves, medical probes; ornaments such as arm-bands and brooches; needles and weapons such as short swords, daggers, sickle-swords, arrows, lance-heads and an axe. Some of this metal, which occurs with tools and weapons of stone and native obsidian, is probably bronze.[15] Copper and bronze also occur both in the old and later (2000 B.C.; after the entry of the Hittites) periods at Carchemish (Jerablus). The older objects include poker-shaped spears, which are sometimes tanged, and bent pins with eyelets; the younger objects are socketed axes, riveted dagger blades, leaf-like spear-heads with the suggestion of central ribs, narrow celts with piercing for rivets, many pins (some with round heads, some with the shanks pierced), ear-rings of twisted bronze, torques with curled ends, beads of paste, rock crystal, carnelian, lapis lazuli and shell, and shell pendants and mother-of-pearl discs.[16] The Hittite chariots had bronze fittings.[17]

The site at Sákjé-Geuzi in the southern foot of the Taurus, dating from the 14th century B.C., with a high standard of culture, is in a metal-producing country in contact with the mines of the Caucasus, Cyprus and the Taurus,

[1] Berthelot, Arch., 81.
[2] Meissner, i, 174.
[3] Strabo, XI, xiv, 10, 530 C.
[4] i, 4.
[5] XII, iii, 19, 548 C.; Charlesworth, Trade Routes, 88, 258.
[6] Leaf, Troy, 1912, 291.
[7] Gowland, Archæologia, 1920, lxix, 157, 159; Perrot-Chipiez, Sardinia and Judæa, ii, 268; von der Osten, Explor., No. 6, 1929, 130—Kara Hissar.
[8] Leaf, Troy, 291; Sayce, q. in J. Hellenic Stud., xxx, 315; Bury, CAH, ii, 492; Evans, Palace of Minos, ii, 169.
[9] Schrader, RL, 766; Orth, PW, Suppl. iv, 113.

[10] Myres, CAH, iii, 661; Lenz, Mineralogie, 3.
[11] Iliad, ii, 857; Strabo, XII, iii, 20, 548 C.
[12] Blümner, PW, iii A, 14.
[13] Woolley, in Garstang, Hittite Empire, 1929, 290; Olmstead, H. Pal. Syr., 80, 117.
[14] Q. by Hall, Oldest Civ. Greece, 193 f.
[15] Von Oppenheim, Tell Halaf, 1931, 52, 75, 173, 188, 194, 196, plates 48, 57.
[16] Garstang, Land of the Hittites, 1910, 81; Olmstead, H. Pal. Syr., 80, 117, Fig. 29 f.; Woolley, Lawrence and Hogarth, Carchemish, 2 vols., 1914–21.
[17] Perrot-Chipiez, Sardinia, etc., ii, 268; Maspero, Struggle, 356.

and able to obtain tin for the composition of bronze.[1] The copper probably came from the rich mines of Diarbekir, which were worked in the Assyrian Period [2], although it also occurs south-west of Aleppo, in the Anti-Lebanon near Petra, and at Chirbet es-Sawra in north Arabia.[3] Copper mines near Cisthene in Mysia were apparently not important.[4] The 14th century Boghazköi texts indicate that the Hittites obtained copper and bronze from Cyprus [5], although Sayce thinks "the City of Alasiya and Mount Taggata" is not Cyprus but a city at the foot of the south-eastern Taurus, and that before 2100 B.C. bronze went from Asia Minor to Assyria and Palestine.[6] The tin for the bronze may have come from the Caucasus, where very old mines are said to be still worked at Ochtschutschai.[7] The Subarians may have traded bronzes of copper and lead (or tin) and alloys of copper, tin and arsenic with Assyria in 3000–2000 B.C. by way of Kültepe and Ursu. A bronze cult waggon (or portable fireplace) decorated with tin inlay and with iron fire-bars, of c. 2000 B.C., was found at Tell Halaf, and a bronze statue of c. 3000 B.C. at Boghazköi.[8] A very archaic bronze figure found by Mecrant in the bed of the Orontes contained 3·9 per cent. of lead and 3·4 of tin.[9]

LEAD

Galena (native lead sulphide), found near Zephyrium in Cilicia, was exported for medical (?) and other purposes.[10] Six rolls of lead with Hittite inscriptions (then still in use) of the 9th–7th centuries B.C. were found at Assur.[11] Lead idols occur in various parts of Asia Minor.[12] Lead was also added to Hittite bronzes (see above).

IRON

The north-east part of the Hittite kingdom was the cradle of a developed technology of iron, and is the place where Jewish and Greek traditions locate the earliest working of that metal.[13] The Cappadocian tablets (2300 B.C.) refer to barzi-ili, "metal of god," the Hebrew parzil, and in this region iron and steel were worked very early[14], perhaps because the copper mines of the Pontos were becoming exhausted.[15] The "northern iron" in Jeremiah[16] may be steel from the Chalybes ($X\acute{a}\lambda\upsilon\beta\epsilon\varsigma$), who were an Asiatic people living in dens and caves on the Black Sea coast between Samsun and Trebizond, and in Greek accounts they are renowned workers of iron ($\sigma\iota\delta\eta\rho\sigma\tau\acute{\epsilon}\kappa\tau\sigma\nu\epsilon\varsigma$). They are mentioned by Aischylos (500 B.C.)[17] and other early Greek poets.[18] Apollonius Rhodius (of Alexandria ; 200 B.C.)[19] says the Chalybes endure heavy toil in the midst of black soot and smoke ; Vergil[20] and Ammianus

[1] Garstang, Land of the Hittites, 1910, 317, 322 ; Belck, Verhl. Berlin Ges. f. Anthropol., 1893, 61, 68.

[2] Von Oppenheim, Tell Halaf, 188 ; Dussaud, La Lydie et ses Voisins, 1930, 80 ; von der Osten, Explor., No. 8, 1930, 161—Ergani Maden.

[3] Thomsen, RV, i, 426.

[4] Strabo, XIII, i, 51, 607 C. ; Charlesworth, Trade Routes, 89, 258.

[5] Knudtzon, Die El Amarna Tafeln, Vorderasiatische Bibliothek, Leipzig, 1915, ii, 1077 ; Köster, Beih. Alt. Orient, i, 19.

[6] Sayce, Man, 1921, xxi, Art. 97 ; ib., in Garstang, Land of the Hittites, 1910, p. ix.

[7] Belck, op. cit., 61, 68 ; Dussaud, La Lydie, 1930, 81.

[8] Von Oppenheim, Tell Halaf, 1931, 52, 75, 172, 190 ; cf. Landsberger, Der Alte Orient, 1925, xxiv, 4, 13 ; Dussaud, La Lydie et ses Voisins, 1930, 23, 80.

[9] Gladstone, J. Anthropol. Inst., 1896–7, xxvi, 313.

[10] Pliny, xxxiv, 18.

[11] CAH, iii, 112 ; Meyer, Alt., II, ii, 364 ; von Oppenheim, Tell Halaf, 53 ; [Hrozný, Les inscriptions 'hittites' hieroglyphiques sur plomb trouvées à Assur, Archiv orientalni, Prag, 1933—not available.]

[12] Dussaud, La Lydie et ses Voisins, 1930, 135.

[13] Otto, 1925, 40.

[14] Sayce, J. Roy. Asiat. Soc., 1929, 895.

[15] Dussaud, La Lydie et ses Voisins, 1930, 84.

[16] xv, 12.

[17] Prometheus Vinctus, 714 ; ed. Weil, Leipzig, 1907, 29.

[18] Lepsius, Met., 54.

[19] Argonautica, ii, 1002.

[20] Georgics, i, 58.

Marcellinus [1], that they were the first workers of iron (per Chalybas erutum et domitum est primus ferrum).[2]

Hamilton found the people of Uniyah Kalah, west of Kotyora, making iron from charcoal and the ore found in small nodules in yellow clay over limestone, and selling it to Constantinople.[3]

The Chalybes, whose original name was perhaps Chaldis ($X\acute{a}\lambda\delta o\iota$), so that old writers often confuse the Chalybes, Alybes and Chaldæans [4], were certainly very early and celebrated workers in iron and steel. They gave their name to that of steel ($\chi\acute{a}\lambda\upsilon\psi$) among the Greeks, although old scholia connect this with Chalyps, son of Ares.[5]

A troop of Chalybes who emigrated to Doliche, a small city in Commagene, introduced the worship of Baal of Doliche, which during the 1st–3rd centuries A.D. was widely spread over the Roman Empire by soldiers, merchants and slaves. Jupiter Dolichenus is called in votive tablets Jupiter optimus Dolichenus natus ubi ferrum nascitur ("born where iron originates," i.e. was first worked).[6] This Chalybian god was Tešub, whose attribute was the double-axe, so that some Roman troops in Germany took him for the god of tree-fellers.[7]

The Etymologicon Magnum (9th–12th centuries) and Suidas (10th century) say the Chalybes were a "Scythian" tribe living "where iron was born."[8] They may be the remains, dispersed over a large area, of the Khaldi of Ararat and Lake Van in Armenia [9], or of the Alarodians, mentioned only by Herodotos[10], or of the people of Meshech and Tubal of the Bible, "sons of Japheth," known to the Assyrians about 1200 B.C. and related to the Tibareni (Tabali) and Mossynœki (Muški) who also lived in the Pontos.[11] Their knowledge of iron and steel was probably derived from the Hittites and they supplied iron and steel to Assyria and Egypt.[12]

.ι Boghazköi text (14th century B.C.) speaks of "black iron of heaven from the sky," which seems to suggest that meteoric iron was first known.[13] Iron smelting was, however, an old industry in Syria[14], Palestine[15] and Asia Minor.[16] The North Syrians (T'nai) sent Thothmes III silver and three iron vessels, and Rameses II obtained iron and steel from Hattusil at Kizwatna (Qisswadna, etc.) which is fairly certainly identified with the Pontos[17], although Sayce[18] objects on the ground that it paid tribute of purple (murex), which is not found there. The association of the place with an iron industry, however, speaks strongly for the Pontos region. In a Boghazköi text, Rameses asks

[1] XXII, viii, 21.
[2] Büchsenschütz, Hauptst. d. Gewerbfleisses, 44.
[3] Long, DG, i, 602 ; Ainsworth, in transl. of Xenophon's Anabasis, Bohn ed., 1875, 335.
[4] Strabo, XII, iii, 19, 548 C.
[5] Lenormant, Prem. Civ., i, 91 ; Meyer, Chetiter, 76 ; Cumont, Études Syriennes, 1917, 199 ; Perrot-Chipiez, H. de l'Art, ix, 58 ; Justin, Hist. Philippicæ, xliv, 3 ; Ruge, PW, iii, 2100.
[6] Cumont, Oriental Religions in Roman Paganism, Chicago, 1911, 113, 147 ; ib., Études Syriennes, 196 f. ; Meyer, Chetiter, 120, 122, 163 ; Thulin, PW, x, 1139 ; Dussaud, La Lydie et ses Voisins, 1930, 82.
[7] Farnell, Greece and Babylon, Edinburgh, 1911, 63 ; Meyer, Chetiter, 67, 90.
[8] Cook, Zeus, i, 631.
[9] Maspero, H. anc., 1904, 764 ; Meyer, Alt. (1909), I, ii, 622 ; Cumont, Études Syr., 199 ; Minns, CAH, iii, 201 ; Myres, ib., 662 ; Lenormant, Prem. Civ., i, 123 ; Much, Kupferzeit, 365.

[10] vii, 79 ; Meyer, Alt. (1909), I, ii, 622 ; II, i, 7 ; Lehmann-Haupt, PW, xi, 400 ; Karst, Grundsteine zu einer mittelländisch-asianischen Urgeschichte, Leipzig, 1928.
[11] Eckstein, in Lenormant, Prem. Civ., i, 91, 102, 112, 122 ; ib., Hist. ancienne de l'Orient, 1869, i, 104 ; Strabo, II, v, 31 ; VII, iv, 3 ; XI, xiv, 1 ; XII, iii, 18 ; Long, DG, i, 602 ; Much, 366 ; Hoernes, Urg., ii, 293 ; Meyer, Alt., II, ii, 362 ; Schrader-Jevons, 207 ; Olmstead, H. Pal. Syr., 533.
[12] Meyer, Chetiter, 90.
[13] Sayce, Man, 1921, xxi, Art. 97 ; cf. MGM, vi, 362 f. ; Feldhaus, Technik, 233.
[14] Cumont, Études Syr., 202.
[15] Petrie, Gerar, 1928, 14.
[16] Hoernes, in Hoops, i, 549.
[17] Meyer, Chetiter, 76 ; ib., Alt., II, i, 130, 158 ; Hrozný, EB[14], xi, 603 ; Bilabel, Geschichte Vorderasiens, 1927, 270.
[18] J. Roy. Asiat. Soc., 1929, 897.

for iron and Hattusil replies : "As for pure iron (chabalkinu, i.e. steel), about which you have written, there is no iron in my warehouse at Kizwatna and to make it at this time is inconvenient, but I have given written orders that it shall be made. As soon as it is ready I will send some : at the moment I send only a dagger."[1] The Egyptian Pharaoh probably knew the value of steel weapons from his wars against the Hittites, and was anxious to obtain them for himself.[2]

A relation between the Chalybes and the inhabitants of the basin of the Kuban, on the other side of the mountains on the southern slopes of the Caucasus, with a later diffusion of iron in Europe[3], is hypothetical and is based on the existence of similar iron spear-heads in both localities.

The old name of Aleppo was probably Chalybon, but this—and Helbon, also the name of the Chalybonians—is perhaps not derived from $\chi \acute{a} \lambda \nu \psi$ but from the Semitic root cheb.[4]

Iron bars on a bronze wagon and many iron weapons of c. 2000 B.C. were found at Tell Halaf.[5] The metal was still rather rare at Carchemish in the Hittite Period beginning about 1200 B.C.[6] Iron was still mined in the Roman Period near the River Cerasus in Cappadocia, in the hills near Pharnacia, and near Andeira, and the district around Cibyra was famous for beaten iron work.[7] Iron ores occur at present in Armenia, the Caucasus and the Pontos region.[8]

MERCURY

Remains of an ancient cinnabar (mercuric sulphide) mine near Iconium, formerly a Hittite town, contained fifty skeletons with *stone* hammers and lamps, evidently entombed by the collapse of a gallery.[9] Although we have no information that mercury was known to the Hittites, Arabic pineapple-shaped mercury pots are found on old Hittite sites[10], so that the cinnabar was still worked by the Arabs. The red miltos ($\mu \acute{\iota} \lambda \tau o s$) of Sinope, which was excelled only by that of Spain (Almaden), was found in Cappadocia and brought from Sinope before the Ephesian trade extended as far as Cappadocia.[11] Pliny calls it minium : the best was from Ephesus, the Colchian was adulterated, and nearly all of it used in Rome came from Almaden (Sisapo).[12] This may have been cinnabar or perhaps red lead.

ARSENIC

Bright red very poisonous sandaraché ($\sigma a \nu \delta a \rho \acute{a} \chi \eta$), i.e. realgar or arsenic disulphide (As_2S_2), was taken from a mine in Paphlagonia by condemned criminals, who died in the noxious atmosphere.[13] Arsenikon ($\mathring{a} \rho \sigma \epsilon \nu \iota \kappa \grave{o} \nu$; orpiment, arsenic trisulphide, As_2S_3) was mined near the River Hypanis (Bug, in the Ukraine).[14] A lump of orpiment was found in Hittite remains at Zinjirli, and it is suggested that it may have been used (with slaked lime ?) as a depilatory.[15] It is supposed[16] that sandaraché and sandyx ($\sigma \acute{a} \nu \delta v \xi$) are

[1] Meyer, Chetiter, 76 ; Meissner, i, 265, 348 ; Götze, 38 ; Hogarth, CAH, ii, 267.
[2] Maspero, Struggle, 356 ; Hall, CAH, ii, 292.
[3] Peake, Bronze Age and the Celtic World, 1922, 118, 161.
[4] Cumont, Études Syriennes, 200 ; Thomsen, RV, i, 426 ; cf. Bilabel, 397.
[5] Von Oppenheim, Tell Halaf, 75, 190.
[6] Garstang, Hittite Empire, 1929, 287, 289, 295 ; Meyer, Alt., II, i, 22.
[7] Strabo, XII, iii, 19, 549 C. ; XIII, i, 56, 610 C. ; XIII, iv, 17, 631 C. ; Pliny, xxxiv, 14 ; Charlesworth, Trade Routes, 89, 258.
[8] Lippert, Kulturgeschichte, 1886-7, ii,

225 ; Blümner, iv, 67 ; von der Osten, Explor., No. 6, 1929, 130—Kara Hissar ; ib., No. 8, 1930, 161—hinterland of Trebizond.
[9] Gowland, Archæologia, 1920, lxix, 157.
[10] Perrot-Chipiez, Sardinia and Judæa, i, 357, Fig. 250 ; ib., ii, 280.
[11] Strabo, XII, ii, 10, 539 C.
[12] Pliny, xxxiii, 7 ; Vitruvius, vii, 8.
[13] Strabo, XII, iii, 40, 562 C. ; Hill, ed. Theophrastus, On Stones, 1774, 174 ; Nies, PW, ii, 1273.
[14] Vitruvius, vii, 7.
[15] Meissner, Babylonien und Assyrien, 1920, i, 244.
[16] Eisler, Weltenmantel und Himmelszelt, Munich, 1910, 167, 178, 227, 285.

connected, through the meaning "red" in σανδ and σαρδ, with the god of purple, Sandan, or Herakles-Sandan, Santas in the Lûish language, a Hittite god of agriculture.[1] Sandyx, of which there was a mine in Lycia [2], was often a kind of ochre, since Vopiscus [3] mentions an Indian kind which becomes purple on burning (purpuram facere si curetur)[4] ; sandyx also meant red lead (minium) [5], or a vegetable dye obtained from madder.[6] Festus (2nd century A.D.) [7] says the Greeks also called sandaraché sandyx. Celsus [8] distinguishes between minium and sandaraché, but also uses the latter name for the gum resin. Hesychios [9] calls sandaraché a mineral substance (εἶδος μεταλλικόν). Philostratos[10] mentions a curious optical illusion caused by sandaraché under the soil in an Indian well, perhaps a hot spring.

PIGMENTS, ETC.

A rouge box for the king's use (men still apply rouge in Kurdistan), also yellow and red ochre, were found at Tell Halaf. The palace walls at Tell Halaf were covered with a smooth plaster, which had probably been coloured with yellow or red ochre, found in lumps in the old remains ; there was no fresco work.[11] There were also glazed bricks[12] (116), gypsum cement floors (75) and burnt tiles (76) ; beads of serpentine, porphyry, onyx, carneol, agate and blue frit, as well as glass (191, 194 f.), and a little ivory, also in the oldest layers ·(75, 191, 195) with coloured pottery, and seal cylinders (197 f.). The vessels were of limestone, alabaster and basalt, a pestle and flat mortar of basalt being found (180, plate 49). The pottery is decorated in red, brown and black, the oldest having a brilliant glaze (Firnismalerei) (181 f., 250, plates 51, 52). The Hittite pottery is essentially that of the regions in which it was made : that of Asia Minor is different from that of north Syria (which resembles generally the pottery of Babylonia and Assyria).[13] A Boghazköi text (14th century) says the Hittite kings brought lapis lazuli from Mount Takniyara, alabaster from the district of Kaneš (Kara-Euyuk), dusu stone from Ilamda, and kunkunuzzi stone from "wine cups" [?].[14]

In a religious ceremony "salt" (? natron), also "soap" from plants, is dissolved in water and used for washing the hands.[15] Specularis lapis, which came from Cappadocia, was probably selenite ; it was not so good as the Spanish as it was rather more opaque.[16]

The Hittites wore scanty red or blue mantles, fringed like those of the Chaldæans, so that they must have practised dyeing. They also wore gloves, and shoes with thick soles turning up distinctly at the toes. Their dress was of better and thicker material than that of the Syrians and Egyptians.[17] They perhaps first kept bees on a large scale, and from them bee-keeping spread over Nearer Asia[18] : beehives are mentioned in 1300 B.C.[19] There was fenugreek (Trigonella fœnum græcum) seed at Tell Halaf (perhaps for use in magic).[20] Intoxicating drinks are mentioned in religious texts.[21]

[1] Strong and Garstang, Syrian Goddess, 1913, 76 ; Meyer, Alt., II, i, 6 ; II, ii, 366.
[2] Strabo, XI, xiv, 9, 529 C.
[3] Aurelian, 29.
[4] Cf. Roscoe and Schorlemmer, Treatise on Chemistry, 1923, ii, 1269.
[5] Nies, PW, ii, 1273.
[6] Lindemann, Corpus Grammaticorum Latinorum, 1832, ii, 146, 254, 693.
[7] Ed. Lindsay, Leipzig, 1913, 434.
[8] De Medicina, vi, 8, 11 ; ed. Ninnin, Paris, 1821, ii, 62, 70.
[9] Lexikon, s.v.
[10] Apollonios of Tyana, iii, 14.
[11] Von Oppenheim, 87, 194.
[12] References in text to von Oppenheim.

[13] Contenau, Manuel, ii, 1010 f. ; Dussaud, La Lydie, 123.
[14] Sayce, Man, 1921, xxi, Art. 97.
[15] Friedrich, Der Alte Orient, 1925, xxv, Heft 2, 12, 14.
[16] Pliny, xxxvi, 22 ; Kidd, Outlines of Mineralogy, Oxford, 1809, i, 66, 70, 183.
[17] Maspero, Struggle, 353 ; von Oppenheim, Tell Halaf, 89.
[18] Otto, 1925, 40.
[19] Friedrich, Der Alte Orient, 1923, xxiii, 19.
[20] Von Oppenheim, Tell Halaf, 197; Lippmann, Gesch. der Rübe, 1925, 21.
[21] Friedrich, Der Alte Orient, 1925, xxv, Heft 2, 11.

THE PHRYGIANS AND LYDIANS

The Hittite kingdom ended with a Phrygian invasion in the 12th century B.C., these people crossing the Hellespont under pressure of an Illyrian invasion of their homes in Macedonia and Thrace. The Phrygians were at least partly Indo-European.[1] They took over the mining and metallurgy of the Hittites and were acquainted with iron, which occurs in old graves [2], and the fabulous gold of King Midas of Phrygia, said to have been obtained from the Pactolus near Mount Tmolos [3], may have been partly obtained from Greece in exchange for iron.[4]

The Phrygian power, which was checked in the west by Lydia from the 12th century, ultimately succumbed to waves of barbaric Cimmerian and Scythian invasion by way of the Caucasus about 750–600.[5]

Lydia, ruled from Sardis, the emblem of the king being the double axe, was formerly thought to be within the bounds of the Hittite Empire, and the Heraklid dynasty was considered as of Hittite origin. The Lydian culture is, however, older than the Hittite.[6] There were relations between Lydia, Assyria and the Ionian Greeks.[7] Ašurbanipal (668–626) reports an embassy from Gyges of Lydia, "a distant region whose name my fathers had not heard."[8] After the fall of Nineveh in 612 B.C., the Lydians divided Asia Minor with the Medes, with the Halys as the boundary. Smyrna, Ephesus and Colophon were important cities before the fall of Phrygia. In 546 B.C. Sardis, with Crœsus as ruler, fell to Cyrus the Persian, but his influence from Susa was weakly felt. In 334 Alexander crossed the Hellespont, and the land became Hellenised after two or three centuries [9]. Small articles of jewellery and pottery have been found on Lydian sites and some iron, but the weapons are late.

Herodotos[10] says Lydia was very rich in silver and in gold, which was washed down from Tmolos by the River Pactolus, thence called Chrysorrhoas. In the time of Strabo[11] "no gold dust was found," and Chikhachev (Tchihatchef)[12] supposes that the gold was an isolated pocket, worked out in the 7th century B.C. At present the stream carries down only a reddish mud.[13] Gold mines in Lydia[14] are no longer known, and may not have existed, but the riches in silver and gold are correct.[15] The Phrygian name for gold was glouros, the origin of the Greek χλωρός, greenish-yellow[16]: the Semitic hârûz, usually thought to be the origin of the Greek χρυσός by Phœnician transmission, may be derived from a word of Lydia or Asia Minor.[17]

The Greeks represented the rulers and officials of Asia Minor as enormously

[1] Garstang, Land of the Hittites, 1910, 58; Meyer, Alt., II, i, 567; Maspero, Empires, 331; Hogarth, CAH, iii, 502; Strabo, XII, iv, 4, 564 C.; Herodotos, ii, 2; Hippolytos, Refutationis Omnium Hæresium, v, 9, ed. Duncker and Schneidewin, Göttingen, 1859, 170; Eisele, Ro., iv, 257; Anrich, Das antike Mysterienwesen, Göttingen, 1894, 76; Inge, The Philosophy of Plotinus, 1918, i, 42; Ramsay, Cities and Bishoprics of Phrygia, Oxford, 1895–97.

[2] Perrot and Chipiez, Art in Phrygia, Lydia, Caria and Lycia, 1892, 19, 214; Schrader, RL, 1016.

[3] Schmitz, DBM, iii, 1084.

[4] Davies, Nature, 1933, cxxx, 985.

[5] Garstang, Land of the Hittites, 1910, 58, 62, 64; Giles, CAH, ii, 16; Herodotos, iv, 11, 12: the Cimmerians are in Homer, Odyssey, xi, 14.

[6] Garstang, Land of the Hittites, 1910, viii, 64 f.; Dussaud, La Lydie, 16, 40.

[7] BMGB, 184; Hogarth, CAH, iii, 507.

[8] Luckenbill, Anct. Rec. of Assyria, 1927, ii, 352.

[9] Garstang, 1910, 65; Giles, CAH, ii, 16; Perrot-Chipiez, Art in Phrygia, 326; Hogarth, Ionia and the East, 1909; Rostovtzeff, i, 129.

[10] i, 93; v, 49, 101.

[11] XIII, iv, 5, 625 C.

[12] Le Bospore et Constantinople, 2 ed., 1866, 232.

[13] Schmitz, DG, ii, 508.

[14] ψ-Aristotle, De mirab. auscult., 52.

[15] Dussaud, La Lydie et ses Voisins, 1930, 7.

[16] Schrader-Jevons, 176; Hesychios, γλούρεα.

[17] Hall, Oldest Civ. Greece, 193; Meyer, Alt. (1909), I, ii, 627; Schrader, RL, 300; Schrader-Jevons, 174.

wealthy. Pythius (480 B.C.) derived over three millions sterling from the gold mines of Kelainai near Apamea in Phrygia, and the presents sent by Crœsus to Delphi must have been worth over a million sterling.[1] The gold of Gyges (χρυσὸς Γυγιδάς or Γυγάδας), cited by Pollux (A.D. 178) [2], was probably electrum from the Pactolus, or from the quartz of Tmolos or Sipylos, containing more than the standard 20 per cent. of silver, usually 27 per cent. Electrum of Asia Minor mostly contains 60 per cent. of silver, that with 50 per cent. being uncommon and that with 40 very rare.[3] Crœsus (560–546) sent to Delphi among other gifts ingots of refined gold (ἄπεφθος χρυσός) and white gold (λευκός χρυσός), and had large quantities of gold dust (ψῆγμα χρυσοῦ) in his treasury.[4]

LYDIAN COINAGE

This wealth suggested coinage [5] and thus, as Aristotle [6] says : "the world was spared perpetual weighing." Herodotos [7], probably correctly, makes the Lydians the first to coin money (as distinct from weighed bars or rings used in Egypt and Babylonia).[8] This early Lydian electrum coinage, about 700 B.C., was stamped, not cast [9], into oval pastilles (φθοῖδες χρυσοῦ) slightly flattened at the sides, having the stamp deeply impressed on one side and striations on the other, and the devices show considerable artistic taste.[10] Xenophanes of Colophon, 6th century B.C., quoted by Pollux, reports that coinage spread from Lydia to Ionia, the Ægean and Greece[11], gold and silver money being unknown in Greece before about 720 B.C.[12] Lydian electrum coins from near Sardis contain 73 of silver to 27 of gold, and the metal is "white gold" (λευκός χρυσός)[13] rather than electrum (ἤλεκτρον) : Sophokles (5th century B.C.) already distinguished between the electrum of Sardis and the native gold (χρυσός) of India.[14] Although electrum is harder than gold, it is of indefinite composition, the percentage of silver varying in specimens from different places from 6–16[15], 20–48·3[16], or 40–60[17], and this makes it unsuitable for coinage. Some differentiation in quality may be disclosed by the touchstone, which as its name "Lydian stone" (λίθος Λυδή) indicates, was probably invented in Lydia. The usual statement that the first Greek mention of the touchstone is by Theognis of Megara (509 B.C.) is based on a misunderstanding.[18] The indications of the touchstone are also less accurate than Theophrastos[19] and Pliny[20] suppose[21]. Aristotle[22] calls the stone βάσανος (basanos), and it was a kind of black basalt.

The use of a coinage of pure gold in place of variable electrum is probably a

[1] Herodotos, vii, 27 ; ib., i, 50 ; Perrot-Chipiez, Phrygia, etc., 249 ; Long, DG, i, 577 ; Mason, DBM, ii, 628.
[2] Onomastikon, iii, 87.
[3] Rawlinson, Mon., i, 407 ; Maspero, H. anc., 1905, 605 ; ib., Empires, 608 ; Forrer, RL, 200 ; Hultsch, Gewichte d. Alt., 1898, III, 167.
[4] Herodotos, i, 50 f. ; vi, 125.
[5] Perrot-Chipiez, Phrygia, 251, 253, 299 ; ib., Hist. de l'Art, ix, 44.
[6] Politics, i, 3.
[7] i, 94.
[8] Rawlinson, note on Herodotos, ad loc. ; Lenormant, Monnaie, i, 125 ; Maspero, H. anc., 1905, 604 ; Forrer, RL, 506 ; Gardner, DA, ii, 447 ; Hill, CAH, iv, 126 ; Perrot-Chipiez, H. de l'Art, ix, 44 ; Madden, Coins of the Jews, 1903, 17 f. ; Dussaud, La Lydie, 9.
[9] Forrer, RL, 506, 924.
[10] Lenormant, Monnaie, i, 134 ; Maspero,

H. anc., 1905, 605, and Fig. ; Gardner, DA, i, 714 and Figs.—Lydian electrum coins in the British Museum ; Rawlinson, Mon., ii, 407.
[11] Perrot-Chipiez, Phrygia, 254 ; Regling, PW, vii, 983 ; Hogarth, CAH, iii, 519.
[12] Pausanias, III, xii, 3.
[13] Herodotos, i, 70.
[14] Perrot-Chipiez, Phrygia, 255, 282.
[15] Humboldt, Asie centrale, 1843, i, 493.
[16] Blümner, PW, v, 2315.
[17] Hultsch, Gewichte, III, 167.
[18] Hoover, Agricola's De Re Metallica, 1912, 399.
[19] De lapidibus, 4, 45 ; Hill's note, Theophrastus on Stones, 1774, 186.
[20] xxxiii, 8—lapis Lydius ; Salmasius, Plinianæ Exercitationes, 1689, 776.
[21] Muspratt, Chemistry, 1853–61, iv, 297 ; Nicholson, First Principles of Chemistry, 1790, 227 ; Hoover, 252, 613.
[22] Hist. Animal., viii, 12.

Lydian invention of the time of Crœsus (550 B.C.), who prohibited the use of electrum[1] : it implies a knowledge of separating gold from silver. The electrum coins were minted in the standard of weights used for silver, not gold, and the ratio of values of gold, electrum and silver was 13½ : 10 : 1. At first only large pieces, later smaller subdivisions, were coined.[2]

The separation of silver from gold, known in Egypt and Babylonia about 1500 B.C. [3], is thought to be established for Lydia by the pure gold jewellery found near Aidin (ancient Tralles), on the border between Lydia and Caria ; various decorative pieces have heads, etc., attached by rivets, but the granulated work is soldered. These objects, however, may be Egyptian, Syrian or Carthaginian, and a serpentine stone mould for casting statuettes is similar to the Babylonian.[4] The artificial production of electrum from gold and silver by the Lydians is possible[5] : they seem to have been especially skilled in gold work, and a Lydian goldsmith among the Helvetii is mentioned in a Roman inscription.[6]

The inventiveness of the Lydians is also evident in other directions and several new games are attributed to them [7], but there is another claimant with good credentials for the invention of coinage. Pollux [8] hesitated to decide between the Lydians and King Pheidon of Argos (650 B.C. ?), and the problem is still where he left it. Probably the first gold coins were issued by the kings of Lydia and the first silver coins by Pheidon in Ægina [9], in the form of a tortoise or turtle, which was the Phœnician emblem of trade.[10] Pheidon is said to have deposited metal bars (ὀβελίσκοι) in the temple of Hera at Argos as a witness of the old means of exchange which his new coinage had set aside[11], and Herodotos[12] says that he established weights and measures throughout the Peloponnesos. Pheidon's silver money was a copy of the Lydian electrum coinage.[13] That he coined copper money[14] is not stated by Herodotos or by Pausanias[15], and Strabo[16] merely says he "stamped money, silver in particular."

POTTERY

Lydian pottery was of good quality, with light red paste of fine texture and parallel band decorations in white or yellow, or in imitation of the wavy glass decoration of Egypt in black lines on red. Carian pottery from Tralles was famous in Roman times.[17] The ornamentation consists of bands and spirals in dull opaque colours—dark violet, brick-red, light green—which are easily rubbed off. Gold, bronze, and iron objects occur.[18]

Pliny[19] describes a peculiar stone, lapis Alabandicus, found at Alabanda in Caria, which was fusible and used for making glass and glazing pottery. Iassos, in Caria, furnished a beautiful blood-red and white ornamental marble.[20]

[1] Blümner, iv, 161 ; ib., PW, v, 2315 ; Bucher, ii, 141 ; Rossignol, Les métaux dans l'antiquité, 1863, 334 ; Perrot and Chipiez, Phrygia, etc., 257, 283 ; Meyer, Alt., II (1893), 552.

[2] Gardner, DA, i, 714 ; Regling, PW, vii, 983.

[3] Pp. 27, 36, 41, 232.

[4] Perrot-Chipiez, Phrygia, 287, 292 ; Maspero, The Passing of the Empires, 1900, 605.

[5] Blümner, iv, 161 ; ib., PW, v, 2315.

[6] Charlesworth, Trade Routes, 96.

[7] Rawlinson, Mon., ii, 408.

[8] Onomastikon, ix, 83.

[9] Perrot-Chipiez, Phrygia, 253, 257; ib., H. de l'Art, ix, 49 ; Schrader, RL, 286; CAH, iii, 761.

[10] Strabo, VIII, vi, 16, 375 C. ; Maspero, H. anc., 1905, 606 ; Lenormant, Monnaie, i, 135.

[11] Perrot-Chipiez, Phrygia, 256 ; Etymologicon Magnum (9th–12th centuries), ed. Gaisford, Oxford, 1848, 6134, ὀβελίσκος.

[12] vi, 127 ; Perrot-Chipiez, Hist. de l'Art, ix, 50.

[13] Perrot-Chipiez, ix, 49.

[14] Elder, DBM, iii, 255.

[15] VI, xxii, 2.

[16] VIII, iii, 33, 358 C. ; from Ephoros of Cumæ, 408 B.C., fragments in Müller-Didot, Fragm. Hist. Græc., i, 234–77.

[17] Pliny, xxxv, 12.

[18] Perrot-Chipiez, Phrygia, 285, 319, 326, 399.

[19] xxxvi, 8.

[20] Schmitz, DG, ii, 5

25

The gem lychnis [1], from near Orthosia in Caria, is perhaps the red ruby, and the carbuncle was also found.[2] Other gems of Asia Minor specified by Pliny are opals and jasper.[3] Jet, a corruption of gagates lapis, was found at the mouth of the River Gagis in Lycia.[4] The Chimæra ($\chi i\mu\alpha\iota\rho\alpha$), a fire-breathing female monster, is probably a personification of the small Lycian volcano of the same name [5], said by Ktesias to have a flame burning on a rock, inextinguishable by water, which only increased it. Beaufort found it to be a stream of inflammable gas from a crevice near the ruins of Olympos, and it is probably the origin of the story in Homer.[6]

Lydia produced textiles, a purple dye [7], saffron [8] and two kinds of vegetable honey, one from wheat (perhaps malt extract) and one from the tamarisk tree (probably a kind of manna).[9] A Lydian necklace and pectoral (8th century B.C.) of granulated gold work and decorated with amber beads was formerly in the Hermitage Museum in Leningrad (St. Petersburg).[10]

[1] Pliny, xxxvii, 7.
[2] Rawlinson, Mon., iii, 162.
[3] xxxvii, 6, 8.
[4] Pliny, xxxvi, 19 ; Rawlinson, Mon., iii, 163.
[5] Schmitz, DBM, i, 694 ; Eisler, Weltenmantel, 512.
[6] Maury, Religions de la Grèce Antique,

iii, 189 ; Long, DG, i, 608 ; Strabo, XIV, iii, 5, 665 C. ; Humboldt, Asie Centrale, ii, 505 ; Lippmann, Isis, 1928, xi, 496.
[7] Perrot-Chipiez, Phrygia, 297.
[8] Schmitz, DG, ii, 228.
[9] Herodotos, vii, 31.
[10] Williamson, Amber, 1932, 76.

PERSIA

PERSIA I

GENERAL

SUSA

Recent excavations on different sites in Persia have disclosed very old remains, dating from about 3000 B.C., in the earliest of which there is no metal, but somewhat later finds include gold, silver, copper and bronze. These are discussed in a later section. The remains at Susa and those at Anau (in Turkestan) have been published in some detail and will first be considered. The town of Susa [1] (O.T. Shushan) is in the territory called by the Jews Elam and by the Greeks Kissia and later Susis or Susiana, on the east bank of the River Choaspes.[2] The town, largely built of unbaked bricks now consolidated by rain and heat, was first excavated by Dieulafoy [3] and since 1885 by de Morgan.[4] The earliest (Proto-Elamite) periods are already in the Copper Age.[5] Although Elam was well known to the old Sumerians (about 3000 B.C.) it exerted very little influence upon Sumerian art and industry. Graves at Hekatompylos in North Persia, with stone vessels and copper weapons very like those found at Ur, show that Sumerian influences had spread into that region at an early date.[6]

The oldest civilisation (Susa I), beginning about 3000 B.C. in the stone-copper period, is different from the Elamite and closely related to the Sumerian.[7] Agriculture, flax growing and spinning, metallurgy, making pottery and glazed ware and stone cutting were all developed in this period, and obsidian and lapis lazuli were obtained by trade. Ground stone axes were used along with flat copper axe-heads. These axes have splayed blades but at first are without shaft-holes, although these are found somewhat later (Susa Ic). Copper was also made into narrow chisels, needles with eyelets and mirrors up to 19 cm. dia. The pottery is described below. Ornaments include beads of carnelian, shell, limestone, lapis lazuli and glazed faience, strung with spacers of the latter material ; also shell finger rings and limestone nose plugs. One perforated bead is engraved on the flat face and is perhaps the earliest known seal.[8]

The second civilisation (Susa II) was that of the peoples of Susa, Awan and Hamasi, afterwards called Elamites, and was only distantly connected with that of Susa I.[9] Susa II belongs to an Elamite culture as contrasted with the Sumerian civilisation. Stamp seals of this period have an undeciphered semi-pictographic script, believed to be Anzanite, the "Japhetic" speech of Elam. The copper work found in Susa II and the neighbouring site of Musyan is of Sumerian type and includes axes with cast or folded sockets, a transverse axe,

---

[1] See p. 217.
[2] Rawlinson, Mon., i, 26.
[3] L'Acropole de Suse, Paris, 1893.
[4] Maspero, New Light on Ancient Egypt, 1908, 10 ; C. Huart, La Perse Antique et la Civilisation Iranienne, 1925, 268 ; V. G. Childe, New Light on the Most Ancient East, 1934, 231 f.
[5] De Morgan, Délégation en Perse, xiii, 3, 66, 69 ; Mecquenem, Revue des Arts Asiatiques, 1929–30, vi, 73–88.

[6] Woolley, Ur Excav., 1934, ii, 397 ; Childe, Most Ancient East, 1934, 232, 250.
[7] Loftus, Travels, 1857, 335 ; Meyer, Alt., I, ii, 499 ; Jastrow, Civ., 112 ; King, S. and A., 339 ; Maspero, Hist. anc., 1905, 188 ; ib., Dawn, 563 ; ib., Struggle, 30.
[8] Childe, Most Ancient East, 233 f., 237.
[9] Peake and Fleure, Priests and Kings, 1927, 42.

forked arrow butts, vessels and pins. The recent dating of Susa II places it about 2750–2500 B.C. The semi-pictographic script was now replaced by cuneiform, and there were relations with the Indus Valley.[1]

CERAMICS AT SUSA

Very characteristic of Susa I is the pottery, in the forms of beakers, plates and jugs of good shape with linear and geometrical painted decorations and stylised figures of plants and animals. Similar painted pottery was prevalent in mountainous Asia, but not in contemporary Egypt. The people wrote on clay tablets in their own script, and made calculations on a decimal system, not sexagesimal as in Babylonia.[2] The pottery of Susa I (3000–2800 B.C.) is made with a fine paste, well mixed, and is often very thin (1 mm.). It is of six main forms and was painted directly on the clay with oxide of iron applied with a brush before firing. Some objects were of obsidian, probably from Ararat.[3]

Although the earliest pottery at Susa has been regarded as peculiar to this site and unrelated to other ceramics of Nearer Asia [4], it has also been supposed to resemble some Sumerian pottery.[5] It is said to be technically related to that of al-'Ubaid ; bowls and tumblers are normally given stability by an incipient ring base (as at al-'Ubaid), implying long practice. The decorations include highly stylised natural objects, perhaps derived from basket work. Technically and artistically, the pottery from Tell Halaf is comparable with the finest products of Susa I.[6]

Analysis of the body of vases of Susa I indicates that limonite (a native oxide of iron) was used for the colour in the non-glazed ware and a "ferruginous glass" for the glazed. Limonite may contain up to 10 per cent. of manganese, but no manganese was found on the vases. The iron oxide was applied with soda and a little potash, perhaps mixed with sand, and the resulting black glaze is similar to that on some Minoan and Greek pottery.[7]

	Painted Vases	Reddish Pottery
Silica, SiO_2	46·3	50·4
Alumina, Al_2O_3	15·1	10·5
Ferrous oxide, FeO	6·4	not dtmd.
Ferric oxide, Fe_2O_3	—	7·2
Lime, CaO	22·1	14·7
Magnesia, MgO	4·9	4·7
Water, H_2O	0·9	not dtmd.
Carbon dioxide, CO_2 (loss on heating) .	4·5	Alkalies 10·5
	100·2	98·0

The pottery of Susa II (which was not analysed) includes cylinder seals in place of the button seals of Susa I and vases with decorations partly geometrical and partly representations of animals less stylised than those of Susa I and

[1] Childe, Most Ancient East, 1934, 243, 248.
[2] Meyer, Alt., I, ii, 501 ; Jastrow, Civ., 114.
[3] De Morgan, Délég., xiii, 5 f., 8, 14, 28, 31, 65 ; Dussaud, Revue de l'Art, 1931, lix, 5.
[4] De Morgan, Délég., xiii, 1 ; cf. Perrot

and Chipiez, History of Art in Persia, 1892, 470—some resemblance to Babylonian.
[5] Mecquenem, Revue d'Assyriol., 1923, xix, 111.
[6] Childe, Most Ancient East, 234, 255.
[7] Couyat-Barthoux, in Délég. en Perse, xiii, 161 ; Harrison, Pots and Pans, 63.

tending towards naturalism. Some decorations resemble those on Minoan pottery.[1] The Susa I pottery is probably much older than that of Susa II and has been considered to be unrelated to it.[2] Pottery found at Persepolis, Nihâwand, Teheran, etc., is very similar to that of Susa I and II, and indicates an influence extending beyond the frontiers of Elam. A curious type of object at Susa is represented by baked clay head-dresses for statues, studded with copper nails, and a lapis lazuli model of a dove with gold nails (2 millen).[3]

The cylinder seals of Susa II were of fine white paste, worked whilst soft and then covered with a very thin layer of slightly greenish glaze ; they occur with cylinders of hard stone, one perhaps of hæmatite.[4] Susa II pottery is coarser and thicker than that of Susa I[5] : probably the fine clay was found naturally and not treated, but in the first period it was perhaps washed.[6] Both types were probably formed on the wheel.[7] A "vermilion" and a purple-red are "less solid" than in Susa I ; the black is the same in both periods. Although it was suggested that the colours in Susa II were applied after firing, this is not probable. In the later period ("Anzanite récent") the pottery is coarser and often polychrome.[8] Porous pottery was made watertight by a layer of bitumen, as in Mesopotamia.[9]

A specimen of early pottery clay from Susa consisted of : combined water 4·05, "humidity" 2·70, limestone ("calcaire") 37·58, clay matter 28·57, sand and rocky débris 27·10, per cent. It is very ferruginous, burns badly and is fusible, melting (like the pottery) in a porcelain furnace at 1400°. The pottery was probably burnt at 900°–1000°. The black designs on the painted pottery showed much iron and a little manganese, as in the Greek pottery analysed by Foster[10] ; perhaps magnetite was used with a flux containing silicates of lime and soda in vases of Susa I : those of Susa II are not glazed, the same iron oxide pigment and body being used. The black, which is less beautiful than the Greek, has a brown cast and is thicker.[11] The metal objects from Susa are considered in the appropriate sections in the next chapter.

ANAU (TURKESTAN)

Apparently related to the Susian is the very old civilisation at Anau, a ruined city in the foothills of Russian Turkestan between Transcaucasia and Iran.[12] This culture existed in the third millennium and was perhaps related to the Sumerian. Three cultures at Anau were distinguished, dated[13] for their commencements by Schmidt as : Anau I, 3000 B.C. ; Anau II, 2000 B.C. ; Anau III, 1000 B.C. Both de Morgan and Pottier think the pottery is a decadent form of the earlier Susian, the result of influence from Elam. Langdon[14] suggests, from the pottery, that the Anau civilisation, which may be as old as the

[1] De Morgan, Déllég. en Perse, xiii, 23, 56, 65.

[2] Frankfort, Studies in Early Pottery, 1924, i, 22 ; Contenau, Manuel, ii, 662 ; Hall and Woolley, Ur Excav., 1927, i, 9, 167 ; Childe, Most Ancient East, 246.

[3] Contenau, Manuel, ii, 645, 929.

[4] De Morgan, Déllég., viii, 6, 9.

[5] Ib., xiii, 28, 41.

[6] Cf. Déllég., i, 184, plates 17–20 ; 188, plates 21–22 ; Contenau, Civ. assyro-babylonienne, 1922, 95.

[7] Déllég., xiii, 30 ; cf. Perrot-Chipiez, Persia, 475—perhaps hand-made.

[8] Ib., xiii, 45, 56, 62.

[9] Ib., i, 66.

[10] Cf. Brongniart, Arts Céram., i, 16 ;

Couyat-Barthoux, Déllég., xiii, 161, found no manganese.

[11] Granger, in Déllég., xiii, 160.

[12] Myres, CAH, i, 69, 85 ; Hall, ib., 579 ; J. M. Tyler, The New Stone Age, 1922, 93 ; R. Pumpelly, Excavations in Turkestan, Carnegie Inst. Publ., No. 26, Washington, 1905 ; ib., Explorations in Turkestan ; Prehistoric Civilisation in Anau, Carnegie Inst. Publ., No. 73, 2 vols., Washington, 1908, i, 16 ; Woolley, Sumerians, 1928, 47.

[13] Myres, CAH, i, 85 ; Pumpelly, Explor., 1908, ii, 50 ; Pottier, in de Morgan, Déllég., xiii, 70 ; Childe, Most Anct. East, 314, says Schmidt's dating is "obsolete" but gives no alternative suggestion ; Huntington had dated I–III as 9000–5200.

[14] CAH, i, 362.

Egyptian [1], was part of a great prehistoric culture spreading from Central Asia to the Iranian plateau, Syria and Egypt before 4000 B.C., the Sumerians being a somewhat later branch of this race, but these hypotheses have been criticised and both the great antiquity of the Anau sites and their relations with Susa have been called in question : the pottery of Anau I is said to be "diametrically opposed to those of Susa," that of Anau II is more related to that of Anato'lia and North Syria, and the relation of that of Anau III to pottery of Elam and Mesopotamia is remote.[2]   Civilisations II and III at Anau are separated by a drought, when the site was abandoned ; after a further drought a culture IV in Persian times brought in the use of iron.[3]

The early pottery at Anau is of smooth grey or red body mostly painted in black, but also bluish-black, violet-black, violet and black-brown.[4]   Very early painted pottery also occurs on sites in Persia.[5]   In Anau III there is blue glaze like that of Egypt.[6]

Some copper and lead, but no gold, silver or tin were found in Anau I.   The copper contains some antimony and arsenic and, with the lead, probably came from Central Asia [7] or Syria, or further west, and some probably came from Babylonia in the later period.[8]   There were turquoise beads but no lapis lazuli.[9]   No bronze, gold or silver was found in Anau II.[10]   In Anau III there is copper and bronze with low and variable tin content, perhaps due to smelting of mixed or stanniferous ores, although some copper seems to have been intentionally alloyed with lead.   Analysis gave :   copper 70·42, lead 21·69, tin 5·57, arsenic 0·49, antimony 0·41.   The tin for the bronze may have come from Khorassan.[11]

OTHER PERSIAN SITES

Many interesting objects have recently been obtained on what are apparently old sites in Iran.[12] In the oldest remains, at Dâmghân (ancient Hekatompylos) and Persepolis, of about 3000 B.C. or soon after (2900–2600 ?), Herzfeld found no metal but stone and flint tools and painted, thin, hand-made pottery.   The remains at Nihâwand (Tepe Giyân) and other sites in western Iran are of the Bronze Age.   The objects have been hastily plundered by the avaricious natives and have come into the market through dealers, so that the authenticity of some has been questioned and some of the bronzes are modern imitations, probably, according to Pope, made in Paris.   The patina of these spurious bronzes is poor and dull and becomes black when oiled.   There was abundance of metal at Nihâwand, including gold, silver, copper and bronze.   These objects are dated about 2000 B.C., but some authorities think the bronzes are as late as A.D. 200.   The dates of the earliest objects have been inferred from objects said to have been found with them : a pottery bowl with the name of Šargali-šarri, king of Akkad (2624–2603), a bronze dagger inscribed with the name of

[1] Cook, CAH, i, 181.
[2] Hall, CAH, i, 579 ; Hall and Woolley, Ur Excav., 1927, i, 156 ; Frankfort, Studies in Early Pottery, 1924, i, 76 ; 1927, ii, 3.
[3] Myres, CAH, i, 88 ; Pumpelly, Explor., 1908, ii, 49.
[4] Explor., 1908, i, 106, 114, 125, 127, 132, 140, 144, 148 ; Excav., 1905, 8, pl. 1.
[5] Peake and Fleure, Merchant Venturers in Bronze, Oxford, 1931, 136.
[6] Myres, 88 ; Explor., 1908, i, 173 ; ii, 46.
[7] Explor., 1908, i, 38, 40, 150–174 ; plates 36–39 ; analyses ; Myres, 85.
[8] Myres, 89, 91 ; Explor., 1908, i, 152.
[9] Explor., 1908, i, 38, plates 40–42 ; Myres, 85, and Peake and Fleure, Priests

and Kings, 1927, 137, say lapis lazuli occurs.
[10] Explor., 1908, ii, 42 f., 238 f.
[11] Ib., 1908, i, 40 ; ii, 43 f., 152, 238 f. ; Myres, 88.
[12] F. Sarre, Die Kunst des alten Persien, 1922, 19, 48, pls. 45 f. ; E. Herzfeld, Archäologische Mitteilungen aus Iran, 1929 f., i, 65, pls. 4–7 ; ib., Illustr. London News, 1929, clxxiv, 892, 942, 982 ; 1930, clxxvii, 1024 ; A. U. Pope, ib., 1930, clxxvii, 389, 444 ; 1932, clxxxi, 613, 666, 1054 ; 1934, clxxxv, 1005 ; ib., An Introduction to Persian Art since the Seventh Century A.D., 1930, 175, 186 ; Contenau and Ghirshman, Syria, 1933, xiv, I — bronze dagger, c. 1100 B.C., Nihâwand ; Contenau, Manuel, iii, 1576.

Gimil-Ištar, of the period of Sulgi (2399–2282), and another with the name of a king of Babylon of 1123–1113. These objects, which were obtained through dealers, may have come from Mesopotamia, although Pope considers this unlikely. Objects from Luristan, also obtained through dealers, are supposed to be of the Iron Age (800 B.C. ?).

The metal from Luristan was both copper and bronze, the latter containing up to 19·7 per cent. of tin. Nickel was present only in traces, except in one specimen, which contained 1·1 per cent.[1] The metal of another specimen, analysed in New York, contained 84·70 of copper, 14·69 of tin, 0·26 of arsenic, about 0·1–0·2 of iron, 0·10 of sand and a trace of silver : it was free from antimony, lead, zinc, nickel and cobalt, and hence is regarded by Pope as a modern imitation. There were also idols of hammered copper sheet, bronze weapons, pins, needles, axe-heads similar to Cretan and a bronze spoon. A large plain goblet, figured cups of Assyrian [Elamite ?] style and a fragment of a mirror have a highly polished curious black finish produced by some unknown technique. A copper or bronze bull's head is perhaps of about 1000–600 B.C. Quite different are the crude "Syro-Hittite" bronze figures of the 2nd millenn. found in a Persian dolmen at Pirawand, north of Taq-i-Bustan.

Other bronze or copper objects from Luristan are socketed axes of curious form, needles, bridles, lance-heads, daggers and tools, sheet diadems and spouted beakers like those from Ur, but no swords. There was a solid silver statuette and gold and silver jewellery like that of Queen Shub-ad at Ur, perhaps older than the other objects, which Herzfeld dates about 2200 B.C. The pottery at Nihâwand is partly like the Cretan. There is a glazed jar from Luristan supposed to be of the 2nd millenn. Some pottery from Nihâwand resembles Anatolian ware and there is also black-topped red ware like Pre-dynastic Egyptian. The Luristan pottery is like that of Susa II. Glass and glass paste are not certainly of Luristan origin. It has been supposed that the remains indicate a homogeneous civilisation stretching from west of Asia Minor across North Syria and Mesopotamia to the Iranian table-land and reaching to the borders of India, with probable relations with China.

At Tepe Hissar near Dâmghân, on the south-eastern slopes of the Elburz, three cultures are recognised, numbered from the top. Hissar I (2000–1600) has silver vessels, ribbon and cord, gold sheet, a copper wand, a copper plate, a small lead flask, bronze daggers inlaid with silver in a check pattern, lapis lazuli, columns and discs of alabaster, pottery and seals like those of Susa II and clay braziers. Hissar II (3000–2000) has pottery resembling that of Susa Ic, and copper or poor bronze spear-heads with a looped tang and pins with double spiral heads—both resembling Early Cycladic Ægean types. Hissar III (to 3000 B.C.) has beak-spouted jugs in grey pottery and silver resembling Early Minoan, copper (or bronze ?) tools similar to those of Copper Age graves in the Kuban Valley, an axe-adze like one found near the surface at Mohenjo-daro and beads of frit or stone resembling those of Early Dynastic Sumerian graves and of Mohenjo-daro.[2]

Bronze pins, knives, spears and axes found at Kákh, Khinámán, south-east Persia, and at Van (Armenia) and Hamadan (ancient Ekbatana) are probably of the later Bronze Age, as are objects from Redkin near Delishan in Russian Armenia and on the southern slope of the Caucasus. The famous cemetery at Kuban, almost midway between the Black and Caspian Seas on the northern slope, supposed to belong to the early Bronze Age from the relative scarcity of iron, is perhaps later, since the form of brooches is that found in Greece and Italy and later than the types at Mycenæ. There was probably Mediterranean

---

[1] Desch, B.A. Rep., 1931, 271.
[2] E. Schmidt, Illustr. London News, 1932, clxxxi, 772 ; A. U. Pope, ib., 1933,

clxxxii, 116 ; V. G. Childe, New Light on The Most Ancient East, 1934, 268.

and Central European influence. Tombs at Lenkoran, on the south-west shore of the Caspian, date from the early Bronze Age : daggers without handles or with cast hilts and long swords show relations with the Ægean.[1] The Caucasian and Scythian antiquities are considered in a later section. Excavations by Sir Aurel Stein in alluvial basins in the valleys of Waziristan and Baluchistan show cultures related to the Indus civilisation but less advanced and probably later. Metallurgy was well known, copper and bronze being apparently freely used, although few metal objects (flat axes, single-bladed knife, stamp seals, saws, etc.) were actually found. Glass is believed to have been known and there are beads of imported lapis lazuli. A copper pin with a head of lapis lazuli found at Kulli is of "Babylonian" form.[2]

The so-called "treasure of Astrabad," found in 1841 at this town, south-east of the Caspian, and described by Bode in 1844, contained gold vases and objects, arms of copper (not bronze or iron) and two small idols in red and yellow stone. It is supposed to be of Sumerian origin. Finds in the Caucasus, including figured silver vases and gold ornaments, are also supposed to be as old as those at Susa.[3]

THE IRANIANS

A cuneiform text of about 1400 B.C. found at Boghazköi [4] contains a version of a treaty of Mattiuza, king of the Mitanni, with Šubbi-luliuma, king of the Hittites, which shows that the Mitanni in Mesopotamia had as gods Mitra (= Mithra), Indra, Varuna and the Nâsâtyâs (or Nashatianna).[5] Although the identifications have been criticised [6], it is now considered that, since Mithra and Indra are found in India and Persia but the Nāsātyās and Varuna are peculiar to India, a common origin of the Hindus and Persians (Iranians), who afterwards separated, is probable.[7]

We touch here, for the first and last time, the highly controversial question of the origin of the Aryan race—if such existed—a subject with a vast, confused and contradictory literature. It is clear that there were in 1400 B.C. people in the Hittite capital of Boghazköi speaking languages containing strong Indo-European ("Aryan") elements. The first Indo-Europeans to reach western Asia may have been the Medes, who are called Manda by Narâm-Sin (c. 2550 B.C.), and afterwards Mada.[8]

According to one theory, the "Aryans" or Indo-Iranians spread into Asia Minor and India from a region on the Danube, bounded on the north by the mountains separating Bohemia from Germany, on the east by the Carpathians, on the west by the Bohemian Forest and on the south by the Balkans. An alternative hypothesis (equally speculative) places the cradle of the "Nordic" Aryans in the steppe beginning in Russia and extending without interruption to the slopes of the Hindu Kush, with certain westward prolongations, especially in the sandy heaths to the north of the Carpathians stretching from the Russian steppe across Galicia to the neighbourhood of Breslau. The movement probably occurred in a period of transition to a Bronze Age civilisation.[9]

[1] BMG Bronze Age, 175.
[2] Childe, Most Ancient East, 1934, 269 f., 274, 276.
[3] Woolley, Sumerians, 1928, 47 ; Contenau, Manuel, iii, 1564 f., 1569 f. ; criticised by Frankfort, Studies in Early Pottery, 1924, i, 85.
[4] Winckler, Die Völker Vorderasiens, Der Alte Orient, 1899, i ; Farnell, Greece and Babylon, Edinburgh, 1911, 46 ; Giles, CAH, ii, 13 ; Cook, ib., 400.
[5] BMGB, 8 ; J. H. Moulton, Early Zoroastrianism, 1913, 115, 139 ; Meyer, Alt., II, i, 34.

[6] E. G. Browne, Literary History of Persia, 1902, 34 ; Keith, in Cambridge History of India, 1922, i, 113 ; Moulton, 6.
[7] Giles, Cambridge Hist. of India, i, 72 ; Hall, Bronze Age, 85.
[8] Giles, CAH, ii, 13, 15, 28 ; Meyer, Alt., I, ii, 891 ; II, i, 33, 35, 40.
[9] Lassen, Indische Altertumskunde, Bonn and Leipzig, 1847–61, i, 527 ; Contenau, Manuel, ii, 882 ; V. G. Childe, Dawn of European Civilisation, 1925 ; ib., The Aryans, 1926 ; Peake, Bronze Age, 132, 139, 153 ; Peake and Fleure, The Steppe and the Sown, 1928, 34 ; S. Feist, RV, vi, 54.

The dispersal and the settlements in the Punjab and in Iran, formerly dated as early as 4000 B.C.[1], probably occurred much later than this, although the dates given by modern authorities vary from 2400 to 1200 B.C.[2] Apparently the Iranian movement was going on from about 1700, and the settlement, which was achieved about 900 B.C., was apparently confined at first to the small region of Persis, the modern Fars. This was at first called Pârsa, later Parsua, and is situated between the desert of central Iran and the Persian Gulf.[3] The old Indus civilisation at Mohenjo-daro, etc., was pre-Aryan and non-Aryan, and was probably the true origin of several elements of Indian religion and culture previously regarded as post-Vedic accretions to Brahmanism. Lower strata at Amri in Lower Sindh contain pottery analogous to that found at Jemdet Nasr in Akkad.[4]

The earlier Iranian civilisation is very obscure, and little is known of it before the time of Cyrus (560 B.C.), although the country was apparently plundered by Šalmaneser III in 838 B.C. and became an Assyrian province under Tiglathpileser III (746–28).[5] Persian authors, such as Firdawsî[6], have much to say of early battles between the peoples of Iran and Turan, but the early history of Persia is shrouded in mystery and the poetical accounts are largely fabulous. The Iranians were strongly influenced by Assyria in the 1st millenn. B.C., but had a civilisation of their own and a comparatively independent art, which was maintained in a very conservative way until the latest periods, so that, as in the case of the archæology of India and China, even quite late sources can give useful indications of the technology of the earlier periods. The country formed a bridge between Asia and the Mediterranean region, and there are elements from Egypt, Assyria and Asia Minor in Persian art.[7] The earliest antiquities are of stone and of little interest from our point of view, but they represent embroideries, and wine carried in leather bags.[8]

In the period of the greatest extension of the Persian Empire, the territory was of enormous extent. The Iranian highland is bounded on the north by the Hindu Kush, which is prolonged through the Paropamisus and other mountain chains towards the west, thus separating Iran from the steppes of western Turkestan, and then by the Elburz, sloping nearly vertically into the Caspian. Iran communicates with Turkestan practically only by roads at Balkh and Merv. To the east it is separated from India by the Sulaiman and Kirthar mountains, although the Indus Valley is geographically a part of Iran. To the south the mountains of Mekran and Luristan slope down to the Persian Gulf, and the periphery is completed by the Zagros mountains bordering on Mesopotamia. Communication to the Far East, to Central Asia and Mongolia, passes over the Pamir plateau. The whole region is intersected with mountain ranges and is rich in metalliferous ores.[9]

Persia, Bactria with the adjoining steppe country, and the north-west corner

[1] Deussen, Allgemeine Geschichte der Philosophie, Leipzig, 1894, I, i, 10, 39.

[2] Peake, Bronze Age, 157 ; Peake and Fleure, Merchant Venturers in Bronze, Oxford, 1931, 132 ; Meyer, Alt., I, ii, 857 ; II, i, 6 ; ib., EB¹¹, xxi, 202 ; Meston, EB¹⁴, xii, 184 ; Rogers, History of Ancient Persia, New York, 1929, 9 ; Charpentier, q. in Isis, 1927, ix, 552 ; Havell, Short History of India, 1924, 9, 19, 35, 59 ; Cambridge Hist. of India, i, 70 ; Marshall and Langdon, Mohenjo-daro, 1931, i, 112, ii, 432 ; O. Schroeder, RV, vi, 67.

[3] F. Schachermeyr, RV, x, 68, 80 ; Maspero, Empires, 89.

[4] V. G. Childe, New Light on the Most Ancient East, 1934, 204 f., 223 f.

[5] Maspero, Empires, 89 ; Schachermeyr, RV, x, 68 ; Laufer, Sino-Iranica, Chicago, 1919, 185.

[6] Le livre des rois, tr. by J. Mohl, 1876–8, iii, 1 f. : the work was completed in A.D. 1010.

[7] Rostovtzeff, Iranians and Greeks in South Russia, Oxford, 1922, 192 ; Sarre, Kunst den alten Persien, p. vi ; de Morgan, Délég. en Perse, viii, 46.

[8] Sarre, 12, 15.

[9] Rawlinson, Mon., iii, 84 f. ; de Gobineau, Histoire des perses, 1869, i, 152.

of India were in 500 B.C.–A.D. 500 in touch not only with the route crossing Chinese Turkestan but also with one far to the north, running from Russia along the steppes and foothills of the central mountains to the borders of China. The steppes connected the opposite ends of Asia and also the continents of Europe and Asia, and thus influences from Mesopotamia penetrated through Bactria to the northern route, whence they reached Europe and China.[1] There were relations between the Iranian plateau and Syria in the 9th century B.C.[2]

THE HISTORY OF PERSIA

The periods of Persian history of interest to us may be divided as follows[3] :

(1) Indo-Iranian, 2400 B.C. ?
(2) Early Iranian, c. 2000 B.C. ?
(3) Period of Assyrian influence, 1000 B.C. or earlier.
(4) Medic Period (800–700 B.C.).
(5) Old Persian (Achæmenian), from 550 B.C.
(6) Interregnum, from the invasion of Alexander to the Sassanian Restoration (330 B.C.–A.D. 226), including the Arsacid (Parthian) rule, 250 B.C.–A.D. 226.
(7) Sassanian Period, A.D. 226–652.
(8) Muḥammadan Period, from A.D. 652.

Seleukos (321–281) founded seventy towns on Greek models in Iran, and the Parthian Period (of which little is known) was also under strong Greek influence. The Sassanian Period, however, represents a reaction against Hellenism and a return to the old Iranian traditions, with the Zoroastrian religion and the importance of the magi or priests.[4]

Iran was inhabited about 800–700 B.C. in the north-west by the Medes and in the south-west by the Persians.[5] The first mention of the Medes by name is in Assyrian texts of 810–781, and after this there is repeated but vague mention of Media as dependent on Assyria.[6] Although the Medes and Persians were closely allied in language, religion and customs, the former were conquered in 560 B.C. by Cyrus (Kai-Khusru) the Persian, who captured Ekbatana (founded as the capital of Media in 720) and established the Achæmenid (Akhmânish) dynasty and the Persian Empire, which lasted until 330 B.C.[7]

Cambyses (529–522), son of Cyrus, engaged in the conquest of Egypt, from whence in 525 he imported much spoil in precious metals, ivory and building materials, and perhaps Egyptian artificers.[8] The Persian kingdom was firmly established and reached the zenith of its prosperity under Darius I (521–485). Xerxes (485–465) built the magnificent palace at Persepolis [9] and had engraved the famous rock inscriptions at Behistun (Besutun). During this period, Aramaic ("Chaldee," the most northern Semitic dialect) was used throughout the western Persian kingdom, including Egypt and Asia Minor.[10] Persian

[1] Dalton, Treasure of the Oxus, 1926, pp. xlvi f., lxxvi ; Aurel Stein, Sand-buried Ruins of Khotan, 1903 ; ib., Ancient Khotan, 2 vols., Oxford, 1907 ; ib., Ruins of Desert Cathay, 2 vols., 1912 ; ib., On Alexander's Track to the Indus, 1929 ; ib., Innermost Asia, 3 vols., Oxford, 1928.
[2] Dussaud, Syria, 1934, xv, 187.
[3] Browne, Literary History of Persia, 1902, 37 ; Meyer, Alt., I, ii, para. 581 ; III, 16 ; Rogers, History of Ancient Persia, 1929 ; Dalton, Treasure of the Oxus, 1926, p. xxii ; A. T. Wilson, A Bibliography of Persia, Oxford, 1930.
[4] Sarre, Kunst d. alt. Persien, 24, 26, 32.

[5] Meyer, Alt., III, 16.
[6] Perrot-Chipiez, Persia, 4.
[7] Browne, Lit. Hist. of Persia, 22 ; Maspero, Empires, 455 ; Rawlinson, Mon., ii, 420.
[8] Perrot-Chipiez, Art in Persia, 113.
[9] Ib., Persia, 21, 38, 277 ; Rawlinson, Mon., iii, 267 ; Sarre, Kunst. d. alt. Persien, plates ; Herzfeld, Illustr. London News, 1933, clxxxii, 207, 401, 406, 453, 488.
[10] E. Meyer, Papyrusfund von Elephantine, 1912, 17 ; Sayce and Cowley, Aramaic Papyri discovered at Assuan, 1906—the 5th century B.C. Aramaic papyri were found at Aswân, not Elephantine.

rule in conquered lands was milder and more tolerant than that of Egypt or Assyria, and the life, religion and customs of the dependent nations were respected [1], but this tolerance probably ceased in Babylonia after Darius.[2] The Sassanian Empire was a dominant power from the 4th century A.D. Sapor I (240–273) defeated and captured the Emperor Valerian at Edessa and Sapor II (309–380) worsted the Emperor Julian at Ctesiphon. It reached its climax under Khusraw Anuschirwan (531–579), to whom Byzantium paid tribute, but Khusraw II was defeated by the Arabs at Nineveh in 627 and the Persian army under Rustum was heavily defeated at Qâdisîya in 635 and at Nihâwand in 639.

PERSIAN LITERATURE

The Persian sacred book, the Avesta, gives the reputed teachings of Zoroaster (Zarathustra), whose dates are variously given from 1000 to 522 B.C.[3] The Avesta contains parts of various dates, mostly before the 5th century B.C. The suggestion [4] that the Avesta and the Indian Rig-Veda are of 1200 B.C. makes the first too old and the second too young, according to general opinion. Parts of the Avesta other than the old Gâthâs (liturgies and prayers) are the Yashts (hymns), of the later Achæmenian Age, and the Prose Avesta, in particular the ritual of the Vendidad, which was probably composed after Alexander's conquest. The material may sometimes be very old even though the form is late, and different chapters and sections of the Yashts, Vendidad and Yasna may vary considerably in age.[5] The Vendidad is probably fairly complete[6] ; the latest part, the code against demons, is of the Arsacid Period (250 B.C.–A.D. 226) [7], perhaps about 147 B.C.[8] The Avesta in its present form is supposed to contain remains of early Persian, Medic or Bactrian literature.[9] Thiele thinks most of it goes back to the Achæmenian Period[10], Gasquet[11] puts it in the 4th century B.C., but the present work is usually supposed to date to the period of the Sassanid renaissance in the 3rd century A.D.[12], with parts, usually differentiable by the language, as late as the 10th–12th centuries A.D.[13] In his later translation Darmesteter[14] doubts the authenticity of the extant text of the Avesta and considers that it has been influenced by Brahmanical, Buddhist and Greek elements, but principally by Neo-Platonism and Judaism, including Philo of Alexandria. This view is strongly rebutted by Carnoy[15] and by Meyer.[16]

THE LEGEND OF GÂYÔMARD

The Bundeheš (more correctly Bundahišn), although of the Sassanian Period[17], contains extracts from old cosmogonies based on lost texts of the

[1] Rostovtzeff, , 153, 251.

[2] Peet, JEA, 1928, xiv, 196.

[3] Meyer, Alt., I, ii, para. 581 ; ib., Christentums, ii, 58 ; Hertel, q. in Isis, 1928, x, 104 ; Browne, Lit. H. Pers., 30 ; Moulton, Early Zoroastrianism, 1913, viii, 8, 78, 87, 103, 197, 204 ; Jackson, Cambridge History of India, 1922, i, 323.

[4] Peake and Fleure, Merchant Venturers in Bronze, Oxford, 1931, 127 f.

[5] Jackson, Cambr. Hist. of India, i, 323 ; Moulton, Early Zoroastrianism, viii f., 8, 22, 78, 87, 103, 197, 204 ; Spiegel, Eranische Altertumskunde, iii, 778.

[6] Horn, Geschichte der persischen Litteratur, Leipzig, 1901, 4.

[7] Meyer, Christentums, ii, 58.

[8] Christensen, Études sur le Zoroastrianisme de la Perse antique, Copenhagen, 1928, 43.

[9] Justi, A. Rel., vi, 252 ; Dieterich, Mithrasliturgie, 189.

[10] Bousset, A. Rel., iv, 157.

[11] Essai sur le culte et les mystères de Mithra, 1899, 16.

[12] Meyer, Alt., I, ii, para. 581 ; Darmesteter, Sacred Books of the East [SBE], iv, pp. xxxviii, xlviii.

[13] West, and Bartholomae, q. by Justi, and Dieterich, opp. cit.

[14] Ann. Musée Guimet, xxiv, p. vi ; cf. SBE, iv, p. li.

[15] ERE, xii, 863.

[16] Christentums, ii, 58.

[17] West, SBE, v, p. xli.

398   PERSIA I

Avesta [1], particularly the Dâmdâd Nask, which is supposed to have been composed in 500–430 B.C. and to contain late Babylonian (Chaldæan) elements.[2] An important legend in the Dâmdâd Nask, that of the primæval divine man Gâyômard, forms the basis of later Iranian gnosis. Gâyômard was also regarded as the cosmos in human form, a direct equality of macrocosm and microcosm and not mere analogy or comparison as in later revisions of the legend.[3] The body of Gâyômard thus contained the elements of the cosmos—the seven metals of the seven planets. On his death (brought about by the planet Saturn and its evil spirit Beelzebub), but not before, the seven metals issued from his various members and flowed into the earth.[4] From the head came lead (srub), from the blood tin (arjiz), from the marrow silver (sim, or asim), from the feet bronze ("brass," West) (asin), from the bones copper (rod), from the fat glass (abginag) ("mercury," West), from the flesh steel (polâd) ("adamant," West), and from the soul ("sperm and life," West), as the sum of the whole, came the eighth most perfect substance, gold (zar). The translation of abginag (avginako) as "mercury" by West is said to be incorrect, since the Persian name for mercury is zivag or simab ("silver water," hydrargyros). Later texts give a different order : gold, silver (asîm), bronze (or brass) (asîn), copper, tin, lead, glass, steel (almas, originally steel, later, by confusion with the Greek adamas, diamond) ; or, by omitting the purely Iranian addition of the metal of the soul, according to value : gold, silver, bronze (or brass), copper, tin, steel and "mixed iron" (crude iron ?).[5] Pollux [6] gives as the metals gold, silver, brass (ὀρείχαλκος), iron, tin (καττίτερος), lead and glass (ὕαλος), and even Isidore of Seville [7] compares glass with "other metals" (aliis metallis). In the modern glass factory the fused glass in the pot is often called "metal."

The translation of avginako as "mercury" is also given by Brandt from Mandæan books, written under Persian influence, in which there are eight (not seven) flat metal "anvils" making up the body of the earth ; one is also of "adamant" (steel ?) and one rests in "mercury."[8]

In the Gâthâs, Scharewar (Khshathra Vairya) is "overlord of metals" or "master of the seven metals," and melts iron, lead and "the commonest metal" (presumably copper). In the Bundahišn he is "lord of weapons and the power of weapons," in the Pahlawî texts (A.D. 226–880) "giver of treasures," i.e. metals, which he will melt on the last day by a meteor. The just, however, will walk through the resulting molten mass as if it were warm milk.[9] Scharewar is one of the seven Amesha Spentas, angels of light or emanations of the great lord Ahura Mazda (Ormuzd), who is opposed to the evil principle, Angra Mainyu (Ahriman) and his powers, which appear in Persian dualism, probably under Chaldean influence, about the 4th century B.C.[10] In the resurrection the bones (copper) of Gâyômard will be the first to arise.[11]

The order of diminishing value of the metals and of the "ages of the world"

[1] Horn, Pers. Lit., 37.

[2] Reitzenstein and Schaeder, Studien zur antiken Synkretistik aus Iran und Griechenland, Leipzig, 1926, 6, 11, 121, 130, 209, 221, 349 ; Eisler, Weltenmantel, 93.

[3] West, SBE, v, 183 ; Reitzenstein and Schaeder, 37, 205, 211, 225, 228, 232.

[4] West, op. cit. ; Reitzenstein and Schaeder, 18, 220, 223, 225 ; Gray, A. Rel., vii, 359 ; Bousset, A. Rel., iv, 206, 218 ; ib., Hauptprobleme der Gnosis, Göttingen, 1907, 206.

[5] Reitzenstein and Schaeder, 228, 232.

[6] Onomastikon, iii, 87.

[7] Etymologiæ, XVI, xvi, 1.

[8] A. Brandt, Die Mandäische Religion, Utrecht, 1889, 60.

[9] SBE, v, p. x f., 10, 125, 359, 365, 373, 375, 401 ; xxiii, 5 ; Spiegel, Eran. Alt., ii, 29, 37 ; Gray, A. Rel., vii, 359 ; Rawlinson, Mon., ii, 334 ; Albiruni, Chronology of Ancient Nations, tr. by Sachau, 1879, 207—written A.D. 1000 ; Moulton, Early Zoroastrianism, 98, 109, 157.

[10] Spiegel, ii, 21, 28, 37, 121 ; Moulton, 74, 98, 126, 157 ; Meyer, Christ., ii, 62 ; ib., Alt., III, 124 ; Cumont, Oriental Religions in Roman Paganism, Chicago, 1911, 190.

[11] Geldner, in Bertholet, Religionsgeschichtliches Lesebuch, Tübingen, 1908, 356 ; from the Bundahišn.

reappears in Hesiod [1] and in Buddhist doctrines.[2] Hesiod's five ages, four distinguished by metals (gold, silver, bronze and iron) are perhaps not of astrological significance, and although the conception may be of Babylonian origin [3], Reitzenstein and Schaeder [4] think it is derived from Persian sources, as are also the four ages in Orphicism and in Plato, who was under its influence especially in the "Timaeus." Iranian teachings came to Greece during the Persian invasion of Asia Minor. The doctrine of the macrocosm and microcosm in the syncretistic Treatise on the Number Seven ($\pi\epsilon\rho\grave{\iota}\ \epsilon\beta\delta o\mu\acute{a}\delta\omega\nu$) is probably derived from an Iranian source. The work is attributed to Hippokrates, but is probably due to a physician of the school of Knidos, and is usually regarded as pre-Hippokratic (6th–5th centuries B.C.), although Reitzenstein and Schaeder date it in the 4th century B.C. The homunculi of gold, silver, copper and lead in the vision of Zosimos (c. A.D. 250–300) are also supposed to be derived from an Iranian source.[5]

A legend similar to that of Gâyômard is contained in the Indian Mahâbhârata [6], in which Narayana, a late Vedic god connected with obscure ideas of a primæval sacrifice and the creation, contains the universe in himself. The Mahâbhârata is a compilation of various dates, some as late as the 3rd century A.D., but some parts go back to the old Vedic Period [7], and the relation of metals to planets goes back in Babylonia to the period of the Aryan Cassites.

LATE PERSIAN LITERATURE

The Dînkart, a kind of Zoroastrian encyclopædia, is a very late Pahlawî work, although an old Pahlawî text, as such go, begun in al-Ma'mûn's time and completed by Âtûrpât about A.D. 881.[8] Later Persian works giving useful information, some from much earlier sources, are the "Pharmacology" (Liber fundamentorum pharmacologiæ) of Abu Mansûr Muwaffaq ibn Alî al-Harawî (A.D. 975) [9], the medical encyclopædia (Zakhîra al-Khwârizmshâhî) of Ismâ'îl al-Jurjânî (d. 1136)[10] and the Tuḥfat ul-Mûminîn, written in 1669 by Muḥammad Mûmin Ḥusaini, available in a European language only in one very rare translation made at Ispahan and published by Joseph Labrosse.[11]

Much old Iranian material is said to be contained in the epic "Book of Kings" (Shâhnâma) of Firdawsî, completed in 1010[12]; the references to technology in it include processes contemporary with the author, which he nevertheless refers to mythical ancient kings. Huscheng discovered the separation of iron from its ore and the way of fashioning it into tools; Jemschid invented bricks, perfumes and medicines, and discovered gems and amber and the way of separating other stones by art.[13]

[1] Works and Days, 110 f.
[2] Meyer, Christ., ii, 190; de Gobineau, Histoire des Perses, 1869, i, 58; Reitzenstein and Schaeder, 61, 67.
[3] Cf. King, H. Bab., 302; Smith, EHA, 165.
[4] Op. cit., 61, 67, 70, 118, 147.
[5] Reitzenstein and Schaeder, 7, 67, 118, 130; Berthelot, Coll., ii, 108, 115.
[6] Oldenberg, Das Mahabharata, Göttingen, 1922, 41.
[7] Winternitz, Geschichte der indischen Literatur, Leipzig, 1908–22, i, 24, 263, 389, 396.
[8] Sarton, Introduction to the Hist. of Science, 1927, i, 583, 591; Isis, 1928, x, 119; SBE, xxxvii.

[9] Transl. by Abdul Chalig Achundow, in Kobert, Historische Studien aus dem Pharmakologischen Institut der Kaiserlichen Universität Dorpat, Band III, Halle, 1893; summary in Lippmann, Abh., i, 81.
[10] Naficy, La médicine en Perse des origines à nos jours, 1933.
[11] Pharmacopœa Persica ex idiomate Persico in Latinum conversa . . . , Paris, 1681; dedicated by Frater Angelus, i.e. Labrosse; preface + 370 pp. + indexes. I have been able to see this work.
[12] Transl., Das Königsbuch, E. Rückert, 3 vols., 1890–95—incomplete; complete tr. by J. Mohl, Le livre des rois, 7 vols., 1876–78.
[13] Mohl, Le livre des rois, i, 25, 35.

Similar stories are given in the "History of the Kings of Persia" by 'Abd al-Malik ibn Muḥammad (Abu Mansûr) al-Tha'âlibî, of Naisapûr (961–1037).[1] Such works cannot safely be used for the ancient period, since they often contain much Greek material, as is the case with Abu Mansûr Muwaffaq, and also information derived from Arabic sources. Both the Greek and Arabic material may ultimately have come from Persian sources, but until the dependence is proved it is uncertain.

[1] Ibn Khallikân, Biographical Dictionary, tr. by Mac Guckin de Slane, 4 vols., 1842–71, ii, 129 ; al-Tha'-âlibî, Histoire des Rois des Perses, tr. by Zotenberg, 1900.

PERSIA II

METALS

In dealing with the archæology of Persia we find that it divides into two parts. The earliest period includes that of the Elamites, who inhabited part of the region before the Persians made their appearance. The later period, as will be seen from the dates given on p. 396, really takes us out of the true province of the present work and brings us into the Iron Age.[1] The finds of metal from later Persia are not numerous : they include decorated gold sheet and a gold jug ; silver statuettes, dishes and vessels ; a bronze bull's head from north-west Persia and other bronze objects. Many of the objects came from the Caucasus.[2] There is a glimpse of ancient Persian metallurgy in the Vendidad [3], but the industrial arts were little esteemed [4] and the Greeks knew of hardly any place in Iran except Ekbatana : city life began much later than in Mesopotamia.[5]

GOLD

Gold statuettes and objects of 1100 B.C. found at Susa were really of electrum. Bars of gold and electrum, and gold leaf, have cuneiform inscriptions of the 12th century B.C.[6] Gold jewellery found in a bronze coffin of the Achæmenid Period (c. 350 B.C.) at Susa included a torque which weighed 385 gm. The ends formed lions' heads encrusted with mother-of-pearl (nacre), turquoise and lapis lazuli. Two bracelets weighed 98·8 gm. and 97 gm. A large gold bead collar weighing 152 gm., with ornaments of pendants, lapis lazuli, turquoise and carnelian (which spoils the effect), the stones being poorly worked, is based on late and decadent Egyptian art. Other gold jewellery comprised ear-rings decorated with lapis and turquoise, buttons incrusted with gems, a fine pearl collar (which probably originally had 400–500 pearls, but many fell in pieces after the discovery), small beads soldered on to a circular collar of gold, and discs.[7]

The so-called "treasure of the Oxus" consists of a number of gold and silver objects found in 1877, perhaps at Kabadian on a tributary of the Oxus, the styles resembling those of objects found at Susa. They are probably Achæmenian of the 5th–4th centuries B.C. and made in Bactria. The objects are cast, beaten or embossed, punched or engraved, and inlaid with flat stones (usually turquoise), this cloisonnée jewellery, probably of Sumerian origin, being especially developed in Achæmenid Persia.[8]

Gold vessels and gold working are mentioned in the Vendidad.[9] Although there is a little gold in the district of Fars,[10] there is little gold in the mountains of Iran.[11] Tradition speaks of gold from mountains near Takht-i-Suleimân,

---

[1] Contenau, Manuel, iii, 1397, 1431.
[2] Sarre, Kunst. d. alt. Persien, 19 ; A. U. Pope, Persian Art, 175 ; ib., Illustr. London News, 1930, clxxvii, 388, 444.
[3] Schrader-Jevons, 157.
[4] Strabo, XV, iii, 19, 733 C. ; Herodotos, ii, 167.
[5] Perrot-Chipiez, Persia, 469.
[6] De Morgan, Délég. en Perse, vii, 63, 68, 131 ; ib., La Délégation en Perse, 1897–1905, 123.

[7] Délég. en Perse, viii, 43, 48, and plates ii, iv, v ; 50, 51, 54, 56.
[8] O. M. Dalton, The Treasure of the Oxus, with other examples of early Oriental Metal Work, 2 ed., 1926, p. xliv.
[9] Rougemont, Âge du bronze, 36.
[10] Malcom, History of Persia, 1829, ii, 369.
[11] Maspero, H. anc., 1905, 556 ; ib., Empires, 453 ; Spiegel, Eranische Altertumskunde, i, 250; Rawlinson, Mon., iii, 146.

one of which is still called Zerreh Sharân or the mountain of the gold washers, and quartz, the matrix of gold, abounds in Kurdistan.[1] Hiuen Tsiang (A.D. 629) says Persia (Po-la-sse) produced gold, and that of Bolor (Po-lo-lo : Turkestan) was fiery red.[2] Strabo [3] says the Persians prized the metal because of its colour of fire [the sacred element].

Possible sources of Persian gold are Tibet, the Altai, India, Lydia, Armenia, Kabul and the district near Meshed (where it is still found).[4] The sand of the River Hyktanis in Carmania was said to be auriferous.[5] An inscription of Darius at Susa says he imported gold from Sardis and Bactria (which probably had an indigenous metallurgy).[6] In his time the alluvial goldfields of Dardistan produced vast quantities of the metal, paying a tribute of 360 talents (20,736 lb. = £1,078,272) to the king. The miners of Dardistan still keep fierce yellow Tibetan mastiffs to guard their houses, and these are perhaps the legendary "ants" which attacked seekers after the "ant gold" ($\chi\rho\upsilon\sigma\grave{o}s$ $\mu\upsilon\rho\mu\eta\kappa\acute{\iota}as$) thrown up in excavating their burrows.[7] In a late form of the legend given by Ibn al-Baitâr, the ants dig red sulphur for King Solomon.[8] Perhaps some resemblance between alluvial gold and the earth of anthills is the basis of the story, which also appears in Indian legends of ants (pipîlaka). Arrian [9], who reports the story from Megasthenes, is sceptical, but later authors profess to have seen the skins and even horns of the "ants."[10] Humboldt[11] and Wecker[12] identified the gold-digging ants with the ant-eaters, which Moorcroft found throwing up the auriferous sand of Tibet, and Blümner[13] also is sceptical as regards the Tibetan mastiff theory.

The Dardistan mines were practically worked out in the Roman Period, when large quantities of gold (nearly £550,000 a year) were sent from Rome to India, China and Arabia for Oriental luxuries.[14] Herodotos[15] says the Massagetæ in Central Asia had gold and bronze in abundance, but no iron or silver. In Northern Asia, beyond the land of the Scythians, there was much gold, which Herodotos[16], perhaps from the poem "Arimaspeia" of Aristeas (c. 550 B.C. ? ; known only in a quotation in Tzetzes), sceptically says was supposed to be guarded by griffins and the Arimaspi, men with only one eye.[17] The land of the Arimaspi, according to Humboldt[18], was the Urals or Altai, where there are remains of the mines of the Chudes, a nation which has disappeared. Finds of gold in the Altai and in Chudic graves have representations of the griffin. According to Tomaschek, Arimâçpô is the Iranian name of the Scythian tribes of the Pontos, and the Arimaspi were the Huns of China, Tibet and the Gobi Desert.[19]

[1] Rawlinson, Mon., ii, 294.
[2] Si-yu-ki, tr. Beal, ii, 278, 298.
[3] XV, iii, 18, 733 C.
[4] Rawlinson, Mon., iii, 158.
[5] Strabo, XV, ii, 14, 726 C.
[6] Kent, J. Amer. Orient. Soc., 1933, liii, 1 ; Rawlinson, Mon., iii, 146 ; Rougemont, 36 ; Dalton, Treasure of the Oxus, 1926, xvii f.
[7] Herodotos, iv, 13 ; Heliodoros, Æthiopica, 10, in Script. Erot. Græc., tr. Smith, 1889, 248 ; Ælian, H. Animal., iii, 4 ; Olympiodoros, in Berthelot, Coll., ii, 95.
[8] Ruska, Das Steinbuch des Aristoteles, Heidelberg, 1912, 161.
[9] Anabasis, v, 4 ; Indika, c. 15.
[10] Nearchos, fragm. 12, ed. Müller-Didot, Script. rerum Alexandri Magnum, Paris, 1846, 61 ; ψ-Kallisthenes, lib. ii, cap. 29, ed. Müller, op. cit., 85 ; Ausfeld, Der griechische Alexanderroman, ed. Kroll, Leipzig, 1907, 184 ;

Tomaschek, PW, iv, 2153 ; Oldenberg, Religion du Véda, tr. Henry, 1903, 58 ; H. G. Rawlinson, India, 1926, 18, 23 ; Hermann, PW, ix, 2236, 2245 ; Ziegler, ib., vii, 1918.
[11] Asie Centrale, 1843, 158, 221.
[12] PW, ix, 1301.
[13] Ib., vii, 1556.
[14] Rawlinson, India, 102.
[15] i, 215.
[16] iv, 13, 27, 116.
[17] Latham, DG, i, 213 ; Schmitz, DBM, i, 293 ; Berger, Erdkunde der Griechen, 48 ; Bethe, PW, ii, 877 ; Prinz, PW, vii, 1904 ; Schroeder, A. Rel., viii, 75 ; Pliny, vii, 2 ; Ælian, H. Animal., iv, 27 ; Lenz, Zoologie, 1856, 8.
[18] Asie Centrale, 1843, 158, 221 f.
[19] Tomaschek and Wernicke, PW, ii, 826 ; iii, 2203 ; Lenz, Zoologie, 8 ; Schrader-Jevons, 176; Carruthers, Unknown Mongolia, 1913, 49, 70 ; E. H. Minns, Scythians and Greeks, 4to., Cambridge, 1913, 7, 112, 244.

The tumuli, abandoned mines and ruined furnaces in the region between the Urals and the Yenissi basin, the home of the Chudes, were discovered about 1550, and were examined by Pallas. The mines contain petrified wood, indicating their great age. Gold is plentiful in the tumuli, but iron, which also occurs in the mine galleries, is rare, and the principal metals are pure copper and bronze with 10 per cent. of tin.[1] Objects in gold, electrum and silver, belonging to the earliest period, were found at Talyche in the Caucasus.[2] The old Persian name for gold, zaranya, is the same as the East Finnish and Sanskrit hiranya ; the Pahlawî and modern Persian name is zar (golden is zarik). The Ural-Altai name altun is Turko-Tartar and the Armenian oski is perhaps Caucasian.[3]

The Persian and Lydian kings possessed large amounts of gold, perhaps exaggerated in the Greek stories. The amount of gold and silver which Pliny says Cyrus took on the conquest of Asia would now be worth over 125 million pounds.[4] Gold was largely used for adorning palaces and for personal ornaments.[5] Arrian [6] reports that gold ear-rings set with jewels, and tapestries, were found in the tomb of Cyrus at Pasargadæ, a structure on seven (or six) stages which still remains in ruins and has been supposed to be modelled on the Babylonian stage tower.[7] Greek authors describe the throne of the Persian king as of gold with silver feet. The sceptre was of gold and the palace walls were plated with gold, but the thrones were probably really of wood covered with gold and silver plates. The palaces of Susa and Ekbatana were said to have been decorated with gold, electrum, bronze and ivory, and both gold and silver were extensively used for plate, furniture, etc.[8] Darius had a statue of his favourite wife in beaten gold, but although he amassed great quantities of gold in the treasury, Xerxes was even richer.[9]

The Persian guard of Xerxes wore scarlet tunics over breastplates of golden scales and carried spears with golden pomegranates at the lower end instead of spikes, and other troops had spears adorned with silver and gold pomegranates and golden apples. The Persian camp had great stores of coined and uncoined gold and silver and drinking cups of precious metals, to be used for bribing the Greeks. After the battle of Platæa (479 B.C.) the Greeks found vast quantities of gold, which they used partly as gifts to the temples of Delphi and Olympia, and for years after gold and silver were picked up on the field of battle.[10]

Alexander took from the royal treasury at Susa 40,000 talents of precious metal in the form of ingots, but only 9,000 in coined gold.[11] Strabo[12] states that most of the gold and silver was "manufactured" and coined money was minted only as required. Herodotos[13] speaks of vast issues of coin from the royal mints, and says the Persian king melted down the gold tribute in an earthen crucible ($\pi i\theta os$), in which he allowed it to cool, afterwards removing it as required.[14] Under Darius gold had 13 times the value of silver, whilst in Greece the ratio was 10 : 1 in Plato's time and long afterwards : the Persian

[1] P. S. Pallas, Voyages en differentes provinces de l'empire de Russie et dans l'Asie septentrionale, 4to., 1788–93 ; i. 425 ; ii, 160, 212 ; iii, 331 ; Stoll, Geschlechtsleben, 391 ; von Bibra, Bronzen, 1869, 18 ; Lenormant, Prem. Civ., i, 116, suggests that the Chudes are the Thyssagetians of Herodotos and his Arimaspi are the miners of the district worked in his time.

[2] De Morgan, Délég. en Perse, viii, 336 f.

[3] Schrader-Jevons, 119, 171.

[4] Blümner, PW, vii, 1555 ; Napier, Arts, 62, 75.

[5] Strabo, XV, iii, 18, 733 C.

[6] Anabasis, VI, xxix, 6.

[7] F. Justi, Geschichte des alten Persiens, 1879, 45.

[8] Rawlinson, Mon., iii, 205, 207, 216, 234 ; ψ-Aristotle, De mundo, cap. 6.

[9] Herodotos, vii, 27 f., 69 ; Perrot-Chipiez, Persia, 22.

[10] Herodotos, vii, 41 ; ix, 22, 41, 80, 83.

[11] Huart, La Perse antique, 1925, 122.

[12] XV, iii, 21, 735 C.

[13] vii, 28, 29.

[14] Herodotos, iii, 96 ; Regling, PW, vii, 979 ; Meyer, Alt., III, 89.

Empire paid Darius as tribute 7,740 Babylonian talents of silver, India 360 talents of gold dust.[1] Although the later Persians attributed the discovery of gold to Jemschid, and of gold and silver coinage to Dahhâk, both mythical ancient kings [2], silver and gold coins actually originated in Lydia. Darius coined money of very pure refined gold containing only 3 per cent. of alloy.[3] His coin, first issued in 516, was the daric (δαρεικός), a name perhaps derived from Darius, or the Persian darah, king[4] : the derivation from the Babylonian dariku [5] is doubtful.[6] The darics were sometimes called τόξοται on account of the device of an archer, still used by Alexander.[7] This was the attribute of Ahura Mazda, as it was still earlier of Assur. The darics were flattened lumps of metal, very thick, irregular and rudely stamped [8]; each contained 124 [9] or 130[10] grains, or 0·2788 oz. Troy = 134 grains of gold and 3,000 of them made a talent.[11] The silver stater, drachma, shekel or Medic siglos, twenty of which were equal to a gold daric, weighed 86 grains.[12]

Coinage was in general use only in Mediterranean countries and was unknown in the interior of the empire before Alexander. The provinces to the west of Lebanon, Amanus and Taurus had used currency before Darius, and Cyrus and Cambyses probably allowed the royal mint at Sardis to continue the issue of gold and silver staters. The mints established in Cilicia and Syria, probably also those at Tarsus and Tyre, seem to have coined only bronze. The double daric (gold tetradrachm) was not issued before Alexander.[13]

Although Persia adopted the Lydian gold and silver coinage, and electrum coins rapidly went out of use, electrum was again issued about 500 B.C. by a few cities, especially Cyzicus, Phocæa and Lampsacus. Their staters and hektæ (sixths), which were current in the Euxine, Asia Minor and Greece, are mentioned in Attic inscriptions from 434 B.C. In Pontos a Cyzicene stater was equivalent to 28 Attic drachmae of silver. In the 4th century B.C. electrum coins were largely used at Carthage and Syracuse.[14] The rise of gold coinage in Greece is supposed by Rostovtzeff[15] to have been due to the fall of Athenian commercial hegemony and the increasing commercial and political influence of Persia, whilst Gardner assumed that Athens took the lead.

SILVER

Silver objects of the Elamite Period (1100 B.C.) at Susa include statuettes, rings and a mask, probably originally covering a wood statuette : the face, of ancient Chaldæan type, perhaps represented the goddess Nana.[16] A fragment of a vase, perhaps of the Elamite Period (before 750 B.C.), was examined by Berthelot. It was disintegrated by salt and largely transformed into chloride, but the interior face had preserved its polish. The principal material was

[1] Herodotos, iii, 90 f., 95.
[2] Al-Tha'âlibî, Histoire des rois des Perses, tr. by Zotenberg, 1900, 12, 22; Firdawsî, Le livre des rois, tr. by Mohl, i, 25, 35.
[3] Herodotos, iv, 166 ; Perrot-Chipiez, Persia, 459 ; Lenormant, Monnaie, i, 187.
[4] Hoefer, Hist. de la Chimie, i, 55 ; cf. English "sovereign."
[5] Gardner, DA, i, 598.
[6] Liddell and Scott, Greek Lexicon, 1926, 370.
[7] Perrot-Chipiez, 460.
[8] Rawlinson, Mon., iii, 428 ; the development of coinage, from the electrum stater of Asia Minor, early 7th century B.C., to coins of Alexander, 325 B.C., is well shown, from

coins in the British Museum, in Plate LIX of Rostovtzeff, i, 196.
[9] Rawlinson, Mon., iii, 427.
[10] Gardner, DA, i, 597.
[11] Rogers, Hist. Persia, 1929, 109.
[12] Gardner, DA, i, 598 : giving the correct ratio 13 : 1 of gold to silver ; Rawlinson's figure, Mon., iii, 428, of 224–230 grains, does not agree.
[13] Perrot-Chipiez, Persia, 458, 462 ; Rogers, 109 ; Schrader, RL, 284 ; Lenormant, Monnaie, i, 89, 149.
[14] Gardner, DA, i, 715.
[15] Iranians and Greeks in South Russia, Oxford, 1922, 229.
[16] De Morgan, Délég. en Perse, 1897–1905, 1905, 69, 123 ; ib., Délég. en Perse, 1905, vii, 43, and plate.

silver chloride, with no arsenic or carbonate, but some iron, gold and copper. Analyses gave the following percentages: total silver 65·27 and 64·14, copper 2·95, gold 1·12, chlorine 16·98 and 16·72, sand 1·44, loss on heating in hydrogen 21·08. A second specimen gave : silver 63·60, copper 1·55, gold 0·34, chlorine 18·85, lime 4·58, iron oxide 0·27, carbon dioxide 5·34, water and loss 3·7. Another Elamite vase contained silver, lead, copper and arsenic ; a white metal vase, probably before the 5th century B.C., was of silver containing one-fifth of gold.[1] A silver ring from Luristan contained both gold and copper and is perhaps of native metal.[2]

A large cast silver dish weighing 589 gm., found in a bronze coffin of the Achæmenid Period (c. 350 B.C.) at Susa [3], is 148 mm. dia. and 44 mm. high. The work is very artistic ; the graver was used on the ornamentation and the circular form trued up on the lathe. Pieces of silver of the shape of crucibles occur on various sites [4], and Herodotos [5] says the metal was melted in this way for storage in the king's treasury. Boiled water from the Choaspes for the king's use was carried in silver flasks.[6] The Iranian name for silver was erezata,[7] and silver vessels and working are mentioned in the Vendidad.[8]

Silver plate was probably made in the Parthian Period, but it is doubtful if any existing piece is of this date. The Sassanian silver plate (bowls, dishes and cups) was an important article of commerce and travelled great distances along the paths of trade, the best examples (4th century A.D. and later) being found in Siberia and in the province of Perm.[9]

Aryandes, the Persian governor of Egypt, minted money of such pure silver that to the day of Herodotos there was none so pure as "the Aryandic": Darius, hearing of this, put Aryandes to death.[10] Coining gold money was a royal prerogative, although the satraps were usually allowed to coin silver.[11] A process of purification, probably cupellation with lead, was apparently in use at that time.

Darius obtained silver as tribute from Egypt[12], Cappadocia, Media, the sea coast and Carmania, where there were mines.[13] A little silver is found in Azerbaijan[14] and Tavernier describes mines at Karven, but the cost of working them was too great.[15] There is some silver in the mountains of Iran[16], and according to tradition it was procured from the mountains near Takht-i-Suleimân. Besides the mines of Kerman (Carmania), Armenia, Asia Minor and the Elburz, much silver was found at Kapan Maden near Kharput, and Denek Maden on the right bank of the Halys between Kaiseriyeh and Angora.[17] Abulfeda[18] describes very extensive mines in Khorassan at Panjhîr, in the country of Ghûr, where the earth was so much worked that it resembled a sieve. When lamps refused to burn the miners ceased work. Men became wealthy and also lost everything in a day : the first to strike a lode became the owner, and "they worked harder than devils to get there first." The Persian silver

[1] Berthelot, Arch., 85, 93.
[2] Desch, B.A. Rep., 1931, 271.
[3] Délég. en Perse, 1905, viii, 43, plates II and III.
[4] Regling, PW, vii, 979.
[5] iii, 96.
[6] Herodotos, i, 188 ; Rawlinson, Mon., iii, 226.
[7] Schrader, RL, 766.
[8] Rougemont, L'âge du bronze, 36.
[9] Dalton, Treasure of the Oxus, 1926, lxix, lxxv f. ; Sarre, Kunst d. alt. Persien, 32, 48.
[10] Herodotos, iv, 166.
[11] Breasted, Anct. Times, 186.

[12] Kent, J. Amer. Orient. Soc., 1933, liii, 1.
[13] Strabo, XI, xiii, 8 ; XV, iii, 14, 21 ; 524 C., 726 C., 735 C. ; Herodotos, v, 49; Rawlinson, Mon., iii, 146.
[14] Malcom, Hist. of Persia, ii, 369.
[15] Huart, Perse, 11 ; Contenau, Manuel, iii, 1510.
[16] Maspero, H. anc., 1905, 556 ; ib., Empires, 453 ; Spiegel, Eranische Altertumskunde, i, 250.
[17] Rawlinson, Mon., ii, 294 ; iii, 146, 159.
[18] Géographie, tr. Reinaud and Guyard, iii, 101, 201.

mines are mentioned by Hiuen Tsiang (A.D. 629) [1], who says the Persians used large silver coins : "they care not for learning, but devote themselves entirely to works of art."

COPPER AND BRONZE

Persian traditions attribute the discovery of copper to the mythical kings Jemschid or Hushank(Husheng).[2] The Aryans appear to have been acquainted with copper before their separation [3], the Sanskrit (áyas) and Iranian (ayah) names for the metal being the same.[4] Bronze is supposed to be mentioned as ayanh in the Avesta, the adjectives attached to the name suiting bronze rather than copper.[5]

Copper occurs (with gold) in the oldest layers at Susa [6], including very thin ($1\frac{1}{2}$–$2\frac{1}{2}$ mm.) mirrors [7], and axes or blades wrapped in linen of very fine texture.[8] Loftus [9] found a large piece of copper with green glazed bricks at Susa. The casting of copper was practised very early in Susa and the metal workers were so expert that Ur Nina (2880 B.C.) sent for them to carry out work in his country.[10]

ANALYSES OF COPPER AND BRONZE

An analysis of a copper axe from Susa gave[11] after separation of the outer crust : copper 98·70, iron 0·95, lead 0·44 in percentages, and traces of tin and antimony. Another, found in a deep layer, contained 99·12 per cent. of copper and a mere trace of nickel[12], the rest being non-metallic. Several copper and bronze objects from Susa of the Elamite, Achæmenid, etc., periods, were analysed by Berthelot[13] and by Desch.[14]

Object and Date	Cu	Sn	Fe	Ni	
1. Hoe ; Sargon I	—	tr.	1·34	0·12	Ag, Sb, As, tr. ; Pb, nil.
2. Axe ; Ḥammurabi . . .	—	tr.	2·9	0·45	As, Sb, tr. ; Pb, Ag, nil.
3. Elamite Period . . .	—	tr.	0·96	tr.	Sb, tr. ; Pb, nil.
4. ,, ,, . . .	—	tr.	1·6	0·3	Pb, Sb, nil.
5. ,, ,, . . .	—	tr.	1·05	tr.	Pb, tr. ; Sb, nil.
6. ,, ,, . . .	—	1·63	0·88	0·35	Pb, tr. ; Sb, nil.
7. Fragment, 1000–700 B.C. .	c. 100	tr.	tr.	—	Pb, tr. ; As, Zn, nil.
8. Wire, 1000–700 B.C. . .	96·91	tr.	—	tr.?	Pb, tr. ; Ag, Zn, As, nil.
9. Nail from temple, 1000–700 B.C. .	c. 100	0	tr.	0	Pb, Ag, Zn, Hg, As, P, Sb, nil.
10. Nail, 1000–700 B.C. . . .	c. 100	0	—	0	Ag, Pb, As, Sb, Zn, nil.
11. Object, 1000–700 B.C. . . .	c. 100	—	—	—	"pure copper."
12. Rod, 1000–700 B.C. . . .	c. 100	—	—	—	"pure copper."

[1] Si-yu-ki, tr. Beal, ii, 278.

[2] Firdawsî, Das Königsbuch, tr. Rückert, i, 9 ; al-Tha'âlibî, tr. Zotenberg, 1900, 12.

[3] Hoernes, Urg., ii, 216.

[4] Schrader, RL, 488.

[5] Lenormant, Prem. Civ., i, 129 ; Schrader-Jevons, 190.

[6] De Morgan, Délég. en Perse, 1897–1905, 1905, 80.

[7] De Morgan, Délég. en Perse, i, 60, 132, 134 ; xiii, 9, 12, 25, Figs. 33–34.

[8] Contenau, Civ. Assyr., 95.

[9] Travels, 352, 396.

[10] King, S. and A., 337 ; ib., H. Bab., 20 ; Langdon, CAH, i, 545 ; Loftus, Travels, 352 ; Delaporte, Mésopotamie, 68.

[11] Granger, in Délég. en Perse, xiii, 612.

[12] Marples, Ancient Egypt, 1929, 100.

[13] Arch., 86–99 ; cf. Petrie, Tools and Weapons ; nos. 7–22 in table.

[14] B.A. Rep., 1931, 271 ; nos. 1–6 in table.

Object and Date	Cu	Sn	Fe	Ni	
13. Shovel, 1000–750 B.C. . . .	98·7	0·3	0·2	0·3	—
14. Foot of statuette, 1000–750 B.C. .	98·6	1·0	0·3	tr.	—
15. Ring, 1000–750 B.C. . . .	89·8	5·7	0·9	tr.	Pb, Ag, Sb, As, Zn, P, nil.
16. Fragment of vase, Persian ?. .	89·46	{8·45 {8·60	0·53	tr.	{SiO$_2$, CaO, tr. ; As, { Ag, Pb, Zn, nil.
17. Bell-shaped door-knocker support ; King Šilkhak (13th century B.C.).	{85·20 {85·56	11·35	—	—	Pb, Zn, As, nil.
18. Javelin point, 1000–750 B.C. ? .	67·3	11·4	tr.	—	Not cast.
19. Irregular sheet, 1000–750 B.C. .	74·5	10·3	0·15	tr.	
20. Fragments of vases and plaques, 1000–750 B.C.	{52·27 {51·83	{19·80 {18·97	tr.	tr.	Pb 7·04, 7·37 ; Cl • 1·39 ; Fe$_2$O$_3$ 0·46; CaO, tr. ; Ag, Zn, As, CO$_2$, nil. A tin-lead bronze.
21. Bronze, Achæmenid . . .	82·7	13·9	—	—	Pb, 3·4 ; As, Ag, Zn, nil. Original composition as calculated from analyses.
22. Nails, 1000–750 B.C. . . .	77·2	16·1	tr.	0 ?	Ag, Pb, Zn, As, Sb, P, nil.

The nickel found in several fragments was probably derived from some mineral. In scoria from No. 7 (nearly pure copper), which were free from zinc and cobalt but contained traces of arsenic, tin and lead, Berthelot found : copper 67·79 and 67·59, nickel oxide 2·30 and 2·43, iron oxide 2·59, chlorine 9·36 and silica 4·21.

Statuettes of Sulgi (2276–2231) and of the 12th century B.C. in foundations of a temple at Susa, also axes and lance-heads of the Second Elamite Period (2555–1955) are described as bronze by de Morgan.[1] A bronze axe of c. 2000 B.C. is similar in shape to a Hittite votive axe in ivory.[2] There are 18th century B.C. "bronze" castings from Susa, which were too large and heavy for Ašurbanipal to remove, also bows, daggers, horns, etc. [3], and axes, etc., of the Elamite Period from Tepe Aly-Abad.[4] These are all of doubtful composition.

Large quantities of "bronze," in all 5,000 kgm., were found at Susa and probably much was removed in ancient times or remains to be discovered. A statue of Queen Napir-Asu (about 1500 B.C.) weighed 1,750 kg. The Susians were very expert at bronze working, and this statue is solid, but was filled with molten metal in portions, a mass of metal in the form of a cylinder being pressed into the middle. The method used in casting large pieces free from flaws cannot be stated definitely, but the metal must have been poured simultaneously from a large number of crucibles. Bronze axes with curved blades and round bars with square ends for fixing in walls are of 1100 B.C., and there are large bronze columns of the 10th century B.C.[5]

Bronze (and gold) plates were used as wall coverings in palaces, and a piece

[1] Déleg. en Perse, 1897–1905, 1905, 123, 126 ; ib., Déleg. en Perse, vii, 37, 73.

[2] Contenau, Manuel, ii, 925.

[3] De Morgan, Déleg. en Perse, 1897–1902,

1902, 88 ; ib., Déleg. en Perse, i, 107, 118, 122, 163, plates 12–13.

[4] Déleg. en Perse, 1905, viii, 144.

[5] Ib., 245 ; King, S. and A., 337 ; Contenau, Manuel, ii, 914, 923.

of bronze from a door covering was found at Susa.[1] Bronze statuettes of
c. 2500 B.C. were plated with gold by burnishing over the surface a thin gold
sheet, the two edges of which were then hammered into a groove made in an
inconspicuous place, i.e. a kind of damascening process, also much used later in
Babylon.[2] Remains of a jeweller's balance (10th century B.C.) with copper
pans and two sets of weights (for buying and selling ?) were found at Susa.[3]

SOURCES OF PERSIAN COPPER

Copper ores are found in the mountains of Iran and in Kurdistan [4], and
Masûdî says copper was produced at Oman [5], which is thought to have been
the source of early Sumerian copper (see p. 243). There is an old mine at
Arghana Ma'adan in Kurdistan, which was worked by the Germans during the
Great War, and one at Sabzawar in Afghanistan.[6] De Morgan considered
that the copper at Susa came from the Caucasus.[7] There were old mines
in Carmania [8], Cyprus and Armenia are possible sources [9], and Darius obtained
copper from Egypt.[10] It is found in the Altai region, where there are remains
of furnaces and very hard bronze of the Chudes.[11] Some copper is still mined
in Khorassan and is found between Astrabad and Schrund, but most of the
metal is imported[12], although some is worked in the region of Zawar and on
the eastern border of the great desert at Kala-Seb-Zarre.[13] The easily reduced
carbonate ores on the plateau near Binamar probably served for the production
of the Perso-Caucasian bronze swords of Talyche.[14] The Persian copper
mines are mentioned by Hiuen Tsiang in 629.[15]

To the south of Trebizond, near Erzerum in Armenia and also at Diarbekir
in the upper basin of the Tigris, there are vast accumulations of mining and
metallurgical refuse, and numerous excavations mark the sites of an important
ancient, probably Subarian, copper industry.[16]

Two fragments of bronze coffins from an old Chudic grave in the Altai contained 80·27 and 73·00 per cent. of copper and 19·66 and 26·74 per cent. of
tin, respectively. A ring from the mouth of the Don (pre-Christian era)
contained 91·00 per cent. of copper and 9·00 per cent. of zinc, with traces of
iron, nickel and antimony. Bronzes from Scythic graves on the Dneiper, from
the 4th century B.C., often contained appreciable amounts of lead, e.g. up to
7 per cent., and occasionally zinc, but are mostly fairly normal tin-bronzes
(up to about 16 per cent. of tin).[17] Môsul bronzes, so called after the place
where they were sold, were an important article of export under the Chalifate.[18]
Copper is now mostly worked in Persia at Kashan.[19]

[1] Huart, Perse antique, 109.
[2] Contenau, Manuel, ii, 858 ; iii, 1171.
[3] De Mecquenem, Revue des arts
asiatiques, 1929–30, vi, 86.
[4] Maspero, H. anc., 1905, 556 ; ib.,
Empires, 453 ; Spiegel, Eranische Altertumskunde, i, 250 ; Rawlinson, Mon., ii,
294 ; iii, 146.
[5] Wilson, Persian Gulf, 1928, 58.
[6] Contenau, Manuel, iii, 1510.
[7] De Mecquenem, op. cit., 80 ; on
minerals and ores in Persia, see A. F. Stahl,
Chem. Zeit., 1894, xviii, 3, 487, 882, 1424,
1568.
[8] Strabo, XV, ii, 14, 726 C.
[9] Rawlinson, Mon., iii, 159.
[10] Kent, J. Amer. Orient. Soc., 1933, liii,
1 f.
[11] Minns, Scythians and Greeks, 1913, 7,
244.

[12] G. N. Curzon, Persia and the Persian
Question, 2 vols., 1892, ii, 510, 517, 522.
[13] Huart, Perse, 11 ; Stahl, op. cit.
[14] De Morgan, Délég. en Perse, viii, 337 ;
BMG Bronze Age, 1920, 175 f.
[15] Si-yu-ki, tr. Beal, ii, 278.
[16] Gowland, J. Anthropol. Inst., 1912, xlii,
245 ; von Oppenheim, Tell Halaf, 1931,
189.
[17] E. von Bibra, Bronzen, 1869, 98, 108,
113, 152 ; also quoting C. C. T. Fr. Göbel,
Ueber den Einfluss der Chemie auf die
Ermittlung der Völker der Vorzeit, Erlangen, 1842.
[18] Woermann, Geschichte der Kunst,
Leipzig, 1915, ii, 388.
[19] R. Murdoch Smith, Persian Art
(South Kensington Museum Handbook),
1876, 28.

TIN

That aonya in the Vendidad means tin is doubtful and the definite mention of tin in this work is a later interpolation.[1] Tin (arjiz) is mentioned in the Dâmdâd-Nask (c. 500 B.C.) [2], which, it is true, is known only in a very late form. Although the supposed deposits of tin in the Caucasus are regarded as fictitious [3], pure tin arm-rings were found in old Caucasian graves [4] and bronzes and alloys rich in tin in Iron Age graves [5]:

	I	II	III		IV
Copper . . .	80·1	91·34	81·37	Lead . . .	85·53
Tin . . .	17·4	5·35	14·05	Tin . . .	6·35
Oxygen, etc. . .	2·5	3·31	4·58	Oxygen, etc. .	8·12

Specimens I, II and III were free from lead, zinc, silver, arsenic and antimony, and IV from copper.

The ancient fame of the Caucasus as a seat of metallurgical invention was partly due to old legends such as that of the Argonauts, partly to its supposed position as one of the boundaries of the world and partly to the neighbourhood of tribes like the Tibareni, Moschi and Chalybes, who worked metal and trafficked with it to foreign countries.[6] A region of Bamian in the Hindu Kush, anciently called Iberia and the modern Georgia, is said to show traces of ancient tin mines, but neither later classical authors nor Moses of Chorene (A.D. 450) mention them, so that if they existed they must have been abandoned from the 1st century A.D.[7] Greek authors, however, call the Paropamisus "the Caucasus" and the Hindu Kush is rich in minerals, especially iron.[8]

Ancient Persian tin was probably obtained from Khorassan, the old Drangiana or Sarangia and modern Seistan [9], although the mines appear to be exhausted and Curzon[10] does not mention tin, but only lead, as a product of Khorassan. This region, south-east of the Caspian and Aral seas, the mountains of Khorassan and Transoxania, the Paropamisus, Drangiana, Eastern Iran and the valley of the Etymandrus or Erymandrus (modern Hilmend; one of the four rivers of Paradise), was formerly very fertile and well watered by canals, attributed to Rustum, which existed to the Middle Ages.[11] The region is so extensive and relatively imperfectly known that it would be better not to say dogmatically that no tin occurs in any part of it. Tin ore is found in the Trans-Baikal provinces, the Urals and other parts of Russia, and in Finland, but is not worked.[12] The objection to Persia as an early source of tin, on the ground that bronze should then be known earlier in Babylonia than in Egypt[13] is invalid, since tin-bronze was extensively used in an early period at Ur (see p. 246).

[1] Lenormant, Prem. Civ., i, 129; Rougemont, 36.
[2] Reitzenstein and Schaeder, 223.
[3] See p. 255.
[4] Belck, Verh. Berlin Ges. f. Anthropol., 1893, 64.
[5] Berthelot, Arch., 13; cf. BMG Bronze Age, 177.
[6] BMG Bronze Age, 176.
[7] James, DG, ii, 9; Lenormant, Prem. Civ., i, 129.
[8] Smith, DG, i, 571; Vaux, ib., ii, 552; Tozer, Hist. of Ancient Geography, 133.
[9] Vaux, DG, i, 787, 850; ii, 552;

Rawlinson, Mon., iii, 158; Hilzheimer, RV, xiv, 183 f.—metals in Asia Minor; V. G. Childe, Most Anct. East, 1934, 187—all relying on Strabo, XI, x, 1, 516 C.; XV, ii, 10, 725 C.
[10] Persia, ii, 510 f.
[11] Lenormant, Prem. Civ., i, 129; Vaux, op. cit.; Tomaschek, PW, v, 1666—who quotes only Strabo for the tin mines; Kiessling, PW, vi, 806.
[12] Mantell, Tin, New York, 1929, 60.
[13] R. Hennig, Rheinisches Museum für Philologie, 1934, lxxxiii, 166—where Persian tin is said to be mentioned by Ktesias; I have been unable to confirm this.

BRASS AND ZINC

Brass was probably made in Persia in the Achæmenian Period, since Darius is said to have had a bowl like gold in appearance but distinguishable by its unpleasant smell.[1] Brass is not named in the Vendidad, but is mentioned in a later commentary.[2] The Persian name for brass, pirin, later birinj (Kurdish pirinjok, Armenian plinj), which has not passed into any other language, is probably not the origin of the word bronze.[3] Birinj also means rice ; the Greek name ὄρυξα is usually regarded as a loan word from Sanskrit vrihi, βρίζα is rye.[4] Laufer thinks brass was probably first made in Persia from the rich deposits of calamine, but brass occurs in Palestine in 1400 B.C.[5]

Hiuen Tsiang (A.D. 629) says the metal plates of a colossal statue of Buddha were made from copper and a stone of golden-yellow colour, found in Persia and elsewhere and called t'ou shi, which was probably calamine.[6] Ibn al-Faqîh in 902 describes the zinc mines at Demawend (Dumbawand) in Kerman, the ore being (as it is still) a government monopoly, and al-Jawbarî in 1225 gives the method of smelting calamine (al qalamî).[7] The "copper" of Damascus which Nasîr ibn Khusraw says [8] is like gold was probably brass.

According to Marco Polo [9] (c. 1300) in the Moslem town of Kobiam in Kerman, Persia, "there is much andanico [steel] and zinc. The inhabitants procure tutia [zinc oxide] which makes an excellent collyrium, together with spodium, by taking the crude ore from a vein and putting it into a heated furnace, over which they place an iron grating of small bars close together. The smoke ascending from the ore in burning attaches itself to the bars, and as it cools becomes hard. This is tutia ; the heavy part, which remains as a cinder in the furnace, becomes spodium." This is merely the old process described for Cyprus by Dioskurides and Galen[10], and there is no mention of metallic zinc. J. Bontius[11] was told by Persian and Armenian merchants that "the earth from which tutia is made is found in great quantities in Kerman." Bontius says that Teixeira in 1610 reported that tutia (tutyah) was used as a depilatory and was made in Kerman from an ore from the mountains, which was kneaded with water, baked in crucibles in a pottery kiln and sent in boxes to Hormuz for sale. It was not the ash of a plant, as Garcia da Orta had reported.[12] The zinc ores of Persia are no longer worked, the metal being imported from Russia, etc., for the manufacture of brass, which is mostly carried out at Ispahan.[13]

General Houtum-Schindler[14] first suggested that tutia, the mediæval Latin name for zinc oxide, is the Arabicised form of the Persian dudhâ, smokes, since it was formed as a sublimate in brass furnaces. Tutia in Kerman, however, now means vitriol, generally blue vitriol (nila-tutia) ; tutia-i-salz, green tutia, is green vitriol ; tutia-i-zard, yellow tutia, is alum (? white vitriol) ; tutia-i-safid, white tutia, is "an argillaceous zinc ore."[15] The meaning vitriol for tutia

[1] ψ-Aristotle, De mirab. auscult., cap. 49 ; cf. J. J. Becher, Physica subterranea, Leipzig, 1738, 279.
[2] Rougemont, L'âge du bronze, 36.
[3] Laufer, Sino-Iranica, 512, 513, 515.
[4] Ib., 373, 581.
[5] See section on Palestine.
[6] Si-yu-ki, tr. Beal, 1884, i, 51, 166, 177, 197 f. ; ii, 45 f., 174, 272.
[7] Laufer, 512 ; Wiedemann, Sitzb. Phys. Med. Soc. Erlangen, 1911, xliii, 219 ; ibn al-Baitâr, Treatise on Simples, cap. 437, tr. Leclerc, Notices et extraits des MSS., xxiii, 322.
[8] Sefer Nameh, tr. Schefer, Paris, 1881, 73, 81, 152.

[9] i, 20 ; Everyman tr., 71 ; cf. Lippmann, Abh., ii, 264.
[10] See p. 368.
[11] Account of the Diseases, Natural History and Medicines of the East Indies, 1769, 180.
[12] Simples and Drugs of India, tr. by Markham, 1913, 412 : Garcia really distinguishes it from tabaschir.
[13] Curzon, Persia, ii, 522 ; Polack, MGM, ii, 152 ; R. M. Smith, Persian Art, 1876, 28.
[14] J. R. Asiat. Soc., 1881, 497 ; Cordier, in Yule's Marco Polo, 1903, i, 127.
[15] Cordier, op. cit. ; Bonnin, Tutenag and Paktong, Oxford, 1924, 3 ; Balfour, Cyclopædia of India, 1885, iii, 963.

is said to be derived from the Sanskrit tuttha [1], which is perhaps not related to the Persian tutiya.[2] Ruska [3] thinks tutia may be a foreign word in Persian and Sanskrit, and Thompson [4] derives it from the Assyrian tuskû, which is probably tin oxide, not zinc oxide. The Tuḥfat ul-Mûminîn (A.D. 1669) mentions the yellow and blue minerals (vitriols), tutia (zinc oxide) obtained from brass furnaces and tutia-i-qalam (carbon black) made from roots, as constituents of collyrium (kohl), and the name tutia seems to have meant collyrium in general.[5] The name tutia first occurs, perhaps with the meaning of vitriol, in the 9th century A.D. Lapidary of ψ-Aristotle.[6] Avicenna (11th century) [7] and Geber (? 13th century) [8] also use the word. A supposed earlier mention (as τυθία) by an anonymous Greek chemist of the 1st century A.D. [9] is doubtful, but the material was, of course, well known at that time.

An alternative place of origin of brass and zinc is China, where there are also deposits of zinc ores.[10] A primitive method of extraction of zinc is described and illustrated in the Chinese work, " Tien kong kaï wu," of 1637.[11] The process is like that obscurely mentioned by Strabo[12] for the production of "mock-silver" (ψευδαργυρος)—the first definite mention of metallic zinc—and the same imperfect process is described by Löhneyss as in use at Goslar in the 17th century.[13] Brass was well known in Rome in the 1st century B.C. ; it could have reached China from this source, if it were not known there before, and Laufer's assumption that knowledge of brass passed from Persia through Samarkand to China is without definite proof. Brass is definitely mentioned in China in the 6th century A.D., but the assumption that the Chinese t'ou shi (t'ou stone) is brass[14] seems doubtful, since the usual meaning is calamine.

A supposed mention of zinc, as roy or ruj, in Abu Mansûr Muwaffaq (975 A.D.) is doubtful, as this word usually means copper.[15] According to Hommel[16] the Persian name for zinc is jast or jasta, the Sanskrit jasada.[17] Al-Dimešqi (c. 1300 A.D.) mentions zinc as coming from China but not well known : Jâbir ibn Hayyân (8th–9th centuries ?), he says, "speaks of it in his writings, but I know of no other author who describes it."[18] Al-Dimešqi gives its Persian name as séfîd-roû, i.e. isfîd-ruj, white copper.[19] Other Persian names are isbadâri and sbiadâr[20], the origin of the European name spiauter, spelter, which occurs in Boyle.[21]

[1] Ray, History of Hindu Chemistry, Calcutta, 1909, ii, 25 ; Garbe, Die indischen Mineralien, Leipzig, 1882, 51, 59 ; with 24 names.
[2] Laufer, Sino-Iranica, Chicago, 1919, 513.
[3] Steinbuch des Aristoteles, 1912, 175.
[4] Chemistry of the Anct. Assyrians, 30, 38.
[5] Cordier, in Yule's Marco Polo, 1903, i, 126 f. ; Frater Angelus, Pharmacopœa persica, Paris, 1681, caps. 40 f., 261, 537, 894–905, 1070 ; pp. 18 f., 93, 177, 276–80, 341.
[6] Ruska, Steinbuch, 175.
[7] Liber canonis, tr. by Gerard of Cremona, Lugduni, 1522, 126 ; Beckmann, H. Invent., ii, 36.
[8] Gebri Regis Arabum Chymia, Dantisci, 1682, 75.
[9] Berthelot, Coll., ii, 424 ; Lagercrantz, Studien zur Geschichte der Chemie, ed. Ruska, 1927, 15.
[10] Hommel, Z. angew. Chem., 1912, 99 ; 1919, 73 ; Frantz, Berg-und-Hüttenmänn. Z., 1881, xl, 231, 251, 337, 376, 387 ; K. B. Hofmann, ib., 1882, xli, 505 ; Neumann, Z. angew. Chem., 1902, xv, 512, 1217 ; Gowland, J. Inst. Metals, 1912, vii, 43.
[11] S. Julien and P. Champion, Industries anciennes et modernes de l'empire Chinois, 1869, 46, plate II ; F. de Mély, Les lapidaires Chinois, 1896, p. xxxii ; G. L. Staunton, Account of an Embassy . . . to the Emperor of China, 4to., 1797, ii, 540 ; J. F. Davis, The Chinese, 1836, ii, 246 ; Partington, Everyday Chemistry, 1929, 73, Fig. 68.
[12] XIII, i, 56, 610 C. ; Leaf, Strabo and the Troad, Cambridge, 1923, 287.
[13] Bericht von Bergwercken, Leipzig, 1690, 83.
[14] Laufer, Sino-Iranica, 512 ; Lippmann, Alchemie, ii, 143.
[15] Diergart, MGM, ii, 150.
[16] Z. angew. Chem., 1912, 99 ; Chem. Zeit., 1912, 905, 918.
[17] Bucher, iii, 46.
[18] Cosmographie, tr. Mehren, Copenhagen, 1874, 60.
[19] Cf. Bonnin, Tutenag and Paktong, Oxford, 1924, pp. x, 6, 18.
[20] Karabacek, q. by Wiedemann, Sitzb. Phys. Med. Soc. Erlangen, 1905, xxxvii, 403.
[21] The Origin of Fluidity and Firmness, Works, ed. Shaw, 1725, i, 316 ; Bonnin, 16.

Metallic zinc was used in later Persia for the preparation of various alloys with copper, previously made by mixed smelting.[1]

	Red	Yellow	White	Malleable Brass	Beidri	Koftgari
Copper	80–97	50–70	20–55	60	3	Composition
Zinc	20–3	50–30	80–45	40	94 (lead and tin 3)	not given.

The alloy haftjus, of one-seventh (ἑπτα) each of gold, silver, copper, iron, tin, lead and zinc, is still used in Persia for beads worn round the neck as amulets, but the amulets of "seven metals" with astrological signs really contain only tin, lead and zinc.[2]

It is possible that zinc compounds were used as pigments in Persia, since a Persian account of the Great Hall of Xerxes says that "in ancient times persons ascended to the summits of these columns, now fallen, and took earth and clay therefrom, which they crushed and found amongst it Indian tutty, useful as a medicine for the eyes."[3] This may, however, have been the green patina of copper (or perhaps of brass ?), which would serve this purpose. The later Persian brass was of a good golden colour.[4] Besides a considerable amount of bronze at Khotan, a brass ornament was found at Kuchâ.[5]

IRON

An iron stiletto found with a copper spoon in a very old grave (3000 B.C.) at Samarra, the metal of which is free from nickel [6], may be an accidental intrusion.[7] Objects found at Susa include a lump of iron [8], an iron bar—both of uncertain date—an oxidised iron ring 30 mm. dia. (1000–750 B.C.) with a small quantity of copper patina[9], and a blade and "armature," with iron nails, etc., in temple foundations not later than 1100 B.C.[10] Iron blades, lance-heads, etc., in old tombs at Talyche, on the borders of the Caspian, perhaps date to 2500 B.C. In this region a Bronze Age came to a sudden end by the irruption of a new race with iron weapons, probably from the Caucasus, where iron was in use as early as 1300 B.C.[11] Extensive remains of early iron workings, which were doubtless the source of the vast number of iron implements and objects found in the plain near Persepolis, occur in Northern Persia near Papa, between Kerman and Shiraz, and great mounds of "prehistoric" iron slag in the Karadagh district. Nineveh may have obtained some iron from these regions.[12]

Neither copper nor iron is mentioned in the Vendidad, but a later commentary refers to iron and steel, including arrow-heads[13], and an iron pot is mentioned in the Yasna.[14] Firdawsî[15] and al-Tha'âlibî[16] refer the discovery of iron and steel to the mythical kings Husheng (Hushank) and Jemschid.

[1] Bucher, iii, 46.
[2] S. Seligmann, Der böse Blick, 1910, ii, 6; ib., Die magische Heil- und Schütz-mittel aus der unbelebten Natur, Stuttgart, 1927, 156, 174, 176, Fig. 50.
[3] Fergusson, q. by Thompson, Chem. of Assyr., 1925, 40.
[4] A. U. Pope, Intr. to Persian Art, 1930, 186.
[5] Aurel Stein, Innermost Asia, 3 vols., Oxford, 1928, ii, 828.
[6] Herzfeld, Die Vorgeschichtliche Töpfereien von Samarra, 1930, pp. vi, 3, 5.
[7] Childe, Most Ancient East, 1934, 313.

[8] De Morgan, Délég. en Perse, i, 107.
[9] Berthelot, Arch., 91, 94.
[10] Délég. en Perse, vii, 45 f.
[11] Ib., viii, 279, 339, 342.
[12] Gowland, J. Anthropol. Inst., 1912, xlii, 282.
[13] Rougemont, L'âge du bronze, 36.
[14] Horn, Pers. Litteratur, 22.
[15] Königsbuch, tr. Rückert, i, 17, 49; ii, 459, 497; Livre des rois, tr. Mohl, i, 25, 35.
[16] Hist. des rois des Perses, tr. Zotenberg, 32, 38.

Firdawsî speaks of swords, lances, etc., of Indian and of Chinese steel and of damascened blades and weapons hardened with blood.[1] Al-Tha'âlibî also mentions the blacksmith Kahweh, who fought the rebellious Afridun in antiquity and whose leather apron was used as a banner for the nation until it was captured by the Arabs at the battle of Qadisîya (A.D. 636).[2] According to legend, Alexander the Great built walls from masses of iron cemented with molten bronze to keep back the savages of Gog and Magog from the far north.[3]

The early Persians probably used iron and steel, a knowledge of which they may have derived from India[4] ; Artaxerxes (464–425 B.C.) is said to have given Ktesias a present of a sword of Indian steel, apparently a "damscened" blade [5], and Roman authors praise the Parthian (Persian) iron as second only to that of the Seres (Chinese ?).[6] Scale and chain armour in China in the Han Period were probably based on Parthian models.[7] The name Seres meant "silk people" and had no ethnographic significance[8] ; the usual meaning is the people of North China.[9] An iron battle-axe plated with gold of the 6th century B.C., found with gold objects at Kuban, is probably Persian.[10] The Persian troops in 480 B.C. had iron scale armour[11] : the spear- and arrow-heads were either of bronze or iron, the javelins had iron tips and the horses were protected by steel armour.[12] Suidas (10th century A.D.) says the Parthian knights were so encased in armour that they seemed "all iron" (ὅλον σιδηροῦν).[13] Heliodoros[14] describes the scale armour and couched lances of the Persian cavalry, the horses also carrying armour, and Persian armour and swords are always praised by mediæval authors. The armour worn in the later period (e.g. Khosroes II) was probably true mail with links riveted together : stirrups are doubtful, and are first definitely mentioned as σκάλαι by the Byzantine Emperor Maurice, 582–602, about which time they were generally known in Europe.[15] The use of mail was confined to a few warriors only.[16]

Iron is frequently mentioned in the post-exilic book of Daniel, which is under strong Persian influences, but "steel" in other parts of the Authorised Version of the Bible usually means copper.[17] In the Avesta, ayanh, originally copper or bronze, gradually assumes the meaning of iron.[18] In modern Persian, steel is pûlâd, from which are derived the Mongolian bolot, the Russian bulatu and names in Tibetan, Armenian, Ossetic, Grunsinian and Turkish[19]; a later name, andun, is found in many Caucasian (Ossetic) languages[20], and the mountainous region of the Caucasus, probably producing iron as early as 1500 B.C., is perhaps the original home of iron.[21] Some iron (mouth-piece of vessel, cauldron) was found

[1] Rückert, i, 105, 129 ; ii, 323, 394, 459.
[2] Justi, Geschichte des alten Persiens, 1879, 30 f. ; de Gobineau, Histoire des perses, 1869, i, 95, 161.
[3] Al-Tha'âlibî, 32, 38, 441 ; al-Dimešqi, tr. Mehren, 30 ; Horovitz, Koranische Untersuchungen, 1926, 150.
[4] Freise, Geschichte der Bergbau, 1908, 120.
[5] Zippe, Geschichte der Metalle, Vienna, 1857, 115.
[6] Pliny, xxxiv, 14 ; Dyer, DG, ii, 968.
[7] Huart, Perse antique, 242 ; cf. Hermann, PW, IIA, ii, 1678.
[8] Krause, Geschichte Ostasiens, Göttingen, 1925, i, 286, 317.
[9] E. H. Warmington, Commerce between the Roman Empire and India, Cambridge, 1928, 71, 129.

[10] Rostovtzeff, Iranians and Greeks in South Russia, Oxford, 1922, 50.
[11] Herodotos, vii, 61, 62 ; Strabo, XV, iii, 19, 733 C.
[12] Rawlinson, Mon., iii, 175, 177, 178.
[13] Lenz, Zoologie, 1856, 34.
[14] Æthiopica, ix ; Script. Erotici Græci, tr. Smith, 1889, 217.
[15] Dalton, Treasure of the Oxus, 1926, p. lxviii.
[16] Ed. Meyer, EB11, xxi, 207.
[17] Smith, Concise Dict. Bible, 1865, 900 ; Beck, Gesch. des Eisens, i, 170.
[18] Schrader-Jevons, 203.
[19] Schrader, RL, 796 ; Laufer, Sino-Iranica, 575.
[20] Schrader-Jevons, 204.
[21] Zippelius, MGM, i, 168 ; Hoops, i, 549 ; Hoernes, Urg., ii, 224—Armenia ; Forrer, RL, 416.

at Khotan.[1]   There are scarce finds of ornamental iron in 13th–11th century B.C. graves at Kuban, near Tiflis.[2]

The name andun is derived from the Persian and Arabic hundwaniy or hundwan, "Indian," so that the names ondanique, andaine, etc., used by mediæval authors, mean "iron of Indian origin." Avicenna (11th century) and al-Idrîsî (12th century) both mention it as hindiah (= Indian [iron]).[3] The Chinese name pin for a variety of iron is derived from the Iranian spaina, which in the Pamir dialect is spin.[4]   The process of making damascened blades, which is now carried out at Ispahan, involves heat treatment and the application of some preparation to the finished blade to bring out the markings.[5]

Iron ore occurs with "native steel" (?) in Kurdistan and in the mountains of Iran, also in huge boulders and nodules in Persian territory.[6]   Masûdî (10th century A.D.) mentions the iron mines on the coast of the Abyssinian sea in the countries about Kerman (Carmania) [7], where the best iron and steel (used for large polished mirrors in Kerman and Cobinan) were made.[8] These mines were not worked after the Arabic Period.[9]   The minium ($\mu\lambda\tau o s$) of Carmania[10] was probably iron ochre.

NICKEL AND COBALT

Nickel and cobalt ores occur at Kamsur[11] and a coin of King Euthydemos (235 B.C.) found in Bactria contains 77·585 of copper, 20·038 of nickel, 0·544 of cobalt, 1·048 of iron, 0·038 of tin, 0·090 of sulphur (all percentages) and a trace of silver.[12] The alloy is the same as the Chinese pai-t'ung, "white copper," pronounced in Canton paktong and usually given, erroneously, as pakfong.[13] It probably came from farther east ; there were Buddhist priests in Bactria in the 1st century B.C., if not earlier.[14]   The cobalt mines of Kashan are really iron mines, which are still worked.[15]

LEAD

Specimens of the Elamite Period from Susa include a cuboid of almost pure lead with a hole in one of the six faces ; a piece of oxidised sheet (1000–750 B.C.) ; statuettes, rings, small serpent heads, etc., of total weight 620 gm. found in a temple foundation, and a piece of lead pipe about 4 in. long, the pommel of a rod found with silver objects, but no gold, at Musyan.[16] Lead as a material for vessels is mentioned in the Vendidad.[17] According to Xenophon the Persian slingers threw stone balls only, whilst the Rhodians threw lead balls twice as far.[18] Nasîr ibn Khusraw, who visited Lahsa (Al-Ahsâ) in A.D. 1051, found that lead in kufs (baskets or sacks) weighing 6,000 dirhems was used as money, but its export was forbidden.[19] A small quantity of lead was found by Aurel Stein[20] near Domoko. Lead ores occur in the mountains

[1] Aurel Stein, Innermost Asia, 3 vols., Oxford, 1928, i, 108, 149.
[2] Hoernes, Urg., ii, 224 ; ib., Urzeit, ii, 116 ; Borovka, Scythian Art, 1928, 75.
[3] Yule, in ed. of Marco Polo, 1903, i, 93, 215.
[4] Laufer, Sino-Iranica, 515.
[5] Beckmann, H. Invent., ii, 331 ; R. Murdoch Smith, Persian Art, 1876, 20, 28, 30 ; Guertler, Chem. Zeit., 1915, Refer. 71 ; Gsell, 88 ; Feldhaus, Technik, 179.
[6] Rawlinson, Mon., ii, 294 ; iii, 146, 159 ; Maspero, H. anc., 1905, 556 ; ib., Empires, 453 ; Spiegel, Eranische Altertums., i, 250 ; cf. Stahl, Chem. Z., 1893, xvii, 1910.
[7] A. T. Wilson, The Persian Gulf, Oxford, 1928, 58.

[8] Yule, Marco Polo, i, 90, 125, 212.
[9] Curzon, Persia, ii, 518.
[10] Strabo, XV, ii, 14, 726 C.
[11] Curzon, ii, 519.
[12] Flight, Pogg. Ann., 1870, cxxxix, 507.
[13] Bonnin, Tutenag and Paktong, Oxford, 1924, 18 f., 25 f.
[14] Spiegel, Eran. Alt., i, 671.
[15] Stahl, Chem. Z., 1893, xvii, 1910.
[16] Berthelot, Arch., 87, 99 ; de Morgan, Délég. en Perse, 1897–1905, 1905, 123 ; ib., Délég. en Perse, vii, 72, 91.
[17] Rougemont, L'âge du bronze, 36.
[18] Huart, Perse antique, 53 ; Rawlinson, Mon., iii, 176.
[19] Wilson, Persian Gulf, 88.
[20] Innermost Asia, 3 vols., Oxford, 1928, i, 133.

of Iran, Bactria, Armenia, Kerman, Afghanistan, Transoxania [1], Fars, near Neyriz, near Murgab, Kurdistan and in Khorassan.[2] Demawend and Rêy supplied the Arabs with lead oxide (litharge) [3] and Ispahan with white lead.[4]

ANTIMONY

Antimony sulphide (stibnite) occurs in Armenia, Afghanistan, Kurdistan and Media.[5] In the Abbâsid Period it was exported from Old Basra (a mile or so east of modern Zubair), the trade with which extended to China.[6] Arabic authors mention stimmi (not necessarily antimony sulphide and possibly galena) as found at Ispahan, Rêy (near Teheran) and Demawend.[7] It was applied to the upper and lower eyelids.[8] The composition of the remains of decomposed eye-paint found in pots in the oldest remains at Susa is not stated.[9] The Tuḥfat (1669) says eye-paint (kohl or surmeh) is either antimony (stibium) of Ispahan or a composition of iron or soot[10], and even the modern product is not always antimony sulphide.[11] The lampblack collyrium, always called surmeh, is obtained in Kerman from the smoke of lamps burning castor oil or goat's fat, deposited on porcelain saucers. In the mountainous districts it is the soot of the Gavan plant, an astragalus which yields gum tragacanth.[12] Albiruni[13] says the Persians had a special "festival of stimmi," and the Sabians of Harran also had special days for painting the eyes.[14]

MERCURY

Cinnabar (zinjifr) came from countries bordering on the "Abyssinian Sea,"[15] and was also found in Persia and the provinces.[16] When mercury (zeibaq)[17] became known in Persia is doubtful. It is prescribed for venereal disease in the Tuḥfat.[18]

ARSENIC

Orpiment (auripigmentum, arsenic trisulphide, As_2S_3) was obtained from Kurdistan, Bactria and the Hazareh country.[19] Onesikritos[20] mentions a "mountain" of it in Carmania. The name arsenic ($ἀρσενικὸν$)[21] is probably derived from the old Persian zaranya, modern Persian zar, zarnich, zarik (gold or golden, from its colour), Armenian zarik, Syriac zarnîkâ[22], Hebrew zarniq.[23]

[1] Maspero, H. anc., 1905, 556; ib., Empires, 453; Spiegel, Eran. Alt., i, 250 f.; A. von Kremer, Culturgeschichte des Orients unter den Chalifen, Vienna, 1875-7, i, 303, 329; Rawlinson, Mon., ii, 294; iii, 146 f., 159; Stahl, Chem. Z., 1894, xviii, 364.

[2] Curzon, Persia, ii, 510, 517.

[3] Kremer, i, 334.

[4] Abu Mansûr Muwaffaq, Liber fundamentorum, tr. Achundow, Halle, 1893, 316; Frater Angelus, Pharmacopœa persica, Paris, 1681, caps. 30, 42, 47, 258 f., 506 f., 901, etc., etc.

[5] Rawlinson, Mon., ii, 294; iii, 159.

[6] Wilson, Persian Gulf, 68.

[7] Avicenna, Canon, tr. Gerard of Cremona, Lugduni, 1522, 424 v.; Abulfeda, Géographie, iii, 170; Kremer, i, 46, 334; ii, 212, 223, 253.

[8] Rawlinson, Mon., iii, 243.

[9] Délég. en Perse, xiii, 9.

[10] Frater Angelus, Pharmacopœa persica, Paris, 1681, caps. 30, 268, 518 f., 535 f., 825, 896 f., etc.; Yule's Marco Polo, 1903, i, 127.

[11] Curzon, ii, 519.

[12] Cordier, in Yule's Marco Polo, 1903, i, 127.

[13] Chronology of Ancient Nations, tr. Sachau, 1879, 317.

[14] Chwolson, Die Ssabier, ii, 36.

[15] Masûdî; Wilson, Persian Gulf, 58.

[16] Abulfeda, Géogr., iii, 100; al-Dimešqi, Cosmographie, tr. Mehren, 1874, 311; in Spain, ib., 345, 349; Kremer, i, 303, 329; Abu Mansûr, tr. Achundow, 214, 363.

[17] Abu Mansûr, 170, 214, 363.

[18] Frater Angelus, Pharmacopœa persica, Paris, 1681, caps. 242, 654, 659, 685, 726, 1071.

[19] Rawlinson, Mon., ii, 294; iii, 146, 159.

[20] In Strabo, XV, ii, 14, 726 C.

[21] Hippokrates, transl. Fuchs, Munich, 1895, iii, 293; Theophrastos, De Lapidibus, 40, 50, 51; ψ-Aristotle, Problemata, c. 38; Salmasius, Plinianæ Exercitationes, 1689, ii, 812.

[22] Schrader, RL, 45; Schrader-Jevons, 119, 171.

[23] Liddell and Scott, Greek Lexicon, 1925, 247.

PERSIA III

NON-METALS

GEMS AND PRECIOUS STONES

The Persians made much use of precious stones and were expert engravers of gems.[1] The stones in an Achæmenid (c. 350 B.C.) necklace found at Susa were identified by de Morgan [2] as turquoise, lapis lazuli, emerald, agate, amethyst, jasper of various colours, felspar, red and yellow carnelians, quartz, hæmatite, breccia, coloured marbles and (doubtfully) jade.

Turquoise, pérôze, is the precious stone *par excellence* of Persia. There are rich mines, irregularly exploited, in Khorassan and turquoises (some of indifferent quality) are still mined at Kerman and Khojend and at Madan near Nishapur—the source of "999 out of every 1,000 turquoises that come into the market."[3] Darius obtained it from Chorasmia.[4]

Lapis lazuli was said to be found near Kashan, but the workings are no longer known. The Persians say it occurs in quantity in the mountainous district of Kuhpa, between Yezd and Ispahan, or Kashan, but the vein is not exploited. It probably came from Bokhara and is found in abundance in Afghanistan. Specimens occur fairly frequently in remains at Susa, often in fragments with no trace of working and probably a stock of material.[5] There are really no mines of it in Persia or Susiana.[6] Assyrian texts of 681–669 say that it came from Demawend [7], and this has been supposed to have been the sapphirus of Pliny [8], found in Mount Bikni (modern Demawend) and described as a purple stone with bright golden specks (probably pyrites) in it which made it unfit for cutting.[9] Lapis lazuli is not found in Demawend and the material may have come through Bactria from farther east. Whole cliffs of lapis lazuli are said to overhang the River Kashkar in Kaferistan, and other sources suggested are Badakshan, Lake Baikal or the upper valley of the Kokcha.[10] Darius obtained lapis lazuli (kapauta) from Sogdiana.[11]

There is no documentary evidence of the existence of emeralds in Persia although they are said to be found there.[12] Pliny refers to Bactrian, Scythian, Median, Cyprian and Egyptian emeralds, and the Persian stone tanos, classed with emerald, but these are mostly other materials. The Median emeralds were the largest, but not of very good quality.[13] The smaragdus ("emerald") of Pliny and other classical authors was never the modern emerald, which is a

---

[1] Sarre, Kunst d. alt. Persien, 1922, 52.

[2] Délég. en Perse, viii, 53, which is also the authority for some of the discussion of the separate stones.

[3] Curzon, Persia, i, 264 ; Rawlinson, Mon., iii, 147, 161 ; Spiegel, Eran. Alt., i, 250.

[4] Kent, J. Amer. Orient. Soc., 1933, liii, 1.

[5] Délég. en Perse, 1897–1905, 1905, 124 ; Délég., i, 49.

[6] Maspero, H. anc., 1905, 556 ; *ib.*,

Empires, 453 ; Rawlinson, Mon., iii, 146, 161 ; Spiegel, Eran. Alt., i, 250.

[7] Huart, Perse, 11, 33.

[8] xxxvii, 9.

[9] Maspero, Empires, 453.

[10] Rawlinson, Mon., ii, 305 ; iii, 162 ; Meissner, i, 350 ; Yule, Marco Polo, 1871, i, 153.

[11] Kent, J. Amer. Orient. Soc., 1933, liii, 1 f.

[12] Maspero, Empires, 453.

[13] Pliny, xxxvii, 5 ; Cæsius, Mineralogia, Lugduni, fol., 1636, 553.

variety of beryl. The "Bactrian emeralds" of Pliny, dark green, very hard and free from flaws, were supposed by King to be green rubies.[1]

Agates are abundant in the alluvia of south-east Persia and in Carmania, Susiana and Armenia and were used as jewellery in the oldest period. The banded agate is called "stone of Solomon" in modern Persia and is used as an amulet. Amethysts, jasper of various colours and red and yellow carnelians occur in Persia, as well as hæmatite, breccia, marbles, topaz, garnet, sapphire and chalcedony—the last (the blue variety of which is now called sapphirine) being much used by gem engravers. The actual seal cylinder of Darius I in the British Museum is of green chalcedony. Garnets and beryls are found in Armenia.[2] Transparent (hyaline) and milky quartz occur abundantly in Kurdistan [3], and Hiuen Tsiang [4] (A.D. 629) says Persia produces crystal (sphâtika). Etched carnelian (with glass) beads at Khotan [5] may be of Indian origin (see p. 289).

The balas rubies, found in and named after Badakshan, are mentioned by Albertus Magnus (d. A.D. 1280) [6], Marco Polo [7] and Hayton (d. 1308).[8] The common ruby is found abundantly in Badakshan [9] and in Ceylon, where balas rubies (paler and tinged with blue) also occur.[10] The present mines of balas rubies (spinels) are at Ghârân.[11]

Jade is found in Khotan.[12] Hard semi-gem stones said to be produced in Persia are iritis, a species of rock crystal; atizoë, a white stone with a pleasant odour ; mithrax, a gem of many hues ; nipparêni, which resembled ivory ; thelycardios or mulc ; and three unknown gems of inferior value : zathene, gassidanes and narcissitis.[13] Obsidian could be obtained from the volcanic ranges overlooking Assyria.[14]

No amber (already known in Egypt) was found in remains of 350 B.C. at Susa, and it was not used there.[15] The Persian name for amber, kahrupâi or kâhrubâ, "that which attracts straws," is used by the Arabs (e.g. Ibn al-Abbâs) and in the Pahlawî Bundahišn. In the 9th–10th centuries the word penetrated from Arabic into Syriac. Another Persian name for amber is šahbarî.[16] Pearls found at Susa[17] (c. 350 B.C.) were from the Persian Gulf, where they were probably fished at least as early as 2000 B.C.[18] They were also known in India in the Vedic Period.[19]

BUILDING MATERIALS

Excellent stone is found all over Media, the most important being Tabriz marble, probably deposited by petrifying springs and found only on the flanks of the hills near Lake Urmia. The slabs are used for tombstones, skirting of

[1] Rawlinson, Mon., iii, 161 ; the ruby and "Oriental" emerald are both alumina, coloured by traces of metallic oxides, whilst the "Peruvian" emerald is beryllium aluminium silicate.

[2] De Morgan, Déllég., viii, 53 ; Rawlinson, Mon., iii, 161 f. ; Maspero, Empires, 453 ; Huart, Perse, 123.

[3] Rawlinson, Mon., ii, 294.

[4] Si-yu-ki, tr. Beal, ii, 278.

[5] Aurel Stein, Innermost Asia, 3 vols., Oxford, 1928, i, 109 f. ; ii, 826.

[6] De Mineralibus, i, 1 ; ii, 2—balagius, palacius ; ed. Venice, 1542, 14, 173 ; d'Herbelot, Bibliothèque orientale, 1776, 152 ; Huart, Perse, 6.

[7] Travels, i, 26 ; ed. Yule, 1871, i, 149.

[8] Fleur des histoires de la terre d'Orient, ed. Omont, Notices et extraits des MSS., 1903, xxxviii, 237, 257.

[9] Rawlinson, Mon., iii, 162.

[10] Ibn Batûta, Travels, tr. by S. Lee, 1829, 187 ; J. Hill, Theophrastus' History of Stones, 1774, 76.

[11] Aurel Stein, Innermost Asia, ii, 877.

[12] Stein, Sand-Buried Ruins of Khotan, 1903, 254 ; ib., Ancient Khotan, 2 vols., Oxford, 1907, i, 132.

[13] Pliny, xxxvii, 9, 10, 11 ; Rawlinson, Mon., ii, 251, 305 ; iii, 147, 161.

[14] Childe, New Light on the Most Ancient East, 1934, 229.

[15] Délég. en Perse, viii, 54.

[16] Laufer, Sino-Iranica, 521.

[17] Délég. en Perse, viii, 51 f.

[18] See p. 289.

[19] Finot, Les Lapidaires Indiens, Bibl. de l'École des hautes Études, Phil.-Hist., 1896, Fasc. cxi, pp. xv, xxxi, xxxiii, 64, 153.

rooms and pavements for baths and palaces, and were cut in thin slices for windows, being semi-transparent. The marble of Iran is white or pale yellow, occasionally streaked with red, green or copper-coloured veins. Gypsum occurs in Kurdistan and talc in the mountains of Koum, near Tabriz. Hornblende, quartz, talc and asbestos are found in various parts of the Taurus. Some of the stone blocks used in building the palace at Persepolis are of great size, 49 to 55 ft. long and 6⅓ to 9⅘ ft. broad, but are laid to form a smooth perpendicular wall.[1]

The Persian palaces, probably based on Babylonian models, were built on a very large scale from blocks of grey or sometimes yellow or brown marble-like limestone, united by iron clamps set in lead, but the inner walls were of sun-dried brick faced with burnt brick.[2] Abd al-Latîf (1161–1231) says the bricks of Khosroes in Irâq were twice as large as those in Egypt.[3] At Susa and perhaps elsewhere bricks were used, cemented with bitumen, also enamelled bricks and coloured reliefs as at Babylon.[4] The roofs were of wood, mainly cedar and often painted, and the art was of high standard : the so-called "Archers' frieze" from Susa is a well-known example. The enamelled bricks, not mere facing tiles as in a much later period, were sometimes coated with gold, silver and electrum.[5] The so-called "bricks" are not formed of clay but of a sand-lime cement (the use of which has recently been resumed) or a kind of artificial stone. This was made by grinding a paste of lime, quartz sand and powdered flints, together with some chopped straw, and the mixture was then forced into moulds. It does not appear to have been fired.[6] The technique was probably derived from Babylon [7] and the actual work may have been carried out by foreign workmen. The Persians set no great store on trade, commerce and industry ; only the poorest were artisans and traders and shops were not allowed in some parts of the towns.[8]

BITUMEN AND PETROLEUM

Bitumen and naphtha are found together, generally with indications of volcanic activity, at the foot of the Caucasus where it drops into the Caspian Sea, in the Deshtistan or low country of Persia proper and in the Bakhtiyari mountains. The bitumen, which was used in ancient times, is of excellent quality ; the naphtha is of two kinds, black (petroleum) and white (which is preferred).[9] Coal is now worked near Teheran and Meshed.[10]

Herodotos[11] describes a well at Arderikka, not far from Susa, which produced bitumen, salt (i.e. brine) and oil (petroleum). The liquid was drawn in a skin and poured into a reservoir, from which it passed into another, where it took three different shapes. The salt and bitumen collected and hardened. The black oil, with an unpleasant smell, and called by the Persians rhadinaké (ῥαδινάκη), was drawn off into casks. This was crude petroleum, now obtained in Persia. Strabo[12] reports that "on digging near the River Ochus [modern Tedjen,

---

[1] Sir R. Ker Porter, Travels in Georgia, Persia, Armenia, Ancient Babylonia, etc., 2 vols., 4to., 1821–2, ii, 527 ; Maspero, H. anc., 1905, 556; Spiegel, Eran. Altertumsk., i, 250; Rawlinson, Mon., ii, 265 f., 293 f.; iii, 272 ; geology of Persia, in Blandford, Eastern Persia, an Account of the Journeys of the Persian Boundary Commission, 1876, ii, 439.

[2] Perrot-Chipiez, Persia, 47, 69 ; Meyer, Alt., III, 117 ; Curzon, Persia, ii, 72.

[3] Reitemeyer, Beschreibung Ägyptens, 135.

[4] Meyer, Alt., III, 118, 121 ; Perrot-Chipiez, 48, 51, 79, 139, 141, 149, 152, 315, 420, 439, 477 ; Curzon, ii, 71, 115, 310 ;

Maspero, Empires, 694 ; Breasted, Illustr. London News, 1933, clxxxii, 488.

[5] Perrot-Chipiez, Persia, 150.

[6] Contenau, Manuel, iii, 1441.

[7] Sarre, Kunst d. alt. Persien, 18.

[8] Perrot-Chipiez, 496 ; Rawlinson, Mon., iii, 242.

[9] Rawlinson, Mon., iii, 160 ; Cæsius, Mineralogia, 1636, 354–65.

[10] Huart, Perse, 11 ; Stahl, Chem. Z., 1893, xvii, 1596.

[11] vi, 119.

[12] XI, xi, 5, 517 C.—perhaps from Poseidonios ; C. Reinhardt, Poseidonios, Munich, 1921, 116.

flowing to the Caspian] a spring of oil was discovered. It is probable that as certain nitrous [alkaline], astringent [containing alum], bituminous and sulphurous liquids permeate the earth, greasy fluids may be found, but the rarity of their occurrence makes their existence almost doubtful."

In the Apocryphal Prayer of Azariah, written after 168 B.C. by various Jewish authors, the furnace in which Ananias [Shadrach], Azarias [Abednego] and Misael [Meshach] enter without injury is heated with naphtha ($\nu\acute{\alpha}\phi\theta\alpha$).[1] In Greek, $\nu\acute{\alpha}\phi\theta\alpha$ means a thick oil; in modern Persian naft is petroleum, which is found in Kurdistan, Baku, Afghanistan, etc.[2] A bronze lamp of the 10th century B.C. found at Susa contains remains of petroleum (analysed), which therefore appears to have been used for illumination at an early date.[3] Bitumen and naphtha occur near Dalaki, which is an ancient source.[4] The naphtha springs at Kerkuk are described by Ker Porter [5] as well as the "sacred fires" of Baku, which were also seen by Humboldt.[6] Dioskurides [7], Pliny [8] and Solinus [9] also mention a fountain of petroleum at Agrigentum in Sicily, which is still in existence.[10]

Bitumen occurs abundantly near Susa and was used in the earliest period.[11] The analysis by Le Chatelier of the black bituminous material of some reliefs and statuettes of the second period at Susa gave : water 2·8, spermaceti [sic : probably mineral wax] 1·6, mineral matter (calcium carbonate, etc.) 71·2, fixed carbon 10·6, volatile matter (gas, tar and water of decomposition of carbonaceous matter) 13·8 : sum 100·0. The mineral matter contained : calcium carbonate 45·2, calcium sulphate 3·5, calcium phosphate 0·8, "clay and iron (SiO_2, Al_2O_3, Fe_2O_3)" 21·7 : sum 71·2. The material is a kind of asphalt composed of rock and bitumen, like that used at Tello (see p. 270). There was also a dish supported on three legs, of bituminous stone, at Susa (c. 2000 B.C.).[12]

GLAZED WARE

Two statues of lions found at Susa, of the 11th century B.C., are of baked clay of reddish colour covered with a glaze which was originally bright blue or green ("vert franc"), but has undergone chemical decomposition in contact with the soil.[13] Although polychrome glazed bricks (11th–9th centuries B.C.) were found at Susa, together with blue and green bricks of one colour[14], the art of "enamelling" bricks was fully developed in the Achæmenid Period, when it was used both for tiles and pottery. The designs show no Greek influence.[15]

The silver and gold tiles and bricks said by Polybios (c. 150 B.C.)[16] to have been found by Antiochos (3rd century B.C.) in the palace of Ekbatana and the seven coloured walls of that city (white, black, scarlet, blue, orange, silver and gold)[17] were probably silvered, gilt and enamelled tiles[18], and although Diodoros Siculus[19] reports that it was after Cambyses took craftsmen from Egypt that the Persians built the palaces of Ekbatana and Persepolis, it is not certain that the manufacture of enamelled tiles in Persia was of Egyptian origin, since they had long been in use in Assyria and Babylonia.[20]

[1] Charles, Apocrypha and Pseudepigrapha of the O.T., Oxford, 1913, i, 628, 632, 634.
[2] Schrader, RL, 51 ; Rawlinson, Mon., ii, 294 ; Spiegel, Eran. Altert., i, 252.
[3] De Mecquenem, Revue des arts asiatiques, 1929–30, vi, 83.
[4] Pliny, vi, 23 ; Rawlinson, Mon., iii, 146.
[5] Travels, ii, 440, 515.
[6] Asie Centrale, 1843, ii, 505.
[7] i, 100.
[8] xxxv, 15.
[9] v, 22.
[10] Bunbury, DG, i, 80.

[11] Délég. en Perse, xiii, 20, 62, 162.
[12] Contenau, Manuel, ii, 807.
[13] De Morgan, Délég. en Perse, viii, 166.
[14] Rawlinson, Mon., iii, 311 ; de Morgan, Délég. en Perse, 1897–1905, 1905, 128, 177 ; ib., Délég. en Perse, i, 126, plate VI.
[15] Perrot-Chipiez, Persia, 470, 475 ; R. Murdoch Smith, Persian Art, 1876, 5, 15 ; Sarre, Kunst d. alt. Persien, 4, 55.
[16] Hist., x, 27; tr. Shuckburgh, 1889, ii, 26.
[17] Herodotos, i, 98.
[18] Rawlinson, Mon., ii, 265, 269.
[19] i, 46, 95.
[20] Franchet, Céram. prim., 103.

An Elamite vase from Susa contained no lead in the glaze, but a frit of litharge, lime and a little sand (? after 750 B.C.) was found.[1] Vases of metallic lustre are usually supposed to have been developed in later Persia, perhaps under Egyptian influence, since Nasîr ibn Khusraw, who visited Egypt in A.D. 1050 [2], says[3] : "in this country they manufacture faience of all kinds ; bowls, cups, plates and other utensils are decorated with colours analogous to the fabric called bûqalemûn," and that the nuances change according to the positions given to the vase. The art passed with the Arabs to Spain, and thence to France (Poictiers and Narbonne) and Italy. It was rediscovered in France about 1870. The Arabic formula is as follows : 91·80 gm. copper, 5·05 gm. silver and 91·80 gm. sulphur are fused until combination occurs. The sulphides are ground with 367·20 gm. of red ochre and the whole, reduced to an impalpable powder, is painted with vinegar on the already glazed and fired ware. The ochre is not really necessary. The ware is again fired at about 600° and reduction is effected by producing an intense smoke in the furnace for three hours. On cooling the ochry deposit is washed off.[4]

The glazed ware, which was described by Chardin in the 17th century, is not "porcelain" but a faience with a light porous body, still made at Shiraz, etc. The ware à reflet metallique was certainly made in A.D. 1250 and the art may be very much older. In some cases a tin glaze may have been used.[5] The later Persian green glaze is supposed to have influenced the green porcelain glaze of the Han dynasty in China.[6]

GLASS

Of the old Persian glass industry practically nothing is known. Glass furnaces are mentioned in a later commentary to the Vendidad.[7] It has been supposed that Persian transparent glass, a late name for which is bulur (Arabic billaur, rock-crystal) is really Alexandrian [8], and that glass, with glazed pottery and fine metal work, was exported through Scythia to China in 200 B.C.–A.D. 200.[9] Aristophanes[10] (c. 400 B.C.) says the drinking vessels at the Persian court were of gold or glass, and we may, therefore, infer that the Persians before Alexander's time liked to drink from glass vessels.[11]

The glass industry, highly developed in the 16th–17th centuries A.D., probably dates to the Roman conquest of Syria, from whence it reached Persia.[12] Persian glass was perhaps transparent from the first.[13] Specimens of early Persian glass are very rare, although numerous fragments were found at Susa and Persepolis and on the site of Rêy (Rhages)—where glass was made in the 10th–11th centuries A.D. There are a magnificent bowl of Khosroes I (A.D. 532) and some coloured glass discs in Paris. Glass of A.D. 900 was found at Samarra (see p. 288), and there were many glass jars, mirrors and coloured glass decorations in the palace of Khosroes at Ctesiphon when it was taken by the Arabs in 637 ; the mosque (12th century) at Tabriz was profusely decorated with glazed tiles.[14] Benjamin of Tudela (12th century A.D.) says the Sinjar of Persia put the body of the prophet Daniel into a glass coffin, an old story

[1] Berthelot, Arch., 85 f., 91, 93.

[2] Ethé, EB[11], xix, 248 ; Rehmân, J. Osmania Univ. Hyderabad, 1933, i, 61.

[3] Sefer Nameh de Nassiri Khosrau, ed. and tr. by C. Schefer, Publ. des langues orientales vivantes, Ser. II, vol. 1, Paris, 1881, 113, 151.

[4] Franchet, Céram. prim., 112.

[5] R. Murdoch Smith, Persian Art, 1876, 5, 11, 13, 15.

[6] Sarre, Kunst d. alt. Persien, 1922, 20 ;

A. U. Pope, Intr. to Persian Art, 1930, 61, 65, 93 : no chemical information.

[7] Rougemont, Âge du bronze, 36.

[8] Kisa, Glas, 104 f. ; A. U. Pope, Persian Art, 191, 195.

[9] Kümmel, q. in MGM, 1929, xxviii, 192.

[10] Acharanians, verse 73.

[11] Kisa, 104 f.

[12] Ib., 105.

[13] Fowler, Archæologia, 1880, xlvi, 86.

[14] Wallace-Dunlop, Glass in the Old World, 1883, 71 ; Kisa, 104 f.

applied to several other celebrities.[1] The magnificent coloured glass windows of a palace near Teheran are mentioned by Ker Porter, who also saw glass factories at Maraga and Shiraz, the coloured glass from the latter being of very good quality.[2] The best eastern glass is now made at Shiraz, "like pieces of gems"[3] ; wine bottles of curious shapes are also made there.[4] According to Franchet [5] and Thompson [6], green Persian glass of the time of Darius was coloured with copper and lead, and tin oxide was used to render glass opaque.

SALT

Salt occurs in various parts of Persia, sometimes as rock salt of various colours. Strabo's [7] "mountain of salt" in Carmania has been seen by modern travellers. In other places the salt covers the ground as an incrustation, is deposited on the shores of lakes and is dissolved in lake and stream waters. It occurs in great abundance in various parts of the Persian Empire : large tracts were covered with it in North Africa, Media, Carmania and Lower Babylonia ; there are salt lakes in Asia Minor, Armenia, Syria, Palestine and other places, and "mountains" of salt in Kerman and Palestine.[8] Salt is abundant on the side of the desert, and near Tabriz there are vast plains which glisten with crystals, whilst brine springs are numerous. Rock salt is quarried and is preferred by the natives. Lake Neyriz, or Kheir, dries up in summer and salt is then obtained in large quantities. Touz-Ghieul, on the upland of Phrygia, is a very saline lake containing in some seasons of the year over 32 per cent. of saline matter (more than the Dead Sea) and salt is manufactured from the water by evaporation. The plain is often incrusted with salt.[9] The lakes in the central desert (Loût or Kawîr) dry up and leave a saline incrustation called sebkha by the Algerians.[10]

SODA, ETC.

Strabo[11] says Lake Spauta (anct. Kapauta, the modern Blue Lake, near Lake Urmia) in Media formed efflorescences of a solid salt (soda) which caused itching and pain. Oil was a cure, and fresh water restored the colour of clothes which appeared burnt after immersion in the lake by ignorant persons for the purpose of washing. He also mentions the soda in Lake Van (see p. 226).

Nitre is found in the Elburz and in Azerbaijan, and both potassium and sodium nitrates are found in Persia.[12] The nitrum of Media, called halmyraga, which was found in dry glens in a very pure state[13], was probably the halmyris (ἀλμυρίς), a saline efflorescence (cf. ἀφρόνιτρον ; ἀφρὸς νίτρου) mentioned by Aristotle[14], and was perhaps natron, although nitre is often assumed.[15]

Sulphur is found in Persia, in Elburz and Azerbaijan, on the high plateau, on the surface of the ground in some places, in mines near Môsul, on the coast of Mekran and near the Dead Sea.[16] The petroleum springs at Kerkuk produce large quantities of sulphur.[17]

Alum is found in Persia, e.g. near Tabriz and also in the Zagros range.[18]

[1] Fowler, 91 ; Kisa, 101.
[2] Travels, i, 303 ; ii, 494.
[3] Kisa, 105 ; Pope, Persian Art, 191, 195.
[4] Curzon, Persia, ii, 100.
[5] Céram. prim., 103.
[6] Chem. Assyr., 1925, 9, 11.
[7] XV, ii, 14, 726 C.
[8] Rawlinson, Mon., iii, 146, 159.
[9] Rawlinson, Mon., ii, 294 ; iii, 89, 125.
[10] Huart, Perse, 5.

[11] XI, xiii, 2, 523 C. ; Rawlinson, Mon., ii, 305.
[12] Rawlinson, Mon., ii, 294, 305 ; Curzon, Persia, ii, 522.
[13] Pliny, xxxi, 10.
[14] Meteor., II, iii, 9 ; here sea-salt.
[15] Cæsius, Mineralogia, 1636, 326, 332.
[16] Curzon, ii, 522 ; Rawlinson, Mon., ii, 294 ; iii, 147, 160.
[17] Ker Porter, Travels, 1822, ii, 441.
[18] Curzon, Persia, ii, 522 ; Rawlinson, Mon., ii, 294.

In the Middle Ages the alum of Jemen and that of Aleppo (which probably came from Roha, or Rocha, near Edessa) were famous.[1]

VEGETABLE PRODUCTS

The mountains of ancient Media were clothed with dense forests of trees, including the pine, oak, cedar and terebinth. The ancients believed that the lemon tree originally came from Persia and the apple of Media (Medicum malum), which would not grow elsewhere, was a powerful antidote to poison.[2] To-day many fruit trees flourish, the olive is easily acclimatised and the vine produces grapes for the table or wine. The papyrus plant was cultivated, and all kinds of cereals and vegetables grow readily where there is water.[3] Media produced the plant silphium, a drug greatly prized in antiquity; although the Median kind was less valuable than the Cyrenaic, Parthian or Syrian, it was exported to Greece and Rome and sold by druggists.[4] Other products (or supposed products) were bdellium, amomum, cardamomum, gum tragacanth (still produced), wild-vine oil and sagapenum (Ferula persica).[5] Rawlinson identifies silphium with the assafœtida, which is still produced to a small extent; it has been doubtfully identified with Ferula assafœtida Linn., Ferula tingitana Linn., Thapsia Silphium Vivian and Laserpitium ferulaceum Linn.[6], but the plant may be extinct.

In Zagros most trees are common. The oak bears large galls and the gum-tragacanth plant frequently covers the mountain-sides. On the plains of the high plateau there is a great scarcity of vegetation; on the waterless plains only the tamarisk and a few other sapless shrubs occur, the soap-wort being the most common shrub between Koum and Teheran. With irrigation, the great plain of Iran can produce good crops of grain, cotton, tobacco, saffron, rhubarb, madder, poppies which give a good opium, senna and assafœtida. The rose grows luxuriantly and of all colours, and there are great numbers of wild flowers.[7]

Vegetable products of Persia proper were[8] the vine, peach (persica, the Persian fruit), citron, assafœtida, walnut ("Persian nut," καρύα περσική, of Theophrastos)[9], many trees and grain crops, madder, henna and cotton; opium, tobacco, indigo and maize are recent introductions.[10] Medicinal and magic herbs mentioned by Pliny include hestiatoris, theombrotios and napy (= σίναπι, mustard).[11] The Avesta (Yasna; not the Gâthâs) mentions two sacred plants: (1) the yellow, golden or golden-eyed haoma, growing on the summits of high mountains[12], and (2) the (mythical) milky white gaokerena, or gokard tree, growing in the middle of the sea and surrounded by 10,000 healing plants.[13] The juices of sacred plants were regarded as gods.[14] The usual identification of haoma is the acid Asclepias Sarcostemma viminalis[15]; according to Havell[16]

---

[1] Cæsius, 334; Lemery, Cours de Chymie, 1747, 550; ib., Traité universel des drogues simples, 1698, 27; Heyd, Commerce du Levant, Leipzig, 1886, ii, 565; Doren, Studien aus der Florentinischen Wirtschaftsgeschichte, Stuttgart, 1901–8, i, 82; ii, 98; Stapleton and Azo, Mem. Asiat. Soc. Bengal, 1927, viii, 373; Dussaud, Deschamps and Seyrig, La Syrie Antique, 1931, 98.

[2] Vergil, Georgics, ii, 126; Dioskurides, i, 166, and note in Saracenus's ed., Frankfurt, 1598, 23; Pliny, xii, 3; Maspero, Empires, 453.

[3] Maspero, Empires, 453 f.—de Morgan says the olive will not grow; Rawlinson, Mon., iii, 140, 157.

[4] Strabo, XI, xiii, 7, 524 C.

[5] Rawlinson, Mon., ii, 304; iii, 156.

[6] Smith, Lat. Dict., 1904, 624.

[7] Rawlinson, Mon., ii, 289, 292 f.

[8] Ib., iii, 140 f.

[9] Hist. plant., III, vi, 2; cf. Lagercrantz, Papyr. græc. Holmiensis, 37, 224, "royal nut," κάρυον βασιλικόν.

[10] Rawlinson, Mon., iii, 140.

[11] Pliny, xxiv, 17; xxvii, 13.

[12] SBE, xxxi, 230 f., 240; Plutarch's omomi, ὤμωμι, Isis and Osiris, c. 46; Moulton, Early Zoroastrianism, 399; Rawlinson, Mon., ii, 329.

[13] SBE, iv, 225; ib., v, 100; ib., xxiii, 141; ib., xxxi, 242.

[14] Spiegel, Eran. Alt., ii, 114.

[15] Rawlinson, Mon., ii, 329; Maspero, Empires, 573; Maury, Magie, 30.

[16] J. Roy. Asiat. Soc., 1920, 349.

it is common millet, the grass Eleusine coracana (Indian ragi), used to make an intoxicating drink called marua in the Eastern Himalayas, but this would not agree at all with the ancient accounts. Whether the juice of the haoma was fermented [1] or not is doubtful; that it was "strongly alcoholic" [2] is improbable, unless it was mead or fermented honey (madhu). Another identification of the haoma plant is Ephedra distachya. The Greek ambrosia was supposed by Bopp to have been derived, through ἀ-μβρότος (immortal), from the Aryan a-mrita, with the same meaning; in Greek legend Melissa and Amaltheia fed the infant Zeus in Crete with milk and honey.[3] Strabo [4] says the Medes made cakes of sliced and dried apples, and bread of roasted almonds, and expressed a wine from some kinds of roots.

Wine was profusely used by Persian kings and their guests, the king drinking wine of Helbon from gold cups and the guests an inferior wine.[5] Helbon wine is the Chalybonian wine which Strabo [6] says was sent to the Persian kings. Chalybon is Aleppo, but he probably meant Damascus.[7] Although the earlier Persians were abstemious, the use of wine increased greatly in the later period and intoxication was very common. Herodotos [8] says that all important decisions in council were made with the members intoxicated, and the king was compelled by law to be intoxicated once a year at the feast of Mithras.[9]

Manna is produced on the oaks of Kurdistan.[10] Common foods were wheat bread, barley cakes and roasted or boiled meat with salt and bruised cress seeds; the poor ate dates, figs, wild pears, acorns and terebinth-tree fruit, and drank water. The meats included those commonly used, together with the flesh of goats, horses, asses and camels. Oysters and other fish were eaten.[11]

TEXTILES

Some portions of tissue wrapped around copper axes at Susa were of linen, so fine as hardly to be surpassed by modern work.[12] The textiles of Persia were famous in antiquity. Herodotos[13] specially mentions Persian embroideries of the time of Xerxes, and carpet-making was an important industry in the Hellenistic Period.[14] Hiuen Tsiang (A.D. 629) refers to the fine brocaded silks, woollen stuffs and carpets made in Persia[15], and some specimens of dyed silk stuffs of A.D. 600 and later are in museums (e.g. Victoria and Albert, London).[16] The silk industry probably began in Khuzistan (Susiana) under Sapor I (240–273) and some fabrics of about this period dyed in red, green and blue are known.[17] Silk fabrics are apparently unknown to Pliny, and the Sassanian Empire was the channel for the silk trade with China. Specimens of early silk have been found on the route of this traffic.[18]

The Medic robe was long and loose-fitting and of many colours, sometimes purple, sometimes scarlet, occasionally dark grey or deep crimson. The robe called sarapis was striped alternately white and purple. Procopius says the robe was of silk, which Justin describes as transparent, but the material

[1] Rawlinson, Mon., ii, 329; Havell, 349.
[2] Moulton, 71.
[3] Lippmann, Z. Deutsch. Zuckerindustr., 1934, lxxxiv, 812, 823, Techn. Teil.
[4] XI, xiii, 11, 526 C.
[5] Rawlinson, Mon., iii, 214, 226.
[6] XV, iii, 22, 735 C.
[7] Note to Strabo ad loc., Falconer and Hamilton's tr., iii, 140.
[8] i, 133.
[9] Rawlinson, Mon., iii, 236.
[10] Ker Porter, Travels, 1822, ii, 471; Sadegh Moghadam, Les mannes de Perse, 1930, q. by Naficy, La Médicine en Perse, 1933, 56.

[11] Rawlinson, Mon., iii, 237.
[12] Lecaisse, in de Morgan, Délég. en Perse, xiii, 163.
[13] v, 49; ix, 80.
[14] Perrot-Chipiez, Persia, 470.
[15] Si-yu-ki, tr. Beal, ii, 278.
[16] Sarre, Kunst d. alt. Persien, 1922, 32, 46, pl. 95 f.; R. Murdoch Smith, Persian Art, 1876, 21.
[17] Pfister, Revue des arts asiatiques, 1929–30, vi, 1.
[18] Aurel Stein, Ruins of Desert Cathay, 1912, i, 380 f.; ii, 204 f.; ib., Innermost Asia, 3 vols., Oxford, 1928, i, 243 f.

probably varied. Silk was used for the Parthian standards, and Media and Persia probably imported it from China, India and Cashmere. The shirt was purple, and embroidered trousers and felt caps dyed in various colours were worn. The native name for the Medic robe (not the tunic) was probably kandys : it was sometimes embroidered with gold, and Plutarch says the entire dress of the king was worth 12,000 talents, or £2,925,000. The trousers were crimson (ὑαγινοβαφεῖς, Xenophon, ? purple) or in stripes of bright colours, and round the base of the cap (kitaris) was a fillet (the diadem) of blue spotted with white (cærulea fascia albo distincta), represented on seal cylinders. Whether this was so dyed directly or made up from blue and white fabrics is not known. The hanging curtains were dyed green and violet, and Persian dyes had some reputation. Tanning was known and the shoes of the king were dyed deep yellow or saffron (κροκόβαπτος).[1]

ANIMAL PRODUCTS

The Medes used cosmetics, false hair, eye-paints ('ὀφθαλμῶν ὑπογραφή)[2] and scented ointments, which Pliny[3] says originated in Persia. They used perfumes lavishly, the king applying an ointment of lion's fat, palm wine, saffron and the (fabulous) plant helianthes, which was considered to beautify the complexion.[4] The royal ointment of Parthian kings, probably a Persian invention, was composed of cinnamon, spikenard, myrrh, cassia, gum storax, saffron, cardamom, ben (myrobalanos), costus, amomum, arbutus (comarus), ladanum, opobalsamum, sweet calamus, wild vine flower (œnanthus), malabathrum, serichatum (otherwise unknown), cyperus, aspalathus, opoponax (panacea), henna, marjoram, clarified honey and wine.[5] Arabia sent to Persia 1,000 talents of frankincense every year and censers are often represented. Alabaster boxes or bottles were used to contain the ointments and some were found at Nimrûd.[6]

Bee-keeping is very old in Persia, Afghanistan, Tibet and China, but was introduced into Siberia and across the Urals only in the 18th century A.D.[7] Wax was used, and Strabo[8] says the Persians covered dead bodies with it before burial, whilst the Magi (priests) were not buried but exposed to be devoured by birds for fear of defiling the sacred elements. There is an ivory statuette of the Elamite Period in the Louvre[9] and this material was used for statuettes in the Parthian Period.[10]

Poisons of some kind were in use, particularly in the time of Artaxerxes Mnemon (405–359)[11], and poisoning by bull's blood is mentioned from the Achæmenid Period ; even in the time of Sapor II (A.D. 309–380) a cowardly general was put to death in this way.[12] Bull's blood was also considered poisonous in Egypt.[13] Herodotos[14] says that Cambyses compelled Psammenitus, king of Egypt, to drink bull's blood and it killed him immediately. The belief in its poisonous properties occurs frequently in older authors[15]

---

[1] Rawlinson, Mon., ii, 316, 364 ; iii, 202, 204, 215, 345 ; Dalton, Treasure of the Oxus, 1926, p. xxx.
[2] Rawlinson, Mon., ii, 317.
[3] xiii, 1.
[4] Pliny, xxiv, 17 ; Smith, Latin Dict., 1904, 494.
[5] Pliny, xiii, 2.
[6] Rawlinson, Mon., iii, 212 f.
[7] Lippmann, Z. Deutsch. Zuckerindustr., 1934, lxxxiv, 811, Techn. Teil.
[8] XV, iii, 20, 733 C.
[9] Contenau, Manuel, ii, 794.
[10] Sarre, Kunst d. alt. Persien, 29.

[11] Rawlinson, Mon., iii, 246 ; Justi, Gesch. d. alt. Persiens, 1879, 136.
[12] Moses of Chorene, c. A.D. 420, q. in Huart, Perse, 196.
[13] Wiedemann, Alte Ägypten, 1920, 294, quoting Apostolides, Bull. Soc. Médicine de Caire, 1909.
[14] iii, 15.
[15] Diodoros Siculus, xi, 12 ; Strabo, I, iii, 21, 61 C. ; Grote, Hist. of Greece, 1869, v, 143 ; W. Smith and Marinden, Hist. of Greece, 1902, 264—from the time of Aristophanes.

THE SCYTHIANS   425

and persisted into mediæval times.[1] Van Helmont [2], who accepts the poisonous property as a well-known fact, gives an explanation on psycho-pathological lines which was used much later in a similar connection.[3] According to Pliny [4] *fresh* bull's blood is poisonous, yet the priestess of the goddess Ops at Ægira in Attica drank it before going into the sacred vault. I do not know any adequate explanation for the belief; the usual one of ptomaines in putrefied blood [5] seems inadequate, since fresh blood is always mentioned: perhaps "bull's blood" is a cover-name. Turpentine was regarded by the Egyptians as the "blood of Osiris," [6] and an Assyrian text says that a condemned criminal was made to drink the "blood of the cedar" (turpentine).[7]

THE SCYTHIANS

The Scythians were originally a nomadic people who were living in South Russia east of the Dneiper and were paramount in this region in the 5th century B.C. They were probably a mixed race, mostly Iranian with some Mongolian elements.[8] They are described by Herodotos [9] and by Hippokrates[10], the former of whom represents them as rich in gold, the first specimens of which fell from heaven in the forms of a plough, a yoke, a battle-axe and a drinking cup—one of the fabulous stories related by the Ionian Greeks who came in contact with the Scythians.[11]

The large region of South Russia was highly civilised and from the second, perhaps the third, millennium B.C. was in relations with adjacent civilisations, receiving and perhaps transmitting artistic and technical influences. Similarities in artistic work with that of the Caucasus, the Black Sea, the Danube, Iran, Greece and even China (from 1100 B.C., through Turkestan and the Altai) have been traced.[12] The original inhabitants of the Kuban region were perhaps Sauromatians, Sindians, Mæotians and Taurians. Successive waves of conquest from the east were perhaps the result of pressure by Mongolian tribes on Iranian tribes in Central Asia and Turkestan, the Cimmerians, the Scythians and the Sarmatians, the first-named establishing a kingdom around Kertch (the Cimmerian Bosporos), the Scythians (who were perhaps Iranians) establishing a state (consolidated in the 7th century B.C.) on the Iranian model in the valley of the Kuban and later in the steppes of the Don and Dneiper, afterwards taken over by the Sarmatians, who were probably only a branch of the Scythians. Scythian influence perhaps extended to Pontos, Cappadocia and Armenia. Central Caucasus and Transcaucasia (in later times the kingdom of Van) had mines of gold, silver, copper, iron and other metals which were worked from the earliest period and probably supplied Mesopotamia. In the 6th century B.C. there was Greek influence from the Ionian and Æolian colonies in Asia Minor, which developed at the expense of the Oriental from the end of the 5th century B.C. The objects of gold, silver and copper in the oldest graves in the Kuban valley (e.g. at Maikop) are like those of Egypt, Mesopotamia and Elam. The legend of the Argonauts is perhaps Carian and corresponds with expeditions of Carians and Achaians

[1] Lippmann, Beiträge zur Gesch. d. Naturwiss., 1923, 187.
[2] Ortus Medicinæ, Leyden, 1652, 143.
[3] Carus, in Liebig, Familiar Letters on Chemistry, tr. by Gregory, 1851, 531.
[4] xxviii, 9.
[5] Hoefer, Hist. de la chimie, i, 217.
[6] Dawson, Magician and Leech, 30.
[7] Delaporte, La Mésopotamie, 1923, 336.
[8] E. H. Minns, Scythians and Greeks, Cambridge, 4to., 1913; ib., CAH, iii, 192; Rostovtzeff, Iranians and Greeks in South Russia, Oxford, 4to., 1922; O. M. Dalton, Treasure of the Oxus, 1926, p. xlvi f.; Gutschmid, Kleine Schriften, 1892, iii, 421; G Borovka, Scythian Art, tr. Childe, 1928.
[9] Book iv.
[10] De Aere, Aquis et Locis; Opera Omnia, ed. van der Linden, Leyden, 1665, i, 352.
[11] Sprengel, H. Med., i, 206.
[12] Rostovtzeff, Iranians and Greeks in South Russia, 7 f., 10, 14, 22, 32, 59, 197 f., 208; G. Borovka, Scythian Art, 7, 16 f., 21, 25 f., 82 f.

about 1000 B.C. to the Black Sea region in search of iron and gold.[1] Borovka [2] assumes that in the Early Copper and Bronze Ages at Maikop and Kuban there was influence from the Ægean, whilst in the full Bronze Age the influence was mainly from Asia Minor and the Caucasus.

Sinope and Trebizond were ancient centres of civilisation and were ports of shipment for copper and iron from the Transcaucasian mines.[3] Polybios [4] describes the export of honey, wax, and dried fish from the Pontos and Euxine, and Pliny mentions hides, salt, timber, precious stones, amber, drugs and gold.[5] The abundance of gold in Scythic graves probably came from mines in the Siberian steppes (Ural-Altai) which are now exhausted.[6]

The Scythians wore gold and bronze jewellery of Greek make, and used arrow- and spear-heads of stone, bronze and iron, with bronze and iron swords, sometimes gilt. They sacrificed on a great pile of brushwood, constantly added to, on the top of which was planted every year "an antique iron sword ($\sigma\iota\delta\eta\rho\epsilon o\varsigma$ $\dot{a}\kappa\iota\nu\dot{a}\kappa\eta\varsigma$), an image of Ares."[7] The Khonds, a primitive tribe of Hindustani, also have a "god of iron" (Loha-Pennu), who is also a god of war and is represented in each village by a buried piece of iron.[8] The early iron work is decorative and the metal is relatively scarce.[9] The Sarmatians (like the Parthians) used conical helmets and stirrups of iron, both unknown to the Scythians.[10] The Scythians used great bronze cauldrons, sometimes made from the metal of arrow-heads; those found contain bones of horses and sheep. Although Herodotos says they used no silver or bronze, both metals have been found in tombs.[11]

Glass vessels either cast or hewn out of solid blocks at Kuban belong to about the 1st century B.C.–1st century A.D. The style prevailed in China after the manufacture of glass began there in the 5th century A.D.[12] At the mouth of the River Borysthenes (Dneiper), second only to the Nile, "salt forms in great plenty without human aid."[13] The salt made at the mouth of the Dneiper and salted fish were exported to Greece and elsewhere.[14]

A kind of soap was used in washing the head ($\sigma\mu\eta\sigma\dot{a}\mu\epsilon\nu o\iota$ $\tau\dot{a}\varsigma$ $\kappa\epsilon\phi a\lambda\dot{a}\varsigma$) and the Scythians took vapour baths made by throwing hempseed on hot stones in a tent, the fumes from which caused them to "shout for joy."[15] Hemp seeds ($\kappa\dot{a}\nu\nu a\beta\iota\varsigma$) or the extract (the modern extract of Cannabis Indica) were perhaps used in the Thracian Dionysos festivals and by the Corybantes.[16] A Scythian tribe living on islands in the Araxes fed on roots and the stored fruits of trees. They had "a tree which bears the strangest produce. When they meet in companies they throw some of the fruit on the fire round which they are sitting, and presently, by the mere smell of the fumes which it gives out on burning, they grow drunk."[17] The thorn-apple (Datura stramonium) may be meant. The women used a cosmetic of cypress, cedar and frankincense wood pounded with water to a paste on a stone, with which they plastered themselves, washing it off next day, when the skin became clean and glossy.[18]

[1] Rostovtzeff, 18 f., 34, 38, 41, 54, 58, 60 f., 113 f., 120, 228 ; Giles, CAH, ii, 32.
[2] Scythian Art, 16 f.
[3] Rostovtzeff, 62.
[4] iv, 38.
[5] xxvii, proem, 4 ; xxxi, 7 ; xxxiii, 4 ; xxxvii, 2 ; cf. Minns, Scythians and Greeks, 1913, 440 ; F. Oertel, CAH, x, 401.
[6] Borovka, 29 ; Minns, 1913, 63 f.
[7] Herodotos, iv, 62 ; Meyer, Alt. (1909), I, ii, 822 ; Minns, CAH, iii, 197 ; Rostovtzeff, Iranians and Greeks, 45.
[8] Maury, Religions de la Grèce Antique, 1857–9, i, 125.
[9] Borovka, 75.
[10] Rostovtzeff, 121.

[11] Herodotos, iv, 71, 81 ; Minns, CAH, iii, 199, 201 ; ib., Scythians and Greeks, 79.
[12] Rostovtzeff, 127, 233.
[13] Herodotos, iv, 53 : these salines are still very important ; Rawlinson, ad loc.
[14] Minns, Scythians and Greeks, 440.
[15] Herodotos, iv, 73 f.; the intoxicating fumes of hemp would be evolved.
[16] Rohde, Psyche, Freiburg i. B., 1898, ii, 17, 51 ; Herodotos, i, 202 ; ib., iv, 74, 75 ; Pomponius Mela, ii, 21 ; Kanngiesser, A. Nat., iii, 84.
[17] Herodotos, i, 202.
[18] Ib., iv, 75 ; hydrogen peroxide would be formed by autoxidation of the essential oils.

A drink was made by pressing in cloths the fruit of a tree called Ponticon [a species of cherry], bearing a fruit like a bean with a stone inside. The juice was black and thick and called by the natives aschy. "They lap it up with their tongues and also mix it with milk. The lees are made into cakes, eaten instead of meat."[1]   The first mention of butter (without name) is in Herodotos [2], who says the Scythians poured mare's milk into deep wooden casks and stirred it, drawing off the part which rises to the top, which was considered best.[3]   Hippokrates [4] gives the same account but first uses the name butyron ($\beta o\acute{v}\tau\upsilon\rho o\nu$).[5]   The suggestion of Dawson [6] that $\beta o\acute{v}\tau\upsilon\rho o\nu$ is not butter but a large juicy fruit, the Egyptian bdd or water melon, because a variant text gives $\sigma\iota\kappa\acute{v}\eta\nu$ $\mathring{\eta}$ $\beta o\acute{v}\tau\rho o\nu$, "cucumber or butyron," and Hesychios says $\beta o\acute{v}\tau\upsilon\rho o\varsigma$ is a kind of plant ($\beta o\tau\acute{a}\nu\eta\varsigma$ $\epsilon\mathring{\iota}\delta o\varsigma$), does not seem very probable. The Scythians may have made an intoxicating drink (kumys, koumiss) by fermenting mare's milk. Koumiss is now made in leather bags, whereas Herodotos refers to wooden casks; it keeps well.[7]

[1] Herodotos, iv, 23 ; Minns, Scythians and Greeks, 1913, 108.
[2] iv, 2.
[3] Schrader, RL, 121 : this could not have been cream, which does not rise when the milk is *stirred*.
[4] De Morbis, lib. iv, cap. 25 ; ed. van der Linden, ii, 144.
[5] Schrader, RL, 122 ; Lippert, Kulturgeschichte, 1886, i, 538.
[6] Magician and Leech, 1929, 142.
[7] Herodotos, iv, 2—he seems to describe cream, not koumiss, as Minns, CAH, iii, 196, and Yule, Marco Polo, 1871, i, 226, claim.

PHŒNICIA

PHŒNICIA I

GENERAL

THE PHŒNICIAN CULTURE

Two generations ago all strange elements in Mediterranean culture were regarded as Phœnician, whilst the last generation virtually denied all Phœnician influence. The Phœnicians can now be traced to nearly 3000 B.C., and the tombs of their kings give indications of great wealth and high civilisation. Until recently the pre-Classical history of Central and North Syria was a blank, but now a continuous narrative can be constructed from Egyptian, Hittite and Assyrian records. The early civilisation at Byblos on the Syrian coast was in relation with XII dyn. Egypt, and although the remains may not, strictly speaking, be "Phœnician" there is no evidence of a change of population in this region [1], which is the part of the Syrian coast afterwards inhabited by the race called by that name.

The Phœnicians passed as great inventors, but in reality much credited to them was discovered elsewhere : glass in Egypt, purple dye in Crete, woollen clothing in Syria and metal working in Egypt and Babylonia. Although not great inventors they were, however, great adapters, using (often without understanding them) ideas and models from Egypt, Mesopotamia, Assyria, Cyprus and the Hittites, and possibly introducing improvements in technical processes.[2] Although no literature has come down to us [3], the Phœnicians probably invented the alphabet about 1250 B.C., but may have been anticipated in Sinai about 1500 B.C. [4], and cuneiform tablets at Râs Shamra on the Syrian coast opposite Cyprus, which were written about 1400 B.C. in a primitive Semitic language resembling Hebrew, contain a rudimentary alphabet.[5] The Phœnician language has been supposed since St. Jerome and St. Augustine to have been closely related to Hebrew [6], but this has been disputed. Phœnician history is that of an immense business house and the position of the country predestined it to become an entrepôt between East and West ; in later times the African colony of Carthage had extensive factories turning out all kinds of products, and the Phœnicians were business men and traders long before the Jews, although both were probably originally pastoral peoples.[7]

The native name for the country was Chna or Canaan [8], the Egyptians called it Zahi and its inhabitants Fenkhu. The name Phœnician has been derived from φοινός, blood red, the colour of the famous purple dye, or perhaps "red

---

[1] Olmstead, H. Pal. Syr., x, 66, 319, 367, 404 ; Eissfeldt, A. Rel., 1934, xxxi, 24 ; Meyer, Alt., II, ii, 61 ; Hall, Bronze Age, 1928, 36, 38 ; the results of the French excavations at Byblos, etc., are reported in the journal Syria, 1920 f., general index to vols. i–x (not very accurate) in ib., 1929, x, 371.

[2] Meyer, Alt., II, ii, 131 ; Otto, 1925, 48 ; Köster, Beih. alt. Orient, i, 16 ; Dussaud, Civilisations préhelleniques, 1914, 303.

[3] Mosso, Medit. Civ., 282.

[4] Contenau, La Civilisation phénicienne,

1926, 309, 322 ; Myres, CAH, iii, 640 ; Meyer, Alt., II, ii, 69, 115, 374 ; Torrey, Isis, 1927, ix, 545 ; Sethe, ib., 546.

[5] C. Marston, New Knowledge about the Old Testament, 1933.

[6] Worrell, Races in the Ancient Near East, 1927, 111.

[7] Lenormant, H. anc., iii, 96, 225 ; Meyer, Alt., II, ii, 131 ; Otto, 1925, 46.

[8] Meyer, Alt. II, i, 83, 97 ; II, ii, 63 ; Olmstead, H. Pal. Syr., 66 ; Bilabel, Gesch. Vorderasiens, 1927, 9, 429.

431

skins," as the Cretans may have called the people[1] ; or from the name of the palm tree (φοίνιξ in Homer)[2] ; or from the Red Sea (Mare Erythræum, really the Persian Gulf). The name of Phœnix, the brother of Cadmus, is connected with a later tradition.[3] The Phœnicians are regarded as a remnant of the Semitic Canaanite tribe cut off on the coast by foreign invasion at some imperfectly known date, certainly before 2000 and perhaps as early as 3000 B.C., the rest going inland.[4] The race may have been formed from various elements : autochthonous non-Semites, Semites of Amurru, Amorites proper, Hittites and Ægeans. The culture of Syria was mixed and full of local differences which became important after 2000 B.C.[5] The Semites probably entered originally in the south from Arabia. Their occupation of the pre-Phœnician sites of Tyre, Sidon, etc., is supposed to date from 2800 B.C.[6] The country was a narrow passage between Africa and Asia, a fertile strip of the Syrian coast about 120 miles long between the mouths of the Nile and the Euphrates and within easy reach overland of the latter, of the frontier of Egypt at Pelusium and of Cyprus. The hills at the back, a few miles from the coast, send several rivers into the sea.[7] The ancient boundaries were from the Mediterranean to Lebanon on the east, Aradus (Arvad) on the north and the River Chorseas (now a mountain stream), near Cæsarea, in the south.

Although a late tradition makes Tyre a colony founded by Sidonians [8], it was probably an early site and was already important in the 15th century B.C. [9] ; the date 2750 B.C. given from native sources for its foundation by Herodotos[10] is probably near the truth.[11] The name of Sidon, a very old town, probably recalls Sid, the hunter god, the derivation as "the fishery" being due to false etymology.[12] Tyre, the Hebrew Tsur, is supposed to mean "the rock."[13] Other old towns on the Syrian coast are Amor (the source of the Babylonian name Amurru for Syria) or Marath, Beruth (Roman Berytus, modern Beyrut)[14], Arvad (Greek Arados), which dominated south Phœnicia as Tyre did the north[15], and the very ancient town Gebal or (more correctly) Gubl, called by the Egyptians Kupna (a non-Semitic name) and by the Greeks Byblos, "the tomb of the god Adonis" and the centre of an important community from the first to the last days of the ancient world. Byblos was an important trade centre before Tyre and Sidon, e.g. for early Egyptian trade in cedarwood, and its ships, which occupied only four days in the voyage with a

[1] Budge, Amulets, 250, 259 ; Meyer, Alt., II, i, 97 ; ib., II, ii, 66 ; Olmstead, 66.
[2] Dyer, DG, ii, 605 ; G. Rawlinson, History of Phœnicia, 1889, 1 f.
[3] Schmitz, DBM, i, 524 ; Meyer, Alt., II, ii, 114.
[4] Maspero, H. anc., 1905, 217 ; Perrot and Chipiez, Art in Phœnicia and Cyprus, 1885, i, 14 ; Cook, CAH, i, 186 ; Köster, Beih. alt. Orient, i, 4, 7 ; Obermaier and Thomsen, RV, x, 7 ; Reche, Thomsen and Bosch-Gimpera, ib., x, 133 ; Roeder, ib., xiii, 153.
[5] Otto, 1925, 44 ; Contenau, Civ. phénic., 351, 359 ; Hall, CAH, ii, 278 ; Cook, ib., 380 ; Meyer, Alt., II, i, 108.
[6] Meyer, Alt., II, ii, 64 f., 300 ; F. C. Eiselen, Sidon, Columbia Univ. Orient. Studies, iv, New York, 1907, 29.
[7] Contenau, Civ. phénic., 37 ; Perrot-Chipiez, Phœnicia, i, 3 ; Grote, History of Greece, 1869, iii, 267 ; Meyer, Alt., II, i, 97 ; Olmstead, H. Pal. Syr., 66 ; Dussaud, Topographie historique de la Syrie antique et médiévale, 1927.

[8] Cook, CAH, ii, 379 ; Eiselen, Sidon, 16.
[9] W. B. Fleming, Tyre, Columbia Univ. Oriental Studies, x, New York, 1915, 8 ; Perrot-Chipiez, Phœnicia, i, 19 ; Strabo, XVI, ii, 22 ; XVI, iii, 4 ; Pliny, vi, 28, calls it Tylos ; Ezekiel, xxvii, gives a detailed account of its industries ; Lenormant, H. anc., iii, 36 ; Smith, CDB, 965 ; Dussaud, Topographie, 18.
[10] ii, 44.
[11] Contenau, Civ. phénic., 43 ; Dyer, DG, ii, 608 ; cf. Meyer, Alt., II, ii, 80, 122.
[12] Olmstead, H. Pal. Syr., 76 ; Eiselen, Sidon, 11 ; Meyer, Alt., II, ii, 64 ; Dussaud, Topographie, 37 ; excavations at Sidon (objects mostly Hellenistic), Contenau, Syria, 1920, i, 16, 108, 198, 287 ; 1923, iv, 261 ; 1924, v, 9, 123.
[13] Fleming, Tyre, p. ix ; Perrot-Chipiez, Phœnicia, i, 19 ; Grote, Hist. of Greece, iii, 268 ; Dyer, DG, ii, 608 ; Contenau, Civ. phénic., 79.
[14] Olmstead, H. Pal. Syr., 74.
[15] Perrot-Chipiez, Phœnicia, i, 20 ; Meyer, Alt., II, i, 99.

favouring south wind, were well known in IV dyn. Egypt.[1] The Cappadocian tablets show that there was a Semitic colony on the Syrian coast at least as early as the 25th century B.C.[2] ; we know something of the towns, mostly dating from before 2000 B.C., from the Egyptian conquest of Syria in the 16th century, and they are frequently mentioned in Assyrian records of the 9th to 7th centuries B.C.[3]

Phœnician ships were shown in an XVIII dyn. Egyptian painting found at Thebes in 1895 but now destroyed[4], which also showed large oil jars, smaller vases (gold or silver ?) and Minoan jars, which hint at direct commerce with Crete. Numerous Cretan potsherds occur in Phœnician towns and Crete perhaps imported ivory, alabaster, faience, glass, real or artificial lapis (kyanos) and (to a small extent) amber, copper, silver and gold from the Syrian coast. Artistic work in these materials was then exported from Crete to Egypt, with copper bars, lapis in baskets and elephant tusks.[5] There were uninterrupted relations, probably by sea, with Egypt from the Thinite Period (3300–3000), a seal cylinder of which, copper idols and many later Egyptian vases, were found in a temple at Byblos. Syria and Palestine were the medium of contact between the earliest Egyptians and the Semites.[6]

The antiquity of Phœnicia in sea-power and commerce has been exaggerated. The Minoan power preceded it by many centuries and direct influence of Phœnicia in the Ægean is not much earlier than the 12th century B.C. [7], Phœnician trade being important only in the period 1500–1000, perhaps 1075–1000 or 1100–800 B.C. Phœnician ships are first mentioned under Thothmes III and were copied from the Egyptian.[8] For the traditional historian, however, the civilisation of the Orient was gradually spread through south Europe and the west by Phœnician trade.[9] The theory which made the Phœnicians the diffusers of cultural relations in the Mediterranean, which lay between Tyre and Carthage, is mostly due to Samuel Bochart[10] and Movers[11], but is now abandoned.[12]

Phœnician sea traffic and trade began to be displaced by Ionian Greek in the 8th–7th centuries.[13] There was a revival under the Achæmenid Persians (from 550 B.C.), when Phœnician ships formed the core of the Persian navy and were used especially against the Greeks, the struggle being very severe in Cyprus. The downfall of Phœnician power was complete and final ("delenda est Carthago"—Cato) when Carthage was sacked by the Romans in 146 B.C. and North Africa became a Roman province. The conquest of Syria by Pompey in 64 B.C. revived the importance of the Phœnician towns, but the nation no longer had any separate existence.[14] It had stood for the commercial ideal in its simplest form, and even after the Roman conquest Tyre and

[1] Perrot-Chipiez, Phœnicia, i, 23 ; Grote, Hist. of Greece, iii, 268 ; Lenormant, H. anc., iii, 17 ; Köster, Seefahrten alt. Ägypter, 15 ; ib., Antike Seewesen, 45 f. ; Olmstead, H. Pal. Syr., 66, 68 ; Meyer, Alt., II, i, 98 ; II, ii, 65 ; Otto, 1925, 44 ; Dussaud, Topographie, 63.
[2] Contenau, Civ. phénic., 44.
[3] Maspero, Struggle, 169 ; Perrot-Chipiez, Phœnicia, i, 19 ; Bilabel, Gesch. Vorderasiens, 1927, 73 ; Hall, NE, 158.
[4] Fimmen, Kret.-myken. Kultur, 1921, 118, Fig. 111 ; Olmstead, H. Pal. Syr., 145.
[5] Meyer, Alt., II, i, 209 ; II, ii, 66.
[6] Contenau, 45, 51, 150 ; Wiedemann, A. Rel., xxvi, 336 ; Olmstead, 68 f.
[7] Cooke, EB[14], xvii, 769 ; Woolley, Syria, 1921, ii, 190.
[8] Otto, 1925, 47 ; Köster, Beih. alt.

Orient, 1924, i, 3, 17, 19 f. ; ib., Seefahrten, 1926, 31 ; Fleming, Tyre, 137 ; Hawes, Crete, 44 ; Glotz, Ægean Civ., p. vi.
[9] Breasted, H., 262.
[10] Geographia Sacra, Phaleg et Chanaan, Caen, 1646 ; edition used in Opera Omnia, fol., Leyden, 1712 ; ib., Hierozoicon, 2 vols., fol., Frankfurt, 1675, both containing much interesting and scholarly material.
[11] Die Phönizier, Bonn, 1841–56 ; ib., "Phönizier" in Ersch-Gruber, Allgemeine Encyclopädie—very detailed and less speculative.
[12] Meyer, Alt., II, ii, 86.
[13] Contenau, Civ. phénic., 84 ; Meyer, Alt., II, ii, 106 ; Grote, Hist. of Greece, 1869, iii, 266, 271, 277.
[14] Perrot-Chipiez, Phœnicia, i, 37, 40 ; Olmstead, H. Pal. Syr., 578.

28

Sidon remained important manufacturing cities, their final downfall coming only with the Turks in 1516.[1]

PHŒNICIAN COLONIES

Phœnician colonisation, probably from Tyre, was extensive. The people settled on the coast of Asia Minor, in Cyprus—where they were later and less important than the Greeks—and in the Ægean, e.g. Crete, from whence they could reach Rhodes, the Sporades and Cyclades. The rich silver mines on Siphnos and Kimolos were exploited, and there were small commercial stations at Thera and Melos and on the Thracian islands, e.g. Thasos, where the gold mines were worked [2], as well as those at Pangaios. The Phœnicians perhaps penetrated to the Euxine, carrying copper, iron and steel from the Chalybes.[3]

Phœnician colonisation was like the Greek, with attempts to trade with local inhabitants, coastal points, outlying islands and easily defended promontories occupied and developed with factories, sometimes with portions of the hinterland for agriculture. The colonisation, however, was on a small scale as compared with Greek settlements in Asia Minor, lower Italy and Sicily.[4]

Although later Phœnician influence on Greece has been over-estimated [5], there is no doubt that Phœnician art and craftsmanship left a marked impression.[6] Late Minoan tombs at Eleusis contain Egyptian glazed ware and Phœnician imitations of Egyptian scarabs, indicating Phœnician trade with Attica, although there was probably not a settlement there.[7] Italy proper was never invaded by the Phœnicians, who were allied with the Etruscans, but their wares appeared in the 8th century B.C.[8]

The late Palæolithic (Capsian ; from Gafsa, Roman Capsa, in south Tunis) culture of North Africa is little known, but seems to be related to the Pre-dynastic Egyptian[9] : it had extensions to Palestine, Sicily and south Italy, also south and east Spain. The traditional date of 1112 or 1101 B.C. for the foundation of Utica, the greatest city of ancient Africa, is too early, but the city is older than Carthage, and Phœnician trading posts on the coast of North Africa were established from about 1100. Sicily was occupied about the same date as the foundation of Carthage,[10] Qartchadašt, "the new city," about 850–800 B.C., which was always a daughter-city of Tyre.[11] Herodotos[12] says the Phœnicians from Carthage traded with the western coast of Africa beyond the Pillars of Hercules (Gibraltar) by a method of barter, the natives offering gold for the goods. Later trade was largely concentrated in Carthage and Gades (Cadiz, in Spain), Tyre being unimportant after the 6th century B.C. This is the period to which Ezekiel[13] refers, and although identifications are doubtful his account is supposed to mean that metals were exported from Tharshish (Tartessos ?)[14], bronze vessels from Javan (Greece ?) and from the Tibareni and Moschi of Pontos, ivory and ebony from Rhodes and other islands, costly stones and coloured stuffs from Edom (Idumæa), agricultural produce from Judah and Israel, wine and wool from Damascus, balsams, gems and gold from Saba (south Arabia) and goods from Harran

[1] Dyer, DG, ii, 613.
[2] Herodotos, vi, 47.
[3] Perrot-Chipiez, Phœnicia, i, 30, 38 ; Lenormant, H. anc., iii, 28, 32 ; W. H. Engel, Kypros, 1841, i, 168 ; Maspero, Struggle, 204 ; Meyer, Alt., II, ii, 81, 86, 117.
[4] Meyer, Alt., II, ii, 84, 87, 106, 113.
[5] Myres, CAH, iii, 639 ; Contenau, Civ. phénic., 340, 345, 350.
[6] Meyer, Alt., II, ii, 115, 117.
[7] Gardner, CAH, iii, 575.
[8] Olmstead, H. Pal. Syr., 405.

[9] Obermaier, RV, ii, 274 ; Scharff, Morgenland, 1927, Heft xii, 14.
[10] Meyer, Alt., II, ii, 83, 106, 108 ; Hackforth, CAH, iv, 347 ; Olmstead, 104, 404.
[11] Contenau, Civ. phénic., 93 ; Meyer, Alt., II, ii, 109 ; Ehrenberg, Morgenland, 1927, xiv ; very various dates—see Smith, DG, i, 531.
[12] iv, 196 ; cf. Hennig, Isis, 1929, xiii, 200.
[13] xxvii.
[14] Smith, CDB, 1865, 921.

and other parts of Mesopotamia as far as Assur.[1] Overland trade to Babylon, Assyria and Egypt, however, was in the hands of Syrians and Bedouin Arabs.[2] Carthage extended its influence to Malta, Sardinia, Sicily and the Balearic Islands, and through the Pillars of Hercules to the west, where its Atlantic trade—a jealously guarded monopoly—reached the Kassiterides.[3] Carthage probably had factories at Massilia or Massalia (Marseilles), a town not founded by Phœnicians but about 600 B.C. by Greek refugees from Phokaia, the most northerly Ionian city in Asia Minor.[4] The circumnavigation of Africa by Phœnician sailors subsidised by King Necho of Egypt (609–593) is described only by Herodotos [5] and has been called into question.[6] A voyage undertaken by Hanno, a sea captain of Tyre, with a commission from the Carthaginian senate [7], at some date not definitely known, perhaps 500 B.C., probably reached the Gulf of Guinea or Sierra Leone.[8] Himilco of Carthage sailed north about the same time, reconnoitring the coasts of Spain and Gaul, touching the British Isles and penetrating to the Sargasso Sea.[9] His lost report is known only from quotations in Rufus Festus Avienus (c. A.D. 400).[10] The extant Greek Periplus of Skylax of Caryanda, in Caria[11], who is supposed to have been sent by Darius I (521–485) to explore the coast of Asia from the Indies to the Red Sea, was edited about 338 B.C., and mentions a colony founded by Hanno, but is of doubtful authenticity.[12]

TARTESSOS

Direct proofs of Phœnician influence in Spain are wanting before the period of Carthage. The rich gold ornaments, silver, gilt silver, rings with imitation Egyptian scarabs, bronze and a glass vessel with a pseudo-Egyptian inscription found at Aliseda belong to the 6th century B.C., although a few ivory plaques with griffins, battle scenes, etc., from Spain are supposed to be pre-Carthaginian.[13] Supposed relations of Tartessos with the east from 3000 B.C. are doubtful, but south Spain in the period 3000–2000 B.C. was a centre of highly developed culture, particularly in ceramics and metal working.[14] The earliest Phœnician colony in Spain was Gadir, "the enclosure,"[15] i.e. Gades, the modern Cadiz. Phœnicians were in the Mediterranean by 1500 B.C.[16] and the date 1100–1050 given by Greek tradition for the foundation of Gades is probably correct.[17]

[1] Rawlinson, Phœnicia, 283.
[2] Meyer, Alt., II, ii, 128.
[3] Perrot-Chipiez, Phœnicia, i, 49, 414 ; Olmstead, 404.
[4] Perrot-Chipiez, Phœnicia, i, 46 ; Long, DG, ii, 290 ; Schmitz, ib., ii, 603 ; Meyer, Alt. (1893), II, 694.
[5] iv, 43 ; Grote, Hist. of Greece, 1869, iii, 284.
[6] Tozer, Hist. of Anct. Geography, 99.
[7] Perrot-Chipiez, Phœnicia, i, 48 ; Cory, Anct. Fragments, 1832, 203 ; text in Müller-Didot, Geogr. Græci Min., i, 1–14, and Aly, Hermes, 1927, lxii, 321 ; tr. by W. H. Schoff, The Periplus of Hanno, Philadelphia, 1913 ; complete transl. in Rawlinson, Phœnicia, 389–392 ; cf. A. Schulten, Tartessos, Hamburg Univ. Abhandl. viii, Hamburg, 1922, 38 ; Tozer, 104.
[8] Meyer, Alt., II, ii, 100 ; Schoff, op. cit. ; Olmstead, H. Pal. Syr., 404 ; Hart, Isis, 1928, x, 105.
[9] Pliny, ii, 67.

[10] Ora Maritima, 117, etc. ; in Pomponius Mela, etc., Bipont ed., Argentorati, 1809, 140 ; Meyer, Alt., II, ii, 99 ; Tozer, Anct. Geogr., 36, 109.
[11] Müller-Didot, Geogr. Græc. Min., i, 15–96.
[12] Gisinger, PW, va, 619 f., 635 ; Tozer, Anct. Geogr., 119.
[13] Meyer, Alt., II, ii, 105; del Castillo, RV, i, 100.
[14] Otto, 1925, 33 ; Hackforth, CAH, iv, 347 ; Meyer, Alt., II, ii, 84, 95, 105 ; Olmstead, H. Pal. Syr., 405 ; Bosch-Gimpera, RV, x, 352 f., 360 ; H. and L. Siret, Les premiers âges du métal dans le sud-est de l'Espagne, Antwerp, 1887.
[15] Maspero, H. anc., 1905, 372.
[16] Schulten, Tartessos, 6, 18 ; denied by Ashby, CAH, ii, 581, who says 750.
[17] Hübner, PW, viii, 446 ; Meyer, Alt., II, ii, 90, 92, 94—but cp. with his date of 800 for Carthage ; Schulten, PW, viii, 2032 ; ib., Tartessos, 6, 181 ; Boyd Dawkins, Early Man in Britain, 454.

Wonderful stories of the great metallic wealth of Spain reached the Greeks when the district was in relations with Gades, especially of the rich silver mines of Sierra Montana.[1]

Tartessos, the Phœnician and Hebrew Tharshish (Tarsisi in Assyrian texts of Asarhaddon, 670 B.C.) [2], was a name given by later Greek and Latin writers to a town situated at the mouth of the River Bætis (the Guadalquivir), to this river itself, to a town Gades (destroyed about 500 B.C.) or to an island Erytheia (also called Gades) in the Bætis delta, with a "mountain" called Kassios.[3] It was identified by Schulten [4] with an island in the mouth of the Guadalquivir, and the site was pre-Phœnician and distinct from Gades. The Phœnician towns Abdera, Sexi and Malaka at the foot of the Sierra Nevada were probably founded in the 5th century B.C. : there is no evidence that Karteja (Algeciras) and Kalpe (Gibraltar) were ever Phœnician.[5]

SARDINIA

In the early period (before 2000 B.C.) Sardinia and Sicily acted both as a barrier and as an entrepôt between East and West. The Sardinians (Shardina) appear in Egypt in Rameses III's time (1200–1090).[6] The Phœnician conquest of Sardinia dates from 500–480, although it had been exploited earlier, ships from Tyre putting into the harbour of Caralis (Cagliari).[7] The Phœnicians established themselves firmly, altering the character and even the population of the island, and remained masters of it until 238 B.C., when it became Roman by treaty.[8] The country had rich mines of metals, especially silver [9], but the iron deposits were not worked.[10] The silver and lead mines, formerly very productive, were almost worked out by the Phœnicians. There are rich veins of copper in the mountains near Iglesias.[11] Metal objects were found in the tombs, the bronze figures being probably native work of 1000–700 B.C.[12] There are lead ingots with the casting pieces still attached ; these, or stones with hollows, were used as supports.[13] Votive swords are mounted upright on pedestals (73 f.)—a unique type, although this is a Scythian custom in Herodotos.[14] Enormous quantities of pig-metal were found near ancient Valentia at Forraxi, old bronze articles being melted down in crucibles (78 f.). Metal working formed an important occupation of the islanders. Native daggers, swords, stilettos, etc., are of bronze, and coats of mail are represented on statuettes. The bronze was often impure : the copper had the composition : copper 78·424, iron 9·640, silica 6, sulphur 2·475, lead 1·800 (85 f., 89). Numerous cakes or pigs of copper are perhaps of Etruscan origin. No tin

[1] Diodoros Siculus, v, 35 ; ψ-Aristotle, De mirab. auscult., 135 ; Justin, Historiæ Philippicæ, xliv, 1 f. ; Meyer, Alt., II, ii, 95, 97, 103 ; Rawlinson, Phœnicia, 313 ; Hübner, PW, iii, 859 ; Blümner, iv, 101.
[2] Meyer, Alt., II, ii, 94, 98, 102 ; Schulten, PW, viii, 2032 ; Hübner, PW, vii, 439.
[3] Dyer, DG, ii, 1106 ; Smith, DG, i, 367, 923 ; Hübner, PW, ii, 2763 ; Skymnos of Chios, in Müller-Didot, Geogr. Græci Minores, i, 196 ; Pomponius Mela, ii, 6, Bipont ed., 1809, 56 ; Festus Avienus, 85, in Poetæ Latini Minores, ed. Lemaire, Paris, 1825, v, 414 ; Rawlinson, Phœnicia, 125.
[4] Tartessos, 37, 53, 59, 81 ; Isis, 1926, viii, 585 ; ib., 1929, xiii, 200 f. ; Meyer, Alt., II, ii, 96, 106.

[5] Meyer, Alt., II, ii, 93.
[6] Otto, 1925, 34 ; Hall, Bronze Age, 240.
[7] Perrot-Chipiez, Phœnicia, i, 30.
[8] Bunbury, DG, ii, 909 ; Olmstead, H. Pal. Syr., 405.
[9] Solinus, Polyhistor, 4.
[10] Perrot-Chipiez, Art in Sardinia, Judæa, etc., 1890, i, 102 ; the article on metals, etc., i, 89–91, is faulty.
[11] Ib., Sardinia, i, 2 ; Rawlinson, Phœnicia, 117 f.
[12] Forrer, Urg., 436, 563 ; Perrot-Chipiez, Sardinia, i, 9, 58 f. ; Dussaud, La Lydie et ses Voisins, 1930, 12.
[13] Perrot-Chipiez, Sardinia, i, 72, 89 ; references in the text are to this.
[14] iv, 62.

occurs and the bronze was probably imported. A piece of mineral which had been called cassiterite (tin oxide) from its fibrous structure was really zinc oxide (89, 90 : text obscure). Enormous heaps of scoria show that the native lead ore was worked (89, 90), but no early iron objects were found, those reported being probably Roman (9, 89, 102), and there was no glass (94). There was a considerable production of purple dye from murex.[1]

[1] Suidas ; Bunbury, DG, ii, 913.

PHŒNICIA II

METALS

METAL WORKING

The substratum of the entire Phœnician region is Jurassic limestone without any important minerals, although the sandstone overlaying it in places is often rich in iron, which is the only metal found in the country. Trap and basalt are plentiful and there are porphyry and greenstone. Quartz and chalcedony occur at Carmel.[1] Metal working in copper, bronze, silver and gold, decorated with more or less complex and elegant designs, was for seven or eight centuries a Phœnician speciality [2], and this was possible only by the use of imported ores. There is in all cases a mixture of designs, Egyptian—which is predominant and perhaps oldest—Assyrian and probably Hittite. The dishes are considered by some as Phœnician, by others as Cypriote, perhaps copied in Phœnicia.[3] The test of genuine Phœnician art, distinguishing it from the north Syrian, is said to be its greater clumsiness, its conscious and careful eclecticism and its use of scenes and motives directly though badly imitated from Egyptian, Assyrian, Syro-Hittite or other Oriental arts side by side on the same object, with occasional reminiscences of Minoan art, derived from Cyprus and perhaps Cilicia. This is seen on the bronze bowls from Nimrûd (Assyria), Cyprus, Olympia, Delphi and Crete, on the silver bowls exported to Italy and in the treasure of Aliseda in Spain. The bowls date from the 9th century (Nimrûd) to the 7th century (Italy). In 6th century Cyprus, Egyptian designs were adopted with more success than by the Phœnicians.[4]

The old supposition [5], partly based on the legend of Cadmus, that the Phœnicians initiated the mining and metallurgy in the lands they visited is probably incorrect. They really in most cases merely developed mining and metallurgical processes already used by the native inhabitants, and tended towards imitation, exploitation and elimination rather than initiation—they were, in fact, the true forerunners of modern "rationalisation" in industry [6] and represented a transitory and unimportant stage in the development of the technical arts. The metallic ores first worked by natives were probably near the surface and the Phœnicians no doubt introduced deeper mining, perhaps learnt from the Egyptians in Sinai.[7]

GOLD

Pliny [8] makes Cadmus (Kadmos) the Phœnician discover the secret of smelting gold. The legend of Kadmos, which has been supposed to indicate

[1] Rawlinson, Phœnicia, 47 ; Dyer, DG, ii, 607.
[2] Perrot-Chipiez, Phœnicia, ii, 338 ; Lenormant, H. anc., iii, 120 ; Beck, Geschichte des Eisens, 1884, i, 171 ; Wroth, DA, i, 324.
[3] Perrot-Chipiez, Phœnicia, ii, 10, 74, 364 ; Contenau, Civ. phénic., 123, Fig. 37, 148–183, 207, 278 ; Meyer, Alt., II, ii, 132, 135 ; Maspero, Struggle, 576 ; Dussaud, Civ. préhellenique, 303, 326.

[4] Hall, CAH, iii, 327.
[5] Lenormant, Prem. Civ., i, 151, 154.
[6] Schrader-Jevons, 156 ; Schrader, RL, 69 f. ; Orth, PW, Suppl. iv, 109 ; Gowland, Archæologia, 1920, lxix, 131.
[7] Rawlinson, Phœnicia, 309 ; Movers, art. "Phönizien," in Ersch-Gruber, Encyclopädie, 1847, 361, 371 ; Freise, Gesch. der Bergbau, 9.
[8] vii, 56.

an early Phœnician colonisation of Greece [1], is really of Bœotian origin, Kadmos being probably the hero eponymos of the town Kadmeia in Thebes.[2] For Kallimachos (d. 240 B.C.) Kadmos was the equivalent of Hermes-Kadmilos (or Kamillos), one of the Kabiri [3], who were mysterious divinities supposed to be of Phœnician origin, and the relation of Kadmos to metallurgy in Greek legend was very close. He is supposed to have first worked the mines of Pangaios in Thrace.[4]

The Sidonians are represented by Homer as very skilled in working the precious metals, and artistic work in gold and silver occurs in the oldest remains at Byblos. Gold objects occur in the remains of 2000 and of 1800–1600 B.C. ; some are soldered, but not the earliest.[5] In the early remains (c. 2000) are a gold horn [6], pectorals in cloisonné with an Egyptian name in hieroglyphics, and pectorals of gold sheet worked in repoussé in the form of a hawk with outstretched wings, of Phœnician, not Egyptian work, since there are faults of design.[7] There is an obsidian bottle with gold setting with the name of Amenhemet III (1849–1801), and remains of 1800–1600 include a pendant of gold incrusted with stones, composed of two gold discs soldered one to the other and provided at the back with a cylindrical piece, the stones or gold beads in the decoration being held in place in small cavities by means of bituminous cement.[8] Gold and other metals occur in remains of 1800–1400 at Qatna in central Syria.[9] There is Phœnician gilt silver and gilt bronze work of 1000–400 B.C.[10] One of the two pillars outside the temple of Herakles at Tyre was said to be of pure gold.[11] The Phœnicians paid Šalmaneser III in 859 B.C. tribute of ingots of gold, silver, lead and copper[12], and Ašurnasirpal received gold as tribute in 916 B.C.[13]

Weighed bars of gold and silver were used as money, the ratio of value being probably 13 : 1, as in Assyria (9th–7th centuries) ; stamped coinage, copied from the Lydian, appeared in the 6th century B.C. Several examples of early Phœnician coinage are known, but trade was essentially by barter.[14] In cuneiform tablets written in a pre-Phœnician alphabet about 1200 B.C., found at Sapuna on the north Syrian coast with objects showing Egyptian, Hittite, Cyprian and Cretan influences, gold has double the value of "good" (? unadulterated) silver.[15] The Phœnicians exploited a gold mine on the island of Thasos[16] and also those on Pangaios in Macedonia, near Philippi[17], where Herodotos[18] says there were the remains of old Phœnician mines, that at Skapte Hyle having furnished yearly 80 talents of gold. The mines at Thasos yielded less than the Skapte Hyle one, but were prolific and a huge mountain had been

[1] Herodotos, i, 56 ; ii, 57 ; v, 57 ; Diodoros Siculus, i, 18 ; Blümner, iv, 3, 173 ; Lenormant, Prem. Civ., ii, 313 ; Giles, CAH, ii, 27 ; Eisler, Orphisch-Dionysische Mysteriengedanken, Leipzig, 1925, 192, 283, 325, 392 ; qadmu, qadmilu, "ancient" ; cf. Weill, Syria, 1921, ii, 120 ; Woolley, ib., 177.

[2] Meyer, Alt., II (1893), 70, 150, 189, 317 ; II, i, 254 ; II, ii, 115 ; Crusius, Ro., ii, 882, 2529 ; cf. Bürchner, PW, x, 1460 ; Latte, ib., 1461, 1470.

[3] Bloch, Ro., ii, 2530 ; Kern, PW, x, 1399, 1430, 1442 ; Meyer, opp. cit. ; Karst, Grundsteine zu einer mittelländisch-asianischen Urgeschichte, Leipzig, 1928, 132 ; Zielinski, A. Rel., ix, 57.

[4] Clement of Alexandria, Stromateis, i, 16 ; Latte, PW, x, 1462 ; Gressmann, PW, ia, 1816.

[5] Contenau, Civ. phénic., 153, 157.

[6] Olmstead, H. Pal. Syr., 70.

[7] Contenau, Civ. phénic., 59, 154, Fig. 15, 388, Fig. 135 ; ib., Manuel, ii, 863.

[8] Ib., Civ. phénic., 69, 152, 157, Fig. 19 ; Olmstead, H. Pal. Syr., 96, Fig. 43.

[9] Garstang, The Hittite Empire, 1929, 325 ; on excavations at Qatna see du Mesnil du Buisson and Dussaud, Syria, 1926, vii, 289 ; 1927, viii, 277, 337 ; 1928, ix, 6, 81 ; 1930, xi, 146.

[10] Perrot-Chipiez, Phœnicia, ii, 341, 343, 359, 362, 372, 373 ; N. Davis, Carthage and her Remains, 1861, 209.

[11] Meyer, Alt., II, ii, 265.

[12] Olmstead, H. Pal. Syr., 375.

[13] Lenormant, H. anc., iii, 70.

[14] Contenau, Civ. phénic., 214 f. ; Lenormant, Prem. Civ., i, 154.

[15] Olmstead, H. Pal. Syr., 234, 238.

[16] Perrot-Chipiez, Phœnicia, i, 30 ; Lenormant, H. anc., iii, 33.

[17] Maspero, Struggle, 204.

[18] vi, 46 f.

turned upside down in search of ore. There is no gold in Thasos at present, and the natives do not know it ever had any.[1] Although the Phœnicians are said to have worked gold and silver mines on the opposite coast [2], only one silver mine there is described by Leake.[3]

Most of the Phœnician gold probably came from Egypt, the Sudan and the interior of Africa [4], which metal later went to Carthage instead of Egypt, gold being obtained from the west coast of Africa by barter.[5] The story told, with scepticism, by Herodotos of the island Kyraunis (Karkenah) off the east coast of Africa, where young girls draw up gold dust by dipping into the mud birds' feathers smeared with pitch, seems to be fabulous. Some gold occurs in the sands of Carthage.[6] That the Phœnicians obtained gold from India [7] is doubtful, as is the statement [8] that they worked the gold mines in Spain, e.g. at Malaga.[9] They may have traded in gold from Colchis, between the Euxine and Caucasus[10] and the fabled land of the Argonauts and the Golden Fleece. It is usually supposed that the Greek name χρυσός for gold is derived from the Babylonian churâsu by way of the Hebrew-Phœnician chârûz.[11]

Both the Hebrews and the Greeks regarded the Phœnicians as expert in the manufacture of jewellery. This included beads of glass, faience, etc., of good quality, as well as work in gold and gems. Work resembling cloisonné enamel was produced by inlaying gems and glass between thin plates of gold. Silver and gold buttons were made ; soldering gold and metals had been learnt from Egypt and much developed.[12] Phœnician jewellery of the later period (c. 500 B.C.) shows individuality and includes gold with pale blue paste incrustations as ear-rings, gold plates and buttons, bracelets, collars of gold or gold with paste and rings with scarabs.[13] Amulets, seals, scarabs, necklaces and personal ornaments were produced in large numbers.[14]

The gold and electrum jewellery with Egyptian and Assyrian styles, including granulated work, found by Salzmann at Kameiros (Camirus) in Rhodes and now in the Louvre and the British Museum, is supposed to be Phœnician work of the 8th-6th centuries B.C.[15], although Egyptian and Greek-Egyptian faience made at Naukratis was exported to Kameiros and farther afield.[16] The jewellery described by Rawlinson[17] is mostly from Cyprus and doubtfully Phœnician.

SILVER

Artistic silver ware of 2000 B.C. at Byblos is perhaps in part imported Minoan, such as the silver blade inlaid with gold and niello, and a fragment of a dish with a spiral decoration in repoussé. The form of a teapot-shaped vase or ewer, the prototype of which is found in the old Egyptian spouted libation vase (gebeh) and in the oldest ceramics of Elam, was generally adopted in

[1] Cf. Grote, Hist. of Greece, 1869, iii, 266.
[2] Herodotos, vii, 112 ; ψ-Aristotle, De mirabilibus auscultat., 42 ; Thukydides, iv, 105 ; Diodoros Siculus, xvi, 8 ; Pliny, vii, 56.
[3] Rawlinson, Phœnicia, 312.
[4] Budge, Amulets, 25 ; Maspero, H. anc., 1905, 374.
[5] Herodotos, iv, 195, 196 ; Rawlinson, Phœnicia, 303 ; Blümner, iv, 12.
[6] Davis, Carthage and her Remains, 1861, 29.
[7] H. G. Rawlinson, India, 1926, 11.
[8] Pliny, vii, 56 ; Strabo, III, iv, 6, 158 C. ; XIV, v, 28, 680 C.
[9] G. Rawlinson, Phœnicia, 103.

[10] Perrot-Chipiez, Phœnicia, i, 30 ; Lenormant, H. anc., iii, 28, 34, 35 ; Strabo, XI, ii, 19, 498 C.
[11] Bochart, Hierozoicon, ii, 9 ; Lenormant, Prem. Civ., ii, 425 ; Hehn, Kulturpflanzen und Haustiere, 1877, 62, 498, from Pott and Renan ; Schrader-Jevons, 174 ; Schrader, RL, 298 f.
[12] Dyer, DG, ii, 616 ; Perrot-Chipiez, Phœnicia, ii, 373 f., 388, 391, 393 ; Grote, Hist. of Greece, vol. iii, ch. 18.
[13] Contenau, Civ. phén., 211 f.
[14] Perrot-Chipiez, Phœnicia, ii, 7, 227 ; Budge, Amulets, 251 ; ib., Mummy², 285.
[15] Wroth, DA, i, 325 ; Forrer, RL, 136.
[16] Hall, CAH, iii, 326.
[17] Phœnicia, 360 f.

Phœnicia in the period 2000–1500 B.C. Among the silver objects of 1800–1600, perhaps 1900, B.C. found in a jar was a curious vase of Phœnician pattern, composed of truncated cones placed together at their bases, on a stand; there were also sandals, a mirror and girdles formed of a ribbon of metal, sometimes ornamented by points in repoussé.[1]

Silver of 1800–1400 B.C. occurs at Qatna in central Syria.[2] From about 1600 B.C. the Phœnicians took large quantities of silver by sea to Babylonia and silver coinage was in use early in Phœnicia and Palestine.[3] Ašurnasirpal in 916 B.C. says the Phœnicians paid him silver as tribute[4], and in Homer[5] the Sidonian artists are the makers of the finest silver craters (bowls). The source of the early Phœnician silver is not certainly known.[6] The Phœnicians are said to have worked the mines of Siphnos and Kimolos[7], and those of Thasos till the 8th century B.C.[8], and they had available from about 500 B.C. the rich silver mines of Sardinia.[9] There are silver mines in Carthage.[10]

Diodoros Siculus says the silver mines in Spain, accidentally made known by forest fires, were actively exploited by the Phœnicians, the natives not knowing the use of the metal and bartering it for goods of little value. There was so much silver that the navigators, when they had filled all the space in the ships, cut off the lead of the anchors and replaced it by silver.[11] The account implies that the natives worked the mines, although the Phœnicians may have owned some of them.[12] These Spanish mines were probably in Andalusia[13]; those near Abdera in south Spain were afterwards worked from Carthagena, a Phœnician colony founded in 243 B.C.[14] The silver in Spain had been worked at an early date, the mines of Sierra Morena perhaps by the natives before 2500 B.C., and the metal was carried in Phœnician ships from about 1200–1100 B.C.[15] Both Pliny[16] and Diodoros Siculus[17] say the Spanish silver, which contained 3 to 12·5 per cent. of gold, was considered the best. The ore was very abundant and occurred largely near the surface, but some veins ran to a great depth.[18]

COPPER

Copper mines in Lebanon, Syria and Palestine are mentioned by old authors[19], and small copper idols (5–8 cm. high) and statuettes of animals (some of which are found only in Syro-Phœnicia) of Phœnician origin occur in the oldest remains (3200–3000 B.C.) at Byblos.[20] Copper of 1800–1400 occurs at Qatna in central Syria.[21] That copper was first brought by Phœnicians to Greece from Cyprus and "Arabia" (?) and afterwards from Spain and Lusitania, the Greeks themselves working the Cyprus mines in Homer's time[22], is doubtful,

[1] Contenau, Civ. phénicienne, 55, 153, 155 f., Figs. 13, 14, 16; ib., Manuel, ii, 865, 867; Olmstead, H. Pal. Syr., 94 f., Figs. 39, 40; A. Evans, Palace of Minos, ii, 825; supposed Elamite influence on Byblos, Frankfort, Studies in Early Pottery, 1924, i, 73.
[2] Garstang, Hittite Empire, 1929, 325.
[3] Movers, Phönizier, II, iii, 28, 36 f.; Blümner, iv, 31.
[4] Lenormant, H. anc., iii, 70.
[5] Iliad, xxiii, 704–757; Odyssey, iv, 615–19; xv, 115–19.
[6] Meyer, Alt., II, ii, 95.
[7] Perrot-Chipiez, Phœnicia, i, 38; Lenormant, H. anc., iii, 70.
[8] Perrot-Chipiez, Histoire de l'Art, vi, 70; ix, 99.
[9] Solinus, c. 4; Perrot-Chipiez, Sardinia, i, 2.
[10] Davis, Carthage, 29.
[11] Diodoros Siculus, v, 35; ψ-Aristotle, De mirab., 135; Rawlinson, Phœnicia, 309; Dyer, DG, ii, 617.
[12] Gowland, Archæologia, 1920, lxix, 131.
[13] Hackforth, CAH, iv, 347.
[14] Rawlinson, Phœnicia, 121; Strabo, III, iv, 6, 158 C.
[15] Meyer, Alt., II, ii, 95, 97; Gowland, op. cit., 131; Schulten, Tartessos, 9, 13, 16, 25.
[16] xxxiii, 4, 6.
[17] v, 36.
[18] Rawlinson, Phœnicia, 313.
[19] Orth, PW, Suppl. iv, 113.
[20] Contenau, Civ. phénic., 150 f.
[21] Garstang, Hittite Empire, 1929, 325.
[22] Cart, Daremberg-Saglio, Dict., I, i, 121.

but the Greek name χαλκός is supposed by Lenormant [1] to be derived from the Phœnician halaq, to polish or work, whilst Eisler [2] derives it from the Aramaic name of copper, hal-hi (chalchi). The copper mines of the interior of Cyprus were probably not worked by Phœnicians, who confined their activities to coastal trade, but by the natives.[3]

The Phœnicians are said to have carried copper from the Chalybes of the Euxine [4]—from whom the Greek name for copper may be derived—and from the Moschi and Tibareni of the central parts of Asia Minor and the country later known as Cappadocia, in which regions the metal occurs abundantly.[5] The principal source was probably Cyprus.[6] There are copper mines in Carthage [7] and very old mines in Ochtschutschai in the Caucasus are still worked.[8] The later Phœnicians obtained copper from the Andalusian mines [9], which may have been worked by natives as early as 2500 B.C., the copper being shipped by Minoans before 1200, then by Phœnicians.[10] Hall[11] suggests that "Phœnicians" [pre-Phœnicians] on the Syrian coast shipped the first known copper to Egypt and Crete in 4000–3000 B.C. from the interior of Asia.

BRONZE

Bronze, which appears to have been exported to Egypt in the 16th century B.C.[12], was used in Syria as early as 2500[13] and has been considered[14] to be a Phœnician discovery, although others thought that a knowledge of the alloy probably came to Syria from Egypt or Babylonia.[15] The Phœnician origin of the early bronzes, both in Europe and in the East, had been doubted by von Bibra[16] on the grounds of their very variable composition and their content of lead.

Bronze cups and weapons made in Phœnicia under Egyptian influence but of Babylonian-Assyrian type, e.g. a scimitar (ἅρπη) with a gold uræus encrusted in the blade, occur in remains of 2000 B.C. at Byblos, and tubes and tubular spirals in those of 1800–1600 B.C.[17] Bronze of 1800–1400 B.C. was found at Qatna in central Syria.[18] Bronze of 1800–1600 B.C. found in a jar at Byblos with scarabs with Hittite inscriptions and cylinder seals resembling those used on the Cappadocian tablets (2300–1800 B.C.) includes rings of two wires zig-zagging from side to side, long pins with large heads, ornaments in the form of pail handles resembling later torques found in Gaul[19], tubes formed of wire spirals in the form of coiled springs, figures of the ibis, baboon or Harpokrates[20], keys and large bronze tridents similar to one of copper found in tomb IV at Mycenæ. The torques, tubes made from spirals, pendant, girdles and per-forated pins resemble those from the Caucasus, which had an early civilisation analogous to those of Elam and Sumer. Similar jewellery from the Caucasus occurs in a fairly late foundation at Susa.[21] That found at Byblos is supposed to date to about 1700–1650 B.C., when the objects in the jar (some as old as 2000 B.C., e.g. the cylinder seals) were collected. The objects may be imported

[1] Prem. Civ., ii, 425.
[2] Z. f. Assyriol., 1926, xxxvii, 113.
[3] Meyer, Alt., II, ii, 87.
[4] Perrot-Chipiez, Phœnicia, i, 30 ; Lenor-mant, H. anc., iii, 28, 34 f.
[5] Rawlinson, Phœnicia, 296 ; Xenophon, Anabasis, IV, i, 8.
[6] Budge, Amulets, 251.
[7] Davis, Carthage, 29.
[8] Belck, Verhl. Berlin. Ges. f. Anthropol., 1893, 61, 68.
[9] Hackforth, CAH, iv, 347.
[10] Schulten, Tartessos, 9, 13.
[11] Bronze Age, 1928, 31, 36, 38 ; cf. Hoernes, Urzeit, ii, 25, 30.

[12] Meyer, Alt. (1909), I, ii, 606.
[13] Hoernes, Urzeit, ii, 25.
[14] Kahlbaum and Hoffmann, in Diergart, 1909, 92.
[15] Perrot-Chipiez, Phœnicia, ii, 2 ; Beck, Gesch. des Eisens, ii, 184.
[16] Die Bronzen, Erlangen, 1869, 158.
[17] Contenau, Civ. phénic., 152 f., 157 f., 229 ; ib., Manuel, ii, 863.
[18] Garstang, Hittite Empire, 1929, 325.
[19] Yates, DA, ii, 857.
[20] Olmstead, H. Pal. Syr., 94 ; Contenau, Civ. phénic., 153, 157 ; ib., Manuel, ii, 866.
[21] De Morgan, Délég. en Perse, vii, 69, 87, plate 12.

or local imitations.[1] Although torques have been regarded as of Caucasian origin, they occur also in XII dyn. Egyptian remains and in tombs in Central Europe, and the type was also found in the tomb of Shub-ad at Ur (3100–3000).[2] The suggestion that they are of Irish origin [3] seems fantastic. The torque is a fairly simple object which may well have been evolved in various places independently of Irish genius.

Thothmes III in a battle near Aleppo captured a bronze spear inlaid with gold from the Syrian royal troops [4], and the Phœnicians in 916 paid Ašurnasirpal bronze as tribute.[5] Sardinian bronzes are dated 1000–700 B.C., and enormous quantities of bronze were cast at Forraxi, probably from imported metal.[6] The Phœnicians cast small objects of bronze at an early period. In the 11th century B.C. they cast immense pillars for the temple at Gades ("pillars of Hercules"), and in 989–982 B.C. the two hollow pillars (Jachin and Boaz) for Solomon's temple were cast in clay moulds in the Jordan Valley under the direction of a Phœnician craftsman.[7] The Phœnician bronze work was generally artistic and of good quality, swords and razors, for example, being made from the alloy.[8] The hollow image of the god Melkart, in which infants are said to have been burnt, was of bronze [9], but life-size statues were not usually produced. Statuettes, sometimes of a crude character, were made. The important series of bronze figures in the Louvre representing the Phœnician pantheon show Syro-Hittite influences, e.g. a statue which resembles Hadad, one in Egyptian style covered with gold leaf and some with two bodies coming from a common base (male-female) as in Hittite examples. Bronze ritual razors with Egyptian divinities engraved on them were found at Carthage.[10] A sacrificial dish has an inscription calling it the "first-born of copper."[11]

TIN—THE KASSITERIDES

The Phœnicians were formerly regarded as the first traders in tin.[12] In later times it was no doubt one of their most profitable commodities, since they were able to transport it by sea and supply it more cheaply than the metal carried overland from Central Asia.[13] Ašurnasirpal obtained tin from them in 916.[14] A late tradition makes King Midas of Phrygia (c. 970 B.C.) the discoverer of tin and lead[15], but tin was known long before Midas, and Midacritus in Pliny may stand for Melkart, the Phœnician god, hero, defender, etc., afterwards identified (e.g. in the "Pillars of Hercules") with the Greek Herakles.[16]

Tin is already known to Homer as kassiteros ($\kappa\alpha\sigma\sigma\acute{\iota}\tau\epsilon\rho\sigma\varsigma$)[17], but the first extant mention of its source as "the islands called Kassiterides" is in Herodotos (450 B.C.)[18], who says : "nor do I know of any islands called the Kassiterides,

---

[1] Rostovtzeff, JEA, 1920, vi, 6, 21 ; Contenau, Civ. phénic., 158, 160 ; Olmstead, H. Pal. Syr., 98, 115, Fig. 46 ; cf. Frankfort, Studies in Early Pottery, 1927, ii, 150.

[2] Contenau, Manuel, ii, 870 ; Dussaud, La Lydie et ses Voisins, 1930, 847.

[3] Petrie, Daily Telegraph, Aug., 1932 ; June, 1933.

[4] Breasted, H., 303.

[5] Lenormant, H. anc., iii, 70.

[6] Perrot-Chipiez, Sardinia, etc., i, 58, 78, 90 ; Forrer, Urg., 436, 563.

[7] Gsell, 35, 78 ; Feldhaus, Technik, 114 ; Rawlinson, Phœnicia, 425.

[8] Lenormant, Prem. Civ., i, 162 ; Perrot-Chipiez, Phœnicia, ii, 415 ; Rawlinson, Phœnicia, 214, 294.

[9] Diodoros Siculus, xx, 14.

[10] Contenau, Civ. phénic., 185, 209 f.

[11] Meyer, Alt., II, ii, 126, 170.

[12] Rawlinson, Phœnicia, 291, 294, 300, 314 ; Lenormant, Prem. Civ., i, 146, 152.

[13] Perrot-Chipiez, Phœnicia, i, 37.

[14] Lenormant, H. anc., iii, 70.

[15] Pliny, vii, 56 ; Hyginus, Fabulæ, 274, in Mythographi Latini, ed. Munckerus, 1681 ; Cassiodorus, Variorum, iii, 31 in Migne, Patrologia latina, lxix, 594 ; cf. Schulten, Tartessos, 26 ; Reinach, Cultes, Mythes et Religions, 1905, iii, 323.

[16] Besnier, Daremberg-Saglio, Dict., IV, ii, 1461 ; Cary, J. Hellenic Studies, 1924, xliv, 169 ; Meyer, Alt., II, ii, 81, 107, 141, 146 ; Bochart, Geographia Sacra, 1712, 651.

[17] Iliad, xi, 25, 34 ; xviii, 474, 565, 574, 613 ; xxi, 592 ; xxiii, 503, 561.

[18] iii, 115.

whence comes the tin we use . . . nevertheless amber and tin do certainly come to us from the ends of the earth." This section of Herodotos is usually supposed to be a compilation from Hekataios of Miletos (c. 520 B.C.) [1], although it occurs in the part retailing the fabulous Eastern stories [2], which may have been derived either from a Phœnician source during the visit made by Herodotos to Tyre or from some collection of travellers' tales.[3] The islands of the Kassiterides (Cassiterides) are usually identified with the British Isles.

The first Greek actually to reach Britain is supposed to have been Pytheas of Marseilles about 325 B.C., who is first mentioned in fragments of Dikaiarchos (d. 285 B.C.). Pytheas is supposed to have reached the tin mines near the promontory Belerion on the extreme western coast of Cornwall (Land's End) and to have taken a cargo of tin back to Marseilles. His account was perhaps used by the superficial Timaios (d. 256 B.C.), whose lost works were drawn upon and criticised by the prosaic and sceptical Polybios (198–117), who transferred the locality to the north-west coast of Spain [4], and so gave rise to the confusion which is found in the descriptions of Diodoros Siculus (c. 8 B.C.) and Strabo (A.D. 7). These two authors are the first since Herodotos whose accounts of the Kassiterides we possess in complete form ; since they are practically identical, they are probably taken from the same source, which is commonly assumed to be Poseidonios (135–51 B.C.) [5], although Pytheas, from Timaios, has also been supposed to have been used.[6]

Diodoros Siculus gives two accounts of the tin traffic, one in the section on Britain and the other in the section on Spain.[7] In the first he reports that the people of the British promontory of Belerion [Land's End, Cornwall] " . . . make the tin which with a great deal of care and labour they dig out of the ground, which is rocky, the metal being found in veins in the inside of the earth. They melt the metal and then refine it. They then form [? cast] it into cubical pieces like dice (astragala) and carry it to a British isle near at hand called Iktis. For at low tide, all being dry between them and the island, they convey over in carts abundance of tin. The merchants transport the tin which they buy from the natives to Gaul (France), and thence by a thirty days' journey they carry it in packs upon horseback through Gaul to the mouth of the River Rhône."

The translators of Diodoros [8] usually adopt Whittaker's identification of Iktis with the Isle of Wight, the Roman Vectis, but Barham's identification with St. Michael's Mount, off Penzance, is much more probable.[9] An old Cornish manuscript brought to Oxford in 1450 and now in the Bodleian Library identifies Iktis with Black Rock Island, off Falmouth.[10] Pliny's[11] statement that Iktis (which he calls Mictis) is an island producing tin (candidum plumbum) and six days' sail in coracles from Britain is obviously confused and of no value.[12]

[1] Meyer, Alt. (1893), II, 692, 759 ; Schmitz, DBM, ii, 362 ; Berger, Erdkunde, 53, 357.

[2] Herodotos, iii, 107, 110, 111.

[3] Latham, DG, i, 432 ; Tozer, Anct. Geography, 7, 153.

[4] T. R. Holmes, Ancient Britain, Oxford, 1907, 217 ; Berger, Erdkunde, 353, 361, 512 ; ib., PW, vi, 1304 ; Lelewel, Pytheas de Marseille, 1836, 14, 30 ; Hübner, PW, iii, 859 ; Tozer, 17, 152 f., 158, 163 ; Cary, J. Hellenic Studies, 1924, xliv, 168, 170, 171 ; Martini, PW, v, 561, 569.

[5] Berger, Erdk., 361, 544, 559, 630.

[6] Holmes, Anct. Britain, 499, 505 ;

Tozer, Anct. Geography, 153, 156 ; cf. R. Hennig, Rheinisches Museum, 1934, lxxxiii, 166.

[7] v, 22, 38, respectively.

[8] Booth, fol., London, 1700, 185 ; Hoefer, Paris, 1865, ii, 23 ; also Ridgeway, and Latham, DG, ii, 12.

[9] Bapst, L'Étain, 18 ; Geo. Smith, The Cassiterides, 1863, 79, 114 ; Tozer, Anct. Geogr., 156 ; Holmes, Anct. Britain, 239, 499, 502, 508.

[10] W. Pryce, Mineralogia Cornubiensis, fol., London, 1778, p. vi.

[11] iv, 16.

[12] Holmes, 499 f., 514.

The ingots shaped like dice, or really knuckle bones (ἀστραγαλα) [1], are known. An ingot of this shape, 2 ft. 8 in. long, weighing 158 lb., was found in the sea at Falmouth and is now in Truro Museum.[2] It is said by Schulten [3] to have the form of Cretan copper ingots; Rawlinson [4], who noted that the shape was different from the Roman, Norman and later ingots, concluded that it was Phœnician. Other ingots (e.g. one from the Thames and another of 29 lb.) and British tin coins are also known.[5]

In his account of Spain [6] Diodoros says: "Tin is found in various places in Iberia (Spain), not on the surface of the ground as some historians have pretended, but in mines from which it is dug to smelt like silver and gold. The richest mines of tin are the Isles of the Ocean, opposite Iberia and above Lusitania and called, for this reason, the Isles of Kassiterides. There is also much tin brought from the island of Britain to Gaul. The merchants pack it on horses and transport it through the Keltic interior to Marseilles and Narbonne. The latter is a Roman colony, the most important entrepôt of the country on account of its situation and opulence." Strabo [7] gives the same account, quoting Poseidonios, and [8] says vaguely that the Kassiterides are in the high seas in nearly the same latitude as Britain, fronting Europe outside the Pillars [Gibraltar] along with Gades and Britain, which is a hopelessly confused statement. He proceeds [7] to say that " tin is produced both in places among the barbarians who dwell beyond the Lusitanians and in the islands of Kassiterides, and from the Britannic Islands it is carried to Marseilles. Among the Artabri [9], who are the last of the Lusitanians towards the north and west, the earth is powdered with silver, tin [i.e. alluvial tin ore] and white gold." At the end of his chapter on Iberia[10] Strabo says: "The Kassiterides are ten in number and lie near each other in the ocean towards the north from the haven of the Artabri. One of them is deserted, but the others are inhabited by men in black cloaks, clad in tunics reaching to the feet, girt about the breast, and walking with staves, thus resembling the furies we see in tragic representa- tions. They subsist by their cattle, leading for the most part a wandering life. Of the metals they have tin and lead, which, with skins, they barter with the merchants for earthenware, salt and bronze vessels. Formerly the Phœnicians alone carried on this traffic from Gades, concealing the passage from every- one; and when the Romans followed a certain ship-master, that they might also find the market, the ship-master of jealousy purposely ran his vessel upon a shoal, leading on those who followed him into the same destructive disaster; he himself escaped by means of a fragment of the ship, and received from the state the value of the cargo he had lost. The Romans nevertheless by frequent efforts discovered the passage, and as soon as Publius Crassus[11], passing over to them, perceived that the metals were dug out at a little depth, and that the men were peaceably disposed, he declared it to those who already wished to traffic in this sea for profit, although the passage was longer than that to Britain." The account is probably fabulous[12] and should be rejected.

Julius Cæsar[13] says that in Britain: "the provinces remote from the sea (in mediterraneis regionibus) produce tin (plumbum album) and those on the

[1] Yates and Wayte, DA, ii, 759 and Fig.
[2] Montelius, Ält. Bronzez., 1900, 209, Fig. 510; Gowland, Archæologia, 1899, lxvi, 300; ib., J. Anthropol. Inst., 1912, xlii, 250.
[3] Tartessos, 7.
[4] Phœnicia, 318.
[5] Gowland, opp. cit.; Besnier, Darem- berg-Saglio, Dict., IV, ii, 1460, 1463.
[6] v, 38.
[7] III, ii, 9, 147 C.
[8] II, v, 15, 30.
[9] In Galicia, in Spain; Smith, DG, i, 226.
[10] III, v, 11, 175 C., after a citation from Poseidonios.
[11] Probably the Governor of Hispania Ulterior, in Julius Cæsar's time; Smith, Cassiterides, 80; Meyer, Alt., II, ii, 101.
[12] Tozer, 38; Hennig, Rhein. Mus., 1934, lxxxiii, 166.
[13] De Bell. Gall., v, 12, c. 40 B.C.

coast (in maritimis) iron, but the latter in no great quantity, and bronze is imported." Cæsar no doubt means by "the sea" the English Channel, Cornwall being, for him, far inland towards the west, whilst the iron was in Kent and Sussex.

Pliny [1] reports from Timaios that : "an island called Mictis [Ictis] is within six days' sail of Britannia, in which tin (plumbum album) is found"; also [2] that "opposite to Celtiberia (Spain) are a number of islands, called by the Greeks Cassiterides, in consequence of their abounding in tin." In his section on lead [3] Pliny says that of the two kinds, black and white (i.e. lead and tin), "the white is the most valuable, called plumbum candidum, by the Greeks cassiteros, and there is a fable of their going to the islands of the Atlantic to find it, and of its being brought in boats made of osiers covered with hides [i.e. coracles]. It is now known to be a product of Lusitania and Gallæcia (Galicia) and is a sand found on the surface of the earth, of a black colour and distinguished from the rest of the soil only by its weight." This last statement refers to alluvial tinstone.

We obviously have in Diodoros Siculus and Strabo a conflation of at least two accounts. The first deals with the tin trade of Britain and is on the whole clear ; the other contains vague and often contradictory references, perhaps from later sources, to other places where tin was, or had been, found. In attempting to extract some concrete information from this second narrative it must be kept in mind that some part of it, at least, may be so confused as to be worthless[4], and that no coherent theory can hope to reconcile all the statements in these authors, both of whom drew on contradictory sources.[5] Both Diodoros and Strabo agree that the continental port of shipment of tin from the Kassiterides was Marseilles, which was an Ionian Greek and not a Phœnician foundation.

Cary [6] points out that the overland trade of the Mediterranean peoples with the Atlantic regions did not pass through the Greek ports of southern France till comparatively late. Holmes had given reasons for supposing that even before Pytheas's voyage there was a regular trade in tin between Cornwall and across Gaul to Marseilles, perhaps also a seaborne trade between Cornwall and the Carthaginian port of Cadiz.[7] Cary maintains that the tin traffic through Marseilles (Massilia) can be traced only in the last three centuries B.C., when Greek merchants penetrated more than half-way through France but did not reach the Atlantic seaboard, and that the direct sea route to the centres of Atlantic trade was probably discovered by a Phokaian captain in the 6th century, and was explored by Pytheas of Marseilles in the 4th century B.C. This hypothesis rests on uncertain foundations, and there is every reason to think that Phœnician trade between Gades (Cadiz) and Britain is much older than any Greek exploitation.

The Phœnicians must have gone all the way by sea from Gades or even Carthage to Cornwall, since it is quite incredible that they should have allowed the tin shipped in Britain to pass out of their hands in an overland transit through Gaul, which was for them a completely unknown and unexplored territory. After the conquest of Gades by the Romans in 206 B.C., however, the jealously guarded Phœnician traffic with Britain probably came to an end, and the sea route was again available to Greek ships carrying south Italian crews.[8] The routes described by Diodoros Siculus and Strabo will thus refer

---

[1] iv, 16; one of his authorities is Poseidonios.
[2] iv, 22.
[3] xxxiv, 16.
[4] Meyer, Alt., II, ii, 101.
[5] Cf. Polybios, iii, 57.
[6] J. Hellenic Stud., 1924, xliv, 166; Isis, 1930, xiv, 507.

[7] Holmes, Anct. Britain, 218, 499 f., with copious references ; Stevens, Ancient Writers on Britain, Antiquity, 1927, i, 189 ; Isis, 1929, xiii, 201.
[8] Bapst, L'Étain, 12, 18 ; Tozer, 36 ; Smith, The Cassiterides, 1863, 57, 77 ; Cary, 171.

to the Roman Period for active commerce [1], which may go back to about 300 B.C. for restricted tin trade among Europeans. This overland transport perhaps lasted till the end of the 1st century B.C. A comparison with earlier routes for the transport of amber is not relevant, since amber went by way of the Rhine and Elbe, not Gaul.[2]

There are different routes possible across Gaul. In one (vaguely mentioned by Diodoros Siculus [3]) Narbonne was the objective. The cargo from Britain, perhaps shipped by sailors from Brittany [4], would go down the coast of Gaul to the mouth of the Garonne, then upstream to the neighbourhood of Tolosa (Toulouse), whence it was transported on pack animals along the slopes of the Pyrenees to the Sinus Gallicus (Gulf of Lyons). This route is also given by Ibn Sa'îd (1214–74 ?) for tin destined for Alexandria [5], and probably gave rise to Pliny's [6] idea that tin was obtained from the Pyrenees. The other route, which Cary [7] thinks older, is given in outline by Strabo. It was along the reaches and canals of the Loire, Seine or Rhine, from the middle or upper reaches of which, by means of slight differences of level, the Saône and Rhône were gained, along which the east coast of Gaul, Massilia, was attained. As an alternative, by a track over the Alps and the region of the Taurini, the Ligurian coast in north Italy and the mouth of the Po could be reached.[8] The city of Corbilon at the mouth of the Loire, which Strabo [9] says was a great emporium, seems to have been the place from which the tin was carried overland to Massilia and Narbo.[10]

The question as to how the Phœnicians became acquainted with British tin is one which, with existing sources of information, cannot be answered except conjecturally. It seems probable that they became acquainted with western tin at Gades and from there, step by step, they traced it to its source by way of Portugal and Brittany, which furnished small amounts of the metal but were never sources of any importance. The assumption that tin was obtained at Tartessos (Gades)[11] is based on a statement by Skymnos of Chios, in a poem of doubtful date[12], in which the River Bætis (Guadalquivir) is said to carry from Keltika "stream-rolled tin" (κασσίτερον ποταμόρρυτον).[13] Gades (Tartessos), however, was merely an entrepôt and no tin occurs there.[14]

It is, however, a common mistake to assert[15] that no tin occurs in Spain, since fairly extensive remains of tin workings exist near Oviedo and at Salabe, the ancient land of the Cantabrians.[16] Tin ores occur both in the Spanish north-west provinces of Salamanca, Zamora, Orense, Pontevedra and Corunna and in the Portuguese provinces of Troz os Montes and Beira Alta and smaller deposits in the provinces of Murcia and Almeria in Spain. The deposits (lodes and alluvial) are still worked on a small scale.[17]

[1] Cf. Berger, Erdkunde, 101, 232, 332 ; Speck, Handelsgeschichte des Altertums, Leipzig, 1900 f., ii, 470 ; Hoernes, Urg., ii, 514, who assume Greek shipment from Marseilles in the 5th century B.C., when transport through Gaul is improbable.

[2] Cary, J. Hellenic Stud., 1924, xliv, 172, 176.

[3] v, 38.

[4] Tozer, Anct. Geography, 36.

[5] Abulfeda, Géogr., tr. Reinaud, ii, 307 ; Gsell, 36.

[6] iv, 20.

[7] Op. cit., 178 f.

[9] H. Genthe, Über den etruskischen Tauschhandel nach dem Norden, Frankfurt, 1874, 65 f., 92 f., 101 f., 105 ; von Sadowski, Handelsstrassen der Griechen und Römer, Jena, 1877, 198 f. ; Haug, PW, ia, 764 ; Strabo, IV, i, 14, 189 C. ;

Lenormant, Prem. Civ., i, 156 ; Berger, Erdkunde, 102, 232, 332.

[9] IV, ii, 1, 189 C.

[10] Tozer, 37.

[11] Schulten, Tartessos, 6 f., 10 f., 25 f.

[12] Gisinger, PW, va, 661, 250–150 B.C. ; Wellmann, Die Georgika des Demokritos, Abh. Preuss. Akad., 1921, 15, c. 185 B.C. ; Meyer, Alt., II, ii, 98, c. 90 B.C.

[13] Skymnos, 165 ; Müller-Didot, Geogr. Græc. Min., i, 201.

[14] Movers, art. "Phönizier," in Ersch-Gruber, Encyclopädie, 1847, 361 ; Cary, 166 f.

[15] Smith, Cassiterides, 45.

[16] Besnier, Daremberg-Saglio, Dict., IV, ii, 1495.

[17] Wilkinson, Anct. Egyptians, ii, 229 ; Boyd Dawkins, 403 ; Lucas, JEA, 1928, xiv, 101.

The tin ores in Portugal, in Galicia, would also become known to the Phœnicians. The Callaïci, the people of Callæcia or Gallæcia (Καλλαικία or Καλαικία), north of Lusitania, the modern Galicia [1], are first mentioned by Poseidonios [2] as the Artabri, but in another place [3] as Galicians, "formerly called Lusitanians." The towns in this region were important in the Roman Period, and the tin mines of Portugal were worked even in the 18th century A.D.[4] They are, however, first mentioned in the 1st century A.D., and Humboldt found them very poor in tin, so that they hardly count as an important ancient source.[5] Justin [6] says Gallæcia produced abundance of copper, lead, minium and gold, the latter being turned up with the plough. The iron was of excellent quality, being tempered in the waters of the Rivers Bilbilis (mod. Bambola) or Chalybs.[7] Strabo seems to mean that tin was found near the coast of Portugal (Lusitania).[8] Pliny's statement [9] that the ore was found there on the surface also agrees with Strabo's that "the earth is powdered with silver, tin and white gold" (electrum). Charlesworth considers that the port of the Artabri was merely an entrepôt for tin from the Kassiterides. The ten or eleven small islands on the coast of Pontevedra, between Cape Folceiro and Silleiro, to which Diodoros and Strabo seem to refer[10] can hardly have been the real Kassiterides. If the Phœnicians obtained tin from these places it probably came from Britain, and they themselves as the demand increased tracked the metal to its true source, even if they did not know it before.

The assumption that the Galician (Portuguese) alluvial deposits were worked only from 95 B.C.[11], or even 24–19 B.C., by Roman engineers, who alone could erect the large hydraulic equipment necessary[12], seems doubtful, since alluvial deposits are notably amenable to small-scale primitive working. That there is no evidence for Phœnician working of these deposits[13], is neither surprising nor important, but the statement[14] that the Phœnicians were unknown in the western Mediterranean lands before 750 B.C. seems questionable, and they may have exploited the tin in Portugal at an early date.[15]

The islands off the Iberian coast appear to have been confused with the Œstrymnides, mentioned only in the poem "Ora Maritima" of Rufus Festus Avienus (A.D. 350–370), supposed to quote fragments of the Periplus (500–400 B.C.) of the Carthaginian Himilco and probably based on a Greek poem, itself compiled from two Greek narratives of different dates, its information being quite antiquated.[16] Avienus[17] says the Phœnicians of Tartessos "of yore" reached the promontory of the Œstrymnides, the islands of which were rich in tin and lead (in quo insulæ sese exserunt Œstrymnides, laxe jacentes et metallo divites : stanni atque plumbi). Apparently islands off Brittany, including Ushant, are meant[18], the supposed mines of which (perhaps really on the

---

[1] Smith, DG, i, 932 ; ii, 218 ; Charlesworth, Trade Routes, 162, 272 ; Hübner, PW, iii, 1356.
[2] In Strabo, III, ii, 9, 147 C.
[3] III, iv, 20, 166 C.
[4] Gmelin, Geschichte der Chemie, Göttingen, 1799, iii, 1140, 1143.
[5] Hennig, Rhein. Mus., 1934, lxxxiii, 167.
[6] Historiæ Philippicæ, xliv, 3.
[7] Smith, DG, i, 402 ; the Chalybs is supposed to have been a river between the Ana and the Tagus.
[8] Smith, DG, ii, 218 ; Berger, 559 ; Meyer, Alt., II, ii, 95.
[9] xxxiv, 16.
[10] Hübner, PW, iii, 859 f. ; Berger, PW, vi, 1307 ; Meyer, Alt., II, ii, 101 ; Schulten, PW, viii, 1987—"off Spain" ; Davis, Carthage, 23.

[11] Haverfield, PW, x, 2328.
[12] Cary, 168.
[13] Lucas, 102.
[14] Ashby, CAH, ii, 581.
[15] Bosch-Gimpera, RV, x, 352, 360 ; Meyer, Alt., II, ii, 95 ; Olmstead, H. Pal. Syr., 405.
[16] Ramsay, DBM, i, 432 ; Holmes, 490 ; Marx, PW, ii, 2387 ; Berger, Erdkunde, 235, 356 ; Meyer, Alt., II, ii, 97 f., c. 300.
[17] Ora Maritima, 94 f. ; in Lemaire, Poetæ Latini Minores, Paris, 1825, v, 415.
[18] Schulten, PW, viii, 1987 ; ib., Tartessos, 67 ; Meyer, Alt., II, ii, 101 f. ; Tozer, 36, 110, 156 ; Cary, 166 f. ; Peake and Fleure, Way of the Sea, 92 ; Lelewel, Pytheas, 24 ; Latham, DG, i, 433 ; Long, ib., ii, 499.

mainland opposite, in the country of the Osismi) were worked about 500 B.C. but rapidly became derelict. The use of bronze in western Gaul spread from Gascony, not Brittany, which seems to imply that it was imported from Spain.[1] Some tin does appear to have been obtained in Brittany and Gaul, where there are traces of ancient mines (e.g. at Montebras, in Limousin), and tin is still worked intermittently in France.[2] A tin traffic with Ireland [3] is quite improbable.[4]

It is possible that the tin which the Phœnicians found at Tartessos (Gades) came partly from Spain and partly by land or sea from Lusitania (Portugal), Galicia and the small islands off the north-west Iberian coast ; somewhat later it may have arrived there from western Brittany and neighbouring islands, and finally it came from Britain. The Phœnicians and Carthaginians, in their endeavour to locate the source of this tin, would then have proceeded by sea first to the west coast of the Iberian peninsula, then to the coast of Gaul and finally to Britain.[5] This hypothesis seems plausible and there is no better alternative to offer.

The Phœnician exploitation of Cornish tin has been denied[6] : no Phœnician remains, it is said, are known in Britain, and the mines were perhaps first worked by Bretons from the Continent or miners from Spain or Portugal.[7] Bronze ingots and moulds occur in Britain, very like those of the Continent [8], and it has been supposed that bronze was made on the island, but this is compatible with the use also of imported bronze ; the absence of pure copper instruments from old tumuli [9] also suggests that bronze was imported and made into the characteristic British objects, and Cæsar[10] expressly states that Britain imported bronze *or copper*. Bronze was first known in the East about 3000 B.C., and in the West in 2000–1000, and tin in the East in 2000–1000, but the argument that tin mining in Britain cannot, therefore, have begun before 1000, or even 500 B.C., when the postulated original supply of ore in the East was becoming exhausted[11], would apply only to Phœnician exploitation and we have no definite indication as to when tin was first produced in Britain. Hennig[12] thinks that British tin was the principal source for the ancient world in the period 2500–100 B.C., that the Bronze Age in Britain began in the 2nd millenn. B.C. and that there were then intensive relations with Spain. The tin bars found in England, he thinks, are as old as 1700 B.C. Other suggested dates are 1500–1000[13], or 500 B.C. for surface workings and 325 B.C. for deep workings[14], or even as late as 300–200 B.C.[15] It is further supposed that the Cornish mines were abandoned when the alluvial deposits of Portugal were exploited by the Romans, and were not worked again before A.D. 250.[16] British tin (ἡ βρεττανικὴ μέταλλος) is mentioned as shipped to Alexandria by Stephanos[17]

[1] Cary, 166 f. ; Boyd Dawkins, Early Man in Britain, 460.
[2] ψ-Aristotle, De mirab., 50, 81 ; Pliny, xxxiv, 17 ; De Launay, L'Anthropologie, 1901, 495 ; Coutil, ib., 624 ; Besnier, Daremberg-Saglio, Dict., IV, ii, 1459 ; Bapst, 3, 15 f.—Cape Pennestain ; Lucas, JEA, 1928, xiv, 102 f.
[3] Abulfeda, Géogr., tr. Reinaud, ii, 307 ; Boyd Dawkins, loc. cit. ; Hackforth, CAH, iv, 348.
[4] Rougemont, L'âge du bronze, 121 ; Cary, 167.
[5] Meyer, Alt., II (1893), 142 f. ; Bapst, 12 f. ; E. Speck, Handelsgeschichte des Altertums, Leipzig, 1900–06; i, 94 f., 103 f., 463, 483, 505 ; ii, 470 f. ; iii, 157, 290 ; W. Götz, Die Verkehrswege des Welthandels, Stuttgart, 1888, 109f., 267, 269, 290, 348, 352.

[6] Hübner, PW, iii, 860 ; Cooke, EB[14], xvii, 769.
[7] Lucas, 103.
[8] Latham, DG, i, 434 ; Holmes, Anct. Brit., 148 ; BMG Bronze Age, 112.
[9] BMG Bronze Age, 1920, 5 f.
[10] De Bell. Gall., v, 12.
[11] Lucas, JEA, 1928, xiv, 104 f.
[12] Rhein. Mus., 1934, lxxxiii, 167 f. ; cf. Oertel, CAH, x, 407.
[13] Pryce, Mineralogia Cornubiensis, 1778, p. ii ; Haverfield, PW, x, 2328 ; Holmes, Anct. Brit., 139 ; Latham, DG, i, 434.
[14] Cary, 167 ; cf. BMG Bronze Age, 21.
[15] Lucas, JEA, 1928, xiv, 103.
[16] Cary, 168 ; Besnier, Daremberg-Saglio, IV, ii, 1460.
[17] Ideler, Physici et Medici Græci Minores, 1841–2, ii, 206.

29

(7th century A.D.), and there was certainly a small Roman settlement at Bodmin in the 1st century A.D.[1] I feel unable to offer any definite suggestion, since the assumption of each authority is made to fit in with some other theory and no very satisfactory arguments are advanced as to the probable age of British tin working, which must have been in existence before the Phœnicians sought the metal in Britain, and hence may well go back to 1500 B.C., if not earlier. A knowledge of British tin probably soon spread to the Continent, since there was apparently active intercourse in the North Sea and Baltic in the later Stone Age.[2]

The later Greek and Roman geographers knew no more than did Diodoros Siculus and Strabo about the Kassiterides. The Spaniard Pomponius Mela (c. A.D. 43) merely says [3] all the Keltic islands which abound in tin (plumbum) are called by one name, Kassiterides. Dionysios Periegetes of Alexandria (A.D. 117–138 ?) [4], in a confused and corrupt account [5], seems to refer to the mines of Drangiana in Khorassan, although he may mean islands off Ushant ; and Ptolemaios (2nd century A.D.) could only say the ten Kassiterides lay in the western ocean (τῷ δυτικῷ ὠκεανῷ).[6]

An old and weak theory [7] supposes that there were no definite Kassiterides as islands, this being merely a general name for the unknown western European tin depôts from which the metal was collected. Holmes [8], after pointing out that Spain and Britain were possible sources of tin, argues that the Kassiterides either comprised Britain and the Scilly Isles or else that the name was later applied to imaginary islands. In the early period, when the tin traffic was Phœnician, all that the Greeks could learn was that the Kassiterides were in the northern sea, by which Britain, with islands off its coast, or Ireland, may have been vaguely meant. In a later period they were identified with islands known to be associated with the tin trade—the Œstrymnides, or the islands off the Galician coast, or the Scillies.[9]

The only important island source of ancient tin was Britain, and the name Kassiterides probably referred to Britain generally and to Cornwall, perhaps also by mistake to the Scilly Isles, which were known to Greek geographers of the 5th century B.C. but were rarely visited and forgotten. Tin was perhaps exchanged from the Scilly Isles, and hence these were thought to contain tin (of which they possess none, or only microscopic amounts)[10] and were called the Kassiterides[11]. Strabo seems to have intended the Kassiterides to mean the Scilly Isles and the Cornish peninsula ; other ancient authors wrongly identified them with Spain, but there are no islands in Spanish waters, except Ons, which is out of the question.[12]

The information on the Phœnician traffic in tin is so vague and scanty that practically nothing beyond the possibility that it reached Cornwall directly at some fairly early date can be deduced from it, and most of the accounts do not

[1] Charlesworth, 216, 282.
[2] Vogel, in Hoops, iv, 156.
[3] De Situ Orbis, iii, 6 ; Bipont ed., 1809, 76.
[4] Leue, Philologos, 1884, xlii, 175.
[5] Müller-Didot, Geogr. Græc. Min., ii.
[6] Haverfield, PW, x, 2328 ; Tozer, 284 ; Gisinger, PW, Suppl. iv, 559, 594, 663, 675 ; Orth, ib., 122 f.
[7] Smith, Cassiterides, 50 f. ; Haverfield, PW, x, 2328 ; Bremer, RV, iv, 545 ; cf. Hennig, Rhein. Mus., 1934, lxxxiii, 166, who rejects it.
[8] Ancient Britain, 483–98.
[9] Tozer, 38.
[10] Cary, 166.
[11] Meyer, Alt. (1893), II, 691 ; ib., II,

ii, 97 f. ; Tozer, 37 ; Lenz, Mineralogie, 1861, 87—"das Zinn kommt aus Britannien selbst" ; Reinach, L'Anthropologie, 1892, iii, 275 ; Hoernes, Urg., ii, 232 ; Besnier, 1460.
[12] Holmes, 497 ; it may be mentioned that Borlase, Observations on the Antiquities of Cornwall, fol., Oxford, 1754 ; ib., Observations on the Isles of Scilly, 4to., Oxford, 1756 ; Smith, Cassiterides, 65 f. ; Gutschmid, EB, 1885, xviii ; Rawlinson, Phœnicia, 301 f. ; Hübner, PW, iii, 860 ; and Cooke, EB[14], xvii, 769, have all identified the Cassiterides with the Scillies ; Pryce, Mineralogia Cornubiensis, 1778, p. iv, thought the name included Land's End.

refer to Phœnician trade at all. The usual accounts, based on Bochart's [1], are now antiquated, but very little positive fact has taken their place. The suggestion that the Coptic name for tin, pitran (cf. p. 81), is connected with britan (Britain) is made by Sethe.[2]

LEAD

Lead is among the metals traded by the Phœnicians of Tarshish [3] and they obtained it from the Kassiterides [4]—there is abundance of lead in Britain— and from the Andalusian mines in Spain.[5] There was so much lead ore in Spain that it was thought to grow again when the mine was closed for a time. Much of the Spanish lead contained silver, which was separated by cupellation, although some was nearly pure.[6] The Tyrian lead ($\mu\acute{o}\lambda\upsilon\beta\delta o\varsigma$ $\acute{\epsilon}\kappa$ $\tau\hat{\omega}\nu$ $T\upsilon\rho\acute{\iota}\omega\nu$) of Aristotle [7] may mean tin.[8] The lead mines of Sardinia were almost certainly worked from about 500 B.C.[9] and there are lead mines in Carthage.[10]

Strabo[11] says the inhabitants of Arvad collected fresh water from a spring which rose in the middle of the sea. They covered the outlet of the spring with a lead bell to which was attached a leather tube, the upper end of which was held in a boat in which vessels were filled with the fresh water. This fountain is still known.[12] The Phœnicians believed they could hold intercourse with the dead by dropping little rolls made of thin sheet lead, with inscriptions, into the tombs.[13] They made less use of lead than of bronze[14], but used it as a solder for bronze, in thin sheets (probably for ornaments), and as coffins (mostly of late date), formed of several sheets placed one over the other and then soldered.[15]

MERCURY

Cinnabar, native mercury sulphide, was exported from Spain and Pliny says 2,000 lb. of it came annually to Rome for preparation as a pigment.[16] Much of the mercury now used comes from Spanish ore. The Phœnicians do not, however, appear to have made any use of metallic mercury.[17]

IRON

Some sandstones of Phœnicia are rich in iron oxide, those in the south of Lebanon containing up to 90 per cent., and ochre occurs near Beyrut.[18] Several finds of prehistoric iron in Lebanon, Syria and Palestine are recorded[19] ; it occurs as ornamental iron at Byblos[20], as arrow-heads, along with bronze, in Phœnician tombs at Rachedieh, near Tyre, of about 1000 B.C.[21], and as iron rings and nails in several Phœnician graves.[22] A little iron of 1800–1400 B.C.

[1] Geographia Sacra, 1712, 648, who suggests that the name Britannia is from the Syriac barat anak, land of tin, etc.— quid quod ex ipso nomine res videtur posse colligi; Boerhaave, Elementa Chemiæ, Leyden, 1732, i, 41.

[2] Q. by Hennig, Rhein. Mus., 1934, lxxxiii, 169, who thinks the name Kassiterides is connected with the Keltic kass, "far," hence "distant islands"—a suggestion due to Reinach, L'Anthropologie, 1892, ii, 275 ; M. C. P. Schmidt, Kulturhistorische Beiträge z. Kunde des griech. u. röm. Altertums, Leipzig, 1906, i, 10, maintains that the name is of Eastern origin.

[3] Ezekiel, xxvii, 12.

[4] Strabo, III, v, 11, 175 C.

[5] Hackforth, CAH, iv, 347.

[6] Polybios, xxxiv, 9 ; Pliny, xxxiv, 16, 17, 18.

[7] Oikonomikon, ii, 36.

[8] Beckmann, History of Inventions, 1846, ii, 216 ; Rougemont, L'âge du bronze, 98.

[9] Perrot-Chipiez, Sardinia, i, 2.

[10] Davis, Carthage, 29.

[11] XVI, ii, 13, 754 C.

[12] Rawlinson, Phœnicia, 77.

[13] Budge, Amulets, 253.

[14] Perrot-Chipiez, Phœnicia, ii, 413.

[15] Rawlinson, 270.

[16] xxxiii, 7; cf. Lenormant, H. anc., iii, 60 ; Lenz, Mineralogie, 105.

[17] Rawlinson, Phœnicia, 315, 317.

[18] Ib., 47, 311.

[19] Orth, PW, Suppl. iv, 114 ; Feldhaus, Geschichtsblätter, 1927, xi, 200.

[20] Dussaud, La Lydie et ses Voisins, 1930, 83.

[21] Contenau, Civ. phénic., 238.

[22] Rawlinson, Phœnicia, 270.

was found at Qatna in central Syria.[1] The iron instruments sent by Phœnicians as tribute to Ašurnasirpal in 916 [2] perhaps came from Cyprus.[3] The iron mines of Sardinia were available from 500 B.C.[4]

The Phœnicians traded the iron and steel of the Chalybes of Armenia [5] and probably did not work the metal themselves, their speciality being bronze [6], although Sanchuniathon (c. 1200 B.C. ?) of Berytos or Tyre, as quoted by Philo of Byblos (A.D. 50–100) [7], says the Phœnicians called the inventor of iron, who was also a magician, Chrysor (perhaps Kharats, or Chusor Chorosch) [8], and that Kronos (Melkart ?) shaped himself a spear and sword of iron on the advice of a goddess whom he calls Athene.[9]

The Phœnicians probably imported iron ore from Spain.[10] There are small quantities of bituminous and also brown coal in the hills above Berytos, but it is sulphurous and was probably not worked.[11] There was a small (Dyer says a considerable) quantity of iron ore in these hills and the iron mines of Germaniceia in Cappadocia have been worked till modern times, but the metal was used only for local purposes. In the Roman Period, Indian iron and steel were imported to Damascus.[12] The Carthaginians appear to have worked iron at an early date ; there were picks of steely iron, together with Carthaginian coins, at a depth of 110 metres in an argentiferous lead mine in Spain.[13] Hiram of Tyre sent skilled iron workers to Solomon[14] and Ezekiel says the Phœnicians of Tyre obtained iron from Uzal (perhaps south Arabia), with cassia and calumus.[15]

---

[1] Garstang, Hittite Empire, 1929, 325 ; Dussaud, Lydie, 83.
[2] Lenormant, H. anc., iii, 70.
[3] Budge, Amulets, 251 ; cf. Perrot-Chipiez, Phœnicia, ii, 421.
[4] Perrot-Chipiez, Sardinia, i, 2.
[5] Cumont, Études Syriennes, 199 f. ; Bultmann, A. Rel., xxiv, 105.
[6] Perrot-Chipiez, Phœnicia, i, 30 ; ii, 413 ; Lenormant, H. anc., iii, 28, 34 f. ; ib., Prem. Civ., i, 162.
[7] Contenau, Civ. phénic., 100 ; Grimm, PW, ia, 2239, 2244 ; Paton, ERE, ix, 887 ; xi, 178 ; Meyer, Alt., II, ii, 177 f. ; Deubner, Ro., iii, 2093 ; Cook, Zeus, ii, 1036.

[8] Cory and Hodges, Anct. Fragments, 1876, 8.
[9] Onka ?, Hesychios, s.v. ; Rawlinson, Phœnicia, 335.
[10] Rawlinson, Phœnicia, 269 ; Ezekiel, xxvii, 12 ; Strabo, III, ii, 8, 145 C.
[11] Dyer, DG, ii, 607.
[12] Charlesworth, Trade Routes, 44, 251 ; Cumont, Études Syriennes, 151 f.
[13] Jagnaux, Hist. de la Chimie, 1891, ii, 218.
[14] II Chron., ii, 14.
[15] Ezek., xxvii, 19, text uncertain ; Fleming, Tyre, 140 ; Rawlinson, Phœnicia, 283, 291 ; Olmstead, H. Pal. Syr., 534 ; Smith, CDB, 974.

PHŒNICIA III—1

NON-METALS—INORGANIC

GEMS

The Phœnicians, who are described in the Bible [1] as expert gem engravers, made cylinder seals and scarabs with Egyptian, Assyrian and Persian designs, but the line engravings are more carefully executed and the work is more delicate than that on the Babylonian or Egyptian originals. The cylinders are in glass, steatite, carnelian, green serpentine, black hæmatite and green jasper, scratched rather than cut.[2] A great number of scarabs and seals came from Sardinia and Phœnicia had no reputation as a country where gems were found, but coral occurs off the Carthaginian site of Hippo-Zaritis near Utica.[3] That the Phœnicians traded in precious stones [4] is very probable, but we have no detailed information in this connection. The supposed emerald pillar in the temple at Tyre [5] was probably of green glass. A stone anthrakion was found at Carthage.[6] The lychnis was collected in the mountains of the Nasamones (in Libya), where it fell like rain, was gathered under the full moon and brought to Carthage.[7]

CERAMIC PRODUCTS

The Phœnicians probably made sun-dried bricks with clay and straw, and claimed the invention of bricks and tiles[8]. On account of the restricted sites, houses were probably tall, with several floors.[9] Phœnicia proper had no very good stone of fine grain, the limestone being shelly, and stone was imported from Egypt, the Taurus and Cyprus.[10] Although the mineral wealth of Syria was not great as compared with the west, there was plenty of good stone.[11] Wall painting, particularly in red and green, occurs in late tombs (2nd century B.C.–2nd century A.D.).[12]

Phœnicia has no good pottery clay and the products exported to native tribes[13] were coarse : earthenware vessels in Phœnicia proper before the Roman Period are very scarce and they do not occur on Phœnician sites in Sardinia, Corsica, Spain, Africa, Sicily, Malta or Gozzo. The pottery of Cyprus has already been fully considered. Some pottery at Jerusalem and at Camirus (in Rhodes) has been supposed to be Phœnician.[14] Cobalt is said to have been used in Persian and Syrian glazes "a few centuries before the Yuan dynasty" in China (A.D. 1280–1368).[15] Faience was copied from Egypt, but the white, blue and green glaze was thinner and less brilliant. Painted terra-cotta was imitated from Assyria, and exported.[16] Herodotos[17] says

[1] II Chron., ii, 14.
[2] Budge, Amulets, 252 ; Rawlinson, Phœnicia, 234, 238, 290.
[3] Rawlinson, 106.
[4] Krause, Pyrgoteles, 1856, 3.
[5] Herodotos, ii, 44.
[6] Theophrastos, De lapidibus, 34 ; garnet, Carchedonius lapis, Hill, Theophrastus on Stones, 1774, 152.
[7] Pliny, xxxvii, 7.
[8] Sanchuniathon, in Cory and Hodges, Ancient Fragments, 1876, 8 ; Birch, i, 155.

[9] Meyer, Alt., II, ii, 78.
[10] Rawlinson, 181.
[11] Charlesworth, Trade Routes, 44.
[12] Contenau, Civ. phénic., 258 f.
[13] Strabo, III, v, 11, 175 C.
[14] Birch, Anct. Pottery, i, 154 ; Rawlinson, 259 : no details except for Cyprus.
[15] Neumann, Chem. Z., 1927, li, 1015.
[16] Perrot-Chipiez, Phœnicia, ii, 5 f., 265 f.
[17] iii, 6.

Syrian wine was exported to Egypt in glazed jars, which were afterwards used as water jars. Yellow clay jars found in a I dyn. tomb at Abydos probably came filled with cedar oil from Lebanon.[1]

GLASS

Although the country is popularly supposed to have been "the great centre of the glass industry in antiquity," no glass factories of any period are established for Phœnicia, nor are Phœnician sites notable for glass deposits.[2]   Although Charlesworth[3] states that the manufacture of glass was not introduced from Egypt into Syria before 50 B.C., this is almost certainly incorrect, the manufacture being adopted and improved long before this date [4], perhaps after 1500 B.C. [5], although 2500 B.C. has been suggested.[6] Glass blowing was probably invented in Sidon, where blown bottles were made in A.D. 100.[7] The story said by Bucher [8] to be given by Josephus that the invention of glass was due to an accidental forest fire is not in that author [9], but is given by Palissy.[10]

Scepticism as to Phœnician glass appeared even with Movers[11] and Dyer[12], who regarded much supposed Phœnician glass as Egyptian[13]; Froehner[14] believed that no single glass object in museums is certainly Phœnician and there are no certain signs for differentiating Syrian from Egyptian glass.[15] Knowledge of glass was probably spread to the West, in part at least, by Phœnician traders, so that the Greek and Roman traditions may be valid as to the immediate provenance of the first glass seen in sub-Mycenæan Hellas : coarse glass beads in Greek and Italian tombs of the 9th–8th centuries B.C. are similar to those found in Cyprus.[16]

Several ancient authors mention glass works on the Syrian coast[17], and traces of old works are said to have been found at Tyre.[18] The glass industry of Venice was probably a continuation of the tradition of Tyre, where there was a Venetian colony in the 14th century A.D.[19] The remains of furnaces, slags, glazed objects and coloured glass pastes found at Tyre by Renan[20] may be those of Jewish workers[21], who were employed at Antioch and Tyre in the 12th century A.D.[22], when the glass factories must have been in full activity.[23] Tyre was noted for its glass (and sugar) until 1291, when it was abandoned by the Crusaders and largely destroyed by the Arabs.[24]

Two pieces of ancient glass, thought to have been made in Tyre, are the "emerald" vase ("Holy Grail") in Genoa Cathedral (examined by Guyton de Morveau) and the supposed "sapphire" in the basilica of Monza, both formerly thought to be precious stones.[25] Pliny[26] mentions glass works at Sidon

[1] Frankfort, Studies in Early Pottery, 1924, i, 105, pl. x ; Köster, Seefahrten der alten Ägypter, 1926, 10.
[2] Myres-Richter, Cat. Cyprus Museum, Oxford, 1899, 100.
[3] Trade Routes, 51, 252.
[4] Perrot-Chipiez, Phœnicia, ii, 326 ; Maspero, Struggle, 580 f. ; Kisa, 33 f., 90.
[5] Petrie, J. Soc. Glass Technology, 1926, x, 229 ; King, in Cesnola, Cyprus, 366 ; Murray, Smith and Walters, Excav. in Cyprus, 24—a Syrian discovery.
[6] Neumann, Chem. Z., 1927, li, 1015.
[7] Contenau, Civ. phénic., 218.
[8] iii, 267 ; also Kisa, Glas, 97.
[9] Trowbridge, Philological Studies in Ancient Glass, 1928, 95.
[10] Des Eaux et Fontaines, Œuvres, ed. Cap, 1844, 156.
[11] Phönizier, II, iii, 322.

[12] DG, ii, 616.
[13] Contenau, Civ. phénic., 217 ; Kisa, 38, 90.
[14] La Verrerie antique, 1879, 19.
[15] Perrot-Chipiez, Phœnicia, ii, 327.
[16] Myres-Richter, 100.
[17] Perrot-Chipiez, ii, 328.
[18] Marquardt, Privatleben, 726.
[19] Merrifield, Original Treatises on the Arts of Painting, 1849, i, p. xciii.
[20] Blümner, iv, 381 ; Merrifield, op. cit.
[21] Kisa, 92.
[22] Benjamin of Tudela, Itinerary, ed. Adler, 1907, 16, 18 ; Merrifield, i, p. xciii ; Kisa, 92.
[23] Perrot-Chipiez, Phœnicia, ii, 328.
[24] Benjamin of Tudela, ed. Adler, 19.
[25] Perrot-Chipiez, Phœnicia, ii, 336 ; Fleming, Tyre, 143 f.
[26] xxvi, 26.

and Lukian transparent glass made there : the Sidonian makers signed their products.[1] At Sarepta there is good sand and large banks of débris of broken glass of all colours, the waste from a large glassworks.[2] The Phœnicians used scrap glass instead of pebbles in making mortar.[3]

The glass industry established on the coast afterwards migrated inland and fine blue glass was made at Hebron in the 19th century A.D.[4] It was also transplanted from Tyre to Carthage.[5] Great quantities of glass of the Augustan Period are found all over Palestine, and this late Phœnician blown glass appears all over the Empire. The West, however, soon drove it out of the market.[6] Coloured semi-transparent vessels, supposed to represent old Phœnician work, and found in Cyprus, Camirus in Rhodes (now in the British Museum), on the Syrian coast near Beyrut and elsewhere, include small flasks or bottles, 3–6 in. long, probably for perfume, small jugs (οἰνοχόη ; œnochoæ), 3–5 in. high, vases about the same size, amphoræ pointed at the lower end and other varieties.[7] Some fragments from the Hathor temple of Sarâbît al-Khâdim in Sinai have been supposed to be Phœnician.[8]

Among finds regarded as old Phœnician glass are : (1) an opaque green glass paste group in the Louvre, found in Phœnicia, of a goddess with two animals, something like the Lion Gate at Mycenæ ; (2) a necklace of beads with two cylindrical yellow pieces with animal masks, and a bearded mask with goggle-eyes, covered with coloured drops of glass, found in a grave in Sardinia. Phœnician blown vessels in Asia Minor and the whole region of Phœnician colonisation are Hellenistic or Græco-Roman, the old art having then disappeared.[9]

The suggestion[10] that Egyptian glass was opaque and Phœnician transparent because of the use of natron instead of plant ashes is incorrect, since there is early transparent glass in Egypt, where natron was used at a very early date.[11] The *extensive* manufacture of transparent glass has been regarded as a Phœnician achievement.[12] The fine quality of Phœnician glass, coloured and colourless, was largely due to the excellent sand used. Strabo[13] mentions the glass sand (ὑαλῖτις ψάμμος) of the coast between Ptolemaios (Acre) and Tyre, which was not used there but sent to Sidon, although "some say the Sidonians have in their own country the vitrifiable sand, others that every kind of sand can be used." Josephus[14] and Tacitus[15] say the hill of sand at the mouth of the River Belus was inexhaustible, probably meaning that the golden yellow sand of Sidon is constantly washed up by the sea from Egypt : "sands of Africa invading Asia."[16]

Herodotos[17] himself saw in the temple of Melkart at Tyre a pillar of "emerald" shining with great brilliancy at night, in all probability a hollow pillar of nearly transparent green glass with a lamp inside.[18] An emerald pillar in a temple of Serapis in the Egyptian labyrinth[19] was probably glass ; two pillars in a temple on the island of Arados were said by Clement of Rome to be of glass.[20]

The specimens of supposed Phœnician coloured glass show the same forms, technique and colours as those of Egypt, so that a very short account of them

[1] Contenau, Civ. phénic., 199, 218, Fig. 76 ; Marquardt, 726.
[2] Rawlinson, Phœnicia, 252 ; Eiselen, Sidon, 122; Perrot-Chipiez, Phœnicia, ii, 328.
[3] Contenau, Civ. phénic., 219.
[4] Perrot-Chipiez, Phœnicia, ii, 328.
[5] Marquardt, Privatleben, 726.
[6] Perrot-Chipiez, Sardinia, i, 358 ; Charlesworth, Trade Routes, 51, 252.
[7] Rawlinson, Phœnicia, 253 f.
[8] Fowler, Archæologia, 1880, xlvi, 81.
[9] Kisa, 42, 43, 93, 95, Figs. 17–20 ; perhaps Egyptian Alexandrian work ? ; Kisa thinks the alabastra with coloured decorations found at Sidon, Tortosa, Cyprus, etc., are "certainly" Egyptian.
[10] Froehner, La Verrerie Antique, 3 ; Contenau, 217.
[11] Kisa, 33 f.
[12] Contenau, Manuel, iii, 1496.
[13] XVI, ii, 25, 758 C. ; Lenz. Mineralogie, 62 f.
[14] War, II, x, 2.
[15] Hist., v, 7.
[16] E. S. Stevens, Cedars, Saints and Sinners in Syria, 108.
[17] ii, 44.
[18] Meyer, Alt., II, ii, 265 ; Perrot-Chipiez, Phœnicia, ii, 336.
[19] Pliny, xxxvii, 5, from Apion.
[20] Kisa, 91, 92 ; Trowbridge, Philological Studies in Ancient Glass, Urbana, 1928, 143.

456 PHŒNICIA III

will suffice. The commonest shape is the alabastron.[1] Rawlinson [2] thought the glass was blown, but moulds (one of hard green stone from Camirus is in the British Museum) were probably used. All the early specimens, of thick glass, are moulded, the decorations being applied in the form of coloured threads to the shape when hot and "dragged" by a pointed tool (the Venetian millefiori technique, used in Egypt at an early date), the protruding threads being afterwards ground off. The colours, white, blue, yellow, green, purplish brown, red (very rare except pale pinkish and this is uncommon) are not vivid, but are pleasing and well contrasted. The opaque white was probably produced by bone ash or tin oxide ; the yellow by iron as ferric oxide ; the brown, black and violet by manganese (pyrolusite) ; the opaque deep red (Pliny's hæmatinum) contained 30 per cent. of copper in the form of cuprous oxide ; the blue was coloured with cobalt, and occasionally with cupric oxide. The rare white pearly variety occurs in Cyprus. Amber and veinings of agate were imitated.[3]

Imitations of Phœnician glass were early made in pottery : glazed vases with waved lines of warm orange and red, highly polished, were found in tumuli of Bin-tepe near Sardis and other sites, and the style was in use in Cyprus down to a comparatively late date.[4]

Imitations of gems, e.g. for beads, were made in coloured glass. In some specimens the beads are partly genuine (agate, onyx, crystal, carnelian) and partly imitations. One necklace from Sardinia, now in the Louvre, with over forty beads, two cylinders and four pendants, is entirely of glass ; another, found in Phœnicia, is partly glass and partly agate and carnelian. Small flat plaques pierced with holes were probably used for sewing on clothes and a supposed optical lens of colourless glass is probably an ornamental button.[5] Pliny [6] praises the black glass mirrors made at Sidon, although in his time the best period was past.[7]

Whereas in Egypt, Babylonia, Crete, Mycenæ and Tiryns the blue glass (kyanos) was used as a substitute for lapis lazuli, and was continued in the Archaic Greek Period (e.g. old Greek scarabs of blue glass), in Phœnicia green glass was used in imitation of the green jasper much in favour there. White glass appears in Sardinia at the end of the Archaic Greek Period, and the use of glass paste (λίθος χυτή) increased in the period just before Alexander.[8]

SALTS

Since the temperature in Phœnicia is lower than would normally correspond with its latitude [9], solar salt was probably not made, and salt was perhaps obtained from the Dead Sea. The salt pans at Hexi (or Sexi) at the foot of the Sierra Nevada were worked and salted provisions exported from Gades to Greece in the 5th century B.C.[10] Pliny[11] mentions a hill of salt near Utica.

Since no analyses of Phœnician glass seem to be available, it is not known whether potash or soda was used in its manufacture. The former could have been used, since the country is well wooded, but Pliny's story (see p. 119) seems to imply that Egyptian soda was used. I have found no notices of other salts (e.g. alum) made in Phœnicia.

[1] Perrot-Chipiez, Phœnicia, ii, 334 f., with excellent coloured plates ; Contenau, Civ. phénic., 200.
[2] Phœnicia, 256, 258.
[3] Perrot-Chipiez, ii, 331 f., 335 f., 338 ; Rawlinson, 254, 256 f. ; Marquardt, Privatleben, 726.
[4] Tubbs, DA, ii, 921.
[5] Rawlinson, Phœnicia, 255 f. ; Perrot-

Chipiez, Phœnicia, ii, 335 f. ; Lenormant, H. anc., iii, 119 ; Kisa, Glas, 95.
[6] xxxvi, 26.
[7] Kisa, 92.
[8] Furtwängler, Antiken Gemmen, 1900, iii, 92, 135.
[9] Dyer, DG, ii, 607.
[10] Strabo, III, iv, 6, 158 C. ; Grote, Hist. of Greece, iii, 274.
[11] xxxi, 7.

PHŒNICIA III—2

NON-METALS—ORGANIC

VEGETABLE PRODUCTS

Modern trees, shrubs and plants of Phœnicia and Syria, which probably correspond with those of the ancient country in the majority of cases, are very numerous and comprise the common trees, olive, palm and date palm, fig and sycamore fig (in the lowlands), cedar (the famous tree), three varieties of oak (one yielding good galls), plane (not common), acacia, sumach, carob or locust tree, lentisk (Pistachia lentiscus), cypress, oleander, myrtle, styrax (S. officinalis : east end of Carmel), rhododendron, caper plant, vine, lavender, rue, wormwood, numerous bulbous plants including blue, yellow and white crocuses, the mandrake ("one of the most striking plants of the country"), melons and other edible plants. Oak wood and grain were imported from Palestine.[1] Ebony was supplied to Ašurnasirpal in 916 B.C.[2] According to Newberry [3] the name "cedar" has been loosely and incorrectly used. The wood of Cedrus libani is of poor quality and the trees growing on the lower spurs of the mountains near the coast, pines, junipers and cypresses, all give better wood than the cedar growing at a height of 2,000 m. The wood of the sd-tree which the Egyptians imported from Syria was probably that of the cypress (Cupressus sempervivens) with horizontal branches. Wine of Helbon (οἶνος Χαλυβώνιος), the favourite drink of the Persian kings [4], was imported from Damascus and the Phœnicians exported wine to Egypt.[5] The olive tree, which never grew well in Egypt, is indigenous to Syria, where it was cultivated at an early period and the products exported to Egypt. King Sahurā (v dyn.) brought olive wood from Syria to Egypt in his ships.[6]

The Phœnicians, like other Eastern nations, made much use of perfumes and their perfume trade was very important. The Greek μύρον (myron, perfume) may be derived from the Semitic môr and βάλσαμον (balsamon, balsam) from bâsâm or besem. Spices, frankincense, cinnamon, balm, cassia, myrrh, calamus and ladanum were imported from Arabia, but some probably came from farther east, e.g. India.[7] In the Periplus of ψ-Skylax [8] the Phœnicians are said to have sold perfume to the natives of the west coast of Africa. Phœnician perfumes were probably made by treating the flowers with oil or fat to form a pomade, since distillation was unknown. They were sold in bottles of Oriental alabaster, glazed earthenware, glass, rock crystal or even gold.[9] Egypt probably imported cedar oil from the Syrian coast in the peculiar foreign yellow clay jars with a white coating which occur already in ɪ dyn. graves at Abydos.[10] Henna was cultivated in Cyprus.[11]

---

[1] Rawlinson, Phœnicia, 18, 32, 41 ; Maspero, Struggle, 188.
[2] Lenormant, H. anc., iii, 70.
[3] B.A. Rep., 1923, 175 ; Der Alte Orient, 1927, xxvii, Heft 1, 26, with notes by Behrens.
[4] Strabo, XV, iii, 22, 735 C.
[5] Rawlinson, Phœnicia, 289, 291.
[6] Köster, Beih. alt. Orient, i, 8.

[7] Ezekiel, xxvii ; Rawlinson, 284, 291.
[8] Cap. 112 ; Müller-Didot, Geogr. Græc. Min., i, 94.
[9] Perrot-Chipiez, Phœnicia, ii, 368, 395 f.; Dyer, DG, ii, 616 ; Contenau, Civ. phénic., 301.
[10] Köster, Beih. alt. Orient, i, 10 ; ib., Seefahrten der alten Agypter, 1926, 10 f.
[11] Perrot-Chipiez, Phœnicia, ii, 89.

The Phœnicians collected amber (doubtfully supposed to be mentioned by Ezekiel [1]) from the north by trade at the mouth of the Po [Eridanos], where the Greeks incorrectly located the "islands of elektron," [2] although remains of amber-digging are said to have been found on the Phœnician coast.[3] Amber and ivory were worked in Carthage.[4] The amber traded by the Phœnicians is mentioned by Homer.[5] There is no evidence that they went to the Baltic, first definitely mentioned, as Sinus Cadanus, by Pomponius Mela (A.D. 43) [6], to collect amber for themselves, although this is possible.[7]

Gum and ivory came from the interior of Africa, the ivory being worked in Carthage [8], papyrus from Egypt as early as 1100 [9] and wax and pitch from Spain.[10] Bee-keeping in horizontal hives of dried mud or clay (as in ancient and modern Egypt) goes back to an early period in Palestine, Syria and Phœnicia, and spread to the Phœnician colonies of Carthage, Sicily, South Italy and Spain[11], although Justin[12] attributes the discovery of honey to Gargoris, a mythical king of Tartessos. Herodotos[13] says the Gyzantians, a Libyan tribe of the African coast, obtained a great amount of honey from bees, but more still "by the skill of men," perhaps by pressing dates.[14] Honey was an important product of Malta.[15]

TEXTILES AND DYEING

Phœnicia was noted for textile work[16], the coloured products being mentioned by Homer.[17] Linen was woven in the Roman Period at Laodicea, Byblos, Tyre and especially Scythopolis (Bethshan), and silk was dyed and worked at Tyre and Berytos from raw silk imported by Persian merchants from China through Damascus.[18] The white wool of Syria was supplied from Damascus, the wool of lambs, rams and goats probably came from distant parts of Arabia and linen yarn from Egypt. Silk was usually mixed with cotton or linen and the Phœnician embroidery was famous.[19]

The Phœnicians sent Ašurnasirpal in 916 stuffs dyed with purple and saffron (? safflower), together with sandalwood, ebony and sealskins, as tribute.[20] They were fond of colours, and their dye industry was famous in antiquity.[21] The "purple of Tyre" (murex), however, was early in use in the Ægean and is probably a Cretan invention[22], although Movers[23] thought it was an Assyrian discovery.

THE PURPLE DYE

There are immense heaps of murex shells at Tyre and Sidon and the dye was certainly made in large quantities in Phœnicia.[24] There are also deposits of shells at Salamis and at Sour on the coast of Attica.[25] Ovid[26] calls the dye

[1] i, 4, 27 ; viii, 2 : Smith, CDB, 38.
[2] Lenormant, H. anc., iii, 99 ; ib., Prem. Civ., 157 ; Bunbury, DG, i, 849 ; Schrader, RL, 72.
[3] Rawlinson, Chambers' Encyclopædia, 1888, art. "Phœnicia."
[4] Perrot-Chipiez, Phœnicia, ii, 403 f. ; 396 f.
[5] Odyss., xv, 460.
[6] De Situ Orbis, 71, 77 ; Schmitz, DG, i, 641.
[7] Rawlinson, Phœnicia, 302 ; Schrader, RL, 72.
[8] Maspero, H. anc., 1905, 374 ; Perrot-Chipiez, Phœnicia, ii, 396 f.
[9] Cook, CAH, iii, 423.
[10] Lenormant, H. anc., iii, 60.
[11] Lippmann, Z. Deutsch. Zuckerindustr., 1934, lxxxiv, 810, Tech. Teil.

[12] Historiæ Philippicæ, xliv, 4.
[13] iv, 194.
[14] Lippmann, Chem. Z., 1924, xlviii, 38.
[15] Rawlinson, Phœnicia, 116.
[16] Orth, PW, xii, 606.
[17] Iliad, vi, 289 ; Odyssey, xv, 417 : "work of the women of Sidon."
[18] Charlesworth, Trade Routes, 50, 252.
[19] Rawlinson, Phœnicia, 243 f., 290.
[20] Lenormant, H. anc., iii, 70.
[21] Contenau, Civ. phénic., 304.
[22] See p. 337.
[23] Art. "Phönizier" in Ersch-Gruber, 368.
[24] Wornum and Middleton, DA, i, 486 ; Contenau, Civ. phénic., 303.
[25] Perrot-Chipiez, Phœnicia, ii, 425 ; Marquardt, Privatleben, 493.
[26] Tristia, iv, 2.

Sidonian purple. Whether the Phœnicians restricted the use of fabrics dyed with purple to persons of the highest rank, as did the Romans ("royal purple") [1], is not known, but the Phœnician chief priest was clothed in purple and wore a gold tiara [2], and legend says the Phœnician king who first saw the purple dye reserved it for royal use.[3] According to Palæphatos, quoted in the so-called Easter Chronicle (Chronicon Paschale, composed about A.D. 630) [4], the art of dyeing began with the discovery of purple. Several authors [5] say the purple was discovered accidentally : a shepherd's dog crunched a shell on the sea-shore and its tongue was dyed purple. In Pollux[6], the Phœnician god Herakles (Melkart) or Herakles-Sandan of Tyre gave the nymph Tyros the first robe dyed with purple.[7] Bochart [8] explained this story by the statement that in Syriac the names for dog (chelab) and dyer (chilab) are the same ; the Hebrew name for purple is těchêleth or argâmân, the Aramaic is thichla or thachla. Eisler [9], who relates argâmân with the Babylonian argamânu, red-purple, thinks it is the name of a Tyrian god also called Morreus, the origin of the name murex.

The best kinds of purple were very expensive : under Diocletian the first quality cost £187, but the poorest only 7s. 6d. per lb.[10] Theopompos, quoted by Athenaios[11], says it was worth its weight in silver, and Pliny[12] that a pound of the best double-dyed Tyrian cloth cost £31 5s., but in his youth the light violet kind only £3 2s. 6d. For this reason cheaper imitations were made, a number of recipes for which are preserved in the Stockholm Chemical Papyrus.[13]

The chief sites where Phœnician purple was made were Tyre, Sidon and Dora (on the southern boundary), but the best was made at Tyre.[14] Clement of Alexandria[15] says the "blood" of the purple fish produces the dye and that it was found at Tyre, Sidon and on the coast of Sparta. The Spartans are said to have had colonies in Crete[16], where they may have learnt purple dyeing, and their purple was the best European variety.[17] In Ezekiel[18] the dye came from Greece (Elisa). Harduin, in his edition of Pliny (Paris, 1685), said the Tarentines of his day reported vast heaps of shells and remains of old purple factories.[19] In Africa purple was found on the rocky coasts in Gætulia ; King Juba had dyeworks on islands off the Moorish coast[20], and there were dyeworks off the coast of the North African Syrtes.[21] The Gætulians in the north-west of Libya were warlike savages who carried on part of the trade with the interior of Africa and their purple was famous.[22] The purple fish is also shown on the coins of Corinth, where Sisyphos (its legendary king) was the father of Porphyrion (the purple industry).[23] The famous fisheries off the Anatolian coast are mentioned in Diocletian's edict and Thyateira was renowned for purple dyeworks.[24] Purple was also fished or dyed at Lakonia[25]

[1] Lucretius, iv, 1118 ; Mommsen, Juristische Schriften, ii, 322, 329.
[2] Rawlinson, Phœnicia, 342.
[3] Goguet, ii, 88 ; Lenormant, H. anc., iii, 117.
[4] Migne, Patrologia græca, xcii, 162 ; p. 43, ed. of du Cange.
[5] Pollux, i, 45 ; Cassiodorus, Variarum, i, 26 ; Palæphatos, op. cit. ; Achilles Tatius, Loves of Clitopho and Leucippus, Scriptores erotici Græci, tr. R. Smith, 1889, 379.
[6] Onomastikon, i, 4.
[7] Hoefer, Hist. de la Chimie, 1866, i, 60 ; Lenz, Zoologie, 629 ; Goguet, ii, 88 ; Maspero, Struggle, 187.
[8] Hierozoicon, II, v, 9, 11 ; 1692 ed., 726, 740.
[9] Weltenmantel, 167, 178, 227, 285.
[10] Marquardt, Privatleben, 493 f.
[11] xii, 31.
[12] ix, 39.
[13] Papyrus Græcus Holmiensis, ed. Lagercrantz, Uppsala, 1913.
[14] Dyer, DG, ii, 607, 616.
[15] Pædagogus, ii, 11 ; Ante-Nicene Library tr., iv, 263.
[16] Smith and Marinden, Hist. of Greece, 1902, 41.
[17] Charlesworth, Trade Routes, 125, 265.
[18] xxvii, 7.
[19] Bailey, Pliny's Chapters, i, 157.
[20] Horace, Odes, II, xvi, 35 ; Charlesworth, 125, 265.
[21] Goguet, ii, 89 ; Charlesworth, 145, 268.
[22] Smith, DG, i, 925.
[23] Tozer, Anct. Geogr., 5.
[24] Charlesworth, Trade Routes, 94, 260.
[25] Horace, Odes, II, xviii, 7.

Bulis [1] and Amyclæ.[2]   The great sacrifice offered by Crœsus of Lydia included purple robes.[3]   Although Scaliger and others assumed that the purple dye was unknown to Homer and that any bright colour was called πορφύρεος (porphyreos) by the Greeks, yet the frequent use in Homer of this word to describe the appearance of clothes makes it probable that a definite colour, no doubt the shellfish purple, is intended.[4]

The literature on Tyrian purple is very extensive.[5]   There is an account of purple fishing in the Ἰωνιά (Ionia; "Bed of Violets"), attributed to Eudokia of Makrembolis, the daughter of Constantine VIII (11th century)[6] : the sources were similar to those used by Suidas (10th century) [7], but the work has been regarded as a forgery of Constantine Palaiokappa (c. 1543).[8]   The ancient accounts of the purple fish and some new observations were published by G. Rondelet [9] and Fabio Colonna.[10] Réaumur[11] made some observations on the coast of Poitou. His memoir, with the older literature on purple, is reprinted by A. Dedekind.[12] E. Bancroft[13] gives a history of purple, its rediscovery in the 17th–18th centuries, and his own experiments. An important memoir on purple by H. Lacaze Duthiers[14] contains specimens of wool dyed with the actual purple, which, in the copy I saw, were still perfect in colour.

Aristotle, whose biology is the best part of his works[15], says[16] the juice is either dark or bright red ; the fish are best caught in spring. "The colour is found between the liver and neck, where there is a kind of white skin, which is removed. The juice colours the hands. The rest is like alum. The large fish are taken from the shells to get the purple, the small are stamped with the shells, which gives a poorer dye. They must be kept alive, as they lose the colour when they die." In actual fact, the colour is contained in a small sac in the body of the fish, the fresh juice of which is a creamy yellowish-white liquid of garlic odour, which becomes yellow, green and finally some shade of violet or purple by oxidation on exposure to air and light.[17]

Davy[18] found that a purple material in a vase from the "Baths of Titus," which had lost its colour where exposed to air but in the centre was carmine,

[1] Pausanias, X, xxxvii, 3.
[2] Charlesworth, 125, 265.
[3] Herodotos, i, 50.
[4] Goguet, ii, 89 ; Hoefer, i, 60, 165 ; Bailey, Pliny's Chapters, 1929, i, 156.
[5] Classical references in Lenz, Zoologie, 1856, 624 ; Goguet, ii, 88 ; Hammer Jensen, PW, Suppl. iii, 465. The two memoirs : W. A. Schmidt, Die Purpurfarberei und der Purpurhandel im Alterthum, Forschungen auf dem Gebiete des Alterthums, i, 96–212, which is praised by Marquardt, 491, and Bischoff, Versuche einer Geschichte der Farberkunst, 1788, favourably mentioned by Berthollet, in Berthollet-Ure, i, 11, were not available to me. Goguet, ii, 88–95, and Movers, art. "Phönizier" in Ersch-Gruber, 373 f., collected most of the references to the older literature.
[6] Published by d'Ansse de Villoison, Anecdota Græca e Regia Parisiensi et e Veneta S. Marci Bibliothecis deprompta, 2 vols., Venice, 1781, i, 42 f.
[7] Means, DBM, ii, 80 ; Berthollet-Ure, i, 18.
[8] Sandys, History of Classical Scholarship, 3rd. ed., Cambridge, 1921, i, 408.
[9] Libri de Piscibus Marini, Lyons, fol., 1554-7, pt. ii, 64–87.

[10] Opusculum de Purpura, Romæ primum An 1616 editum, ed. J. D[aniel] Major, Kiel, 1675, sm. 4to., text 44 pp., notes 114 pp.—a poor edition, but the notes are useful. The work, which is illustrated, deals only with zoology.
[11] Mém. Acad. des Sciences, 1736 : Quelques expériences sur la liqueur colorante qui fournit la pourpre, q. by Hoefer, i, 164.
[12] Ein Beitrag zur Purpurfrage, 4 vols., Berlin, 1898–1906–08–11.
[13] Experimental Researches concerning the Philosophy of Permanent Colours, 2 ed., 2 vols., London, 1813, i, 120–164.
[14] Annales des Sciences Naturelles, Zoologie, 4th series, xii, 1–84 ; cf. Perrot and Chipiez, ii, 425, and Lenormant, H. anc., iii, 118.
[15] Singer, Greek Biology and Greek Medicine, 1922, 21 ; Aristotle, History of Animals, tr. D'Arcy Thompson, Oxford, 1910.
[16] H. Animal., v, 12, 15 ; cf. Vitruvius, vii, 13, and note in the Elzevir ed., 1649, 147.
[17] Rawlinson, Phœnicia, 46 f., 248 f., chemistry out of date ; Dyer, DG, ii, 616.
[18] Phil. Trans., 1815 ; Works, vi, 147.

was mainly calcium carbonate and clay, but the colour was organic, and Davy thought it was the ancient purple. The colouring matter of murex was isolated in a nearly pure state by Schunck in 1879 and called by him punicin ; its constitution as 6,6'-dibromoindigo was discovered by Friedländer in 1906, who also found that Murex brandaris contains another colour of bluer shade, the constitution of which is not yet established.[1]

According to Lacaze Duthiers the species of shellfish most used [2] were (1) Murex trunculus, found in deep water, giving the best colour and probably used at Tyre, although the shells in rubbish heaps there are of (2) Murex brandaris, the commonest variety on the Phœnician coast, also found in Greek seas at Nisyros, Cythere, the Peloponnesos and the Lakonian coast. In addition to these two principal varieties (3) the Buccinum lapillus of Pliny, the modern Purpura lapillus, found in shallow water, and (4) the Helix ianthina were probably also used. The Purpura lapillus, found off the English coast, was at one time used in Ireland for marking initials, etc., on linen.[3] The Helix ianthina is completely different from the Murex, the shell being smooth, not spiny ; although it is found abundantly in the Eastern Mediterranean it is doubtful if it was used by the Phœnicians.[4] Shellfish giving a purple dye have also been found in South American waters.[5] The Murex was fished up from deep water in a contrivance like a lobster trap, baited with mussels or frogs.[6]

The Greek name for the purple was ἁλουργός (halurgos) or more commonly ἁλουργής (halurges), "sea purple" or genuine purple as opposed to imitations [7]; the Latin name was purpura. Another name was sarranus, after Sar [Sur], the old name for Tyre.[8] The Romans called robes dyed with genuine purple conchiliatæ vestes.[9] A Byzantine name for purple was blatta (βλάττα or βλάττη ; dim. βλαττίον)[10], blattam, blatteam, blatellam, blattea[11], meaning a clot of blood ; thence purpura quæ blatta vel oxyblatta, vel hyacinthina dicitur.[12] The Byzantine purple was made until the Turkish conquest.[13]

The ancient purple was really a blackish-violet or bluish-black colour, as found on the oldest garments from Akhmîm. Under the later Roman Empire the shade was brighter and in the Byzantine Period the colour we now call purple was frequent.[14] The violet tint was not greatly prized, though it was fashionable for a time under Augustus ; redder hues were preferred, the favourite being "a rich dark purple, the colour of coagulated blood" ; purpureus or πορφύρεος probably means a deep crimson.[15] Homer and Vergil call the colour of blood "purple,"[16] and the British dye was described as "black-purple." [17] The specimens in Lacaze Duthiers vary from a very dark colour, which seemed to me almost brown, to a pale heliotrope. The colour must have

[1] A. G. Perkin and A. E. Everest, Natural Organic Colouring Matters, 1918, 525 f. ; J. F. Thorpe and C. K. Ingold, Vat Colours, 1923, 7 f., 13.
[2] Seneca, Quæstionum Naturalium, i, 2, says there were several kinds.
[3] Cole, Phil. Trans., 1685 ; abridged ed., iii, 252 ; Rawlinson, Phœnicia, 246 ; long extract in Thorpe and Ingold, Vat Colours, 1923, 13.
[4] Rawlinson, 247.
[5] Berthollet-Ure, i, 17 f. ; Goguet, ii, 90.
[6] Pliny, ix, 36 f. ; cf. Pollux, i, 45 f.
[7] Lacaze Duthiers, 58 ; Liddell and Scott, Greek Lexicon, s.v.
[8] Meyer, Alt., II, ii, 74 ; Salmasius, Plin. exerc., 1689, 936 : purpura optima quæ sarrana Latines quia Sar, olim Tyrus ;

Bochart, Geogr. sacra, 735 ; Thorpe and Ingold, 8 ; Diefenbach, Glossarium Latino-Germanicum, 1857, 513.
[9] Goguet, ii, 89.
[10] Diocletian's edict ; Lydius, de mensibus, i, 21—a Phœnician name for Aphrodite ; Salmasius, Plin. exerc., 1689, 810 ; note to Vitruvius, 1649, 147.
[11] Vopiscus, Aurelian, c. 46.
[12] Codex Theodosios, 1736–45, iii, 541 ; Schrader, RL, 645 ; Ducange, Gloss. med. et infim. Latinitatis, Niort, 1883–87, i, 678 ; W. Smith, Latin Dict., 130.
[13] Marquardt, 492 ; Dyer, DG, ii, 616.
[14] Forrer, RL, 150 f., 641.
[15] Pliny, ix, 36 ; Rawlinson, Phœnicia, 250.
[16] Goguet, ii, 93.
[17] Lenormant, H. anc., iii, 118.

been permanent, since Plutarch [1] says Alexander found in the treasury of the kings of Persia a vast quantity of purple stuffs which were 190 years old and yet were fresh "because they had been treated with honey."[2]

In Pliny's time the whole reputation of Tyre rested on its dye industry [3], which was the chief source of its wealth[4] ; the superintendent of the dye houses was a crown official, although the workmen were probably Phœnician.[5] Tyre then had a monopoly of purple.[6] It was so crowded with dyeworks that it was difficult to walk through the town, and the smell of the dye made it unpleasant.[7]

The description of purple fish and purple dyeing given by Pliny [8] is very long and detailed. There were several varieties of purple fish, which are characterised by trivial names (lutense, algense, tæniense, calculosæ, dialetæ), but the purple was principally obtained from two kinds : (1) buccinum (trumpet-formed), murex or κῆρυξ (keryx), and (2) purpura, pelagia or πορφύρα (porphyra). The buccinum, which gave a fugitive colour alone, was combined with the purpura in the proportion of 200 to 111 for a pound of wool, producing a rich amethyst colour. The purple differed according to the locality and food of the fish. Vitruvius [9] says in the north it was black, in the east and west violet ; other kinds were blue-black (lividum) and red. It was obtained by crushing the shells with an iron rod and since the liquor "from its saltness soon dries, it is mixed with a little honey." Pliny says the fish was either taken out of the shell to get the colour sac or small fish were crushed whole in mills. Vitruvius calls the colour ostrum.[10] Marquardt[11] suggests that Vitruvius's four varieties were perhaps reducible to two principal colours, black and red.

The method of applying the dye[12] was to add to the murex juice about 1 per cent. of salt[13], macerate for three days and evaporate to one-sixth in a lead boiler (fervere in plumbo). Mosso thought some lead vessels found in Crete were used for this purpose.[14] The liquor was tested by putting in it a piece of washed wool (vellus elutriatum mergitur in experimentum) and the concentration continued until the right colour was obtained, which was a very beautiful dark red.[15] The wool was then put in and left for five hours, taken out and immersed anew.

The art of getting a good colour consisted in combining the buccinum and pelagium : one thus obtained two true purples, the Ianthin or amethyst purple, and the Tyrian purple with its varieties[16], the former by one coloration with the mixture and the latter by dipping in pelagium and then in buccinum. The Tyrian (or Lakonian) purple was hence called "double dyed" or "twice dipped."[17] Although Pliny says this purpura dibapha was a new invention of the Tyrians in his day, a similar expression (bis tinctum) is used in the Bible[18], but perhaps for scarlet twice dyed with the same colour, as was still done with cochineal dyeing in the 18th century.[19] The purpura dibapha was a very

[1] Alexander, 36.

[2] Goguet, ii, 95 : Pliny, ix, 36, says the dye was mixed with honey.

[3] Pliny, v, 19 : nunc omnis ejus nobilitas conchylio et purpura constat.

[4] Strabo, XVI, ii, 23, 756 C.

[5] Fleming, Tyre, 145.

[6] Dyer, DG, ii, 616.

[7] Strabo, XVI, ii, 23, 756 C. ; Pliny, ix, 36 ; Martial says fabrics dyed with purple had a disagreeable odour, which is probably correct ; Goguet, ii, 92.

[8] ix, 36–40 ; perhaps from Demokritos, who is mentioned as an authority ; Hammer Jensen, PW, Suppl. iii, 465.

[9] vii, 13.

[10] ὄστρεον or ὄστρειον ; in Plato's

Republic, Liddell and Scott, Lexicon, 1933, 1264 ; Blümner, iv, 496.

[11] Privatleben, 491.

[12] Pliny, ix, 38.

[13] Perhaps sea-salt, Hoefer, Hist. Chim., i, 60, or an alkaline carbonate, ib., 163 ; alum, suggested by Lenz, Zoologie, 624, is not mentioned.

[14] Bailey, Pliny's Chapters, i, 155.

[15] Pliny, ix, 36.

[16] Pliny, ix, 38 ; Marquardt, 491.

[17] διβαφος, Cicero, Epist. ad Atticum—purpura dibapha, Pliny, ix, 38 ; xxi, 8.

[18] Exodus, xxv, 4.

[19] Goguet, ii, 87 ; Berthollet-Ure, ii, 155 f.

costly dye of a deep blackish blood-red colour (color sanguinis concreti, nigricans adspectu ; idemque suspectu refulgens) [1], which probably corresponds with Homer's αἷμα πορφύρεον [2] and Vergil's purpura nigra [3] or purpura anima.[4] Ovid [5] and Martial [6] also refer to it.

Lighter purples were obtained by adding to the juice of the purpura (without buccinum) water, urine and fucus (φῦκος ; a seaweed).[7] If a fabric first dyed scarlet with coccus (kermes) was then dyed with Tyrian purple one obtained hysginum purple, named after the plant ὕσγη (hysge) [8], which is mentioned by Xenophon [9]: in Diocletian's time it is called ἰσγένη (hisgenê). Plutarch[10] even speaks of a whitish shade obtained with white olive oil, which may have been a pale heliotrope.[11] Diocletian's tariff[12] refers to several kinds of purple. The best is metaxablatta (μεταξαβλάττα), which is perhaps purple raw silk. There were four kinds of purple wool, all varieties of Tyrian purple, the best being oxytyria (ὀξυτυρία). In later accounts, blatta means black-purple ; in Diocletian's tariff, hypoblattê (ὑποβλάττη) is probably the pale purple. The genuine sea-purple is θαλασσια πορφύρα, ἁλιπορφύρα or ἀλουργίς. The substitutes were kermes or the Galatian colour (Galaticus rubor)[13], probably used in the dyeworks of Nicæa or shipped from its harbour. A cheaper substitute was ἰσγένη or ἰσγένη ἀλγενησία, corrupted from ὕσγινον, the Latin hisginum from the φῦκος θαλὰσσιον, fucus marinus, or alga (i.e. litmus or archil), which is described by Theophrastos.[14] Quintilian says this kind lost its colour through sulphur[15], perhaps meaning a reduction by theion hydor, a solution of calcium polysulphide known in the Roman Period[16], or bleaching by fumes of burning sulphur.

The φῦκος, fucus, added to the purple, was archil, a species of lichen growing on rocks adjoining the sea and yielding a red colour[17], and in the Alexandrian Period was much used as a substitute for the genuine purple.[18] Fucus is said[19] to be a Phœnician word : it is probably the puch of the Bible.[20] The Hebrew tolaat scheni, translated coccus in the Greek and Vulgate, is probably scarlet, not crimson, the latter being a dark colour made with cochineal, "absolutely unknown to the ancients."[21] The scarlet is a bright colour made with small "grains" collected on the leaves and bark of a sort of holm-oak common in Palestine, Crete and several other countries, and found as little eggs or bladders about the size of juniper berries, formerly thought to be caused by the sting of small grubs, but really the immobile female kermes insects (Coccus ilicis).[22] These are the Arabic al qirmiz or kermes, called in Europe grains of scarlet or vermilion.[23] The grains are actually called "fruits of the holm-oak" (πρίνου καρπόν) by Plutarch : Theophrastos[24], Pliny[25] and Dioskurides[26] all say they were collected from that tree.[27] Pausanias[28] says the colour is formed by a

[1] Pliny, ix, 38, 39 ; Marquardt, Privatleben, 492.
[2] Iliad, xvii, 361.
[3] Georg., iv, 275.
[4] Ænid, ix, 349.
[5] De Arte Amandi, iii, 170 : nec quæ bis Tyrio murice tincta rubra.
[6] Lib. iv, epig. 4 : quod bis murice vellus inquinatum.
[7] Pliny, ix, 39 ; xxvi, 10 ; Horace, Odes, III, v, 28, lana medicata fuco.
[8] Pliny, ix, 41 ; Marquardt, 492, 495.
[9] Cyropæd., VIII, iii, 3.
[10] Alex., 36.
[11] Hoefer, i, 60.
[12] Mommsen, Juristiche Schriften, ii, 318f., 327 f.
[13] Beckmann, Hist. Invent., i, 400.

[14] Hist. plant., IV, vii, 3 ; Salmasius, Plin. exerc., 804.
[15] Mommsen, 321.
[16] Gmelin, Handbook of Chemistry, tr. Watts, 1858, xii, 360.
[17] Berthollet-Ure, ii, 183.
[18] Lagercrantz, Papyrus Græcus Holmiensis, 1913, 206, 219.
[19] Cooke, EB[14], xvii, 769
[20] Löw, Flora der Juden, i, 20 f.; see Index.
[21] Goguet, ii, 96 f.
[22] Berthollet-Ure, ii, 173.
[23] Goguet, ii, 96 ; Schrader, RL, 420.
[24] Hist. plant., III, xvi, 1.
[25] xvi, 12.
[26] iv, 48.
[27] Goguet, ii, 97.
[28] x, 36.

grub and the tree is called kokkos by the Ionians and Greeks, but the Galatians above Phrygia call it in their native tongue hys (ὗς ; hysginum).

The kermes (coccus) was made from the coccus insect of Central Europe and the East, dried and stamped. In the later Roman Empire the preparation of the colour was not known in Europe and it was imported from Persia and the Orient. Flavius Vopiscus [1] says the king of Persia sent to Aurelian purple which was much superior to the Roman [2], and Ktesias speaks of an Indian worm-bearing tree from which scarlet was made.[3] Isidore of Seville (d. A.D. 636) [4] first mentions it in the West as "Κόκκον Græci, nos rubrum seu vermiculum dicimus. Est enim vermiculus ex silvestribus frondibus. Russata quam Græci phœniceam vocant, nos coccinam," but Charlesworth [5] still speaks of the coccus "plant" as a "vegetable dye." The dye is called σκώληξ (worm) by the Greek alchemists.[6]

[1] Aurelian, c. 29.
[2] Forrer, RL, 641.
[3] Schrader, RL, 420.
[4] Etymol., XIX, xxii, 10, xxviii, 1.
[5] Trade Routes, 1924, 144, 165—an otherwise useful book.
[6] ψ-Demokritos, Berthelot, Coll., ii, 42 ; Salmasius, Plin. exercitat., 1689, 195.

PALESTINE

PALESTINE I

GENERAL

The Holy Land

Palestine, the Pelesheth of the Bible and the Greek Syria-Palestina, is properly the long and broad strip of maritime plain later inhabited by the Philistines. In Roman and later Greek authors it meant the whole country of the Jews, east and west of the Jordan, and what we now call the Holy Land.[1] The country was early, perhaps in prehistoric times, occupied by Semites [2], the oldest branch being called Canaanites, although this name (which is probably not Semitic) first appears after 2000 B.C. It is one of many names adopted by the Semitic inhabitants of Syria and perhaps came from Asia Minor. It is improbable that the Canaanites, or Western Semites, were nationally distinct from the Amorites, by which name they are called in the Bible.[3] The Edomites (edom = red) lived in the "rose-red mountains" of Se'îr and the Jebusites were merely a local tribe of Hittites.[4]

Palestine and Syria were early influenced by Egypt, Babylonia and probably the old Ægean culture (from about 2000 B.C.) ; Ægean influence became important as "Philistine" from about 1200 B.C. in Palestine, this name meaning "land of the Philistines," which the Egyptians called Rezenu.[5] The earliest foreign influence was that from XII dyn. Egypt.[6] In the coastal towns of Syria (Phœnicia) Egyptian influence was much earlier. Rezenu (Palestine) sent gold, silver (including ornamental vessels), copper, lead in blocks and rings, and ivory as annual tribute to Thothmes III.[7] An Egyptian name for South Palestine, Kharu, was sometimes applied to the whole of Syria.[8] In Syria and Palestine about 1500 B.C. there were many small states, partly Semitic and partly under dynasties from Asia Minor. There was influence from Egypt and Babylonia, also from the north, especially Asia Minor. The towns were surrounded by brick walls with many towers, as in Assur, and stone glacis, the houses being built round a courtyard and the dead buried in them, mostly in earthenware jars.[9] Imports from Egypt and Babylonia (Shinar) and earthenware from Cyprus were important, and they were imitated. The upper classes were clothed especially in Babylonian fashions, e.g. with blue and red woollen shawls. Bronze weapons, gilt chariots with horses, silver and gold vessels, ivory and ebony furniture, precious stones and statues, etc., were in use.[10] Amulets as magic apparatus, which were freely

[1] Smith, CDB, 666 ; Williams, DG, ii, 516 f. ; Benzinger, JE, ix, 479.
[2] Rostovtzeff, i, 69, 131.
[3] Otto, 1925, 43 ; Meyer, Alt., II, i, 88 ; II, ii, 220.
[4] Sayce, Patriarchal Palestine, 1912, 33, 43.
[5] Otto, 1925, 43 f. ; Olmstead, H. Pal. Syr., viii, 44, 257 f. ; Meyer, Alt., II, i, 83 ; Bilabel, Geschichte Vorderasiens, 1927, 241, 429.
[6] Macalister, A History of Civilisation in Palestine, Cambridge, 1912, 33.

[7] Meyer, Alt., II, i, 136.
[8] Bilabel, Gesch. Vorderasiens, 1927, 9, 138.
[9] L. H. Vincent, Canaan d'après l'exploration récente, 1907, 16, 426 ; Meyer, Alt., I, ii, 680 ; Maspero, Struggle, 111, 127, 138, 154 ; Rawlinson, Mon., i, 370 ; Hall, in A. S. Peake, The People and the Book, Oxford, 1925, 21 ; Jastrow, Civ., 56 ; King, H. Bab., 126.
[10] Meyer, Alt., I, ii, 681 f. ; Perrot-Chipiez, Hist. de l'Art, i, p. v.

imported from Egypt and modified in imitations, although their use was not understood, included the cross of life, the solar disc with wings, the sphinx, etc. They are still worn by natives of Palestine. The goddesses, derived from Egypt, but always represented full-face and not in profile, were combined with those of Babylonia.[1] According to Cook there was no considerable Egyptian or Babylonian influence in *early* Palestine, but later on Babylonian influence was very strong. Elamite influence, though suspected, is uncertain, as the Elamites could rarely have penetrated there, and the assumption of an "Aryan" element in the population in 3000–2000 B.C., on the basis of megalithic remains, is very improbable.[2]

Newcomers in the land of Canaan were the Philistines, called Pulsatha in Egyptian texts of Rameses III, who fought and defeated them on his frontiers and drove them in great numbers into Palestine and Syria, where they settled in 1200–1190.[3] They were non-Semitic and are said in the Bible [4] to have come from Caphtor, i.e. Crete. Their Cretan origin is generally accepted [5], although the Cretans were perhaps joined by neighbouring islanders.[6] The Philistines, who probably carried the Minoan traditions to the Phœnicians, are shown with Sardinians on the walls of the tomb of Rameses III at Medinet Habu.[7] They perhaps formed only a small but leading part of the population, dominating the Canaanites, whose speech and gods they adopted, in the period 1075–1010 B.C.[8] They occupied Gaza, Askalon and Ashdod on the coast, Edom and Gath farther inland, and seem (doubtfully) to have possessed iron weapons, although the Cretans did not. It is difficult to explain how the Philistines became familiar with, and obtained supplies of, this metal in their flight from the Ægean unless they had come under Hittite influences, which seems probable. The Philistines gradually became Semitised, but not completely.[9]

The coming of the Philistines united the various tribes of another people who had entered Canaan before them, the Hebrews[10]. The Khabiru of the Amarna letters (14th century B.C.), who invaded and conquered Palestine in Ikhnaton's time and are mentioned in Hittite and earlier (c. 2000 B.C.) Babylonian texts, are generally identified with the Hebrews[11], although Friedrich[12] maintains that the Khabiru were Syrian Bedouins, not Hebrews. The Israelites (the name being perhaps Arabic and appearing as asrar on a stela of the pharaoh Merenptah, c. 1240 B.C.) certainly appeared in Palestine in the 13th century B.C.[13], or perhaps the 14th.[14] The usual theory is that the Khabiru were Semitic (Aramæan) nomads from Arabia who first settled in north-west Arabia (Midian) near Mount Sinai. The occupation of Palestine by the Israelites and the diffusion of the Aramæans in Syria and Mesopotamia are two aspects of the same movement. The newcomers spoke Aramaic and probably adopted the Hebrew

[1] Vincent, Canaan, 152 f.; Macalister, Excavations of Gezer, 1912, ii, 331, 449; Meyer, Alt., I, ii, 683.
[2] Cook, CAH, i, 54; Hall, in Peake, 1925, 20; Sayce, Patriarchal Palestine, 1912, 71, 78; Otto, 1925, 43.
[3] Maspero, Struggle, 462; Fimmen, 1921, 191, 194.
[4] Jer. xlvii, 4, 5; cf. Tacitus, Hist., v, 2.
[5] Otto, 1925, 43; Meyer, Alt., II, i, 560; II, ii, 239; Olmstead, H. Pal. Syr., x, 257, 264, 271; Köster, Beih. alt. Orient, i, 37.
[6] Hall, Bronze Age, 137, 203, 240; ib., CAH, ii, 285, 289; ib., NE, 72; ib., in Peake, The People and the Book, 24, maintains that the Philistines were at least partly Lycians and Carians—an old theory, cf. Maspero, Struggle, 462.

[7] Quibell, 125; Hall, in Peake, The People and the Book, 1925, 24.
[8] Meyer, Alt., II, ii, 240, 242.
[9] Olmstead, H. Pal. Syr., 265, 268; Maspero, Struggle, 464; Macalister, Civ. in Palestine, 50; Rostovtzeff, i, 109.
[10] Macalister, Civ., 52 f.
[11] Hall, in Peake, The People and the Book, 2, 7, 14; Meyer, Alt., II, i, 344; Burney, Judges, 1918, lxxiv; Jirqu, Alte Orient, 1924, xxiv, 14; 1925, xxv, Heft 4, 13; Bilabel, Gesch. Vorderasiens, 1927, 121.
[12] Alte Orient, 1924, xxiv, Heft 5, 18; 1925, xxv, Heft 2, 30.
[13] Meyer, Alt., II, ii, 214, 215; cf. Jirku, Alte Orient, 1925, xxv, Heft 4, 21 f.; Bilabel, 119, 121.
[14] Olmstead, H. Pal. Syr., 215; Cook, CAH, iv, 167 f.

language from the Canaanites. The Aramæans were great traders over the land, as the Phœnicians were over the sea, and their language was widely diffused from about 1000 B.C.[1]

The history of Syria and Palestine begins after the best days of Egypt and of the Hittites in Asia Minor. In the time of Amenhotep III and IV those parts of Syria called Canaan and Amurru were under Egyptian influence, but were trying to shake it off with the aid of the Hittites.[2] The Hebrews migrated from the North Arabian desert not long before 1500 B.C.[3] : they appeared in Babylonia, crossed the Euphrates into Mesopotamia and penetrated into Central Syria and Canaan in separate tribes. These invaders were called Khabiru (old Babylonian for "nomad"), then Ibrim ("Hebrews"), now interpreted as "those who have passed over." Abraham is said to have come from Ur of the Chaldees, and the Hebrews never forgot their Aramæan origin.[4] A Joshua (Iashuia) who crossed the Jordan is in the Amarna letters and Jericho, a Hyksos town, is known to have been taken and burnt by an invader about 1400 B.C., the walls falling by an earthquake.[5] The first settlement in Palestine was on the ridge of Mount Ephraim and the valleys on either side, with a sacred place at Bethel, the old Canaanite population partly remaining. Various tribes then spread north and south, from the 13th and first half of the 12th century, when Palestine was still a part of the Egyptian empire, though weakly held. The Hebrews largely adopted the Canaanite culture. The tribe of Judah and the Edomites penetrated farther south into Palestine, the first centred around Bethlehem and Mount Se'îr and the second in the fertile valley of the Hebron, in Negeb and adjoining places. Jerusalem, formerly occupied by Jebusites and separating Judah and Israel, first became important under David.[6]

The early Hebrews were still in the Bronze Age and their civilisation, as compared with those of Egypt, Mesopotamia and the Ægean, was mediocre.[7] They were not acquainted with iron, which they probably adopted from the Philistines and afterwards avoided in religious ceremonies.[8] Their pottery even in Solomon's time is an inferior imitation of Cretan.[9] It has been supposed that the old Hebrew script, which soon gave way to Aramaic[10], may have been adopted from the Philistines.[11] It was formerly believed that this script was not known before about 1000 B.C., but an inscription on a sarcophagus at Byblos shows that it was fully developed and even deteriorating about 1250 B.C. Most of the old documents were probably on papyrus, which was largely imported to Byblos from Egypt, and have disappeared in the damp soil.[12]

The cuneiform tablets recently found at Râs Shamra, a site in Syria near the coast opposite Cyprus which was under Mycenæan influences, date back to 1400 B.C., but are written in a primitive Semitic dialect resembling ancient Hebrew and contain a rudimentary alphabet. They contain descriptions of sacrifices and ritual and mythical poems very like the Mosaic writings of the Pentateuch. The people were perhaps related to those who used a primitive

[1] Meyer, Alt., II, i, 344; Olmstead, H. Pal. Syr., 194 ; Worrell, Races in the Anct. Near East, 1927, 113 f. ; Jirqu, *Alte Orient*, 1925, xxv, Heft 4, 14, regards the Israelites as a mixed race of Khabiru and Aramæans.

[2] Cook, CAH, i, 183 ; Delaporte, La Mésopotamie, 49.

[3] Macalister's date, Civ. Palest., 29, of 2500 is too early.

[4] Olmstead, H. Pal. Syr., 194, 196 ; Burney, Judges, 1918, lv f.

[5] Olmstead, 197, 201 ; Garstang, *Illustr. London News*, 1933, clxxxiii, 994.

[6] Meyer, Alt., II, ii, 218, 223, 226, 236, 250.

[7] Macalister, Civ. in Palest., 30 f., 54, 58 ; *ib.*, A Century of Excavation in Palestine, 1925, 208 ; Otto, 1925, 45.

[8] I Sam., xiii, 19–22 ; I Kings, vi, 7 ; Ex., xx, 25.

[9] Macalister, Civ. Pal., 58 f., 66.

[10] Lidzbarsky, JE, i, 439 ; ii, 68.

[11] Macalister, Civ., 58 f.

[12] *Ib.*, Century, 246, 250.

alphabet in the temple of Sarâbît al-Khâdim in Sinai (see p. 11). The supreme deity is El (Kronos-Chronos), with the plural Elohim, and other gods include Môt, who is mentioned by Sanchuniathon as the Phœnician "cosmic egg," produced by an epicene being which arose from chaos and a wind of "dark air."[1] There are various kinds of sacrifices (burnt, whole burnt, of the new moon, bread of the gods, etc.) ; a table of gold in the sanctuary and a sacred object (ark ?) are mentioned, and the number seven occurs in various connections.[2] The finds at Râs Shamra include many bronze objects of the 14th–13th centuries B.C., e.g. a tripod, engraved axe- and spear-heads (13th century), statuettes inlaid with gold and silver and large fire shovels like those found at Cyprus (q.v.), from whence copper was probably imported ; also bronze like XII dyn. Egyptian and that of Byblos, with Egyptian work. There is gold and silver in bars, vases, jewellery and statuettes, and ivory with Mycenæan carving. Iron was very rare and was used only for ornaments, in the 13th century B.C. A fine "porcelain" head is of the earliest period. Râs Shamra probably exported Asiatic products to the Ægean islands and the Greek mainland.[3]

The Captivity of the Jews in Egypt and the Exodus are obscure events, since there is no mention of them in Egyptian records. The Pharaoh of the Oppression is usually identified with Rameses II (1292–30) and the Exodus is considered to have occurred under his successor Merenptah (1230–1215).[4] Yahuda [5] and Marston [6] date the Exodus 1450–1447 B.C., with a sojourn in Egypt of 430 years [7], which was a formative period for Israel, and Hall [8] even accepts the story of Josephus that the Exodus was the expulsion of the Hyksos in the 16th century, a theory which, although also accepted by Budge and Gardiner [9], rests on no ancient evidence and is generally rejected.[10] Yahuda claims that the Bible narratives go back to an early period of direct contact with an Egyptian environment, but his theories (although containing much suggestive material) are not generally accepted and it is supposed that only about half the Jews went to Egypt.[11]

The Assyrians came in contact with the Hebrews in 853 B.C., when Šalmaneser III captured and sacked the town of Qarqar in Syria, and in 841 Israel paid him tribute. Adadnirari III (810–782) actually entered Palestine, and Israel ('Omri), Edom and Philistia submitted to him—Judah not being mentioned. Samaria and Israel fell to Sargon II in 722. In 586 the captivity of Judah by Nebuchadnezzar destroyed both Israel and Judah and carried them captive to Babylon. Israel disappeared and Judah remained in Babylon until the more tolerant Persians allowed a remnant to return about 520 B.C. to Jerusalem and rebuild the Temple. The Nabatæans had by then occupied the land of Judah and the Samaritans that of Israel, and the Diaspora (Dispersion) had begun. Most of the Jews, however, remained in Babylonia, which until A.D. 1000 was the "second land of Israel."[12]

[1] Cory, Ancient Fragments, 1832, 3 ; Paton, ERE, xi, 177 ; Deubner, Ro., iii, 2093 ; Cook, Zeus, ii, 1036 ; Contenau, Civ. phénic., 100 ; W. Scott, Hermetica, Oxford, 1925, ii, 113.

[2] Meyer, Alt., II, ii, 69 ; Sir C. Marston, New Knowledge about the Old Testament, 1933, 37, 59, 63, 139 ; Montgomerey, J. Amer. Orient. Soc., 1933, liii, 97, 283 ; Sarton, Isis, 1934, xx, 478.

[3] Schaeffer, Illustr. London News, 1931, clxxix, 806 ; 1932, clxxx, 382 ; ib., Syria, 1929, x, 285 ; 1931, xii, 1 ; 1932, xiii, 1 ; 1933, xiv, 93 ; 1934, xv, 105.

[4] Maspero, Struggle, 442 ; cf. Jirqu, Alte Orient, 1925, xxv, Heft 4, 15.

[5] Die Sprache des Pentateuch in ihre Beziehung zum Aegyptischen, 1929.

[6] New Knowledge about the Old Testament, 1933, 35, 87, 116.

[7] Ex., xii, 40.

[8] In Peake, 1925, 3, 6, 15, 19.

[9] JEA, 1924, x, 88.

[10] Burney, Judges, cxiv f., cxvi.

[11] W. F. Albright, The Archæology of Palestine and the Bible, New York, 1933, 214.

[12] Maspero, Empires, 71, 101, 216 ; Hall, in Peake, The People and the Book, 34, 39.

THE ARCHÆOLOGY OF CANAAN

Archæological research in Palestine is, compared with that in Egypt and Mesopotamia, relatively late [1] and incomplete. Important sites are Gezer (mentioned in the Amarna letters, c. 1450–1400) near Jaffa [2], Tell Gemmeh (ancient Gerar), 8–9 miles south of Gaza and almost in the desert on the road from Egypt to the Judæan hills [3], and Beth-pelet, now Tell Fâr'ah, 18 miles from Gaza and on the Egyptian border, regarded by Petrie as a Hyksos fortress.[4] Three periods of culture in Canaan may be recognised[5]:

I	3000–1550 (2500–2000; Vincent gives 3000–2500)	Ancient Canaan	Contemporary with the transition from Copper to Bronze Ages in Cyprus. Babylonian influence predominant.
II	1550–1100 (2000–1600)	Middle Canaan	Contemporary with the Second (Mycenæan) Bronze Age in Cyprus. Initiated by the Egyptian XVIII dyn. conquest in Syria.
III	1100–332 (1600–1200)	Recent Canaan	Contemporary with the Græco-Phœnician Iron Age in Cyprus and Phœnician domination in the Mediterranean. New classification of second part of period: 1. Ancient Palestine, 1200–600: (a) Philistine (b) Ancient Jewish 2. Middle Palestine or middle Jewish, 600–300 3. Recent Palestine, 300–50: (a) Recent Jewish (b) Hellenistic

The remains as a whole in Palestine are commonplace and poor, indicating a much lower level of civilisation than those of contemporary Egypt or Babylonia.

Rudely chipped flints which occur in great numbers in Palestine have been associated with a Palæolithic culture, which, like those on all other sites, is dated as ending "about 10,000 B.C."[6] This passed into a Neolithic culture, with pottery.[7] A more recent view is that all the known remains correspond with some knowledge of metal, as is probably the case in Persia, Mesopotamia, Susiana and Egypt.[8] Recent excavations in Palestine are said to have disclosed extremely ancient Palæolithic remains. The pre-Semitic Gezer

[1] Summary of excavations in Palestine: A. Bertholet and A. K. Dallas, A History of Hebrew Civilisation, 1926, 31 f.; Handcock, Archæology of the Holy Land, 1916; Sayce, Patriarchal Palestine, 1912; Sir G. A. Smith, Syria and the Holy Land, 1918; W. F. Albright, The Archæology of Palestine and the Bible, New York, 1933, 60 f.—summary of excav. and bibliography; R. A. S. Macalister, History of Civilisation in Palestine, Cambridge, 1912; ib., A Century of Excavation in Palestine, 1925; J. G. Duncan, Digging up Biblical History, 2 vols., 1931, includes recent work at Gerar and good index; Dussaud, Deschamps and Seyrig, La Syrie antique et médiévale illustrée, 4to., 1933; L. H. Vincent, Canaan d'après l'exploration récente, 1907—somewhat out of date; Olmstead, H. Pal. Syr., 4 f.; Nature, 1931, cxxvii, 301; detailed accounts in: R. A. S. Macalister, The Excavations of Gezer, 3 vols., 1912; Petrie, Gerar, Brit. School of Arch. in Egypt, 1928; Reisner, Fisher and Lyon, Harvard Excavations in Samaria, 2 vols., Cambridge,

Mass., 1924; on excavations at Neirab (Jerusalem) see Abel, Carrière and Barrois, Syria, 1927, viii, 201; 1928, ix, 187, 303—bronze and iron weapons, Neo-Babylonian Period.

[2] Macalister, Excav., i, 9; BMGE, 133.

[3] Petrie, Nature, 1927, cxx, 56; ib., Gerar, 1 f.

[4] Brit. Assoc. Journ., 1930, 63; Macdonald, Starkey and Harding, Beth-Pelet II, 1932; Peake and Fleure, Merchant Venturers in Bronze, Oxford, 1931, 118; Albright, 53; Marston, New Knowledge, 68 f., 76.

[5] Contenau, Civ. phénic., 203; ib., Manuel, ii, 1038, who gives the dates in brackets as a new classification of 1923; Dussaud, Civ. préhellen., 290; Pottier, in de Morgan, Délég. en Perse, xiii, 81; Vincent, Canaan, 426.

[6] Macalister, Civ. Palest., 8; Bertholet-Dallas, 35, 45.

[7] Macalister, Civ., 11.

[8] De Morgan, Syria, 1923, iv, 23; Macalister, Century, 224 f.

cave-dwellers, dated about 3000 B.C.[1], made rude pottery without the wheel [2], cremated their dead, sacrificed pigs and apparently had no relations with Egypt or Mesopotamia. There was another non-Semitic (Armenoid) race living in caves in south Palestine from Hebron to Ashdod, who made hand-painted pottery and used a little copper : they were later displaced by the Hebrew invaders.[3] In Hebrew Gezer there is Cypriote art, flint, bronze and two arrow-heads of Egyptian iron.[4] Period dates at Gezer adopted by Macalister [5] are :

B.C.	B.C.		
To 2000	(2500)	Pre-Semitic	—
2000–1800	} (2500–1600)	I Semitic	—
1800–1400		II Semitic (Philistine)	Egyptian, Cretan and especially Cypriote influences
1400–1000	(1600–1200)	III Semitic	Influences as in Semitic II but "rather reminiscent"
1000–550	(1200–800)	IV Semitic	Influences of Semitic II failing, but fresh imports from Cyprus.
550–100	(800–500)	Hellenistic	Greek influences, with a transitional Persian Period.

The contents of the earlier graves [6] are not very systematically dated, and a lack of precision in the following account is unavoidable. Woolley [7] thinks most of the objects ascribed to Gezer III probably date to about 1200 B.C.

Non-Metals

In remains at Gezer and elsewhere are ruins of houses with walls of stones of moderate size set in mud mortar, a fine lime cement being reserved for water cisterns. Bricks, generally soft-baked and rather large, were also used and much of the wall at Jericho is brick.[8] Stone and pottery seals and many scarabs (from XII dyn.) and amulets (including also silver) [9] occur ; hæmatite beads in early graves (i, 293 f.), stone mortars of various periods (ii, 35 f. and Figs.), and stones for the fire drill (ii, 44) ; an apparatus for separating oil from water (ii, 69) ; and very inaccurate stone weights (ii, 278 f.), of which the gold-smith had two sets, one for buying and the other for selling, as in Babylonia.[10]

Good pottery made on the wheel appears in the II Semitic Period (ii, 131 ; iii, plates 140, 159) in red, brown, yellow, grey and black ; in the III Semitic Period these colours occur with blue. The pottery[11] found on early sites is unglazed, though sometimes burnished (ii, 138 f.), and has red decoration, red ochre used for this purpose occurring on the site. The potter's wheel appears in the I Semitic Period. A Venetian red and a pale cream clay were used.[12] The red pigment decoration was perhaps a technique adopted from the Hittites north of the Halys.[13] The quality of the pottery suddenly deteriorates with the Israelite immigration.[14] Some of the early pottery (2500–1200 ?) is glazed

[1] Macalister, Civ., 11, 15 ; ib., Excav., i, 6 ; Bertholet-Dallas, 36 f.
[2] Various types of Palestinian pottery in Contenau, Civ. Phénic., 203 f. ; Duncan, passim.
[3] Macalister, Civ., 13, 23 ; Olmstead, H. Pal. Syr., 16 f., 20.
[4] Macalister, Civ., 43.
[5] Excav., ii, 131, 308 ; "a little vacillating," Dussaud, 302 ; Contenau, Manuel, ii, 1039, who gives the older dates in brackets.
[6] Macalister, Excav., i, 265 f., 293, 297, 303 ; Olmstead, H. Pal. Syr., 61 f. ; some

of Macalister's "Philistine" graves are probably of the Persian Period : Dussaud, 301.
[7] Syria, 1921, ii, 177.
[8] Macalister, Century, 216, 221.
[9] Excav., ii, 293, 314—references in the text are to this work ; Bertholet-Dallas, 62 f. ; Vincent, Canaan, 177.
[10] Macalister, Civ., 44.
[11] Cf. Macalister, Century, 237 f.
[12] Handcock, Holy Land, 132, 215, 219 f., 238.
[13] Sayce, Patriarchal Palestine, 1895, 239.
[14] Macalister, Century, 238.

("enamelled") (ii, 128 f., 131).[1] Pottery blowing-tuyères, pottery saucers, early stone and earthenware moulds for casting (ii, 260, 265, 266) [2], a funnel (III Semitic) (ii, 184, Fig. 343) and some "stone or porcelain" [!] crucibles of the IV Semitic Period (ii, 258 f., 260) occur at Gezer, but crucibles and moulds are rare. Lamps, of pottery in the "Amorite" Period, were perhaps derived from Egypt.[3] Palaces dated 3100 and 2500 at Gaza have bathrooms and stucco floors.[4] All the best pottery is foreign : Minoan (LM, rarely MM) or Cypriote in the earlier period, Greek in the Maccabæan Period.[5]

Glass, both coloured and clear, occurs on all the sites examined. Ornamental coloured glass beads and vases, which begin to appear in the III Semitic Period at Gezer, were imported from Egypt (ii, 239 f. ; iii, plate 123), and include blue, violet, green, brown and yellow, but not murrhine glass ; the base of a square vessel was probably Egyptian (i, 108, 310). Clear glass begins in the early IV Semitic Period, from 1000 B.C. A glass lamp, nearly intact, was found in a tomb, but was probably early Christian (i, 363, Fig. 189). At Gerar there was a blue and white bead and blue glass of 1194 B.C. ; blue glass of 930, 660, etc. ; beads of black, blue, dirty brown and lemon yellow glass from 960 to 650, and patterned glass of the 8th century B.C. like that made at Cumæ in 800–500.[6] Glass of uncertain date was found in Samaria.[7] "Opaque green glass covered with white glaze" in the forms of Astarte idols occurred at Beth Shan with remains showing Egyptian, Babylonian, Cyprian and Cretan influences.[8]

Stones found on early sites include carnelian, agate, jasper, chalcedony and hæmatite.[9] At Gerar there was lapis lazuli of 810 B.C., amber of the same date, a gypsum ring and malachite as a "weight"[10] ; a "resinous amber-like paste" as beads in an early grave at Gezer is doubtful, and a jade scarab there is "probably not Egyptian."[11]

Pigments at Gezer were red iron oxide, yellow limonite and rarely blue and green (azurite and malachite, or copper frits ?). They were in cakes and for use were spread on a stone palette and laid on with a brush.[12] A large dye vat was found at Tell Beit Mersim (Bibl. Debir).[13] The fuel used was wood, charcoal and dried dung.[14] Excavators do not always distinguish clearly between charred wood from burnt sites and stocks of prepared charcoal.[15] Ivory of various dates (930, 660, etc., B.C.) occurs at Gerar[16] and on old sites at Gezer, and was used for inlay, etc. ; some of it was perhaps Egyptian.[17]

GOLD

Metals used in Philistine Gezer were gold, silver, copper, bronze, brass, lead and iron.[18] Gold is rare and was probably not much used ; it is doubtful if any true Bronze Age specimens are known.[19] It was imported in ingots, two of which, of the IV Semitic Period (1000–550 B.C.) were found on analysis to

[1] Vincent, 219, 223.
[2] Handcock, Holy Land, 175.
[3] Ex., xxv, 37 ; Bertholet-Dallas, 77 ; Bellwood, J. Soc. Chem. Ind., 1922, 213 R.
[4] Petrie, Fathoming the Ages, a pamphlet issued in 1933.
[5] Macalister, Civ., 66 ; ib., Century, 239 f.
[6] Petrie, Gerar, 10 ; 13 ; 24, 30.
[7] Reisner, Fisher and Lyon, Excav. in Samaria, 1924, 27.
[8] Olmstead, H. Pal. Syr., 153.
[9] Handcock, Arch. of Holy Land, 157 f. ; on bead materials see Duncan, ii, 234 f., jewellery, ib., 231 f.

[10] Petrie, Gerar, 10, 13, 20, 26.
[11] Excav., i, 293 f.
[12] Olmstead, H. Pal. Syr., 63.
[13] Albright, Arch. of Palest., 63, 120.
[14] Excav., ii, 45.
[15] Marples, J. Inst. Metals, 1930, xliii, 358.
[16] Petrie, Gerar, 11, 16 f.
[17] Handcock, Holy Land, 153 f.
[18] Excav., ii, 258, and index ; Vincent, Canaan, 177, 219, 223, 234, 344, 448 ; Olmstead, H. Pal. Syr., 20 ; Duncan, Digging, ii, 236 ; Handcock, Holy Land, 179.
[19] Macalister, Century, 258.

be "very pure gold," beaten into shape.[1] They weighed *approximately* 27·6 oz.
or 13,248 grains and 50 Babylonian shekels of 260 grains each make 13,000
grains. One is tongue-shaped, like the "tongue" ("wedge" in Authorised Version)
which Achan stole from Jericho [2], and the silver tongues at Troy.[3] Beating out
to leaf, casting, hammering, welding and drawing (?) into wire were all well
known from the earliest Semitic Period.[4] The gold found at Gezer (2500–
1700) includes a gold tube and gilt bronze.[5] Gold objects from Gaza (some
c. 2300 B.C. ?) include two pins (1–2 in. long) with remarkably worked heads,
ten heavy armlets and an exquisitely wrought pair of ear-rings (? 1700 B.C.)
which show the curious twisting and characteristic shapes of torques (also
found at Susa, p. 401), and hence, with another crude pair of ear-rings of
twisted sheet or "four bladed rod," *c.* 1500 B.C., are thought by Petrie to have
come from Ireland. A pendant of spread falcon pattern is put in the Hyksos
Period (2000 B.C. ; *sic*). Gold and silver torn to shreds and melted were found
in a layer dated 2500 B.C., with smashed copper and pottery, perhaps the result
of invasion and fire [6], or deliberately broken from religious motives. At Gerar
Petrie found evidence of "abundant use of gold," including foil from 1190 B.C.[7]

SILVER

Silver includes vessels, wire, a seal ring and an amulet of the early period (?)
and objects and vases of the Israelite Period (dated 1200–600 by Vincent) at
Gezer.[8] The amulet is a circular ¾-in. piece of metal (no date given), enamelled
deep blue and white, "the only example found of enamel on silver."[9] The
two silver vases, "probably imported," are really 8th century B.C., and all
the silver at Gerar[10] is also late[11], although Petrie puts it down as 1190 B.C.,
etc. Silver and gold jewellery was found by Selin at Ta'anach[12], and a silver
crescent from Beth Shemish (middle Bronze Age) contained : silver 57·23,
lead 2·9, copper 0·1, gold trace.[13]

COPPER ; BRONZE ; BRASS

Copper was used sparingly by the non-Semitic Neolithic race living in south
Palestine[14], and there is very little pure copper on the sites.[15] Petrie found
copper rings and beads in remains of 2000 and 1180 B.C. at Gerar.[16] Bronze
was used for arrow-heads (sparingly—mostly stone), knives, chisels, daggers,
axes, saws, spoons, pins, cooking pots, mirrors, ear-rings, buckles and statuettes,
but hammers were usually of stone ; from the earliest "Amorite" Period to
1350 B.C. it was the dominant metal in Palestine and Samaria. The bronze
axes are of early European or Egyptian form, not socketed. The sickles were
of flints mounted in wood.[17] Some of the "bronze" is probably copper, such as
the "very archaic red bronze" (2500–1200) of Vincent.[18] A magnificent curved
bronze sword found at Gezer is probably a copy of that of Adadnirari I of
Babylon.[19]

[1] Excav., ii, 258 f., Fig. 405 ; Handcock, Arch. Holy Land, 211 f. ; Duncan, ii, 229.
[2] Joshua, vii, 24.
[3] Excav., ii, 259 ; Forrer, RL, 851.
[4] Excav., ii, 260 ; wire-drawing is doubtful, but see p. 229.
[5] *Ib.*, i, 293 f. ; Vincent, 219, 223.
[6] Fathoming the Ages ; *Daily Telegraph*, 3 Aug., 1932 ; June, 1933 ; *Illustr. London News*, 1932, clxxxi, 57 ; cf. Albright, Arch. of Palestine, 1933, 53.
[7] Gerar, 1928, 10, 29 ; Olmstead, H. Pal. Syr., Fig. 118.
[8] Excav. at Gezer, i, 293 f., 303 f. ; Vincent, Canaan, 177, 234 f.
[9] Excav., ii, 263.
[10] Petrie, Gerar, 1928, 10, 12.
[11] Duncan, ii, 231.
[12] Sayce, Patriarchal Palestine, 243.
[13] Meldrum and Palmer, *J. Chem. Education*, 1931, viii, 2171.
[14] Olmstead, H. Pal. Syr., 20.
[15] Duncan, ii, 236 f. ; Handcock, Arch. of the Holy Land, 179 f.
[16] Gerar, 1928, 10, 12, 13.
[17] Macalister, Excav., i, 265, 293, 303 ; *ib.*, Century, 228 f., 232, 236 f.; Handcock, Holy Land, 179 f. ; Duncan, ii, 236 f. ; Reisner, Fisher and Lyon, Excav. in Samaria, 1924, 26, 346.
[18] Canaan, 224.
[19] Bertholet-Dallas, 92.

The most interesting feature of the Gezer bronzes is the presence in them, along with tin, of considerable amounts (up to 23·4 per cent.) of zinc[1] :

	Semitic I 2000–1800	End of Semitic II 1400	Semitic III 1400–1000	Beginning of Semitic IV 1000
Copper	77·90	63·5	66·40	90·00
Zinc	2·89	3·7	23·40	1·86
Tin	11·20	33·7	10·17	7·73
	91·99	100·9	99·97	99·59

The considerable deficit from 100 in the earliest specimen may be due to corrosion products, or perhaps to the presence of lead, which may not have been sought. Analyses of copper and bronze articles found by Petrie at Tell el-Hesy in Palestine, the ruins of ancient Lachiš, gave[2] :

	Copper	Tin	Iron	Lead	Antimony	Oxygen
Amorite axe or adze	94·9	?	0·77	0·68	?	2·7
1500 B.C. knife	97·0	0·0	0·15	tr.	tr.	tr.
Israelite needle	63·4	7·5	—	0·0	—	—
Israelite needle	45·8	10·3	—	—	—	—

The axe contained 25 per cent. of cuprous oxide, which made it red, very hard and brittle. The needle with 10·3 per cent. of tin was much corroded and had perhaps lost a proportion of copper. Flint tools occurred with all these metal objects, the latest being of excellent manufacture. In the upper Israelite layers at Lachiš bronze as a useful metal is replaced by iron, but there were many bronze arrow-heads of 1400–800 B.C. At Gerar Petrie found bronze of various dates from 1480 B.C. (knife, chisels, tools, chain, axe, bolt, etc.).[3] Bronze of 1700 occurs at Jericho.[4] Analyses of corroded bronzes from Beth Shemish gave[5] :

	Copper	Tin	Lead	Miscellaneous
1. Spear-head	63·8	5·2	0·5	Gold 0·28 ; silver tr.
2. Fragments	70·8	13·9	1·3	Zinc 0·20.
3. Fragment of small crescent	55·4	9·6	0·2	——
4. Needle	85·9	2·7	0·6	——
5. Arrow-head	70·9	8·8	tr.	——
6. Bar (late), 107 gm., inner core	88·7	9·7	tr.	——
7. Corroded lump, lowest level, ? imported ore	50·6	12·7	2·8	Sulphur 5·6, silver tr.
8. Arrow-head, lowest level	66·8	5·2	tr.	——
9. Mirror ?, wholly corroded	47·0	18·5	tr.	Silver tr.
10. Needle or punch ?	67·7	12·5	0·1	——
11. Ear-rings, wholly corroded	54·5	6·2	0·3	——
12. Curved needle, early Iron Age	91·6	5·6	0·5	——
13. Lump, corroded	90·9	5·3	tr.	Silver tr.

[1] Excav., ii, 265, 293 f., 303 f. ; analyses by J. E. Purvis.
[2] Gladstone, PSBA, 1894, xvi, 95 ; ib., B.A. Rep., 1893, 715 ; ib., J. Anthropol. Inst., 1897, xxvi, 310 f., 316 ; Montelius, Ält. Bronzezeit, 1900, 140.
[3] Gerar, 10, 12, 13, 14.
[4] Garstang, Illustr. London News, 1933, clxxxiii, 994.
[5] Meldrum and Palmer, J. Chem. Education, 1931, viii, 2171.

The very varying amounts of tin are noteworthy and do not indicate a skilful metallurgy. The bronze daggers found at Râs Shamra (which was perhaps established on the opposite coast by the Mycenæan population of Salamis, in Cyprus) seem to have been imported in an unfinished form from Cyprus and worked locally. The numerous bronze statuettes found in Syria of about 1000 B.C. are generally called Syro-Hittite.[1]

LEAD

Lead was not much used at Gezer, but occurs as a small ring in Semitic III ; other forms, mostly in the Hellenistic Period, are weights, wire, a buckle, a perforated disc and a gate socket.[2] There was a late (? 1200 B.C.) net sinker at Gerar.[3]

IRON

Information on the early use of iron and steel in Palestine has been much modified by recent archæological discoveries. Vincent [4] considered that iron occurs in Palestine in the Israelite Period (1200–600), that at Gezer being 10th–9th century, but Petrie's excavations at Gerar have disclosed abundance of much earlier iron and even steel in Palestine. Small iron work began there at least as early as 1300 B.C. ; there is a ring of 1180, and massive tools of 1170.[5] Handcock [6] did not think steel was known in Palestine, as Belck [7] had assumed, but a broken steel dagger of 1350 B.C. was found at Gerar [8], and a steel dagger "before 1300" at Beth-pelet.[9] The earliest iron at Gezer consists of knives (sometimes riveted with bronze), probably late Semitic II, before 1250 or even 1300 B.C.[10], and two wedges (now lost) at the bottom of a sloping water passage which had been sealed up before 1250.[11] Useful iron appears at Gezer in Semitic III–IV, about 1000 B.C.[12], whilst its use reached a maximum at Gerar about 1200 B.C.[13] Bronze was still used along with iron at Gezer for tools and nails.[14] Some iron "before 300 B.C." was found in Samaria.[15] Fragments of fluted iron mail from Tell Duweir are probably of about 1200 B.C.[16] A smithy with iron dross and lumps of clay ironstone at Tell el-Mutesellin[17] is of uncertain date. A little ornamental iron was found at Râs Shamra.[18] A structure found by Bliss at Lachiš (Tell el-Hesy ; 15th century B.C.) is supposed to be a blast furnace, although Vincent[19] thinks it may be a pottery kiln. The example of such a kiln (c. 2300) found in Babylonia at Nippur[20], however, is quite different, being like a normal modern eastern kiln with nine arches. Iron was found in the Fourth City (c. 1400) of Lachiš.[21]

[1] Dussaud, La Lydie et ses Voisins, 1930, 86, 100 f. ; Contenau, Manuel, ii, 1063, 1069 f.
[2] Excav., i, 293 f. ; ii, 263 f. ; Duncan, ii, 240 ; Handcock, Arch. Holy Land, 211.
[3] Petrie, Gerar, 13.
[4] Canaan, 234 f., 344, 448 ; Macalister, Century, 230—axes and adzes of modern form.
[5] Petrie, Gerar, 1928, 10, 14 f., 29.
[6] Holy Land, 203.
[7] Z. f. Ethnologie, 1910, xlii, 15.
[8] Petrie, Nature, 1929, cxxiii, 838 ; ib., Brit. Assoc. J., 1930, 63.
[9] Petrie, Brit. Assoc. J., 1930, 63.
[10] Macalister, Excav., i, 299 ; cf. ib., ii, 269 ; ib., Century, 230 ; Duncan, ii, 239.
[11] Macalister, Excav., ii, 269 f. ; in ib., Civ.

Palest., 43, there are "arrow-heads," but in Century, 237, Macalister says these are late, just before the Roman Period ; Friend, Iron in Antiquity, 1926, 167 f.
[12] Excav., ii, 243, 269 f. ; Handcock, Holy Land, 203.
[13] Duncan, ii, 239.
[14] Excav., ii, 270.
[15] Reisner, Fisher and Lyon, 26 f., 346 f.
[16] Nature, 1933, cxxxii, 167.
[17] Bertholet-Dallas, 91.
[18] Dussaud, La Lydie et ses Voisins, 1930, 147.
[19] Canaan, 76, Figs. 48–50.
[20] Hilprecht, Explor., 489 ; Vincent, Canaan, 77 ; Meissner, i, 235, two figures.
[21] Hall, Oldest Civ. Greece, 1901, 193 f.

At Gerar Petrie [1] found iron furnaces of about 1200 B.C. and very large tools, such as a pick of 7 lb. weight, hoes and plough points. The ore was probably hæmatite, resulting from decomposition of pyrites in the Beersheba basin. Iron knives were made from 1350 B.C., probably the earliest manufactured iron known. Petrie's photographs [2] of a "large iron furnace" and two "sword furnaces" do not convey the impression of metallurgical furnaces, but Petrie reconstructs the latter and dates the "three iron furnaces" as 1195 [3], 1100 and 870 B.C. "The flues had been violently heated and therefore must have been covered in with movable tile tops, as there is a regular edge without break. It seems plain that they were for heating bars of metal about three feet long. Such bars could only be needed for swords [?], and such a flue-bed, with charcoal along it, would serve well both for smithy heat and also for tempering." Of the three furnaces "the earliest is the largest and best preserved, with the draught hole complete. All the furnaces have recesses at the side, sloping wider upwards, to allow of a draught without being so much encumbered with the charge. The openings all faced west to catch the wind. The earliest of these is as old as any dated iron known in the Mediterranean, excepting the knives here of 1300 and 1250 B.C. It proves that the smelting was done on the spot, and that the metal was not imported." This last conclusion is not quite certain, since Petrie does not describe analysed scoria and slags, and the furnaces could have been used merely for heating and tempering imported iron bars—for which purpose their shape would seem better adapted than for smelting.

The iron objects found at Gerar [4] included large picks and tools (930 B.C.), the largest pick oxidised through to $9\frac{1}{2}$ lb. of "hydrated oxide, equal to 6 lb. of metallic iron," which is incorrect, as iron rust is usually $2Fe_2O_3,3H_2O$, $9\frac{1}{2}$ lb. of which correspond with $4\frac{1}{2}$ lb. of iron; hoes of 1100, 1000 and 932 B.C.; adzes, spear-heads and daggers; "one of the earliest pieces of iron more than a century below the building level of 1194 B.C."; a nail, harpoon and pruning hook of 810 B.C.; sickles, mostly of a uniform degree of bend, of 1250 (?)–600 B.C.; spear-heads and daggers, with forms like those at Nineveh and Anau, of 1200, 1100, 1000, 960, 900–600; large square heads, probably from armour-piercing arrows, of 900–600; arrow- and lance-heads (the latter, unusually, more common) of peculiar Central Asiatic (Tomsk, Perm, Caspian) type, 932 B.C., with some from 1250, one (XXVI dyn.) supposed to be Scythian; a finger ring of c. 800; hooks, borers, chisels, gouge and poker (?); many knives, including a riveted one of 1300 B.C. ("one of the earliest pieces of iron"), others (rusted through) of various forms of 1250, 1200, 1150, 950, used for different purposes, and curved knives of the XXVI dyn. period "like those from Adelsburg and Idria, doubtless Noric steel imported" [?].

It is generally supposed that iron working originated outside Palestine, where there is practically no ore [5], and came with the Philistines in the 12th century [6]; the Philistines were expert in the metallurgy of iron, knowledge of which they kept secret from the hill tribes of Palestine [7], and they probably learnt the use of iron from the Hittites. Egyptian documents of 1209–1205 B.C. mention three places in Canaan as famous for chariots, perhaps because they were of iron. In David's time the Hebrew weapons and tools were of iron, which his conquests in the north had made familiar.[8] A corrupt passage in the Bible [9] speaking of iron of the north (ab Aquilone in the Vulgate) may refer to the

[1] Nature, 1927, cxx, 56.
[2] Gerar, 1928, plate xxv.
[3] So on the plate vi; 1175 in text, ib., 14.
[4] Petrie, Gerar, 14 f., 29 f.; plates xxvii, lxvi; Olmstead, H. Pal. Syr., Fig. 119.
[5] Bertholet-Dallas, 90 f.

[6] Macalister, Civ. Pal., 44, 59; Petrie, Gerar, 29, common use.
[7] I Sam., xiii, 19; Macalister, Excav., i, 299 f.; Olmstead, H. Pal. Syr., 268.
[8] Bertholet-Dallas, 90, 211, 249.
[9] Jer., xv, 12.

Philistines ; Ezekiel [1] by Dan may mean south Arabia.[2]   The Kenites, who are supposed to be represented by Tubal Cain [3], were a nomadic tribe of smiths associated with Sinai, where iron ore occurs.[4]   The statements in the Old Testament show that the Jews had no clear idea as to how the use of iron originated in Palestine ; the true source was Asia Minor.

[1] xxvii, 19.
[2] CDB, 187.

[3] Gen., iv, 22.
[4] Bertholet-Dallas, 90.

PALESTINE II

THE BIBLE

THE OLD TESTAMENT

The Old Testament is a collection of works of very different dates, some components going back to about 1100 B.C. and others being as late as 165 B.C., the earliest actual books being not older than the 8th century B.C. It is beyond our scope to give a critical survey of the dates of these books, many of which are of composite authorship and much interpolated, but it may be stated that the prophets from Amos to Samuel cover the period of about 750–600 ; Proverbs is 8th–6th centuries ; Psalms (very composite) are mostly 536–333 ; Ezekiel is about 580 ; Judges and Kings are 6th century ; the Pentateuch (containing various components, principally Jahvistic J and Elohist E, but also Deuteronomic D, Priestly Document P or Priestly Codex PC) and Joshua are 7th century (J is sometimes dated 9th century and E 8th century), but in the present form about 450, if not later ; Job is about 400 or perhaps later ; Chronicles, Ezra and Nehemiah are about 300 ; Ecclesiastes is 350–200 ; Ecclesiasticus (Wisdom of Jesus the Son of Sirach, abbrev. Ecclus. ; extra-canonical) is before 200 ; Daniel is 167.[1] Needless to say, older traditions are contained in many parts of the Old Testament, some ultimately of Egyptian, Babylonian and Syrian origin, and some transmitted through Persian sources. Persian influences on the post-exilic books have long been recognised. They include views on the last day and the ages of the world, in which successive "kingdoms" are associated at first with four (gold, silver, copper, iron) and then with seven metals. The repeated references to the purification of metals in the furnace and the refining from dross, as well as the drink which gives health and long life, are probably also of Persian origin. Persian elements in the New Testament, which have been assumed by Reitzenstein on the basis of comparisons with late Mandæan and other similar works, are doubtful.[2]

The standard Hebrew Old Testament (Text of the Sopherim) was not established till the 2nd century A.D., vowel points and accents being added in the 7th century to form the Massoretic Text. The Targums ("Chaldee paraphrase") is an Aramaic version, first written out in the 1st–4th centuries A.D. The Mishnâh, or instruction of the rabbis, was fairly well developed about A.D. 200. The Gĕmârâ is a similar work in Aramaic. The Talmûd, including Mishnâh

[1] Driver, Introduction to the Literature of the Old Testament, 9 ed., 1913 ; G. B. Gray, A Critical Introduction to the Old Testament, 1913 ; W. F. Albright, Archæol. of Palestine, 1933, 146 ; A. S. Peake, The People and the Book, Oxford, 1925 ; Meyer, Alt., II, ii, 198, 288, 297 ; Jastrow, Hebrew and Babylonian Traditions, 1914 ; Cook, CAH, i, 166 ; iii, 378, 417 ; vi, 167 f. ; Encyclopædia Biblica [E. Bibl.], ed. Cheyne and Black, 4 vols., 1899–1903 ; Dictionary of the Bible, ed. Hastings [HDB], 5 vols. ; Jewish Encyclopædia [JE], 12 vols., New York, 1901–06 ; Sarton, Introd. to the Hist. of Science, i, 107, 151, 531, 401. In all quotations the chapters and verses of the Authorised Version are given ; those in the Hebrew text are either the same or differ only by one unit in the verses—the end of one chapter sometimes being the beginning of the next.

[2] E. Meyer, Ursprunge und Anfänge des Christentums, 1921–23, ii, 179, 183, 189, 192, 198, 339, 352 ; Reitzenstein, Das iranische Erlösungsmysterium, Bonn, 1921 ; F. C. Burkitt, Religion of the Manichees, Cambridge, 1925 ; Wesendonk, Urmensch und Seele in der iranischen Überlieferung, Hannover, 1924.

and Gĕmârâ, is a chaotic work which exists in two forms : the Palestinian ("Jerusalem") Talmûd completed about A.D. 425, and the more elaborate Babylonian Talmûd (the usual "Talmûd") about A.D. 500 in Sura, and first printed by Daniel Bomberg (a Christian) at Venice in 1520–1523, whose text is that generally adopted.[1] The 'Arûk, a great Talmudic dictionary completed in 1101 by Nathan ben Yehiel in Rome, contains a great amount of information, as do also the writings of the Geonim (singl. gaon), or heads of the Babylonian schools from A.D. 589 to A.D. 1040.[2] The Midrâsh is a body of explanatory tradition auxiliary to the Talmûd and partly contemporary with it.[3] The Greek Old Testament (Septuagint, LXX, or 𝔊) was made for Egyptian Jews in Alexandria from about 250 to the 1st century B.C.[4] The Tôsâfôt (Additions) are glosses on the Talmûd of the 12th–13th centuries. The commentaries of Rashi (Rabbi Solomon ben Isaac, 1040–1105) are also frequently referred to by writers on Jewish antiquities.

The Jewish Apocryphal and Pseudepigraphic literature was composed between about 200 B.C. and A.D. 150, much of it now existing only in Greek, Syriac or other translations.[5] The most interesting work from our point of view is Enoch, known in Ethiopic and (fragmentary) Greek, and in another form (Enoch II ; Secrets of Enoch) in Slavonic, translations. It was probably first written in Hebrew or Aramaic in Palestine from different sources in the period 167 to 64 B.C., was translated into Greek in Egypt in the 1st–2nd centuries A.D., and from this translation the Ethiopic version was made in A.D. 500–600.[6] The New Testament Apocrypha [7] contain little of interest to us. Magic [8] and astrology [9] frequently occur in the Talmûd and are mentioned in the Old Testament.

CHEMISTRY IN THE OLD TESTAMENT

The information of chemical interest in the Old Testament[10] is not very great and the industrial arts were probably not held in very great esteem by the

[1] Mischna, sive totius Hebræorum juris, etc., systema, cum Maimonides et Bartenoræ comment., Lat. donavit, etc., G. Surenhusius, 6 vols., fol., Amsterdam, 1698–1703—not available to me ; H. Danby, The Mishnah, Oxford, 1933 ; M. Schwab, Le Talmud de Jérusalem, 11 vols., 1871–89 ; M. L. Rodkinson, The Babylonian Talmud, 20 vols., New York, 1896–1903—incomplete and unsatisfactory ; L. Goldschmidt, Der Babylonische Talmud neu übertragen, 8 vols., 1930–33—still appearing ; several separate works in numerous editions and translations, e.g. The Babylonian Talmud, Tractate Bĕrâkôt, tr. by A. Cohen, Cambridge, 1921 ; W. O. E. Oesterley, Tractate Shabbath, 1927 ; The Talmud, J. Barclay, 1878 ; criticism, etc., in Cohen, op. cit. ; Danby, op. cit. ; I. Bloch and E. Lévy, Histoire de la Littérature juive d'après G. Karpelès, 1901, Mischnâh, 80, Talmûd, 190, Midrâsch, 212 ; H. L. Strack, Introduction to the Talmud and Midrash, Philadelphia, 1931—copious bibliography ; H. Graetz, History of the Jews, 5 vols., 1891-2, ii, 639 ; Bacher, JE, xii, 1 ; Richtmann, ib., 27. The separate treatises in the Mishnâh and Talmûd are quoted by name in our references.
[2] JE, v, 567.
[3] Strack, 201.
[4] Hautsch, PW, iia, 1586.

[5] Kautzsch, Apokryphen und Pseudepigraphen des alten Testaments, 2 vols., Tübingen, 1900 ; Charles, Apocrypha and Pseudepigrapha of the Old Testament, 2 vols., Oxford, 1913—much better than Kautzsch ; Fabricius, Codex Pseudepigraphus Veteris Testamenti, 2 vols., Hamburg, 1713–23—still useful ; Migne, Dictionnaire des Apocryphes, 2 vols., 1856–8.
[6] Charles, The Book of Enoch, Oxford, 1893; Morfill and Charles, The Book of the Secrets of Enoch, Oxford, 1896 ; Charles, Apocrypha, 1913, ii, 164 ; Beer, in Kautzsch, ii, 217 f. ; Lods, Le Livre d'Hénoch, fragments grecs découverts à Akhmîm, 1892.
[7] Tischendorf, Evangelia apocrypha, Leipzig, 1853 ; ib., Acta apostolorum apocrypha, Leipzig, 1851 ; Fabricius, Codex Apocryphi Novi Testamenti, 3 parts, Hamburg, 1703–19 ; A. Walker, Apocryphal Gospels, Acts and Revelations, Ante-Nicene Library, xvi, Edinburgh, 1870 ; B. H. Cowper, Apocryphal Gospels, etc., 1867.
[8] CDB, 499 ; L. Blau, Das altjüdische Zauberwesen, Strasburg, 1898, 43 f., 79, 85, 117, 127 f., 156 f. ; R. C. Thompson, Semitic Magic, its Origins and Development, 1908.
[9] Sabbath, xxiv ; Berakoth, v, 1 ; ix, 1–5 ; ib., tr. Cohen, 216, 386, 424.
[10] Pinner, Chemisches aus der Bibel, in

Hebrews, who were essentially at first a pastoral and agricultural people. The Book of Sirach contrasts the comparative well-being of the scribe with the manual labour of the potter, smith and stone-worker of Jerusalem [1] in a manner reminiscent of the Egyptian Sallier Papyrus.[2] Such knowledge as the Hebrews possessed of technical processes probably came from Egyptian, Babylonian, Hittite and Phœnician sources. The poetical description in Job [3], the only account of mining which has come from the ancient Hebrews [4], is obscure, but seems to follow the sequence of actual operations and would apply to the account of the Egyptian mines as given by Agatharchides.[5] There is also a resemblance of the account in Job to that of the mines in south Spain in Diodoros Siculus, Strabo and Pliny, which may be from a Phœnician source also used by Job, although the Egyptian mines of Ethiopia would be nearer Palestine.[6] The Jews were not miners or metallurgists, but they may have been acquainted with the processes used in Lebanon.[7] The Hebrew names for various artisans in the Old Testament and Talmûd are collected by Lévias.[8] In the towns the various trades were collected in shops and bazaars, locked at night, and the streets sometimes had names.[9] There is no general name for metal in the Bible: matteket (plural mattakot) is post-Biblical[10], and in later works many Greek, Persian and other names appear, e.g. asimon ($\alpha\sigma\eta\mu o\nu$) for unstamped silver coin in the Midrâsh and Talmûd, kastiterion (tin) and zarnikh (arsenic sulphide) in the Talmûd; ba'az is probably tin, since abaza is the Targum for bedil.[11]

The identification of the metals named in Hebrew works[12] is difficult; e.g. in one case[13] it is uncertain whether bedil (tin) or barzel (iron) is meant, and whether another word means "lead" or "dust." The metals are named in a fixed order[14], viz. gold, silver, copper [or bronze] and iron; also lead and perhaps tin. In Numbers[15] the order is gold, silver, copper (or bronze), tin and lead; in Ezekiel[16] it is copper (or bronze), tin, iron and lead, regarded as "drosses" and perhaps in the order of ease of oxidation.[17] Mercury is not mentioned.[18] The name for metal refiners is zorefim; unrefined [metals] are sigim (sig meaning slag), refined is muzuqqaq.[19] Gold, copper and iron are

Diergart, 1909, 195–200—superficial; Napier, Manufacturing Arts in Ancient Times, Paisley, 1879—antiquated; V. E. Johnson, Chaldean Science, 1890—trifling; Hoefer, Hist. de la Chimie, 1866, i, 33 f., 233 f.—incomplete, but better than other accounts.
[1] Ecclus., xxxviii, 24 f.
[2] Maspero, Dawn, 311, 454.
[3] xxviii, 1–11.
[4] P. Haupt, J. Amer. Orient. Soc., 1923, xliii, 116.
[5] CDB, 558; Hull, HDB, iii, 374; E. Bibl., iii, 3097.
[6] Tozer, Anct. Geography, 188.
[7] Benzinger, JE, v, 531; Krauss, ib., viii, 513; Nowack, ib., viii, 596; Vigouroux, Dict. de la Bible, iv, 1045, 1099.
[8] JE, ii, 152; cf. Franz Delitzsch, Jewish Artisan Life in the time of Christ, 1902.
[9] Jer., xxxvii, 21; I Kings, xx, 34— in Syria and Damascus; Cant., iii, 3; Acts, ix, 11; on Jewish weights and measures see I. Benzinger, Hebräische Archäologie, Leipzig, 1907, 188; W. Nowack, Lehrbuch der hebräischen Archäologie, Freiburg i. B. and Leipzig, 1894, 198.

[10] Krauss, JE, viii, 513, 515.
[11] Bochart, Geographia sacra, 1712, 648, 650.
[12] E. F. C. Rosenmüller, Handbuch der biblischen Altertumskunde, 4 vols., Leipzig, 1823–31; IV, i, 47 f.
[13] Sirach; Ecclus., xlvii, 18; Charles, Apocrypha, 1913, i, 498.
[14] Wiegleb, Kritische Untersuchung der Alchemie, Weimar, 1777, 26 f.; C. C. W. F. Bähr, Symbolik des Mosaischen Cultus, 2 vols., Heidelberg, 1837–39, 256, 281; Krauss, JE, viii, 518; Kopp, Geschichte der Chemie, 1843, i, 24; Pinner, in Diergart, 1909, 196; Hoefer, Hist. de la Chimie, i, 45; Napier, Arts, 29; Beck, Gesch. d. Eisens, i, 142 f.; Movers, Phönizier, II, iii, 27–69; Benzinger, Hebräische Archäologie, 149 f.; P. Volz, Die biblische Altertümer, Calw and Stuttgart, 1914, 387.
[15] xxxi, 22.
[16] xxii, 18.
[17] Wiegleb, 32.
[18] Roscoe and Schorlemmer, Treatise on Chemistry, 1913, ii, 671.
[19] JE, viii, 516.

31

mentioned as known before the Flood [1]—gold from the River Pison, encompassing the land of Havilah (perhaps Bahrain on the Persian Gulf, one of the islands being called Aval).[2]

In Genesis [3] Tubal Cain (Qain) is a worker in copper and iron, a legend which has been thought to suggest that the early Israelites were miners and metal workers [4], learning the art from the Egyptians or Phœnicians [5], which art they are supposed to have lost as the race of metal workers died out. Tubal has been supposed to mean the hero eponymos of the Tabali or Tibareni, iron-working tribes of Pontos and Asia Minor[6] : Gesenius's hybrid etymology from Pers. tûbâl, tûpâl, slag or dross, or iron or copper, and Arabic qain, Aramaic qainaya, smith, is uncertain. Tubal Cain is perhaps "a pale form of Gibil [the Babylonian fire-god] . . . in the earliest form of Hebrew legend the instructor of men in the art of getting fire," a humanised god like Chusôr, the Phœnician Hephaistos, the name Tubal being perhaps of Babylonian origin.[7] Although Palestine is not rich in minerals [8], old non-Biblical sources speak of copper mines in Lebanon, Syria and Palestine, and iron mines in Lebanon.[9]

There is no information in the Bible as to the means of getting fire, but stone fire-drills were found at Gezer.[10] Timber was scarce; some unusual source of supply is usually given[11], and fires were usually of dung and hay.[12] For building, both the wood and workmen were imported from Syria.[13] Coal in the Authorised Version represents five different Hebrew words : (1) gacheleth, a live ember[14]; (2) pechâm, fuel not kindled[15]—in both cases charcoal[16]; (3) retseph or ritspâh, properly a hot stone for baking[17]; (4) rešeph[18], burning ; (5) šĕkhôr[19], black.[20] True coal has not been found in Palestine.[21] Bellows occur only once[22], where they are used to blow a lead smelting furnace.[23] Various types of furnace are mentioned[24] : (1) tannûr[25], generally a baker's oven, fixed or portable ; (2) kibšân, a smelting or calcining furnace[26] or lime-kiln[27] ; (3) kûr, a refining furnace [28]; (4) attûn, a large furnace built like a brick-kiln[29] ; (5) the potter's furnace[30]; (6) the blacksmith's furnace[31]. The word 'ălîl means crucible.[32]

[1] Gen., iv, 22 ; ii, 11.
[2] Williams, DG, i, 1032 ; speculations on the localities of Pison, Eden, etc., by Gen. Gordon—who had a good sense of locality, in Strand Magazine, 1899, xvii, 314.
[3] iv, 22.
[4] Orth, PW, Suppl. iv, 114.
[5] Hoefer, i, 34, 52 ; Napier, 12 f., 21 ; Pinner, in Diergart, 1909, 196 ; Tharshish in Authorised Version of Ezek., xxvii, 12, is Carthage in the Vulgate.
[6] Cook, CAH, vi, 185 ; Meyer, Alt. II, ii, 300 ; Dussaud, La Lydie et ses Voisins, 1930, 81, 83—Gog = Strabo's Gogarene, in Armenia ; Gomer = Niphates, in Armenia.
[7] CDB, 964 ; Cheyne, E. Bibl., i, 626 f. ; cf. Bertholet-Dallas, 90 f.
[8] Bertholet-Dallas, 22 ; Deut., viii, 9, and Job, xxviii, 1-11, mention mines.
[9] Orth, PW, Suppl. iv, 114 ; Thomsen, RV, i, 425.
[10] Excav., ii, 44.
[11] I Sam., vi, 14 ; II Sam., xxiv, 22 ; I Kings, xix, 21 ; Gen., xxii, 3, 6, 7.
[12] Ez., iv, 12, 15 ; Matth., vi, 30 ; cf. Kennedy, HDB, ii, 72, opp. view.
[13] I Kings, v, 6, 8 ; CDB, 20.
[14] Prov., xxv, 22 ; xxvi, 21 ; II Sam., xxii, 9, 13 ; Ps., xviii, 8, 12, 13 ; cxl, 10.
[15] Prov., xxvi, 21.

[16] CDB, 168.
[17] I Kings, xix, 6 ; Is., vi, 6.
[18] Hab., iii, 5 ; Cant., viii, 6.
[19] Lam., iv, 8.
[20] Patrick, HDB, i, 451 ; Kennedy, E. Bibl., i, 854.
[21] Patrick, op. cit., 452.
[22] Jer., vi, 29.
[23] CDB, 104 ; Kennedy, HDB, i, 269 ; E. Bibl., i, 527.
[24] Cf. Roscoe and Schorlemmer, Treatise on Chemistry, 1913, ii, 1162.
[25] Gen., xv, 17 ; Is., xxxi, 9 ; Neh., iii, 11 ; xii, 38 ; Hos., vii, 4 ; Ex., viii, 3 ; Lev., xxvi, 26.
[26] Gen., xix, 28 ; Ex., ix, 8, 10 ; xix, 18.
[27] Is., xxxiii, 12 ; Amos, ii, 1.
[28] Is., i, 25 ; Numb., xxxi, 23 ; Prov., xvii, 3 ; xxvii, 21 ; Ez., xxii, 18 f.; Malachi, iii, 2, 3.
[29] Dan., iii, 22, 23 ; Jer., xxix, 22 ; Hosea, vii, 7.
[30] Ecclus., xxvii, 5 ; xxxviii, 30.
[31] Jer., xi, 4 ; Deut., iv, 20 ; Ecclus., xxxviii, 28 ; CDB, 275, 662 ; Benzinger, JE, v, 531 : Hebrews did not smelt metals ; E. Bibl., ii, 1575 ; Wortabet, HDB, ii, 72.
[32] E. Bibl., ii, 1576.

GOLD 483

The Book of Enoch [1], in the early part written before 161 B.C. [2], describes how Enoch told Noah that men were to be destroyed by a Flood because they had learnt all the secrets of the angels and devils, including witchcraft, casting images and "how silver is produced from the dust of the earth and also soft metal"; tin and lead were not so produced, but by a fountain with an angel in it. In an obscure chapter [3] Enoch is carried by a whirlwind to the west and sees mountains of iron, copper, silver, gold, soft metal and lead, which "will be as wax before the fire" in the presence of the Elect One. The curious legend of the production of tin from a fountain, with some further embellishments, appears in the Syriac version of Zosimos.[4] We should have expected seven metals in Enoch, as in Celsus's ladder of Mithra.[5] The original text for "soft metal" is "easily fusible metal."[6] Charles [7] thinks the chapter is made up of two independent sources with a part in common. In another chapter [8] the unrighteous are shut in a burning valley in the west among mountains of gold, silver, iron, soft metal and tin, with a convulsion of fiery molten metal and a smell of sulphur; hot springs with healing properties are also said to be produced and thus features of the Deluge and volcanic eruptions are combined. In the Apocryphal literature the end of the world is to be achieved by a fiery comet, which will fuse down the metals of the earth into a river in which the condemned sinners will be destroyed, but which will feel to the pious only "like a bath of warm milk."[9] It should be noted that Enoch gives two different orders of these metals: (1) iron, copper, silver, gold, "soft metal" [? tin] and lead[10]; (2) gold, silver, iron, "soft metal" [? lead][11] and tin, lead being named in one and tin in the other.

GOLD

Gold is frequently mentioned[12] in the Old Testament, both as such and figuratively as an emblem of purity[13] and nobility.[14] The Hebrew name for gold, zâhâb, is always derived from a root meaning to glitter or shine, to be clear or bright, or to be yellow.[15] A Hebrew name ketem for gold occurs in the Egyptian Harris Papyrus (Rameses III).[16] Chârûz (ḥârûz) is a poetical name for gold[17] derived from the Assyrian ḥurâsu; the Hebrew bezer, used twice[18], means bars or rings of gold. A derivation of the Greek χρυσός from the Hebrew chârûz, proposed by Bochart[19], who thought chârûz meant "dug from the ground," is accepted by Schrader[20]; another derivation from the Sanskrit hiranya, "to glitter or flame," is less probable. Gold is represented as abundant in ancient times[21] and much used by the wealthy for decorating

[1] lxv; Charles, 1893, 170 f.; 1913, ii, 230.
[2] Charles, 1913, ii, 163, 170, 180, 191; Sel, JE, ix, 321.
[3] R. Laurence, The Book of Enoch, 1821, 191; Charles, 1893, 141.
[4] Berthelot, Mâ., II, 244; the angel has become a demon, who must first be attracted from the spot.
[5] Origen, Contra Celsum, vi, 22; Keim, Celsus' Wahres Wort, Zürich, 1874, 84.
[6] Beer, in Kautzsch, ii, 266, "weiches Metall," no note; Charles, 1893, 141, perhaps "a general name for tin and lead"; Bousset, A. Rel., iv, 243, mercury?
[7] 1913, ii, 219.
[8] lxvii; Charles, 1893, 174; 1913, ii, 232.
[9] Justi, A. Rel., vi, 251.
[10] liii; Kautzsch, ii, 265.
[11] lxvii; Kautzsch, ii, 274.

[12] E. Bibl., ii, 1749 f.; Wiegleb, Alchemie, 1777, 26 f.; Vigouroux, Dict. de la Bible, iv, 1836.
[13] Job, xxiii, 10.
[14] Lam., iv, 1.
[15] Roscoe and Schorlemmer, 1913, ii, 491; Petrie, HDB, ii, 225; E. Bibl., ii, 1749; P. Haupt, J. Amer. Orient. Soc., 1923, xliii, 116; Hoefer, Hist. de la Chimie, 1866, i, 45.
[16] Breasted, Anct. Rec., iv, 117, 166.
[17] Prov., viii, 10; xii, 27.
[18] Job, xxii, 24, 25.
[19] Hierozoicon, ii, 9.
[20] RL, 298; cf. Roscoe and Schorlemmer, Treatise on Chemistry, 1913, ii, 491.
[21] I Chron., xxii, 14; II Chron., i, 15; ix, 9; Nah., ii, 9; Dan., iii, 1; Josephus, Antiq., viii, 7.

furniture, etc.[1] It was handled in the form of "tongues."[2] Among the spoils taken from the Midianites was jewellery to the amount of 16,750 shekels of gold[3]; 1700 shekels of gold in nose-jewels (A.V. ear-rings) alone were taken.[4] The amount of gold accumulated by David is probably exaggerated, [5] and the account of Solomon's temple (probably built in 967–957 B.C.) [6] is partly interpolated with later additions, including the prodigal use of gold [7], e.g. the 120 talents (5893·2 kg.) obtained from Hiram of Tyre. The story of the annual tribute of 666 talents of gold [8] is also exaggerated.[9] Payment in gold is mentioned only once[10] and probably by mistake, since the amount is too large and the parallel passage[11] specifies silver, the usual metal.[12] Various kinds of jewellery are mentioned[13]—the decoration of Joseph with the gold neck-chain[14] corresponds with the Egyptian investiture of a high official by the bestowal of "the gold of praise."[15]

Gold comes "from the North,"[16] as it did from "the mountains of Arallu" in Assyrian texts, a mythical land[17], or "the heaviest part of the earth"[18]; there is "gold of Parvaim,"[19] gold of Ophir[20] and of Uphaz[21], gold of Havilah[22] and "dust of gold"[23] (alluvial gold). The Talmûd[24] mentions seven kinds of gold : gold, good gold, gold of Ophir, best, beaten, pure and gold of Parvaim. In the Qabbâlâh these are : concealed, beaten, gold of Tharshish, of Ophir, of Saba and pure gold.[25]

Ophir, from whence came gold, silver, ivory, gems, precious woods and curious animals, has been very variously placed.[26] Its locality cannot really be identified.[27] India, Ceylon, South-East Africa, South Arabia, Persia, Java and even America have been suggested[28], and in ψ-Kallisthenes[29] India and China are named. The identification with India is improbable,[30] although the Hebrew kôph, ape, is like the Sanskrit kapi, and the Hebrew tukki, peacock, is nearly the same as the Tamil tôka.[31] The usual identification of Ophir is

[1] I Kings, vi, 22 ; x, *passim* ; Cant., iii, 9, 10 ; Esth., i, 6 ; Jerem., x, 9 ; Olmstead, H. Pal. Syr., 346 f. ; Bertholet-Dallas, 213 f.
[2] Josh., vii, 21, A.V., "wedge" ; CDB, 550.
[3] Num., xxxi, 48 f. ; on supposed gold in Midian, see Burton, Gold Mines of Midian, 1878, 242 f.
[4] Judg., viii, 26.
[5] CDB, 550.
[6] Marston, New Knowledge, 120.
[7] Meyer, Alt., II, ii, 263 ; Flint, HDB, iv, 566 ; Davies, *Nature*, 1932, cxxx, 985, thinks some of the gold came from the Ægean.
[8] I Kings, x, 14.
[9] Meyer, 268.
[10] I Chron., xxi, 25.
[11] II Sam., xxiv, 24.
[12] CDB, 735.
[13] Is., iii, 18–21, etc.
[14] Gen., xli, 42 ; xlv, 8.
[15] Yahuda, *Daily Telegraph*, 14–15 Sept., 1933.
[16] Job, xxxvii, 22 ; Aquilone in the Vulgate.
[17] Karppe, Zohar, 516 ; Maspero, Dawn, 690.
[18] CDB, 215.
[19] II Chron., iii, 6 ; Sanskrit pūrva, easterly, "Eastern gold," Napier, Arts, 56 ; CDB, 295, 684 ; cf. E. Bibl., ii, 1751 ; iii, 3588.

[20] I Kings, ix, 28 ; Job, xxviii, 16.
[21] Jer., x, 9 ; Dan., x, 5 ; cf. E. Bibl., iv, 5231.
[22] Gen., ii, 11, before the Flood ; CDB, 222, E. or S. Arabia ? ; Petrie, HDB, ii, 311, N. Arabia ; E. Bibl., ii, 1973.
[23] Job, xxviii, 6.
[24] Yomah, iv ; Bähr, Symbolik, i, 256.
[25] Christian Knorr von Rosenroth, Kabbala denudata seu doctrina Hebræorum transcendentalis . . ., 4to., Sulzbach, 1677, i, 298 f.
[26] E. Bibl., iii, 3514 f. ; Napier, Arts, 56.
[27] Meyer, Alt., II, ii, 124.
[28] Lenormant, H. anc., iii, 68 ; Wilkinson, i, 151 ; Davis, Carthage, 1861, 12, 27 f. ; James, DG, ii, 484 ; Blümner, iv, 14 ; Cook, CAH, iii, 357 ; Maspero, H. anc., 1905, 393 ; *ib.*, Struggle, 742 ; Moritz, Arabien, Hannover, 1923, criticised by Meyer, Alt., II, ii, 124—not seen ; Belck, Z. f. Ethnol., 1910, xlii, 23 ; Hoernes, Urg., ii, 515 ; Oppert, in Diergart, 1909, 132 ; Lassen, Indische Altertumskunde, 1847, i, 538 ; good summary in CDB, 656.
[29] Ed. Müller, Didot, Paris, 1846, 102 ; Budge, Life and Exploits of Alexander the Great, being a series of translations of the Ethiopic Histories of Alexander, Cambridge, 1896, pp. xxxii, 179 ; cf. Thorndike, History of Magic, 1923, i, 551.
[30] Bevan, Cambridge History of India, 1922, i, 391.
[31] CDB, 922.

south or south-east Arabia, perhaps Afar, a little outside the Straits of Bâb al-Mandib [1], or near Debæ (the Red Sea) and the land of the Nabatæans.[2]

Bochart [3] thought there were two Ophirs, an old one on the east coast of Arabia or the Persian Gulf, and a new one in Solomon's time in Ceylon. Schoff [4] does not consider that Ophir was a definite place, but that it is the disguised name [gold of] ophir = [χρυσὸς] ἄπυρος, i.e. apyros, native [gold].[5] Rawlinson [6], who points out that the name is Σωφάρα in the Septuagint, and Sophir is Coptic for South India, is also inclined to think Ophir was an entrepôt on the shores of Arabia, perhaps on the mouth of the Persian Gulf off the coast of Oman, where Indians and Phœnicians met for trade, a theory which is really due to Maspero.[7] Dahse [8] concludes that Uphaz was the West African Gold Coast : he also distinguishes it from Ophir and says Solomon had two sources of gold. The first, Ophir, was Zimbabwe in South Africa.[9] There is now a tendency to identify Ophir with the Egyptian Punt, from which gold was obtained in early times, and this in turn with Nubia.[10] According to Nugel[11], the source of Solomon's gold was not South-East Africa, but Mashonaland, between Zambesi and Limpopo.

Gold is named first in the lists as the most valuable metal.[12] It is worked by the goldsmith[13] in all varieties of ways : cast[14], beaten into plates and leaves (the gold-beater is "carpenter" in the Authorised Version)[15] and made into wire[16], both being used for ornaments[17] as in ancient Egypt.[18] It is "purified" in furnaces[19], as in the Hittite-Egyptian letters[20], and both "gold" and "pure (tahor) gold" are mentioned.[21]

The method of disposal of the golden calf[22] occupied more attention among the older chemists[23] than it does at the present day. Bochart[24] suggested that the calf was cut, ground and filed to powder ; Napier that it was cast into bars, these beaten into leaves, and the leaves crumbled to powder and suspended in water. "Dissolved" may mean "fused,"[25] and the meaning of the Hebrew is "he absorbed it in the fire."[26] Wiegleb[27] supposed that the calf was of wood covered with gold sheet, like the ark of the Tabernacle[28], that the wood was

[1] Wiegleb, Alchemie, 1777, 29 ; CDB, 656 ; Meyer, Alt., II, ii, 268 ; Price, HDB, iii, 626 ; Rawlinson, Phœnicia, 307, 432 ; Cook, CAH, iii, 357 ; Cheyne, E. Bibl., iii, 3515 ; Albright, JEA, 1921, vii, 83 ; cf. Strabo, XVI, iv, 26, 784 C. ; C. Craufurd, Treasure of Ophir, n.d.
[2] Blümner, iv, 15.
[3] Geographia sacra, 1712, 18, 137 f., 691 ; Adler, Benjamin of Tudela, 1907, 63.
[4] Periplus of the Erythræan Sea, 1912, 160.
[5] Diodoros Siculus, iii, 13 ; Hoefer, Hist. de la Chimie, 1866, i, 115.
[6] India, 1926, 11.
[7] H. anc., 1905, 393.
[8] Z. f. Ethnol., 1911, xliii, 1–79, with older references.
[9] "King Solomon's Mines" ; Frobenius, Das unbekannte Afrika, Munich, 1923.
[10] Albright, JEA, 1921, vii, 83 ; Clarke, q. in Olmstead, H. Pal. Syr., 340.
[11] In Ullmann, Enzyklopädie der techn. Chemie, vi, 291.
[12] Numb., xxxi, 22 ; Ex., xxxi, 4 ; Rosenmüller, IV, i, 47 ; Haupt, J. Amer. Orient. Soc., 1923, xliii, 92 f. ; Isis, 1925, vii, 275.
[13] Neh., iii, 8.
[14] Ex., xxxii, 4.

[15] Is., xli, 7 ; Jerem., x, 9.
[16] Ex., xxxix, 3.
[17] Gen., xxix, 22.
[18] CDB, 306 ; Petrie, Arts and Crafts in Ancient Egypt, 90.
[19] Prov., xvii, 3 ; xxvii, 21 ; Is., xiii, 12 ; Malachi, iii, 2, 3.
[20] See p. 38.
[21] Ex., xi, 2 ; xxv, 11 f., 17 f., 24 f. ; xxvi, 19, 21 ; xxvii, 10 f., 17 ; xxviii, 8 ; xxxii, 2, etc.
[22] Ex., xxxii, 20 ; Deut., ix, 21.
[23] Wiegleb, Alchemie, 1777, 110 f. ; Napier, Arts, 77 f.
[24] Hierozoicon, I, ii, 34 ; vol. i, 330 f., 350 ; CDB, 128.
[25] Spielmann, Institutiones Chemiæ, Strasburg, 1766, 110 ; Wiegleb's Chemistry, tr. Hopson, 1789, 107 : "the former of these two modes of solution, viâ siccâ, is otherwise by chemists called fusion."
[26] Hoefer, H. Chim., i, 44.
[27] Geschichte des Wachsthums und der Erfindungen in der Chemie, Berlin and Stuttgart, 1792, 96 ; ib., Alchemie, 1777, 110 f., 119, 121 ; Bergman, Physical and Chemical Essays, Edinburgh, 1791, ii, 72 ; CDB, 128.
[28] Ex., xxvi, 29 ; Is., xliv, 10–20 ; Josephus, Antiq., III, vi, 5.

burnt and the gold thus melted into granules, and then the ashes were stamped and thrown into water. The oldest form of image (ephod, "cover" ; pesel and massêkâh, "carved and cast work") was probably of wood covered with gold (or silver). The teraphîm were similar idols in human form protecting the house.[1] A 13th century gloss on the passage says "those who worshipped the calf had their beards gilded" (? by drinking the suspension) [2], and Josephus that the cavalry of Solomon powdered their hair with gold dust.[3] This theory of mechanical subdivision, supported by Michaelis [4], was rejected by the fanciful F. J. W. Schröder [5], who was convinced that Moses used some chemical process, and points out that the calf is expressly said to have been cast.[6] Chemical methods have, therefore, been proposed. Aqua regia (a mixture of nitric and hydrochloric acids, which dissolves gold), suggested by Herapath [7], is, of course, impossible. Stahl [8] suggested that the gold was got into solution by fusing 1 part of gold with 3 of potash and 3 of sulphur, when a soluble compound is formed.[9] In all probability, however, some mechanical method was intended.

The "water of separation" used to treat things other than metals (gold, silver, copper, iron, tin and lead), which were purified by fire to make them clean—"all that abideth not the fire ye shall make go through the water[10]"— although supposed to have been mercury, was probably the same as the "water of purification" used for ceremonial purification and prepared by burning a red heifer whole, mixing the ashes with water and allowing to stand.[11] That "purely purge away thy dross"[12] meant "by the aid of solvents such as alkali"[13] is doubtful. So far as is known, all ancient processes for the separation of gold from silver or other metals depended on a dry method of heating with salt, misy (impure iron sulphate) and other materials, still used in the 18th century.[14]

SILVER

Silver is frequently mentioned[15], including "refined silver,"[16] etc., and comes second after gold in the lists.[17] It was weighed in the balance for money[18] and its name késeph, from a root "to be pale,"[19] is also used for money,[20] for trumpets[21] and for ornaments.[22] It was also cast.[23] Foreign idols were of silver [or gold] or covered with it[24], and the manufacture of silver shrines for

[1] Meyer, Alt., II, ii, 228, 321, 323; CDB, 340, 932.
[2] Napier, Arts, 86.
[3] Antiq., VIII, vii, 3 ; Bochart, Hierozoicon, i, 351.
[4] Anmerkungen zur Deutschen Uebersetzung des alten Testaments, in Exod., xxxii, 20 ; q. by Wiegleb, 1792, 96.
[5] Geschichte der ältesten Chemie, Marburg, 1775, 310 f.
[6] Ex., xxxii, 4 : a "molten calf."
[7] Phil. Mag., 1852, iii, 528.
[8] Vitulus aureus igne combustus, in Opusculum Chymico-Physico-Medicum, Halle, 1715, 585 f. ; Thomson, History of Chemistry, 1830, i, 256 ; Napier, Arts, 82.
[9] Roscoe and Schorlemmer, Chemistry, 1913, ii, 519.
[10] Numb., xxxi, 22 f.
[11] Numb., xix, 5 ; Napier, Arts, 11 f. ; Kennedy, HDB, iv, 208.
[12] Is., i, 25.
[13] CDB, 785.
[14] Berthelot, Intr., 14.
[15] Gen., xiii, 2 ; Job, xxviii, 1 ; Ps., lxvi, 10.

[16] Ps., cix, 140.
[17] Num., xxxi, 22 ; I Chron., xxii, 16.
[18] Is., xlvi, 6 ; Gen., xxiii, 16 ; xlvii, 13, 14 ; Ex., xxi, 21 ; Deut., xxiii, 19 ; on weights and measures, see Lauterbach, JE, xii, 483.
[19] E. Bibl., iv, 4523 ; Haupt, J. Amer. Orient. Soc., 1923, xliii, 116, thinks "to be fusible."
[20] Hoefer, Hist. de la Chimie, i, 45, 55 ; Thomson, Hist. of Chemistry, 1830, i, 53 ; Roscoe and Schorlemmer, Chemistry, 1913, ii, 445 ; CDB, 882 ; Patrick, HDB, iv, 516 ; Krauss, JE, viii, 513 f. ; Rosenmüller, IV, i, 53 ; E. Bibl., iv, 4523 ; Vigouroux, Dict. de la Bible, i, 945 : cf. ἄργυρος, silver metal, ἀργύριον, silver money, French argent, Scots siller ; Gen., xxiii, 16 ; xvii, 12 ; xx, 16 ; xxxvii, 28, "pieces of silver," CDB, 573, 735.
[21] II Chron., v, 12 ; CDB, 176.
[22] Gen., xxiv, 53.
[23] Judg., xvii, 4.
[24] Ex., xx, 23 ; Hos., xiii, 2 ; Hab., ii, 19 ; Bar., iii, 17.

Diana was a trade in Ephesus.[1] Vast quantities of metals are said to have been used in the construction of the Ark of the Covenant and its furniture : 3,007 lb. of gold, 10,346 lb. of silver and 7,282 lb. of bronze.[2] In Solomon's time (950 B.C.) silver was plentiful in Palestine.[3] Josephus [4] says the Temple at Jerusalem had ten gates, nine covered with gold and silver and an outer, more excellent gate, covered with Corinthian brass.[5] The proportions of gold and silver collected by David were 1 to 9.[6] These accounts are probably all exaggerated. In Solomon's time silver came from Arabia [7] and in plates from Tharshish [8], which also supplied Tyre.[9] Tharshish was perhaps Spain, the silver mines of which are mentioned in Maccabees I (100–70 B.C.)[10], or perhaps Assur, the metal coming by way of the Hittites.[11] Some silver may have been found in Palestine.[12]

The refining of silver by fire is frequently mentioned or implied, and a knowledge of cupellation is evident.[13] The "fining pot for silver"[14] may be a cupel, or a crucible for the process used by the later Egyptians in separating the gold and silver from the alloy asem. That it was a pot for treatment with a mineral acid[15] is impossible. The cupellation of silver in presence of lead is no doubt the process usually described. In James[16] ἰός means the tarnish on silver ; βρῶσις in Matthew[17], also translated "rust," probably means the larva of some moth, perhaps Tinea granella.[18] The "refuse" or "reprobate" silver[19] was perhaps an alloy, as opposed to "choice silver."[20] "Silverlings"[21] is merely a translation of késeph, silver.[22]

COINAGE

The 100 qesitah paid by Jacob[23] are often taken to be coins or sheep[24]; probably, however, they were weighed minas of silver.[25] The first coins mentioned in the Old Testament are Persian darics (darkemôn ; adarkôn)[26]; the first mention of Jewish coins is in the Apocrypha[27], viz. in the permission given to Simon the Maccabee by Antiochos VII to stamp coin in 140 (or 138) B.C. The average weight of the Jewish silver coins is 220 grains troy (1 shekel) and 110 grains troy (½ shekel). Various copper coins were issued from the time of Judas Maccabæus (d. 161 B.C.), those supposed to be of Simon being doubtful.[28] The inscriptions on Jewish coins are in a peculiar character, resembling Samaritan or Phœnician letters, and many spurious coins are extant.[29]

[1] Acts, xix, 24.
[2] Napier, Arts, 17 f., 19 ; 1 talent = 49·11 kgm., Meyer, Alt., II, ii, 264.
[3] I Kings, x, 21, 27 ; Ridgeway, Early Age of Greece, i, 78.
[4] War, V, v, 3.
[5] Cf. Acts, iii, 2 ; III Baruch, proleg., in Charles, Apocrypha, 1913, ii, 533.
[6] CDB, 550.
[7] II Chron., ix, 14.
[8] II Chron., ix, 21 ; I Kings, x, 22 ; Jer., x, 9.
[9] Ez., xxvii, 12.
[10] viii, 3.
[11] E. Bibl., iv, 4524.
[12] CDB, 882.
[13] Napier, Arts, 47, 52 ; Num., xxxi, 23 ; Prov., xxvi, 23 ; xviii, 3 ; xxvii, 21 ; Malachi, iii, 2, 3 ; Is., i, 22, 25—"purge thy dross" ; Ez., xxii, 18 f.—"dross of silver," perhaps litharge from cupellation ; Ps., xii, 6 ; lxvi, 10 ; Jer., vi, 28–30— "silver tried in a furnace of earth purified

many times" ; Ez., xxii, 18–22—"I will gather you and blow upon you in the fire . . . as silver is melted in the midst of the furnace."
[14] Prov., xvii, 3 ; xxvii, 21.
[15] Napier, Arts, 54 f.
[16] v, 3.
[17] vi, 19, 20.
[18] CDB, 882.
[19] Jer., vi, 30.
[20] Prov., viii, 19.
[21] Is., vii, 23.
[22] CDB, 882.
[23] Gen., xxxiii, 18, 19.
[24] CDB, 573.
[25] E. Bibl., ii, 2659.
[26] Ezra, ii, 69 ; viii, 27 ; Neh., vii, 70.
[27] Maccabees I, xv, 6.
[28] CDB, 574 ; Madden, Coins of the Jews, 1903, 43, 61, 71 ; Reinach, JE, ix, 350 ; Benzinger, 1907, 196.
[29] Reinach, JE, ix, 353 ; cf. ib., 439 ; ii, 68.

The passages in the Old Testament relating to payment are of two kinds :
(1) those in which the Authorised Version uses "pieces" for a word understood in
the Hebrew, which gives "a thousand," etc., [pieces] "of silver"[1] ; in similar
passages the word "shekels" occurs in the Hebrew ; (2) passages in which the
Authorised Version supplies the word "shekels" instead of "pieces."[2] An
exceptional case is where the word "rats" occurs in the Hebrew[3], meaning a piece
broken off [? from a notched bar].[4] In the Authorised Version of the New
Testament "piece of silver" translates : (1) δραχμή [5], the Greek silver coin
equivalent to the Roman denarius at that time ; (2) ἀργύριον [6], perhaps a
tetradrachm rather than denarius[7] ; (3) στατήρ [8], etc.[9]

ELECTRUM

The "shining ḥašmal" (chashmal) mentioned thirteen times in the Bible[10],
where its colour is compared with a cloud of fire, translated ἤλεκτρος in the
Septuagint and electrum in the Vulgate, perhaps sometimes means the alloy
of gold and silver electrum, the name ḥašmal being perhaps derived from the
Egyptian asem.[11] Another suggestion is that it was the Assyrian alloy called
ešmarû[12], and a common identification is amber.[13] Baltic amber was found at
Lachiš with XVIII dyn. scarabs and red amber is common in Lebanon.[14]
Josephus[15] says the musical instruments of David were of brass ; that those of
Solomon were of electrum (or "finest brass") ; that some vessels in Solomon's
Temple were of brass like gold, and in rebuilding the Temple after the Babylonian
Captivity "brass [aurichalcum] more precious than gold" was used. Daniel[16]
also speaks of "shining brass" and the χαλκολιβάνος of the Apocalypse[17] is
perhaps electrum or brass.

COPPER ; BRONZE ; BRASS

Copper, the name for which as well as bronze and brass, is něchôšeth (nâchûš,
něchâš, něchûšâh, etc.), perhaps derived from nahâš, to glisten, Arabic nahas,
Moorish-Spanish farass[18], is described[19] as known and worked by Tubal Cain
before the Flood. It is always uncertain whether copper, bronze or brass is
the correct translation of the Hebrew word, and the χαλκός and æs of the
Septuagint and Vulgate are equally ambiguous. "Brass" in the Authorised
Version usually means copper or bronze.[20] "(Fine) copper" is mentioned only
once in the Authorised Version[21] and there means "shining (or yellow) brass."[22]
Copper (or bronze) is frequently mentioned[23]; it comes third, after silver, in

[1] Gen., xx, 16 ; xxxvii, 28 ; xlv, 22 ;
Judg., ix, 4 ; xvi, 5 ; II Kings, vi, 25 ;
Hos., iii, 2 ; Zech., xi, 12, 13.
[2] Deut., xxii, 19, 29 ; Judg., xvii, 2, 3, 4,
10 ; II Sam., xviii, 11, 12.
[3] Ps. lxviii, 30.
[4] CDB, 735 ; Madden, Coins of the
Jews, 11, 15, 43.
[5] Luke, xv, 8, 9.
[6] Matt., xxvi, 15 ; xxvii, 3, 5, 6, 9.
[7] Zech., xi, 12, 13, suggests that shekels
must be understood.
[8] Matt., xvii, 27
[9] CDB, 735, 860 ; Madden, 289 f.
[10] Ez., i, 4 ; viii, 2.
[11] Bochart, Hierozoicon, ii, 870, 881 ;
CDB, 38 ; Schrader, RL, 541 ; Lepsius,
Met., 13, 60.,
[12] Levi, JE, i, 487.
[13] Bochart, Hierozoicon, i, unpaged pref.,
sign. c 4 ; ii, 870 f. ; Levi, op. cit ;
Williamson, Book of Amber, 1932, 52 f. ;
Ridgeway, E. Bibl., i, 134.
[14] Sayce, Patriarchal Palestine, 73 ; E.

Bibl., i, 135 ; Williamson, Amber, 58,
says Syrian amber is yellow.
[15] Antiq., VII, xii, 3 ; VIII, iii, 7, 8 ;
IX, v, 2.
[16] x, 6.
[17] i, 15.
[18] Rosenmüller, IV, i, 56 ; CDB, 173 ;
Roscoe and Schorlemmer, Chemistry, 1913,
ii, 399 ; Nowack, Lehrbuch der hebräischen
Archäologie, 1894, 243 f. ; E. Bibl., i, 603 ;
Wilkinson, ii, 351 ; cf. Hoefer, H. de la
Chimie, i, 47.
[19] Gen., iv, 22.
[20] Gladstone, J. Anthropol. Inst., 1896–7,
xxvi, 313 ; cf. "mines of brasse," Holland's
tr. of Pliny, Historie of the World, 1634, ii,
486.
[21] Ezra, viii, 27.
[22] Margin ; CDB, 173.
[23] Cheyne, E. Bibl., i, 893 f. ; e.g. Numb.,
xxi, 9 ; Ex., xxvi, 11 ; xxviii, 10, 11, 17 ;
xxx, 18 ; Deut., viii, 9 ; xxviii, 23 ; Lev.,
xxvi, 19 ; I Kings, vii, 14 ; II Sam., xxi,
16 ; Job, xxviii, 2.

order.[1] Its ore was dug from the mountains.[2] Beyond the brief statement [3] that "copper is melted out of the stone [ore]," perhaps by piling lumps of ore in heaps with wood and setting fire to the mass [4], there is no information about the extraction of copper, which the Hebrews probably learnt from the Egyptians.[5] They perhaps never actually smelted the metal themselves [6], but the tradition that Solomon opened copper mines in Lebanon is not improbable [7] and copper ore occurs in the hills of Palestine as well as in Syria and Lebanon.[8] Old copper mines at Arabah on the Persian Gulf are supposed to have been worked in Solomon's time.[9] Some passages[10] seem to imply that the Hebrews worked their own mines. That the copper ore in Sinai was unknown to the Israelites[11] seems doubtful. Copper was probably imported from the Tibarenians and Moschians of the Pontos or the region of Colchis.[12] St. Jerome mentions the copper mines of Phæno in Idumæa, between Zoar and Petra, as worked in Diocletian's time by Christian slaves.[13]

Copper was very cheap and abundant[14]—Solomon used so much for the Temple that it could not be estimated[15]—and it was used to make chains[16], pillars[17], lavers (including the "copper sea," cf. Babylonian apsu)[18] and other temple vessels. Although the Jews were not ignorant of metallurgy[19], the large pieces for the Temple were made under Phœnician supervision.[20] Such large castings were probably beyond the skill of the native smiths and even Sennacherib (705–681) in Assyria speaks of their production as a remarkable achievement, revealed to him by his god (see p. 252).

Bronze, known from about 2000 B.C.[21], was probably made by fusing together copper and tin, since Philo Judæus (1st century A.D.), probably from Hebrew sources, refers to definite proportions of these metals which were used.[22] Most of the metal was probably imported ready made.[23] The great bronze "sea" which was cast ("molten") for Solomon was made from copper taken by David from Tibhath and Chun, cities of the king of Zobah, in Syria.[24] The dimensions[25] were 5 cubits ($7\frac{1}{2}$ ft.) high, 10 cubits (15 ft.) in dia. and 30 cubits (45 ft.) in circumference [thus making $\pi = 3 \cdot 0$ instead of $3 \cdot 1416$], and one hand-breadth ($3 \cdot 17$ in.)[26] thick. Its capacity is given as 2,000 or[27] 3,000 baths : Rabbinists make 1 bath or ephah $4 \cdot 4286$ gallons ; Josephus, $8 \cdot 6696$ gallons.[28] The discrepancy between the dimensions and cubic capacity of the vessel has led to several guesses as to its form : it was perhaps cauldron-shaped, bulging below and contracting to the mouth.[29] The vessel was mutilated by Ahaz and broken up by the Assyrians.[30] There were also shovels, pots, pans, etc. The bronze serpent, Nechuštan, "made by Moses," which protected against bites of serpents, was in the Temple at Jerusalem.[31] The mirrors (maroth ;

[1] Numb., xxxi, 22 ; Ex., xxxi, 4.
[2] Deut., viii, 9 ; "brass" in Authorised Version ; cf. Holland, q. in note 20, p. 488.
[3] Job, xxviii, 2.
[4] Napier, Arts, 113.
[5] Ib., 12 f. ; CDB, 560.
[6] Benzinger, JE, v, 531.
[7] Bertholet-Dallas, 213.
[8] CDB, 550 ; Cheyne, E. Bibl., i, 893 ; Orth, PW, Suppl. iv, 113.
[9] Illustr. London News, 1934, clxxxv, 26.
[10] Deut., viii, 9 ; Is., li, 1.
[11] Petrie, Sinai, p. ix.
[12] Ez., xxvii, 13 ; CDB, 173 ; Nowack, 243.
[13] CDB, 559.
[14] II Chron., iv, 18.
[15] I Kings, vii, 47.
[16] Judg., xvi, 21, "fetters."

[17] I Kings, vii, 15–21.
[18] II Kings, xxv, 13 ; I Chron., xviii, 8.
[19] Ez., xxii, 18 ; Deut., iv, 20, etc.
[20] I Kings, vii, 13.
[21] Volz, Bibl. Altert., 1914, 389.
[22] Gsell, 76.
[23] Meldrum and Turner, J. Chem. Education, 1931, viii, 2171.
[24] CDB, 1038.
[25] I Kings, vii, 23 f. ; I Chron., xviii, 8 ; 1 cubit = $1\frac{1}{2}$ ft.
[26] CDB, 1003.
[27] II Chron., iv, 5.
[28] CDB, 1004.
[29] Thenius, q. in CDB, 834 ; Olmstead, H. Pal. Syr., 347.
[30] II Kings, xvi, 14, 17 ; xxv, 13.
[31] Meyer, Alt., II, ii, 323 ; Lofthouse, in Peake, 1925, 248 ; Oesterley, ib., 323, 333.

Egyptian maau, from maa, to see or show) [1] were of polished bronze.[2]   Besides
arms and armour [3], the furniture and ornaments of the Tabernacle, fetters and
city gates, and bows were made from bronze.[4]   There was much elaborate
bronze work in Solomon's temple.[5]   The two ornamental pillars Jachin and
Boaz [6] which stood in the porch were cast hollow, with a thickness of four
fingers (3 in. ?), a circumference of 18 ft. and a height of 27 ft., with capitals
cast in bronze $7\frac{1}{2}$ ft. high, and covered with trellis work and two rows of
pomegranates. They were cast by Huram-abi, the son of a Tyrian bronze
worker, who imitated in them the two pillars of gold and "emerald" in the
temple of Bêl-Melkart at Tyre. The dimensions in II Chron., iii, 15, are dis-
crepant from those in I Kings, vii, 15, and probably incorrect.[7]   Benjamin
of Tudela (d. A.D. 1173) [8] saw these pillars, with the name of Solomon engraved
on them, in the Porta Latina of St. John in the Lateran at Rome. On the day of
the destruction of Jerusalem every year they sweated so that water ran down
them. The pillars, however, were said to have been broken up by the Chaldæans
and carried off to Babylon, but they are shown, one on each side of the temple,
on a gilded glass of A.D. 250–350 [9]; those seen at Rome must have been from
Herod's temple, not Solomon's : later Jewish works describe the pillars as of
silver, gold, copper and tin.[10]

Brass was known early in Palestine[11] and was probably used for cymbals and
bells. Zinc as such was probably quite unknown[12], the brass being made
by smelting a zinc ore with copper and charcoal. That χαλκολιβάνος was copper,
not brass, the latter being unknown to the Israelites[13], is a mere opinion.
Bochart[14] and Rosenmüller[15] think ḥašmal may be an alloy of copper and
gold (χαλκοχρυσίον).[16]

IRON

Iron (Hebrew barzel ; Aramaic parzila ; Assyrian parzillu) is often mentioned in the Old Testament as applied to a variety of uses[17], and also figuratively[18], although copper and bronze are more frequently mentioned than iron
in the ratio of 83 : 4, and in the oldest parts of the Pentateuch copper occurs
forty times for iron's twice.[19]   Iron appears in the lists of metals after copper
but before lead and tin (?).[20]   It is said to be known before the Flood[21], in
which the art of working it was lost.[22]   Hall[23] thinks the name barzel is non-
Semitic, and derived from the Chalybes, Tubal and other Armenian races who
supplied iron to the Semites ; another derivation[24] is from the Hebrew bazal,

[1] Ex., xxxviii, 8 ; Job, xxvii, 13 ;
CDB, 566.
[2] Yahuda, *Daily Telegraph*, 22 Sept.,
1932 ; J. Jahn, Biblische Archäologie,
Vienna, 1817–18, I, ii, 155 ; *ib.*, Biblical
Archæology, tr. by Upham, Oxford, 1836.
[3] I Sam., xvii, 5, 6, 38 ; II Sam., xxi, 16.
[4] II Sam., xxii, 35 ; Job, xx, 24, Revised Version ; "steel" in Authorised
Version ; Gladstone, *J. Anthropol. Inst.*,
1896–7, xxvi, 313.
[5] Bertholet-Dallas, 212.
[6] The meanings of the names are unknown : Meyer, Alt., II, ii, 265 ; Olmstead,
H. Pal. Syr., 347.
[7] Fleming, Tyre, 19 ; cf. Rawlinson,
Phœnicia, 210 ; Dyer, DG, ii, 616.
[8] Itinerary, ed. A. Asher, London and
Berlin, 1840, i, 10 ; ed. Adler, 1907, 7 ;
Eisler, Weltenmantel, 48.
[9] Cook, Zeus, ii, 427, Fig. 331.
[10] JE, viii, 517.
[11] See p. 475 ; CDB, 182, 550.

[12] CDB, 550.
[13] Neumann, *Z. angew. Chem.*, 1902, xv,
512.
[14] Hierozoicon, i, pref., sign. c 4 ; ii, 870 f.
[15] IV, i, 57, 59.
[16] Cf. Madden, Coins of the Jews, 1903,
12 f., and refs.
[17] Rosenmüller, IV, i, 60 ; Goguet, i,
172 ; Beck, Gesch. d. Eisens, i, 142 f. ;
Patrick, HDB, ii, 481 ; E. Bibl., ii, 2172.
[18] Deut., xxviii, 23—"earth shall be as
iron" ; Lev., xxvi, 19—"heaven as iron" ;
"rods of iron" in Ps., ii 9 ; a hard heart, like
"chains of iron," in Is., xlviii, 4.
[19] Schrader-Jevons, 203 ; Hoernes, Urg.,
i, 374.
[20] Numb., xxxi, 22.
[21] Gen., iv, 22.
[22] Hoernes, Urg., i, 377.
[23] Oldest Civ. Greece, 1901, 193 f.
[24] Roscoe and Schorlemmer, Chemistry,
1913, ii, 1162 ; cf. Wilkinson, ii, 247 ;
Hoefer, Hist. de la Chimie, i, 45.

"to be hard" ; a derivation from ber, "bright" and nezel, "to melt," [1] is improbable, since cast iron was unknown to the Hebrews.[2] An African origin has
been suggested : the Abyssinians believe that a smith has an evil spirit, and
the Ethiopic name of a female evil spirit is Werzelyâ.[3] Lemm [4] thinks the
Ethiopic Werzelyâ, Coptic Berzelia [cf. $\beta = v$ in Greek] was a goddess of
iron. Berzeleia is a proper name in Josephus.[5] Hellenistic treatises of Jewish-
Egyptian origin refer to iron and steel as $\beta a\lambda\lambda a\theta\grave{a}$ of the Jews.[6]

Iron ore is plentiful in Syria and Palestine[7] : Canaan was "a land whose
stones are iron."[8] The Lebanon iron mines were probably worked by
Phœnicians from the 10th century B.C., after the 6th century by the Hebrews.[9]
There were also mines in the mountains and lands bordering on Moab, in north
Edom (south of Palestine), near Beyrut on the coast of Syria, at Jabal Mirad
and Rabbath Ammon.[10] The ore was probably worked in the "iron mountain"
in Transjordania, described by Josephus, where traces of old mines exist,
and it is at present mined in Kefr Hûneh in the south of the valley Zaharâni.[11]
Iron ore is mentioned in the Old Testament[12], and iron furnaces, for "preparing"
iron.[13]

Sirach[14] describes a blacksmith's shop and says the conduit through the
rock for the Pool of Siloam was cut with iron tools in the time of Hezekiah
(c. 700 B.C.). The blacksmith (charaš barzel) was traditionally called nappach
(user of bellows) or pechami (user of charcoal).[15] In Solomon's time the
Phœnician was a more skilful metal worker than the Jew.[16] Nebuchadnezzar II
in 597 B.C. sent thousands of smiths and carpenters into exile in Babylon[17],
so that in his time the Jews were familiar with blacksmith's work. Iron is
frequently mentioned in the post-Exilic Daniel, under strong Persian influences.[18]

Iron objects specially named are rods[19], chains[20], pens[21], swords[22], axe-heads[23],
knives[24], yokes[25], chariots[26], vessels[27], tools for cutting stone, etc.[28], and sheet
iron cooking pans[29], but the "iron bedstead" of Og, king of Bashan[30], was
probably a natural rock formation of basalt, used as a sarcophagus.[31] Saws,
axes and arrows of iron are mentioned for David's time[32], when the metal was
plentiful.[33] In the building of Solomon's temple 100,000 talents of iron
were used.[34] Hoefer[35] criticises the statement of Goguet[36] that in the time of

[1] Thomson, Hist. of Chem., 1830, i, 63.
[2] CDB, 347.
[3] Littmann, in Brockelmann, Geschichte
der christlichen Litteratur des Orients,
Leipzig, 1907, 236, 239.
[4] MGM, vii, 485.
[5] Antiq., VII, ix, 5 ; E. Bibl., i, 494 ;
cf. the name of the chemist Berzelius,
Lemm, MGM, vii, 485.
[6] Dieterich, Abraxas, Leipzig, 1891, 191.
[7] Gsell, 20.
[8] Deut., viii, 9.
[9] Albright, Archæol. of Palest., 155, 215 ;
cf. J. Kenrick, Phœnicia, 1855, 37.
[10] Josephus, War, IV, viii, 2 ; Orth,
PW, Suppl. iv, 114 ; Bertholet-Dallas, 22 f.
[11] CDB, 550, 560 ; Vigouroux, Dict. de la
Bible, iv, 1099.
[12] Deut., viii, 9, "stones" ; Job, xxviii, 2,
"dust" ; Napier, Arts, 8.
[13] Jer., xi, 4 ; Deut., iv, 20 ; I Kings, viii,
51 ; Roscoe and Schorlemmer, Chemistry,
1913,ii,1162–not"melting,"Hoefer,i,49,53.
[14] Ecclus., xxxviii, 28 ; xlviii, 17.
[15] Lévias, JE, ii, 152.
[16] Judg., viii, 24 ; xvii, 4 ; I Kings, vii,
13, 45, 46 ; Is., xli, 7 ; Wisd., xv, 9 ;
Ecclus., xxxviii, 28.

[17] II Kings, xxiv, 14 ; Maspero, Struggle,
537 ; Faulmann, Culturgeschichte, 1881,
470.
[18] Dan., ii, 33 f., 40 f. ; iv, 15, 23 ; v, 5, 23.
[19] Ps. ii, 9.
[20] Is., xlviii, 4.
[21] Job, xix, 24—a stylus ?
[22] Numb., xxxv, 16.
[23] Deut., xix, 5.
[24] Lev., i, 17.
[25] Deut., xxviii, 48.
[26] Josh., xvii, 18 ; Judges, iv, 3.
[27] Josh., vi, 19.
[28] Deut., xxvii, 5 ; Josh., viii, 31—not
to be used in building altars.
[29] Lev., vii, 9 ; Ez., iv, 3.
[30] Deut., iii, 11 ; Og was an Amorite
chief of the pre-Israelite inhabitants of the
hill country east of Jordan : Cook, CAH,
ii, 366, 376.
[31] Bertholet-Dallas, 23 ; yet Josephus,
Antiq., IV, v, 3, says it was of iron.
[32] II Sam., xii, 31.
[33] I Chron., xxii, 13.
[34] I Chron., xxix, 7 ; Josephus, Antiq.,
VII, xiv, 9.
[35] Hist. de la Chimie, i, 49 f.
[36] Origine des lois, 1809, i, 172.

Moses swords, knives, axes and implements for working stone were of steel. Knives [1] are of stone ; an "instrument of iron" (barzel) [2] is probably a bar or rod [3], not a sword, and the iron weapons [4] and instruments [5] are bars or hammers, not swords. The Canaanites, unlike the early Hebrews, used war chariots which were plated or studded with iron after the Hittite rather than the Egyptian pattern.[6] In Joshua's time iron was in the temple treasury.[7] Goliath's spear-head (c. 1000 B.C.) was of iron, but the defensive armour was of bronze.[8] Literary references presume the use of iron from 1100 B.C. ; finds of weapons at Lachiš and other Philistine sites go back to 1200 [9], and those at Gerar[10] to the 14th century B.C.

The prejudice against the use of iron in religious ceremonies, prevalent among many ancient peoples[11], was shared by the Jews. The metal instruments for hewing the stones of altars must not be iron[12]; it was then, about 1100 B.C., new to the Israelites and a "foreign" metal. Josephus[13] says the altar in the Temple at Jerusalem was "formed without any iron tool" and Jewish tradition states that the stones of the first temple were not cut with iron.[14] In an eleventh commandment (really "word" [of God] in Hebrew) added to the usual ten in the Samaritan Pentateuch (5th–4th centuries B.C.)[15], two stones covered with plaster are to be erected and the commandments written on them, also a stone altar set up, and "thou shalt not lift up any iron thereon."[16] In the Talmûd an iron tube is said to protect a powder contained in it from demons.[17] Although no iron was used in the construction of the Tabernacle, the metal was then well known to the Jews, and the Canaanites of Palestine and Syria used it freely.[18]

When the Jews were established in Canaan, the occupation of blacksmith was recognised as a distinct employment[19], and smith's work is often mentioned.[20] From the time of Amos (760 B.C.) iron was in general use by the Hebrews and Syrians and smelting furnaces are known to later Hebrew authors.[21] Although the Bible[22] makes the Jews, who had no blacksmiths, apply to the Philistines for the repair of their iron ploughshares, etc., the text is not clear[23], and the Philistines had no iron ore in south Phœnicia.[24] The account is perhaps legendary.[25] In an account of the journey of an ambassador of Rameses II to Jaffa in the Papyrus Anastasi I, his sword is stolen in his sleep, his armour cut off and his chariot broken. The chariot, however, was repaired by iron workers, who must, therefore, have existed at that time (c. 1250 B.C.) in Palestine.[26] It has been supposed[27] that until 1000

[1] Jos., v, 2, 3 ; Ex., iv, 25 ; Ps. lxxxix, 43.
[2] Numb., xxxv, 16.
[3] As in Ex., ii, 11, 30 ; vii, 16 ; Deut., xxv, 3.
[4] Job, xx, 24.
[5] Job, xix, 24 ; Deut., xix, 5 ; xxvii, 5 ; Josh., viii, 31.
[6] Whitehouse, E. Bibl., i, 725 ; Friend, Iron in Antiquity, 171.
[7] Lepsius, Met., 54.
[8] Ib., 54 ; Hall, CAH, ii, 292.
[9] Myres, CAH, i, 109.
[10] See p. 476 f.
[11] Stoll, Geschlechtsleben, 395 f.
[12] Josh., viii, 31 ; Meyer, Alt., II, ii, 227.
[13] War, V, v, 6.
[14] Kuttner, Jüdische Sagen, 4 vols., Frankfurt, 1902–6, i, 18 ; Delitzsch, Jewish Artisan Life, 1902, 12.
[15] Written in Samaritan character, a development of the old Hebrew-Phœnician

script ; Sarton, Intr., i, 151 ; Isis, 1928, xi, 157.
[16] CDB, 931.
[17] Berakoth, i, 1 ; tr. Cohen, 27.
[18] Ex., xx, 25 ; xxv, 3 ; xxvii, 3, 19 ; Numb., xxxv, 16 ; Deut., iii, 11 ; iv, 20 ; viii, 9 ; Josh., viii, 31 ; xvii, 16, 18.
[19] I Sam., xiii, 19.
[20] II Sam., xii, 31 ; I Kings, vi, 7 ; II Chron., xxvi, 14 ; Is., xliv, 12 ; liv, 16.
[21] Cheyne, E. Bibl., ii, 2172 ; Friend, 173.
[22] I Sam., xiii, 19 f.
[23] Giebeler, Z. f. Ethnol., 1909, xli, 99.
[24] Hoernes, Urg., ii, 223, 295 ; Olmstead, H. Pal. Syr., 296.
[25] Meyer, Alt. II, ii, 241.
[26] Sayce, Patriarchal Palestine, 1912, 184 f.
[27] Nowack, Hebr. Archäol., 1894, 244— "time of Deuteronomy" ; Hoernes, Urg., ii, 223 ; Meyer, Alt., II, ii, 227.

B.C. the Jews knew only foreign iron, which was perhaps brought by caravans from Mesopotamia to the Jordan Valley [1], but more probably from the iron-working tribes of Asia Minor : Meshech and Tubal have, since Bochart [2], been identified with the Moschi and Tibareni of Pontos ; the Arabic version of Genesis, x, 2, gives Khorassan and China, respectively ; Eusebius, Illyria and Thessaly.[3] Since the native Canaanites and the newly settled Phœnicians had iron before the 9th century B.C., its introduction and spread are conjecturally associated with the extensive movements involving Philistines from Crete (or Lycia and Caria) in which "Dorians" and others were involved from shortly before 1000 B.C.[4]

The expression "iron (barzel) sharpens iron (barzel)" [5] has been thought to mean "steel sharpens iron."[6] "Steel" in the Authorised Version should frequently read "copper" or "bronze," [7] and it has been supposed that steel was unknown to the ancient Hebrews.[8] The paldâh, which occurs only once [9], has, however, been thought to mean the flashing steel scythes of the Assyrian chariots[10], and the "northern iron" (ferrum ab Aquilone in the Vulgate, i.e. the same as that made in Aquileia)[11] is quite possibly steel from the Chalybes of Pontos.[12] The "bright iron" taken by the merchants of Dan (South Arabia ?) and Javan (Greece ?) to Tyre[13] may also have been steel from the Ionians.[14] India[15] or South Arabia[16] are less probable sources.

There is no reason to suppose that the Jews or any other nation of antiquity were acquainted with cast iron, although passages in Job[17] and Hesiod[18] have been thought to suggest it. The stock of "cast iron" in Whiston's translation of Josephus[19], which was accumulated in the fortress of Masada in the Roman Period, is in the original ἀργός σίδηρος, ferrum infectum, "unworked iron," i.e. bars of wrought iron. The Septuagint version of Job[20] certainly suggests the casting of iron.[21] The very obscure passage in Diodoros Siculus[22] probably refers to the production of "blooms" of spongy malleable iron, as in the primitive methods of Africa and India.[23]

LEAD

Lead is called 'ôphereth,[24] perhaps[25] from aphar, grey, although Hoefer[26] thinks 'ôphereth is copper, as aphar is red or red earth.[27] The Aramaic name is abar or abra.[28] Lead is mentioned in a few places in the Old Testament.[29]

[1] Perrot and Chipiez, Art in Sardinia and Judæa, i, 345.
[2] Geographia Sacra, 1712, 179.
[3] CDB, 964.
[4] Cook, CAH, iii, 417 ; Hogarth, ib., iii, 162 ; Hall, ib., ii, 292, 294.
[5] Prov., xxvii, 17.
[6] Gsell, 49.
[7] II Sam., xxii, 35 ; Job, xx, 24—with iron ; Ps. xviii, 34 ; CDB, 173, 550, 900 ; Beck, Gesch. des Eisens, i, 170.
[8] Gsell, 49 ; CDB, 900.
[9] Nahum, ii, 3—"torches" in A.V.
[10] CDB, 900 ; Rosenmüller, IV, i, 63 ; Munk, Palestine, in L'Universe, histoire et description de tous les Peuples, 1856, 390 ; Cheyne, E. Bibl., ii, 2174—doubtful.
[11] Jer., xv, 12—written after 600 B.C. and probably interpolated, Driver, Introd. to the Lit. of the Old Testament, 1913, 273.
[12] Wilkinson, ii, 248 ; Thomson, History of Chemistry, i, 64 ; Rosenmuller, IV, i, 63 ; CDB, 347, 550.
[13] Ez., xxvii, 19.

[14] Bochart, Geographia Sacra, 1712, 153.
[15] Dyer, DG, ii, 617.
[16] CDB, 187, 364.
[17] xl, 18, in LXX.
[18] Theog., 864.
[19] War, VII, viii, 4 ; Josephus, Opera, ed. Dindorf, Paris, 1847, ii, 325.
[20] xl, 18 ; αὐτοῦ σίδηρος χυτός.
[21] CDB, 347—"bars of iron" in A.V.
[22] v, 13 ; καίουσιν, liquefactus, is given as "la fonte" in Hoefer's translation, 1865, ii, 13.
[23] CDB, 560 ; Johannsen, Geschichte des Eisens, Düsseldorf, 1925, 75.
[24] Rosenmüller, IV, i, 64.
[25] Roscoe and Schorlemmer, Chemistry, 1913, ii, 872.
[26] Hist. de la Chimie, i, 45, 53.
[27] Job, xxviii, 6 ; Prov., viii, 26.
[28] Krauss, JE, viii, 513.
[29] Job, xix, 24 ; xxii, 14 ; Ex., xv, 10—"heavy" ; Zach., v, 7, 8 ; Ecclus., xxii, 14 ; xlvii, 18—based on I Kings, x, 27, which has "stone" ; E. Bibl., iii, 2751 ; Petrie, HDB, iii, 88.

In the list of metals it comes after tin [1] or iron.[2] In IV Ezra (A.D. 100–135) [3] the list is : gold, silver, copper (or brass), iron, lead and clay.[4] In Job the metal is part of the spoil taken from the Midianites (in the peninsula of Sinai) [5], and this metal may have come from ore from the mines of Gebel Rusâs, near the Red Sea [6], which supplied the Egyptians (see p. 81).

In the Septuagint 'ôphereth is translated $\mu\acute{o}\lambda\iota\beta$os, i.e. $\mu\acute{o}\lambda\upsilon\beta\delta$os, but molybdaina later came to mean black minerals like stibnite, graphite (plumbago) and molybdenum sulphide.[7] Isaiah [8] distinctly mentions the cupellation of argentiferous lead and also soldering [9], perhaps with tin and lead.[10] Cupellation is also described by Ezekiel.[11] In Zechariah (6th century B.C.)[12] a lead disc is used as a heavy cover for a basket. Josephus[13] speaks of stocks of lead ($\mu\acute{o}\lambda\iota\beta$os) in the fortress of Masada (the modern Sebbeh).[14] According to tradition the river bed near Jerusalem was lined with lead.[15]

White lead occurs only in late Hebrew works, in which it has the Persian name sapidag ; the Syriac is uspedka. It appears as alsefidag in the works of the Geonim.[16] The bedil in Isaiah[17] may mean litharge.[18] The sinoper (šâšêr)[19], translated vermilion in the Authorised Version, is probably a red iron ochre[20], although šâšêr is usually given as red lead.[21]

TIN

The mention of tin in the Old Testament is rather doubtful. In the passage[22] "I will purge thy dross ("lye" in the Revised Version)[23] and take away all thy tin [bedil]," no metal is mentioned in the Greek LXX, and "alloy" is in the Revised Version, but tin is in the Talmûd[24] and (doubtfully) in the Mishnâh.[25] The word is plural in the Hebrew.[26] Bedil would ouit lead, removed or separated as litharge by cupellation from other metal, such as silver.[27] Gesenius derived bedil from hedal, to separate or eliminate[28] ; others suggest an origin from the Arabic badal, "substituted" [for silver]. Although bedil is usually given as $\kappa\alpha\sigma\sigma\acute{\iota}\tau\epsilon\rho$os in the LXX, and so adopted by translators[29], the Greek translation is vacillating, and sometimes bedil is given as $\mu\acute{o}\lambda\iota\beta$os, i.e. "lead" (plumbum in the Vulgate), whilst 'ôphereth (usually translated lead) is $\kappa\alpha\sigma\sigma\acute{\iota}\tau\epsilon\rho$os (stannum in the Vulgate).[30] The translation "tin" is, therefore, very doubtful[31], although Munk[32] asserts that tin is meant. In the Talmûd the

[1] Numb., xxxi, 22.
[2] Ez., xxii, 18 ; xxvii, 12, copper, tin, iron, lead.
[3] Charles, Apocrypha, 1913, ii, 585.
[4] Dross ? ; cf. Is., i, 25.
[5] CDB, 555.
[6] Ib., 560.
[7] Kopp, Gesch. d. Chemie, iv, 131 ; Roscoe and Schorlemmer, Chemistry, 1913, ii, 872, 1059 ; Rössing, Geschichte der Metalle, Ergänzungsheft Verhl. Verein z. Beförderung des Gewerbfleisses, 1901, 51, thought $\mu\acute{o}\lambda\upsilon\beta$os was from the "old Indian" mulva : the Sanskrit name is sîśa, Schrader, RL, 97.
[8] i, 22 f.
[9] xli, 7.
[10] CDB, 460.
[11] xxii, 18 f.
[12] v, 7.
[13] War, VII, viii, 4.
[14] Williams, DG, ii, 288.
[15] Letter of Aristeas, cap. 90 ; Kautzsch, ii, 13 ; the date is 1st century B.C. ; Wendland, JE, ii, 92 ; Charles, 1913, ii, 83 f.
[16] Krauss, JE, viii, 513 f.
[17] i, 25.
[18] Nowack, Hebr. Archäol., 1894, 245.
[19] Ez., xxiii, 14 ; Jer., xxii, 14.
[20] Hendrie, Theophilus, 104 ; Vigouroux, Dict. de la Bible, ii, 1065.
[21] Canney, E. Bibl., i, 875 ; Thatcher, HDB, i, 458 ; Bertholet-Dallas, 171, 306.
[22] Is., i, 25.
[23] Shipley, E. Bibl., iii, 2840.
[24] Sabbath, xx ; tr. Rodkinson, ii, 313.
[25] Kelim, x, 2 ; "tin or lead" in Danby's tr., 1933, 619.
[26] Beckmann, H. Inv., ii, 208.
[27] Hoefer, H. de la Chimie, i, 52 f.; Nowack, Hebr. Archäol., 1894, 245.
[28] Tychsen, in Beckmann, ii, 208 ; Krauss, JE, viii, 513 f. ; CDB, 948 ; Hoefer, op. cit.
[29] Kopp, Gesch. d. Chemie, iv, 125 ; Krauss, JE, viii, 513.
[30] E.g. in Numb., xxxi, 22, where both metals are mentioned together ; Kopp, Gesch. d. Chemie, iv, 126 ; Beckmann, ii, 208 ; Bapst, L'Étain, 5, 14.
[31] Rosenmüller, IV, i, 63 ; Roscoe and Schorlemmer, 1913, ii, 841 ; Kopp, iv, 125.
[32] Palestine, 391.

Greek name kastiterion, gasteron, is used.[1] The Babylonian name anak (adopted by the Syrians and Arabs), originally meaning lead, then tin, occurs once only in the Old Testament, viz. in Amos [2], and there perhaps in the sense of lead.[3]

Jehu, king of Israel, sent Šalmaneser III in 842 B.C. gold, silver and "bars of anak," which may mean either lead [4] or (less probably) tin.[5] The "stone of bedil" [6] is probably a lead plummet, as in the Authorised Version [7], and the "tin" given as a common metal in Sirach [8] is also lead.[9] Tin (bedil) and lead ('ôphereth) are sometimes mentioned together, e.g. in Ezekiel[10]: "all they are copper and tin and iron and lead, in the midst of the furnace, they are the dross of silver," and[11] "they shall gather silver and copper and iron and lead and tin into the midst of the furnace, to blow the fire upon it, to melt it" [? each metal separately]. In Numbers bedil is taken as spoil from the Midianites, the inhabitants of Midian (Madian ; $Μαδιαμ$ in the LXX), perhaps the Arabs of the peninsula of Sinai, or northern Arabs.[12] Midian is mentioned only once in the Bible : in Maqrîzî's time the cities were in ruins and the inhabitants gone.[13]

It is possible that tin from Cornwall reached the Hebrews, since they seem to have obtained Baltic amber.[14] In Ezekiel tin and other metals came from Tharshish[15], which is Tarsisi in an Assyrian text of Asarhaddon. The description in which "Ships of Tharshish" brought "gold of Ophir" for Solomon has given rise to endless discussion, but the voyage probably refers to the Red Sea[16], although the tin of Portugal and Galicia shipped from Gades[17], or perhaps the Taurus[18], may be meant. The tinning of copper[19] is not mentioned in the Bible.[20]

MERCURY

There is no indication that the metal mercury is ever mentioned in the Bible or was known to the old Hebrews, although Haupt[21] conjectured that the "stone of Tharshish"[22] was cinnabar (red mercuric sulphide) from the mines of Almaden in Spain. Since Phœnician traffic with Spain was fairly early, and the mines of Almaden were worked in antiquity, this is quite possible. Mercury is said to be mentioned (with sulphur and a crucible) in the Talmûd[23]: it is, apparently, not in the Mishnâh.[24]

ANTIMONY

Although metallic antimony was known to Sumerians (see p. 256) it does not seem to be mentioned in the Bible (unless it is bedil), but occurs as a tin-white metal in the Midrâsh.[25] The sulphide, stibnite, is commonly understood

[1] Bochart, Geographia sacra, 648, 650.
[2] vii, 7, 8.
[3] A.V. "plumbline," Beckmann, ii, 209 ; Bochart, Geogr. sacra, 649—either tin or lead.
[4] Rogers, H. Bab. Assyr., ii, 76 ; BMGB, 46.
[5] Hommel, q. in Lippmann, Alch., i, 578.
[6] Zech., iv, 10 ; $τοῦ λίθου κασσιτερίου$ in Septuagint.
[7] Beckmann, ii, 207.
[8] Ecclus., xlvii, 18 ; CDB, 948,
[9] Kautzsch, Apokryphen, i, 461.
[10] xxii, 18.
[11] xxii, 20.
[12] CDB, 555 ; R. F. Burton, The Gold Mines of Midian, 1878 ; ib., The Land of Midian Revisited, 2 vols., 1879.
[13] CDB, 556.

[14] CDB, 948 ; Sayce, Patriarchal Palestine, 73, 209.
[15] Ez., xxvii, 12 ; Bochart, Geographia sacra, 1712, 165, 170 ; Bapst, L'Étain, 14 ; Lucas, JEA, 1928, xiv, 98.
[16] Meyer, Alt., II, ii, 102.
[17] Hall, in Peake, 1925, 29.
[18] Olmstead, H. Pal. Syr., 341, 406.
[19] Dioskurides, i, 38.
[20] Napier, Arts, 138 ; Beckmann, H. Inv., ii, 221.
[21] MGM, i, 386.
[22] Ex., xxviii, 20, etc. ; Ez., i, 16 ; x, 9 ; xxviii, 13 ; Dan, x, 6 ; cf. JE, v, 593 f.
[23] Sabbath, viii ; tr. Rodkinson, i, 151, 154.
[24] Danby, The Mishnâh, 1933—index insufficient.
[25] JE, v, 593 f.

to be mentioned [1], always as an eye-paint, as the Hebrew pûkh.[2] In the Septuagint the passage in Ezekiel is : ἐστιβίζου τοὺς ὀφθαλμους συο, that in Kings : ἐστιμμίσατο τοὺς ὀφθαλμους αὐτῆς, and the word pûkh is thus given as στίμμι, antimony. The Vulgate has stibium, and the Aramaic and Syriac versions also give antimony.[3] In the Greek fragment of Enoch [4] found at Akhmîm, pûkh is translated as λίθου φουκά, usually taken to mean stibium [5], although Gesenius (1786–1842) had emphasised the resemblance between pûkh and φῦκος, fucus, a red colour (archil) made from certain seaweeds and used as a cosmetic for the face and eyebrows by the Greeks, who probably got its use from the Ionians.[6] In the Talmûd [7] pûkh is said to be applied to the eyes to dry tears and cause the lashes to grow, which corresponds with the use of eye-paint (koḥl) rather than a red colour, and in the later period at least it had the same meaning as koḥl (kuḥl).[8] It is also a red paint for the cheeks, otherwise called siqra and formerly used for ink (rouge ?).[9]

Löw[10] adopts the suggestion (apparently without mentioning Gesenius) that pûkh (or fuqus) was not stibium but a red cosmetic made from archil (φῦκος) for the face and cheeks, and sometimes used to surround the eyes, as Josephus[11] confirms for a much later period. The translation "thou rentest thy face with painting" in Jeremiah[12] is, however, said in the margin to be wrong, "eyes" not "face" being in the Hebrew. In the Apocryphal Wisdom of Solomon[13] an image is painted with minium and its "eyes coloured red with paint." Sirach[14] regards painting the eyelids as the sign of an abandoned woman, yet the use of black koḥl (perhaps galena, or soot) for painting the eyes has always been regarded as legitimate and medicinal in the East.[15] Schröder[16] suggested that pûkh was galena, viz. Pliny's[17] platyophthalmon or stibium fœmina, not antimony sulphide, and a small pot of galena was found in a tomb (late Maccabæan ?) at Beit Jibrin.[18]

A late Hebrew name koḥl is probably the Arabic word,[19] and the Spanish translation of Ezekiel[20] is : alcoholaste tuo ojos.[21] The modern koḥl of Arabia and Persia is usually antimony sulphide, that of Egypt a soot from burning frankincense or charred almond shells. It is moistened with oil and kept in a small jar (keren ḥappûkh, "horn for paint")[22], being then applied in a special way.[23] The references to eye-paint in the Mishnâh[24] are vague and do not convey any information as to the nature of the material.

[1] Ez., xxiii, 40 ; Jer., iv, 30 ; II Kings, ix, 30.

[2] Pinner, in Diergart, 1909, 199 ; Rosenmüller, IV, i, 67 ; Roscoe and Schorlemmer, Chemistry, 1913, ii, 963, say the Hebrew name was koḥl.

[3] Thomson, Hist. of Chem., i, 75 ; Kopp, iv, 100 ; Pinner, 199—in LXX, pûkh = στίβι ; Hamerus Poppius, Basilica Antimonii, Geneva, 1647, 597, gives Hebrew zadadah, Aramaic zedidah, Spanish alcohol, i.e. al koḥl.

[4] xviii, 8.

[5] Beer, in Kautzsch, ii, 218 ; Charles, Book of Enoch, 1893, 354.

[6] Smith and Wayte, DA, i, 880.

[7] Sabbath, xiii.

[8] Albright, Arch. of Palest., 123, 203.

[9] Oesterley, Tractate Shabbath, 1927, 41, 47.

[10] Flora der Juden, Vienna, 1926, i, 20 f. ; ib., JE, x, 76.

[11] War, IV, ix, 10.

[12] iv, 30, A.V.

[13] xiii, 14 ; Kautzsch, i, 498, 479 : written 100 to 50 B.C. at Alexandria.

[14] Ecclus., xxvi, 9 ; Charles, Apocrypha, 1913, i, 403.

[15] Bertholet-Dallas, 179 ; Kopp, Beiträge zur Geschichte der Chemie, Brunswick, 1869, i, 9.

[16] Geschichte der ältesten Chemie, Marburg, 1775, 354, 361 f. ; Jahn, Biblische Archäologie, 1817–18, I, ii, 158.

[17] xxxiii, 6.

[18] Macalister, Century, 262 : "powdered lead."

[19] Salmasius, Homonymis Hyles Iatricæ, Utrecht, 1689, 177—"cochal."

[20] xxiii, 40.

[21] Kopp, Gesch. der Chemie, iv, 100.

[22] Job, xlii, 14.

[23] CDB, 664.

[24] Sabbath, viii, 3 ; x, 6 ; Makoth, iii, 6 ; Berakoth, vii, 3 ; Kelim, xiii, 2 ; xvi, 8.

NON-METALS (INORGANIC)

SALT

Salt (melach) in the Bible symbolised durability, fidelity and purity, on account of its antiseptic properties [1], but it was also a material of sterility.[2] It was used with food, including bread.[3] Rock salt occurs in large amounts, particularly the salt mound of Kashm Usdum at the southern end of the Dead Sea, a pile 5 miles long, $2\frac{1}{2}$ miles wide and some hundred feet high.[4] According to Jewish authors, the Dead Sea salt was used in ancient times for the Temple service. The Talmûdists and Galen [5] call it "Sodom salt" ($\H{a}\lambda\epsilon\varsigma\ \Sigma o\delta o\mu\eta\nu o\acute{\iota}$), and Zephaniah (c. 630 B.C.) [6] mentions salt pits in the neighbourhood.[7] Julius Sextus Africanus (born at Jerusalem ; died about A.D. 300) [8] says [9] the Dead Sea produces alum and salt differing somewhat from the common kinds, as they are pungent and transparent. The connection of the Dead Sea with Sodom goes back to II Esdras (IV Ezra)[10]. The remains of Sodom were described by Strabo[11] and 17th century travellers were told that they were beneath the Dead Sea.[12] Benjamin of Tudela (1173) saw the pillar of salt about seven miles from the Dead Sea ; he says that sheep licked it, but it regained its original shape.[13] Actually, Sodom and Gomorrah were probably destroyed by bitumen or naphtha fires and not by submergence[14], and Talmûdists suggest[15] that the "salt which has lost its savour"[16] is liquid bitumen, said to have been used under the name "salt" in Jewish sacrifices. The phenomenon of the pillar of salt is caused by salt springs which deposit solid salt and seal up the aperture until it is licked off by animals, when a new deposit forms, and so on continually. Such pillars are found at Jabal Usdum (one of which is traditionally that of Lot's wife), and also at Hammam Meskutim in Algeria.[17]

THE DEAD SEA

Palestine was celebrated among the Romans for the balsam tree and the Dead Sea.[18] The Dead Sea is one of the most remarkable as well as one of the oldest lakes in the world ; at the end of the Tertiary Period it stood 106 m. higher in level than at present and an enormous amount of evaporation has taken place.[19] Its name in the Bible[20] is "the salt sea" ($\dot{\eta}\ \theta\acute{a}\lambda a\sigma\sigma a\ \tau\hat{\omega}\nu\ \dot{a}\lambda\hat{\omega}\nu$ in the Greek ; mare salis in the Vulgate), sometimes "the sea of the Arabah." The first classical author to say much about it is Diodoros Siculus[21], who calls it "the asphalt lake" ($\dot{a}\sigma\phi a\lambda\hat{\iota}\tau\iota\varsigma\ \lambda\acute{\iota}\mu\nu\eta$, or $\dot{\eta}\ a\sigma\phi a\lambda\hat{\iota}\tau\iota\varsigma$), a name adopted

[1] CDB, 811 ; Macalister, HDB, ii, 38 ; Hull, ib., iv, 355 ; E. Bibl., iv, 4247 ; Sel and Benzinger, JE, x, 660.
[2] Judg., ix, 45.
[3] Job, vi, 6 ; CDB, 272 ; Mishnâh and Talmûd, frequently, e.g. Berakoth, i, 1 ; vi, 1, 7.
[4] CDB, 675, 811, 839 ; Rosenmüller, IV, i, 5.
[5] De Simpl. Fac. Med., iv, 20.
[6] ii, 9.
[7] Rawlinson, Mon., ii, 488 ; CDB, 811.
[8] Kroll, PW, x, 116 f.
[9] Quoted by Synkellos, in Ante-Nicene Library tr. of Hippolytos, 1867, vi, 175.
[10] v, 7 ; 50–30 B.C., but much edited in A.D. 90–120 ; Charles, Apocrypha, 1913, ii, 542 ; CDB, 834, 890.
[11] XVI, ii, 44.
[12] Lenz, Mineralogie, 64 ; Maundrell,

Journey from Aleppo to Jerusalem, 1697, in Pinkerton, Voyages, 1811, x, 348.
[13] Itinerary, ed. Asher, i, 37 ; ib., ed. Adler, 23.
[14] C. Geikie, The Holy Land and the Bible [1898], 625 ; Sayce, Patriarchal Palestine, 153.
[15] J. D. Michaelis, Commentationes Societati Regiæ Scient. Gœttingensi, 1760, iv, de Natura et Origene Maris Mortui, 70 f., 91 f.—Sodom salt is common salt, not bitumen ; Thomson, Hist. of Chemistry, 1830, i, 106.
[16] Matt., v, 13.
[17] Geikie, 626 ; Adler, Benjamin of Tudela, 23.
[18] Tozer, Anct. Geogr., 260.
[19] CDB, 835 ; Williams, DG, ii, 522 ; Benzinger, PW, ii, 1730.
[20] Gen., xiv, 3.
[21] ii, 48 ; xix, 98.

32

by Josephus[1]; the name Dead Sea ($\theta\acute{a}\lambda a\sigma\sigma a$ $\mathring{\eta}$ $\nu\epsilon\kappa\rho\acute{a}$; mortuum mare) dates only to 2nd century A.D. authors such as Justin [2] and Pausanias.[3] Strabo's name [4], Lake Sirbonis, is based on some mistake.[5] Diodoros says the lake is very salt and fish will not live in it, and it is practically free from life, even corals and mussels [6], yet some low forms of life exist in it.[7] Diodoros says that even dense materials like lead, silver and gold sink in it more slowly than in common water, and a man cannot sink in it. Lynch [8] gives its specific gravity as 1·13 ; it is not quite saturated, although in places salt crystals form on the surface. Some old analyses are[9] :

	Marcet 1807	Gay Lussac 1818	Gmelin 1826	Apjohn 1839
Specific Gravity . .	1·211	1·228	1·212	1·153
$CaCl_2$ . . .	3·920	3·98	3·2141	2·438
$MgCl_2$ . . .	10·246	15·31	11·7734	7·370
$MgBr_2$ . . .	—	—	0·4393	0·201
KCl	—	—	1·6738	0·852
NaCl . . .	10·360	6·95	7·0777	7·839
$MnCl_2$ . . .	—	—	0·2117	0·005
$AlCl_3$ . . .	—	—	0·0896	—
NH_4Cl . . .	—	—	0·0075	—
$CaSO_4$ . . .	0·054	—	0·0527	0·075
Total solids per cent. .	24·580	26·24	24·5398	18·780

In recent years a company has been promoted to work the Dead Sea for salts, including bromides.[10] Diodoros Siculus and Strabo mention noxious odours or vapours rising from the lake : the former says they tarnish gold, silver and copper ; the latter that copper, silver and other brilliant things except [or "even"] gold are tarnished ($\kappa a\tau\iota\acute{o}\epsilon\sigma\theta a\iota$).[11] There are sulphur springs near the Dead Sea[12], and a strong smell of hydrogen sulphide is said to hang over the lake ; the water itself is inodorous but corrodes iron strongly.[13]

ALKALIES

The alkalies, represented by sodium carbonate (nether) and (probably) potassium carbonate (borith), were known, but do not seem to have been causticised. The words nether and borith[14] are translated "nitre" and "soap" in the Authorised Version. Luther translated the first "chalk," since it effervesced with vinegar[15], but St. Jerome (4th century) already stated in his commentary on Proverbs that it was soda (nitrum), which he knew was made in Egypt in the province of Nitria by the drying up of lakes by the heat of the sun and was used as a detergent (crepitat autem in aqua quomodo calx viva et ipsum quidem disperit, sed aquam lavatione habilem reddit ; cujus natura cui sit apta figuræ, cernens Solomon ait : acetum in nitro, qui cantat

[1] Antiq., I, ix.
[2] Hist. Philipp., xxvi, 3.
[3] V, vii, 4.
[4] XVI, ii, 42, 764 C.
[5] Lenz, Mineralogie, 63 ; Falconer and Hamilton, tr. of Strabo, iii, 182.
[6] Benzinger, PW, ii, 1730.
[7] CDB, 839.
[8] Williams, DG, ii, 524 f.
[9] Robinson, Researches in Palestine, 1841, ii, 224 ; Vigouroux, Dict. de la Bible, iv, 1289.

[10] Melamede, Discovery, 1929, x, 128, 165 ; Chemical Age [London], 1933, xxix, 163.
[11] Lenz, Mineralogie, 63.
[12] Geikie, Holy Land, 622.
[13] Williams, DG, ii, 525.
[14] Jerem., ii, 22 ; Rosenmüller, IV, i, 9, 112.
[15] Prov., xxv, 20 ; Roscoe and Schorlemmer, Chemistry, 1913, ii, 237 ; Hoefer, Hist. de la Chimie, i, 58, derives nether from natar, to effervesce.

carmina cordi pessimo. Acetum quippe si mittatur in nitrum, protinus ebullit).[1] Although some travellers report nitre (saltpetre) in the Dead Sea, this is doubtful [2], and saltpetre [as a soil efflorescence ?] is rare in Palestine.[3] Since nether is mentioned only twice it was perhaps not much used.

"Cinders of borith" [4] probably mean alkaline plant ash ; Pinner's [5] translation as "borax, potash or soda" is almost certainly incorrect in the first item ; potash (potassium carbonate) is the most probable translation of borith, and soda of nether.[6]

The soda plant (Salsola kali), called gasûl in later commentaries, modern hubeibeh, grows near the Dead Sea and is very common in Syria, Judæa, Egypt and Arabia, the ash being called kali.[7] In modern times Egyptian soda was made from the ash of Mesembrianthemum Copticum and M. nodiflorum (which also grows in Palestine) ; the Turks introduced the Mesembrianthemum crystallinum into Greece.[8] The ash of such plants contains both potash and soda (see p. 147).

The alkaline materials were used for washing.[9] The "fuller's soap"[10] was probably a saponaceous plant, more commonly used, as well as fuller's earth (creta cimolia), than alkali[11]; the plant was probably the soap-weed (Struthium).[12] The saponaceous plant called ajram grows near Sinai and is used as a substitute for soap, and the Saponaria officinalis grows in Palestine.[13] Soap was unknown to the ancient Hebrews[14], but is now made at Nablus by the Arabs from plant ashes.[15]

Alum (seriph) is mentioned in the Mishnâh, Talmûd[16] and Midrâsh[17]; Josephus[18] mentions the alum of Machairus in Peræa.

GLASS

The name zĕkûkîth, generally understood to mean glass, is from a root meaning "clear," not necessarily transparent.[19] The use of glass came from Egypt to Palestine in the III Semitic Period, but clear glass was not made till the IV Semitic Period[20], and the common use of glass is supposed to date only from the Ptolemaic Period.[21] Although it has been supposed that glass was made by the Jews "before the separation of the ten tribes,"[22] in reality practically nothing is known of early Jewish glass-working. Glass was made in Palestine and Syria in the Roman and later periods[23], and the abundance and good quality of the natural soda, now extracted by the Arabs of the

[1] Beckmann, H. of Invent., ii, 487 ; Jahn, Biblische Archäologie, Vienna, 1817–18, I, i, 579 f. ; Hoefer, I, 58.
[2] Rawlinson, Mon., ii, 488.
[3] CDB, 675.
[4] Mal., iii, 2.
[5] In Diergart, 1909, 199 f.
[6] Hoefer, i, 58 ; Shipley, E. Bibl., iii, 2840, "lye" in R.V. ; Löw, Flora der Juden, i, 637 f., 644, 647 ; Wiedemann, A. Rel., 1928, xxvi, 332 ; E. Bibl., iii, 3425 ; Olaf Celsius, Hierobotanicon sive de plantis Sacræ Scripturæ dissertationes breves, 2 vols., Uppsala, 1745–47, i, 449 ; C. B. Michaelis, Epistola [to D. F. Hoffmann] de herba borith, Halle, 1728—supposes borith is struthion, saponaria, lanaria ; J. D. Michaelis, Commentationes Societate Regiæ Scientiarum Gœttingensi per annos 1758–68, 1769, vii, 151 f., 155, de nitro Hebræorum sive bŏrith—supposes it is the kali plant, used for making glass ; cf. ib., vi, de nitro Plinii, 134 f. ; R. P. M. Jullien, Sinaï et Syrie, Lille, 1893, 83, says the Anabasis

articulata (Forsk.) is at present burnt for soda.
[7] Bochart, Hierozoicon, i, 874 ; Rosenmüller, IV, i, 112 ; Goguet, i, 144.
[8] Löw, i, 644, 647 ; CDB, 889.
[9] Jer., ii, 22 ; Job, ix, 30, washing with "snow water," A.V.
[10] Mal., iii, 2.
[11] Smith, CDB, 274 ; cf. Michaelis, ref. 6, above.
[12] Dioskurides, ii, 193 ; Pliny, xxv, 5, radicula lavandis ; Löw, i, 640.
[13] CDB, 889.
[14] Löw, i, 640.
[15] Geikie, Holy Land, 703.
[16] Kelim, ii, 1 ; x, 1 ; Sabbath, xiv.
[17] Bochart, Geographia Sacra, 1712, 27.
[18] War, VII, vi, 3.
[19] E. Bibl., ii, 1737; iii, 3821; HDB, ii, 180.
[20] Handcock, Holy Land, 271.
[21] E. Bibl., iii, 3821.
[22] Michaelis, q. by Fowler, Archæologia, 1880, xlvi, 82.
[23] Kisa, Glas, 96, 99.

desert from the ashes of saline plants (ice plant) growing in clefts of the rocks, especially around the shores of the Dead Sea [1], make it probable that glass-making was practised in Judæa at an early date. A white opaque glass flagon with six panels decorated in high relief, "picked up somewhere in Syria," now in the Louvre, is supposed to be of Jewish origin, although certainly "long after the exile" in date, and Phœnician glass has been found in Palestine.[2] A few balsamaria found in Palestine were perhaps imported from Egypt a few centuries B.C., and much Roman glass was dug up in laying the Bagdad railway, especially from Jaffa to Jerusalem.[3]

Glass is often mentioned in the Mishnâh and Talmûd [4] and it was very well known to Josephus. Arabic authors report a tradition that the throne of Solomon stood on a pavement of glass.[5] Kisa thinks all pre-Roman glass was imported and that the supposed references to glass in the Old Testament [6] are doubtful, but most authorities follow Michaelis and Gesenius in supposing that zĕkûkîth in Job (ΰαλος in the Septuagint ; Arabic zadjadj, glass ; the Hebrew chol is sand) meant glass.[7] This is sometimes [8] called the only mention in the Old Testament, but it occurs again in Proverbs.[9] All the Jewish interpreters say the "treasures hid in the sands"[10] mean glass.[11]

In the later Roman Empire the Phœnician glass-works at Tyre were operated by Jews, and there were perhaps Jewish glass-workers in Constantinople in 531–565 ; they may have been glass-blowers or perhaps only produced glass blocks for mosaics.[12] Mosaics appear to be mentioned in the Old Testament[13], but Bucher's suggestion that the name is derived from the Hebrew maskith is improbable : it comes from the Muses, and there are fragments of a 5th century B.C. mosaic pavement of a temple of Zeus at Olympia.[14] Greek glass-workers are said to have gone from Constantinople to France in 687 to make Jewish or lead glass (plumbeum vitrum, Judeum scilicet), which was not well known in Europe till long after the 13th century[15], although lead glaze was known early in Mesopotamia (see p. 282). Fusible lead glass, for painting and imitation gems, was called Jewish glass in the Middle Ages, and ruby glass was called "Jews' glass" in Birmingham in the 19th century[16], when the Jews made glass, including coloured bracelets, at Hebron. The Jewish workers from Tyre and Hebron founded the Venice works in the 9th century, the sand at first being brought from Belus and from the desert between Cairo and Alexandria.[17] Nasîr ibn Khusraw (11th century) says the portraits in the Church of the Holy Sepulchre at Jerusalem were varnished and also covered with sheets of transparent glass.[18] Benjamin of Tudela (A.D. 1173) found two Jewish glass-workers in factories at Antioch ; he and al-Idrîsî

[1] Robinson, Res. in Palestine, 1841, ii, 211.
[2] Perrot and Chipiez, Art in Sardinia and Judæa, 1890, i, 358 ; Wallace-Dunlop, Glass in the Old World, [1883], 62 f.
[3] Kisa, Glas, 97.
[4] Aboth, iv, vi ; Derech Eretz, x ; Sabbath, xvi, xvii, xviii, xxi ; Baba Bathra, v, 10 ; Berakoth, ix, 1–5 ; Mishnâh, Kelim, viii, 9—"glass-blower's furnace"; no details of glass-making are given.
[5] Wallace-Dunlop, 62, 67 f., 96 ; Kisa, 96 f.
[6] Job, xxviii, 17 ; Prov., xxiii, 31.
[7] Bochart, Hierozoicon, ii, 723 ; Fowler, Archæologia, 1880, xlvi, 86.
[8] Pinner, in Diergart, 1909, 195 f. ; CDB, 294 ; Patrick, HDB, ii, 180.
[9] xxiii, 31 ; Wiegleb, Alchemie, 1777, 4 ; Thomson, Hist. of Chemistry, i, 82.

[10] Deut., xxxiii, 19.
[11] Geikie, Holy Land, 774 ; Wallace-Dunlop, 63.
[12] Bucher, iii, 281 ; Kisa, 99 ; Merrifield, Anct. Practice of Painting, i, pp. xxxviii, xcii.
[13] Esther, i, 6 ; Cant. iii, 10 ; Ez., xl, 17 f. ; Bucher, i, 97, 101.
[14] Bucher, i, 97, 102, Fig. 19 ; Seyffert, Dict. of Antiquities, 1891, 398 f.
[15] Merrifield, i, p. xciii, from Filiasi ; Eraclius, iii, 49, in Merrifield, i, 243 ; Kisa, 100.
[16] Bucher, iii, 280 ; Wallace-Dunlop, 70 ; Merrifield, i, p. xcii.
[17] Kisa, 100 ; Merrifield, i, p. xciii ; Geikie, Holy Land and the Bible, 299, 307 ; Wallace-Dunlop, 69.
[18] Sefer Nameh, tr. Schefer, Paris, 1881, 108 ; Wallace-Dunlop, 70.

say that Tyrian glass was good and esteemed in all countries and Benjamin also mentions a wall of glass (? mosaics) in the Gomah synagogue at Damascus.[1]

An analysis of a specimen of glass from the Temple at Jerusalem gave[2] : silica 69·30 ; soda (Na_2O) 13·79 ; potash (K_2O) 1·49 ; lime (CaO) 8·50 ; magnesia (MgO) 0·55 ; lead, trace ; iron oxide (Fe_2O_3) 2·00 ; alumina 3·30 ; phosphoric anhydride (P_2O_5) 0·80. The date of the specimen is doubtful : it was probably mediæval. The marâh [3] and rěî [4], translated "looking glass" in the Authorised Version, were (as the texts show) polished metal mirrors, which had to be frequently polished to prevent tarnish.[5] The obscure word gilyônîm [6] probably means "transparent" fine linen garments.[7]

BUILDING MATERIALS

The whole mountain chain of Palestine is of limestone, with eruptions of basalt.[8] For masonry and buildings the stones used in the early periods included diorite, sandstone, basalt, quartzites, serpentine, various kinds of marble, limestone and Egyptian "alabaster" ; there are stone mace-heads and weights. Gypsum is probably the Hebrew sîd.[9] "Marble" (šeš), used by Solomon, is called "white stone" ($\lambda i\theta os \lambda \epsilon v \kappa \acute{o} s$) by Josephus,[10] and may have been limestone from near Jerusalem or Lebanon (Jura limestone) as used for the temple at Baalbec, or else white marble from Arabia : the marble pillars, etc., in the palace at Susa were probably Persian.[11] Stones were used for the common purposes.[12] Alabaster, which occurs only in the New Testament[13] as the material of the precious box of ointment, is really a variety of calcite[14], as in Egypt (q.v.). The name is from a city in Egypt, Alabastra, with quarries of the stone.[15]

The Canaanite houses are of sun-dried bricks, also pinkish pure clay bricks, with mortar of sandy clay without straw. The Israelite bricks sometimes contain chopped straw, and bricks with straw were found at Megiddo, with an oven.[16] In the Bible[17] it is said only that Pharaoh refused to supply straw to the Hebrews for making bricks (lebênâh), not that they made them without straw. Bricks without straw were found by Naville in 1883 in the "store city" of Pithom built by the Israelites for Pharaoh.[18] The Jews learnt brick-making in Egypt, and the brick kiln was used in David's time[19]: Isaiah[20] complains that altars were built of brick instead of unhewn stone as directed by the law.[21] Teben[22], Aramaic 'ûr[23], is chopped wheat or barley straw used for mixing with the clay ; chăšaš is hay and môts is chaff.[24]

The usual name for potter's clay is chômer ; other names are tît and chăsaph ; melet is doubtful : most of the modern pottery is made at Gaza

[1] Fowler, *Archæologia*, 1880, xlvi, 93.
[2] Muspratt, *Chemie*[4], iii, 1366.
[3] Ex., xxxviii, 8.
[4] Job, xxxvii, 18.
[5] I Cor., xiii, 12 ; Wisd., vii, 26 ; Ecclus., xii, 11 ; Wallace-Dunlop, 67 ; CDB, 566.
[6] Is., iii, 23 ; Vulgate specula, A.V. glasses.
[7] CDB, 566.
[8] *Ib.*, 674 ; Williams, DG, ii, 518 ; Hull, HDB, ii, 150 ; Blankenhorn, *Z. f. Deutsch. Palestina Vereins*, 1914, xxxvi, 20–44, on geology, not seen.
[9] Handcock, Holy Land, 65, 157, 168 ; Myres, E. Bibl., iv, 4800 ; Patrick, HDB, iv, 617 ; Rosenmüller, IV, i, 18.
[10] Antiq., VIII, iii, 2.
[11] Esth., i, 6 ; Rosenmüller, IV, i, 19 ;

Bochart, Hierozoicon, ii, 708 ; CDB, 513.
[12] CDB, 901.
[13] Matt., xxvi, 7 ; Mark, xiv, 3, where "breaking" means of the seal ; Luke, vii, 37 ; $\dot{a}\lambda\dot{a}\beta a\sigma\tau\rho o\nu$.
[14] CDB, 28.
[15] Donne, DG, i, 81.
[16] Handcock, 66, 85, 96, 122, 124.
[17] Ex., i, 7.
[18] Yahuda, *Daily Telegraph*, 17 Sept., 1932 ; cf. Vigouroux, Dict. de la Bible, i, 1929.
[19] II Sam., xii, 31.
[20] lxv, 3.
[21] Ex., xx, 25.
[22] Ex., v, 7, etc. ; Job, xxi, 18.
[23] Only in Dan., ii, 35.
[24] CDB, 147, 903 ; Petrie, HDB, i, 326 ; Müller, E. Bibl., i, 608.

in dark blue clay.[1] Clay was also used for sealing.[2] Various types of pots are named[3] : asûk, an amphora [4], cheres, an earthen pot for stewing [5], dûd, a small culinary vessel [6], and sîr, combined with other words to denote special uses.[7] The potter's trade was learnt in Egypt.[8] The clay was trodden [9], and formed on the wheel, perhaps also adopted from Egypt.[10] There was a royal pottery at Jerusalem.[11] That the pottery was glazed[12], beyond the use of a clay slip, is doubtful ; that it was lead glazed[13] is improbable for the early pottery. The "potsherd covered with silver dross,"[14] which might (as Munk suggests) mean glazed with litharge, is translated "silver dishonestly given is to be considered as a potsherd" in the Septuagint ; the Hebrew seems to imply intentional overlaying with some dross or slag, and since cupellation was known, the use of a lead glaze could have been discovered accidentally at a fairly early period. Glazed pottery, occasionally found in Palestine, was probably imported.[15]

MORTAR ; BITUMEN

Lime (gir, sîd ; Arabic gayyârum, quicklime) is named only three times in the Old Testament[16], including burnt lime, the later Hebrew name for which was kelšâ, from the Greek kalx.[17] Quicklime is a constituent of a depilatory (with orpiment ?) in the Mishnâh.[18] Various mentions of mortar include (probably) lime mortar, bitumen and mud, as well as plaster.[19] Bitumen or asphalt (chêmâr ; Arabic al hummar)[20] occurs only in the valley of the Jordan and in and about the Dead Sea.[21] It is obtained at Hasbeiya, the most remote source of the Jordan, from pits sunk through the bituminous earth to a depth of about 180 ft.[22] The three Hebrew words for bitumen in the Old Testament are zépheth (perhaps from Aramaic ziphtâ ; Arabic zift) for the liquid kind, chêmâr for the solid kind, and kópher for a wood varnish, the first two being probably what the Greeks called asphaltos (ἀσφαλτος ; bitumen durum) and naphtha (νάφθα ; bitumen liquidum), respectively.[23] What is called rosin in the Authorised Version is naphtha[24], which is mentioned in the Talmûd[25] as used (with other oils) for lamps.

The bitumen pits near the Dead Sea, beĕrôth chêmâr in the valley of Siddim in Genesis[26], are what the Arabs call biâret hummar[27], and the asphalt from them was called Jew's pitch (bitumen Judaicum). The asphalt, of specific gravity 1·07–1·16, also floats on the surface of the Dead Sea, of specific gravity 1·19–1·21.[28] Dioskurides,[29] Serapion[30] and other medical authors

[1] CDB, 167 ; E. Bibl., i, 835 ; iii, 3819 ; Hull, HDB, i, 447.
[2] Job, xxxviii, 14 ; Jer., xxxii, 14.
[3] E. Bibl., iii, 3815, 3820 f.
[4] II Kings, iv, 2.
[5] Ez., iv, 9 ; Lev., vi, 28.
[6] I Sam., ii, 14.
[7] Ex., xvi, 3 ; Ps. lx, 8 ; Prov., xxvii, 21 ; Jer., xxxv, 5.
[8] Ps. lxxxi, 6.
[9] Is., xli, 25 ; Wisd., xv, 7.
[10] Is., xlv, 9 ; Jer., xviii, 3 ; Brongniart, i, 8 ; ii, 94.
[11] I Chron., iv, 23 ; cf. Is., xxx, 14.
[12] CDB, 747.
[13] Birch, Anct. Pottery, i, 153 ; Munk, Palestine, 1856, 389 ; J. Jahn, Biblische Archäologie, I, i, 1817–18, 442, who merely says glaze, not lead.
[14] Prov., xxvi, 23 ; a later part of the book, written c. 350–250 B.C., Driver, 404 f. ; some parts use Egyptian sources, Isis, 1928, xi, 157.

[15] See p. 282 ; E. Bibl., iii, 3821 ; Brongniart, ii, 106.
[16] Deut., xxvii, 2, 4 ; Is., xxxiii, 12 ; Am., ii, 1 ; CDB, 473 ; E. Bibl., iii, 2799.
[17] E. Bibl., iii, 2799.
[18] Sabbath, viii, 4.
[19] CDB, 579, 741.
[20] Gen., xi, 3.
[21] Robinson, Res. in Palestine, 1841, ii, 228.
[22] CDB, 675.
[23] Ib., 738, 888 ; E. Bibl., iii, 3782, 3819 ; Bochart, Geographia Sacra, 1712, 38 ; Nies, PW, ii, 1726.
[24] CDB, 799.
[25] Sabbath, iii.
[26] xiv, 10 ; "slime pits," Authorised Version.
[27] Bochart, Geographia Sacra, 38 ; CDB, 888.
[28] Nies, op. cit.
[29] i, 99.
[30] Practica, de temperamentis simplicium, lib. ii, cap. 177 ; ed. Venice, 1550, 147.

mention a red or purple Jew's pitch, and some authors confused it with sulphur.[1] Julius Africanus [2] mentions the bitumen springs of the Dead Sea region. Although Diodoros Siculus [3] describes an island of solid asphalt which rises every year in the lake, to which the natives go out on rafts and "cut the asphalt like soft stone," modern travellers found no islands but numerous fragments of solid bitumen among the pebbles on the west shore.[4] The account in Strabo [5] is similar : the asphalt is liquid when forced up by the underground fire, but solidifies in the cold water of the lake. Poseidonios (in Strabo) and Josephus [6] also say that magicians solidify the asphalt by pouring urine over it, pressing out the liquids and uttering incantations; Tacitus [7] that it is solidified by [the very cold] vinegar. Josephus says the asphalt, mixed with various drugs, was used medicinally. The Jews used asphalt or pitch as mortar and for caulking wood, and the Egyptians for waterproofing papyrus boats.[8] Nasîr ibn Khusraw [9] says bitumen from the Dead Sea was put round roots of trees to preserve them from worms (see p. 269).

Maccabees I[10] refers to instruments for throwing fire, and Maccabees II[11] to naphtha ("nephthor, which some call nephtha"), the Persian naptar, Greek νέφθαι or νέφθαι (naptar apanm is Zend for "water of purification" or "sacred water," used in connection with the sacred fire)[12], which took fire "miraculously" when thrown on the hot stones of the altar. This seems to be a remnant of the Persian ceremony at the fire-altar.

SULPHUR

The Hebrew name for sulphur, gophrîth (Aramaic kophrîth)[13] is probably connected with that for bitumen, gôpher or kópher, and is not a Bactrian word as Lagarde suggested.[14] Sulphur does not appear to have been put to any use by the Jews, except perhaps in the topheth.[15] The name gophrîth has also been connected with gôpher in Genesis[16], perhaps a resin from a tree[17], this name being extended to other inflammable materials, and particularly to the "balls of nearly pure sulphur" found plentifully on the south, south-east and west shores of the Dead Sea.[18] Josephus[19] mentions the sulphur mines of Machairus in Peræa, and sulphur is now mined in Palestine at Gaza.[20] There are many sulphur springs in Palestine.[21] Sulphur is mentioned in the Mishnâh.[22]

PRECIOUS STONES

The precious stone (yĕqârâh) was early in use and several kinds are frequently named in the Old Testament, although their identification is very uncertain.[23]

[1] Bochart, Geogr. Sacra, 40, 42 ; Adams, The Seven Books of Paulus Ægineta, 1844–47, i, 60.
[2] Quoted by Synkellos ; Ante-Nicene Libr. tr. of Hippolytos, ii, 175.
[3] xix, 98.
[4] Williams, DG, ii, 526.
[5] XVI, ii, 42 f., 764 C.
[6] War, IV, viii, 4.
[7] Hist., v, 6.
[8] Gen., vi, 14 ; xi, 3 ; Ex., ii, 3.
[9] Sefer Nameh, tr. Schefer, Paris, 1881, 57.
[10] vi, 52 ; Charles, Apocrypha, 1913, i, 89 ; c. 100 B.C.
[11] i, 36 ; Charles, 1913, i, 133 ; c. 106 B.C.
[12] Charles, ib.; Kautzsch, Apokryphen, i, 88.

[13] Rosenmüller, IV, i, 11 ; E. Bibl., i, 611.
[14] E. Bibl., i, 611.
[15] Ib., i, 611 ; iii, 3186.
[16] vi, 14.
[17] Ursinus, Arboretum Biblicum, Norimbergæ, 1663, 76.
[18] Gen., xix, 24 ; CDB, 123 ; Rawlinson, Mon., ii, 488 ; Geikie, Holy Land, 625.
[19] War, VII, vi, 3.
[20] Chemical Age [London], 1933, xxix, 519.
[21] Hull, HDB, i, 328.
[22] Sabbath, viii, 4.
[23] Hoefer, Hist. de la Chimie, i, 63 ; CDB, 902 ; Petrie, HDB, iv, 619; Myres, E. Bibl., iv, 4799 ; Bertholet-Dallas, 177 ; JE, v, 593 ; Bochart, Geographia Sacra, 19 ; Rosenmüller, IV, i, 28 f.

The information on gems in the Septuagint is probably based on Greek knowledge as summarised in Theophrastos ; that in Revelations [1] reflects Pliny.[2] Josephus [3], who agrees with the Vulgate, gives his own interpretation, and in his time and in that of St. Jerome the breastplate could still be seen in the Temple of Concord.[4] The Book of Sirach [5] contents itself with a general account without detail.[6] The identifications of the twelve stones in the breastplate given in the treatise of Epiphanios (A.D. 315–403) [7] also preserve the traditional names.[8]

The gem engraver (châraš eben) [9] probably made use of emery, which is the stone šâmîr, translated "adamant," "diamond," etc., in the Authorised Version[10], the Egyptian àsmer ($\sigma\mu\acute{v}\rho\iota s$), since diamond was unknown to the old Hebrews.[11] Šâmîr is connected in legends with a (boring ?) worm.[12] Flint, challâmîš[13], or tzor[14], was probably also used for softer stones.[15]

The identifications of precious stones in the Bible, like those of classical authors[16], are, as stated, very conjectural, and the twelve stones in Aaron's breastplate[17] have been very variously described.[18] The word hôšen translated "breastplate" is really of unknown meaning ; it denoted a sacred receptacle attached to the ephod of the High Priest and contained the 'Ûrîm and Thummîm, which were perhaps small plaques of stone, wood or ivory, or perhaps two natural precious or semi-precious stones of different colours[19], although the usual interpretation is small statuettes.[20] The twelve stones were engraved with the names of the twelve tribes[21] and gem cutting was probably known early among the Jews. Josephus[22] says the twelve stones denoted either the signs of the zodiac or the months, but this is a late tradition, probably derived from Philo Judæus, who was fond of such allegories.

The following is the usual identification of the stones in the breastplate[23], the names from the Vulgate being in brackets :

Row I . .	ôdhem carnelian or sard [sardius]	pitdhâh topaz or peridot [topazius]	bêrĕqeth emerald [smaragdus]
Row II . .	nôphek ruby or carbuncle [carbunculus]	sappîr sapphire or lapis lazuli [sapphirus]	yâšĕphâh jasper or onyx [jaspis]
Row III . .	lésem jacinth [ligurius]	šêbô agate [achates]	achlâmâh amethyst [amethystus]
Row IV . .	taršîš beryl or yellow jasper [chrysolithus]	šôham chrysolite [onychinus]	yahălôm jasper [beryllus]

[1] xxi, 9–11.

[2] xxxvii.

[3] Antiq., III, vii, 5 ; War, V, v, 7.

[4] CDB, 902.

[5] Ecclus., xlv, 11.

[6] JE, v, 593.

[7] Περὶ τῶν ιβ′ λίθων ἐν τῷ λογίῳ τοῦ ἱερέως ἐμπεπηγμένων ; Liber de xii gemmis rationali summi sacerdotis Hebræorum infixis, text in de Mély, Les lapidaires grecs, 1898, i, 193 ; a longer old Latin version, in Epiphanii Opera, ed. Dindorf, Leipzig, 1862, iv, 141–224.

[8] Rosenmüller, IV, i, 28 f. ; Lipsius, in Smith and Wace, Dict. of Christian Biography, 1877–87, ii, 154.

[9] Ex., xxviii, 11, 21, 36.

[10] Ez., iii, 9 ; Zech., vii, 12—adamant ; Jer., xvii, 1—diamond.

[11] CDB, 14 ; JE, v, 593 f. ; Bochart, Hierozoicon, ii, 841.

[12] Blau, JE, xi, 229.

[13] Deut., viii, 15 ; xxxii, 13 ; Ps. cxiv, 8 ; Is., l, 7 ; Job, xxviii, 9.

[14] Ez., iii, 9.

[15] CDB, 271.

[16] Furtwängler, Antiken Gemmen, 1900, iii, 384 f. ; Krause, Pyrgoteles, 1856.

[17] Ex., xxviii, 17–21.

[18] Petrie, HDB, iv, 619 ; Myres, E. Bibl., iv, 4810 ; Budge, Amulets, 327 ; British Museum of Natural History [South Kensington], Special Guide No. 5 to the Exhibition of Animals, Plants and Minerals mentioned in the Bible, 1911, 68 f.

[19] Budge, Amulets, 328.

[20] Hoefer, Hist. de la Botanique, de la Minéralogie et de la Géologie, 1872, 290 f., 300.

[21] Gen., xxxviii, 18.

[22] Antiq., III, vii, 7.

[23] Budge, Amulets, 327 f. ; E. Bibl., iv, 4799 f. ; Bertholet-Dallas, 177.

Ôdhem or ôdem [1] is probably, as the Septuagint and Josephus say, the sard, a superior kind of agate.[2] It was a red stone, the name being from a root meaning "to be red" (as in the name Adam), and the margin gives (improbably) "ruby": the name may also indicate merely its Edomite origin.[3] The stone may also have been the opaque blood-red jasper, or the carnelian.[4]

It is generally supposed that the ancient and modern topaz (Hebrew pitdhâh) [5] and chrysolith are interchanged, since Pliny [6] says the topaz was a green stone; chrysolite, also called olivine and peridot, is a soft green magnesium and iron silicate.[7] The word pitdhâh is now generally derived from the Assyrian hipindu, flashing stone, and the material is regarded as yellow quartz or, more probably, chrysolite [8], although a relation with the Sanskrit pita, yellow, has been suggested.[9]

Bârĕqath, bâreqeth, or bĕrĕqeth,[10] also found in Tyre,[11] is unanimously identified by the Septuagint, Vulgate and Josephus with the σμάραγδος, smaragdus[12], but this is not the modern emerald, as is often supposed.[13] The Massoretic texts give yahălôm and the identification is impossible; perhaps rock crystal or beryl is meant.[14] The name bĕrĕqeth is probably derived from bâraq, to flash.[15]

Nôphek is probably the red garnet (Septuagint carbuncle or ruby)[16]; in Ezekiel[17] the Authorised Version text has emerald and the margin chrysoprase, and the identification is very doubtful.[18] W. Max Müller's derivation of nôphek from the Egyptian mafek (malachite), the lupaaku stone of the Amarna tablets, is doubted by Knudtzon.[19] Buffon supposed it to be the ruby.[20] Kĕrûm in the Talmûd is the Greek χρῶμα and means a precious red stone.[21] The carbuncle (ανθρ[αξ] ; carbunculus lapis) and "stone of Ophir" are in Tobit (225–200 B.C., revised A.D. 150).[22] Chalcedon (χαλκηδών)[23] is probably not chalcedony but malachite.[24]

Sappîr[25] was a very precious stone, apparently bright blue. The Hebrew name is the origin of the Greek σάπφειρος and the Latin sapphirus, and all these have been identified with the sapphire, a very hard species of corundum[26], although the true sapphire was unknown before the Roman Period and first occurs in Rabbinical works.[27] The identification with lapis lazuli (Babylonian uknû, Egyptian chesbet)[28] is possible, although this is a very soft stone.

Yâšĕphâh (jâšepe ; ἴασπις ; perhaps not a Semitic word)[29], from a root šapah, to be smooth, is usually translated beryl, although jasper seems more

---

[1] Ex., xxvii, 17 ; xxxix, 10 ; Ez., xxviii, 13 ; Rev., iv, 3 ; xxi, 20.
[2] CDB, 824 ; Hoefer, H. Min., 290.
[3] JE, v, 593 f.
[4] Rosenmüller, IV, i, 28 f.
[5] Ex., xxviii, 17 ; xxxix, 10 ; Job, xxviii, 19—"of Ethiopia" ; Ez., xxviii, 13 ; Rev., xxi, 20.
[6] xxxvii, 8.
[7] CDB, 960 ; JE, v, 593 f. ; cf. Krause, Pyrgoteles, 62 f. ; Hill, Theophrastus on Stones, 1774, 73—the "topaz of the Antients, now called the Chrysolite . . . has always an admixture of green with the yellow."
[8] JE, v, 593 f.
[9] Hoefer, H. Min., 292.
[10] Ex., xxviii, 17 ; xxxix, 10—"carbuncle," A.V.
[11] Ez., xxviii, 13.
[12] Rosenmüller, IV, i, 28 f.
[13] CDB, 142 ; Hoefer, H. Min., 293.
[14] JE, v, 593 f.
[15] Hoefer, H. Min., 293 ; Lewy, Die

Semitischen Fremdwörter im Griechischen, 1895, 57.
[16] Rosenmüller, IV, i, 28 f. ; Hoefer, H. Min., 293 ; JE, v, 593 f.
[17] xxviii, 13.
[18] CDB, 19.
[19] JE, v, 593 f.
[20] Hoefer, H. Min., 295.
[21] Bĕrâkôt, tr. Cohen, 32.
[22] Charles, Apocrypha, 1913, i, 174 f., 238.
[23] Rev., xxi, 19, only.
[24] Hill, Theophrastus on Stones, 1774, 116 ; CDB, 147.
[25] Ex., xxiv, 10 ; xxviii, 18 ; Job, xxviii, 6, 16 ; Ez., xxviii, 13.
[26] Rosenmüller, IV, i, 36 f.
[27] JE, v, 593 f.
[28] Wiegleb, Alchemie, 1777, 28, quoting Michaelis's translation of Job, and suggesting that "dust of gold" mentioned with sappîr in the same verse, xxviii, 6, is gold-spotted lapis lazuli ; CDB, 824 ; JE, v, 593 ; xi, 52.
[29] Lewy, Fremdwörter, 56.

obvious.[1]   Onyx, also suggested, was not known till a later period, although its name may be derived from the Assyrian unqu, ring, hence elemêšu, ringstone.[2]

Léšem (Septuagint λιγύριον ; Authorised Version ligure) [3] is perhaps the jacinth or hyacinth, a white, grey, yellow, red, reddish-brown or pale green variety of zircon [4], but the ligurium has been supposed to be amber, belemnite, opal, rubellite or tourmaline ; Beckmann suggested hyacinth, which (like amber) becomes electric when rubbed.[5]   The identification of léšem is very doubtful, since all that is known is that it was a golden-yellow lustrous stone (the Midrâsh gives it the tin-white colour of antimony), and both amber and hyacinth would suit the case.[6]

Šĕbô [7] is probably agate[8] ; the translation of kadkôd [9] as agate is probably wrong, and this stone may have been ruby, although the marginal reading is chrysoprase.[10]   The name šĕbô may be derived from the Assyrian šubû, shining, if it is not from some place-name, Ψεφω.[11]

Achlâmâh[12] is probably the amethyst.[13]   The name may be derived from the word chalom, meaning dreams, unless it is from Achlamu, the place where the stone was found.[14]

Taršîš[15], rendered beryl in the Authorised Version without any real reason, is probably the ancient chrysolite or the modern yellow topaz.[16]   Amber has also been suggested, or pale green topaz, but the best rendering is "stone of Tharshish,"[17] which Haupt[18] thought was Spanish cinnabar (mercuric sulphide) from Almaden, shipped from Tartessos.

Sôham in the Septuagint is given as onyx, although emerald has been suggested.[19]   Apparently it was a leek-green stone, and Myres suggests malachite.[20]   The usual identification of sôham[21] is onyx or its variety sardonyx[22], which Josephus[23] distinguishes from onyx, although beryl[24] has been suggested. The name sardonyx is now thought to have the same origin as σανδαράχη (see p. 381), although Kohut[25] derived it from the Persian name for ruby, and others relate it with the Greek σάρξ (flesh).[26]

Yahâlôm, translated diamond on the authority of Ibn Ezra (Abraham ben Meïr ibn Ezra, of Toledo, c. 1090–1107, the author of influential commentaries on the Old Testament)[27], was a precious stone of the king of Tyre[28] and has been supposed to be onyx[29] or emerald[30], the latter being often mentioned in the Authorised Version and Apocrypha.[31]   The stone translated jasper[32]

[1] Hoefer, H. Min., 299.
[2] JE, v, 593 f.
[3] Ex., xxviii, 19 ; Rev., xxi, 20 ; ib., ix, 17, means a dark purple colour.
[4] CDB, 357.
[5] Ib., 472, 902 ; Hoefer, H. Min., 296.
[6] JE, v, 593 f. ; Rosenmüller, IV, i, 28 f.
[7] Ex., xxviii, 19 ; xxxix, 12.
[8] CDB, 19 ; Rosenmüller, IV, i, 28 f.
[9] Is., liv, 12 ; Ez., xxvii, 16.
[10] CDB, 18.
[11] JE, i, 230 ; v, 593 f.
[12] Ex., xxviii, 19 ; ἀμέθυστος in N.T., Rev., xxi, 20.
[13] Rosenmüller, IV, i, 28 f. ; CDB, 38.
[14] Hoefer, H. Min., 298 ; JE, v, 593 f.
[15] Ex., xxviii, 20 ; xxxix, 13 ; Cant., v, 14 ; Ez., i, 16 ; x, 9 ; xxviii, 13 ; Dan., x, 6.
[16] Hoefer, H. Min., 298 ; Rosenmüller, IV, i, 28 f. ; CDB, 108, 162 ; A.V. margin.
[17] JE, v, 593 f.
[18] MGM, i, 386.
[19] CDB, 656.
[20] JE, v, 593 f. ; cf. Bochart, Geographia sacra, 19.
[21] Gen., ii, 12 ; Ex., xxviii, 9, 10 ; I Chron., xxix, 2 ; Job, xxviii, 16 ; Ez., xxviii, 13.
[22] Rev., xxi, 20.
[23] War, V, v, 7 ; Hoefer, H. Min., 298.
[24] Rosenmüller, IV, i, 28 f. ; CDB, 656, 825 ; HDB, iii, 624.
[25] JE, v, 593 f.; on the derivations of names of gems, see H. Lewy, Semitischen Fremdwörter, 1895, 53–62.
[26] Hoefer, H. Min., 292.
[27] Sarton, Intr. to the Hist. of Science, 1931, ii, 187.
[28] Ez., xxviii, 13.
[29] Rosenmüller, IV, i, 28 f.
[30] JE, v, 593 f. ; CDB, 204, 902—jasper.
[31] Ex., xxviii, 18 ; xxxix, 11 ; Ez., xxvii, 16 ; xxviii, 13 ; Ecclus., xxxii, 6 ; Judith, x, 21 ; Rev., iv, 3 ; xxi, 19 ; Tobit, xiii, 16.
[32] Ex., xxviii, 20 ; xxxix, 13 ; Rev., iv, 3 ; xxi, 11, 19.

has also been supposed to be diamond.[1] The diamond was probably unknown in antiquity and it has been suggested that yahălôm should be read yâšĕphâh, of Egyptian origin, meaning jasper[2] ; kadkôd in Isaiah [3] is rendered ἴασπις in the Septuagint. According to Fraas, the jasper of the Bible was opal, which is found in Egyptian tombs, including statuettes.[4] Diamonds are mentioned in the Mishnâh and Talmûd [5] and were perhaps known to the earlier Jews.[6]

Chrysoprase [7] is not the true chrysoprase but a similar stone found in Egyptian jewellery.[8] The stone 'ekdâch [9], "carbuncle" in the Authorised Version, is some unknown brilliant gem.[10] The turquoise is not in the Old Testament, although it is said to occur in Sinai[11], but is mentioned in the Targums.[12] The word kerach, often used for ice and frost, is once used, apparently, for rock crystal[13], which is more probable than glass.[14] The pearl probably occurs once in the Old Testament[15] as gâbîš[16], which literally means ice, e.g. in gâbîšel abnî (hailstones, "ice-stones").[17] In the New Testament pearls often occur[18] : the "pearl of great price" is probably a true pearl from the pearl oyster (Avicula margaritifera) of the Persian Gulf.[19] Bochart[20] supposed penîyyim, penînîm[21], to mean pearls, but others read coral[22] ("pearls" and "rubies" occur together in the Authorised Version of Job). Bochart supposed that bedôlach[23] (usually given as bdellium) and dar were names for pearls. Pearls are frequently mentioned in the Talmûd.[24] Coral is supposed to be râmôth, only in Job[25] and Ezekiel [26], and the translation is supported by etymology[27]; râmôth may be black, and penînîm red, coral.[28] Coral (almug), is mentioned in the Mishnâh.[29]

Enoch[30] saw in a vision "towards the south . . . seven mountains of magnificent stones" (λίθων πολυτελῶν, very costly, in Akhmîm text), which occur also in the Book of Jubilees.[31] One was of coloured stone (λίθου χρώματος, Akhmîm), one of pearls and one of jacinth (in the Akhmîm text λίθου ταθεν, formerly thought to be ἀχάτης, agate, but now regarded as a corruption of ἰακίνθου).[32] The Ethiopic text gives "stone of healing," perhaps an emendation of ταθεν into ἰατρου ; ταθεν is usually translated "antimony," but the Akhmîm text gives λίθου φουκά, Hebrew pûkh[33], a red cosmetic (see p. 496). The mountains towards the south were of red stone (λίθου πυρροῦ, Akhmîm) ; the middle one, like a throne, was of alabaster, with a summit of "sapphire" (λίθου σαπφείρου, Akhmîm).

There are no references to the imitation of gems by coloured glass in the Bible, but the art of making false gems was well known to the later Jews.[34]

---

[1] CDB, 364, 902.
[2] Rosenmüller, IV, i, 28 f.
[3] liv, 12.
[4] JE, v, 593 f.
[5] Aboth, v, 6 : šâmîr.
[6] Bertholet-Dallas, 177 f.
[7] χρυσόπρασος, only in Rev., xxi, 20.
[8] C. W. King, Natural Hist. of Precious Stones and Gems, etc., 1865, 163 ; see p. 49.
[9] Only in Is., liv, 12.
[10] CDB, 142.
[11] R. P. M. Jullien, Sinaï et Syrie, Lille, 1893, 75—an inferior kind and perhaps malachite ; see pp. 50, 60.
[12] JE, v, 593 f.
[13] Ez., i, 22 ; κρύσταλλος in Rev., iv, 6 ; xxii, 1.
[14] CDB, 180.
[15] Dakin, Pearls, Cambridge, 1913, 1.
[16] Job, xxviii, 18, with râmôth, coral ?
[17] Ez., xiii, 11, 13 ; xxxviii, 22.
[18] Matt., xiii, 45 ; I Tim., ii, 9 ; Rev., xvii, 4 ; xxi, 21.
[19] CDB, 703 ; Dakin, 1.

[20] Hierozoicon, ii, 681 f. ; Rosenmüller, IV, ii, 456.
[21] Job, xxviii, 18 ; Prov., iii, 15 ; viii, 11 ; xxxi, 10 ; Lam., iv, 7 ; A.V. "rubies."
[22] CDB, 799 ; Bochart, Hierozoicon, i, pref. sign. g 4 verso ; ii, 673 f., 693, 708 ; ib., Geographia Sacra, 30 ; Hirsch, JE, iv, 261 ; Nowack and Sel, ib., ix, 569.
[23] Num., xi, 7.
[24] Aboth, v, vi; Sabbath, xvi; Taanith, iii.
[25] xxviii, 18.
[26] xxvii, 16.
[27] CDB, 173 ; Hirsch, JE, iv, 261 ; Nowack and Sel, JE, ix, 569 ; E. Bibl., i, 895.
[28] Bochart, Hierozoicon, ii, 681 f., 702 f.
[29] Kelim, xiii, 6.
[30] xviii, xxiv ; Charles, 1893, 89, 97, 353.
[31] Cap. viii ; 135–105 B.C. ; Charles, Apocrypha, 1913, ii, 26.
[32] Charles, Apocrypha, 1913, ii, 200.
[33] Charles, Enoch, 1893, 354.
[34] JE, v, 595 ; S. Krauss, Griechische und lateinische Lehnwörter im Talmud, Midrasch und Targum, 1898–9, ii, 132.

ORGANIC MATERIALS

HEBREW AGRICULTURE

Before Solomon's time (c. 970 B.C.) the Jews were purely agricultural and pastoral ; trade slowly commenced about then. Even Ezekiel (6th century B.C.) speaks of agricultural exports to Tyre [1] and the Jews became important traders only after the Babylonian exile (c. 550 B.C.).[2] The flora of Palestine is very extensive ; 126 orders, 850 genera and 3500 species have been listed.[3] The Mishnâh gives 230 plant names, 180 old Hebrew and 40 from the Greek ; the Jerusalem Talmûd gives 100, and the Babylonian Talmûd 175, of which 20 are from the Greek.[4] Varieties of oaks are very common in Palestine, and other common trees are pistacia, carob (Ceratonia Siliqua), plane, sycamore fig, poplar, walnut (more common in Syria than Palestine), arbutus, cypress, styrax officinalis, tamarisk, many kinds of shrubs, vine (especially in south Palestine, and producing large grapes, especially at Eshcol), olive, fig, quince, apple, almond, apricot, pomegranate, orange, prickly pear, cistus, etc. There are over 2,000 flowering plants, and the papyrus plant grows abundantly in western Syria.[5] In the following account the trees and plants are arranged according to their uses and properties ; the modern botanical classification was, of course, quite unknown to the ancients, and although it is adopted by Löw it leads to some irregularities in the historical survey.

EDIBLE CROPS

Fallows [6] and perhaps rotation of crops were used.[7] Cereal crops were wheat and barley, more rarely rye and millet (?) [8], the first two, with the vine, olive and fig, and the use of the plough, harrow and irrigation, being mentioned in Job.[9] The Jews were important producers of wine and oil in antiquity.[10] Staple produce were two kinds of cummin (the black called "fitches" in the Authorised Version)[11], beans and lentils ; later writers mention peas, kidney beans, lettuce, endive, leeks, garlic, onions, melons, cucumber, cabbage, etc. Animal manure was used.[12] Lentils ('ădâšîm), bean (pol), cucumbers (qiššûîm), melons (ăbattichîm), wild gourds (pl. pĕqâ'îm), onions (bĕzâlîm), leek (hâzîr), garlic (šûm), bitter herbs (mĕrôrîm), almond (šâqêd), quince ("apple" in Authorised Version) (tappûach), date (dĕbaš), fig (tĕênâh), sycamore figs (šiqmîm), mulberry (baca), nuts (botnîm), olive (zayîth), pomegranate (rimmôn), vine (gephen), walnut ('ĕgôz), mallow (mallûah) and juniper root (rôthem) are named as vegetable foods.[13] ; truffles are in the Mishnâh and Talmûd.[14]

[1] Otto, 1925, 46.
[2] Meyer, Alt., II, ii, 268 ; J. Newman, The Agricultural Life of the Jews in Babylonia between the Years 200 C.E. and 500 C.E., 1933—not seen.
[3] CDB, 675 f. ; Löw, JE, x, 72 f. ; G. E. Post, Flora of Syria, Palestine and Sinai, Beirut [1896]—an enlarged edition of 1932 I have not seen.
[4] Löw, JE, x, 77.
[5] H. B. Tristram, Flora and Fauna of Palestine, Survey of Western Palestine, 4to, 1884—lists only ; ib., Natural History of the Bible, 1867 ; Post, Flora, 1896 ; British Museum (Nat. Hist.) Guide to the Exhibition of Animals, Plants and Minerals mentioned in the Bible, Special Guide No. 5, 1911, 23 f. ; Rosenmüller, IV, 71–329 ; Olaf Celsius, Hierobotanicon, sive de plantis sacræ scripturæ dissertationes breves, 2 vols., Uppsala, 1745–7 ; Ursinus, Arboretum

Biblicum, Norimbergæ, 1663 ; ib., Historiæ Plantarum Biblicæ, sive (1) De Sacra Phytologia, (2) Herbarius sacer, (3) Hortus Aromaticus, Norimbergæ, 1665 ; ib., Silva Theologiæ Symbolicæ, Norimbergæ, 1665 ; I. Löw, Flora der Juden, Vienna, I, i and ii (1926–8), II, III (1924), IV (1934) ; ib., JE, x, 72 f.—an excellent summary.
[6] Jer. iv, 3 ; Hos., x, 12.
[7] CDB, 22.
[8] Hoefer, H. Bot., 1872, 2, says this is really Holcus Sorghum L.
[9] xxxi, 40 ; xv, 33 ; xxiv, 6 ; xxix, 19 ; xxxix, 10.
[10] JE, i, 262 ; ii, 152 ; viii, 596 ; Benzinger, Hebr. Archäol., 1907, 139 f.
[11] Is., xxviii, 27.
[12] CDB, 21, 22, 96.
[13] Macalister, HDB, ii, 27 f. ; E. Bibl., i, 965 ; ii, 1540 f.
[14] Uktzin, iii, 2 ; Berakoth, vi, 3.

The most important crop was wheat (chittâh ; Aramaic chintîn), then barley (se'ôrâh), spelt (kussemet) [1], oats (šîfôn ; not much grown), millet (dôchan), bean (pol), lentils ('ădâšîm) and flax (pištah) ; cotton (karpas) was only later cultivated.[2] The curious identification of Serapis with Joseph [3] arose from the fact that wheat shipped from Alexandria to the Serapaion in Ostia was under the tutelage of Serapis, the god with the corn measure on his head, who suggested Joseph, the seller of corn.[4] Sesame is šumšum.[5] Emmer (split) wheat grows wild in Palestine.[6] Zêr'ôîm and zêr'ônîm [7] mean any uncooked grain ("seeds").[8] Rice is not in the Old Testament and was cultivated only in a later period ; it is mentioned in the Mishnâh and Talmûd.[9]

BREAD

Bread-making, which was practised very early in Egypt, implies a certain level of culture. The inhabitants of the Canary Islands ate flour cooked with meat or butter, not as bread, and the Romans before they used bread made the grain into a kind of boiled soup (pulmentaria).[10] Besides being made into bread by the Jews, corn was eaten green, or the picked grains were roasted into parched corn in a pan over a fire.[11] Otherwise the grain was bruised and dried in the sun, and eaten or offered mixed with oil[12], or made into a soft cake.[13]

Lechem[14] really means "food," but is often used for bread or cakes[15], a staple food[16], the best being made from wheat ground to flour or "meal,"[17] and also sifted as "fine flour,"[18] generally used in the sacred offerings or in meals of the rich.[19] This fine flour was also made up in a pan with honey and oil[20], perhaps as in Babylonia[21], the "marzipan" so formed being called châbîz or rihătă in the Talmûd.[22] Barley was used only by the poor[23] or in times of scarcity[24] : it was made into bread after mixing with wheat, beans, lentils, millet, etc.[25], and also baked into cakes[26], and used as fodder for horses,[27] and it had a low value as compared with wheat.[28] The mixed bread is called "barley cakes."[29] Spelt (rye, fitches or spelt, in the Authorised Version) was used in Egypt[30] and Palestine.[31] From Solomon's time Palestine

[1] Fitches (vetches) in A.V. of Ex., ix, 32 ; Is., xxviii, 25 ; Ez., iv, 9 ; "rye, oats, or spelt (Triticum spelta)"—CDB, 271, 800 ; E. Bibl., ii, 1540 f. ; HDB, ii, 28.
[2] CDB, 96, 463, 558, 1005 ; JE, i, 267 ; Löw, Flora, i, 658 f., 686, 707 ; Macalister, HDB, i, 316.
[3] Rollin, Anct. Hist., tr. Bell, Glasgow, 1845, ii, 303.
[4] JE, i, 264.
[5] Löw, Flora, iii, 1 f.
[6] Breasted, Ancient Times, 1916, 38.
[7] Plurals in Dan., i, 12, 16—"pulse," A.V.
[8] CDB, 767.
[9] Mishnâh, tr. Danby, 19, 21, 41, 83, 86, 87, 774 ; Barclay, The Talmûd, 1878, 65 ; Löw, Flora, i, 730 f.
[10] Pliny, xviii, 8 ; Hoefer, H. de la Chimie, i, 40.
[11] Lev., ii, 14 ; xxiii, 14 ; Deut., xxiii, 25 ; II Kings, iv, 42 ; Ruth, ii, 14 ; I Sam., xvii, 17 ; xxv, 18 ; II Sam., xvii, 28 ; Benzinger, Hebr. Archäol., 62 f. ; Macalister, HDB, ii, 27 ; E. Bibl., ii, 1538.
[12] Lev., ii, 14 f.
[13] Num., xv, 20 ; Neh., x, 37 ; Ez., xliv, 30.

[14] Gen., iii, 19 ; A.V., bread ; Vulgate, pane.
[15] Gen., xviii, 6.
[16] CDB, 272 ; Macalister, HDB, i, 315 ; Kennedy, E. Bibl., i, 604.
[17] Judg., vi, 19 ; I Sam., i, 24 ; I Kings, iv, 22 ; xvii, 12, 14.
[18] Gen., xviii, 6 ; Ex., xxix, 2 ; I Kings, iv, 22.
[19] Ex., xxix, 40 ; Lev., ii, 1 ; I Kings, iv, 22 ; II Kings, vii, 1, 18 ; Ez., xvi, 13, 19 ; xlvi, 14 ; cf. N.T., Rev., xviii, 13.
[20] Ez., xvi, 13, 19 ; Lev., ii, 7 ; vii, 9.
[21] Meissner, Bab. und Assyr., i, 415 ; Kennedy, E. Bibl., i, 460.
[22] Berakoth, vi, 1 ; tr. Cohen, 247, 250.
[23] Lev., xxvii, 16 ; Deut., viii, 8 ; Ruth, ii, 17 ; Judg., vii, 13 ; II Kings, iv, 42 ; John, vi, 9, 13.
[24] Ruth, iii, 15, cf. i, 1 ; II Kings, iv, 38, 42 ; Rev., vi, 6.
[25] Ez., iv, 9 ; II Sam., xvii, 28.
[26] Ez., iv, 12.
[27] I Kings, iv, 28.
[28] CDB, 96.
[29] Ez., iv, 12.
[30] Ex., ix, 32.
[31] Is., xxviii, 25 ; Ez., iv, 9.

exported grain to Tyre.[1] The grain was ground in the house by means of saddle-querns (rechaîm) consisting of two stones [2], or in a stone (sometimes basalt) mortar (mĕdôkâh ; maktĕš) and pestle (ĕlî).[3] In spite of frequent descriptions and representations to the contrary, the mill of two circular stones, the upper pivoted on the lower, was not used, the arrangement being a large lower saddle-shaped stone with a plane or slightly concave upper surface, and a riding stone about 12 in. long and 4–6 in. wide, with a flat oval base and round back, rubbed backwards and forwards by one woman whilst another woman sprinkled the grain on the upper surface of the saddle-stone.[4] The ground grain was then passed through a sieve (nâpâh).[5]

Baking at home was done by women, in bakeries by men ; in Palestine the bakeries were in one quarter of the town, and no doubt had fixed ovens [6], but in domestic baking a pot was used, heated inside by a fire of wood, dry grass or dung, and on the surface of the heated pot the flat cakes were spread, as in Egypt.[7] The cakes could also be laid on hot stones or on embers [8] or on a dung fire.[9] The flour was kneaded to dough, leaven (sour dough) being usually added ; the dough was then divided into flat cakes, sometimes punctured and mixed with oil ; or it was rolled into wafers and coated with oil (to prevent sticking to the oven ?).[10] The leaven (sĕôr) was usually a piece of old fermenting dough, the use of which, on account of its "corruption" (also mentioned by Greek authors) was omitted in offerings or, on account of the time necessary for its formation, in hasty baking.[11]

Cooking pans were the flat frying pan (mǎchabath ; masrêth) and the cauldron (sîr) for boiling, set on three stones.[12]

Some Edible Plants

The cabbage (kĕrûb ; Sanskrit karambhâ ; Greek κράμβη) is in the Mishnâh and Talmûd[13] ; Egyptian cucumbers (qiššuîm)[14], now growing round Cairo, the gherkin (Cucumis sativus), the melon (pl. ǎbattichîm[14], melopepon in the Mishnâh) and the water-melon (Arabic batêkh) all now grow in Palestine.[15] The beet was a foreign plant in Syria, where it appears about 800 B.C. ; it is not mentioned in the Old Testament, but (with beet juice) is in the Mishnâh[16] (tĕrâdin) and Talmûd (silqa) and now grows in Palestine.[17] Châtsîr may mean the leek, grass or an edible plant like clover (Trigonella fœnum Græcum)[18] ; 'ădâšîm may mean lentils (Arabic adas)[19] ; betsâlîm are onions (Assyrian bisru) and shuni is garlic (Allium sativum), long known in China, where it is called suan.[20] Mallûach[21] is probably some species of Orache, say the Atriplex halimus, although nettles and some species of mallow (Jew's mallow, Corchorus

[1] CDB, 176.
[2] Ib., 558; Macalister, HDB, i, 317; Carslaw, ib., iii, 369 ; Kennedy, E. Bibl., iii, 3091.
[3] CDB, 579 ; Kennedy, E. Bibl., iii, 3201 ; Handcock, Holy Land, 132, 165.
[4] Macalister, Century, 233.
[5] Kennedy, E. Bibl., iii, 3095.
[6] Neh., iii, 11 ; xii, 38 ; "tower of the furnaces," A.V.
[7] CDB, 122 ; Kennedy, E. Bibl., i, 459, 604.
[8] I Kings, xix, 6.
[9] Ez., iv, 12, 15.
[10] CDB, 122 and refs. ; Macalister, HDB, i, 318.
[11] Ex., xii, 20 ; xiii, 3 ; CDB, 122, 461 ; White, HDB, iii, 90 ; E. Bibl., iii, 2752 ; Aristotle, De animal. gen., iii, 4.

[12] CDB, 680 ; Macalister, HDB, i, 318 ; Kennedy, E. Bibl., i, 887, on cooking.
[13] Löw, Flora, i, 483 ; ib., JE, x, 77 ; Mishnâh, tr. Danby, 28, 49, 64, etc.
[14] Only in Numb., xi, 5.
[15] CDB, 181, 541 ; Löw, Flora, i, 530, 535 ; Post, HDB, i, 531 ; iii, 337 ; E. Bibl., i, 965 ; iii, 3018.
[16] Tr. by Danby, 28, 64, 92, 786.
[17] Berakoth, vi, 1 ; ix, 1 ; tr. Cohen, 240, 260, 378 ; Lippmann, Gesch. d. Rübe, 1925, 35 ; Löw, Flora, i, 346 ; ib., JE, ii, 638 ; x, 77 ; Post, Flora, 679.
[18] CDB, 462.
[19] Gen., xxv, 34 ; II Sam., xvii, 28 xxiii, 11 ; Ez., iv, 9 ; CDB, 463.
[20] Numb., xi, 5 ; CDB, 655 ; Löw, Flora, ii, 126 ; Vigouroux, Dict. de la Bible, iv, 1761.
[21] In Job, xxx, 4, only.

olitorius) have been suggested.[1] Neither the Greek ἀσπάραγος nor the asparagus in the Talmûd [2] means the modern plant, but is the name given to young shoots of various plants. Ašišah [3] means a cake of compressed raisins.[4] Grapes (usually dried, as raisins), lentils, beans, leeks, onions, garlic [5] and numerous spices and condiments were used as food and many plants were grown in kitchen gardens.[6]

TREES YIELDING EDIBLE PRODUCTS

The word šâqêd means the almond tree or its fruit.[7] The word lûz [8], "hazel" in the Authorised Version, also means the almond [9], as in Arabic ; the name šâqêd is from a root meaning "to hasten," as the tree flowered very early.[10] The "almond bowls" of the golden candlestick[11] were really of rock crystal.[12] The tappûach tree and its fruit are given as apple in the Authorised Version[13], and this translation has been supported[14], although the quince, apricot and citron are sometimes preferred.[15] The apricot was not known before Roman times, when it was introduced, perhaps from China, and the quince is the best identification.[16] The orange, which now grows in Palestine (and on the lower slopes of the Himalayas), would suit the text[17], but was probably not known in ancient times ; ětrôg is the citrus or citron[18], which is in the Mishnâh.[19]

The pomegranate (rimmôn ; Arabic rummân) is frequently mentioned in the Old Testament ; the Hebrews probably got to know it in Egypt and it is now common in Aleppo.[20] The "sycamine" and "sycamore," only in Luke[21], mean the mulberry tree (morus), both the black and white varieties of which are common in Syria and Palestine.[22] It is not mentioned in the Old Testament, beka'[23] not being the mulberry[24], the post-Biblical name for which is tût.[25]

Nuts (botnîm) are probably of the Pistacia vera, a tree which grows in Syria and Palestine ; ěgôz is probably the walnut[26], which Josephus[27] says was formerly a common tree, i.e. the Juglans regia, Linn., the Aramaic gauzâ. Walnuts are used to prepare a hair-dye in the Mishnâh, which also mentions nut oil.[28]

The only oil used by the ancient Hebrews was that of the olive[29], the wild olive being probably called šemen.[30] The tree and olive oil (also called šemen) were well known. The olives were crushed in stone presses, smaller than wine presses[31], with heavy stone rollers or with pyramidal stone weights tied to a

[1] CDB, 506 ; Bochart, Hierozoicon, i, pref. sign. c 4.
[2] Berakoth, vii, 4 ; tr. Cohen, 326.
[3] II Sam., vi, 19 ; I Chron., xvi, 3 ; Cant., ii, 5 ; Hos., iii, 1—A.V. "flagon."
[4] CDB, 271 ; Löw, Flora, i, 82.
[5] Num., xi, 5.
[6] CDB, 272, 280 ; Rosenmüller, IV, i, 211 ; Löw, Flora, i, 48 f.
[7] Gen., xliii, 11 ; Ex., xxv, 33, 34 ; xxxvii, 19, 20 ; Num., xvii, 8 ; Eccles., xii, 5 ; Jer., i, 11 ; Celsius, i, 297 ; E. Bibl., i, 116.
[8] Only in Gen., xxx, 37.
[9] Celsius, i, 253.
[10] Post, HDB, i, 67 ; ii, 313.
[11] Ex., xxv, 33.
[12] CDB, 33.
[13] Cant., ii, 3, 5 ; vii, 8 ; viii, 5 ; Joel, i, 12 ; Prov., xxv, 11.
[14] Celsius, i, 254 ; E. Bibl., i, 267 ; Kohler, JE, ii, 23 ; Löw, Flora, iii, 212 ; ib., JE, x, 73.
[15] CDB, 55.
[16] E. Bibl., i, 267 ; cf. Adams, Paulus Ægineta, 1844–7, i, 134, 137 ; iii, 250.

[17] Rohde, Garden-Craft in the Bible, 1927, 36.
[18] CDB, 55 ; Löw, Flora, iii, 278 ; ib., JE, v, 261.
[19] Sukkah, iii, 4 f. ; iv, 7, 9 ; etc.
[20] CDB, 745 ; pomum granatum = "grained apple"; malus punica ; Celsius, i, 271 ; Löw, Flora, iii, 80 f.
[21] xvii, 6 ; xix, 4.
[22] CDB, 906.
[23] Sam., v, 23.
[24] Benzinger, JE, ix, 104.
[25] Löw, Flora, i, 266.
[26] Only in Cant., vi, 11.
[27] War, III, x, 8.
[28] CDB, 637 ; Benzinger, JE, ix, 363 ; Löw, Flora, ii, 29, 46.
[29] Löw, JE, ix, 392.
[30] Neh., viii, 15 ; "pine" in A.V. ; CDB, 737 ; Celsius, ii, 271.
[31] Handcock, Holy Land, 62 ; Macalister, Century, 234 ; Bellwood, J. Soc. Chem. Ind., 1922, 213 R.

wooden lever engaged in a hole in a wooden standard, and the oil was used as food, for lamps, for anointing the body, medicinally (externally) and for dressing leather.[1] Oils said to be used for lamps in the Mishnâh and Talmûd are olive oil, nut oil, fish oil and oils of gourds and coloquinta, as well as naphtha.[2] The oil tree (êts šemen) [3] is perhaps the Balanites Ægyptiaca, the Arabic zaqqûm (the oil being zûq).[4] Qiqâiôn (qiqâyôn) [5] is probably the castor oil plant (Ricinus communis), the Egyptian qiqi [6]; the oil (qiqšemen, the plant being qiq) appears first in the Mishnâh and Talmûd.[7]

The fig tree and fig (têênâh, Aramaic tînâ, Arabic ti'n, Assyrian tittu, Egyptian teb) are frequently mentioned in the Old Testament and the tree is very common in Palestine, figs being still gathered on Mount Olivet. Figs, as distinct from the tree, are often named in the plural (têênîm) ; other names are bikkûrah (first ripe fruit of the fig tree), pag (unripe fig) and dĕbêlâh, a cake of compressed figs.[8] Caprification [9] was early in use.[10] According to one tradition in the Talmûd, the fig tree grew in Paradise and its fruit was gathered by Eve, although another made this tree the vine.[11] The šiqmâh[12] is really the fig-mulberry or sycamore fig (Ficus sycamorus).[13] Abd al-Latîf (A.D. 1203) says the "male" sycamore grows in the Jordan valley and has a black astringent fruit.[14] The suggestion[15] that "fig leaves" were banana (a Malay plant, unknown to the Egyptians) is improbable.[16]

The palm (tâmâr ; Aramaic diqla) included the date palm (Phœnix dactylifera, Linn.), regarded by the ancients as peculiarly characteristic of Palestine and neighbouring regions, and often mentioned in the Bible, but now comparatively rare except in the Philistine plain and old Phœnicia near Beyrut.[17] There is no clear allusion in the Bible to the production of date honey or date wine (lakbî), but the dates were used as food and the "honey" frequently mentioned may include date honey.[18] Josephus says dates yield excellent "honey" when pressed.[19] Date honey is mentioned in the Mishnâh[20] and Talmûd[21]; it is the Arabic dibs.[22] The date palm is called the "tree of life" in the Apocrypha[23], but not in Genesis.[24]

Although the sugar cane is now extensively cultivated in Syria[25], the supposed references to it ("sweet cane") and to sugar in the Old Testament[26] are misinterpreted, since châsâb is not the sugar cane.[27] Two places in the Talmûd are supposed to mention sugar (sîqôrâ), which otherwise does not occur in the Bible, Apocrypha or Mishnâh[28], and first appears definitely among the

[1] Ex., xxv, 6 ; xxx, 24 ; etc. ; CDB, 641, 651 ; E. Bibl., iii, 3466 ; Macalister, HDB, iii, 590 ; Post, ib., 616.
[2] Sabbath, ii, 1 f.
[3] Neh., viii, 15 ; I Kings, vi, 23 ; Is., xli, 19 ; Celsius, i, 309 ; Post, HDB, iii, 592.
[4] CDB, 641, 676 ; cf. E. Bibl., iii, 3471.
[5] Only in Jonah, iv, 6 f. ; "gourd" in A.V.
[6] Bochart, Geographia Sacra, 918 ; Celsius, i, 273 ; CDB, 299 ; Löw, Flora, i, 219.
[7] Löw, op. cit. ; JE, x, 73, 82.
[8] II Kings, xx, 7.
[9] Theophrastos, De causis plant., V, ii, 4.
[10] CDB, 269 ; Löw, Flora, i, 224 f., 233 f.
[11] Berakoth, vi, 2.
[12] I Kings, x, 27 ; I Chron., xxvii, 28 ; Ps. lxxviii, 47 ; Am., vii, 14, bôtês šiqmin ; sycamine in the Septuagint.
[13] CDB, 906 ; Löw, Flora, i, 274 ; Ursinus, Arboretum Biblicum, 1663, 455 ; E. Bibl., ii, 1519 ; iv, 4831 ; Post, HDB, ii, 5.
[14] Reitemeyer, Beschreibung Ägyptens, 67.

[15] Hoefer, H. Bot., 7.
[16] E. Bibl., ii, 1520.
[17] CDB, 679, and refs. ; Levi, JE, ix, 505.
[18] II Chron., xxxi, 5—margin "dates" ; Bochart, Hierozoicon, pref. sign. c 3 verso ; CDB, 680.
[19] War, IV, viii, 3 ; CDB, 331, 680 ; Bochart, Hierozoicon, ii, 518 ; Lippmann, Gesch. des Zuckers, 1929, 662.
[20] Terumoth, xi, 2, 3 ; Nedarim, vi, 8, 9.
[21] Berakoth, vi, 1.
[22] Bochart, Hierozoicon, i, pref. sign. c 3 verso, g 1 verso.
[23] Enoch, xxiv, 4.
[24] Rohde, Garden-Craft in the Bible, 1927, 21 ; cf. Sarton, Isis, 1934, xxi, 8.
[25] Post, Flora, 849.
[26] Is., xliii, 24 ; Jer., vi, 20 ; Hoefer, H. Bot., 4 ; Falconer, Mem. Manchester Lit. and Phil. Soc., 1793, iv, 291.
[27] Löw, Flora, i, 746, 751 ; ii, 690 f.
[28] The σάκχαρον in Josephus, Antiq., III, vii, 6, is said to be the Aramaic šakrma, hyoscyamus ; Löw, JE, x, 73.

Babylonian Jews of the 8th–9th centuries A.D. The sugar cane (kanjâ dešak-kar) and sugar (šakkar) are in the Hălâchôth gĕdôlôth of Rabbi Simon of Cairo (c. A.D. 900) as novelties. From this time they are repeatedly mentioned as sugar-cane juice, solid sugar and preserved ginger (not known in antiquity), and sugar for medicinal use is often mentioned by Maimonides (d. 1204).[1] Preserved ginger is in the Talmûd.[2] Manna (mân) [3] has already been fully considered.[4] Pilter [5] considered that the manna of the Israelites was a mixture of tamarisk gum and the manna lichen, both of which have been separately identified with it.[6] According to Jullien [7] the Sinai manna is tamarisk manna produced by a coccus ; not much (600 lb. per annum) is collected as compared with the manna from Alhagi (a leguminous plant about 2 ft. high, almost without leaves), of which 2,000 lb. are annually collected in Herat and Kandahar and sent to North India.

WINE

The invention of wine was regarded by the Jews as one made in the remote past, since wine was used by Noah and bread and wine were primitive offerings.[8] In the Talmûd [9] wine must be at least forty days old in order to be "strong drink" as required by the law[10], although when drunk it was diluted with water. The crushing of seeds and fruits to produce, according to their character, flour, oil and must probably dates from the same period.[11] Besides grape juice, the juice of the pomegranate (rimmôn, Arabic rummân, Punica granatum) was also fermented.[12]

The usual name in the Old Testament for intoxicating drink is šêkar (schek-kar ; σίκερα in the Septuagint, sicera in the Vulgate), which usually means wine, but sometimes perhaps beer.[13] Wine from grapes is commonly yáyin (jajin), perhaps a foreign word, either related to the Armenian voino (Greek oîνος), i.e. non-Semitic[14], or from the Assyrian înu (Arabic wain).[15] Eight other names are also used[16], e.g. chémer (Aramaic hamra), sobî, tîrôš and mések (in the Mishnâh).[17]

In later Hebrew literature many varieties of wine and other intoxicating drinks and mixtures of wines are described.[18] St. Jerome and other writers say the Jews of later times drank beer made from barley and flavoured with lupin and skirret, also cider (apple wine in the Mishnâh)[19], two kinds of honey

---

[1] Löw, Flora, i, 740, 754 ; iv, 148 f., 177, 185, 208, 227 f., 518 ; ib., JE, x, 81, 84 ; ib., Chem. Zeit., 1927, li, 15 ; E. Bibl., ii, 1543 ; Lippmann, Gesch. des Zuckers, 1929, 216 ; ib., Z. d. Vereins d. Deutsch. Zucker-Ind., 1934, lxxxiv, 846.
[2] Sabbath, vi ; Berakoth, vi, 1, tr. Cohen, 246—"from India."
[3] Ex., xvi, 14–36 ; Num., xi, 7–9 ; Deut., viii, 3, 16 ; Jos., v, 12 ; Ps. lxxviii, 24, 25 ; Wisd., xvi, 20, 21 ; Josephus, Antiq., III, i, 6 ; Rosenmüller, IV, 316 f. ; CDB, 512 ; JE, viii, 292 ; E. Bibl., iii, 2929 ; Macalister, HDB, iii, 236.
[4] See p. 162, also Löw, Flora, iv, 97, 415, 452, and Vigouroux, Dict. de la Bible, i, 367 ; iv, 656.
[5] PSBA, 1917, xxxix, 155, 187.
[6] See also Bochart, De variis mannæ speciebus, Opera Omnia, 1712 ; Geographia sacra, 871 f. ; ib., Hierozoicon, ii, 519 ; Salmuth, in Pancirollius, Nova Reperta sive Rerum Memorabilium, Ambergæ, 1602, 297 f. ; C. Niebuhr, Description de l'Arabie, Copenhagen, 1776, 128 f.

[7] Sinaï et Syrie, 68.
[8] Gen., ix, 21 ; xiv, 18.
[9] Berakoth, iv, 1 ; vii, 4 ; tr. Cohen, 177, 322.
[10] Num., xxviii, 7.
[11] Hoefer, H. de la Chimie, i, 43 ; Goguet, i, 108, 118, 125.
[12] Cant., vii, 2 ; CDB, 213, 745 ; Löw, Flora, i, 94 f. ; V. Zapletal, Der Wein in der Bibel, Biblische Studien, Bd. xx, Heft 1, Freiburg i. B., 1920, 51.
[13] Zapletal, 50 f. ; CDB, 213 ; Bertholet-Dallas, 184 f.
[14] Zapletal, 1 f. ; Macalister, HDB, ii, 33 ; E. Bibl., iv, 5306 ; JE, xii, 532.
[15] Richter, A. Nat., iv, 438 ; cf. Hoefer, H. de la Chimie, i, 42, who connected it with a Hebrew word meaning effervescence.
[16] Tristram, Nat. Hist. of the Bible, 411 ; Benzinger, Hebr. Archäol., 71.
[17] Zapletal, 8 f. ; E. Bibl., iv, 5306 ; Tristram, op. cit.
[18] Eisenstein, JE, xii, 532.
[19] Berakoth, vi, 1 ; tr. Cohen, 253 ; Terumoth, xi, 2 ; Nedarim, vi, 9.

33

wine (mixtures of wine, honey and pepper—pilpêl) [1], decoctions of grape juice or must (tîrôš or 'asîs) [2] called děbas (Syrian dibs), date wine made by macerating the fruit in water and fermenting, and various other drinks fermented from fruits and vegetables, perhaps including raisins buried in jars with water, as is done by the Arabs. A kind of sherbet was made from fig cake and water [3], and drinks may have been cooled by snow from Mount Hermon.[4]

In making wine the grapes were pressed, either by treading or in deep stone presses, very old specimens of which, for use with grapes and olives, were found at Gezer.[5] The lees (šemer) or dregs of wine [6], from which it was strained or "well refined" [7] before use, might be yeast [8], or perhaps tartar. "Wine on the lees" could mean old wine which had deposited tartar, or perhaps well-fermented wine. The wine was strained or filtered before use.[9] Beer is mentioned in the Mishnâh, e.g. Median beer of wheat or barley, Edomite "barley wine" (?) and Egyptian zythum, which is said in the Talmûd to be made by fermenting barley, salt and wild saffron[10], and in later Palestine large quantities of date wine were made.[11] The distillation of alcoholic spirit from wines (including date and raisin wines) was begun at some time not accurately known, but was familiar to Jewish authors of the 16th century and was practised by Jews in Italy in the "late middle ages"[12]. Wine was used medicinally[13], and alcohol was no doubt also first prepared for medical purposes, as it was by European physicians.

Vinegar (chômets), made from fermented (chamets) liquor turned sour, or made artificially from wine and barley[14], was not drunk[15], the "vinegar" used by labourers for dipping bread being a thin sour wine like the acetum used by Roman soldiers. Vinegar is frequently mentioned in the Mishnâh and Talmûd.[16] Among the Talmûdists, wine and vinegar are often interchanged[17]: in the text of Matthew[18] "vinegar and gall" is more correctly[19] "wine and myrrh," reading οἶνος instead of ὄξος[20]: ἐσμυρνισμένον οἶνον in Mark replaces οἶνον μετὰ χολῆς μεμιγμένον in Matthew.

BITTER OR FLAVOURING PLANTS

"Bitter herbs" (měôrîm)[21] were probably various bitter plants, such as bitter cresses, hawkweeds, sow-thistles and the wild lettuce (chass ; Lactuca sativa), which is abundant in Sinai, Palestine and Egypt (where it was used as a salad).[22] The usual translation of perâgûm as poppy is incorrect.[23] Babûnêg is camomile (Anthemis).[24] Coriander (gad ; Coriandrum sativum), found in

[1] Sabbath, xx, 2 ; cf. Uktzin, iii, 5.
[2] Löw, Flora, i, 90 ; Zapletal, 8 f.
[3] CDB, 213, 272, 1010 ; Bertholet-Dallas, 184 f. ; Zapletal, 52 ; see p. 304.
[4] Prov., xxv, 13 ; Geikie, Holy Land, 859.
[5] Zapletal, 30 f. ; Handcock, Holy Land, 59, 62 ; Macalister, Century, 233 f.— grapes pressed in stone vats by treading only ; Benzinger, Hebr. Archäol., 144, Fig. 72.
[6] Ps., lxxv, 8.
[7] Is., xxv, 6.
[8] Löw, Flora, i, 95 f. ; Zapletal, 8 f. ; Macalister, HDB, ii, 33 f.
[9] Nowack, Hebr. Archäol., 1894, 237 ; filter-cloth in the Mishnâh, Sabbath, xx, 1.
[10] Pesachim, iii, 1, tr. Rodkinson, vol. v, p. 68 ; Berakoth, vii, 4–5, tr. Cohen, 326.
[11] Talmûd ; Sabbath, xx ; Berakoth, vi, 1 ; Löw, Flora, i, 114, 130, 245 ; iii, 348.
[12] Löw, op. cit.

[13] W. Ebstein, Die Medizin im alten Testament, Stuttgart, 1901, 37.
[14] Ruth, ii, 14, in A.V. ; Num., vi, 3, sour wine ; Prov., x, 26 ; Ps. lxix, 21 ; CDB, 991 ; Macalister, HDB, ii, 34 ; iv, 870.
[15] Zapletal, 37, contradicting CDB, 991 ; cf. Löw, JE, xii, 439.
[16] Aboth, iii ; Sabbath, xiii, xiv, xx, xxii ; Berakoth, i, 1 ; vi, 1, grape vinegar: Pesachim, iii, 3, Edomite vinegar from barley and wine ; Rodkinson, Talmud, i (ix), 73 ; ii, 224, 228, 317, 330 ; v, 68 ; Cohen, Běrâkot, 25, 253.
[17] Berakoth, i, 1 ; Cohen, 25.
[18] xxvii, 34.
[19] As in Mark, xv, 23.
[20] Westcott and Hort, New Testament in Greek, Matt., xxvii, 34.
[21] Ex., xii, 8.
[22] CDB, 118 ; Löw, Flora, i, 425.
[23] Löw, Flora, ii, 364.
[24] Ib., i, 376.

Egypt, Persia and India, is twice mentioned as a white seed.[1] Cummin was cultivated in Palestine.[2] Anise (ἄνηθον), mentioned only once in the New Testament, is probably dill (Anethum graveolens ; šebet in the Mishnâh) rather than true anise (Pimpinella anisum).[3] Dill-water (perhaps distilled) is mentioned in the Talmûd as used for sweetening (the breath ?).[4] Mint (ἡδύοσμον) occurs only in the New Testament [5] and in late Hebrew texts. It was used both medicinally and in cookery by the Romans. Wild mint (Mentha sylvestris) grows in Ṣyria.[6] Chicory is in the Mishnâh.[7] Mustard [8] is probably the ordinary plant (Sinapis nigra), which grows to a large size in Palestine, although Royle had suggested the Salvadora persica, the seeds of which (if it is the same as the tree called khardal—the Arabic for mustard ; Egyptian shuft ?) are used in Syria as a substitute for mustard.[9] Senna (šěnimuth) is mentioned in the Mishnâh.[10] Some twenty different Hebrew words are translated "thorns" (perhaps including blackthorn) and thistles[11], and the nettle is (doubtfully) represented by the words chârûl[12], qîmôš or qîmôš[13] and qimměšônîm (Authorised Version, thorns).[14]

AROMATIC MATERIALS, INCENSE AND PERFUMES

The Hebrews anointed the hair or body profusely with scented oil or ointments[15] and made much use of perfumes.[16] The aromatics (sammîm)[17] were carried by Arabs from Gilead to Egypt by camels.[18] Josephus[19] says that olive oil was boiled with myrrh, cassia, cinnamon and calamus to make the holy oil, which Moses was commanded to make "after the art of the apothecary" (perfumer)[20]; also that incense trees were first brought by the Queen of Sheba to Palestine, where one was still growing in his time. Although the point is much disputed, it is said that incense was not used by the Jews before the Babylonian exile.[21] A dry yellow powder found in a jar at Gezer, which became semi-fluid and resinous on exposure, was supposed to be incense, from an analysis. An earthenware altar (7th–8th centuries B.C.), probably for "burning" incense by evaporation in a heated basin, was found at Ta'anach.[22]

Incense was also (as in Egypt) used in profane life, but not in times of mourning.[23] That for use in the Temple was ordained to be composed of equal weights of materials translated as stacte, onycha, galbanum and frankincense, and all incense not so made was forbidden.[24] Rashi (d. 1105)[25] in his commentary

[1] Ex., xvi, 31 ; Num., xi, 7; Löw, Flora, iii, 441 ; HDB, ii, 38 ; E. Bibl., i, 897.

[2] Is., xxviii, 25, 27 ; Matt., xxiii, 23 ; CDB, 181.

[3] Matt., xxiii, 23 ; Bochart, Hierozoicon, i, pref. sign. c 4 ; CDB, 46, from Royle ; Löw, Flora, i, 518 ; ib., JE, x, 80 ; Post, HDB, i, 99 ; Macalister, E. Bibl., i, 170 ; ii, 38.

[4] Berakoth, vi, 1 ; tr. Cohen, 260.

[5] Matt., xxiii, 23 ; Luke, xi, 42.

[6] CDB, 561 ; Macalister, HDB, ii, 38 ; Post, ib., iii, 379 ; E. Bibl., iii, 3151 ; Löw, Flora, ii, 75.

[7] Kilaim, i, 2 ; Shebiith, vii, 1.

[8] Matt., xiii, 31 ; xvii, 20 ; Mark, iv, 31 ; Luke, xiii, 19 ; xvii, 6.

[9] CDB, 587, 676 f. ; Celsius, ii, 253 ; Macalister, HDB, ii, 38 ; Post, ib., iii, 463 ; E. Bibl., iii, 3244 ; Löw, Flora, i, 516.

[10] Löw, JE, x, 79.

[11] CDB, 940.

[12] Job, xxx, 7 ; Prov., xxiv, 31 ; Löw, JE, x, 72—"vetchling," perhaps wild mustard.

[13] Is., xxxiv, 13 ; Hos., ix, 6.

[14] Prov., xxiv, 31 ; CDB, 608 ; Löw, JE, x, 73.

[15] Ruth, iii, 3 ; II Sam., xiv, 2 ; Ps. xxiii, 5 ; xlv, 7 ; xcii, 10 ; Eccl., ix, 8 ; Is., iii, 24 ; Matt., vi, 17 ; xxvi, 7 ; Luke, vii, 46.

[16] Prov., vii, 17 ; xxvii, 9 ; Ps. xlv, 8 ; Cant., iv, 11 ; Dan., ii, 46 ; CDB, 711; Macalister, HDB, iii, 593 ; E. Bibl., iii, 3471.

[17] Rosenmüller, IV, i, 146 f. ; CDB, 898; Price, JE, ii, 475 ; Ursinus, Hist. plant. Bibl., 1665, 184 f., 192 f.

[18] Gen., xxxvii, 25 ; Whiston, in Josephus, Antiq., VIII, vi, 6, thinks this was turpentine.

[19] Antiq., III, viii, 3 ; VIII, vi, 6.

[20] Ex., xxx, 24, 25, 35 ; xxxvii, 29.

[21] Albright, Archæol. of Palestine, 108.

[22] Macalister, Century, 286, 290.

[23] Cant., iii, 6 ; CDB, 711 ; Selbie, HDB, ii, 467 ; E. Bibl., ii, 2165.

[24] Ex., xxx, 9, 34 ; CDB, 345 ; E. Bibl., ii, 2167.

[25] Sarton, Intr. to the Hist. of Science, i, 752 ; Isis, 1928, xi, 422.

gave 11 ingredients, Josephus 13 ; Maimonides' recipe is 16 manehs (mina =
560 gm.) each of myrrh, cassia, spikenard and saffron, 12 manehs of costus,
9 of cinnamon and 3 of sweet bark [?]. To the mixture was added a quarter of a
qab (qab = ½ or ¼ gallon ; 2197 c.c.) of salt of Sodom and Jordan amber [? bitu-
men] and a herb which made the smoke ascend vertically, called "smoke-
raiser," known to secret tradition [? saltpetre]. Salt was the symbol of incor-
ruption, and Maimonides says nothing was offered without it except wine,
blood and wood. Philo Judæus thought the four ingredients of the old incense
represented the four elements.[1]

Libneh [2] may be storax (Styrax officinalis, Linn.), which is mentioned in
Sirach [3] with other aromatic substances : the white flowers would agree.[4]
The styrax gum is now unknown in Arabia and none has been available in trade
since the 19th century.[5]   Stacte (nâtâf ; šorî kâtâf) [6] is identified by Rosenmüller and Löw with storax gum (from Styrax officinalis), but many other
materials have been suggested.[7]

The frankincense, lěbônâ ("shining" ? ; Assyrian lubbânu), a brittle, glittering
vegetable resin of a bitter taste used for fumigation[8], was obtained by successive
incisions in the bark of the incense tree (Arbor thuris), the first incision giving
a pure white variety (laban, white) and the others a kind spotted with yellow ;
when old it loses its whiteness.[9]   The Hebrews imported it from Arabia,
particularly Saba[10], but that now growing there (olibanum) only in Yemen[11]
is inferior. The best incense now comes through Arabia from the islands of
the Indian Archipelago and is produced in India by the Boswellia serrata
(Roxburgh) or Boswellia thurifera (Colebrooke). The tree producing the
Arabian olibanum is the Boswellia Carteri (Birdw.)[12]; it has been thought to
be the Amyris Gileadensis.[13] Galbanum[14], used in making the sacred incense[15],
is now brought from India and the Levant and has been regarded as the product
of the Ferula ferulago (Linn.), growing in North Africa, Crete and Asia Minor ;
or of the Bubon galbanum (Linn.), a native of the Cape of Good Hope ; or of
the Opoidia galbanifera.[16]

Nêrd[17] is spikenard[18], the Arabic sunbul, the Greek nardos (νάρδος) and the
Hindu jatamansee, growing on the mountains overhanging the Ganges and
Jumna, the Nardostachys jatamansi of De Candolle.[19] The Syrian nard
mentioned by Greek physicians is supposed to be some species of valerian.[20]
Dioskurides says the spikenard ointment was composed of amomum, balsam,
costus, myrrh, nard and schœnus in nut oil, but it probably varied in composition, since the Coptic version of St. John calls the alabaster box of ointment "genuine" as well as "precious"[21]—a reminiscence of the old Egyptian
terminology (p. 46 f.).

[1] CDB, 345 ; E. Bibl., ii, 2167 f.
[2] Poplar in A.V. of Gen., xxx, 37 ;
Hos., iv, 13 ; "white," hence Populus
alba according to Celsius, i, 292.
[3] Ecclus., xxiv, 15.
[4] CDB, 746 ; Ursinus, Arboretum Biblicum, 158.
[5] Löw, Flora, iii, 388 f.
[6] Ex., xxx, 34.
[7] Celsius, i, 529 ; CDB, 899 ; Löw,
Flora, i, 195 ; iii, 388 f., 393—"tears" ;
cf. Bochart, Geographia sacra, 119, 713 ;
Hierozoicon, ii, 532, někôth.
[8] Ex., xxx, 34 f.
[9] CDB, 274 ; Celsius, i, 231 ; Bochart,
Geographia sacra, 103 f., 713, and refs. ;
Löw, Flora, i, 312.
[10] Is., lx, 6 ; Jer., vi, 20.
[11] Bochart, Geographia sacra, 129 f. ;

Niebuhr, Descr. de l'Arabie, 1776, 126 f. ;
Hasselquist, Voyages and Travels in the
Levant, 1766, 297.
[12] Löw, Flora, i, 312.
[13] CDB, 274 ; Price, JE, ii, 475.
[14] Löw, Flora, iii, 455.
[15] Ex., xxx, 34.
[16] CDB, 277.
[17] Only in Cant., i, 12 ; iv, 13, 14.
[18] Mark, xiv, 3–5 ; John, xii, 3.
[19] CDB, 898 ; Bochart, Geographia
sacra, 713 ; E. Balfour, Cyclopædia of
India, 1885, ii, 1062 ; Löw, Flora, iii, 483 ;
ib., JE, ix, 170 ; Dioskurides, i, 6—νάρδος
γαγγῖτις ; any Indian aromatic was called
nard.
[20] Adams, Paulus Ægineta, 1847, iii, 264.
[21] E. S. Rohde, Garden-Craft in the Bible,
1927, 33.

Môr [1] is myrrh, the product of the Balsamodendron myrrha of Arabia.[2] The lôt, also translated myrrh [3], is probably ladanum, from the Cistus Creticus.[4] Löw [5] thinks there was also a liquid variety of myrrh (šemen hamôr), but this was no doubt a perfumed oil; also that the source of myrrh was the Commiphora Abyssinia (Engel.) and C. Schimperi in Arabia and Abyssinia. According to Schweinfurth [6] môr is not myrrh but balsam, but this has a separate name. He says it is obtained by incision from the Commiphora Abyssinia or C. Schimperi, mostly the former, growing in Yemen and Assir. The Balsamodendron myrrha yields practically no exudation on incision. The words bâsâm, besem or bôsem (balsam ; Arabic bishâm ; "spice" in the Authorised Version) [7] probably mean the balm of Gilead from the Balsamodendron opobalsamum or Amyris Gileadensis, which grows in some parts of Arabia (Mecca) and Africa and yields the balsam by incision or from the green or ripe berries.[8]   In later times the very costly balm was obtained from trees which grew only in Judæa in two royal gardens, and was the best and most costly balsam.[9] Even in the time of Theophrastos no true balsam of Gilead was exported[10], and Pliny describes how it was adulterated. The name balsam (βάλσαμον ; balsamum) comes from Hebrew words meaning "king of oils." The gum, also called ὀποβάλσαμον, was obtained by incision of the tree as "tears" (δάκρυα ; lachrimæ) for the best quality. There were tests to distinguish the best from the poorer qualities, such as emulsification in water, which is still used by dealers.[11] Justin[12] says that opobalsamum is produced only in Judæa in a shut-in valley called Ericus [Jericho], and in fact in this locality a day temperature over 100° F. may be reached, so that a sub-tropical vegetation out of keeping with the latitude flourishes.[13] Josephus[14] says it was transplanted, for they still possessed a root from a gift from the Queen of Sheba. The balsam tree was grown in the 13th–14th centuries at Matarîya, just south of 'Ain Schems (Heliopolis) by Christians, where Nasîr ibn Khusraw[15] saw the gum collected in glass bottles. Another variety, bašâm (Amyris opobalsamum) grew in Arabia. Later the true balsam was transplanted to the Hedjaz and disappeared in Egypt about 1615.[16] The material now sold as balm of Gilead is made from the Balanites Ægyptiaca.[17] Schweinfurth (op. cit.) says the balm of Gilead is a product of the Commiphora opobalsamum Engl., which grows on the coasts of Arabia and also in Nubia, but the balsam is collected only in valleys near Mecca, only a few pounds coming into the market. The tree would not yield sufficient balsam by incision, so that perhaps a process of crushing and boiling is used. In a 12th century Persian MS., however, the Judæan balm (duhn balsân) is represented as produced by incision near the ground.[18] The incense plant does grow in Arabia, in spite of frequent assertions to the contrary.[19]

[1] Ex., xxx, 23 ; Esth., ii, 12 ; Ps. xlv, 8 ; Prov., vii, 17 ; Matt., ii, 11, σμύρνα; Mark, xv, 23 ; John, xix, 39—for embalming.

[2] CDB, 588 ; Bochart, Geographia Sacra, 113, 119, 123 ; Celsius, i, 520 ; Schrader, RL, 566.

[3] Gen., xxvii, 25 ; xliii, 11.

[4] CDB, 580 ; see p. 168.

[5] Flora, i, 305 f.

[6] Q. in Pharm. J., 1894, 897 ; Schmidt, Drogen, 1924, 27 ; Wainwright, Balabish, 1920, 15.

[7] Cant., i, 13 ; v, 1 ; vi, 2.

[8] CDB, 898 ; Löw, Flora, i, 299 ; Ursinus, Hist. Plant., 259 f. ; Hasselquist, Travels in the Levant, 1766, 293 f. ; Post, HDB, iv, 610 ; E. Bibl., i, 465 ; iv, 4746.

[9] Pliny, xii, 25.

[10] Hist. plantarum, IX, vi, 2.

[11] Dioskurides, i, 18 ; Hasselquist, Travels, 293 ; Schmidt, Drogen, 34 ; Wagler, PW, ii, 2836, 2838.

[12] Hist. Phillipicæ, xxxvi, 3.

[13] Wainwright, Balabish, 1920, 15.

[14] Antiq., VIII, vi, 6.

[15] Sefer Nameh, tr. Schefer, Paris, 1881, 143.

[16] Reitemeyer, Beschr. Ägyptens, 66 f., 68, 71.

[17] Price, JE, ii, 476.

[18] Rimmel, Le livre des parfums, [1870], 61 f., 68 f.

[19] Tkač, PW, i a, 1311, 1419, 1495 f. ; Art. "Saba," cols. 1298–1515 ; on Punt, ib., 1312 ; cf. Senn, q. by Bilabel, Gesch. Vorderasiens, 1927, 431—myrrh now grows north of the Gulf of Aden, incense in Africa only on the Somali coast. No two writers on this subject are in agreement, unless one simply copies from the other.

Another name for the balm or gum of the Balsamodendron opobalsamum was probably tzôrî or tzêrî.[1] Although tzôrî has been supposed to mean mastic, the gum of the Pistacia lentiscus, this occurs only in the Apocrypha.[2]

Bedôlach [3] has been translated bdellium, an odoriferous exudation from the tree Borassus flabelliformis (Linn.) of Arabia.[4] The identification of bedôlach is uncertain, since it is merely called "honey-coloured" and might be animal, vegetable or mineral.[5] Amber has been suggested [6] and Bochart [7] proposed pearls from the Persian Gulf. Saadia Gaon (882–928) in his Arabic translation of the Old Testament [8] regarded bedôlach as meaning pearls, and was followed by Benjamin of Tudela (A.D. 1170), who says it was obtained at Katifa (El Khatif) on the Persian Gulf.[9]

Qiddâh, an ingredient of the holy oil[10], or qetzîôth[11], is translated cassia in the Authorised Version and probably correctly.[12] Cassia bark is from various kinds of Cinnamomum growing in different parts of India, but the name is used by Greek and Latin authors for different products, and the material is not obtained from any tree now growing in Arabia. In Exodus[13] cinnamon is separately mentioned; in other places[14] it is an aromatic plant; the New Testament[15] makes it a Babylonian article of trade.[16] What is called cinnamon (qinnâmôn) is really the Chinese cassia and not true cinnamon.[17] ʻCinnamon in "reeds" (the quill-like form of true cinnamon) is mentioned in the Harris Papyrus (Rameses III).[18]

Mace is a product of the Myristica fragrans Houtt., the muscat nut, which is mentioned by Simeon Seth (11th century A.D.) as karyon aromaticus : it was not known early.[19] Costus (Sanskrit kushta, kushtum) is first mentioned in the Mishnâh[20]—with amomum and pepper—but must have been known at least as early as the 4th century B.C.[21] Někôth or nâkôth[22] is probably gum tragacanth, Arabic naka'at, from the Astragalus[23], although storax is a common identification.[24]

Pepper (pipêl, pipělâ, through Greek from the Sanskrit pippali), from a shrub, is mentioned as something new in Enoch[25], and is in the Mishnâh and Talmûd.[26] The Jews were the great pepper merchants in the Roman and later periods.[27] Cloves are not mentioned in Hebrew works till the 11th century A.D.[28], but are apparently mentioned by Aëtios[29] in the 6th century.

[1] Gen., xxxvii, 25 ; xliii, 11 ; Jer., viii, 22 ; xlvi, 11 ; li, 8 ; Ez., xxvii, 17, marg. "rosin" ; mastic is also proposed ; CDB, 92 ; Post, HDB, i, 235.

[2] Hist. of Susanna, 54, in Septuagint ; margin A.V. "lentisk" ; CDB, 526 ; Löw, Flora, i, 195.

[3] Gen., ii, 12 ; Num., xi, 7 ; Josephus, Antiq., III, i, 6.

[4] Celsius, i, 324 ; CDB, 100 ; Ursinus, H. Plant., 202 ; Löw, Flora, i, 304 ; E. Bibl., i, 504.

[5] CDB, 100.

[6] Hoefer, Hist. de la Chimie, i, 62.

[7] Hierozoicon, ii, 674 f. ; cf. ib., Geographia sacra, 19, 713 ; Hull, HDB, i, 259.

[8] Sarton, Intr., i, 627 ; Isis, 1929, xiv, 153.

[9] Itinerary, ed. Adler, 1907, 63.

[10] Ex., xxx, 24 ; Ez., xxvii, 19.

[11] Only in Ps. xlv, 8.

[12] CDB, 143 ; Post, HDB, i, 358 ; E. Bibl., i, 708 ; Löw, Flora, ii, 114 ; ib., JE, x, 73.

[13] xxx, 23.

[14] Prov., vii, 17 ; Cant., iv, 14.

[15] Rev., xviii, 13.

[16] CDB, 165.

[17] Löw, Flora, ii, 107, 407 ; E. Bibl., i, 828 ; cf. Post, HDB, i, 442—true cinnamon.

[18] Breasted, Anct. Rec., iv, 132, 137, 153, etc.

[19] Löw, Flora, ii, 60 f.

[20] Sabbath, xiv, 3 ; Uktzin, iii, 5.

[21] Löw, Flora, i, 391 ; ii, 114 ; Salmasius, Hyles iatricæ, 128 ; Garcia da Orta, tr. Markham, 148 f. ; Gorræus, Definitionum Medicarum, 1578, 242.

[22] Gen., xxxvii, 25 ; xliii, 11 ; II Kings, xx, 13 ; Is., xxxix, 2 ; "spice" in the A.V. ; θυμίαμα in the Septuagint.

[23] CDB, 898 ; Löw, Flora, ii, 419.

[24] Ursinus, H. plant., 206 ; Celsius, i, 548 ; Bochart, Hierozoicon, ii, 532 ; E. Bibl., iv, 4746 ; Post, HDB, iv, 611—not tragacanth.

[25] xxxii, 1 ; Kautzsch, ii, 256.

[26] Uktzin, iii, 5 ; Talmûd, Sabbath, vi ; Berakoth, vi, 1, tr. Cohen, 246.

[27] Löw, Flora, iii, 49.

[28] Ib., ii, 277.

[29] Tetrabiblos, I, i ; ed. Basle, 1549, 35 ; caroum ; cf. Garcia da Orta, tr. Markham, 213.

Hyssop (êzôb ; ὕσσωπος) was some detergent plant very variously identified [1], e.g. as the ὕσσωπος of Dioskurides [2], Satureia Græca, Satureia Juliana (neither being appropriate), or Origanum Ægyptiacum in Egypt, Origanum Syriacum in Palestine (O. Smyrnæum in Dioskurides) ; Bochart makes it marjoram or some similar plant [3], and according to Celsius and Post old tradition supports this. The monks on Jabal Musa call the fragrant plant, ja'deh, which grows there hyssop. Celsius gives Hyssopus Officinalis ; Royle concluded that êzôb was the caper (Capparis spinosa, Linn., Arabic asuf).[4] Jullien [5], who refers to identifications with Origanum maru, Micromeria græca (B.) and Satureia græca (L.)—called hyssop by Beyrut pharmacists—says that true hyssop does not grow in Sinai or Palestine. The caper, however, is usually identified with sělâf, which was cultivated in ancient Palestine [6] and is mentioned in the Mishnâh and Talmûd.[7] The ăbîyyônah [8] is supposed to be the caper.[9]

The kôpher, translated camphor (camphire) in the Authorised Version[10], is really henna, Egyptian puker, called khofreh in Nubia, and the Greek κύπρος (hence "cypress" in the margin of the Authorised Version and cyprus in the Vulgate).[11] The Lawsonia alba, which grows in Egypt, Syria, Arabia and North India, was probably used by the ancient Hebrews. The young shrub is without thorns, but acquires them when older, hence Linnæus thought (incorrectly) that there were two varieties, Lawsonia inermis and L. spinosa.[12] Henna is in the Mishnâh.[13] Camphor is not mentioned in the Old Testament ; the first Oriental author to mention it is Imru' al-Qais in the 6th century.[14]

The "thyine wood"[15] is probably that of the Thuya articulata, Desfont., mod. Callitris quadrivalvis, yielding sandarach resin : it is a cypress and was called citrus by the Romans.[16] The algum or almug tree, brought from Ophir[17], sometimes thought to be Brazil wood, was probably the red sandalwood (Pterocarpus santalinus), hard, heavy and fine-grained, a native of India and Ceylon, imported from Tyre[18], although many other identifications, including coral, have been proposed.[19] The precious aromatic wood ahâlîm, ahâlôth[20], is probably the aloes-wood (lign aloes), agallochum, Arabic ûd[21], a pathological product of the large tree Aquilaria agallochum, the wood of which, penetrated with oil and resin, is used in India for fumigation and incense. It may have been cedar-wood.[22] Barton[23] says the references to ahâlîm, except in Numbers, refer not to the wood but to the fragrant gum of the Aloexylon and Aquilaria ovata of Malabar, and the Aquilaria agallochum of Bengal. True aloes ('albaj) are in the Mishnâh, but not in the Old Testament.[24]

[1] Ps. li, 7 ; CDB, 339 ; Macalister, HDB, ii, 38 ; Post, ib., 442 ; Celsius, i, 407 f. ; Hoefer, H. Bot., 13.
[2] iii, 30.
[3] Hierozoicon, i, 587 f., 589, and refs.
[4] CDB, 339 ; Tristram, Nat. Hist. of the Bible, 457 ; Post, HDB, ii, 442 ; E. Bibl., ii, 2142.
[5] Sinaï et Syrie, 150.
[6] Löw, Flora, i, 322.
[7] Demai, i, 1 ; Maaseroth, iv, 6 ; Talmûd, Berakoth, vi, 1, 3 ; tr. Cohen, 242, 269.
[8] Eccles., xii, 5, R.V. ; "desire" in A.V.
[9] E. Bibl., i, 695 ; Post, HDB, i, 350.
[10] Cant., i, 14 ; iv, 13.
[11] Bochart, Geographia sacra, 713, 916.
[12] CDB, 130 and Fig., 664 ; Celsius, i, 223 ; Post, HDB, i, 346 ; E. Bibl., i, 637 ; Löw, Flora, ii, 114 f., 218 f. ; ib., JE, x, 72 ; Vigouroux, Dict. de la Bible, iii, 590.
[13] Shebiith, vi, 6.

[14] Löw, Flora, ii, 116.
[15] Rev., xviii, 12.
[16] CDB, 942.
[17] II Chron., ii, 8 ; ix, 9, 10, 11 ; I Kings, x, 11, 12.
[18] Ursinus, Arboretum Biblicum, 1663, 565 f. ; Celsius, i, 171 ; CDB, 31, 658.
[19] Post, HDB, i, 63 ; E. Bibl., i, 119.
[20] Num., xxiv, 6 ; Ps. xlv, 8 ; Prov., vii, 17 ; Cant., iv, 14 ; once in N.T., John, xix, 39, 40.
[21] Ursinus, Arboret. Bibl., 559 ; Hist. Plant., 194 ; Post, HDB, i, 69.
[22] Celsius, i, 135 ; CDB, 34, 898 ; Balfour, Cyclopædia of India, 38, i, 80, 122, etc. ; E. Bibl., i, 120 ; Niebuhr, Descr. de l'Arabie, 1776, 127.
[23] JE, i, 437 ; cf. Vigouroux, Dict. de la Bible, i, 398.
[24] Löw, Flora, ii, 150 ; Post, HDB, i, 69.

The fir (běrôš ; běrôth) [1] probably did not mean exclusively the cedar, but also one or other of the trees Pinus sylvestris, larch and cypress (Cupressus sempervivens), although the cypress is properly saru [2], all of which trees still grow in Lebanon. The turpentine tree [3] is supposed to be the Pistacia terebinthus, Hebrew êlâh, but the latter name may mean the oak.[4]   The terebinth (bûtum) grows in Palestine but is not tapped for turpentine.[5] Gôpher wood [6] is probably the wood of some resinous tree (pine, fir, etc., or cypress) [7]: acacia [8] has also been suggested. Books were sprinkled with cedar oil to preserve them from insects.[9]

The ôren[10] is called "ash" in the Authorised Version, but some species of pine in the Septuagint and Vulgate ; it is perhaps the larch (Larix Europæa)[11] or laurel[12]; the Assyrian irin is a fir.[13] Erez, always given as the cedar in the Authorised Version, is probably this tree, particularly Cedrus Libani[14], but the name was also used for other conifers and firs, and perhaps also the juniper (Juniperus Sabina).[15] The 'ărô'êr[16] and 'ar'âr[17], "heath" in the Authorised Version, is probably the Arabic arar, some species of juniper, probably Juniperus Sabina or savin.[18] What the Authorised Version calls juniper[19], on the other hand, is supposed to be a kind of broom (Genista monosperma ; Arabic rethem).[20] The tirzâh[21], translated cypress in the Authorised Version, is doubtful : holm oak, beech and fir have been proposed.[22] Hadas[23], growing on hills round Jerusalem, is correctly given in the Authorised Version as myrtle, Myrtus communis, still growing in Jerusalem, Samaria and Galilee[24]; Löw[25] gives âsâ as the name of the myrtle, which is mentioned only in later books of the Old Testament.

TREES YIELDING USEFUL WOOD

The oak probably appears under various names, the usual being êl, êlâh, êlôn, îlân, allâh, alon and allôn[26]; tidhâr[27] may be oak or Indian plane, larch or elm.[28] Ezrâch[29] does not mean "bay tree" (Authorised Version) but a tree which has grown in its own soil and not transplanted, or indigenous (margin).[30] Safsâfâ[31] is the willow, the usual name being 'ârâb (Arabic gharab)[32]. The translation box-tree for těassûr[33] given by Talmûdical authors is probably correct[34], although the Syriac and Arabic versions give sherbin, a kind of cedar

[1] Celsius, i, 74 f. ; M'Lean, E. Bibl., ii, 1522, fir or pine.
[2] Vigouroux, Dict. de la Bible, ii, 1171 ; Löw, Flora, iii, 13 f., 26, 40 f. ; CDB, 269; Post, HDB, ii, 8.
[3] Only in Ecclus., xxiv, 16, and doubtful ; Kautzsch, Apokryphen, i, 354.
[4] CDB, 638, 964.
[5] Ib., 676, 965 ; Löw, Flora, i, 191 f. ; E. Bibl., i, 466.
[6] Gen., vi, 14.
[7] Bochart, Geographia Sacra, 22 ; CDB, 296—conjectures.
[8] Ursinus, Arboretum Biblicum, 310 f.
[9] Assumpt. Mosis, i, 17 ; early 1st century A.D. ; Kautzsch, Apokryphen, ii, 320.
[10] Is., xliv, 14.
[11] CDB, 70 ; Celsius, i, 185 ; Post, HDB, i, 163.
[12] Löw, Flora, ii, 120.
[13] E. Bibl., i, 325.
[14] Ib., i, 716 ; Post, HDB, i, 364.
[15] CDB, 145 ; Löw, Flora, iii, 33.
[16] Jer., xlviii, 6.
[17] Jer., xvii, 6.

[18] CDB, 314 ; E. Bibl., ii, 1982 ; Post, HDB, ii, 319 ; Robinson, Researches in Palestine, 1841, ii, 506.
[19] I Kings, xix, 4, 5 ; Ps. cxx, 4 ; Job, xxx, 4.
[20] CDB, 434.
[21] Only in Is., xliv, 14.
[22] CDB, 182.
[23] Neh., viii, 15 ; Is., xli, 19 ; lv, 13 ; Zech., i, 8, 10, 11.
[24] CDB, 589.
[25] Flora, ii, 258 ; ib., JE, ix, 136 ; on oleander, see Löw, Flora, i, 206 f.
[26] Jos., xix, 33, with meaning of places ; Judg., iv, 11 ; CDB, 32, 638, deciduous and evergreen ; Casanowicz, JE, ix, 364; Löw, Flora, i, 622, doubtful.
[27] Is., xli, 19 ; lx, 13 ; A.V. "pine."
[28] CDB, 737 ; Celsius, ii, 271.
[29] Ps. xxxvii, 35.
[30] CDB, 100.
[31] Ez., xvii, 5.
[32] CDB, 1009 ; Löw, Flora, iii, 325.
[33] Is., xli, 19 ; lx, 13.
[34] CDB, 217 ; M'Lean and Thiselton-Dyer, E. Bibl., i, 600.

or cypress.[1] The wood was probably used for writing tablets.[2] Hobnîm [3] is perhaps ebony from Ethiopia, India or Ceylon.[4]

Šittâh (plural šittîm, "shittim wood") [5] is perhaps the acacia (Egyptian sont), and possibly the variety Acacia Seyal, which grows in some parts of Palestine and, like the Acacia Arabica, yields gum arabic, although there is no indication that the old Jews used this gum [6], which is mentioned, as kûmôs, in the Mishnâh.[7]

Atâd [8] is sometimes supposed to be the box-thorn, Lycium Europæum, or L. Afrum, both of which grow in Palestine.[9] Hasselquist[10] calls it the Rhamnus spina Christi (buckthorn) : other thorny or prickly plants are chôach (which Celsius thought was buckthorn), chêdeq, dardar and šâmîr.[11] The 'armôn[12] is probably not the chestnut, given in the Authorised Version, but the plane, a sacred tree among the Greeks and Persians : Herodotos mentions one of enormous size near Sardes, to which Xerxes paid homage.[13]

TEXTILES

Linen is the translation of at least five different words in the Old Testament[14]: (1) šeš, fine linen of Egypt[15]; (2) bad, a spun thread used only for religious garments[16]; flax was spun by women, whose dress was largely of linen[17]; (3) bûts, "fine linen"[18], $\beta\acute{v}\sigma\sigma\sigma\varsigma$ in the Septuagint and apparently a late Hebrew word[19]; (4) êtûn, used once only in connection with Egypt[20] and probably a kind of thread used for tapestry work ; (5) pešeth, pišteh or pištâh, including all kinds of linen, also flax and the plant from which it was made[21], is frequently mentioned.[22] The word mikvêh or mikvê, translated "linen yarn" in the Authorised Version[23], following Junius and Tremellius, came from Egypt, which was long famous for linen. The name sâdîn, used[24] for linen garments, is perhaps the origin of $\sigma\iota\nu\delta\acute{\omega}\nu$.[25] The dress of the poor was probably unbleached flax.[26] Flax is certainly mentioned in Exodus[27] and doubtfully in Joshua[28]: it was grown very early in Palestine[29] and its preparation is described.[30] The words šeš (the older form) and bûts meant linen[31], although bûts has been translated cotton.[32]

Although cotton is now grown with flax and hemp in Syria and Palestine[33], the meaning of the word kittan (? cotton) is disputed[34] and there is no evidence

[1] CDB, 217 ; Post, HDB, i, 213 ; Löw, Flora, i, 316 ; ib., JE, iv, 389.
[2] II Esdras, xiv, 24.
[3] Ez., xxvi, 15, only.
[4] CDB, 217.
[5] Ex., xxv f.
[6] Celsius, i, 498 ; Robinson, Res. in Palestine, ii, 210 ; Tristram, Nat. History of the Bible, 391 ; CDB, 876 ; Löw, Flora, ii, 377, 388 f.
[7] Sabbath, ⌜xii, 4 ; Tract. Shabbath, tr. Oesterley, 1927, 47 ; Egyptian qemi.
[8] Judg., xi, 14, 15 ; Ps. lviii, 9 ; "bramble," "thorns."
[9] CDB, 940.
[10] Voyages and Travels, 1766, 276 f., 282, 288.
[11] CDB, 940.
[12] Gen., xxx, 37 ; Ez., xxxi, 8.
[13] Herodotos, vii, 31 ; CDB, 153 ; Tristram, Nat. Hist. of the Bible, 345.
[14] Goguet, i, 140 ; Wilkinson, ii, 159 ; CDB, 210, 474 ; Ewing, HDB, iii, 124 ; E. Bibl., iii, 2799 ; iv, 5059.
[15] Gen., xli, 42 ; Ex., xxv, 4 ; xxxv, 6 ; Ez., xxvii, 7.

[16] Ex., xxviii, 42 ; Lev., vi, 10 ; xvi, 4 ; cf. Ex. xxxix, 28 ; Celsius, i, 50.
[17] Ex., xxxv, 25 ; Prov., xxxi, 22, A.V. " silk" ; Ez., xvi, 10, 13 ; cf. Rev., xviii, 16.
[18] Except in II Chron., v, 12, "white linen."
[19] Used in I Chron., xv, 27 ; II Chron., iii, 14 ; Esth., viii, 15 ; Luke, xvi, 19 ; Rev., xviii, 12.
[20] Prov., vii, 16.
[21] Judg., xv, 14 ; Josh., ii, 6.
[22] CDB, 474 ; Post, HDB, ii, 13 ; E. Bibl., ii, 1533.
[23] I Kings, x, 28 ; II Chron., i, 16.
[24] Judg., xiv, 12, 13.
[25] CDB, 474.
[26] Ecclus., xl, 4, "hairy mantle"; CDB, 471.
[27] ix, 31.
[28] ii, 6.
[29] Hos., 5, 9.
[30] Is., xix, 9 ; Löw, Flora, ii, 208 f.
[31] Celsius, ii, 169 ; CDB, 177.
[32] Thomson, Hist. Chem., i, 92.
[33] CDB, 676.
[34] Löw, Flora, ii, 208 f.

that the Hebrews were acquainted with cotton as distinct from linen, or with its name, until they came in contact with Persia. There is no name for cotton in Hebrew until the Talmûd [1], and karpas (Sanskrit karpâsa ; in the Periplus) in Esther [2] is a colour (Vulgate, carbasini coloris ; Authorised Version, green). Ancient Asiatic cotton was produced by the *plant* Gossypium herbaceum L., indigenous to India but growing in Bactriana in Alexander's time. The cotton *shrub* (G. arboreum L.), was little known to the ancients, but is once mentioned by Pliny [3], and it grew originally in Upper Guinea, Abyssinia, Sennâr and Upper Egypt, was brought down from the Sudan and was probably the first cotton grown in Lower Egypt, where Alpinus saw it in the 16th century.[4] "Indian linen" in the Mishnâh [5] is perhaps cotton.

Wool is repeatedly mentioned in the Old Testament and was the staple material for clothing [6], that of Damascus being highly prized at Tyre.[7] Rams "with the wool" are specified as tribute.[8] The names tsemer and gêz imply the act of shearing, gêz referring to the fleece.[9] A mixture (shaatnêz) of wool and linen in garments was forbidden.[10]

Silk was introduced only in the Roman Period : the words meši[11] and demešek[12], supposed to mean silk, are very doubtful, and the first *certain* mention is in the New Testament[13], although most commentators support the meaning "silk" for meši.[14] Silk is in the Mishnâh and Talmûd.[15]

DYES AND DYEING

The translation "coat of many colours"[16] is doubtful ; white was probably most worn and dyed robes were imported, probably from Phœnicia, and were expensive[17]: blue, scarlet and purple are mentioned in the places given. Rams' skins were dyed red (adom)[18], and skins are still dyed red for tanning in Syria.[19] They were probably dyed with kermes, called coccum in the Vulgate[20]: Hoefer[21] suggested that adom in the Hebrew meant yellow, perhaps from adamah, earth or yellow ochre ; red ochre, however, is just as plausible, and adom is now usually taken to mean red.[22] Pinner[23] suggested that only three colours are mentioned in the Old Testament, viz., purple, kermes and madder, the last of which he thinks is the Hebrew argâmân. The latter word, however, means purple.[24] Dyeing and tanning are not specifically mentioned in the Old Testament, but dyers are mentioned in the Mishnâh[25] and Benjamin of Tudela (A.D. 1173) mentions Jewish dyers at

[1] Bochart, Geographia sacra, 690 ; Löw, Flora, ii, 235 ; cf. Mishnâh, Kilaim, vii, 2— "cotton-tree."
[2] i, 6.
[3] xix, 1 ; gossipion ; xylon.
[4] De Candolle, q. by M'Lean and Thiselton-Dyer, E. Bibl., i, 915 ; cf. Anon., EB[14], vi, 524; Chambers's Encyclopædia, 1923, iii, 503—G. arboreum is also cultivated in India and China.
[5] Yoma, iii, 7.
[6] Lev., xiii, 47 ; Deut., xxii, 11 ; Job, xxxi, 20 ; Prov., xxxi, 13 ; Ez., xxxiv, 3 ; Hos., ii, 5.
[7] Ez., xxvii, 18.
[8] II Kings, iii, 4.
[9] Deut., xviii, 4 ; Job, xxxi, 20 ; CDB, 1014 ; E. Bibl., iv, 5353.
[10] Lev., xix, 19 ; Deut., xxii, 11 ; Josephus, Antiq., IV, viii, 11.
[11] Ez., xvi, 10, 13 ; A.V. "silk."
[12] Amos, iii, 12 ; A.V. "Damascus."
[13] Rev., xviii, 12, σιρικος ; Josephus, War, VII,[v, 3, in Rome ; CDB, 210, 880.

[14] E. Bibl., iv, 4521.
[15] Kilaim, ix, 2 ; Talmûd, Aboth, iv, "Roman."
[16] Gen., xxxvii, 3, 23.
[17] Zeph., i, 8 ; Prov., xxxi, 22 ; Luke, xvi, 19 ; II Sam., i, 24 ; Ez., xxvii, 7 ; xxiii, 6 ; Judg., viii, 26 ; Esth., viii, 15 ; Goguet, ii, 87 f. ; Jahn, Biblische Archäologie, 1817–18, I, ii, 54 ; CDB, 210.
[18] Ex., xxv, 5 ; pelles arietum rubricatas.
[19] Pinner, in Diergart, 1909, 198 f. ; Porter, HDB, i, 632.
[20] Ex., xxv, 4.
[21] H. de la Chimie, i, 59.
[22] Bochart, Hierozoicon, i, pref. signs. c 3 verso and g 3 recto, 986, 991 ; Thatcher, HDB, i, 457 ; Canney, E. Bibl., i, 874 f.
[23] In Diergart, 1909, 198 f.
[24] Bochart, Hierozoicon, i, pref. sign. c 3 verso ; ii, 734 f.
[25] Porter, HDB, i, 631 ; Benzinger, Hebr. Archäol., 153 ; Shebiith, vii, 3 ; Sabbath, i, 6 ; Baba Kamma, ix, 4 ; Eduyoth, vii, 8 ; Kelim, xiv, 6 ; Mikwaoth, vii, 3 ; etc.

Brindisi, Thebes (Greece ; purple and silk) and various parts of Syria and Palestine.[1] The translation "scarlet thread"[2] is probably correct, the dye (color coccineus, Vulgate) being almost certainly kermes. There were, apparently, four species of scarlet, named by Arabic, Talmûdic, late Greek and Portuguese authors, and all were species of coccus.[3] The identification with kermes (Arabic qirmiz ; Sanskrit krimi ; Greek κόκκος) is confirmed by the name tôlâ (worm) sometimes used for it, generally as tôla'ath, with šânî (brilliant) before or after.[4]

The notices concerning purple are confused. Pinner follows Bochart[5] in considering, probably with reason, that těchêleth (Septuagint πορφύρα ; Vulgate purpura) was the famous purple dye of the purple fish (chilzon)[6], and Bochart considered that the translation of Exodus[7], "blue, scarlet and purple" in the Authorised Version, should be "hyacinth-red, scarlet and purple," the first being hyacinthum in the Vulgate (Syriac sasgona).[8] Maccabees I[9] refers to "blue and sea-purple" (i.e. purple dye). Bochart considered that těchêleth was the old name for the true purple, hyacinthus, from shell-fish ; argâmân was a reddish-purple, not from the shell-fish but from a snail, of which there were several varieties (non ex conchylio facta sed ex illius cochleæ sanie quæ proprie purpura dicitur). It is more likely, perhaps, that different shades of colour produced by special processes, with admixture in some cases with other dyes, are meant. Canney[10] suggests that těchêleth (Assyrian ta-kil-tu) was a purplish-blue and argâmân (Assyrian argamannu) a purplish-red, and this would bring some order into the confused information given by old authors. Bochart's long dissertation, in particular, plentifully bestrewn with Hebrew, Aramaic, Syriac, and Arabic words, I have found difficult to follow.[11]

Although it is often stated that indigo (Indigofera tinctoria), which is now said to be cultivated in Palestine[12], was known to the old Hebrews, this is not correct, since the plant was woad (Isatis tinctoria), nine kinds of which grow in the country.[13] Its use was first learnt from the Greeks and is in the Mishnâh.[14] The commentaries on the Mishnâh put, by mistake, nîl (indigo) for woad (stis), which was often confused with indigo in the Middle Ages, and so a mistaken idea arose that indigo was cultivated by the Jews. The oldest mention of indigo (called lîlaq by the Jews, from the Persian nîlag, lîlag)[15] is by the Geonim. It was well known to Maimonides and the Jews grew indigo at Palermo.[13]

According to Post[16] varieties of indigo (ssbâgh ; nîl) now growing in Palestine are Indigofera Arabica, I. pauciflora and I. argentea, the last cultivated for indigo in the Jordan Valley and as far south as Egypt. Kalailan, the Greek καλλάϊνος, from the Sanskrit kaljâna, is generally supposed to be the turquoise, a greenish-blue stone in Pliny.[17] Following a hint given by Bochart[18], who collected references to show that it was a green or blue material,

[1] Itinerary, ed. Adler, 1907, 9, 10, 18, 20, 22, 25, 27, 28, 31.
[2] Gen., xxviii, 28.
[3] Bochart, Hierozoicon, ii, 734 f., 620 f. ; Hoefer, H. de la Chimie, i, 59 ; Schrader, RL, 420.
[4] Canney, E. Bibl., i, 874 ; iv, 4316.
[5] Hierozoicon, ii, 719, 727 f.
[6] Rosenmüller, IV, ii, 451.
[7] xxv, 4, 5 ; xxvi, 1, 14.
[8] Bochart, Hierozoicon, i, 985 ; ii, 727 f., 731 f.
[9] iv, 23 ; Charles, Apocrypha, 1913, i, 80.
[10] E. Bibl., i, 875.
[11] See the discussion of purple under Phœnicia, p. 458, with bibliography ; also J. Kenrick, Phœnicia, 1855, 237 f., 255 f.,

who refers to Q. Amati, De Restitutione Purpurarum, 3rd ed., Cæsenæ, 1784, not available to me.
[12] CDB, 676.
[13] Löw, Flora, i, 493 f., 497 f., 503 ; ib., JE, x, 78, 84.
[14] Shebiith, vii, 1 ; Sabbath, ix, 5 ; Megillah, iv, 7.
[15] Partington, Isis, 1934, xxii, 145 ; lulakin in mediæval treatises on painting ; cf. H. Hedfors, Compositiones ad tingenda musiva, Uppsala, 1932, 103.
[16] Flora, 252.
[17] xxxvii, 8 ; callaïs, καλάïς ; Periplus mar. erythr., c. 39, ed. Müller, p. 288, λίθος καλλεανος [καλλαϊνος].
[18] Hierozoicon, ii, 730.

Löw [1] claims that kalailan is indigo, the name coming from the Sanskrit Kaljâna, a place (the modern Kaljan) north-east of Bombay, so that the material is really indigo, which was called kalailan in the 2nd century. The Indian black (μέλαν ᾿Ινδικόν) of the Periplus [2] is not indigo, as has sometimes been supposed, but black Chinese ink, as Salmasius knew.[3]

Löw [4] maintains, with what seems to me suspicious over-emphasis, that madder (Rubia tinctorum), fû'â [5], Aramaic pûthâ, Arabic fuwwâ, was known to the ancient Hebrews, and that there were three kinds (all doubtful). Dioskurides [6] says madder (ἐρυθρόδανον ; rubia) was grown in Galilee. Warša (Arabic wars) is not henna or saffron but a red colour from the glands of the Moghania (Flemingia) rhodocarpa.[7] Dyer's weld (Reseda luteola) is a very old dye plant growing wild in Palestine ; it has no name in ancient Hebrew, šagarat being late.[8]

Carthamus tinctorius (safflower) is a very old dye, named with woad and madder in the Mishnâh [9] as qôs or qôsâ, in Aramaic kurtĕmâ, from the Arabic qartum (dye), ultimately from the Indian kurkum. It is cultivated in Syria and Palestine.[10] Löw[11] has argued at length that chârîâ' is an inermous safflower. It has been supposed that red was the only dye actually manufactured by the Jews[12], the others being imported. Six varieties of safflower are found at present, C. tinctorius being cultivated and the flowerets used in cookery and as a dye.[13] Hasselquist[14] describes the preparation (in Egypt). The flowers are pressed, washed several times, squeezed and dried exposed to the air and to dew at night, when they turn deep yellow, and are exported and known as saffranon. The powdered leaves are used to coagulate milk into cheese.

The karkôm[15], Syriac kurkemâ, Arabic qurqum, is the saffron crocus, Crocus sativus, not safflower.[16] Löw[17] states that two kinds of karkôm meant the Crocus sativus[18] and the Curcuma longa (Sanskrit haridra), still used as a dye (turmeric), but the latter is quite different from saffron.

Algummim has been supposed to mean Brazil wood[19], but there is no information that the latter was used as a dye by the Jews. Archil (fuqus) and henna have already been mentioned.[20] Challamut (alkanet) is said by Löw[21] to be mentioned in the Old Testament.

POISONOUS PLANTS

Bošâh[22], "cockle" in the Authorised Version, thought by Celsius[23] to be aconite, is probably any bad weeds or fruit, here perhaps bad or smutted barley.[24] Poisons are rarely and vaguely mentioned in the 'Bible.[25] Rôs is twice translated hemlock[26] and twice serpent venom[27], but otherwise usually gall. The ancients believed that the poison of a reptile was contained in its

[1] Flora, i, 500 f.
[2] Cap. 39; Schrader, RL, 398.
[3] Plin. exerc., 180, 810.
[4] Flora, iii, 325 ; cf. ib., JE, x, 79.
[5] Φουά in the Septuagint is a proper name ; Littmann, in Kautzsch, Apokryphen, ii, 38.
[6] iii, 160 ; G. R. Cameron, The History of Madder (rubia tinctorum), in *Annals of Medical History*, 1932, iv, 466–73 ; not seen.
[7] Löw, Flora, ii, 26, "carthamus" ; Berthelot, Mâ., ii, 138, "henna."
[8] Löw, Flora, iii, 127.
[9] Shebiith, vii, 1 ; Kilaim, ii, 8 ; Uktzin, iii, 5—"lozenges of safflower."
[10] CDB, 676, 809 ; Löw, JE, x, 77 ; saffron and safflower are often confused.

[11] Flora, i, 396.
[12] Cohen, JE, v, 24.
[13] Post, Flora, 473.
[14] Travels, 1766, 252.
[15] Cant., iv, 14.
[16] Celsius, ii, 11 ; CDB, 809 ; E. Bibl., iv, 4240.
[17] Flora, ii, 7 ; ib., JE, x, 72, 77.
[18] Saffron, in the Talmûd, Sabbath, xiv.
[19] Ursinus, Arboretum Bibl., 565 f.
[20] See pp. 496, 519.
[21] JE, x, 72.
[22] Only in Job, xxxi, 40.
[23] Hierobotanicon, ii, 199.
[24] CDB, 168.
[25] Ebstein, Medizin, 1901, 102.
[26] Hos., x, 4 ; Amos, vi, 12 ; A.V.
[27] Deut., xxxii, 33 ; Job, xx, 16.

gall.[1] Rôš was probably some bitter poisonous plant, perhaps colocynth [2], but Post [3] suggests an infusion of poppy heads or opium. Rôš cannot mean hops, as these (homlon) were first used by the Franco-German Jews in the 11th century A.D.[4] Měrêrâh or měrôrâh [5] means gall or bile ; chêmâh [6] is some arrow poison of unknown nature.[7] Suicide by poison is in the Apocrypha.[8]

Ketsach is the Nigella sativa, an herbaceous annual of the order Ranunculaceæ and sub-order Helleboreæ, growing in South Europe and North Africa.[9] Paqqû'ôth and pĕqâ'îm, "wild gourds"[10], were some species of gourd, Cucurbitaceæ, which tribe contains some bitter and poisonous plants with leaves and tendrils resembling vines. The etymology, from pâka, to split or burst, has been supposed to favour the identification with the squirting cucumber, Ecbalium elaterium, but old texts understood the colocynth (Colocynthis vulgaris, Arabic faqqa', also meaning mushrooms), with a poisonous fruit the size of an orange, which when dry bursts with a crashing noise. It is called handal by the Bedouins.[11] Ivy (Hedera helix) is mentioned only in the Apocrypha[12] and the Mishnâh.[13] Dûdâîm[14] is generally supposed to mean the mandrake, Mandragora officinalis L.[15] A great mass of superstition gathered in antiquity about this plant, with its curious forked root, and some of it still lingers on even in Europe.[16] The account of the ritual of gathering the root (baara) given by Josephus[17], which agrees with that in Ælian[18], is an old one[19], although Theophrastos[20]—who is usually free from superstition—mentions digging the root after drawing three circles around it with a sword, pulling it out whilst looking towards the west and "speaking of erotic matters."[21] The mandrake (Cucumis dûdâîm, love apples ; Atropa mandragora)[22] is still used medicinally in Cairo and in Bagdad small figures of it could be bought for use as amulets. The roots assume very curious shapes and the whole plant has a fetid and somewhat nauseating smell[23], although Dioskurides[24] says its odour is pleasant, and it may have been so to Orientals.

It has also been supposed[25] that dûdâîm meant the fruits of the Dead Sea apple (Arabic ôsher), the Asclepias gigantea, the fruit of which is a large hollow ball which crashes with an explosive puff when crushed, giving the appearance of escaping smoke[26]; or the globular yellow fruit of the Solanum Sodomæum.[27] Josephus[28] and Julius Africanus[29] say the fruit near the Dead

---

[1] Job, xx, 14 ; Bochart, Hierozoicon, i, 23 f.
[2] CDB, 278, 319, 745 ; Löw, Flora, ii, 364 ; E. Bibl., ii, 1636 ; W. Ebstein, Die Medizin im alten Testament, Stuttgart, 1901, 102.
[3] HDB, ii, 104, 348.
[4] Löw, Flora, i, 263 ; ib., JE, x, 83.
[5] Job, xvi, 13 ; xx, 25.
[6] Job, vi, 4.
[7] CDB, 278, 319, 745 ; Post, HDB, ii, 104 ; E. Bibl., ii, 1636.
[8] II Maccabees, x, 13 ; φαρμακ[ε]ία in Gal., v, 20, is witchcraft or the preparation of philtres, not poisoning ; CDB, 745.
[9] CDB, 271.
[10] II Kings, iv, 39 ; I Kings, vi, 18 ; vii, 24.
[11] Celsius, i, 402 ; CDB, 299 ; E. Bibl., ii, 1899 ; Löw, Flora, i, 537 ; Vigouroux, Dict. de la Bible, ii, 1899.
[12] II Maccabees, vi, 7 ; CDB, 356.
[13] Sukkah, i, 4, etc. ; Löw, JE, x, 73.
[14] Gen., xxx, 14 f. ; Cant., vii, 13—the word is always in the plural.
[15] Ursinus, Hist. Plantarum Biblicæ, 84, 136 ; Bochart, Geographia sacra, 866 ;

Celsius, i, 1 f. ; Rosenmüller, IV, i, 128 ; CDB, 511 ; Post, HDB, iii, 233 ; E. Bibl., iii, 2927 ; Lippmann, Abh., i, 190 ; Löw, Flora, iii, 364.
[16] Gubernatis, Mythologie des plantes, 1878–82, ii, 213 ; H. H. Ewers, Alraune, Berlin, 1919.
[17] War, VII, vi, 3.
[18] Hist. animal., xiv, 27.
[19] Steier, PW, xiv, 1028 f., 1037.
[20] Hist. plant., VI, ii, 9.
[21] Cf. Pliny, xxv, 13 ; J. Schmidel, Disputatis de Mandragora, Leipzig, 1669 ; K. Sprengel, Geschichte der Botanik, 1817, i, 23, 63 ; Keimer, Gartenpflanzen im alten Ägypten, i, 20, 87, 136, 173.
[22] Bontius, Account of the . . . East Indies, 1769, 33 ; Löw, JE, x, 76.
[23] Eisler, Weltenmantel, 32 ; Budge, Amulets, 490.
[24] iv, 76.
[25] Hoefer, H. Bot., 11.
[26] Kohler, JE, ii, 25.
[27] CDB, 676 ; Löw, JE, x, 74.
[28] War, IV, viii, 4.
[29] Synkellos' Chronicle, in Ante-Nicene Library tr. of Hippolytos, ii, 175.

Sea is "found full of foul smoke." E. Robinson [1] found the Dead Sea apple in Palestine only on the shores of the Dead Sea, although it grows abundantly in Upper Egypt, Nubia and Arabia Felix. The fruit of the Asclepias gigantea vel procera, he says, yellow and delicious to the eye, explodes when pressed, leaving only the shreds of the thin rind and a few fibres. Hasselquist says the fruit is hollow only when it has been attacked by an insect ; he found the mandrake growing in Galilee but not in Judæa, and considered it to be dûdâim.[2]

JEWISH MEDICINE

The information on Jewish medicine is scanty. The description of embalming in Genesis refers to Egypt and the "physicians" (rofeim) correspond with the Egyptian sewenu or seyen, physician, who was also an embalmer.[3] Josephus [4] and the Talmûd [5] say that Solomon composed a book of prescriptions or incantations for curing almost all ailments, but the magic and medicine of the Hebrews were probably both borrowed from other nations with whom they were in contact.[6] Works on Hebrew medicine which I have been able to see contain nothing of chemical interest.[7] The influence of Greek medicine appears only in the period of the Geonim.[8]

FLOWERS

Flowering plants grow in great variety in Palestine.[9] The lily, šûšân or šôšannah, is incorrectly given as the rose in the Aramaic Targum and by Maimonides, and as the violet by Kimchi and Ben Melech[10], but the correct reading is preserved by the Septuagint (κρίνον) and by the Arabic, Persian (susan), Syriac and Coptic versions. The flower was probably the scarlet Lilium Chalcedonicum, which grows profusely in the Levant, although the white lily, lotus, wild artichoke and even turnip have been proposed.[11] The arum lily is called luf : the modern šûšan is the iris, which was called qiddah by the old Jews.[12] The Authorised Version " rose " for chabatstseleth[13] is incorrect: the narcissus (Polyanthus narcissus)[14] and autumn crocus (Colchicum autumnale, by Gesenius) have been suggested and the former is probable.[15] The rose is mentioned in the Apocrypha[16] and Mishnâh,[17] but not in the Old Testament, and was introduced at a later date.[18] Mâmînân is chelidonium.[19]

TANNING ; INK AND WRITING

Galls are not produced by the oaks of Palestine and the famous galls (Aleppo galls) of Syria are not mentioned as used for tanning until the time of the

[1] Biblical Researches in Palestine, Mount Sinai, and Arabia Petræa, 3 vols., 1841, ii, 237.
[2] Voyages and Travels in the Levant, 1766, 288 ; 160.
[3] Gen., 1, 2 ; CDB, 534 ; Yahuda, Daily Telegraph, 15th Sept., 1932.
[4] Antiq., VIII, ii, 5.
[5] Berakoth, i, 2 ; tr. Cohen, 62.
[6] CDB, 499, 534.
[7] R. Mead, Medica sacra, sive de morbis insignioribus qui in Bibliis memorantur commentarius, London, 1749 ; Sprengel, H. Med., i, 65 ; S. Krauss, Geschichte der jüdischen Ärzte vom frühesten Mittelalter bis zur Gleichberechtigung, Vienna, 1930 ; W. Ebstein, Die Medizin im alten Testament, Stuttgart, 1901 ; Benzinger, Hebr. Archäol., 185 ; Jahn, Bibl. Archäol., I, i, 566 ; Vigouroux, Dict. de la Bible, iv, 911 ; Bertholet-Dallas, 298 ; I have not seen J. Bergel, Medicin d. Talmudisten, Leipzig and Berlin, 1885, nor Kagan, Medicine according to the ancient Hebrew literature, Medical Life, 1930, xxxvii, 309–39.
[8] Löw, JE, x, 82 ; bibliography in Strack, Introduction to the Talmud, 1931, 193.
[9] CDB, 675 f. ; Löw, JE, x, 72 f.
[10] In I Kings, vii, 19.
[11] CDB, 473 ; Bochart, Geographia sacra, 916 ; Celsius, i, 383 ; Löw, Flora, ii, 165.
[12] Löw, Flora, i, 214 ; ii, 1 f. ; ib., JE, x, 74.
[13] Cant., ii, 1 ; Is., xxxv, 1.
[14] Celsius, i, 488.
[15] CDB, 799.
[16] Ecclus., xxiv, 14.
[17] Shebiith, vii, 6 ; Sabbath, xiv, 4— "rose-oil."
[18] Löw, Flora, iii, 195 ; ib., JE, x, 73, 471.
[19] Ib., Flora, ii, 374.

Geonim.[1] The sumach (Rhus coriaria L.), which yields a tannin, grows in Palestine.[2] Notices of leather are very few in the Bible—it is named only twice in the Authorised Version [3]—but "skin" probably sometimes means leather.[4] The Talmûdists say leather was used for the soles of sandals.[5] The dressing of skins was practised by the Hebrews.[6] Parchment rolls were used for writing[7] ; tablets and books are also mentioned.[8] Josephus [9] says the sacred books were written in gold letters on thin parchments, but the note states that the Talmûdists made it unlawful to write with gold letters.[10] Ink (děyô)[11], although it has been regarded as Egyptian carbon ink[12], was perhaps not this, which does not write well on parchment, but an iron-gall ink.[13] "Bitter waters" are used for blotting out a name, perhaps in ink.[14] Ink is often mentioned in the Talmûd[15], also graphite (?) and galls.[16] What is called "gum" (kûmôs) is a thick fluid for writing.[17] Danby translates as "copperas" a constituent of ink mentioned in the Mishnâh.[18]

There are no references to papyrus paper in the Old Testament, where writing is always said to be on skins or on wood tablets[19]—apart, of course, from tablets of stone—but papyrus paper is in the Apocrypha[20] and New Testament.[21] Gôme[22], an Egyptian word, means the papyrus plant, which now grows in Syria.[23] Other "reeds" mentioned are : agmon, which is also an Egyptian plant, and also some aquatic reed ; 'arôth is wrongly "paper reed" in the Authorised Version, the word meaning open grassland. Qâneh is one of various kinds of reed[24], including fragrant varieties (qenêh). Various identifications are Acorus calamus ; the κάλαμος ἀρωματικός of Dioskurides[25]; some species of Andropogon ; or lemon grass. The word âchû[26], which Jerome says is of Egyptian origin, is usually supposed to mean reeds or rushes, or Cyperus esculentus, or Butomus umbellatus[27]; it is also used for sûph, an Egyptian word for the fertile, inundated banks of the Nile, or the reed thickets on them, or for a papyrus swamp[28], or for water weeds or algæ.[29]

HONEY AND WAX

Bee-keeping is not mentioned in the Old or New Testaments, and only superficially in the Mishnâh—e.g. a "bee smoker"[30]—and Talmûd.[31] The

[1] CDB, 675 ; Löw, Flora, i, 633 ; Nierenstein, Incunabula of Tannin Chemistry, 1932, 159 ; ib., Natural Organic Tannins, 1934, 175.
[2] Löw, Flora, i, 200 f.
[3] II Kings, i, 8 ; Matt., iii, 4.
[4] Lev., xi, 32 ; xiii, 48 ; Num., xxxi, 20 ; CDB, 460 ; E. Bibl., iii, 2751.
[5] Kelim, xxvi, 4 ; CDB, 823.
[6] Ex., xxv, 5 ; Lev., xiii, 48.
[7] CDB, 795, 1019.
[8] Is., viii, 1 ; xxx, 8.
[9] Antiq., XII, ii, 11.
[10] Cf. Hoefer, Hist. de la Chimie, i, 61.
[11] Jer., xxxvi, 18 ; II Cor., iii, 2 ; II John, 12 ; III John, 13—μέλαν.
[12] Hendrie, Theophilus, 1847, 74 ; Vigouroux, Dict. de la Bible, ii, 1780.
[13] Budge, Mummy², 176 f. ; Selbie, HDB, ii, 473 ; Kenyon, ib., iv, 944 f. ; E. Bibl., ii, 2169 ; Bochart, Hierozoicon, ii, 197 ; Quandt, De Atramento Hebræorum, 4to ; Königsberg, 1713, I have not seen.
[14] Numb., v, 23 ; Hoefer, i, 62.
[15] Aboth, iv ; Sabbath, viii, xii, incl. black ink.

[16] Sabbath, xii ; Rodkinson.
[17] Oesterley, Tract. Shabbath, 1927, 47 ; cf. Egy. qemi, gum arabic.
[18] Sabbath, xii, 4 ; Megillah, ii, 2 ; Sotah, ii, 4 ; Gittin, ii, 3 ; Parah, ix, 1 ; Yadaim, i, 3 ; Danby, The Mishnah, 1933, 203¹⁵.
[19] Num., xvii, 3 ; II Esd., xiv, 24.
[20] III Maccabees, iv, 20—χαρτήρια.
[21] II John, 12—χάρτης.
[22] Ex., ii, 3 ; Is., xviii, 2 ; xxxv, 7 ; Job, viii, 11 ; rush or bulrush in the A.V.
[23] CDB, 784 ; Celsius, ii, 137 ; Bochart, Hierozoicon, i, 342 ; Ursinus, Hist. plant., 157 ; E. Bibl., iii, 3556 ; Löw, Flora, i, 562.
[24] CDB, 784 f.
[25] i, 17.
[26] Job, viii, 11 ; Gen., xli, 2, 18, "meadow."
[27] CDB, 271 : A.V. "flag."
[28] Bochart, Geographia sacra, 1104.
[29] CDB, 271 ; Ursinus, Hist. plant., 85.
[30] Kelim, xvi, 6 ; honey frequently, Danby, Mishnah, 1933, index, 825.
[31] Lippmann, Z. Deutsch. Zuckerindustr., 1934, lxxxiv, 810, Techn. Teil.

528   PALESTINE II

bee (debôrâh) [1] was very common in Palestine, as it now is in Moab (which was populous in Byzantine times) [2], although apiculture is first mentioned by Philo Judæus as much practised by the Essenes.[3] The "honey out of the stony rock" [4] refers to the nesting of the bees in clefts in the rocks, as at present.[5] It has been supposed that the bee of Palestine (Apies fasciata) is distinct from the European honey-bee (Apies mellifica), but it is probably a sub-species of the latter.[6]

Various kinds of honey (děbaš) were probably used[7] ; besides bee honey, there was date honey [8] and inspissated grape juice (modern dibs), which was perhaps the kind sent by Jacob to Joseph [9] and purchased from Palestine by Phœnicians of Tyre.[10] The honey which Jonathan ate in the wood[11] and the wild honey of John the Baptist[12], although supposed to have been manna from the Tamarix mannifera, were probably honey of wild bees.[13]

If the Jews did not keep bees they would not make beeswax, so that the wax (dônag, sometimes mum ; Aramaic š'uah), which is frequently mentioned as something fusible[14], was probably mostly mineral wax[15], although beeswax was perhaps imported ; the seals which are often mentioned may have been impressed on clay.[16]

SOME ANIMAL PRODUCTS

Milk (châlâb), curdled milk (chem'âh) and possibly once butter[17] were well known, the butter-like milk-curd now used in Palestine and called leban or samn being almost like butter.[18] "Cheese" (chârîz)[19] is probably dried and powdered buttermilk curd.[20] Eggs were used only sparingly[21]: the hen (now common in-Palestine) is mentioned only twice and in the New Testament.[22] Löw[23] maintains that the reading "egg" in Job[24] is wrong, since the Syriac version shows that the material is the Anchusa officinalis, a slimy plant eaten as a salad.

Musk is mentioned in the Talmûd.[25] Onyx, or onycha (šechêleth), a constituent of incense, was some fragrant material[26] and probably the operculum of some species of Strombus, a genus of gasteropodous mollusca.[27] A gloss in the Talmûd[28] says onycha is costus root. The word tachaš ("badger" in the Authorised Version)[29] has been supposed to mean a blue or sky-blue colour, but is probably a skin, perhaps of the seal (the Arabic tuchaš is the dolphin)[30].

[1] Deut., i, 44 ; Judg., xiv, 8 ; Ps. cxviii, 12 ; Is., vii, 18 ; Bochart, Hierozoicon, i, pref. sign. g 1 verso, 69 ; ii, 502 f.
[2] Sir G. A. Smith, Syria and the Holy Land, 1918, 31.
[3] E. Bibl., ii, 2105.
[4] Deut., xxxii, 13 ; Ps. lxxxi, 16.
[5] CDB, 102.
[6] Bochart, Hierozoicon, ii, 502 f., 517 f.—a long dissertation ; CDB, 102, 272, 331 ; E. Bibl., i, 512 ; Post, HDB, i, 263.
[7] Bochart, Hierozoicon, ii, 517 f. ; Macalister, HDB, ii, 37 ; Post, ib., i, 264 ; E. Bibl., ii, 2104 ; CDB, 331 ; Zapletal, Wein in der Bibel, 35 ; Benzinger, Hebr. Archäol., 68.
[8] Mishnâh, Terumoth, xi, 2, 3 ; Nedarim, vi, 8, 9 ; Talmûd, Berakoth, vi, 1.
[9] Gen., xliii, 11.
[10] Ez., xxviii, 17.
[11] I Sam., xiv, 25.
[12] Matt., iii, 4.
[13] CDB, 272, 331.
[14] Ps. xxii, 14 ; lxviii, 2, etc.

[15] Bochart, Hierozoicon, i, pref. sign. c 3 verso ; ii, 531 ; E. Bibl., iv, 5274.
[16] CDB, 840.
[17] Prov., xxx, 33.
[18] CDB, 125, 272, 557 ; Macalister, HDB, ii, 36 ; E. Bibl., iii, 3088.
[19] Job, x, 10 ; I Sam., xvii, 18 ; II Sam., xvii, 29.
[20] CDB, 151, 557 ; E. Bibl., i, 734; iii, 3091.
[21] Is., x, 14 ; lix, 5 ; Job, vi, 6 ; CDB, 272.
[22] Matt., xxiii, 37 ; Luke, xiii, 34 ; CDB, 319 ; E. Bibl., ii, 1202 ; Macalister, HDB, ii, 37.
[23] Flora, i, 292.
[24] vi, 6.
[25] Berakoth, vi, 5 ; tr. Cohen, 280.
[26] Ecclus., xxiv, 15.
[27] Dioskurides, ii, 10, ὄνυξ ; Pliny, xxxii, 10, onyx ; CDB, 656 ; Bochart, Hierozoicon, i, pref. sign. c 4 recto ; ii, 803 ; Rosenmüller, IV, ii, 454.
[28] Sabbath, xiv.
[29] Ex., xxv, 5, etc.
[30] CDB, 90.

The sponge (σπόγγος) is mentioned only in the New Testament [1], but may have been known to the ancient Hebrews, who could have obtained good specimens from the Persian Gulf.[2]

Ivory is the translation of šên and šenhabbîm[3] ; it means literally "tooth" and the name of the elephant does not occur, except as the marginal for behemoth [4], where it may mean hippopotamus.[5] Elephants' teeth is the marginal for ivory.[6] Elephants are repeatedly mentioned in the Apocrypha.[7] Carved (Indian ?) ivory came to Tyre from Chittim [8], usually regarded as Cyprus.[9] The Phœnicians probably carved Solomon's ivory throne and overlaid it with gold[10], the (Indian ?) ivory coming by the caravans of Dedan[11], probably from the Persian Gulf.[12] Ivory also came by sea from Tharshish[13], i.e. by Phœnician trade.[14] Ahab had a palace panelled in ivory[15] and beds were inlaid with this material.[16]

[1] Matt., xxvi, 48, etc.
[2] CDB, 899 ; Bochart, Hierozoicon, i, 53.
[3] I Kings, x, 22 ; II Chron., ix, 21.
[4] Job, xl, 15.
[5] CDB, 104, 234, 355 ; Post, HDB, ii, 522 ; E. Bibl., ii, 2297.
[6] I Kings, x, 22 ; II Chron., ix, 41.
[7] I Maccabees, vi ; Kautzsch, Apokryphen, i, 51.
[8] Ez., xxvii, 6.
[9] CDB, 154, 356 ; but see p. 357.
[10] I Kings, x, 18 ; II Chron., ix, 17.
[11] Is., xxi, 13 ; Ez., xxvii, 15.
[12] CDB, 198, 967.
[13] I Kings, x, 22.
[14] Bochart, Hierozoicon, i, 254.
[15] I Kings, xxii, 39.
[16] Amos, vi, 4 ; Cant., vii, 4.

ADDITIONAL NOTES

(These have *not* been included in the Indexes)

P. 5, n. 1 : the correct reading of the hieroglyphic name is Qem or Qemt, the crocodile's tail, etc., being a determinative sign.

P. 11, n. 2 : further on alphabet on p. 469.

P. 19 : Taubach (near Weimar) has remains of the Mousterian Period (? 50,000–35,000 B.C.) : Forrer, Urgesch. des Europäers, Stuttgart, 1908, 61.

P. 20, n. 8 : the title of Freise's book has been abbreviated as Gesch. der Bergbau, although the last word (masculine as a noun) is the adjective qualifying the (feminine) noun Technik.

P. 33, n. 6 : Lepsius, Les métaux, 7, refers to Linant, Carte de l'Etbaye ou pays habité par les Arabes Bisharis, 1854, giving a map of the mines.

P. 84, n. 5 : the modern Arabic kohl is galena : Niebuhr, in Pinkerton, Travels, 1811, x, 157.

P. 88, n. 3, and p. 380, n. 17 : Meyer, Alt. II, i, 529, says the pharaoh is Merenptah and the Hittite king is Arnuanda IV ; Hogarth, CAH, ii, 272, thinks Kizwatna is in the Issus Gulf in the Anti-Taurus.

P. 89, n. 24 : on the very varying dates proposed for Hero, see Hammer-Jensen, *Hermes*, 1913, xlviii, 224 ; Heath, History of Greek Mathematics, Oxford, 1921, ii, 298 ; J. L. Heiberg, Science and Mathematics in Classical Antiquity, 1922, 85 f. ; *ib.*, MGM, 1925, xxiv, 23 ; *ib.*, *Archeion*, 1925, vi, 202 ; Diels, Antike Technik, 1924, 57 ; Tittel, PW, viii, 992 ; Weinleitner, *Archeion*, 1925, vi, 202 ; *ib.*, MGM, 1926, xxv, 156, 159 ; *ib.*, *A. Nat.*, 1928, x, 239 ; Tropfke, *A. Nat.*, 1928, x, 450 (100 B.C.). Hero of Alexandria, The Pneumatics, transl. by J. G. Greenwood and ed. by B. Woodcroft, 4to, 1851 ; Hero's *Automata* is contained in Thevenot's Veterum Mathematicorum Opera, Paris, 1693, 243 f.

P. 103 : Calcite (common) and aragonite (rather rare) are two different crystalline forms of calcium carbonate, $CaCO_3$, calcite crystallising in forms belonging to the hexagonal system and aragonite in the rhombic system. Alabaster is a compact, translucent crystalline variety of gypsum, or hydrated calcium sulphate, $CaSO_4,2H_2O$, a transparent form of which, easily splitting into thin sheets, is selenite. Common mica is quite a different mineral from selenite, being a potassium hydrogen aluminium ortho-silicate, $KH_2Al_3(SiO_4)_3$.

P. 141, n. 11 : in the Harris Papyrus wod or uaz is a green mineral : Breasted, Ancient Records, iv, 175 ; another name for green ink or paint was reâ or ri-t : Budge, Dict., 417, 419.

P. 147, n. 4 : Blount found the weeds suhit and gazull burnt in Alexandria and the ash exported to Venice : Pinkerton, Travels, x, 235.

P. 157, n. 11 : castor oil, suggested by Lucas, is improbable, as this contains practically no stearic acid.

P. 161, n. 8 : the lotus is called sšn in the Harris Papyrus : Breasted, Ancient Records, iv, 139

P. 161, n. 14 : "pure olive oil" is mentioned in the Harris Papyrus : Breasted, Ancient Records, iv, 175 ; the name for olive oil was beq or bak : Budge, Hierogl. Dict., 207, 224, 903.

P. 165, n. 16 : both "incense" and "inflammable incense" are named in the Harris Papyrus : Breasted, Ancient Records, iv, 136.

P. 173, n. 2 : the whole mummy was sometimes bogus, the body being replaced by rubbish : Petrie, Hawara, 1889, 14 ; on mummifying, K. Sethe, Zur Geschichte der Einbalsamierung bei den Aegyptern, 1934.

P. 180, n. 12 : Egyptian names for prescriptions are khes and tep red ; the spell or spells are âakhu, ḥeka, thes, skhenu, šemu : Budge, Dict., 22, 339, 515, 563, 693, 831, 860.

P. 195 : Blount says the Nile water tasted like new milk, but was somewhat nitrous and purging unless first treated with bruised almonds : Pinkerton, Travels, x, 244.

P. 197, n. 7 : Niebuhr, however, reports that plum-tree lichen was used by the Arabs for fermenting beer and was exported to Alexandria : Pinkerton, Travels, x, 192.

P. 202, n. 15 : What Aretaios actually says (De Curat. diut. Morb., ii, 13 ; in Medicæ Artis Principes, Paris, 1567, i, 96) is that soap is an invention of the Gauls, and Aëtios (Tetrabiblos, lib. II, sermo iv, 6 ; lib. IV, sermo iii, 14 ; ed. Basle, 1549, 398, 826) says the same.

P. 204 : honey was known in the Old Kingdom, wax (menech) is mentioned in the Middle Kingdom (c. 2000 B.C.) and bee-culture and the collection of honey are represented in 1450 B.C. and in 600 B.C. Honey was not a monopoly in the Hellenistic Period (cf. p. 17). The curious legend of bugonia (generation of bees in putrid carcases of cattle) appears to go back to Bolos of Mendes (200 B.C.) : Lippmann, Z. Deutsch. Zuckerindustr., 1934, lxxxiv, 808.

P. 207, n. 1 : a block stamp was used in the Ptolemaic Period : Petrie, Hawara, 1889, 29.

P. 211, n. 6 : Articles on excavations at Warkâ (Uruk) in Abh. Preuss. Akad. Wiss., Phil.-hist. Kl. :

(1) Jordan, 1931, pp. 1–55 : ziqqurat, clay cones as mosaic, clay flowers and rosettes.

(2) Jordan, Nöldeke and Heinrich, 1932, 1–37 : plate of wall mosaic ; obsidian ; sexagesimal system.

(3) Leuzen and v. Haller, 1932, 1–47 : mosaic, ceramics.

(4) Nöldeke, Heinrich and Schott, 1933 : copper, iron (forked throwing stick, sword and weapons).

Summary in Childe, Most Ancient East, 1934, 134 f.

P. 212 : Woolley's dates for the "predynastic" tombs at Ur, which may not be royal tombs at all, have been contested and the remains supposed later than the 1 dyn. : Speiser, Antiquity, 1934, viii, 451. This would agree better with the truly remarkable results of the analyses of the bronzes (see p. 245 f.).

P. 224, n. 3 : horse in Mesopotamia : Rostovtzeff, Syria, 1931, xi, 48.

P. 226, n. 6 : A silver box found at Van contained powdered silver sulphide, which is still used in the district to produce the so-called Tula-work. The powder is applied to places on polished silver and forms a darker colour. What was possibly a blowpipe was found. This niello-technique seems to be very old in this region. Lehmann-Haupt, Materialen zur älteren Geschichte Armeniens und Mesopotamiens, Abh. Kgl. Ges. Wiss. zu Göttingen, Phil.-hist. Kl., 1907, ix, 90 f. : on bronze work, ib., 93 f., iron weapons, ib., 101 f.

P. 239, n. 4 : decorative copper objects from Kiš : Watelin, Revue des arts asiatiques, 1929–30, vi, 148.

P. 254 : Niebuhr mentions a very rich lead mine in Oman : Pinkerton, Travels, x, 199.

P. 257, n. 12 f. : iron dagger of 2700 B.C. at Tell Asmar (anct. Ešnunna) with copper knives with silver foil hilts ; also silver and lapis lazuli amulets strung with blue faience and carnelian beads on silver wire, c. 2500 B.C. : Frankfort, Illustr. London News, 1933, clxxxiii, 97, 124 ; on Khafaje, ib., 1932, clxxxi, 526.

P. 281, n. 2 : Babylonian pottery kiln at Nippur, Babylonian Exped. of Univ. of Pennsylvania, Excav. at Nippur, Plans . . . with descriptive text by C. S. Fisher, Part 1, Philadelphia, 1905, 40, and Fig. 16.

P. 289, n. 6 : bleaching carnelian with soda is apparently mentioned by Pliny, was well known in Italy in the Renaissance Period, and is described by Barbot : King, Nat. Hist. of Precious Stones, etc., 1865, 300.

P. 302 : Hoefer, Hist. de la botanique, 5, says artificial fertilisation of the date palm was known in ancient Egypt. The text on p. 302 is perhaps a little ambiguous owing to the introduction of the treatment of figs (caprification) in the description of the palm. The process of caprification is described by Theophrastos (Hist. plant., II, viii) and Pliny (xv, 19 ; xvii, 27) and was seen in use in the Greek Archipelago by Tournefort : Hoefer, *op. cit.*, 56.

P. 316, n. 3 : R. C. Thompson, JRAS, 1933, 855 : šamaîtum = blue vitriol ; peš = ætites (botryoidal hæmatite) (? ; *see* p. 317³ and Blancardus, Lexicon Novum Medicum Græco-Latinum, Lugduni Batavorum, 1702, 14).

P. 323 : pottery from Phraté (Crete) ; D. Levi, *Illustr. London News*, 1931, clxxix, 1042.

P. 334 : bronze shield from Phraté (Crete) ; D. Levi, *Illustr. London News*, 1931, clxxix, 1042.

P. 356 f. : Bronze Age (3000–2100 B.C.) remains in Cyprus (red polished ware ; white-painted polished ware of 2100–1600 B.C. ; gypsum idols—rare and a new type—very composite jar, black red-topped ware, glass paste beads, gold ornaments, bronze dagger) : see P. Dikaios, *Illustr. London News*, 1931, clxxix, 678, 891 ; 1933, clxxxiii, 29 ; I. Gjerstad, *ib.*, 1933, clxxxiii, 29.

P. 368 : Arabic authors mention copper, vitriol, asbestos and alum mines in Cyprus (al-Idrîsî, Dimešqî, ibn al-Wardî) : Hâjjî Khalîfa also mentions sulphur and the soda plant ; the sugar cane was cultivated from A.D. 1200, and madder is also a recent introduction : Oberhummer, Die Insel Cypern, i, 46, 54, 56, 75, 180, 282 f., 291 ; Dandini (17th century) speaks of mines of gold (mentioned in a fragment attributed to Aristotle, V. Rose, Aristoteles Pseudepigraphus, Leipzig, 1863, 260), copper, marcasite, lattin (? zinc ore), iron, rock alum, pitch, sulphur and saltpetre in Cyprus, also the occurrence of "scarlet berry" (kermes—see p. 464), coral, emerald, "crystal diamonds" and other precious stones (Pinkerton, Travels, x, 281) : Pococke mentions madder, asbestos and "seed of alkermes" (*ib.*, 592) : on "Cyprian diamonds" see King, Nat. Hist. of Precious Stones, etc., 1865, 23.

P. 373 : K. Bittel, Prähistorische Forschungen in Kleinasien, *Istanbuler Forschungen*, Bd. 6, 1934.

P. 375, n. 5 : Mallowan, *Illustr. London News*, 1933, clxxxiii, 436—double axe of *c.* 4000 B.C. at Tell Arpachiyah, near Nineveh.

P. 420, n. 8 : glass from Susa, C. J. Lamm, *Syria*, 1931, xi, 358, who is mistaken in thinking old Babylonian glass is Egyptian : see p. 285, n. 15.

P. 462, n. 14 : the liquor was boiled until 100 amphoræ (8,000 lb.) were reduced to 500 lb. : Kenrick, Phœnicia, 241.

P. 512, n. 3 : various identifications : Post suggests some resinous pine and the E. Bibl. the Eleagnus angustifolia.

P. 513, n. 1 and n. 4 : I have failed to find a copy of vol. iv of Löw's work and quote it from the paper by Lippmann mentioned in n. 1. There is a brief section on manna in vol. iii of Löw, p. 401 f.

P. 525, n. 15 : Maundrell was told that the "love apples" (dûdâîm) were not mandrakes : Pinkerton, Travels, x, 337.

P. 528, n. 5 : Maundrell found evidence of wild bees in the desert near the Red Sea : Pinkerton, Travels, x, 338.

INDEX OF AUTHORS AND PUBLICATIONS

The place of publication is London for English books, Paris for French books, and Berlin for German books, and the format is octavo, unless the contrary is stated.

35

INDEX OF PERSONS AND NATIONS

(See also Index of Places for corresponding nations)

INDEX OF PLACES

36

SUBJECT INDEX

basalt, 5, 18, 73, 91, 101, 103, 262–5, 325, 354, 359, 382, 384, 438, 491, 501, 510 ; green —, 104 ; — syenite, 103
bâsâm, 517
bašâm, 517
basanos, 384
base-ring, 390
basin, 36, 64, 76–7, 179, 253, 317, 335, 515
basket, 47, 64, 70, 109, 177, 201, 217, 223, 269, 279, 414, 433 ; — work, 4, 109, 266, 390
bašlu, 237
bas-reliefs, 268
bassam, 457
bât, 204
batêkh, 510
baths, 17, 125, 148, 270, 326, 350, 418, 426, 473, 489 ; — of Titus, 117, 137, 460
batrachion, 291
battering rams, 225
baurach, 147
bay tree, 520
bazal, 490
bdd, 427
bdellium, 306, 315, 422, 507, 518
beads, 5–7, 24–7, 29, 39, 42, 48–50, 53, 55, 78, 80–1, 83–4, 89–91, 97, 102, 104, 111, 113, 116, 118, 120–4, 126–8, 134, 141, 170, 192, 201–3, 223, 230–1, 264–5, 279, 281, 285–6, 289–91, 293–4, 308, 324–5, 327–9, 338, 342–3, 345, 354–5, 359 –61, 369, 374, 376, 378, 382, 386, 389, 392, 394, 401, 412, 417, 439–40, 454 –6, 472–4 ; bleached carnelian —, 289, 417 ; gilt glass —, 134 ; millefiore —, 128 ; — materials, 49, 473
beakers, 4, 124, 234, 241, 378, 390, 393
beams, 213, 377
beans, 160, 178, 301, 315, 427, 508–9, 511 ; kidney —, 508
beards, 216–7 ; blue —, 88, 156, 277, 290, 297 ; goats', 168
beaten work, 25, 82, 231, 240–1, 251, 352, 381, 401, 403, 474, 484–5
bed, 221, 491 ; — (salt), 149
bedet, 187, 190
bedil, 481, 494–5
bedôlach, 507, 518
bee, 160, 181, 302, 338, 382, 424, 458, 527–8 ; — hives, 382
beech tree, 520
beeswax, 153, 204, 528

beef, 201
beer, 19, 145, 181, 185, 190, 192–3, 195–7, 200, 266, 301, 303–4, 315, 338, 513– 4 ; — bread, 304
beěrôth chêmâr, 502
beet, 32, 301, 315, 510
behemoth, 529
beidri, 412
beka', 511
belemnite, 506
bell, 75, 235, 245, 250, 252, 259, 490
belladonna (Atropa belladonna), 315
bellows, 21–2, 64, 79, 97–9, 228, 240, 332, 368, 482, 491
belt, 234
ben, 186, 424 ; — oil, 158
benà, 198
benipe, 86–7
ben-i-pet, 185
benzoin, 167–8
ber, 491
bêrěqeth, 504
běroš, 520
běrôth, 520
berries, 186, 517
beryl, 48–9, 288, 290, 325, 417, 504–6
beryllus, 504
besbes, 160, 186
besem, 457, 517
Beta vulgaris, 302
betsâlîm, 510
bet-teser, 197
bězâlîm, 508
bezer, 483
bezmen, 222
biâret-hummar, 502
bikkûrah, 512
bile, 525
billaur, 420
bill-hook, 242
biltu, 220
bin, 200
binder, 112
binding, 234
bînu, 316
birch bark, 204
bird, 144, 176, 203, 318, 325, 440
birinj, 410
birth-charms, 295
biscuit, 200
bishâm, 517
bismuth, 27, 54, 56, 62, 68, 352, 353, 365
bisru, 316, 510
bis tinctum, 462
bistre, 208
bit, 94
bitter herbs, 508 ; materials, 186
bitumen, 73, 137, 143, 157, 170, 173, 175, 198, 213, 217, 219–20, 229, 231, 239, 241, 244, 250, 263, 266–8,

270–2, 297, 391, 418–9, 439, 497, 502–3, 516
bituminous shale, 231
bj-ni-pet, 86
black calf or cow, 187, 189 ; — cummin, 315 ; — finish on bronze, 393 ; —smith, 85–6, 95, 228, 482, 491–2 ; —thorn, 515 ; —wood, 303
blade, 93, 96, 102, 241–2, 257, 355, 389, 406, 412, 440, 442 ; damascened —, 413–4
blatta, blattam, blattea, blatteam, blattellam, 461, 463
Blättermergel, 150
bleaching, 140, 308, 463
block, 11, 20, 59, 61, 64–5, 76, 105, 111, 116, 207, 262– 3, 270–1, 325–6, 333, 348, 354, 467, 500
blood, 65, 109, 138, 181, 187, 189–91, 308, 398, 413, 431, 459, 461, 463, 516 ; bull's —, 424 ; — of the cedar, 306, 425 ; circulation of the —, 189 ; — of death, 189 ; — of Herakles, 189 ; — letting, 178 ; — of the lizard, 189 ; — of Osiris, 425
bloodstone, 359
bloom (iron), 493
blowing drugs, 314 ; — glass, 124–5, 127, 133
blowpipe, 8, 15, 21, 25, 128, 228–9, 287
blue, natural, 327 (see azurite, lapis lazuli)
boat, 5, 9, 11, 75, 111, 178, 202, 205, 220, 223, 234, 252, 268–9, 272, 290, 347, 446, 451, 503
bodet, 199
boiler, 75, 305, 462 ; boiling, 201–2, 237, 313, 315, 510, 517
bole, 137
bolot, 413
bolt, 8, 54, 67, 75, 251, 475
bone(s), 6, 20, 65, 90, 135, 173, 181, 213, 257, 270, 295, 323, 336, 342, 369, 398, 426 ; — ash, 282, 284, 456 ; — black, 136–7, 191, 268 ; burnt —, 94, 138 ; calcined —, 213 ; — dust, 185 ; — of Horus, 86, 90 ; — of Set, 86, 89–90
book(s), 18, 178–80, 182, 189, 206, 220, 272, 301, 520, 527
boomerang, 116
booths, 17, 172
Borassus flabelliformis, Linn., 518
borates (borax), 112, 130, 193, 295, 298, 317, 351, 499

GREEK INDEX

PRINTED IN ENGLAND BY
WILLIAM CLOWES AND SONS, LIMITED, BECCLES.

HISTORY, PHILOSOPHY AND
SOCIOLOGY OF SCIENCE

Classics, Staples and Precursors

An Arno Press Collection

Aliotta, [Antonio]. **The Idealistic Reaction Against Science.** 1914

Arago, [Dominique François Jean]. **Historical Eloge of James Watt.** 1839

Bavink, Bernhard. **The Natural Sciences.** 1932

Benjamin, Park. **A History of Electricity.** 1898

Bennett, Jesse Lee. **The Diffusion of Science.** 1942

[Bronfenbrenner], Ornstein, Martha. **The Role of Scientific Societies in the Seventeenth Century.** 1928

Bush, Vannevar. **Endless Horizons.** 1946

Campanella, Thomas. **The Defense of Galileo.** 1937

Carmichael, R. D. **The Logic of Discovery.** 1930

Caullery, Maurice. **French Science and its Principal Discoveries Since the Seventeenth Century.** [1934]

Caullery, Maurice. **Universities and Scientific Life in the United States.** 1922

Debates on the Decline of Science. 1975

de Beer, G. R. **Sir Hans Sloane and the British Museum.** 1953

Dissertations on the Progress of Knowledge. [1824]. 2 vols. in one

Euler, [Leonard]. **Letters of Euler.** 1833. 2 vols. in one

Flint, Robert. **Philosophy as Scientia Scientiarum and a History of Classifications of the Sciences.** 1904

Forke, Alfred. **The World-Conception of the Chinese.** 1925

Frank, Philipp. **Modern Science and its Philosophy.** 1949

The Freedom of Science. 1975

George, William H. **The Scientist in Action.** 1936

Goodfield, G. J. **The Growth of Scientific Physiology.** 1960

Graves, Robert Perceval. **Life of Sir William Rowan Hamilton.** 3 vols. 1882

Haldane, J. B. S. **Science and Everyday Life.** 1940

Hall, Daniel, et al. **The Frustration of Science.** 1935

Halley, Edmond. **Correspondence and Papers of Edmond Halley.** 1932

Jones, Bence. **The Royal Institution.** 1871

Kaplan, Norman. **Science and Society.** 1965

Levy, H. **The Universe of Science.** 1933

Marchant, James. **Alfred Russel Wallace.** 1916

McKie, Douglas and Niels H. de V. Heathcote. **The Discovery of Specific and Latent Heats.** 1935

Montagu, M. F. Ashley. **Studies and Essays in the History of Science and Learning.** [1944]

Morgan, John. **A Discourse Upon the Institution of Medical Schools in America.** 1765

Mottelay, Paul Fleury. **Bibliographical History of Electricity and Magnetism Chronologically Arranged.** 1922

Muir, M. M. Pattison. **A History of Chemical Theories and Laws.** 1907

National Council of American-Soviet Friendship. **Science in Soviet Russia: Papers Presented at Congress of American-Soviet Friendship.** 1944

Needham, Joseph. **A History of Embryology.** 1959

Needham, Joseph and Walter Pagel. **Background to Modern Science.** 1940

Osborn, Henry Fairfield. **From the Greeks to Darwin.** 1929

Partington, J[ames] R[iddick]. **Origins and Development of Applied Chemistry.** 1935

Polanyi, M[ichael]. **The Contempt of Freedom.** 1940

Priestley, Joseph. **Disquisitions Relating to Matter and Spirit.** 1777

Ray, John. **The Correspondence of John Ray.** 1848

Richet, Charles. **The Natural History of a Savant.** 1927

Schuster, Arthur. **The Progress of Physics During 33 Years (1875-1908).** 1911

Science, Internationalism and War. 1975

Selye, Hans. **From Dream to Discovery: On Being a Scientist.** 1964

Singer, Charles. **Studies in the History and Method of Science.** 1917/1921. 2 vols. in one

Smith, Edward. **The Life of Sir Joseph Banks.** 1911

Snow, A. J. **Matter and Gravity in Newton's Physical Philosophy.** 1926

Somerville, Mary. **On the Connexion of the Physical Sciences.** 1846

Thomson, J. J. **Recollections and Reflections.** 1936

Thomson, Thomas. **The History of Chemistry.** 1830/31

Underwood, E. Ashworth. **Science, Medicine and History.** 2 vols. 1953

Visher, Stephen Sargent. **Scientists Starred 1903-1943 in American Men of Science.** 1947

Von Humboldt, Alexander. **Views of Nature: Or Contemplations on the Sublime Phenomena of Creation.** 1850

Von Meyer, Ernst. **A History of Chemistry from Earliest Times to the Present Day.** 1891

Walker, Helen M. **Studies in the History of Statistical Method.** 1929

Watson, David Lindsay. **Scientists Are Human.** 1938

Weld, Charles Richard. **A History of the Royal Society.** 1848. 2 vols. in one

Wilson, George. **The Life of the Honorable Henry Cavendish.** 1851

Partington. James Riddick. 1886-
 Origins and development of applied chemistry James Riddick Partington. — New York : Arno Press. 1975. c1935.

 xv, 597 p. ; 24 cm — (History. philosophy. and sociology of science)

 Reprint of the ed. published by Longmans. Green. London.
 Includes bibliographical references and indexes.
 ISBN 0-405-06611-2

75